Introduce something extra

It may sometimes be necessary to introduce something new, an auxiliary aid, to help make the connection between the given and the unknown. For instance, in a problem where a diagram is useful the auxiliary aid could be a new line drawn in a diagram. In a more algebraic problem it could be a new unknown that is related to the original unknown.

Take cases

We may sometimes have to split a problem into several cases and give a different argument for each of the cases. For instance, we often have to use this strategy in dealing with absolute value.

Work backward

Sometimes it is useful to imagine that your problem is solved and work backward, step by step, until you arrive at the given data. Then you may be able to reverse your steps and thereby construct a solution to the original problem. This procedure is commonly used in solving equations. For instance, in solving the equation $3x - 5 = 7$, we suppose that x is a number that satisfies $3x - 5 = 7$ and work backward. We add 5 to each side of the equation and then divide each side by 3 to get $x = 4$. Since each of these steps can be reversed, we have solved the problem.

Establish subgoals

In a complex problem it is often useful to set subgoals (in which the desired situation is only partially fulfilled). If we can first reach these subgoals, then we may be able to build on them to reach our final goal.

Indirect reasoning

Sometimes it is appropriate to attack a problem indirectly. In using proof by contradiction to prove that P implies Q we assume that P is true and Q is false and try to see why this cannot happen. Somehow we have to use this information and arrive at a contradiction to what we absolutely know is true.

Mathematical induction

In proving statements that involve a positive integer n, it is frequently helpful to use the Principle of Mathematical Induction, which is discussed in Appendix A.

3. Carry Out the Plan

In Step 2 a plan was devised. In carrying out that plan we have to check each stage of the plan and write the details that prove that each stage is correct.

4. Look Back

Having completed our solution, it is wise to look back over it, partly to see if there are errors in the solution and partly to see if there is an easier way to solve the problem. Another reason for looking back is that it will familiarize us with the method of solution and this may be useful for solving a future problem. Descartes said, "Every problem that I solved became a rule which served afterwards to solve other problems."

See page xi for further discussion and an example of these problem-solving principles and references to books with more thorough treatments of methods of problem solving.

Second Edition

Single Variable Calculus

Early Transcendentals

Second Edition

Single Variable Calculus

Early Transcendentals

James Stewart

McMaster University

Brooks/Cole Publishing Company
Pacific Grove, California

To Dorothy

Brooks/Cole Publishing Company
A Division of Wadsworth, Inc.

Maple is a registered trademark of Waterloo Maple Software.
Derive is a registered trademark of the Soft Warehouse, Inc.
Mathematica is a registered trademark of Wolfram Research, Inc.
Calculus-Pad is a registered trademark of Queen's University.

Printed in the United States of America
10   9   8   7   6   5   4

Library of Congress Cataloging-in-Publication Data
Stewart, James [date]
    Single variable calculus, second edition: early
transcendentals / James Stewart.
    Rev. ed. of: Calculus / James Stewart. 2nd ed. 1991.
    Includes index.
    ISBN 0-534-16414-5 (hardcover)
    ISBN 0-534-16410-2 (softcover)
    1. Calculus.   I. Stewart, James, [date]   Calculus.
II. Title.
QA303.S8826   1991b   91-4474
515—dc20   CIP

Sponsoring Editor: *Jeremy Hayhurst*
Editorial Assistant: *Nancy Champlin*
Permissions Editor: *Carline Haga*
Cover Design: *Katherine Minerva*
Cover Photo: *Lee Hocker*
Interior Illustration: *TECHarts; Vantage Art, Inc.*
Cover Printing: *Lehigh Press Lithographers*
Printing: *Courier/Kendallville, Inc.*
Binding: *Courier/Kendallville/Westford, Inc.*

Cover: Illustrates the remarkable resemblance between the sound hole of a violin viewed from this angle and the elongated "S" of the integral sign, a notation introduced by Leibniz.

Preface

A great discovery solves a great problem but there is a grain of discovery in the solution of any problem. Your problem may be modest; but if it challenges your curiosity and brings into play your inventive faculties, and if you solve it by your own means, you may experience the tension and enjoy the triumph of discovery.

George Polya

The art of teaching, Mark Van Doren said, is the art of assisting discovery. I have tried to write a book that assists students in discovering calculus—both for its practical power and its surprising beauty. I aim to convey to the students a sense of the utility of calculus and develop their technical competence, but I also strive to give them some appreciation for the intrinsic beauty of the subject. Newton undoubtedly experienced a sense of triumph when he made his great discoveries. I want students to share some of that excitement.

The emphasis is on understanding. Enough mathematical detail is presented so that the treatment is precise, but without allowing formalism to become obtrusive. The instructor can follow an appropriate course between intuition and rigor by choosing to include or exclude optional sections and proofs. Section 1.4, for example, on the precise definition of the limit is an optional section. Although a majority of theorems are proved in the text, some of the more difficult proofs are given in Appendix C.

Early Transcendentals *Single Variable Calculus, Second Edition: Early Transcendentals* differs from *Single Variable Calculus, Second Edition,* in that the contents of Chapter 6 of the first and second editions (exponential and logarithmic functions, inverse trigonometric functions, hyperbolic functions, and l'Hospital's Rule) are presented in Chapter 3. This means that there is a plentiful supply of interesting functions at an early stage of the course. They give variety to the curve sketching in Chapter 4, integration in Chapter 5, and applications of integration in Chapter 6. It also means that some of the most substantial applications of differential calculus, those to exponential growth and decay, are given early (see Section 3.6). The other main difference from the second edition is that limits at infinity and infinite limits are in Chapter 1. The rest of the text is essentially unchanged from *Single Variable Calculus, Second Edition.*

Other Changes in the Second Edition

While teaching from the first edition for three years, I (and my students) have had ideas, some major, some minor, for improving the exposition and organization and for adding new and better examples and exercises. I have also had the benefit of some valuable suggestions from colleagues, both friends and strangers, which have been incorporated into the second edition. Here is a summary of some of the principal changes:

• At the request of several users, some of the applications of integration are introduced earlier. In fact, applications of integration now occur in two chapters. The applications, such as volume and work, that ordinarily require only basic techniques of integration are in Chapter 6. Those applications for which it is profitable to have studied further techniques (separable differential equations, arc length, surface area) are now in Chapter 8, together with centers of mass, hydrostatic pressure, and a new section on applications to economics and biology (consumer's surplus, present value of an income stream, blood flow, cardiac output).

• The chapter Limits and Rates of Change (now Chapter 1) has been substantially reorganized. In particular, the properties of limits are introduced earlier.

• In Chapter 2 the section Rates of Change in the Natural and Social Sciences is placed earlier and there is increased emphasis on linear approximation.

• In Chapter 3 the method for differentiating the exponential and logarithmic functions has been changed so that the exponential function is differentiated first.

• The Midpoint Rule for approximate integration is now covered. It first occurs in Section 5.3, where the definite integral is introduced. The treatment of approximate integration in Section 7.8 is expanded to include comparison of errors for the Left Endpoint, Right Endpoint, Trapezoidal, Midpoint, and Simpson's rules.

• In Chapter 10 there are more graphs of Taylor Series approximations, and Taylor's Formula is now proved in the text instead of in the exercises. Multiplication and division of power series are now covered.

• There is a new appendix on complex numbers.

Problem-Solving Emphasis

My educational philosophy was strongly influenced by attending lectures of George Polya and Gabor Szego when I was a student at Stanford University. Both Polya and Szego consistently introduced topics with an intuitive geometrical or physical description and have attempted to tie mathematical concepts to the students' experiences.

I found Polya's lectures on problem solving highly inspirational and his books *How To Solve It, Mathematical Discovery,* and *Mathematics and Plausible Reasoning* have become the core text material for a mathematical problem-solving course that I instituted and teach at McMaster University. I have adapted these problem-solving strategies to the study of calculus both explicitly, by outlining strategies, and implicitly, by illustration and example.

Students usually have difficulties in situations where there is no single, well-defined procedure for obtaining an answer. I think nobody has improved very much on Polya's four-stage problem-solving strategy, and accordingly I have presented a version of this strategy on the front endpapers of this book together with a discussion of it and an example of its use in To the Student. I also urge my students to read Polya's *How To Solve It* for a more leisurely exposition of the principles. I often find myself nagging them into using these principles.

The classic calculus situations where problem-solving skills are especially important are related rates problems, maximum and minimum problems, integration, testing series, and solving differential problems. In these and other situations I have adapted Polya's strategies to the matter at hand. In particular two special sections are devoted to problem solving: 7.6 (Strategy for Integration) and 10.7 (Strategy for Testing Series).

Problems Plus and Applications Plus

In this edition I have added what I call *Problems Plus* after even-numbered chapters. These are problems that go beyond the usual exercises in one way or another and require a higher level of problem-solving ability. The very fact that they do not occur in the context of any particular chapter makes them a little more challenging. For instance, a problem that occurs after Chapter 10 need not have anything to do with Chapter 10. I particularly value problems in which a student has to combine methods from two or three different chapters. In recent years I have been testing these Problems Plus on my own students by putting them on assignments, tests, and exams. Because of their challenging nature I grade these problems in a different way. Here I reward a student significantly for ideas toward a solution and for recognizing which problem-solving principles are relevant. My aim is to teach my students to be unafraid to tackle a problem the likes of which they have never seen before.

A counterpart to the Problems Plus are the *Applications Plus,* which occur after odd-numbered chapters (starting with Chapter 3) and which tend to be challenging because they involve related concepts from science that are usually outside students' experiences. Again the idea is to combine concepts and techniques from different parts of the book. These problems are helpful in demonstrating the sheer variety of the applications of calculus but also in focusing the students' attention on the essential mathematical similarities in diverse situations in science. By solving a wide variety of concrete problems, I hope that they will come to appreciate the power of abstraction. I am grateful to Garrett Etgen for amassing such a wide-ranging collection of applied problems.

Exercises and Examples

I believe that one of the reasons that the first edition has been successful at a range of educational institutions is that there has been such a wide range of abilities among my own calculus students and my goal has been not to lose any of them. In particular I did not want to lose the interest of my very best students and so I have sought to challenge them with stimulating exercises. There is nothing in any text that can compensate for a deficiency of good problems. I have selected my exercises from those used in 20 years of calculus classes and have expressly chosen examples for their instructional value. I have added 750 new exercises to this edition, making a total of about 5,500 exercises that range from the essential, routine ones to those that will challenge your best students. I have made a special effort to include unusual problems at both ends of the spectrum of difficulty. Many of the new exercises are thought provoking and occur toward the end of exercise sets.

Another reason for the first edition's success may be the heuristic flavor of many of the text examples. Examples can, and should, be more than exercise-solving "templates." Carefully constructed examples can be one of the most effective ways of leading students into more advanced material. Many examples herein are designed to promote careful thinking about the problems and ideas behind calculus while giving the students insight into why theorems and proofs are necessary.

Design Pedagogy

From repeatedly correcting the same mistakes I have come to recognize consistent pitfalls that trap many students. I believe that it is best to alert students to them and in many cases have done so with the "caution" symbol. For example, on page 244 the symbol is used to warn the student against a common misuse of l'Hospital's Rule and on page 573 it is used to emphasize that if $\lim_{n \to \infty} a_n = 0$, then we cannot conclude that Σa_n is convergent.

The calculator symbol indicates an exercise, or group of exercises, that requires the use of a calculator or computer. Although most of these exercises are designed to illustrate the power of algorithmic computation, some of them are designed to show machine limitations. See for instance Exercise 18 in Section 1.2. Appendix D, Lies My Calculator or Computer Told Me, alerts the student to many of the situations in which a calculator or computer can give unreliable answers.

Appendixes In addition to the usual review of precalculus topics in the Review and Preview I have included an algebra review appendix, because so many of the errors made by students writing calculus exams are not errors in calculus itself, or even in precalculus material, but rather in very basic algebra. Appendix A contains a substantial review of elementary algebra and drill exercises. Similarly, a review of trigonometry appears in Appendix B, because a review thorough enough for students who really need it would be intrusive in the text per se.

Acknowledgments It took me some eight years to write and test the first edition. I never suspected that it would take nearly two years to revise that same material. Much of that time was spent considering reasoned (but sometimes contradictory) advice from a large number of astute reviewers. I have learned something from each of them.

First Edition Reviewers:

John Alberghini,
Manchester Community College
Daniel Anderson, *University of Iowa*
David Berman,
University of New Orleans
Richard Biggs,
University of Western Ontario
Stephen Brown
David Buchthal, *University of Akron*
James Choike, *Oklahoma State University*
Carl Cowen, *Purdue University*
Daniel Cyphert, *Armstrong State College*
Robert Dahlin
Daniel DiMaria,
Suffolk Community College
Daniel Drucker, *Wayne State University*
Dennis Dunninger,
Michigan State University
Bruce Edwards, *University of Florida*
Garrett Etgen, *University of Houston*
Frederick Gass, *Miami University of Ohio*
Bruce Gilligan, *University of Regina*
Stuart Goldenberg,
California Polytechnic State University

Michael Gregory,
University of North Dakota
Charles Groetsch,
University of Cincinnati
D.W. Hall, *Michigan State University*
Allen Hesse,
Rochester Community College
Matt Kaufmann
David Leeming, *University of Victoria*
Mark Pinsky, *Northwestern University*
Lothar Redlin,
The Pennsylvania State University
Eric Schreiner,
Western Michigan University
Wayne Skrapek,
University of Saskatchewan
William Smith,
University of North Carolina
Richard St. Andre,
Central Michigan University
Steven Willard, *University of Alberta*

Second Edition Reviewers:

Michael Albert,
Carnegie-Mellon University
Jorge Cassio,
Miami-Dade Community College
Jack Ceder,
University of California, Santa Barbara
Seymour Ditor,
University of Western Ontario
Ken Dunn, *Dalhousie University*
John Ellison, *Grove City College*
William Francis,
Michigan Technological University
Gerald Goff, *Oklahoma State University*
Stuart Goldenberg,
California Polytechnic State University

Richard Grassl,
University of New Mexico
Melvin Hausner,
Courant Institute, New York University
Clement Jeske,
University of Wisconsin, Platteville
Jerry Johnson, *Oklahoma State University*
Virgil Kowalik, *Texas A & I University*
Sam Lesseig,
Northeast Missouri State University
Phil Locke, *University of Maine*
Phil McCartney,
Northern Kentucky University
Mary Martin, *Colgate University*

Igor Malyshev, *San Jose State University*   David Ryeburn, *Simon Fraser University*
Richard Nowakowski, *Dalhousie University*   Ricardo Salinas, *San Antonio College*
Vincent Panico, *University of the Pacific*   Stan Ver Nooy, *University of Oregon*
Tom Rishel, *Cornell University*   Jack Weiner, *University of Guelph*

In addition I would like to thank:

Dan Anderson, *University of Iowa,* Dan Drucker, *Wayne State University,* and Barbara Frank, *The Pennsylvania State University,* for their careful working of the solutions to all the problems in the text and their authoring of Volumes 1 and 2 of the **Student Solutions Manual.** Student contributors were Aaron Childs, Loris Corazza, Michael Gmell, Brett Goodwin, and Tony di Silvestro.

Bruce Char, *Oak Ridge National Laboratory, University of Tennessee,* David Hayes, *San Jose State University,* and Vincent Panico, *University of the Pacific,* for their helpful contributions to the appendix material.

Zdislav Kovarik, my colleague at *McMaster University,* for the ideas of Appendix D.

Garrett Etgen, *University of Houston,* for researching, writing, and solving the Applications Plus exercises, and Dan Drucker, *Wayne State University,* David McKay and Saleem Watson, *California State University, Long Beach,* and Lothar Redlin, *The Pennsylvania State University,* for checking the solutions.

Supplement Authors: Richard St. Andre, *Central Michigan University,* for the **Study Guide;** Joan Thomas and *Engineering Press, Inc.,* for their individual contributions to the **Test Item** file; J.S. Devitt, *University of Waterloo,* for his **Maple**® **and Calculus;** K. Heuvers, J. Kuisti, W. Francis, G. Ortner, D. Moak, and D. Lockhart, *Michigan Technological University,* for their **Linear Algebra for Calculus;** J. Douglas Child, *Rollins College,* Chris D'Arcy, *St. George's College,* Robert Kowalczyk and Adam Hausknecht, *Southeastern Massachusetts University,* Phoebe T. Judson, *Trinity University,* Vivian Y. Kraines, *Meredith College,* and Roy E. Myers, *The Pennsylvania State University,* for their individual contributions to **Calculus Laboratories for Brooks/Cole Software Tools.**

Most of these lab authors have also written the software that the labs require. Thanks to Ian Bell, Jon Davis, and Steve Rice, *Queen's University,* for **Calculus-Pad**®; Roy E. Myers, *The Pennsylvania State University,* for **Surface Plotter;** Chris D'Arcy, *St. George's College,* for **Image-Calculus;** J. Douglas Child, *Rollins College,* for **Calculus T/L;** Robert Kowalczyk and Adam Hausknecht, *Southeastern Massachusetts University,* for **TEMATH.**

Finally, I thank Kathi Townes for her production coordination and interior illustration for this text, Lee Hocker for the cover photograph, and the following Brooks/Cole staff: William T. Bokermann, vice president, production; Maureen Allaire, product manager; Margaret Parks, advertising manager; and Nancy Champlin, editorial assistant. Special thanks are due to the mathematics editor, Jeremy Hayhurst.

JAMES STEWART

To the Student

Reading a calculus textbook is different from reading a newspaper or a novel, or even a physics book. Don't be discouraged if you have to read a passage more than once in order to understand it. You should have pencil and paper at hand to make a calculation or sketch a diagram.

Some students start by trying their homework problems and only read the text if they get stuck on an exercise. I suggest that a far better plan is to read and understand a section of the text before attempting the exercises. In particular, you should study the definitions to see the exact meanings of the terms.

Part of the aim of this course is to train you to think logically. Learn to write the solutions of the exercises in a connected step-by-step fashion with explanatory words or symbols—not just a string of disconnected equations or formulas.

The answers to the odd-numbered exercises appear at the back of the book, in Appendix H. There are often several different forms in which to express an answer, so if your answer differs from mine, don't immediately assume that you are wrong. There many be an algebraic or trigonometric identity that connects the answers. For example, if the answer given in the back of the book is $\sqrt{2} - 1$ and you obtain $1/(1 + \sqrt{2})$, then you are right and rationalizing the denominator will show you that the expressions are equivalent.

The symbol ▦ means that a calculator (or computer) is required to do a calculation in an example or exercise. The symbol ● indicates the end of a proof or an example. You will also encounter the symbol ⊘, which warns you against committing an error. I have placed this symbol in the margin in situations where I have observed that a large proportion of my students tend to make the same mistake.

Calculus is an exciting subject; I hope you find it both useful and interesting in its own right.

A Note on Logic

In understanding the theorems it is important to know the meaning of certain logical terms and symbols. If P and Q are mathematical statements, then $P \Rightarrow Q$ is read as "P implies Q" and means the same as "If P is true, then Q is true." The *converse* of a theorem of the form $P \Rightarrow Q$ is the statement $Q \Rightarrow P$. (The converse of a theorem may or may not be true. For example, the converse of the statement "If it rains, then I take my umbrella" is "If I take my umbrella, then it rains.") The symbol $\Leftrightarrow$ indicates that two statements are equivalent. Thus $P \Leftrightarrow Q$ means that both $P \Rightarrow Q$ and $Q \Rightarrow P$. The phrase "if and only if" is also used in this situation. Thus "P is true if and only if Q is true" means the same as $P \Leftrightarrow Q$. The *contrapositive* of a theorem $P \Rightarrow Q$ is the statement that $\sim Q \Rightarrow \sim P$, where $\sim P$ means *not P*. So the contrapositive says "If Q is false, then P is false." Unlike converses, the contrapositive of a theorem is always true.

Methods of Problem Solving

The way to master calculus is to solve problems—lots of problems. The exercise sets at the end of each section begin with straightforward exercises designed to test your mastery of fundamental skills. Later exercises are less straightforward because they involve more complex calculations, or because they make you think harder about what the concepts really mean. The problems at the end of the exercise sets are often quite challenging and may require more than one attempt. The Problems Plus, which can be found after the even-numbered chapters, are also challenging. Often the challenge is to recognize that a problem requires combining methods from two or more different chapters. For instance, the Problems Plus after Chapter 8 don't necessarily relate to Chapter 8 but might require using knowledge from any of the chapters from 1 to 8.

If you have trouble solving any of the more challenging problems, you might find it useful to look at the Principles of Problem Solving that are printed on the front endpapers of this book. For a more thorough discussion of the problem-solving process I recommend that you consult the following books:

Polya, G. *How To Solve It* (2nd ed.). Princeton University Press, 1957.

Polya, G. *Mathematical Discovery.* New York: John Wiley and Sons, 1962.

Wickelgren, W. *How To Solve Problems.* San Francisco: W. H. Freeman, 1974.

Solow, D. *How To Read and Do Proofs.* New York: John Wiley and Sons, 1982.

Here is a problem that involves only precalculus mathematics:

Express the hypotenuse h of a right triangle in terms of its area A and its perimeter P.

After reading the Principles of Problem Solving, try to solve this problem. Then look at the solution on the following page.

EXAMPLE Express the hypotenuse h of a right triangle in terms of its area A and its perimeter P.

Understand the problem **Solution**   Let us first sort out the information by identifying the unknown quantity and the data:

$$\text{Unknown: } h \qquad \text{Given quantities: } A, P$$

Draw a diagram It helps to draw a diagram and we do so in Figure 1.

Connect the given with the unknown   In order to connect the given quantities to the unknown, we introduce two extra var-

Introduce something extra iables a and b, which are the lengths of the other two sides of the triangle. This enables us to express the given condition, which is that the triangle is right-angled, by the Pythagorean Theorem:

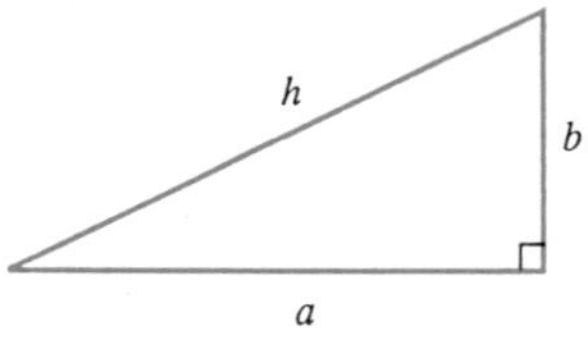

$$h^2 = a^2 + b^2$$

The other connections among the variables come by writing expressions for the area and perimeter:

$$A = \tfrac{1}{2}ab \qquad P = a + b + h$$

Figure 1

Since A and P are given, notice that we now have three equations in the three unknowns a, b, and h:

$$h^2 = a^2 + b^2 \qquad (1)$$
$$A = \tfrac{1}{2}ab \qquad (2)$$
$$P = a + b + h \qquad (3)$$

Although we have the correct number of equations, they are not easy to solve in a straightforward fashion. But if we can use the problem-solving strategy of trying to

Relate to the familiar recognize something familiar, then we can solve these equations by an easier method. Look at the right sides of Equations 1, 2, and 3. Do these expressions remind you of anything familiar? Notice that they contain the ingredients of the familiar formula:

$$(a + b)^2 = a^2 + 2ab + b^2$$

Using this idea, we express $(a + b)^2$ in two ways. From Equations 1 and 2 we have

$$(a + b)^2 = (a^2 + b^2) + 2ab = h^2 + 4A$$

From Equation 3 we have

$$(a + b)^2 = (P - h)^2 = P^2 - 2Ph + h^2$$

Thus
$$h^2 + 4A = P^2 - 2Ph + h^2$$
$$2Ph = P^2 - 4A$$
$$h = \frac{P^2 - 4A}{2P}$$

This is the required expression.

Contents

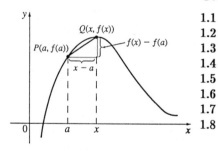

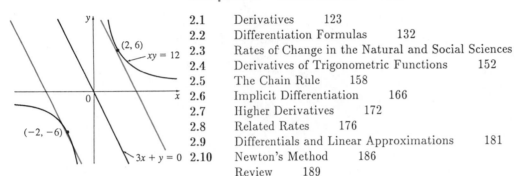

Chapter 2   Derivatives   122

Problems Plus   192

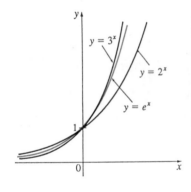

Applications Plus   252

Chapter 4   The Mean Value Theorem and   256
Curve Sketching   256

Problems Plus   312

Chapter 5   Integrals   314

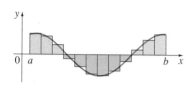

Chapter 6   Applications of Integration   372

Chapter 7   Techniques of Integration   406

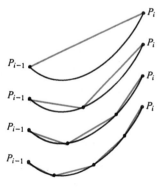

Chapter 8   Further Applications of Integration   470

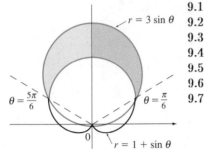
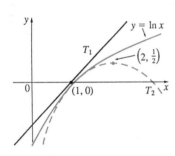
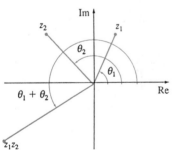

Second Edition

Single Variable Calculus

Early Transcendentals

Review and Preview

In most sciences one generation
tears down what another has built
and what one has established
another undoes. In Mathematics
alone each generation builds a new
story to the old structure.

Hermann Hankel

In the first six sections we review the mathematical topics—algebra, analytic geometry, and functions—that are needed for the study of calculus. Then in the last section we give a preview of some of the main ideas in calculus.

Numbers, Inequalities, and Absolute Values

Calculus is based on the real number system. We start with the **integers:**

$$\ldots, -3, -2, -1, 0, 1, 2, 3, 4, \ldots$$

Then we construct the **rational numbers,** which are ratios of integers. Thus any rational number r can be expressed as

$$r = \frac{m}{n} \qquad \text{where } m \text{ and } n \text{ are integers and } n \neq 0$$

Examples are

$$\frac{1}{2} \qquad -\frac{3}{7} \qquad 46 = \frac{46}{1} \qquad 0.17 = \frac{17}{100}$$

(Recall that division by 0 is always ruled out, so expressions like $\frac{3}{0}$ and $\frac{0}{0}$ are undefined.) There are also real numbers, such as $\sqrt{2}$, that cannot be expressed as a ratio of integers and are therefore called **irrational numbers.** It can be shown, with varying degrees of difficulty, that the following are also irrational numbers:

$$\sqrt{3} \qquad \sqrt{5} \qquad \sqrt[3]{2} \qquad \pi \qquad \sin 1° \qquad \log_{10} 2$$

The set of all real numbers is usually denoted by the symbol R. When we use the word *number* without qualification, we mean "real number."

Every number has a decimal representation. If the number is rational, then the corresponding decimal is repeating. For example,

$$\frac{1}{2} = 0.5000\ldots = 0.5\overline{0} \qquad\qquad \frac{2}{3} = 0.66666\ldots = 0.\overline{6}$$

$$\frac{157}{495} = 0.3171717\ldots = 0.3\overline{17} \qquad\qquad \frac{9}{7} = 1.285714285714\ldots = 1.\overline{285714}$$

3

(The bar indicates that the sequence of digits repeats forever.) On the other hand, if the number is irrational, the decimal is nonrepeating:

$$\sqrt{2} = 1.414213562373095\ldots \qquad \pi = 3.141592653589793\ldots$$

If we stop the decimal expansion of any number at a certain place, we get an approximation to the number. For instance, we can write

$$\pi \approx 3.14159265$$

where the symbol $\approx$ is read "is approximately equal to." The more decimal places we retain, the better the approximation we get.

The real numbers can be represented by points on a line as in Figure 1. The positive direction (to the right) is indicated by an arrow. We choose an arbitrary reference point O, called the **origin**, which corresponds to the real number 0. Given any convenient unit of measurement, each positive number x is represented by the point on the line a distance of x units to the right of the origin, and each negative number $-x$ is represented by the point x units to the left of the origin. Thus every real number is represented by a point on the line, and every point P on the line corresponds to exactly one real number. The number associated with the point P is called the **coordinate** of P and the line is then called a **coordinate line**, or a **real number line**, or simply a **real line**. Often we identify the point with its coordinate and think of a number as being a point on the real line.

Figure 1

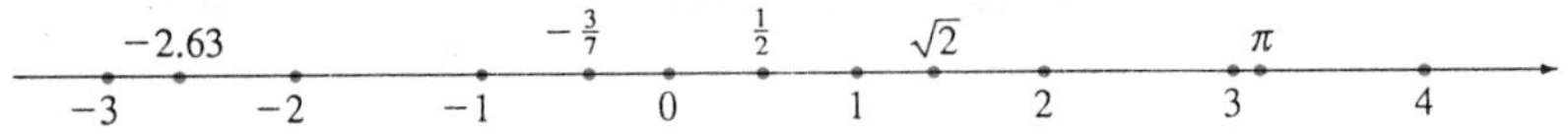

The real numbers are ordered. We say a *is less than* b and write $a < b$ if $b - a$ is a positive number. Geometrically this means that a lies to the left of b on the number line. (Equivalently, we say b *is greater than* a and write $b > a$.) The symbol $a \leq b$ (or $b \geq a$) means that either $a < b$ or $a = b$ and is read "a is less than or equal to b." For instance, the following are true inequalities:

$$7 < 7.4 < 7.5 \qquad -3 > -\pi \qquad \sqrt{2} < 2 \qquad \sqrt{2} \leq 2 \qquad 2 \leq 2$$

In what follows we need to use set notation. A **set** is a collection of objects, and these objects are called the **elements** of the set. If S is a set, the notation $a \in S$ means that a is an element of S, and $a \notin S$ means that a is not an element of S. For example, if Z represents the set of integers, then $-3 \in Z$ but $\pi \notin Z$. If S and T are sets, then their **union** $S \cup T$ is the set consisting of all elements that are in S or T (or in both S and T). The **intersection** of S and T is the set $S \cap T$ consisting of all elements that are in both S and T. In other words, $S \cap T$ is the common part of S and T. The empty set, denoted by $\emptyset$, is the set contains no elements.

Some sets can be described by listing their elements between braces. For instance, the set A consisting of all positive integers less than 7 can be written as

$$A = \{1, 2, 3, 4, 5, 6\}$$

We could also write A in set-builder notation as

$$A = \{x \mid x \text{ is an integer and } 0 < x < 7\}$$

which is read "A is the set of x such that x is an integer and $0 < x < 7$."

Certain sets of real numbers, called **intervals**, occur frequently in calculus and correspond geometrically to line segments. For instance, if $a < b$, the **open interval** from a to b

consists of all numbers between a and b and is denoted by the symbol (a, b). Using set-builder notation, we can write

$$(a,b) = \{x \mid a < x < b\}$$

Notice that the endpoints of the interval—namely, a and b—are excluded. This is indicated by the round brackets ( ) and by the open dots in Figure 2. The **closed interval** from a to b is the set

Figure 2

Open interval (a, b)

$$[a,b] = \{x \mid a \leq x \leq b\}$$

Here the endpoints of the interval are included. This is indicated by the square brackets [ ] and by the solid dots in Figure 3. It is also possible to include only one endpoint in an interval, as shown in Table 1.

Figure 3

Closed interval $[a, b]$

We also need to consider infinite intervals such as

$$(a, \infty) = \{x \mid x > a\}$$

This does not mean that ∞ ("infinity") is a number. The notation (a, ∞) stands for the set of all numbers that are greater than a, so the symbol ∞ simply indicates that the interval extends indefinitely far in the positive direction.

Table 1 lists the nine possible types of intervals. When these intervals are discussed, it is always assumed that $a < b$.

Table of Intervals (1)

Notation	Set Description	Picture
(a, b)	$\{x \mid a < x < b\}$	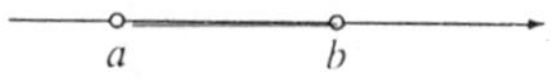
$[a, b]$	$\{x \mid a \leq x \leq b\}$	
$[a, b)$	$\{x \mid a \leq x < b\}$	
$(a, b]$	$\{x \mid a < x \leq b\}$	
(a, ∞)	$\{x \mid x > a\}$	
$[a, \infty)$	$\{x \mid x \geq a\}$	
$(-\infty, b)$	$\{x \mid x < b\}$	
$(-\infty, b]$	$\{x \mid x \leq b\}$	
$(-\infty, \infty)$	R (set of all real numbers)	

When working with inequalities, note the following rules:

Rules for Inequalities (2)

1. If $a < b$, then $a + c < b + c$.
2. If $a < b$ and $c < d$, then $a + c < b + d$.
3. If $a < b$ and $c > 0$, then $ac < bc$.
4. If $a < b$ and $c < 0$, then $ac > bc$.
5. If $0 < a < b$, then $1/a > 1/b$.

Rule 1 says that we can add any number to both sides of an inequality, and Rule 2 says that two inequalities can be added. However, we have to be careful with multiplication. Rule 3 says that we can multiply both sides of an inequality by a *positive* number, but Rule 4 says that *if we multiply both sides of an inequality by a negative number, then we reverse the direction of the inequality.* For example, if we take the inequality $3 < 5$ and multiply by 2, we get $6 < 10$, but if we multiply by -2, we get $-6 > -10$. Finally, Rule 5 says that if we take reciprocals, then we reverse the direction of an inequality (provided the numbers are positive).

EXAMPLE 1   Solve the inequality $1 + x < 7x + 5$.

Solution   The given inequality is satisfied by some values of x but not by others. To *solve* an inequality means to determine the set of numbers x for which the inequality is true. This is called the *solution set*.

First we subtract 1 from each side of the inequality (using Rule 1 with $c = -1$):

$$x < 7x + 4$$

Then we subtract $7x$ from both sides (Rule 1 with $c = -7x$):

$$-6x < 4$$

Now we divide both sides by -6 (Rule 4 with $c = -\frac{1}{6}$):

$$x > -\frac{4}{6} = -\frac{2}{3}$$

These steps can all be reversed, so the solution set consists of all numbers greater than $-\frac{2}{3}$. In other words, the solution of the inequality is the interval $\left(-\frac{2}{3}, \infty\right)$.   ●

EXAMPLE 2   Solve the inequalities $4 \leq 3x - 2 < 13$.

Solution   Here the solution set consists of all values of x that satisfy both inequalities. Using the rules given in (2), we see that the following inequalities are equivalent:

$$4 \leq 3x - 2 < 13$$
$$6 \leq 3x < 15 \qquad \text{(add 2)}$$
$$2 \leq x < 5 \qquad \text{(divide by 3)}$$

Therefore the solution set is $[2, 5)$.   ●

EXAMPLE 3   Solve $2x + 1 \leq 4x - 3 \leq x + 7$.

Solution   This time we first solve the inequalities separately:

$$2x + 1 \leq 4x - 3 \qquad\qquad 4x - 3 \leq x + 7$$
$$4 \leq 2x \qquad\qquad\qquad 3x \leq 10$$
$$2 \leq x \qquad\qquad\qquad x \leq \frac{10}{3}$$

Since x must satisfy both inequalities, we have

$$2 \le x \le \tfrac{10}{3}$$

Thus the solution set is the closed interval $\left[2, \tfrac{10}{3}\right]$.

EXAMPLE 4   Solve the inequality $x^2 - 5x + 6 \le 0$.

Solution   First we factor the left side:

$$(x - 2)(x - 3) \le 0$$

We know that the corresponding equation $(x - 2)(x - 3) = 0$ has the solutions 2 and 3. The numbers 2 and 3 divide the real line into three intervals:

$$(-\infty, 2) \qquad (2, 3) \qquad (3, \infty)$$

On each of these intervals we determine the signs of the factors. For instance,

$$x \in (-\infty, 2) \quad \Rightarrow \quad x < 2 \quad \Rightarrow \quad x - 2 < 0$$

Then we record these signs in the following chart:

Interval	$x - 2$	$x - 3$	$(x - 2)(x - 3)$
$x < 2$	$-$	$-$	$+$
$2 < x < 3$	$+$	$-$	$-$
$x > 3$	$+$	$+$	$+$

Another method for obtaining the information in the chart is to use *test values*. For instance, if we use the test value $x = 1$ for the interval $(-\infty, 2)$, then substitution in $x^2 - 5x + 6$ gives

$$1^2 - 5(1) + 6 = 2$$

The polynomial $x^2 - 5x + 6$ does not change sign inside any of the three intervals, so we conclude that it is positive on $(-\infty, 2)$.

Then we read from the chart that $(x - 2)(x - 3)$ is negative when $2 < x < 3$. Thus the solution of the inequality $(x - 2)(x - 3) \le 0$ is

$$\{x \mid 2 \le x \le 3\} = [2, 3]$$

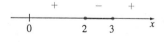

Notice that we have included the endpoints 2 and 3 because we seek values of x such that the product is either negative or zero. The solution is illustrated in Figure 4.

Figure 4

EXAMPLE 5   Solve $\dfrac{1 + x}{1 - x} > 1$.

Solution 1   One method is to take all nonzero terms to the left side and use a common denominator:

$$\frac{1 + x}{1 - x} > 1$$

$$\frac{1+x}{1-x} - 1 > 0$$

$$\frac{1+x-1+x}{1-x} > 0$$

$$\frac{2x}{1-x} > 0$$

The numerator is zero when $x = 0$ and the denominator is zero when $x = 1$. As before, we can set up a chart to determine the sign on each of the intervals $(-\infty, 0)$, $(0, 1)$, and $(1, \infty)$.

Interval	$2x$	$1 - x$	$\dfrac{2x}{1-x}$
$x < 0$	$-$	$+$	$-$
$0 < x < 1$	$+$	$+$	$+$
$x > 1$	$+$	$-$	$-$

From the chart we see that the solution set is $\{x \mid 0 < x < 1\} = (0, 1)$.

Solution 2   Another method is to multiply both sides by $1 - x$, but in view of Rules 3 and 4 we must consider separately the cases where $1 - x$ is positive and negative.

CASE I   If $1 - x > 0$, that is, $x < 1$, then multiplying the given inequality by $1 - x$ gives

$$1 + x > 1 - x$$

which becomes $2x > 0$, that is, $x > 0$. So we have

$$0 < x < 1$$

CASE II   If $1 - x < 0$, that is, $x > 1$, then multiplying the given inequality by $1 - x$ gives

$$1 + x < 1 - x$$

which becomes $2x < 0$, that is, $x < 0$. But the conditions $x > 1$ and $x < 0$ are incompatible, so there is no solution in Case II.

Therefore the solution set is the open interval $(0, 1)$.

EXAMPLE 6   Solve $x^3 + 3x^2 > 4x$.

Solution   First we take all nonzero terms to one side of the inequality sign and factor the resulting expression. (See Appendix A for a review of factoring.)

$$x^3 + 3x^2 - 4x > 0 \qquad \text{or} \qquad x(x-1)(x+4) > 0$$

As in Example 4 we solve the corresponding equation $x(x-1)(x+4) = 0$ and use the solutions $x = -4$, $x = 0$, and $x = 1$ to divide the real line into four intervals $(-\infty, -4)$, $(-4, 0)$, $(0, 1)$, and $(1, \infty)$. On each interval the product keeps a constant sign as shown in the following chart:

Interval	x	$x-1$	$x+4$	$x(x-1)(x+4)$
$x < -4$	$-$	$-$	$-$	$-$
$-4 < x < 0$	$-$	$-$	$+$	$+$
$0 < x < 1$	$+$	$-$	$+$	$-$
$x > 1$	$+$	$+$	$+$	$+$

Then we read from the chart that the solution set is

$$\{x \mid -4 < x < 0 \text{ or } x > 1\} = (-4,0) \cup (1,\infty)$$

Figure 5   The solution is illustrated in Figure 5.

Absolute Value

The **absolute value** of a number a, denoted by $|a|$, is the distance from a to 0 on the real number line. Distances are always positive or 0, so we have

$$|a| \geq 0 \qquad \text{for every number } a$$

For example,

$$|3| = 3 \qquad |-3| = 3 \qquad |0| = 0 \qquad |\sqrt{2}-1| = \sqrt{2}-1 \qquad |3-\pi| = \pi - 3$$

In general, we have

(3)

$$\begin{aligned} |a| &= a && \text{if } a \geq 0 \\ |a| &= -a && \text{if } a < 0 \end{aligned}$$

(Remember that if a is negative, then $-a$ is positive.)

EXAMPLE 7   Express $|3x-2|$ without using the absolute value symbol.

Solution

$$|3x-2| = \begin{cases} 3x-2 & \text{if } 3x-2 \geq 0 \\ -(3x-2) & \text{if } 3x-2 < 0 \end{cases}$$

$$= \begin{cases} 3x-2 & \text{if } x \geq \frac{2}{3} \\ 2-3x & \text{if } x < \frac{2}{3} \end{cases}$$

Let us recall that the symbol $\sqrt{}$ means "the positive square root of." Thus $\sqrt{r} = s$ means $s^2 = r$ and $s \geq 0$. Therefore the equation $\sqrt{a^2} = a$ is not always true. It is true only when $a \geq 0$. If $a < 0$, then $-a > 0$, so we have $\sqrt{a^2} = -a$. In view of (3), we then have the equation

(4)

$$\sqrt{a^2} = |a|$$

which is true for all values of a.

Hints for the proofs of the following properties are given in the exercises.

Properties of Absolute Values (5)

> Suppose a and b are any real numbers and n is an integer. Then
>
> 1. $|ab| = |a|\,|b|$
>
> 2. $\left|\dfrac{a}{b}\right| = \dfrac{|a|}{|b|} \qquad b \neq 0$
>
> 3. $|a^n| = |a|^n$

For solving equations or inequalities involving absolute values, it is often very helpful to use the following statements:

(6)

> Suppose $a > 0$. Then
>
> 4. $|x| = a$ if and only if $x = \pm a$
>
> 5. $|x| < a$ if and only if $-a < x < a$
>
> 6. $|x| > a$ if and only if $x > a$ or $x < -a$

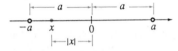

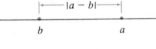

Figure 6

For instance, the inequality $|x| < a$ says that the distance from x to the origin is less than a, and you can see from Figure 6 that this is true if and only if x lies between $-a$ and a.

If a and b are any real numbers, then the distance between a and b is the absolute value of the difference, namely, $|a - b|$, which is also equal to $|b - a|$ (see Figure 7).

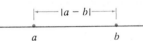

Figure 7
Length of a line segment $= |a - b|$

EXAMPLE 8   Solve $|2x - 5| = 3$.

Solution   By Property 4 of (6), $|2x - 5| = 3$ is equivalent to

$$2x - 5 = 3 \qquad \text{or} \qquad 2x - 5 = -3$$

So $2x = 8$ or $2x = 2$. Thus $x = 4$ or $x = 1$.

EXAMPLE 9   Solve $|x - 5| < 2$.

Solution 1   By Property 5 of (6), $|x - 5| < 2$ is equivalent to

$$-2 < x - 5 < 2$$

Therefore, adding 5 to each side, we have

$$3 < x < 7$$

and the solution set is the open interval $(3, 7)$.

Solution 2   Geometrically the solution set consists of all numbers x whose distance from 5 is less than 2. From Figure 8 we see that this is the interval $(3, 7)$.

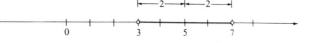

Figure 8

EXAMPLE 10 Solve $|3x + 2| \geq 4$.

Solution By Properties 4 and 6 of (6), $|3x + 2| \geq 4$ is equivalent to

$$3x + 2 \geq 4 \qquad \text{or} \qquad 3x + 2 \leq -4$$

In the first case $3x \geq 2$, which gives $x \geq \frac{2}{3}$. In the second case $3x \leq -6$, which gives $x \leq -2$. So the solution set is

$$\left\{ x \mid x \leq -2 \text{ or } x \geq \frac{2}{3} \right\} = (-\infty, -2] \cup \left[\frac{2}{3}, \infty \right)$$

Another important property of absolute value, called the Triangle Inequality, is used frequently not only in calculus but throughout mathematics in general.

The Triangle Inequality (7)

> If a and b are any real numbers, then
>
> $$|a + b| \leq |a| + |b|$$

Observe that if the numbers a and b are both positive or both negative, then the two sides in the Triangle Inequality are actually equal. But if a and b have opposite signs, the left side involves a subtraction and the right side does not. This makes the Triangle Inequality seem reasonable, but we can prove it as follows.

Notice that

$$-|a| \leq a \leq |a|$$

is always true because a equals either $|a|$ or $-|a|$. The corresponding statement for b is

$$-|b| \leq b \leq |b|$$

Adding these inequalities, we get

$$-(|a| + |b|) \leq a + b \leq |a| + |b|$$

If we now apply Properties 4 and 5 (with x replaced by $a + b$ and a by $|a| + |b|$), we obtain

$$|a + b| \leq |a| + |b|$$

which is what we wanted to show.

EXAMPLE 11 If $|x - 4| < 0.1$ and $|y - 7| < 0.2$, use the Triangle Inequality to estimate $|(x + y) - 11|$.

Solution In order to use the given information, we use the Triangle Inequality with $a = x - 4$ and $b = y - 7$:

$$|(x + y) - 11| = |(x - 4) + (y - 7)|$$

$$\leq |x - 4| + |y - 7|$$

$$< 0.1 + 0.2 = 0.3$$

Thus $|(x + y) - 11| < 0.3$ ●

SECTION 1 **Exercises**

In Exercises 1–12 rewrite each expression without using the absolute value symbol.

1. $|5 - 23|$

2. $|5| - |-23|$

3. $|-\pi|$

4. $|\pi - 2|$

5. $|\sqrt{5} - 5|$

6. $||-2| - |-3||$

7. $|x - 2|$ if $x < 2$

8. $|x - 2|$ if $x > 2$

9. $|x + 1|$

10. $|2x - 1|$

11. $|x^2 + 1|$

12. $|1 - 2x^2|$

In Exercises 13–46 solve the given inequalities in terms of intervals and illustrate the solution sets on the real number line.

13. $2x + 7 > 3$

14. $3x - 11 < 4$

15. $1 - x \leq 2$

16. $4 - 3x \geq 6$

17. $2x + 1 \leq 5x - 8$

18. $1 + 5x > 5 - 3x$

19. $-1 < 2x - 5 < 7$

20. $1 < 3x + 4 \leq 16$

21. $0 \leq 1 - x < 1$

22. $-5 \leq 3 - 2x \leq 9$

23. $4x < 2x + 1 \leq 3x + 2$

24. $2x - 3 < x + 4 < 3x - 2$

25. $1 - x \geq 3 - 2x \geq x - 6$

26. $x > 1 - x \geq 3 + 2x$

27. $(x - 1)(x - 2) > 0$

28. $(2x + 3)(x - 1) \geq 0$

29. $2x^2 + x \leq 1$

30. $x^2 < 2x + 8$

31. $x^2 + x + 1 > 0$

32. $x^2 + x > 1$

33. $x^2 < 3$

34. $x^2 \geq 5$

35. $x^3 - x^2 \leq 0$

36. $(x + 1)(x - 2)(x + 3) \geq 0$

37. $x^3 > x$

38. $x^3 + 3x < 4x^2$

39. $\frac{1}{x} < 4$

40. $-3 < \frac{1}{x} \leq 1$

41. $\frac{4}{x} < x$

42. $\frac{x}{x + 1} > 3$

43. $\frac{2x + 1}{x - 5} < 3$

44. $\frac{2 + x}{3 - x} \leq 1$

45. $\frac{x^2 - 1}{x^2 + 1} \geq 0$

46. $\frac{x^2 - 2x}{x^2 - 2} > 0$

47. The relationship between the Celsius and Fahrenheit temperature scales is given by $C = \frac{5}{9}(F - 32)$, where C is the temperature in degrees Celsius and F is the temperature in degrees Fahrenheit. What interval on the Celsius scale corresponds to the temperature range $50 \leq F \leq 95$?

48. Use the relationship between C and F given in Exercise 47 to find the interval on the Fahrenheit scale corresponding to the temperature range $20 \leq C \leq 30$.

49. As dry air moves upward, it expands and in so doing cools at a rate of about 1°C for each 100-m rise, up to about 12 km.
    (a) If the ground temperature is 20°C, write a formula for the temperature at height h.
    (b) What range of temperature can be expected if a plane takes off and reaches a maximum height of 5 km?

50. If a ball is thrown upward from the top of a building 128 ft high with an initial velocity of 16 ft/s, then the height h above the ground t seconds later will be

$$h = 128 + 16t - 16t^2$$

During what time interval will the ball be at least 32 ft above the ground?

In Exercises 51–54 solve the given equation for x.

51. $|2x| = 3$

52. $|3x + 5| = 1$

53. $|x + 3| = |2x + 1|$

54. $\left|\frac{2x - 1}{x + 1}\right| = 3$

In Exercises 55–68 solve the given inequalities.

55. $|x| < 3$

56. $|x| \geq 3$

57. $|x - 4| < 1$

58. $|x - 6| < 0.1$

59. $|x + 5| \geq 2$

60. $|x + 1| \geq 3$

61. $|2x - 3| \leq 0.4$

62. $|5x - 2| < 6$

63. $1 \leq |x| \leq 4$

64. $0 < |x - 5| < \frac{1}{2}$

65. $|x| > |x - 1|$

66. $|2x - 5| \leq |x + 4|$

67. $\left|\frac{x}{2 + x}\right| < 1$

68. $\left|\frac{2 - 3x}{1 + 2x}\right| \leq 4$

In Exercises 69 and 70 solve for x assuming a, b, and c are positive constants.

69. $a(bx - c) \geq bc$

70. $a \leq bx + c < 2a$

In Exercises 71 and 72 solve for x assuming a, b, and c are negative constants.

71. $ax + b < c$

72. $\frac{ax + b}{c} \leq b$

73. Suppose that $|x - 2| < 0.01$ and $|y - 3| < 0.04$. Use the Triangle Inequality to show that $|(x + y) - 5| < 0.05$.

74. Show that if $|x + 3| < \frac{1}{2}$, then $|4x + 13| < 3$.

75. Show that if $a < b$, then $a < \frac{a + b}{2} < b$.

76. Use Rule 3 to prove Rule 5 of (2).

77. Prove that $|ab| = |a||b|$. (*Hint:* Use Equation 4.)

78. Prove that $\left|\frac{a}{b}\right| = \frac{|a|}{|b|}$.

79. Show that if $0 < a < b$, then $a^2 < b^2$.

80. Prove that $|x - y| \geq |x| - |y|$. (*Hint:* Use the Triangle Inequality with $a = x - y$ and $b = y$.)

81. Show that the sum, difference, and product of rational numbers are rational numbers.

82. (a) Is the sum of two irrational numbers always an irrational number?

(b) Is the product of two irrational numbers always an irrational number?

SECTION 2

Coordinate Geometry and Lines

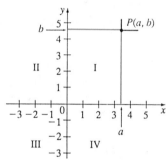

Figure 9

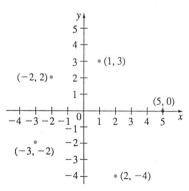

Figure 10

Just as the points on a line can be identified with real numbers by assigning them coordinates, as described in Section 1, so the points in a plane can be identified with ordered pairs of real numbers. We start by drawing two perpendicular coordinate lines that intersect at the origin O on each line. Usually one line is horizontal with positive direction to the right and is called the *x*-axis; the other line is vertical with positive direction upward and is called the *y*-axis.

Any point P in the plane can be located by a unique ordered pair of numbers as follows. Draw lines through P perpendicular to the *x*- and *y*-axes. These lines intersect the axes in points with coordinates a and b as shown in Figure 9. Then the point P is assigned the ordered pair (a, b). The first number a is called the **x-coordinate** (or **abscissa**) of P; the second number b is called the **y-coordinate** (or **ordinate**) of P. We say that P is the point with coordinates (a, b), and we denote the point by the symbol $P(a, b)$. Several points are labeled with their coordinates in Figure 10.

By reversing the preceding process we can start with an ordered pair (a, b) and arrive at the corresponding point P. Often we identify the point P with the ordered pair (a, b) and refer to "the point (a, b)." [Although the notation used for an open interval (a, b) is the same as the notation used for a point (a, b), you will be able to tell from the context which meaning is intended.]

This coordinate system is called the **rectangular coordinate system** or the **Cartesian coordinate system** in honor of the French mathematician René Descartes (1596–1650), even though another Frenchman, Pierre Fermat (1601–1665), invented the principles of analytic geometry at about the same time as Descartes. The plane supplied with this coordinate system is called the **coordinate plane** or the **Cartesian plane** and is denoted by R^2.

The *x*- and *y*-axes are called the **coordinate axes** and divide the Cartesian plane into four quadrants, which are labeled I, II, III, and IV in Figure 9. Notice that the first quadrant consists of those points whose *x*- and *y*-coordinates are both positive.

EXAMPLE 1   Describe and sketch the regions given by the following sets:
(a) $\{(x,y) \mid x \geq 0\}$ (b) $\{(x,y) \mid y = 1\}$ (c) $\{(x,y) \mid |y| < 1\}$

Solution
(a)   The points whose x-coordinates are 0 or positive lie on the y-axis or to the right of it [see Figure 11(a)].

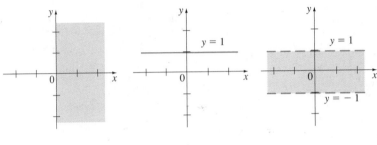

Figure 11   (a) $x \geq 0$ (b) $y = 1$ (c) $|y| < 1$

(b)   The set of all points with y-coordinate 1 is a horizontal line one unit above the x-axis [see Figure 11(b)].
(c)   Recall from Section 1 that

$$|y| < 1 \qquad \text{if and only if} \qquad -1 < y < 1$$

The given region consists of those points in the plane whose y-coordinates lie between -1 and 1. Thus the region consists of all points that lie between (but not on) the horizontal lines $y = 1$ and $y = -1$. [These lines are shown as broken lines in Figure 11(c) to indicate that the points on these lines do not lie in the set.]   ●

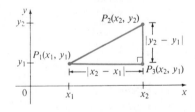

Figure 12

Recall from Section 1 that the distance between points a and b on a number line is $|a - b| = |b - a|$. Thus the distance between points $P_1(x_1, y_1)$ and $P_3(x_2, y_1)$ on a horizontal line must be $|x_2 - x_1|$ and the distance between $P_2(x_2, y_2)$ and $P_3(x_2, y_1)$ on a vertical line must be $|y_2 - y_1|$ (see Figure 12).
To find the distance $|P_1 P_2|$ between any two points $P_1(x_1, y_1)$ and $P_2(x_2, y_2)$, we note that triangle $P_1 P_2 P_3$ in Figure 12 is a right triangle, and so by the Pythagorean Theorem we have

$$|P_1 P_2| = \sqrt{|P_1 P_3|^2 + |P_2 P_3|^2} = \sqrt{|x_2 - x_1|^2 + |y_2 - y_1|^2}$$
$$= \sqrt{(x_2 - x_1)^2 + (y_2 - y_1)^2}$$

Distance Formula (8)

> The distance between the points $P_1(x_1, y_1)$ and $P_2(x_2, y_2)$ is
> $$|P_1 P_2| = \sqrt{(x_2 - x_1)^2 + (y_2 - y_1)^2}$$

EXAMPLE 2   The distance between $(1, -2)$ and $(5, 3)$ is

$$\sqrt{(5 - 1)^2 + [3 - (-2)]^2} = \sqrt{4^2 + 5^2} = \sqrt{41}$$ ●

Lines

We want to find an equation of a given line L; such an equation is satisfied by the coordinates of the points on L and by no other points. To find the equation of L we use its *slope*, which is a measure of the steepness of the line.

Definition (9)

The **slope** of a nonvertical line that passes through the points $P_1(x_1, y_1)$ and $P_2(x_2, y_2)$ is

$$m = \frac{y_2 - y_1}{x_2 - x_1}$$

The slope of a vertical line is not defined.

Thus the slope of a line is the ratio of the change in y to the change in x. From the similar triangles in Figure 13 we see that the slope is independent of which two points are chosen on the line:

$$\frac{y_2 - y_1}{x_2 - x_1} = \frac{y_2' - y_1'}{x_2' - x_1'}$$

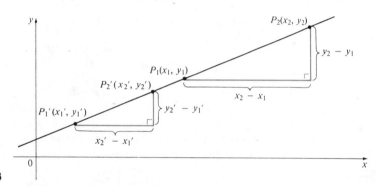

Figure 13

Figure 14 shows several lines labeled with their slopes. Notice that lines with positive slope slant upward to the right, whereas lines with negative slope slant downward to the right. Notice also that the steepest lines are the ones for which the absolute value of the slope is largest, and a horizontal line has slope 0.

Now let us find the equation of the line that passes through a given point $P_1(x_1, y_1)$ and has slope m. A point $P(x, y)$ with $x \neq x_1$ lies on this line if and only if the slope of the line through P_1 and P is equal to m; that is,

$$\frac{y - y_1}{x - x_1} = m$$

This equation can be rewritten in the form

$$y - y_1 = m(x - x_1)$$

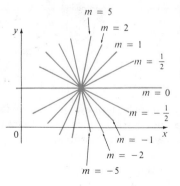

Figure 14

and we observe that this equation is also satisfied when $x = x_1$ and $y = y_1$. Therefore it is the equation of the given line.

Point-Slope Form of the Equation of a Line (10)

An equation of the line passing through the point $P_1(x_1, y_1)$ and having slope m is

$$y - y_1 = m(x - x_1)$$

EXAMPLE 3   Find an equation of the line through $(1, -7)$ with slope $-\frac{1}{2}$.

Solution   Using (10) with $m = -\frac{1}{2}$, $x_1 = 1$, and $y_1 = -7$, we obtain an equation of the line as

$$y + 7 = -\tfrac{1}{2}(x - 1)$$

which we can rewrite as

$$2y + 14 = -x + 1 \qquad \text{or} \qquad x + 2y + 13 = 0$$

     ●

EXAMPLE 4   Find an equation of the line through the points $(-1, 2)$ and $(3, -4)$.

Solution   By Definition 9 the slope of the line is

$$m = \frac{-4 - 2}{3 - (-1)} = -\frac{3}{2}$$

Using the point-slope form with $x_1 = -1$ and $y_1 = 2$, we obtain

$$y - 2 = -\tfrac{3}{2}(x + 1)$$

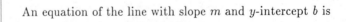

which simplifies to $\qquad\qquad\qquad 3x + 2y = 1$

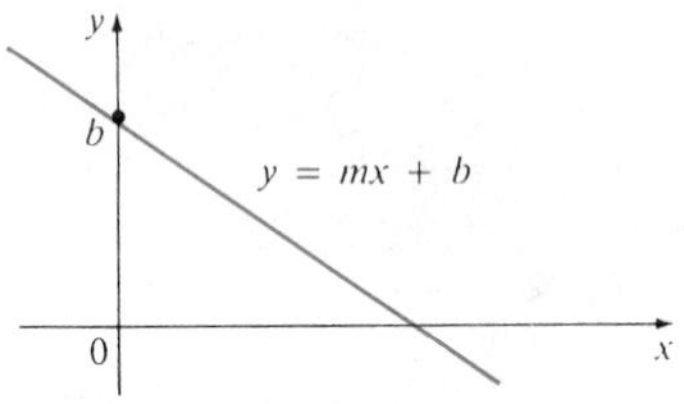

     ●

Figure 15

     Suppose a nonvertical line has slope m and y-intercept b (see Figure 15). This means it intersects the y-axis at the point $(0, b)$, so the point-slope form of the equation of the line, with $x_1 = 0$ and $y_1 = b$, becomes

$$y - b = m(x - 0)$$

This simplifies as follows:

Slope-Intercept Form of the Equation of a Line (11)

An equation of the line with slope m and y-intercept b is

$$y = mx + b$$

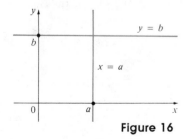

Figure 16

     In particular, if a line is horizontal, its slope is $m = 0$, so its equation is $y = b$, where b is the y-intercept (see Figure 16). A vertical line does not have a slope, but we can write its equation as $x = a$, where a is the x-intercept, because the x-coordinate of every point on the line is a.

     Observe that the equation of every line can be written in the form

(12)

$$Ax + By + C = 0$$

because a vertical line has the equation $x = a$ or $x - a = 0$ ($A = 1$, $B = 0$, $C = -a$) and a nonvertical line has the equation $y = mx + b$ or $-mx + y - b = 0$ ($A = -m$, $B = 1$, $C = -b$). Conversely, if we start with a general first-degree equation, that is, an equation of the form (12) where A, B, and C are constants and A and B are not both 0, then we can show that it is the equation of a line. If $B = 0$, it becomes $Ax + C = 0$ or $x = -C/A$, which represents a vertical line with x-intercept $-C/A$. If $B \neq 0$, the equation can be rewritten by solving for y:

$$y = -\frac{A}{B}x - \frac{C}{B}$$

and we recognize this as being the slope-intercept form of the equation of a line ($m = -A/B$, $b = -C/B$). Therefore an equation of the form (12) is called a **linear equation** or the **general equation of a line**. For brevity, we often refer to "the line $Ax + By + C = 0$" instead of "the line whose equation is $Ax + By + C = 0$."

EXAMPLE 5   Sketch the graph of the equation $3x - 5y = 15$.

Solution   Since the equation is linear, its graph is a line. To draw the graph it suffices to find two points on the line. It is easiest to find the intercepts. Substituting $y = 0$ (the equation of the x-axis) in the given equation, we get $3x = 15$, so $x = 5$ is the x-intercept. Substituting $x = 0$ in the equation, we see that the y-intercept is -3. This allows us to sketch the graph as in Figure 17.

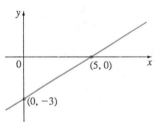

Figure 17

EXAMPLE 6   Graph the inequality $x + 2y > 5$.

Solution   We are asked to sketch the graph of the set $\{(x, y) \mid x + 2y > 5\}$ and we do so by solving the inequality for y:

$$x + 2y > 5$$
$$2y > -x + 5$$
$$y > -\tfrac{1}{2}x + \tfrac{5}{2}$$

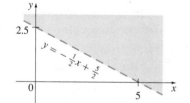

Figure 18

Compare this inequality with the equation $y = -\frac{1}{2}x + \frac{5}{2}$, which represents a line with slope $-\frac{1}{2}$ and y-intercept $\frac{5}{2}$. We see that the given graph consists of points whose y-coordinates are *larger* than those on the line $y = -\frac{1}{2}x + \frac{5}{2}$. Thus the graph is the region that lies above the line, as illustrated in Figure 18.

Parallel and Perpendicular Lines

Slopes can be used to show that lines are parallel or perpendicular. The following facts are proved, for instance, in *Mathematics for Calculus* by Stewart, Redlin, and Watson (Brooks/Cole Publishing Co., Pacific Grove, CA, 1989). Another proof of part (b) is outlined in Exercise 65.

(13)

> (1)   Two nonvertical lines are parallel if and only if they have the same slope.
>
> (2)   Two lines with slopes m_1 and m_2 are perpendicular if and only if $m_1 m_2 = -1$; that is, their slopes are negative reciprocals:
> $$m_2 = -\frac{1}{m_1}$$

EXAMPLE 7   Find an equation of the line through the point $(5, 2)$ that is parallel to the line $4x + 6y + 5 = 0$.

Solution   The given line can be written in the form

$$y = -\tfrac{2}{3}x - \tfrac{5}{6}$$

which is in slope-intercept form with $m = -\tfrac{2}{3}$. Parallel lines have the same slope, so the required line has slope $-\tfrac{2}{3}$ and its equation in point-slope form is

$$y - 2 = -\tfrac{2}{3}(x - 5)$$

This simplifies to $2x + 3y = 16$.   ●

EXAMPLE 8   Show that the lines $2x + 3y = 1$ and $6x - 4y - 1 = 0$ are perpendicular.

Solution   The equations can be written as

$$y = -\tfrac{2}{3}x + \tfrac{1}{3} \qquad \text{and} \qquad y = \tfrac{3}{2}x - \tfrac{1}{4}$$

from which we see that the slopes are

$$m_1 = -\tfrac{2}{3} \qquad \text{and} \qquad m_2 = \tfrac{3}{2}$$

Since $m_1 m_2 = -1$, the lines are perpendicular.   ●

EXAMPLE 9
(a)   As dry air moves upward, it expands and cools. If the ground temperature is 20°C and the temperature at a height of 1 km is 10°C, express the temperature T (in °C) in terms of the height h (in kilometers), assuming the expression is linear.
(b)   Draw the graph of the linear equation. What does the slope represent?
(c)   What is the temperature at a height of 2.5 km?

Solution
(a)   Because we are assuming a linear relationship between h and T, the equation must be of the form

$$T = mh + b$$

where m and b are constants. When $h = 0$, we are given that $T = 20$, so

$$20 = m(0) + b \qquad \text{or} \qquad b = 20$$

Thus we have

$$T = mh + 20$$

When $h = 1$, we have $T = 10$ and so

$$10 = m(1) + 20$$

$$m = 10 - 20 = -10$$

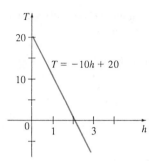

Figure 19

The required expression is

$$T = -10h + 20$$

(b)   The graph is sketched in Figure 19. The slope is $m = -10°/\text{km}$ and represents the rate of change of temperature with respect to height.

(c)   At a height of $h = 2.5$ km, the temperature is

$$T = -10(2.5) + 20 = -25 + 20 = -5°C$$

●

SECTION 2 **Exercises**

In Exercises 1–6 find the distance between the given points.

1. $(1,1)$, $(4,5)$ **2.** $(1,-3)$, $(5,7)$

3. $(6,-2)$, $(-1,3)$ **4.** $(1,-6)$, $(-1,-3)$

5. $(2,5)$, $(4,-7)$ **6.** (a,b), (b,a)

In Exercises 7–10 find the slope of the line through P and Q.

7. $P(1,5)$, $Q(4,11)$ **8.** $P(-1,6)$, $Q(4,-3)$

9. $P(-3,3)$, $Q(-1,-6)$ **10.** $P(-1,-4)$, $Q(6,0)$

11.   Show that the triangle with vertices $A(0,2)$, $B(-3,-1)$, and $C(-4,3)$ is isosceles.

12.   (a)   Show that the triangle with vertices $A(6,-7)$, $B(11,-3)$, and $C(2,-2)$ is a right triangle using the converse of the Pythagorean Theorem.
    (b)   Use slopes to show that ABC is a right triangle.
    (c)   Find the area of the triangle.

13.   Show that the points $(-2,9)$, $(4,6)$, $(1,0)$, and $(-5,3)$ are the vertices of a square.

14.   (a)   Show that the points $A(-1,3)$, $B(3,11)$, and $C(5,15)$ are collinear by showing that $|AB| + |BC| = |AC|$.
    (b)   Use slopes to show that A, B, and C are collinear.

15.   Show that $A(1,1)$, $B(7,4)$, $C(5,10)$, and $D(-1,7)$ are vertices of a parallelogram.

16.   Show that $A(1,1)$, $B(11,3)$, $C(10,8)$, and $D(0,6)$ are vertices of a rectangle.

In Exercises 17–20 sketch the graph of the given equation.

17. $x = 3$ **18.** $y = -2$

19. $xy = 0$ **20.** $|y| = 1$

In Exercises 21–36 find an equation of the line that satisfies the given conditions.

21.   Through $(2,-3)$, slope 6

22.   Through $(-1,4)$, slope -3

23.   Through $(1,7)$, slope $\frac{2}{3}$

24.   Through $(-3,-5)$, slope $-\frac{7}{2}$

25.   Through $(2,1)$ and $(1,6)$

26.   Through $(-1,-2)$ and $(4,3)$

27.   Slope 3, y-intercept -2

28.   Slope $\frac{2}{5}$, y-intercept 4

29. x-intercept 1, y-intercept -3

30. x-intercept -8, y-intercept 6

31.   Through $(4,5)$, parallel to the x-axis

32.   Through $(4,5)$, parallel to the y-axis

33.   Through $(1,-6)$, parallel to the line $x + 2y = 6$

34. y-intercept 6, parallel to the line $2x + 3y + 4 = 0$

35.   Through $(-1,-2)$, perpendicular to the line $2x + 5y + 8 = 0$

36.   Through $(\frac{1}{2}, -\frac{2}{3})$, perpendicular to the line $4x - 8y = 1$

In Exercises 37–42 find the slope and y-intercept of the line and draw its graph.

37. $x + 3y = 0$ **38.** $2x - 5y = 0$

39. $y = -2$ **40.** $2x - 3y + 6 = 0$

41. $3x - 4y = 12$ **42.** $4x + 5y = 10$

In Exercises 43–52 sketch the given region in the xy-plane.

43. $\{(x,y) \mid x < 0\}$ 44. $\{(x,y) \mid y > 0\}$

45. $\{(x,y) \mid xy < 0\}$

46. $\{(x,y) \mid x \geq 1 \text{ and } y < 3\}$

47. $\{(x,y) \mid |x| \leq 2\}$

48. $\{(x,y) \mid |x| < 3 \text{ and } |y| < 2\}$

49. $\{(x,y) \mid 0 \leq y \leq 4 \text{ and } x \leq 2\}$

50. $\{(x,y) \mid y > 2x - 1\}$

51. $\{(x,y) \mid 1 + x \leq y \leq 1 - 2x\}$

52. $\left\{(x,y) \mid -x < y < \dfrac{x+3}{2}\right\}$

53. Find a point on the *y*-axis that is equidistant from $(5, -5)$ and $(1, 1)$.

54. Show that the midpoint of the line segment from $P_1(x_1, y_1)$ to $P_1(x_2, y_2)$ is
$$\left(\frac{x_1 + x_2}{2}, \frac{y_1 + y_2}{2}\right)$$

55. Find the midpoint of the line segment joining the points
     (a) $(1, 3)$ and $(7, 15)$
     (b) $(-1, 6)$ and $(8, -12)$

56. Find the lengths of the medians of the triangle with vertices $A(1, 0)$, $B(3, 6)$ and $C(8, 2)$. (A median is a line segment from a vertex to the midpoint of the opposite side.)

57. Show that the lines $2x - y = 4$ and $6x - 2y = 10$ are not parallel and find their point of intersection.

58. Show that the lines $3x - 5y + 19 = 0$ and $10x + 6y - 50 = 0$ are perpendicular and find their point of intersection.

59. Find the equation of the perpendicular bisector of the line segment joining the points $A(1, 4)$ and $B(7, -2)$.

60. (a)   Find equations for the sides of the triangle with vertices $P(1, 0)$, $Q(3, 4)$, and $R(-1, 6)$.
     (b)   Find equations for the medians of this triangle. Where do they intersect?

61. (a)   Show that if the *x*- and *y*-intercepts of a line are nonzero numbers a and b, then the equation of the line can be put in the form
$$\frac{x}{a} + \frac{y}{b} = 1$$

This equation is called the **two-intercept form** of an equation of a line.

     (b)   Use part (a) to find an equation of the line whose *x*-intercept is 6 and whose *y*-intercept is -8.

62. The monthly cost of driving a car depends on the number of miles driven. Lynn found that in May it cost her \$380 to drive 480 mi and in June it cost her \$460 to drive 800 mi.
     (a)   Express the monthly cost C in terms of the distance driven d, assuming that a linear relationship gives a suitable model.
     (b)   Use part (a) to predict the cost of driving 1500 mi per month.
     (c)   Draw the graph of the linear equation. What does the slope of the line represent?
     (d)   What does the *y*-intercept of the graph represent?
     (e)   Why is a linear relationship a suitable model in this situation?

63. Jason and Debbie leave Detroit at 2:00 P.M. and drive at a constant speed west along I-90. They pass Ann Arbor, 40 mi from Detroit, at 2:50 P.M.
     (a)   Express the distance traveled in terms of the time elapsed.
     (b)   Draw the graph of the equation in part (a).
     (c)   What is the slope of this line? What does it represent?

64. (a)   If L is a line in the plane and ϕ is the angle formed by the line and the *x*-axis as shown in the figure, show that the slope m of the line is given by
$$m = \tan \phi$$

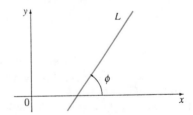

     (b)   Let L_1 and L_2 be two nonparallel lines in the plane with slopes m_1 and m_2, respectively. Let θ be the acute angle formed by the two lines as shown in the figure. Show that
$$\tan \theta = \frac{m_2 - m_1}{1 + m_1 m_2}$$

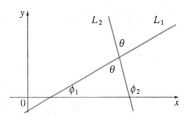

(c)   Find the point of intersection of the lines $3x - y = 3$ and $2x + 3y = 13$.

(d)   Use part (b) to find the acute angle between the lines in part (c) correct to the nearest degree.

65.   Use part (b) of Exercise 64 to prove that if L_1 and L_2 are perpendicular, then $m_1 m_2 = -1$.

SECTION 3

Graphs of Second-Degree Equations

In Section 2 we saw that a first-degree, or linear, equation $Ax + By + C = 0$ represents a line. In this section we discuss second-degree equations such as

$$x^2 + y^2 = 1 \qquad y = x^2 + 1 \qquad \frac{x^2}{9} + \frac{y^2}{4} = 1 \qquad x^2 - y^2 = 1$$

which represent a circle, a parabola, an ellipse, and a hyperbola, respectively.

The graph of such an equation in x and y is the set of all points (x, y) that satisfy the equation; it gives a visual representation of the equation. Conversely, given a curve in the xy-plane, we may have to find an equation that represents it, that is, an equation satisfied by the coordinates of the points on the curve and by no other points. This is the other half of the basic principle of analytic geometry as formulated by Descartes and Fermat. The idea is that if a geometric curve can be represented by an algebraic equation, then the rules of algebra can be used to analyze the geometric problem.

Circles

As an example of this type of problem, let us find the equation of a circle with radius r and center (h, k). By definition, the circle is the set of all points $P(x, y)$ whose distance from the center $C(h, k)$ is r (see Figure 20). Thus P is on the circle if and only if $|PC| = r$. From the distance formula (8) we have

$$\sqrt{(x - h)^2 + (y - k)^2} = r$$

or equivalently, squaring both sides, we get

$$(x - h)^2 + (y - k)^2 = r^2$$

Figure 20

This is the desired equation.

Equation of a Circle (14)

The equation of a circle with center (h, k) and radius r is

$$(x - h)^2 + (y - k)^2 = r^2$$

In particular, if the center is the origin $(0, 0)$, the equation is

$$x^2 + y^2 = r^2$$

EXAMPLE 1   Find the equation of the circle with radius 3 and center $(2, -5)$.

Solution   From Equation 14 with $r = 3$, $h = 2$, and $k = -5$, we obtain

$$(x - 2)^2 + (y + 5)^2 = 9$$

EXAMPLE 2   Sketch the graph of the equation $x^2 + y^2 + 2x - 6y + 7 = 0$ by first showing that it represents a circle and then finding its center and radius.

Solution   We first group the x-terms and y-terms as follows:

$$(x^2 + 2x) + (y^2 - 6y) = -7$$

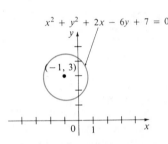

$x^2 + y^2 + 2x - 6y + 7 = 0$

Then we complete the square within each grouping, adding the appropriate constants to both sides of the equation:

$$(x^2 + 2x + 1) + (y^2 - 6y + 9) = -7 + 1 + 9$$

or $$(x + 1)^2 + (y - 3)^2 = 3$$

(See Appendix A for a review of completing the square.) Comparing this equation with the standard equation of a circle (14), we see that $h = -1$, $k = 3$, and $r = \sqrt{3}$, so the given equation represents a circle with center $(-1, 3)$ and radius $\sqrt{3}$. It is sketched in Figure 21.

Figure 21

Parabolas

The geometric properties of parabolas are reviewed in Section 9.6. Here we regard a parabola as a graph of an equation of the form $y = ax^2 + bx + c$.

EXAMPLE 3   Draw the graph of the parabola $y = x^2$.

Solution   We set up a table of values, plot points, and join them by a smooth curve to obtain the graph in Figure 22.

x	$y = x^2$
0	0
$\pm\frac{1}{2}$	$\frac{1}{4}$
±1	1
±2	4
±3	9

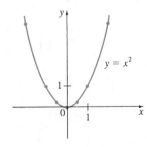

Figure 22

Figure 23 shows the graphs of several parabolas with equations of the form $y = ax^2$ for various values of the number a. In each case the *vertex*, the point where the parabola

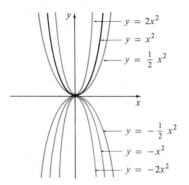

Figure 23

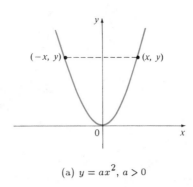

(a) $y = ax^2$, $a > 0$

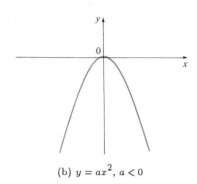

(b) $y = ax^2$, $a < 0$

Figure 24

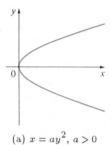

(a) $x = ay^2$, $a > 0$

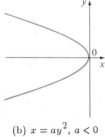

(b) $x = ay^2$, $a < 0$

Figure 25

changes direction, is the origin. We see that the parabola $y = ax^2$ opens upward if $a > 0$ and downward if $a < 0$ (as in Figure 24).

Notice that if (x, y) satisfies the equation $y = ax^2$, then so does $(-x, y)$. This corresponds to the geometric fact that if the right half of the graph is reflected in the y-axis, then the left half of the graph is obtained. We say that the graph is **symmetric with respect to the y-axis**.

> The graph of an equation is symmetric with respect to the y-axis if the equation is unchanged when x is replaced by $-x$.

If we interchange x and y in the equation $y = ax^2$, the result is $x = ay^2$, which also represents a parabola. (Interchanging x and y amounts to reflecting about the diagonal line $y = x$.) The parabola $x = ay^2$ opens to the right if $a > 0$ and to the left if $a < 0$ (see Figure 25). This time the parabola is symmetric with respect to the x-axis because if (x, y) satisfies $x = ay^2$, then so does $(x, -y)$.

> The graph of an equation is symmetric with respect to the x-axis if the equation is unchanged when y is replaced by $-y$.

EXAMPLE 4   Sketch the region bounded by the parabola $x = y^2$ and the line $y = x - 2$.

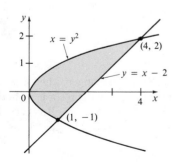

Figure 26

Solution   First we find the points of intersection by solving the two equations. Substituting $x = y + 2$ into the equation $x = y^2$, we get $y + 2 = y^2$, which gives

$$0 = y^2 - y - 2 = (y - 2)(y + 1)$$

so $y = 2$ or -1. Thus the points of intersection are $(4, 2)$ and $(1, -1)$ and we draw the line $y = x - 2$ passing through these points. We then sketch the parabola $x = y^2$ by referring to Figure 25(a) and having the parabola pass through $(4, 2)$ and $(1, -1)$. The region bounded by $x = y^2$ and $y = x - 2$ means the finite region whose boundaries are these curves. It is sketched in Figure 26.   ●

Ellipses

The curve with equation

(15)

$$\frac{x^2}{a^2} + \frac{y^2}{b^2} = 1$$

where a and b are positive numbers, is called an **ellipse** in standard position. (Geometric properties of ellipses are discussed in Section 9.6.) Observe that Equation 15 is unchanged if x is replaced by $-x$ or y is replaced by $-y$, so the ellipse is symmetric with respect to both axes. As a further aid to sketching the ellipse we find its intercepts.

> The **x-intercepts** of a graph are the x-coordinates of the points where the graph intersects the x-axis. They are found by setting $y = 0$ in the equation of the graph. The **y-intercepts** are the y-coordinates of the points where the graph intersects the y-axis. They are found by setting $x = 0$ in its equation.

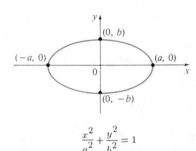

$$\frac{x^2}{a^2} + \frac{y^2}{b^2} = 1$$

Figure 27

If we set $y = 0$ in Equation 15 we get $x^2 = a^2$ and so the x-intercepts are $\pm a$. Setting $x = 0$, we get $y^2 = b^2$, so the y-intercepts are $\pm b$. Using this information together with symmetry, we sketch the ellipse in Figure 27. If $a = b$, then the ellipse is a circle with radius a.

EXAMPLE 5   Sketch the graph of $9x^2 + 16y^2 = 144$.

Solution   We divide both sides of the equation by 144:

$$\frac{x^2}{16} + \frac{y^2}{9} = 1$$

The equation is now in the standard form for an ellipse (15), so we have $a^2 = 16$, $b^2 = 9$, $a = 4$, and $b = 3$. The x-intercepts are ± 4; the y-intercepts are ± 3. The graph is sketched in Figure 28. •

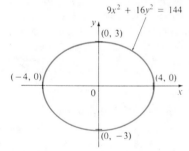

Figure 28

Hyperbolas

The curve with equation

(16)

$$\frac{x^2}{a^2} - \frac{y^2}{b^2} = 1$$

is called a **hyperbola** in standard position. Again, Equation 16 is unchanged when x is replaced by $-x$ or y is replaced by $-y$, so the hyperbola is symmetric with respect to both axes. To find the x-intercepts we set $y = 0$ and obtain $x^2 = a^2$ and $x = \pm a$. However, if we put $x = 0$ in Equation 16, we get $y^2 = -b^2$, which is impossible, so there is no y-intercept. In fact, from Equation 16 we obtain

$$\frac{x^2}{a^2} = 1 + \frac{y^2}{b^2} \geq 1$$

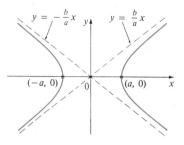

Figure 29

The hyperbola $\dfrac{x^2}{a^2} - \dfrac{y^2}{b^2} = 1$

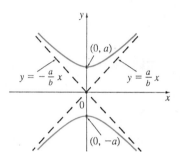

Figure 30

The hyperbola $\dfrac{y^2}{a^2} - \dfrac{x^2}{b^2} = 1$

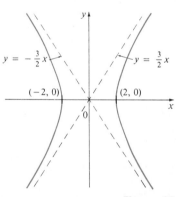

Figure 31

The hyperbola $9x^2 - 4y^2 = 36$

Figure 32

Equilateral hyperbolas

which shows that $x^2 \geq a^2$ and so $|x| = \sqrt{x^2} \geq a$. Therefore we have $x \geq a$ or $x \leq -a$. This means that the hyperbola consists of two parts, called its branches. It is sketched in Figure 29.

In drawing a hyperbola it is useful to draw first its *asymptotes,* which are the lines $y = (b/a)x$ and $y = -(b/a)x$ shown in Figure 29. Both branches of the hyperbola approach the asymptotes; that is, they come arbitrarily close to the asymptotes. (This involves the idea of a limit, which is discussed in Chapter 1. See Exercise 57 in Section 4.5.)

By interchanging the roles of x and y we get an equation of the form

$$\frac{y^2}{a^2} - \frac{x^2}{b^2} = 1$$

which also represents a hyperbola and is sketched in Figure 30.

EXAMPLE 6   Sketch the curve $9x^2 - 4y^2 = 36$.

Solution   Dividing both sides by 36, we obtain

$$\frac{x^2}{4} - \frac{y^2}{9} = 1$$

which is in the standard form of the equation of a hyperbola (Equation 16). Since $a^2 = 4$, the x-intercepts are ± 2. Since $b^2 = 9$, we have $b = 3$ and the asymptotes are $y = \pm(\frac{3}{2})x$. The hyperbola is sketched in Figure 31.   •

If $b = a$, a hyperbola has the equation $x^2 - y^2 = a^2$ (or $y^2 - x^2 = a^2$) and is called an *equilateral hyperbola* [see Figure 32(a)]. Its asymptotes are $y = \pm x$, which are perpendicular. If an equilateral hyperbola is rotated by 45°, the asymptotes become the x- and y-axes, and it can be shown that the new equation of the hyperbola is $xy = k$, where k is a constant [see Figure 32(b)].

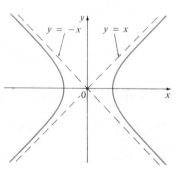

(a) $x^2 - y^2 = a^2$

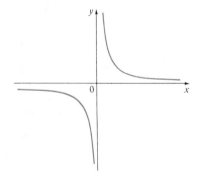

(b) $xy = k \ (k > 0)$

Shifted Conics

Recall that the equation of a circle with center the origin and radius r is $x^2 + y^2 = r^2$, but if the center is the point (h, k), then the equation of the circle becomes

$$(x - h)^2 + (y - k)^2 = r^2$$

Similarly, if we take the ellipse with equation

(17)
$$\frac{x^2}{a^2} + \frac{y^2}{b^2} = 1$$

and translate it (shift it) so that its center is the point (h, k), then its equation becomes

(18)
$$\frac{(x-h)^2}{a^2} + \frac{(y-k)^2}{b^2} = 1$$

(See Figure 33.)

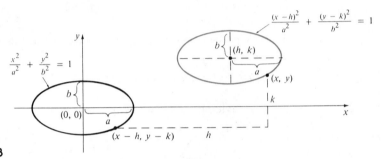

Figure 33

Notice that in shifting the ellipse, we replaced x by $x - h$ and y by $y - k$ in Equation 17 to obtain Equation 18. We use the same procedure to shift the parabola $y = ax^2$ so that its vertex (the origin) becomes the point (h, k) as in Figure 34. Replacing x by $x - h$ and y by $y - k$, we see that the new equation is

$$y - k = a(x - h)^2 \qquad \text{or} \qquad y = a(x - h)^2 + k$$

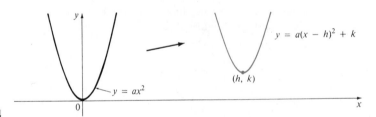

Figure 34

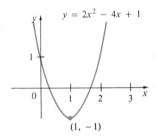

Figure 35

EXAMPLE 7   Sketch the graph of the equation $y = 2x^2 - 4x + 1$.

Solution   First we complete the square:

$$y = 2(x^2 - 2x) + 1 = 2(x - 1)^2 - 1$$

In this form we see that the equation represents the parabola obtained by shifting $y = 2x^2$ so that its vertex is at the point $(1, -1)$. The graph is sketched in Figure 35.   ●

EXAMPLE 8   Sketch the curve $x = 1 - y^2$.

Solution   This time we start with the parabola $x = -y^2$ (as in Figure 25 with $a = -1$) and shift one unit to the right to get the graph of $x = 1 - y^2$ (see Figure 36).

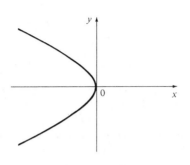
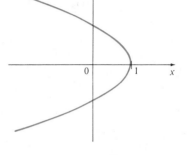

Figure 36   (a) $x = -y^2$ (b) $x = 1 - y^2$ ●

SECTION 3 **Exercises**

In Exercises 1–4 find an equation of a circle that satisfies the given conditions.

1.   Center $(3, -1)$; radius 5

2.   Center $(-2, -8)$; radius 10

3.   Center at the origin; passes through $(4, 7)$

4.   Center $(-1, 5)$; passes through $(-4, -6)$

In Exercises 5–10 show that the given equation represents a circle and find the center and radius.

5. $x^2 + y^2 - 4x + 10y + 13 = 0$

6. $x^2 + y^2 + 6y + 2 = 0$

7. $x^2 + y^2 + x = 0$

8. $16x^2 + 16y^2 + 8x + 32y + 1 = 0$

9. $2x^2 + 2y^2 - x + y = 1$

10.   Under what condition on the coefficients a, b, and c does the equation $x^2 + y^2 + ax + by + c = 0$ represent a circle? When that condition is satisfied, find the center and radius of the circle.

In Exercises 11–32 identify the type of curve and sketch the graph. Do not plot points. Just use the standard graphs given in Figures 24, 25, 27, 29, and 30 and shift if necessary.

11. $y = -x^2$

12. $y^2 - x^2 = 1$

13. $x^2 + 4y^2 = 16$

14. $x = -2y^2$

15. $16x^2 - 25y^2 = 400$

16. $25x^2 + 4y^2 = 100$

17. $4x^2 + y^2 = 1$

18. $y = x^2 + 2$

19. $x = y^2 - 1$

20. $9x^2 - 25y^2 = 225$

21. $9y^2 - x^2 = 9$

22. $2x^2 + 5y^2 = 10$

23. $xy = 4$

24. $y = x^2 + 2x$

25. $9(x - 1)^2 + 4(y - 2)^2 = 36$

26. $16x^2 + 9y^2 - 36y = 108$

27. $y = x^2 - 6x + 13$

28. $x^2 - y^2 - 4x + 3 = 0$

29. $x = 4 - y^2$

30. $y^2 - 2x + 6y + 5 = 0$

31. $x^2 + 4y^2 - 6x + 5 = 0$

32. $4x^2 + 9y^2 - 16x + 54y + 61 = 0$

In Exercises 33 and 34 sketch the region bounded by the given curves.

33. $y = 3x$, $y = x^2$

34. $y = 4 - x^2$, $x - 2y = 2$

35.   Find the equation of the parabola with vertex $(1, -1)$ that passes through the points $(-1, 3)$ and $(3, 3)$.

36.   Find the equation of the ellipse with center at the origin that passes through the points $(1, -10\sqrt{2}/3)$ and $(-2, 5\sqrt{5}/3)$.

In Exercises 37–40 sketch the graph of the given set.

37. $\{(x, y) \mid x^2 + y^2 \le 1\}$

38. $\{(x, y) \mid x^2 + y^2 > 4\}$

39. $\{(x, y) \mid y \ge x^2 - 1\}$

40. $\{(x, y) \mid x^2 + 4y^2 \le 4\}$

Functions and Their Graphs

The area A of a circle depends on the radius r of the circle. The rule that connects r and A is given by the equation $A = \pi r^2$. With each positive number r there is associated one value of A, and we say that A is a *function* of r.

The number N of bacteria in a culture depends on the time t. If the culture starts with 5000 bacteria and the population doubles every hour, then after t hours the number of bacteria will be $N = (5000)2^t$. This is the rule that connects t and N. For each value of t there is a corresponding value of N, and we say that N is a function of t.

The cost C of mailing a first-class letter depends on the mass m of the letter. Although there is no single neat formula that connects m and C, the post office has a rule for determining C when m is known.

In each of these examples there is a rule whereby, given a number (r, t, or m), another number (A, N, or C) is assigned. In each case we say that the second number is a function of the first number.

Definition (19)

> A **function** f is a rule that assigns to each element x in a set A exactly one element, called $f(x)$, in a set B.

We usually consider functions for which the sets A and B are sets of real numbers. The set A in Definition 19 is called the **domain** of the function. The symbol $f(x)$ is read "f of x" and is called the **value of f at x**, or the **image of x under f**. The **range** of f is the set of all possible values of the number $f(x)$ as x varies throughout the domain, that is, $\{f(x) \mid x \in A\}$.

It is helpful to think of a function as a **machine** (see Figure 37). If x is in the domain of the function f, then when x enters the machine, it is accepted as an input and the machine produces an output $f(x)$ according to the rule of the function. Thus we can think of the domain as the set of all possible inputs and the range as the set of all possible outputs.

The preprogrammed functions in a calculator are good examples of a function as a machine. For example, the $\sqrt{x}$ key on your calculator is such a function. First you input x into the display. Then you press the key labeled $\sqrt{x}$. If $x < 0$, then x is not in the domain of this function; that is, x is not an acceptable input, and the calculator will indicate an error. If $x \geq 0$, then an approximation to $\sqrt{x}$ will appear in the display. Thus the $\sqrt{x}$ key on your calculator is not quite the same as the exact mathematical function f defined by $f(x) = \sqrt{x}$. (See Appendix D.)

Another way to picture a function is by an **arrow diagram** as in Figure 38. Each arrow connects an element of A to an element of B. The arrow indicates that $f(x)$ is associated with x, $f(a)$ is associated with a, and so on.

x (input) → f → $f(x)$ (output)

Figure 37

Machine diagram for a function f

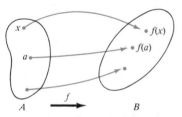

Figure 38

Arrow diagram for f

EXAMPLE 1   The squaring function assigns to each real number x its square x^2. It is defined by the equation

$$f(x) = x^2$$

The values of f are found by substituting for x in this equation. For example,

$$f(3) = 3^2 = 9 \qquad f(-2) = (-2)^2 = 4$$

The domain of f is the set R of all real numbers. The range of f consists of all values of $f(x)$, that is, all numbers of the form x^2. But $x^2 \geq 0$ for all numbers x, and any non-negative number c is a square since $c = (\sqrt{c})^2 = f(\sqrt{c})$. Therefore the range of f is $\{y \mid y \geq 0\} = [0, \infty)$. The machine diagram and arrow diagram for this function are shown in Figure 39.

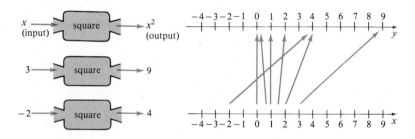

Figure 39   (a) Machine diagram   (b) Arrow diagram

EXAMPLE 2   If we define a function g by

$$g(x) = x^2 \qquad 0 \leq x \leq 3$$

then the domain of g is given as the closed interval $[0, 3]$. This is different from the function f given in Example 1 because in considering g we are restricting our attention to those values of x between 0 and 3. The range of g is

$$\{x^2 \mid 0 \leq x \leq 3\} = \{y \mid 0 \leq y \leq 9\} = [0, 9]$$

In Examples 1 and 2 the domain of the function was given explicitly. But *if a function is given by a formula and the domain is not stated explicitly, the convention is that the domain is the set of all numbers for which the formula makes sense and defines a real number.*

We should distinguish between a function f and the number $f(x)$, which is the value of f at x. Nonetheless, it is common to abbreviate an expression such as

"the function f defined by $f(x) = x^2 + x$"

to   "the function $f(x) = x^2 + x$"

EXAMPLE 3   Find the domain of the function $f(x) = \dfrac{1}{x^2 - x}$.

Solution   Since

$$f(x) = \frac{1}{x^2 - x} = \frac{1}{x(x - 1)}$$

and division by 0 is not allowed, we see that $f(x)$ is not defined when $x = 0$ or $x = 1$. Thus the domain of f is

$$\{x \mid x \neq 0,\ x \neq 1\}$$

which could also be written in interval notation as

$$(-\infty, 0) \cup (0, 1) \cup (1, \infty)$$

EXAMPLE 4   Find the domain of $h(x) = \sqrt{2 - x - x^2}$.

Solution   Since the square root of a negative number is not defined (as a real number), the domain of h consists of all values of x such that

$$2 - x - x^2 \geq 0$$

We solve this inequality using the methods of Section 1. Since $2 - x - x^2$ can be factored as $(2 + x)(1 - x)$, the product will change sign when $x = -2$ or $x = 1$ as indicated in the following chart:

Interval	$2 + x$	$1 - x$	$(2 + x)(1 - x)$
$x \leq -2$	$-$	$+$	$-$
$-2 \leq x \leq 1$	$+$	$+$	$+$
$x \geq 1$	$+$	$-$	$-$

Therefore the domain of h is

$$\{x \mid -2 \leq x \leq 1\} = [-2, 1]$$

●

A symbol that represents an arbitrary number in the *domain* of a function f is called an **independent variable**. A symbol that represents a number in the *range* of f is called a **dependent variable**. For example, the squaring function of Example 1 could be defined by saying that each number x is assigned the number y by the rule $y = x^2$. Then x is the independent variable and y is the dependent variable. In the example on bacteria at the beginning of this section, t is the independent variable, N is the dependent variable, and they are connected by the equation $N = (5000)2^t$. In general, a function describes how one quantity depends on another. For instance, we say that population is a function of time and that pressure is a function of temperature.

EXAMPLE 5   A rectangular storage container with an open top has a volume of $10\,\mathrm{m}^3$. The length of its base is twice its width. Material for the base costs \$10 per square meter; material for the sides costs \$6 per square meter. Express the cost of materials as a function of the width of the base.

Solution   We draw a diagram as in Figure 40 and introduce notation by letting w and $2w$ be the width and length of the base and h be the height.

The area of the base is $(2w)w = 2w^2$, so the cost, in dollars, of the material for the base is $10(2w^2)$. Two of the sides have area wh and the other two have area $2wh$, so the cost of the material for the sides is $6[2(wh) + 2(2wh)]$. The total cost is therefore

$$C = 10(2w^2) + 6[2(wh) + 2(2wh)] = 20w^2 + 36wh$$

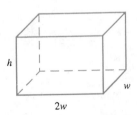

Figure 40   To express C as a function of w alone, we need to eliminate h and we do so by using the fact that the volume is $10\,\mathrm{m}^3$. Thus

$$w(2w)h = 10$$

which gives

$$h = \frac{10}{2w^2} = \frac{5}{w^2}$$

Substituting this into the expression for C, we have

$$C = 20w^2 + 36w\left(\frac{5}{w^2}\right) = 20w^2 + \frac{180}{w}$$

Therefore the equation

$$C(w) = 20w^2 + \frac{180}{w} \qquad w > 0$$

expresses C as a function of w.

In setting up applied functions as in Example 5, it may be useful to review the principles of problem solving as discussed on the inside front cover of this book, particularly Step 1: Understanding the Problem.

Graphs of Functions

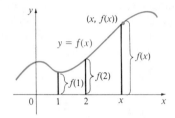

We have seen how to picture functions using machine diagrams and arrow diagrams. A third method for visualizing a function is its graph. If f is a function with domain A, its **graph** is the set of ordered pairs

$$\{(x, f(x)) \mid x \in A\}$$

Figure 41

In other words, the graph of f consists of all points (x, y) in the coordinate plane such that $y = f(x)$ and x is in the domain of f. Thus the graph of a function f is the same as the graph of the equation $y = f(x)$ as discussed in Section 3.

The graph of a function f gives us a useful picture of the behavior or "life history" of a function. Since the y-coordinate of any point (x, y) on the graph is $y = f(x)$, we can read the value of $f(x)$ from the graph as being the height of the graph above the point x (see Figure 41). The graph of f also allows us to picture the domain and range of f on the x-axis and y-axis as in Figure 42.

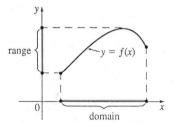

Figure 42

EXAMPLE 6   Sketch the graph of the function $f(x) = 2x - 1$.

Solution   The equation of the graph is $y = 2x - 1$, and we recognize this as being the equation of a line with slope 2 and y-intercept -1. This enables us to sketch the graph of f in Figure 43.

EXAMPLE 7   Sketch the graph of $f(x) = x^2$.

Solution   The equation of the graph is $y = x^2$, which we recognize from Section 3 as a parabola. The graph is shown in Figure 44.

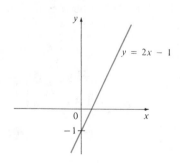

Figure 43

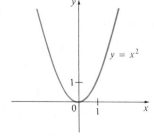

Figure 44

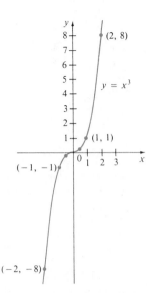

Figure 45

EXAMPLE 8   Sketch the graph of $f(x) = x^3$.

Solution   We list some functional values and the corresponding points on the graph in the following table:

x	0	$\frac{1}{2}$	1	2	$-\frac{1}{2}$	-1	-2
$f(x) = x^3$	0	$\frac{1}{8}$	1	8	$-\frac{1}{8}$	-1	-8
(x, x^3)	$(0,0)$	$\left(\frac{1}{2}, \frac{1}{8}\right)$	$(1,1)$	$(2,8)$	$\left(-\frac{1}{2}, -\frac{1}{8}\right)$	$(-1,-1)$	$(-2,-8)$

Then we plot these points and join them by a smooth curve to obtain the graph shown in Figure 45. At the present stage we cannot be absolutely certain that the graph is exactly as shown, but we will later develop calculus techniques that will confirm the picture.   ●

EXAMPLE 9   Sketch the graph of $f(x) = \sqrt{x+2}$.

Solution 1   We first observe that $\sqrt{x+2}$ is defined when $x + 2 \geq 0$, so the domain of f is $\{x \mid x \geq -2\} = [-2, \infty)$. Then we plot the points given by the following table and use them to produce the sketch in Figure 46.

x	-2	-1	0	1	2	3
$f(x) = \sqrt{x+2}$	0	1	$\sqrt{2}$	$\sqrt{3}$	2	$\sqrt{5}$

Figure 46

Solution 2   The equation of the graph of f is $y = \sqrt{x+2}$ and we note that this implies that $y \geq 0$. Squaring, we obtain $y^2 = x + 2$, or $x = y^2 - 2$, which represents a shifted parabola as discussed in Section 3. [See Figure 47(a).] But since $y \geq 0$, the graph of $y = \sqrt{x+2}$ is just the top half of this parabola [Figure 47(b)]. [The bottom half of the parabola would be the graph of the function $y = -\sqrt{x+2}$ shown in Figure 47(c).]

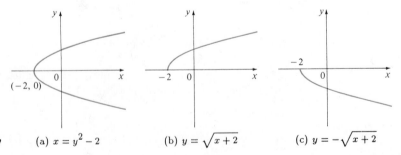

Figure 47   (a) $x = y^2 - 2$ (b) $y = \sqrt{x+2}$ (c) $y = -\sqrt{x+2}$ ●

The graph of a function is a curve in the xy-plane. But the question arises: Which curves in the xy-plane are graphs of functions? This is answered by the following test.

The Vertical Line Test (20)   A curve in the plane is the graph of a function if and only if no vertical line intersects the curve more than once.

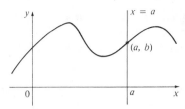

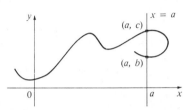

Figure 48

The reason for the truth of the Vertical Line Test can be seen in Figure 48. If each vertical line $x = a$ intersects a curve only once, at (a, b), then exactly one functional value is defined by $f(a) = b$. But if a line $x = a$ intersects the curve twice, at (a, b) and (a, c), then the curve cannot represent a function because a function cannot assign two different values to a.

For example, the parabola $x = y^2 - 2$ shown in Figure 47(a) is not the graph of a function of x because, as you can see, there are vertical lines that intersect the parabola twice. The parabola, however, does represent *two* functions of x; the upper and lower halves of the parabola are the graphs of the functions $f(x) = \sqrt{x + 2}$ and $g(x) = -\sqrt{x + 2}$ [Figure 47(b) and (c)]. We observe that if we reverse the roles of x and y, then the equation $x = h(y) = y^2 - 2$ does define x as a function of y (with y as the independent variable and x as the dependent variable) and the parabola now appears as the graph of the function h.

The functions we have looked at so far have been defined by means of simple formulas. But many functions are not given by such formulas. Here are some examples: the cost of mailing a first-class letter as a function of its mass, the population of New York City as a function of time, and the cost of a taxi ride as a function of distance. The following examples give further illustrations.

EXAMPLE 10   A function f is defined by

$$f(x) = \begin{cases} 1 - x & \text{if } x \leq 1 \\ x^2 & \text{if } x > 1 \end{cases}$$

Evaluate $f(0)$, $f(1)$, and $f(2)$ and sketch the graph.

Solution   Remember that a function is a rule. For this particular function the rule is the following: First look at the value of the input x. If it happens that $x \leq 1$, then the value of $f(x)$ is $1 - x$. On the other hand, if $x > 1$, then the value of $f(x)$ is x^2.

Since $0 \leq 1$, we have $f(0) = 1 - 0 = 1$.

Since $1 \leq 1$, we have $f(1) = 1 - 1 = 0$.

Since $2 > 1$, we have $f(2) = 2^2 = 4$.

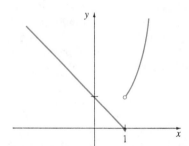

Figure 49   How do we draw the graph of f? We observe that if $x \leq 1$, then $f(x) = 1 - x$, so the part of the graph of f that lies to the left of the vertical line $x = 1$ must coincide with the line $y = 1 - x$, which has slope -1 and y-intercept 1. If $x > 1$, then $f(x) = x^2$, so the part of the graph of f that lies to the right of the line $x = 1$ must coincide with the graph of $y = x^2$, which is a parabola. This enables us to sketch the graph in Figure 49. The solid dot indicates that the point is included on the graph; the open dot indicates that the point is excluded from the graph.

EXAMPLE 11   Sketch the graph of the absolute value function $f(x) = |x|$.

Solution   Recall from Equation 3 that $|x| = \begin{cases} x & \text{if } x \geq 0 \\ -x & \text{if } x < 0 \end{cases}$

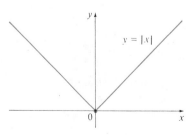

Figure 50

Using the same method as in Example 10, we see that the graph of f coincides with the line $y = x$ to the right of the y-axis and coincides with the line $y = -x$ to the left of the y-axis (see Figure 50).

EXAMPLE 12   The cost of a long-distance daytime phone call from Toronto to New York City is 69 cents for the first minute and 58 cents for each additional minute (or part or a minute). Draw the graph of the cost C (in dollars) of the phone call as a function of the time t (in minutes).

Solution   Let $C(t)$ be the cost for t minutes. Since $t > 0$, the domain of the function is $(0, \infty)$. From the given information we have

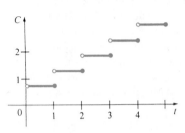

Figure 51

$$C(t) = 0.69 \qquad\qquad \text{if } 0 < t \leq 1$$

$$C(t) = 0.69 + 0.58 = 1.27 \qquad \text{if } 1 < t \leq 2$$

$$C(t) = 0.69 + 2(0.58) = 1.85 \qquad \text{if } 2 < t \leq 3$$

$$C(t) = 0.69 + 3(0.58) = 2.43 \qquad \text{if } 3 < t \leq 4$$

and so on. The graph is shown in Figure 51.

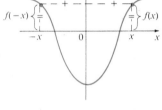

Figure 52

An even function

Symmetry

If a function f satisfies $f(-x) = f(x)$ for every number x in its domain, then f is called an **even function**. For instance, the function $f(x) = x^2$ is even because

$$f(-x) = (-x)^2 = x^2 = f(x)$$

The geometric significance of an even function is that its graph is symmetric with respect to the y-axis (see Figure 52). This means that if we have plotted the graph of f for $x \geq 0$, we obtain the entire graph simply by reflecting about the y-axis.

If f satisfies $f(-x) = -f(x)$ for every number x in its domain, then f is called an **odd function**. For example, the function $f(x) = x^3$ is odd because

$$f(-x) = (-x)^3 = -x^3 = -f(x)$$

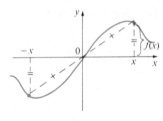

Figure 53

An odd function

The graph of an odd function is symmetric about the origin (see Figure 53). If we already have the graph of f for $x \geq 0$, we can obtain the entire graph by rotating through $180°$ about the origin. For instance, in Example 8, we need only have plotted the graph of $y = x^3$ for $x \geq 0$ and then rotated that part about the origin.

SECTION 4 **Exercises**

1. If $f(x) = x^2 - 3x + 2$, find $f(1)$, $f(-2)$, $f(\frac{1}{2})$, $f(\sqrt{5})$, $f(a)$, and $f(-a)$.

2. If $f(x) = x^3 + 2x^2 - 3$, find $f(0)$, $f(3)$, $f(-3)$, $f(-x)$, and $f(1/a)$.

3. If $g(x) = (1 - x)/(1 + x)$, find $g(2)$, $g(-2)$, $g(\pi)$, $g(a)$, $g(a - 1)$, and $g(-a)$.

4. If $h(t) = t + (1/t)$, find $h(1)$, $h(\pi)$, $h(t + 1)$, $h(t) + h(1)$, and $h(x)$.

In Exercises 5 and 6 find $f(2 + h)$, $f(x + h)$, and
$$\frac{f(x + h) - f(x)}{h}, \text{ where } h \neq 0.$$

5. $f(x) = x - x^2$ 6. $f(x) = \dfrac{x}{x + 1}$

In Exercises 7 and 8 draw a machine diagram, an arrow diagram, and a graph for the given function.

7. $f(x) = \sqrt{x}$, $0 \leq x \leq 4$

8. $f(x) = 2/x$, $1 \leq x \leq 4$

9. The domain of f is $A = \{1, 2, 3, 4, 5, 6\}$ and $f(1) = 2$, $f(2) = 1$, $f(3) = 0$, $f(4) = 1$, $f(5) = 2$, and $f(6) = 4$. What is the range of f? Draw an arrow diagram and a graph for f.

In Exercises 10–18 find the domain and range of the given function.

10. $f(x) = 2x + 7, \ -1 \le x \le 6$

11. $f(x) = 6 - 4x, \ -2 \le x \le 3$

12. $g(x) = \dfrac{1}{x + 4}$

13. $g(x) = \dfrac{2}{3x - 5}$

14. $h(x) = \sqrt[4]{7 - 3x}$

15. $h(x) = \sqrt{2x - 5}$

16. $F(x) = 1 - \sqrt{x}$

17. $F(x) = \sqrt{1 - x^2}$

18. $G(x) = \sqrt{x^2 - 9}$

In Exercises 19–26 find the domain of the given function.

19. $f(x) = \dfrac{x + 2}{x^2 - 1}$

20. $f(x) = \dfrac{x^4}{x^2 + x - 6}$

21. $g(x) = \sqrt[4]{x^2 - 6x}$

22. $g(x) = \sqrt{x^2 - 2x - 8}$

23. $\phi(x) = \sqrt{\dfrac{x}{\pi - x}}$

24. $\phi(x) = \sqrt{\dfrac{x^2 - 2x}{x - 1}}$

25. $f(t) = \sqrt[3]{t - 1}$

26. $f(t) = \sqrt{t^2 + 1}$

In Exercises 27–62 find the domain and sketch the graph of the function.

27. $f(x) = 2$

28. $f(x) = -3$

29. $f(x) = 3 - 2x$

30. $f(x) = \dfrac{x + 3}{2}, \ -2 \le x \le 2$

31. $f(x) = -x^2$

32. $f(x) = x^2 - 4$

33. $f(x) = x^2 + 2x - 1$

34. $f(x) = -x^2 + 6x - 7$

35. $g(x) = x^4$

36. $g(x) = x^5$

37. $g(x) = \sqrt{-x}$

38. $g(x) = \sqrt{6 - 2x}$

39. $h(x) = \sqrt{4 - x^2}$

40. $h(x) = \sqrt{x^2 - 4}$

41. $F(x) = \dfrac{1}{x}$

42. $F(x) = \dfrac{2}{x + 4}$

43. $G(x) = |x| + x$

44. $G(x) = |x| - x$

45. $H(x) = |2x|$

46. $H(x) = |2x - 3|$

47. $f(x) = x/|x|$

48. $f(x) = |x^2 - 1|$

49. $f(x) = \dfrac{x^2 - 1}{x - 1}$

50. $f(x) = \dfrac{x^2 + 5x + 6}{x + 2}$

51. $f(x) = \begin{cases} x + 1 & \text{if } x \ne 1 \\ 1 & \text{if } x = 1 \end{cases}$

52. $f(x) = \begin{cases} x + 3 & \text{if } x \ne -2 \\ 4 & \text{if } x = -2 \end{cases}$

53. $f(x) = \begin{cases} 0 & \text{if } x < 2 \\ 1 & \text{if } x \ge 2 \end{cases}$

54. $f(x) = \begin{cases} -1 & \text{if } x < -1 \\ 1 & \text{if } -1 \le x \le 1 \\ -1 & \text{if } x > 1 \end{cases}$

55. $f(x) = \begin{cases} x & \text{if } x \le 0 \\ x + 1 & \text{if } x > 0 \end{cases}$

56. $f(x) = \begin{cases} 2x + 3 & \text{if } x < -1 \\ 3 - x & \text{if } x \ge -1 \end{cases}$

57. $f(x) = \begin{cases} -1 & \text{if } x < -1 \\ x & \text{if } -1 \le x \le 1 \\ 1 & \text{if } x > 1 \end{cases}$

58. $f(x) = \begin{cases} |x| & \text{if } |x| \le 1 \\ 1 & \text{if } |x| > 1 \end{cases}$

59. $f(x) = \begin{cases} x + 2 & \text{if } x \le -1 \\ x^2 & \text{if } x > -1 \end{cases}$

60. $f(x) = \begin{cases} 1 - x^2 & \text{if } x \le 2 \\ 2x - 7 & \text{if } x > 2 \end{cases}$

61. $f(x) = \begin{cases} -1 & \text{if } x \le -1 \\ 3x + 2 & \text{if } |x| < 1 \\ 7 - 2x & \text{if } x \ge 1 \end{cases}$

62. $f(x) = \begin{cases} \sqrt{-x} & \text{if } x < 0 \\ x & \text{if } 0 \le x \le 2 \\ \sqrt{x - 2} & \text{if } x > 2 \end{cases}$

In Exercises 63–66 state whether the given curve is the graph of a function of x. If it is, state the domain and range of the function.

63.

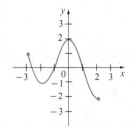

64.

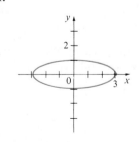

65. 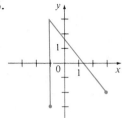   **66.**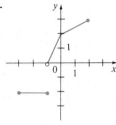

In Exercises 67–70 find a function whose graph is the given curve.

67. The line segment joining the points $(-2, 1)$ and $(4, -6)$

68. The line segment joining the points $(-3, -2)$ and $(6, 3)$

69. The bottom half of the parabola $x + (y - 1)^2 = 0$

70. The top half of the circle $(x - 1)^2 + y^2 = 1$

In Exercises 71–75 find a formula for the described function and state its domain.

71. A rectangle has a perimeter of 20 m. Express the area of the rectangle as a function of the length of one of its sides.

72. A rectangle has an area of $16\,\text{m}^2$. Express the perimeter of the rectangle as a function of the length of one of its sides.

73. Express the area of an equilateral triangle as a function of the length of a side.

74. Express the surface area of a cube as a function of its volume.

75. An open rectangular box with a volume of $2\,\text{m}^3$ has a square base. Express the surface area of the box as a function of the length of a side of the base.

76. A Norman window has the shape of a rectangle surmounted by a semicircle. If the perimeter of the window is $30\,\text{ft}$, express the area A of the window as a function of the width x of the window.

77. A box with an open top is to be constructed from a rectangular piece of cardboard with dimensions 12 in. by 20 in. by cutting out equal squares of side x at each corner and then folding up the sides as in the figure. Express the volume V of the box as a function of x.

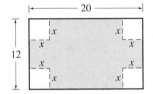

78. A taxi company charges two dollars for the first mile (or part of a mile) and 20 cents for each succeeding tenth of a mile (or part). Express the cost C (in dollars) of a ride as a function of the distance x traveled (in miles) for $0 < x < 2$ and sketch the graph of this function.

In Exercises 79–84 determine whether f is even, odd, or neither. If f is even or odd, use symmetry to sketch its graph.

79. $f(x) = x^{-2}$ **80.** $f(x) = x^{-3}$

81. $f(x) = x^2 + x$ **82.** $f(x) = x^4 - 4x^2$

83. $f(x) = x^3 - x$ **84.** $f(x) = 3x^3 + 2x^2 + 1$

SECTION 5

Combinations of Functions

Two functions f and g can be combined to form new functions $f + g$, $f - g$, fg, and f/g in a manner similar to the way we add, subtract, multiply, and divide real numbers. If we define the sum $f + g$ by the equation

$$(f + g)(x) = f(x) + g(x)$$

(21)

then the right side of Equation 21 makes sense if both $f(x)$ and $g(x)$ are defined, that is, if x belongs to the domain of f and also to the domain of g. If the domain of f is A and the domain of g is B, then the domain of $f + g$ is the intersection of these domains, that is, $A \cap B$.

Notice that the $+$ sign on the left side of Equation 21 stands for the operation of addition of *functions*, but the $+$ sign on the right side of the equation stands for addition of the *numbers* $f(x)$ and $g(x)$.

Similarly we can define the difference $f - g$ and the product fg, and their domains are also $A \cap B$. But in defining the quotient f/g we must remember not to divide by 0.

Algebra of Functions (22)

Let f and g be functions with domains A and B. Then the functions $f + g$, $f - g$, fg, and f/g are defined as follows:

$$(f + g)(x) = f(x) + g(x) \qquad \text{domain} = A \cap B$$

$$(f - g)(x) = f(x) - g(x) \qquad \text{domain} = A \cap B$$

$$(fg)(x) = f(x)g(x) \qquad \text{domain} = A \cap B$$

$$\left(\frac{f}{g}\right)(x) = \frac{f(x)}{g(x)} \qquad \text{domain} = \{x \in A \cap B \mid g(x) \neq 0\}$$

EXAMPLE 1   If $f(x) = \sqrt{x}$ and $g(x) = \sqrt{4 - x^2}$, find the functions $f + g$, $f - g$, fg, and f/g.

Solution   The domain of $f(x) = \sqrt{x}$ is $[0, \infty)$. The domain of $g(x) = \sqrt{4 - x^2}$ consists of all numbers x such that $4 - x^2 \geq 0$, that is, $x^2 \leq 4$. Taking square roots of both sides, we get $|x| \leq 2$, or $-2 \leq x \leq 2$, so the domain of g is the interval $[-2, 2]$. The intersection of the domains of f and g is

$$[0, \infty) \cap [-2, 2] = [0, 2]$$

Using the definitions in (22), we have

$$(f + g)(x) = \sqrt{x} + \sqrt{4 - x^2} \qquad\qquad 0 \leq x \leq 2$$

$$(f - g)(x) = \sqrt{x} - \sqrt{4 - x^2} \qquad\qquad 0 \leq x \leq 2$$

$$(fg)(x) = \sqrt{x}\,\sqrt{4 - x^2} = \sqrt{4x - x^3} \qquad 0 \leq x \leq 2$$

$$\left(\frac{f}{g}\right)(x) = \frac{\sqrt{x}}{\sqrt{4 - x^2}} = \sqrt{\frac{x}{4 - x^2}} \qquad 0 \leq x < 2$$

Notice that the domain of f/g is the interval $[0, 2)$ because we must exclude the points where $g(x) = 0$, that is, $x = \pm 2$.   ●

The graph of the function $f + g$ is obtained from the graphs of f and g by **graphical addition**. This means that we add corresponding y-coordinates as in Figure 54. Figure 55 shows the result of using this procedure to graph the function $f + g$ from Example 1.

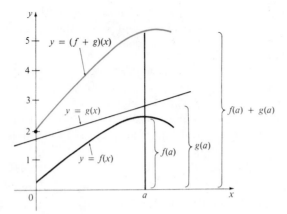

Figure 54

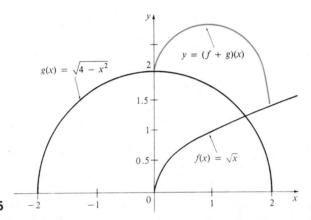

Figure 55

Composition of Functions

There is another way of combining two functions to get a new function. For example, suppose that $y = f(u) = \sqrt{u}$ and $u = g(x) = x^2 + 1$. Since y is a function of u and u is, in turn, a function of x, it follows that y is ultimately a function of x. We compute this by substitution:

$$y = f(u) = f(g(x)) = f(x^2 + 1) = \sqrt{x^2 + 1}$$

The procedure is called **composition** because the new function is composed of the two given functions f and g.

In general, given any two functions f and g, we start with a number x in the domain of g and find its image $g(x)$. If this number $g(x)$ is in the domain of f, then we can calculate the value of $f(g(x))$. The result is a new function $h(x) = f(g(x))$ obtained by substituting g into f. It is called the **composition** (or **composite**) of f and g and is denoted by $f \circ g$ ("f circle g").

Definition (23)

Given two functions f and g, the **composite function** $f \circ g$ (also called the **composition** of f and g) is defined by

$$(f \circ g)(x) = f(g(x))$$

The domain of $f \circ g$ is the set of all x in the domain of g such that $g(x)$ is in the domain of f. In other words, $(f \circ g)(x)$ is defined whenever both $g(x)$ and $f(g(x))$ are defined. The best way to picture $f \circ g$ is by a machine diagram (Figure 56) or an arrow diagram (Figure 57).

Figure 56

The $f \circ g$ machine is composed of the g machine (first) and then the f machine

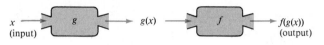

Figure 57

Arrow diagram for $f \circ g$

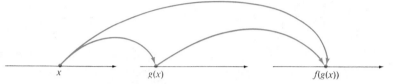

EXAMPLE 2 If $f(x) = x^2$ and $g(x) = x - 3$, find the composite functions $f \circ g$ and $g \circ f$ and their domains.

Solution We have

$$(f \circ g)(x) = f(g(x)) = f(x - 3) = (x - 3)^2$$

$$(g \circ f)(x) = g(f(x)) = g(x^2) = x^2 - 3$$

The domains of both $f \circ g$ and $g \circ f$ are R (the set of all real numbers). ●

  Note: You can see from Example 2 that, in general, $f \circ g \neq g \circ f$. Remember, the notation $f \circ g$ means that the function g is applied first and then f is applied second. In Example 2, $f \circ g$ is the function that *first* subtracts 3 and *then* squares; $g \circ f$ is the function that *first* squares and *then* subtracts 3.

EXAMPLE 3 If $f(x) = \sqrt{x}$ and $g(x) = \sqrt{2 - x}$, find the functions $f \circ g$, $g \circ f$, $f \circ f$, and $g \circ g$ and their domains.

Solution $$(f \circ g)(x) = f(g(x)) = f(\sqrt{2 - x}) = \sqrt{\sqrt{2 - x}} = \sqrt[4]{2 - x}$$

The domain of $f \circ g$ is $\{x \mid 2 - x \geq 0\} = \{x \mid x \leq 2\} = (-\infty, 2]$.

$$(g \circ f)(x) = g(f(x)) = g(\sqrt{x}) = \sqrt{2 - \sqrt{x}}$$

For $\sqrt{x}$ to be defined we must have $x \geq 0$. For $\sqrt{2 - \sqrt{x}}$ to be defined we must have $2 - \sqrt{x} \geq 0$, that is, $\sqrt{x} \leq 2$, or $x \leq 4$. Thus we have $0 \leq x \leq 4$, so the domain of $g \circ f$ is the closed interval $[0, 4]$.

$$(f \circ f)(x) = f(f(x)) = f(\sqrt{x}) = \sqrt{\sqrt{x}} = \sqrt[4]{x}$$

The domain of $f \circ f$ is $[0, \infty)$.

$$(g \circ g)(x) = g(g(x)) = g(\sqrt{2 - x}) = \sqrt{2 - \sqrt{2 - x}}$$

This expression is defined when $2 - x \geq 0$, that is, $x \leq 2$ and $2 - \sqrt{2 - x} \geq 0$. This latter inequality is equivalent to $\sqrt{2 - x} \leq 2$, or $2 - x \leq 4$, that is, $x \geq -2$. Thus $-2 \leq x \leq 2$, so the domain of $g \circ g$ is the closed interval $[-2, 2]$. ●

It is possible to take the composition of three or more functions. For instance, the composite function $f \circ g \circ h$ is found by first applying h, then g, and then f as follows:

$$(f \circ g \circ h)(x) = f(g(h(x)))$$

EXAMPLE 4   Find $f \circ g \circ h$ if $f(x) = x/(x+1)$, $g(x) = x^{10}$, and $h(x) = x+3$.

Solution

$$(f \circ g \circ h)(x) = f(g(h(x))) = f(g(x+3)) = f((x+3)^{10}) = \frac{(x+3)^{10}}{(x+3)^{10}+1}$$ ●

So far we have used composition to build complicated functions from simpler ones. But in calculus it is sometimes useful to be able to decompose a complicated function into simpler ones, as in the following example.

EXAMPLE 5   Given $F(x) = \sqrt[4]{x+9}$, find functions f and g such that $F = f \circ g$.

Solution   Since the formula for F says to first add 9 and then take the fourth root, we let

$$g(x) = x+9 \qquad \text{and} \qquad f(x) = \sqrt[4]{x}$$

Then $$(f \circ g)(x) = f(g(x)) = f(x+9) = \sqrt[4]{x+9} = F(x)$$ ●

SECTION 5   **Exercises**

In Exercises 1–6 find $f+g$, $f-g$, fg, and f/g and their domains.

1. $f(x) = x^2 - x$, $g(x) = x+5$

2. $f(x) = x^3 + 2x^2$, $g(x) = 3x^2 - 1$

3. $f(x) = \sqrt{1+x}$, $g(x) = \sqrt{1-x}$

4. $f(x) = \sqrt{9-x^2}$, $g(x) = \sqrt{x^2-1}$

5. $f(x) = \sqrt{x}$, $g(x) = \sqrt[3]{x}$

6. $f(x) = \sqrt[4]{x+1}$, $g(x) = \sqrt{x+2}$

In Exercises 7 and 8 find the domain of the given function.

7. $F(x) = \dfrac{\sqrt{4-x} + \sqrt{3+x}}{x^2 - 2}$

8. $F(x) = \sqrt{1-x} + \sqrt{x-2}$

In Exercises 9–12 use the graphs of f and g and the method of graphical addition to sketch the graph of $f+g$.

9. $f(x) = x^3$, $g(x) = 1$ 10. $f(x) = \sqrt{x}$, $g(x) = 3$

11. $f(x) = x$, $g(x) = \frac{1}{x}$ 12. $f(x) = x^3$, $g(x) = -x^2$

In Exercises 13–22 find the functions $f \circ g$, $g \circ f$, $f \circ f$, and $g \circ g$ and their domains.

13. $f(x) = 2x+3$, $g(x) = 4x - 1$

14. $f(x) = 6x - 5$, $g(x) = x/2$

15. $f(x) = 2x^2 - x$, $g(x) = 3x + 2$

16. $f(x) = \sqrt{x-1}$, $g(x) = x^2$

17. $f(x) = 1/x$, $g(x) = x^3 + 2x$

18. $f(x) = \dfrac{1}{x-1}$, $g(x) = \dfrac{x-1}{x+1}$

19. $f(x) = \sqrt[3]{x}$, $g(x) = 1 - \sqrt{x}$

20. $f(x) = \sqrt{x^2 - 1}$, $g(x) = \sqrt{1-x}$

21. $f(x) = \dfrac{x+2}{2x+1}$, $g(x) = \dfrac{x}{x-2}$

22. $f(x) = \dfrac{1}{\sqrt{x}}$, $g(x) = x^2 - 4x$

In Exercises 23–26 find $f \circ g \circ h$.

23. $f(x) = x - 1$, $g(x) = \sqrt{x}$, $h(x) = x - 1$

24. $f(x) = 1/x$, $g(x) = x^3$, $h(x) = x^2 + 2$

25. $f(x) = x^4 + 1$, $g(x) = x - 5$, $h(x) = \sqrt{x}$

26. $f(x) = \sqrt{x}$, $g(x) = \frac{x}{x-1}$, $h(x) = \sqrt[3]{x}$

In Exercises 27–30 express the given function in the form $f \circ g$.

27. $F(x) = (x-9)^5$

28. $F(x) = \sqrt{x} + 1$

29. $G(x) = \frac{x^2}{x^2 + 4}$

30. $G(x) = \frac{1}{x+3}$

In Exercises 31 and 32 express the given function in the form $f \circ g \circ h$.

31. $H(x) = \frac{1}{x^2 + 1}$

32. $H(x) = \sqrt[3]{\sqrt{x} - 1}$

33. A stone is dropped into a lake, creating a circular ripple that travels outward at a speed of $60 \, \text{cm/s}$. Express the area of this circle as a function of time t (in seconds).

34. A spherical balloon is being inflated. If the radius of the balloon is increasing at a rate of $1 \, \text{cm/s}$, express the volume of the balloon as a function of time t (in seconds).

35. If $f(x) = 3x + 5$ and $h(x) = 3x^2 + 3x + 2$, find a function g such that $f \circ g = h$.

36. If $f(x) = x + 4$ and $h(x) = 4x - 1$, find a function g such that $g \circ f = h$.

37. Let $f(x) = 1/x$ and $g(x) = x$. How does $f \circ f$ differ from g?

SECTION 6

Types of Functions; Shifting and Scaling

In solving calculus problems you will find that it is helpful to be familiar with the graphs of some commonly occurring functions. We classify various types of functions as follows.

Constant Functions. The constant function $f(x) = c$ has domain R and its range consists of the single number c. Its graph is a horizontal line and is illustrated in Figure 58 for $c = 2$.

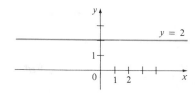

Figure 58

Power Functions. A function of the form $f(x) = x^a$, where a is a constant, is called a **power function**. A general study of such functions will have to wait until Chapter 3, but we can graph the following special cases now.

(a) $a = n$, *n a positive integer.* The graphs of $f(x) = x^n$ for $n = 1, 2, 3, 4,$ and 5 are shown in Figure 59. We have already seen the graphs of $y = x$ (a line through the origin with slope 1), $y = x^2$ (a parabola), and $y = x^3$ (Example 8 in Section 4).

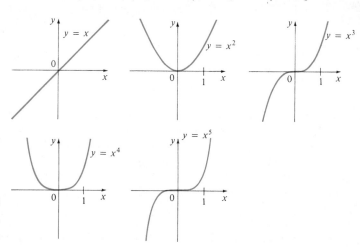

Figure 59
Graphs of $f(x) = x^n$,
$n = 1, 2, 3, 4, 5$

The general shape of the graph of $f(x) = x^n$ depends on whether n is even or odd. If n is even, then $f(x) = x^n$ is an even function and its graph is similar to the parabola $y = x^2$. If n is odd, then $f(x) = x^n$ is an odd function and its graph is similar to that of $y = x^3$. Notice from Figure 60, however, that as n increases, the graph of $y = x^n$ becomes flatter near 0 and steeper when $|x| \geq 1$. (If x is small, then x^2 is smaller, x^3 is even smaller, x^4 is smaller still, and so on.)

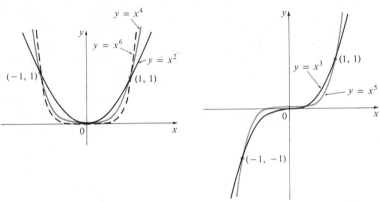

Figure 60

(b) $a = -1$. The graph of the reciprocal function $f(x) = x^{-1} = 1/x$ is shown in Figure 61. Its graph has the equation $y = 1/x$ or $xy = 1$. This is an equilateral hyperbola with the coordinate axes as its asymptotes. (See Section 3.)

(c) $a = 1/n$, n *a positive integer*. The function $f(x) = x^{1/n} = \sqrt[n]{x}$ is a **root function**. For $n = 2$ it is the square root function $f(x) = \sqrt{x}$ whose domain is $[0, \infty)$ and whose graph is the upper half of the parabola $x = y^2$ [see Figure 62(a)]. For other even values of n, the graph of $y = \sqrt[n]{x}$ is similar to that of $y = \sqrt{x}$. For $n = 3$ we have the cube root function $f(x) = \sqrt[3]{x}$ whose domain is R (recall that every real number has a cube root) and whose graph is shown in Figure 62(b). The graph of $y = \sqrt[n]{x}$ for n odd ($n > 3$) is similar to that of $y = \sqrt[3]{x}$.

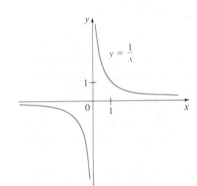

Figure 61

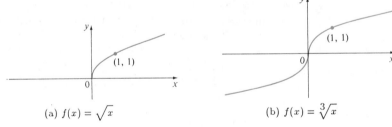

(a) $f(x) = \sqrt{x}$ (b) $f(x) = \sqrt[3]{x}$

Figure 62

Graphs of root functions

Polynomials.   A function P is called a **polynomial** if

$$P(x) = a_n x^n + a_{n-1} x^{n-1} + \cdots + a_2 x^2 + a_1 x + a_0$$

where n is a nonnegative integer and the numbers a_0, a_1, a_2, ... , a_n are constants called the **coefficients** of the polynomial. The domain of any polynomial is $R = (-\infty, \infty)$. If the leading coefficient $a_n \neq 0$, then the **degree** of the polynomial is n. For example, the function

$$P(x) = 2x^6 - x^4 + \tfrac{2}{5}x^3 + \sqrt{2}$$

is a polynomial of degree 6.

A polynomial of degree 1 is of the form $P(x) = ax + b$ and is called a **linear function** because its graph is the line $y = ax + b$ (slope a, y-intercept b).

A polynomial of degree 2 is of the form $P(x) = ax^2 + bx + c$ and is called a **quadratic function**. The graph of a quadratic function is always a parabola obtained by shifting the parabola $y = ax^2$. (See Example 2.)

A polynomial of degree 3 is of the form

$$P(x) = ax^3 + bx^2 + cx + d$$

and is called a **cubic function**. The graphs of cubic functions and higher-degree polynomials will be discussed in Chapter 4.

Rational Functions. A **rational function** f is a ratio of two polynomials:

$$f(x) = \frac{P(x)}{Q(x)}$$

where P and Q are polynomials. The domain consists of all values of x such that $Q(x) \neq 0$. For example, the function

$$f(x) = \frac{2x^4 - x^2 + 1}{x^2 - 4}$$

is a rational function with domain $\{x \mid x \neq \pm 2\}$. We will learn how to graph rational functions in Chapter 4.

Algebraic Functions. A function f is called **algebraic** if it can be constructed using algebraic operations (addition, subtraction, multiplication, division, and taking roots) starting with polynomials. Any rational function is automatically an algebraic function. Here are two more examples:

$$f(x) = \sqrt{x^2 + 1} \qquad g(x) = \frac{x^4 - 16x^2}{x + \sqrt{x}} + (x - 2)\sqrt[3]{x + 1}$$

Transcendental Functions. Functions that are not algebraic are called **transcendental**. Some examples of transcendental functions are the following.

(a)   *Exponential functions.* These are functions of the form $f(x) = a^x$, where a is a positive constant. They will be studied in Chapter 3.

(b)   *Logarithmic functions.* These are functions $f(x) = \log_a x$, where the base a is a positive constant. We will study logarithmic functions in Chapter 3.

(c)   *Trigonometric functions.* Trigonometry and the trigonometric functions are reviewed in Appendix B. In calculus the convention is that radian measure is always used (except when otherwise indicated). For example, when we use the function $f(x) = \sin x$, it is understood that $\sin x$ means the sine of the angle whose radian measure is x. Thus the graphs of the sine and cosine functions are as shown in Figures 63 and 64.

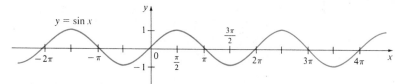

Figure 63

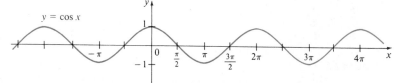

Figure 64

Transformations of Functions

By applying certain transformations to the graph of a given function we can obtain the graphs of certain related functions and thereby reduce the amount of work in graphing. Let us first consider **translations.** By adding the constant function $g(x) = c > 0$ to a given function f by graphical addition, we see that the graph of $y = f(x) + c$ is just the graph of $y = f(x)$ shifted upward a distance of c units. Likewise, if $g(x) = f(x - c)$, where $c > 0$, then the value of g at x is the same as the value of f at $x - c$ (c units to the left of x). Therefore the graph of $y = f(x - c)$ is just the graph of $y = f(x)$ shifted c units to the right (see Figure 65).

Vertical and Horizontal Shifts (24)

Suppose $c > 0$. To obtain the graph of

$y = f(x) + c$, shift the graph of $y = f(x)$ c units upward

$y = f(x) - c$, shift the graph of $y = f(x)$ c units downward

$y = f(x - c)$, shift the graph of $y = f(x)$ c units to the right

$y = f(x + c)$, shift the graph of $y = f(x)$ c units to the left

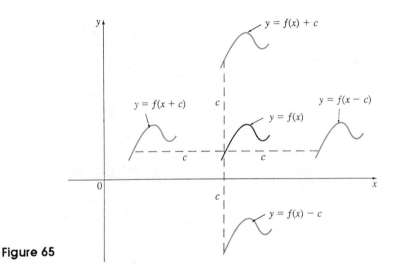

Figure 65

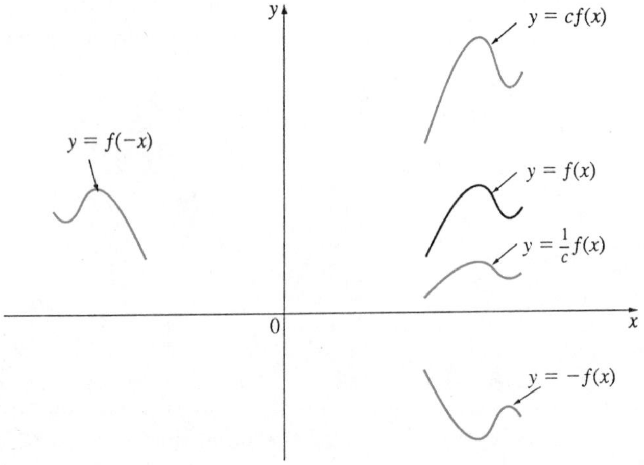

Figure 66

Now let us consider the **stretching** and **reflecting** transformations. By multiplying the given function f by the constant function $g(x) = c$, where $c > 1$, we see that the graph of $y = cf(x)$ is the graph of $y = f(x)$ stretched by a factor of c in the vertical direction. The graph of $y = -f(x)$ is the graph of $y = f(x)$ reflected about the x-axis because the point (x, y) is replaced by the point $(x, -y)$. [See Figure 66 and (25), where the results of other stretching, compressing, and reflecting transformations are also given.]

Vertical and Horizontal Stretching and Reflecting

(25)

Suppose $c > 1$. To obtain the graph of

$y = cf(x)$, stretch the graph of $y = f(x)$ vertically by a factor of c

$y = (1/c)f(x)$, compress the graph of $y = f(x)$ vertically by a factor of c

$y = f(cx)$, compress the graph of $y = f(x)$ horizontally by a factor of c

$y = f(x/c)$, stretch the graph of $y = f(x)$ horizontally by a factor of c

$y = -f(x)$, reflect the graph of $y = f(x)$ about the x-axis

$y = f(-x)$, reflect the graph of $y = f(x)$ about the y-axis

EXAMPLE 1   Given the graph of $y = \sqrt{x}$, use transformations to graph $y = \sqrt{x} - 2$, $y = \sqrt{x-2}$, $y = -\sqrt{x}$, $y = 2\sqrt{x}$, and $y = \sqrt{-x}$.

Solution   The graph of the square root function $y = \sqrt{x}$, obtained from Figure 62, is shown in Figure 67(a). In the other parts of the figure we sketch $y = \sqrt{x} - 2$ by shifting 2 units down, $y = \sqrt{x-2}$ by shifting 2 units to the right, $y = -\sqrt{x}$ by reflecting about the x-axis, $y = 2\sqrt{x}$ by stretching vertically by a factor of 2, and $y = \sqrt{-x}$ by reflecting about the y-axis.

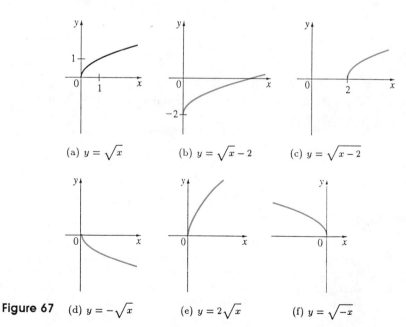

Figure 67   (a) $y = \sqrt{x}$ (b) $y = \sqrt{x} - 2$ (c) $y = \sqrt{x-2}$ (d) $y = -\sqrt{x}$ (e) $y = 2\sqrt{x}$ (f) $y = \sqrt{-x}$

EXAMPLE 2   Sketch the graph of the function $f(x) = x^2 + 6x + 10$.

Solution   Completing the square, we write the equation of the graph as

$$y = x^2 + 6x + 10 = (x+3)^2 + 1$$

According to (24) this means we obtain the desired graph by starting with the parabola $y = x^2$ and shifting 3 units to the left and then 1 unit upward (see Figure 68).

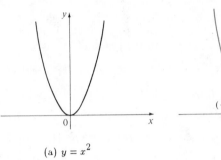

Figure 68   (a) $y = x^2$ (b) $y = (x+3)^2 + 1$

EXAMPLE 3   Sketch the graphs of the functions:

(a) $y = \sin 2x$ (b) $y = 1 - \sin x$

Solution

(a)   According to (25) we obtain the graph of $y = \sin 2x$ from that of $y = \sin x$ by compressing horizontally by a factor of 2 (see Figure 69).

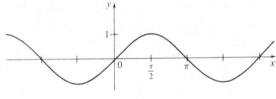
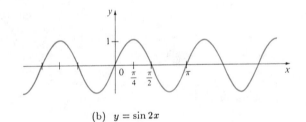

Figure 69   (a) $y = \sin x$ (b) $y = \sin 2x$

(b)   To obtain the graph of $y = 1 - \sin x$, we again start with $y = \sin x$. We reflect about the x-axis to get $y = -\sin x$ and then we shift 1 unit upward to get $y = 1 - \sin x$ (see Figure 70).

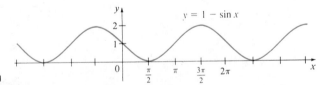

Figure 70

Another transformation of some interest is taking the absolute value of a function. If $y = |f(x)|$, then according to the definition of absolute value, $y = f(x)$ when $f(x) \geq 0$ and $y = -f(x)$ when $f(x) < 0$. This tells us how to get the graph of $y = |f(x)|$ from the graph of $y = f(x)$: The part of the graph that lies above the x-axis remains the same; the part that lies below the x-axis is reflected about the x-axis.

EXAMPLE 4   Sketch the graph of the function $y = |x^2 - 1|$.

Solution   We first graph the parabola $y = x^2 - 1$ in Figure 71(a) by shifting the parabola $y = x^2$ down 1 unit. We see that the graph lies below the x-axis when $-1 < x < 1$, so we reflect that part of the graph about the x-axis to obtain the graph of $y = |x^2 - 1|$ in Figure 71(b).

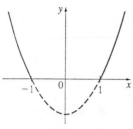

Figure 71   (a) $y = x^2 - 1$ (b) $y = |x^2 - 1|$

SECTION 6 **Exercises**

Graph the functions in Exercises 1–24, not by plotting points, but by starting with the graphs of the standard functions given in this section and then applying the appropriate transformations.

1. $y = x^8$

2. $y = \sqrt[4]{x}$

3. $y = -1/x$

4. $y = -x^3$

5. $y = 2\sin x$

6. $y = 1 + \sqrt{x}$

7. $y = (x-1)^3 + 2$

8. $y = -\cos x$

9. $y = 1 + \sqrt[6]{-x}$

10. $y = \sqrt[3]{x+2}$

11. $y = \cos(x/2)$

12. $y = x^2 + x + 1$

13. $y = \dfrac{1}{x-3}$

14. $y = -2\sin \pi x$

15. $y = \frac{1}{3}\sin\left(x - \frac{\pi}{6}\right)$

16. $y = 2 + \dfrac{1}{x+1}$

17. $y = 1 + 2x - x^2$

18. $y = \frac{1}{2}\sqrt{x+4} - 3$

19. $y = 2 - \sqrt{x+1}$

20. $y = 1 - (x-8)^6$

21. $y = |x^2 - 2x|$

22. $y = |\cos x|$

23. $y = ||x| - 1|$

24. $y = |||x| - 2| - 1|$

25. (a)   How is the graph of $y = f(|x|)$ related to the graph of f?

 (b)   Sketch the graph of $y = \sin|x|$.

26. Sketch the graph of $y = \sqrt{|x|}$.

SECTION 7

A Preview of Calculus

Calculus is fundamentally different from the mathematics that you have previously studied. Calculus is less static and more dynamic. It is concerned with change and motion; it deals with quantities that approach other quantities.

For that reason it may be useful to have an overview of the subject before beginning its intensive study. In this section we give a glimpse of some of the main ideas of calculus by showing how limits arise when we attempt to solve a variety of problems.

The Area Problem

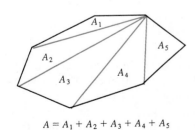

$A = A_1 + A_2 + A_3 + A_4 + A_5$

Figure 72

The origins of calculus go back at least 2500 years to the ancient Greeks, who found areas using the "method of exhaustion." They knew how to find the area A of any polygon by dividing it into triangles as in Figure 72 and adding the areas of these triangles.

It is a much more difficult problem to find the area of a curved figure. Their method of exhaustion was to inscribe polygons in the figure and circumscribe polygons about the figure and then let the number of sides of the polygons increase. Figure 73 illustrates this process for the special case of a circle with inscribed regular polygons.

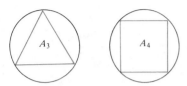   

Figure 73

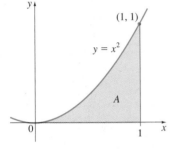

Figure 74

Let A_n be the area of the inscribed polygon with n sides. As n increases, it appears that A_n becomes closer and closer to the area of the circle. We say that the area of the circle is the *limit* of the areas of the inscribed polygons, and we write

$$A = \lim_{n \to \infty} A_n$$

The Greeks themselves did not explicitly use limits. However, by indirect reasoning, Eudoxus (fifth century B.C.) used exhaustion to prove the familiar formula for the area of a circle: $A = \pi r^2$.

We will use a similar idea in Chapter 5 to find areas of regions of the type shown in Figure 74. We will approximate the desired area A by areas of rectangles (as in Figure 75), let the width of the rectangles decrease, and then calculate A as the limit of these sums of the areas of rectangles.

The area problem is the central problem in the branch of calculus called *integral calculus*. The techniques that we will develop in Chapter 5 for finding areas will also enable us to compute the volume of a solid, the length of a curve, the force against a dam, the mass and center of gravity of a rod, and the work done in pumping water out of a tank.

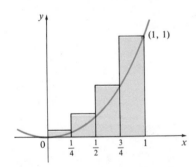 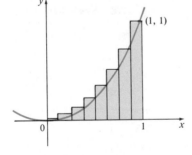 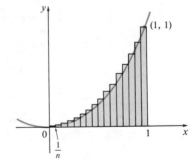

Figure 75

The Tangent Problem

Consider the problem of trying to find the equation of the tangent line t to a curve with equation $y = f(x)$ at a given point P. (We will give a precise definition of a tangent line in Chapter 1. For now you can think of it as a line that touches the curve at P as in Figure 76.) Since we know that the point P lies on the tangent line, we can find the equation of t if we know its slope m. The problem is that we need two points to compute the slope and we know only one point, P, on t. To get around this problem we first find an approximation to m by taking a nearby point Q on the curve and computing the slope m_{PQ} of the secant line PQ. From Figure 77 we see that

(26)
$$m_{PQ} = \frac{f(x) - f(a)}{x - a}$$

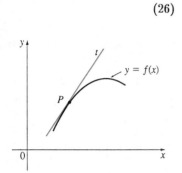

Figure 76

The tangent line at P

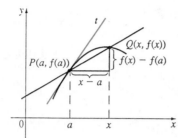

Figure 77

The secant line PQ

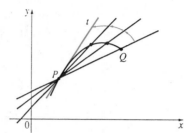

Figure 78

Now imagine that Q moves along the curve toward P as in Figure 78. You can see that the secant line rotates and approaches the tangent line as its limiting position. This means that the slope m_{PQ} of the secant line becomes closer and closer to the slope m of the tangent line. We write

$$m = \lim_{Q \to P} m_{PQ}$$

and we say that m is the limit of m_{PQ} as Q approaches P along the curve. Since x approaches a as Q approaches P, we could also use Equation 26 to write

(27)
$$m = \lim_{x \to a} \frac{f(x) - f(a)}{x - a}$$

Specific examples of this procedure will be given in Chapter 1.

The tangent problem has given rise to the branch of calculus called *differential calculus*, which was not invented until more than 2000 years after integral calculus. The main ideas behind differential calculus are due to the French mathematician Pierre Fermat (1601–1665) and were developed by the English mathematicians John Wallis (1616–1703), Isaac Barrow (1630–1677), and Isaac Newton (1642–1723) and the German mathematician Gottfried Leibniz (1648–1716).

The two branches of calculus and their chief problems, the area problem and the tangent problem, appear to be very different, but it turns out that there is a very close connection between them. The tangent problem and the area problem are inverse problems in a sense that will be described in Chapter 5.

Velocity

When we look at the speedometer of a car and read that the car is traveling at $48 \, \text{mi/h}$, what does that indicate to us? We know that if our velocity remains constant, then after an hour we will have traveled 48 mi. But if the velocity of the car varies, what does it mean to say that the velocity at a given instant is $48 \, \text{mi/h}$?

In order to analyze this question, let us analyze the motion of a car that travels along a straight road and assume that we can measure the distance traveled by the car (in feet) at 1-second intervals as in the following chart:

t = time elapsed (s)	0	1	2	3	4	5
d = distance (ft)	0	2	10	25	43	78

As a first step toward finding the velocity after 2 seconds have elapsed, let us find the average velocity during the time interval $2 \leq t \leq 4$:

$$\text{average velocity} = \frac{\text{distance traveled}}{\text{time elapsed}}$$

$$= \frac{43 - 10}{4 - 2}$$

$$= 16.5 \text{ ft/s}$$

Similarly the average velocity in the time interval $2 \leq t \leq 3$ is

$$\text{average velocity} = \frac{25 - 10}{3 - 2} = 15 \text{ ft/s}$$

We have the feeling that the velocity at the instant $t = 2$ cannot be much different from the average velocity during a short time interval starting at $t = 2$. So let us imagine that the distance traveled has been measured at 0.1-second time intervals as in the following chart:

t	2.0	2.1	2.2	2.3	2.4	2.5
d	10.00	11.02	12.16	13.45	14.96	16.80

Then we can compute, for instance, the average velocity over the time interval $[2, 2.5]$:

$$\text{average velocity} = \frac{16.80 - 10.00}{2.5 - 2} = 13.6 \text{ ft/s}$$

The results of such calculations are shown in the following chart:

time interval	$[2,3]$	$[2,2.5]$	$[2,2.4]$	$[2,2.3]$	$[2,2.2]$	$[2,2.1]$
average velocity (ft/s)	15.0	13.6	12.4	11.5	10.8	10.2

The average velocities over successively smaller intervals appear to be getting closer to a number near 10, and so we expect that the velocity at exactly $t = 2$ is about 10 ft/s. In Chapter 1 we will define the instantaneous velocity of a moving object as the limiting value of the average velocities over smaller and smaller time intervals.

In Figure 79 we show a graphical representation of the motion of the car by plotting the distance traveled as a function of time. If we write $d = f(t)$, then $f(t)$ is the number of feet traveled after t seconds. The average velocity in the time interval $[2, t]$ is

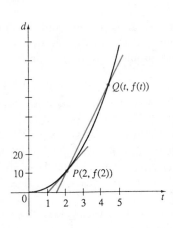

Figure 79

$$\text{average velocity} = \frac{\text{distance traveled}}{\text{time elapsed}} = \frac{f(t) - f(2)}{t - 2}$$

which is the same as the slope of the secant line PQ in Figure 79. The velocity v when $t = 2$ is the limiting value of this average velocity as t approaches 2; that is,

$$v = \lim_{t \to 2} \frac{f(t) - f(2)}{t - 2}$$

and we recognize from Equation 27 that this is the same as the slope of the tangent line to the curve at P.

Thus when we solve the tangent problem in differential calculus, we are also solving problems concerning velocities. The same techniques also enable us to solve problems involving rates of change in all of the natural and social sciences.

The Limit of a Sequence

In the fifth century B.C. the Greek philosopher Zeno of Elea posed four problems, now known as *Zeno's paradoxes,* that were intended to challenge some of the ideas concerning space and time that were held in his day. Zeno's second paradox concerns a race between the Greek hero Achilles and a tortoise that has been given a head start. Zeno argued as follows that Achilles could never pass the tortoise: Suppose that Achilles starts at position a_1 and that the tortoise starts at position t_1 (see Figure 80). When Achilles reaches the point $a_2 = t_1$, the tortoise is farther ahead at position t_2. When Achilles reaches $a_3 = t_2$, the tortoise is at t_3. This process continues indefinitely and so it appears that the tortoise will always be ahead! But this defies common sense.

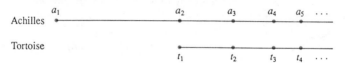

Figure 80

One way of explaining this paradox is with the idea of a *sequence.* The successive positions of Achilles $(a_1, a_2, a_3, \ldots)$ or the successive positions of the tortoise $(t_1, t_2, t_3, \ldots)$ form what is known as a sequence.

In general, a sequence $\{a_n\}$ is a set of numbers written in a definite order. For instance, the sequence

$$\left\{ 1, \tfrac{1}{2}, \tfrac{1}{3}, \tfrac{1}{4}, \tfrac{1}{5}, \ldots \right\}$$

can be described by giving the following formula for the nth term:

$$a_n = \frac{1}{n}$$

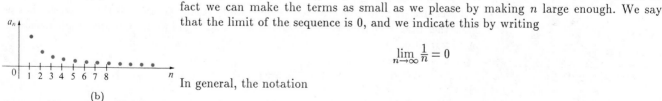

We can visualize this sequence by plotting its terms on a number line as in Figure 81(a) or by drawing its graph as in Figure 81(b). Observe from either picture that the terms of the sequence $a_n = 1/n$ are becoming closer and closer to 0 as n increases. In fact we can make the terms as small as we please by making n large enough. We say that the limit of the sequence is 0, and we indicate this by writing

$$\lim_{n \to \infty} \frac{1}{n} = 0$$

In general, the notation

Figure 81

$$\lim_{n \to \infty} a_n = L$$

is used if the terms a_n approach the number L as n becomes large. This means that the numbers a_n can be made as close as we like to the number L by taking n sufficiently large.

The concept of the limit of a sequence occurs whenever we use the decimal representation of a real number. For instance, if

$$a_1 = 3.1$$
$$a_2 = 3.14$$
$$a_3 = 3.141$$
$$a_4 = 3.1415$$
$$a_5 = 3.14159$$
$$a_6 = 3.141592$$
$$a_7 = 3.1415926$$
$$\vdots$$

then
$$\lim_{n \to \infty} a_n = \pi$$

The terms in this sequence are rational approximations to π.

Let us return to Zeno's paradox. The successive positions of Achilles and the tortoise form sequences $\{a_n\}$ and $\{t_n\}$, where $a_n < t_n$ for all n. But using the more precise treatment of sequences given in Chapter 10, it can be shown that both sequences have the same limit:

$$\lim_{n \to \infty} a_n = p = \lim_{n \to \infty} t_n$$

It is precisely at this point p that Achilles overtakes the tortoise.

The Sum of a Series

Another of Zeno's paradoxes, as passed on to us by Aristotle, is the following: "A man standing in a room cannot walk to the wall. In order to do so, he would first have to go half the distance, then half the remaining distance, and then again half of what still remains. This process can always be continued and can never be ended." (See Figure 82.)

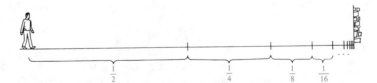

Figure 82

Of course we know that the man can actually reach the wall, so this suggests that perhaps the total distance can be expressed as the sum of infinitely many smaller distances as follows:

(28)
$$1 = \frac{1}{2} + \frac{1}{4} + \frac{1}{8} + \frac{1}{16} + \cdots + \frac{1}{2^n} + \cdots$$

Zeno was arguing that it does not make sense to add infinitely many numbers together. But there are other situations in which we implicitly use infinite sums. For instance, in decimal notation, the symbol $0.\overline{3} = 0.33333\ldots$ means

$$\frac{3}{10} + \frac{3}{100} + \frac{3}{1000} + \frac{3}{10,000} + \cdots$$

and so, in some sense, it must be true that

$$\frac{3}{10} + \frac{3}{100} + \frac{3}{1000} + \frac{3}{10,000} + \cdots = \frac{1}{3}$$

More generally, if d_n denotes the nth digit in the decimal representation of a number, then

$$0.d_1 d_2 d_3 d_4 \ldots = \frac{d_1}{10} + \frac{d_2}{10^2} + \frac{d_3}{10^3} + \cdots + \frac{d_n}{10^n} + \cdots$$

Therefore some infinite sums, or infinite series as they are called, have a meaning. But we must define carefully what the sum of an infinite series is.

Returning to the series in Equation 28, we denote by s_n the sum of the first n terms of the series. Thus

$$s_1 = \frac{1}{2} = 0.5$$

$$s_2 = \frac{1}{2} + \frac{1}{4} = 0.75$$

$$s_3 = \frac{1}{2} + \frac{1}{4} + \frac{1}{8} = 0.875$$

$$s_4 = \frac{1}{2} + \frac{1}{4} + \frac{1}{8} + \frac{1}{16} = 0.9375$$

$$s_5 = \frac{1}{2} + \frac{1}{4} + \frac{1}{8} + \frac{1}{16} + \frac{1}{32} = 0.96875$$

$$s_6 = \frac{1}{2} + \frac{1}{4} + \frac{1}{8} + \frac{1}{16} + \frac{1}{32} + \frac{1}{64} = 0.984375$$

$$s_7 = \frac{1}{2} + \frac{1}{4} + \frac{1}{8} + \frac{1}{16} + \frac{1}{32} + \frac{1}{64} + \frac{1}{128} = 0.9921875$$

$$\vdots$$

$$s_{10} = \frac{1}{2} + \frac{1}{4} + \cdots + \frac{1}{1024} \approx 0.99902344$$

$$\vdots$$

$$s_{16} = \frac{1}{2} + \frac{1}{4} + \cdots + \frac{1}{2^{16}} \approx 0.99998474$$

Observe that as we add more and more terms, the partial sums become closer and closer to 1. In fact, it can be shown that by taking n large enough (that is, by adding sufficiently many terms of the series), we can make the partial sum s_n as close as we please to the number 1. It therefore seems reasonable to say that the sum of the infinite series is 1 and to write

$$\frac{1}{2} + \frac{1}{4} + \frac{1}{8} + \cdots + \frac{1}{2^n} + \cdots = 1$$

In other words, the reason the sum of the series is 1 is that

$$\lim_{n \to \infty} s_n = 1$$

In Chapter 10 we will make these ideas precise and we will use Newton's idea of combining infinite series with differential and integral calculus.

Summary

We have seen that the concept of a limit arises in trying to find the area of a region, the slope of a tangent to a curve, the velocity of a car, or the sum of an infinite series. In each case the common theme is the calculation of a quantity as the limit of other, easily calculated, quantities. It is this basic idea of a limit that sets calculus apart from other areas of mathematics. In fact, we could define calculus as the part of mathematics that deals with limits.

Sir Isaac Newton invented his version of calculus in order to explain the motion of the planets around the sun. Today calculus is used not only in calculating the orbits of satellites and spacecraft and in the study of astronomy, nuclear physics, electricity, thermodynamics, acoustics, design of machines, chemical reactions, growth of organisms, weather prediction, and the calculation of life insurance premiums, but also in such everyday concerns as fencing off a field so as to enclose the maximum area or computing the most economical speed for driving a car. Some of these uses of calculus will be explored throughout this book.

REVIEW AND PREVIEW

Review

Key Topics

Define, state, or discuss the following.

1. Rational and irrational numbers
2. Real number line
3. Open interval; closed interval
4. Rules for inequalities
5. Absolute value of a number
6. Properties of absolute values
7. Triangle Inequality
8. Rectangular coordinate system
9. Distance formula
10. Slope of a line
11. Point-slope form of the equation of a line
12. Slope-intercept form of the equation of a line
13. Slope relationships for parallel and perpendicular lines
14. Equation of a circle
15. Equations of parabolas, ellipses, and hyperbolas
16. Function
17. Domain and range of a function
18. Independent and dependent variables
19. Graph of a function
20. Vertical Line Test
21. Even function; odd function
22. Sum, difference, product, and quotient of functions
23. Composition of functions
24. Constant function
25. Power function
26. Polynomial
27. Linear function
28. Quadratic function
29. Rational function
30. Algebraic and transcendental functions
31. Rules for vertical and horizontal shifts of functions
32. Rules for stretching and reflecting functions

Exercises

Solve the inequalities in Exercises 1–6 in terms of intervals.

1. $2 + 5x \leq 9 - 2x$

2. $-5 \leq 1 - 2x \leq 3$

3. $|x + 3| < 7$

4. $|2x - 7| > 1$

5. $x^2 - 5x + 6 > 0$

6. $\dfrac{2x^2 + x}{x^2 + 2} \leq 1$

7. Find the distance between the points $(2, 4)$ and $(-4, -4)$.

8. Sketch the region $\{(x,y)\,|\,|x| \leq 2,\ |y| > 1\}$.

9. Find an equation of the circle with center $(2,1)$ and radius 3.

10. Find an equation of the circle that passes through the origin and has center $(-6,4)$.

11. Find the center and radius of the circle
$$x^2 + y^2 + 2x - 8y + 8 = 0.$$

In Exercises 12–17 find an equation of the line that satisfies the given conditions.

12. Through $(2,1)$, with slope -3

13. Through $(-1,-6)$ and $(2,-1)$

14. With slope $-\frac{1}{3}$, y-intercept 2

15. Through $(2,3)$, parallel to $x + 2y = 1$

16. Through $(-1,1)$, perpendicular to $3x - 4y = 6$

17. x-intercept $\frac{3}{2}$, y-intercept $-\frac{5}{2}$

18. Find the slope and y-intercept of the line
$3x + 5y = 10.$

Identify the curves in Exercises 19–24 and sketch their graphs.

19. $y = 8 - 2x^2$ 20. $x^2 + 4y^2 = 16$

21. $x^2 - 4y^2 = 4$ 22. $x + 4y^2 = 0$

23. $2x^2 + y^2 - 16x + 30 = 0$

24. $y^2 - x^2 = 4$

25. If $f(x) = 1 + \sqrt{x-1}$, find $f(5)$, $f(9)$, $f(-x)$, $f(x^2)$, and $[f(x)]^2$.

In Exercises 26–28 find the domain of the function.

26. $f(x) = \dfrac{\sqrt[3]{2x+1}}{\sqrt[3]{2x+2}}$ 27. $g(x) = \dfrac{x^2 + x + 1}{x^2 + x - 1}$

28. $h(x) = \sqrt{5 - 4x - x^2}$

In Exercises 29–38 sketch the graph of the function.

29. $f(x) = -1$ 30. $f(x) = 2 + 3x$

31. $g(x) = x^2 + 2$ 32. $g(x) = 4x - x^2$

33. $h(x) = \sqrt{x-5}$ 34. $h(x) = 1 - x^5$

35. $y = -\sin 2x$ 36. $y = |16 - x^4|$

37. $y = x^2 - 2|x|$

38. $F(x) = \begin{cases} 1 - \dfrac{x}{2} & \text{if } x < 2 \\ x - 3 & \text{if } x \geq 2 \end{cases}$

In Exercises 39 and 40 find the following functions and their domains: (a) $f + g$, (b) f/g, (c) $f \circ g$, (d) $g \circ f$.

39. $f(x) = x^2$, $g(x) = x + 2$

40. $f(x) = x^2 + 3x$, $g(x) = \sqrt{x+2}$

41. If $h(x) = \sqrt{x^2 + x + 9}$, find functions f and g such that $h = f \circ g$.

42. If $F(x) = 1/\sqrt[3]{x^2 + 3}$, find functions f, g, and h such that $F = f \circ g \circ h$.

43. Express the area A of a circle as a function of the circumference C.

44. Two ships leave a port at the same time. One sails south at $15\,\text{mi/h}$ and the other sails east at $20\,\text{mi/h}$. Express the distance d between the ships as a function of t, the time (in hours) after their departure.

1

Limits and Rates of Change

The calculus was the first achievement of modern mathematics and it is difficult to overestimate its importance. I think it defines more unequivocally than anything else the inception of modern mathematics; and the system of mathematical analysis, which is its logical development, still constitutes the greatest technical advance in exact thinking.

John von Neumann

In A Preview of Calculus (Section 7 in Review and Preview) we saw how the idea of a limit underlies the various branches of calculus. Thus it is appropriate to begin our study of calculus by investigating limits and their properties.

The Tangent and Velocity Problems

In this section we see how limits arise when we attempt to find the tangent to a curve or the velocity of an object.

The Tangent Problem

The word *tangent* is derived from the Latin word *tangens*, which means "touching." Thus a tangent to a curve is a line that touches the curve. How can this idea be made precise?

For a circle we could simply follow Euclid and say that a tangent is a line that intersects the circle once and only once as in Figure 1.1(a).

Figure 1.1   (a)   (b)

For more complicated curves this definition is inadequate. Figure 1.1(b) shows two lines l and t passing through a point P on a curve C. The line l intersects C only once, but it certainly does not look like what we think of as a tangent. The line t, on the other hand, looks like a tangent but it intersects C twice.

To be specific, let us look at the problem of trying to find a tangent line t to the parabola $y = x^2$ in the following example.

57

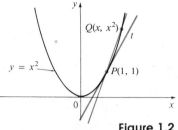

Figure 1.2

EXAMPLE 1   Find an equation of the tangent line to the parabola $y = x^2$ at the point $P(1,1)$.

Solution   We will be able to find the equation of the tangent line t as soon as we know its slope m. The difficulty is that we know only one point, P, on t, whereas we need two points to compute the slope. But observe that we can compute an approximation to m by choosing a nearby point $Q(x, x^2)$ on the parabola (as in Figure 1.2) and computing the slope m_{PQ} of the secant line PQ.

We choose $x \neq 1$ so that $Q \neq P$. Then

(1.1)
$$m_{PQ} = \frac{x^2 - 1}{x - 1}$$

For instance, for the point $Q(1.5, 2.25)$ we have

$$m_{PQ} = \frac{2.25 - 1}{1.5 - 1} = \frac{1.25}{0.5} = 2.5$$

The tables in (1.2) show the values of m_{PQ} for several values of x close to 1.

(1.2)

x	m_{PQ}	x	m_{PQ}
2	3	0	1
1.5	2.5	0.5	1.5
1.1	2.1	0.9	1.9
1.01	2.01	0.99	1.99
1.001	2.001	0.999	1.999

The closer Q is to P, the closer x is to 1 and, it appears from (1.2), the closer m_{PQ} is to 2. This suggests that the slope of the tangent line t should be $m = 2$.

We say that the slope of the tangent line is the *limit* of the slopes of the secant lines, and we express this symbolically by writing

$$\lim_{Q \to P} m_{PQ} = m$$

and

$$\lim_{x \to 1} \frac{x^2 - 1}{x - 1} = 2$$

Assuming that the slope of the tangent line is indeed 2, we use the point-slope form of the equation of a line (10) from Review and Preview to write the equation of the tangent line through $(1,1)$ as

$$y - 1 = 2(x - 1) \qquad \text{or} \qquad y = 2x - 1 \qquad\qquad \bullet$$

Figure 1.3 illustrates the limiting process of Example 1. As Q approaches P along the parabola, the corresponding secant lines rotate about P and approach the tangent line.

The Velocity Problem

If you watch the speedometer of a car while traveling in city traffic, you see that the needle does not stay still for very long; that is, the velocity of the car is not constant.

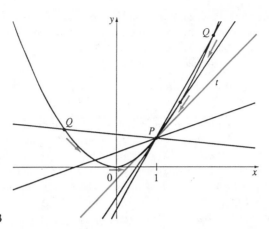

Figure 1.3

We assume from watching the speedometer that the car has a definite velocity at each moment, but how is the "instantaneous" velocity defined? Let us investigate the example of a falling ball.

EXAMPLE 2   Suppose that a ball is dropped from the upper observation deck of the CN Tower in Toronto, 450 m above the ground. Find the velocity of the ball after 5 s.

Solution   In trying to solve this problem we use the fact, discovered by Galileo almost four centuries ago, that the distance fallen by any freely falling body is proportional to the square of the time it has been falling. (This neglects air resistance.) If the distance fallen after t seconds is denoted by $s(t)$ and measured in meters, then Galileo's law is expressed by the equation

$$s(t) = 4.9t^2$$

The difficulty in finding the velocity after 5 s is that we are dealing with a single instant of time ($t = 5$) so there is no time interval involved. However, we can approximate the desired quantity by computing the average velocity over the brief time interval of a tenth of a second from $t = 5$ to $t = 5.1$:

$$\text{average velocity} = \frac{\text{distance traveled}}{\text{time elapsed}}$$

$$= \frac{s(5.1) - s(5)}{0.1}$$

$$= \frac{4.9(5.1)^2 - 4.9(5)^2}{0.1} = 49.49 \, \text{m/s}$$

The following table shows the results of similar calculations of the average velocity over successively smaller time periods.

The CN Tower is currently the highest freestanding building in the world. Courtsey of CN Tower.

Time interval	Average velocity (m/s)
$5 \leq t \leq 6$	53.9
$5 \leq t \leq 5.1$	49.49
$5 \leq t \leq 5.05$	49.245
$5 \leq t \leq 5.01$	49.049
$5 \leq t \leq 5.001$	49.0049

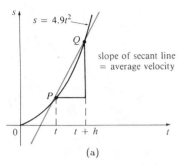

(a)

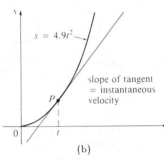

slope of tangent
= instantaneous
velocity

(b)

Figure 1.4

It appears that as we shorten the time period, the average velocity is becoming closer to 49 m/s. The **instantaneous velocity** when $t = 5$ is defined to be the limiting value of these average velocities over shorter and shorter time periods that start at $t = 5$. Thus the (instantaneous) velocity after 5 s is

$$v = 49\,\text{m/s}$$

You may have the feeling that the calculations used in solving this problem are very similar to those used earlier in this section to find tangents. In fact, there is a close connection between the tangent problem and the problem of finding velocities. If we draw the graph of the distance function of the ball (as in Figure 1.4) and we consider the points $P(a, 4.9a^2)$ and $Q(a + h, 4.9(a + h)^2)$ on the graph, then the slope of the secant line PQ is

$$m_{PQ} = \frac{4.9(a + h)^2 - 4.9a^2}{(a + h) - a}$$

which is the same as the average velocity over the time interval $[a, a + h]$. Therefore the velocity at time t (the limit of these average velocities as h approaches 0) must be equal to the slope of the tangent line at P (the limit of the slopes of these secant lines).

Examples 1 and 2 show that in order to solve tangent and velocity problems we must be able to find limits. After studying methods for computing limits in the next four sections, we will return to the problems of finding tangents and velocities in Section 1.8.

SECTION 1.1   **Exercises**

1. The point $P(1, 3)$ lies on the curve $y = 1 + x + x^2$.
   (a) If Q is the point $(x, 1 + x + x^2)$, find the slope of the secant line PQ for the following values of x:

(i) 2	(ii) 1.5	(iii) 1.1
(iv) 1.01	(v) 1.001	(vi) 0
(vii) 0.5	(viii) 0.9	(ix) 0.99
(x) 0.999		

   (b) Using the results of part (a), guess the value of the slope of the tangent line to the curve at $P(1, 3)$.
   (c) Using the slope from part (b), find the equation of the tangent line to the curve at $P(1, 3)$.

2. The point $P(-1, 3)$ lies on the curve $y = 1 - 2x^3$.
   (a) If Q is the point $(x, 1 - 2x^3)$, find the slope of the secant line PQ for the following values of x:

(i) −2	(ii) −1.5	(iii) −1.1
(iv) −1.01	(v) −1.001	(vi) 0
(vii) −0.5	(viii) −0.9	(ix) −0.99
(x) −0.999		

   (b) Using the results of part (a), guess the value of the slope of the tangent line to the curve at $P(-1, 3)$.

(c) Using the slope from part (b), find the equation of the tangent line to the curve at $P(-1, 3)$.

3. The point $P(4, 2)$ lies on the curve $y = \sqrt{x}$.
   (a) If Q is the point $(x, \sqrt{x})$, use your calculator to find the slope of the secant line PQ (correct to six decimal places) for the following values of x:

(i) 5	(ii) 4.5	(iii) 4.1	(iv) 4.01
(v) 4.001	(vi) 3	(vii) 3.5	(viii) 3.9
(ix) 3.99	(x) 3.999		

   (b) Using the results of part (a), guess the value of the slope of the tangent line to the curve at $P(4, 2)$.
   (c) Using the slope from part (b), find the equation of the tangent line to the curve at $P(4, 2)$.

4. The point $P(0.5, 2)$ lies on the curve $y = 1/x$.
   (a) If Q is the point $(x, 1/x)$, use your calculator to find the slope of the secant line PQ (correct to six decimal places) for the following values of x:

(i) 2	(ii) 1	(iii) 0.9	(iv) 0.8
(v) 0.7	(vi) 0.6	(vii) 0.55	(viii) 0.51
(ix) 0.45	(x) 0.49		

(b)   Using the results of part (a), guess the value of the slope of the tangent line to the curve at $P(0.5, 2)$.

(c)   Using the slope from part (b), find the equation of the tangent line to the curve at $P(0.5, 2)$.

(d)   Sketch the curve, two of the secant lines, and the tangent line.

5.   If a ball is thrown into the air with a velocity of 40 ft/s, its height in feet after t seconds is given by $y = 40t - 16t^2$.

(a)   Find the average velocity for the time period beginning when $t = 2$ and lasting

    (i)   0.5 s   (ii)   0.1 s

    (iii) 0.05 s   (iv) 0.01 s

(b)   Find the instantaneous velocity when $t = 2$.

6.   If an arrow is shot upward on the moon with a velocity of 58 m/s, its height in meters after t seconds is given by $h = 58t - 0.83t^2$.

(a)   Find the average velocity over the given time intervals:

    (i)   [1, 2]   (ii)   [1, 1.5]   (iii) [1, 1.1]

    (iv)   [1, 1.01] (v)   [1, 1.001]

(b)   Find the instantaneous velocity after 1 s.

7.   The displacement in meters of a particle moving in a straight line is given by $s = t^2 + t$, where t is measured in seconds.

(a)   Find the average velocity over the following time periods:

    (i)   [0, 2]   (ii)   [0, 1]

    (iii) [0, 0.5] (iv) [0, 0.1]

(b)   Find the instantaneous velocity when $t = 0$.

(c)   Draw the graph of s as a function of t and draw the secant lines whose slopes are the average velocities in part (a).

(d)   Draw the tangent line whose slope is the instantaneous velocity in part (b).

8.   The displacement in feet of a particle moving in a straight line is given by $s = t^3/6$, where t is measured in seconds.

(a)   Find the average velocity over the following time periods:

    (i)   [1, 3]   (ii)   [1, 2]

    (iii) [1, 1.5] (iv) [1, 1.1]

(b)   Find the instantaneous velocity when $t = 1$.

(c)   Draw the graph of s as a function of t and draw the secant lines whose slopes are the average velocities in part (a).

(d)   Draw the tangent line whose slope is the instantaneous velocity in part (b).

SECTION 1.2

The Limit of a Function

Having seen in the preceding section how limits arise when we want to find the tangent to a curve or the velocity of an object, we now turn our attention to limits in general and methods for computing them.

    Let us investigate the behavior of the function f defined by $f(x) = x^2 - x + 2$ for values of x near 2. The following table gives values of $f(x)$ for values of x close to 2 but not equal to 2.

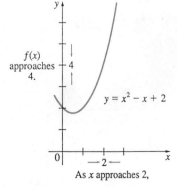

$f(x)$ approaches 4.

$y = x^2 - x + 2$

As x approaches 2,

Figure 1.5

x	$f(x)$	x	$f(x)$
1.0	2.000000	3.0	8.000000
1.5	2.750000	2.5	5.750000
1.8	3.440000	2.2	4.640000
1.9	3.710000	2.1	4.310000
1.95	3.852500	2.05	4.152500
1.99	3.970100	2.01	4.030100
1.995	3.985025	2.005	4.015025
1.999	3.997001	2.001	4.003001

From the table and graph of f (a parabola) shown in Figure 1.5 you can see that when x is close to 2 (on either side of 2), $f(x)$ is close to 4. In fact, it appears that we can make the values of $f(x)$ as close as we like to 4 by taking x sufficiently close to 2. We express this by saying "the limit of the function $f(x) = x^2 - x + 2$ as x approaches 2 is equal to 4." The notation for this is

$$\lim_{x \to 2} (x^2 - x + 2) = 4$$

In general, we use the following notation.

Definition (1.3)

> We write
>
> $$\lim_{x \to a} f(x) = L$$
>
> and say   "the limit of $f(x)$, as x approaches a, equals L"
>
> if we can make the values of $f(x)$ arbitrarily close to L (as close to L as we like) by taking x to be sufficiently close to a but not equal to a.

Roughly speaking, this says that the values of $f(x)$ get closer and closer to the number L as x gets closer and closer to the number a (from either side of a) but $x \neq a$. A more precise definition will be given in Section 1.4.

An alternative notation for

$$\lim_{x \to a} f(x) = L$$

is $f(x) \to L \quad \text{as} \quad x \to a$

which is usually read "$f(x)$ approaches L as x approaches a."

Notice the phrase "but $x \neq a$" in the definition of limit. This means that in finding the limit of $f(x)$ as x approaches a, we never consider $x = a$. In fact $f(x)$ need not even be defined when $x = a$. The only thing that matters is how f is defined *near a*.

Figure 1.6 shows the graphs of three functions. Note that in part (c), $f(a)$ is not defined and in part (b), $f(a) \neq L$. But in each case, regardless of what happens at a, $\lim_{x \to a} f(x) = L$.

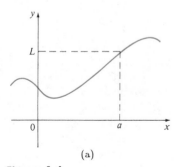
(a)

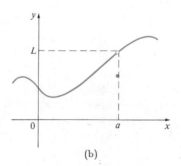
(b)

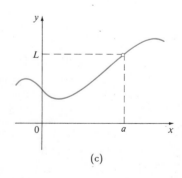
(c)

Figure 1.6

$\lim_{x \to a} f(x) = L$ in all three cases

EXAMPLE 1   Guess the value of $\lim_{x \to 1} \dfrac{x - 1}{x^2 - 1}$.

Solution   Notice that the function $f(x) = (x-1)/(x^2-1)$ is not defined when $x = 1$, but that doesn't matter because the definition of $\lim_{x \to a} f(x)$ says that we consider values of x that are close to a but not equal to a. The following table gives values of $f(x)$ (correct to six decimal places) for values of x that approach 1 (but are not equal to 1).

$x < 1$	$f(x)$	$x > 1$	$f(x)$
0.5	0.666667	1.5	0.400000
0.9	0.526316	1.1	0.476190
0.99	0.502513	1.01	0.497512
0.999	0.500250	1.001	0.499750
0.9999	0.500025	1.0001	0.499975

On the basis of the values in the table, we make the guess that

$$\lim_{x \to 1} \frac{x-1}{x^2-1} = 0.5$$

Example 1 is illustrated by the graph of f in Figure 1.7. Now let us change f slightly by giving it the value 2 when $x = 1$ and calling the resulting function g:

$$g(x) = \begin{cases} \dfrac{x-1}{x^2-1} & \text{if } x \neq 1 \\ 2 & \text{if } x = 1 \end{cases}$$

This new function g still has the same limit as x approaches 1 (see Figure 1.8).

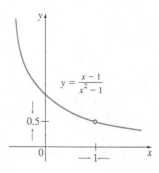

Figure 1.7

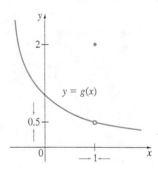

Figure 1.8

EXAMPLE 2   Find $\lim\limits_{t \to 0} \dfrac{\sqrt{t+9}-3}{t}$.

Solution   The accompanying table lists values of the function for several values of t near 0. As t approaches 0, the values of the function seem to approach 0.1666666... and so we guess that

$$\lim_{t \to 0} \frac{\sqrt{t+9}-3}{t} = \frac{1}{6}$$

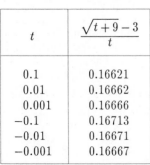

t	$\dfrac{\sqrt{t+9}-3}{t}$
0.1	0.16621
0.01	0.16662
0.001	0.16666
−0.1	0.16713
−0.01	0.16671
−0.001	0.16667

In Example 2 what would have happened if we had taken even smaller values of t? Here are the results from my calculator; you can see that something strange seems to be happening:

t	0.000001	0.0000001	0.00000001	0.000000001
$\dfrac{\sqrt{t+9}-3}{t}$	0.16670	0.17000	0.20000	0.00000

Does this mean that the answer is really 0, instead of $\frac{1}{6}$? No, the value of the limit is $\frac{1}{6}$, as we will show in the next section. The problem is that the calculator gave false values because $\sqrt{t+9}$ is very close to 3 when t is small. For a further explanation, see Appendix D (Lies my Calculator or Computer Told Me) and, in particular, the section called The Perils of Subtraction.

EXAMPLE 3   Find $\lim\limits_{x\to 0}\dfrac{\sin x}{x}$.

Solution   Again the function $f(x)=(\sin x)/x$ is not defined when $x=0$. Using a calculator (and remembering that, if $x\in R$, $\sin x$ means the sine of the angle whose *radian* measure is x), we construct the accompanying table of values correct to eight decimal places. From the table and Figure 1.9 (drawn with the aid of the table) we guess that

$$\lim_{x\to 0}\frac{\sin x}{x}=1$$

x	$f(x)=\dfrac{\sin x}{x}$
± 1.0	0.84147098
± 0.5	0.95885108
± 0.4	0.97354586
± 0.3	0.98506736
± 0.2	0.99334665
± 0.1	0.99833417
± 0.05	0.99958339
± 0.01	0.99998333
± 0.005	0.99999583
± 0.001	0.99999983

This guess is in fact correct, as will be proved in Chapter 2 using a geometric argument.

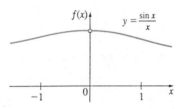

Figure 1.9

EXAMPLE 4   Find $\lim\limits_{x\to 0}\sin\dfrac{\pi}{x}$.

Solution   Once again the function $f(x)=\sin(\pi/x)$ is undefined at 0. Evaluating the function for some small values of x, we get

$$f(1)=\sin\pi=0 \qquad\qquad f(\tfrac{1}{2})=\sin 2\pi=0$$

$$f(\tfrac{1}{3})=\sin 3\pi=0 \qquad\qquad f(\tfrac{1}{4})=\sin 4\pi=0$$

$$f(0.1)=\sin 10\pi=0 \qquad f(0.01)=\sin 100\pi=0$$

Similarly, $f(0.001)=f(0.0001)=0$. On the basis of this information we might be tempted to guess that

$$\lim_{x\to 0}\sin\frac{\pi}{x}=0$$

but this time our guess is wrong. Note that although $f(1/n)=\sin n\pi=0$ for any integer n, it is also true that $f(x)=1$ for infinitely many values of x that approach 0. [In fact, $\sin(\pi/x)=1$ when

$$\frac{\pi}{x}=\frac{\pi}{2}+2n\pi$$

and, solving for x, we get $x=2/(4n+1)$.] The graph of f is given in Figure 1.10. The values of $\sin(\pi/x)$ oscillate between 1 and -1 infinitely often as x approaches 0. Since

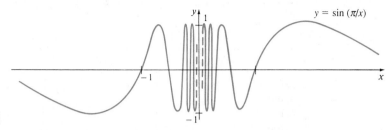

Figure 1.10

the values of $f(x)$ do not approach a fixed number as x approaches 0,

$$\lim_{x \to 0} \sin \frac{\pi}{x} \text{ does not exist}$$

EXAMPLE 5   Find $\lim_{x \to 0} \left(x^3 + \frac{\cos 5x}{10,000} \right)$.

Solution   As before, we construct a table of values.

x	$x^3 + \dfrac{\cos 5x}{10,000}$
1	1.000028
0.5	0.124920
0.1	0.001088
0.05	0.000222
0.01	0.000101

From the first table it appears that

$$\lim_{x \to 0} \left(x^3 + \frac{\cos 5x}{10,000} \right) = 0$$

But if we persevere with smaller values of x, the second table suggests that

$$\lim_{x \to 0} \left(x^3 + \frac{\cos 5x}{10,000} \right) = 0.000100 = \frac{1}{10,000}$$

x	$x^3 + \dfrac{\cos 5x}{10,000}$
0.005	0.00010009
0.001	0.00010000

Later we will see that $\lim_{x \to 0} \cos 5x = 1$ and then it follows that the limit is 0.0001.

    Examples 4 and 5 illustrate some of the pitfalls in guessing the value of a limit. It is easy to guess the wrong value if we use inappropriate values of x, but it is difficult to know when to stop calculating values. And, as the discussion after Example 2 shows, sometimes our calculators give the wrong values. In the next two sections, however, we develop foolproof methods for calculating limits.

EXAMPLE 6   Find $\lim_{x \to 0} \dfrac{1}{x^2}$ if it exists.

Solution   We see from the table that as x becomes close to 0, x^2 also becomes close to 0, so $1/x^2$ becomes very large. In fact, you can see from the graph of the function $f(x) = 1/x^2$ shown in Figure 1.11 that the values of $f(x)$ can be made arbitrarily large by taking x close enough to 0. Thus the values of $f(x)$ do not approach a number, so $\lim_{x \to 0} (1/x^2)$ does not exist.

x	$\dfrac{1}{x^2}$
± 1	1
± 0.5	4
± 0.2	25
± 0.1	100
± 0.05	400
± 0.01	10,000
± 0.001	1,000,000

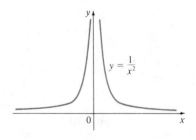

Figure 1.11

EXAMPLE 7   The Heaviside function H is defined by

$$H(t) = \begin{cases} 0 & \text{if } t < 0 \\ 1 & \text{if } t \geq 0 \end{cases}$$

[This function is named after the electrical engineer Oliver Heaviside (1850–1925) and can be used to describe an electric current that is switched on at time $t = 0$.] Its graph is shown in Figure 1.12.

At t approaches 0 from the left, $H(t)$ approaches 0. As t approaches 0 from the right, $H(t)$ approaches 1. There is no single number that $H(t)$ approaches as t approaches 0. Therefore $\lim_{t \to 0} H(t)$ does not exist.

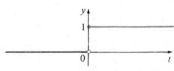

Figure 1.12

One-Sided Limits

We noticed in Example 7 that $H(t)$ approaches 0 as t approaches 0 from the left and $H(t)$ approaches 1 as t approaches 0 from the right. We indicate this situation symbolically by writing

$$\lim_{t \to 0^-} H(t) = 0 \qquad \text{and} \qquad \lim_{t \to 0^+} H(t) = 1$$

The symbol "$t \to 0^-$" indicates that we consider only values of t that are less than 0. Likewise "$t \to 0^+$" indicates that we consider only values of t that are greater than 0.

Definition (1.4)

> We write
>
> $$\lim_{x \to a^-} f(x) = L$$
>
> and say the **left-hand limit of $f(x)$ as x approaches a** (or the **limit of $f(x)$ as x approaches a from the left**) is equal to L if we can make the values of $f(x)$ arbitrarily close to L by taking x to be sufficiently close to a and $x < a$.

Notice that Definition 1.4 differs from Definition 1.3 only in that we require x to be less than a. Similarly, if we require that x be greater than a, we get "the **right-hand limit of $f(x)$ as x approaches a** is equal to L" and we write

$$\lim_{x \to a^+} f(x) = L$$

Thus the symbol "$x \to a^+$" means that we consider only $x > a$. These definitions are illustrated in Figure 1.13.

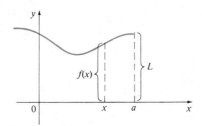

Figure 1.13   (a) $\lim\limits_{x \to a^-} f(x) = L$ (b) $\lim\limits_{x \to a^+} f(x) = L$

By comparing Definition 1.3 with the definitions of one-sided limits, we see that the following is true.

(1.5)

$$\lim_{x \to a} f(x) = L \quad \text{if and only if} \quad \lim_{x \to a^-} f(x) = L \quad \text{and} \quad \lim_{x \to a^+} f(x) = L$$

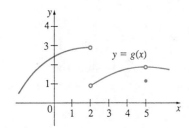

EXAMPLE 8   The graph of a function g is shown in Figure 1.14. Use it to state the values (if they exist) of the following:

(a) $\lim\limits_{x \to 2^-} g(x)$ (b) $\lim\limits_{x \to 2^+} g(x)$ (c) $\lim\limits_{x \to 2} g(x)$

(d) $\lim\limits_{x \to 5^-} g(x)$ (e) $\lim\limits_{x \to 5^+} g(x)$ (f) $\lim\limits_{x \to 5} g(x)$

Solution   From the graph we see that

Figure 1.14   (a) $\lim\limits_{x \to 2^-} g(x) = 3$ (b) $\lim\limits_{x \to 2^+} g(x) = 1$

(c)   Since the left and right limits are different, we conclude from (1.5) that $\lim\limits_{x \to 2} g(x)$ does not exist. The graph also shows that

(d) $\lim\limits_{x \to 5^-} g(x) = 2$ (e) $\lim\limits_{x \to 5^+} g(x) = 2$

(f)   This time the left and right limits are the same and so, by (1.5), we have

$$\lim_{x \to 5} g(x) = 2$$

●

SECTION 1.2 **Exercises**

1. For the function f whose graph is given, state the value of the given quantity, if it exists.

(a) $\lim\limits_{x \to 1} f(x)$ (b) $\lim\limits_{x \to 3^-} f(x)$ (c) $\lim\limits_{x \to 3^+} f(x)$

(d) $\lim\limits_{x \to 3} f(x)$ (e) $f(3)$ (f) $\lim\limits_{x \to -2^-} f(x)$

(g) $\lim\limits_{x \to -2^+} f(x)$ (h) $\lim\limits_{x \to -2} f(x)$ (i) $f(-2)$

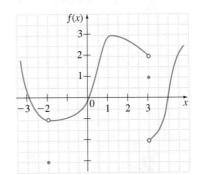

2. For the function g whose graph is given, state the value of the given quantity, if it exists.

(a) $\lim\limits_{x \to -2^-} g(x)$ (b) $\lim\limits_{x \to -2^+} g(x)$

(c) $\lim\limits_{x \to -2} g(x)$ (d) $g(-2)$

(e) $\lim\limits_{x \to 2^-} g(x)$ (f) $\lim\limits_{x \to 2^+} g(x)$

(g) $\lim\limits_{x \to 2} g(x)$ (h) $g(2)$

(i) $\lim\limits_{x \to 4^+} g(x)$ (j) $\lim\limits_{x \to 4^-} g(x)$

(k) $g(0)$ (l) $\lim\limits_{x \to 0} g(x)$

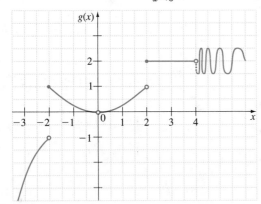

3. State the value of the limit, if it exists, from the given graph.

(a) $\lim\limits_{x \to 3} f(x)$ (b) $\lim\limits_{x \to 1} f(x)$

(c) $\lim\limits_{x \to -3} f(x)$ (d) $\lim\limits_{x \to 2^-} f(x)$

(e) $\lim\limits_{x \to 2^+} f(x)$ (f) $\lim\limits_{x \to 2} f(x)$

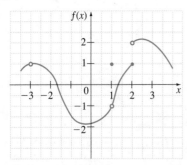

4. State the value of the limit, if it exists, from the given graph.

(a) $\lim\limits_{x \to 1} g(x)$ (b) $\lim\limits_{x \to 0} g(x)$

(c) $\lim\limits_{x \to 2} g(x)$ (d) $\lim\limits_{x \to -2} g(x)$

(e) $\lim\limits_{x \to -1^-} g(x)$ (f) $\lim\limits_{x \to -1} g(x)$

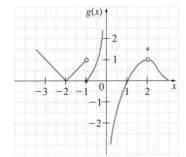

5. (a) Sketch the graph of the function
$$f(x) = \begin{cases} 1 - x & \text{if } x \le 0 \\ 1 & \text{if } x > 0 \end{cases}$$

(b) Use the graph from part (a) to state the values of the following limits, if they exist.

(i) $\lim\limits_{x \to 0^-} f(x)$ (ii) $\lim\limits_{x \to 0^+} f(x)$

(iii) $\lim\limits_{x \to 0} f(x)$

6. (a) Sketch the graph of the function
$$g(x) = \begin{cases} 2 - x & \text{if } x < -1 \\ x & \text{if } -1 \le x < 1 \\ 4 & \text{if } x = 1 \\ 4 - x & \text{if } x > 1 \end{cases}$$

(b) Use the graph from part (a) to state the values of the following limits, if they exist.

(i) $\lim\limits_{x \to -1^-} g(x)$ (ii) $\lim\limits_{x \to -1^+} g(x)$

(iii) $\lim\limits_{x \to -1} g(x)$ (iv) $\lim\limits_{x \to 1^-} g(x)$

(v) $\lim\limits_{x \to 1^+} g(x)$ (vi) $\lim\limits_{x \to 1} g(x)$

⊞ *In Exercises 7–14 evaluate the function at the given numbers (correct to six decimal places) and then use the results to guess the value of the limit.*

7. $f(x) = 1 - 2x^3$; $x = 1, 1.5, 1.9, 1.99, 1.999, 3, 2.5,$ $2.1, 2.01, 2.001$; $\lim\limits_{x \to 2} (1 - 2x^3)$

8. $f(x) = \dfrac{x^2 + 3x + 2}{x + 2}$; $x = -3, -2.5, -2.1, -2.01,$

  $-2.001, -1, -1.5, -1.9, -1.99, -1.999$;

  $\displaystyle\lim_{x \to -2} \dfrac{x^2 + 3x + 2}{x + 2}$

9. $g(x) = \dfrac{x - 1}{x^3 - 1}$; $x = 0.2, 0.4, 0.6, 0.8, 0.9, 0.99, 1.8,$

  $1.6, 1.4, 1.2, 1.1, 1.01$; $\displaystyle\lim_{x \to 1} \dfrac{x - 1}{x^3 - 1}$

10. $g(x) = \sqrt{x}$; $x = 1, 0.8, 0.6, 0.4, 0.2, 0.1, 0.01, 0.001,$

  $0.0001, 0.00001$; $\displaystyle\lim_{x \to 0^+} \sqrt{x}$

11. $F(x) = \dfrac{(1/\sqrt{x}) - \frac{1}{5}}{x - 25}$; $x = 26, 25.5, 25.1, 25.05, 25.01,$

  $24, 24.5, 24.9, 24.95, 24.99$; $\displaystyle\lim_{x \to 25} \dfrac{(1/\sqrt{x}) - \frac{1}{5}}{x - 25}$

12. $F(t) = \dfrac{\sqrt[3]{t} - 1}{\sqrt{t} - 1}$; $t = 1.5, 1.2, 1.1, 1.01, 1.001$;

  $\displaystyle\lim_{t \to 1} \dfrac{\sqrt[3]{t} - 1}{\sqrt{t} - 1}$

13. $f(x) = \dfrac{1 - \cos x}{x^2}$; $x = 1, 0.5, 0.4, 0.3, 0.2, 0.1, 0.05,$

  0.01; $\displaystyle\lim_{x \to 0} \dfrac{1 - \cos x}{x^2}$

14. $g(x) = \dfrac{\cos x - 1}{\sin x}$; $x = 1, 0.5, 0.4, 0.3, 0.2, 0.1, 0.05,$

  0.01; $\displaystyle\lim_{x \to 0} \dfrac{\cos x - 1}{\sin x}$

15. The existence of the limit

$$\lim_{x \to 0} (1 + x)^{1/x}$$

will be established in Chapter 10. Estimate the value of the limit to five decimal places by first evaluating the function for $x = 1, 0.1, 0.01, 0.001, 0.0001, 0.00001, 0.000001, 0.0000001, 0.00000001,$ and 0.000000001.

16. The slope of the tangent line to the graph of the exponential function $y = 2^x$ at the point $(0, 1)$ is $\lim_{x \to 0} (2^x - 1)/x$. Estimate the slope to three decimal places by first evaluating $(2^x - 1)/x$ for $x = 0.5, 0.1, 0.05, 0.01, 0.005, 0.001, 0.0005,$ and 0.0001.

17. (a) Evaluate the function $f(x) = x^2 - (2^x/1000)$ for $x = 1, 0.8, 0.6, 0.4, 0.2, 0.1,$ and 0.05, and guess the value of

$$\lim_{x \to 0} \left(x^2 - \dfrac{2^x}{1000} \right)$$

  (b) Evaluate $f(x)$ for $x = 0.04, 0.02, 0.01, 0.005, 0.003,$ and 0.001. Guess again.

18. (a) Evaluate $h(x) = (\tan x - x)/x^3$ for $x = 1, 0.5, 0.1, 0.05, 0.01,$ and 0.005.

  (b) Guess the value of $\displaystyle\lim_{x \to 0} \dfrac{\tan x - x}{x^3}$.

  (c) Evaluate h for $x = 0.001, 0.0005, 0.0004, 0.0003, 0.0002, 0.0001, 0.00005, 0.00001, 0.000001,$ and 0.0000001. Are you still confident that your guess in part (b) is correct? In Section 3.9 a method for evaluating the limit will be explained. See Appendix D for an explanation of what went wrong with the calculator computations.

SECTION 1.3

Calculating Limits using the Limit Laws

In Section 1.2 we used calculators and graphs to guess the values of limits, but we saw that such methods don't always lead to the correct answer. In this section we use the following properties of limits, called the *Limit Laws*, to calculate limits.

Limit Laws (1.6)

Suppose that c is a constant and the limits

$$\lim_{x \to a} f(x) \qquad \text{and} \qquad \lim_{x \to a} g(x)$$

exist. Then

1. $\lim\limits_{x \to a} [f(x) + g(x)] = \lim\limits_{x \to a} f(x) + \lim\limits_{x \to a} g(x)$

2. $\lim\limits_{x \to a} [f(x) - g(x)] = \lim\limits_{x \to a} f(x) - \lim\limits_{x \to a} g(x)$

3. $\lim\limits_{x \to a} [c f(x)] = c \lim\limits_{x \to a} f(x)$

4. $\lim\limits_{x \to a} [f(x) g(x)] = \lim\limits_{x \to a} f(x) \cdot \lim\limits_{x \to a} g(x)$

5. $\lim\limits_{x \to a} \dfrac{f(x)}{g(x)} = \dfrac{\lim\limits_{x \to a} f(x)}{\lim\limits_{x \to a} g(x)} \qquad \text{if } \lim\limits_{x \to a} g(x) \neq 0$

These five laws can be stated verbally as follows:

Sum Law   1.   The limit of a sum is the sum of the limits.

Difference Law   2.   The limit of a difference is the difference of the limits.

Constant Multiple Law   3.   The limit of a constant times a function is the constant times the limit of the function.

Product Law   4.   The limit of a product is the product of the limits.

Quotient Law   5.   The limit of a quotient is the quotient of the limits (provided that the limit of the denominator is not 0).

It is easy to believe that these properties are true. For instance, if $f(x)$ is close to L and $g(x)$ is close to M, it is reasonable to conclude that $f(x) + g(x)$ is close to $L + M$. This gives us an intuitive basis for believing that Law 1 is true. In Section 1.4 we give a precise definition of a limit and use it to prove this law. The proofs of the remaining laws are given in Appendix C.

If we use the Product Law repeatedly with $g(x) = f(x)$, we obtain the following law.

Power Law   6. $\lim\limits_{x \to a} [f(x)]^n = \left[\lim\limits_{x \to a} f(x)\right]^n \qquad$ where n is a positive integer

In applying these six limit laws we need to use two special limits.

7. $\lim\limits_{x \to a} c = c$ 8. $\lim\limits_{x \to a} x = a$

These limits are obvious from an intuitive point of view (state them in words or draw graphs of $y = c$ and $y = x$), but proofs based on the precise definition are requested in the exercises for Section 1.4.

If we now put $f(x) = x$ in Law 6 and use Law 8, we get another useful special limit.

> 9. $\lim\limits_{x \to a} x^n = a^n$ where n is a positive integer

A similar limit holds for roots as follows. (For square roots the proof is outlined in Exercise 25 in Section 1.4.)

> 10. $\lim\limits_{x \to a} \sqrt[n]{x} = \sqrt[n]{a}$ where n is a positive integer
>
> (If n is even, we assume that $a > 0$.)

More generally, we have the following law, which is proved as a consequence of Law 10 in Section 1.5.

Root Law
> 11. $\lim\limits_{x \to a} \sqrt[n]{f(x)} = \sqrt[n]{\lim\limits_{x \to a} f(x)}$ where n is a positive integer
>
> (If n is even, we assume that $\lim\limits_{x \to a} f(x) > 0$.)

EXAMPLE 1   Find $\lim\limits_{x \to 5} (2x^2 - 3x + 4)$ and justify each step.

Solution

$$\lim_{x \to 5} (2x^2 - 3x + 4) = \lim_{x \to 5} (2x^2) - \lim_{x \to 5} (3x) + \lim_{x \to 5} 4 \qquad \text{(by Laws 2 and 1)}$$

$$= 2 \lim_{x \to 5} x^2 - 3 \lim_{x \to 5} x + \lim_{x \to 5} 4 \qquad \text{(by 3)}$$

$$= 2\,(5^2) - 3\,(5) + 4 \qquad \text{(by 9, 8, and 7)}$$

$$= 39$$

EXAMPLE 2   Find $\lim\limits_{x \to -2} \dfrac{x^3 + 2x^2 - 1}{5 - 3x}$ and justify each step.

Solution   We start by using Law 5, but its use is only fully justified at the final stage when we see that the limits of the numerator and denominator exist and the limit of the denominator is not 0.

$$\lim_{x \to -2} \frac{x^3 + 2x^2 - 1}{5 - 3x} = \frac{\lim\limits_{x \to -2} (x^3 + 2x^2 - 1)}{\lim\limits_{x \to -2} (5 - 3x)} \qquad \text{(by Law 5)}$$

$$= \frac{\lim\limits_{x \to -2} x^3 + 2 \lim\limits_{x \to -2} x^2 - \lim\limits_{x \to -2} 1}{\lim\limits_{x \to -2} 5 - 3 \lim\limits_{x \to -2} x} \qquad \text{(by Laws 1, 2, and 3)}$$

$$= \frac{(-2)^3 + 2\,(-2)^2 - 1}{5 - 3\,(-2)} \qquad \text{(by 9, 8, and 7)}$$

$$= -\frac{1}{11}$$

EXAMPLE 3   Calculate $\lim\limits_{x \to 1}\left[\sqrt[5]{x^2 - x} + (x^3 + x)^9\right]$ and justify each step.

Solution

$$\lim_{x \to 1}\left[\sqrt[5]{x^2 - x} + (x^3 + x)^9\right]$$

$$= \lim_{x \to 1}\sqrt[5]{x^2 - x} + \lim_{x \to 1}(x^3 + x)^9 \qquad \text{(by Law 1)}$$

$$= \sqrt[5]{\lim_{x \to 1}(x^2 - x)} + \left[\lim_{x \to 1}(x^3 + x)\right]^9 \qquad \text{(by 11 and 6)}$$

$$= \sqrt[5]{\lim_{x \to 1}x^2 - \lim_{x \to 1}x} + \left[\lim_{x \to 1}x^3 + \lim_{x \to 1}x\right]^9 \qquad \text{(by 2 and 1)}$$

$$= \sqrt[5]{1^2 - 1} + [1^3 + 1]^9 \qquad \text{(by 9 and 8)}$$

$$= 2^9 = 512 \qquad\qquad\qquad \bullet$$

Note: If we let $f(x) = 2x^2 - 3x + 4$, then $f(5) = 39$. In other words, we would have gotten the correct answer in Example 1 by substituting 5 for x. Similarly, direct substitution provides the correct answers in Examples 2 and 3. The functions in Examples 1 and 2 are a polynomial and a rational function, respectively, and similar use of the Limit Laws proves that direct substitution always works for such functions (see Exercises 75 and 76). We state this fact as follows.

(1.7) | If f is a polynomial or a rational function and a is in the domain of f, then
$$\lim_{x \to a}f(x) = f(a)$$

Functions with the direct substitution property stated in (1.7) are called *continuous at a* and will be studied in Section 1.5. However, not all limits can be evaluated by direct substitution, as the following examples show.

EXAMPLE 4   Find $\lim\limits_{x \to 1}\dfrac{x^2 - 1}{x - 1}$.

Solution   Let $f(x) = (x^2 - 1)/(x - 1)$. We cannot find the limit by substituting $x = 1$ because $f(1)$ is not defined. Nor can we apply the Quotient Law because the limit of the denominator is 0. Instead we need to do some preliminary algebra. We factor the numerator as a difference of squares:

$$\frac{x^2 - 1}{x - 1} = \frac{(x - 1)(x + 1)}{(x - 1)}$$

The numerator and denominator have a common factor of $x - 1$. When we take the limit as x approaches 1, we have $x \neq 1$ and so $x - 1 \neq 0$. Therefore we can cancel the common factor and compute the limit as follows:

$$\lim_{x \to 1}\frac{x^2 - 1}{x - 1} = \lim_{x \to 1}\frac{(x - 1)(x + 1)}{x - 1}$$

$$= \lim_{x \to 1}(x + 1)$$

$$= 1 + 1 = 2$$

The limit in this example arose in Section 1.1 when we were trying to find the tangent to the parabola $y = x^2$ at the point $(1, 1)$.

EXAMPLE 5   Find $\lim\limits_{x \to 1} g(x)$ where

$$g(x) = \begin{cases} x + 1 & \text{if } x \neq 1 \\ \pi & \text{if } x = 1 \end{cases}$$

Solution   Here g is defined at $x = 1$ and $g(1) = \pi$, but the value of a limit as x approaches 1 does not depend on the value of the function at 1. Since $g(x) = x + 1$ for $x \neq 1$, we have

$$\lim\limits_{x \to 1} g(x) = \lim\limits_{x \to 1} (x + 1) = 2$$

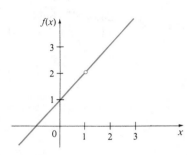

Note that the values of the functions in Examples 4 and 5 are identical except when $x = 1$ (see Figure 1.15) and so they have the same limit as x approaches 1.

EXAMPLE 6   Evaluate $\lim\limits_{h \to 0} \dfrac{(3 + h)^2 - 9}{h}$.

Solution   If we define

$$F(h) = \frac{(3 + h)^2 - 9}{h}$$

then, as in Example 4, we cannot compute $\lim_{h \to 0} F(h)$ by letting $h = 0$ since $F(0)$ is undefined. But if we simplify $F(h)$ algebraically, we find

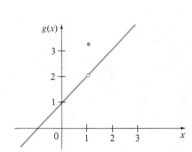

$$F(h) = \frac{(9 + 6h + h^2) - 9}{h} = \frac{6h + h^2}{h} = 6 + h$$

Figure 1.15

(Recall that we consider only $h \neq 0$ when letting h approach 0.) Thus

$$\lim\limits_{h \to 0} \frac{(3 + h)^2 - 9}{h} = \lim\limits_{h \to 0} (6 + h) = 6$$

EXAMPLE 7   Find $\lim\limits_{t \to 0} \dfrac{\sqrt{t + 9} - 3}{t}$.

Solution   We cannot apply the Quotient Law immediately, since the limit of the denominator is 0. Here the preliminary algebra consists of rationalizing the numerator:

$$\lim\limits_{t \to 0} \frac{\sqrt{t + 9} - 3}{t} = \lim\limits_{t \to 0} \frac{\sqrt{t + 9} - 3}{t} \cdot \frac{\sqrt{t + 9} + 3}{\sqrt{t + 9} + 3}$$

$$= \lim\limits_{t \to 0} \frac{(t + 9) - 9}{t(\sqrt{t + 9} + 3)} = \lim\limits_{t \to 0} \frac{t}{t(\sqrt{t + 9} + 3)}$$

$$= \lim\limits_{t \to 0} \frac{1}{\sqrt{t + 9} + 3} = \frac{1}{\sqrt{\lim\limits_{t \to 0} (t + 9)} + 3} = \frac{1}{3 + 3} = \frac{1}{6}$$

This calculation confirms the guess that we made in Example 2 in Section 1.2.

Some limits are best calculated by first finding the left- and right-hand limits. The following theorem is a reminder of what we discovered in Section 1.2. It says that a two-sided limit exists if and only if both of the one-sided limits exist and are equal. (For the proof, see Exercise 28 in Section 1.4.)

Theorem (1.8)

$$\lim_{x \to a} f(x) = L \qquad \text{if and only if} \qquad \lim_{x \to a^-} f(x) = L = \lim_{x \to a^+} f(x)$$

When computing one-sided limits we use the fact that the Limit Laws also hold for one-sided limits.

EXAMPLE 8   Show that $\lim_{x \to 0} |x| = 0$.

Solution   Recall that

$$|x| = \begin{cases} x & \text{if } x \geq 0 \\ -x & \text{if } x < 0 \end{cases}$$

Since $|x| = x$ for $x > 0$, we have

$$\lim_{x \to 0^+} |x| = \lim_{x \to 0^+} x = 0$$

For $x < 0$, we have $|x| = -x$ and so

$$\lim_{x \to 0^-} |x| = \lim_{x \to 0^-} (-x) = 0$$

Therefore by Theorem 1.8,

$$\lim_{x \to 0} |x| = 0$$

●

EXAMPLE 9   Prove that $\lim_{x \to 0} \dfrac{|x|}{x}$ does not exist.

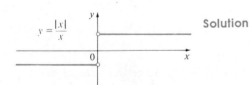

Solution
$$\lim_{x \to 0^+} \frac{|x|}{x} = \lim_{x \to 0^+} \frac{x}{x} = \lim_{x \to 0^+} 1 = 1$$

$$\lim_{x \to 0^-} \frac{|x|}{x} = \lim_{x \to 0^-} \frac{-x}{x} = \lim_{x \to 0^-} (-1) = -1$$

Since the right- and left-hand limits are different, it follows from Theorem 1.8 that **Figure 1.16** $\lim_{x \to 0} |x|/x$ does not exist. The graph of $f(x) = |x|/x$ is shown in Figure 1.16.   ●

EXAMPLE 10   If

$$f(x) = \begin{cases} \sqrt{x - 4} & \text{if } x > 4 \\ 8 - 2x & \text{if } x < 4 \end{cases}$$

determine whether $\lim_{x \to 4} f(x)$ exists.

Solution   Since $f(x) = \sqrt{x-4}$ for $x > 4$, we have

$$\lim_{x \to 4^+} f(x) = \lim_{x \to 4^+} \sqrt{x-4} = \sqrt{4-4} = 0$$

Since $f(x) = 8 - 2x$ for $x < 4$, we have

$$\lim_{x \to 4^-} f(x) = \lim_{x \to 4^-} (8 - 2x) = 8 - 2 \cdot 4 = 0$$

The right- and left-hand limits are equal. Thus the limit exists and

$$\lim_{x \to 4} f(x) = 0$$

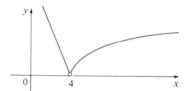

Figure 1.17   The graph of f is shown in Figure 1.17.

EXAMPLE 11   The *greatest integer function* is defined by $[\![x]\!] =$ the largest integer that is less than or equal to x. (For instance, $[\![4]\!] = 4$, $[\![4.8]\!] = 4$, $[\![\pi]\!] = 3$, $[\![\sqrt{2}]\!] = 1$, $[\![-\frac{1}{2}]\!] = -1$.) Show that $\lim_{x \to 3} [\![x]\!]$ does not exist.

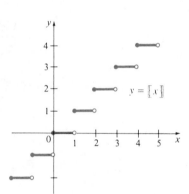

Solution   The graph of the greatest integer function is shown in Figure 1.18. Since $[\![x]\!] = 3$ for $3 \le x < 4$, we have

$$\lim_{x \to 3^+} [\![x]\!] = \lim_{x \to 3^+} 3 = 3$$

Since $[\![x]\!] = 2$ for $2 \le x < 3$, we have

$$\lim_{x \to 3^-} [\![x]\!] = \lim_{x \to 3^-} 2 = 2$$

Because these one-sided limits are not equal, $\lim_{x \to 3} [\![x]\!]$ does not exist by Theorem 1.8.

Figure 1.18

Greatest integer function

Note: If you have a calculator with an "integer part" key INT, note that $\text{INT}(x) = [\![x]\!]$ for $x \ge 0$, but be warned that in some calculators, $\text{INT}(x) = -[\![-x]\!]$ for $x < 0$. For instance, these calculators would compute $\text{INT}(-3.6) = -3$, whereas $[\![-3.6]\!] = -4$.

The next two theorems give two additional properties of limits. Their proofs can be found in Appendix C.

Theorem (1.9)

If $f(x) \le g(x)$ for all x in an open interval that contains a (except possibly at a) and the limits of f and g both exist as x approaches a, then

$$\lim_{x \to a} f(x) \le \lim_{x \to a} g(x)$$

The Squeeze Theorem (1.10)

If $f(x) \le g(x) \le h(x)$ for all x in an open interval that contains a (except possibly at a) and

$$\lim_{x \to a} f(x) = \lim_{x \to a} h(x) = L$$

then

$$\lim_{x \to a} g(x) = L$$

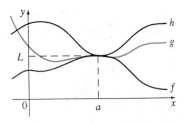

Figure 1.19

The Squeeze Theorem, sometimes called the Sandwich Theorem or the Pinching Theorem, is illustrated by Figure 1.19. It says that if $g(x)$ is squeezed between $f(x)$ and $h(x)$ near a, and if f and h have the same limit L at a, then g is forced to have the same limit L at a.

EXAMPLE 12   Show that $\lim\limits_{x \to 0} x \sin\frac{1}{x} = 0$.

Solution   First note that we *cannot* use

$$\lim_{x \to 0} x \sin\frac{1}{x} = \lim_{x \to 0} x \cdot \lim_{x \to 0} \sin\frac{1}{x}$$

because $\lim_{x \to 0} \sin(1/x)$ does not exist. (See Example 4 in Section 1.2.) However, since

$$-1 \le \sin\frac{1}{x} \le 1$$

we have, as illustrated by Figure 1.20,

$$-|x| \le x \sin\frac{1}{x} \le |x|$$

We know that

$$\lim_{x \to 0} |x| = 0 \qquad \text{and} \qquad \lim_{x \to 0} -|x| = 0$$

from Example 8. Taking $f(x) = -|x|$, $g(x) = x \sin(1/x)$ and $h(x) = |x|$ in the Squeeze Theorem, we obtain

$$\lim_{x \to 0} x \sin\frac{1}{x} = 0$$

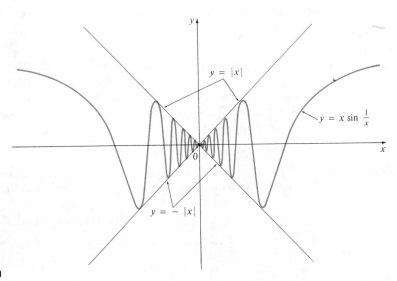

Figure 1.20

SECTION 1.3 **Exercises**

In Exercises 1–14 evaluate the given limit and justify each step by indicating the appropriate Limit Laws.

1. $\lim_{x \to 4} (5x^2 - 2x + 3)$

2. $\lim_{x \to -3} (x^3 + 2x^2 + 6)$

3. $\lim_{x \to 2} (x^2 + 1)(x^2 + 4x)$

4. $\lim_{x \to -2} (x^2 + x + 1)^5$

5. $\lim_{x \to -1} \dfrac{x - 2}{x^2 + 4x - 3}$

6. $\lim_{t \to -2} \dfrac{t^3 - t^2 - t + 10}{t^2 + 3t + 2}$

7. $\lim_{x \to -1} \sqrt{x^3 + 2x + 7}$

8. $\lim_{x \to 64} (\sqrt[3]{x} + 3\sqrt{x})$

9. $\lim_{t \to -2} (t + 1)^9 (t^2 - 1)$

10. $\lim_{r \to 3} (r^4 - 7r + 4)^{2/3}$

11. $\lim_{w \to -2} \sqrt[3]{\dfrac{4w + 3w^3}{3w + 10}}$

12. $\lim_{y \to 3} \dfrac{3(8y^2 - 1)}{2y^2(y - 1)^4}$

13. $\lim_{h \to 1/2} \dfrac{2h}{h + (1/h)}$

14. $\lim_{t \to 16} t^{-1/2} (t^2 - 14t)^{3/5}$

15. Given that
$$\lim_{x \to a} f(x) = -3 \qquad \lim_{x \to a} g(x) = 0 \qquad \lim_{x \to a} h(x) = 8$$
find the limits that exist.

(a) $\lim_{x \to a} [f(x) + h(x)]$ (b) $\lim_{x \to a} [f(x)]^2$

(c) $\lim_{x \to a} \sqrt[3]{h(x)}$

(d) $\lim_{x \to a} \dfrac{1}{f(x)}$

(e) $\lim_{x \to a} \dfrac{f(x)}{h(x)}$

(f) $\lim_{x \to a} \dfrac{g(x)}{f(x)}$

(g) $\lim_{x \to a} \dfrac{f(x)}{g(x)}$

(h) $\lim_{x \to a} \dfrac{2f(x)}{h(x) - f(x)}$

In Exercises 16–41 evaluate the limits, if they exist.

16. $\lim_{x \to 3} \dfrac{x^2 - x + 12}{x + 3}$

17. $\lim_{x \to -3} \dfrac{x^2 - x + 12}{x + 3}$

18. $\lim_{x \to -3} \dfrac{x^2 - x - 12}{x + 3}$

19. $\lim_{x \to 0} \dfrac{x + x^2}{x}$

20. $\lim_{x \to 1} \dfrac{x^2 - x - 2}{x + 1}$

21. $\lim_{x \to -1} \dfrac{x^2 - x - 2}{x + 1}$

22. $\lim_{x \to -1} \dfrac{x^2 - x - 3}{x + 1}$

23. $\lim_{t \to 1} \dfrac{t^3 - t}{t^2 - 1}$

24. $\lim_{x \to 1} \dfrac{x^3 - 1}{x^2 - 1}$

25. $\lim_{h \to 0} \dfrac{(h - 5)^2 - 25}{h}$

26. $\lim_{h \to 0} \dfrac{(2 + h)^3 - 8}{h}$

27. $\lim_{h \to 0} \dfrac{(1 + h)^4 - 1}{h}$

28. $\lim_{x \to 1} \dfrac{x^2 + x - 2}{x^2 - 3x + 2}$

29. $\lim_{x \to -2} \dfrac{x + 2}{x^2 - x - 6}$

30. $\lim_{t \to 2} \dfrac{t^2 + t - 6}{t^2 - 4}$

31. $\lim_{x \to -2} \left[\dfrac{x^2}{x + 2} + \dfrac{2x}{x + 2} \right]$

32. $\lim_{x \to 2} \dfrac{x^4 - 16}{x - 2}$

33. $\lim_{t \to 9} \dfrac{9 - t}{3 - \sqrt{t}}$

34. $\lim_{t \to 1} \dfrac{\sqrt[3]{t} - 1}{t - 1}$

35. $\lim_{t \to 0} \dfrac{\sqrt{2 - t} - \sqrt{2}}{t}$

36. $\lim_{x \to 1} \left[\dfrac{1}{x - 1} - \dfrac{2}{x^2 - 1} \right]$

37. $\lim_{x \to 9} \dfrac{x^2 - 81}{\sqrt{x} - 3}$

38. $\lim_{h \to 0} \dfrac{(3 + h)^{-1} - 3^{-1}}{h}$

39. $\lim_{t \to 0} \left[\dfrac{1}{t\sqrt{1 + t}} - \dfrac{1}{t} \right]$

40. $\lim_{x \to 2} \dfrac{\frac{1}{x} - \frac{1}{2}}{x - 2}$

41. $\lim_{x \to 0} \dfrac{x}{\sqrt{1 + 3x} - 1}$

42. If $1 \le f(x) \le x^2 + 2x + 2$ for all x, find $\lim_{x \to -1} f(x)$.

43. If $3x \le f(x) \le x^3 + 2$ for $0 \le x \le 2$, evaluate $\lim_{x \to 1} f(x)$.

44. Prove that $\lim_{x \to 0} x^2 \sin \frac{1}{x} = 0$.

45. Prove that $\lim_{x \to 0} \sqrt[3]{x} \sin \dfrac{1}{\sqrt[3]{x}} = 0$.

In Exercises 46–65 find the limit if it exists. If the limit does not exist, explain why. The symbol $[\![\ ]\!]$ denotes the greatest integer function, defined in Example 11.

46. $\lim_{x \to 0^+} (\sqrt[4]{x} - 1)$

47. $\lim_{x \to 5^+} (\sqrt{x - 5} + \sqrt{5x})$

48. $\lim_{x \to 4^-} \sqrt{16 - x^2}$

49. $\lim_{x \to -1.5^+} (\sqrt{3 + 2x} + x)$

50. $\lim_{x \to -4} |x + 4|$

51. $\lim_{x \to -4^-} \dfrac{|x + 4|}{x + 4}$

52. $\lim_{x \to 2} \dfrac{|x - 2|}{x - 2}$

53. $\lim_{x \to 2^+} \dfrac{1}{x - 2}$

54. $\lim_{x \to 9^-} [\![x]\!]$

55. $\lim_{x \to 9^+} [\![x]\!]$

56. $\lim_{x \to -2^+} [\![x]\!]$

57. $\lim_{x \to -2} [\![x]\!]$

58. $\lim_{x \to -2.4} [\![x]\!]$

59. $\lim_{x \to 8^+} (\sqrt{x - 8} + [\![x + 1]\!])$

60. $\lim_{x \to 1^+} \sqrt{x^2 + x - 2}$

61. $\lim_{x \to -2^-} \sqrt{x^2 + x - 2}$

62. $\displaystyle\lim_{x\to 0^+}\frac{2+x^{3/2}}{\sqrt{x+2}}$

63. $\displaystyle\lim_{x\to 5^+} x\sqrt[6]{x^2-25}$

64. $\displaystyle\lim_{x\to 0^-}\left(\frac{1}{x}-\frac{1}{|x|}\right)$

65. $\displaystyle\lim_{x\to 0^+}\left(\frac{1}{x}-\frac{1}{|x|}\right)$

66. The *signum* (or *sign*) *function*, denoted by sgn, is defined by
$$\text{sgn } x = \begin{cases} -1 & \text{if } x<0 \\ 0 & \text{if } x=0 \\ 1 & \text{if } x>0 \end{cases}$$
(a) Sketch the graph of this function.
(b) Find the following limits or explain why they do not exist.
    (i) $\displaystyle\lim_{x\to 0^+}\text{sgn } x$ (ii) $\displaystyle\lim_{x\to 0^-}\text{sgn } x$

    (iii) $\displaystyle\lim_{x\to 0}\text{sgn } x$ (iv) $\displaystyle\lim_{x\to 0}|\text{sgn } x|$

67. The Heaviside function H was defined in Example 7 in Section 1.2. Use Theorem 1.8 to show that $\lim_{t\to 0} H(t)$ does not exist.

68. Let
$$f(x) = \begin{cases} x^2-2x+2 & \text{if } x<1 \\ 3-x & \text{if } x\geq 1 \end{cases}$$
(a) Find $\lim_{x\to 1^-} f(x)$ and $\lim_{x\to 1^+} f(x)$.
(b) Does $\lim_{x\to 1} f(x)$ exist?
(c) Sketch the graph of f.

69. Let
$$g(x) = \begin{cases} -x^3 & \text{if } x<-1 \\ (x+2)^2 & \text{if } x>-1 \end{cases}$$
(a) Find $\lim_{x\to -1^-} g(x)$ and $\lim_{x\to -1^+} g(x)$.
(b) Does $\lim_{x\to -1} g(x)$ exist?
(c) Sketch the graph of g.

70. Let
$$h(x) = \begin{cases} x & \text{if } x<0 \\ x^2 & \text{if } 0<x\leq 2 \\ 8-x & \text{if } x>2 \end{cases}$$
(a) Evaluate the following limits if they exist.
    (i) $\displaystyle\lim_{x\to 0^+} h(x)$ (ii) $\displaystyle\lim_{x\to 0} h(x)$ (iii) $\displaystyle\lim_{x\to 1} h(x)$

    (iv) $\displaystyle\lim_{x\to 2^-} h(x)$ (v) $\displaystyle\lim_{x\to 2^+} h(x)$ (vi) $\displaystyle\lim_{x\to 2} h(x)$

(b) Sketch the graph of h.

71. (a) If n is an integer, evaluate
    (i) $\displaystyle\lim_{x\to n^-}[\![x]\!]$ (ii) $\displaystyle\lim_{x\to n^+}[\![x]\!]$
(b) For what values of a does $\lim_{x\to a}[\![x]\!]$ exist?

72. Let $f(x) = x - [\![x]\!]$.
(a) Sketch the graph of f.
(b) If n is an integer, evaluate
    (i) $\displaystyle\lim_{x\to n^-} f(x)$ (ii) $\displaystyle\lim_{x\to n^+} f(x)$
(c) For what values of a does $\lim_{x\to a} f(x)$ exist?
(d) If you have a calculator with a "fractional part" key FRAC, how is FRAC related to f? (See the note after Example 11.) Sketch the graphs of INT and FRAC.

73. Let $F(x) = \dfrac{x^2-1}{|x-1|}$.
(a) Find (i) $\displaystyle\lim_{x\to 1^+} F(x)$ (ii) $\displaystyle\lim_{x\to 1^-} F(x)$
(b) Does $\lim_{x\to 1} F(x)$ exist?
(c) Sketch the graph of F.

74. Let $g(x) = [\![x/2]\!]$.
(a) Sketch the graph of g.
(b) Evaluate the following limits if they exist.
    (i) $\displaystyle\lim_{x\to 1^+} g(x)$ (ii) $\displaystyle\lim_{x\to 1^-} g(x)$ (iii) $\displaystyle\lim_{x\to 1} g(x)$

    (iv) $\displaystyle\lim_{x\to 2^+} g(x)$ (v) $\displaystyle\lim_{x\to 2^-} g(x)$ (vi) $\displaystyle\lim_{x\to 2} g(x)$

(c) For what values of a does $\lim_{x\to a} g(x)$ exist?

75. If p is a polynomial, show that $\lim_{x\to a} p(x) = p(a)$.

76. If r is a rational function, use Exercise 75 to show that $\lim_{x\to a} r(x) = r(a)$ for every number a in the domain of r.

77. If
$$f(x) = \begin{cases} x^2 & \text{if } x \text{ is rational} \\ 0 & \text{if } x \text{ is irrational} \end{cases}$$
prove that $\lim_{x\to 0} f(x) = 0$.

78. Show by means of an example that $\lim_{x\to a}[f(x)+g(x)]$ may exist even though neither $\lim_{x\to a} f(x)$ nor $\lim_{x\to a} g(x)$ exists.

79. Show by means of an example that $\lim_{x\to a}[f(x)g(x)]$ may exist even though neither $\lim_{x\to a} f(x)$ nor $\lim_{x\to a} g(x)$ exists.

80. Evaluate $\displaystyle\lim_{x\to 2}\frac{\sqrt{6-x}-2}{\sqrt{3-x}-1}$.

81. Evaluate $\displaystyle\lim_{x\to 0}\frac{\sqrt[3]{1+cx}-1}{x}$.

82. Evaluate $\displaystyle\lim_{x\to 1}\frac{\sqrt[3]{x}-1}{\sqrt{x}-1}$.

SECTION 1.4

The Precise Definition of a Limit

The intuitive definition of a limit given in Section 1.2 is inadequate for some purposes because such phrases as "x is close to 2" and "$f(x)$ gets closer and closer to L" are vague. In order to be able to prove conclusively that

$$\lim_{x \to 0}\left(x^3 + \frac{\cos 5x}{10,000}\right) = 0.0001 \qquad \text{or} \qquad \lim_{x \to 0}\frac{\sin x}{x} = 1$$

we must make the definition of a limit more precise.

To motivate the precise definition of a limit, let us consider the function

$$f(x) = \begin{cases} 2x - 1 & \text{if } x \neq 3 \\ 6 & \text{if } x = 3 \end{cases}$$

Intuitively it is clear that when x is close to 3 but $x \neq 3$, $f(x)$ is close to 5, and so $\lim_{x \to 3} f(x) = 5$.

To obtain more detailed information about how $f(x)$ varies when x is close to 3, let us ask the following question.

How close to 3 does x have to be so that $f(x)$ differs from 5 by less than 0.1?

The distance from x to 3 is $|x - 3|$ and the distance from $f(x)$ to 5 is $|f(x) - 5|$, so our problem is to find a number δ such that

$$|f(x) - 5| < 0.1 \qquad \text{if} \qquad |x - 3| < \delta \text{ but } x \neq 3$$

[It is traditional to use the Greek letter δ (delta) in this situation.] If $|x - 3| > 0$, then $x \neq 3$, so an equivalent formulation of our problem is to find a number δ such that

$$|f(x) - 5| < 0.1 \qquad \text{if} \qquad 0 < |x - 3| < \delta$$

Notice that if $0 < |x - 3| < (0.1)/2 = 0.05$, then

$$|f(x) - 5| = |(2x - 1) - 5| = |2x - 6| = 2|x - 3| < 0.1$$

that is,

$$|f(x) - 5| < 0.1 \qquad \text{if} \qquad 0 < |x - 3| < 0.05$$

Thus an answer to the problem is given by $\delta = 0.05$; that is, if x is within a distance of 0.05 from 3, then $f(x)$ will be within a distance of 0.1 from 5.

If we change the number 0.1 in our problem to the smaller number 0.01, then by using the same method we find that $f(x)$ will differ from 5 by less than 0.01 provided that x differs from 3 by less than $(0.01)/2 = 0.005$:

$$|f(x) - 5| < 0.01 \qquad \text{if} \qquad 0 < |x - 3| < 0.005$$

Similarly,

$$|f(x) - 5| < 0.001 \qquad \text{if} \qquad 0 < |x - 3| < 0.0005$$

If, instead of tolerating an error of 0.1 or 0.01 or 0.001, we want accuracy to within a tolerance of an arbitrary positive number ε (the Greek letter epsilon), then we find as before that

(1.11)
$$|f(x) - 5| < \varepsilon \qquad \text{if} \qquad 0 < |x - 3| < \delta = \frac{\varepsilon}{2}$$

This is a precise way of saying that $f(x)$ is close to 5 when x is close to 3 because (1.11) says that we can make the values of $f(x)$ within an arbitrary distance ε from 5 by taking the values of x within a distance $\varepsilon/2$ from 3 (but $x \neq 3$).

Note that (1.11) can be rewritten as

$$5 - \varepsilon < f(x) < 5 + \varepsilon \qquad \text{whenever} \qquad 3 - \delta < x < 3 + \delta \qquad (x \neq 3)$$

and this is illustrated in Figure 1.21. By taking the values of x ($\neq 3$) to lie in the interval $(3 - \delta, 3 + \delta)$ we can make the values of $f(x)$ lie in the interval $(5 - \varepsilon, 5 + \varepsilon)$.

Using (1.11) as a model, we give a precise definition of a limit.

Definition (1.12)

Let f be a function defined on some open interval that contains the number a, except possibly at a itself. Then we say that the **limit of $f(x)$ as x approaches a is L**, and we write

$$\lim_{x \to a} f(x) = L$$

if for every number $\varepsilon > 0$ there is a corresponding number $\delta > 0$ such that

$$|f(x) - L| < \varepsilon \qquad \text{whenever} \qquad 0 < |x - a| < \delta$$

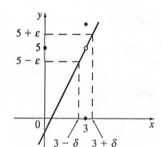

Figure 1.21

Another way of writing the last line of this definition is

$$\text{if } 0 < |x - a| < \delta \qquad \text{then} \qquad |f(x) - L| < \varepsilon$$

Another notation for $\lim_{x \to a} f(x) = L$ is

$$f(x) \to L \text{ as } x \to a$$

Since $|x - a|$ is the distance from x to a and $|f(x) - L|$ is the distance from $f(x)$ to L, and since ε can be arbitrarily small, the definition of a limit can be expressed in words as follows:

$\lim_{x \to a} f(x) = L$ means that the distance between $f(x)$ and L can be made arbitrarily small by taking the distance from x to a sufficiently small (but not 0).

Alternatively,

$\lim_{x \to a} f(x) = L$ means that the values of $f(x)$ can be made as close as we please to L by taking x close enough to a (but not equal to a).

We can also reformulate Definition 1.12 in terms of intervals by observing that the inequality $|x - a| < \delta$ is equivalent to $-\delta < x - a < \delta$, which in turn can be written as $a - \delta < x < a + \delta$. Also $0 < |x - a|$ is true if and only if $x - a \neq 0$, that is, $x \neq a$. Similarly, the inequality $|f(x) - L| < \varepsilon$ is equivalent to $L - \varepsilon < f(x) < L + \varepsilon$. Therefore, in terms of intervals, Definition 1.12 can be stated as follows:

$\lim_{x \to a} f(x) = L$ means that for every $\varepsilon > 0$ (no matter how small ε is) we can find $\delta > 0$ such that if x lies in the open interval $(a - \delta, a + \delta)$ and $x \neq a$, then $f(x)$ lies in the open interval $(L - \varepsilon, L + \varepsilon)$.

We interpret this statement geometrically by representing a function by an arrow diagram as in Figure 1.22, where f maps a subset of R onto another subset of R.

Figure 1.22

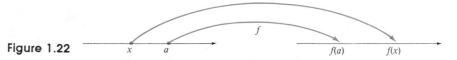

The definition of limit says that if any small interval $(L - \varepsilon, L + \varepsilon)$ is given around L, then we can find an interval $(a - \delta, a + \delta)$ around a such that f maps all the points in $(a - \delta, a + \delta)$ (except possibly a) into the interval $(L - \varepsilon, L + \varepsilon)$ (see Figure 1.23).

Figure 1.23

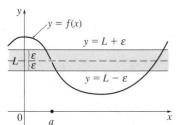

Figure 1.24

Another geometric interpretation of limits can be given in terms of the graph of a function. If $\varepsilon > 0$ is given, then we draw the horizontal lines $y = L + \varepsilon$ and $y = L - \varepsilon$ and the graph of f (see Figure 1.24). If $\lim_{x \to a} f(x) = L$, then we can find a number $\delta > 0$ such that if we restrict x to lie in the interval $(a - \delta, a + \delta)$ and take $x \neq a$, then the curve $y = f(x)$ lies between the lines $y = L - \varepsilon$ and $y = L + \varepsilon$ (see Figure 1.25). You can see that if such a δ has been found, then any smaller δ will also work.

It is important to realize that the process illustrated in Figures 1.24 and 1.25 must work for *every* positive number ε no matter how small it is chosen. Figure 1.26 shows that if a smaller ε is chosen, then a smaller δ may be required.

In proving limit statements it may be helpful to think of the definition of limit as a challenge. First it challenges you with a number ε. Then you must be able to produce a suitable δ. You have to be able to do this for *every* $\varepsilon > 0$, not just a particular ε.

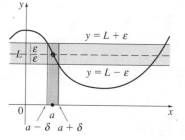

Figure 1.25

Imagine a contest between two people, A and B, and imagine yourself to be B. A stipulates that the fixed number L should be approximated by the values of $f(x)$ to within a degree of accuracy ε (say, 0.01). B then responds by finding a number δ such that $|f(x) - L| < \varepsilon$ whenever $0 < |x - a| < \delta$. Then A may become more exacting and challenge B with a smaller value of ε (say, 0.0001). Again B has to respond by finding a corresponding δ. Usually the smaller the value of ε, the smaller the corresponding value of δ must be. If B always wins, no matter how small A makes ε, then $\lim_{x \to a} f(x) = L$.

EXAMPLE 1   Prove that $\lim_{x \to 3} (4x - 5) = 7$.

Solution

1.   *Preliminary analysis of the problem (guessing a value for δ).* Let ε be a given positive number. We want to find a number δ such that

$$|(4x - 5) - 7| < \varepsilon \qquad \text{whenever} \qquad 0 < |x - 3| < \delta$$

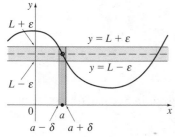

Figure 1.26

But $|(4x - 5) - 7| = |4x - 12| = |4(x - 3)| = 4|x - 3|$. Therefore we want

$$4|x - 3| < \varepsilon \qquad \text{whenever} \qquad 0 < |x - 3| < \delta$$

that is, $|x - 3| < \frac{\varepsilon}{4}$ whenever $0 < |x - 3| < \delta$

This suggests that we should choose $\delta = \varepsilon/4$.

   2.   *Proof (showing that the δ works).* Given $\varepsilon > 0$, choose $\delta = \varepsilon/4$. If $0 < |x - 3| < \delta$, then

$$|(4x - 5) - 7| = |4x - 12| = 4|x - 3| < 4\delta = 4\left(\frac{\varepsilon}{4}\right) = \varepsilon$$

Thus $|(4x - 5) - 7| < \varepsilon$ whenever $0 < |x - 3| < \delta$

Therefore, by the definition of a limit,

$$\lim_{x \to 3}(4x - 5) = 7$$

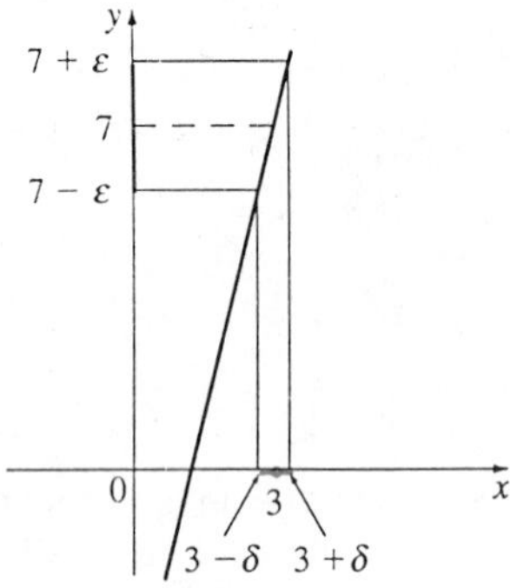

Figure 1.27   This example is illustrated by Figure 1.27.   ●

   Note that in the solution of Example 1 there were two stages—guessing and proving. We made a preliminary analysis that enabled us to guess a value for δ. But then in the second stage we had to go back and prove in a careful logical fashion that we had made a correct guess. This procedure is typical of much of mathematics. Sometimes it is necessary to first make an intelligent guess about the answer to a problem and then prove that the guess is correct.

EXAMPLE 2   Show that $\lim_{x \to 0}\dfrac{1}{x^2}$ does not exist.

Solution   The function $f(x) = 1/x^2$ does not have a value at $x = 0$, but that does not matter. [Recall that $\lim_{x \to a} f(x)$ can exist even if $f(a)$ is not defined.]

   To show that a limit does not exist we use an indirect proof. Suppose that

$$\lim_{x \to 0}\frac{1}{x^2} = L$$

for some number L. Then, if we take $\varepsilon = 1$, for example, the definition of limit says that there exists a number δ such that

$$\left|\frac{1}{x^2} - L\right| < 1   \text{whenever}   -\delta < x < \delta   (x \neq 0)$$

that is, $L - 1 < \dfrac{1}{x^2} < L + 1$ whenever $-\delta < x < \delta$ $(x \neq 0)$

This is impossible, since $1/x^2$ takes on arbitrarily large values in any interval $(-\delta, \delta)$ (see Figure 1.28). In fact,

$$\frac{1}{x^2} > L + 1   \text{if}   0 < x < \frac{1}{\sqrt{L + 1}}$$

Therefore $\lim_{x \to 0} 1/x^2$ does not exist.   ●

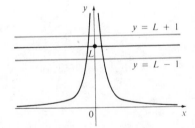

Figure 1.28

   The intuitive definitions of one-sided limits that were given in Section 1.2 can be precisely reformulated as follows.

Definition of Left-Hand Limit (1.13)	$$\lim_{x \to a^-} f(x) = L$$ if for every number $\varepsilon > 0$ there is a corresponding number $\delta > 0$ such that $$|f(x) - L| < \varepsilon \qquad \text{whenever} \qquad a - \delta < x < a$$

Definition of Right-Hand Limit (1.14)	$$\lim_{x \to a^+} f(x) = L$$ if for every number $\varepsilon > 0$ there is a corresponding number $\delta > 0$ such that $$|f(x) - L| < \varepsilon \qquad \text{whenever} \qquad a < x < a + \delta$$

Notice that Definition 1.13 is the same as Definition 1.12 except that x is restricted to lie in the *left* half $(a - \delta, a)$ of the interval $(a - \delta, a + \delta)$. In Definition 1.14 x is restricted to lie in the *right* half $(a, a + \delta)$ of the interval $(a - \delta, a + \delta)$.

EXAMPLE 3   Use Definition 1.14 to prove that $\lim_{x \to 0^+} \sqrt{x} = 0$.

Solution

1. *Guessing a value for δ.* Let ε be a given positive number. Here $a = 0$ and $L = 0$, so we want to find a number δ such that

$$|\sqrt{x} - 0| < \varepsilon \qquad \text{whenever} \qquad 0 < x < \delta$$

that is,
$$\sqrt{x} < \varepsilon \qquad \text{whenever} \qquad 0 < x < \delta$$

or, squaring both sides of the inequality $\sqrt{x} < \varepsilon$, we get

$$x < \varepsilon^2 \qquad \text{whenever} \qquad 0 < x < \delta$$

This suggests that we should choose $\delta = \varepsilon^2$.

2. *Showing that δ works.*   Given $\varepsilon > 0$, let $\delta = \varepsilon^2$. If $0 < x < \delta$, then

$$\sqrt{x} < \sqrt{\delta} = \sqrt{\varepsilon^2} = \varepsilon$$

so
$$|\sqrt{x} - 0| < \varepsilon$$

According to Definition 1.14, this shows that $\lim_{x \to 0^+} \sqrt{x} = 0$.

EXAMPLE 4   Prove that $\lim_{x \to 3} x^2 = 9$.

Solution

1. *Guessing a value for δ.*   Let $\varepsilon > 0$ be given. We have to find a number $\delta > 0$

such that

$$|x^2 - 9| < \varepsilon \qquad \text{whenever} \qquad 0 < |x - 3| < \delta$$

To connect $|x^2 - 9|$ with $|x - 3|$ we write $|x^2 - 9| = |(x - 3)(x + 3)|$. Then we want

$$|x + 3||x - 3| < \varepsilon \qquad \text{whenever} \qquad 0 < |x - 3| < \delta$$

Notice that if we can find a positive constant C such that $|x + 3| < C$, then

$$|x + 3||x - 3| < C|x - 3|$$

and we can make $C|x - 3| < \varepsilon$ by taking $|x - 3| < \varepsilon/C = \delta$.

We can find such a number C if we restrict x to lie in some interval centered at 3. In fact, since we are interested only in values of x that are close to 3, it is reasonable to assume that x is within a distance 1 from 3, that is, $|x - 3| < 1$. Then $2 < x < 4$, so $5 < x + 3 < 7$. Thus we have $|x + 3| < 7$ and so $C = 7$ is a suitable choice for the constant.

But now there are two restrictions on $|x - 3|$, namely,

$$|x - 3| < 1 \qquad \text{and} \qquad |x - 3| < \frac{\varepsilon}{C} = \frac{\varepsilon}{7}$$

To make sure that both of these inequalities are satisfied, we take δ to be the smaller of the two numbers 1 and $\varepsilon/7$. The notation for this is $\delta = \min\{1, \varepsilon/7\}$.

2.   *Showing that δ works.*   Given $\varepsilon > 0$, let $\delta = \min\{1, \varepsilon/7\}$. If $0 < |x - 3| < \delta$, then $|x - 3| < 1 \Rightarrow 2 < x < 4 \Rightarrow |x + 3| < 7$ (as in part 1). Also $|x - 3| < \varepsilon/7$, so

$$|x^2 - 9| = |x + 3||x - 3| < 7 \cdot \frac{\varepsilon}{7} = \varepsilon$$

This shows that $\lim_{x \to 3} x^2 = 9$.   ●

As Example 4 shows, it is not always easy to prove that limiting statements are true using the ε, δ definition. In fact for a more complicated function such as $f(x) = (6x^2 - 8x + 9)/(2x^2 - 1)$, it would require a great deal of ingenuity. Fortunately this is not necessary because the Limit Laws stated in Section 1.3 can be proved using Definition 1.12, and then the limits of complicated functions can be found rigorously from the Limit Laws without resorting to the definition directly.

As a sample, we prove the Sum Law: If $\lim_{x \to a} f(x) = L$ and $\lim_{x \to a} g(x) = M$ both exist, then

$$\lim_{x \to a} [f(x) + g(x)] = L + M$$

The remaining laws are proved in the exercises and in Appendix C.

Proof of the Sum Law   Let $\varepsilon > 0$ be given. We must find $\delta > 0$ such that

$$|f(x) + g(x) - (L + M)| < \varepsilon \qquad \text{whenever} \qquad 0 < |x - a| < \delta$$

Using the Triangle Inequality [(7) in Section 1 of Review and Preview] we can write

(1.15)
$$|f(x) + g(x) - (L + M)| = |(f(x) - L) + (g(x) - M)|$$
$$\leq |f(x) - L| + |g(x) - M|$$

We make $|f(x) + g(x) - (L + M)|$ less than ε by making each of the terms $|f(x) - L|$ and $|g(x) - M|$ less than $\varepsilon/2$.

Since $\varepsilon/2 > 0$ and $\lim_{x \to a} f(x) = L$, there exists a number $\delta_1 > 0$ such that

$$|f(x) - L| < \frac{\varepsilon}{2} \qquad \text{whenever} \qquad 0 < |x - a| < \delta_1$$

Similarly, since $\lim_{x \to a} g(x) = M$, there exists a number $\delta_2 > 0$ such that

$$|g(x) - M| < \frac{\varepsilon}{2} \qquad \text{whenever} \qquad 0 < |x - a| < \delta_2$$

Let $\delta = \min\{\delta_1, \delta_2\}$. Notice that

$$\text{if} \quad 0 < |x - a| < \delta \quad \text{then} \quad 0 < |x - a| < \delta_1 \quad \text{and} \quad 0 < |x - a| < \delta_2$$

and so

$$|f(x) - L| < \frac{\varepsilon}{2} \quad \text{and} \quad |g(x) - M| < \frac{\varepsilon}{2}$$

Therefore, by (1.15),

$$|f(x) + g(x) - (L + M)| \leq |f(x) - L| + |g(x) - M|$$
$$< \frac{\varepsilon}{2} + \frac{\varepsilon}{2} = \varepsilon$$

To summarize,

$$|f(x) + g(x) - (L + M)| < \varepsilon \qquad \text{whenever} \qquad 0 < |x - a| < \delta$$

Thus, by the definition of a limit,

$$\lim_{x \to a} [f(x) + g(x)] = L + M$$

SECTION 1.4 **Exercises**

1. How close to 3 do we have to take x so that $6x + 1$ is within a distance of (a) 0.1 and (b) 0.01 from 19?

2. How close to 2 do we have to take x so that $8x - 5$ is within a distance of (a) 0.01, (b) 0.001, and (c) 0.0001 from 11?

In Exercises 3-6 prove the statements using the ε, δ definition of limit and illustrate with a diagram like Figure 1.27.

3. $\lim_{x \to 2} (3x - 2) = 4$

4. $\lim_{x \to 4} (5 - 2x) = -3$

5. $\lim_{x \to -1} (5x + 8) = 3$

6. $\lim_{x \to -1} (3 - 4x) = 7$

In Exercises 7-20 prove the statements using the ε, δ definition of limit.

7. $\lim_{x \to 2} \frac{x}{7} = \frac{2}{7}$

8. $\lim_{x \to 4} \left(\frac{x}{3} + 1\right) = \frac{7}{3}$

9. $\lim_{x \to -5} \left(4 - \frac{3x}{5}\right) = 7$

10. $\lim_{x \to 3} x = 3$

11. $\lim_{x \to a} x = a$

12. $\lim_{x \to 1} \pi = \pi$

13. $\lim_{x \to 2} c = c$

14. $\lim_{x \to a} c = c$

15. $\lim_{x \to 0} x^2 = 0$

16. $\lim_{x \to 0} x^3 = 0$

17. $\lim_{x \to 0} |x| = 0$

18. $\lim_{x \to 9^-} \sqrt[4]{9 - x} = 0$

19. $\lim_{x \to -2} (x^2 - 1) = 3$

20. $\lim_{x \to 2} x^3 = 8$

21. Prove that the following limits do not exist.

(a) $\lim_{x \to 0} \frac{1}{x}$

(b) $\lim_{x \to -4} \frac{1}{(x + 4)^2}$

22. Draw a diagram like Figure 1.25 to illustrate Example 4 by showing graphically that $\lim_{x \to 3} x^2 = 9$.

23. Verify that another possible choice of δ for showing that $\lim_{x \to 3} x^2 = 9$ in Example 4 is $\delta = \min\{2, \varepsilon/8\}$.

24. Prove that $\lim_{x \to 2} \frac{1}{x} = \frac{1}{2}$.

25. Prove that $\lim_{x \to a} \sqrt{x} = \sqrt{a}$ if $a > 0$.

$$\left[\textit{Hint: Use } |\sqrt{x} - \sqrt{a}| = \frac{|x - a|}{\sqrt{x} + \sqrt{a}}. \right]$$

26. If H is the Heaviside function defined in Example 7 in Section 1.2, prove, using Definition 1.12, that $\lim_{t \to 0} H(t)$ does not exist.

27. If the function f is defined by

$$f(x) = \begin{cases} 0 & \text{if } x \text{ is rational} \\ 1 & \text{if } x \text{ is irrational} \end{cases}$$

prove that $\lim_{x \to 0} f(x)$ does not exist.

28. By comparing Definition 1.12, 1.13, and 1.14, prove Theorem 1.8.

Continuity

We noticed in Section 1.3 that the limit of a function as x approaches a can often be found simply by calculating the value of the function at a. Functions with this property are called *continuous* at a. We will see that the mathematical definition of continuity corresponds closely with the meaning of the word *continuity* in everyday language. (A continuous process is one that takes place gradually, without interruption or abrupt change.)

Definition (1.16)

A function f is **continuous at a number a** if

$$\lim_{x \to a} f(x) = f(a)$$

If f is not continuous at a, we say f is **discontinuous at a**, or f has a **discontinuity** at a. Notice that Definition 1.16 implicitly requires three things if f is continuous at a:

1. $f(a)$ is defined (that is, a is in the domain of f).
2. $\lim_{x \to a} f(x)$ exists (so f must be defined on an open interval that contains a).
3. $\lim_{x \to a} f(x) = f(a)$.

Intuitively speaking, f is continuous at a if $f(x)$ gets closer and closer to $f(a)$ as x gets closer and closer to a. Thus a continuous function f has the property that a small change in x produces only a small change in $f(x)$. In fact, the change in $f(x)$ can be kept as small as we please by keeping the change in x sufficiently small.

Physical phenomena are usually continuous. For instance, the displacement or velocity of a vehicle varies continuously with time, as does a person's height. But discontinuities do occur in such situations as electric currents. (See Example 7 in Section 1.2, where the Heaviside function is discontinuous at 0 because $\lim_{t \to 0} H(t)$ does not exist.)

Geometrically, you can think of a function that is continuous at every number in an interval as a function whose graph has no breaks in it. The graph can be drawn without removing your pen from the paper.

EXAMPLE 1 $f(x) = \dfrac{x^2 - x - 2}{x - 2}$

This function is discontinuous at $a = 2$ since $f(2)$ is not defined.

EXAMPLE 2 $f(x) = \begin{cases} \dfrac{1}{x^2} & \text{if } x \neq 0 \\ 1 & \text{if } x = 0 \end{cases}$

This function is discontinuous at $a = 0$. Here $f(0) = 1$ is defined but

$$\lim_{x \to 0} f(x) = \lim_{x \to 0} \frac{1}{x^2}$$

does not exist. (See Example 6 in Section 1.2 or Example 2 in Section 1.4)

EXAMPLE 3 $f(x) = \begin{cases} \dfrac{x^2 - x - 2}{x - 2} & \text{if } x \neq 2 \\ 1 & \text{if } x = 2 \end{cases}$

This function has a discontinuity at $a = 2$. Here $f(2) = 1$ is defined and

$$\lim_{x \to 2} f(x) = \lim_{x \to 2} \frac{x^2 - x - 2}{x - 2} = \lim_{x \to 2} (x + 1) = 3$$

exists. But

$$\lim_{x \to 2} f(x) \neq f(2)$$

so f is not continuous at 2.

EXAMPLE 4   The greatest integer function $f(x) = [\![x]\!]$ has discontinuities at all of the integers because $\lim_{x \to n} [\![x]\!]$ does not exist if n is an integer. (See Example 11 and Exercise 71 in Section 1.3.)

Figure 1.29 shows the graphs of the functions in Examples 1–4. In each case the graph cannot be drawn without lifting the pen from the paper because there is a hole or break or jump in the graph. The kind of discontinuity illustrated in parts (a) and (c) is called **removable** because we could remove the discontinuity by redefining f at 2. [The function $g(x) = x + 1$ is continuous.] The discontinuity in part (b) is called an **infinite discontinuity**. The discontinuities in part (d) are called **jump discontinuities** because the function "jumps" from one value to another.

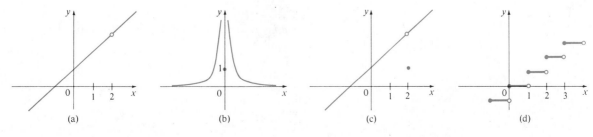

(a)   (b)   (c)   (d)

Figure 1.29

Definition (1.17)

A function f is **continuous from the right at a number a** if

$$\lim_{x \to a^+} f(x) = f(a)$$

and f is **continuous from the left at a** if

$$\lim_{x \to a^-} f(x) = f(a)$$

EXAMPLE 5   At each integer n, the function $f(x) = [\![x]\!]$ [see Figure 1.29(d)] is continuous from the right but discontinuous from the left because

$$\lim_{x \to n^+} f(x) = \lim_{x \to n^+} [\![x]\!] = n = f(n)$$

but

$$\lim_{x \to n^-} f(x) = \lim_{x \to n^-} [\![x]\!] = n - 1 \neq f(n)$$

Definition (1.18)

A function f is **continuous on the open interval** (a, b) [or (a, ∞) or $(-\infty, b)$ or $(-\infty, \infty)$] if it is continuous at every number in the interval. It is **continuous on the closed interval** $[a, b]$ if it is continuous on (a, b) and is also continuous from the right at a and continuous from the left at b.

Similarly, one can define continuity of functions on other types of intervals such as $[a, b)$, $(a, b]$, $[a, \infty)$, and $(-\infty, b]$. For instance, f is continuous on $[a, b)$ if it is continuous on (a, b) and continuous from the right at a.

EXAMPLE 6   Show that the function $f(x) = 1 - \sqrt{1 - x^2}$ is continuous on the interval $[-1, 1]$.

Solution   If $-1 < a < 1$, then using the Limit Laws, we have

$$\lim_{x \to a} f(x) = \lim_{x \to a} \left(1 - \sqrt{1 - x^2}\right)$$
$$= 1 - \lim_{x \to a} \sqrt{1 - x^2} \qquad \text{(by Laws 2 and 7)}$$
$$= 1 - \sqrt{\lim_{x \to a} (1 - x^2)} \qquad \text{(by 11)}$$
$$= 1 - \sqrt{1 - a^2} \qquad \text{(by 2, 7, and 9)}$$
$$= f(a)$$

Thus, by Definition 1.16, f is continuous at a if $-1 < a < 1$. We must also calculate the right-hand limit at -1 and the left-hand limit at 1.

$$\lim_{x \to -1^+} f(x) = \lim_{x \to -1^+} \left(1 - \sqrt{1 - x^2}\right)$$
$$= 1 - \sqrt{\lim_{x \to -1^+} (1 - x^2)} \qquad \text{(as above)}$$

$$= 1 - \sqrt{1 - 1^2}$$

$$= 1 = f(-1)$$

So f is continuous from the right at -1. Similarly,

$$\lim_{x \to 1^-} f(x) = \lim_{x \to 1^-} \left(1 - \sqrt{1 - x^2}\right)$$

$$= 1 - \sqrt{\lim_{x \to 1^-}(1 - x^2)} = 1 - 0 = 1 = f(1)$$

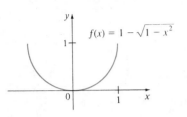

So f is continuous from the left at 1. Therefore, according to Definition 1.18, f is continuous on $[-1, 1]$.

The graph of f is sketched in Figure 1.30. It is the lower half of the circle given by

Figure 1.30 $x^2 + (y - 1)^2 = 1$.   ●

Instead of always using Definitions 1.16, 1.17, and 1.18 to verify the continuity of a function as we did in Example 6, it is often convenient to use the next theorem, which shows how to build up complicated continuous functions from simple ones.

Theorem (1.19)

> If f and g are continuous at a and c is a constant, then the following functions are also continuous at a:
>
> 1. $f + g$
>
> 2. $f - g$
>
> 3. cf
>
> 4. fg
>
> 5. $\dfrac{f}{g}$ if $g(a) \neq 0$

Proof   Each of the five parts of this theorem follows from the corresponding Limit Law in (1.6). For instance, we give the proof of part 1. Since f and g are continuous at a, we have

$$\lim_{x \to a} f(x) = f(a) \qquad \text{and} \qquad \lim_{x \to a} g(x) = g(a)$$

Therefore $\lim_{x \to a} (f + g)(x) = \lim_{x \to a} [f(x) + g(x)]$

$$= \lim_{x \to a} f(x) + \lim_{x \to a} g(x) \qquad \text{(by Law 1)}$$

$$= f(a) + g(a)$$

$$= (f + g)(a)$$

This shows that $f + g$ is continuous at a.   ●

It follows from Theorem 1.19 and Definition 1.18 that if f and g are continuous on an interval, then so are the functions $f + g$, $f - g$, cf, fg, and (if g is never 0) f/g. The following theorem was stated as (1.7) in Section 1.3.

Theorem (1.20)

(a)   Any polynomial is continuous everywhere; that is, it is continuous on $R = (-\infty, \infty)$.

(b)   Any rational function is continuous wherever it is defined; that is, it is continuous on its domain.

Proof

(a)   A polynomial is a function of the form

$$P(x) = c_n x^n + c_{n-1} x^{n-1} + \cdots + c_1 x + c_0$$

where $c_0, c_1, \dots, c_n$ are constants. We know that

$$\lim_{x \to a} c_0 = c_0 \qquad \text{(by Law 7)}$$

(1.21)   and

$$\lim_{x \to a} x^m = a^m \qquad m = 1, 2, \dots, n \qquad \text{(by 9)}$$

Equation 1.21 is precisely the statement that the function $f(x) = x^m$ is a continuous function. Thus, by Theorem 1.19 (3), the function $g(x) = cx^m$ is continuous. Since P is a sum of functions of this form and a constant function, it follows from Theorem 1.19(1) that P is continuous.

(b)   A rational function is a function of the form

$$f(x) = \frac{P(x)}{Q(x)}$$

where P and Q are polynomials. The domain of f is $D = \{x \in R \mid Q(x) \neq 0\}$. We know from part (a) that P and Q are continuous everywhere. Thus by Theorem 1.19(5), f is continuous at every number in D. ●

As an illustration of Theorem 1.20, observe that the volume of a sphere varies continuously with its radius because the formula

$$V(r) = \tfrac{4}{3}\pi r^3$$

shows that V is a polynomial function of r. Likewise, if a ball is thrown vertically into the air with a velocity of 50 ft/s, then the height of the ball in feet after t seconds is given by the formula $h = 50t - 16t^2$. Again this is a polynomial function, so the height is a continuous function of the elapsed time.

Knowledge of which functions are continuous enables us to evaluate some limits very quickly, as the following example shows. Compare it with Example 2 in Section 1.3.

EXAMPLE 7   Find $\displaystyle \lim_{x \to -2} \frac{x^3 + 2x^2 - 1}{5 - 3x}$.

Solution   The function

$$f(x) = \frac{x^3 + 2x^2 - 1}{5 - 3x}$$

is rational, so by Theorem 1.20 it is continuous on its domain. Therefore

$$\lim_{x \to -2} \frac{x^3 + 2x^2 - 1}{5 - 3x} = \lim_{x \to -2} f(x) = f(-2)$$

$$= \frac{(-2)^3 + 2(-2)^2 - 1}{5 - 3(-2)} = -\frac{1}{11}$$

●

The following theorem is a consequence of Law 10 of limits.

Theorem (1.22)

> If n is a positive even integer, then $f(x) = \sqrt[n]{x}$ is continuous on $[0, \infty)$.
>
> If n is a positive odd integer, then f is continuous on $(-\infty, \infty)$.

EXAMPLE 8   On what intervals are the following functions continuous?

(a) $f(x) = x^{100} - 2x^{37} + 75$ (b) $g(x) = \dfrac{x^2 + 2x + 17}{x^2 - 1}$

(c) $h(x) = \sqrt{x} + \dfrac{x+1}{x-1} - \dfrac{x+1}{x^2+1}$

Solution

(a) f is a polynomial, so it is continuous on $(-\infty, \infty)$ by Theorem 1.20 (a).

(b) g is a rational function, so by Theorem 1.20 (b) it is continuous on its domain, which is $D = \{x \mid x^2 - 1 \neq 0\} = \{x \mid x \neq \pm 1\}$. Thus g is continuous on the intervals $(-\infty, -1)$, $(-1, 1)$, and $(1, \infty)$.

(c) We can write $h(x) = F(x) + G(x) - H(x)$, where

$$F(x) = \sqrt{x} \qquad G(x) = \frac{x+1}{x-1} \qquad H(x) = \frac{x+1}{x^2+1}$$

F is continuous on $[0, \infty)$ by Theorem 1.22. G is a rational function, so it is continuous everywhere except when $x - 1 = 0$, that is, $x = 1$. H is also a rational function, but its denominator is never 0, so H is continuous everywhere. Thus, by parts 1 and 2 of Theorem 1.19, h is continuous on the intervals $[0, 1)$ and $(1, \infty)$.   ●

Another way of combining continuous functions f and g to get a new continuous function is to form the composite function $f \circ g$. This fact is a consequence of the following theorem.

Theorem (1.23)

> If f is continuous at b and $\lim_{x \to a} g(x) = b$, then
>
> $$\lim_{x \to a} f(g(x)) = f(b) = f\left(\lim_{x \to a} g(x)\right)$$

Intuitively this theorem is reasonable because if x is close to a, then $g(x)$ is close to b, and since f is continuous at b, if $g(x)$ is close to b, then $f(g(x))$ is close to $f(b)$. A proof of Theorem 1.23 is given in Appendix C.

Let us now apply Theorem 1.23 in the special case where $f(x) = \sqrt[n]{x}$, with n being a positive integer. Then

$$f(g(x)) = \sqrt[n]{g(x)}$$

and

$$f\left(\lim_{x \to a} g(x)\right) = \sqrt[n]{\lim_{x \to a} g(x)}$$

If we put these expressions into Theorem 1.23 we get

$$\lim_{x \to a} \sqrt[n]{g(x)} = \sqrt[n]{\lim_{x \to a} g(x)}$$

and so Limit Law 11 has now been proved. (We assume that the roots exist.)

Theorem (1.24)

> If g is continuous at a and f is continuous at $g(a)$, then $(f \circ g)(x) = f(g(x))$ is continuous at a.

This theorem is often expressed informally by saying "a continuous function of a continuous function is a continuous function."

Proof   Since g is continuous at a, we have

$$\lim_{x \to a} g(x) = g(a)$$

Since f is continuous at $b = g(a)$, we can apply Theorem 1.23 to obtain

$$\lim_{x \to a} f(g(x)) = f(g(a))$$

which is precisely the statement that the function $h(x) = f(g(x))$ is continuous at a; that is, $f \circ g$ is continuous at a.   ●

EXAMPLE 9   Where are the following functions continuous?

(a) $h(x) = |x|$ (b) $F(x) = \dfrac{1}{\sqrt{x^2 + 7} - 4}$

Solution

(a)   Since $|x| = \sqrt{x^2}$ for all x, we have $h(x) = f(g(x))$, where

$$g(x) = x^2 \qquad \text{and} \qquad f(x) = \sqrt{x}$$

Now g is continuous on R since it is a polynomial and f is continuous on $[0, \infty)$ by Theorem 1.22. Thus $h = f \circ g$ is continuous on R by Theorem 1.24.

(b)   Notice that F can be broken up as the composition of four continuous functions:

$$F = f \circ g \circ h \circ k \qquad \text{or} \qquad F(x) = f(g(h(k(x))))$$

where $f(x) = \dfrac{1}{x}$ $g(x) = x - 4$ $h(x) = \sqrt{x}$ $k(x) = x^2 + 7$

We know that each of these functions is continuous on its domain (by Theorems 1.20 and 1.22), so by Theorem 1.24 F is continuous on its domain, which is

$$\{x \in R \mid \sqrt{x^2 + 7} \neq 4\} = \{x \mid x \neq \pm 3\} = (-\infty, -3) \cup (-3, 3) \cup (3, \infty) \qquad ●$$

An important property of continuous functions is expressed by the following theorem, whose proof is found in more advanced books on calculus.

The Intermediate Value Theorem (1.25)

> Suppose that f is continuous on the closed interval $[a, b]$ and let N be any number strictly between $f(a)$ and $f(b)$. Then there exists a number c in (a, b) such that $f(c) = N$.

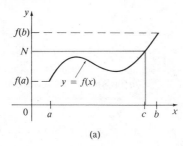

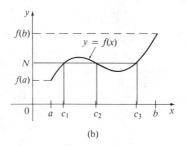

Figure 1.31   (a)   (b)

The Intermediate Value Theorem states that a continuous function takes on every intermediate value between the function values $f(a)$ and $f(b)$. It is illustrated by Figure 1.31. Note that the value N can be taken on once [as in part (a)] or more than once [as in part (b)].

If we think of a continuous function as a function whose graph has no holes or breaks, then it is easy to believe that the Intermediate Value Theorem is true. In geometric terms it says that if any horizontal line $y = N$ is given between $y = f(a)$ and $y = f(b)$ as in Figure 1.32, then the graph of f cannot jump over the line. It must intersect $y = N$ somewhere.

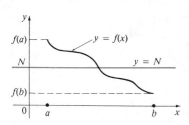

Figure 1.32

It is important that the function f in the theorem be continuous. The Intermediate Value Theorem is not true in general for discontinuous functions (see Exercise 47).

One use of the Intermediate Value Theorem is in locating roots of equations as in the following example.

EXAMPLE 10   Show that there is a root of the equation

$$4x^3 - 6x^2 + 3x - 2 = 0$$

between 1 and 2.

Solution   Let $f(x) = 4x^3 - 6x^2 + 3x - 2$. We are looking for a solution of the given equation, that is, a number c between 1 and 2 such that $f(c) = 0$. Therefore we take $a = 1$, $b = 2$, and $N = 0$ in Theorem 1.25. We have

$$f(1) = 4 - 6 + 3 - 2 = -1 < 0 \quad \text{and} \quad f(2) = 32 - 24 + 6 - 2 = 12 > 0$$

Thus $f(1) < 0 < f(2)$; that is, $N = 0$ is a number between $f(1)$ and $f(2)$. Now f is continuous since it is a polynomial, so the Intermediate Value Theorem says there is a number c between 1 and 2 such that $f(c) = 0$. In other words, the equation $4x^3 - 6x^2 + 3x - 2 = 0$ has a root c in the interval $(1, 2)$.

In fact, we can locate a root more precisely by using the Intermediate Value Theorem again. Since

$$f(1.2) = -0.128 < 0 \qquad \text{and} \qquad f(1.3) = 0.548 > 0$$

a root must lie between 1.2 and 1.3. A calculator gives, by trial and error,

$$f(1.22) = -0.007008 < 0 \qquad \text{and} \qquad f(1.23) = 0.056068 > 0$$

so a root lies in the interval $(1.22, 1.23)$.

SECTION 1.5 **Exercises**

1. (a) From the graph of f, state the numbers at which f is discontinuous.
    (b) For each of the numbers stated in part (a), state whether f is continuous from the right or from the left or neither.

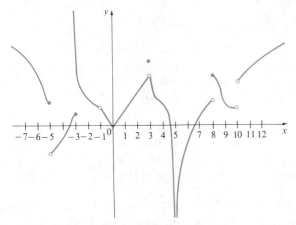

2. From the graph of g, state the intervals on which g is continuous.

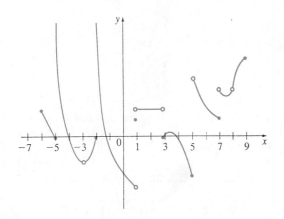

In Exercises 3–7 use the definition of continuity and the properties of limits to show that the given function is continuous at the given number.

3. $f(x) = x^4 - 5x^3 + 6$, $a = 3$

4. $f(x) = x^2 + (x-1)^9$, $a = 2$

5. $f(x) = 1 + \sqrt{x^2 - 9}$, $a = 5$

6. $g(x) = \dfrac{x+1}{2x^2 - 1}$, $a = 4$

7. $g(t) = \dfrac{\sqrt[3]{t}}{(t+1)^4}$, $a = -8$

In Exercises 8–11 use the definition of continuity and the properties of limits to show that the given function is continuous on the given interval.

8. $f(x) = x + \sqrt{x - 1}$, $[1, \infty)$

9. $f(x) = x\sqrt{16 - x^2}$, $[-4, 4]$

10. $F(x) = \dfrac{x+1}{x-3}$, $(-\infty, 3)$

11. $f(x) = (x^2 - 1)^8$, $(-\infty, \infty)$

In Exercises 12–20 explain why the given function is discontinuous at the given point. In each case sketch the graph of the function.

12. $f(x) = \dfrac{x^2 - 1}{x + 1}$, $a = -1$

13. $f(x) = \dfrac{3x^2 - 5x - 2}{x - 2}$, $a = 2$

14. $f(x) = \dfrac{1}{x - 1}$, $a = 1$

15. $f(x) = -\dfrac{1}{(x-1)^2}$, $a = 1$

16. $f(x) = \begin{cases} \dfrac{x^2 - 1}{x + 1} & \text{if } x \neq -1 \\ 6 & \text{if } x = -1 \end{cases}$ $a = -1$

17. $f(x) = \begin{cases} -\dfrac{1}{(x-1)^2} & \text{if } x \neq 1 \\ 0 & \text{if } x = 1 \end{cases}$ $a = 1$

18. $f(x) = \begin{cases} \dfrac{x^2 - 2x - 8}{x - 4} & \text{if } x \neq 4 \\ 3 & \text{if } x = 4 \end{cases}$ $a = 4$

19. $f(x) = \begin{cases} x^2 - 2 & \text{if } x \neq -3 \\ 5 & \text{if } x = -3 \end{cases}$ $a = -3$

20. $f(x) = \begin{cases} 1 - x & \text{if } x \leq 2 \\ x^2 - 2x & \text{if } x > 2 \end{cases}$ $a = 2$

In Exercises 21–33 use Theorems 1.19, 1.20, 1.22, and 1.24 to show that the given functions are continuous on their domains. State these domains.

21. $f(x) = (x + 1)(x^3 + 8x + 9)$

22. $g(x) = (x^6 - x^4 + 8x^2 - 7)^{10}$

23. $h(x) = \dfrac{x^2 + 2x - 1}{x + 1}$ 24. $G(x) = \dfrac{x^4 + 17}{6x^2 + x - 1}$

Definition (1.26)

Let f be a function defined on some interval (a, ∞). Then

$$\lim_{x \to \infty} f(x) = L$$

means that the values of $f(x)$ can be made arbitrarily close to L by taking x sufficiently large.

Another notation for $\lim_{x \to \infty} f(x) = L$ is

$$f(x) \to L \qquad \text{as } x \to \infty$$

The symbol ∞ does not represent a number. However the expression $\lim_{x \to \infty} f(x) = L$ is often read as

"the limit of $f(x)$, as x approaches infinity, is L"

or   "the limit of $f(x)$, as x becomes infinite, is L"

or   "the limit of $f(x)$, as x increases without bound, is L"

The meaning of such phrases is given by Definition 1.26. A more precise definition, similar to the ε, δ definition of Section 1.4, is given at the end of this section.

Geometric illustrations of Definition 1.26 are shown in Figure 1.34. Notice that there are many ways for the graph of f to approach the line $y = L$ (which is called a *horizontal asymptote*).

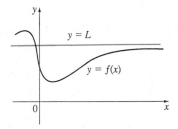
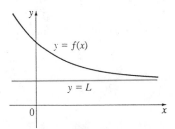
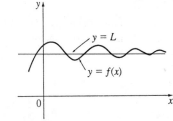

Figure 1.34

Examples illustrating $\lim\limits_{x \to \infty} f(x) = L$

Referring back to Figure 1.33 we see that for numerically large negative values of x, the values of $f(x)$ are close to 1. By letting x decrease through negative values without bound, we can make $f(x)$ as close as we like to 1. This is expressed by writing

$$\lim_{x \to -\infty} \frac{x^2 - 1}{x^2 + 1} = 1$$

The general definition is as follows:

Definition (1.27)

Let f be a function defined on some interval $(-\infty, a)$. Then

$$\lim_{x \to -\infty} f(x) = L$$

means that the values of $f(x)$ can be made arbitrarily close to L by taking x sufficiently large negative.

Again, the symbol $-\infty$ does not represent a number, but the expression $\lim_{x \to -\infty} f(x) = L$ is often read as

"the limit of $f(x)$, as x approaches negative infinity, is L"

Definition 1.27 is illustrated in Figure 1.35.

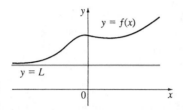   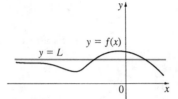

Figure 1.35

Examples illustrating $\lim_{x \to -\infty} f(x) = L$

Definition (1.28)

The line $y = L$ is called a **horizontal asymptote** of the curve $y = f(x)$ if either

$$\lim_{x \to \infty} f(x) = L \qquad \text{or} \qquad \lim_{x \to -\infty} f(x) = L$$

For instance, the curve illustrated in Figure 1.33 has the line $y = 1$ as a horizontal asymptote because

$$\lim_{x \to \infty} \frac{x^2 - 1}{x^2 + 1} = 1$$

The curve $y = f(x)$ sketched in Figure 1.36 has both $y = -1$ and $y = 2$ as horizontal asymptotes because

$$\lim_{x \to \infty} f(x) = -1 \qquad \text{and} \qquad \lim_{x \to -\infty} f(x) = 2$$

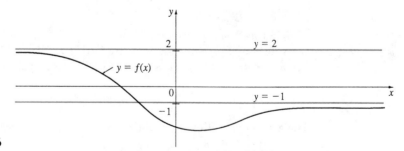

Figure 1.36

EXAMPLE 1   Find $\lim_{x \to \infty} \frac{1}{x}$ and $\lim_{x \to -\infty} \frac{1}{x}$.

Solution   Observe that when x is large, $1/x$ is small. For instance,

$$\frac{1}{100} = 0.01 \qquad \frac{1}{10,000} = 0.0001 \qquad \frac{1}{1,000,000} = 0.000001$$

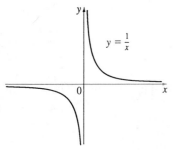

Figure 1.37

$\lim\limits_{x \to \infty} (1/x) = 0, \quad \lim\limits_{x \to -\infty} (1/x) = 0$

In fact, by taking x large enough, we can make $1/x$ as close to 0 as we please. Therefore, according to Definition 1.26, we have

$$\lim_{x \to \infty} \frac{1}{x} = 0$$

Similar reasoning shows that when x is large negative, $1/x$ is small negative, so we also have

$$\lim_{x \to -\infty} \frac{1}{x} = 0$$

It follows that the line $y = 0$ (the x-axis) is a horizontal asymptote of the curve $y = 1/x$. (This is an equilateral hyperbola; see Figure 1.37.)

Most of the Limit Laws that were given in Section 1.3 also hold for limits at infinity. It can be proved that the *Limit Laws listed Section 1.3 (with the exception of Laws 8, 9, and 10) are also valid if "$x \to a$" is replaced by "$x \to \infty$" or "$x \to -\infty$."* In particular, if we combine Laws 6 and 11 with the results of Example 1 we obtain the following important rule for calculating limits.

Theorem (1.29)

If $r > 0$ is a rational number, then

$$\lim_{x \to \infty} \frac{1}{x^r} = 0$$

If $r > 0$ is a rational number such that x^r is defined for all x, then

$$\lim_{x \to -\infty} \frac{1}{x^r} = 0$$

EXAMPLE 2   Evaluate

$$\lim_{x \to \infty} \frac{3x^2 - x - 2}{5x^2 + 4x + 1}$$

and indicate which properties of limits are used at each stage.

Solution   To evaluate the limit at infinity of a rational function, we first divide both the numerator and denominator by the highest power of x that occurs. (We may assume that $x \neq 0$ since we are interested only in large values of x.) In this case the highest power of x is x^2, so we have

$$\lim_{x \to \infty} \frac{3x^2 - x - 2}{5x^2 + 4x + 1} = \lim_{x \to \infty} \frac{3 - \frac{1}{x} - \frac{2}{x^2}}{5 + \frac{4}{x} + \frac{1}{x^2}}$$

$$= \frac{\lim_{x \to \infty} \left(3 - \frac{1}{x} - \frac{2}{x^2}\right)}{\lim_{x \to \infty} \left(5 + \frac{4}{x} + \frac{1}{x^2}\right)} \quad \text{(by Law 5)}$$

$$= \frac{\lim_{x \to \infty} 3 - \lim_{x \to \infty} \frac{1}{x} - 2 \lim_{x \to \infty} \frac{1}{x^2}}{\lim_{x \to \infty} 5 + 4 \lim_{x \to \infty} \frac{1}{x} + \lim_{x \to \infty} \frac{1}{x^2}} \quad \text{(by 1, 2, and 3)}$$

$$= \frac{3-0-0}{5+0+0} \qquad \text{(by 7 and Theorem 1.29)}$$

$$= \frac{3}{5}$$

EXAMPLE 3   Find the horizontal asymptotes of the graph of the function

$$f(x) = \frac{\sqrt{2x^2+1}}{3x-5}$$

Solution   Dividing both numerator and denominator by x and using the properties of limits, we have

$$\lim_{x \to \infty} \frac{\sqrt{2x^2+1}}{3x-5} = \lim_{x \to \infty} \frac{\sqrt{2+\dfrac{1}{x^2}}}{3-\dfrac{5}{x}} \qquad \text{(since } \sqrt{x^2} = x \text{ for } x > 0 \text{)}$$

$$= \frac{\displaystyle\lim_{x \to \infty} \sqrt{2+\dfrac{1}{x^2}}}{\displaystyle\lim_{x \to \infty} \left(3-\dfrac{5}{x}\right)}$$

$$= \frac{\sqrt{\displaystyle\lim_{x \to \infty} 2 + \lim_{x \to \infty} \dfrac{1}{x^2}}}{\displaystyle\lim_{x \to \infty} 3 - 5 \lim_{x \to \infty} \dfrac{1}{x}}$$

$$= \frac{\sqrt{2+0}}{3-5 \cdot 0} = \frac{\sqrt{2}}{3}$$

Therefore the line $y = \sqrt{2}/3$ is a horizontal asymptote of the graph of f.

In computing the limit as $x \to -\infty$, we must remember that for $x < 0$, we have $\sqrt{x^2} = |x| = -x$, so when we divide the numerator by x, when $x < 0$, we get

$$\frac{1}{x}\sqrt{2x^2+1} = -\frac{1}{\sqrt{x^2}}\sqrt{2x^2+1} = -\sqrt{2+\frac{1}{x^2}}$$

Therefore

$$\lim_{x \to -\infty} \frac{\sqrt{2x^2+1}}{3x-5} = \lim_{x \to -\infty} \frac{-\sqrt{2+\dfrac{1}{x^2}}}{3-\dfrac{5}{x}}$$

$$= \frac{-\sqrt{2+\displaystyle\lim_{x \to -\infty}\dfrac{1}{x^2}}}{3-5\displaystyle\lim_{x \to -\infty}\dfrac{1}{x}} = -\frac{\sqrt{2}}{3}$$

Thus the line $y = -\sqrt{2}/3$ is also a horizontal asymptote. The graph of f is sketched in Figure 1.38.

EXAMPLE 4   Compute $\displaystyle\lim_{x \to \infty} \left(\sqrt{x^2+1} - x\right)$.

Solution   We first multiply numerator and denominator by the conjugate radical:

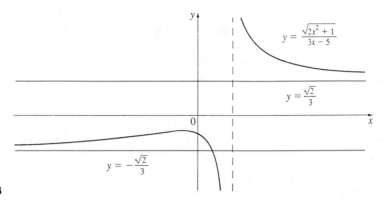

Figure 1.38

$$\lim_{x \to \infty} \left(\sqrt{x^2 + 1} - x \right) = \lim_{x \to \infty} \left(\sqrt{x^2 + 1} - x \right) \frac{\sqrt{x^2 + 1} + x}{\sqrt{x^2 + 1} + x}$$

$$= \lim_{x \to \infty} \frac{(x^2 + 1) - x^2}{\sqrt{x^2 + 1} + x} = \lim_{x \to \infty} \frac{1}{\sqrt{x^2 + 1} + x}$$

The Squeeze Theorem could be used to show that this limit is 0. But an easier method is to divide numerator and denominator by x. Doing this and using the Limit Laws, we obtain

$$\lim_{x \to \infty} \left(\sqrt{x^2 + 1} - x \right) = \lim_{x \to \infty} \frac{\frac{1}{x}}{\sqrt{1 + \frac{1}{x^2}} + 1}$$

$$= \frac{0}{\sqrt{1 + 0} + 1} = 0$$

EXAMPLE 5   Evaluate $\lim_{x \to \infty} \sin x$.

Solution   As x increases, the values of $\sin x$ oscillate between 1 and -1 infinitely often. Thus $\lim_{x \to \infty} \sin x$ does not exist.

Precise Definitions

Definition 1.26 can be stated precisely as follows:

Definition (1.30)

Let f be a function defined on some interval (a, ∞). Then

$$\lim_{x \to \infty} f(x) = L$$

means that for every $\varepsilon > 0$ there is a corresponding number N such that

$$|f(x) - L| < \varepsilon \qquad \text{whenever} \qquad x > N$$

In words, this says that the values of $f(x)$ can be made arbitrarily close to L (within a distance ε, where ε is any positive number) by taking x sufficiently large (larger than

N, where N depends on ε). Graphically it says that by choosing x large enough (larger than some number N) we can make the graph of f lie between the given horizontal lines $y = L - \varepsilon$ and $y = L + \varepsilon$ as in Figure 1.39. This must be true no matter how small we choose ε. Figure 1.40 shows that if a smaller value of ε is chosen, then a larger value of N may be required.

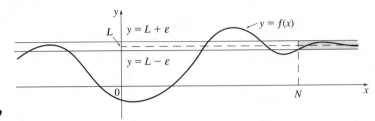

Figure 1.39

$\lim\limits_{x \to \infty} f(x) = L$

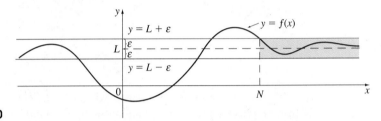

Figure 1.40

$\lim\limits_{x \to \infty} f(x) = L$

Similarly, a precise version of Definition 1.27 is given by Definition 1.31, which is illustrated in Figure 1.41.

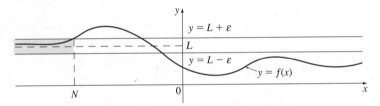

Figure 1.41

$\lim\limits_{x \to -\infty} f(x) = L$

Definition (1.31)

> Let f be a function defined on some interval $(-\infty, a)$. Then
>
> $$\lim\limits_{x \to -\infty} f(x) = L$$
>
> means that for every $\varepsilon > 0$ there is a corresponding number N such that
>
> $$|f(x) - L| < \varepsilon \qquad \text{whenever} \qquad x < N$$

EXAMPLE 6   Use Definition 1.30 to prove that $\lim\limits_{x \to \infty} \frac{1}{x} = 0$.

Solution

1. *Preliminary analysis of the problem: guessing a value for N.* Given $\varepsilon > 0$, we want

to find N such that

$$\left|\tfrac{1}{x}-0\right|<\varepsilon \qquad \text{whenever} \qquad x>N$$

In computing the limit we may assume $x>0$, in which case

$$\left|\tfrac{1}{x}-0\right|=\left|\tfrac{1}{x}\right|=\tfrac{1}{x}$$

Therefore we want

$$\tfrac{1}{x}<\varepsilon \qquad \text{whenever} \qquad x>N$$

that is, $\qquad\qquad\quad x>\tfrac{1}{\varepsilon} \qquad \text{whenever} \qquad x>N$

This suggests that we should take $N=1/\varepsilon$.

 2. *Proof* (*showing that N works*).   Given $\varepsilon>0$, we choose $N=1/\varepsilon$. Let $x>N$. Then

$$\left|\tfrac{1}{x}-0\right|=\tfrac{1}{|x|}=\tfrac{1}{x}<\tfrac{1}{N}=\varepsilon$$

Thus $\qquad\qquad\qquad \left|\tfrac{1}{x}-0\right|<\varepsilon \qquad \text{whenever} \qquad x>N$

Therefore, by Definition 1.30,

$$\lim_{x\to\infty}\tfrac{1}{x}=0$$

Figure 1.42 illustrates the proof by showing some values of ε and the corresponding values of N.

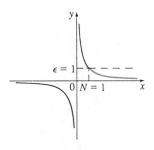
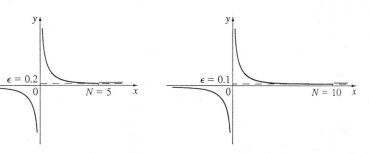

Figure 1.42

SECTION 1.6 **Exercises**

In Exercises 1–10 evaluate the given limit and justify each step by indicating the appropriate properties of limits.

1. $\displaystyle\lim_{x\to\infty}\frac{1}{x\sqrt{x}}$

2. $\displaystyle\lim_{x\to-\infty}5x^{-2/3}$

3. $\displaystyle\lim_{x\to\infty}\frac{5+2x}{3-x}$

4. $\displaystyle\lim_{x\to\infty}\frac{x+4}{x^2-2x+5}$

5. $\displaystyle\lim_{x\to-\infty}\frac{2x^2-x-1}{4x^2+7}$

6. $\displaystyle\lim_{t\to\infty}\frac{7t^3+4t}{2t^3-t^2+3}$

7. $\displaystyle\lim_{x\to-\infty}\frac{(1-x)(2+x)}{(1+2x)(2-3x)}$

8. $\displaystyle\lim_{x\to\infty}\sqrt{\frac{2x^2-1}{x+8x^2}}$

9. $\displaystyle\lim_{x\to\infty}\frac{1}{3+\sqrt{x}}$

10. $\displaystyle\lim_{x\to\infty}\frac{\sin^2 x}{x^2}$

In Exercises 11–24 find the given limits.

11. $\displaystyle\lim_{r\to\infty}\frac{r^4-r^2+1}{r^5+r^3-r}$

12. $\displaystyle\lim_{t\to-\infty}\frac{6t^2+5t}{(1-t)(2t-3)}$

13. $\displaystyle\lim_{x\to\infty}\frac{\sqrt{1+4x^2}}{4+x}$

14. $\displaystyle\lim_{x\to-\infty}\frac{\sqrt{x^2+4x}}{4x+1}$

15. $\displaystyle\lim_{x\to\infty}\frac{\sqrt[3]{x^2+8}}{x+2}$

16. $\displaystyle\lim_{x\to\infty}\frac{\sqrt[3]{x^3+8}}{x+2}$

17. $\displaystyle\lim_{x\to\infty}\frac{1-\sqrt{x}}{1+\sqrt{x}}$

18. $\displaystyle\lim_{x\to\infty}\left(\sqrt{x^2+3x+1}-x\right)$

19. $\displaystyle\lim_{x\to\infty}\left(\sqrt{x^2+1}-\sqrt{x^2-1}\right)$

20. $\displaystyle\lim_{x\to-\infty}\left(x+\sqrt{x^2+2x}\right)$

21. $\displaystyle\lim_{x\to\infty}\left(\sqrt{1+x}-\sqrt{x}\right)$

22. $\displaystyle\lim_{x\to\infty}\left(\sqrt[3]{1+x}-\sqrt[3]{x}\right)$

23. $\displaystyle\lim_{x\to-\infty}\left(\sqrt{x^2+x+1}+x\right)$

24. $\displaystyle\lim_{x\to\infty}\cos x$

In Exercises 25–28 find the horizontal asymptotes of the graphs of the given functions.

25. $f(x)=\dfrac{1+2x}{2+x}$

26. $g(x)=\dfrac{3x^4+6x^2-x}{2x^4-5x^3+9}$

27. $h(x)=\dfrac{x}{\sqrt[4]{x^4+1}}$

28. $F(x)=\dfrac{x-9}{\sqrt{4x^2+3x+2}}$

29. Find $\lim_{x\to\infty}f(x)$ if
$$\frac{4x-1}{x}<f(x)<\frac{4x^2+3x}{x^2}$$
for all $x>5$.

30. (a) A tank contains $5000\,\text{L}$ of pure water. Brine that contains $30\,\text{g}$ of salt per liter of water is pumped into the tank at a rate of $25\,\text{L/min}$. Show that the concentration of salt after t minutes (in grams per liter) is
$$C(t)=\frac{30t}{200+t}$$

(b) What happens to the concentration as $t\to\infty$?

31. Guess the value of the limit
$$\lim_{x\to\infty}\frac{x^2}{2^x}$$
by evaluating the function $f(x)=x^2/2^x$ for $x=0,1,2,3,4,5,6,7,8,9,10,20,50$, and 100.

32. How large do we have to take x so that $1/x^2<0.0001$?

33. Taking $r=2$ in Theorem 1.29, we have the statement
$$\lim_{x\to\infty}\frac{1}{x^2}=0$$
Prove this directly using Definition 1.30.

34. How large do we have to take x so that $1/\sqrt{x}<0.0001$?

35. Taking $r=\frac{1}{2}$ in Theorem 1.29, we have the statement
$$\lim_{x\to\infty}\frac{1}{\sqrt{x}}=0$$
Prove this directly using Definition 1.30.

36. Use Definition 1.31 to prove that $\lim_{x\to-\infty}\frac{1}{x}=0$.

37. Prove that
$$\lim_{x\to\infty}f(x)=\lim_{t\to0^+}f\!\left(\frac{1}{t}\right)$$
and
$$\lim_{x\to-\infty}f(x)=\lim_{t\to0^-}f\!\left(\frac{1}{t}\right)$$
if these limits exist.

SECTION 1.7

Infinite Limits; Vertical Asymptotes

In Example 6 in Section 1.2 and Example 2 in Section 1.4 we saw that $\lim_{x\to0}\left(1/x^2\right)$ does not exist by examining the behavior of the function $f(x)=1/x^2$ near 0. The closer we take x to 0, the larger $1/x^2$ becomes, and in fact we can make $f(x)$ as large as we like by taking x close enough to 0 (see Figure 1.43).

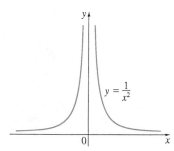

Figure 1.43

To indicate this kind of behavior of the function we use the notation

$$\lim_{x \to 0} \frac{1}{x^2} = \infty$$

In general, we write symbolically

$$\lim_{x \to a} f(x) = \infty$$

to indicate that the values of $f(x)$ become larger and larger (or "increase without bound") as x becomes closer and closer to a.

Definition (1.32)

Let f be a function defined on both sides of a, except possibly at a itself. Then

$$\lim_{x \to a} f(x) = \infty$$

means that the values of $f(x)$ can be made arbitrarily large (as large as we please) by taking x sufficiently close to a $(x \neq a)$.

Another notation for $\lim_{x \to a} f(x) = \infty$ is

$$f(x) \to \infty \qquad \text{as} \qquad x \to a$$

Again the symbol ∞ is not a number, but the expression $\lim_{x \to a} f(x) = \infty$ is often read as

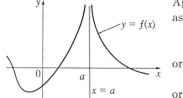

"the limit of $f(x)$, as x approaches a, is infinity"

or   "$f(x)$ becomes infinite as x approaches a"

or   "$f(x)$ increases without bound as x approaches a"

Figure 1.44
$$\lim_{x \to a} f(x) = \infty$$

This definition is illustrated in Figure 1.44. (A more precise definition is given at the end of this section.)

A similar sort of limit, for functions that become large negative as x gets close to a, is defined in Definition 1.33 and is illustrated in Figure 1.45.

Definition (1.33)

Let f be defined on both sides of a, except possibly at a itself. Then

$$\lim_{x \to a} f(x) = -\infty$$

means that the values of $f(x)$ can be made arbitrarily large negative by taking x sufficiently close to a $(x \neq a)$.

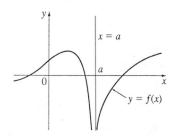

Figure 1.45
$$\lim_{x \to a} f(x) = -\infty$$

The symbol $\lim_{x \to a} f(x) = -\infty$ can be read as "the limit of $f(x)$, as x approaches a, is negative infinity" or "$f(x)$ decreases without bound as x approaches a." As an example we have

$$\lim_{x \to 0} \left(-\frac{1}{x^2} \right) = -\infty$$

Similar definitions can be given for the one-sided infinite limits

$$\lim_{x \to a^-} f(x) = \infty \qquad\qquad \lim_{x \to a^+} f(x) = \infty$$

$$\lim_{x \to a^-} f(x) = -\infty \qquad\qquad \lim_{x \to a^+} f(x) = -\infty$$

remembering that "$x \to a^-$" means that we consider only values of x that are less than a, and similarly "$x \to a^+$" means that we consider only $x > a$. Illustrations of these four cases are given in Figure 1.46.

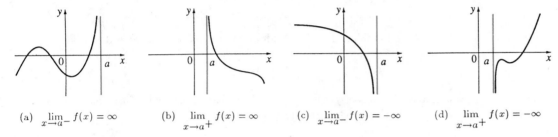

(a) $\lim_{x \to a^-} f(x) = \infty$ (b) $\lim_{x \to a^+} f(x) = \infty$ (c) $\lim_{x \to a^-} f(x) = -\infty$ (d) $\lim_{x \to a^+} f(x) = -\infty$

Figure 1.46

Definition (1.34)

The line $x = a$ is called a **vertical asymptote** of the curve $y = f(x)$ if at least one of the following statements is true:

$$\lim_{x \to a} f(x) = \infty \qquad \lim_{x \to a^-} f(x) = \infty \qquad \lim_{x \to a^+} f(x) = \infty$$

$$\lim_{x \to a} f(x) = -\infty \qquad \lim_{x \to a^-} f(x) = -\infty \qquad \lim_{x \to a^+} f(x) = -\infty$$

For instance, the y-axis is a vertical asymptote of the curve $y = 1/x^2$ because $\lim_{x \to 0}(1/x^2) = \infty$. In Figure 1.46 the line $x = a$ is a vertical asymptote in each of the four cases.

EXAMPLE 1   Find $\lim_{x \to 3^+} \dfrac{2}{x - 3}$ and $\lim_{x \to 3^-} \dfrac{2}{x - 3}$.

Solution   If x is close to 3 but larger than 3, then $x - 3$ is a small positive number and so $2/(x - 3)$ is a large positive number. Thus intuitively we see that

$$\lim_{x \to 3^+} \frac{2}{x - 3} = \infty$$

Likewise if x is close to 3 but smaller than 3, then $x - 3$ is a small negative number and so $2/(x - 3)$ is a numerically large negative number. Thus

$$\lim_{x \to 3^-} \frac{2}{x - 3} = -\infty$$

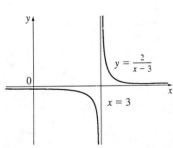

The graph of the curve $y = 2/(x - 3)$ is given in Figure 1.47. The line $x = 3$ is a vertical

Figure 1.47 asymptote.

The following theorem can be established intuitively by arguing as in Example 1 or proved from Definition 1.36.

Theorem (1.35)

(a)   If n is a positive even integer, then

$$\lim_{x \to a} \frac{1}{(x-a)^n} = \infty$$

(b)   If n is a positive odd integer, then

$$\lim_{x \to a^+} \frac{1}{(x-a)^n} = \infty \qquad \text{and} \qquad \lim_{x \to a^-} \frac{1}{(x-a)^n} = -\infty$$

EXAMPLE 2   Find the vertical asymptotes and horizontal asymptotes of the function

$$f(x) = \frac{x}{x^2 + x - 2}$$

Solution   Write

$$f(x) = \frac{x}{(x-1)(x+2)}$$

The vertical asymptotes are likely to occur when the denominator is 0, that is, when $x = 1$ or -2. If x is close to 1 but $x > 1$, then the denominator is close to 0, but $x > 0$, $x - 1 > 0$, and $x + 2 > 0$, so $f(x) > 0$. Thus

$$\lim_{x \to 1^+} \frac{x}{(x-1)(x+2)} = \infty$$

Similarly we see that
$$\lim_{x \to 1^-} \frac{x}{(x-1)(x+2)} = -\infty$$

If x is close to -2 but $x < -2$, then the denominator is close to 0, but $x < 0$, $x - 1 < 0$, and $x + 2 < 0$, so $f(x) < 0$. Thus

$$\lim_{x \to -2^-} \frac{x}{(x-1)(x+2)} = -\infty$$

Similarly we see that
$$\lim_{x \to -2^+} \frac{x}{(x-1)(x+2)} = \infty$$

Thus the vertical asymptotes are $x = 1$ and $x = -2$. Since

$$\lim_{x \to \infty} \frac{x}{x^2 + x - 2} = \lim_{x \to \infty} \frac{\frac{1}{x}}{1 + \frac{1}{x} - \frac{2}{x^2}}$$

$$= \frac{\lim_{x \to \infty} \frac{1}{x}}{1 + \lim_{x \to \infty} \frac{1}{x} - 2 \lim_{x \to \infty} \frac{1}{x^2}} = \frac{0}{1 + 0 - 0} = 0$$

the horizontal asymptote is $y = 0$. A similar computation shows that

$$\lim_{x \to -\infty} \frac{x}{x^2 + x - 2} = 0$$

Infinite Limits at Infinity

The notation

$$\lim_{x \to \infty} f(x) = \infty$$

is used to indicate that the values of $f(x)$ become large as x becomes large. Similar meanings are attached to the following symbols:

$$\lim_{x \to -\infty} f(x) = \infty \qquad \lim_{x \to \infty} f(x) = -\infty \qquad \lim_{x \to -\infty} f(x) = -\infty$$

EXAMPLE 3   Find $\lim_{x \to \infty} x^3$ and $\lim_{x \to -\infty} x^3$.

Solution   When x becomes large, x^3 also becomes large. For instance,

$$10^3 = 1000 \qquad 100^3 = 1,000,000 \qquad 1000^3 = 1,000,000,000$$

In fact we can make x^3 as big as we like by taking x large enough. Therefore we can write

$$\lim_{x \to \infty} x^3 = \infty$$

Similarly, when x is large negative, so is x^3. Thus

$$\lim_{x \to -\infty} x^3 = -\infty$$

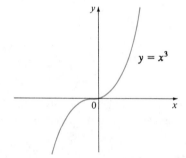

Figure 1.48

$$\lim_{x \to \infty} x^3 = \infty, \ \lim_{x \to -\infty} x^3 = -\infty$$

These limit statements can also be seen from the graph of $y = x^3$ in Figure 1.48.

EXAMPLE 4   Find $\lim_{x \to \infty} (x^2 - x)$.

Solution   Note that we *cannot* write

$$\lim_{x \to \infty} (x^2 - x) = \lim_{x \to \infty} x^2 - \lim_{x \to \infty} x = \infty - \infty$$

The Limit Laws cannot be applied to infinite limits because ∞ is not a number ($\infty - \infty$ cannot be defined). However we can write

$$\lim_{x \to \infty} (x^2 - x) = \lim_{x \to \infty} x(x - 1) = \infty$$

because both x and $x - 1$ become arbitrarily large.

EXAMPLE 5   Find $\lim_{x \to \infty} \dfrac{x^2 + x}{3 - x}$.

Solution 1   As in Example 2 in Section 1.6 we can evaluate the limit at infinity of a rational function by dividing numerator and denominator by the highest power of x that occurs:

$$\lim_{x \to \infty} \frac{x^2 + x}{3 - x} = \lim_{x \to \infty} \frac{1 + \frac{1}{x}}{\frac{3}{x^2} - \frac{1}{x}} = -\infty$$

because $1 + 1/x \to 1$ and $3/x^2 - 1/x \to 0$ through negative values as $x \to \infty$. (Note that $3/x^2 - 1/x < 0$ for $x > 3$.)

Solution 2   It is also possible to find the given limit by dividing numerator and denominator by x instead of x^2:

$$\lim_{x \to \infty} \frac{x^2 + x}{3 - x} = \lim_{x \to \infty} \frac{x + 1}{\frac{3}{x} - 1} = -\infty$$

because $x + 1 \to \infty$ and $3/x - 1 \to -1$ as $x \to \infty$.   ●

The next example shows that by using infinite limits at infinity, together with intercepts, we can get a rough idea of the graph of a factored polynomial.

EXAMPLE 6   Sketch the graph of $y = (x - 2)^4 (x + 1)^3 (x - 1)$ by finding its intercepts and its limits as $x \to \infty$ and as $x \to -\infty$.

Solution   The y-intercept is $f(0) = (-2)^4 (1)^3 (-1) = -16$ and the x-intercepts are found by setting $y = 0$: $x = 2, -1, 1$. Notice that since $(x - 2)^4$ is positive, the function does not change sign at 2; thus the graph does not cross the x-axis at 2. The graph crosses the axis at -1 and 1.

When x is large, all three factors are large, so

$$\lim_{x \to \infty} (x - 2)^4 (x + 1)^3 (x - 1) = \infty$$

When x is large negative, the first factor is large positive and the second and third factors are both large negative, so

$$\lim_{x \to -\infty} (x - 2)^4 (x + 1)^3 (x - 1) = \infty$$

Combining this information we give a rough sketch of the graph in Figure 1.49.   ●

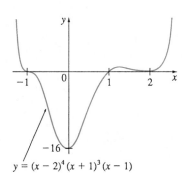

$y = (x - 2)^4 (x + 1)^3 (x - 1)$

Figure 1.49

Precise Definitions

Definition 1.32 can be stated more precisely as follows:

Definition (1.36)

Let f be a function defined on some open interval that contains the number a, except possibly at a itself. Then

$$\lim_{x \to a} f(x) = \infty$$

means that for every positive number M there is a corresponding number $\varepsilon > 0$ such that

$$f(x) > M \qquad \text{whenever} \qquad 0 < |x - a| < \delta$$

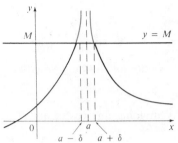

Figure 1.50

This says that the value of $f(x)$ can be made arbitrarily large (larger than any given number M) by taking x close enough to a (within a distance δ, where δ depends on M). A geometric illustration is shown in Figure 1.50.

Given any horizontal line $y = M$, we can find a number $\delta > 0$ such that if we restrict x to lie in the interval $(a - \delta, a + \delta)$, then the curve $y = f(x)$ lies above the line $y = M$. You can see that if a larger M is chosen, then a smaller δ may be required.

EXAMPLE 7   Use Definition 1.36 to prove that

$$\lim_{x \to 0} \frac{1}{x^2} = \infty$$

Solution

1. *Preliminary analysis of the problem: guessing a value for δ.* Given $M > 0$, we want to find $\delta > 0$ such that

$$\frac{1}{x^2} > M \qquad \text{whenever} \qquad 0 < |x - 0| < \delta$$

that is,

$$x^2 < \frac{1}{M} \qquad \text{whenever} \qquad 0 < |x| < \delta$$

or

$$|x| < \frac{1}{\sqrt{M}} \qquad \text{whenever} \qquad 0 < |x| < \delta$$

This suggests that we should take $\delta = 1/\sqrt{M}$.

2. *Proof: showing that this δ works.* If $M > 0$ is given, let $\delta = 1/\sqrt{M}$. If $0 < |x - 0| < \delta$, then

$$|x| < \delta \quad \Rightarrow \quad x^2 < \delta^2$$

$$\Rightarrow \quad \frac{1}{x^2} > \frac{1}{\delta^2} = M$$

Thus

$$\frac{1}{x^2} > M \qquad \text{whenever} \qquad 0 < |x - 0| < \delta$$

Therefore, by Definition 1.36,

$$\lim_{x \to 0} \frac{1}{x^2} = \infty$$

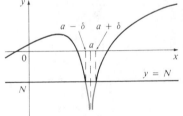

Figure 1.51

Similarly, a precise version of Definition 1.33 is given by Definition 1.37, which is illustrated by Figure 1.51.

Defintion (1.37)

Let f be a function defined on some open interval that contains the number a, except possibly at a itself. Then

$$\lim_{x \to a} f(x) = -\infty$$

means that for every negative number N there is a corresponding number $\delta > 0$ such that

$$f(x) < N \qquad \text{whenever} \qquad 0 < |x - a| < \delta$$

Finally we note that an infinite limit at infinity can be defined as follows. The geometric illustration is given in Figure 1.52.

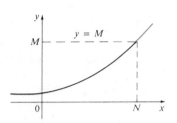

Definition (1.38)

Let f be a function defined on some open interval (a, ∞). Then

$$\lim_{x \to \infty} f(x) = \infty$$

means that for every positive number M there is a corresponding number $N > 0$ such that

$$f(x) > M \qquad \text{whenever} \qquad x > N$$

Figure 1.52

$$\lim_{x \to \infty} f(x) = \infty$$

SECTION 1.7 **Exercises**

1. For the function f whose graph is shown, state the following.

   (a) $\lim\limits_{x \to 3} f(x)$ (b) $\lim\limits_{x \to 7} f(x)$

   (c) $\lim\limits_{x \to -4} f(x)$ (d) $\lim\limits_{x \to -9^-} f(x)$

   (e) $\lim\limits_{x \to -9^+} f(x)$ (f) $\lim\limits_{x \to \infty} f(x)$

   (g) $\lim\limits_{x \to -\infty} f(x)$

   (h) The equations of the vertical asymptotes.
   (i) The equations of the horizontal asymptotes.

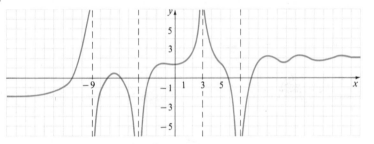

In Exercises 2–29 find each limit.

2. $\lim\limits_{x \to 0} \dfrac{2}{x^4}$

3. $\lim\limits_{x \to -1} \dfrac{-2}{(x+1)^6}$

4. $\lim\limits_{x \to 3} \dfrac{1}{(x-3)^8}$

5. $\lim\limits_{x \to 5^+} \dfrac{6}{x-5}$

6. $\lim\limits_{x \to 5^-} \dfrac{6}{x-5}$

7. $\lim\limits_{x \to -1^-} \dfrac{x^2}{x+1}$

8. $\lim\limits_{x \to -6^+} \dfrac{x}{x+6}$

9. $\lim\limits_{t \to 3^-} \dfrac{t+3}{t^2-9}$

10. $\lim\limits_{x \to 0^+} \dfrac{3}{x^2+2x}$

11. $\lim\limits_{x \to -3^-} \dfrac{x^2}{x^2-9}$

12. $\lim\limits_{x \to 0} \dfrac{x-1}{x^2(x+2)}$

13. $\lim\limits_{x \to -2^+} \dfrac{x-1}{x^2(x+2)}$

14. $\lim\limits_{x \to 0^+} \left(\dfrac{1}{x} - \dfrac{1}{x^2} \right)$

15. $\lim\limits_{x \to -3^+} \dfrac{x^3}{x^2+5x+6}$

16. $\lim\limits_{x \to 5^-} \dfrac{2}{\sqrt{5-x}}$

17. $\lim\limits_{x \to 2^+} \dfrac{-3}{\sqrt[3]{x-2}}$

18. $\lim\limits_{x \to \infty} \sqrt{x}$

19. $\lim\limits_{x \to -\infty} \sqrt[3]{x}$

20. $\lim\limits_{x \to \infty} (x - \sqrt{x})$

21. $\lim\limits_{x \to \infty} (x + \sqrt{x})$

22. $\lim\limits_{x \to -\infty} (x^3 - 5x^2)$

23. $\lim\limits_{x \to \infty} (x^2 - x^4)$

24. $\lim\limits_{x \to \infty} \dfrac{x^7-1}{x^6+1}$

25. $\lim\limits_{x \to \infty} \dfrac{x^3-1}{x^4+1}$

26. $\lim\limits_{x \to \infty} \dfrac{\sqrt{x}+3}{x+3}$

27. $\lim\limits_{x \to \infty} \dfrac{x}{\sqrt{x-1}}$

28. $\lim\limits_{x \to \infty} \dfrac{\sqrt{x}}{\sqrt[3]{x}}$

29. $\lim\limits_{x \to -\infty} \dfrac{x^8+3x^4+2}{x^5+x^3}$

In Exercises 30–33 find the vertical and horizontal asymptotes of the given curves.

30. $y = \dfrac{x}{x+4}$

31. $y = \dfrac{x^2+4}{x^2-1}$

32. $y = \dfrac{x^3}{x^2+3x-10}$

33. $y = \dfrac{x^3+1}{x^3+x}$

34. Make a rough sketch of the curve $y = x^n$ (n an integer) for the following five cases: (i) $n = 0$, (ii) $n > 0$, n odd, (iii) $n > 0$, n even, (iv) $n < 0$, n odd, and (v) $n < 0$, n even. Then use these sketches to find the following limits.

(a) $\displaystyle\lim_{x \to 0^+} x^n$ (b) $\displaystyle\lim_{x \to 0^-} x^n$

(c) $\displaystyle\lim_{x \to \infty} x^n$ (d) $\displaystyle\lim_{x \to -\infty} x^n$

In Exercises 35–38 find the limits as $x \to \infty$ and $x \to -\infty$. Use this information, together with intercepts, to give a rough sketch of the graph as in Example 6.

35. $y = x^2 (x - 2)(1 - x)$

36. $y = (2 + x)^3 (1 - x)(3 - x)$

37. $y = (x + 4)^5 (x - 3)^4$

38. $y = (1 - x)(x - 3)^2 (x - 5)^2$

39. How close to -3 do we have to take x so that

$$\frac{1}{(x + 3)^4} > 10{,}000$$

40. Prove, using Definition 1.36, that

$$\lim_{x \to -3} \frac{1}{(x + 3)^4} = \infty$$

41. Prove that

$$\lim_{x \to -1^-} \frac{5}{(x + 1)^3} = -\infty$$

42. Prove, using Definition 1.38, that $\lim_{x \to \infty} x^3 = \infty$.

43. If $\lim_{x \to a} f(x) = \infty$ and $\lim_{x \to a} g(x) = c$, where c is a real number, prove that

$$\lim_{x \to a} [f(x) + g(x)] = \infty$$

44. Suppose $\lim_{x \to a} f(x) = \infty$ and $\lim_{x \to a} g(x) = c$, where c is a real number.

(a) If $c > 0$, prove that $\lim_{x \to a} [f(x)g(x)] = \infty$.

(b) If $c < 0$, prove that $\lim_{x \to a} [f(x)g(x)] = -\infty$.

SECTION 1.8

Tangents, Velocities, and Other Rates of Change

In Section 1.1 we guessed the values of slopes of tangent lines and velocities on the basis of numerical evidence. Now that we have defined limits and have learned techniques for computing them, we return to the tangent and velocity problems with the ability to calculate slopes of tangents, velocities, and other rates of change.

Tangents

If a curve C has equation $y = f(x)$ and we want to find the tangent to C at the point $P(a, f(a))$, then we consider a nearby point $Q(x, f(x))$, where $x \neq a$, and compute the slope of the secant line PQ:

$$m_{PQ} = \frac{f(x) - f(a)}{x - a}$$

Then we let Q approach P along the curve C by letting x approach a. If m_{PQ} approaches a number m, then we define the *tangent* t to be the line through P with slope m. (This amounts to saying that the tangent line is the limiting position of the secant line PQ as Q approaches P. See Figure 1.53.)

Definition (1.39) | The **tangent line** to the curve $y = f(x)$ at the point $P(a, f(a))$ is the line through P with slope

$$m = \lim_{x \to a} \frac{f(x) - f(a)}{x - a}$$

provided that this limit exists.

In our first example we confirm the guess we made in Example 1 in Section 1.1.

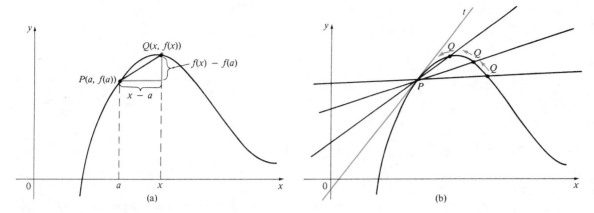

Figure 1.53

EXAMPLE 1   Find the equation of the tangent line to the parabola $y = x^2$ at the point $P(1, 1)$.

Solution   Here we have $a = 1$ and $f(x) = x^2$, so the slope is

$$m = \lim_{x \to 1} \frac{f(x) - f(1)}{x - 1} = \lim_{x \to 1} \frac{x^2 - 1}{x - 1}$$

$$= \lim_{x \to 1} \frac{(x - 1)(x + 1)}{x - 1} = \lim_{x \to 1} (x + 1) = 1 + 1 = 2$$

Using the point-slope form of the equation of a line, we find that the equation of the tangent line at $(1, 1)$ is

$$y - 1 = 2(x - 1) \qquad \text{or} \qquad y = 2x - 1 \qquad \qquad \bullet$$

There is another expression for the slope of a tangent line that is sometimes easier to use. Let

$$h = x - a$$

Then

$$x = a + h$$

so the slope of the secant line PQ is

$$m_{PQ} = \frac{f(a + h) - f(a)}{h}$$

(See Figure 1.54 where the case $h > 0$ is illustrated and Q is to the right of P. If $h < 0$, however, Q would be to the left of P.) Notice that as x approaches a, h approaches 0 and so the expression for the slope of the tangent line in Definition 1.39 becomes

(1.40)

$$m = \lim_{h \to 0} \frac{f(a + h) - f(a)}{h}$$

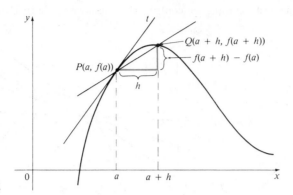

Figure 1.54

EXAMPLE 2   Find the equation of the tangent line to the hyperbola $y = 3/x$ at the point $(3, 1)$.

Solution   Let $f(x) = 3/x$. Then the slope of the tangent at $(3, 1)$ is

$$m = \lim_{h \to 0} \frac{f(3 + h) - f(3)}{h}$$

$$= \lim_{h \to 0} \frac{\dfrac{3}{3 + h} - 1}{h} = \lim_{h \to 0} \frac{\dfrac{3 - (3 + h)}{3 + h}}{h}$$

$$= \lim_{h \to 0} \frac{-h}{h(3 + h)} = \lim_{h \to 0} -\frac{1}{3 + h} = -\frac{1}{3}$$

Therefore the equation of the tangent at the point $(3, 1)$ is

$$y - 1 = -\tfrac{1}{3}(x - 3)$$

which simplifies to $\qquad\qquad x + 3y - 6 = 0$

Figure 1.55   The hyperbola and its tangent are shown in Figure 1.55.

EXAMPLE 3   Find the slopes of the tangent lines to the graph of the function $f(x) = \sqrt{2x + 3}$ at the points $(\tfrac{1}{2}, 2)$, $(3, 3)$, and $(5, \sqrt{13})$.

Solution   Since three slopes are requested, it is efficient to start by finding the slope at the general point $(a, \sqrt{2a + 3})$:

$$m = \lim_{h \to 0} \frac{f(a + h) - f(a)}{h} = \lim_{h \to 0} \frac{\sqrt{2(a + h) + 3} - \sqrt{2a + 3}}{h}$$

Rationalize the numerator

$$= \lim_{h \to 0} \frac{\sqrt{2(a + h) + 3} - \sqrt{2a + 3}}{h} \cdot \frac{\sqrt{2(a + h) + 3} + \sqrt{2a + 3}}{\sqrt{2(a + h) + 3} + \sqrt{2a + 3}}$$

$$= \lim_{h \to 0} \frac{(2a + 2h + 3) - (2a + 3)}{h(\sqrt{2a + 2h + 3} + \sqrt{2a + 3})} = \lim_{h \to 0} \frac{2h}{h(\sqrt{2a + 2h + 3} + \sqrt{2a + 3})}$$

Continuous function of h

$$= \lim_{h \to 0} \frac{2}{\sqrt{2a+2h+3}+\sqrt{2a+3}} = \frac{2}{\sqrt{2a+3}+\sqrt{2a+3}} = \frac{1}{\sqrt{2a+3}}$$

At the point $(\frac{1}{2}, 2)$, $a = \frac{1}{2}$, so the slope of the tangent is $m = 1/\sqrt{2(\frac{1}{2})+3} = \frac{1}{2}$. At $(3,3)$, we have $m = 1/\sqrt{2(3)+3} = \frac{1}{3}$; at $(5, \sqrt{13})$, $m = 1/\sqrt{2(5)+3} = 1/\sqrt{13}$. ●

Velocities

In Section 1.1 we investigated the motion of a ball dropped from the CN Tower and defined its velocity to be the limiting value of average velocities over shorter and shorter time periods.

In general, suppose an object moves along a straight line according to an equation of motion $s = f(t)$, where s is the displacement (directed distance) of the object from the origin at time t. The function f that describes the motion is called the **position function** of the object. In the time interval from $t = a$ to $t = a + h$ the change in position is $f(a+h) - f(a)$ (see Figure 1.56). The average velocity over this time interval is

$$\text{average velocity} = \frac{\text{displacement}}{\text{time}} = \frac{f(a+h) - f(a)}{h}$$

which is the same as the slope of the secant line PQ in Figure 1.57.

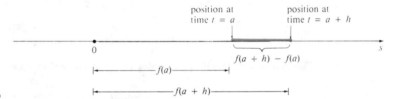

Figure 1.56

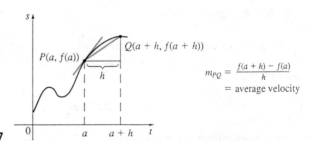

Figure 1.57

Now suppose we compute the average velocities over shorter and shorter time intervals $[a, a+h]$. In other words, we let h approach 0. As in the example of the falling ball, we define the **velocity** (or **instantaneous velocity**) $v(a)$ at time $t = a$ to be the limit of these average velocities:

(1.41)

$$v(a) = \lim_{h \to 0} \frac{f(a+h) - f(a)}{h}$$

This means that the velocity at time $t = a$ is equal to the slope of the tangent line at P. (Compare Equations 1.40 and 1.41.)

Now that we know how to compute limits, let us reconsider the problem of the falling ball.

EXAMPLE 4   Suppose that a ball is dropped from the upper observation deck of the CN Tower, 450 m above the ground.
(a) What is the velocity of the ball after 5 s?
(b) How fast is it traveling when it hits the ground?

Solution   We first use the equation of motion $s = f(t) = 4.9t^2$ to find the velocity $v(a)$ after a seconds:

$$v(a) = \lim_{h \to 0} \frac{f(a+h) - f(a)}{h} = \lim_{h \to 0} \frac{4.9(a+h)^2 - 4.9a^2}{h}$$

$$= \lim_{h \to 0} \frac{4.9(a^2 + 2ah + h^2 - a^2)}{h} = \lim_{h \to 0} \frac{4.9(2ah + h^2)}{h}$$

$$= \lim_{h \to 0} 4.9(2a + h) = 9.8a$$

(a)   The velocity after 5 s is $v(5) = (9.8)(5) = 49$ m/s.
(b)   Since the observation deck is 450 m above the ground, the ball will hit the ground at the time t_1 when $s(t_1) = 450$, that is,

$$4.9t_1^2 = 450$$

This gives $t_1^2 = \dfrac{450}{4.9}$ and $t_1 = \sqrt{\dfrac{450}{4.9}} \approx 9.6$ s

The velocity of the ball as it hits the ground is therefore

$$v(t_1) = 9.8t_1 = 9.8\sqrt{\frac{450}{4.9}} \approx 94 \text{ m/s}$$

Other Rates of Change

Suppose y is a quantity that depends on another quantity x. Thus y is a function of x and we write $y = f(x)$. If x changes from x_1 to x_2, then the change in x (also called the **increment** of x) is

$$\Delta x = x_2 - x_1$$

and the corresponding change in y is

$$\Delta y = f(x_2) - f(x_1)$$

The difference quotient

$$\frac{\Delta y}{\Delta x} = \frac{f(x_2) - f(x_1)}{x_2 - x_1}$$

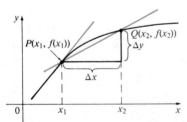

average rate of change $= m_{PQ}$
instantaneous rate of change $=$ slope of tangent at P

Figure 1.58

is called the **average rate of change of y with respect to x** over the interval $[x_1, x_2]$ and can be interpreted as the slope of the secant line PQ in Figure 1.58. By analogy with velocity, we consider the average rate of change over smaller and smaller intervals by letting x_2 approach x_1 and therefore letting Δx approach 0. The limit of these average rates of change is called the **(instantaneous) rate of change of y with respect to x** at

$x = x_1$, which is interpreted as the slope of the tangent to the curve $y = f(x)$ at $P((x)_1, f(x_1))$:

(1.42)

$$\text{instantaneous rate of change} = \lim_{\Delta x \to 0} \frac{\Delta y}{\Delta x}$$

$$= \lim_{x_2 \to x_1} \frac{f(x_2) - f(x_1)}{x_2 - x_1}$$

EXAMPLE 5   Temperature readings T (in degrees Celsius) were recorded every hour starting at midnight on a day in April in Whitefish, Montana, as in the following table. The time x is measured in hours from midnight.

x (h)	0	1	2	3	4	5	6	7	8	9	10	11	12
T (°C)	6.5	6.1	5.6	4.9	4.2	4.0	4.0	4.8	6.1	8.3	10.0	12.1	14.3

x (h)	13	14	15	16	17	18	19	20	21	22	23	24
T (°C)	16.0	17.3	18.2	18.8	17.6	16.0	14.1	11.5	10.2	9.0	7.9	7.0

(a)   Find the average rate of change of temperature with respect to time
    (i)   from noon to 3 P.M.   (ii)   from noon to 2 P.M.
    (iii)   from noon to 1 P.M.
(b)   Estimate the instantaneous rate of change at noon.

Solution
  (a) (i)   From noon to 3 P.M. the temperature changes from 14.3° to 18.2°, so

$$\Delta T = T(15) - T(12) = 18.2 - 14.3 = 3.9°$$

while the change in time is $\Delta x = 3$ h. Therefore the average rate of change of temperature with respect to time is

$$\frac{\Delta T}{\Delta x} = \frac{3.9}{3} = 1.3°/\text{h}$$

    (ii)   From noon to 2 P.M. the average rate of change is

$$\frac{\Delta T}{\Delta x} = \frac{T(14) - T(12)}{14 - 12} = \frac{17.3 - 14.3}{2} = 1.5°/\text{h}$$

    (iii)   From noon to 1 P.M. the average rate of change is

$$\frac{\Delta T}{\Delta x} = \frac{T(13) - T(12)}{13 - 12} = \frac{16.0 - 14.3}{1} = 1.7°/\text{h}$$

(b)　We plot the given data in Figure 1.59 and use them to sketch a smooth curve that approximates the graph of the temperature function. Then we draw the tangent at point P where $x = 12$ and, after measuring the sides of triangle ABC, we estimate that the slope of the tangent line is

$$\frac{|BC|}{|AC|} = \frac{10.3}{5.5} \approx 1.9$$

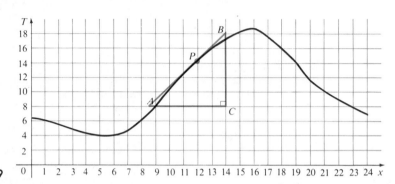

Figure 1.59

Therefore the instantaneous rate of change of temperature with respect to time at noon is about 1.9°/h.　　　　　　　　　　　　　　　　　　　　　　　　　　　　　●

The velocity of a particle is the rate of change of displacement with respect to time. Physicists are interested in other rates of change as well—for instance, the rate of change of work with respect to time (which is called power). Chemists who study a chemical reaction are interested in the rate of change in the concentration of a reactant with respect to time (called the rate of reaction). A steel manufacturer is interested in the rate of change of the cost of producing x tons of steel per day with respect to x (called the marginal cost). A biologist is interested in the rate of change of the population of a colony of bacteria with respect to time. In fact, the computation of rates of change is important in all of the natural sciences, engineering, and even the social sciences. Further examples will be given in Section 2.3.

All these rates of change can be interpreted as slopes of tangents. This gives added significance to the solution of the tangent problem. Whenever we solve a problem involving tangent lines, we are not just solving a problem in geometry. We are also implicitly solving a great variety of problems involving rates of change in science and engineering.

SECTION 1.8　**Exercises**

1. (a)　Find the slope of the tangent line to the parabola $y = x^2 + 2x$ at the point $(-3, 3)$
   (i)　using Definition 1.39
   (ii)　using Equation 1.40
   (b)　Find the equation of the tangent line in part (a).
   (c)　Graph the parabola and the tangent line.

2. (a)　Find the slope of the tangent line to the curve $y = x^3$ at the point $(-1, -1)$
   (i)　using Definition 1.39
   (ii)　using Equation 1.40
   (b)　Find the equation of the tangent line in part (a).
   (c)　Graph the curve and the tangent line.

In Exercises 3–6 find the equation of the tangent line to the given curve at the given point.

3. $y = 1 - 2x - 3x^2$, $(-2, -7)$

4. $y = 1/\sqrt{x}$, $(1, 1)$

5. $y = 1/x^2$, $(-2, \frac{1}{4})$

6. $y = x/(1 - x)$, $(0, 0)$

7. (a) Find the slope of the tangent to the curve $y = 2/(x + 3)$ at the point where $x = a$.
    (b) Find the slopes of the tangent lines at the points whose x-coordinates are given.
        (i) -1 (ii) 0 (iii) 1

8. (a) Find the slope of the tangent to the parabola $y = 1 + x + x^2$ at the point where $x = a$.
    (b) Find the slopes of the tangent lines at the points whose x-coordinates are given.
        (i) -1 (ii) $-\frac{1}{2}$ (iii) 1
    (c) Graph the curve and the three tangents.

9. If a ball is thrown into the air with a velocity of 40 ft/s, its height in feet after t seconds is given by $y = 40t - 16t^2$. Find the velocity when $t = 2$.

10. If an arrow is shot upward on the moon with a velocity of 58 m/s, its height in meters after t seconds is given by $h = 58t - 0.83t^2$.
    (a) Find the velocity of the arrow after 1 s.
    (b) Find the velocity of the arrow when $t = a$.
    (c) When will the arrow hit the moon?
    (d) With what velocity will the arrow hit the moon?

11. The displacement in meters of a particle moving in a straight line is given by the equation of motion $s = 4t^3 + 6t + 2$, where t is measured in seconds. Find the velocity of the particle at times $t = a$, $t = 1$, $t = 2$, and $t = 3$.

12. The displacement in meters of a particle moving in a straight line is given by $s = t^2 - 8t + 18$, where t is measured in seconds.
    (a) Find the average velocities over the following time intervals:
        (i) $[3, 4]$ (ii) $[3.5, 4]$
        (iii) $[4, 5]$ (iv) $[4, 4.5]$
    (b) Find the instantaneous velocity when $t = 4$.
    (c) Draw the graph of s as a function of t and draw the secant lines whose slopes are the average velocities in part (a) and the tangent line whose slope is the instantaneous velocity in part (b).

13. (a) Use the data in Example 5 to find the average rate of change of temperature with respect to time
        (i) from 8 P.M. to 11 P.M.
        (ii) from 8 P.M. to 10 P.M.
        (iii) from 8 P.M. to 9 P.M.
    (b) Estimate the instantaneous rate of change of T with respect to x at 8 P.M. by measuring the slope of a tangent.

14. The population P (in thousands) of a city from 1980 to 1986 is given in the accompanying table.
    (a) Find the average rate of growth
        (i) from 1982 to 1986
        (ii) from 1982 to 1985
        (iii) from 1982 to 1984
        (iv) from 1982 to 1983
    (b) Estimate the instantaneous rate of growth in 1982 by measuring the slope of a tangent.

Year	1980	1981	1982	1983	1984	1985	1986
P (in thousands)	105	110	117	126	137	150	164

15. The cost (in dollars) of producing x units of a certain commodity is $C(x) = 5000 + 10x + 0.05x^2$.
    (a) Find the average rate of change of C with respect to x when the production level is changed (i) from $x = 100$ to $x = 105$ and (ii) from $x = 100$ to $x = 101$.
    (b) Find the instantaneous rate of change of C with respect to x when $x = 100$. (This is called the *marginal cost*. Its significance will be explained in Sections 2.3 and 4.6.)

16. If a cylindrical tank holds 100,000 gallons of water, which takes 1 h to drain from the bottom of the tank, then Torricelli's Law gives the volume V of water remaining in the tank after t minutes as

$$V(t) = 100,000\left(1 - \frac{t}{60}\right)^2 \qquad 0 \le t \le 60$$

Find the rate at which the water is flowing out of the tank (the instantaneous rate of change of V with respect to t) after 20 min.

CHAPTER 1

Review

Key Topics

Define, state, or discuss the following.

1. Limit of $f(x)$ as x approaches a
2. One-sided limits
3. Limit Laws
4. The Squeeze Theorem
5. Continuous at a number a
6. Discontinuities of a function
7. Continuous on an interval
8. Intermediate Value Theorem
9. Limit at infinity
10. Horizontal asymptote
11. Infinite limit
12. Vertical asymptote
13. Infinite limit at infinity
14. Tangent line
15. Velocity
16. Average and instantaneous rate of change

Exercises

In Exercises 1–10 determine whether the statement is true or false.

1. $\displaystyle \lim_{x \to 4}\left(\frac{2x}{x-4} - \frac{8}{x-4}\right) = \lim_{x \to 4}\frac{2x}{x-4} - \lim_{x \to 4}\frac{8}{x-4}$

2. $\displaystyle \lim_{x \to 1}\frac{x^2+6x-7}{x^2+5x-6} = \frac{\lim_{x \to 1}(x^2+6x-7)}{\lim_{x \to 1}(x^2+5x-6)}$

3. $\displaystyle \lim_{x \to 1}\frac{x-3}{x^2+2x-4} = \frac{\lim_{x \to 1}(x-3)}{\lim_{x \to 1}(x^2+2x-4)}$

4. If $\lim_{x \to 5}f(x) = 2$ and $\lim_{x \to 5}g(x) = 0$, then $\lim_{x \to 5}[f(x)/g(x)]$ does not exist.

5. If $f(x) > 1$ for all x and $\lim_{x \to 0}f(x)$ exists, then $\lim_{x \to 0}f(x) > 1$.

6. Let f be a function such that $\lim_{x \to 4}f(x) = 7$. If $3.9 < x < 4.1$, then $6.9 < f(x) < 7.1$.

7. Let f be a function such that $\lim_{x \to 0}f(x) = 6$. Then there exists a number δ such that if $0 < |x| < \delta$, then $|f(x) - 6| < 1$.

8. If $\lim_{x \to 6}f(x)g(x)$ exists, then the limit must be $f(6)\,g(6)$.

9. If f is continuous at 5 and $f(5) = 2$ and $f(4) = 3$, then $\lim_{x \to 2}f(4x^2 - 11) = 2$.

10. If f is continuous on $[-1, 1]$, and $f(-1) = 4$ and $f(1) = 3$, then there exists a number r such that $|r| < 1$ and $f(r) = \pi$.

In Exercises 11–38 find the given limits.

11. $\displaystyle \lim_{x \to 4}\sqrt{x + \sqrt{x}}$

12. $\displaystyle \lim_{x \to 0^-}\sqrt{-x}$

13. $\displaystyle \lim_{t \to -1}\frac{t+1}{t^3-t}$

14. $\displaystyle \lim_{t \to 4}\frac{t-4}{t^2-3t-4}$

15. $\displaystyle \lim_{h \to 0}\frac{(1+h)^2 - 1}{h}$

16. $\displaystyle \lim_{h \to 0}\frac{(1+h)^{-2} - 1}{h}$

17. $\displaystyle \lim_{x \to -1}\frac{x^2-x-2}{x^2+3x-2}$

18. $\displaystyle \lim_{x \to -1}\frac{x^2-x-2}{x^2+3x+2}$

19. $\displaystyle \lim_{s \to 16}\frac{4-\sqrt{s}}{s-16}$

20. $\displaystyle \lim_{v \to 2}\frac{v^2+2v-8}{v^4-16}$

21. $\displaystyle \lim_{x \to 8^-}\frac{|x-8|}{x-8}$

22. $\displaystyle \lim_{x \to 9^+}\left(\sqrt{x-9} + [\![x+1]\!]\right)$

23. $\displaystyle \lim_{x \to 0}\frac{1-\sqrt{1-x^2}}{x}$

24. $\displaystyle \lim_{x \to 2}\frac{\sqrt{x+2}-\sqrt{2x}}{x^2-2x}$

25. $\displaystyle \lim_{t \to 6}\frac{17}{(t-6)^2}$

26. $\displaystyle \lim_{x \to 9^-}\left(\frac{-2}{x+9}\right)$

27. $\displaystyle \lim_{x \to 2^-}\frac{3}{\sqrt{2-x}}$

28. $\displaystyle \lim_{x \to \infty}\frac{1+2x-x^2}{1-x+2x^2}$

29. $\displaystyle \lim_{x \to \infty}\frac{1+x}{1-x^2}$

30. $\displaystyle \lim_{x \to \infty}x\tan\frac{1}{x}$

31. $\displaystyle \lim_{x \to \infty}\frac{\sqrt{x^2-9}}{2x-6}$

32. $\displaystyle \lim_{x \to 3^+}\frac{\sqrt{x^2-9}}{2x-6}$

33. $\displaystyle \lim_{x \to -\infty}\frac{4x^5+x^2+3}{2x^5+x^3+1}$

34. $\displaystyle \lim_{x \to -\infty}\frac{\cos^2 x}{x^2}$

35. $\displaystyle \lim_{x \to \infty}\left(\sqrt[3]{x} - \frac{x}{3}\right)$

36. $\displaystyle \lim_{x \to 2\pi^-}\frac{\cot x}{1+x^2}$

37. $\displaystyle \lim_{x \to -2^+}\frac{x}{x^2-4}$

38. $\displaystyle \lim_{x \to \infty}\left(\sqrt{x^2+x+1} - \sqrt{x^2-x}\right)$

In Exercises 39–41 prove the statements using the ε, δ definition of a limit.

39. $\displaystyle \lim_{x \to 5}(7x - 27) = 8$

40. $\displaystyle \lim_{x \to 0}\sqrt[3]{x} = 0$

41. $\lim_{x \to 2} (x^2 - 3x) = -2$

42. If $2x - 1 \le f(x) \le x^2$ for $0 < x < 3$, find $\lim_{x \to 1} f(x)$.

43. Let
$$f(x) = \begin{cases} \sqrt{-x} & \text{if } x < 0 \\ 3 - x & \text{if } 0 \le x < 3 \\ (x-3)^2 & \text{if } x > 3 \end{cases}$$

(a) Evaluate the following limits if they exist.
   (i) $\lim_{x \to 0^+} f(x)$ (ii) $\lim_{x \to 0^-} f(x)$

   (iii) $\lim_{x \to 0} f(x)$ (iv) $\lim_{x \to 3^-} f(x)$

   (v) $\lim_{x \to 3^+} f(x)$ (vi) $\lim_{x \to 3} f(x)$

(b) Where is f discontinuous?
(c) Sketch the graph of f.

44. Let
$$g(x) = \begin{cases} 2x - x^2 & \text{if } 0 \le x \le 2 \\ 2 - x & \text{if } 2 < x \le 3 \\ x - 4 & \text{if } 3 < x < 4 \\ \pi & \text{if } x \ge 4 \end{cases}$$

(a) For each of the numbers 2, 3, and 4, discover whether g is continuous from the left, continuous from the right, or continuous at the number.
(b) Sketch the graph of g.

Show that the functions in Exercises 45 and 46 are continuous on their domains. State these domains.

45. $f(x) = \dfrac{x+1}{x^2 + x + 1}$ **46.** $g(x) = \dfrac{\sqrt{x^2 - 9}}{x^2 - 2}$

In Exercises 47 and 48 use the Intermediate Value Theorem to show that there is a root of the given equation in the given interval.

47. $2x^3 + x^2 + 2 = 0$, $(-2, -1)$

48. $x^4 + 1 = 1/x$, $(0.5, 1)$

49. (a) Find the slope of the tangent line to the curve $y = 9 - 2x^2$ at the point $(2, 1)$.
   (b) Find the equation of this tangent line.

50. Find the equations of the tangent lines to the curve $y = 2/(1 - 3x)$ at the points with x-coordinates 0 and -1.

51. The displacement (in meters) of an object moving in a straight line is given by $s = 1 + 2t + t^2/4$, where t is measured in seconds.
(a) Find the average velocity over the following time periods.
   (i) $[1, 3]$ (ii) $[1, 2]$
   (iii) $[1, 1.5]$ (iv) $[1, 1.1]$
(b) Find the instantaneous velocity when $t = 1$.

52. According to Boyle's Law, if the temperature of a confined gas is held fixed, then the product of the pressure P and the volume V is constant. Suppose that for a certain gas, $PV = 800$, where P is measured in pounds per square inch and V is measured in cubic inches.
(a) Find the average rate of change of P as V increases from 200 in^3 to 250 in^3.
(b) Express V as a function of P and show that the instantaneous rate of change of V with respect to P is proportional to the inverse square of P.

53. Suppose that $|f(x)| \le g(x)$ for all x, where $\lim_{x \to a} g(x) = 0$. Find $\lim_{x \to a} f(x)$.

2

Derivatives

Mathematics compares the most diverse
phenomena and discovers the secret
analogies that unite them.

Joseph Fourier

In this chapter we begin our study of differential calculus, which is concerned with how one quantity changes in relation to another quantity. The central concept of differential calculus is the *derivative*. After learning how to calculate derivatives, we use them to solve problems involving rates of change.

Derivatives

In Section 1.8 we defined the slope of the tangent to a curve with equation $y = f(x)$ at the point where $x = a$ to be

(2.1)
$$m = \lim_{h \to 0} \frac{f(a + h) - f(a)}{h}$$

We also saw that the velocity of an object with position function $s = f(t)$ at time $t = a$ is

$$v(a) = \lim_{h \to 0} \frac{f(a + h) - f(a)}{h}$$

In fact, limits of the form

$$\lim_{h \to 0} \frac{f(a + h) - f(a)}{h}$$

arise whenever we calculate a rate of change in any of the sciences or engineering, such as a rate of reaction in chemistry or a marginal cost in economics. Since this type of limit occurs so widely, it is given a special name and notation.

Definition (2.2)

The **derivative of a function f at a number a,** denoted by $f'(a)$, is

$$f'(a) = \lim_{h \to 0} \frac{f(a+h) - f(a)}{h}$$

if this limit exists.

If we write $x = a + h$, then $h = x - a$ and h approaches 0 if and only if x approaches a. Therefore an equivalent way of stating the definition of the derivative, as we saw in finding tangent lines, is

(2.3)

$$f'(a) = \lim_{x \to a} \frac{f(x) - f(a)}{x - a}$$

EXAMPLE 1   Find the derivative of the function $f(x) = x^2 - 8x + 9$ at the number a.

Solution   From Definition 2.2 we have

$$f'(a) = \lim_{h \to 0} \frac{f(a+h) - f(a)}{h}$$

$$= \lim_{h \to 0} \frac{[(a+h)^2 - 8(a+h) + 9] - [a^2 - 8a + 9]}{h}$$

$$= \lim_{h \to 0} \frac{a^2 + 2ah + h^2 - 8a - 8h + 9 - a^2 + 8a - 9}{h}$$

$$= \lim_{h \to 0} \frac{2ah + h^2 - 8h}{h} = \lim_{h \to 0} (2a + h - 8)$$

$$= 2a - 8$$

Interpretations of the Derivative

The Interpretation of the Derivative as the Slope of a Tangent.   In Section 1.8 we defined the tangent line to the curve $y = f(x)$ at the point $P(a, f(a))$ to be the line that passes through P and has slope m given by Equation 2.1. Since, by Definition 2.2, this is the same as the derivative $f'(a)$, we can now say that *the tangent line to $y = f(x)$ at $(a, f(a))$ is the line through $(a, f(a))$ whose slope is equal to $f'(a)$,   the derivative of f at a.* Thus the geometric interpretation of a derivative [as defined by either (2.2) or (2.3)] is as shown in Figure 2.1.

Using the point-slope form of the equation of a line (Equation 10 in Review and Preview), we have the following:

(2.4)

If $f'(a)$ exists, then the equation of the tangent line to the curve $y = f(x)$ at the point $(a, f(a))$ is

$$y - f(a) = f'(a)(x - a)$$

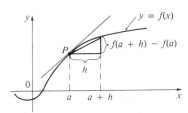

Figure 2.1

Geometric interpretation of
the derivative

(a) $f'(a) = \lim\limits_{h \to 0} \dfrac{f(a+h) - f(a)}{h}$

= slope of tangent at P

(b) $f'(a) = \lim\limits_{x \to a} \dfrac{f(x) - f(a)}{x - a}$

= slope of tangent at P

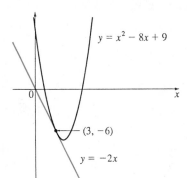

Figure 2.2

EXAMPLE 2   Find the equation of the tangent line to the parabola $y = x^2 - 8x + 9$ at the point $(3, -6)$.

Solution   From Example 1 we know that the derivative of $f(x) = x^2 - 8x + 9$ at the number a is $f'(a) = 2a - 8$. Therefore the slope of the tangent line at $(3, -6)$ is $f'(3) = 2(3) - 8 = -2$. Thus the equation of the tangent line, shown in Figure 2.2, is

$$y - (-6) = (-2)(x - 3) \qquad \text{or} \qquad y = -2x$$

The Interpretation of the Derivative as a Rate of Change.   In Section 1.8 we defined the instantaneous rate of change of $y = f(x)$ with respect to x at $x = x_1$ as the limit of the average rate of change over smaller and smaller intervals. If the interval is $[x_1, x_2]$, then the change in x is $\Delta x = x_2 - x_1$, the corresponding change in y is

$$\Delta y = f(x_2) - f(x_1)$$

and

(2.5)   instantaneous rate of change $= \lim\limits_{\Delta x \to 0} \dfrac{\Delta y}{\Delta x} = \lim\limits_{x_2 \to x_1} \dfrac{f(x_2) - f(x_1)}{x_2 - x_1}$

From Equation 2.3 we recognize this limit as being the derivative of f at x_1, that is, $f'(x_1)$.

This gives a second interpretation of the *derivative $f'(a)$ as the instantaneous rate of change of $y = f(x)$ with respect to x when $x = a$.* [The connection with the first interpretation is that if we sketch the curve $y = f(x)$, then the instantaneous rate of change is the slope of the tangent to this curve at the point where $x = a$.]

In particular, if $s = f(t)$ is the position function of a particle that moves along a straight line, then $f'(a)$ is the rate of change of the displacement s with respect to the time t. In other words, $f'(a)$ *is the velocity of the particle at time $t = a$.* (See Section 1.8.) The *speed* of the particle is the absolute value of the velocity, that is, $|f'(a)|$.

EXAMPLE 3   The position of a particle is given by the equation of motion $s = f(t) = t^2 - 8t + 9$, where t is measured in seconds and s in meters. Find the velocity and the speed after 2 seconds.

Solution   Again from Example 1 we know that the derivative of the position function is $f'(t) = 2t - 8$. Thus the velocity after 2 s is $f'(2) = 2(2) - 8 = -4\,\text{m/s}$ and the speed is $|f'(2)| = |-4| = 4\,\text{m/s}$.

The Derivative as a Function

If we replace a by x in Definition 2.2, we obtain

(2.6)
$$f'(x) = \lim_{h \to 0} \frac{f(x+h) - f(x)}{h}$$

Given a function f, we associate with it a new function f', called the **derivative of f**, defined by Equation 2.6. We know that the value of f' at x, $f'(x)$, can be interpreted geometrically as the slope of the tangent line to the graph of f at the point $(x, f(x))$.

The function f' is called the derivative of f because it has been "derived" from f by the limiting operation in Equation 2.6. The domain of f' is the set $\{x \mid f'(x) \text{ exists}\}$ and may be smaller than the domain of f.

EXAMPLE 4   If $f(x) = \sqrt{x - 5}$ for $x \geq 5$, find the derivative of f. What is the domain of f'?

Solution   When using Equation 2.6 to compute a derivative, we must remember that the variable is h and that x is temporarily regarded as a constant during the calculation of the limit.

$$f'(x) = \lim_{h \to 0} \frac{f(x+h) - f(x)}{h}$$

$$= \lim_{h \to 0} \frac{\sqrt{x+h-5} - \sqrt{x-5}}{h}$$

$$= \lim_{h \to 0} \frac{\sqrt{x+h-5} - \sqrt{x-5}}{h} \cdot \frac{\sqrt{x+h-5} + \sqrt{x-5}}{\sqrt{x+h-5} + \sqrt{x-5}}$$

$$= \lim_{h \to 0} \frac{(x+h-5) - (x-5)}{h(\sqrt{x+h-5} + \sqrt{x-5})}$$

$$= \lim_{h \to 0} \frac{1}{\sqrt{x+h-5} + \sqrt{x-5}}$$

$$= \frac{1}{\sqrt{x-5} + \sqrt{x-5}} = \frac{1}{2\sqrt{x-5}}$$

We see that $f'(x)$ exists if $x > 5$. Thus the domain of f' is $(5, \infty)$, which is smaller than the domain of f.   ●

EXAMPLE 5   Find f' if $f(x) = \dfrac{1-x}{2+x}$.

Solution $$f'(x) = \lim_{h \to 0} \frac{f(x+h) - f(x)}{h}$$

$$= \lim_{h \to 0} \frac{\dfrac{1-(x+h)}{2+(x+h)} - \dfrac{1-x}{2+x}}{h}$$

$$\frac{\frac{a}{b} - \frac{c}{d}}{e} = \frac{ad - bc}{bd} \cdot \frac{1}{e}$$

$$= \lim_{h \to 0} \frac{(1 - x - h)(2 + x) - (1 - x)(2 + x + h)}{h(2 + x + h)(2 + x)}$$

$$= \lim_{h \to 0} \frac{(2 - x - 2h - x^2 - xh) - (2 - x + h - x^2 - xh)}{h(2 + x + h)(2 + x)}$$

$$= \lim_{h \to 0} \frac{-3h}{h(2 + x + h)(2 + x)}$$

$$= \lim_{h \to 0} \frac{-3}{(2 + x + h)(2 + x)} = -\frac{3}{(2 + x)^2} \qquad \bullet$$

Other Notations

If we use the traditional notation $y = f(x)$ to indicate that the independent variable is x and the dependent variable is y, then some common alternative notations for the derivative are as follows:

$$f'(x) = y' = \frac{dy}{dx} = \frac{df}{dx} = \frac{d}{dx} f(x) = Df(x) = D_x f(x)$$

The symbols D and d/dx are called **differentiation operators** because they indicate the operation of **differentiation,** which is the process of calculating a derivative.

The symbol dy/dx, which was introduced by Leibniz, should not be regarded as a ratio (for the time being); it is simply a synonym for $f'(x)$. Nonetheless, it is a very useful and suggestive notation, especially when used in conjunction with increment notation. Referring to Equation 2.5, we can rewrite the definition of derivative in Leibniz notation in the form

$$\frac{dy}{dx} = \lim_{\Delta x \to 0} \frac{\Delta y}{\Delta x}$$

If we want to indicate the value of a derivative dy/dx in Leibniz notation at a specific number a, we use the notation

$$\frac{dy}{dx}\bigg|_{x=a}$$

which is a synonym for $f'(a)$.

Definition (2.7)

A function f is **differentiable at a** if $f'(a)$ exists. It is **differentiable on an open interval** (a, b) [or (a, ∞) or $(-\infty, a)$ or $(-\infty, \infty)$] if it is differentiable at every number in the interval.

EXAMPLE 6   Where is the function $f(x) = |x|$ differentiable?

Solution   If $x > 0$, then $|x| = x$ and we can choose h small enough that $x + h > 0$ and hence $|x + h| = x + h$. Therefore for $x > 0$ we have

$$f'(x) = \lim_{h \to 0} \frac{|x + h| - |x|}{h}$$

$$= \lim_{h \to 0} \frac{(x + h) - x}{h} = \lim_{h \to 0} \frac{h}{h} = \lim_{h \to 0} 1 = 1$$

and so f is differentiable for any $x > 0$.

Similarly, for $x < 0$ we have $|x| = -x$ and h can be chosen small enough that $|x + h| = -(x + h)$. Therefore

$$f'(x) = \lim_{h \to 0} \frac{|x + h| - |x|}{h}$$

$$= \lim_{h \to 0} \frac{-(x + h) - (-x)}{h} = \lim_{h \to 0} \frac{-h}{h} = \lim_{h \to 0} (-1) = -1$$

and so f is differentiable for any $x < 0$.

For $x = 0$ we have to investigate

$$f'(0) = \lim_{h \to 0} \frac{f(0 + h) - f(0)}{h} = \lim_{h \to 0} \frac{|0 + h| - |0|}{h} \qquad \text{(if it exists)}$$

Let us compute the left and right limits separately:

$$\lim_{h \to 0^+} \frac{|0 + h| - |0|}{h} = \lim_{h \to 0^+} \frac{|h|}{h} = \lim_{h \to 0^+} \frac{h}{h} = \lim_{h \to 0^+} 1 = 1$$

and

$$\lim_{h \to 0^-} \frac{|0 + h| - |0|}{h} = \lim_{h \to 0^-} \frac{|h|}{h} = \lim_{h \to 0^-} \frac{-h}{h} = -1$$

Since these limits are different, $f'(0)$ does not exist. Thus f is differentiable at all x except 0.

A formula for f' is given by

$$f'(x) = \begin{cases} 1 & \text{if } x > 0 \\ -1 & \text{if } x < 0 \end{cases}$$

and its graph is shown in Figure 2.3(b). The fact that $f'(0)$ does not exist is reflected geometrically in the fact that the curve $y = |x|$ does not have a tangent line at $(0, 0)$.

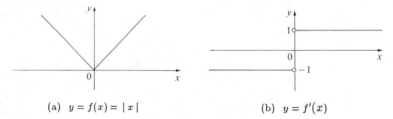

Figure 2.3   (a) $y = f(x) = |x|$ (b) $y = f'(x)$

In general if the graph of a function f has corners or kinks in it, then the graph has no tangent at those points and f is not differentiable there. If, on the other hand, f is differentiable on an interval, we expect its graph to be "smooth."

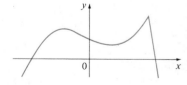

Figure 2.4

EXAMPLE 7   Sketch the graph of f' from the graph of f given in Figure 2.4.

Solution   Using the interpretation of $f'(x)$ as the slope of the tangent line at $(x, f(x))$, we can sketch the graph of f' directly beneath the graph of f (see Figure 2.5). Notice that the vertical coordinate of P' is the slope at P, and so on. Remember that a horizontal line has slope 0 and there is no tangent at a corner point.

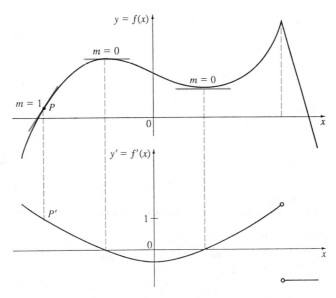

Figure 2.5

The technique for graphing f' shown in Example 7 is useful because many functions are not defined by formulas but can be represented graphically from scientific data. From the graph of f' we can see the rate of change of y at any value of x and determine where the rate of change itself is increasing or decreasing.

Theorem (2.8)

> If f is differentiable at a, then f is continuous at a.

Proof   To prove that f is continuous at a, we have to show that $\lim_{x \to a} f(x) = f(a)$. We do this by showing that the difference $f(x) - f(a)$ approaches 0.

For $x \neq a$ we can divide and multiply by $x - a$:

$$f(x) - f(a) = \frac{f(x) - f(a)}{x - a}(x - a)$$

We did this in order to involve the difference quotient. (We somehow have to use the fact that f is differentiable at a, that is, $f'(a)$ exists.) Thus we can use the Product Law and Equation 2.3 to write

$$\lim_{x \to a} [f(x) - f(a)] = \lim_{x \to a} \frac{f(x) - f(a)}{x - a}(x - a)$$

$$= \lim_{x \to a} \frac{f(x) - f(a)}{x - a} \lim_{x \to a}(x - a)$$

$$= f'(a) \cdot 0 = 0$$

Therefore

$$\lim_{x \to a} f(x) = \lim_{x \to a} [f(a) + (f(x) - f(a))]$$

$$= \lim_{x \to a} f(a) + \lim_{x \to a} [f(x) - f(a)]$$

$$= f(a) + 0 = f(a)$$

and so f is continuous at a.

Note: The converse of Theorem 2.8 is false; that is, there are functions that are continuous but not differentiable. For instance, the function $f(x) = |x|$ is continuous at 0 because

$$\lim_{x \to 0} f(x) = \lim_{x \to 0} |x| = 0 = f(0)$$

(See Example 8 in Section 1.3.) But in Example 6 we showed that f is not differentiable at 0.

SECTION 2.1 **Exercises**

1. If $f(x) = 3x^2 - 5x$, find $f'(2)$ and use it to find the equation of the tangent line to the parabola $y = 3x^2 - 5x$ at the point $(2, 2)$.

2. If $g(x) = 1 - x^3$, find $g'(0)$ and use it to find the equation of the tangent line to the curve $y = 1 - x^3$ at the point $(0, 1)$.

3. If $F(x) = \sqrt{x + 7}$, find $F'(a)$ and use it to find the equation of the tangent line to the curve $y = \sqrt{x + 7}$ at the point $(2, 3)$.

4. If $G(x) = 1/(2x - 1)$, find $G'(a)$ and use it to find the equation of the tangent line to the curve $y = 1/(2x - 1)$ at the point $(-1, -\frac{1}{3})$.

In Exercises 5 and 6 a particle moves along a straight line with equation of motion $s = f(t)$, where s is measured in meters and t in seconds. Find the velocity when $t = 2$.

5. $f(t) = t^2 - 6t - 5$ 6. $f(t) = 2t^3 - t + 1$

In Exercises 7–12 find $f'(a)$.

7. $f(x) = 1 + x - 2x^2$ 8. $f(x) = x^3 + 3x$

9. $f(x) = \dfrac{x}{2x - 1}$ 10. $f(x) = \dfrac{x}{x^2 - 1}$

11. $f(x) = \dfrac{2}{\sqrt{3 - x}}$ 12. $f(x) = \sqrt{x - 1}$

13. If $f(x) = \sqrt[3]{x}$, find $f'(a)$ using (a) Equation 2.3 and (b) Definition 2.2.

14. If $f(x) = x^{2/3}$, show that $f'(0)$ does not exist. Sketch the curve $y = x^{2/3}$.

In Exercises 15–20 each limit represents the derivative of some function f at some number a. State f and a in each case.

15. $\displaystyle\lim_{h \to 0} \frac{\sqrt{1 + h} - 1}{h}$ 16. $\displaystyle\lim_{h \to 0} \frac{(2 + h)^3 - 8}{h}$

17. $\displaystyle\lim_{x \to 1} \frac{x^9 - 1}{x - 1}$ 18. $\displaystyle\lim_{x \to 3\pi} \frac{\cos x + 1}{x - 3\pi}$

19. $\displaystyle\lim_{t \to 0} \frac{\sin\left(\frac{\pi}{2} + t\right) - 1}{t}$ 20. $\displaystyle\lim_{x \to 0} \frac{3^x - 1}{x}$

In Exercises 21–32 find the derivative of the given function using the definition of derivative. State the domain of the function and the domain of its derivative.

21. $f(x) = 5x + 3$ 22. $f(x) = 18$

23. $f(x) = x^3 - x^2 + 2x$ 24. $f(x) = \sqrt{6 - x}$

25. $f(x) = x - \frac{2}{x}$ 26. $f(x) = \frac{x + 1}{x - 1}$

27. $g(x) = \sqrt{1 + 2x}$ 28. $g(x) = \dfrac{1}{x^2}$

29. $G(x) = \dfrac{4 - 3x}{2 + x}$ 30. $F(x) = \dfrac{1}{\sqrt{x - 1}}$

31. $f(x) = x^4$ 32. $f(t) = \dfrac{6}{1 + t^2}$

33. Compute the derivatives of the functions $f(x) = x$, $f(x) = x^2$, and $f(x) = x^3$ and observe the result of Exercise 31. On the basis of these results, guess the derivative of $f(x) = x^n$ for n a positive integer. Test your guess by computing the derivative of $f(x) = x^5$.

34. Match the graph of each function in (a)–(d) with the graph of its derivative in (i)–(iv).

(a)   (b)

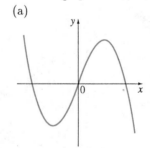

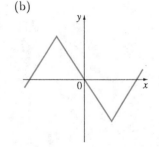

(c)
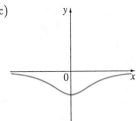

(d)
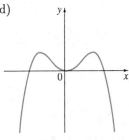

39.
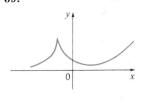

40.
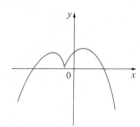

(i)
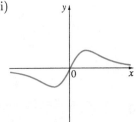

(ii)
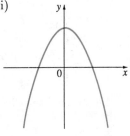

41.
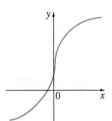

42.
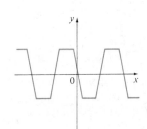

(iii)
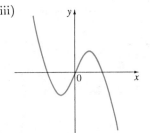

(iv)
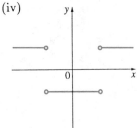

43.
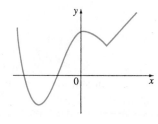

In Exercises 35–43 trace or copy the graph of the given function f. Then use the method of Example 7 to sketch the graph of f' below it.

35.
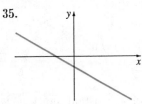

36.
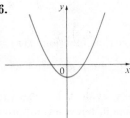

37.
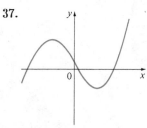

38.
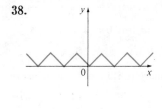

44. Make a careful sketch of the graph of the sine function and below it sketch of the graph of its derivative in the same manner as in Exercises 35–43. Can you guess what the derivative of the sine function is from its graph?

45. Show that the function $f(x) = |x - 6|$ is not differentiable at 6. Find a formula for f' and sketch its graph.

46. Where is the greatest integer function $f(x) = [\![x]\!]$ not differentiable? Find a formula for f' and sketch its graph.

47. (a) Sketch the graph of the function $f(x) = x|x|$.
    (b) For what values of x is f differentiable?
    (c) Find a formula for f'.

48. Where (and why) is the following function discontinuous? Where is it not differentiable? Sketch the graph.

$$g(x) = \begin{cases} \dfrac{x^3 - x}{x^2 + x} & \text{if } x < 1 \quad (x \neq 0) \\ 0 & \text{if } x = 0 \\ 1 - x & \text{if } x \geq 1 \end{cases}$$

49. The **left-hand** and **right-hand derivatives of** f at a are defined by

$$f'_-(a) = \lim_{h \to 0^-} \frac{f(a+h) - f(a)}{h}$$

and

$$f'_+(a) = \lim_{h \to 0^+} \frac{f(a+h) - f(a)}{h}$$

if these limits exist. Then $f'(a)$ exists if and only if these one-sided derivatives exist and are equal.

(a) Find $f'_-(0.6)$ and $f'_+(0.6)$ for the function $f(x) = |5x - 3|$.

(b) Show that $f'(0.6)$ does not exist.

50. (a) Use the definitions in Exercise 49 to compute $f'_-(4)$ and $f'_+(4)$ for the function

$$f(x) = \begin{cases} 0 & \text{if } x \leq 0 \\ 5 - x & \text{if } 0 < x < 4 \\ 1/(5 - x) & \text{if } x \geq 4 \end{cases}$$

(b) Sketch the graph of f.

(c) Where is f discontinuous?

(d) Where is f not differentiable?

In Exercises 51 and 52 determine whether or not $f'(0)$ exists.

51. $f(x) = \begin{cases} x \sin\frac{1}{x} & \text{if } x \neq 0 \\ 0 & \text{if } x = 0 \end{cases}$

52. $f(x) = \begin{cases} x^2 \sin\frac{1}{x} & \text{if } x \neq 0 \\ 0 & \text{if } x = 0 \end{cases}$

53. A function f is called *even* if $f(-x) = f(x)$ for all x in its domain and *odd* if $f(-x) = -f(x)$ for all such x. Prove that

(a) the derivative of an even function is an odd function, and

(b) the derivative of an odd function is an even function.

SECTION 2.2

Differentiation Formulas

If it were always necessary to compute derivatives directly from the definition, as we did in the preceding section, such computations would be tedious and the evaluation of some limits would require ingenuity. Fortunately there are several rules for finding derivatives without having to use the definition directly. These formulas greatly simplify the task of differentiation.

Theorem (2.9)

> If f is a constant function, $f(x) = c$, then $f'(x) = 0$.

This result is geometrically evident because the graph of a constant function is a horizontal line, which has slope 0, but a formal proof is also easy.

Proof
$$f'(x) = \lim_{h \to 0} \frac{f(x+h) - f(x)}{h} = \lim_{h \to 0} \frac{c - c}{h}$$
$$= \lim_{h \to 0} 0 = 0$$

In Leibniz notation, Theorem 2.9 can be written as

$$\frac{d}{dx} c = 0$$

The next theorem gives a formula for differentiating the power function $f(x) = x^n$. In the preceding section (Exercise 33) we found that $D(x) = 1$, $D(x^2) = 2x$, $D(x^3) = 3x^2$, $D(x^4) = 4x^3$, and so it seems reasonable to make the guess that $D(x^n) = nx^{n-1}$. We give two proofs of this fact; the second uses the Binomial Theorem.

The Power Rule (2.10)

> If n is a positive integer and $f(x) = x^n$, then
>
> $$f'(x) = nx^{n-1}$$

First Proof   The formula

(2.11)
$$x^n - a^n = (x - a)(x^{n-1} + x^{n-2}a + \cdots + xa^{n-2} + a^{n-1})$$

can be verified simply by multiplying out the right-hand side (or by summing the second factor as a geometric series). Thus, by using Equation 2.3 for $f'(a)$ and then using Equation 2.11, we get

$$f'(a) = \lim_{x \to a} \frac{f(x) - f(a)}{x - a} = \lim_{x \to a} \frac{x^n - a^n}{x - a}$$

$$= \lim_{x \to a} (x^{n-1} + x^{n-2}a + \cdots + xa^{n-2} + a^{n-1})$$

$$= a^{n-1} + a^{n-2}a + \cdots + aa^{n-2} + a^{n-1}$$

$$= na^{n-1}$$

Second Proof

$$f'(x) = \lim_{h \to 0} \frac{f(x+h) - f(x)}{h} = \lim_{h \to 0} \frac{(x+h)^n - x^n}{h}$$

Expanding $(x + h)^n$ by the Binomial Theorem (see Appendix A), we get

$$f'(x) = \lim_{h \to 0} \frac{\left[x^n + nx^{n-1}h + \dfrac{n(n-1)}{2} x^{n-2}h^2 + \cdots + nxh^{n-1} + h^n \right] - x^n}{h}$$

$$= \lim_{h \to 0} \frac{nx^{n-1}h + \dfrac{n(n-1)}{2} x^{n-2}h^2 + \cdots + nxh^{n-1} + h^n}{h}$$

$$= \lim_{h \to 0} \left[nx^{n-1} + \frac{n(n-1)}{2} x^{n-2}h + \cdots + nxh^{n-2} + h^{n-1} \right]$$

$$= nx^{n-1}$$

because every term except the first has h as a factor and therefore approaches 0.   ●

The Power Rule can be written in Leibniz notation as

> $$\frac{d}{dx}(x^n) = nx^{n-1}$$

We illustrate the Power Rule using various notations in Example 1.

EXAMPLE 1

(a) If $f(x) = x^6$, then $f'(x) = 6x^5$.   (b) If $y = x^{1000}$, then $y' = 1000x^{999}$.

(c) If $y = t^4$, then $\dfrac{dy}{dt} = 4t^3$.   (d) $\dfrac{d}{dr}(r^3) = 3r^2$

(e) $D_u(u^m) = mu^{m-1}$

The next differentiation formulas tell us that *the derivative of a constant times a function is the constant times the derivative of the function*, and *the derivative of a sum (or difference) of functions is the sum (or difference) of the derivatives* (assuming these derivatives exist).

Theorem (2.12)

Suppose c is a constant and $f'(x)$ and $g'(x)$ exist.

(a)   If $F(x) = cf(x)$, then $F'(x)$ exists and $F'(x) = cf'(x)$.

(b)   If $G(x) = f(x) + g(x)$, then $G'(x)$ exists and $G'(x) = f'(x) + g'(x)$.

(c)   If $H(x) = f(x) - g(x)$, then $H'(x)$ exists and $H'(x) = f'(x) - g'(x)$.

In short:

(a) $(cf)' = cf'$ (b) $(f + g)' = f' + g'$ (c) $(f - g)' = f' - g'$

Proof

(a)
$$F'(x) = \lim_{h \to 0} \frac{F(x+h) - F(x)}{h} = \lim_{h \to 0} \frac{cf(x+h) - cf(x)}{h}$$

$$= \lim_{h \to 0} c\left[\frac{f(x+h) - f(x)}{h}\right]$$

$$= c \lim_{h \to 0} \frac{f(x+h) - f(x)}{h} \qquad \text{(by Limit Law 3)}$$

$$= cf'(x)$$

(b)
$$G'(x) = \lim_{h \to 0} \frac{G(x+h) - G(x)}{h}$$

$$= \lim_{h \to 0} \frac{[f(x+h) + g(x+h)] - [f(x) + g(x)]}{h}$$

$$= \lim_{h \to 0} \left[\frac{f(x+h) - f(x)}{h} + \frac{g(x+h) - g(x)}{h}\right]$$

$$= \lim_{h \to 0} \frac{f(x+h) - f(x)}{h} + \lim_{h \to 0} \frac{g(x+h) - g(x)}{h} \quad \text{(by Limit Law 1)}$$

$$= f'(x) + g'(x)$$

Part (c) can be proved similarly using Limit Law 2 from Section 1.3.

Using Leibniz notation, the results of Theorem 2.12 can be summarized as follows.

$$\text{(a)} \quad \frac{d}{dx}(cf) = c\frac{df}{dx} \qquad\qquad \text{(b)} \quad \frac{d}{dx}(f+g) = \frac{df}{dx} + \frac{dg}{dx}$$

$$\text{(c)} \quad \frac{d}{dx}(f-g) = \frac{df}{dx} - \frac{dg}{dx}$$

The result of Theorem 2.12(b) can be extended to the sum of any number of functions. For instance, using this theorem twice, we get

$$(f + g + h)' = [(f + g) + h]' = (f + g)' + h' = f' + g' + h'$$

Theorem 2.12 can be combined with the Power Rule to differentiate any polynomial, as the following examples demonstrate.

EXAMPLE 2

$$\frac{d}{dx}(x^8 + 12x^5 - 4x^4 + 10x^3 - 6x + 5)$$
$$= \frac{d}{dx}(x^8) + 12\frac{d}{dx}(x^5) - 4\frac{d}{dx}(x^4) + 10\frac{d}{dx}(x^3) - 6\frac{d}{dx}(x) + \frac{d}{dx}(5)$$
$$= 8x^7 + 12(5x^4) - 4(4x^3) + 10(3x^2) - 6(1) + 0$$
$$= 8x^7 + 60x^4 - 16x^3 + 30x^2 - 6$$

EXAMPLE 3
If $f(x) = x^4 - x^3 + x^2 - x + 1$, find the equation of the tangent to the graph of f at the point $(1,1)$.

Solution   The slope is $f'(1)$, which we calculate as follows:

$$f'(x) = 4x^3 - 3x^2 + 2x - 1$$
$$f'(1) = 4 - 3 + 2 - 1 = 2$$

Therefore the equation of the tangent at $(1,1)$ is

$$y - 1 = 2(x - 1)$$
or
$$2x - y - 1 = 0$$

Next we need a formula for the derivative of a product of two functions. By analogy with Theorem 2.12(b) and (c), one might be tempted to guess, as Leibniz did three centuries ago, that the derivative of a product is the product of the derivatives. We can see, however, that this guess is wrong by looking at a particular example. Let $f(x) = x$ and $g(x) = x^2$. Then the Power Rule gives $f'(x) = 1$ and $g'(x) = 2x$. But $(fg)(x) = x^3$, so $(fg)'(x) = 3x^2$. Thus $(fg)' \neq f'g'$. The correct formula was discovered by Leibniz (soon after his false start) and is called the Product Rule.

The Product Rule (2.13)

> If $F(x) = f(x)g(x)$ and $f'(x)$ and $g'(x)$ both exist, then
>
> $$F'(x) = f(x)g'(x) + g(x)f'(x)$$
>
> In short: $(fg)' = fg' + gf'$

Proof

$$F'(x) = \lim_{h \to 0} \frac{F(x+h) - F(x)}{h}$$
$$= \lim_{h \to 0} \frac{f(x+h)g(x+h) - f(x)g(x)}{h}$$

In order to evaluate this limit, we would like to separate the functions f and g as in the proof of Theorem 2.12(b). We can achieve this separation by adding and subtracting the term $f(x+h)g(x)$ in the numerator:

$$F'(x) = \lim_{h \to 0} \frac{f(x+h)g(x+h) - f(x+h)g(x) + f(x+h)g(x) - f(x)g(x)}{h}$$
$$= \lim_{h \to 0} \left[f(x+h) \frac{g(x+h) - g(x)}{h} + g(x) \frac{f(x+h) - f(x)}{h} \right]$$
$$= \lim_{h \to 0} f(x+h) \cdot \lim_{h \to 0} \frac{g(x+h) - g(x)}{h} + \lim_{h \to 0} g(x) \cdot \lim_{h \to 0} \frac{f(x+h) - f(x)}{h}$$
$$= f(x)g'(x) + g(x)f'(x)$$

Note that $\lim_{h \to 0} g(x) = g(x)$ because $g(x)$ is a constant with respect to the variable h. Also, since f is differentiable at x, it is continuous at x by Theorem 2.8, and so $\lim_{h \to 0} f(x+h) = f(x)$. (See Exercise 61 in Section 1.5.)   ●

When expressed in Leibniz notation, the Product Rule becomes

$$\frac{d}{dx}(fg) = f \frac{dg}{dx} + g \frac{df}{dx}$$

In words, this says that *the derivative of a product of two functions is the first function times the derivative of the second function plus the second function times the derivative of the first function.*

EXAMPLE 4   Find $F'(x)$ if $F(x) = (6x^3)(7x^4)$.

Solution   By the Product Rule, we have

$$F'(x) = (6x^3) \frac{d}{dx}(7x^4) + (7x^4) \frac{d}{dx}(6x^3)$$
$$= (6x^3)(28x^3) + (7x^4)(18x^2)$$
$$= 168x^6 + 126x^6 = 294x^6$$ ●

Notice that we could verify the answer to Example 4 directly by first multiplying the factors:

$$F(x) = (6x^3)(7x^4) = 42x^7 \quad \Rightarrow \quad F'(x) = 42(7x^6) = 294x^6$$

But later we will meet functions, such as $y = x^2 \sin x$, for which the Product Rule is the only possible method.

The Quotient Rule (2.14)

> If $F(x) = f(x)/g(x)$ and $f'(x)$ and $g'(x)$ exist, then $F'(x)$ exists and
>
> $$F'(x) = \frac{g(x)f'(x) - f(x)g'(x)}{[g(x)]^2}$$
>
> In short:
> $$\left(\frac{f}{g}\right)' = \frac{gf' - fg'}{g^2}$$

Proof

$$F'(x) = \lim_{h \to 0} \frac{F(x+h) - F(x)}{h} = \lim_{h \to 0} \frac{\dfrac{f(x+h)}{g(x+h)} - \dfrac{f(x)}{g(x)}}{h}$$

$$= \lim_{h \to 0} \frac{f(x+h)g(x) - f(x)g(x+h)}{hg(x+h)g(x)}$$

We can separate f and g in this expression by adding and subtracting the term $f(x)g(x)$ in the numerator:

$$F'(x) = \lim_{h \to 0} \frac{f(x+h)g(x) - f(x)g(x) + f(x)g(x) - f(x)g(x+h)}{hg(x+h)g(x)}$$

$$= \lim_{h \to 0} \frac{g(x)\dfrac{f(x+h) - f(x)}{h} - f(x)\dfrac{g(x+h) - g(x)}{h}}{g(x+h)g(x)}$$

$$= \frac{\displaystyle\lim_{h \to 0} g(x) \cdot \lim_{h \to 0} \frac{f(x+h) - f(x)}{h} - \lim_{h \to 0} f(x) \cdot \lim_{h \to 0} \frac{g(x+h) - g(x)}{h}}{\displaystyle\lim_{h \to 0} g(x+h) \cdot \lim_{h \to 0} g(x)}$$

$$= \frac{g(x)f'(x) - f(x)g'(x)}{[g(x)]^2}$$

Again g is continuous by Theorem 2.8, so $\lim_{h \to 0} g(x+h) = g(x)$.   ●

The Quotient Rule, in Leibniz notation, becomes

$$\frac{d}{dx}\left(\frac{f(x)}{g(x)}\right) = \frac{g(x)\dfrac{d}{dx}f(x) - f(x)\dfrac{d}{dx}g(x)}{[g(x)]^2}$$

In words, this says that the *derivative of a quotient is the denominator times the derivative of the numerator minus the numerator times the derivative of the denominator, all divided by the square of the denominator.*

The theorems of this section show that any polynomial is differentiable on R and any rational function is differentiable on its domain. Furthermore, the Quotient Rule and the other differentiation formulas enable us to compute the derivative of any rational function, as the next example illustrates.

EXAMPLE 5   Let $y = \dfrac{x^2 + x - 2}{x^3 + 6}$.

Then
$$y' = \frac{(x^3 + 6)D(x^2 + x - 2) - (x^2 + x - 2)D(x^3 + 6)}{(x^3 + 6)^2}$$

$$= \frac{(x^3 + 6)(2x + 1) - (x^2 + x - 2)(3x^2)}{(x^3 + 6)^2}$$

$$= \frac{(2x^4 + x^3 + 12x + 6) - (3x^4 + 3x^3 - 6x^2)}{(x^3 + 6)^2}$$

$$= \frac{-x^4 - 2x^3 + 6x^2 + 12x + 6}{(x^3 + 6)^2}$$

The Quotient Rule can also be used to extend the Power Rule to the case where the exponent is a negative integer.

Theorem (2.15)

> If $f(x) = x^{-n}$, where n is a positive integer, then
>
> $$f'(x) = -nx^{-n-1}$$

Proof
$$f'(x) = \frac{d}{dx}(x^{-n}) = \frac{d}{dx}\left(\frac{1}{x^n}\right)$$

$$= \frac{x^n D(1) - 1 \cdot D(x^n)}{(x^n)^2}$$

$$= \frac{x^n \cdot 0 - 1 \cdot nx^{n-1}}{x^{2n}}$$

$$= \frac{-nx^{n-1}}{x^{2n}} = -nx^{n-1-2n} = -nx^{-n-1}$$

EXAMPLE 6

(a)   If $y = \frac{1}{x}$, then $\frac{dy}{dx} = \frac{d}{dx}(x^{-1}) = -x^{-2} = -\frac{1}{x^2}$.

(b) $\frac{d}{dt}\left(\frac{6}{t^3}\right) = 6\frac{d}{dt}(t^{-3}) = 6(-3)t^{-4} = -\frac{18}{t^4}$

By (2.10) and (2.15) the Power Rule holds if the exponent n is a positive or negative integer. If $n = 0$, then $x^0 = 1$, which we know has a derivative of 0. Thus the Power Rule holds for any integer n. In fact, it also holds for *any real number n*, as we will prove in Chapter 3. (A proof for rational values of n is indicated in Exercise 36 in Section 2.6.) In the meantime we state the general version and use it in the examples and exercises.

**The Power Rule
(General Version)
(2.16)**

> Let n be any real number. If $f(x) = x^n$, then $f'(x) = nx^{n-1}$.
>
> Or, in Leibniz notation, $\frac{d}{dx}(x^n) = nx^{n-1}$.

EXAMPLE 7

(a) If $f(x) = x^{\pi}$, then $f'(x) = \pi x^{\pi - 1}$.

(b) $\dfrac{d}{dx}\sqrt{x} = \dfrac{d}{dx}(x^{1/2}) = \frac{1}{2}x^{(1/2)-1} = \frac{1}{2}x^{-1/2} = \dfrac{1}{2\sqrt{x}}$

(c) Let

$$y = \dfrac{1}{\sqrt[3]{x^2}}$$

Then

$$\dfrac{dy}{dx} = \dfrac{d}{dx}(x^{-2/3}) = -\frac{2}{3}x^{-(2/3)-1}$$

$$= -\frac{2}{3}x^{-5/3}$$

EXAMPLE 8   Differentiate the function $f(t) = \sqrt{t}(1 - t)$.

Solution 1   Using the Product Rule, we have

$$f'(t) = \sqrt{t}\,\dfrac{d}{dt}(1 - t) + (1 - t)\dfrac{d}{dt}\sqrt{t}$$

$$= \sqrt{t}(-1) + (1 - t)\cdot\frac{1}{2}t^{-1/2}$$

$$= -\sqrt{t} + \dfrac{1 - t}{2\sqrt{t}} = \dfrac{1 - 3t}{2\sqrt{t}}$$

Solution 2   If we first use the laws of exponents, then we can proceed directly without using the Product Rule.

$$f(t) = \sqrt{t} - t\sqrt{t} = t^{1/2} - t^{3/2}$$

$$f'(t) = \frac{1}{2}t^{-1/2} - \frac{3}{2}t^{1/2}$$

which is equivalent to the answer given in Solution 1.

EXAMPLE 9   At what points on the hyperbola $xy = 12$ is the tangent line parallel to the line $3x + y = 0$?

Solution   Since $xy = 12$ can be written as $y = 12/x$, we have

$$\dfrac{dy}{dx} = 12\dfrac{d}{dx}(x^{-1}) = 12(-x^{-2}) = -\dfrac{12}{x^2}$$

Let the x-coordinate of one of the points in question be a. Then the slope of the tangent line at that point is $-12/a^2$. This tangent line will be parallel to the line $3x + y = 0$, or $y = -3x$, if it has the same slope, that is, -3. Equating slopes we get

$$-\dfrac{12}{a^2} = -3 \qquad \text{or} \qquad a^2 = 4 \qquad \text{or} \qquad a = \pm 2$$

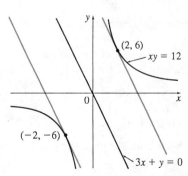

Figure 2.6

Therefore the required points are $(2, 6)$ and $(-2, -6)$. The hyperbola and the tangents are shown in Figure 2.6.

We summarize the differentiation formulas we have learned so far in the following table.

Table of Differentiation
Formulas (2.17)

$$(cf)' = cf' \qquad\qquad (f+g)' = f' + g'$$

$$(f-g)' = f' - g' \qquad (fg)' = f'g + fg'$$

$$\left(\frac{f}{g}\right)' = \frac{f'g - fg'}{g^2} \qquad \frac{d}{dx}c = 0$$

$$\frac{d}{dx}(x^n) = nx^{n-1}$$

SECTION 2.2 **Exercises**

Differentiate the functions given in Exercises 1–36.

1. $f(x) = x^2 - 10x + 100$ 2. $g(x) = x^{100} + 50x + 1$

3. $V(r) = \frac{4}{3}\pi r^3$

4. $s(t) = t^8 + 6t^7 - 18t^2 + 2t$

5. $F(x) = (16x)^3$ 6. $G(y) = (y^2 + 1)(2y - 7)$

7. $Y(t) = 6t^{-9}$ 8. $R(x) = \frac{\sqrt{10}}{x^7}$

9. $g(x) = x^2 + \frac{1}{x^2}$ 10. $f(t) = \sqrt{t} - \frac{1}{\sqrt{t}}$

11. $h(x) = \frac{x+2}{x-1}$ 12. $f(u) = \frac{1-u^2}{1+u^2}$

13. $G(s) = (s^2 + s + 1)(s^2 + 2)$

14. $H(t) = \sqrt[3]{t}(t+2)$

15. $y = \frac{x^2 + 4x + 3}{\sqrt{x}}$ 16. $y = \frac{\sqrt{x} - 1}{\sqrt{x} + 1}$

17. $y = \sqrt{5x}$ 18. $y = x^{4/3} - x^{2/3}$

19. $y = \frac{1}{x^4 + x^2 + 1}$ 20. $y = x^2 + x + x^{-1} + x^{-2}$

21. $y = ax^2 + bx + c$ 22. $y = A + \frac{B}{x} + \frac{C}{x^2}$

23. $y = \frac{3t - 7}{t^2 + 5t - 4}$ 24. $y = \frac{4t + 5}{2 - 3t}$

25. $y = x + \sqrt[5]{x^2}$ 26. $y = x^4 - \sqrt[4]{x}$

27. $u = x^{\sqrt{2}}$ 28. $u = \sqrt[3]{t^2} + 2\sqrt{t^3}$

29. $v = x\sqrt{x} + \frac{1}{x^2\sqrt{x}}$ 30. $v = \frac{6}{\sqrt[3]{t^5}}$

31. $f(x) = \frac{x}{x + \frac{c}{x}}$ 32. $f(x) = \frac{ax + b}{cx + d}$

33. $f(x) = \frac{x^5}{x^3 - 2}$

34. $g(x) = \sqrt[5]{x} - \frac{6}{x^{1.8}} + 0.1(x^{1.8})$

35. $s = \frac{2 - \frac{1}{t}}{t + 1}$ 36. $s = \sqrt{t}(t^3 - \sqrt{t} + 1)$

37. The general polynomial of degree n has the form

$$P(x) = a_n x^n + a_{n-1}x^{n-1} + \cdots + a_2 x^2 + a_1 x + a_0$$

where $a_n \neq 0$. Find the derivative of P.

In Exercises 38–41 find the equation of the tangent line to the given curve at the given point.

38. $y = \frac{x}{x - 3}$, $(6, 2)$ 39. $y = x + \frac{4}{x}$, $(2, 4)$

40. $y = x^{5/2}$, $(4, 32)$ 41. $y = \frac{1}{x^2 + 1}$, $(-1, \frac{1}{2})$

42. Find the equations of the tangent lines to the curve $y = (x - 1)/(x + 1)$ that are parallel to the line $x - 2y = 1$.

43. At what point on the curve $y = x\sqrt{x}$ is the tangent line parallel to the line $3x - y + 6 = 0$?

44. For what values of x does the graph of $f(x) = 2x^3 - 3x^2 - 6x + 87$ have a horizontal tangent?

45. Find the points on the curve $y = x^3 - x^2 - x + 1$ where the tangent is horizontal.

46. Draw a diagram to show that there are two tangent lines to the parabola $y = x^2$ that pass through the point $(0, -4)$. Find the coordinates of the points where these tangent lines intersect the parabola.

47. How many tangent lines to the curve $y = x/(x + 1)$ pass through the point $(1, 2)$? At which points do these tangent lines touch the curve?

48. Find the equations of both lines through the point $(2, -3)$ that are tangent to the parabola $y = x^2 + x$.

49. Show that the curve $y = 6x^3 + 5x - 3$ has no tangent lines with slope 4.

50. A manufacturer of cartridges for stereo systems has designed a stylus with parabolic cross-section as shown in the figure. The equation of the parabola is $y = 16x^2$, where x and y are measured in millimeters. If the stylus sits in a record groove whose sides make an angle of θ with the horizontal direction, where $\tan \theta = 1.75$, find the points of contact P and Q of the stylus with the groove.

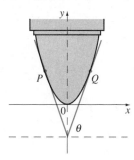

*In Exercises 51–54 find the equation of the normal line to the given curve at the given point. (The **normal line** to a curve C at a point P is, by definition, the line that passes through P and is perpendicular to the tangent line to C at P.) Also sketch the curve and the normal line in Exercises 51–53.*

51. $y = 1 - x^2$, $(2, -3)$ **52.** $y = \dfrac{1}{x - 1}$, $(2, 1)$

53. $y = \sqrt[3]{x}$, $(-8, -2)$ **54.** $y = f(x)$, $(a, f(a))$

55. At what point on the curve $y = x^4$ does the normal line have slope 16?

56. If $f(3) = 4$, $g(3) = 2$, $f'(3) = -6$, and $g'(3) = 5$, find the following numbers:
(a) $(f + g)'(3)$ (b) $(fg)'(3)$
(c) $\left(\dfrac{f}{g}\right)'(3)$ (d) $\left(\dfrac{f}{f - g}\right)'(3)$

57. Suppose that $f(5) = 1$, $f'(5) = 6$, $g(5) = -3$, and $g'(5) = 2$. Find the values of (a) $(fg)'(5)$, (b) $(f/g)'(5)$, and (c) $(g/f)'(5)$.

58. If f is a differentiable function, find expressions for the derivatives of the following functions.
(a) $y = x^2 f(x)$ (b) $y = \dfrac{f(x)}{x^2}$
(c) $y = \dfrac{x^2}{f(x)}$ (d) $y = \dfrac{1 + x f(x)}{\sqrt{x}}$

59. (a) Use the Product Rule twice to prove that if f, g, and h are differentiable, then
$$(fgh)' = f'gh + fg'h + fgh'$$
(b) Taking $f = g = h$ in part (a), show that
$$\frac{d}{dx}[f(x)]^3 = 3[f(x)]^2 f'(x)$$

Use Exercise 59 to differentiate the functions in Exercises 60–62.

60. $y = (x + 5)(x^2 + 7)(x - 3)$

61. $y = \sqrt{x}(x^4 + x + 1)(2x - 3)$

62. $y = (x^4 + 3x^3 + 17x + 82)^3$

63. Let
$$f(x) = \begin{cases} 2 - x & \text{if } x \le 1 \\ x^2 - 2x + 2 & \text{if } x > 1 \end{cases}$$
Is f differentiable at 1? Sketch the graphs of f and f'.

64. At what numbers is the function g given by
$$g(x) = \begin{cases} -1 - 2x & \text{if } x < -1 \\ x^2 & \text{if } -1 \le x \le 1 \\ x & \text{if } x > 1 \end{cases}$$
differentiable? Give a formula for g' and sketch the graphs of g and g'.

65. (a) For what values of x is the function $f(x) = |x^2 - 9|$ differentiable? Find a formula for f'.
(b) Sketch the graphs of both f and f'.

66. Where is the function $h(x) = |x - 1| + |x + 2|$ differentiable? Give a formula for h' and sketch the graphs of h and h'.

67. An easy proof of the Quotient Rule (2.14) can be given if we make the prior assumption that $F'(x)$ exists, where $F = f/g$. Write $f = Fg$; then differentiate using the Product Rule and solve the resulting equation for F'.

68. A tangent line is drawn to the hyperbola $xy = c$ at a point P. Show that the midpoint of the line segment cut off this tangent line by the coordinate axes is P.

Rates of Change in the Natural and Social Sciences

Recall from Section 2.1 that if $y = f(x)$, then the derivative dy/dx can be interpreted as the rate of change of y with respect to x. In this section we examine some of the applications of this idea to physics, chemistry, biology, economics, and other sciences.

Let us recall from Section 1.8 the basic idea behind rates of change. If x changes from x_1 to x_2, then the change in x is

$$\Delta x = x_2 - x_1$$

and the corresponding change in y is

$$\Delta y = f(x_2) - f(x_1)$$

The difference quotient

$$\frac{\Delta y}{\Delta x} = \frac{f(x_2) - f(x_1)}{x_2 - x_1}$$

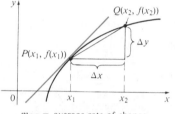

m_{PQ} = average rate of change
$m = f'(x_1)$ = instantaneous rate of change

Figure 2.7

is the **average rate of change of y with respect to x** over the interval $[x_1, x_2]$ and can be interpreted as the slope of the secant line PQ in Figure 2.7. Its limit as $\Delta x \to 0$ is the derivative $f'(x_1)$, which can therefore be interpreted as the instantaneous rate of change of y with respect to x or the slope of the tangent line at $P(x_1, f(x_1))$. Using the Leibniz notation, we write the process in the form

$$\frac{dy}{dx} = \lim_{\Delta x \to 0} \frac{\Delta y}{\Delta x}$$

Whenever the function $y = f(x)$ has a specific interpretation in one of the sciences, its derivative will have a specific interpretation as a rate of change. We now look at some of these interpretations in the natural and social sciences.

Physics

If $s = f(t)$ is the position function of a particle that is moving in a straight line, then $\Delta s/\Delta t$ represents the average velocity over a time period Δt, and $v = ds/dt$ represents the instantaneous **velocity** (the rate of change of displacement with respect to time). This was discussed in Sections 1.8 and 2.1, but now that we know the differentiation formulas, we are in a position to solve velocity problems more easily.

EXAMPLE 1   The position of a particle is given by the equation

$$s = f(t) = t^3 - 6t^2 + 9t$$

where t is measured in seconds and s in meters.
(a)   Find the velocity at time t.
(b)   What is the velocity after $2\,\text{s}$? after $4\,\text{s}$?
(c)   When is the particle at rest?
(d)   When is the particle moving in the positive direction?
(e)   Draw a diagram to represent the motion of the particle.
(f)   Find the total distance traveled by the particle during the first $5\,\text{s}$.

Solution

(a) The velocity function is the derivative of the position function.

$$s = f(t) = t^3 - 6t^2 + 9t$$

$$v(t) = \frac{ds}{dt} = 3t^2 - 12t + 9$$

(b) The velocity after $2\,\mathrm{s}$ means the instantaneous velocity when $t = 2$, that is,

$$v(2) = \frac{ds}{dt}\bigg|_{t=2} = 3(2)^2 - 12(2) + 9 = -3\,\mathrm{m/s}$$

The velocity after $4\,\mathrm{s}$ is

$$v(4) = 3(4)^2 - 12(4) + 9 = 9\,\mathrm{m/s}$$

(c) The particle is at rest when $v(t) = 0$, that is,

$$3t^2 - 12t + 9 = 3(t^2 - 4t + 3) = 3(t-1)(t-3) = 0$$

and this is true when $t = 1$ or $t = 3$. Thus the particle is at rest after $1\,\mathrm{s}$ and after $3\,\mathrm{s}$.

(d) The particle moves in the positive direction when $v(t) > 0$, that is,

$$3t^2 - 12t + 9 = 3(t-1)(t-3) > 0$$

This inequality is true when both factors are positive $(t > 3)$ or when both factors are negative $(t < 1)$. Thus the particle moves in the positive direction in the time intervals $t < 1$ and $t > 3$. It moves in the negative direction when $1 < t < 3$.

(e) The motion of the particle is illustrated schematically in Figure 2.8.

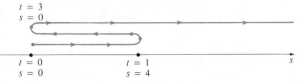

Figure 2.8

(f) The distance traveled in the first second is

$$|\,f(1) - f(0)\,| = |\,4 - 0\,| = 4\,\mathrm{m}$$

From $t = 1$ to $t = 3$ the distance traveled is

$$|\,f(3) - f(1)\,| = |\,0 - 4\,| = 4\,\mathrm{m}$$

From $t = 3$ to $t = 5$ the distance traveled is

$$|\,f(5) - f(3)\,| = |\,20 - 0\,| = 20\,\mathrm{m}$$

The total distance is $4 + 4 + 20 = 28\,\mathrm{m}$.

EXAMPLE 2   If a rod or piece of wire is homogeneous, then its linear density is uniform and is defined as the mass per unit length $(\rho = m/l)$ and measured in kilograms per meter. Suppose, however, that the rod is not homogeneous but that its mass measured from its left end to a point x is $m = f(x)$ as shown in Figure 2.9.

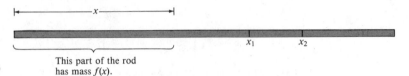

Figure 2.9

This part of the rod
has mass $f(x)$.

The mass of the part of the rod that lies between $x = x_1$ and $x = x_2$ is given by $\Delta m = f(x_2) - f(x_1)$, so the average density of that part of the rod is

$$\text{average density} = \frac{\Delta m}{\Delta x} = \frac{f(x_2) - f(x_1)}{x_2 - x_1}$$

If we now let $\Delta x \to 0$ (that is, $x_2 \to x_1$), we are computing the average density over a smaller and smaller interval. The **linear density** ρ at x_1 is the limit of these average densities as $\Delta x \to 0$; that is, the linear density is the rate of change of mass with respect to length. Symbolically,

$$\rho = \lim_{\Delta x \to 0} \frac{\Delta m}{\Delta x} = \frac{dm}{dx}$$

Thus the linear density of the rod is the derivative of mass with respect to length.

For instance, if $m = f(x) = \sqrt{x}$, where x is measured in meters and m in kilograms, then the average density of the part of the rod given by $1 \le x \le 1.2$ is

$$\frac{\Delta m}{\Delta x} = \frac{f(1.2) - f(1)}{1.2 - 1} = \frac{\sqrt{1.2} - 1}{0.2} \approx 0.48 \, \text{kg/m}$$

while the density right at $x = 1$ is

$$\rho = \frac{dm}{dx}\bigg|_{x=1} = \frac{1}{2\sqrt{x}}\bigg|_{x=1} = 0.50 \, \text{kg/m}$$ •

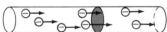

Figure 2.10

EXAMPLE 3   A current exists whenever electric charges move. Figure 2.10 shows part of a wire and electrons moving through a shaded plane surface. If ΔQ is the net charge that passes through this surface during a time period Δt, then the average current during this time interval is defined as

$$\text{average current} = \frac{\Delta Q}{\Delta t} = \frac{Q_2 - Q_1}{t_2 - t_1}$$

If we take the limit of this average current over smaller and smaller time intervals, we get what is called the **current** I at a given time t_1:

$$I = \lim_{\Delta t \to 0} \frac{\Delta Q}{\Delta t} = \frac{dQ}{dt}$$

Thus the current is the rate at which charge flows through a surface.   •

Velocity, density, and current are not the only rates of change that are important in physics. Others include power (the rate at which work is done), the rate of heat flow, temperature gradient (the rate of change of temperature with respect to position), and the rate of decay of a radioactive substance in nuclear physics.

Chemistry

EXAMPLE 4   A chemical reaction results in the formation of one or more substances (called products) from one or more starting materials (called reactants). For instance, the "equation"

$$2H_2 + O_2 \rightarrow 2H_2O$$

indicates that two molecules of hydrogen and one molecule of oxygen form two molecules of water. Let us consider the reaction

$$A + B \rightarrow C$$

where A and B are the reactants and C is the product. The **concentration** of a reactant A is the number of moles (6.022×10^{23} molecules) per liter and is denoted by $[A]$. The concentration varies during a reaction, so $[A]$, $[B]$, and $[C]$ are all functions of time (t). The average rate of reaction of the product C over a time interval $t_1 \leq t \leq t_2$ is

$$\frac{\Delta[C]}{\Delta t} = \frac{[C](t_2) - [C](t_1)}{t_2 - t_1}$$

But chemists are more interested in the **instantaneous rate of reaction,** which is obtained by taking the limit of the average rate of reaction as the time interval Δt approaches 0:

$$\text{rate of reaction} = \lim_{\Delta t \to 0} \frac{\Delta[C]}{\Delta t} = \frac{d[C]}{dt}$$

Since the concentration of the product increases as the reaction proceeds, the derivative $d[C]/dt$ will be positive. (You can see intuitively that the slope of the tangent to the graph of an increasing function is positive.) Thus the rate of reaction of C is positive. The concentrations of the reactants, however, decrease during the reaction, so, to make the rates of reaction of A and B positive numbers, we put minus signs in front of the derivatives $d[A]/dt$ and $d[B]/dt$. Since $[A]$ and $[B]$ each decrease at the same rate $[C]$ increases, we have

$$\text{rate of reaction} = \frac{d[C]}{dt} = -\frac{d[A]}{dt} = -\frac{d[B]}{dt}$$

More generally, it turns out that for a reaction of the form

$$aA + bB \rightarrow cC + dD$$

we have

$$-\frac{1}{a}\frac{d[A]}{dt} - \frac{1}{b}\frac{d[B]}{dt} = \frac{1}{c}\frac{d[C]}{dt} + \frac{1}{d}\frac{d[D]}{dt}$$

The rate of reaction can be determined by graphical methods (See Exercise 20). In some cases we can use the rate of reaction to find explicit formulas for the concentrations as functions of time (see Exercise 7 in Section 3.6 and Exercise 27 in Section 8.1).   ●

EXAMPLE 5   One of the quantities of interest in thermodynamics is compressibility. If a given substance is kept at a constant temperature, then its volume V depends on its pressure P. We can consider the rate of change of volume with respect to pressure—namely, the derivative dV/dP. As P increases, V decreases, so $dV/dP < 0$. The compressibility is defined by introducing a minus sign and dividing this derivative by the volume V:

$$\text{isothermal compressibility} = \beta = -\frac{1}{V}\frac{dV}{dP}$$

Thus β measures how fast, per unit volume, the volume of a substance decreases as the pressure on it increases at constant temperature.

For instance, the volume V (in cubic meters) of a sample of air at 25°C was found to be related to the pressure P (in kilopascals) by the equation

$$V = \frac{5.3}{P}$$

The rate of change of V with respect to P when $P = 50\,\text{kPa}$ is

$$\frac{dV}{dP}\bigg|_{P=50} = -\frac{5.3}{P^2}\bigg|_{P=50}$$

$$= -\frac{5.3}{2500} = -0.00212\,\text{m}^3/\text{kPa}$$

The compressibility at that pressure is

$$\beta = -\frac{1}{V}\frac{dV}{dP}\bigg|_{P=50}$$

$$= \frac{0.00212}{\frac{5.3}{50}} = 0.02\,(\text{m}^3/\text{kPa})/\text{m}^3 \qquad \bullet$$

Biology

EXAMPLE 6   Let $n = f(t)$ be the number of individuals in an animal or plant population at time t. The change in the population size between the times $t = t_1$ and $t = t_2$ is $\Delta n = f(t_2) - f(t_1)$, and so the average rate of growth during the time period $t_1 \leq t \leq t_2$ is

$$\text{average rate of growth} = \frac{\Delta n}{\Delta t} = \frac{f(t_2) - f(t_1)}{t_2 - t_1}$$

The **instantaneous rate of growth** is obtained from this average rate of growth by letting the time period Δt approach 0:

$$\text{growth rate} = \lim_{\Delta t \to 0} \frac{\Delta n}{\Delta t} = \frac{dn}{dt}$$

Strictly speaking, this is not quite accurate because the actual graph of a population function $n = f(t)$ would be a step function that is discontinuous whenever a birth or death occurs and therefore not differentiable. However, for a large animal or plant population we can replace the graph by a smooth approximating curve as in Figure 2.11.

To be more specific, consider a population of bacteria in a homogeneous nutrient medium. Suppose that by sampling the population at certain intervals it is determined that the population doubles every hour. If the initial population is n_0 and the time t is measured in hours, then

$$f(1) = 2n_0$$

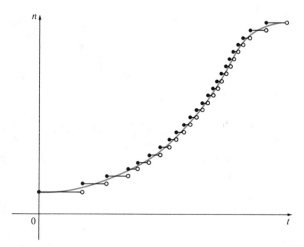

Figure 2.11

A smooth curve approximating
a growth function

$$f(2) = 2f(1) = 2^2 n_0$$

$$f(3) = 2f(2) = 2^3 n_0$$

and, in general,

$$f(t) = 2^t n_0$$

The population function is $n = n_0 2^t$.

This is an example of an exponential function. In Chapter 3 we will discuss exponential functions in general and at that time we will be able to compute their derivatives and thereby determine the rate of growth of the bacteria population. ●

EXAMPLE 7   When we consider the flow of blood through a blood vessel, such as a vein or artery, we can take the shape of the blood vessel to be a cylindrical tube with radius R and length l as illustrated in Figure 2.12.

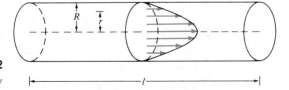

Figure 2.12

Blood flow in an artery

Because of friction at the walls of the tube, the velocity v of the blood is greatest along the central axis of the tube and decreases as the distance r from the axis increases until v becomes 0 at the wall. The relationship between v and r is given by the **law of laminar flow** discovered by the French physician Poiseuille in 1840. This states that

(2.18)
$$v = \frac{P}{4\eta l}(R^2 - r^2)$$

where η is the viscosity of the blood and P is the pressure difference between the ends of the tube. If P and l are constant, then v is a function of r with domain $[0, R]$. [For more detailed information see D.A. McDonald, *Blood Flow in Arteries* (London: Arnold, 1960).]

The average rate of change of the velocity as we move from $r = r_1$ outward to $r = r_2$ is

$$\frac{\Delta v}{\Delta r} = \frac{v(r_2) - v(r_1)}{r_2 - r_1}$$

and if we let $\Delta r \rightarrow 0$, we obtain the instantaneous rate of change of velocity with respect to r:

$$\text{velocity gradient} = \lim_{\Delta r \rightarrow 0} \frac{\Delta v}{\Delta r} = \frac{dv}{dr}$$

Using Equation 2.18, we obtain

$$\frac{dv}{dr} = \frac{P}{4\eta l}(0 - 2r) = -\frac{Pr}{2\eta l}$$

In a typical human artery we can take $\eta = 0.027$, $R = 0.008$ cm, $l = 2$ cm, and $P = 4000$ dynes/cm^2, which gives

$$v = \frac{4000}{4(0.027)2}(0.000064 - r^2)$$

$$\approx 1.85 \times 10^4(6.4 \times 10^{-5} - r^2)$$

At $r = 0.002$ cm the blood is flowing at a speed of

$$v(0.002) \approx 1.85 \times 10^4(64 \times 10^{-6} - 4 \times 10^{-6})$$

$$= 1.11 \text{ cm/s}$$

and the velocity gradient at that point is

$$\left.\frac{dv}{dr}\right|_{r=0.002} = -\frac{4000(0.002)}{2(0.027)2} \approx -74 \text{ (cm/s)/cm} \qquad \bullet$$

Economics

EXAMPLE 8   Suppose $C(x)$ is the total cost that a company incurs in producing x units of a certain commodity. The function C is called a **cost function**. If the number of items produced is increased from x_1 to x_2, the additional cost is $\Delta C = C(x_2) - C(x_1)$, and the average rate of change of the cost is

$$\frac{\Delta C}{\Delta x} = \frac{C(x_2) - C(x_1)}{x_2 - x_1} = \frac{C(x_1 + \Delta x) - C(x_1)}{\Delta x}$$

The limit of this quantity as $\Delta x \rightarrow 0$, that is, the instantaneous rate of change of cost with respect to the number of items produced, is called the **marginal cost** by economists:

$$\text{marginal cost} = \lim_{\Delta x \rightarrow 0} \frac{\Delta C}{\Delta x} = \frac{dC}{dx}$$

[Since x can usually take on only integer values, it may not make literal sense to let Δx approach 0, but we always replace $C(x)$ by a smooth approximating function as in Example 6.]

Taking $\Delta x = 1$ and n large, we have

$$C'(n) \approx C(n+1) - C(n)$$

Thus the marginal cost of producing n units is approximately equal to the cost of producing one more unit [the $(n+1)$st unit].

It is often appropriate to represent a total cost function by a polynomial

$$C(x) = a + bx + cx^2 + dx^3$$

where a represents the overhead cost (rent, heat, maintenance) and the other terms represent the cost of raw materials, labor, and so on. (The cost of raw materials may be proportional to x, but labor costs might depend partly on higher powers of x because of overtime costs and inefficiencies involved in large-scale operations.)

For instance, suppose a company has estimated that the cost (in dollars) of producing x items is

$$C(x) = 10,000 + 5x + 0.01x^2$$

Then the marginal cost function is

$$C'(x) = 5 + 0.02x$$

The marginal cost at the production level of 500 items is

$$C'(500) = 5 + (0.02)500 = \$15/\text{item}$$

This gives the rate at which costs are increasing with respect to the production level when $x = 500$.

The cost of producing the 501st item is

$$C(501) - C(500) = [10,000 + 5(501) + 0.01(501)^2]$$

$$- [10,000 + 5(500) + 0.01(500)^2]$$

$$= \$15.01$$

Notice that $C'(500) \approx C(501) - C(500)$.   •

Economists also study marginal demand, marginal revenue, and marginal profit, which are the derivatives of the demand, revenue, and profit functions. These will be considered in Chapter 4 after we have developed techniques for finding the maximum and minimum values of functions.

Other Sciences

Rates of change occur in all the sciences. A geologist is interested in knowing the rate at which an intruded body of molten rock cools by conduction of heat into surrounding rocks. An engineer wants to know the rate at which water flows into or out of a reservoir. An urban geographer is interested in the rate of change of the population density in a city as the distance from the city center increases. A meteorologist is concerned with the rate of change of atmospheric pressure with respect to height. (See Exercise 17 in Section 3.6.)

In psychology, those interested in learning theory study the so-called learning curve, which graphs the performance $P(t)$ of someone learning a skill as a function of the training time t. Of particular interest is the rate at which performance improves as time passes, that is, dP/dt.

In sociology, differential calculus is used in analyzing the spread of rumors (or innovations or fads or fashions). If $p(t)$ denotes the proportion of a population that knows a rumor by time t, then the derivative dp/dt represents the rate of spread of the rumor. (See Exercise 30 in Section 3.2.)

Summary

Velocity, density, current, power, and temperature gradient in physics, rate of reaction and compressibility in chemistry, rate of growth and blood velocity gradient in biology, marginal cost and marginal profit in economics, rate of heat flow in geology, rate of improvement of performance in psychology, rate of spread of a rumor in sociology—these are all special cases of a single mathematical concept, the derivative.

    This is an illustration of the fact that part of the power of mathematics lies in its abstractness. A single abstract mathematical concept (such as the derivative) can have different interpretations in each of the sciences. When we develop the properties of the mathematical concept once and for all, we can then turn around and apply these results to all of the sciences. This is much more efficient than developing properties of special concepts in each separate science. The French mathematician Joseph Fourier (1768–1830) put it succinctly: "Mathematics compares the most diverse phenomena and discovers the secret analogies that unite them."

SECTION 2.3   **Exercises**

In Exercises 1–6 a particle moves according to a law of motion $s = f(t)$, $t \geq 0$, where t is measured in seconds and s in feet.

(a) *Find the velocity at time t.*
(b) *What is the velocity after 2 s?*
(c) *When is the particle at rest?*
(d) *When is the particle moving in the positive direction?*
(e) *Find the total distance traveled during the first 4 s.*
(f) *Draw a diagram like Figure 2.8 to illustrate the motion of the particle.*

1. $f(t) = t^2 - 6t + 9$
2. $f(t) = 4t^3 - 9t^2 + 6t + 2$
3. $f(t) = 2t^3 - 9t^2 + 12t + 1$
4. $f(t) = t^4 - 4t + 1$
5. $s = \dfrac{t}{t^2 + 1}$ 6. $s = \sqrt{t}\,(5 - 5t + 2t^2)$

7.   The position function of a particle is given by $s = t^3 - 4.5t^2 - 7t$, $t \geq 0$. When does the particle reach a velocity of 5 m/s?

8.   If a ball is thrown vertically upward with a velocity of 80 ft/s, then its height after t seconds is $s = 80t - 16t^2$.
    (a) What is the maximum height reached by the ball?
    (b) What is the velocity of the ball when it is 96 ft above the ground on its way up? on its way down?

9.   (a) Find the average rate of change of the volume of a cube with respect to its edge length x as x changes from
      (i) 5 to 6   (ii) 5 to 5.1   (iii) 5 to 5.01
    (b) Find the instantaneous rate of change when $x = 5$.
    (c) Show that the rate of change of the volume of a cube with respect to its edge length (at any x) is equal to half the surface area of the cube.

10.   (a) Find the average rate of change of the area of a circle with respect to its radius r as r changes from
      (i) 2 to 3   (ii) 2 to 2.5   (iii) 2 to 2.1
    (b) Find the instantaneous rate of change when $r = 2$.
    (c) Show that the rate of change of the area of a circle with respect to its radius (at any r) is equal to the circumference of the circle.

11.   A stone is dropped into a lake, creating a circular ripple that travels outward at a speed of 60 cm/s. Find the rate at which the area within the circle is increasing after (a) 1 s, (b) 3 s, and (c) 5 s.

12.   (a) The volume of a growing spherical cell is $V = \frac{4}{3}\pi r^3$, where the radius r is measured in micrometers $(1\,\mu m = 10^{-6}\,m)$. Find the

average rate of change of V with respect to r when r changes from

(i) 5 to 8 μm   (ii) 5 to 6 μm

(iii) 5 to 5.1 μm

(b) Find the instantaneous rate of change of V with respect to r when $r = 5\,\mu$m.

13. A spherical balloon is being inflated. Find the rate of increase of the surface area $(S = 4\pi r^2)$ with respect to the radius r when r is (a) 1 ft, (b) 2 ft, and (c) 3 ft.

14. Show that the rate of change of the volume of a sphere with respect to its radius is equal to its surface area.

15. The mass of the part of a metal rod that lies between its left end and a point x meters to the right is $3x^2$ kg. Find the linear density (see Example 2) when x is (a) 1 m, (b) 2 m, and (c) 3 m.

16. If a tank holds 5000 gallons of water that drain from the bottom of the tank in 40 min, then Torricelli's Law gives the volume V of water remaining in the tank after t minutes as

$$V = 5000\left(1 - \frac{t}{40}\right)^2 \qquad 0 \le t \le 40$$

Find the rate at which water is draining from the tank after (a) 5 min, (b) 10 min, and (c) 20 min.

17. The quantity of charge Q in coulombs (C) that passes through a surface at time t (measured in seconds) is given by $Q(t) = t^3 - 2t^2 + 6t + 2$. Find the current when (a) $t = 0.5$ s and (b) $t = 1$ s. [See Example 3. The unit of current is an ampere $(1\,\text{A} = 1\,\text{C/s})$.]

18. Newton's Law of Gravitation says that the magnitude F of the force exerted by a body of mass m on a body of mass M is

$$F = \frac{GmM}{r^2}$$

where G is the gravitational constant and r is the distance between the bodies. If the bodies are moving, find the rate of change of F with respect to r.

19. Boyle's Law states that when a sample of gas is compressed at a constant temperature, the product of the pressure and the volume remains constant: $PV = C$.

(a) Find the rate of change of volume with respect to pressure.

(b) Prove that the isothermal compressibility (see Example 5) is given by $\beta = 1/P$.

20. The data in the following table concern the lactonization of hydroxyvaleric acid at 25°C. They give the concentration $C(t)$ of this acid in moles per liter after t minutes.

t	0	2	4	6	8
$C(t)$	0.0800	0.0570	0.0408	0.0295	0.0210

(a) Find the average rate of reaction for the following time intervals:

(i) $2 \le t \le 6$ (ii) $2 \le t \le 4$

(iii) $0 \le t \le 2$

(b) Plot the points from the table and draw a smooth curve through them as an approximation to the graph of the concentration function. Then draw the tangent at $t = 2$ and use it to estimate the instantaneous rate of reaction when $t = 2$.

21. If, in Example 4, one molecule of the product C is formed from one molecule of the reactant A and one molecule of the reactant B, and the initial concentrations of A and B have a common value $[A] = [B] = a$ moles/L, then $[C] = a^2kt/(akt + 1)$, where k is a constant.

(a) Find the rate of reaction at time t.

(b) Show that if $x = [C]$, then

$$\frac{dx}{dt} = k(a - x)^2$$

22. If f is the focal length of a convex lens and an object is placed at a distance p from the lens, then its image will be at a distance q from the lens, where f, p, and q are related by the *lens equation*

$$\frac{1}{f} = \frac{1}{p} + \frac{1}{q}$$

Find the rate of change of p with respect to q.

23. The mass of glucose in a metabolic experiment decreased according to the equation $m = 5 - (0.02)t^2$, where t is measured in hours. Find the rate of change of the amount of glucose after 1 h.

24. The population of a slowly growing bacterial colony after t hours is given by

$$n = 100 + 24t + 2t^2$$

Find the growth rate after 2 h.

25. Use the law of laminar flow (see Example 7) to find the velocity gradient at point 0.005 cm from the central axis of a blood vessel with radius 0.01 cm, length 3 cm, pressure difference 3000 dynes/cm^2, and viscosity $\eta = 0.027$.

26. The population of a bacterial culture is 5000 initially and doubles every 20 min. Find an expression for the population function $n = f(t)$ where t is measured in hours.

In Exercises 27–30 a cost function is given for a certain commodity. Find the marginal cost function. Then compare the marginal cost at the production level of 100 units with the cost of producing the 101st item.

27. $C(x) = 420 + 1.5x + 0.002x^2$

28. $C(x) = 1200 + \frac{x}{10} + \frac{x^2}{10,000}$

29. $C(x) = 2000 + 3x + 0.01x^2 + 0.0002x^3$

30. $C(x) = 2500 + 2\sqrt{x}$

SECTION 2.4

Derivatives of Trigonometric Functions

Before starting this section, you might need to review the trigonometric functions in Appendix B. In particular it is important to remember that when we talk about the function f defined for all real numbers x by

$$f(x) = \sin x$$

it is understood that $\sin x$ means the sine of the angle whose *radian* measure is x. A similar convention holds for the other trigonometric functions cos, tan, csc, sec, and cot.

In order to compute the derivatives of these functions, we first have to evaluate some trigonometric limits.

Theorem (2.19)

$$\lim_{\theta \to 0} \sin \theta = 0$$

If we let $f(\theta) = \sin \theta$, then $f(0) = \sin 0 = 0$ and so Theorem 2.19 tells us that $\lim_{\theta \to 0} f(\theta) = f(0)$. This is precisely the statement that the sine function is continuous at 0.

Proof In calculating $\lim_{\theta \to 0^+} \sin \theta$, we may assume that $0 < \theta < \pi/2$. Figure 2.13 shows a sector of a circle with center O, central angle θ, and radius 1. BC is drawn perpendicular to OA. By the definition of radian measure, we have arc $AB = \theta$. Also $|BC| = |OB| \sin \theta = \sin \theta$. From the diagram we see that

$$|BC| < \text{arc } AB$$

(2.20) Therefore $0 < \sin \theta < \theta$

Since we know that $\lim_{\theta \to 0^+} 0 = 0$ and $\lim_{\theta \to 0^+} \theta = 0$, it follows from the Squeeze Theorem that

$$\lim_{\theta \to 0^+} \sin \theta = 0$$

If $-\pi/2 < \theta < 0$, then $0 < -\theta < \pi/2$, so by (2.20) we have

$$0 < \sin(-\theta) < -\theta$$

or $0 < -\sin \theta < -\theta$

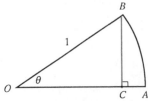

Figure 2.13

which implies $\qquad\qquad\qquad\qquad\qquad \theta < \sin\theta < 0$

This inequality, together with the fact that $\lim_{\theta\to 0^-}\theta = 0$ and the Squeeze Theorem, shows that $\lim_{\theta\to 0^-}\sin\theta = 0$. Thus

$$\lim_{\theta\to 0}\sin\theta = 0$$

Corollary (2.21)

$$\boxed{\lim_{\theta\to 0}\cos\theta = 1}$$

Again, by the definition of continuity and the fact that $\cos 0 = 1$, Corollary 2.21 says that the cosine function is continuous at 0.

Proof   Using $\sin^2\theta + \cos^2\theta = 1$ together with $\cos\theta \geq 0$ for $-\pi/2 \leq \theta \leq \pi/2$, we have $\cos\theta = \sqrt{1 - \sin^2\theta}$ for those values of θ. Thus, using the properties of limits, we have

$$\lim_{\theta\to 0}\cos\theta = \lim_{\theta\to 0}\sqrt{1 - \sin^2\theta}$$
$$= \sqrt{\lim_{\theta\to 0}(1 - \sin^2\theta)}$$
$$= \sqrt{1 - 0} \qquad \text{(by Theorem 2.19)}$$
$$= 1$$

In Example 3 in Section 1.2 we made the guess that $\lim_{x\to 0}(\sin x)/x = 1$. We are now in a position to prove that the guess is correct.

Theorem (2.22)

$$\boxed{\lim_{\theta\to 0}\frac{\sin\theta}{\theta} = 1}$$

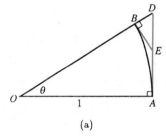

(a)

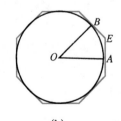

(b)

Figure 2.14

Proof   First suppose that $0 < \theta < \pi/2$. Again, Figure 2.14 shows a sector of a circle with center O, central angle θ, and radius 1. Let the tangents at A and B intersect at E. You can see from Figure 2.14(b) that the circumference of a circle is smaller than the length of a circumscribed polygon, so $\mathrm{arc}\,AB < |AE| + |EB|$. Thus

$$\theta = \mathrm{arc}\,AB < |AE| + |EB|$$
$$< |AE| + |ED|$$
$$= |AD| = |OA|\tan\theta$$
$$= \tan\theta$$

(In Appendix C it is proved that $\theta \leq \tan\theta$ directly from the definition of the length of an arc without resorting to geometric intuition as we did above.) Therefore we have

$$\theta < \frac{\sin\theta}{\cos\theta}$$

or $\qquad\qquad\qquad\qquad \cos\theta < \frac{\sin\theta}{\theta}$

From the proof of Theorem 2.19 we have $\sin\theta < \theta$, so

(2.23)
$$\cos\theta < \frac{\sin\theta}{\theta} < 1$$

If $-\pi/2 < \theta < 0$, then $0 < -\theta < \pi/2$, so (2.23) gives

(2.24)
$$\cos(-\theta) < \frac{\sin(-\theta)}{-\theta} < 1$$

But $\cos(-\theta) = \cos\theta$ and $\sin(-\theta) = -\sin\theta$, so (2.24), when simplified, becomes (2.23). Thus (2.23) holds whenever $\theta \in (-\pi/2, \pi/2)$ and $\theta \neq 0$. Also, we know that $\lim_{\theta\to0} 1 = 1$ and $\lim_{\theta\to0} \cos\theta = 1$ by Corollary 2.21. Hence, by the Squeeze Theorem,

$$\lim_{\theta\to0} \frac{\sin\theta}{\theta} = 1$$

Corollary (2.25)

$$\boxed{\lim_{\theta\to0} \frac{\cos\theta - 1}{\theta} = 0}$$

Proof   To put the function in a form in which we can use the limits we know, we multiply numerator and denominator by $\cos\theta + 1$:

$$\lim_{\theta\to0} \frac{\cos\theta - 1}{\theta} = \lim_{\theta\to0} \left[\frac{\cos\theta - 1}{\theta} \cdot \frac{\cos\theta + 1}{\cos\theta + 1}\right] = \lim_{\theta\to0} \frac{\cos^2\theta - 1}{\theta(\cos\theta + 1)}$$

$$= \lim_{\theta\to0} \frac{-\sin^2\theta}{\theta(\cos\theta + 1)} = -\lim_{\theta\to0} \frac{\sin\theta}{\theta} \cdot \frac{\sin\theta}{\cos\theta + 1}$$

$$= -\lim_{\theta\to0} \frac{\sin\theta}{\theta} \cdot \lim_{\theta\to0} \frac{\sin\theta}{\cos\theta + 1} = -1 \cdot \left(\frac{0}{1+1}\right) = 0$$

Here we have used Theorem 2.22 together with Theorem 2.19 and Corollary 2.21.

Another method for proving Corollary 2.25 is outlined in Exercise 41.

EXAMPLE 1   Find $\lim_{x\to0} \frac{\sin 7x}{4x}$.

Solution   In order to apply Theorem 2.22, we first rewrite the function as follows:

$$\frac{\sin 7x}{4x} = \frac{7}{4}\left(\frac{\sin 7x}{7x}\right)$$

Notice that as $x\to0$, we have $7x\to0$, and so, by Theorem 2.22 with $\theta = 7x$,

$$\lim_{x\to0} \frac{\sin 7x}{7x} = \lim_{7x\to0} \frac{\sin(7x)}{7x} = 1$$

Thus
$$\lim_{x\to0} \frac{\sin 7x}{4x} = \lim_{x\to0} \frac{7}{4}\left(\frac{\sin 7x}{7x}\right)$$

$$= \frac{7}{4} \lim_{x\to0} \frac{\sin 7x}{7x} = \frac{7}{4} \cdot 1 = \frac{7}{4}$$

EXAMPLE 2   Calculate $\lim_{x\to0} x\cot x$.

Solution

$$\lim_{x \to 0} x \cot x = \lim_{x \to 0} x \frac{\cos x}{\sin x}$$

$$= \lim_{x \to 0} \frac{\cos x}{\frac{\sin x}{x}} = \frac{\lim\limits_{x \to 0} \cos x}{\lim\limits_{x \to 0} \frac{\sin x}{x}}$$

$$= \frac{1}{1} \qquad \text{(by Corollary 2.21 and Theorem 2.22)}$$

$$= 1$$

Derivatives

If you sketch the graph of the function $f(x) = \sin x$ and use the interpretation of $f'(x)$ as the slope of the tangent to the sine curve in order to sketch the graph of f' (see Exercise 44 in Section 2.1), then it looks as if the graph of f' may be the same as the cosine curve (see Figure 2.15). This is confirmed in the next theorem by using the definition of derivative to calculate $f'(x)$.

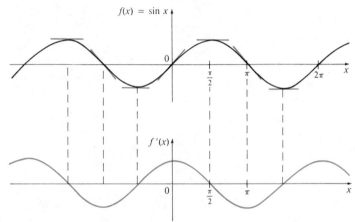

Figure 2.15

Theorem (2.26)

$$\boxed{\frac{d}{dx} \sin x = \cos x}$$

Proof

If $f(x) = \sin x$, then

$$f'(x) = \lim_{h \to 0} \frac{f(x+h) - f(x)}{h}$$

$$= \lim_{h \to 0} \frac{\sin(x+h) - \sin x}{h}$$

$$= \lim_{h \to 0} \frac{\sin x \cos h + \cos x \sin h - \sin x}{h}$$

$$= \lim_{h \to 0} \left[\sin x \left(\frac{\cos h - 1}{h} \right) + \cos x \left(\frac{\sin h}{h} \right) \right]$$

$$= \lim_{h \to 0} \sin x \cdot \lim_{h \to 0} \frac{\cos h - 1}{h} + \lim_{h \to 0} \cos x \cdot \lim_{h \to 0} \frac{\sin h}{h}$$

$$= \sin x \cdot 0 + \cos x \cdot 1 \qquad \text{(by Corollary 2.25 and Theorem 2.22)}$$

$$= \cos x$$

EXAMPLE 3   Differentiate $y = x^2 \sin x$.

Solution   Using the Product Rule and Theorem 2.26, we have

$$\frac{dy}{dx} = x^2 \frac{d}{dx} \sin x + \sin x \frac{d}{dx} (x^2)$$

$$= x^2 \cos x + 2x \sin x$$

Using the same methods as in the proof of Theorem 2.26, one can prove (see Exercise 42) that

(2.27)

$$\boxed{\frac{d}{dx} \cos x = -\sin x}$$

The tangent function can also be differentiated from first principles, but it is easier to use the Quotient Rule together with Theorem 2.26 and Equation 2.27:

$$\frac{d}{dx} \tan x = \frac{d}{dx} \left(\frac{\sin x}{\cos x} \right)$$

$$= \frac{\cos x D \sin x - \sin x D \cos x}{\cos^2 x}$$

$$= \frac{\cos x \cdot \cos x - \sin x(-\sin x)}{\cos^2 x}$$

$$= \frac{\cos^2 x + \sin^2 x}{\cos^2 x}$$

$$= \frac{1}{\cos^2 x} = \sec^2 x$$

(2.28)

$$\boxed{\frac{d}{dx} \tan x = \sec^2 x}$$

The derivatives of the remaining trigonometric functions csc, sec, and cot, can also be easily found using the Quotient Rule (see Exercises 21–23). We collect all of the differentiation formulas for trigonometric functions in Table 2.29.

Table of Derivatives of Trigonometric Functions (2.29)

$\dfrac{d}{dx}(\sin x) = \cos x$	$\dfrac{d}{dx}(\csc x) = -\csc x \cot x$
$\dfrac{d}{dx}(\cos x) = -\sin x$	$\dfrac{d}{dx}(\sec x) = \sec x \tan x$
$\dfrac{d}{dx}(\tan x) = \sec^2 x$	$\dfrac{d}{dx}(\cot x) = -\csc^2 x$

In memorizing Table 2.29 it is helpful to notice that the minus signs go with the derivatives of the "cofunctions," that is, cosine, cosecant, and cotangent.

These differentiation formulas show that each trigonometric function is differentiable at every number in its domain. It follows from Theorem 2.8 that *all of the trigonometric functions are continuous on their domains.*

EXAMPLE 4   Differentiate $f(x) = \dfrac{\sec x}{1 + \tan x}$.

Solution   The Quotient Rule gives

$$f'(x) = \frac{(1 + \tan x)D \sec x - \sec x\, D(1 + \tan x)}{(1 + \tan x)^2}$$

$$= \frac{(1 + \tan x)\sec x \tan x - \sec x \cdot \sec^2 x}{(1 + \tan x)^2}$$

$$= \frac{\sec x[\tan x + \tan^2 x - \sec^2 x]}{(1 + \tan x)^2}$$

$$= \frac{\sec x(\tan x - 1)}{(1 + \tan x)^2}$$

In simplifying the answer we have used the identity $\tan^2 x + 1 = \sec^2 x$.

SECTION 2.4 **Exercises**

Find the limits in Exercises 1–20.

1. $\displaystyle \lim_{x \to 0} (x^2 + \cos x)$

2. $\displaystyle \lim_{x \to 0} \cos(\sin x)$

3. $\displaystyle \lim_{x \to \pi/3} (\sin x - \cos x)$

4. $\displaystyle \lim_{x \to \pi} x^2 \sec x$

5. $\displaystyle \lim_{x \to \pi/4} \frac{\sin x}{3x}$

6. $\displaystyle \lim_{x \to 0} \frac{\sin x}{3x}$

7. $\displaystyle \lim_{t \to -3\pi} t^3 \sin^4 t$

8. $\displaystyle \lim_{t \to \pi/6} \csc t \cot^2 t$

9. $\displaystyle \lim_{t \to 0} \frac{\sin 5t}{t}$

10. $\displaystyle \lim_{t \to 0} \frac{\sin 8t}{\sin 9t}$

11. $\displaystyle \lim_{\theta \to 0} \frac{\sin(\cos \theta)}{\sec \theta}$

12. $\displaystyle \lim_{\theta \to 0} \frac{\cos \theta - 1}{\sin \theta}$

13. $\displaystyle \lim_{x \to \pi/4} \frac{\tan x}{4x}$

14. $\displaystyle \lim_{x \to 0} \frac{\tan x}{4x}$

15. $\displaystyle \lim_{\theta \to 0} \frac{\sin^2 \theta}{\theta}$

16. $\displaystyle \lim_{h \to 0} \frac{\sin 5h}{\tan 3h}$

17. $\displaystyle \lim_{x \to 0} \frac{\tan 3x}{3 \tan 2x}$

18. $\displaystyle \lim_{x \to 0} \frac{\sec x}{1 - \sin x}$

19. $\displaystyle \lim_{t \to 0} \frac{\sin^2 3t}{t^2}$

20. $\displaystyle \lim_{t \to 0} \frac{t^3}{\tan^3 2t}$

21. Prove that $\dfrac{d}{dx}(\csc x) = -\csc x \cot x$.

22. Prove that $\dfrac{d}{dx}(\sec x) = \sec x \tan x$.

23. Prove that $\dfrac{d}{dx}(\cot x) = -\csc^2 x$.

In Exercises 24–35 find $\dfrac{dy}{dx}$.

24. $y = \cos x - 2 \tan x$

25. $y = \sin x + \cos x$

26. $y = x \csc x$

27. $y = \csc x \cot x$

28. $y = \dfrac{\sin x}{1 + \cos x}$

29. $y = \dfrac{\tan x}{x}$

30. $y = \dfrac{\tan x - 1}{\sec x}$

31. $y = \dfrac{x}{\sin x + \cos x}$

32. $y = 2x(\sqrt{x} - \cot x)$

33. $y = x^{-3} \sin x \tan x$

34. $y = x \sin x \cos x$

35. $y = \dfrac{x^2 \tan x}{\sec x}$

In Exercises 36–38 find the equation of the tangent line to the given curve at the given point.

36. $y = 2 \sin x$, $(\pi/6, 1)$ 37. $y = \tan x$, $(\pi/4, 1)$

38. $y = \sec x - 2 \cos x$, $(\pi/3, 1)$

39. For what values of x does the graph of $f(x) = x + 2 \sin x$ have a horizontal tangent?

40. Find the points on the curve $y = (\cos x)/(2 + \sin x)$ where the tangent is horizontal.

41. If we put $\theta = 2x$ in the identity $\cos 2x = 1 - 2 \sin^2 x$, it becomes $\cos \theta = 1 - 2 \sin^2(\theta/2)$. Use this identity to give an alternative proof of Corollary 2.25.

42. Prove, using the definition of derivative, that if $f(x) = \cos x$, then $f'(x) = -\sin x$.

Find the limits in Exercises 43–50.

43. $\displaystyle\lim_{x\to 0} \frac{\cot 2x}{\csc x}$

44. $\displaystyle\lim_{x\to 0} \frac{1-\cos x}{2x^2}$

45. $\displaystyle\lim_{x\to \pi} \frac{\tan x}{\sin 2x}$

46. $\displaystyle\lim_{x\to \pi/4} \frac{\sin x - \cos x}{\cos 2x}$

47. $\displaystyle\lim_{\theta\to 0} \frac{\sin \theta}{\theta + \tan \theta}$

48. $\displaystyle\lim_{x\to 0} \frac{x}{\sin(x/2)}$

49. $\displaystyle\lim_{x\to 0^+} \sqrt{x}\,\csc\sqrt{x}$

50. $\displaystyle\lim_{x\to 1} \frac{\sin(x-1)}{x^2+x-2}$

51. Differentiate the following trigonometric identities to obtain new (or familiar) identities.

   (a) $\tan x = \dfrac{\sin x}{\cos x}$ (b) $\sec x = \dfrac{1}{\cos x}$

   (c) $\sin x + \cos x = \dfrac{1 + \cot x}{\csc x}$

52. Find $\displaystyle\lim_{x\to\infty} \cos\frac{1}{x}$.

53. Find $\displaystyle\lim_{x\to\infty} \left(x - x\cos\frac{1}{x}\right)$.

SECTION 2.5

The Chain Rule

Suppose that you were asked to differentiate the function

$$F(x) = \sqrt{x^2 + 1}$$

The differentiation formulas you learned in the preceding section do not enable you to calculate $F'(x)$.

Observe that F is a composite function. If we let $y = f(u) = \sqrt{u}$ and $u = g(x) = x^2 + 1$, then we can write $y = F(x) = f(g(x))$, that is, $F = f \circ g$. (See Section 5 in Review and Preview for a review of composite functions.) We know how to differentiate both f and g, so it would be useful to have a rule that tells us how to find the derivative of $F = f \circ g$ in terms of the derivatives of f and g.

It turns out that the derivative of the composite function $f \circ g$ is the product of the derivatives of f and g. This fact is one of the most important of the differentiation rules and is called the Chain Rule. It seems plausible if we interpret derivatives as rates of change. Regard du/dx as the rate of change of u with respect to x, dy/du as the rate of change of y with respect to u, and dy/dx as the rate of change of y with respect to x. If u changes twice as fast as x and y changes three times as fast as u, then it seems reasonable that y changes six times as fast as x, and so we expect that

$$\frac{dy}{dx} = \frac{dy}{du}\frac{du}{dx}$$

The Chain Rule (2.30)

> If the derivatives $g'(x)$ and $f'(g(x))$ both exist, and $F = f \circ g$ is the composite function defined by $F(x) = f(g(x))$, then $F'(x)$ exists and is given by the product
>
> $$F'(x) = f'(g(x))g'(x)$$
>
> In Leibniz notation, if $y = f(u)$ and $u = g(x)$ are both differentiable functions, then
>
> $$\frac{dy}{dx} = \frac{dy}{du}\frac{du}{dx}$$

Comments on the Proof of the Chain Rule   Let Δu be the change in u corresponding to a change of Δx in x, that is,

$$\Delta u = g(x + \Delta x) - g(x)$$

Then the corresponding change in y is

$$\Delta y = f(u + \Delta u) - f(u)$$

It is tempting to write

(2.31)

$$\frac{dy}{dx} = \lim_{\Delta x \to 0} \frac{\Delta y}{\Delta x}$$

$$= \lim_{\Delta x \to 0} \frac{\Delta y}{\Delta u} \cdot \frac{\Delta u}{\Delta x}$$

$$= \lim_{\Delta x \to 0} \frac{\Delta y}{\Delta u} \cdot \lim_{\Delta x \to 0} \frac{\Delta u}{\Delta x}$$

$$= \lim_{\Delta u \to 0} \frac{\Delta y}{\Delta u} \cdot \lim_{\Delta x \to 0} \frac{\Delta u}{\Delta x} \qquad \text{(Note that } \Delta u \to 0 \text{ as } \Delta x \to 0 \text{ since } g \text{ is continuous.)}$$

$$= \frac{dy}{du}\frac{du}{dx}$$

The only thing wrong with this reasoning is that in (2.31) it might happen that $\Delta u = 0$ for infinitely many values of Δx as $\Delta x \to 0$. Most functions g have the property that there is an open interval I containing x such that $g(t) \neq g(x)$ for all t in I ($t \neq x$). For such functions we have $\Delta u \neq 0$, so division by Δu causes no problem in (2.31). The above is a perfectly good proof of the Chain Rule for such functions. However, some differentiable functions do not have this property (for example, constant functions and functions like the one in Exercise 52 in Section 2.1), and so the above reasoning does not work. But if $\Delta u = 0$ for infinitely many values of Δx as $\Delta x \to 0$, then the same is true of Δy and it can be shown that for such functions we have both $dy/dx = 0$ and $du/dx = 0$. Thus the Chain Rule is still true for such functions.

Another way to prove the Chain Rule is given at the end of this section.

The Chain Rule can be written either in the prime notation

(2.32)

$$(f \circ g)'(x) = f'(g(x))g'(x)$$

or, if $y = f(u)$ and $u = g(x)$, in Leibniz notation:

(2.33)

$$\frac{dy}{dx} = \frac{dy}{du}\frac{du}{dx}$$

Equation 2.33 is easy to remember because if dy/du and du/dx were quotients, then we could cancel du. Remember, however, that du has not been defined and du/dx should not be thought of as an actual quotient.

EXAMPLE 1   Find $F'(x)$ if $F(x) = \sqrt{x^2 + 1}$.

Solution 1   (using Equation 2.32): At the beginning of this section we expressed F as $F(x) = (f \circ g)(x) = f(g(x))$ where $f(u) = \sqrt{u}$ and $g(x) = x^2 + 1$. Since

$$f'(u) = \tfrac{1}{2} u^{-1/2} = \frac{1}{2\sqrt{u}} \qquad \text{and} \qquad g'(x) = 2x$$

we have

$$F'(x) = f'(g(x))g'(x)$$

$$= \frac{1}{2\sqrt{x^2+1}} \cdot 2x = \frac{x}{\sqrt{x^2+1}}$$

Solution 2   (using Equation 2.33): If we let $u = x^2 + 1$ and $y = \sqrt{u}$, then

$$F'(x) = \frac{dy}{du}\frac{du}{dx} = \frac{1}{2\sqrt{u}}(2x)$$

$$= \frac{1}{2\sqrt{x^2+1}}(2x) = \frac{x}{\sqrt{x^2+1}}$$ •

Note: In using the Chain Rule we work from the outside to the inside. Formula 2.32 says that *we differentiate the outer function f [at the inner function $g(x)$] and then we multiply by the derivative of the inner function.*

When using Formula 2.33, we should bear in mind that dy/dx refers to the derivative of y when y is considered as a function of x (called the *derivative of y with respect to x*), whereas dy/du refers to the derivative of y when considered as a function of u (the derivative of y with respect to u). For instance, in Example 1, y can be considered as a function of x ($y = \sqrt{x^2+1}$) and also as a function of u ($y = \sqrt{u}$). Note that

$$\frac{dy}{dx} = F'(x) = \frac{x}{\sqrt{x^2+1}} \qquad \text{whereas} \qquad \frac{dy}{du} = f'(u) = \frac{1}{2\sqrt{u}}$$

EXAMPLE 2   If $y = u^3 + u^2 + 1$, where $u = 2x^2 - 1$, find

$$\frac{dy}{dx}\bigg|_{x=2}$$

Solution 1   Using the Chain Rule, we have

$$\frac{dy}{dx} = \frac{dy}{du}\frac{du}{dx} = (3u^2 + 2u)(4x)$$

When $x = 2$, $u = 2(2)^2 - 1 = 7$, so

$$\frac{dy}{dx}\bigg|_{x=2} = (3 \cdot 7^2 + 2 \cdot 7)(4 \cdot 2) = 161 \cdot 8 = 1288$$

Solution 2   We could solve this problem without using the Chain Rule by expressing y explicitly as a function of x:

$$y = u^3 + u^2 + 1 = (2x^2 - 1)^3 + (2x^2 - 1)^2 + 1$$

$$= (8x^6 - 12x^4 + 6x^2 - 1) + (4x^4 - 4x^2 + 1) + 1$$

$$= 8x^6 - 8x^4 + 2x^2 + 1$$

Therefore

$$\frac{dy}{dx} = 48x^5 - 32x^3 + 4x$$

$$\frac{dy}{dx}\bigg|_{x=2} = 48 \cdot 2^5 - 32 \cdot 2^3 + 4 \cdot 2 = 1288$$

Solution 1 is clearly preferable.   •

Let us make explicit the special case of the Chain Rule where the outer function f is a power function. If $y = [g(x)]^n$, then we can write $y = f(u) = u^n$ where $u = g(x)$. By using the Chain Rule and then the Power Rule, we get

$$\frac{dy}{dx} = \frac{dy}{du}\frac{du}{dx} = nu^{n-1}\frac{du}{dx} = n[g(x)]^{n-1} \; g'(x)$$

The Power Rule Combined with The Chain Rule (2.34)

If n is any real number and $u = g(x)$ is differentiable, then

$$\frac{d}{dx}\,(u^n) = nu^{n-1}\frac{du}{dx}$$

Alternatively,

$$\frac{d}{dx}\,[g(x)]^n = n[g(x)]^{n-1} \cdot g'(x)$$

Notice that the derivative in Example 1 could be calculated by taking $n = \frac{1}{2}$ in (2.34).

EXAMPLE 3   Differentiate $y = (x^3 - 1)^{100}$.

Solution   Taking $u = g(x) = x^3 - 1$ and $n = 100$ in (2.34), we have

$$\frac{dy}{dx} = \frac{d}{dx}\,(x^3 - 1)^{100} = 100(x^3 - 1)^{99}\frac{d}{dx}\,(x^3 - 1)$$

$$= 100(x^3 - 1)^{99} \cdot 3x^2$$

$$= 300x^2(x^3 - 1)^{99}$$

EXAMPLE 4   Find $f'(x)$ if

$$f(x) = \frac{1}{\sqrt[3]{x^2 + x + 1}}$$

Solution   First rewrite f: $f(x) = (x^2 + x + 1)^{-1/3}$. Thus

$$f'(x) = -\tfrac{1}{3}(x^2 + x + 1)^{-4/3}\frac{d}{dx}\,(x^2 + x + 1)$$

$$= -\tfrac{1}{3}(x^2 + x + 1)^{-4/3}(2x + 1)$$

EXAMPLE 5   Find the derivative of the function

$$g(t) = \left(\frac{t - 2}{2t + 1}\right)^9$$

Solution   Combining the Power Rule, Chain Rule, and Quotient Rule, we get

$$g'(t) = 9\left(\frac{t - 2}{2t + 1}\right)^8\frac{d}{dt}\left(\frac{t - 2}{2t + 1}\right)$$

$$= 9\left(\frac{t - 2}{2t + 1}\right)^8\frac{(2t + 1)\cdot 1 - 2(t - 2)}{(2t + 1)^2}$$

$$= \frac{45(t - 2)^8}{(2t + 1)^{10}}$$

EXAMPLE 6   Differentiate $y = (2x + 1)^5(x^3 - x + 1)^4$.

Solution   In this example we must use the Product Rule before using the Chain Rule:

$$\frac{dy}{dx} = (2x + 1)^5 \frac{d}{dx}(x^3 - x + 1)^4 + (x^3 - x + 1)^4 \frac{d}{dx}(2x + 1)^5$$

$$= (2x + 1)^5 \cdot 4(x^3 - x + 1)^3 \frac{d}{dx}(x^3 - x + 1)$$

$$+ (x^3 - x + 1)^4 \cdot 5(2x + 1)^4 \frac{d}{dx}(2x + 1)$$

$$= 4(2x + 1)^5(x^3 - x + 1)^3(3x^2 - 1) + 5(x^3 - x + 1)^4(2x + 1)^4 \cdot 2$$

By using common factors, we could write the answer as

$$\frac{dy}{dx} = 2(2x + 1)^4(x^3 - x + 1)^3(17x^3 + 6x^2 - 9x + 3)$$ •

EXAMPLE 7   Differentiate (a) $y = \sin(x^2)$ and (b) $y = \sin^2 x$.

Solution
(a)   If $u = x^2$, then $y = \sin u$, and the Chain Rule gives

$$\frac{dy}{dx} = \frac{dy}{du}\frac{du}{dx} = \cos u \cdot 2x = 2x \cos(x^2)$$

(b)   Note that $y = \sin^2 x = (\sin x)^2$. Here $u = \sin x$ and $y = u^2$, so the Chain Rule gives

$$\frac{dy}{dx} = \frac{dy}{du}\frac{du}{dx} = 2u \cdot \cos x = 2 \sin x \cos x$$

The answer can be left as above or written as

$$\frac{dy}{dx} = \sin 2x$$ •

In Example 7(a) we combined the Chain Rule with Theorem 2.26. In general, if $y = \sin u$, where u is a differentiable function of x, then, by the Chain Rule,

$$\frac{dy}{dx} = \frac{dy}{du}\frac{du}{dx} = \cos u \frac{du}{dx}$$

Thus

$$\frac{d}{dx}(\sin u) = \cos u \frac{du}{dx}$$

In a similar fashion, all of the formulas for differentiating trigonometric functions can be combined with the Chain Rule.

The reason for the name "Chain Rule" becomes clear when we make a longer chain by adding another link. Suppose that $y = f(u)$, $u = g(x)$, and $x = h(t)$, where f, g, and h are differentiable functions. Then to compute the derivative of y with respect to t, we use the Chain Rule twice:

$$\frac{dy}{dt} = \frac{dy}{dx}\frac{dx}{dt} = \frac{dy}{du}\frac{du}{dx}\frac{dx}{dt}$$

EXAMPLE 8   If $f(x) = \sin(\cos(\tan x))$, then

$$f'(x) = \cos(\cos(\tan x))\frac{d}{dx}\cos(\tan x)$$

$$= \cos(\cos(\tan x)) \left[-\sin(\tan x)\right] \frac{d}{dx}(\tan x)$$

$$= -\cos(\cos(\tan x)) \sin(\tan x) \sec^2 x$$

Notice that the Chain Rule has been used twice.

EXAMPLE 9   Let

$$y = \left(x^4 + \sqrt{x + \sqrt[3]{x-1}}\right)^{\sqrt{2}}$$

Then

$$\frac{dy}{dx} = \sqrt{2}\left(x^4 + \sqrt{x + \sqrt[3]{x-1}}\right)^{\sqrt{2}-1} \frac{d}{dx}\left(x^4 + \sqrt{x + \sqrt[3]{x-1}}\right)$$

$$= \sqrt{2}\left(x^4 + \sqrt{x + \sqrt[3]{x-1}}\right)^{\sqrt{2}-1}\left[4x^3 + \frac{1}{2}\left(x + \sqrt[3]{x-1}\right)^{-1/2} \frac{d}{dx}\left(x + \sqrt[3]{x-1}\right)\right]$$

$$= \sqrt{2}\left(x^4 + \sqrt{x + \sqrt[3]{x-1}}\right)^{\sqrt{2}-1}\left\{4x^3 + \frac{1}{2\sqrt{x + \sqrt[3]{x-1}}}\left[1 + \frac{1}{3}(x-1)^{-2/3}\frac{d}{dx}(x-1)\right]\right\}$$

$$= \sqrt{2}\left(x^4 + \sqrt{x + \sqrt[3]{x-1}}\right)^{\sqrt{2}-1}\left[4x^3 + \frac{1}{2\sqrt{x + \sqrt[3]{x-1}}}\left(1 + \frac{1}{3(x-1)^{2/3}}\right)\right]$$

Notice that (2.34) has been used three times, with $n = \sqrt{2}$, $\frac{1}{2}$, and $\frac{1}{3}$.

Proof of the Chain Rule   If f is differentiable at a given number u and we define a function E by

(2.35) $$E(t) = \frac{f(u+t) - f(u)}{t} - f'(u)$$

then by the definition of derivative we have

$$\lim_{t \to 0} E(t) = \lim_{t \to 0} \frac{f(u+t) - f(u)}{t} - \lim_{t \to 0} f'(u)$$

$$= f'(u) - f'(u) = 0$$

In Equation 2.35 if we multiply both sides by t we get

(2.36) $$f(u+t) - f(u) = [f'(u) + E(t)]t$$

If we define $E(0) = 0$, then Equation 2.36 remains true even for $t = 0$. Replacing t by Δu in Equation 2.36, we get

$$\Delta y = f(u + \Delta u) - f(u)$$

$$= [f'(u) + E(\Delta u)]\Delta u$$

We divide both sides of this equation by Δx:

(2.37) $$\frac{\Delta y}{\Delta x} = [f'(u) + E(\Delta u)] \frac{\Delta u}{\Delta x}$$

Since g is differentiable at x, it must be continuous there (by Theorem 2.8) and so

$$\lim_{\Delta x \to 0} \Delta u = \lim_{\Delta x \to 0} [g(x + \Delta x) - g(x)] = 0$$

Therefore $$\lim_{\Delta x \to 0} E(\Delta u) = \lim_{\Delta x \to 0} E(\Delta u) = 0$$

Thus Equation 2.37 gives

$$\frac{dy}{dx} = \lim_{\Delta x \to 0} \frac{\Delta y}{\Delta x} = \lim_{\Delta x \to 0} [f'(u) + E(\Delta u)] \frac{\Delta u}{\Delta x}$$

$$= [f'(u) + \lim_{\Delta x \to 0} E(\Delta u)] \lim_{\Delta x \to 0} \frac{\Delta u}{\Delta x}$$

$$= [f'(u) + 0] \frac{du}{dx} = \frac{dy}{du} \frac{du}{dx}$$

SECTION 2.5 **Exercises**

In Exercises 1–4 find dy/dx and $dy/dx|_{x=1}$ in two ways:
(a) using the Chain Rule and (b) without using the Chain Rule, as in Example 2.

1. $y = u^2$, $u = x^2 + 2x + 3$

2. $y = u^2 - 2u + 3$, $u = 5 - 6x$

3. $y = u^3$, $u = x + (1/x)$

4. $y = u - u^2$, $u = \sqrt{x} + \sqrt[3]{x}$

In Exercises 5–52 find the derivatives of the given functions.

5. $F(x) = (x^2 + 4x + 6)^5$ 6. $F(x) = (x^3 - 5x)^4$

7. $G(x) = (3x - 2)^{10}(5x^2 - x + 1)^{12}$

8. $g(t) = (6t^2 + 5)^3(t^3 - 7)^4$

9. $f(t) = (2t^2 - 6t + 1)^{-8}$

10. $f(t) = \dfrac{1}{(t^2 - 2t - 5)^4}$ 11. $g(x) = \sqrt{x^2 - 7x}$

12. $k(x) = \sqrt[3]{1 + \sqrt{x}}$ 13. $h(t) = \left(t - \dfrac{1}{t}\right)^{3/2}$

14. $F(s) = \sqrt{s^3 + 1}(s^2 + 1)^4$

15. $F(y) = \left(\dfrac{y - 6}{y + 7}\right)^3$ 16. $s(t) = \sqrt[4]{\dfrac{t^3 + 1}{t^3 - 1}}$

17. $f(z) = \dfrac{1}{\sqrt[5]{2z - 1}}$ 18. $f(x) = \dfrac{x}{\sqrt{7 - 3x}}$

19. $y = (2x - 5)^4(8x^2 - 5)^{-3}$

20. $y = (x^2 + 1)\sqrt[3]{x^2 + 2}$

21. $y = \tan 3x$ 22. $y = 4\sec 5x$

23. $y = \cos(x^3)$

24. $y = \cos^3 x$

25. $y = (1 + \cos^2 x)^6$

26. $y = \tan(x^2) + \tan^2 x$

27. $y = \cos(\tan x)$

28. $y = \sin(\sin x)$

29. $y = \sec^2 2x - \tan^2 2x$

30. $y = \sqrt{1 + 2\tan x}$ 31. $y = \csc \dfrac{x}{3}$

32. $y = \cot \sqrt[3]{1 + x^2}$ 33. $y = \sin^3 x + \cos^3 x$

34. $y = \sin^2(\cos 4x)$ 35. $y = \sin \dfrac{1}{x}$

36. $y = \dfrac{\sin^2 x}{\cos x}$ 37. $y = \dfrac{1 + \sin 2x}{1 - \sin 2x}$

38. $y = x \sin \dfrac{1}{x}$ 39. $y = \tan^2(x^3)$

40. $y = \left(\sin \sqrt{x^2 + 1}\right)^{\sqrt{2}}$ 41. $y = \cos^2\left(\dfrac{1 - \sqrt{x}}{1 + \sqrt{x}}\right)$

42. $y = \sqrt{1 + \tan\left(x + \dfrac{1}{x}\right)}$

43. $y = \cos^2(\cos x) + \sin^2(\cos x)$

44. $y = \sin(\sin(\sin x))$ 45. $y = \sqrt{x + \sqrt{x}}$

46. $y = \sqrt{x + \sqrt{x + \sqrt{x}}}$

47. $f(x) = [x^3 + (2x - 1)^3]^3$

48. $g(t) = \sqrt[4]{(1 - 3t)^4 + t^4}$

49. $p(t) = \left[\left(1 + \dfrac{2}{t}\right)^{-1} + 3t\right]^{-2}$

50. $N(y) = \left(y + \sqrt[3]{y + \sqrt{2y-9}}\right)^8$

51. $y = \sin(\tan\sqrt{\sin x})$ **52.** $y = \sqrt{\cos(\sin^2 x)}$

In Exercises 53–58 find the equations of the tangent lines to the given curves at the given points.

53. $y = (x^3 - x^2 + x - 1)^{10}$, $(1,0)$

54. $y = \sqrt{x + \frac{1}{x}}$, $(1, \sqrt{2})$

55. $y = \dfrac{8}{\sqrt{4 + 3x}}$, $(4,2)$

56. $y = \dfrac{x}{(3 - x^2)^5}$, $(2,-2)$

57. $y = \cot^2 x$, $(\pi/4, 1)$

58. $y = \sin x + \cos 2x$, $(\pi/6, 1)$

59. Find all points on the graph of the function $f(x) = 2\sin x + \sin^2 x$ where the tangent line is horizontal.

60. Find the x-coordinates of all points on the curve $y = \sin 2x - 2\sin x$ where the tangent line is horizontal.

61. Suppose that $F(x) = f(g(x))$ and $g(3) = 6$, $g'(3) = 4$, $f'(3) = 2$, and $f'(6) = 7$. Find $F'(3)$.

62. Suppose that $w = u \circ v$ and $u(0) = 1$, $v(0) = 2$, $u'(0) = 3$, $u'(2) = 4$, $v'(0) = 5$, and $v'(2) = 6$. Find $w'(0)$.

In Exercises 63–66 find f' and state the domains of f and f'.

63. $f(x) = x^2 \sec^2 3x$ **64.** $f(x) = \sin\sqrt{2x + 1}$

65. $f(x) = \sqrt{\cos\sqrt{x}}$

66. $f(x) = \cos\sqrt{x} + \sqrt{\cos x}$

67. The displacement of a particle on a vibrating string is given by the equation
$$s(t) = 10 + \tfrac{1}{4}\sin(10\pi t)$$
where s is measured in centimeters and t in seconds. Find the velocity of the particle after t seconds.

68. If the equation of motion of a particle is given by $s = A\cos(\omega t + \delta)$, the particle is said to undergo *simple harmonic motion*.
(a) Find the velocity of the particle at time t.
(b) When is the velocity 0?

69. A Cepheid variable star is a star whose brightness alternately increases and decreases. The most easily

visible such star is Delta Cephei, for which the interval between times of maximum brightness is 5.4 days. The average brightness of this star is 4.0 and its brightness changes by ± 0.35. In view of these data, the brightness of Delta Cephei at time t, where t is measured in days, has been modeled by the function
$$B(t) = 4.0 + 0.35\sin(2\pi t/5.4)$$
(a) Find the rate of change of the brightness after t days.
(b) Find, correct to two decimal places, the rate of increase after 1 day.

70. The frequency of vibrations of a vibrating violin string is given by
$$f = \frac{1}{2L}\sqrt{\frac{T}{\rho}}$$
where L is the length of the string, T is its tension, and ρ is its linear density. [See Chapter 11 in D.E. Hall, *Musical Acoustics,* 2nd ed. (Pacific Grove, CA: Brooks/Cole, 1991).] Find the rate of change of the frequency with respect to
(a) the length (when T and ρ are constant),
(b) the tension (when L and ρ are constant), and
(c) the linear density (when L and T are constant).

71. Let h be differentiable on $[0, \infty)$ and define G by $G(x) = h(\sqrt{x})$.
(a) Where is G differentiable?
(b) Find an expression for $G'(x)$.

72. Suppose f is differentiable on R and α is a real number. Let $F(x) = f(x^\alpha)$ and $G(x) = [f(x)]^\alpha$. Find expressions for (a) $F'(x)$ and (b) $G'(x)$.

73. Suppose f is differentiable on R. Let $F(x) = f(\cos x)$ and $G(x) = \cos(f(x))$. Find expressions for (a) $F'(x)$ and (b) $G'(x)$.

74. If $g(t) = [f(\sin t)]^2$, where f is a differentiable function, find $g'(t)$.

75. Use the Chain Rule to give easier proofs of Exercise 53 in Section 2.1.

76. Use the Chain Rule and the Product Rule to give an alternative proof of the Quotient Rule. [*Hint:* Write $f(x)/g(x) = f(x)[g(x)]^{-1}$.]

77. If n is a positive integer, prove that
$$\frac{d}{dx}(\sin^n x \cos nx) = n\sin^{n-1} x \cos(n+1)x$$

78. Find a formula for $D(\cos^n x \cos nx)$ that is similar to the one in Exercise 77.

Find the derivatives of the functions in Exercises 79–81.
(Hint: $|x| = \sqrt{x^2}$.)

79. $f(x) = |x|$ 80. $g(x) = |x|/x$

81. $h(x) = x|2x - 1|$

82.   (a)   Sketch the graph of the function $f(x) = |\sin x|$.
      (b)   At what points is f not differentiable?
      (c)   Give a formula for f' and sketch its graph.

(d)   Do the same for $g(x) = \sin|x|$.

83.   Use the Chain Rule to show that

$$\frac{d}{d\theta}(\sin \theta°) = \frac{\pi}{180}\cos \theta°$$

(This gives one reason for the convention that radian measure is always used when dealing with trigonometric functions in calculus: the differentiation formulas would not be as simple if we used degrees.)

SECTION 2.6

Implicit Differentiation

The functions that we have met so far can be described by expressing one variable explicitly in terms of another variable—for example,

$$y = \sqrt{x^3 + 1} \qquad \text{or} \qquad y = x \sin x$$

or in general $y = f(x)$. Some functions, however, are defined implicitly by a relation between x and y such as

(2.38) $x^2 + y^2 = 25$

(2.39)   or $x^3 + y^3 = 6xy$

Sometimes it may be possible to solve such an equation for y as an explicit function (or several functions) of x. For instance, if we solve Equation 2.38 for y, we get $y = \pm\sqrt{25 - x^2}$, so two functions determined by the implicit Equation 2.38 are $f(x) = \sqrt{25 - x^2}$ and $g(x) = -\sqrt{25 - x^2}$. The graphs of f and g are the upper and lower semicircles of the circle $x^2 + y^2 = 25$ (see Figure 2.16).

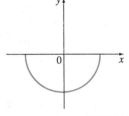

(a) $x^2 + y^2 = 25$ (b) $f(x) = \sqrt{25 - x^2}$ (c) $g(x) = -\sqrt{25 - x^2}$

Figure 2.16

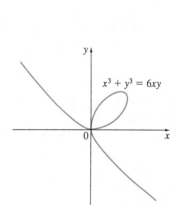

$x^3 + y^3 = 6xy$

Figure 2.17

The folium of Descartes

It would be very difficult (though possible) to solve Equation 2.39 for y explicitly as a function of x. Nonetheless, it is the equation of a curve called the *folium of Descartes* shown in Figure 2.17 and it implicitly defines y as several functions of x. The graphs of three such functions are shown in Figure 2.18. When we say that f is a function defined implicitly by Equation 2.39, we mean that the equation

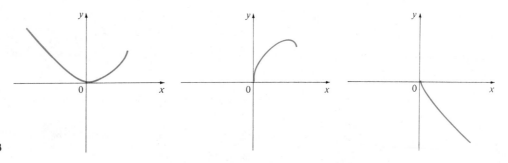

Figure 2.18

$$x^3 + [f(x)]^3 = 6xf(x)$$

is true for all values of x in the domain of f.

Fortunately it is not necessary to solve an equation for y in terms of x in order to find the derivative of y. Instead we can use the method of **implicit differentiation**. This consists of differentiating both sides of the relation with respect to x and then solving the resulting equation for y'. In the examples and exercises of this section it is always assumed that the given equation determines y implicitly as a differentiable function of x so that the method of implicit differentiation can be applied.

EXAMPLE 1

(a)   If $x^2 + y^2 = 25$, find $\dfrac{dy}{dx}$.

(b)   Find the equation of the tangent to the circle $x^2 + y^2 = 25$ at the point $(3, 4)$.

Solution 1

(a)   Differentiate both sides of the equation $x^2 + y^2 = 25$:

$$\frac{d}{dx}(x^2 + y^2) = \frac{d}{dx}(25)$$

$$\frac{d}{dx}(x^2) + \frac{d}{dx}(y^2) = 0$$

Remembering that y is a function of x and using the Chain Rule, we have

$$\frac{d}{dx}(y^2) = 2y\frac{dy}{dx}$$

Thus $$2x + 2y\frac{dy}{dx} = 0$$

Now we solve this equation for dy/dx:

$$\frac{dy}{dx} = -\frac{x}{y}$$

(b)   At the point $(3, 4)$ we have $x = 3$ and $y = 4$, so

$$\frac{dy}{dx} = -\frac{3}{4}$$

An equation of the tangent to the circle at $(3, 4)$ is therefore

$$y - 4 = -\tfrac{3}{4}(x - 3)$$

or $$3x + 4y = 25$$

Solution 2

(b)   Solving the equation $x^2 + y^2 = 25$ we get $y = \pm\sqrt{25 - x^2}$. The point $(3, 4)$ lies on the upper semicircle $y = \sqrt{25 - x^2}$, and so we consider the function $f(x) = \sqrt{25 - x^2}$. Differentiating f using the Chain Rule, we have

$$f'(x) = \tfrac{1}{2}(25 - x^2)^{-1/2}\,\tfrac{d}{dx}\,(25 - x^2)$$

$$= \tfrac{1}{2}(25 - x^2)^{-1/2}(-2x)$$

$$= -\frac{x}{\sqrt{25 - x^2}}$$

$$f'(3) = -\frac{3}{\sqrt{25 - 3^2}} = -\frac{3}{4}$$

and, as in Solution 1, the equation of the tangent is $3x + 4y = 25$.   •

Note 1:   Example 1 illustrates that even when it is possible to solve an equation explicitly for y in terms of x, it may be easier to use implicit differentiation.

Note 2:   The expression $dy/dx = -x/y$ gives the derivative in terms of both x and y. It is correct no matter which function y is determined by the given equation. For instance, for $y = f(x) = \sqrt{25 - x^2}$ we have

$$\frac{dy}{dx} = -\frac{x}{y} = -\frac{x}{\sqrt{25 - x^2}}$$

whereas for $y = g(x) = -\sqrt{25 - x^2}$ we have

$$\frac{dy}{dx} = -\frac{x}{y} = -\frac{x}{-\sqrt{25 - x^2}} = \frac{x}{\sqrt{25 - x^2}}$$

EXAMPLE 2

(a) Find y' if $x^3 + y^3 = 6xy$.

(b) Find the tangent to the folium of Descartes $x^3 + y^3 = 6xy$ at the point $(3, 3)$.

Solution

(a)   Differentiating both sides of $x^3 + y^3 = 6xy$ with respect to x, regarding y as a function of x, and using the Chain Rule on the y^3 term and the Product Rule on the $6xy$ term, we get

$$3x^2 + 3y^2 y' = 6y + 6xy'$$

or $$x^2 + y^2 y' = 2y + 2xy'$$

We now solve for y':

$$(y^2 - 2x)\,y' = 2y - x^2$$

or $$y' = \frac{2y - x^2}{y^2 - 2x}$$

(b)   When $x = y = 3$,

$$y' = \frac{2 \cdot 3 - 3^2}{3^2 - 2 \cdot 3} = -1$$

so an equation of the tangent to the folium at $(3,3)$ is

$$y - 3 = -1(x - 3) \qquad \text{or} \qquad x + y = 6$$

Note: There is a formula for the three roots of a cubic equation that is like the quadratic formula but much more complicated. If this formula is used to solve the equation $x^3 + y^3 = 6xy$ for y in terms of x, then three functions determined by the equation are

$$y = f(x) = \sqrt[3]{-\frac{x^3}{2} + \sqrt{\frac{x^6}{4} - 8x^3}} + \sqrt[3]{-\frac{x^3}{2} - \sqrt{\frac{x^6}{4} - 8x^3}}$$

and

$$y = \frac{1}{2}\left[-f(x) \pm \sqrt{-3}\left(\sqrt[3]{-\frac{x^3}{2} + \sqrt{\frac{x^6}{4} - 8x^3}} - \sqrt[3]{-\frac{x^3}{2} - \sqrt{\frac{x^6}{4} - 8x^3}} \right) \right]$$

You can see that the method of implicit differentiation saves a lot of work in cases such as this. Moreover, implicit differentiation works just as easily for equations such as

$$y^5 + 3x^2y^2 + 5x^4 = 12$$

which are *impossible* to solve for y in terms of x. [The Norwegian mathematician Niels Abel proved in 1824 that there is no general formula for the roots of a fifth-degree equation. Later the French mathematician Evariste Galois proved that it is impossible to find a general formula for the roots of an nth-degree equation (in terms of algebraic operations on the coefficients) if n is any integer larger than 4.]

EXAMPLE 3   Find y' if $\sin(x + y) = y^2 \cos x$.

Solution   Differentiating implicitly with respect to x and remembering that y is a function of x, we get

$$\cos(x + y) \cdot (1 + y') = 2yy' \cos x + y^2(-\sin x)$$

(Note that we have used the Chain Rule on the left side and the Product Rule and Chain Rule on the right side.) Solving for y' gives

$$y' = \frac{y^2 \sin x + \cos(x + y)}{2y \cos x - \cos(x + y)}$$

Two curves are called **orthogonal** if at each point of intersection their tangent lines are perpendicular. In the next example we use implicit differentiation to show that two families of curves are **orthogonal trajectories** of each other; that is, every curve in one family is orthogonal to every curve in the other family. Orthogonal families arise in several areas of physics. For example, the lines of force in an electrostatic field are orthogonal to the lines of constant potential.

EXAMPLE 4   The equation

(2.40)
$$xy = c \qquad c \neq 0$$

represents a family of hyperbolas. (Different values of the constant c give different hyperbolas. See Figure 2.19.) The equation

(2.41)
$$x^2 - y^2 = k \qquad k \neq 0$$

represents another family of hyperbolas with asymptotes $y = \pm x$. Show that every curve in the family (2.40) is orthogonal to every curve in the family (2.41); that is, the families are orthogonal trajectories.

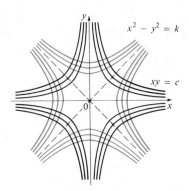

Figure 2.19

Solution   Implicit differentiation of Equation 2.40 gives

(2.42)
$$y + x\frac{dy}{dx} = 0 \qquad \text{so} \qquad \frac{dy}{dx} = -\frac{y}{x}$$

Implicit differentiation of Equation 2.41 gives

(2.43)
$$2x - 2y\frac{dy}{dx} = 0 \qquad \text{so} \qquad \frac{dy}{dx} = \frac{x}{y}$$

From (2.42) and (2.43) we see that at any point of intersection of curves from each family, the slopes of the tangents are negative reciprocals of each other. Therefore the curves intersect at right angles.   ●

SECTION 2.6 **Exercises**

In Exercises 1–6, (a) find y' by implicit differentiation, (b) solve the given equation explicitly for y and differentiate to get y' in terms of x, and (c) check that your solutions to parts (a) and (b) are consistent by substituting the expressions for y into your solution for part (a).

1. $x^2 + 3x + xy = 5$ 2. $\dfrac{x^2}{2} + \dfrac{y^2}{4} = 1$

3. $\dfrac{1}{x} + \dfrac{1}{y} = 3$ 4. $x^2 + xy - y^2 = 3$

5. $2y^2 + xy = x^2 + 3$ 6. $\sqrt{x} + \sqrt{y} = 4$

In Exercises 7–20 find dy/dx by implicit differentiation.

7. $x^2 - xy + y^3 = 8$ 8. $\sqrt{xy} - 2x = \sqrt{y}$

9. $2y^2 + \sqrt[3]{xy} = 3x^2 + 17$

10. $y^5 + 3x^2y^2 + 5x^4 = 12$

11. $x^4 + y^4 = 16$ 12. $\sqrt{x + y} + \sqrt{xy} = 6$

13. $2xy = (x^2 + y^2)^{3/2}$ 14. $x^2 = \dfrac{y^2}{y^2 - 1}$

15. $\dfrac{y}{x - y} = x^2 + 1$

16. $x\sqrt{1 + y} + y\sqrt{1 + 2x} = 2x$

17. $\cos(x - y) = y\sin x$

18. $x\sin y + \cos 2y = \cos y$

19. $xy = \cot(xy)$

20. $x\cos y + y\cos x = 1$

In Exercises 21 and 22 regard y as the independent variable and x as the dependent variable, and use implicit differentiation to find dx/dy.

21. $y^4 + x^2y^2 + yx^4 = y + 1$

22. $(x^2 + y^2)^2 = ax^2y$

23. If $x[f(x)]^3 + xf(x) = 6$ and $f(3) = 1$, find $f'(3)$.

24. If $[g(x)]^2 + 12x = x^2g(x)$ and $g(4) = 12$, find $g'(4)$.

In Exercises 25–30 find an equation of the tangent line to the given curve at the given point.

25. $\dfrac{x^2}{16} - \dfrac{y^2}{9} = 1$, $\left(-5, \frac{9}{4}\right)$ (hyperbola)

26. $\dfrac{x^2}{9} + \dfrac{y^2}{36} = 1$, $(-1, 4\sqrt{2})$ (ellipse)

27. $y^2 = x^3(2 - x)$, $(1, 1)$ (piriform)

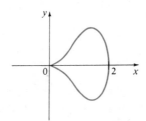

28. $x^{2/3} + y^{2/3} = 4$, $(-3\sqrt{3}, 1)$ (astroid)

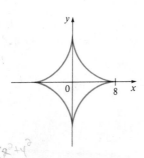

29. $2(x^2 + y^2)^2 = 25(x^2 - y^2)$, $(3, 1)$ (lemniscate)

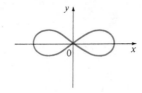

30. $x^2y^2 = (y + 1)^2(4 - y^2)$, $(0, -2)$
(conchoid of Nicomedes)

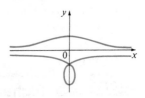

31. Find the points on the lemniscate in Exercise 29 where the tangent is horizontal.

32. Show by implicit differentiation that the tangent to the ellipse

$$\frac{x^2}{a^2} + \frac{y^2}{b^2} = 1$$

at the point (x_0, y_0) is $\dfrac{x_0x}{a^2} + \dfrac{y_0y}{b^2} = 1$.

33. Find the equation of the tangent line to the hyperbola

$$\frac{x^2}{a^2} - \frac{y^2}{b^2} = 1$$

at the point (x_0, y_0).

34. Show that the sum of the x- and y-intercepts of any tangent line to the curve $\sqrt{x} + \sqrt{y} = \sqrt{c}$ is equal to c.

35. Show, using implicit differentiation, that any tangent line at a point P to a circle with center O is perpendicular to the radius OP.

36. The Power Rule (2.16) can be proved using implicit differentiation for the case where n is a rational number, $n = p/q$, and $y = f(x) = x^n$ is assumed beforehand to be a differentiable function. If $y = x^{p/q}$, then $y^q = x^p$. Use implicit differentiation to show that

$$y' = \frac{p}{q}x^{(p/q)-1}$$

In Exercises 37 and 38 show that the given curves are orthogonal.

37. $2x^2 + y^2 = 3$, $x = y^2$

38. $x^2 - y^2 = 5$, $4x^2 + 9y^2 = 72$

In Exercises 39–42 show that the given families of curves are orthogonal trajectories of each other. Sketch both families of curves.

39. $x^2 + y^2 = r^2$, $ax + by = 0$

40. $x^2 + y^2 = ax$, $x^2 + y^2 = by$

41. $y = cx^2, \quad x^2 + 2y^2 = k$

42. $y = ax^3, \quad x^2 + 3y^2 = b$

43. Find all points on the curve $x^2y^2 + xy = 2$ where the slope of the tangent line is -1.

44. Find the equations of both the tangent lines to the ellipse $x^2 + 4y^2 = 36$ that pass through the point $(12, 3)$.

SECTION 2.7

Higher Derivatives

If f is a differentiable function, then its derivative f' is also a function, so f' may have a derivative of its own, denoted by $(f')' = f''$. This new function f'' is called the **second derivative** of f because it is the derivative of the derivative of f. Thus

$$f''(x) = \frac{d}{dx}(f'(x)) = \frac{d}{dx}\left(\frac{d}{dx} f(x)\right)$$

For instance, if $f(x) = x^8$, then $f'(x) = 8x^7$, so

$$f''(x) = \frac{d}{dx}(f'(x)) = \frac{d}{dx}(8x^7) = 56x^6$$

Notation: If $y = f(x)$, then

$$y'' = f''(x) = \frac{d}{dx}\left(\frac{dy}{dx}\right) = \frac{d^2y}{dx^2} = D^2f(x) = D_x^2f(x)$$

The symbol D^2 indicates that the operation of differentiation is performed twice.

Similarly the *third derivative* f''' is the derivative of the second derivative: $f''' = (f'')'$. If $y = f(x)$, then alternative notations are

$$y''' = f'''(x) = \frac{d}{dx}\left(\frac{d^2y}{dx^2}\right) = \frac{d^3y}{dx^3} = D^3f(x) = D_x^3f(x)$$

The process can be continued. The fourth derivative f'''' is usually denoted by $f^{(4)}$. In general the nth derivative of f is denoted by $f^{(n)}$ and is obtained from f by differentiating n times. If $y = f(x)$, we write

$$y^{(n)} = f^{(n)}(x) = \frac{d^ny}{dx^n} = D^nf(x) = D_x^nf(x)$$

EXAMPLE 1   If

$$y = x^3 - 6x^2 - 5x + 3$$

then

$$y' = 3x^2 - 12x - 5$$

$$y'' = 6x - 12$$

$$y''' = 6$$

$$y^{(4)} = 0$$

and in fact $y^{(n)} = 0$ for all $n \geq 4$.

EXAMPLE 2   If $f(x) = \frac{1}{x}$, find $f^{(n)}(x)$.

Solution

$$f(x) = \frac{1}{x} = x^{-1}$$

$$f'(x) = -x^{-2} = \frac{-1}{x^2}$$

$$f''(x) = (-2)(-1)x^{-3} = \frac{2}{x^3}$$

$$f'''(x) = -3 \cdot 2 \cdot 1 \cdot x^{-4}$$

$$f^{(4)}(x) = 4 \cdot 3 \cdot 2 \cdot 1 \cdot x^{-5}$$

$$f^{(5)}(x) = -5 \cdot 4 \cdot 3 \cdot 2 \cdot 1 \cdot x^{-6} = -5! x^{-6}$$

$$\vdots$$

$$f^{(n)}(x) = (-1)^n n(n-1)\,(n-2) \cdots 2 \cdot 1 \cdot x^{-(n+1)}$$

or

$$f^{(n)}(x) = \frac{(-1)^n n!}{x^{n+1}}$$

Here we have used the factorial symbol $n!$ for the product of the first n positive integers.

$$\boxed{n! = 1 \cdot 2 \cdot 3 \cdots (n-1)n}$$

If $s = f(t)$ is the position function of an object that moves in a straight line, we know that its first derivative represents the velocity $v(t)$ of the object as a function of time:

$$v(t) = f'(t) = \frac{ds}{dt}$$

The instantaneous rate of change of velocity with respect to time is called the **acceleration** $a(t)$ of the object. Thus the acceleration function is the derivative of the velocity function and is therefore the second derivative of the position function:

$$a(t) = v'(t) = f''(t)$$

or, in Leibniz notation,

$$a = \frac{dv}{dt} = \frac{d^2 s}{dt^2}$$

EXAMPLE 3   The equation of motion of a particle is $s = 2t^3 - 5t^2 + 3t + 4$, where s is measured in centimeters and t in seconds. Find the acceleration as a function of time. What is the acceleration after $2\,\text{s}$?

Solution   The velocity and acceleration are

$$v(t) = \frac{ds}{dt} = 6t^2 - 10t + 3$$

$$a(t) = \frac{dv}{dt} = 12t - 10$$

The acceleration after $2\,\text{s}$ is $a(2) = 14\,\text{cm/s}^2$.

Another application of second derivatives occurs in curve sketching (Section 4.4). Second and higher derivatives will be used in Chapter 10 to represent functions as sums of infinite series.

The following example shows how to find second derivatives of functions that are defined implicitly.

EXAMPLE 4   Find y'' if $x^4 + y^4 = 16$.

Solution   Differentiating the equation implicitly with respect to x, we get

$$4x^3 + 4y^3 y' = 0$$

Solving for y' gives

(2.44)
$$y' = -\frac{x^3}{y^3}$$

To find y'' we differentiate this expression for y' using the Quotient Rule and remembering that y is a function of x:

$$y'' = \frac{d}{dx}\left(-\frac{x^3}{y^3}\right) = -\frac{y^3 D(x^3) - x^3 D(y^3)}{(y^3)^2}$$

$$= -\frac{y^3 \cdot 3x^2 - x^3(3y^2 y')}{y^6}$$

If we now substitute Equation 2.44 into this expression for y'' we get

$$y'' = -\frac{3x^2 y^3 - 3x^3 y^2\left(\dfrac{-x^3}{y^3}\right)}{y^6}$$

$$= -\frac{3(x^2 y^4 + x^6)}{y^7} = -\frac{3x^2(y^4 + x^4)}{y^7}$$

But the values of x and y must satisfy the original equation $x^4 + y^4 = 16$. So the answer simplifies to

$$y'' = -\frac{3x^2(16)}{y^7} = -48\frac{x^2}{y^7}$$

$\bullet$

EXAMPLE 5   Find $D^{27} \cos x$.

Solution   The first few derivatives of $\cos x$ are as follows:

$$D \cos x = -\sin x$$

$$D^2 \cos x = -\cos x$$

$$D^3 \cos x = \sin x$$

$$D^4 \cos x = \cos x$$

$$D^5 \cos x = -\sin x$$

We see that the successive derivatives occur in a cycle of length 4 and, in particular, $D^n \cos x = \cos x$ whenever n is a multiple of 4. Therefore

$$D^{24} \cos x = \cos x$$

and, differentiating three more times, we have

$$D^{27} \cos x = \sin x$$

$\bullet$

SECTION 2.7 **Exercises**

In Exercises 1–14 find the first and second derivatives of the given function.

1. $f(x) = x^4 - 3x^3 + 16x$

2. $f(t) = t^{10} - 2t^7 + t^4 - 6t + 8$

3. $h(x) = \sqrt{x^2 + 1}$ 4. $G(r) = \sqrt{r} + \sqrt[3]{r}$

5. $F(s) = (3s + 5)^8$ 6. $g(u) = \dfrac{1}{\sqrt{1-u}}$

7. $y = \dfrac{x}{1-x}$ 8. $y = x^\pi$

9. $y = (1 - x^2)^{3/4}$ 10. $y = \dfrac{x^2}{x+1}$

11. $H(t) = \tan^3(2t - 1)$ 12. $f(x) = \csc^2(5x)$

13. $F(r) = \sec\sqrt{r}$ 14. $g(s) = s^2 \cos s$

In Exercises 15–18 find y'''.

15. $y = ax^2 + bx + c$ 16. $y = \dfrac{1-x}{1+x}$

17. $y = \sqrt{5t - 1}$ 18. $y = \dfrac{1}{1+x^2}$

19. If $f(x) = (2 - 3x)^{-1/2}$, find $f(0)$, $f'(0)$, $f''(0)$, and $f'''(0)$.

20. If $g(t) = (2 - t^2)^6$, find $g(0)$, $g'(0)$, $g''(0)$, and $g'''(0)$.

21. If $f(\theta) = \cot\theta$, find $f'''(\pi/6)$.

22. If $g(x) = \sec x$, find $g^{(4)}(\pi/4)$.

In Exercises 23–26 find y'' by implicit differentiation.

23. $x^3 + y^3 = 1$ 24. $\sqrt{x} + \sqrt{y} = 1$

25. $x^2 + 6xy + y^2 = 8$ 26. $\dfrac{x^2}{a^2} - \dfrac{y^2}{b^2} = 1$

27. Find the first 73 derivatives of
$$f(x) = x - x^2 + x^3 - x^4 + x^5 - x^6$$

In Exercises 28–31 find a formula for $f^{(n)}(x)$.

28. $f(x) = \sqrt{x}$ 29. $f(x) = x^n$

30. $f(x) = \dfrac{1}{(1-x)^2}$ 31. $f(x) = \dfrac{1}{3x^3}$

In Exercises 32–34 find the given derivative by finding the first few derivatives and observing the pattern that occurs.

32. $D^{99} \sin x$

33. $D^{50} \cos 2x$ 34. $D^{35} x \sin x$

In Exercises 35–38 the equation of motion is given for a particle, where s is in meters and t is in seconds. Find

(a) *the velocity and acceleration as functions of t,*
(b) *the acceleration after 1 s, and*
(c) *the acceleration at the instants when the velocity is 0.*

35. $s = t^3 - 3t$ 36. $s = t^2 - t + 1$

37. $s = At^2 + Bt + C$

38. $s = 2t^3 - 7t^2 + 4t + 1$

In Exercises 39 and 40 an equation of motion is given, where s is in meters and t in seconds.
(a) *At what time is the acceleration 0?*
(b) *Find the displacement and velocity at these times.*

39. $s = t^4 - 4t^3 + 2$ 40. $s = 2t^3 - 9t^2$

41. A mass attached to a vertical spring has position function given by $y(t) = A\sin\omega t$, where A is the amplitude of its oscillations and ω is a constant.
   (a) Find the velocity and acceleration as functions of time.
   (b) Show that the acceleration is proportional to the displacement y.
   (c) Show that the speed is a maximum when the acceleration is 0.

42. An object moves in such a way that its velocity v is related to its displacement s by the equation
$$v = \sqrt{2gs + c}$$
where c and g are constants. Show that the acceleration is constant.

43. Find a second-degree polynomial P such that $P(2) = 5$, $P'(2) = 3$, and $P''(2) = 2$.

44. Find a third-degree polynomial Q such that $Q(1) = 1$, $Q'(1) = 3$, $Q''(1) = 6$, and $Q'''(1) = 12$.

45. If P is a polynomial of degree n, show that $P^{(m)}(x) = 0$ for $m > n$.

46. If $f(x) = |x^2 - x|$, find f' and f''. What are their domains? Sketch the graphs of all three functions.

In Exercises 47–49 g is a twice differentiable function. Find f'' in terms of g, g', and g''.

47. $f(x) = xg(x^2)$ 48. $f(x) = \dfrac{g(x)}{x}$

49. $f(x) = g(\sqrt{x})$

50. (a) If $F(x) = f(x)g(x)$, where f and g have derivatives of all orders, show that
$$F'' = f''g + 2f'g' + fg''$$

(b) Find similar formulas for F''' and $F^{(4)}$.

(c) Guess a formula for $F^{(n)}$.

$$\frac{d^2y}{dx^2} = \frac{d^2y}{du^2}\left(\frac{du}{dx}\right)^2 + \frac{dy}{du}\frac{d^2u}{dx^2}$$

51. If $y = f(u)$ and $u = g(x)$, where f and g are twice differentiable functions, show that

52. If $y = f(u)$ and $u = g(x)$, where f and g possess third derivatives, find a formula for d^3y/dx^3 similar to the one in Exercise 51.

SECTION 2.8

Related Rates

In a related rates problem the idea is to compute the rate of change of one quantity in terms of the rate of change of another quantity (which may be more easily measured). The procedure is to find an equation that relates the two quantities and then use the Chain Rule to differentiate both sides with respect to time.

EXAMPLE 1   Air is being pumped into a spherical balloon so that its volume increases at a rate of $100\,\text{cm}^3/\text{s}$. How fast is the radius of the balloon increasing when the diameter is 50 cm?

Solution   If we let V be the volume of the balloon and r the radius, then V and r are related by the equation

(2.45)
$$V = \tfrac{4}{3}\pi r^3$$

We are given that

$$\frac{dV}{dt} = 100\,\text{cm}^3/\text{s}$$

and we are asked to find dr/dt when $r = 25\,\text{cm}$. We use the Chain Rule to differentiate each side of Equation 2.45:

$$\frac{dV}{dt} = \frac{dV}{dr}\frac{dr}{dt} = 4\pi r^2 \frac{dr}{dt}$$

and so
$$\frac{dr}{dt} = \frac{1}{4\pi r^2}\frac{dV}{dt}$$

If we put $r = 25$ and $dV/dt = 100$ in this equation, we obtain

$$\frac{dr}{dt} = \frac{1}{4\pi(25)^2}100 = \frac{1}{25\pi}$$

The radius of the balloon is increasing at the rate of $1/(25\pi)\,\text{cm/s}$.   ●

EXAMPLE 2   A ladder 10 ft long rests against a vertical wall. If the bottom of the ladder slides away from the wall at a rate of 1 ft/s, how fast is the top of the ladder sliding down the wall when the bottom of the ladder is 6 ft from the wall?

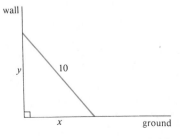

Figure 2.20

Solution   We first draw a diagram and label it as in Figure 2.20. Let x meters be the distance from the bottom of the ladder to the wall and y meters the distance from the top of the ladder to the ground. Note that x and y are both functions of t (time).

We are given that $dx/dt = 1\,\text{ft/s}$ and we are asked to find dy/dt when $x = 6$ ft. In this question, the relationship between x and y is given by the Pythagorean Theorem:

$$x^2 + y^2 = 100$$

Differentiating each side with respect to t using the Chain Rule, we have

$$2x\frac{dx}{dt} + 2y\frac{dy}{dt} = 0$$

and solving this equation for the desired rate, we obtain

$$\frac{dy}{dt} = -\frac{x}{y}\frac{dx}{dt}$$

When $x = 6$, the Pythagorean Theorem gives $y = 8$ and so, substituting these values and $dx/dt = 1$, we have

$$\frac{dy}{dt} = -\tfrac{6}{8}(1) = -\tfrac{3}{4}\,\text{ft/s} \qquad\qquad \bullet$$

EXAMPLE 3   A water tank has the shape of an inverted circular cone with base radius $2\,\text{m}$ and height $4\,\text{m}$. If water is being pumped into the tank at a rate of $2\,\text{m}^3/\text{min}$, find the rate at which the water level is rising when the water is $3\,\text{m}$ deep.

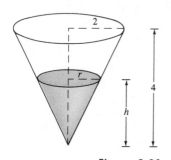

Figure 2.21

Solution   We first sketch the cone and label it as in Figure 2.21. Let V, r, and h be the volume of the water, the radius of the surface, and the height at time t, where t is measured in minutes.

We are given that $dV/dt = 2\,\text{m}^3/\text{min}$ and we are asked to find dh/dt when h is $3\,\text{m}$. The quantities V and h are related by the equation

$$V = \tfrac{1}{3}\pi r^2 h$$

but we must express V as a function of h alone. In order to eliminate r we use the similar triangles in Figure 2.21 to write

$$\frac{r}{h} = \frac{2}{4} \qquad r = \frac{h}{2}$$

and the expression for V becomes

$$V = \tfrac{1}{3}\pi\left(\frac{h}{2}\right)^2 h = \tfrac{\pi}{12}h^3$$

Now we can differentiate each side with respect to t:

$$\frac{dV}{dt} = \tfrac{\pi}{4}h^2\frac{dh}{dt}$$

so

$$\frac{dh}{dt} = \frac{4}{\pi h^2}\frac{dV}{dt}$$

Substituting $h = 3\,\text{m}$ and $dV/dt = 2\,\text{m}^3/\text{min}$, we have

$$\frac{dh}{dt} = \frac{4}{\pi(3)^2}\cdot 2 = \frac{8}{9\pi} \approx 0.28\,\text{m/min} \qquad\qquad \bullet$$

Strategy: It is useful to recall some of the problem-solving principles from the inside front cover of this book and adapt them to related rates in light of our experience in Examples 1–3:

1. Read the problem carefully.
2. Draw a diagram if possible.
3. Introduce notation. Assign symbols to all quantities that are functions of time.
4. Express the given information and the required rate in terms of derivatives.
5. Write an equation that relates the various quantities of the problem. If necessary, use the geometry of the situation to eliminate one of the variables by substitution (as in Example 3).
6. Use the Chain Rule to differentiate each side of the equation with respect to t.
7. Substitute the given information into the resulting equation and solve for the unknown rate.

Warning: A common error is to substitute the given numerical information (for quantities that vary with time) too early. This should be done only *after* the differentiation. (Step 7 follows Step 6.) For instance, in Example 3 we dealt with general values of h until we finally substituted $h = 3$ at the last stage. (If we had put $h = 3$ earlier, we would have gotten $dV/dt = 0$, which is clearly wrong.)

The following examples are further illustrations of the strategy.

EXAMPLE 4   Car A is going west at $50 \, \text{mi/h}$ and car B is going north at $60 \, \text{mi/h}$. Both are headed for the intersection of the two roads. At what rate are the cars approaching each other when car A is 0.3 mi and car B is 0.4 mi from the intersection?

Solution   We draw Figure 2.22 where C is the intersection of the roads. At a given time t, let x be the distance from car A to C, let y be the distance from car B to C, and let z be the distance between the cars, where x, y, and z are measured in miles.

We are given that $dx/dt = -50 \, \text{mi/h}$ and $dy/dt = -60 \, \text{mi/h}$. (We take the derivatives to be negative since x and y are decreasing.) We are asked to find dz/dt. The equation that relates x, y, and z is given by the Pythagorean Theorem:

$$z^2 = x^2 + y^2$$

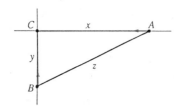

Figure 2.22   Differentiating each side with respect to t, we have

$$2z \frac{dz}{dt} = 2x \frac{dx}{dt} + 2y \frac{dy}{dt}$$

$$\frac{dz}{dt} = \frac{1}{z} \left(x \frac{dx}{dt} + y \frac{dy}{dt} \right)$$

When $x = 0.3$ mi and $y = 0.4$ mi, the Pythagorean Theorem gives $z = 0.5$ mi, so

$$\frac{dz}{dt} = \frac{1}{0.5} \, [0.3(-50) + 0.4(-60)]$$

$$= -78 \, \text{mi/h}$$

The cars are approaching each other at a rate of 78 mi/h.

EXAMPLE 5   A man walks along a straight path with a speed of 4 ft/s. A searchlight is located on the ground 20 ft from the path and is kept focused on the man. At what rate is the searchlight rotating when the man is 15 ft from the point on the path closest to the searchlight?

Solution   We draw Figure 2.23 and let x be the distance from the point on the path closest to the searchlight to the man. We let θ be the angle between the beam of the searchlight and the perpendicular to the path.

We are given that $dx/dt = 4$ ft/s and are asked to find $d\theta/dt$ when $x = 15$. The equation that relates x and θ can be written from Figure 2.23:

$$\frac{x}{20} = \tan\theta \qquad x = 20\tan\theta$$

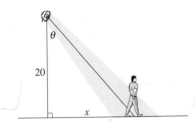

Figure 2.23   Differentiating each side with respect to t, we get

$$\frac{dx}{dt} = 20\sec^2\theta\,\frac{d\theta}{dt}$$

so
$$\frac{d\theta}{dt} = \frac{1}{20}\cos^2\theta\,\frac{dx}{dt} = \frac{1}{20}\cos^2\theta(4) = \frac{1}{5}\cos^2\theta$$

When $x = 15$, the length of the beam is 25, so $\cos\theta = \frac{4}{5}$ and

$$\frac{d\theta}{dt} = \frac{1}{5}\left(\frac{4}{5}\right)^2 = \frac{16}{125} = 0.128$$

The searchlight is rotating at a rate of 0.128 rad/s.

SECTION 2.8 **Exercises**

1. If V is the volume of a cube with edge length x, find dV/dt in terms of dx/dt.

2. If A is the area of a circle with radius r, find dA/dt in terms of dr/dt.

3. If $xy = 1$ and $dx/dt = 4$, find dy/dt when $x = 2$.

4. If $x^2 + 3xy + y^2 = 1$ and $dy/dt = 2$, find dx/dt when $y = 1$.

5. A spherical snowball is melting in such a way that its volume is decreasing at a rate of $1\,\text{cm}^3/\text{min}$. At what rate is the diameter decreasing when the diameter is $10\,\text{cm}$?

6. If a snowball melts so that its surface area decreases at a rate of $1\,\text{cm}^2/\text{min}$, find the rate at which the diameter decreases when the diameter is $10\,\text{cm}$.

7. A street light is at the top of a 15-ft pole. A man 6 ft tall walks away from the pole with a speed of 5 ft/s along a straight path.
    (a) How fast is the tip of his shadow moving when he is 40 ft from the pole?
    (b) How fast is his shadow lengthening at that point?

8. A spotlight on the ground shines on a wall 12 m away. If a man 2 m tall walks from the spotlight toward the building at a speed of 1.6 m/s, how fast is his shadow on the building decreasing when he is 4 m from the building?

9. A plane flying horizontally at an altitude of 1 mi and a speed of 500 mi/h passes directly over a radar station. Find the rate at which the distance from the plane to the station is increasing when it is 2 mi away from the station.

10. A baseball diamond is a square with side 90 ft. A batter hits the ball and runs toward first base with a speed of 24 ft/s.

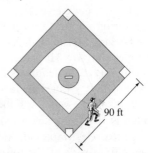

90 ft

(a) At what rate is his distance from second base decreasing when he is halfway to first base?

(b) At what rate is his distance from third base increasing at the same moment?

11. Two cars start from the same point. One travels south at 60 mi/h and the other travels west at 25 mi/h. At what rate is the distance between them increasing two hours later?

12. At noon, ship A is 150 km west of ship B. Ship A is sailing east at 35 km/h and ship B is sailing north at 25 km/h. How fast is the distance between them changing at 4:00 P.M.?

13. At noon, ship A is 100 km west of ship B. Ship A is sailing south at 35 km/h and ship B is sailing north at 25 km/h. How fast is the distance between them changing at 4:00 P.M.?

14. A man starts walking north at 4 ft/s from a point P. Five minutes later a woman starts walking south at 5 ft/s from a point 500 ft due east of P. At what rate are they separating 15 min after the woman starts?

15. The altitude of a triangle is increasing at a rate of 1 cm/min while the area of the triangle is increasing at a rate of 2 cm²/min. At what rate is the base of the triangle changing when the altitude is 10 cm and the area is 100 cm²?

16. A boat is pulled into a dock by a rope attached to the bow of the boat and passing through a pulley on the dock that is 1 m higher than the bow of the boat. If the rope is pulled in at a rate of 1 m/s, how fast is the boat approaching the dock when it is 8 m from the dock?

17. Water is leaking out of an inverted conical tank at a rate of 10,000 cm³/min at the same time that water is being pumped into the tank at a constant rate. The tank has height 6 m and the diameter at the top is 4 m. If the water level is rising at a rate of 20 cm/min when the height of the water is 2 m, find the rate at which water is being pumped into the tank.

18. A trough is 10 ft long and its ends have the shape of isosceles triangles that are 3 ft across at the top and have a height of 1 ft. If the trough is filled with water at a rate of 12 ft³/min, how fast does the water level rise when the water is 6 in. deep?

19. A water trough is 10 m long and a cross-section has the shape of an isosceles trapezoid that is 30 cm wide at the bottom, 80 cm wide at the top, and has height 50 cm. If it is being filled with water at the rate of

0.2 m³/min, how fast is the water level rising when the water is 30 cm deep?

20. A swimming pool is 20 ft wide, 40 ft long, 3 ft deep at the shallow end, and 9 ft deep at its deepest point. A cross-section is shown in the figure. If the pool is being filled at a rate of 0.8 ft³/min, how fast is the water level rising when the depth at the deepest point is 5 ft?

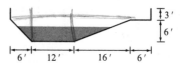

21. Gravel is being dumped from a conveyor belt at a rate of 30 ft³/min and its coarseness is such that it forms a pile in the shape of a cone whose base diameter and height are always equal. How fast is the height of the pile increasing when the pile is 10 ft high?

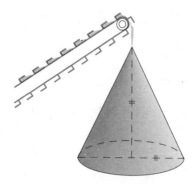

22. A kite 100 ft above the ground moves horizontally at a speed of 8 ft/s. At what rate is the angle between the string and the horizontal decreasing when 200 ft of string have been let out?

23. Two sides of a triangle are 4 m and 5 m in length and the angle between them is increasing at a rate of 0.06 rad/s. Find the rate at which the area of the triangle is increasing when the angle between the sides of fixed length is $\pi/3$.

24. Two sides of a triangle have lengths 12 m and 15 m. The angle between them is increasing at a rate of 2°/min. How fast is the length of the third side increasing when the angle between the sides of fixed length is 60°?

25. Boyle's Law states that when a sample of gas is compressed at a constant temperature, the pressure P and volume V satisfy the equation $PV = C$, where C is a constant. Suppose that at a certain instant the volume is 600 cm³, the pressure is 150 kPa, and the pressure is increasing at a rate of 20 kPa/min. At what rate is the volume decreasing at this instant?

26. When air expands adiabatically (without gaining or losing heat), its pressure P and volume V are related by the equation $PV^{1.4} = C$, where C is a constant. Suppose that at a certain instant the volume is $400 \, \text{cm}^3$ and the pressure is $80 \, \text{kPa}$ and is decreasing at a rate of $10 \, \text{kPa/min}$. At what rate is the volume increasing at this instant?

27. A plane flying with a constant speed of $300 \, \text{km/h}$ passes over a ground radar station at an altitude of $1 \, \text{km}$ and climbs at an angle of $30°$. At what rate is the distance from the plane to the radar station increasing $1 \, \text{min}$ later?

28. Two people start from the same point. One walks east at $3 \, \text{mi/h}$ and the other walks northeast at $2 \, \text{mi/h}$. How fast is the distance between them changing after $15 \, \text{min}$?

29. A ladder $10 \, \text{ft}$ long rests against a vertical wall. If the bottom of the ladder slides away from the wall at a speed of $2 \, \text{ft/s}$, how fast is the angle between the top of the ladder and the wall changing when the angle is $\pi/4$ radians?

30. A lighthouse is on a small island $3 \, \text{km}$ away from the nearest point P on a straight shoreline and its light makes four revolutions per minute. How fast is the beam of light moving along the shoreline when it is $1 \, \text{km}$ from P?

SECTION 2.9

Differentials and Linear Approximations

We have used the Leibniz notation dy/dx to denote the derivative of y with respect to x, but we have regarded it as a single entity and not a ratio. In this section we give the quantities dy and dx separate meanings in such a way that their ratio is equal to the derivative. We also see that these quantities, called differentials, are useful in finding approximate values of functions.

Definition (2.46)

> Let $y = f(x)$, where f is a differentiable function. Then the **differential** dx is an independent variable; that is, dx can be given the value of any real number. The **differential** dy is then defined in terms of dx by the equation
>
> $$dy = f'(x) \, dx$$

Note 1: The differentials dx and dy are both variables, but dx is an independent variable whereas dy is a dependent variable—it depends on the values of x and dx. If dx is given a specific value and x is taken to be some specific number in the domain of f, then the numerical value of dy is determined.

Note 2: If $dx \neq 0$, we can divide both sides of the equation in Definition 2.46 by dx to obtain

$$\frac{dy}{dx} = f'(x)$$

We have seen similar equations before, but now the left side can genuinely be interpreted as a ratio of differentials.

EXAMPLE 1
(a) Find dy if $y = x^3 + 2x^2$.
(b) Find the value of dy when $x = 2$ and $dx = 0.1$.

Solution

(a) If $f(x) = x^3 + 2x^2$, then $f'(x) = 3x^2 + 4x$, so

$$dy = (3x^2 + 4x) \, dx$$

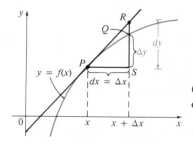

Figure 2.24

(b)   Substituting $x = 2$ and $dx = 0.1$ in the expression for dy, we have

$$dy = (3 \cdot 2^2 + 4 \cdot 2)0.1 = 2$$

The geometric meaning of differentials is shown in Figure 2.24. Let $P(x, f(x))$ and $Q(x + \Delta x, f(x + \Delta x))$ be points on the graph of f and set $dx = \Delta x$. The corresponding change in y is

$$|QS| = \Delta y = f(x + \Delta x) - f(x)$$

The slope of the tangent line PR is the derivative $f'(x)$. But from triangle PRS you can see that the slope of the tangent line can also be written as $|RS|/|PS|$. Thus

$$|RS| = f'(x)|PS| = f'(x)\,dx = dy$$

Therefore dy represents the amount that the tangent line rises or falls, whereas Δy represents the amount that the curve $y = f(x)$ rises or falls when x changes by an amount dx.

Since

$$\frac{dy}{dx} = \lim_{\Delta x \to 0} \frac{\Delta y}{\Delta x}$$

we have

(2.47) $$\frac{\Delta y}{\Delta x} \approx \frac{dy}{dx}$$

when Δx is small. (Geometrically, this says that the slope of the secant line PQ is very close to the slope of the tangent line at P when Δx is small.) If we take $dx = \Delta x$, then (2.47) becomes

(2.48) $$\Delta y \approx dy$$

which says that if Δx is small, then the actual change in y is approximately equal to the differential dy. (Again this is geometrically evident in the case illustrated by Figure 2.24.)

The approximation given by (2.48) can be used in computing approximate values of functions. Suppose that $f(x_1)$ is a known number and it is desired to calculate an approximate value for $f(x_1 + \Delta x)$ where Δx is small. Since

$$f(x_1 + \Delta x) = f(x_1) + \Delta y$$

(2.48) gives

(2.49) $$f(x_1 + \Delta x) \approx f(x_1) + dy$$

EXAMPLE 2   Compare the values of Δy and dy if $y = f(x) = x^3 + x^2 - 2x + 1$ and x changes (a) from 2 to 2.05 and (b) from 2 to 2.01.

Solution
(a)   We have

$$f(2) = 2^3 + 2^2 - 2(2) + 1 = 9$$

$$f(2.05) = (2.05)^3 + (2.05)^2 - 2(2.05) + 1 = 9.717625$$

$$\Delta y = f(2.05) - f(2) = 0.717625$$

In general, $$dy = f'(x)\,dx = (3x^2 + 2x - 2)\,dx$$

When $x = 2$ and $dx = \Delta x = 0.05$, this becomes

$$dy = [3(2)^2 + 2(2) - 2]0.05 = 0.7$$

(b) $$f(2.01) = (2.01)^3 + (2.01)^2 - 2(2.01) + 1 = 9.140701$$

$$\Delta y = f(2.01) - f(2) = 0.140701$$

When $dx = \Delta x = 0.01$,

$$dy = [3(2)^2 + 2(2) - 2]0.01 = 0.14$$

    ●

Notice that the approximation $\Delta y \approx dy$ becomes better as Δx becomes smaller in Example 2. Notice also that dy was easier to compute than Δy. For more complicated functions it may be impossible to compute Δy exactly. In such cases the approximation by differentials is especially useful.

EXAMPLE 3   Use differentials to find an approximate value for $\sqrt[3]{65}$.

Solution   Let $y = f(x) = \sqrt[3]{x} = x^{1/3}$. Then

$$dy = \tfrac{1}{3}x^{-2/3}dx$$

Since $f(64) = 4$, we take $x_1 = 64$ and $dx = \Delta x = 1$. This gives

$$dy = \tfrac{1}{3}(64)^{-2/3}(1) = \frac{1}{3 \cdot 16} = \frac{1}{48}$$

Therefore (2.49) gives

$$\sqrt[3]{65} = f(64 + 1) \approx f(64) + dy$$

$$= 4 + \tfrac{1}{48} \approx 4.021$$

    ●

Note: The actual value of $\sqrt[3]{65}$ is 4.0207257.... Thus the approximation by differentials in Example 3 is accurate to three decimal places even when $\Delta x = 1$.

EXAMPLE 4   The radius of a sphere was measured and found to be 21 cm with a possible error in measurement of at most 0.05 cm. What is the maximum error in using this value of the radius to compute the volume?

Solution   If the radius of the sphere is r, then its volume is $V = \tfrac{4}{3}\pi r^3$. If the error in the measured value of r is denoted by $dr = \Delta r$, then the corresponding error in the calculated value of V is ΔV, which can be approximated by the differential

$$dV = 4\pi r^2 dr$$

When $r = 21$ and $dr = 0.05$, this becomes

$$dV = 4\pi(21)^2 0.05 \approx 277$$

The maximum error in the calculated volume is about 277 cm^3.   ●

Note: Although the possible error in Example 4 may appear to be rather large, a better picture of the error is given by the **relative error,** which is computed by dividing the error by the total volume:

$$\frac{\Delta V}{V} \approx \frac{dV}{V} \approx \frac{277}{38,792} \approx 0.00714$$

Thus a relative error of $dr/r = 0.05/21 \approx 0.0024$ in the radius produces a relative error of about 0.007 in the volume. The errors could also be expressed as **percentage errors** of 0.24% in the radius and 0.7% in the volume.

Linear Approximations

The equation of the tangent line to the curve $y = f(x)$ at $(x_1, f(x_1))$ is

$$y = f(x_1) + f'(x_1)(x - x_1)$$

and the right side of this equation is just $f(x_1) + f'(x_1)\,dx$, or $f(x_1) + dy$, so when we use the approximation in (2.49) we are, in effect, using the tangent line at $P(x_1, f(x_1))$ as an approximation to the curve $y = f(x)$ when x is near x_1. For this reason, the approximation

(2.50) $$f(x) \approx f(x_1) + f'(x_1)(x - x_1)$$

is called the **linear approximation** or **tangent line approximation** of f at x_1, and the function

(2.51) $$L(x) = f(x_1) + f'(x_1)(x - x_1)$$

(whose graph is the tangent line) is called the **linearization** of f at x_1.

EXAMPLE 5   Find the linearization of the function $f(x) = \sqrt{x + 3}$ at $x_1 = 1$ and use it to approximate $\sqrt{3.98}$ and $\sqrt{4.05}$.

Solution   The derivative of $f(x) = (x + 3)^{1/2}$ is

$$f'(x) = \tfrac{1}{2}(x + 3)^{-1/2} = \frac{1}{2\sqrt{x + 3}}$$

and so we have $f(1) = 2$ and $f'(1) = 1/4$. Putting these values into Equation 2.51, we see that the linearization is

$$L(x) = f(1) + f'(1)(x - 1) = 2 + \tfrac{1}{4}(x - 1) = \frac{7}{4} + \frac{x}{4}$$

The corresponding linear approximation (2.50) is

$$\sqrt{x + 3} \approx \frac{7}{4} + \frac{x}{4}$$

In particular, we have:

$$\sqrt{3.98} \approx \frac{7}{4} + \frac{0.98}{4} = 1.995$$

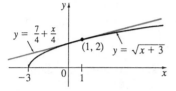

and

$$\sqrt{4.05} \approx \frac{7}{4} + \frac{1.05}{4} = 2.0125$$

The linear approximation in Example 5 is illustrated in Figure 2.25. You can see that, indeed, the tangent line approximation is a good approximation to the given function when x is near 1. Of course, a calculator could give us approximations for $\sqrt{3.98}$ and $\sqrt{4.05}$, but the linear approximation gives an approximation over an entire interval.

Figure 2.25

SECTION 2.9 **Exercises**

In Exercises 1–8 find the differential of the given function.

1. $y = x^5$

2. $y = \sqrt[4]{x}$

3. $y = \sqrt{x^4 + x^2 + 1}$

4. $y = (x^2 - 2x - 3)^{10}$

5. $y = \dfrac{x - 2}{2x + 3}$

6. $y = \sqrt{x + \sqrt{2x - 1}}$

7. $y = \sin 2x$

8. $y = x \tan x$

In Exercises 9–16, (a) find the differential dy and (b) evaluate dy for the given values of x and dx.

9. $y = 1 - x^2$, $x = 5$, $dx = \frac{1}{2}$

10. $y = x^4 - 3x^3 + x - 1$, $x = 2$, $dx = 0.1$

11. $y = (x^2 + 5)^3$, $x = 1$, $dx = 0.05$

12. $y = \sqrt{1 - x}$, $x = 0$, $dx = 0.02$

13. $y = \dfrac{1}{\sqrt[3]{3x + 2}}$, $x = 2$, $dx = -0.04$

14. $y = \dfrac{x - 1}{x^2 + 1}$, $x = 1$, $dx = -0.3$

15. $y = \cos x$, $x = \frac{\pi}{6}$, $dx = 0.05$

16. $y = \sin x$, $x = \frac{\pi}{6}$, $dx = -0.1$

In Exercises 17–20 compute Δy and dy for the given values of x and $dx = \Delta x$. Then sketch a diagram like Figure 2.24 showing the line segments with lengths dx, dy, and Δy.

17. $y = x^2$, $x = 1$, $\Delta x = 0.5$

18. $y = \sqrt{x}$, $x = 1$, $\Delta x = 1$

19. $y = 6 - x^2$, $x = -2$, $\Delta x = 0.4$

20. $y = 16/x$, $x = 4$, $\Delta x = -1$

In Exercises 21 and 22 compute Δy, dy, and $\Delta y - dy$ for the given values of x and for each of the following values of $dx = \Delta x$: 1, 0.5, 0.1, and 0.01.

21. $y = 2x^2 + 3x - 4$, $x = 3$

22. $y = x^4 + x^2 + 1$, $x = 1$

In Exercises 23–32 use differentials to find an approximate value for the given number.

23. $\sqrt{36.1}$

24. $\sqrt{99}$

25. $\sqrt[3]{218}$

26. $\sqrt[3]{1.02} + \sqrt[4]{1.02}$

27. $\dfrac{1}{10.1}$

28. $(1.97)^6$

29. $\sin 59°$

30. $\cos 31.5°$

31. $\tan 47°$

32. $\sec 62°$

In Exercises 33–36 find the linearization $L(x)$ of the given function at x_1.

33. $f(x) = x^3$, $x_1 = 1$

34. $f(x) = 1/\sqrt{2 + x}$, $x_1 = 0$

35. $f(x) = 1/x$, $x_1 = 4$

36. $f(x) = \sqrt[3]{x}$, $x_1 = -8$

In Exercises 37–40, verify the given linear approximation at $x_1 = 0$.

37. $\sqrt{1 + x} \approx 1 + \frac{1}{2}x$

38. $\sin x \approx x$

39. $\dfrac{1}{(1 + 2x)^4} \approx 1 - 8x$

40. $\dfrac{1}{\sqrt{4 - x}} \approx \frac{1}{2} + \frac{1}{16}x$

41. Find the linear approximation of the function $f(x) = \sqrt{1 - x}$ at $x_1 = 0$ and use it to approximate the numbers $\sqrt{0.9}$ and $\sqrt{0.99}$. Illustrate with a sketch like Figure 2.25.

42. Find the linear approximation of the function $g(x) = \sqrt[3]{1+x}$ at $x_1 = 0$ and use it to approximate the numbers $\sqrt[3]{0.95}$ and $\sqrt[3]{1.1}$. Illustrate with a sketch like Figure 2.25.

43. The edge of a cube was found to be 30 cm with a possible error in measurement of 0.1 cm. Use differentials to estimate the maximum possible error in computing (a) the volume of the cube and (b) the surface area of the cube.

44. The radius of a circular disk is given as 24 cm with a maximum error in measurement of 0.2 cm.
   (a) Use differentials to estimate the maximum error in the calculated area of the disk.
   (b) What is the relative error?

45. The circumference of a sphere was measured to be 84 cm with a possible error of 0.5 cm.
   (a) Use differentials to estimate the maximum error in the calculated surface area.
   (b) What is the relative error?

46. Do Exercise 45 with "surface area" replaced by "volume."

47. (a) Use differentials to find a formula for the approximate volume of a thin cylindrical shell with height h, inner radius r, and thickness Δr.
   (b) What is the error involved in using this formula?

48. Use differentials to estimate the amount of paint needed to apply a coat of paint 0.05 cm thick to a hemispherical dome with diameter 50 m.

49. Establish the following rules for working with differentials (where c denotes a constant and u and v are functions of x).
   (a) $dc = 0$ (b) $d(cu) = c\,du$
   (c) $d(u+v) = du + dv$
   (d) $d(uv) = u\,dv + v\,du$
   (e) $d\!\left(\dfrac{u}{v}\right) = \dfrac{v\,du - u\,dv}{v^2}$
   (f) $d(x^n) = nx^{n-1}\,dx$

SECTION 2.10

Newton's Method

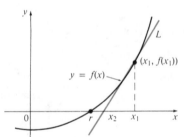

Figure 2.26

Many problems in science, engineering, and mathematics lead to the problem of finding the roots of an equation of the form $f(x) = 0$ where f is a differentiable function. For a quadratic equation $ax^2 + bx + c = 0$ there is a well-known formula for the roots. For third- and fourth-degree equations there are also formulas for the roots but they are extremely complicated. If f is a polynomial of degree 5 or higher, there is no such formula (see the note after Example 2 in Section 2.6). Likewise there is no formula that will enable us to find the exact roots of a transcendental equation such as $\cos x = x$. However there are methods that give *approximations* to the roots of such equations.

One such method is called **Newton's method** or the **Newton–Raphson method**. The idea behind it is shown in Figure 2.26 where the root that we are trying to find is labeled r. We start with a first approximation x_1, which is obtained by guessing or from a rough sketch of the graph of f. Consider the tangent line L to the curve $y = f(x)$ at the point $(x_1, f(x_1))$ and look at the x-intercept of L, labeled x_2. If x_1 is close to r, it appears that x_2 is even closer to r and we use it as the second approximation to r. To find a formula for x_2 in terms of x_1 we use the fact that the slope of L is $f'(x_1)$, so its equation is

$$y - f(x_1) = f'(x_1)(x - x_1)$$

Since the x-intercept of L is x_2, we set $y = 0$ and obtain

$$0 - f(x_1) = f'(x_1)(x_2 - x_1)$$

If $f'(x_1) \neq 0$, we can solve this equation for x_2:

$$x_2 = x_1 - \frac{f(x_1)}{f'(x_1)}$$

Next we repeat this procedure with x_1 replaced by x_2, using the tangent line at $(x_2, f(x_2))$. This gives a third approximation:

$$x_3 = x_2 - \frac{f(x_2)}{f'(x_2)}$$

If we keep repeating this process we obtain a sequence of approximations x_1, x_2, x_3, x_4, ... as shown in Figure 2.27. In general, if the nth approximation is x_n and $f'(x_n) \neq 0$, then the next approximation is given by

(2.52)

$$x_{n+1} = x_n - \frac{f(x_n)}{f'(x_n)}$$

If the numbers x_n become closer and closer to r as n becomes large, then we say that the sequence converges to r and we write

$$\lim_{n \to \infty} x_n = r$$

(See Section 10.1 for a discussion of general sequences.) Although the sequence of successive approximations converges to the desired root for functions of the type illustrated in Figure 2.27, in certain circumstances the sequence may not converge. For example, consider the situation shown in Figure 2.28. You can see that x_2 is a worse approximation than x_1. This is likely to be the case when $f'(x_1)$ is close to 0. It might even happen that an approximation (such as x_3 in Figure 2.28) falls outside the domain of f. Then Newton's method breaks down and a better initial approximation x_1 should be chosen. See Exercises 21–24 for specific examples in which Newton's method works very slowly or does not work at all.

Figure 2.27

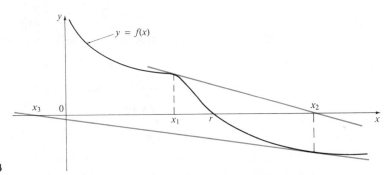

Figure 2.28

EXAMPLE 1   Starting with $x_1 = 2$, find the third approximation x_3 to the root of the equation $x^3 - 2x - 5 = 0$.

Solution   We apply Newton's method with

$$f(x) = x^3 - 2x - 5 \qquad f'(x) = 3x^2 - 2$$

Newton himself used this equation to illustrate his method and he chose $x_1 = 2$ after some experimentation because $f(1) = -6$, $f(2) = -1$, and $f(3) = 16$. Equation 2.52 becomes

$$x_{n+1} = x_n - \frac{x_n^3 - 2x_n - 5}{3x_n^2 - 2}$$

With $n = 1$ we have

$$x_2 = x_1 - \frac{x_1^3 - 2x_1 - 5}{3x_1^2 - 2}$$

$$= 2 - \frac{2^3 - 2(2) - 5}{3(2)^2 - 2} = 2.1$$

Then with $n = 2$ we obtain $\qquad x_3 = x_2 - \dfrac{x_2^3 - 2x_2 - 5}{3x_2^2 - 2}$

$$= 2.1 - \frac{(2.1)^3 - 2(2.1) - 5}{3(2.1)^2 - 2} \approx 2.0946$$

It turns out that this third approximation $x_3 \approx 2.0946$ is accurate to four decimal places. ●

Suppose that we want to achieve a given accuracy, say to eight decimal places, using Newton's method. How do we know when to stop? The rule of thumb that is generally used is that we can stop when successive approximations x_n and x_{n+1} agree to eight decimal places. (A precise statement concerning accuracy in Newton's method will be given as Exercise 51 in Section 10.11.)

Notice that the procedure in going from n to $n+1$ is the same for all values of n. (It is called an *iterative* process.) This means that Newton's method is particularly convenient for use with a programmable calculator or a computer.

EXAMPLE 2   Use Newton's method to find $\sqrt[6]{2}$ correct to eight decimal places.

Solution   First we observe that finding $\sqrt[6]{2}$ is equivalent to finding the positive root of the equation

$$x^6 - 2 = 0$$

so we take $f(x) = x^6 - 2$. Then $f'(x) = 6x^5$ and Formula 2.52 (Newton's method) becomes

$$x_{n+1} = x_n - \frac{x_n^6 - 2}{6x_n^5}$$

If we choose $x_1 = 1$ as the initial approximation, then we obtain

$$x_2 \approx 1.166666667 \qquad x_3 \approx 1.126443678$$

$$x_4 \approx 1.122497067 \qquad x_5 \approx 1.122462051$$

$$x_6 \approx 1.122462048$$

Since x_5 and x_6 agree to eight decimal places, we conclude that

$$\sqrt[6]{2} \approx 1.12246205$$

to eight decimal places. ●

EXAMPLE 3   Find, correct to six decimal places, the root of the equation $\cos x = x$.

Solution   We first rewrite the equation in standard form:

$$\cos x - x = 0$$

Therefore we let $f(x) = \cos x - x$. Then $f'(x) = -\sin x - 1$, so Formula 2.52 becomes

$$x_{n+1} = x_n - \frac{\cos x_n - x_n}{-\sin x_n - 1} = x_n + \frac{\cos x_n - x_n}{\sin x_n + 1}$$

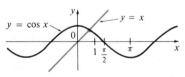

Figure 2.29

In order to guess a suitable value for x_1 we sketch the graphs of $y = \cos x$ and $y = x$ in Figure 2.29. It appears that they intersect at a point whose x-coordinate is somewhat less than 1, so let us take $x_1 = 1$ as a convenient first approximation. Then

$$x_2 \approx 0.75036387 \qquad x_3 \approx 0.73911289$$

$$x_4 \approx 0.73908513 \qquad x_5 \approx 0.73908513$$

Since x_4 and x_5 agree to six decimal places (eight, in fact), we conclude that the root of the equation, correct to six decimal places, is 0.739085. ●

▦ SECTION 2.10 **Exercises**

In Exercises 1–4 use Newton's method with the given initial approximation x_1 to find x_3, the third approximation to the root of the given equation. (Give your answer to four decimal places.)

1. $x^3 + x + 1 = 0$, $x_1 = -1$

2. $x^3 + x^2 + 2 = 0$, $x_1 = -2$

3. $x^5 - 10 = 0$, $x_1 = 1.5$

4. $x^7 - 100 = 0$, $x_1 = 2$

In Exercises 5 and 6 use Newton's method to approximate the given number correct to eight decimal places.

5. $\sqrt[4]{22}$ 6. $\sqrt[10]{100}$

In Exercises 7–12 use Newton's method to approximate the indicated root of the equation correct to six decimal places.

7. The root of $x^3 - 2x - 1 = 0$ in the interval $[1, 2]$

8. The root of $x^3 + x^2 + x - 2 = 0$ in the interval $[0, 1]$

9. The root of $x^4 + x^3 - 22x^2 - 2x + 41 = 0$ in the interval $[3, 4]$

10. The root of $x^4 + x^3 - 22x^2 - 2x + 41 = 0$ in the interval $[1, 2]$

11. The positive root of $2 \sin x = x$

12. The root of $\tan x = x$ in the interval $(\pi/2, 3\pi/2)$

In Exercises 13–18 use Newton's method to find all roots of the given equation correct to six decimal places.

13. $x^3 = 4x - 1$ 14. $x^5 = 5x - 2$

15. $x^4 = 1 + x - x^2$ 16. $(x - 2)^4 = \frac{x}{2}$

17. $2 \cos x = 2 - x$ 18. $\sin \pi x = x$

19. (a) Apply Newton's method to the equation $x^2 - a = 0$ to derive the following square-root

algorithm (used by the ancient Babylonians to compute $\sqrt{a}$):

$$x_{n+1} = \tfrac{1}{2}\left(x_n + \tfrac{a}{x_n}\right)$$

(b) Use part (a) to compute $\sqrt{1000}$ correct to six decimal places.

20. (a) Apply Newton's method to the equation $1/x - a = 0$ to derive the following reciprocal algorithm:

$$x_{n+1} = 2x_n - ax_n^2$$

(This algorithm enables a computer to find reciprocals without actually dividing.)

(b) Use part (a) to compute $1/1.6984$ correct to six decimal places.

21. Explain why Newton's method does not work for finding the root of the equation $x^3 - 3x + 6 = 0$ if the initial approximation is chosen to be $x_1 = 1$.

22. (a) Use Newton's method with $x_1 = 1$ to find the root of the equation $x^3 - x = 1$ correct to six decimal places.

(b) Solve the equation in part (a) using $x_1 = 0.6$ as the initial approximation.

(c) Solve the equation in part (a) using $x_1 = 0.57$. Sketch the graph of $f(x) = x^3 - x - 1$ to show why x_2 is such a poor approximation.

23. Show that Newton's method fails when applied to the equation $\sqrt[3]{x} = 0$ with any initial approximation $x_1 \neq 0$.

24. If
$$f(x) = \begin{cases} \sqrt{x} & \text{if } x \geq 0 \\ -\sqrt{-x} & \text{if } x < 0 \end{cases}$$

then the root of the equation $f(x) = 0$ is $x = 0$. Explain why Newton's method fails to find the root no matter which initial approximation $x \neq 0$ is used. Illustrate your explanation with a sketch.

CHAPTER 2

≡≡≡≡≡≡   **Review**

Key Topics

Define, state, or discuss the following.

1. Derivative of a function
2. Interpretations of the derivative
3. Equation of the tangent line to $y = f(x)$ at $(a, f(a))$
4. Differentiable function

5. Relation between differentiability and continuity
6. Power Rule
7. Product Rule
8. Quotient Rule

9. Derivatives of trigonometric functions
10. Chain Rule
11. Implicit differentiation
12. Orthogonal curves
13. Second derivative; higher derivatives
14. Position function, velocity, acceleration

15. Average and instantaneous rate of change
16. Related rates
17. Differentials
18. Linear approximation
19. Linearization of f at x_1
20. Newton's method

Exercises

In Exercises 1–10 determine whether the statement is true or false.

1. If f is continuous at a, then f is differentiable at a.

2. If f and g are differentiable, then
$$\frac{d}{dx}[f(x) + g(x)] = f'(x) + g'(x)$$

3. If f and g are differentiable, then
$$\frac{d}{dx}[f(x)g(x)] = f'(x)g'(x)$$

4. If f and g are differentiable, then
$$\frac{d}{dx}[f(g(x))] = f'(g(x))g'(x)$$

5. If f is differentiable, then $\dfrac{d}{dx}\sqrt{f(x)} = \dfrac{f'(x)}{2\sqrt{f(x)}}$.

6. If f is differentiable, then $\dfrac{d}{dx}f(\sqrt{x}) = \dfrac{f'(x)}{2\sqrt{x}}$.

7. $\dfrac{d}{dx}|x^2 + x| = |2x + 1|$

8. If $f'(r)$ exists, then $\lim\limits_{x \to r} f(x) = f(r)$.

9. If $g(x) = x^5$, then $\lim\limits_{x \to 2} \dfrac{g(x) - g(2)}{x - 2} = 80$.

10. $\dfrac{d^2 y}{dx^2} = \left(\dfrac{dy}{dx}\right)^2$

In Exercises 11–14 find $f'(x)$ from first principles, that is, directly from the definition of a derivative.

11. $f(x) = x^3 + 5x + 4$ 12. $f(x) = \dfrac{4 - x}{3 + x}$

13. $f(x) = \sqrt{3 - 5x}$ 14. $f(x) = x \sin x$

In Exercises 15–42 calculate y'.

15. $y = (x + 2)^8 (x + 3)^6$ 16. $y = \sqrt[3]{x} + \dfrac{1}{\sqrt[3]{x}}$

17. $y = \dfrac{x}{\sqrt{9 - 4x}}$ 18. $y = \left(x + \dfrac{1}{x^2}\right)^{\sqrt{7}}$

19. $x^2 y^3 + 3y^2 = x - 4y$ 20. $y = (1 - x^{-1})^{-1}$

21. $y = \sqrt{x\sqrt{x\sqrt{x}}}$ 22. $y = -2/\sqrt[4]{x^3}$

23. $y = x/(8 - 3x)$ 24. $y\sqrt{x - 1} + x\sqrt{y - 1} = xy$

25. $y = \sqrt[5]{x \tan x}$ 26. $y = \sin(\cos x)$

27. $x^2 = y(y + 1)$ 28. $y = 1/\sqrt[3]{x + \sqrt{x}}$

29. $y = \dfrac{(x - 1)(x - 4)}{(x - 2)(x - 3)}$ 30. $y = \sqrt{\sin\sqrt{x}}$

31. $y = \tan\sqrt{1 - x}$ 32. $y = \dfrac{1}{\sin(x - \sin x)}$

33. $y = \sin\left(\tan\sqrt{1 + x^3}\right)$ 34. $y = \dfrac{(x + \lambda)^4}{x^4 + \lambda^4}$

35. $y = \cot(3x^2 + 5)$ 36. $y = \dfrac{\cos^2 x}{1 + \tan x} + \dfrac{\sin^2 x}{1 + \cot x}$

37. $y = \dfrac{\sin mx}{x}$ 38. $y = \sqrt[10]{4 + 7\sqrt[8]{x^5}}$

39. $y = \cos^2(\tan x)$ 40. $x \tan y = y - 1$

41. $y = \sqrt{7 - x^2}(x^3 + 7)^5$ 42. $y = \sqrt{x}\sec\sqrt{x}$

43. If $f(x) = \dfrac{1}{(2x - 1)^5}$, find $f''(0)$.

44. If $g(t) = \csc 2t$, find $g'''(-\pi/8)$.

45. Find y'' if $x^6 + y^6 = 1$.

46. Find $f^{(n)}(x)$ if $f(x) = 1/(2 - x)$.

In Exercises 47–50 find the equation of the tangent to the given curve at the given point.

47. $y = \dfrac{x}{x^2 - 2}$, $(2, 1)$ 48. $\sqrt{x} + \sqrt{y} = 3$, $(4, 1)$

49. $y = \tan x$, $(\pi/3, \sqrt{3})$ 50. $y = x\sqrt{1 + x^2}$, $(1, \sqrt{2})$

51. At what points on the curve $y = \sin x + \cos x$, $0 \le x \le 2\pi$, is the tangent line horizontal?

52. Find the points on the ellipse $x^2 + 2y^2 = 1$ where the tangent line has slope 1.

53. A particle moves on a vertical line so that its coordinate at time t is $y = t^3 - 12t + 3$, $t \ge 0$.
    (a) Find the velocity and acceleration functions.
    (b) When is the particle moving upward and when is it moving downward?
    (c) Find the distance that it travels in the time interval $0 \le t \le 3$.

54. A particle moves along a horizontal line so that its coordinate at time t is $x = \sqrt{b^2 + c^2 t^2}$, $t \ge 0$, where b and c are positive constants.
    (a) Find the velocity and acceleration functions.
    (b) Show that the particle always moves in the positive direction.

55. If $f(x) = (x - a)(x - b)(x - c)$, show that
$$\frac{f'(x)}{f(x)} = \frac{1}{x - a} + \frac{1}{x - b} + \frac{1}{x - c}$$

56. (a) By differentiating the double-angle formula $\cos 2x = \cos^2 x - \sin^2 x$, obtain the double-angle formula for the sine function.

(b) By differentiating the addition formula $\sin(x + a) = \sin x \cos a + \cos x \sin a$, obtain the addition formula for the cosine function.

57. Suppose that $h(x) = f(x)g(x)$ and $F(x) = f(g(x))$, where $f(2) = 3$, $g(2) = 5$, $g'(2) = 4$, $f'(2) = -2$, and $f'(5) = 11$. Find (a) $h'(2)$ and (b) $F'(2)$.

58. The graph of a function f is given here. Trace or copy it and use slopes of tangents to sketch the graph of f' beneath it.

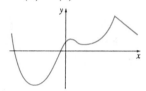

In Exercises 59–64 find f' in terms of g'.

59. $f(x) = x^2 g(x)$ **60.** $f(x) = g(x^2)$

61. $f(x) = [g(x)]^2$ **62.** $f(x) = x^a g(x^b)$

63. $f(x) = g(g(x))$ **64.** $f(x) = g(\tan \sqrt{x})$

In Exercises 65–67 find h' in terms of f' and g'.

65. $h(x) = \dfrac{f(x)g(x)}{f(x) + g(x)}$ **66.** $h(x) = \sqrt{\dfrac{f(x)}{g(x)}}$

67. $h(x) = f(g(\sin 4x))$

68. The volume of a right circular cone is $V = \pi r^2 h/3$, where r is the radius of the base and h is the height.
(a) Find the rate of change of the volume with respect to the height if the radius is constant.
(b) Find the rate of change of the volume with respect to the radius if the height is constant.

69. The mass of part of a wire is $x(1 + \sqrt{x})$ kilograms, where x is measured in meters from one end of the wire. Find the linear density of the wire when $x = 4$ m.

70. The cost, in dollars, of manufacturing x units of a given item is $950 + 12x + 0.01x^2$. Find the marginal cost at a production level of 200 units. Compare this with the cost of producing the 201st unit.

71. The volume of a cube is increasing at a rate of $10 \text{ cm}^3/\text{min}$. How fast is the surface area increasing when the length of an edge is 30 cm?

72. A paper cup has shape of a cone with height 10 cm and radius 3 cm (at the top). If water is poured into the cup at a rate of $2 \text{ cm}^3/\text{s}$, how fast is the water level rising when the water is 5 cm deep?

73. A balloon is rising at a constant speed of 5 ft/s. A boy is cycling along a straight road at a speed of 15 ft/s. When he passes under the balloon it is 45 ft above him. How fast is the distance between the boy and the balloon increasing 3 s later?

74. A waterskier skis over the ramp shown in the figure at a speed of 30 ft/s. How fast is she rising as she leaves the ramp?

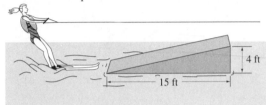

75. The angle of elevation of the sun is decreasing at a rate of 0.25 rad/h. How fast is the shadow cast by a 400-ft-tall building increasing when the angle of elevation of the sun is $\pi/6$?

76. Find dy if $y = (4 - x^2)^{3/2}$.

77. Evaluate dy if $y = x^3 - 2x^2 + 1$, $x = 2$, and $dx = 0.2$.

78. Use differentials to approximate $8 + \sqrt{143.6}$.

79. Find the linearization of $f(x) = \sqrt[3]{1 + 3x}$ at $x_1 = 0$. State the corresponding linear approximation and use it to give an approximate value for $\sqrt[3]{1.03}$.

80. A window has the shape of a square surmounted by a semicircle. The base of the window is measured as having width 60 cm with a possible error in measurement of 0.1 cm. Use differentials to estimate the maximum possible error in computing the area of the window.

81. Use Newton's method to find the root of the equation $x^4 + x - 1 = 0$ in the interval $[0, 1]$ correct to six decimal places.

82. Use Newton's method to find all roots of the equation $6 \cos x = x$ correct to six decimal places.

83. Find the coordinates (correct to four decimal places) of the point on the curve $y = x^6 + 2x^2 - 8x + 3$ where the tangent is horizontal.

In Exercises 84–86 express the limit as a derivative and evaluate.

84. $\displaystyle \lim_{x \to 1} \frac{x^{17} - 1}{x - 1}$ **85.** $\displaystyle \lim_{h \to 0} \frac{(2 + h)^6 - 64}{h}$

86. $\displaystyle \lim_{\theta \to \pi/3} \frac{\cos \theta - 0.5}{\theta - \pi/3}$

87. Show that the triangle formed by any tangent line to the hyperbola $xy = 1$ and the coordinate axes has area 2.

88. Show that the length of the portion of any tangent line to the astroid $x^{2/3} + y^{2/3} = a^{2/3}$ cut off by the coordinate axes is constant.

Problems Plus

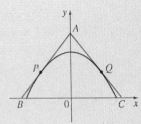

1. Find points P and Q on the parabola $y = 1 - x^2$ so that the triangle ABC formed by the x-axis and the tangent lines at P and Q is an equilateral triangle.

2. (a) Find the sum of the finite geometric series $1 + x + x^2 + \cdots + x^n$.
   (b) Find a formula for the sum $1 + 2x + 3x^2 + \cdots + nx^{n-1}$.

3. Solve the inequality $1 + x + x^2 + \cdots + x^{100} \geq 0$.

4. (a) Solve the inequality $|x + 1| + |x - 2| < 7$.
   (b) Sketch the graph of the function $f(x) = |x + 1| + |x - 2|$.
   (c) Where is f continuous?
   (d) Where is f differentiable?
   (e) Sketch the graph of $g(x) = |x| + |x + 1| + |x - 1|$.
   (f) Where is g differentiable?

5. Sketch the graph of the set $\{(x, y) \mid |x| + |y| \leq 1\}$.

6. Sketch the graph of the set $\{(x, y) \mid |x - y| + |x| - |y| \leq 2\}$.

7. Evaluate $\lim\limits_{x \to 0} \dfrac{|2x - 1| - |2x + 1|}{x}$.

8. If f is differentiable at a, where a > 0, evaluate the following limit in terms of $f'(a)$:

$$\lim_{x \to a} \frac{f(x) - f(a)}{\sqrt{x} - \sqrt{a}}$$

9. Prove that $\dfrac{d^n}{dx^n}\left(\sin^4 x + \cos^4 x\right) = 4^{n-1} \cos(4x + n\pi/2)$.

10. Draw a diagram to show that there are two lines that are tangent to both of the parabolas $y = -1 - x^2$ and $y = 1 + x^2$. Find the coordinates of the points at which these tangents touch the parabolas.

11. (a) Find the domain of the function $f(x) = \sqrt{1 - \sqrt{2 - \sqrt{3 - x}}}$.
    (b) Find $f'(x)$.

12. (a) Sketch the graph of the function $g(x) = |x^2 - 4| - |x^2 - 9|$.
    (b) Find $g'(x)$.

13. (a) Sketch the graph of the function $h(x) = |x^2 - 6|x| + 8|$.
    (b) At what values of x is h not differentiable?

14. If $[\![x]\!]$ denotes the greatest integer function, sketch the regions in the plane defined by the following equations.
    (a) $[\![x]\!]^2 + [\![y]\!]^2 = 1$
    (b) $[\![x]\!]^2 - [\![y]\!]^2 = 3$
    (c) $[\![x + y]\!]^2 = 1$
    (d) $[\![x]\!] + [\![y]\!] = 1$

15. Let P be a point on the curve $y = x^3$ and suppose the tangent line at P intersects the curve again at Q. Prove that the slope at Q is four times the slope at P.

16. Evaluate $\lim\limits_{x \to 0} \dfrac{\sin(3 + x)^2 - \sin 9}{x}$.

17. Suppose that f is differentiable at 0 and

$$\lim_{x \to 0} \frac{f(x)}{x} = 4 \qquad \lim_{x \to 0} \frac{g(x)}{x} = 2$$

(a) Find $f(0)$. (b) Find $f'(0)$. (c) Find $\lim_{x \to 0} \frac{g(x)}{f(x)}$.

18. (a) Use a calculator to draw carefully the graph of the exponential function
    $f(x) = 2^x$.
    (b) Sketch the graph of f' below the graph of f by measuring slopes.
    (c) Write an expression for $f'(0)$ as a limit and use your calculator to estimate the value to two decimal places.
    (d) Show that $f'(x) = f'(0)f(x)$.
    (e) Use part (d) to check that your graph in part (b) is reasonable.

19. Suppose f is a function with the property that $|f(x)| \le x^2$ for all x. Show that f is differentiable at 0.

20. (a) Draw a diagram and interpret the quotient

$$\frac{f(x+h) - f(x-h)}{2h}$$

as the slope of a secant line.

(b) If f is differentiable at x, show that

$$\lim_{h \to 0} \frac{f(x+h) - f(x-h)}{2h} = f'(x)$$

(*Hint:* In the proof of the Product Rule we used the trick of subtracting and adding the same quantity in a numerator. A similar trick works here.)

(c) Show that it is possible for the limit

$$\lim_{h \to 0} \frac{f(x+h) - f(x-h)}{2h}$$

to exist but for $f'(x)$ not to exist. [*Hint:* Consider $f(x) = |x|$.]

21. (a) Use the identity for $\tan(x - y)$ (see Equation 14b in Appendix B) to show that if two lines L_1 and L_2 intersect at an angle α, then

$$\tan \alpha = \frac{m_2 - m_1}{1 + m_1 m_2}$$

where m_1 and m_2 are the slopes of L_1 and L_2, respectively.

(b) The **angle between the curves** C_1 and C_2 at a point of intersection P is defined to be the angle between the tangent lines to C_1 and C_2 at P (if these tangent lines exist). Use part (a) to find, correct to the nearest degree, the angle between each of pair of curves at each point of intersection.
(i) $y = x^2$ and $y = (x - 2)^2$ (ii) $x^2 - y^2 = 3$ and $x^2 - 4x + y^2 + 3 = 0$

22. There are two points on the curve $y = x^4 - 2x^2 - x$ that have a common tangent line. Find those points.

3

Inverse Functions:
Exponential, Logarithmic, and Inverse Trigonometric Functions

Each problem that I solved became
a rule which served afterwards to
solve other problems.

René Descartes

Two of the most important functions that occur in mathematics and its applications are the exponential function $f(x) = a^x$ and its inverse function, the logarithmic function $g(x) = \log_a x$. In this chapter we investigate their properties, compute their derivatives, and use them to describe exponential growth and decay in such sciences as chemistry, physics, biology, and economics. We also study the inverses of trigonometric and hyperbolic functions. Finally, we look at a method (l'Hospital's Rule) for computing difficult limits and apply it to sketching curves.

SECTION 3.1

Exponential Functions

An **exponential function** is a function of the form

$$f(x) = a^x$$

where a is a positive constant. It is defined in five stages:

1. If $x = n$, a positive integer, then

$$a^n = \underbrace{a \cdot a \cdot \cdots \cdot a}_{n \text{ factors}}$$

2. If $x = 0$, $a^0 = 1$.
3. If $x = -n$, n a positive integer, then

$$a^{-n} = \frac{1}{a^n}$$

4. If x is a rational number, $x = p/q$, where p and q are integers and $q > 0$, then

$$a^x = a^{p/q} = \sqrt[q]{a^p}$$

5. If x is an irrational number, we wish to define a^x so as to fill in the holes of the graph of the function $y = a^x$, where x is rational. In other words, we want to make $f(x) = a^x$, $x \in R$, a continuous function. Since any irrational number can be approximated as closely as we like by a rational number (see Section 1 in Review and Preview), we define

(3.1)

$$a^x = \lim_{r \to x} a^r \qquad r \text{ rational}$$

It can be shown that this definition uniquely specifies a^x and makes the function $f(x) = a^x$ continuous. [See Chapter 9 in M.P. Dolciani et al., *Modern Introductory Analysis* (Boston: Houghton Mifflin, 1964).]

To illustrate Equation 3.1 let us take $x = \sqrt{2}$. The number $\sqrt{2}$ is an irrational number that has a nonrepeating decimal representation:

$$\sqrt{2} = 1.414213562373095 \ldots$$

This means that $\sqrt{2}$ is the limit of the sequence of rational numbers

$$1, \quad 1.4, \quad 1.41, \quad 1.414, \quad 1.4142, \quad 1.41421, \quad 1.414213, \quad \ldots$$

According to Equation 3.1, $a^{\sqrt{2}}$ is the limit of the sequence

$$a^1, \quad a^{1.4}, \quad a^{1.41}, \quad a^{1.414}, \quad a^{1.4142}, \quad a^{1.41421}, \quad a^{1.414213}, \quad \ldots$$

and each term of this sequence has been defined in stage 4. In fact, if $a > 1$, $a^{\sqrt{2}}$ is the unique number that satisfies all of the following inequalities:

$$a^1 < a^{\sqrt{2}} < a^2$$
$$a^{1.4} < a^{\sqrt{2}} < a^{1.5}$$
$$a^{1.41} < a^{\sqrt{2}} < a^{1.42}$$
$$a^{1.414} < a^{\sqrt{2}} < a^{1.415}$$
$$a^{1.4142} < a^{\sqrt{2}} < a^{1.4143}$$
$$\vdots$$

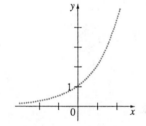

(a) Representation of $y = 2^x$, x rational

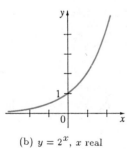

(b) $y = 2^x$, x real

Figure 3.1

For example, if $a = 2$, then this approximation process allows us to compute $2^{\sqrt{2}}$ correct to six decimal places:

$$2^{\sqrt{2}} \approx 2.665144$$

Figure 3.1 illustrates the definition of the exponential function with base $a = 2$. Part (a) is a representation of the graph of $y = 2^x$ up to stage 4, that is, for x rational. Part (b) shows how stage 5 completes the picture by filling in all the holes to give the graph of the continuous function $y = 2^x$, $x \in R$.

The graphs of the function $y = a^x$ are shown in Figure 3.2 for various values of the base a. Notice that all of these graphs pass through the same point $(0,1)$ because $a^0 = 1$ for $a \neq 0$.

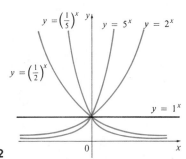

Figure 3.2

You can see from Figure 3.2 that there are basically three kinds of exponential functions $y = a^x$. If $0 < a < 1$, the exponential function decreases rapidly; if $a = 1$, it is a constant; and if $a > 1$, it increases rapidly. These three cases are illustrated in Figure 3.3. Since $(1/a)^x = 1/a^x = a^{-x}$, the graph of $y = (1/a)^x$ is just the reflection of the graph of $y = a^x$ about the y-axis.

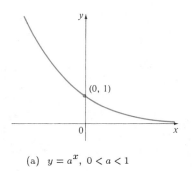
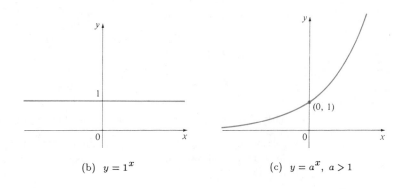

(a) $y = a^x$, $0 < a < 1$ (b) $y = 1^x$ (c) $y = a^x$, $a > 1$

Figure 3.3

The properties of the exponential function are summarized in the following theorem.

Theorem (3.2)

> If $a > 0$ and $a \neq 1$, then $f(x) = a^x$ is a continuous function with domain R and range $(0, \infty)$. In particular, $a^x > 0$ for all x. If $0 < a < 1$, $f(x) = a^x$ is a decreasing function; if $a > 1$, f is an increasing function. If $a, b > 0$ and $x, y \in R$, then
>
> (a) $a^{x+y} = a^x a^y$ (b) $a^{x-y} = \dfrac{a^x}{a^y}$
>
> (c) $(a^x)^y = a^{xy}$ (d) $(ab)^x = a^x b^x$

The reason for the importance of the exponential function lies in properties (a)–(d), which are called the **Laws of Exponents**. If x and y are rational numbers, then these laws are well known from elementary algebra (see Appendix A). For arbitrary real numbers x and y these laws can be deduced from the special case where the exponents are rational by using (3.1).

The following limits can be read from the graphs shown in Figure 3.3 or proved from the definition of a limit at infinity. (See Exercise 85 in Section 3.4.)

(3.3)

> If $a > 1$, then
>
> $$\lim_{x \to \infty} a^x = \infty \qquad \text{and} \qquad \lim_{x \to -\infty} a^x = 0$$
>
> If $0 < a < 1$, then
>
> $$\lim_{x \to \infty} a^x = 0 \qquad \text{and} \qquad \lim_{x \to -\infty} a^x = \infty$$

In particular, if $a \neq 1$, then the x-axis is a horizontal asymptote of the graph of the exponential function $y = a^x$.

EXAMPLE 1

(a) Find $\lim\limits_{x \to \infty} (2^{-x} - 1)$.

(b) Sketch the graph of the function $y = 2^{-x} - 1$.

Solution

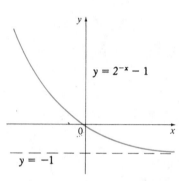

$y = 2^{-x} - 1$

$y = -1$

(a)
$$\lim_{x \to \infty} (2^{-x} - 1) = \lim_{x \to \infty} \left[\left(\tfrac{1}{2}\right)^x - 1 \right]$$
$$= 0 - 1 \qquad \text{[by (3.3) with } a = \tfrac{1}{2} < 1]$$
$$= -1$$

(b) We write $y = \left(\tfrac{1}{2}\right)^x - 1$ as above. The graph of $y = \left(\tfrac{1}{2}\right)^x$ is shown in Figure 3.2, so we shift it down one unit to obtain the graph of $y = \left(\tfrac{1}{2}\right)^x - 1$ shown in Figure 3.4. (For a review of shifting graphs, see Section 6 and, in particular, Table 24 in Review and Preview.) ●

Figure 3.4   **EXAMPLE 2**   Find $\lim\limits_{x \to 0^-} 2^{1/x}$.

Solution   As $x \to 0^-$, $1/x \to -\infty$, so by (3.3),

$$\lim_{x \to 0^-} 2^{1/x} = 0 \qquad \qquad ●$$

You might prefer to calculate the limit in Example 2 by explicitly introducing a new variable for the exponent. If we let $t = 1/x$, then as $x \to 0^-$ we have $t \to -\infty$ and so

$$\lim_{x \to 0^-} 2^{1/x} = \lim_{t \to -\infty} 2^t = 0$$

SECTION 3.1 **Exercises**

(Note: For a review of the Laws of Exponents, see the exercises in Appendix A.)

1. Sketch, using the same axes, the graphs of the following functions.

(a) $y = 3^x$

(b) $y = 10^x$

(c) $y = \left(\tfrac{1}{3}\right)^x$

(d) $y = \left(\tfrac{1}{10}\right)^x$

2. Sketch, using the same axes, the graphs of the following functions.

    (a) $y = 4^x$ (b) $y = 4^{-x}$ (c) $y = 1^x$

In Exercises 3–16 make a rough sketch of the graph of the given function. Do not use a calculator. Just use the graphs given in Figures 3.2 and 3.3 and, if necessary, the transformations of Section 6 of Review and Preview.

3. $y = 100^x$

4. $y = (1.1)^x$

5. $y = (0.9)^x$

6. $y = (0.1)^x$

7. $y = 2^x + 1$

8. $y = 2^{x+1}$

9. $y = 3^{-x}$

10. $y = -3^x$

11. $y = -3^{-x}$

12. $y = 2^{|x|}$

13. $y = 5^{-3x}$

14. $y = 5^{x-3}$

15. $y = 3 - 2^{x-1}$

16. $y = 2 + 5(1 - 10^{-x})$

Find the limits in Exercises 17–36.

17. $\lim_{x \to \infty} (1.1)^x$

18. $\lim_{x \to -\infty} (1.1)^x$

19. $\lim_{x \to \infty} \pi^{-x}$

20. $\lim_{x \to -\infty} \pi^{-x}$

21. $\lim_{x \to \infty} 5^{2x+5}$

22. $\lim_{x \to -\infty} 2^{3x+1}$

23. $\lim_{x \to \infty} (2^{-0.8x} + 1/x)$

24. $\lim_{x \to \infty} (0.8)^{x+1}$

25. $\lim_{x \to -\infty} (\pi/4)^x$

26. $\lim_{x \to \infty} (2\pi/7)^x$

27. $\lim_{x \to (\pi/2)^-} 2^{\tan x}$

28. $\lim_{x \to -(\pi/2)^+} 2^{\tan x}$

29. $\lim_{x \to \infty} 3^{1/x}$

30. $\lim_{x \to -\infty} 3^{1/x}$

31. $\lim_{x \to 0^+} 3^{1/x}$

32. $\lim_{x \to 0^-} 3^{1/x}$

33. $\lim_{x \to \infty} \dfrac{10^x}{10^x + 1}$

34. $\lim_{x \to \infty} \dfrac{2^x - 2^{-x}}{2^x + 3 \cdot 2^{-x}}$

35. $\lim_{t \to 0^+} \pi^{\csc t}$

36. $\lim_{t \to 0^-} (3 - 2^{\csc t})$

37. Evaluate:

    (a) $\lim_{x \to 2^+} 4^{x/(2-x)}$ (b) $\lim_{x \to 2^-} 4^{x/(2-x)}$

    (c) $\lim_{x \to \infty} 4^{x/(2-x)}$

38. Evaluate: (a) $\lim_{x \to \infty} 2^{-3^x}$ (b) $\lim_{x \to \infty} 2^{3^{-x}}$

39. If $f(x) = 10^x$, show that

$$\frac{f(x+h) - f(x)}{h} = 10^x \left(\frac{10^h - 1}{h} \right)$$

40. (a) Compare the functions $f(x) = x^2$ and $g(x) = 2^x$ by evaluating each of them for $x = 0, 1, 2, 3, 4,$ $5, 6, 7, 8, 9, 10, 15,$ and 20. Then draw the graphs of f and g on the same set of axes for $-4 \le x \le 6$.

    (b) Show that if the graphs of f and g are drawn on a coordinate grid where the unit of measurement is 1 inch, then at a distance 2 ft to the right of the origin the height of the graph of f is 48 ft but the height of the graph of g is about 265 mi.

SECTION 3.2

Derivatives of Exponential Functions

Let us try to compute the derivative of the exponential function $f(x) = a^x$ using the definition of a derivative:

$$f'(x) = \lim_{h \to 0} \frac{f(x+h) - f(x)}{h} = \lim_{h \to 0} \frac{a^{x+h} - a^x}{h}$$

$$= \lim_{h \to 0} \frac{a^x a^h - a^x}{h} = \lim_{h \to 0} \frac{a^x (a^h - 1)}{h}$$

$$f'(x) = a^x \lim_{h \to 0} \frac{a^h - 1}{h}$$

Notice that the limit is the value of the derivative of f at 0, that is,

$$\lim_{h \to 0} \frac{a^h - 1}{h} = f'(0)$$

Therefore we have shown that if the exponential function $f(x) = a^x$ is differentiable at

0, then it is differentiable everywhere and

(3.4)
$$f'(x) = f'(0)a^x$$

Numerical evidence for the existence of $f'(0)$ is given in the accompanying table for the cases $a = 2$ and $a = 3$. (Values are stated correct to four decimal places.) It appears that the limits exist and

h	$\dfrac{2^h - 1}{h}$	$\dfrac{3^h - 1}{h}$
0.1	0.7177	1.1612
0.01	0.6956	1.1047
0.001	0.6934	1.0992
0.0001	0.6932	1.0987

for $a = 2$, $\displaystyle f'(0) = \lim_{h \to 0} \frac{2^h - 1}{h} \approx 0.69$

for $a = 3$, $\displaystyle f'(0) = \lim_{h \to 0} \frac{3^h - 1}{h} \approx 1.10$

In fact, it can be proved that the limits exist and, correct to six decimal places, the values are

(3.5)
$$\frac{d}{dx} 2^x \bigg|_{x=0} \approx 0.693147 \qquad \frac{d}{dx} 3^x \bigg|_{x=0} \approx 1.098612$$

Thus, from Equation 3.4, we have

$$\frac{d}{dx}(2^x) \approx (0.69)2^x \qquad \frac{d}{dx}(3^x) \approx (1.10)3^x$$

Of all possible choices for the base a in Equation 3.4, the simplest differentiation formula occurs when $f'(0) = 1$. In view of the estimates of $f'(0)$ for $a = 2$ and $a = 3$, it seems reasonable that there is a number a between 2 and 3 for which $f'(0) = 1$. It is traditional to denote this value by the letter e.* Thus

(3.6)
$$\boxed{\;e \text{ is the number such that } \lim_{h \to 0} \frac{e^h - 1}{h} = 1\;}$$

Geometrically, this means that of all the possible exponential functions $y = a^x$, $f(x) = e^x$ is the one whose tangent line at $(0, 1)$ has a slope $f'(0)$ that is exactly 1. (See Figures 3.5 and 3.6.)

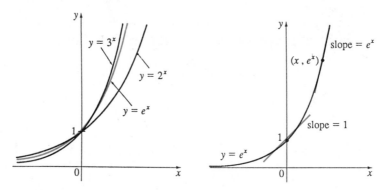

Figure 3.5   Figure 3.6

If we put $a = e$ and therefore $f'(0) = 1$ in Equation 3.4, it becomes the following important differentiation formula:

---

*The notation e for this number was chosen by the Swiss mathematician Leonhard Euler in 1727, probably because it is the first letter of the word *exponential*.

(3.7)
$$\frac{d}{dx} e^x = e^x$$

Thus the exponential function $f(x) = e^x$ has the property that it is its own derivative. The geometrical significance of this fact is that the slope of a tangent line to the curve $y = e^x$ is equal to the y-coordinate of the point (see Figure 3.6).

EXAMPLE 1   Differentiate the function $y = e^{\tan x}$.

Solution   To use the Chain Rule, we let $u = \tan x$. Then we have $y = e^u$, so

$$\frac{dy}{dx} = \frac{dy}{du}\frac{du}{dx} = e^u \frac{du}{dx} = e^{\tan x} \sec^2 x \qquad \bullet$$

In general if we combine Formula 3.7 with the Chain Rule, as in Example 1, we get

(3.8)
$$\frac{d}{dx} e^u = e^u \frac{du}{dx}$$

EXAMPLE 2   Find y' if $y = e^{-4x} \sin 5x$.

Solution   Using Formula 3.8 and the Product Rule, we have

$$y' = e^{-4x}(-4)\sin 5x + e^{-4x}(\cos 5x)(5) = e^{-4x}(5\cos 5x - 4\sin 5x) \qquad \bullet$$

We have seen that e is a number that lies somewhere between 2 and 3, but we can use Equation 3.4 to estimate the numerical value of e more accurately. Let $e = 2^c$. Then $e^x = 2^{cx}$. If $f(x) = 2^x$, then from Equation 3.4 we have $f'(x) = k2^x$, where $k = f'(0) \approx 0.693147$. Thus, by the Chain Rule,

$$e^x = \frac{d}{dx} e^x = \frac{d}{dx} 2^{cx} = k2^{cx}\frac{d}{dx}(cx) = ck2^{cx}$$

Putting $x = 0$, we have $1 = ck$, so $c = 1/k$ and

$$e = 2^{1/k} \approx 2^{1/0.693147} \approx 2.71828$$

It can be shown that the approximate value to 20 decimal places is

$$e \approx 2.71828182845904523536$$

The decimal expansion of e is nonrepeating because e is an irrational number (see Exercise 53 in Section 10.11).

EXAMPLE 3   In Example 6 in Section 2.3 we considered a population of bacteria in a homogeneous nutrient medium. We showed that if the population doubles every hour, then the population after t hours is

$$n = n_0 2^t$$

where n_0 is the initial population. Now we can use (3.4) and (3.5) to compute the growth rate:

$$\frac{dn}{dt} \approx n_0 \, (0.691347) \, 2^t$$

For instance, if the initial population is $n_0 = 1000$ cells, then the growth rate after $2\,\text{h}$ is

$$\frac{dn}{dt}\bigg|_{t=2} \approx (1000)(0.693147)2^t\big|_{t=2}$$

$$= (4000)(0.693147) \approx 2773 \, \text{cells/h}$$ •

The exponential function $f(x) = e^x$ is one of the most frequently occurring functions in calculus and its applications, so it is important to be familiar with its graph and properties. We summarize these properties as follows, using the fact that this function is just a special case of the exponential functions considered in Section 3.1 but with base $a = e > 1$.

Properties of Exponential Functions (3.9)

> The exponential function $f(x) = e^x$ is an increasing continuous function with domain R and range $(0, \infty)$. Thus $e^x > 0$ for all x. Also
>
> $$\lim_{x \to -\infty} e^x = 0 \qquad\qquad \lim_{x \to \infty} e^x = \infty$$

EXAMPLE 4   Find $\lim\limits_{x \to \infty} \dfrac{e^{2x}}{e^{2x}+1}$.

Solution   We divide numerator and denominator by e^{2x}:

$$\lim_{x \to \infty} \frac{e^{2x}}{e^{2x}+1} = \lim_{x \to \infty} \frac{1}{1+e^{-2x}} = \frac{1}{1+\lim\limits_{x\to\infty} e^{-2x}}$$

$$= \frac{1}{1+0} = 1$$

We have used the fact that $t = -2x \to -\infty$ as $x \to \infty$ and so

$$\lim_{x \to \infty} e^{-2x} = \lim_{t \to -\infty} e^t = 0$$ •

SECTION 3.2   **Exercises**

▦ 1. (a) Sketch the curve $y = 4^x$.
   (b) Use a calculator to evaluate the quantity
   $$\frac{4^h - 1}{h}$$
   for $h = 0.1, 0.01, 0.001,$ and 0.0001.
   What does the quantity represent?
   (c) Estimate the value of the limit
   $$\lim_{h \to 0} \frac{4^h - 1}{h}$$
   correct to two decimal places.
   (d) What does the limit in part (c) represent?

▦ 2. Use a calculator to estimate the values of the limits to two decimal places.
   (a) $\lim\limits_{h \to 0} \dfrac{2.7^h - 1}{h}$ (b) $\lim\limits_{h \to 0} \dfrac{2.8^h - 1}{h}$

Differentiate the functions in Exercises 3–18.

3. $f(x) = e^{\sqrt{x}}$

4. $f(x) = xe^{-x^2}$

5. $y = xe^{2x}$

6. $g(x) = e^{-5x} \cos 3x$

7. $h(t) = \sqrt{1 - e^t}$

8. $h(\theta) = e^{\sin 5\theta}$

9. $y = e^{x \cos x}$

10. $y = \dfrac{e^{-x^2}}{x}$

11. $y = e^{-1/x}$

12. $y = e^{x+e^x}$

13. $y = \tan(e^{3x-2})$

14. $y = \sqrt[3]{2x + e^{3x}}$

15. $y = \dfrac{e^{3x}}{1 + e^x}$

16. $y = \dfrac{e^x + e^{-x}}{e^x - e^{-x}}$

17. $y = x^e$

18. $y = \sec\!\left(e^{\tan x^2}\right)$

In Exercises 19 and 20 find the equation of the tangent line to the given curve at the given point.

19. $y = e^{-x} \sin x$, $(\pi, 0)$ 20. $y = x^2 e^{-x}$, $(1, 1/e)$

21. Find y' if $\cos(x - y) = xe^x$.

22. Find the equation of the tangent line to the curve $2e^{xy} = x + y$ at the point $(0, 2)$.

23. Show that the function $y = e^{2x} + e^{-3x}$ satisfies the differential equation $y'' + y' - 6y = 0$.

24. Show that the function $y = Ae^{-x} + Bxe^{-x}$ satisfies the differential equation $y'' + 2y' + y = 0$.

25. For what values of r does the function $y = e^{rx}$ satisfy the equation $y'' + 5y' - 6y = 0$?

26. Find the values of λ for which $y = e^{\lambda x}$ satisfies the equation $y + y' = y''$.

27. If $f(x) = e^{-2x}$, find $f^{(8)}(x)$.

28. Find the thousandth derivative of $f(x) = xe^{-x}$.

29. The velocity of a particle that moves in a straight line under the influence of viscous forces is $v(t) = ce^{-kt}$ where c and k are positive constants.

(a) Show that the acceleration is proportional to the velocity.
(b) What is the significance of the number c?
(c) At what time is the velocity equal to half the initial velocity?

30. Under certain circumstances a rumor spreads according to the equation

$$p(t) = \frac{1}{1 + ae^{-kt}}$$

where $p(t)$ is the proportion of the population that knows the rumor at time t and a and k are positive constants. Find the rate of spread of the rumor.

In Exercises 31 and 32 use the graph of $y = e^x$ to draw the graph of the given function.

31. $y = e^{-x}$

32. $y = 3 - e^x$

In Exercises 33–38 find the limit.

33. $\displaystyle\lim_{x \to \infty} \frac{e^{3x} - e^{-3x}}{e^{3x} + e^{-3x}}$

34. $\displaystyle\lim_{x \to -\infty} \frac{e^{3x} - e^{-3x}}{e^{3x} + e^{-3x}}$

35. $\displaystyle\lim_{x \to 1^-} e^{2/(x-1)}$

36. $\displaystyle\lim_{x \to 1^+} e^{2/(x-1)}$

37. $\displaystyle\lim_{x \to (\pi/2)^-} \frac{2}{1 + e^{\tan x}}$

38. $\displaystyle\lim_{x \to 0^-} \frac{2}{1 + e^{\cot x}}$

39. Use the Intermediate Value Theorem to show that there is a root of the equation $e^x + x = 0$.

40. Use Newton's method to find the root of the equation in Exercise 39 correct to six decimal places.

SECTION 3.3

Inverse Functions

Let us compare the functions f and g whose arrow diagrams are shown in Figure 3.7. Note that f never takes on the same value twice (any two numbers in A have different images), whereas g does take on the same value twice (both 2 and 3 have the same image, 4). In symbols,

$$g(2) = g(3)$$

but $f(x_1) \neq f(x_2)$ whenever $x_1 \neq x_2$

Figure 3.7

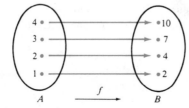

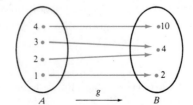

Functions that have this latter property are called *one-to-one.*

Definition (3.10)

> A function f with domain A is called a **one-to-one function** if no two elements of A have the same image; that is
>
> $$f(x_1) \neq f(x_2) \qquad \text{whenever } x_1 \neq x_2$$

If a horizontal line intersects the graph of f in more than one point, then we see from Figure 3.8 that there are numbers x_1 and x_2 such that $f(x_1) = f(x_2)$.

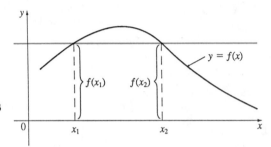

Figure 3.8

This function is not one-to-one because $f(x_1) = f(x_2)$

This means that f is not one-to-one. Therefore we have the following geometric method for determining whether or not a function is one-to-one.

Horizontal Line Test (3.11)

> A function is one-to-one if and only if no horizontal line intersects its graph more than once.

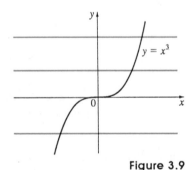

Figure 3.9

$f(x) = x^3$ is one-to-one

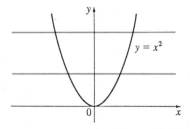

Figure 3.10

$g(x) = x^2$ is not one-to-one

EXAMPLE 1   Is the function $f(x) = x^3$ one-to-one?

Solution 1   If $x_1 \neq x_2$, then $x_1^3 \neq x_2^3$ (two different numbers cannot have the same cube). Therefore, by Definition 3.10, $f(x) = x^3$ is one-to-one.

Solution 2   From Figure 3.9 we see that no horizontal line intersects the graph of $f(x) = x^3$ more than once. Therefore, by the Horizontal Line Test, f is one-to-one.   ●

EXAMPLE 2   Is the function $g(x) = x^2$ one-to-one?

Solution 1   This function is not one-to-one because, for instance,

$$g(1) = 1 = g(-1)$$

and so 1 and -1 have the same image.

Solution 2   From Figure 3.10 we see that there are horizontal lines that intersect the graph of g more than once. Therefore, by the Horizontal Line Test, g is not one-to-one.   ●

Notice that the function f of Example 1 is increasing and is also one-to-one. More generally we have the following theorem (see Exercise 35):

Theorem (3.12)

> Every decreasing function and every increasing function are one-to-one functions.

One-to-one functions are important because they are precisely the functions that possess inverse functions according to the following definition.

Definition (3.13)

> Let f be a one-to-one function with domain A and range B. Then its **inverse function** f^{-1} has domain B and range A and is defined by
>
> $$f^{-1}(y) = x \quad \Leftrightarrow \quad f(x) = y$$
>
> for any y in B.

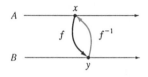

Figure 3.11

This definition says that if f maps x into y, then f^{-1} maps y back into x. (If f was not one-to-one, then f^{-1} would not be uniquely defined.) The arrow diagram in Figure 3.11 indicates that f^{-1} reverses the effect of f. Note that

$$\text{domain of } f^{-1} = \text{range of } f$$

$$\text{range of } f^{-1} = \text{domain of } f$$

For example, the inverse function of $f(x) = x^3$ is $f^{-1}(x) = x^{1/3}$ because if $y = x^3$, then

$$f^{-1}(y) = f^{-1}(x^3) = (x^3)^{1/3} = x$$

⊘ *Caution:* Do not mistake the -1 in f^{-1} for an exponent. Thus

$$f^{-1}(x) \quad \text{does } not \text{ mean} \quad \frac{1}{f(x)}$$

The reciprocal $1/f(x)$ could, however, be written as $[f(x)]^{-1}$.

The letter x is traditionally used as the independent variable, so when we concentrate on f^{-1} rather than on f we usually reverse the roles of x and y in Definition 3.13 and write

(3.14)

$$f^{-1}(x) = y \quad \Leftrightarrow \quad f(y) = x$$

By substituting for y in Definition 3.13 and substituting for x in (3.14) we get the following **cancellation equations:**

(3.15)

$$f^{-1}(f(x)) = x \qquad \text{for every } x \text{ in } A$$

$$f(f^{-1}(x)) = x \qquad \text{for every } x \text{ in } B$$

The first cancellation equation says that if we start with x, apply f, and then apply f^{-1}, we arrive back at x, where we started. Thus f^{-1} undoes what f does. The second equation says that f undoes what f^{-1} does.

For example, if $f(x) = x^3$, then $f^{-1}(x) = x^{1/3}$ and the cancellation equations become

$$f^{-1}(f(x)) = (x^3)^{1/3} = x$$

$$f(f^{-1}(x)) = (x^{1/3})^3 = x$$

These equations simply say that the cube function and the cube root function cancel each other out.

Let us now see how to compute inverse functions. If we have a function $y = f(x)$ and are able to solve this equation for x in terms of y, then according to Definition 3.13 we must have $x = f^{-1}(y)$. If we then interchange x and y, we have $y = f^{-1}(x)$, which is the desired equation.

How to Find the Inverse Function of a One-to-One Function f (3.16)

(a)   Write $y = f(x)$.

(b)   Solve this equation for x in terms of y (if possible).

(c)   Interchange x and y. The resulting equation is $y = f^{-1}(x)$.

EXAMPLE 3   Find the inverse function of $f(x) = x^3 + 2$.

Solution   According to (3.16) we first write

$$y = x^3 + 2$$

Then we solve this equation for x:

$$x^3 = y - 2$$
$$x = \sqrt[3]{y - 2}$$

Finally we interchange x and y: $y = \sqrt[3]{x - 2}$

Therefore the inverse function is $f^{-1}(x) = \sqrt[3]{x - 2}$.   ●

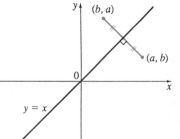

Figure 3.12

The principle of interchanging x and y to find the inverse function also gives us the method for obtaining the graph of f^{-1} from the graph of f. If $f(a) = b$, then $f^{-1}(b) = a$. Thus the point (a, b) is on the graph of f if and only if the point (b, a) is on the graph of f^{-1}. But we get the point (b, a) from (a, b) by reflecting about the line $y = x$ (see Figure 3.12). Therefore, as illustrated by Figure 3.13:

The graph of f^{-1} is obtained by reflecting the graph of f about the line $y = x$.

EXAMPLE 4   Sketch the graphs of $f(x) = \sqrt{-1 - x}$ and its inverse function using the same coordinate axes.

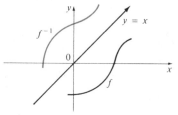

Figure 3.13

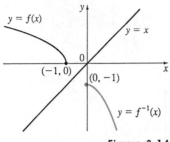

Figure 3.14

Solution   First we sketch the curve $y = \sqrt{-1-x}$ (the top half of the parabola $y^2 = -1 - x$ or $x = -y^2 - 1$) and then we reflect about the line $y = x$ to get the graph of f^{-1} (see Figure 3.14).   ●

Now let us look at inverse functions from the point of view of calculus. Suppose that f is both one-to-one and continuous. We think of a continuous function as one whose graph has no breaks in it. (It consists of just one piece.) Since the graph of f^{-1} is obtained from the graph of f by reflecting about the line $y = x$, the graph of f^{-1} has no breaks in it either (see Figure 3.13). Thus we would expect that f^{-1} is also a continuous function.

This geometrical argument does not prove the following theorem but at least it makes the theorem plausible. A proof can be found in Appendix C.

Theorem (3.17)

> If f is a one-to-one continuous function defined on an interval, then its inverse function f^{-1} is also continuous.

Now suppose that f is a one-to-one differentiable function. Geometrically we can think of a differentiable function as one whose graph has no corners or kinks in it. We get the graph of f^{-1} by reflecting the graph of f about the line $y = x$, so the graph of f^{-1} has no corners or kinks in it either. We therefore expect that f^{-1} is also differentiable (except where its tangents are vertical). In fact, we can predict the value of the derivative of f^{-1} at a given point by a geometric argument. In Figure 3.15 the graphs of f and its inverse $g = f^{-1}$ are shown. If $f(b) = a$, then $g(a) = f^{-1}(a) = b$ and $g'(a)$ is the slope of the tangent to the graph of g at (a, b), which is $\tan \phi$. Likewise $f'(b) = \tan \theta$, From Figure 3.15 we see that $\theta + \phi = \pi/2$, so

$$g'(a) = \tan \phi = \tan\left(\frac{\pi}{2} - \theta\right) = \frac{1}{\tan \theta} = \frac{1}{f'(b)}$$

that is,

$$g'(a) = \frac{1}{f'(g(a))}$$

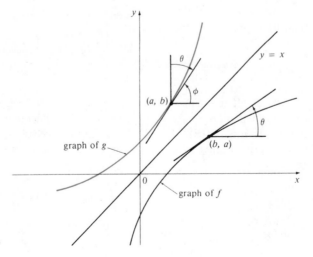

Figure 3.15

Theorem (3.18)

> If f is a one-to-one differentiable function with inverse function $g = f^{-1}$ and $f'(g(a)) \neq 0$, then the inverse function is differentiable at a and
>
> $$g'(a) = \frac{1}{f'(g(a))}$$

Proof   Write the definition of derivative in the form given by (2.3):

$$g'(a) = \lim_{x \to a} \frac{g(x) - g(a)}{x - a}$$

By (3.14) we have

$$g(x) = y \quad \Leftrightarrow \quad f(y) = x$$

and

$$g(a) = b \quad \Leftrightarrow \quad f(b) = a$$

Since f is differentiable, it is continuous, so $g = f^{-1}$ is continuous by Theorem 3.17. Thus if $x \to a$, then $g(x) \to g(a)$, that is, $y \to b$. Therefore

$$g'(a) = \lim_{x \to a} \frac{g(x) - g(a)}{x - a}$$

$$= \lim_{y \to b} \frac{y - b}{f(y) - f(b)}$$

$$= \lim_{y \to b} \frac{1}{\dfrac{f(y) - f(b)}{y - b}} = \frac{1}{\lim\limits_{y \to b} \dfrac{f(y) - f(b)}{y - b}}$$

$$= \frac{1}{f'(b)} = \frac{1}{f'(g(a))} \qquad \bullet$$

Note 1:   Replacing a by the general number x in the formula of Theorem 3.18, we get

(3.19)

$$g'(x) = \frac{1}{f'(g(x))}$$

If we write $y = g(x)$, then $f(y) = x$, so Equation 3.19, when expressed in Leibniz notation, becomes

$$\frac{dy}{dx} = \frac{1}{\dfrac{dx}{dy}}$$

Note 2:   If it is known in advance that f^{-1} is differentiable, then its derivative can be computed more easily than in the proof of Theorem 3.18 by using implicit differentiation. If $y = f^{-1}(x)$, then $f(y) = x$. Differentiating the equation $f(y) = x$ implicitly with respect to x, remembering that y is a function of x, and using the Chain Rule, we get

$$f'(y)\frac{dy}{dx} = 1$$

Therefore

$$\frac{dy}{dx} = \frac{1}{f'(y)} = \frac{1}{\dfrac{dx}{dy}}$$

EXAMPLE 5   Although the function $y = x^2$, $x \in R$, is not one-to-one and therefore does not have an inverse function, we can make it one-to-one by restricting its domain. For instance, the function $f(x) = x^2$, $0 \le x \le 2$, is one-to-one (by the Horizontal Line Test) and has domain $[0,2]$ and range $[0,4]$ (see Figure 3.16). Thus f has an inverse function $g = f^{-1}$ with domain $[0,4]$ and range $[0,2]$.

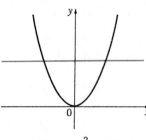

Figure 3.16   (a) $y = x^2$, $x \in R$ (b) $f(x) = x^2$, $0 \le x \le 2$

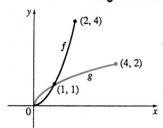

Figure 3.17

Without computing a formula for g' we can still calculate $g'(1)$. Since $f(1) = 1$, we have $g(1) = 1$. Also $f'(x) = 2x$. So by Theorem 3.18 we have

$$g'(1) = \frac{1}{f'(g(1))} = \frac{1}{f'(1)} = \tfrac{1}{2}$$

In this case it is easy to find g explicitly. In fact, $g(x) = \sqrt{x}$, $0 \le x \le 4$. [In general we could use the method given by (3.16).] Then $g'(x) = 1/(2\sqrt{x})$, so $g'(1) = \tfrac{1}{2}$, which agrees with the preceding computation. The functions f and g are graphed in Figure 3.17.   ●

EXAMPLE 6   If $f(x) = 2x + \cos x$ and $g = f^{-1}$, find $g'(1)$.

Solution   To use Theorem 3.18 we need to know $g(1)$ and we can find it by inspection:

$$f(0) = 1 \quad \Rightarrow \quad g(1) = 0$$

Therefore

$$g'(1) = \frac{1}{f'(g(1))} = \frac{1}{f'(0)} = \frac{1}{2 - \sin 0} = \tfrac{1}{2} \qquad ●$$

SECTION 3.3 **Exercises**

In Exercises 1–6 the graph of a function f is shown.
Determine whether or not f is one-to-one.

1. 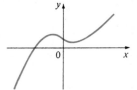   **2.** 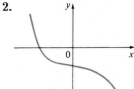   **3.** 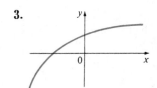   **4.**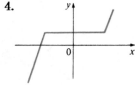

5.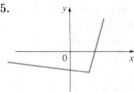

6.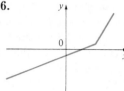

In Exercises 7–12 determine whether or not the given function is one-to-one.

7. $f(x) = 7x - 3$

8. $f(x) = x^2 - 2x + 5$

9. $g(x) = \sqrt{x}$

10. $g(x) = |x|$

11. $h(x) = x^4 + 5$

12. $h(x) = x^4 + 5, \; 0 \le x \le 2$

In Exercises 13–18 show that f is one-to-one and find its inverse function.

13. $f(x) = 4x + 7$

14. $f(x) = \dfrac{x-2}{x+2}$

15. $f(x) = \dfrac{1+3x}{5-2x}$

16. $f(x) = 5 - 4x^3$

17. $f(x) = \sqrt{2 + 5x}$

18. $f(x) = x^2 + x, \; x \ge -\frac{1}{2}$

In Exercises 19–24, (a) show that f is one-to-one, (b) use Theorem 3.18 to find $g'(a)$, where $g = f^{-1}$, (c) calculate $g(x)$ and state the domain and range of g, (d) calculate $g'(a)$ from the formula in part (c) and check that it agrees with the result of part (b), and (e) sketch the graphs of f and g.

19. $f(x) = 2x + 1, \quad a = 3$
20. $f(x) = 6 - x, \quad a = 2$

21. $f(x) = x^3, \quad a = 8$
22. $f(x) = \sqrt{x - 2}, \quad a = 2$

23. $f(x) = 9 - x^2, \quad 0 \le x \le 3, \quad a = 8$

24. $f(x) = 1/(x - 1), \quad x > 1, \quad a = 2$

In Exercises 25–30 find $g'(a)$, where g is the inverse function of the given function.

25. $f(x) = x^3 + x + 1, \quad a = 1$

26. $f(x) = x^5 - x^3 + 2x, \quad a = 2$

27. $f(x) = 3 + x^2 + \sin \pi x, \quad -0.4 < x < 0.4, \quad a = 3$

28. $f(x) = \sin x + \cos x, \quad -\pi/4 < x < \pi/4, \quad a = 1$

29. $f(x) = e^x, \quad a = 1$

30. $f(x) = 3 + x + e^x, \quad a = 4$

31. Suppose g is the inverse function of f and $f(4) = 5$, $f'(4) = \frac{2}{3}$. Find $g'(5)$.

32. Suppose g is the inverse function of a differentiable function f and let $G(x) = 1/g(x)$. If $f(3) = 2$ and $f'(3) = \frac{1}{9}$, find $G'(2)$.

33. If n is a positive integer, then $y = \sqrt[n]{x}$ has domain R if n is odd and domain $[0, \infty)$ if n is even. It is the inverse function of the power function. Use the method of Note 2 to find dy/dx.

34. Show that $h(x) = \sin x$, $x \in R$, is not one-to-one, but its restriction $f(x) = \sin x$, $-\pi/2 \le x \le \pi/2$ is one-to-one. Compute the derivative of $f^{-1} = \sin^{-1}$ by the method of Note 2.

35. Prove that every increasing function is one-to-one.

36. If f is a one-to-one, twice differentiable function with inverse function g, show that

$$g''(x) = -\frac{f''(g(x))}{[f'(g(x))]^3}$$

SECTION 3.4

Logarithmic Functions

If $a > 0$ and $a \ne 1$, Theorem 3.2 tells us that the exponential function $f(x) = a^x$ is either increasing or decreasing. Thus, by Theorem 3.12, f is one-to-one. It therefore has an inverse function f^{-1} which is called the **logarithmic function to the base a** and is denoted by $\log_a$. If we use the formulation of an inverse function given by (3.14),

$$f^{-1}(x) = y \quad \Leftrightarrow \quad f(y) = x$$

then we have

(3.20)

$$\log_a x = y \quad \Leftrightarrow \quad a^y = x$$

Thus, if $x > 0$, $\log_a x$ is the exponent to which the base a must be raised to give x.

EXAMPLE 1   Evaluate (a) $\log_3 81$, (b) $\log_{25} 5$, and (c) $\log_{10} 0.001$.

Solution

(a) $\log_3 81 = 4$ because $3^4 = 81$

(b) $\log_{25} 5 = \frac{1}{2}$ because $25^{1/2} = 5$

(c) $\log_{10} 0.001 = -3$ because $10^{-3} = 0.001$ •

The cancellation equations (3.15), when applied to $f(x) = a^x$ and $f^{-1}(x) = \log_a x$, become

(3.21)

$$\log_a(a^x) = x \qquad \text{for every } x \in R$$

$$a^{\log_a x} = x \qquad \text{for every } x > 0$$

The logarithmic function, $\log_a$, has domain $(0, \infty)$ and range R and is continuous by Theorem 3.17 since it is the inverse of the exponential function. Its graph is the reflection of the graph of $y = a^x$ about the line $y = x$.

Figure 3.18 shows the case where $a > 1$. (The most important logarithmic functions have base $a > 1$.) The fact that $y = a^x$ is a very rapidly increasing function for $x > 0$ is reflected in the fact that $y = \log_a x$ is a very slowly increasing function for $x > 1$. Figure 3.19 shows the graphs of $y = \log_a x$ for various values of the base a. Since $\log_a 1 = 0$, the graphs of all logarithmic functions pass through the point $(1, 0)$.

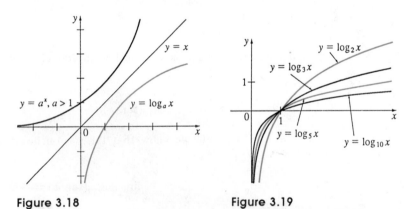

Figure 3.18   **Figure 3.19**

The following theorem summarizes the properties of logarithmic functions.

Theorem (3.22)

If $a > 1$, the function $f(x) = \log_a x$ is a one-to-one, continuous, increasing function with domain $(0, \infty)$ and range R. If $x, y > 0$, then

(a) $\log_a(xy) = \log_a x + \log_a y$

(b) $\log_a\left(\frac{x}{y}\right) = \log_a x - \log_a y$

(c) $\log_a(x^y) = y \log_a x$

Properties (a), (b), and (c) follow from the corresponding properties in Theorem 3.2 (see Exercise 57).

EXAMPLE 2   Use the properties of logarithms in Theorem 3.22 to evaluate the following: (a) $\log_4 2 + \log_4 32$ (b) $\log_2 80 - \log_2 5$

Solution
(a)   Using (a) in Theorem 3.22, we have

$$\log_4 2 + \log_4 32 = \log_4 (2 \cdot 32) = \log_4 64 = 3$$

since $4^3 = 64$.
(b)   Using (b) we have

$$\log_2 80 - \log_2 5 = \log_2 \left(\frac{80}{5}\right) = \log_2 16 = 4$$

since $2^4 = 16$.   ●

The limits of exponential functions given in (3.3) are reflected in the following limits of logarithmic functions.

(3.23)

> If $a > 1$, then
>
> $$\lim_{x \to \infty} \log_a x = \infty \qquad \text{and} \qquad \lim_{x \to 0^+} \log_a x = -\infty$$

In particular, the y-axis is a vertical asymptote of the curve $y = \log_a x$.

EXAMPLE 3   Find $\lim\limits_{x \to 0^+} \log_{10}(\tan^2 x)$.

Solution   As $x \to 0$, we know that $\tan x \to \tan 0 = 0$, so by (3.23) with $a = 10 > 1$, we have

$$\lim_{x \to 0^+} \log_{10}(\tan^2 x) = -\infty$$ ●

Natural Logarithms

Of all possible bases a for logarithms, we will see in the next section that the most convenient choice of a base is the number e, which was defined in Section 3.2. The logarithm with base e is called the **natural logarithm** and has a special notation:

$$\log_e x = \ln x$$

(Another notation that is sometimes used is $\log_e x = \log x$; that is, the base is omitted because e is the most frequently used base.)

If we put $a = e$ and $\log_e = \ln$ in (3.20) and (3.21), then the defining properties of the natural logarithmic function become

(3.24)

$$\ln x = y \quad \Leftrightarrow \quad e^y = x$$

(3.25) and

$$\ln(e^x) = x \qquad x \in R$$
$$e^{\ln x} = x \qquad x > 0$$

In particular, if we set $x = 1$, we get

$$\ln e = 1$$

EXAMPLE 4   Find x if $\ln x = 5$.

Solution 1   From (3.24) we see that

$$\ln x = 5 \qquad \text{means} \qquad e^5 = x$$

Therefore $x = e^5$.

(If you have trouble working with the "ln" notation, just replace it by $\log_e$. Then the equation becomes $\log_e x = 5$; so by the definition of logarithm, $e^5 = x$.)

Solution 2   Start with the equation

$$\ln x = 5$$

and apply the exponential function to both sides of the equation:

$$e^{\ln x} = e^5$$

But the second cancellation equation in (3.25) says that $e^{\ln x} = x$. Therefore $x = e^5$.   ●

EXAMPLE 5   Solve the equation $e^{5-3x} = 10$.

Solution   We take natural logarithms of both sides of the equation and use (3.25):

$$\ln(e^{5-3x}) = \ln 10$$

$$5 - 3x = \ln 10$$

$$3x = 5 - \ln 10$$

$$x = \tfrac{1}{3}(5 - \ln 10)$$

Since the natural logarithm is found on scientific calculators, we can approximate the solution to four decimal places: $x \approx 0.8991$.   •

EXAMPLE 6   Express $\ln a + \tfrac{1}{2}\ln b$ as a single logarithm.

Solution   Using properties (c) and (a) of logarithms, we have

$$\ln a + \tfrac{1}{2}\ln b = \ln a + \ln b^{1/2}$$

$$= \ln a + \ln \sqrt{b}$$

$$= \ln(a\,\sqrt{b})$$ •

The graphs of the exponential function $y = e^x$ and its inverse function, the natural logarithm function, are shown in Figure 3.20.

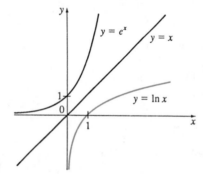

Figure 3.20

If we put $a = e$ in (3.23), then we have the following limits:

(3.26)

$$\lim_{x\to\infty} \ln x = \infty \qquad \lim_{x\to 0^+} \ln x = -\infty$$

EXAMPLE 7   Sketch the graph of the function $y = \ln(x - 2) - 1$.

Solution   We start with the graph of $y = \ln x$ as given in Figure 3.20. Using the transformations of Section 6 in Review and Preview, we shift it two units to the right to get the graph of $y = \ln(x - 2)$ and then we shift it one unit downward to get the graph of $y = \ln(x - 2) - 1$ (see Figure 3.21). Notice that the line $x = 2$ is a vertical asymptote since

$$\lim_{x\to 2^+} [\ln(x - 2) - 1] = -\infty$$

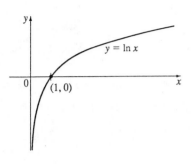
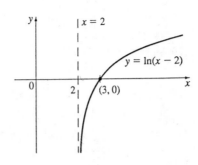
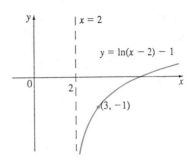

Figure 3.21

SECTION 3.4 **Exercises**

In Exercises 1–14 evaluate the given expression.

1. $\log_2 64$
2. $\log_6 \frac{1}{36}$
3. $\log_8 2$
4. $\log_8 4$
5. $\log_3 \frac{1}{27}$
6. $e^{\ln 6}$
7. $\ln e^{\sqrt{2}}$
8. $\log_3 3^{\sqrt{5}}$
9. $\log_{10} 1.25 + \log_{10} 80$
10. $\log_3 108 - \log_3 4$
11. $\log_8 6 - \log_8 3 + \log_8 4$
12. $\log_5 10 + \log_5 20 - 3 \log_5 2$
13. $2^{(\log_2 3 + \log_2 5)}$
14. $e^{3 \ln 2}$

In Exercises 15–20 express the given quantity as a single logarithm.

15. $\log_5 a + \log_5 b - \log_5 c$
16. $\log_2 x + 5 \log_2(x+1) + \frac{1}{2} \log_2(x-1)$
17. $2 \ln 4 - \ln 2$
18. $\ln 10 + \frac{1}{2} \ln 9$
19. $\frac{1}{3} \ln x - 4 \ln(2x+3)$
20. $\ln x + a \ln y - b \ln z$

In Exercises 21 and 22 sketch, using the same axes, the graphs of the following functions.

21. (a) $y = \log_3 x$ (b) $y = \log_{10} x$
22. (a) $y = \log_2 x$ (b) $y = \ln x$
    (c) $y = \log_{10} x$

In Exercises 23–34 make a rough sketch of the graph of the given function. Do not use a calculator. Just use the graphs given in Figures 3.18, 3.19, and 3.20 and, if necessary, the transformations of Section 6 in Review and Preview.

23. $y = \log_{1.1} x$
24. $y = \log_{100} x$
25. $y = \log_{10}(x+5)$
26. $y = 1 + \log_5(x-1)$
27. $y = -\ln x$
28. $y = \ln(-x)$
29. $y = -\ln(-x)$
30. $y = \ln|x|$
31. $y = \ln(x^2)$
32. $y = \ln(1/x)$
33. $y = \ln(x+3)$
34. $y = \ln|x+3|$

In Exercises 35–52 solve the given equations for x.

35. $\log_2 x = 3$
36. $2^{x-5} = 3$
37. $e^x = 16$
38. $\ln x = -1$
39. $\ln(2x-1) = 3$
40. $e^{3x-4} = 2$
41. $3^{x+2} = m$
42. $5^{\log_5(2x)} = 6$
43. $\ln x = \ln 5 + \ln 8$
44. $\ln x^2 = 2 \ln 4 - 4 \ln 2$
45. $\ln(e^{2x-1}) = 5$
46. $\ln x + \ln(x-1) = 1$
47. $\ln(\ln x) = 1$
48. $e^{e^x} = 10$
49. $2^{3^x} = 5$
50. $\log_2(\log_3(\log_4 x)) = C$
51. $\ln(x+6) + \ln(x-3) = \ln 5 + \ln 2$
52. $\ln\left(\frac{x-2}{x-1}\right) = 1 + \ln\left(\frac{x-3}{x-1}\right)$

In Exercises 53–56 find the solution of the equation correct to four decimal places.

53. $\ln(x-5) = 3$
54. $e^{5x-1} = 12$
55. $e^{2-3x} = 20$
56. $2^{-x} = 5$

57. Prove properties (a), (b), and (c) of logarithms in Theorem 3.22 by deducing them from properties (a), (b), and (c) of exponential functions in Theorem 3.2.

58. (a) Prove that $(\log_a b)(\log_b c) = \log_a c$.

    (b) Deduce that $\log_a b = \dfrac{1}{\log_b a}$.

    (c) Deduce that $\log_a e = \dfrac{1}{\ln a}$.

59. Deduce from Exercise 58(a) that $\log_b c = \dfrac{\ln c}{\ln b}$.

60. Use Exercise 59 to evaluate the following logarithms correct to six decimal places.

    (a) $\log_2 5$ (b) $\log_5 26.05$

    (c) $\log_3 e$ (d) $\log_{0.7} 14$

61. The geologist C. F. Richter defined the magnitude of an earthquake to be $\log_{10}(I/S)$, where I is the intensity of the earthquake (measured by the amplitude of a seismograph 100 km from the earthquake) and S is the intensity of a "standard" earthquake (where the amplitude is only 1 micron $= 10^{-4}$ cm). The 1989 earthquake in San Francisco had a magnitude of 6.9 on the Richter scale. The 1906 San Francisco earthquake was 25 times as intense. What was its magnitude on the Richter scale?

62. A sound so faint that it can just be heard has intensity $I_0 = 10^{-12}$ watt/m^2 at a frequency of 1000 hertz (Hz). The loudness, in decibels (dB), of a sound with intensity I is then defined to be $L = 10 \log_{10}(I/I_0)$. Rock music with amplifiers is measured at 120 dB. The noise from a power mower is measured at 106 dB. Find the ratio of the intensity of the rock music to that of the power mower.

Find the limits in Exercises 63–70.

63. $\displaystyle\lim_{x \to 5^+} \ln(x-5)$

64. $\displaystyle\lim_{x \to 0^+} \log_{10}(4x)$

65. $\displaystyle\lim_{x \to \infty} \log_2(x^2 - x)$

66. $\displaystyle\lim_{x \to 0^+} \ln(\sin x)$

67. $\displaystyle\lim_{x \to (\pi/2)^-} \log_{10}(\cos x)$

68. $\displaystyle\lim_{x \to \infty} \dfrac{\ln x}{1 + \ln x}$

69. $\displaystyle\lim_{x \to \infty} \ln(1 + e^{-x^2})$

70. $\displaystyle\lim_{x \to \infty} [\ln(2+x) - \ln(1+x)]$

In Exercises 71–74 find the domain and range of the given function.

71. $f(x) = \log_{10}(1 - x)$ 72. $g(x) = \ln(4 - x^2)$

73. $F(t) = \sqrt{t}\,\ln(t^2 - 1)$ 74. $G(t) = \ln(t^3 - t)$

In Exercises 75–80 find the inverse functions of the given functions.

75. $y = \ln(x + 3)$ 76. $y = 2^{10^x}$

77. $y = e^{\sqrt{x}}$ 78. $y = (\ln x)^2, \quad x \geq 1$

79. $y = \dfrac{10^x}{10^x + 1}$ 80. $y = \dfrac{1 + e^x}{1 - e^x}$

81. (a) Show that the function $f(x) = \ln\left(x + \sqrt{x^2 + 1}\right)$ is an odd function.

    (b) Find the inverse function of f.

82. Find the equation of the tangent to the curve $y = e^{-x}$ that is perpendicular to the line $2x - y = 8$.

83. Without using a calculator, determine which of the numbers $\log_{10} 99$ or $\log_9 82$ is larger.

84. Any function of the form $f(x) = [g(x)]^{h(x)}$, where $g(x) > 0$, can be analyzed as a power of e by writing $g(x) = e^{\ln g(x)}$ so that $f(x) = e^{h(x)\ln g(x)}$. Using this device, calculate:

    (a) $\displaystyle\lim_{x \to \infty} x^{\ln x}$ (b) $\displaystyle\lim_{x \to 0^+} x^{-\ln x}$

    (c) $\displaystyle\lim_{x \to 0^+} x^{1/x}$ (d) $\displaystyle\lim_{x \to \infty} (\ln 2x)^{-\ln x}$

85. Let $a > 1$. Prove, using Definitions 1.31 and 1.38, that

    (a) $\displaystyle\lim_{x \to -\infty} a^x = 0$ (b) $\displaystyle\lim_{x \to \infty} a^x = \infty$

86. Solve the inequality $\ln(x^2 - 2x - 2) \leq 0$.

87. Solve the equation $4^x - 2^{x+3} + 12 = 0$.

SECTION 3.5

Derivatives of Logarithmic Functions

In this section we find the derivatives of the logarithmic functions $y = \log_a x$ and the exponential functions $y = a^x$. We start with the natural logarithmic function $y = \ln x$. We know that it is differentiable by Theorem 3.18 because it is the inverse of the differentiable function $y = e^x$.

(3.27)

$$\frac{d}{dx}(\ln x) = \frac{1}{x}$$

Proof   Let $y = \ln x$. Then

$$e^y = x$$

Differentiating this equation implicitly with respect to x, we get

$$e^y \frac{dy}{dx} = 1$$

and so

$$\frac{dy}{dx} = \frac{1}{e^y} = \frac{1}{x}$$

EXAMPLE 1   Differentiate $y = \ln(x^3 + 1)$.

Solution   To use the Chain Rule we let $u = x^3 + 1$. Then $y = \ln u$, so

$$\frac{dy}{dx} = \frac{dy}{du}\frac{du}{dx} = \frac{1}{u}\frac{du}{dx} = \frac{1}{x^3 + 1}(3x^2) = \frac{3x^2}{x^3 + 1}$$

In general, if we combine Formula 3.27 with the Chain Rule as in Example 1, we get

(3.28)
$$\boxed{\frac{d}{dx}\ln u = \frac{1}{u}\frac{du}{dx}} \quad \text{or} \quad \boxed{\frac{d}{dx}\ln g(x) = \frac{g'(x)}{g(x)}}$$

EXAMPLE 2   Find $\frac{d}{dx}\ln(\sin x)$.

Solution   Using (3.28), we have

$$\frac{d}{dx}\ln(\sin x) = \frac{1}{\sin x}\frac{d}{dx}\sin x = \frac{1}{\sin x}\cos x = \cot x$$

EXAMPLE 3   Differentiate $f(x) = \sqrt{\ln x}$.

Solution   This time the logarithm is the inner function, so the Chain Rule gives

$$f'(x) = \frac{1}{2}(\ln x)^{-1/2}\frac{d}{dx}(\ln x) = \frac{1}{2\sqrt{\ln x}} \cdot \frac{1}{x} = \frac{1}{2x\sqrt{\ln x}}$$

EXAMPLE 4   Find $\frac{d}{dx}\ln\frac{x+1}{\sqrt{x-2}}$.

Solution 1
$$\frac{d}{dx}\ln\frac{x+1}{\sqrt{x-2}} = \frac{1}{\frac{x+1}{\sqrt{x-2}}}\frac{d}{dx}\frac{x+1}{\sqrt{x-2}}$$

$$= \frac{\sqrt{x-2}}{x+1}\frac{1\cdot\sqrt{x-2} - (x+1)(\frac{1}{2})(x-2)^{-1/2}}{x-2}$$

$$= \frac{x-2-(\frac{1}{2})(x+1)}{(x+1)(x-2)} = \frac{x-5}{2(x+1)(x-2)}$$

Solution 2   If we first simplify the given function using the laws of logarithms (3.22), then the differentiation becomes easier:

$$\frac{d}{dx}\ln\frac{x+1}{\sqrt{x-2}} = \frac{d}{dx}\Big[\ln(x+1) - \tfrac{1}{2}\ln(x-2)\Big]$$

$$= \frac{1}{x+1} - \tfrac{1}{2}\Big(\frac{1}{x-2}\Big)$$

(This answer can be left as it is, but if we used a common denominator we would see that it gives the same answer as in Solution 1.)   ●

EXAMPLE 5   Find $f'(x)$ if $f(x) = \ln|x|$.

Solution   Since

$$f(x) = \begin{cases} \ln x & \text{if } x > 0 \\ \ln(-x) & \text{if } x < 0 \end{cases}$$

it follows that

$$f'(x) = \begin{cases} \dfrac{1}{x} & \text{if } x > 0 \\ \dfrac{1}{-x}(-1) = \dfrac{1}{x} & \text{if } x < 0 \end{cases}$$

Thus $f'(x) = 1/x$ for all $x \neq 0$.   ●

The result of this example is worth remembering:

(3.29)
$$\boxed{\frac{d}{dx}\ln|x| = \frac{1}{x}}$$

General Exponential and Logarithmic Functions

In Section 3.2 we showed that the derivative of the general exponential function $f(x) = a^x$, $a > 0$, is a constant multiple of itself:

$$f'(x) = f'(0)a^x \qquad \text{where} \qquad f'(0) = \lim_{h \to 0} \frac{a^h - 1}{h}$$

We are now in a position to show that the value of the constant is $f'(0) = \ln a$.

(3.30)
$$\boxed{\frac{d}{dx}a^x = a^x \ln a}$$

Proof   We use the fact that $e^{\ln a} = a$:

$$\frac{d}{dx}a^x = \frac{d}{dx}(e^{\ln a})^x = \frac{d}{dx}e^{(\ln a)x} = e^{(\ln a)x}\frac{d}{dx}(\ln a)x$$

$$= (e^{\ln a})^x(\ln a) = a^x \ln a \qquad ●$$

In Example 6 in Section 2.3 we considered a population of bacteria that doubles every hour and saw that the population after t hours is $n = n_0 2^t$, where n_0 is the initial population. Formula 3.30 enables us to find the growth rate:

$$\frac{dn}{dt} = n_0 2^t \ln 2$$

EXAMPLE 6 $\frac{d}{dx} 10^{x^2} = 10^{x^2}(\ln 10)\frac{d}{dx} x^2 = (2\ln 10)x10^{x^2}$ ●

Since the general logarithmic function $y = \log_a x$ is the inverse function of $y = a^x$, we can find its derivative from Formula 3.30 as follows.

(3.31)

$$\boxed{\frac{d}{dx}\log_a x = \frac{1}{x\ln a}}$$

Proof   Let $y = \log_a x$. Then $a^y = x$, so implicit differentiation gives

$$a^y(\ln a)y' = 1$$

$$y' = \frac{1}{a^y \ln a} = \frac{1}{x\ln a}$$ ●

EXAMPLE 7 $\frac{d}{dx}\log_{10}(2+\sin x) = \frac{1}{(2+\sin x)\ln 10}\frac{d}{dx}(2+\sin x) = \frac{\cos x}{(2+\sin x)\ln 10}$ ●

From Formula 3.31 we see one of the main reasons that natural logarithms (logarithms with base e) are used in calculus: The differentiation formula is simplest when $b = e$ because $\ln e = 1$.

Logarithmic Differentiation

The calculation of derivatives of complicated functions involving products, quotients, or powers can often be simplified by taking logarithms. The method used in the following example is called **logarithmic differentiation.**

EXAMPLE 8   Differentiate

$$y = \frac{x^{3/4}\sqrt{x^2+1}}{(3x+2)^5}$$

Solution   We take logarithms of both sides of the equation:

$$\ln y = \tfrac{3}{4}\ln x + \tfrac{1}{2}\ln(x^2+1) - 5\ln(3x+2)$$

Differentiating implicitly with respect to x gives

$$\frac{1}{y}\frac{dy}{dx} = \frac{3}{4}\cdot\frac{1}{x} + \frac{1}{2}\cdot\frac{2x}{x^2+1} - 5\cdot\frac{3}{3x+2}$$

Solving for dy/dx, we get

$$\frac{dy}{dx} = y\left(\frac{3}{4x} + \frac{x}{x^2+1} - \frac{15}{3x+2}\right)$$

$$= \frac{x^{3/4}\sqrt{x^2+1}}{(3x+2)^5}\left(\frac{3}{4x}+\frac{x}{x^2+1}-\frac{15}{3x+2}\right)$$

Steps in Logarithmic Differentiation

1. Take logarithms of both sides of an equation $y = f(x)$.

2. Differentiate implicitly with respect to x.

3. Solve the resulting equation for y'.

If $f(x) < 0$ for some values of x, then $\ln f(x)$ is not defined, but we can write $|y| = |f(x)|$ and use Equation 3.29. We illustrate this procedure by proving the general version of the Power Rule, as promised in Section 2.2.

Power Rule (3.32)

If n is any real number and $f(x) = x^n$, then

$$f'(x) = nx^{n-1}$$

Proof   Let $y = x^n$ and use logarithmic differentiation:

If $x = 0$, we can show that $f'(0) = 0$ directly from the definition of a derivative.

$$\ln|y| = \ln|x|^n = n\ln|x| \qquad x \neq 0$$

Therefore

$$\frac{y'}{y} = \frac{n}{x}$$

Hence

$$y' = n\frac{y}{x} = n\frac{x^n}{x} = nx^{n-1}$$

 You should distinguish carefully between the Power Rule ($Dx^n = nx^{n-1}$), where the base is variable and the exponent is constant, and the rule for differentiating exponential functions ($Da^x = a^x \ln a$), where the base is constant and the exponent is variable. In general there are four cases for exponents and bases:

1. $\dfrac{d}{dx}(a^b) = 0$ (a and b are constants)

2. $\dfrac{d}{dx}[f(x)]^b = b[f(x)]^{b-1}f'(x)$

3. $\dfrac{d}{dx}[a^{g(x)}] = a^{g(x)}(\ln a)g'(x)$

4. To find $(d/dx)[f(x)]^{g(x)}$, logarithmic differentiation can be used as in the next example.

EXAMPLE 9   Differentiate $y = x^{\sqrt{x}}$.

Solution 1   Using logarithmic differentiation, we have

$$\ln y = \ln x^{\sqrt{x}} = \sqrt{x}\ln x$$

$$\frac{y'}{y} = \frac{1}{2\sqrt{x}}\ln x + \sqrt{x}\cdot\frac{1}{x}$$

$$y' = y\left(\frac{\ln x}{2\sqrt{x}} + \frac{1}{\sqrt{x}}\right)$$

$$= x^{\sqrt{x}}\left(\frac{\ln x + 2}{2\sqrt{x}}\right)$$

Solution 2   Another method is to write $x^{\sqrt{x}} = \left(e^{\ln x}\right)^{\sqrt{x}}$:

$$\frac{d}{dx}\left(x^{\sqrt{x}}\right) = \frac{d}{dx}\left(e^{\sqrt{x}\ln x}\right)$$

$$= e^{\sqrt{x}\ln x}\frac{d}{dx}\left(\sqrt{x}\ln x\right)$$

$$= x^{\sqrt{x}}\left(\frac{\ln x + 2}{2\sqrt{x}}\right) \quad \text{(as above)}$$

The Number e as a Limit

We have shown that if $f(x) = \ln x$, then $f'(x) = 1/x$. Thus $f'(1) = 1$. We now use this fact to express the number e as a limit.

From the definition of a derivative as a limit, we have

$$f'(1) = \lim_{h\to0}\frac{f(1+h)-f(1)}{h} = \lim_{x\to0}\frac{f(1+x)-f(1)}{x}$$

$$= \lim_{x\to0}\frac{\ln(1+x)-\ln 1}{x} = \lim_{x\to0}\frac{1}{x}\ln(1+x)$$

$$= \lim_{x\to0}\ln(1+x)^{1/x} = \ln\left[\lim_{x\to0}(1+x)^{1/x}\right] \quad \text{(since ln is continuous)}$$

Because $f'(1) = 1$ we have

$$\ln\left[\lim_{x\to0}(1+x)^{1/x}\right] = 1$$

Therefore

(3.33)

$$\boxed{\lim_{x\to0}(1+x)^{1/x} = e}$$

Formula 3.33 is illustrated by the graph of the function $y = (1+x)^{1/x}$ in Figure 3.22 and a table of values for small values of x.

x	$(1+x)^{1/x}$
0.1	2.59374246
0.01	2.70481383
0.001	2.71692393
0.0001	2.71814593
0.00001	2.71826824
0.000001	2.71828047
0.0000001	2.71828169
0.00000001	2.71828182

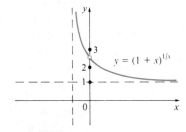

Figure 3.22

If we put $n = 1/x$ in Formula 3.33, then $n \to \infty$ as $x \to 0^+$ and so an alternative expression for e is

(3.34)
$$e = \lim_{n \to \infty} \left(1 + \frac{1}{n}\right)^n$$

SECTION 3.5 Exercises

In Exercises 1–10 find f' and state the domains of f and f'.

1. $f(x) = \ln(x + 1)$ 2. $f(x) = \ln \ln x$

3. $f(x) = \ln(\cos x)$ 4. $f(x) = \cos(\ln x)$

5. $f(x) = \ln(2 - x - x^2)$ 6. $f(x) = \ln \ln \ln x$

7. $f(x) = x^2 \ln(1 - x^2)$

8. $f(x) = \ln(\sqrt{x} - \sqrt{x - 1})$

9. $f(x) = \log_3(x^2 - 4)$ 10. $f(x) = \sqrt{3 - 2^x}$

In Exercises 11–14 find y' and y''.

11. $y = x \ln x$ 12. $y = \ln(ax)$

13. $y = \log_{10} x$ 14. $y = \ln(\sec x + \tan x)$

Differentiate the functions in Exercises 15–56.

15. $f(x) = \sqrt{x} \ln x$ 16. $f(x) = \log_{10}\left(\frac{x}{x-1}\right)$

17. $g(x) = \ln\frac{a-x}{a+x}$ 18. $h(x) = \ln\left(x + \sqrt{x^2 - 1}\right)$

19. $F(x) = \ln\sqrt{x}$ 20. $G(x) = \sqrt{\ln x}$

21. $f(t) = \log_2(t^4 - t^2 + 1)$

22. $g(t) = \sin(\ln t)$

23. $h(y) = \ln(y^3 \sin y)$ 24. $k(r) = r \sin r \ln r$

25. $g(u) = \dfrac{1 - \ln u}{1 + \ln u}$ 26. $G(u) = \ln\sqrt{\dfrac{3u + 2}{3u - 2}}$

27. $y = (\ln \sin x)^3$ 28. $y = \ln(x + \ln x)$

29. $y = \dfrac{\ln x}{1 + x^2}$ 30. $y = \ln(x\sqrt{1 - x^2}\sin x)$

31. $y = \ln\left(\dfrac{x+1}{x-1}\right)^{3/5}$ 32. $y = \ln|\tan 2x|$

33. $y = \ln|x^3 - x^2|$ 34. $y = \tan[\ln(ax + b)]$

35. $F(x) = e^x \ln x$ 36. $G(x) = 5^{\tan x}$

37. $f(t) = \pi^{-t}$ 38. $g(x) = 1.6^x + x^{1.6}$

39. $h(t) = t^3 - 3^t$ 40. $h(\theta) = 10^{\sec \theta}$

41. $y = 2^{3^x}$

42. $y = \ln\left(e^{2x} + \sqrt{e^{4x} + 1}\right)$

43. $y = \ln(e^{-x} + xe^{-x})$ 44. $y = x^x$

45. $y = x^{\sin x}$ 46. $y = (\sin x)^x$

47. $y = x^{e^x}$ 48. $y = (x^e)^x$

49. $y = (\ln x)^x$ 50. $y = x^{\ln x}$

51. $y = x^{1/\ln x}$ 52. $y = (\sin x)^{\cos x}$

53. $y = x^{1/x}$ 54. $y = x^{x^x}$

55. $y = (x^x)^x$ 56. $y = \cos\left(x^{\sqrt{x}}\right)$

57. If $f(x) = \dfrac{x}{\ln x}$, find $f'(e)$.

58. If $f(x) = x^2 \ln x$, find $f'(1)$.

In Exercises 59–62 find the equation of the tangent line to the given curve at the given point.

59. $y = \ln \ln x$, $(e, 0)$ 60. $y = 10^x$, $(1, 10)$

61. $y = \ln(1 + e^x)$, $(0, \ln 2)$

62. $y = (x + 1)^x$, $(1, 2)$

63. Find y' if $y = \ln(x^2 + y^2)$.

64. Find y' if $x^y = y^x$.

65. Find a formula for $f^{(n)}(x)$ if $f(x) = \ln(x - 1)$.

66. Find $\dfrac{d^9}{dx^9}(x^8 \ln x)$.

In Exercises 67–72 use logarithmic differentiation to find the derivatives of the given functions.

67. $y = (3x - 7)^4(8x^2 - 1)^3$

68. $y = x^{2/5}(x^2 + 8)^4 e^{x^2 + x}$

69. $y = \dfrac{(x+1)^4(x-5)^3}{(x-3)^8}$ 70. $y = \sqrt{\dfrac{x^2 + 1}{x + 1}}$

71. $y = \dfrac{e^x\sqrt{x^5 + 2}}{(x+1)^4(x^2+3)^2}$ 72. $y = \dfrac{(x^3+1)^4\sin^2 x}{\sqrt[3]{x}}$

73. If g is the inverse function of $f(x) = 2x + \ln x$, find $g'(2)$.

74. If $f(x) = e^x - \ln x$ and $h(x) = f^{-1}(x)$, find $h'(e)$.

75. Show that $\lim\limits_{n \to \infty} \left(1 + \dfrac{x}{n}\right)^n = e^x$ for any $x > 0$.

76. Use the definition of derivative to prove that

$$\lim_{x \to 0} \frac{\ln(1 + x)}{x} = 1$$

SECTION 3.6

Exponential Growth and Decay

In many natural phenomena, quantities grow or decay at a rate proportional to their size. For instance, if $y = f(t)$ is the number of individuals in a population of animals or bacteria at time t, then it seems reasonable to expect that the rate of growth $f'(t)$ is proportional to the population $f(t)$; that is, $f'(t) = kf(t)$ for some constant k. Indeed, under ideal conditions (unlimited environment, adequate nutrition, freedom from disease) that is what happens. Another example occurs in nuclear physics where the mass of a radioactive substance decays at a rate proportional to the mass. In chemistry, the rate of a unimolecular first-order reaction is proportional to the concentration of the substance. In finance, the value of a savings account with continuously compounded interest increases at a rate proportional to that value.

In general, if $y(t)$ is the value of a quantity y at time t and if the rate of change of y with respect to t is proportional to its size $y(t)$ at any time, then

(3.35)

$$\frac{dy}{dt} = ky$$

where k is a constant. Equation 3.35 is sometimes called the **law of natural growth** (if $k > 0$) or the **law of natural decay** (if $k < 0$). It is called a **differential equation** because it involves an unknown function y and its derivative dy/dt. (This and other differential equations will be studied in Section 8.1.)

It is not hard to think of a solution of Equation 3.35. This equation asks us to find a function whose derivative is a constant multiple of itself. We have met such functions in this chapter. Any exponential function of the form $y(t) = Ce^{kt}$, where C is a constant, satisfies

$$y'(t) = Cke^{kt} = k(Ce^{kt}) = ky(t)$$

Conversely we will show in Section 8.1 that *any* function that satisfies $dy/dt = ky$ must be of the form $y = Ce^{kt}$. Also $y(0) = Ce^{k0} = C$, so C is the initial value of y.

Theorem (3.36)

The only solutions of the differential equation $dy/dt = ky$ are the exponential functions

$$y(t) = y(0)e^{kt}$$

EXAMPLE 1　　A bacteria culture starts with 1000 bacteria and after 2 h there are 2500 bacteria. Assuming that the culture grows at a rate proportional to its size, find the population after 6 h.

Solution　　Let $y(t)$ be the number of bacteria after t hours. Then $y(0) = 1000$ and $y(2) = 2500$. Since we are assuming $dy/dt = ky$, Theorem 3.36 gives

$$y(t) = y(0)e^{kt} = 1000e^{kt}$$

$$y(2) = 1000e^{2k} = 2500$$

Therefore　　　　　　　　$e^{2k} = 2.5$　　and　　$2k = \ln 2.5$

Substituting the value $k = \frac{1}{2}\ln 2.5$ back into the expression for $y(t)$, we have

(3.37)　　　　　　　　　　$y(t) = 1000e^{\ln 2.5(t/2)}$

Since $e^{\ln 2.5} = 2.5$, an alternative expression for Equation 3.37 is

$$y(t) = 1000(2.5)^{t/2}$$

and so　　　　　　　　　　$y(6) = 1000(2.5)^3 = 15,625$　　　　　　　　　●

EXAMPLE 2　　The *half-life* of radium-226 $\left(^{226}_{88}\text{Ra}\right)$ is 1590 years. This means that the rate of decay is proportional to the amount present, and half of any given quantity will disintegrate in 1590 years.
(a)　A sample of radium-226 has a mass of 100 mg. Find a formula for the mass of $^{226}_{88}\text{Ra}$ that remains after t years.
(b)　Find the mass after 1000 years correct to the nearest milligram.
(c)　When will the mass be reduced to 30 mg?

Solution
　　(a)　Let $y(t)$ be the mass of radium (in milligrams) that remains after t years. Then $dy/dt = ky$ and $y(0) = 100$, so Theorem 3.36 gives

$$y(t) = y(0)e^{kt} = 100e^{kt}$$

In order to determine the value of k, we use the fact that $y(1590) = 50$. Thus

$$100e^{1590k} = 50 \quad \text{so} \quad e^{1590k} = \tfrac{1}{2}$$

and　　　　　　　　　　$1590k = \ln\left(\tfrac{1}{2}\right) = -\ln 2$

$$k = -\frac{\ln 2}{1590}$$

Therefore　　　　　　　　$y(t) = 100e^{-(\ln 2/1590)t}$

As in Example 1 we could use the fact that $e^{\ln 2} = 2$ to write the expression for $y(t)$ in the alternative form

$$y(t) = 100 \times 2^{-t/1590}$$

(b)   The mass after 1000 years is

$$y(1000) = 100e^{-(\ln 2/1590)1000} \approx 65\,\text{mg}$$

(c)   We want to find the value of t such that $y(t) = 30$, that is,

$$100e^{-(\ln 2/1590)t} = 30$$

or

$$e^{-(\ln 2/1590)t} = 0.3$$

We solve this equation for t by taking the natural logarithm of both sides:

$$-\frac{\ln 2}{1590}\, t = \ln 0.3$$

Thus

$$t = -1590\,\frac{\ln 0.3}{\ln 2} \approx 2762 \text{ years}$$

EXAMPLE 3   Newton's Law of Cooling states that the rate of cooling of an object is proportional to the temperature difference between the object and its surroundings, provided that this difference is not too large. Suppose the object takes 40 min to cool from 30°C to 24°C in a room that is kept at 20°C.
(a) What was the temperature of the object 15 minutes after it was 30°C?
(b) How long would it take the object to cool down to 21°C?

Solution
(a)   Let $y(t)$ be the temperature t minutes after it was 30°C. Then Newton's Law of Cooling states that

$$\frac{dy}{dt} = k(y - 20)$$

This differential equation is not quite the same as Equation 3.35, so we introduce a new function $u(t) = y(t) - 20$. Then $du/dt = dy/dt$, so

$$\frac{du}{dt} = ku$$

Therefore, by Theorem 3.36, we have

$$u(t) = u(0)e^{kt} = 10e^{kt}$$

We are given that $y(40) = 24$, so $u(40) = 4$ and

$$10e^{40k} = 4 \qquad e^{40k} = 0.4$$

Taking logarithms, we have

$$k = \frac{\ln 0.4}{40}$$

Thus

$$u(t) = 10e^{(\ln 0.4)t/40}$$

(3.38)

$$y(t) = 20 + 10e^{(\ln 0.4)t/40}$$

$$y(15) = 20 + 10e^{(\ln 0.4)15/40} \approx 27.1°C$$

(b)   We have $y(t) = 21$ when

$$20 + 10e^{(\ln 0.4)t/40} = 21$$

$$e^{(\ln 0.4)t/40} = \tfrac{1}{10}$$

$$\frac{(\ln 0.4)t}{40} = \ln \tfrac{1}{10} = -\ln 10$$

$$t = -40\,\frac{\ln 10}{\ln 0.4} \approx 100.5$$

The object will cool down to 21°C after 1 hour and 41 minutes.

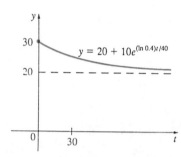

Notice that, in Example 3, $\ln(0.4) < 0$ since $0.4 < 1$. Thus, from Equation 3.38 we have

$$\lim_{t\to\infty} y(t) = 20 + \lim_{t\to\infty} 10e^{(\ln 0.4)t/40}$$

$$= 20 + 0 = 20$$

Figure 3.23   which is to be expected. The graph of the temperature function is shown in Figure 3.23.

EXAMPLE 4   If $1000 is invested at 12% interest, compounded annually, then after 1 year the investment is worth $1000(1.12) = $1120; after 2 years it is worth $[1000(1.12)]1.12 = $1254.40; and after t years it is worth $1000(1.12)^t. In general, if an amount A_0 is invested at an interest rate i ($i = 0.12$ in this example), then after t years it is worth $A_0(1 + i)^t$. Usually, however, interest is compounded more frequently, say n times a year. Then in each compounding period the interest rate is i/n and there are nt compounding periods in t years, so the investment is worth

$$A_0\left(1 + \frac{i}{n}\right)^{nt}$$

For instance, after 3 years at 12% interest a $1000 investment will be worth

$$\$1000(1.12)^3 = \$1404.93 \qquad \text{with annual compounding}$$

$$\$1000(1.06)^6 = \$1418.52 \qquad \text{with semiannual compounding}$$

$$\$1000(1.03)^{12} = \$1425.76 \qquad \text{with quarterly compounding}$$

$$\$1000(1.01)^{36} = \$1430.77 \qquad \text{with monthly compounding}$$

$$\$1000\left(1 + \frac{0.12}{365}\right)^{365 \cdot 3} = \$1433.24 \qquad \text{with daily compounding}$$

You can see that the interest paid increases as the number of compounding periods (n) increases. If we let $n \to \infty$, then we will be compounding the interest continuously and the value of the investment will be

$$A(t) = \lim_{n\to\infty} A_0\left(1 + \frac{i}{n}\right)^{nt} = \lim_{n\to\infty} A_0\left[\left(1 + \frac{i}{n}\right)^{n/i}\right]^{it}$$

$$= A_0\left[\lim_{n\to\infty}\left(1 + \frac{i}{n}\right)^{n/i}\right]^{it}$$

$$= A_0 \left[\lim_{m \to \infty} \left(1 + \frac{1}{m} \right)^m \right]^{it} \quad \text{(where } m = n/i \text{)}$$

$$= A_0 e^{it} \quad \text{(by Equation 3.33)}$$

This formula for continuous compounding $[A(t) = A_0 e^{it}]$ should be compared with the expression for exponential growth in Theorem 3.36: $y(t) = y(0)e^{kt}$. When differentiated, it becomes

$$\frac{dA}{dt} = iA_0 e^{it} = iA(t)$$

which says that, with continuous compounding of interest, the rate of increase of an investment is proportional to its size.

Returning to the example of $1000 invested for 3 years at 12% interest, we see that with continuous compounding of interest the value of the investment will be

$$A(3) = \$1000 e^{(0.12)3}$$

$$= \$1000 e^{0.36} = \$1433.33$$

SECTION 3.6   Exercises

In Exercises 1–6 assume that the population grows at a rate proportional to its size.

1. A cell of the bacterium *Escherichia coli* in a nutrient broth medium divides into two cells every 20 min. If there are initially 100 cells, find (a) an expression for the number of cells after t hours, (b) the number of cells after 10 hours, and (c) the time required for the population to reach 10,000 cells.

2. A bacteria culture starts with 4000 bacteria and the population triples every half-hour. (a) Find an expression for the number of bacteria after t hours. (b) Find the number of bacteria after 20 min. (c) When will the population reach 20,000?

3. A bacteria culture starts with 500 bacteria and after 3 hours there are 8000 bacteria. (a) Find an expression for the number of bacteria after t hours. (b) Find the number of bacteria after 4 hours. (c) When will the population reach 30,000?

4. The count in a bacteria culture was 400 after 2 hours and 25,600 after 6 hours. (a) What was the initial size of the culture? (b) Find an expression for the population after t hours. (c) In what period of time does the population double? (d) When will the population be 100,000?

5. The population of a certain city grows at a rate of 5% per year. The population in 1985 was 307,000. What will the population be (a) in 1995 and (b) in 2001?

6. A city had a population of 450,000 in 1980 and 500,000 in 1985. What will its population be (a) in 1995 and (b) in 2001?

7. Experiments show that if the chemical reaction

$$N_2O_5 \to 2NO_2 + \tfrac{1}{2}O_2$$

takes place at 45°C, the rate of reaction of dinitrogen pentoxide is proportional to its concentration as follows:

$$-\frac{d[N_2O_5]}{dt} = 0.0005[N_2O_5]$$

(See Example 4 in Section 2.3) (a) Find an expression for the concentration $[N_2O_5]$ after t seconds if the initial concentration is C. (b) How long will the reaction take to reduce the concentration of N_2O_5 to 90% of its original value?

8. Polonium-210 has a half-life of 140 days (see Example 2). (a) If a sample has a mass of 200 mg, find a formula for the mass that remains after t days. (b) Find the mass after 100 days. (c) When will the mass be reduced to 10 mg? (d) Sketch the graph of the mass function.

9. Polonium-214 has a very short half-life of 1.4×10^{-4} s. (a) If a sample has a mass of 50 mg, find a formula for the mass that remains after t seconds. (b) Find the mass that remains after a hundredth of a second. (c) How long would it take for the mass to decay to 40 mg?

10. After 3 days a sample of radon-222 decayed to 58% of its original amount. (a) What is the half-life of radon-222? (b) How long would it take the sample to decay to 10% of its original amount?

11. Scientists can determine the age of ancient objects by a method called *radiocarbon dating*. The bombardment of the upper atmosphere by cosmic rays converts nitrogen to a radioactive isotope of carbon, ^{14}C, with a half-life of about 5730 years. Vegetation absorbs carbon dioxide through the atmosphere and animal life assimilates ^{14}C through food chains. When a plant or animal dies it stops replacing its carbon and the amount of ^{14}C begins to decrease through radioactive decay. Therefore the level of radioactivity must also decay exponentially. A parchment fragment was discovered that had about 74% as much ^{14}C radioactivity as does plant material on earth today. Estimate the age of the parchment.

12. A curve passes through the point $(0,5)$ and has the property that the slope of the curve at every point P is twice the y-coordinate of P. What is the equation of the curve?

13. A certain object cools at a rate (in degrees Celsius per minute) equal to one-tenth of the difference between its temperature and that of the surrounding air. If a room is kept at 21°C and the temperature of the object is 33°C, find an expression for the temperature of the object t minutes later.

14. A thermometer is taken from a room where the temperature is 20°C to the outdoors where the temperature is 5°C. After 1 min the thermometer reads 12°C. Use Newton's Law of Cooling to answer the following questions: (a) What will the reading on the thermometer be after one more minute? (b) When will the thermometer read 6°C?

15. A roast turkey is taken from an oven when its temperature has reached 185°F and is placed on a table in a room where the temperature is 75°F. (a) If the temperature of the turkey is 150°F after half an hour, what is the temperature after 45 min? (b) When will the turkey have cooled to 100°F?

16. On a hot day a thermometer is taken outside from an air-conditioned room where the temperature is 21°C. After 1 min it reads 27°C and after 2 min it reads 30°C. (a) What is the outdoor temperature? (b) Sketch the graph of the temperature function.

17. The rate of change of atmospheric pressure P with respect to altitude h is proportional to P, provided that the temperature is constant. At 15°C the pressure is 101.3 kPa at sea level and 87.14 kPa at $h = 1000$ m. (a) What is the pressure at an altitude of 3000 m? (b) What is the pressure at the top of Mount McKinley (6187 m)?

18. If $500 is borrowed at 14% interest, find the amount due at the end of 2 years if the interest is compounded (a) annually, (b) quarterly, (c) monthly, (d) daily, (e) hourly, and (f) continuously.

19. If $3000 is invested at 9% interest, find the value of the investment at the end of 5 years if the interest is compounded (a) annually, (b) semiannually, (c) monthly, (d) weekly, (e) daily, and (f) continuously.

20. How long will it take an investment to double in value if the interest rate is 10% compounded continuously?

21. A tank contains 1500 L of brine with a concentration of 0.3 kg of salt per liter. In order to dilute the solution, pure water is run into the tank at a rate of 20 L/min and the resulting solution, which is continuously stirred, runs out at same rate. (a) How many kilograms of salt will remain after half an hour? (b) When will the concentration be reduced to 0.2 kg of salt per liter?

22. Do Exercise 21 if, instead of pure water, brine with a concentration of 0.1 kg of salt per liter is used.

SECTION 3.7

Inverse Trigonometric Functions

In this section we apply the ideas of Section 3.3 to find the derivatives of the so-called inverse trigonometric functions. There is a slight difficulty here because the trigonometric functions are not one-to-one so they do not have inverse functions. The difficulty is overcome by restricting the domains of these functions so that they are one-to-one.

You can see from Figure 3.24 that the sine function $y = \sin x$ is not one-to-one (use the Horizontal Line Test). But the function $f(x) = \sin x$, $-\pi/2 \le x \le \pi/2$ (see Figure

3.25), *is* one-to-one. The inverse function of this restricted sine function f exists and is denoted by $\sin^{-1}$ or arcsin. It is called the **inverse sine function** or the **arcsine function**.

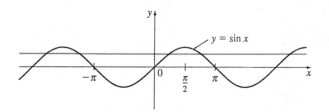

Figure 3.24   **Figure 3.25**

Using the definition of inverse function given by (3.14),

$$f^{-1}(x) = y \quad \Leftrightarrow \quad f(y) = x$$

we have

(3.39)

$$\sin^{-1} x = y \quad \Leftrightarrow \quad \sin y = x \quad \text{and} \quad -\frac{\pi}{2} \le y \le \frac{\pi}{2}$$

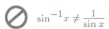 $\sin^{-1} x \ne \dfrac{1}{\sin x}$

Thus if $-1 \le x \le 1$, $\sin^{-1} x$ is the number between $-\pi/2$ and $\pi/2$ whose sine is x.

EXAMPLE 1 Evaluate (a) $\sin^{-1}\frac{1}{2}$ and (b) $\tan\!\left(\arcsin\frac{1}{3}\right)$.

Solution
(a) We have

$$\sin^{-1}\tfrac{1}{2} = \frac{\pi}{6}$$

because $\sin(\pi/6) = \frac{1}{2}$ and $\pi/6$ lies between $-\pi/2$ and $\pi/2$.

(b) Let $\theta = \arcsin\frac{1}{3}$. Then we can draw a right triangle with angle θ as shown in Figure 3.26 and deduce from the Pythagorean Theorem that the third side has length $\sqrt{9-1} = 2\sqrt{2}$. This enables us to read from the triangle that

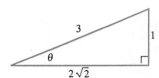

Figure 3.26

$$\tan\!\left(\arcsin\tfrac{1}{3}\right) = \tan\theta = \frac{1}{2\sqrt{2}}$$

The cancellation equations (3.15) for inverse functions become, in this case,

(3.40)

$$\sin^{-1}(\sin x) = x \quad \text{for } -\frac{\pi}{2} \le x \le \frac{\pi}{2}$$

$$\sin(\sin^{-1} x) = x \quad \text{for } -1 \le x \le 1$$

  We must be careful when using the first cancellation equation because it is valid only when x lies in the interval $[-\pi/2, \pi/2]$. The following example shows how to proceed when x lies outside this interval.

EXAMPLE 2   Evaluate:

(a) $\sin(\sin^{-1} 0.6)$ (b) $\sin^{-1}(\sin(\pi/12))$ (c) $\sin^{-1}(\sin(2\pi/3))$

Solution

(a) Since 0.6 lies between -1 and 1, the second cancellation equation in (3.40) gives

$$\sin(\sin^{-1} 0.6) = 0.6$$

(b) Since $\pi/12$ lies between $-\pi/2$ and $\pi/2$, the first cancellation equation gives

$$\sin^{-1}\left(\sin \tfrac{\pi}{12}\right) = \tfrac{\pi}{12}$$

(c) Since $2\pi/3$ does not lie in the interval $[-\pi/2, \pi/2]$, we cannot use the cancellation equation. Instead we note that $\sin(2\pi/3) = \sqrt{3}/2$ and $\sin^{-1}(\sqrt{3}/2) = \pi/3$ because $\pi/3$ lies between $-\pi/2$ and $\pi/2$. Therefore

$$\sin^{-1}\left(\sin \tfrac{2\pi}{3}\right) = \sin^{-1}\left(\tfrac{\sqrt{3}}{2}\right) = \tfrac{\pi}{3} \qquad \bullet$$

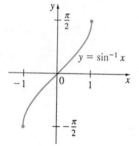

Figure 3.27

The inverse sine function, $\sin^{-1}$, has domain $[-1, 1]$ and range $[-\pi/2,\ \pi/2]$, and its graph, shown in Figure 3.27, is obtained from that of the restricted sine function (Figure 3.25) by reflection about the line $y = x$.

We know that the sine function f is continuous, so by Theorem 3.17 the inverse sine function is also continuous. We also know from Section 2.4 that the sine function is differentiable, so by Theorem 3.18 the inverse sine function is differentiable. We could calculate the derivative of $\sin^{-1}$ by the formula in Theorem 3.18, but since we know that $\sin^{-1}$ is differentiable we can just as easily calculate it by implicit differentiation as follows.

Let $y = \sin^{-1}x$. Then $\sin y = x$ and $-\pi/2 \le y \le \pi/2$. Differentiating the latter equation implicitly with respect to x, we obtain

$$\cos y \, \frac{dy}{dx} = 1$$

and

$$\frac{dy}{dx} = \frac{1}{\cos y}$$

Now $\cos y \ge 0$ since $-\pi/2 \le y \le \pi/2$, so

$$\cos y = \sqrt{1 - \sin^2 y} = \sqrt{1 - x^2}$$

Therefore

$$\frac{dy}{dx} = \frac{1}{\cos y} = \frac{1}{\sqrt{1 - x^2}}$$

(3.41)

$$\frac{d}{dx}(\sin^{-1}x) = \frac{1}{\sqrt{1 - x^2}} \qquad -1 < x < 1$$

EXAMPLE 3   If $f(x) = \sin^{-1}(x^2 - 1)$, find (a) the domain of f, (b) $f'(x)$, and (c) the domain of f'.

Solution

(a)   Since the domain of the inverse sine function is $[-1, 1]$, the domain of f is

$$\{x \mid -1 \leq x^2 - 1 \leq 1\} = \{x \mid 0 \leq x^2 \leq 2\}$$

$$= \left\{x \mid |x| \leq \sqrt{2}\right\} = \left[-\sqrt{2}, \sqrt{2}\right]$$

(b)   Combining Formula 3.41 with the Chain Rule, we have

$$f'(x) = \frac{1}{\sqrt{1 - (x^2 - 1)^2}} \frac{d}{dx}(x^2 - 1)$$

$$= \frac{1}{\sqrt{1 - (x^4 - 2x^2 + 1)}} 2x = \frac{2x}{\sqrt{2x^2 - x^4}}$$

(c)   The domain of f' is

$$\{x \mid -1 < x^2 - 1 < 1\} = \{x \mid 0 < x^2 < 2\}$$

$$= \left\{x \mid 0 < |x| < \sqrt{2}\right\} = (-\sqrt{2}, 0) \cup (0, \sqrt{2})$$ ●

The inverse cosine function is handled similarly. The restricted cosine function $f(x) = \cos x$, $0 \leq x \leq \pi$, is one-to-one [see Figure 3.28(a)] and so it has an inverse function $\cos^{-1} = \arccos$.

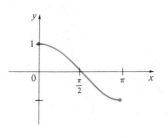

Figure 3.28   (a) $y = \cos x$, $0 \leq x \leq \pi$ (b) $y = \cos^{-1} x$

(3.42)

$$\cos^{-1} x = y \quad \Leftrightarrow \quad \cos y = x \quad \text{and} \quad 0 \leq y \leq \pi$$

The cancellation equations are

(3.43)

$$\cos^{-1}(\cos x) = x \quad \text{for } 0 \leq x \leq \pi$$

$$\cos(\cos^{-1} x) = x \quad \text{for } -1 \leq x \leq 1$$

The inverse cosine function, $\cos^{-1}$, has domain $[-1, 1]$ and range $[0, \pi]$ and is a continuous function whose graph is shown in Figure 3.28(b). Its derivative is given by

(3.44)

$$\frac{d}{dx}(\cos^{-1}x) = -\frac{1}{\sqrt{1-x^2}} \qquad -1 < x < 1$$

Formula 3.44 can be proved by the same method as for Formula 3.41 and is left as Exercise 31.

The tangent function can be made one-to-one by restricting it to the interval $(-\pi/2, \pi/2)$. Thus the **inverse tangent function** is defined as the inverse of the function $f(x) = \tan x$, $-\pi/2 < x < \pi/2$ [see Figure 3.29(a)].

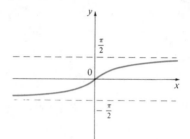

Figure 3.29   (a) $y = \tan x$, $-\pi/2 < x < \pi/2$ (b) $y = \tan^{-1}x = \arctan x$

(3.45)

$$\tan^{-1}x = y \quad \Leftrightarrow \quad \tan y = x \quad \text{and} \quad -\frac{\pi}{2} < y < \frac{\pi}{2}$$

EXAMPLE 4   Simplify the expression $\cos(\tan^{-1}x)$.

Solution 1   Let $y = \tan^{-1}x$. Then $\tan y = x$ and $-\pi/2 < y < \pi/2$. We want to find $\cos y$ but, since $\tan y$ is known, it is easier to find $\sec y$ first:

$$\sec^2 y = 1 + \tan^2 y = 1 + x^2$$

$$\sec y = \sqrt{1 + x^2} \qquad \text{(since } \sec y > 0 \text{ for } -\pi/2 < y < \pi/2\text{)}$$

Thus $$\cos(\tan^{-1}x) = \cos y = \frac{1}{\sec y} = \frac{1}{\sqrt{1+x^2}}$$

Solution 2   Instead of using trigonometric identities as in Solution 1, it is perhaps easier to use a diagram. If $y = \tan^{-1}x$, then $\tan y = x$, and we can read from Figure 3.30 (which illustrates the case where $y > 0$) that

$$\cos(\tan^{-1}x) = \cos y = \frac{1}{\sqrt{1+x^2}}$$

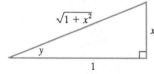

Figure 3.30

The inverse tangent function, $\tan^{-1} = \arctan$, has domain R and range $(-\pi/2, \pi/2)$. Its graph is shown in Figure 3.29(b).

We showed in Example 3 in Section 1.7 that

$$\lim_{x \to (\pi/2)^-} \tan x = \infty \qquad \text{and} \qquad \lim_{x \to -(\pi/2)^+} \tan x = -\infty$$

and so the lines $x = \pm\pi/2$ are vertical asymptotes of the graph of tan. Since the graph of $\tan^{-1}$ is obtained by reflecting the graph of the restricted tangent function about the line $y = x$, it follows that the lines $y = \pi/2$ and $y = -\pi/2$ are horizontal asymptotes of the graph of $\tan^{-1}$. This fact is expressed by the following limits:

(3.46)

$$\lim_{x \to \infty} \tan^{-1} x = \frac{\pi}{2} \qquad \lim_{x \to -\infty} \tan^{-1} x = -\frac{\pi}{2}$$

EXAMPLE 5 Evaluate $\lim\limits_{x \to 2^+} \arctan\left(\dfrac{1}{x - 2}\right)$.

Solution   Since

$$\frac{1}{x - 2} \to \infty \qquad \text{as } x \to 2^+$$

(3.46) gives

$$\lim_{x \to 2^+} \arctan\left(\frac{1}{x - 2}\right) = \frac{\pi}{2}$$

Since $\tan$ is differentiable, $\tan^{-1}$ is also differentiable. To find its derivative, let $y = \tan^{-1} x$. Then $\tan y = x$. Differentiating this latter equation implicitly with respect to x, we have

$$\sec^2 y \frac{dy}{dx} = 1$$

and so

$$\frac{dy}{dx} = \frac{1}{\sec^2 y} = \frac{1}{1 + \tan^2 y} = \frac{1}{1 + x^2}$$

(3.47)

$$\frac{d}{dx}(\tan^{-1} x) = \frac{1}{1 + x^2}$$

The remaining inverse trigonometric functions are not used as frequently and are summarized here.

(3.48)

$$y = \csc^{-1} x \ (|x| \geq 1) \quad \Leftrightarrow \quad \csc y = x \ \text{ and } \ y \in \left(0, \frac{\pi}{2}\right] \cup \left(\pi, \frac{3\pi}{2}\right]$$

$$y = \sec^{-1} x \ (|x| \geq 1) \quad \Leftrightarrow \quad \sec y = x \ \text{ and } \ y \in \left[0, \frac{\pi}{2}\right) \cup \left[\pi, \frac{3\pi}{2}\right)$$

$$y = \cot^{-1} x \ (x \in R) \quad \Leftrightarrow \quad \cot y = x \ \text{ and } \ y \in (0, \pi)$$

The choice of intervals for y in the definitions of $\csc^{-1}$ and $\sec^{-1}$ is not universally accepted. For instance, some authors use $y \in [0, \pi/2) \cup (\pi/2, \pi]$ in the definition of $\sec^{-1}$. The reason for the choice in (3.48) is that the differentiation formulas turn out to be simpler (see Exercise 89).

We collect in Table 3.49 the differentiation formulas for all of the inverse trigonometric functions. The proofs of the formulas for the derivatives of $\csc^{-1}$, $\sec^{-1}$, and $\cot^{-1}$ are left as Exercises 33–35.

Table of Derivatives of Inverse Trigonometric Functions (3.49)

$$\frac{d}{dx}\left(\sin^{-1}x\right) = \frac{1}{\sqrt{1-x^2}} \qquad \frac{d}{dx}\left(\csc^{-1}x\right) = -\frac{1}{x\sqrt{x^2-1}}$$

$$\frac{d}{dx}\left(\cos^{-1}x\right) = -\frac{1}{\sqrt{1-x^2}} \qquad \frac{d}{dx}\left(\sec^{-1}x\right) = \frac{1}{x\sqrt{x^2-1}}$$

$$\frac{d}{dx}\left(\tan^{-1}x\right) = \frac{1}{1+x^2} \qquad \frac{d}{dx}\left(\cot^{-1}x\right) = -\frac{1}{1+x^2}$$

All of these formulas can be combined with the Chain Rule. For instance, if u is a differentiable function of x, then

$$\frac{d}{dx}\left(\sin^{-1}u\right) = \frac{1}{\sqrt{1-u^2}}\frac{du}{dx} \qquad \text{and} \qquad \frac{d}{dx}\left(\tan^{-1}u\right) = \frac{1}{1+u^2}\frac{du}{dx}$$

EXAMPLE 6   Differentiate (a) $y = \dfrac{1}{\sin^{-1}x}$ and (b) $f(x) = x\tan^{-1}\sqrt{x}$.

Solution

(a)
$$\frac{dy}{dx} = \frac{d}{dx}\left(\sin^{-1}x\right)^{-1} = -(\sin^{-1}x)^{-2}\frac{d}{dx}\left(\sin^{-1}x\right)$$

$$= -\frac{1}{(\sin^{-1}x)^2\sqrt{1-x^2}}$$

(b)
$$f'(x) = \tan^{-1}\sqrt{x} + x\frac{1}{1+(\sqrt{x})^2}\frac{1}{2}x^{-1/2}$$

$$= \tan^{-1}\sqrt{x} + \frac{\sqrt{x}}{2(1+x)}$$

●

SECTION 3.7 **Exercises**

Find the exact value of the expressions in Exercises 1–26.

1. $\cos^{-1}(-1)$
2. $\sin^{-1}(0.5)$
3. $\tan^{-1}\sqrt{3}$
4. $\arctan(-1)$
5. $\csc^{-1}\sqrt{2}$
6. $\arcsin 1$
7. $\cot^{-1}(-\sqrt{3})$
8. $\cos^{-1}\left(\dfrac{\sqrt{3}}{2}\right)$
9. $\sin^{-1}\left(-\dfrac{1}{\sqrt{2}}\right)$
10. $\sec^{-1}2$
11. $\arctan\left(-\dfrac{\sqrt{3}}{3}\right)$
12. $\arccos(-0.5)$
13. $\sin(\sin^{-1}0.7)$
14. $\sin^{-1}(\sin 1)$
15. $\tan(\tan^{-1}10)$
16. $\tan^{-1}\left(\tan\dfrac{4\pi}{3}\right)$
17. $\cos\left(\sin^{-1}\dfrac{\sqrt{3}}{2}\right)$
18. $\tan(\cos^{-1}0.5)$

19. $\sin(\cos^{-1}\frac{4}{5})$
20. $\sec(\arctan 2)$
21. $\arcsin\left(\sin\dfrac{5\pi}{4}\right)$
22. $\sin(2\sin^{-1}\frac{3}{5})$
23. $\cos(2\sin^{-1}\frac{5}{13})$
24. $\cos(2\sin^{-1}x),\ |x|\le 1$
25. $\sin\left[\sin^{-1}\frac{1}{3} + \sin^{-1}\frac{2}{3}\right]$
26. $\cos\left[\sin^{-1}\frac{3}{4} + \cos^{-1}\frac{1}{4}\right]$

27. Prove that $\cos(\sin^{-1}x) = \sqrt{1-x^2}$ for $-1 \le x \le 1$.

In Exercises 28–30 simplify each expression as in Exercise 27.

28. $\tan(\sin^{-1}x)$
29. $\sin(\tan^{-1}x)$
30. $\sin(2\cos^{-1}x)$

31. Prove Formula 3.44 for the derivative of $\cos^{-1}$ by the same method as for Formula 3.41.

32. (a) Prove that $\sin^{-1}x + \cos^{-1}x = \pi/2$ for all x.

(b) Use part (a) to prove Formula 3.44.

33. Prove that $D \cot^{-1} x = -\dfrac{1}{1+x^2}$.

34. Prove that $D \sec^{-1} x = \dfrac{1}{x\sqrt{x^2-1}},\ |x| > 1$.

35. Prove that $D \csc^{-1} x = -\dfrac{1}{x\sqrt{x^2-1}},\ |x| > 1$.

In Exercises 36–59 find the derivatives of the given functions. Simplify where possible.

36. $f(x) = \sin^{-1}(2x-1)$ 37. $g(x) = \tan^{-1}(x^3)$

38. $y = (\sin^{-1} x)^2$ 39. $y = \sin^{-1}(x^2)$

40. $G(x) = \sin^{-1}\left(\dfrac{x}{a}\right),\ a > 0$

41. $F(x) = \tan^{-1}\left(\dfrac{x}{a}\right)$

42. $h(x) = (\arcsin x)\ln x$

43. $H(x) = (1+x^2)\arctan x$

44. $f(t) = (\cos^{-1} t)/t$ 45. $g(t) = \sin^{-1}(4/t)$

46. $F(t) = \sqrt{1-t^2} + \sin^{-1} t$

47. $G(t) = \cos^{-1}\sqrt{2t-1}$

48. $y = \tan^{-1}\left(\dfrac{x}{a}\right) + \ln\sqrt{\dfrac{x-a}{x+a}}$

49. $y = \sec^{-1}\sqrt{1+x^2}$

50. $y = x\cos^{-1} x - \sqrt{1-x^2}$

51. $y = \tan^{-1}(\sin x)$

52. $y = \sin^{-1}\left(\dfrac{\cos x}{1+\sin x}\right)$ 53. $y = (\sin^{-1} x)/\cos^{-1} x$

54. $y = \sqrt[4]{\arcsin\sqrt{x^2+2x}}$

55. $y = (\tan^{-1} x)^{-1}$

56. $y = \tan^{-1}\left(x - \sqrt{1+x^2}\right)$

57. $y = x^2 \cot^{-1}(3x)$

58. $y = x\sin x\csc^{-1} x$

59. $y = \arccos\left(\dfrac{b+a\cos x}{a+b\cos x}\right),\ 0 \le x \le \pi,\ a > b > 0$

In Exercises 60–67 find the derivative of the given function. State the domain of the function and the domain of its derivative.

60. $f(x) = \cos^{-1}(\sin^{-1} x)$ 61. $g(x) = \sin^{-1}(3x+1)$

62. $h(x) = x\sec^{-1}(3x^2)$ 63. $S(x) = \sin^{-1}(\tan^{-1} x)$

64. $F(x) = \sqrt{\sin^{-1}(2/x)}$ 65. $G(x) = \sqrt{\csc^{-1} x}$

66. $R(t) = \arcsin(2^t)$ 67. $U(t) = 2^{\arctan t}$

68. If $f(x) = x\tan^{-1} x$, find $f'(1)$.

69. If $g(x) = x\sin^{-1}(x/4) + \sqrt{16-x^2}$, find $g'(2)$.

70. If $h(x) = (3\tan^{-1} x)^4$, find $h'(3)$.

In Exercises 71–80 find the given limits.

71. $\lim\limits_{x \to -1^+} \sin^{-1} x$ 72. $\lim\limits_{x \to \infty} \sin^{-1}\left(\dfrac{x+1}{2x+1}\right)$

73. $\lim\limits_{x \to -\infty} \cot^{-1} x$ 74. $\lim\limits_{x \to \infty} \sec^{-1} x$

75. $\lim\limits_{x \to -\infty} \csc^{-1} x$ 76. $\lim\limits_{x \to \infty} \tan^{-1}(x^2)$

77. $\lim\limits_{x \to \infty} (\tan^{-1} x)^2$ 78. $\lim\limits_{x \to \infty} \tan^{-1}(x - x^2)$

79. $\lim\limits_{x \to 1^-} \dfrac{\arcsin x}{\tan(\pi x/2)}$ 80. $\lim\limits_{x \to \infty} \dfrac{\tan^{-1} x}{x}$

81. A ladder 10 ft long leans against a vertical wall. If the bottom of the ladder slides away from the base of the wall at a speed of 2 ft/s, how fast is the angle between the ladder and the wall changing when the bottom of the ladder is 6 ft from the base of the wall?

82. A lighthouse is on a small island 3 km away from the nearest point P on a straight shoreline and its light makes four revolutions per minute. How fast is the beam of light moving along the shoreline when it is 1 km from P?

83. Sketch the graph of $y = \sec^{-1} x$.

84. Sketch the graph of $y = \arctan(\tan x)$.

85. Prove that, for $xy \ne 1$,
$$\arctan x + \arctan y = \arctan \frac{x+y}{1-xy}$$
if the left side lies between $-\pi/2$ and $\pi/2$.

86. Use the result of Exercise 85 to prove the following:
(a) $\arctan\frac{1}{2} + \arctan\frac{1}{3} = \pi/4$
(b) $2\arctan\frac{1}{3} + \arctan\frac{1}{7} = \pi/4$

87. Let $f(x) = x\arctan(1/x)$ if $x \ne 0$ and $f(0) = 0$.
(a) Is f continuous at 0? (b) Is f differentiable at 0?

88. (a) Sketch the graph of $f(x) = \sin(\sin^{-1} x)$.
(b) Sketch the graph of $g(x) = \sin^{-1}(\sin x),\ x \in R$.
(c) Show that $g'(x) = \dfrac{\cos x}{|\cos x|}$.

(d) Sketch the graph of $h(x) = \cos^{-1}(\sin x)$, $x \in R$, and find its derivative.

89. Some authors define $y = \sec^{-1} x \Leftrightarrow \sec y = x$ and $y \in [0, \pi/2) \cup (\pi/2, \pi]$. Show that if this definition is adopted, then, instead of the formula given in Exercise 34, we have

$$\frac{d}{dx}(\sec^{-1} x) = \frac{1}{|x|\sqrt{x^2 - 1}} \qquad |x| > 1$$

SECTION 3.8

Hyperbolic Functions

Certain combinations of the exponential functions e^x and e^{-x} arise so frequently in mathematics and its applications that they deserve to be given special names. In many ways they are analogous to the trigonometric functions, and they have the same relationship to the hyperbola that the trigonometric functions have to the circle. For this reason they are collectively called **hyperbolic functions** and individually called **hyperbolic sine, hyperbolic cosine,** and so on.

Definition of the Hyperbolic Functions (3.50)

$$\sinh x = \frac{e^x - e^{-x}}{2} \qquad \operatorname{csch} x = \frac{1}{\sinh x}$$

$$\cosh x = \frac{e^x + e^{-x}}{2} \qquad \operatorname{sech} x = \frac{1}{\cosh x}$$

$$\tanh x = \frac{\sinh x}{\cosh x} \qquad \coth x = \frac{\cosh x}{\sinh x}$$

The graphs of hyperbolic sine and cosine can be sketched using graphical addition as in Figure 3.31.

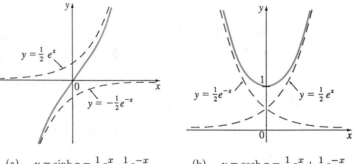

(a) $y = \sinh x = \frac{1}{2}e^x - \frac{1}{2}e^{-x}$ (b) $y = \cosh x = \frac{1}{2}e^x + \frac{1}{2}e^{-x}$

Figure 3.31

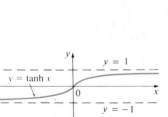

Figure 3.32

Note that sinh has domain R and range R, while cosh has domain R and range $[1, \infty)$. The graph of tanh is shown in Figure 3.32. It has the horizontal asymptotes $y = \pm 1$. (See Exercise 19.)

Some of the mathematical uses of hyperbolic functions will be seen in Chapter 7. Applications to science and engineering occur whenever an entity such as light, velocity, electricity, or radioactivity is gradually absorbed or extinguished, for the decay can be represented by hyperbolic functions. The most famous application is the use of hyperbolic cosine to describe the shape of a hanging wire. It can be proved that if a heavy flexible cable (such as a telephone or power line) is suspended between two points at the same height, then it takes the shape of a curve with equation $y = a\cosh(x/a)$ called a

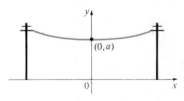

Figure 3.33

A catenary

catenary (see Figure 3.33). (The Latin word *catena* means "chain.")

The hyperbolic functions satisfy a number of identities that are analogues of well-known trigonometric identities. We list some of them here and leave most of the proofs to the exercises.

Hyperbolic Identities (3.51)

$$\sinh(-x) = -\sinh x \qquad \cosh(-x) = \cosh x$$

$$\cosh^2 x - \sinh^2 x = 1 \qquad 1 - \tanh^2 x = \text{sech}^2 x$$

$$\sinh(x + y) = \sinh x \cosh y + \cosh x \sinh y$$

$$\cosh(x + y) = \cosh x \cosh y + \sinh x \sinh y$$

EXAMPLE 1   Prove (a) $\cosh^2 x - \sinh^2 x = 1$ and (b) $1 - \tanh^2 x = \text{sech}^2 x$.

Solution

(a)
$$\cosh^2 x - \sinh^2 x = \left(\frac{e^x + e^{-x}}{2}\right)^2 - \left(\frac{e^x - e^{-x}}{2}\right)^2$$

$$= \frac{e^{2x} + 2 + e^{-2x}}{4} - \frac{e^{2x} - 2 + e^{-2x}}{4}$$

$$= \tfrac{4}{4} = 1$$

(b)   If we start with the identity proved in part (a):

$$\cosh^2 x - \sinh^2 x = 1$$

and we divide both sides by $\cosh^2 x$, we get

$$1 - \frac{\sinh^2 x}{\cosh^2 x} = \frac{1}{\cosh^2 x}$$

or
$$1 - \tanh^2 x = \text{sech}^2 x \qquad \bullet$$

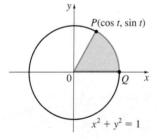

Figure 3.34

The identity proved in Example 1(a) gives a clue to the reason for the name "hyperbolic" functions.

If t is any real number, then the point $P(\cos t, \sin t)$ lies on the unit circle $x^2 + y^2 = 1$ because $\cos^2 t + \sin^2 t = 1$. In fact, t can be interpreted as the radian measure of $\angle POQ$ in Figure 3.34. For this reason the trigonometric functions are sometimes called *circular* functions.

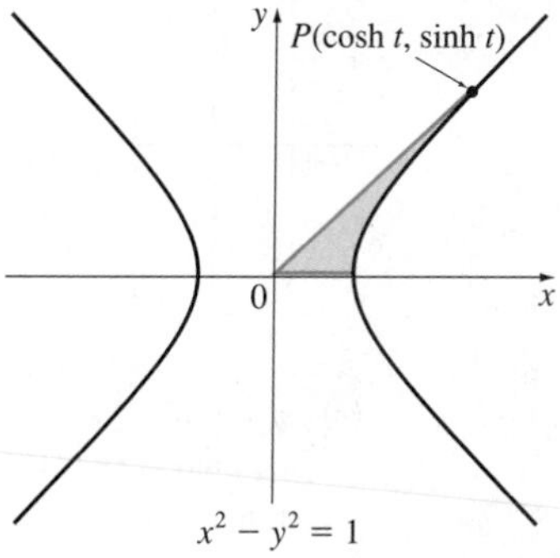

Likewise if t is any real number, then the point $P(\cosh t, \sinh t)$ lies on the right branch of the hyperbola $x^2 - y^2 = 1$ because $\cosh^2 t - \sinh^2 t = 1$ and $\cosh t \geq 1$. This time, t does not represent the measure of an angle. However it turns out that t represents twice the area of the shaded hyperbolic sector in Figure 3.35 (see Exercise 77 in the Review for Chapter 5), just as in the trigonometric case t represents twice the area of the shaded circular sector in Figure 3.34.

The derivatives of the hyperbolic functions are easily computed. For example,

$$\frac{d}{dx}(\sinh x) = \frac{d}{dx}\left(\frac{e^x - e^{-x}}{2}\right) = \frac{e^x + e^{-x}}{2} = \cosh x$$

Figure 3.35   We list the differentiation formulas for the hyperbolic functions as Table 3.52. The remaining proofs are left as exercises. Note the analogy with the differentiation formulas for trigonometric functions, but beware that the signs are sometimes different.

Table of Derivatives of Hyperbolic Functions (3.52)

$$\frac{d}{dx}\sinh x = \cosh x \qquad \frac{d}{dx}\operatorname{csch} x = -\operatorname{csch} x \coth x$$

$$\frac{d}{dx}\cosh x = \sinh x \qquad \frac{d}{dx}\operatorname{sech} x = -\operatorname{sech} x \tanh x$$

$$\frac{d}{dx}\tanh x = \operatorname{sech}^2 x \qquad \frac{d}{dx}\coth x = -\operatorname{csch}^2 x$$

EXAMPLE 2   Any of these differentiation rules can be combined with the Chain Rule. For instance,

$$\frac{d}{dx}\left(\cosh\sqrt{x}\right) = \sinh\sqrt{x} \cdot \frac{d}{dx}\sqrt{x} = \frac{\sinh\sqrt{x}}{2\sqrt{x}}$$

●

Inverse Hyperbolic Functions

You can see from Figures 3.31(a) and 3.32 that sinh and tanh are one-to-one functions and so they have inverse functions denoted by $\sinh^{-1}$ and $\tanh^{-1}$. Figure 3.31(b) shows that cosh is not one-to-one, but when restricted to the domain $[0,\infty)$ it becomes one-to-one. The inverse hyperbolic cosine function is defined as the inverse of this restricted function.

(3.53)

$$y = \sinh^{-1}x \quad \Leftrightarrow \quad \sinh y = x$$

$$y = \cosh^{-1}x \quad \Leftrightarrow \quad \cosh y = x \quad \text{and} \quad y \geq 0$$

$$y = \tanh^{-1}x \quad \Leftrightarrow \quad \tanh y = x$$

The remaining inverse hyperbolic functions are defined similarly (see Exercise 24).

We can sketch the graphs of $\sinh^{-1}$, $\cosh^{-1}$, and $\tanh^{-1}$ in Figure 3.36 by using Figures 3.31 and 3.32.

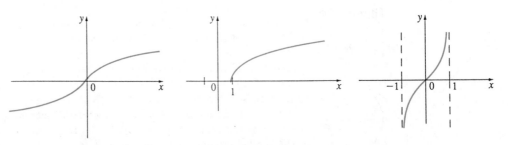

(a) $y = \sinh^{-1}x$
domain $= R$
range $= R$

(b) $y = \cosh^{-1}x$
domain $= [1,\infty)$
range $= [0,\infty)$

(c) $y = \tanh^{-1}x$
domain $= (-1, 1)$
range $= R$

Figure 3.36

Since the hyperbolic functions are defined in terms of exponential functions, it is not surprising to learn that the inverse hyperbolic functions can be expressed in terms of logarithms. In particular we have:

(3.54)
$$\sinh^{-1}x = \ln\left(x + \sqrt{x^2+1}\right) \qquad x \in R$$

(3.55)
$$\cosh^{-1}x = \ln\left(x + \sqrt{x^2-1}\right) \qquad x \geq 1$$

(3.56)
$$\tanh^{-1}x = \tfrac{1}{2}\ln\left(\frac{1+x}{1-x}\right) \qquad -1 < x < 1$$

EXAMPLE 3 Show that $\sinh^{-1}x = \ln\left(x + \sqrt{x^2+1}\right)$.

Solution Let $y = \sinh^{-1}x$. then

$$x = \sinh y = \frac{e^y - e^{-y}}{2}$$

so
$$e^y - 2x - e^{-y} = 0$$

or, multiplying by e^y,

$$e^{2y} - 2xe^y - 1 = 0$$

This is really a quadratic equation in e^y:

$$(e^y)^2 - 2x(e^y) - 1 = 0$$

Solving by the quadratic formula, we get

$$e^y = \frac{2x \pm \sqrt{4x^2+4}}{2} = x \pm \sqrt{x^2+1}$$

Note that $e^y > 0$, but $x - \sqrt{x^2+1} < 0$ (because $x < \sqrt{x^2+1}$). Thus the minus sign is inadmissible and we have

$$e^y = x + \sqrt{x^2+1}$$

Therefore
$$y = \ln(e^y) = \ln\left(x + \sqrt{x^2+1}\right)$$

(See Exercise 21 for another method.) ●

Table of Derivatives of Inverse Hyperbolic Functions (3.57)

$$\frac{d}{dx}\sinh^{-1}x = \frac{1}{\sqrt{1+x^2}} \qquad\qquad \frac{d}{dx}\operatorname{csch}^{-1}x = -\frac{1}{|x|\sqrt{x^2+1}}$$

$$\frac{d}{dx}\cosh^{-1}x = \frac{1}{\sqrt{x^2-1}} \qquad\qquad \frac{d}{dx}\operatorname{sech}^{-1}x = -\frac{1}{x\sqrt{1-x^2}}$$

$$\frac{d}{dx}\tanh^{-1}x = \frac{1}{1-x^2} \qquad\qquad \frac{d}{dx}\operatorname{coth}^{-1}x = \frac{1}{1-x^2}$$

The inverse hyperbolic functions are all differentiable by Theorem 3.18 because the hyperbolic functions are differentiable. The formulas in Table 3.57 can be proved either by the method for inverse functions or by differentiating Formulas 3.54, 3.55, and 3.56.

EXAMPLE 4   Prove that $\dfrac{d}{dx}\sinh^{-1}x = \dfrac{1}{\sqrt{1+x^2}}$.

Solution 1   Let $y = \sinh^{-1}x$. Then $\sinh y = x$. If we differentiate this equation implicitly with respect to x, we get

$$\cosh y\,\frac{dy}{dx} = 1$$

Since $\cosh^2 y - \sinh^2 y = 1$ and $\cosh y \geq 0$, we have $\cosh y = \sqrt{1 + \sinh^2 y}$, so

$$\frac{dy}{dx} = \frac{1}{\cosh y} = \frac{1}{\sqrt{1+\sinh^2 y}} = \frac{1}{\sqrt{1+x^2}}$$

Solution 2   From Equation 3.54 (proved in Example 3), we have

$$\begin{aligned}
\frac{d}{dx}\sinh^{-1}x &= \frac{d}{dx}\ln\!\left(x + \sqrt{x^2+1}\right) \\[2mm]
&= \frac{1}{x+\sqrt{x^2+1}}\,\frac{d}{dx}\!\left(x + \sqrt{x^2+1}\right) \\[2mm]
&= \frac{1}{x+\sqrt{x^2+1}}\left(1 + \frac{x}{\sqrt{x^2+1}}\right) \\[2mm]
&= \frac{\sqrt{x^2+1}+x}{\left(x+\sqrt{x^2+1}\right)\sqrt{x^2+1}} \\[2mm]
&= \frac{1}{\sqrt{x^2+1}}
\end{aligned}$$

EXAMPLE 5   Find $\dfrac{d}{dx}\tanh^{-1}(\sin x)$.

Solution   Using Table 3.57 and the Chain Rule, we have

$$\frac{d}{dx}\tanh^{-1}(\sin x) = \frac{1}{1-(\sin x)^2}\,\frac{d}{dx}(\sin x)$$

$$= \frac{1}{1-\sin^2 x}\cos x = \frac{\cos x}{\cos^2 x} = \sec x$$

SECTION 3.8 **Exercises**

Prove the identities in Exercises 1–15.

1. $\sinh(-x) = -\sinh x$
    (This shows that sinh is an odd function.)

2. $\cosh(-x) = \cosh x$
    (This shows that cosh is an even function.)

3. $\cosh x + \sinh x = e^x$

4. $\cosh x - \sinh x = e^{-x}$

5. $\sinh(x+y) = \sinh x \cosh y + \cosh x \sinh y$

6. $\cosh(x+y) = \cosh x \cosh y + \sinh x \sinh y$

7. $\coth^2 x - 1 = \operatorname{csch}^2 x$

8. $\tanh(x+y) = \dfrac{\tanh x + \tanh y}{1 + \tanh x \tanh y}$

9. $\sinh 2x = 2\sinh x \cosh x$

10. $\cosh 2x = \cosh^2 x + \sinh^2 x$

11. $\sinh \frac{x}{2} = \pm\sqrt{\dfrac{\cosh x - 1}{2}}$

12. $\cosh \frac{x}{2} = \sqrt{\dfrac{\cosh x + 1}{2}}$

13. $\tanh(\ln x) = \dfrac{x^2 - 1}{x^2 + 1}$

14. $\dfrac{1 + \tanh x}{1 - \tanh x} = e^{2x}$

15. $(\cosh x + \sinh x)^n = \cosh nx + \sinh nx$
    (n any real number)

16. If $\sinh x = \frac{3}{4}$, find the values of the other hyperbolic functions at x.

17. If $\tanh x = \frac{4}{5}$, find the values of the other hyperbolic functions at x.

18. Use the graphs of sinh, cosh, and tanh in Figures 3.31 and 3.32 to draw the graphs of csch, sech, and coth.

19. Use the definitions of the hyperbolic functions to find the following limits:
    (a) $\lim\limits_{x\to\infty} \tanh x$ (b) $\lim\limits_{x\to-\infty} \tanh x$ (c) $\lim\limits_{x\to\infty} \sinh x$
    (d) $\lim\limits_{x\to-\infty} \sinh x$ (e) $\lim\limits_{x\to\infty} \operatorname{sech} x$ (f) $\lim\limits_{x\to\infty} \coth x$
    (g) $\lim\limits_{x\to 0^+} \coth x$ (h) $\lim\limits_{x\to 0^-} \coth x$ (i) $\lim\limits_{x\to-\infty} \operatorname{csch} x$

20. Prove the formulas given in Table 3.52 for the derivatives of the following functions: (a) cosh, (b) tanh, (c) csch, (d) sech, and (e) coth.

21. Give an alternative solution to Example 3 by letting $y = \sinh^{-1} x$ and then using Exercise 3 and Example 1(a) (with x replaced by y).

22. Prove Equation 3.55.

23. Prove Equation 3.56 (a) using the method of Example 3, and (b) using Exercise 14 (with x replaced by y).

24. For each of the following functions (i) give a definition like those in (3.53), (ii) sketch the graph, and

(iii) find a formula similar to Equation 3.54.
    (a) csch^{-1} (b) sech^{-1} (c) $\coth^{-1}$

25. Prove the formulas given in Table 3.57 for the derivatives of the following functions.
    (a) $\cosh^{-1}$ (b) $\tanh^{-1}$ (c) csch^{-1}
    (d) sech^{-1} (e) $\coth^{-1}$

In Exercises 26–45 find the derivatives of the given functions.

26. $f(x) = e^x \sinh x$ 27. $f(x) = \tanh 3x$

28. $g(x) = \cosh^4 x$ 29. $h(x) = \cosh(x^4)$

30. $F(x) = e^{\coth 2x}$ 31. $G(x) = x^2 \operatorname{sech} x$

32. $f(t) = \ln(\sinh t)$ 33. $H(t) = \tanh(e^t)$

34. $y = \cos(\sinh x)$ 35. $y = x^{\cosh x}$

36. $y = e^{\tanh x} \cosh(\cosh x)$

37. $y = \cosh^{-1}(x^2)$ 38. $y = \sqrt{x}\sinh^{-1}\sqrt{x}$

39. $y = x \ln(\operatorname{sech} 4x)$

40. $y = x\tanh^{-1}x + \ln\sqrt{1 - x^2}$

41. $y = \tanh^{-1}\left(\dfrac{x}{a}\right)$

42. $y = x\sinh^{-1}(x/3) - \sqrt{9 + x^2}$

43. $y = \operatorname{csch}^{-1}(x^4)$

44. $y = \operatorname{sech}^{-1}\sqrt{1 - x^2}$, $x > 0$

45. $y = \coth^{-1}\sqrt{x^2 + 1}$

46. Evaluate $\lim\limits_{x\to\infty} \dfrac{\sinh x}{e^x}$.

47. At what point of the curve $y = \cosh x$ does the tangent have slope 1?

48. (a) Show that any function of the form $y = A\sinh mx + B\cosh mx$ satisfies the differential equation $y'' = m^2 y$.
    (b) Find $y = y(x)$ such that $y'' = 9y$, $y(0) = -4$, and $y'(0) = 6$.

49. If $x = \ln(\sec\theta + \tan\theta)$, show that $\sec\theta = \cosh x$.

SECTION 3.9

Indeterminate Forms and L'Hospital's Rule

Suppose we are trying to sketch the graph of the function

$$F(x) = \frac{2^x - 1}{x}$$

Although F is not defined when $x = 0$, we need to know how F behaves *near* 0. In particular, we would like to know the value of the limit

(3.58)
$$\lim_{x \to 0} \frac{2^x - 1}{x}$$

But we cannot apply Limit Law 5 (the limit of a quotient is the quotient of the limits) to (3.58) because the limit of the denominator is 0. In fact, although the limit in (3.58) exists, its value is not obvious because both numerator and denominator approach 0 and $\frac{0}{0}$ is not defined.

In general, if we have a limit of the form

$$\lim_{x \to a} \frac{f(x)}{g(x)}$$

where both $f(x) \to 0$ and $g(x) \to 0$ as $x \to a$, then this limit may or may not exist and is called an **indeterminate form of type $\frac{0}{0}$**. We met some limits of this type in Chapter 1. For rational functions, we can cancel common factors:

$$\lim_{x \to 1} \frac{x^2 - x}{x^2 - 1} = \lim_{x \to 1} \frac{x(x-1)}{(x+1)(x-1)} = \lim_{x \to 1} \frac{x}{x+1} = \frac{1}{2}$$

We used a geometric argument to show that

$$\lim_{x \to 0} \frac{\sin x}{x} = 1$$

But these methods do not work for limits such as (3.58), so in this section we introduce a systematic method, known as l'Hospital's Rule, for the evaluation of indeterminate forms.

Another situation in which a limit is not obvious occurs when we try to sketch the curve $y = (\ln x)/x$. In searching for horizontal asymptotes or other aspects of the curve for large values of x, we need to evaluate the limit

(3.59)
$$\lim_{x \to \infty} \frac{\ln x}{x}$$

It is not obvious how to evaluate this limit because both numerator and denominator become large as $x \to \infty$. There is a struggle between numerator and denominator. If the numerator wins, the limit will be ∞; if the denominator wins, the answer will be 0. Or there may be some compromise, in which case the answer may be some finite positive number.

In general, if we have a limit of the form

$$\lim_{x \to a} \frac{f(x)}{g(x)}$$

where both $f(x) \to \infty$ (or $-\infty$) and $g(x) \to \infty$ (or $-\infty$), then the limit may or may not exist and is called an **indeterminate form of type ∞/∞**. We saw in Section 1.6 that this type of limit can be evaluated for certain functions, including rational functions, by dividing numerator and denominator by the highest power of x that occurs. For instance,

$$\lim_{x \to \infty} \frac{x^2 - 1}{2x^2 + 1} = \lim_{x \to \infty} \frac{1 - \dfrac{1}{x^2}}{2 + \dfrac{1}{x^2}} = \frac{1 - 0}{2 + 0} = \frac{1}{2}$$

This method does not work for limits such as (3.59) but l'Hospital's Rule also applies to this type of indeterminate form.

L'Hospital's Rule is named after a French nobleman, the Marquis de l'Hospital (1661–1704), but was discovered by a Swiss mathematician, John Bernoulli (1667–1748).

L'Hospital's Rule (3.60)

Suppose f and g are differentiable and $g'(x) \neq 0$ on an open interval I that contains a (except possibly at a). Suppose that

$$\lim_{x \to a} f(x) = 0 \quad \text{and} \quad \lim_{x \to a} g(x) = 0$$

or that

$$\lim_{x \to a} f(x) = \pm\infty \quad \text{and} \quad \lim_{x \to a} g(x) = \pm\infty$$

(In other words, we have an indeterminate form of type $\frac{0}{0}$ or ∞/∞.) Then

$$\lim_{x \to a} \frac{f(x)}{g(x)} = \lim_{x \to a} \frac{f'(x)}{g'(x)}$$

if the limit on the right side exists (or is ∞ or $-\infty$).

Note 1: L'Hospital's Rule says that the limit of a quotient of functions is equal to the limit of the quotient of their derivatives, provided that the conditions stated in (3.60) are satisfied. It is especially important to verify the conditions regarding the limits of f and g before using l'Hospital's Rule.

Note 2: L'Hospital's Rule is also valid for one-sided limits and for limits at infinity or negative infinity; that is, in Theorem 3.60, "$x \to a$" can be replaced by any of the following symbols: $x \to a^+$, $x \to a^-$, $x \to \infty$, $x \to -\infty$.

Note 3: The full proof of l'Hospital's Rule for the indeterminate form $\frac{0}{0}$ is given in Appendix C, but if we make assumptions about the continuity of f, g, and their derivatives, there is a much simpler proof, which we now present.

Proof of l'Hospital's Rule for a Special Case Suppose that $f(a) = 0 = g(a)$, $g'(a) \neq 0$, and f' and g' are continuous. Then, using the alternate form of the definition of a derivative (2.3), we have

$$\lim_{x \to a} \frac{f(x)}{g(x)} = \lim_{x \to a} \frac{f(x) - f(a)}{g(x) - g(a)}$$

$$= \lim_{x \to a} \frac{\dfrac{f(x) - f(a)}{x - a}}{\dfrac{g(x) - g(a)}{x - a}} = \frac{\lim\limits_{x \to a} \dfrac{f(x) - f(a)}{x - a}}{\lim\limits_{x \to a} \dfrac{g(x) - g(a)}{x - a}}$$

$$= \frac{f'(a)}{g'(a)} = \lim_{x \to a} \frac{f'(x)}{g'(x)}$$

EXAMPLE 1   Find $\lim\limits_{x \to 0} \dfrac{2^x - 1}{x}$.

Solution   Since $\lim\limits_{x \to 0} (2^x - 1) = 0$ and $\lim\limits_{x \to 0} x = 0$, we can apply l'Hospital's Rule:

$$\lim_{x \to 0} \frac{2^x - 1}{x} = \lim_{x \to 0} \frac{\dfrac{d}{dx}(2^x - 1)}{\dfrac{d}{dx}(x)} = \lim_{x \to 0} \frac{2^x \ln 2}{1} = \ln 2$$

EXAMPLE 2   Calculate $\lim\limits_{x \to \infty} \dfrac{e^x}{x^2}$.

Solution   We have $\lim\limits_{x \to \infty} e^x = \infty$ and $\lim\limits_{x \to \infty} x^2 = \infty$, so l'Hospital's Rule gives

$$\lim_{x \to \infty} \frac{e^x}{x^2} = \lim_{x \to \infty} \frac{e^x}{2x}$$

Since $e^x \to \infty$ and $2x \to \infty$ as $x \to \infty$, the limit on the right side is also indeterminate, but a second application of l'Hospital's Rule gives

$$\lim_{x \to \infty} \frac{e^x}{x^2} = \lim_{x \to \infty} \frac{e^x}{2x} = \lim_{x \to \infty} \frac{e^x}{2} = \infty \qquad \bullet$$

EXAMPLE 3   Calculate $\lim_{x \to \infty} \dfrac{\ln x}{\sqrt[3]{x}}$.

Solution   Since $\ln x \to \infty$ and $\sqrt[3]{x} \to \infty$ as $x \to \infty$, l'Hospital's Rule applies:

$$\lim_{x \to \infty} \frac{\ln x}{\sqrt[3]{x}} = \lim_{x \to \infty} \frac{\frac{1}{x}}{\frac{1}{3}x^{-2/3}}$$

Notice that the limit on the right side is now indeterminate of type $\frac{0}{0}$. But instead of applying l'Hospital's Rule a second time as we did in Example 2, we simplify the expression and see that a second application is unnecessary:

$$\lim_{x \to \infty} \frac{\ln x}{\sqrt[3]{x}} = \lim_{x \to \infty} \frac{\frac{1}{x}}{\frac{1}{3}x^{-2/3}} = \lim_{x \to \infty} \frac{3}{\sqrt[3]{x}} = 0 \qquad \bullet$$

EXAMPLE 4   Find $\lim_{x \to 0} \dfrac{\tan x - x}{x^3}$. (See Exercise 18 in Section 1.2.)

Solution   Noting that both $\tan x - x \to 0$ and $x^3 \to 0$ as $x \to 0$, we use l'Hospital's Rule:

$$\lim_{x \to 0} \frac{\tan x - x}{x^3} = \lim_{x \to 0} \frac{\sec^2 x - 1}{3x^2}$$

Since the limit on the right side is still indeterminate of type $\frac{0}{0}$, we apply l'Hospital's Rule again:

$$\lim_{x \to 0} \frac{\sec^2 x - 1}{3x^2} = \lim_{x \to 0} \frac{2 \sec^2 x \tan x}{6x}$$

Again both numerator and denominator approach 0, so a third application of l'Hospital's Rule is necessary. Putting together all three steps, we get

$$\lim_{x \to 0} \frac{\tan x - x}{x^3} = \lim_{x \to 0} \frac{\sec^2 x - 1}{3x^2} = \lim_{x \to 0} \frac{2 \sec^2 x \tan x}{6x}$$

$$= \lim_{x \to 0} \frac{4 \sec^2 x \tan^2 x + 2 \sec^4 x}{6} = \frac{2}{6} = \frac{1}{3} \qquad \bullet$$

EXAMPLE 5   Find $\lim_{x \to \pi^-} \dfrac{\sin x}{1 - \cos x}$.

Solution   If we blindly attempted to use l'Hospital's Rule, we would get

$$\lim_{x \to \pi^-} \frac{\sin x}{1 - \cos x} = \lim_{x \to \pi^-} \frac{\cos x}{\sin x} = -\infty$$

This is *wrong!* Although the numerator $\sin x \to 0$ as $x \to \pi^-$, notice that the denominator $(1 - \cos x)$ does not approach 0, so l'Hospital's Rule cannot be applied here.

The required limit is, in fact, easy to find because the function is continuous and the denominator is nonzero at π:

$$\lim_{x \to \pi^-} \frac{\sin x}{1 - \cos x} = \frac{\sin \pi}{1 - \cos \pi} = \frac{0}{1 - (-1)} = 0$$

Example 5 shows what can go wrong if you use l'Hospital's Rule without thinking. Other limits *can* be found using l'Hospital's Rule but are more easily found by other methods. (See Examples 4, 6, and 7 in Section 1.3, Examples 2 and 3 in Section 1.6, Examples 1 and 2 in Section 2.4, and the discussion at the beginning of this section.) So when evaluating any limit, you should try other methods before using l'Hospital's Rule.

Indeterminate Products

If $\lim_{x \to a} f(x) = 0$ and $\lim_{x \to a} g(x) = \infty$ (or $-\infty$), then it is not clear what the value of $\lim_{x \to a} f(x)g(x)$, if any, will be. There is a struggle between f and g. If f wins, the answer will be 0; if g wins, the answer will be ∞ (or $-\infty$). Or there may be a compromise where the answer is a finite nonzero number. This kind of limit is called an **indeterminate form of type $0 \cdot \infty$**. We can deal with it by writing the product fg as a quotient:

$$fg = \frac{f}{1/g} \qquad \text{or} \qquad fg = \frac{g}{1/f}$$

This converts the given limit into an indeterminate form of type $\frac{0}{0}$ or ∞/∞ so that we can use l'Hospital's Rule.

EXAMPLE 6   Evaluate $\lim\limits_{x \to 0^+} x \ln x$.

Solution   The given limit is indeterminate because $x \to 0^+$ while $\ln x \to -\infty$. Writing $x = 1/(1/x)$, we have $1/x \to \infty$ as $x \to 0^+$, so l'Hospital's Rule gives

$$\lim_{x \to 0^+} x \ln x = \lim_{x \to 0^+} \frac{\ln x}{\frac{1}{x}} = \lim_{x \to 0^+} \frac{\frac{1}{x}}{\frac{-1}{x^2}}$$

$$= \lim_{x \to 0^+} (-x) = 0$$

Indeterminate Differences

If $\lim_{x \to a} f(x) = \infty$ and $\lim_{x \to a} g(x) = \infty$, then the limit

$$\lim_{x \to a} [f(x) - g(x)]$$

is called an **indeterminate form of type $\infty - \infty$**. Again there is a contest between f and g. Will the answer be ∞ (f wins) or will it be $-\infty$ (g wins) or will they compromise on a finite number? To find out, we convert the difference into a quotient (for instance, by using a common denominator or rationalization, or factoring out a common factor) so that we have an indeterminate form of type $\frac{0}{0}$ or ∞/∞.

EXAMPLE 7   Compute $\lim\limits_{x \to (\pi/2)^-} (\sec x - \tan x)$.

Solution   First notice that $\sec x \to \infty$ and $\tan x \to \infty$ as $x \to (\pi/2)^-$, so the limit is indeterminate. Here we use a common denominator:

$$\lim_{x \to \pi/2^-} (\sec x - \tan x) = \lim_{x \to \pi/2^-} \left(\frac{1}{\cos x} - \frac{\sin x}{\cos x} \right)$$

$$= \lim_{x \to \pi/2^-} \frac{1 - \sin x}{\cos x} = \lim_{x \to \pi/2^-} \frac{-\cos x}{-\sin x} = 0$$

Note that the use of l'Hospital's Rule was justified because $1 - \sin x \to 0$ and $\cos x \to 0$ as $x \to (\pi/2)^-$.

Indeterminate Powers

Several indeterminate forms arise from the limit

$$\lim_{x \to a} [f(x)]^{g(x)}$$

1. $\lim_{x \to a} f(x) = 0$ and $\lim_{x \to a} g(x) = 0$ type 0^0

2. $\lim_{x \to a} f(x) = \infty$ and $\lim_{x \to a} g(x) = 0$ type ∞^0

3. $\lim_{x \to a} f(x) = 1$ and $\lim_{x \to a} g(x) = \pm\infty$ type 1^∞

Each of these three cases can be treated either by taking the natural logarithm:

$$\text{let } y = [f(x)]^{g(x)}, \text{ then } \ln y = g(x) \ln f(x)$$

or by writing the function as an exponential:

$$[f(x)]^{g(x)} = e^{g(x) \ln f(x)}$$

(Recall that both of these methods were used in differentiating such functions.) In either method we are led to the indeterminate product $g(x) \ln f(x)$, which is of type $0 \cdot \infty$.

EXAMPLE 8   Calculate $\lim_{x \to 0^+} (1 + \sin 4x)^{\cot x}$.

Solution   First notice that as $x \to 0^+$, we have $1 + \sin 4x \to 1$ and $\cot x \to \infty$, so the given limit is indeterminate. Let

$$y = (1 + \sin 4x)^{\cot x}$$

Then

$$\ln y = \ln\left[(1 + \sin 4x)^{\cot x} \right] = \cot x \ln(1 + \sin 4x)$$

so l'Hospital's Rule gives

$$\lim_{x \to 0^+} \ln y = \lim_{x \to 0^+} \frac{\ln(1 + \sin 4x)}{\tan x}$$

$$= \lim_{x \to 0^+} \frac{\dfrac{4 \cos 4x}{1 + \sin 4x}}{\sec^2 x} = 4$$

So far we have computed the limit of $\ln y$, but what we want is the limit of y. To find this we use the fact that $y = e^{\ln y}$:

$$\lim_{x \to 0^+} (1 + \sin 4x)^{\cot x} = \lim_{x \to 0^+} y$$

$$= \lim_{x \to 0^+} e^{\ln y}$$

$$= e^4$$

EXAMPLE 9   Find $\lim\limits_{x \to 0^+} x^x$.

Solution   Notice that this limit is indeterminate since $0^x = 0$ for any $x > 0$ but $x^0 = 1$ for any $x \neq 0$. We could proceed as in Example 8 or by writing the function as an exponential:

$$x^x = (e^{\ln x})^x = e^{x \ln x}$$

Now l'Hospital's Rule gives

$$\lim_{x \to 0^+} x \ln x = \lim_{x \to 0^+} \frac{\ln x}{\frac{1}{x}} = \lim_{x \to 0^+} \frac{\frac{1}{x}}{\frac{-1}{x^2}}$$

$$= \lim_{x \to 0^+} (-x) = 0$$

Therefore

$$\lim_{x \to 0^+} x^x = \lim_{x \to 0^+} e^{x \ln x} = e^0 = 1$$

SECTION 3.9   **Exercises**

Find the limits in Exercises 1–78.

1. $\lim\limits_{x \to 2} \dfrac{x - 2}{x^2 - 4}$

2. $\lim\limits_{x \to 1} \dfrac{x^2 + 3x - 4}{x - 1}$

3. $\lim\limits_{x \to -1} \dfrac{x^6 - 1}{x^4 - 1}$

4. $\lim\limits_{x \to 1} \dfrac{x^a - 1}{x^b - 1}$

5. $\lim\limits_{x \to 0} \dfrac{e^x - 1}{\sin x}$

6. $\lim\limits_{x \to 1} \dfrac{\ln x}{x - 1}$

7. $\lim\limits_{x \to 0} \dfrac{\sin x}{x^3}$

8. $\lim\limits_{x \to \pi} \dfrac{\tan x}{x}$

9. $\lim\limits_{x \to 0} \dfrac{\tan x}{x + \sin x}$

10. $\lim\limits_{x \to 3\pi/2} \dfrac{\cos x}{x - (3\pi/2)}$

11. $\lim\limits_{x \to \infty} \dfrac{\ln x}{x}$

12. $\lim\limits_{x \to 0^+} \dfrac{\ln x}{\sqrt{x}}$

13. $\lim\limits_{x \to \infty} \dfrac{e^x}{x^3}$

14. $\lim\limits_{x \to \infty} \dfrac{(\ln x)^3}{x^2}$

15. $\lim\limits_{x \to a} \dfrac{\sqrt[3]{x} - \sqrt[3]{a}}{x - a}$, $a \neq 0$

16. $\lim\limits_{x \to 0} \dfrac{6^x - 2^x}{x}$

17. $\lim\limits_{x \to 0} \dfrac{e^x - 1 - x}{x^2}$

18. $\lim\limits_{x \to 0} \dfrac{e^x - 1 - x - (x^2/2)}{x^3}$

19. $\lim\limits_{x \to 0} \dfrac{\sin x}{e^x}$

20. $\lim\limits_{x \to 0} \dfrac{\sin^2 x}{\tan(x^2)}$

21. $\lim\limits_{x \to 0} \dfrac{1 - \cos x}{x^2}$

22. $\lim\limits_{x \to 0} \dfrac{\sin x - x}{x^3}$

23. $\lim\limits_{x \to 2^-} \dfrac{\ln x}{\sqrt{2 - x}}$

24. $\lim\limits_{x \to 0} \dfrac{\sin x}{\sinh x}$

25. $\lim\limits_{x \to \infty} \dfrac{\ln \ln x}{\sqrt{x}}$

26. $\lim\limits_{x \to \infty} \dfrac{\ln(1 + e^x)}{5x}$

27. $\lim\limits_{x \to 0} \dfrac{\tan^{-1}(2x)}{3x}$

28. $\lim\limits_{x \to 0} \dfrac{x}{\sin^{-1}(3x)}$

29. $\lim\limits_{x \to 0} \dfrac{\tan \alpha x}{x}$

30. $\lim\limits_{x \to 0} \dfrac{\sin mx}{\sin nx}$

31. $\lim\limits_{x \to 0} \dfrac{\tan 2x}{\tanh 3x}$

32. $\lim\limits_{x \to 0} \dfrac{\sin^{10} x}{\sin(x^{10})}$

33. $\lim\limits_{x \to 0} \dfrac{x + \sin 3x}{x - \sin 3x}$

34. $\lim\limits_{x \to 0} \dfrac{2x - \sin^{-1} x}{2x + \cos^{-1} x}$

35. $\lim\limits_{x \to 0} \dfrac{e^{4x} - 1}{\cos x}$

36. $\lim\limits_{x \to 0} \dfrac{2x - \sin^{-1} x}{2x + \tan^{-1} x}$

37. $\lim\limits_{x \to 0} \dfrac{\tan x - \sin x}{x^3}$

38. $\lim\limits_{x \to 0} \dfrac{\cos mx - \cos nx}{x^2}$

39. $\lim\limits_{x \to 0^+} \sqrt{x} \ln x$

40. $\lim\limits_{x \to -\infty} x e^x$

41. $\lim\limits_{x \to \infty} e^{-x} \ln x$

42. $\lim\limits_{x \to \pi/2^-} \sec 7x \cos 3x$

43. $\displaystyle\lim_{x\to\infty} x^3 e^{-x^2}$

44. $\displaystyle\lim_{x\to 0^+} \sqrt{x}\,\sec x$

45. $\displaystyle\lim_{x\to\pi} (x-\pi)\cot x$

46. $\displaystyle\lim_{x\to 1^+} (x-1)\tan(\pi x/2)$

47. $\displaystyle\lim_{x\to 0}\left(\frac{1}{x^4}-\frac{1}{x^2}\right)$

48. $\displaystyle\lim_{x\to 0}(\csc x-\cot x)$

49. $\displaystyle\lim_{x\to 0}\left(\frac{1}{x}-\csc x\right)$

50. $\displaystyle\lim_{x\to 1}\left(\frac{1}{\ln x}-\frac{1}{x-1}\right)$

51. $\displaystyle\lim_{x\to\infty}\left(x-\sqrt{x^2-1}\right)$

52. $\displaystyle\lim_{x\to\infty}\left(\sqrt{x^2+x+1}-\sqrt{x^2-x}\right)$

53. $\displaystyle\lim_{x\to\infty}\left(\frac{x^3}{x^2-1}-\frac{x^3}{x^2+1}\right)$

54. $\displaystyle\lim_{x\to\infty}(xe^{1/x}-x)$

55. $\displaystyle\lim_{x\to 0^+} x^{\sin x}$

56. $\displaystyle\lim_{x\to 0^+} (\sin x)^{\tan x}$

57. $\displaystyle\lim_{x\to 0^+} x^{1/\ln x}$

58. $\displaystyle\lim_{x\to 0^+} x^{-1/\sqrt{-\ln x}}$

59. $\displaystyle\lim_{x\to 0} (1-2x)^{1/x}$

60. $\displaystyle\lim_{x\to\infty}\left(1+\frac{a}{x}\right)^{bx}$

61. $\displaystyle\lim_{x\to\infty}\left(1+\frac{3}{x}+\frac{5}{x^2}\right)^{x}$

62. $\displaystyle\lim_{x\to\infty}\left(1+\frac{1}{x^2}\right)^{x}$

63. $\displaystyle\lim_{x\to\infty} x^{1/x}$

64. $\displaystyle\lim_{x\to\infty} (e^x+x)^{1/x}$

65. $\displaystyle\lim_{x\to 0^+} (\cot x)^{\sin x}$

66. $\displaystyle\lim_{x\to\infty}\left(1+\frac{1}{x}\right)^{x^2}$

67. $\displaystyle\lim_{x\to\infty}\left(\frac{x}{x+1}\right)^{x}$

68. $\displaystyle\lim_{x\to 0} (\cos 3x)^{5/x}$

69. $\displaystyle\lim_{x\to 0^+} x^{x^x}$

70. $\displaystyle\lim_{x\to 0^+} (-\ln x)^{x}$

71. $\displaystyle\lim_{x\to\infty} \frac{x}{2x+3\sin x}$

72. $\displaystyle\lim_{x\to\infty}\left(\frac{2x-3}{2x+5}\right)^{2x+1}$

73. $\displaystyle\lim_{x\to 0^+} \frac{x+1-e^x}{x^3}$

74. $\displaystyle\lim_{x\to\infty} x[\ln(x+5)-\ln x]$

75. $\displaystyle\lim_{x\to 0} \frac{2x\sin x}{\sec x-1}$

76. $\displaystyle\lim_{x\to 0} \frac{1-\cos 2x+\tan^2 x}{x\sin x}$

77. $\displaystyle\lim_{x\to 0} \frac{\cos x-1+\dfrac{x^2}{2}}{x^4}$

78. $\displaystyle\lim_{x\to-\infty}\left(\sqrt{x^2+1}-\sqrt{x^2-4x}\right)$

79. If f' is continuous, use l'Hospital's Rule to show that
$$\lim_{h\to 0}\frac{f(x+h)-f(x-h)}{2h}=f'(x)$$

80. If f'' is continuous, show that
$$\lim_{h\to 0}\frac{f(x+h)-2f(x)+f(x-h)}{h^2}=f''(x)$$

81. Prove that
$$\lim_{x\to\infty}\frac{e^x}{x^n}=\infty$$
for any integer n. This shows that the exponential function approaches infinity faster than any power of x.

82. Prove that
$$\lim_{x\to\infty}\frac{\ln x}{x^p}=0$$
for any number $p>0$. This shows that the logarithmic function approaches ∞ more slowly than any power of x.

83. Prove that $\displaystyle\lim_{x\to 0^+} x^\alpha \ln x=0$ for any $\alpha>0$.

84. Let
$$f(x)=\begin{cases} e^{-1/x^2} & \text{if } x\neq 0 \\ 0 & \text{if } x=0 \end{cases}$$

   (a) Use the definition of derivative to compute $f'(0)$.

   (b) Show that f has derivatives of all orders that are defined on R. [*Hint:* First show by induction that there is a polynomial $p_n(x)$ and a nonnegative integer k_n such that
$$f^{(n)}(x)=p_n(x)f(x)/x^{k_n}\text{ for }x\neq 0.]$$

CHAPTER 3

========
====== **Review**
========

Key Topics

Define, state, or discuss the following.

1. Exponential functions
2. Graphs of exponential functions
3. Properties of exponential functions
4. Limits of exponential functions
5. One-to-one function
6. Horizontal Line Test
7. Inverse function
8. Cancellation equations
9. Procedure for finding an inverse function
10. Graph of an inverse function
11. Continuity and differentiability of an inverse function
12. Formula for the derivative of an inverse function
13. Logarithmic function to the base a
14. Graphs of logarithmic functions
15. Properties of logarithms
16. Limits of logarithmic functions
17. The number e
18. Natural logarithm
19. Derivatives of logarithmic and exponential functions
20. Logarithmic differentiation
21. Law of natural growth or decay
22. Inverse trigonometric functions
23. Derivatives of inverse trigonometric functions
24. Hyperbolic functions
25. Hyperbolic identities
26. Derivatives of hyperbolic functions
27. Inverse hyperbolic functions
28. L'Hospital's Rule
29. Indeterminate quotients, products, and powers

Exercises

In Exercises 1–8 sketch the graph of the given function.

1. $y = 7^x$
2. $y = (0.7)^x$
3. $y = -e^{-x}$
4. $y = \pi^{-x}$
5. $y = \log_5 x$
6. $y = \ln(x - 1)$
7. $y = 2 - \ln x$
8. $y = e^x \cos x$

In Exercises 9–16 solve the given equation for x.

9. $e^x = 5$
10. $\ln x = 2$
11. $\log_{10}(e^x) = 1$
12. $e^{e^x} = 2$
13. $\ln(x^\pi) = 2$
14. $\ln(x + 1) - \ln x = 1$
15. $\tan x = 4$
16. $\sin^{-1} x = 1$

In Exercises 17–45 calculate y'.

17. $y = \log_{10}(x^2 - x)$
18. $y = \sqrt{2}^{\,x}$
19. $y = \dfrac{\sqrt{x+1}\,(2-x)^5}{(x+3)^7}$
20. $y = \ln(\csc 5x)$
21. $y = e^{cx}(c \sin x - \cos x)$
22. $y = \sin^{-1}(e^x)$
23. $y = \ln(\sec^2 x)$
24. $y = \ln(x^2 e^x)$
25. $y = xe^{-1/x}$
26. $y = \ln|\csc 3x + \cot 3x|$
27. $y = (\cos^{-1} x)^{\sin^{-1} x}$
28. $y = x^r e^{sx}$
29. $y = e^{e^x}$
30. $y = 5^{x \tan x}$
31. $y = \ln\left(\dfrac{1}{x}\right) + \dfrac{1}{\ln x}$
32. $xe^y = y - 1$
33. $y = 7^{\sqrt{2x}}$
34. $y = e^{\cos x} + \cos(e^x)$
35. $y = xe^{\sec^{-1} x}$
36. $y = \ln\left|\dfrac{x^2 - 4}{2x + 5}\right|$
37. $y = \ln(\cosh 3x)$
38. $y = \log_3 \sqrt{(1 + cx^a)^s}$
39. $y = \cosh^{-1}(\sinh x)$
40. $y = x \tanh^{-1} \sqrt{x}$
41. $y = \ln \sin x - \frac{1}{2}\sin^2 x$
42. $y = (c/x)^x$
43. $y = \sin^{-1}\left(\dfrac{x-1}{x+1}\right)$
44. $y = \arctan(\arcsin \sqrt{x}\,)$

45. $y = \ln \sqrt[4]{\dfrac{x^2 + x + 1}{x^2 - x + 1}}$

$\qquad + \dfrac{1}{2\sqrt{3}}\left[\tan^{-1}\left(\dfrac{2x+1}{\sqrt{3}}\right) + \tan^{-1}\left(\dfrac{2x-1}{\sqrt{3}}\right)\right]$

46. If

$$y = \dfrac{x}{\sqrt{a^2 - 1}} - \dfrac{2}{\sqrt{a^2 - 1}}\arctan\dfrac{\sin x}{a + \sqrt{a^2 - 1} + \cos x}$$

show that $y' = \dfrac{1}{a + \cos x}$.

Find $f^{(n)}(x)$ in Exercises 47 and 48.

47. $f(x) = 2^x$ 48. $f(x) = \ln(2x)$

49. Use mathematical induction to show that if $f(x) = xe^x$, then $f^{(n)}(x) = (x + n)e^x$.

50. Find y' if $y = x + \arctan y$.

In Exercises 51–52 find the equation of the tangent to the given curve at the given point.

51. $y = \ln(e^x + e^{2x})$, $(0, \ln 2)$

52. $y = x \ln x$, (e, e)

53. At what point on the curve $y = [\ln(x + 4)]^2$ is the tangent horizontal?

54. Find the equation of the tangent to the curve $y = e^x$ that is parallel to the line $x - 4y = 1$.

Evaluate the limits in Exercises 55–76.

55. $\lim\limits_{x \to -\infty} 10^{-x}$

56. $\lim\limits_{x \to \infty} \dfrac{4^x}{4^x - 1}$

57. $\lim\limits_{x \to 0^+} \ln(\tan x)$

58. $\lim\limits_{x \to -\infty} e^{x/2} \cos x$

59. $\lim\limits_{x \to -4^+} e^{1/(x+4)}$

60. $\lim\limits_{x \to -1^+} e^{\tanh^{-1} x}$

61. $\lim\limits_{x \to \infty} \dfrac{e^x}{e^{2x} + e^{-x}}$

62. $\lim\limits_{x \to 0} (1 + x)^{2/x}$

63. $\lim\limits_{x \to 1} \cos^{-1}\left(\dfrac{x}{x+1}\right)$

64. $\lim\limits_{x \to -\infty} \tan^{-1}(x^4)$

65. $\lim\limits_{x \to \pi} \dfrac{\sin x}{x^2 - \pi^2}$

66. $\lim\limits_{x \to 0} \dfrac{e^{ax} - e^{bx}}{x}$

67. $\lim\limits_{x \to \infty} \dfrac{\ln(\ln x)}{\ln x}$

68. $\lim\limits_{x \to 0} \dfrac{1 + \sin x - \cos x}{1 - \sin x - \cos x}$

69. $\lim\limits_{x \to 0} \dfrac{\ln(1 - x) + x + \dfrac{x^2}{2}}{x^3}$

70. $\lim\limits_{x \to \pi/2} \left(\dfrac{\pi}{2} - \pi\right)\tan x$

71. $\lim\limits_{x \to 0^+} (\sin x)(\ln x)^2$

72. $\lim\limits_{x \to 0} (\csc^2 x - x^{-2})$

73. $\lim\limits_{x \to 1} (\ln x)^{\sin x}$

74. $\lim\limits_{x \to 1} x^{1/(1-x)}$

75. $\lim\limits_{x \to 0^+} \dfrac{\sqrt[3]{x} - 1}{\sqrt[4]{x} - 1}$

76. $\lim\limits_{x \to \infty} \tan^{-1}\left(\dfrac{\sqrt{x}}{\ln x}\right)$

77. A bacteria culture starts with 1000 bacteria and the growth rate is proportional to the number of bacteria. After 2 h the population is 9000.
    (a) Find an expression for the number of bacteria after t hours.
    (b) Find the number of bacteria after 3 hours.
    (c) In what period of time does the number of bacteria double?

78. An isotope of strontium, Sr^{90}, has a half-life of 25 years. (a) Find the mass of Sr^{90} that remains from a sample of 18 mg after t years. (b) How long would it take for the mass to decay to 2 mg?

79. An amount of $10,000 is invested at an interest rate of 10% per year. Find the value of the investment at the end of 4 years if the interest is compounded (a) annually, (b) semiannually, (c) quarterly, (d) monthly, (e) daily, and (f) continuously.

80. A cup of coffee has a temperature of 200°F and is in a room that has a temperature of 70°F. After 10 min the temperature of the coffee is 150°F.
  (a) What is the temperature of the coffee after 15 min?
  (b) When will the coffee have cooled to 100°F?

81. Let $C(t)$ be the concentration of a drug in the bloodstream. As the body eliminates the drug, $C(t)$ decreases at a rate that is proportional to the amount of the drug that is present at the time. Thus $C'(t) = -kC(t)$, where k is a positive number called the *elimination constant* of the drug.
  (a) If C_0 is the concentration at time $t = 0$, find the concentration at time t.
  (b) If the body eliminates half the drug in 30 h, how long does it take to eliminate 90% of the drug?

82. (a) Show that there is exactly one root of the equation $\ln x = 3 - x$ and that it lies between 2 and e.
  (b) Find the root of the equation in part (a) correct to four decimal places.

83. An equation of motion of the form $s = Ae^{-ct}\cos(\omega t + \delta)$ represents damped oscillation of an object. Find the velocity and acceleration of the object.

In Exercises 84–89 find f' in terms of g'.

84. $f(x) = g(e^x)$
85. $f(x) = e^{g(x)}$

86. $f(x) = g(\ln x)$
87. $f(x) = \ln|g(x)|$

88. $f(x) = xe^{g(\sqrt{x})}$
89. $f(x) = \ln g(e^x)$

90. If $f(x) = x + x^2 + e^x$ and $g(x) = f^{-1}(x)$, find $g'(1)$.

91. If g is the inverse function of $f(x) = \ln x + \tan^{-1}x$, find $g'(\pi/4)$.

Applications Plus

1. A car is traveling at night along a highway shaped like a parabola with its vertex at the origin. The car starts at a point 100 m west and 100 m north of the origin and travels in an easterly direction. There is a statue located 100 m east and 50 m north of the origin. At what point on the highway will the car's headlights illuminate the statue?

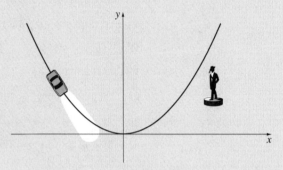

2. The figure shows a rotating wheel with radius 40 cm and a connecting rod AP with length 1.2 m. The pin P slides back and forth along the x-axis as the wheel rotates counterclockwise at a rate of 360 revolutions per minute.
   (a) Find the angular velocity of the connecting rod, $d\alpha/dt$, in radians per second, when $\theta = \pi/3$.
   (b) Express the distance $x = |OP|$ in terms of θ.
   (c) Find an expression for the velocity of the pin P in terms of θ.

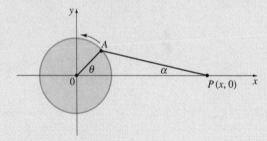

3. A glucose solution is administered intravenously into the bloodstream at a constant rate r. As the glucose is added, it is converted into other substances and removed from the bloodstream at a rate that is proportional to the concentration at that time. The mathematical model for the concentration $C = C(t)$ of the glucose solution in the bloodstream is

$$\frac{dC}{dt} = r - kC = k\left(\frac{r}{k} - C\right)$$

where k is a positive constant.
   (a) Suppose that the concentration at time $t = 0$ is C_0. Determine the concentration at any time t by solving the differential equation. [*Hint:* Consider the function $u(t) = (r/k) - C(t)$.]
   (b) Assuming that $C_0 < r/k$, find $\lim_{t \to \infty} C(t)$ and interpret this result.

4. The annual sales of a new company are expected to grow at a rate that is proportional to the difference between the sales at time t and an upper limit of $25 million. The sales are $0 initially and are projected to be $3.5 million at the end of three years.
   (a) Determine a differential equation that expresses the rate of growth of sales for the company.
   (b) Determine the annual sales of the company at any time t by solving the differential equation in part (a).
   (c) What will the annual sales be at the end of the eighth year?
   (d) How long will it take for the annual sales to be $12.5 million?

5. Assume that a snowball melts so that its volume decreases at a rate proportional to its surface area. If it takes the snowball 3 h for its volume to decrease to half the original volume, how much longer will it take for the snowball to melt completely?

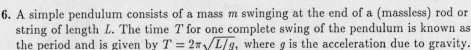

6. A simple pendulum consists of a mass m swinging at the end of a (massless) rod or string of length L. The time T for one complete swing of the pendulum is known as the period and is given by $T = 2\pi\sqrt{L/g}$, where g is the acceleration due to gravity.
   (a) Show that a small change dL in the length produces a change in the period dT satisfying

$$\frac{dT}{T} = \frac{dL}{2L}$$

   (b) Suppose that a pendulum clock loses 15 seconds per hour. How should the length of the pendulum be adjusted?
   (c) The formula for the period given above can be used to measure the acceleration due to gravity. Assuming that the error in measuring the length L is negligible, express the error dg in the acceleration of gravity in terms of the error dT in measuring the period.

7. A planning engineer for a new alum plant must present some estimates to his company regarding the capacity of a silo designed to contain bauxite ore until it can be processed into alum. The ore resembles pink talcum powder and is poured from a conveyor at the top of the silo. The silo is a cylinder 100 ft high with a radius of 200 ft. The conveyor carries $60,000\pi$ ft^3/h and the ore maintains a conical shape whose radius is 1.5 times its height.
   (a) If, at a certain time t, the pile is 60 ft high, how long will it take for the pile to reach the top of the silo?

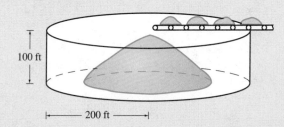

100 ft

200 ft

(b) Management wants to know how much room will be left in the floor area of the silo when the pile is 60 ft high. How fast is the floor area of the pile growing at that height?

(c) Suppose a loader starts removing the ore at the rate of $20{,}000\pi$ ft^3/h when the height of the pile reaches 90 ft. Suppose, also, that the pile continues to maintain its shape. How long will it take for the pile to reach the top of the silo under these conditions?

8. Starting from $0°$ longitude and proceeding in the counterclockwise direction, let $T(x)$ denote the temperature at the point x at any given time. Assume that T is a continuous function of x.

(a) Use the function T to show that at any fixed time there are at least two diametrically opposite points on the equator that have exactly the same temperature.

(b) Does the result in (a) hold for points lying on any circle on the earth's surface?

(c) Does the result in (a) also hold for barometric pressure and for altitude above sea level?

9. Let $P_1(x_1, y_1)$ be a point on the parabola $y^2 = 4px$ with focus $F(p, 0)$. Let α be the angle between the parabola and the line segment FP and let β be the angle between the horizontal line $y = y_1$ and the parabola as in the figure. Prove that $\alpha = \beta$. (Thus, by a principle of geometrical optics, light from a source placed at F will be reflected along a line parallel to the x-axis. This explains why paraboloids, the surfaces obtained by rotating parabolas about their axes, are used as the shape of some automobile headlights and mirrors for telescopes.)

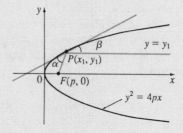

4

The Mean Value Theorem and Curve Sketching

The greatest mathematicians, as Archimedes, Newton, and Gauss, always united theory and applications in equal measure.

Felix Klein

For since the fabric of the universe is most perfect and the work of a most wise Creator, nothing at all takes place in the universe in which some rule of maximum or minimum does not appear.

Leonhard Euler

The Mean Value Theorem, which is stated and proved in Section 4.2, is one of the most important theorems in calculus. It enables us to develop the tools that are needed to graph functions using derivatives. In particular we learn how to find maximum and minimum values of functions. This is a useful skill because many practical problems require us to minimize a cost or maximize an area or somehow find the best possible outcome of a situation.

Maximum and Minimum Values

Some of the most important applications of differential calculus are *optimization problems* in which we are required to find the optimal (best) way of doing something. In many cases these problems can be reduced to finding the maximum or minimum values of a function. Let us first explain exactly what we mean by maximum and minimum values.

Definition (4.1)

A function f has an **absolute maximum** at c if $f(c) \geq f(x)$ for all x in D, where D is the domain of f, and the number $f(c)$ is called the **maximum value** of f on D. Similarly, f has an **absolute minimum** at c if $f(c) \leq f(x)$ for all x in D and the number $f(c)$ is called the **minimum value** of f on D. The maximum and minimum values of f are called the **extreme values** of f.

Figure 4.1 shows the graph of a function f with absolute maximum at d and absolute minimum at a. Note that $(d, f(d))$ is the highest point on the graph and $(a, f(a))$ is the lowest point.

In Figure 4.1 if we consider only values of x near b [for instance, if we restrict our attention to the interval (a, c)], then $f(b)$ is the largest of those values of $f(x)$ and is called a *local maximum value* of f. Likewise $f(c)$ is called a *local minimum value* of f because $f(c) \leq f(x)$ for x near c [in the interval (b, d), for instance]. The function f also has a local minimum at e. In general we have the following definition:

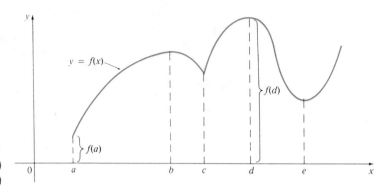

Figure 4.1

Minimum value $f(a)$

Maximum value $f(d)$

Definition (4.2)

A function f has a **local maximum** (or **relative maximum**) at c if there is an open interval I containing c such that $f(c) \geq f(x)$ for all x in I. Similarly, f has a **local minimum** at c if there is an open interval I containing c such that $f(c) \leq f(x)$ for all x in I.

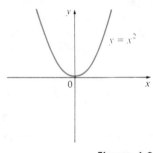

Figure 4.2

Minimum value 0, no maximum

EXAMPLE 1   The function $f(x) = \cos x$ takes on its (local and absolute) maximum value of 1 infinitely many times, since $\cos 2n\pi = 1$ for any integer n and $-1 \leq \cos x \leq 1$ for all x. Likewise $\cos(2n+1)\pi = -1$ is its minimum value, where n is any integer.   ●

EXAMPLE 2   If $f(x) = x^2$, then $f(x) \geq f(0)$ because $x^2 \geq 0$ for all x. Therefore $f(0) = 0$ is the absolute (and local) minimum value of f. This corresponds to the fact that the origin is the lowest point on the parabola $y = x^2$ (see Figure 4.2). However there is no highest point on the parabola and so this function has no maximum value.   ●

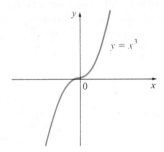

Figure 4.3

No minimum, no maximum

EXAMPLE 3   From the graph of the function $f(x) = x^3$, shown in Figure 4.3, we see that this function has neither an absolute maximum value nor an absolute minimum value. In fact, it has no local extreme values either.   ●

EXAMPLE 4   The graph of the function

$$f(x) = 3x^4 - 16x^3 + 18x^2 \qquad -1 \leq x \leq 4$$

is shown in Figure 4.4. You can see that $f(1) = 5$ is a local maximum, whereas the absolute maximum is $f(-1) = 37$. Also $f(0) = 0$ is a local minimum and $f(3) = -27$ is both a local and an absolute minimum.   ●

We have seen that some functions have extreme values, while some do not. The following theorem gives conditions under which a function is guaranteed to possess extreme values.

The Extreme Value Theorem (4.3)

If f is continuous on a closed interval $[a, b]$, then f attains an absolute maximum value $f(c)$ and an absolute minimum value $f(d)$ at some numbers c and d in $[a, b]$.

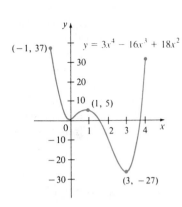

Figure 4.4

$$y = 3x^4 - 16x^3 + 18x^2$$

The Extreme Value Theorem is illustrated in Figure 4.5. Note that an extreme value can be taken on more than once. Although the Extreme Value Theorem is intuitively very plausible, it is difficult to prove and so we omit the proof.

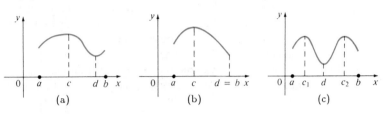

Figure 4.5

The next two examples show that a function need not possess extreme values if either hypothesis (continuity or closed interval) is omitted from the Extreme Value Theorem.

EXAMPLE 5   The function

$$f(x) = \begin{cases} x^2 & \text{if } 0 \le x < 1 \\ 0 & \text{if } 1 \le x \le 2 \end{cases}$$

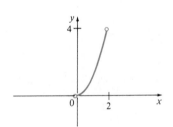

Figure 4.6

is defined on the closed interval $[0,2]$ but has no maximum value. Notice that the range of f is the interval $[0,1)$. The function takes on values arbitrarily close to 1 but never actually attains the value 1. This does not contradict the Extreme Value Theorem because f is not continuous on $[0,2]$. In fact, it has a discontinuity at $x = 1$ (see Figure 4.6).   •

EXAMPLE 6   The function $f(x) = x^2$, $0 < x < 2$, is continuous on the finite interval $(0,2)$ but has neither a maximum nor a minimum value. The range of f is the interval $(0,4)$; the values 0 and 4 are never taken on by f. This does not contradict the Extreme Value Theorem since the interval $(0,2)$ is not closed.

If we alter the function by including either endpoint of the interval $(0,2)$, then we get the situations shown in Figure 4.7. In particular the function $k(x) = x^2$, $0 \le x \le 2$, is continuous on the closed interval $[0,2]$, so the Extreme Value Theorem says that there will be an absolute maximum and an absolute minimum.

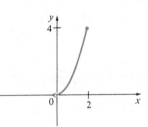
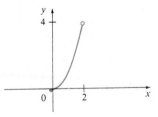
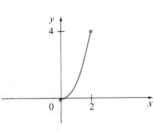

(a) $f(x) = x^2$, $0 < x < 2$
no maximum,
no minimum

(b) $g(x) = x^2$, $0 < x \le 2$
maximum $g(2) = 4$,
no minimum

(c) $h(x) = x^2$, $0 \le x < 2$
no maximum,
minimum $h(0) = 0$

(d) $k(x) = x^2$, $0 \le x \le 2$
maximum $k(2) = 4$,
minimum $k(0) = 0$

Figure 4.7

In spite of Example 6 we point out that a continuous function, like the one whose graph is shown in Figure 4.8, *could* have a maximum or minimum value even when defined on an open interval. Likewise a discontinuous function *might* have maximum and minimum values (see Exercise 69), but there is no guarantee.

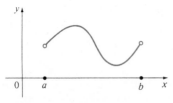

Figure 4.8

The Extreme Value Theorem says that a continuous function on a closed interval has a maximum value and a minimum value, but it does not tell us how to find these extreme values. We start by looking for local extreme values.

Figure 4.9 shows the graph of a function f with a local maximum at c and a local minimum at d. It appears that at the maximum and minimum points the tangent line is horizontal and therefore has slope 0. We know that the derivative is the slope of the tangent line, so it appears that $f'(c) = 0$ and $f'(d) = 0$. The following theorem shows that this is always true for differentiable functions. It is named after the French mathematician Pierre Fermat (1601–1665).

Fermat's Theorem (4.4)   If f has a local extremum (that is, maximum or minimum) at c, and if $f'(c)$ exists, then $f'(c) = 0$.

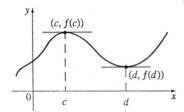

Figure 4.9

Proof   Suppose, for the sake of definiteness, that f has a local maximum at c. Then, according to Definition 4.2, $f(c) \geq f(x)$ if x is sufficiently close to c. This implies that if h is sufficiently close to 0, with h being positive or negative, then

$$f(c) \geq f(c + h)$$

and therefore

(4.5)
$$f(c + h) - f(c) \leq 0$$

We can divide both sides of an inequality by a positive number. Thus if $h > 0$ and h is sufficiently small, we have

$$\frac{f(c + h) - f(c)}{h} \leq 0$$

Taking the right-hand limit of both sides of this inequality (using Theorem 1.9), we get

$$\lim_{h \to 0^+} \frac{f(c + h) - f(c)}{h} \leq \lim_{h \to 0^+} 0 = 0$$

But since $f'(c)$ exists, we have

$$f'(c) = \lim_{h \to 0} \frac{f(c + h) - f(c)}{h} = \lim_{h \to 0^+} \frac{f(c + h) - f(c)}{h}$$

and so we have shown that $f'(c) \leq 0$.

If $h < 0$, then the direction of the inequality (4.5) is reversed when we divide by h:

$$\frac{f(c + h) - f(c)}{h} \geq 0 \qquad h < 0$$

So, taking the left-hand limit, we have

$$f'(c) = \lim_{h \to 0} \frac{f(c + h) - f(c)}{h} = \lim_{h \to 0^-} \frac{f(c + h) - f(c)}{h} \geq 0$$

We have shown that $f'(c) \geq 0$ and also that $f'(c) \leq 0$. Since both of these inequalities must be true, the only possibility is that $f'(c) = 0$.

We have proved Fermat's Theorem for the case of a local maximum. The case of a local minimum can be proved in a similar manner, or it can be deduced from the case that we have proved by using Exercise 70 (see Exercise 71).   ●

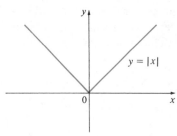

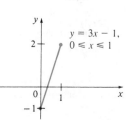

The following examples caution us against reading too much into Fermat's Theorem. We cannot expect to locate extreme values simply by setting $f'(x) = 0$ and then solving for x.

EXAMPLE 7   The function $f(x) = |x|$ has its minimum value (local and absolute) at 0, but it cannot be found by setting $f'(x) = 0$ because, as was shown in Example 6 in Section 2.1, $f'(0)$ does not exist. (See Figure 4.10.)   ●

Figure 4.10

EXAMPLE 8   The function $f(x) = 3x - 1$, $0 \leq x \leq 1$, whose graph is shown in Figure 4.11, has its maximum value when $x = 1$, but $f'(1) = 3 \neq 0$. This does not contradict Fermat's Theorem, since $f(1) = 2$ is not a *local* maximum. (Note that the number 1 is not contained in an *open* interval in the domain of f.)   ●

EXAMPLE 9   If $f(x) = x^3$, then $f'(x) = 3x^2$, so $f'(0) = 0$. But f has no maximum or minimum at 0 as you can see from its graph in Figure 4.3. (Or observe that $x^3 > 0$ for $x > 0$ but $x^3 < 0$ for $x < 0$.) The fact that $f'(0) = 0$ simply means that the curve $y = x^3$ has a horizontal tangent at $(0, 0)$. Instead of having a maximum or minimum at $(0,0)$, the curve crosses its horizontal tangent there.   ●

Figure 4.11

 Warning: Examples 7–9 show that we must be careful when using Fermat's Theorem. Example 9 demonstrates that even when $f'(c) = 0$ there need not be a maximum or minimum at c. (In other words, the converse of Fermat's Theorem is false in general.) Furthermore there may be an extreme value even when $f'(c) \neq 0$ (as in Example 8) or when $f'(c)$ does not exist (as in Example 7).

Fermat's Theorem does suggest that we should at least start looking for extreme values of f at the numbers c where $f'(c) = 0$ or where $f'(c)$ does not exist. Such numbers are given a special name.

Definition (4.6)

> A **critical number** of a function f is a number c in the domain of f such that either $f'(c) = 0$ or $f'(c)$ does not exist.

EXAMPLE 10   Find the critical numbers of $f(x) = x^{3/5}(4 - x)$.

Solution   The Product Rule gives

$$f'(x) = \tfrac{3}{5}x^{-2/5}(4 - x) + x^{3/5}(-1)$$
$$= \frac{3(4 - x) - 5x}{5x^{2/5}} = \frac{12 - 8x}{5x^{2/5}}$$

[The same result could be obtained by first writing $f(x) = 4x^{3/5} - x^{8/5}$.] Therefore $f'(x) = 0$ if $12 - 8x = 0$, that is, $x = \tfrac{3}{2}$, and $f'(x)$ does not exist when $x = 0$. Thus the critical numbers are $\tfrac{3}{2}$ and 0.   ●

In terms of critical numbers, Fermat's Theorem can be rephrased as follows (compare Definition 4.6 with Theorem 4.4):

(4.7)

> If f has a local extremum at c, then c is a critical number of f.

To find an absolute maximum or minimum of a continuous function on a closed interval, we note that either it is a local extremum [in which case it occurs at a critical number by (4.7)] or it occurs at an endpoint of the interval. Thus the following three-step procedure always works.

(4.8)   To find the *absolute* maximum and minimum values of a continuous function f on a closed interval $[a, b]$:

1.   Find the values of f at the critical numbers of f in (a, b).

2.   Find the values $f(a)$ and $f(b)$.

3.   The largest of the values from steps 1 and 2 is the absolute maximum value; the smallest of these values is the absolute minimum value.

EXAMPLE 11   Find the absolute maximum and minimum values of the function

$$f(x) = x^3 - 3x^2 + 1 \qquad -\tfrac{1}{2} \le x \le 4$$

Solution   Since f is continuous on $[-\tfrac{1}{2}, 4]$, we can use the procedure outlined in (4.8):

$$f(x) = x^3 - 3x^2 + 1$$

$$f'(x) = 3x^2 - 6x = 3x(x - 2)$$

Since $f'(x)$ exists for all x, the only critical numbers of f occur when $f'(x) = 0$, that is, $x = 0$ or $x = 2$. Notice that each of these critical numbers lies in the interval $[-\tfrac{1}{2}, 4]$. The values of f at these critical numbers are

$$f(0) = 1 \qquad f(2) = -3$$

The values of f at the endpoints of the interval are

$$f\left(-\tfrac{1}{2}\right) = \tfrac{1}{8} \qquad f(4) = 17$$

Comparing these four numbers, we see that the absolute maximum value is $f(4) = 17$ and the absolute minimum value is $f(2) = -3$.

Note that in this example the absolute maximum occurs at an endpoint, whereas the absolute minimum occurs at a critical number. The graph of f is sketched in Figure 4.12.   ●

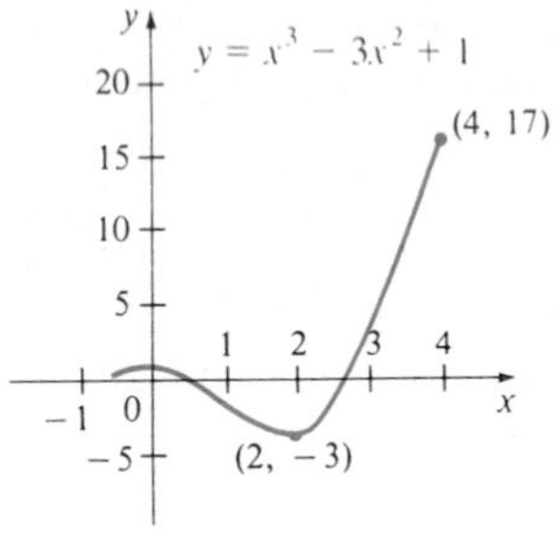

Figure 4.12

SECTION 4.1   **Exercises**

For the functions whose graphs are shown in Exercises 1 and 2, state whether the function has an absolute or local maximum or minimum at the numbers a, b, c, d, e, r, s, and t.

1.

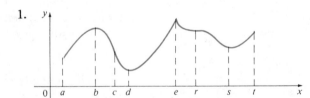

2.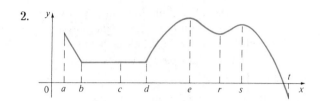

In Exercises 3–22 find the absolute and local maximum and minimum values of the given function. Sketch the graph.

3. $f(x) = 1 + 2x, \quad x \geq -1$

4. $f(x) = 4x - 1, \quad x \leq 8$

5. $f(x) = |x|, \quad -2 \leq x \leq 1$

6. $f(x) = |4x - 1|, \quad 0 \leq x \leq 2$

7. $f(x) = 1 - x^2, \quad 0 < x < 1$

8. $f(x) = 1 - x^2, \quad 0 < x \leq 1$

9. $f(x) = 1 - x^2, \quad 0 \leq x < 1$

10. $f(x) = 1 - x^2, \quad 0 \leq x \leq 1$

11. $f(x) = 1 - x^2, \quad -2 \leq x \leq 1$

12. $f(x) = 1 + (x + 1)^2, \quad -2 \leq x < 5$

13. $f(t) = 1/t, \quad 0 < t < 1$

14. $f(t) = 1/t, \quad 0 < t \leq 1$

15. $f(\theta) = \sin \theta, \quad -2\pi \leq \theta \leq 2\pi$

16. $f(\theta) = \tan \theta, \quad -\pi/4 \leq \theta < \pi/2$

17. $f(\theta) = \cos(\theta/2), \quad -\pi < \theta < \pi$

18. $f(\theta) = \sec \theta, \quad -\pi/2 < \theta \leq \pi/3$

19. $f(x) = x^5$ 20. $f(x) = 2 - x^4$

21. $f(x) = \begin{cases} 2x & \text{if } 0 \leq x < 1 \\ 2 - x & \text{if } 1 \leq x \leq 2 \end{cases}$

22. $f(x) = \begin{cases} x^2 & \text{if } -1 \leq x < 0 \\ 2 - x^2 & \text{if } 0 \leq x \leq 1 \end{cases}$

In Exercises 23–44 find the critical numbers of the given function.

23. $f(x) = 2x - 3x^2$ 24. $f(x) = 5 + 8x$

25. $f(x) = x^3 - 3x + 1$

26. $f(x) = 4x^3 - 9x^2 - 12x + 3$

27. $f(t) = 2t^3 + 3t^2 + 6t + 4$

28. $f(t) = t^3 + 6t^2 + 3t - 1$

29. $s(t) = 2t^3 + 3t^2 - 6t + 4$

30. $s(t) = t^4 + 4t^3 + 2t^2$

31. $g(x) = \sqrt[9]{x}$ 32. $g(x) = |x + 1|$

33. $g(t) = 5t^{2/3} + t^{5/3}$ 34. $g(t) = \sqrt{t}(1 - t)$

35. $f(r) = \dfrac{r}{r^2 + 1}$ 36. $f(z) = \dfrac{z + 1}{z^2 + z + 1}$

37. $F(x) = x^{4/5}(x - 4)^2$ 38. $G(x) = \sqrt[3]{x^2 - x}$

39. $V(x) = x\sqrt{x - 2}$ 40. $T(x) = x^2(2x - 1)^{2/3}$

41. $f(\theta) = \sin^2(2\theta)$ 42. $g(\theta) = \theta + \sin \theta$

43. $f(x) = x \ln x$ 44. $f(x) = xe^{2x}$

In Exercises 45–62 find the absolute maximum and absolute minimum values of f on the given interval.

45. $f(x) = x^2 - 2x + 2, \quad [0, 3]$

46. $f(x) = 1 - 2x - x^2, \quad [-4, 1]$

47. $f(x) = x^3 - 12x + 1, \quad [-3, 5]$

48. $f(x) = 4x^3 - 15x^2 + 12x + 7, \quad [0, 3]$

49. $f(x) = 2x^3 + 3x^2 + 4, \quad [-2, 1]$

50. $f(x) = 18x + 15x^2 - 4x^3, \quad [-3, 4]$

51. $f(x) = x^4 - 4x^2 + 2, \quad [-3, 2]$

52. $f(x) = 3x^5 - 5x^3 - 1, \quad [-2, 2]$

53. $f(x) = x^2 + 2/x, \quad \left[\frac{1}{2}, 2\right]$

54. $f(x) = \sqrt{9 - x^2}, \quad [-1, 2]$

55. $f(x) = x^{4/5}, \quad [-32, 1]$

56. $f(x) = \dfrac{x}{x + 1}, \quad [1, 2]$

57. $f(x) = |x - 1| - 1, \quad [-1, 2]$

58. $f(x) = |x^2 + x|, \quad [-2, 1]$

59. $f(x) = \sin x + \cos x, \quad [0, \pi/3]$

60. $f(x) = x - 2\cos x, \quad [-\pi, \pi]$

61. $f(x) = xe^{-x}, \quad [0, 1]$

62. $f(x) = (\ln x)/x, \quad [1, 3]$

63. Show that every real number is a critical number of the greatest integer function $f(x) = [\![x]\!]$.

64. (a) Use Newton's method to find the critical numbers of the function

$$f(x) = 3x^4 - 28x^3 + 6x^2 + 24x$$

correct to three decimal places.

(b) Find the absolute minimum value of the function $f(x) = 3x^4 - 28x^3 + 6x^2 + 24x$, $-1 \le x \le 7$, correct to two decimal places.

65. Show that 0 is a critical number of the function $f(x) = x^5$ but f does not have a local extremum at 0.

66. Show that 5 is a critical number of the function $g(x) = 2 + (x - 5)^3$ but g does not have a local extremum at 5.

67. Prove that the function $f(x) = x^{101} + x^{51} + x + 1$ has neither a local maximum nor a local minimum.

68. A cubic function is a polynomial of degree 3; that is, it has the form $f(x) = ax^3 + bx^2 + cx + d$, where $a \ne 0$.

(a) Show that a cubic function can have two, one, or no critical numbers. Give examples and sketches to illustrate the three possibilities.

(b) How many local extreme values can a cubic function have?

69. Sketch the graph of a discontinuous function on $[0, 1]$ that has both an absolute maximum and an absolute minimum.

70. If f has a minimum value at c, show that the function $g(x) = -f(x)$ has a maximum value at c.

71. Prove Fermat's Theorem for the case where f has local minimum at c.

SECTION 4.2

The Mean Value Theorem

We will see that many of the results of this chapter depend on one central fact, which is called the Mean Value Theorem. But to arrive at the Mean Value Theorem we first need a result that is named after the French mathematician Michel Rolle (1652–1719).

Rolle's Theorem (4.9)

Let f be a function that satisfies the following three hypotheses:

1. f is continuous on the closed interval $[a, b]$.

2. f is differentiable on the open interval (a, b).

3. $f(a) = f(b)$

Then there is a number c in (a, b) such that $f'(c) = 0$.

Before giving the proof let us take a look at the graphs of some typical functions that satisfy the three hypotheses. Figure 4.13 shows the graphs of four such functions. In each case it appears that there is at least one point $(c, f(c))$ on the graph where the tangent is horizontal and therefore $f'(c) = 0$. Thus Rolle's Theorem is plausible.

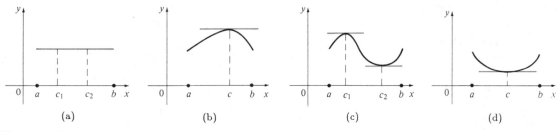

(a)   (b)   (c)   (d)

Figure 4.13

Proof   There are three cases:

CASE I $f(x) = k$, a constant: Then $f'(x) = 0$, so the number c can be taken to be *any* number in (a, b).

CASE II $f(x) > f(a)$ for some x in (a, b) [as in Figure 4.13(b) or (c)]: By the Extreme Value Theorem (which we can apply by hypothesis 1) f has a maximum value somewhere in $[a, b]$. Since $f(a) = f(b)$, it must attain this maximum value at a number c in the open interval (a, b). Then f has a *local* maximum at c and, by hypothesis 2, f is differentiable at c. Therefore $f'(c) = 0$ by Fermat's Theorem.

CASE III $f(x) < f(a)$ for some x in (a, b) [as in Figure 4.13(c) or (d)]: By the Extreme Value Theorem, f has a minimum value in $[a, b]$ and, since $f(a) = f(b)$, it attains this minimum value at a number c in (a, b). Again $f'(c) = 0$ by Fermat's Theorem.   ●

EXAMPLE 1   Let us apply Rolle's Theorem to the position function $s = f(t)$ of a moving object. If the object is in the same place at two different instants $t = a$ and $t = b$, then $f(a) = f(b)$. Rolle's Theorem says that there is some instant of time $t = c$ between a and b when $f'(c) = 0$; that is, the velocity is 0. (In particular you can see that this is true when a ball is thrown directly upward.)   ●

EXAMPLE 2   Prove that the equation $x^3 + x - 1 = 0$ has exactly one real root.

Solution   First we use the Intermediate Value Theorem (1.25) to show that there is a root. Let $f(x) = x^3 + x - 1$. Then $f(0) = -1 < 0$ and $f(1) = 1 > 0$. Since f is a polynomial, it is continuous, so the Intermediate Value Theorem states that there is a number c between 0 and 1 such that $f(c) = 0$. Thus the given equation has a root.

To show that there is no other real root we use Rolle's Theorem and argue by contradiction. Suppose that the equation had two roots a and b. Then $f(a) = 0 = f(b)$ and, since f is a polynomial, it is differentiable on (a, b) and continuous on $[a, b]$. Thus, by Rolle's Theorem, there is a number c between a and b such that $f'(c) = 0$. But

$$f'(x) = 3x^2 + 1 \geq 1 \qquad \text{for all } x$$

(since $x^2 \geq 0$) so $f'(x)$ can never be 0. This gives a contradiction. Therefore the equation cannot have two roots.   ●

Our main use of Rolle's Theorem is in proving the following important theorem, which was first stated by another French mathematician, Joseph–Louis Lagrange (1736–1813).

The Mean Value Theorem (4.10)

Let f be a function that satisfies the following hypotheses:

1. f is continuous on the closed interval $[a, b]$.

2. f is differentiable on the open interval (a, b).

Then there is a number c in (a, b) such that

(4.11)
$$f'(c) = \frac{f(b) - f(a)}{b - a}$$

or, equivalently,

(4.12)
$$f(b) - f(a) = f'(c)(b - a)$$

Before proving this theorem, we can see that it is reasonable by interpreting it geometrically. Figure 4.14 shows the points $A(a, f(a))$ and $B(b, f(b))$ on the graphs of two differentiable functions. The slope of the secant line AB is

(4.13)
$$m_{AB} = \frac{f(b) - f(a)}{b - a}$$

which is the same expression as on the right side of Equation 4.11. Since $f'(c)$ is the slope of the tangent line at the point $(c, f(c))$, the Mean Value Theorem, in the form given by Equation 4.11, says that there is at least one point $P(c, f(c))$ on the graph where the slope of the tangent line is the same as the slope of the secant line AB. In other words, there is a point P where the tangent line is parallel to the secant line AB.

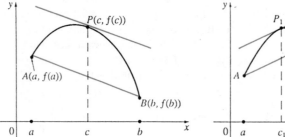

Figure 4.14

Proof   We apply Rolle's Theorem to a new function h defined by

(4.14)
$$h(x) = f(x) - f(a) - \frac{f(b) - f(a)}{b - a}(x - a)$$

[See the note after this proof for a geometric interpretation of $h(x)$.] First we must verify that h satisfies the three hypotheses of Rolle's Theorem.

1.   The function h is continuous on $[a, b]$ because it is the sum of f and a first-degree polynomial, both of which are continuous.
2.   The function h is differentiable on (a, b) because both f and the first-degree polynomial are differentiable. In fact, we can compute h' directly from Equation 4.14:

$$h'(x) = f'(x) - \frac{f(b) - f(a)}{b - a}$$

{Note that $f(a)$ and $[f(b) - f(a)]/(b - a)$ are constants.}

3. $h(a) = f(a) - f(a) - \dfrac{f(b) - f(a)}{b - a}(a - a) = 0$

$$h(b) = f(b) - f(a) - \frac{f(b) - f(a)}{b - a}(b - a)$$
$$= f(b) - f(a) - [f(b) - f(a)] = 0$$

Therefore $h(a) = h(b)$.

Since h satisfies the hypotheses of Rolle's Theorem, that theorem says there is a number c in (a, b) such that $h'(c) = 0$. Therefore

$$0 = h'(c) = f'(c) - \frac{f(b) - f(a)}{b - a}$$

and so

$$f'(c) = \frac{f(b) - f(a)}{b - a}$$

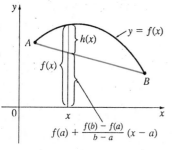

Figure 4.15

Near the figure:
A

$f(x)$

$h(x)$

$y = f(x)$

B

x

$f(a) + \dfrac{f(b) - f(a)}{b - a}\,(x - a)$

Note: The function h used in the proof of the Mean Value Theorem can be interpreted as follows. Using Equation 4.13, we see that the equation of the line AB can be written as

$$y - f(a) = \frac{f(b) - f(a)}{b - a}(x - a)$$

or as

$$y = f(a) + \frac{f(b) - f(a)}{b - a}(x - a)$$

Thus $|h(x)|$ can be interpreted as the vertical distance between the line AB and the curve $y = f(x)$ (see Figure 4.15).

EXAMPLE 3   To illustrate the Mean Value Theorem with a specific function, let us consider $f(x) = x^3 - x$, $a = 0$, $b = 2$. Since f is a polynomial, it is continuous and differentiable for all x, so it is certainly continuous on $[0, 2]$ and differentiable on $(0, 2)$. Therefore, by the Mean Value Theorem, there is a number c in $(0, 2)$ such that

$$f(2) - f(0) = f'(c)(2 - 0)$$

Now $f(2) = 6$, $f(0) = 0$, and $f'(x) = 3x^2 - 1$, so this equation becomes

$$6 = (3c^2 - 1)2 = 6c^2 - 2$$

which gives $c^2 = \frac{4}{3}$, that is, $c = \pm 2/\sqrt{3}$. But c must lie in $(0, 2)$, so $c = 2/\sqrt{3}$.   ●

EXAMPLE 4   If an object moves in a straight line with position function $s = f(t)$, then the average velocity between $t = a$ and $t = b$ is

$$\frac{f(b) - f(a)}{b - a}$$

and the velocity at $t = c$ is $f'(c)$. Thus the Mean Value Theorem tells us that at some time $t = c$ between a and b the instantaneous velocity $f'(c)$ is equal to that average velocity. For instance, if a car traveled 180 km in 2 h, then the speedometer must have read 90 km/h at least once.   ●

The main significance of the Mean Value Theorem is that it can be used to establish some of the basic facts of differential calculus. One of these basic facts is the following theorem. Others will be found in the following sections.

Theorem (4.15) | If $f'(x) = 0$, for all x in an interval (a, b), then f is constant on (a, b).

Proof   Let x_1 and x_2 be any two numbers in (a, b) with $x_1 < x_2$. Since f is differentiable on (a, b) it must be differentiable on (x_1, x_2) and continuous on $[x_1, x_2]$. By applying the Mean Value Theorem to f on the interval $[x_1, x_2]$, we get a number c such that $x_1 < c < x_2$ and

(4.16) $$f(x_2) - f(x_1) = f'(c)(x_2 - x_1)$$

Since $f'(x) = 0$ for all x, we have $f'(c) = 0$, and so Equation 4.16 becomes

$$f(x_2) - f(x_1) = 0 \quad \text{or} \quad f(x_2) = f(x_1)$$

Therefore f has the same value at *any* two numbers x_1 and x_2 in (a, b). This means that f is constant on (a, b).   ●

Corollary (4.17)

> If $f'(x) = g'(x)$ for all x in an interval (a, b), then $f - g$ is constant on (a, b); that is, $f(x) = g(x) + c$ where c is a constant.

Proof   Let $F(x) = f(x) - g(x)$. Then

$$F'(x) = f'(x) - g'(x) = 0$$

for all x in (a, b). Thus by Theorem 4.15, F is constant; that is, $f - g$ is constant.   ●

Note: Care must be taken in applying Theorem 4.15. Let

$$f(x) = \frac{x}{|x|} = \begin{cases} 1 & \text{if } x > 0 \\ -1 & \text{if } x < 0 \end{cases}$$

The domain of f is $D = \{x \mid x \neq 0\}$ and $f'(x) = 0$ for all x in D. But f is obviously not a constant function. This does not contradict Theorem 4.15 because D is not an interval. Notice that f is constant on the interval $(0, \infty)$ and also on the interval $(-\infty, 0)$.

EXAMPLE 5   Prove the identity $\tan^{-1}x + \cot^{-1}x = \pi/2$.

Solution   Although calculus is not needed to prove this identity, the proof using calculus is quite simple. If $f(x) = \tan^{-1}x + \cot^{-1}x$, then

$$f'(x) = \frac{1}{1 + x^2} - \frac{1}{1 + x^2} = 0$$

for all values of x. Therefore, by Theorem 4.15, $f(x) = C$, a constant. To determine the value of C, we put $x = 1$. Then

$$C = f(1) = \tan^{-1}1 + \cot^{-1}1 = \frac{\pi}{4} + \frac{\pi}{4} = \frac{\pi}{2}$$

Thus $\tan^{-1}x + \cot^{-1}x = \pi/2$.   ●

SECTION 4.2 **Exercises**

In Exercises 1–4 verify that the given function satisfies the three hypotheses of Rolle's Theorem on the given interval. Then find all numbers c that satisfy the conclusion of Rolle's Theorem.

1. $f(x) = x^3 - x$, $[-1, 1]$

2. $f(x) = x^3 + x^2 - 2x + 1$, $[-2, 0]$

3. $f(x) = \cos 2x$, $[0, \pi]$

4. $f(x) = \sin x + \cos x$, $[0, 2\pi]$

5. Let $f(x) = 1 - x^{2/3}$. Show that $f(-1) = f(1)$ but there is no number c in $(-1, 1)$ such that $f'(c) = 0$. Why does this not contradict Rolle's Theorem?

6. Let $f(x) = (x - 1)^{-2}$. Show that $f(0) = f(2)$ but there is no number c in $(0, 2)$ such that $f'(c) = 0$. Why does this not contradict Rolle's Theorem?

In Exercises 7–13 verify that the given function satisfies the hypotheses of the Mean Value Theorem on the given interval. Then find all numbers c that satisfy the conclusion of the Mean Value Theorem.

7. $f(x) = 1 - x^2$, $[0,3]$

8. $f(x) = x^2 - 4x + 5$, $[1,5]$

9. $f(x) = x^3 - 2x + 1$, $[-2,3]$

10. $f(x) = 2x^3 + x^2 - x - 1$, $[0,2]$

11. $f(x) = 1/x$, $[1,2]$ 12. $f(x) = \sqrt{x}$, $[1,4]$

13. $f(x) = 1 + \sqrt[3]{x-1}$, $[2,9]$

14. Verify that the function $f(x) = x^4 - 6x^3 + 4x - 1$ satisfies the hypotheses of the Mean Value Theorem on the interval $[0,1]$. Then use Newton's method to find, correct to two decimal places, the numbers c that satisfy the conclusion of the Mean Value Theorem.

15. Let $f(x) = |x - 1|$. Show that there is no value of c such that $f(3) - f(0) = f'(c)(3 - 0)$. Why does this not contradict the Mean Value Theorem?

16. Let $f(x) = (x + 1)/(x - 1)$. Show that there is no value of c such that $f(2) - f(0) = f'(c)(2 - 0)$. Why does this not contradict the Mean Value Theorem?

17. Show that the equation $x^5 + 10x + 3 = 0$ has exactly one real root.

18. Show that the equation $x^7 + 5x^3 + x - 6 = 0$ has exactly one real root.

19. Show that the equation $x^5 - 6x + c = 0$ has at most one root in the interval $[-1,1]$.

20. Show that the equation $x^4 + 4x + c = 0$ has at most two real roots.

21. (a) Show that a polynomial of degree 3 has at most three real roots.
    (b) Show that a polynomial of degree n has at most n real roots.

22. (a) Suppose that f is differentiable on R and has two roots. Show that f' has at least one root.

(b) Suppose that f is twice differentiable on R and has three roots. Show that f'' has at least one real root.

(c) Can you generalize parts (a) and (b)?

23. Does there exist a function f such that $f(0) = -1$, $f(2) = 4$, and $f'(x) \le 2$ for all x?

24. Suppose f is continuous on $[2,5]$ and $1 \le f'(x) \le 4$ for all x in $(2,5)$. Show that $3 \le f(5) - f(2) \le 12$.

25. Use the Mean Value Theorem to prove the inequality

$$|\sin a - \sin b| \le |a - b| \qquad \text{for all } a \text{ and } b$$

26. If $f'(x) = c$, a constant, for all x, use Corollary 4.17 to show that $f(x) = cx + d$ for some constant d.

27. Let $f(x) = 1/x$ and

$$g(x) = \begin{cases} \frac{1}{x} & \text{if } x > 0 \\ 1 + \frac{1}{x} & \text{if } x < 0 \end{cases}$$

Show that $f'(x) = g'(x)$ for all x in their domains. Can we conclude from Corollary 4.17 that $f - g$ is constant?

28. Use Corollary 4.17 to prove the identity

$$\arcsin \frac{x-1}{x+1} = 2 \arctan \sqrt{x} - \frac{\pi}{2}$$

29. Use Corollary 4.17 to prove the identity

$$2 \sin^{-1} x = \cos^{-1}(1 - 2x^2) \qquad x \ge 0$$

30. At 2:00 P.M. a car's speedometer reads $30 \, \text{mi/h}$. At 2:10 P.M. it reads $50 \, \text{mi/h}$. Show that at some time between 2:00 and 2:10 the acceleration was exactly $120 \, \text{mi/h}^2$.

31. Two runners start a race at the same time and finish in a tie. Prove that at some time during the race they had the same velocity. [*Hint:* Consider $f(t) = g(t) - h(t)$, where g and h are the position functions of the two runners.

SECTION 4.3

Monotonic Functions and the First Derivative Test

In sketching the graph of a function it is very useful to know where it rises and where it falls. The graph shown in Figure 4.16 rises from A to B, falls from B to C, and rises again from C to D. The function f is said to be increasing on the interval $[a,b]$, decreas-

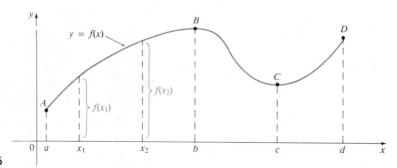

Figure 4.16

ing on $[b,c]$, and increasing again on $[c,d]$. Notice that if x_1 and x_2 are any two numbers between a and b with $x_1 < x_2$, then $f(x_1) < f(x_2)$. We use this as the defining property of an increasing function.

Definition (4.18)

A function f is called **increasing** on an interval I if

$$f(x_1) < f(x_2) \qquad \text{whenever } x_1 < x_2 \text{ in } I$$

It is called **decreasing** on I if

$$f(x_1) > f(x_2) \qquad \text{whenever } x_1 < x_2 \text{ in } I$$

It is called **monotonic** on I if it is either increasing or decreasing on I.

In the definition of an increasing function it is important to realize that the inequality $f(x_1) < f(x_2)$ must be satisfied for *every* pair of numbers x_1 and x_2 in I with $x_1 < x_2$.

The function $f(x) = x^2$ is decreasing on $(-\infty, 0]$ and increasing on $[0, \infty)$. (Make a sketch.) Therefore f is monotonic on $(-\infty, 0]$ and on $[0, \infty)$, but it is not monotonic on $(-\infty, \infty)$.

To see how the derivative can help us determine where a function is increasing or decreasing, look at Figure 4.17. It appears that where the slope of the tangent is positive the function is increasing, and where the slope of the tangent is negative the function is decreasing. We know that $f'(x)$ is the slope of the tangent at $(x, f(x))$. So it appears that the function increases when $f'(x) > 0$ and decreases when $f'(x) < 0$. To prove that this is always the case we use the Mean Value Theorem.

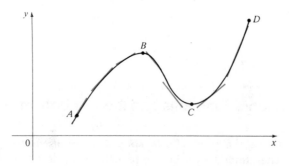

Figure 4.17

Test for Monotonic Functions (4.19)

Suppose f is continuous on $[a,b]$ and differentiable on (a,b).

(a)   If $f'(x) > 0$ for all x in (a,b), then f is increasing on $[a,b]$.

(b)   If $f'(x) < 0$ for all x in (a,b), then f is decreasing on $[a,b]$.

Proof

(a)   Let x_1 and x_2 be any two numbers in $[a,b]$ with $x_1 < x_2$. Then f is continuous on $[x_1, x_2]$ and differentiable on (x_1, x_2), so by the Mean Value Theorem there is a number c between x_1 and x_2 such that

(4.20)
$$f(x_2) - f(x_1) = f'(c)(x_2 - x_1)$$

Now $f'(c) > 0$ by assumption and $x_2 - x_1 > 0$ because $x_1 < x_2$. Thus the right side of Equation 4.20 is positive, and so

$$f(x_2) - f(x_1) > 0 \qquad \text{or} \qquad f(x_1) < f(x_2)$$

This shows that f is increasing on $[a,b]$.

Part (b) is proved similarly (see Exercise 51).   ●

EXAMPLE 1   Find where the function $f(x) = 3x^4 - 4x^3 - 12x^2 + 5$ is increasing and where it is decreasing.

Solution
$$f'(x) = 12x^3 - 12x^2 - 24x = 12x(x-2)(x+1)$$

To use Theorem 4.19 we have to know where $f'(x) > 0$ and where $f'(x) < 0$. This depends on the signs of the three factors of $f'(x)$—namely, $12x$, $x - 2$, and $x + 1$. We divide the real line into intervals whose endpoints are the critical numbers -1, 0, and 2 and arrange our work in a chart. A plus sign indicates that the given expression is positive, and a minus sign indicates that it is negative. The last column of the chart gives the conclusion based on Theorem 4.19. From this information and the values of f at the critical numbers, the graph of f is sketched in Figure 4.18.

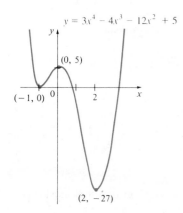

$y = 3x^4 - 4x^3 - 12x^2 + 5$

$(0, 5)$

$(-1, 0)$

$(2, -27)$

Figure 4.18

Interval	$12x$	$x - 2$	$x + 1$	$f'(x)$	f
$x < -1$	$-$	$-$	$-$	$-$	decreasing on $(-\infty, -1]$
$-1 < x < 0$	$-$	$-$	$+$	$+$	increasing on $[-1, 0]$
$0 < x < 2$	$+$	$-$	$+$	$-$	decreasing on $[0, 2]$
$x > 2$	$+$	$+$	$+$	$+$	increasing on $[2, \infty)$

●

Recall from Section 4.1 that if f has a local maximum or minimum at c, then c must be a critical number of f (by Fermat's Theorem), but not every critical number gives rise to an extremum. We therefore need a test that will tell us whether or not f has a local extremum at a critical number.

You can see from Figure 4.18 that $f(0) = 5$ is a local maximum value of f because f increases on $[-1, 0]$ and decreases on $[0, 2]$. Or, in terms of derivatives, $f'(x) > 0$ for $-1 < x < 0$ and $f'(x) < 0$ for $0 < x < 2$. In other words, the sign of $f'(x)$ changes from positive to negative at 0. This observation is the basis of the following test.

The First Derivative Test (4.21)

Suppose that c is a critical number of a function f that is continuous on $[a, b]$.

(a) If $f'(x) > 0$ for $a < x < c$ and $f'(x) < 0$ for $c < x < b$ (that is, f' changes from positive to negative at c), then f has a local maximum at c.

(b) If $f'(x) < 0$ for $a < x < c$ and $f'(x) > 0$ for $c < x < b$ (that is, f' changes from negative to positive at c), then f has a local minimum at c.

(c) If f' does not change sign at c, then f has no local extremum at c.

Proof

(a) Let $x \in (a, b)$. If $a < x < c$, then $f(x) < f(c)$ since $f' > 0$ implies that f is increasing on $[a, c]$. If $c < x < b$, then $f(c) > f(x)$ since $f' < 0$ implies that f is decreasing on $[c, b]$. Therefore $f(c) \geq f(x)$ for every x in (a, b). Thus, by Definition 4.2, f has a local maximum at c.

Parts (b) and (c) are proved similarly and are left as Exercise 52.   •

It is easy to remember the First Derivative Test by visualizing diagrams such as those in Figure 4.19.

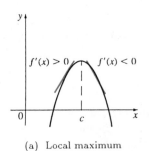

(a) Local maximum

(b) Local minimum

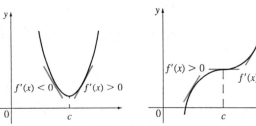

(c) No extremum

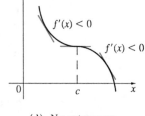

(d) No extremum

Figure 4.19

EXAMPLE 2   Find the local extrema of $f(x) = x(1 - x)^{2/5}$ and sketch its graph.

Solution   First we find the critical numbers of f:

$$f'(x) = (1 - x)^{2/5} + x \cdot \tfrac{2}{5}(1 - x)^{-3/5}(-1)$$
$$= \frac{5(1 - x) - 2x}{5(1 - x)^{3/5}} = \frac{5 - 7x}{5(1 - x)^{3/5}}$$

The derivative $f'(x) = 0$ when $5 - 7x = 0$, that is, $x = \frac{5}{7}$. Also $f'(x)$ does not exist when $x = 1$. So the critical numbers are $\frac{5}{7}$ and 1.

Next we set up a chart as in Example 1, dividing the real line into intervals with the critical numbers as endpoints.

Interval	$5 - 7x$	$(1 - x)^{3/5}$	$f'(x)$	f
$x < \frac{5}{7}$	$+$	$+$	$+$	increasing on $(-\infty, \frac{5}{7}]$
$\frac{5}{7} < x < 1$	$-$	$+$	$-$	decreasing on $[\frac{5}{7}, 1]$
$x < 1$	$-$	$-$	$+$	decreasing on $[1, \infty)$

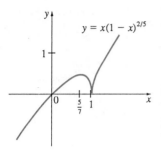

Figure 4.20

You can see from the chart that $f'(x)$ changes from positive to negative at $\frac{5}{7}$. So, by the First Derivative Test, f has a local maximum at $\frac{5}{7}$ and the local maximum value is

$$f\left(\tfrac{5}{7}\right) = \tfrac{5}{7}\left(\tfrac{2}{7}\right)^{2/5}$$

Since $f'(x)$ changes from negative to positive at 1, $f(1) = 0$ is a local minimum. We use these maximum and minimum values and the information in the chart to sketch the graph in Figure 4.20. Note that the curve is not smooth at $(1,0)$ but has a "corner" there (called a *cusp*). This is because $f'(1)$ does not exist. •

EXAMPLE 3   Find the local and absolute extreme values of the function $f(x) = x^3(x-2)^2$, $-1 \le x \le 3$. Sketch its graph.

Solution   Using the Product Rule, we have

$$f'(x) = 3x^2(x-2)^2 + x^3 \cdot 2(x-2) = x^2(x-2)(5x-6)$$

To find the critical numbers we set $f'(x) = 0$ and obtain $x = 0,\ 2,\ \frac{6}{5}$. To determine whether these give rise to extreme values we set up the following chart.

Interval	x^2	$x-2$	$5x-6$	$f'(x)$	f
$-1 \le x < 0$	$+$	$-$	$-$	$+$	increasing on $[-1,0]$
$0 < x < \frac{6}{5}$	$+$	$-$	$-$	$+$	increasing on $\left[0,\frac{6}{5}\right]$
$\frac{6}{5} < x < 2$	$+$	$-$	$+$	$-$	decreasing on $\left[\frac{6}{5},2\right]$
$2 < x \le 3$	$+$	$+$	$+$	$+$	increasing on $[2,3]$

Notice that $f'(x)$ does not change sign at 0, so by part (c) of the First Derivative Test, f has neither a maximum nor a minimum at 0. [The significance of $f'(0) = 0$ is just that the tangent is horizontal there.] Since f' changes from positive to negative at $\frac{6}{5}$, $f\left(\frac{6}{5}\right) = (1.2)^3(-0.8)^2 = 1.10592$ is a local maximum. Since f' changes from negative to positive at 2, $f(2) = 0$ is a local minimum.

To find the absolute extreme values we evaluate f at the endpoints of the interval:

$$f(-1) = -9 \qquad f(3) = 27$$

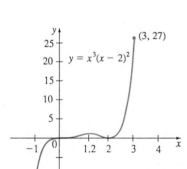

Figure 4.21

Using the procedure in (4.8) we compare these values with the values of f at $\frac{6}{5}$ and 2 and find that the absolute maximum value is $f(3) = 27$ and the absolute minimum value is $f(-1) = -9$. The graph of f is sketched in Figure 4.21. •

EXAMPLE 4   Prove that the inequality $(1+x)^n > 1 + nx$ is true whenever $x > 0$ and $n > 1$.

If n is a positive integer, the inequality follows from the Binomial Theorem or by mathematical induction. But, as this solution shows, n does not have to be an integer; it can be any real number greater than 1.

Solution   Consider the difference

$$f(x) = (1+x)^n - (1+nx)$$

Then

$$f'(x) = n(1+x)^{n-1} - n$$

$$= n[(1+x)^{n-1} - 1]$$

Since $x > 0$ and $n - 1 > 0$, we have $(1 + x)^{n-1} > 1$, so $f'(x) > 0$. Therefore f is increasing on $[0, \infty)$. In particular, $f(0) < f(x)$ when $0 < x$. But $f(0) = 0$, so

$$0 < (1 + x)^n - (1 + nx)$$

and therefore $$(1 + x)^n > 1 + nx$$

when $x > 0$.   •

SECTION 4.3 Exercises

In Exercises 1–22 (a) find the intervals on which f is increasing or decreasing, (b) find the local maximum and minimum values of f, and (c) sketch the graph of f.

1. $f(x) = 20 - x - x^2$ 2. $f(x) = x^3 - x + 1$

3. $f(x) = x^3 + x + 1$

4. $f(x) = 4x^3 - 3x^2 - 18x + 5$

5. $f(x) = x^3 - 2x^2 + x$

6. $f(x) = x^4 - 4x^3 - 8x^2 + 3$

7. $f(x) = 2x^2 - x^4$ 8. $f(x) - x^2(1 - x)^2$

9. $f(x) = x^4 + 4x + 1$

10. $f(x) = 3x^5 - 25x^3 + 60x$

11. $f(x) = x^3(x - 4)^4$ 12. $f(x) = x\sqrt{1 - x^2}$

13. $f(x) = x\sqrt{6 - x}$ 14. $f(x) = x^{2/3}(x - 2)^2$

15. $f(x) = x^{1/5}(x + 1)$ 16. $f(x) = \sqrt[3]{x} - \sqrt[3]{x^2}$

17. $f(x) = x\sqrt{x - x^2}$ 18. $f(x) = x^2\sqrt[3]{6x - 7}$

19. $f(x) = x - 2\sin x,\quad 0 \le x \le 2\pi$

20. $f(x) = x + \cos x,\quad 0 \le x \le 2\pi$

21. $f(x) = \sin^4 x + \cos^4 x,\quad 0 \le x \le 2\pi$

22. $f(x) = x\sin x + \cos x,\quad -\pi \le x \le \pi$

In Exercises 23–30 find the intervals on which the given function is increasing or decreasing.

23. $f(x) = x^3 + 2x^2 - x + 1$

24. $f(x) = x^5 + 4x^3 - 6$

25. $f(x) = x^6 + 192x + 17$

26. $f(x) = 2\tan x - \tan^2 x$

27. $f(x) = xe^x$ 28. $f(x) = x^2 e^x$

29. $f(x) = (\ln x)/\sqrt{x}$ 30. $f(x) = x\ln x$

In Exercises 31–36 find the local and absolute extreme values of the given function and sketch its graph.

31. $f(x) = x^3 - 3x^2 + 6x - 2,\quad -1 \le x \le 1$

32. $f(x) = x + 1/x,\quad 0.5 \le x \le 3$

33. $f(x) = x + \sqrt{1 - x},\quad 0 \le x \le 1$

34. $f(x) = x^3 + 6x^2 + 9x + 2,\quad -4 \le x \le 0$

35. $g(x) = \dfrac{x}{x^2 + 1},\quad -5 \le x \le 5$

36. $g(x) = \sin x - \cos x,\quad -\pi/2 \le x \le \pi/2$

37. Prove that

$$a + \frac{1}{a} < b + \frac{1}{b}\qquad \text{whenever } 1 < a < b$$

[*Hint:* Show that the function $f(x) = x + 1/x$ is increasing on $[1, \infty)$.]

38. Prove that

$$\frac{\tan b}{\tan a} > \frac{b}{a}\qquad \text{whenever } 0 < a < b < \pi/2$$

In Exercises 39–42 prove the given inequality using the method of Example 4.

39. $2\sqrt{x} > 3 - \frac{1}{x},\quad x > 1$

40. $\cos x > 1 - \dfrac{x^2}{2},\quad x > 0$

41. $\sin x > x - \dfrac{x^3}{6},\quad x > 0$

42. $\tan x > x,\quad 0 < x < \pi/2$

43. (a) Use the methods of this section to show that if $x > 0$, then

$$x + \frac{1}{x} \ge 2$$

   (b) Give another proof of this inequality using the fact that $(\sqrt{x} - \sqrt{y})^2 \ge 0$.

44. For what values of a and b will the function $f(x) = x^3 + ax^2 + bx + 2$ have a local maximum when $x = -3$ and a local minimum when $x = -1$?

45. Find a cubic function $f(x) = ax^3 + bx^2 + cx + d$ that has a local maximum value of 3 at -2 and a local minimum value of 0 at 1.

46. Find the local maximum and minimum values of the function f defined by

$$f(x) = \begin{cases} x - 39 & \text{if } x \le -4 \\ x^3 + 3x^2 - 9x & \text{if } -4 < x < 3 \\ 30 - x & \text{if } x \ge 3 \end{cases}$$

47. Sketch the graph of a function that satisfies all of the following conditions.
(a) $f(1) = 5$, $f(4) = 2$ (b) $f'(1) = f'(4) = 0$
(c) $f'(x) > 0$ for $x < 1$ (d) $f'(x) \le 0$ for $x > 1$

48. Sketch the graph of a function that satisfies all of the following conditions.

(a) $f'(4) = f'(10) = 0$
(b) $f'(x) < 0$ if $|x| < 2$, $f'(x) \ge 0$ if $2 < x < 10$, $f'(x) < 0$ if $x > 10$
(c) $f'(x) = 1$ if $x < -2$, $\displaystyle\lim_{x \to -2^+} f'(x) = -1$

49. If f and g are increasing on an interval I, show that $f + g$ is increasing on I.

50. If f and g are positive increasing functions on I, show that their product fg is increasing on I.

51. Prove Theorem 4.19(b).

52. (a) Prove Theorem 4.21(b) .
(b) Prove Theorem 4.21(c).

53. (a) Show that $e^x \ge 1 + x$ for $x \ge 0$.
(b) Show that $e^x \ge 1 + x + \frac{1}{2}x^2$ for $x \ge 0$.
(c) Use mathematical induction to prove that for $x \ge 0$ and any positive integer n,

$$e^x \ge 1 + x + \frac{x^2}{2!} + \cdots + \frac{x^n}{n!}$$

SECTION 4.4

Concavity and Points of Inflection

We have seen that knowledge of the first derivative of a function is useful in sketching its graph. In this section we see, with the help of the Mean Value Theorem, that the second derivative gives additional information that enables us to sketch a better picture of the graph.

Figure 4.22 shows the graphs of two increasing functions on $[a, b]$. Both graphs join point A to point B but they look different because they bend in different directions. How can we distinguish between these two types of behavior? In Figure 4.23 tangents to these curves have been drawn at several points. In (a) the curve lies above the tangents and f is called *concave upward* on $[a, b]$. In (b) the curve lies below the tangents and g is called *concave downward* on $[a, b]$.

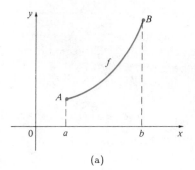
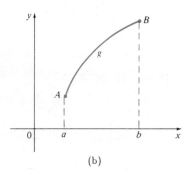

(a)   (b)

Figure 4.22

Definition (4.22)

If the graph of f lies above all of its tangents on an interval I, then it is called **concave upward** on I. If the graph of f lies below all of these tangents, it is called **concave downward** on I.

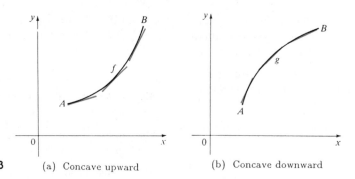

Figure 4.23   (a) Concave upward   (b) Concave downward

Figure 4.24 shows the graph of a function that is concave upward (abbreviated CU) on the intervals $[b,c]$, $[d,e)$, and $(e,p]$, and concave downward (CD) on the intervals $[a,b]$, $[c,d]$, and $[p,q]$.

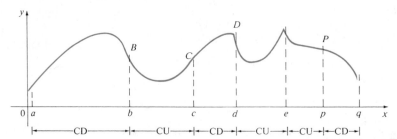

Figure 4.24

Let us see how the second derivative helps determine the intervals of concavity. Looking at Figure 4.23(a) you can see that, going from left to right, the slope of the tangent increases. This means that the derivative $f'(x)$ is an increasing function and therefore its derivative $f''(x)$ is positive. Likewise in Figure 4.23(b) the slope of the tangent decreases from left to right, so $f'(x)$ decreases and therefore $f''(x) < 0$. This reasoning can be reversed and suggests that the following theorem is true.

The Test for Concavity (4.23)

> Suppose f is twice differentiable on the interval I.
>
> (a)   If $f''(x) > 0$ for all x in I, then the graph of f is concave upward on I.
>
> (b)   If $f''(x) < 0$ for all x in I, then the graph of f is concave downward on I.

Proof of (a)   Let a be any number in I. We must show that curve $y = f(x)$ lies above the tangent line at the point $(a, f(a))$. The equation of this tangent is

$$y = f(a) + f'(a)(x - a)$$

So we must show that

$$f(x) > f(a) + f'(a)(x - a)$$

whenever $x \in I$ $(x \neq a)$. (See Figure 4.25.)

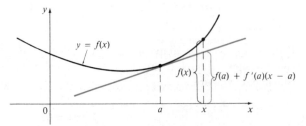

Figure 4.25

First let us take the case where $x > a$. Applying the Mean Value Theorem to f on the interval $[a, x]$, we get a number c, with $a < c < x$, such that

(4.24)
$$f(x) - f(a) = f'(c)(x - a)$$

Since $f'' > 0$ on I we know from Theorem 4.19 that f' is increasing on I. Thus, since $a < c$, we have

$$f'(a) < f'(c)$$

and so, multiplying this inequality by the positive number $x - a$, we get

(4.25)
$$f'(a)(x - a) < f'(c)(x - a)$$

Now we add $f(a)$ to both sides of this inequality:

$$f(a) + f'(a)(x - a) < f(a) + f'(c)(x - a)$$

But from Equation 4.24 we have $f(x) = f(a) + f'(c)(x - a)$. So this inequality becomes

(4.26)
$$f(x) > f(a) + f'(a)(x - a)$$

which is what we wanted to prove.

For the case where $x < a$ we have $f'(c) < f'(a)$, but multiplication by the negative number $x - a$ reverses the inequality, so we get (4.25) and (4.26) as before.   ●

Definition (4.27)

A point P on a curve is called a **point of inflection** if the curve changes from concave upward to concave downward or from concave downward to concave upward at P.

For instance, in Figure 4.24, B, C, D, and P are the points of inflection. Notice that if a curve has a tangent at a point of inflection, then the curve crosses its tangent there.

In view of the Test for Concavity, there is a point of inflection at any point where the second derivative changes sign.

EXAMPLE 1   Determine where the curve $y = x^3 - 3x + 1$ is concave upward and where it is concave downward. Find the inflection points and sketch the curve.

Solution   If $f(x) = x^3 - 3x + 1$, then

$$f'(x) = 3x^2 - 3 = 3(x^2 - 1)$$

Since $f'(x) = 0$ when $x^2 = 1$, the critical numbers are ± 1. Also

$$f'(x) < 0 \quad \Leftrightarrow \quad x^2 - 1 < 0 \quad \Leftrightarrow \quad x^2 < 1 \quad \Leftrightarrow \quad |x| < 1$$

$$f'(x) > 0 \quad \Leftrightarrow \quad x^2 > 1 \quad \Leftrightarrow \quad x > 1 \quad \text{or} \quad x < -1$$

Therefore f is increasing on the intervals $(-\infty, -1]$ and $[1, \infty)$ and is decreasing on $[-1, 1]$. By the First Derivative Test, $f(-1) = 3$ is a local maximum value and $f(1) = -1$ is a local minimum value.

To determine the concavity we compute the second derivative:

$$f''(x) = 6x$$

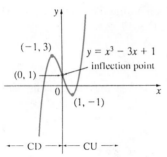

$y = x^3 - 3x + 1$
inflection point

$(-1, 3)$

$(0, 1)$

$(1, -1)$

—— CD —+— CU —·

Figure 4.26

Thus $f''(x) > 0$ when $x > 0$ and $f''(x) < 0$ when $x < 0$. The Test for Concavity then tells us that the curve is concave downward on $(-\infty, 0)$ and concave upward on $(0, \infty)$. Since the curve changes from concave downward to concave upward when $x = 0$, the point $(0, 1)$ is a point of inflection. We use this information to sketch the curve in Figure 4.26.   ●

Another application of the second derivative is in finding maximum and minimum values of a function.

The Second Derivative Test (4.28)

Suppose f'' is continuous on an open interval that contains c.

(a) If $f'(c) = 0$ and $f''(c) > 0$, then f has a local minimum at c.

(b) If $f'(c) = 0$ and $f''(c) < 0$, then f has a local maximum at c.

Proof

(a) If $f''(c) > 0$, then $f''(x) > 0$ on some open interval I that contains c because f'' is continuous. So, by the Test for Concavity, f is concave upward on I. Therefore the graph of f lies *above* its tangent at $P(c, f(c))$. But since $f'(c) = 0$, the tangent at P is horizontal. This shows that $f(x) \geq f(c)$ whenever x is in I [see Figure 4.27(a)] and so f has a local minimum at c.

Part (b) is proved similarly [see Figure 4.27(b)].

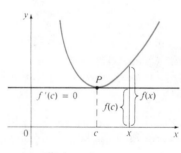

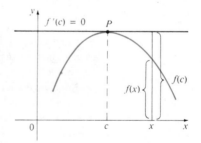

Figure 4.27   (a) $f''(c) > 0$, concave upward   (b) $f''(c) < 0$, concave downward   ●

EXAMPLE 2   Discuss the curve $y = x^4 - 4x^3$ with respect to concavity, points of inflection, and local extrema. Use this information to sketch the curve.

Solution   If $f(x) = x^4 - 4x^3$, then

$$f'(x) = 4x^3 - 12x^2 = 4x^2(x - 3)$$

$$f''(x) = 12x^2 - 24x = 12x(x - 2)$$

To find the critical numbers we set $f'(x) = 0$ and obtain $x = 0$ and $x = 3$. To use the Second Derivative Test we evaluate f'' at these critical numbers:

$$f''(0) = 0 \qquad f''(3) = 36 > 0$$

Since $f'(3) = 0$ and $f''(3) > 0$, $f(3) = -27$ is a local minimum. Since $f''(0) = 0$, the Second Derivative Test gives no information about the critical number 0. But since $f'(x) < 0$ for $x < 0$ and for $0 < x < 3$, the First Derivative Test tells us that f does not have a local extremum at 0.

Since $f''(x) = 0$ when $x = 0$ or 2, we divide the real line into intervals with these numbers as endpoints and complete the following chart.

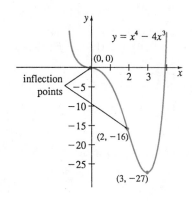

Interval	$f''(x) = 12x(x - 2)$	Concavity
$(-\infty, 0)$	$+$	upward
$(0, 2)$	$-$	downward
$(2, \infty)$	$+$	upward

The point $(0, 0)$ is an inflection point since the curve changes from concave upward to concave downward there. Also $(2, -16)$ is an inflection point since the curve changes from concave downward to concave upward there.

Using the local minimum, the intervals of concavity, and the inflection points, we sketch the curve in Figure 4.28.

Figure 4.28

Note: Example 2 illustrates the fact that the Second Derivative Test gives no information when $f''(c) = 0$. This test also fails when $f''(c)$ does not exist. In such cases the First Derivative Test must be used.

EXAMPLE 3   Sketch the graph of the function $f(x) = x^{2/3}(6 - x)^{1/3}$.

Solution   Calculation of the first two derivatives gives

$$f'(x) = \frac{4 - x}{x^{1/3}(6 - x)^{2/3}} \qquad f''(x) = \frac{-8}{x^{4/3}(6 - x)^{5/3}}$$

Since $f'(x) = 0$ when $x = 4$ and $f'(x)$ does not exist when $x = 0$ or $x = 6$, the critical numbers are 0, 4, and 6.

Interval	$4 - x$	$x^{1/3}$	$(6 - x)^{2/3}$	$f'(x)$	f
$x < 0$	$+$	$-$	$+$	$-$	decreasing on $(-\infty, 0]$
$0 < x < 4$	$+$	$+$	$+$	$+$	increasing on $[0, 4]$
$4 < x < 6$	$-$	$+$	$+$	$-$	decreasing on $[4, 6]$
$x > 6$	$-$	$+$	$+$	$-$	decreasing on $[6, \infty)$

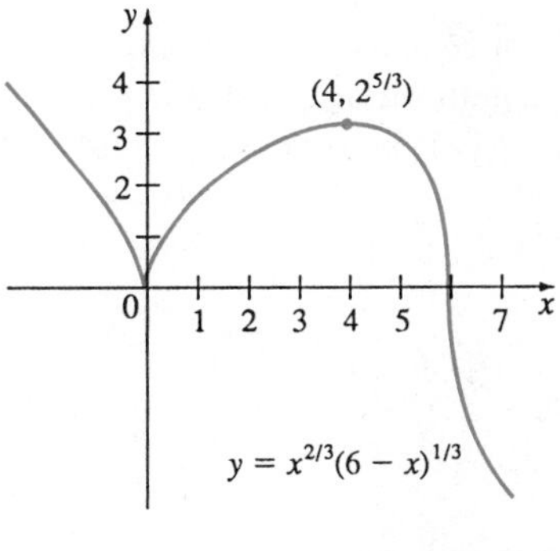

To find the local extreme values we use the First Derivative Test. Since f' changes from negative to positive at 0, $f(0) = 0$ is a local minimum. Since f' changes from positive to negative at 4, $f(4) = 2^{5/3}$ is a local maximum. The sign of f' does not change at 6, so there is no extremum there. (The Second Derivative Test could be used at 4, but not at 0 or 6 since f'' does not exist there.)

Looking at the expression for $f''(x)$ and noting that $x^{4/3} \geq 0$ for all x, we have $f''(x) < 0$ for $x < 0$ and for $0 < x < 6$ and $f''(x) > 0$ for $x > 6$. So f is concave downward on $(-\infty, 0)$ and $(0, 6)$ and concave upward on $(6, \infty)$, and the only inflection point is $(6, 0)$. The graph is sketched in Figure 4.29. Note that the curve has vertical tangents at $(0, 0)$ and $(6, 0)$.

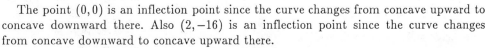

Figure 4.29

SECTION 4.4   **Exercises**

In Exercises 1–18 find (a) the intervals of increase or decrease, (b) the local maximum and minimum values, (c) the intervals of concavity, and (d) the x-coordinates of the points of inflection. Then use this information to sketch the graph.

1. $f(x) = x^3 - x$

2. $f(x) = 2x^3 + 5x^2 - 4x$

3. $f(x) = x^3 - x^2 - x + 1$

4. $f(x) = x^4 - 6x^2$

5. $g(x) = x^4 - 3x^3 + 3x^2 - x$

6. $g(x) = 4 + 72x - 3x^2 - x^3$

7. $h(x) = 3x^5 - 5x^3 + 3$ 8. $h(x) = (x^2 - 1)^3$

9. $G(x) = 8 - \sqrt[3]{x}$ 10. $G(x) = x^{4/3} - 4x^{1/3}$

11. $P(x) = x\sqrt{x^2 + 1}$ 12. $P(x) = x\sqrt{x + 1}$

13. $Q(x) = x^{1/3}(x + 3)^{2/3}$ 14. $Q(x) = x - 3x^{1/3}$

15. $f(\theta) = \sin^2\theta$ 16. $f(\theta) = \cos^2\theta$

17. $f(t) = t + \sin t$ 18. $f(t) = t + \cos t$

In Exercises 19–26 find the intervals on which the curve is concave upward.

19. $y = 6x^2 - 2x^3 - x^4$ 20. $y = x^2/\sqrt{1 + x}$

21. $y = x/(1 + x)^2$ 22. $y = x^3/(x^2 - 3)$

23. $y = xe^x$ 24. $y = x^2 e^x$

25. $y = (\ln x)/\sqrt{x}$ 26. $y = x \ln x$

In Exercises 27–30 sketch the graph of a function that satisfies all of the given conditions.

27. $f'(-1) = f'(1) = 0$, $f'(x) < 0$ if $|x| < 1$, $f'(x) > 0$ if $|x| > 1$, $f(-1) = 4$, $f(1) = 0$, $f''(x) < 0$ if $x < 0$, $f''(x) > 0$ if $x > 0$

28. $f'(-1) = 0$, $f'(1)$ does not exist, $f'(x) < 0$ if $|x| < 1$, $f'(x) > 0$ if $|x| > 1$, $f(-1) = 4$, $f(1) = 0$, $f''(x) < 0$ if $x \neq 1$

29. $f'(-1) = 0$, $f'(2) = 0$, $f(-1) = f(2) = -1$, $f(-3) = 4$, $f'(x) = 0$ if $x < -3$, $f'(x) < 0$ on

(-3, -1) and (0, 2), $f'(x) > 0$ on (-1, 0) and (2, ∞), $f''(x) > 0$ on (-3, 0) and (0, 5), $f''(x) < 0$ on (5, ∞)

30. $f(0) = 0$, $f(-1) = 1$, $f'(-1) = 0$, $f''(x) > 0$ on $(-\infty, -1)$, $f''(x) < 0$ on $(-1, 0)$ and $(0, \infty)$, $f'(x) > 0$ for $x > 0$

31.   Show that a cubic function

$$f(x) = ax^3 + bx^2 + cx + d \quad a \neq 0$$

has exactly one point of inflection.

32.   Show that if a function f is concave upward on an interval I, then the function $-f$ is concave downward on I.

33.   Prove that if $(c, f(c))$ is a point of inflection of the graph of f and f'' exists in an open interval that contains c, then $f''(c) = 0$. (*Hint:* Apply the First Derivative Test and Fermat's Theorem to the function $g = f'$.)

34.   Show that if $f(x) = x^4$, then $f''(0) = 0$, but $(0, 0)$ is not an inflection point of the graph of f.

35.   Show that the function $g(x) = x|x|$ has an inflection point at $(0, 0)$ but $g''(0)$ does not exist.

36.   Use Newton's method to find the coordinates of the inflection point of the curve $y = x^3 + \cos x$ correct to three decimal places.

In Exercises 37–40 assume that all of the functions are twice differentiable.

37.   If f and g are concave upward on I, show that $f + g$ is concave upward on I.

38.   If f is positive and concave upward on I, show that the function $g(x) = [f(x)]^2$ is concave upward on I.

39.   If f and g are positive increasing concave upward functions on I, show that the product function fg is concave upward on I.

40.   Suppose f and g are both concave upward on $(-\infty, \infty)$. Under what condition on f will the composite function $h(x) = f(g(x))$ be concave upward?

SECTION 4.5

Curve Sketching

The graph of a function tells us a great deal about the behavior of the function. One of the main uses of calculus is to provide a procedure for sketching curves with reasonable accuracy.

So far we have been concerned with some particular aspects of curve sketching: domain, range, and symmetry in the Review and Preview; limits, continuity, and asymptotes in Chapter 1; derivatives and tangents in Chapter 2; and extreme values, monotonicity, concavity, and points of inflection in this chapter. It is now time to put all of this information together to sketch graphs that reveal the important features of functions.

You may ask: What is wrong with just using a calculator to plot points and then joining these points with a smooth curve? To see the pitfalls of this approach, suppose you have used a calculator to produce the table of values and corresponding points in Figure 4.30.

x	$f(x)$	x	$f(x)$
-5	22	1	7
-4	7	2	10
-3	-2	3	11
-2	-4	4	10
-1	-2	5	8
0	3	6	-8

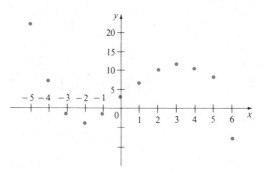

Figure 4.30

 You might then join these points to produce the curve shown in Figure 4.31, but the correct graph might be the one shown in Figure 4.32. You can see the drawbacks of the method of plotting points. Certain essential features of the graph may be missed, such as the maximum and minimum values between -2 and -1 or between 2 and 5. If you

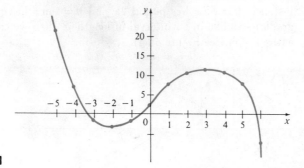

Figure 4.31

just plot points, you don't know when to stop. (How far should you go to the left or right?) But the use of calculus ensures that all the important aspects of the curve are illustrated.

Before sketching the curve it is wise to consider the following items.

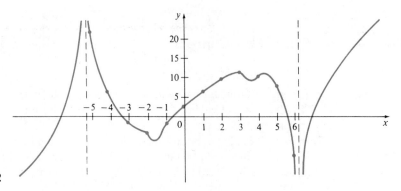

Figure 4.32

Checklist of Information for Sketching a Curve $y = f(x)$

A. **Domain.** The first step is to determine the domain D of f, that is, the set of values of x for which $f(x)$ is defined.

B. **Intercepts.** The y-intercept is $f(0)$ and tells where the curve intersects the y-axis. To find the x-intercepts, we set $y = 0$ and solve for x, if this can be done easily. Sometimes it is difficult, and not worthwhile, to find the x-intercepts. For instance, to find the x-intercept of $f(x) = x^3 - x^2 + 2x - 1$ we would have to solve the equation $x^3 - x^2 + 2x - 1 = 0$. This is not easy, so we do not bother with it.

C. **Symmetry.**

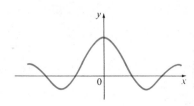

(a) Even function: reflectional symmetry

(i) If $f(-x) = f(x)$ for all x in D, that is, the equation of the curve is unchanged when x is replaced by $-x$, then f is an **even function** and the curve is symmetric about the y-axis. This means that our work is cut in half. If we know what the curve looks like for $x \geq 0$, then we need only reflect about the y-axis to obtain the complete curve [see Figure 4.33(a)]. Here are some examples: $y = x^2$, $y = x^4$, $y = |x|$, and $y = \cos x$.

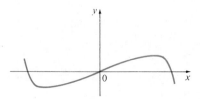

(b)   Odd function: rotational symmetry

Figure 4.33

(ii) If $f(-x) = -f(x)$ for all x in D, then f is an **odd function** and the curve is symmetric about the origin. Again we can obtain the complete curve if we know what it looks like for $x \geq 0$ [see Figure 4.33(b)]. Some simple examples of odd functions are: $y = x$, $y = x^3$, $y = x^5$, and $y = \sin x$.

(iii) If $f(x + p) = f(x)$ for all x in D, where p is a positive constant, then f is called a **periodic function** and the smallest such number p is called the **period**. For instance, $y = \sin x$ has period 2π and $y = \tan x$ has period π. If we know what the graph looks like in an interval of length p, then we can use translation to sketch the entire graph (see Figure 4.34).

Figure 4.34

Periodic function: translational symmetry

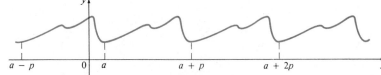

$a - p$ 0 a $a + p$ $a + 2p$ x

D.   **Asymptotes.**
(i)   *Horizontal Asymptotes.*   Recall from Section 1.6 that if either

$$\lim_{x \to \infty} f(x) = L \qquad \text{or} \qquad \lim_{x \to -\infty} f(x) = L$$

then the line $y = L$ is a horizontal asymptote of the curve $y = f(x)$. If it turns out that $\lim_{x \to \infty} f(x) = \infty$ (or $-\infty$), then we do not have an asymptote to the right, but that is still useful information for sketching the curve.

(ii) *Vertical Asymptotes.* Recall from Section 1.7 that the line $x = a$ is a vertical asymptote if at least one of the following statements is true:

(4.29)
$$\lim_{x \to a^+} f(x) = \infty \qquad \lim_{x \to a^-} f(x) = \infty \qquad \lim_{x \to a^+} f(x) = -\infty \qquad \lim_{x \to a^-} f(x) = -\infty$$

(For rational functions you can locate the vertical asymptotes by equating the denominator to 0 after canceling any common factors. But for other functions this method does not apply.) Furthermore, in sketching the curve it is very useful to know exactly which of the statements in (4.29) is true. If $f(a)$ is not defined but a is an endpoint of the domain of f, then you should compute $\lim_{x \to a^-} f(x)$ or $\lim_{x \to a^+} f(x)$, whether or not this limit is infinite.

(iii) *Slant asymptotes.* These are discussed at the end of this section.

E. **Intervals of Increase or Decrease.** Use the Test for Monotonic Functions (4.19). Compute $f'(x)$ and find the intervals on which $f'(x)$ is positive (f is increasing) and the intervals on which $f'(x)$ is negative (f is decreasing).

F. **Local Maximum and Minimum Values.** Find the critical numbers of f [the numbers c where $f'(c) = 0$ or $f'(c)$ does not exist]. Then use the First Derivative Test (4.21). If f' changes from positive to negative at a critical number c, then $f(c)$ is a local maximum. If f' changes from negative to positive at c, then $f(c)$ is a local minimum. Although it is usually preferable to use the First Derivative Test, you can use the Second Derivative Test (4.28) if c is a critical number such that $f''(c) \neq 0$. Then $f''(c) > 0$ implies that $f(c)$ is a local minimum, whereas $f''(c) < 0$ implies that $f(c)$ is a local maximum.

G. **Concavity and Points of Inflection.** Compute $f''(x)$ and use the Test for Concavity (4.23). The curve is concave upward where $f''(x) > 0$ and concave downward where $f''(x) < 0$. Inflection points occur where the direction of concavity changes.

H. **Sketch the Curve.** Using the information in the items A–G, draw the graph. Draw in the asymptotes as broken lines. Plot the intercepts, maximum and minimum points, and inflection points. Then make the curve pass through these points, rising and falling according to E, with concavity according to G, and approaching the asymptotes. If additional accuracy is desired near any point, you could compute the value of the derivative there. The tangent indicates the direction in which the curve proceeds.

EXAMPLE 1   Discuss the curve $y = \dfrac{2x^2}{x^2 - 1}$ under the headings A–H.

Solution

A. The domain is

$$\{x \mid x^2 - 1 \neq 0\} = \{x \mid x \neq \pm 1\} = (-\infty, -1) \cup (-1, 1) \cup (1, \infty)$$

B. The x- and y-intercepts are both 0.

C. Since $f(-x) = f(x)$, f is even. The curve is symmetric about the y-axis.

D.
$$\lim_{x \to \pm\infty} \frac{2x^2}{x^2 - 1} = \lim_{x \to \pm\infty} \frac{2}{1 - 1/x^2} = 2$$

Therefore the line $y = 2$ is a horizontal asymptote. Since the denominator is 0 when $x = \pm 1$, we compute the following limits:

$$\lim_{x \to 1^+} \frac{2x^2}{x^2 - 1} = \infty \qquad \lim_{x \to 1^-} \frac{2x^2}{x^2 - 1} = -\infty$$

$$\lim_{x \to -1^+} \frac{2x^2}{x^2 - 1} = -\infty \qquad \lim_{x \to -1^-} \frac{2x^2}{x^2 - 1} = \infty$$

Therefore the lines $x = 1$ and $x = -1$ are vertical asymptotes.

E.
$$f'(x) = \frac{4x(x^2 - 1) - 2x^2 \cdot 2x}{(x^2 - 1)^2} = \frac{-4x}{(x^2 - 1)^2}$$

Since $f'(x) > 0$ when $x < 0$ $(x \neq -1)$ and $f'(x) < 0$ when $x > 0$ $(x \neq 1)$, f is increasing on $(-\infty, -1)$ and $(-1, 0]$ and decreasing on $[0, 1)$ and $(1, \infty)$.

F. The only critical number is $x = 0$. Since f' changes from positive to negative at 0, $f(0) = 0$ is a local maximum by the First Derivative Test.

G.
$$f''(x) = \frac{-4(x^2 - 1)^2 + 4x \cdot 2(x^2 - 1)2x}{(x^2 - 1)^4} = \frac{12x^2 + 4}{(x^2 - 1)^3}$$

Since $12x^2 + 4 > 0$ for all x, we have

$$f''(x) > 0 \qquad \Leftrightarrow \qquad x^2 - 1 > 0 \qquad \Leftrightarrow \qquad |x| > 1$$

and $f''(x) < 0 \Leftrightarrow |x| < 1$. Thus the curve is concave upward on the intervals $(-\infty, -1)$ and $(1, \infty)$ and downward on $(-1, 1)$. There is no point of inflection since 1 and -1 are not in the domain of f.

H. Using the information in A–G, we sketch the curve in Figure 4.35.

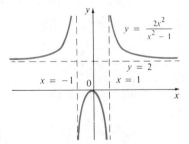

Figure 4.35

EXAMPLE 2   Sketch the graph of $f(x) = \dfrac{x^2}{\sqrt{x + 1}}$.

A. Domain $= \{x \mid x + 1 > 0\} = \{x \mid x > -1\} = (-1, \infty)$

B. The x- and y-intercepts are both 0.

C. Symmetry: none.

D. Since

$$\lim_{x \to \infty} \frac{x^2}{\sqrt{x+1}} = \infty$$

there is no horizontal asymptote. Since $\sqrt{x+1} \to 0$ as $x \to -1^+$ and $f(x)$ is always positive, we have

$$\lim_{x \to -1^+} \frac{x^2}{\sqrt{x+1}} = \infty$$

and so the line $x = -1$ is a vertical asymptote.

E. $$f'(x) = \frac{2x\sqrt{x+1} - x^2 \cdot 1/(2\sqrt{x+1})}{x+1} = \frac{x(3x+4)}{2(x+1)^{3/2}}$$

Note that $f'(x) = 0$ when $x = 0$ or $-\frac{4}{3}$. But $-\frac{4}{3}$ is not in the domain of f, so the only critical number is 0. Since $f'(x) < 0$ when $-1 < x < 0$ and $f'(x) > 0$ when $x > 0$, f is decreasing on $(-1, 0]$ and increasing on $[0, \infty)$.

F. Since $f'(0) = 0$ and f' changes from negative to positive at 0, $f(0) = 0$ is a local (and absolute) minimum by the First Derivative Test.

G. $$f''(x) = \frac{2(x+1)^{3/2}(6x+4) - (3x^2 + 4x)3(x+1)^{1/2}}{4(x+1)^3}$$

$$= \frac{3x^2 + 8x + 8}{4(x+1)^{5/2}}$$

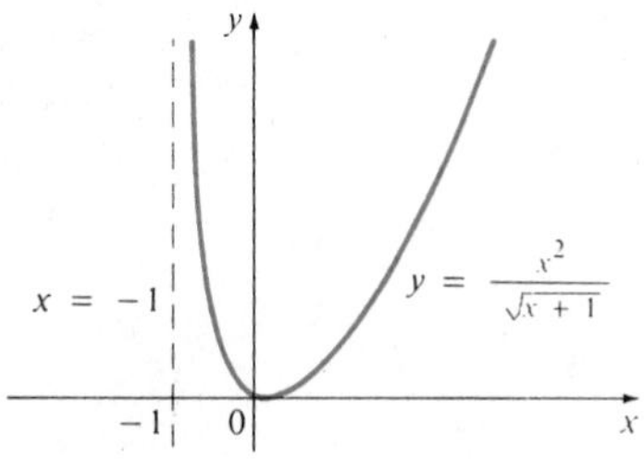

Figure 4.36

Note that the denominator is always positive. The numerator is the quadratic $3x^2 + 8x + 8$, which is always positive because its discriminant is $b^2 - 4ac = -32$ which is negative, and the coefficient of x^2 is positive. Thus $f''(x) > 0$ for all x, which means that f is concave upward on $(-1, \infty)$ and there is no point of inflection.

H. The curve is sketched in Figure 4.36.

EXAMPLE 3   Sketch the graph of $f(x) = 2\cos x + \sin 2x$.

A. The domain is R.

B. The y-intercept is $f(0) = 2$. The x-intercepts occur when

$$2\cos x + \sin 2x = 2\cos x + 2\sin x \cos x = 2\cos x(1 + \sin x) = 0$$

that is, when $\cos x = 0$ or $\sin x = -1$. Thus, in the interval $[0, 2\pi]$, the x-intercepts are $\pi/2$ and $3\pi/2$.

C. f is neither even nor odd, but $f(x + 2\pi) = f(x)$ for all x and so f is periodic and has period 2π. Thus in what follows we need to consider only $0 \le x \le 2\pi$ and then extend the curve by translation in H.

D. Asymptotes: none.

E. $$f'(x) = -2\sin x + 2\cos 2x = -2\sin x + 2(1 - 2\sin^2 x)$$

$$= -2(2\sin^2 x + \sin x - 1) = -2(2\sin x - 1)(\sin x + 1)$$

Thus $f'(x) = 0$ when $\sin x = \frac{1}{2}$ or $\sin x = -1$, so in $[0, 2\pi]$, $x = \pi/6,\ 5\pi/6,\ 3\pi/2$.

Interval	$f'(x)$	f
$0 < x < \pi/6$	$+$	increasing on $[0, \pi/6]$
$\pi/6 < x < 5\pi/6$	$-$	decreasing on $[\pi/6, 5\pi/6]$
$5\pi/6 < x < 3\pi/2$	$+$	increasing on $[5\pi/6, 3\pi/2]$
$3\pi/2 < x < 2\pi$	$+$	increasing on $[3\pi/2, 2\pi]$

F. From the chart in E the First Derivative Test says that $f(\pi/6) = 3\sqrt{3}/2$ is a local maximum and $f(5\pi/6) = -3\sqrt{3}/2$ is a local minimum, but f has no extremum at $3\pi/2$, only a horizontal tangent.

G.
$$f''(x) = -2\cos x - 4\sin 2x = -2\cos x(1 + 4\sin x)$$

Thus $f''(x) = 0$ when $\cos x = 0$ (so $x = \pi/2$ or $3\pi/2$) and when $\sin x = -\frac{1}{4}$. From Figure 4.37 we see that there are two values of x between 0 and 2π for which $\sin x = -\frac{1}{4}$. Let us call them α_1 and α_2. Then $f''(x) > 0$ on $(\pi/2, \alpha_1)$ and $(3\pi/2, \alpha_2)$, so f is concave upward there. Also $f''(x) < 0$ on $(0, \pi/2)$, $(\alpha_1, 3\pi/2)$, and $(\alpha_2, 2\pi)$, so f is concave downward there. Inflection points occur when $x = \pi/2$, α_1, $3\pi/2$, and α_2.

$\alpha_1 = \pi + \arcsin\frac{1}{4}$

$\alpha_2 = 2\pi - \arcsin\frac{1}{4}$

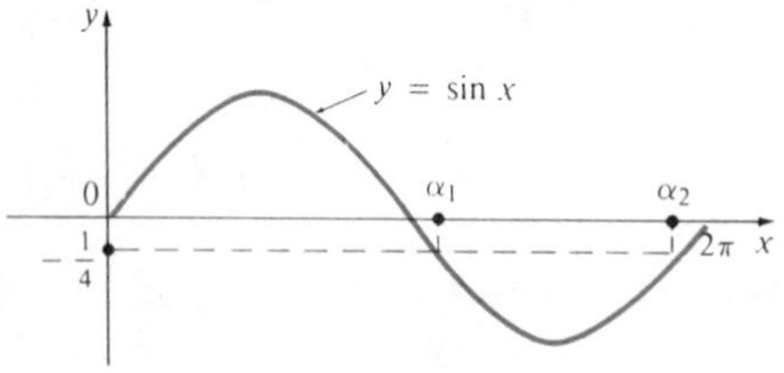

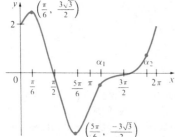

Figure 4.37

H. The graph of the function restricted to $0 \le x \le 2\pi$ is shown in Figure 4.38. Then it is extended, using periodicity, to the complete graph in Figure 4.39.

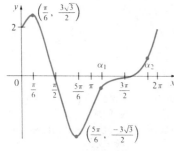

Figure 4.38

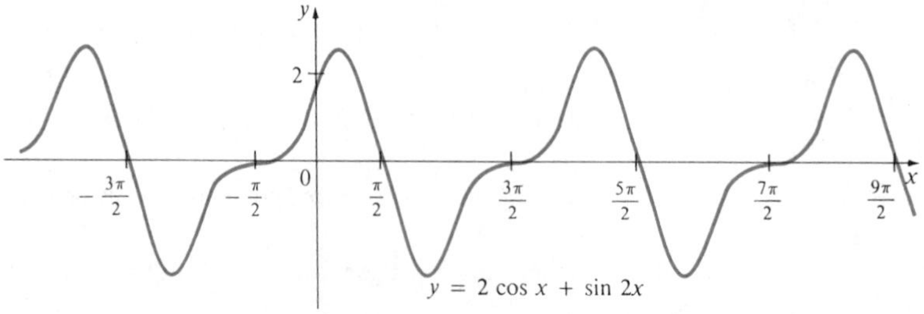

Figure 4.39

EXAMPLE 4   Discuss the curve $y = e^{1/x}$.

A.   The domain is $\{x \mid x \neq 0\} = (-\infty,\ 0) \cup (0, \infty)$.

B. There is no y-intercept since $f(0)$ is undefined. There is no x-intercept since $e^{1/x} > 0$ for all x.
C. Symmetry: none.
D. Since $1/x \to 0$ as $x \to \pm\infty$, we have

$$\lim_{x \to \pm\infty} e^{1/x} = 1$$

so $y = 1$ is a horizontal asymptote. Since $1/x \to \infty$ as $x \to 0^+$, we have

$$\lim_{x \to 0^+} e^{1/x} = \infty$$

so $x = 0$ is a vertical asymptote. Also

$$\lim_{x \to 0^-} e^{1/x} = 0$$

since $1/x \to -\infty$ as $x \to 0^-$.

E. $$f'(x) = -\frac{e^{1/x}}{x^2}$$

Since $e^{1/x} > 0$ and $x^2 > 0$ for all $x \neq 0$, we have $f'(x) < 0$ for all $x \neq 0$. Thus f is decreasing on $(-\infty, 0)$ and on $(0, \infty)$.
F. There is no critical number, so there is no extremum.

G. $$f''(x) = \frac{e^{1/x} \frac{1}{x^2} x^2 + e^{1/x}(2x)}{x^4} = \frac{e^{1/x}(2x+1)}{x^4}$$

Since $e^{1/x} > 0$ and $x^4 > 0$ we have $f''(x) > 0$ when $x > -\frac{1}{2}$ $(x \neq 0)$ and $f''(x) < 0$ when $x < -\frac{1}{2}$. So the curve is concave downward on $(-\infty, -\frac{1}{2})$ and concave upward on $(-\frac{1}{2}, 0)$ and on $(0, \infty)$. The inflection point is $(-\frac{1}{2}, e^{-2})$.
H. The curve is sketched in Figure 4.40. ●

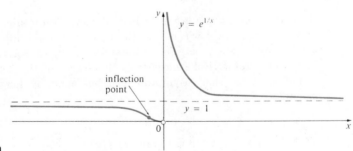

$y = e^{1/x}$

inflection point

$y = 1$

Figure 4.40

EXAMPLE 5   Discuss the curve $y = \ln(4 - x^2)$.
A. The domain is

$$\{x \mid 4 - x^2 > 0\} = \{x \mid x^2 < 4\} = \{x \mid |x| < 2\} = (-2, 2)$$

B. The y-intercept is $f(0) = \ln 4$. To find the x-intercept we set

$$y = \ln(4 - x^2) = 0$$

We know that $\ln 1 = \log_e 1 = 0$ (since $e^0 = 1$), so we have $4 - x^2 = 1 \Rightarrow x^2 = 3$ and therefore the x-intercepts are $\pm \sqrt{3}$.

C. Since $f(-x) = f(x)$, f is even and the curve is symmetric about the y-axis.

D. We look for vertical asymptotes at the endpoints of the domain. Since $4 - x^2 \to 0^+$ as $x \to 2^-$ and also as $x \to -2^+$, we have

$$\lim_{x \to 2^-} \ln(4 - x^2) = -\infty \qquad \lim_{x \to -2^+} \ln(4 - x^2) = -\infty$$

by (3.26). Thus the lines $x = 2$ and $x = -2$ are vertical asymptotes.

E.
$$f'(x) = \frac{-2x}{4 - x^2}$$

Since $f'(x) > 0$ when $-2 < x < 0$ and $f'(x) < 0$ when $0 < x < 2$, f is increasing on $(-2, 0]$ and decreasing on $[0, 2)$.

F. The only critical number is $x = 0$. Since f' changes from positive to negative at 0, $f(0) = \ln 4$ is a local maximum by the First Derivative Test.

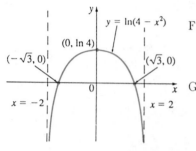

G.
$$f''(x) = \frac{(4 - x^2)(-2) + 2x(-2x)}{(4 - x^2)^2} = \frac{-8 - 2x^2}{(4 - x^2)^2}$$

Since $f''(x) < 0$ for all x, the curve is concave downward on $(-2, 2)$ and there is no inflection point.

Figure 4.41   H. Using this information, we sketch the curve in Figure 4.41.   ●

Slant Asymptotes

Some curves have asymptotes that are oblique, that is, neither horizontal nor vertical. If

$$\lim_{x \to \infty} [f(x) - (mx + b)] = 0$$

then the line $y = mx + b$ is called a **slant asymptote** because the vertical distance between the curve $y = f(x)$ and the line $y = mx + b$ approaches 0. (A similar situation exists if we let $x \to -\infty$.) For rational functions, slant asymptotes occur when the degree of the numerator is one more than the degree of the denominator. In such a case the equation of the slant asymptote can be found by long division as in the following example.

EXAMPLE 6   Sketch the graph of $f(x) = \dfrac{x^3}{x^2 + 1}$.

A. The domain is $R = (-\infty, \infty)$.

B. The x- and y-intercepts are both 0.

C. Since $f(-x) = -f(x)$, f is odd and its graph is symmetric about the origin.

D. Since $x^2 + 1$ is never 0, there is no vertical asymptote. Since $f(x) \to \infty$ as $x \to \infty$ and $f(x) \to -\infty$ as $x \to -\infty$, there is no horizontal asymptote. But long division gives

$$f(x) = \frac{x^3}{x^2 + 1} = x - \frac{x}{x^2 + 1}$$

$$f(x) - x = -\frac{x}{x^2+1} = -\frac{\frac{1}{x}}{1+\frac{1}{x^2}} \to 0 \qquad \text{as } x \to \pm\infty$$

So the line $y = x$ is a slant asymptote.

E.
$$f'(x) = \frac{3x^2(x^2+1) - x^3 \cdot 2x}{(x^2+1)^2} = \frac{x^2(x^2+3)}{(x^2+1)^2}$$

Since $f'(x) > 0$ for all x (except 0), f is increasing on $(-\infty, \infty)$.

F. Although $f'(0) = 0$, f' does not change sign at 0, so there is no local maximum or minimum.

G.
$$f''(x) = \frac{(4x^3+6x)(x^2+1)^2 - (x^4+3x^2)\cdot 2(x^2+1)2x}{(x^2+1)^4}$$

E.
$$= \frac{2x(3-x^2)}{(x^2+1)^3}$$

Since $f''(x) = 0$ when $x = 0$ or $x = \pm\sqrt{3}$, we set up the following chart:

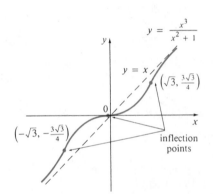

$$y = \frac{x^3}{x^2+1}$$
$$y = x$$
$$\left(\sqrt{3}, \frac{3\sqrt{3}}{4}\right)$$
$$\left(-\sqrt{3}, -\frac{3\sqrt{3}}{4}\right)$$
inflection points

Interval	x	$3-x^2$	$(x^2+1)^3$	$f''(x)$	f
$x < -\sqrt{3}$	$-$	$-$	$+$	$+$	CU on $(-\infty, -\sqrt{3})$
$-\sqrt{3} < x < 0$	$-$	$+$	$+$	$-$	CD on $(-\sqrt{3}, 0)$
$0 < x < \sqrt{3}$	$+$	$+$	$+$	$+$	CU on $(0, \sqrt{3})$
$x > \sqrt{3}$	$+$	$-$	$+$	$-$	CD on $(\sqrt{3}, \infty)$

The points of inflection are $(-\sqrt{3}, -3\sqrt{3}/4)$, $(0,0)$, and $(\sqrt{3}, 3\sqrt{3}/4)$.

Figure 4.42 H. The graph of f is sketched in Figure 4.42.

SECTION 4.5 **Exercises**

In Exercises 1–58 discuss the curves under the headings A–H given in this section.

1. $y = 1 - 3x + 5x^2 - x^3$

2. $y = 2x^3 - 6x^2 - 18x + 7$

3. $y = x^4 - 6x^2$

4. $y = 4x^3 - x^4$

5. $y = \dfrac{1}{x-1}$

6. $y = \dfrac{x}{x-1}$

7. $y = \dfrac{1}{x^2-9}$

8. $y = \dfrac{x}{x^2-9}$

9. $y = \dfrac{x}{(2x-3)^2}$

10. $y = \dfrac{4}{(x-5)^2}$

11. $y = \dfrac{x-3}{x+3}$

12. $y = \dfrac{1}{x^2(x+3)}$

13. $y = \dfrac{1}{(x-1)(x+2)}$

14. $y = \dfrac{1}{4x^3-9x}$

15. $y = \dfrac{1+x^2}{1-x^2}$

16. $y = \dfrac{x^3-1}{x}$

17. $y = \dfrac{1}{x^3-x}$

18. $y = \dfrac{1-x^2}{x^3}$

19. $y = \sqrt{2-x}$

20. $y = x\sqrt{x+3}$

21. $y = x + \sqrt{x}$

22. $y = \sqrt{x} - \sqrt{x-1}$

23. $y = \sqrt{x^2 + 1} - x$

24. $y = \sqrt{\dfrac{x}{x-5}}$

25. $y = \sqrt[4]{x^2 - 25}$

26. $y = \sqrt{x - \sqrt{x}}$

27. $y = x\sqrt{x^2 - 9}$

28. $y = \dfrac{x^2}{\sqrt{1-x^2}}$

29. $y = \dfrac{\sqrt{1-x^2}}{x}$

30. $y = \dfrac{x+1}{\sqrt{x^2+1}}$

31. $y = x + 3x^{2/3}$

32. $y = x^{5/3} - 5x^{2/3}$

33. $y = \cos x - \sin x$

34. $y = \sin x + \cos x$

35. $y = \sin x - \tan x$

36. $y = \sin x + \sqrt{3}\cos x$

37. $y = x\tan x,\ -\pi/2 < x < \pi/2$

38. $y = 2x + \cot x,\ 0 < x < \pi$

39. $y = \dfrac{x}{2} - \sin x,\ 0 < x < 3\pi$

40. $y = 2\sin x + \sin^2 x$

41. $y = 2\cos x + \sin^2 x$

42. $y = \sin x - x$

43. $y = \sin 2x - 2\sin x$

44. $y = \dfrac{\cos x}{2 + \sin x}$

45. $y = e^{-x^2}$

46. $y = xe^{x^2}$

47. $y = e^{-1/(x+1)}$

48. $y = e^{x/(x-2)}$

49. $y = e^{1/x^2}$

50. $y = e^{1/(1-x^2)}$

51. $y = \ln(\cos x)$

52. $y = \ln(\sin x)$

53. $y = \ln\left(x + \sqrt{1+x^2}\right)$

54. $y = \ln(\tan^2 x)$

55. $y = \ln(1 + x^2)$

56. $y = \ln(x^2 - x)$

57. $y = \sin^{-1}\left(\dfrac{x}{x+1}\right)$

58. $y = \tan^{-1}\left(\dfrac{x-1}{x+1}\right)$

In Exercises 59–72 discuss each curve under the headings A–H using l'Hospital's Rule where necessary.

59. $y = xe^{-x}$

60. $y = x^2 e^{-x}$

61. $y = x\ln x$

62. $y = \dfrac{\ln x}{x}$

63. $y = x^2\ln x$

64. $y = x(\ln x)^2$

65. $y = xe^{-x^2}$

66. $y = x^2 e^{-x^2}$

67. $y = \dfrac{e^x}{x}$

68. $y = \dfrac{e^x}{x^2}$

69. $y = xe^{1/x}$

70. $y = x^2 e^{-1/x}$

71. $y = x - \ln(1 + x)$

72. $y = x^x$

In Exercises 73–80 discuss each curve under the headings A–H. In D find an equation of the slant asymptote.

73. $y = \dfrac{x^3}{x^2 - 1}$

74. $y = x - \dfrac{1}{x}$

75. $xy = x^2 + 4$

76. $xy = x^2 + x + 1$

77. $y = \dfrac{1}{x-1} - x$

78. $y = \dfrac{x^2}{2x+5}$

79. $y = e^x + x$

80. $y = e^x - x$

81. Show that the lines $y = (b/a)x$ and $y = -(b/a)x$ are slant asymptotes of the hyperbola $(x^2/a^2) - (y^2/b^2) = 1$.

82. Let $f(x) = (x^3 + 1)/x$. Show that
$$\lim_{x \to \pm\infty} [f(x) - x^2] = 0$$

This shows that the graph of f approaches the graph of $y = x^2$, and we say that the curve $y = f(x)$ is *asymptotic* to the parabola $y = x^2$. Use this fact to help sketch the graph of f.

Applied Maximum and Minimum Problems

The methods we have learned in this chapter for finding extreme values have practical applications in many areas of life. A businessperson wants to minimize costs and maximize profits. Fermat's Principle in optics states that light follows the path that takes the least time. In this section and the next we solve such problems as maximizing areas, volumes, and profits and minimizing distances, times, and costs.

In solving such practical problems the greatest challenge is often to convert the word problem into a maximum-minimum problem by setting up the function that is to be maximized or minimized. Let us recall the problem-solving principles discussed on the inside front cover of this book and adapt them to this situation:

Steps in Solving Applied Maximum and Minimum Problems

1. *Understand the problem.* The first step is to read the problem carefully until it is clearly understood. Ask yourself: What is the unknown? What are the given quantities? What are the given conditions?

2. *Draw a diagram.* In most problems it is useful to draw a diagram and identify the given and required quantities on the diagram.

3. *Introduce notation.* Assign a symbol to the quantity that is to be maximized or minimized (let us call it Q for now). Also select symbols $(a, b, c, \ldots, x, y)$ for other unknown quantities and label the diagram with these symbols. It may help to use initials as suggestive symbols—for example, A for area, h for height, t for time.

4. Express Q in terms of some of the other symbols from Step 3.

5. If Q has been expressed as a function of more than one variable in Step 4, use the given information to find relationships (in the form of equations) among these variables. Then use these equations to eliminate all but one of the variables in the expression for Q. Thus Q will be given as a function of *one* variable x, say $Q = f(x)$. Write the domain of this function.

6. Use the methods of Sections 4.1, 4.3, and 4.4 to find the *absolute* maximum or minimum value of f. In particular, if the domain of f is a closed interval, then the procedure outlined in (4.8) can be used.

EXAMPLE 1   A farmer has 2400 ft of fencing and wants to fence off a rectangular field that borders a straight river. He needs no fence along the river. What are the dimensions of the field that has the largest area?

Solution   First we draw a diagram (Figure 4.43). We wish to maximize the area A of the rectangle. Let x and y be the width and length of the rectangle (in feet).

Then we express A in terms of x and y:

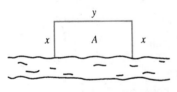

Figure 4.43

$$A = xy$$

We want to express A as a function of just one variable, so we eliminate y by expressing it in terms of x. To do this we use the given information that the total length of fencing is 2400 ft. Thus

$$2x + y = 2400$$

From this equation we have $y = 2400 - 2x$, which gives

$$A = x(2400 - 2x) = 2400x - 2x^2$$

Note that $x \geq 0$ and $x \leq 1200$ (otherwise $A < 0$). So the function that we wish to maximize is

$$A(x) = 2400x - 2x^2 \qquad 0 \leq x \leq 1200$$

To find the critical numbers we solve the equation

$$A'(x) = 2400 - 4x = 0$$

which gives $x = 600$. The maximum value of A must occur either at this critical number or at an endpoint of the interval. Since $A(0) = 0$, $A(600) = 720,000$, and $A(1200) = 0$, the procedure of (4.8) gives the maximum value as $A(600) = 720,000$.

[Alternatively, we could have observed that $A''(x) = -4 < 0$ for all x, so A is always concave downward and the local maximum at $x = 600$ must be an absolute maximum.]

Thus the rectangular field should be 600 ft wide and 1200 ft long.   ●

EXAMPLE 2   A can is to be made to hold 1 L of oil. Find the dimensions that will minimize the cost of the metal to make the can.

Figure 4.44

Solution   Draw the diagram as in Figure 4.44 where r is the radius and h the height (in centimeters). In order to minimize the cost of the metal, we minimize the total surface area of the cylinder (top, bottom, and sides), which is

$$A = 2\pi r^2 + 2\pi r h$$

To eliminate h we use the fact that the volume is given as 1 L, which we take to be $1000 \, \text{cm}^3$. Thus

$$\pi r^2 h = 1000$$

which gives $h = 1000/(\pi r^2)$. Substitution of this into the expression for A gives

$$A = 2\pi r^2 + 2\pi r\left(\frac{1000}{\pi r^2}\right) = 2\pi r^2 + \frac{2000}{r}$$

Therefore the function that we want to minimize is

$$A(r) = 2\pi r^2 + \frac{2000}{r} \qquad r > 0$$

To find the critical numbers, we differentiate:

$$A'(r) = 4\pi r - \frac{2000}{r^2} = \frac{4(\pi r^3 - 500)}{r^2}$$

Then $A'(r) = 0$ when $\pi r^3 = 500$, so the only critical number is $r = \sqrt[3]{500/\pi}$.

Since the domain of A is $(0, \infty)$, we cannot use the argument of Example 1 concerning endpoint extrema. But we can observe that $A'(r) < 0$ for $r < \sqrt[3]{500/\pi}$ and $A'(r) > 0$ for $r > \sqrt[3]{500/\pi}$, so A is decreasing for *all* r to the left of the critical number and increasing for *all* r to the right. Thus $r = \sqrt[3]{500/\pi}$ must give rise to an *absolute* minimum.

[Alternatively, we could argue that $A(r) \to \infty$ as $r \to 0^+$ and $A(r) \to \infty$ as $r \to \infty$, so there must be a minimum value of $A(r)$, which must occur at the critical number.]

The value of h corresponding to $r = \sqrt[3]{500/\pi}$ is

$$h = \frac{1000}{\pi r^2} = \frac{1000}{\pi (500/\pi)^{2/3}} = 2\sqrt[3]{\frac{500}{\pi}} = 2r$$

Thus to minimize the cost of the can, the radius should be $\sqrt[3]{500/\pi}$ cm and the height should be equal to twice the radius, namely, the diameter.   ●

Note: The argument used in Example 2 to justify the absolute minimum is a variant of the First Derivative Test (which applies only to local extrema) and is stated here for future reference.

First Derivative Test for Absolute Extrema (4.30)

Suppose that c is a critical number of a continuous function f defined on an interval.
(a) If $f'(x) > 0$ for all $x < c$ and $f'(x) < 0$ for all $x > c$, then $f(c)$ is the absolute maximum value of f.
(b) If $f'(x) < 0$ for all $x < c$ and $f'(x) > 0$ for all $x > c$, then $f(c)$ is the absolute minimum value of f.

EXAMPLE 3   Find the point on the parabola $y^2 = 2x$ that is closest to the point $(1, 4)$.

Solution   The distance between the point $(1, 4)$ and the point (x, y) is

$$d = \sqrt{(x-1)^2 + (y-4)^2}$$

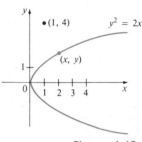

Figure 4.45

(see Figure 4.45). But if (x, y) lies on the parabola, then $x = y^2/2$, so the expression for d becomes

$$d = \sqrt{\left(\frac{y^2}{2} - 1\right)^2 + (y-4)^2}$$

(Alternatively, we could have substituted $y = \sqrt{2x}$ to get d in terms of x alone.) Instead of minimizing d, we minimize its square:

$$d^2 = f(y) = \left(\frac{y^2}{2} - 1\right)^2 + (y-4)^2$$

(You should convince yourself that the minimum of d occurs at the same point as the minimum of d^2, but d^2 is easier to work with.) Differentiating, we obtain

$$f'(y) = 2\left(\frac{y^2}{2} - 1\right)y + 2(y-4) = y^3 - 8$$

so $f'(y) = 0$ when $y = 2$. Observe that $f'(y) < 0$ when $y < 2$ and $f'(y) > 0$ when $y > 2$, so by the First Derivative Test for Absolute Extrema (4.30), the absolute minimum occurs when $y = 2$. (Or we could simply say that because of the geometric nature of problem, it is obvious that there is a closest point but not a farthest point.) The corresponding value of x is $x = y^2/2 = 2$. Thus the point on $y^2 = 2x$ closest to $(1, 4)$ is $(2, 2)$. ●

EXAMPLE 4   A man is at point A on a bank of a straight river, $3\,\text{km}$ wide, and wants to reach point B, $8\,\text{km}$ downstream on the opposite bank, as quickly as possible (see Figure 4.46). He could row his boat directly across the river to point C and then run to B, or he could row directly to B, or he could row to some point D between C and B and then run to B. If he can row at $6\,\text{km/h}$ and run at $8\,\text{km/h}$, where should he land to reach B as soon as possible?

Solution   Let x be the distance from C to D. Then the running distance is $|DB| = 8 - x$ and the Pythagorean Theorem gives the rowing distance as $|AD| = \sqrt{x^2 + 9}$. We assume the speed of the water is $0\,\text{km/h}$ and use the equation

$$\text{time} = \frac{\text{distance}}{\text{rate}}$$

Then the rowing time is $\sqrt{x^2 + 9}/6$ and the running time is $(8 - x)/8$, so the total time T as a function of x is

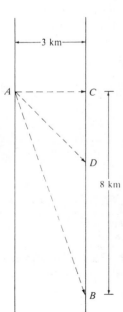

Figure 4.46

$$T(x) = \frac{\sqrt{x^2 + 9}}{6} + \frac{8 - x}{8}$$

The domain of this function T is $[0, 8]$. Notice that if $x = 0$ he rows to C and if $x = 8$ he rows directly to B. The derivative of T is

$$T'(x) = \frac{x}{6\sqrt{x^2 + 9}} - \frac{1}{8}$$

Thus, using the fact that $x \geq 0$, we have

$$T'(x) = 0 \quad \Leftrightarrow \quad \frac{x}{6\sqrt{x^2 + 9}} = \frac{1}{8} \quad \Leftrightarrow \quad 4x = 3\sqrt{x^2 + 9}$$

$$\Leftrightarrow \quad 16x^2 = 9(x^2 + 9) \quad \Leftrightarrow \quad 7x^2 = 81$$

$$\Leftrightarrow \quad x = \frac{9}{\sqrt{7}}$$

The only critical number is $x = 9/\sqrt{7}$. To see whether the minimum occurs at this critical number or at an endpoint of the domain $[0, 8]$, we evaluate T at all three points:

$$T(0) = 1.5 \qquad T\left(\frac{9}{\sqrt{7}}\right) = 1 + \frac{\sqrt{7}}{8} \approx 1.33 \qquad T(8) = \frac{\sqrt{73}}{6} \approx 1.42$$

Since the smallest of these values of T occurs when $x = 9/\sqrt{7}$, the absolute minimum value of T must occur there by (4.8).

Thus the man should land the boat at a point $9/\sqrt{7}\,\text{km}$ ($\approx 3.4\,\text{km}$) downstream from his starting point.   •

EXAMPLE 5   Find the area of the largest rectangle that can be inscribed in a semicircle of radius r.

Solution   Let us take the semicircle to be the upper half of the circle $x^2 + y^2 = r^2$ with center the origin. Then the word *inscribed* means that the rectangle has two vertices on the semicircle and two vertices on the x-axis as shown in Figure 4.47.

Let (x, y) be the vertex that lies in the first quadrant. Then the rectangle has sides of lengths $2x$ and y, so its area is

$$A = 2xy$$

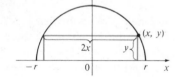

Figure 4.47 To eliminate y we use the fact that (x, y) lies on the circle $x^2 + y^2 = r^2$ and so $y = \sqrt{r^2 - x^2}$. Thus

$$A = 2x\sqrt{r^2 - x^2}$$

The domain of this function is $0 \leq x \leq r$. Its derivative is

$$A' = 2\sqrt{r^2 - x^2} - \frac{2x^2}{\sqrt{r^2 - x^2}} = \frac{2(r^2 - 2x^2)}{\sqrt{r^2 - x^2}}$$

which is 0 when $2x^2 = r^2$, that is, $x = r/\sqrt{2}$ (since $x \geq 0$). This value of x gives a maximum value of A since $A(0) = 0$ and $A(r) = 0$. Therefore the area of the largest inscribed rectangle is

$$A\left(\frac{r}{\sqrt{2}}\right) = 2\frac{r}{\sqrt{2}}\sqrt{r^2 - \frac{r^2}{2}} = r^2$$ •

SECTION 4.6 **Exercises**

1. Find two numbers whose sum is 100 and whose product is a maximum.

2. Find two numbers whose difference is 100 and whose product is a minimum.

3. Find two positive numbers whose product is 100 and whose sum is a minimum.

4. Show that of all the rectangles with a given area, the one with smallest perimeter is a square.

5. Show that of all the rectangles with a given perimeter, the one with greatest area is a square.

6. A farmer wants to fence an area of 1.5 million square feet in a rectangular field and then divide it in half with a fence parallel to one of the sides of the rectangle. How can he do this so as to minimize the cost of the fence?

7. A farmer with 750 ft of fencing wants to enclose a rectangular area and then divide it into four pens with fencing parallel to one side of the rectangle. What is the largest possible total area of the four pens?

8. A box with a square base and open top must have a volume of $32,000 \, cm^3$. Find the dimensions of the box that minimize the amount of material used.

9. If $1200 \, cm^2$ of material is available to make a box with a square base and open top, find the largest possible volume of the box.

10. A rectangular storage container with an open top is to have a volume of $10 \, m^3$. The length of its base is twice the width. Material for the base costs $10 per square meter. Material for the sides costs $6 per square meter. Find the cost of materials for the cheapest such container.

11. Do Exercise 10 assuming the container has a lid that is made from the same material as the sides.

12. A box with an open top is to be constructed from a square piece of cardboard, 3 ft wide, by cutting out a square from each of the four corners and bending up the sides. Find the largest volume that such a box can have.

13. Find the point on the line $y = 2x - 3$ that is closest to the origin.

14. Find the point on the line $2x + 3y + 5 = 0$ that is closest to the point $(-1, -2)$.

15. Find the points on the hyperbola $y^2 - x^2 = 4$ that are closest to the point $(2, 0)$. $y^2 = 4 + x^2$

16. Find the point on the parabola $x + y^2 = 0$ that is closest to the point $(0, -3)$.

17. Find the dimensions of the rectangle of largest area that can be inscribed in a circle of radius r.

18. Find the area of the largest rectangle that can be inscribed in the ellipse $x^2/a^2 + y^2/b^2 = 1$.

19. Find the dimensions of the rectangle of largest area that can be inscribed in an equilateral triangle of side L if one side of the rectangle lies on the base of the triangle.

20. Find the dimensions of the rectangle of largest area that has its base on the x-axis and its other two vertices above the x-axis and lying on the parabola $y = 8 - x^2$.

21. Find the dimensions of the isosceles triangle of largest area that can be inscribed in a circle of radius r.

22. Find the area of the largest rectangle that can be inscribed in a right triangle with legs of lengths $3 \, cm$ and $4 \, cm$ if two sides of the rectangle lie along the legs.

23. A right circular cylinder is inscribed in a sphere of radius r. Find the largest possible volume of such a cylinder.

24. A right circular cylinder is inscribed in a cone with height h and base radius r. Find the largest possible volume of such a cylinder.

25. A right circular cylinder is inscribed in a sphere of radius r. Find the largest possible surface area of such a cylinder.

26. A Norman window has the shape of a rectangle surmounted by a semicircle. (Thus the diameter of the semicircle is equal to the width of the rectangle.) If the perimeter of the window is $30 \, ft$, find the dimensions of the window so that the greatest possible amount of light is admitted.

27. The top and bottom margins of a poster are each $6 \, cm$ and the side margins are each $4 \, cm$. If the area of printed material on the poster is fixed at $384 \, cm^2$, find the dimensions of the poster with the smallest area.

28. A poster is to have an area of $180 \, in.^2$ with 1-in. margins at the bottom and sides and a 2-in. margin

at the top. What dimensions will give the largest printed area?

29. A piece of wire 10 m long is cut into two pieces. One piece is bent into a square and the other is bent into an equilateral triangle. How should the wire be cut so that the total area enclosed is (a) a maximum? (b) a minimum?

30. Answer Exercise 29 if one piece is bent into a square and the other into a circle.

31. A cylindrical can without a top is made to contain V cm^3 of liquid. Find the dimensions that will minimize the cost of the metal to make the can.

32. A fence 8 ft tall runs parallel to a tall building at a distance of 4 ft from the building. What is the length of the shortest ladder that will reach from the ground over the fence to the wall of the building?

33. A conical drinking cup is made from a circular piece of paper of radius R by cutting out a sector and joining the edges CA and CB. Find the maximum capacity of such a cup.

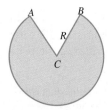

34. The velocity of a wave of length L in deep water is

$$v = K\sqrt{\frac{L}{C} + \frac{C}{L}}$$

where K and C are known positive constants. What is the length of the wave that gives the minimum velocity?

35. An object with weight W is dragged along a horizontal plane by a force acting along a rope attached to the object. If the rope makes an angle θ with the plane, then the magnitude of the force is

$$F = \frac{\mu W}{\mu \sin \theta + \cos \theta}$$

where μ is the coefficient of friction. Show that F is minimized when $\tan \theta = \mu$.

36. A boat leaves a dock at 2:00 P.M. and travels due south at a speed of 20 km/h. Another boat has been heading due east at 15 km/h and reaches the same dock at 3:00 P.M. At what time were the two boats closest together?

37. Solve the problem in Example 4 if the river is 5 km wide and point B is only 5 km downstream from A.

38. A woman at a point A on the shore of a circular lake with radius 2 mi wants to be at the point C diametrically opposite A on the other side of the lake in the shortest possible time. She can walk at the rate of 4 mi/h and row a boat at 2 mi/h. At what angle θ to the diameter should she row?

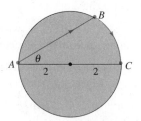

39. The illumination of an object by a light source is directly proportional to the strength of the source and inversely proportional to the square of the distance from the source. If two light sources, one three times as strong as the other, are placed 10 ft apart, where should an object be placed on the line between the sources so as to receive the least illumination?

40. Find the equation of the line through the point $(3, 5)$ that cuts off the least area from the first quadrant.

41. Show that of all the isosceles triangles with a given perimeter, the one with greatest area is equilateral.

42. The frame for a kite is to be made from six pieces of wood. The four exterior pieces have been cut with the lengths indicated in the figure. To maximize the area of the kite, how long should the diagonal pieces be?

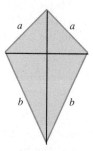

43. Let v_1 be the velocity of light in air and v_2 the velocity of light in water. According to Fermat's Principle, a ray of light will travel from a point A in the air to a point B in the water by a path ACB that minimizes the time taken. Show that

$$\frac{\sin \theta_1}{\sin \theta_2} = \frac{v_1}{v_2}$$

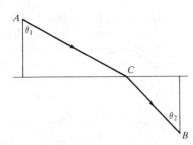

where θ_1 (the angle of incidence) and θ_2 (the angle of refraction) are shown in the figure. This equation is known as Snell's Law.

44. Two vertical poles PQ and ST are secured by a rope PRS going from the top of the first pole to a point R on the ground between the poles and then to the top of the second pole as in the figure. Show that the shortest length of such a rope occurs when $\theta_1 = \theta_2$.

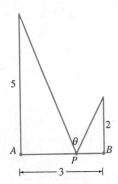

45. Where should the point P be chosen on the line segment AB so as to maximize the angle θ?

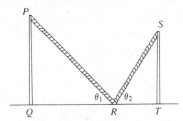

46. A painting in an art gallery has height h and is hung so that its lower edge is a distance d above the eye of an observer (as in the figure). How far from the wall should the observer stand to get the best view? (In other words, where should he stand so as to maximize the angle θ subtended at his eye by the painting?)

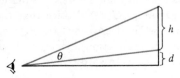

47. The upper left-hand corner of a piece of paper 8 in. wide by 12 in. long is folded over to the right-hand edge as in the figure. How would you fold it so as to minimize the length of the fold? In other words, how would you choose x to minimize y?

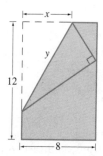

48. A steel pipe is being carried down a hallway 9 ft wide. At the end of the hall there is a right-angle turn into a narrower hallway 6 ft wide. What is the length of the longest pipe that can be carried horizontally around the corner?

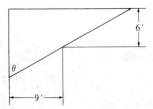

49. An observer stands at a point P 1 unit away from a track. Two runners start at the point S in the figure and run along the track. One runner runs three times as fast as the other. Find the maximum value of the observer's angle of sight θ between the runners. [*Hint:* Maximize $\tan\theta$.]

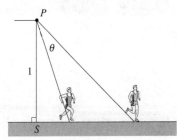

50. A rain gutter is to be constructed from a metal sheet of width 30 cm by bending up one-third of the sheet on each side through an angle θ. How should θ be chosen so that the gutter will carry the maximum amount of water?

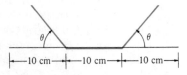

51. Find the maximum area of a rectangle that can be circumscribed about a given rectangle with length L and width W.

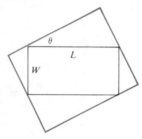

52. The blood vascular system consists of blood vessels (arteries, arterioles, capillaries, and veins) that convey blood from the heart to the organs and back to the heart. This system should work so as to minimize the energy expended by the heart in pumping the blood. In particular, this energy is reduced when the resistance of the blood is lowered. One of Poiseuille's Laws gives the resistance R of the blood as

$$R = C\frac{L}{r^4}$$

© Manfred Kage/Peter Arnold

where L is the length of the blood vessel, r is the radius, and C is a positive constant determined by the viscosity of the blood. (Poiseuille established this law experimentally but it also follows from Equation 8.42 in Section 8.6.) The figure shows a main blood vessel with radius r_1 branching at an angle θ into a smaller vessel with radius r_2.

(a) Use Poiseuille's Law to show that the total resistance of the blood along the path ABC is

$$R = C\left(\frac{a - b\cot\theta}{r_1^4} + \frac{b\csc\theta}{r_2^4}\right)$$

where a and b are the distances shown in the figure.

(b) Prove that the resistance is minimized when

$$\cos\theta = \frac{r_2^4}{r_1^4}$$

(c) Find the optimal branching angle (correct to the nearest degree) when the radius of the smaller blood vessel is two-thirds the radius of the larger vessel.

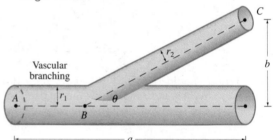

SECTION 4.7

Applications to Economics

In Section 2.3 we introduced the idea of marginal cost. Recall that if $C(x)$, the **cost function**, is the cost of producing x units of a certain product, then the **marginal cost** is the rate of change of C with respect to x. In other words, the marginal cost function is the derivative, $C'(x)$, of the cost function.

The graph of a typical cost function is shown in Figure 4.48. The marginal cost $C'(x)$ is the slope of the tangent to the cost curve at $(x, C(x))$. Notice that the cost curve is initially concave downward (the marginal cost is decreasing) because of economies of scale (more efficient use of the fixed costs of production). But eventually there is an inflection point and the cost curve becomes concave upward (the marginal cost is increas-

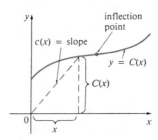

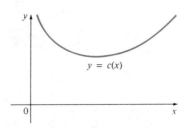

Figure 4.48

Cost function

Figure 4.49

Average cost function

ing) perhaps because of overtime costs or the inefficiencies of a large-scale operation.

The **average cost function**

(4.31)
$$c(x) = \frac{C(x)}{x}$$

represents the cost per unit when x units are produced. We sketch a typical average cost function in Figure 4.49 by noting that $C(x)/x$ is the slope of the line that joins the origin to the point $(x, C(x))$ in Figure 4.48. It appears that there will be an absolute minimum. To find it we locate the critical point of c by using the Quotient Rule to differentiate Equation 4.31:

$$c'(x) = \frac{xC'(x) - C(x)}{x^2}$$

Now $c'(x) = 0$ when $xC'(x) - C(x) = 0$ and this gives

$$C'(x) = \frac{C(x)}{x} = c(x)$$

Therefore:

> If the average cost is a minimum, then
>
> marginal cost = average cost

EXAMPLE 1   A company estimates that the cost (in dollars) of producing x items is $C(x) = 2600 + 2x + 0.001x^2$. (See Example 8 in Section 2.3 for an explanation of why it is reasonable to model a cost function by a polynomial.)

(a) Find the cost, the average cost, and the marginal cost of producing 1000 items, 2000 items, and 3000 items.

(b) At what production level will the average cost be lowest, and what is this minimum average cost?

Solution

(a)   The average cost function is

$$c(x) = \frac{C(x)}{x} = \frac{2600}{x} + 2 + 0.001x$$

The marginal cost function is

$$C'(x) = 2 + 0.002x$$

We use these expressions to fill in the following table giving the cost, average cost, and marginal cost (in dollars, or dollars per item, rounded to the nearest cent).

x	$C(x)$	$c(x)$	$C'(x)$
1000	5,600.00	5.60	4.00
2000	10,600.00	5.30	6.00
3000	17,600.00	5.87	8.00

(b)   To minimize the average cost we must have

$$\text{marginal cost} = \text{average cost}$$

$$C'(x) = c(x)$$

$$2 + 0.002x = \frac{2600}{x} + 2 + 0.001x$$

This equation simplifies to

$$0.001x = \frac{2600}{x}$$

so

$$x^2 = \frac{2600}{0.001} = 2{,}600{,}000$$

and

$$x = \sqrt{2{,}600{,}000} \approx 1612$$

To see that this production level actually gives a minimum we note that $c''(x) = 5200/x^3 > 0$, so c is concave upward on its entire domain. The minimum average cost is

$$c(1612) = \frac{2600}{1612} + 2 + 0.001(1612) = \$5.22/\text{item} \qquad \bullet$$

Now let us consider marketing. Let $p(x)$ be the price per unit that the company can charge if it sells x units. Then p is called the **demand function** (or **price function**) and we would expect it to be a decreasing function of x. If x units are sold and the price per unit is $p(x)$, then the total revenue is

$$R(x) = xp(x)$$

and R is called the **revenue function** (or **sales function**). The derivative R' of the revenue function is called the **marginal revenue function** and is the rate of change of revenue with respect to the number of units sold.

If x units are sold, then the total profit is

$$P(x) = R(x) - C(x)$$

and P is called the **profit function**. The **marginal profit function** is P', the derivative of the profit function. In order to maximize profit we look for the critical numbers of P, that is, the numbers where the marginal profit is 0. But if

$$P'(x) = R'(x) - C'(x) = 0$$

then

$$R'(x) = C'(x)$$

Therefore:

If the profit is a maximum, then

marginal revenue = marginal cost

To ensure that this condition gives a maximum we could use the Second Derivative Test. Note that

$$P''(x) = R''(x) - C''(x) < 0$$

when

$$R''(x) < C''(x)$$

and this condition says that the rate of increase of marginal revenue is less than the rate of increase of marginal cost. Thus the profit will be a maximum when

$$R'(x) = C'(x) \qquad \text{and} \qquad R''(x) < C''(x)$$

EXAMPLE 2   What production level will maximize profits for a company with cost and demand functions

$$C(x) = 3800 + 5x - \frac{x^2}{1000} \qquad p(x) = 50 - \frac{x}{100}$$

Solution   The revenue function is

$$R(x) = xp(x) = 50x - \frac{x^2}{100}$$

so the marginal revenue function is

$$R'(x) = 50 - \frac{x}{50}$$

and the marginal cost function is

$$C'(x) = 5 - \frac{x}{500}$$

Thus marginal revenue is equal to marginal cost when

$$50 - \frac{x}{50} = 5 - \frac{x}{500}$$

Solving, we get

$$x = 2500$$

To check that this gives a maximum we compute the second derivatives:

$$R''(x) = -\frac{1}{50} \qquad C''(x) = -\frac{1}{500}$$

Thus $R''(x) < C''(x)$ for all values of x. Therefore a production level of 2500 units will maximize profits.   ●

For each of the cost functions in Exercises 1–6 find (a) the cost, average cost, and marginal cost of producing 1000 units; (b) the production level that will minimize the average cost; and (c) the minimum average cost.

1. $C(x) = 10,000 + 25x + x^2$

2. $C(x) = 1600 + 8x + 0.01x^2$

3. $C(x) = 45 + \frac{x}{2} + \frac{x^2}{560}$

4. $C(x) = 2000 + 10x + 0.001x^3$

5. $C(x) = 2\sqrt{x} + x^2/8000$

6. $C(x) = 1000 + 96x + 2x^{3/2}$

For each of the cost and demand functions in Exercises 7–12 find the production level that will maximize profits.

7. $C(x) = 680 + 4x + 0.01x^2$, $p(x) = 12$

8. $C(x) = 680 + 4x + 0.01x^2$, $p(x) = 12 - x/500$

9. $C(x) = 1200 + 25x - 0.0001x^2$, $p(x) = 55 - x/1000$

10. $C(x) = 900 + 110x - 0.1x^2 + 0.02x^3$,
    $p(x) = 260 - 0.1x$

11. $C(x) = 1450 + 36x - x^2 + 0.001x^3$, $p(x) = 60 - 0.01x$

12. $C(x) = 10,000 + 28x - 0.01x^2 + 0.002x^3$,
    $p(x) = 90 - 0.02x$

In Exercises 13 and 14 find the production level at which the marginal cost function starts to increase.

13. $C(x) = 0.001x^3 - 0.3x^2 + 6x + 900$

14. $C(x) = 0.0002x^3 - 0.25x^2 + 4x + 1500$

15. A baseball team plays in a stadium that holds 55,000 spectators. With ticket prices at \$10, the average attendance has been 27,000. When ticket prices were lowered to \$8, the average attendance rose to 33,000. (a) Find the demand function, assuming that it is linear. (b) How should ticket prices be set to maximize revenue?

16. During the summer months Terry makes and sells necklaces on the beach. Last summer he sold the necklaces for \$10 each and his sales averaged 20 per day. When he increased the price by \$1, he found that he lost two sales per day. (a) Find the demand function, assuming that it is linear. (b) If the material for each necklace costs Terry \$6, what should the selling price be to maximize profits?

17. A manufacturer has been selling 1000 television sets a week at \$450 each. A market survey indicates that for each \$10 rebate offered to the buyer, the number of sets sold will increase by 100 per week. (a) Find the demand function. (b) How large a rebate should the company offer the buyer in order to maximize its revenue? (c) If its weekly cost function is $C(x) = 68,000 + 150x$, how should it set the size of the rebate in order to maximize its profits?

18. The manager of 100-unit apartment complex knows from experience that all units will be occupied if the rent is \$400 per month. A market survey suggests that, on the average, one additional unit will remain vacant for each \$5 increase in rent. What rent should the manager charge to maximize revenue?

Antiderivatives

We know how to solve the derivative problem: given a function, find its derivative. But many problems in mathematics and its applications require us to solve the inverse of the derivative problem: given a function f, find a function F whose derivative is f. If such a function F exists, it is called an *antiderivative* of f.

Definition (4.32)

A function F is called an **antiderivative** of f on an interval I if $F'(x) = f(x)$ for all x in I.

For instance, let $f(x) = x^2$. It is not difficult to discover an antiderivative of f if we keep the Power Rule in mind. In fact, if $F(x) = \frac{1}{3}x^3$, then $F'(x) = x^2 = f(x)$. But the function $G(x) = \frac{1}{3}x^3 + 100$ also satisfies $G'(x) = x^2$. Therefore both F and G are antiderivatives of f. Indeed any function of the form $H(x) = \frac{1}{3}x^3 + C$, where C is a constant, is an antiderivative of f. The question arises: Are there any others?

To answer this question, recall that in Section 4.2 we used the Mean Value Theorem to prove that if two functions have identical derivatives on an interval, then they must differ by a constant (Corollary 4.17). Thus if F and G are any two antiderivatives of f, then

$$F'(x) = f(x) = G'(x)$$

so $G(x) - F(x) = C$, where C is a constant. We can write this as $G(x) = F(x) + C$, so we have the following result:

Theorem (4.33)

> If F is an **antiderivative** of f on an interval I, then the most general antiderivative of f on I is
>
> $$F(x) + C$$
>
> where C is an arbitrary constant.

Going back to the function $f(x) = x^2$, we see that the general antiderivative of f is $x^3/3 + C$. By assigning specific values to the constant C we obtain a family of functions whose graphs are vertical translates of one another (see Figure 4.50).

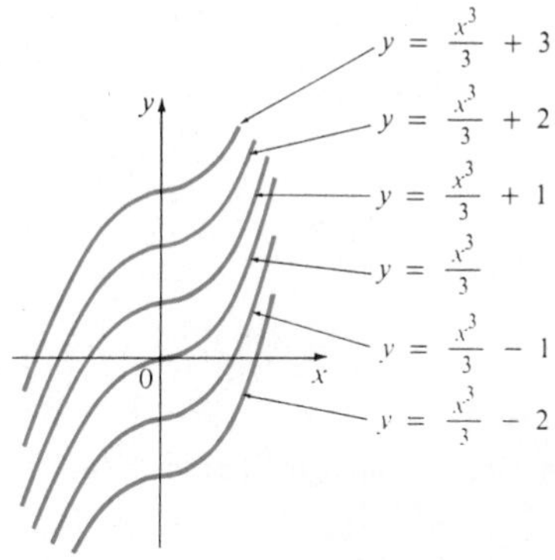

EXAMPLE 1   Find the most general antiderivatives of the following functions:
(a) $f(x) = \sin x$ (b) $f(x) = x^n$, $n \geq 0$ (c) $f(x) = x^{-3}$ (d) $f(x) = 1/x$

Solution

(a)   If $F(x) = -\cos x$, then $F'(x) = \sin x$, so an antiderivative of sine is $-$cosine. By Theorem 4.33, the most general antiderivative is $G(x) = -\cos x + C$.

(b)

$$\frac{d}{dx}\left(\frac{x^{n+1}}{n+1}\right) = \frac{(n+1)x^n}{n+1} = x^n$$

Thus the general antiderivative of $f(x) = x^n$ is

$$F(x) = \frac{x^{n+1}}{n+1} + C$$

Figure 4.50

The antiderivatives of $f(x) = x^2$

($y = \frac{x^3}{3} + 3$, $y = \frac{x^3}{3} + 2$, $y = \frac{x^3}{3} + 1$, $y = \frac{x^3}{3}$, $y = \frac{x^3}{3} - 1$, $y = \frac{x^3}{3} - 2$)

This is valid for $n \geq 0$ since then $f(x) = x^n$ is defined on the interval $(-\infty, \infty)$.

(c)   If we put $n = -3$ in part (b) we get the particular antiderivative $F(x) = x^{-2}/(-2)$ by the same calculation. But notice that $f(x) = x^{-3}$ is not defined at $x = 0$. Thus Theorem 4.33 tells us only that the general antiderivative of f is $x^{-2}/(-2) + C$ on any interval that does not contain 0. So the general antiderivative of $f(x) = 1/x^3$ is

$$F(x) = \begin{cases} -\dfrac{1}{2x^2} + C_1 & \text{if } x > 0 \\[2mm] -\dfrac{1}{2x^2} + C_2 & \text{if } x < 0 \end{cases}$$

(d)   Formula 3.29 shows that a particular antiderivative of $1/x$ is $\ln|x|$. But since

f is not defined where $x = 0$, Theorem 4.33 tells us that the most general antiderivative is

$$F(x) = \begin{cases} \ln x + C_1 & \text{if } x > 0 \\ \ln(-x) + C_2 & \text{if } x < 0 \end{cases}$$

    As in Example 1, every differentiation formula, when read from right to left, gives rise to an antidifferentiation formula. In Table 4.34 we list some particular antiderivatives. Each formula in the table is true because the derivative of the function in the right column appears in the left column. In particular, the first formula says that the antiderivative of a constant times a function is the constant times the antiderivative of the function. The second formula says that the antiderivative of a sum is the sum of the antiderivatives. (We use the notation $F' = f$, $G' = g$.)

Table of Antidifferentiation Formulas (4.34)

Function	Particular Antiderivative
$cf(x)$	$cF(x)$
$f(x) + g(x)$	$F(x) + G(x)$
$x^n \quad (n \neq -1)$	$\dfrac{x^{n+1}}{n+1}$
$1/x$	$\ln|x|$
e^x	e^x
$\cos x$	$\sin x$
$\sin x$	$-\cos x$
$\sec^2 x$	$\tan x$
$\dfrac{1}{\sqrt{1-x^2}}$	$\sin^{-1} x$
$\dfrac{1}{1+x^2}$	$\tan^{-1} x$

    To obtain the most general antiderivative from the particular ones in Table 4.34 we have to add a constant (or constants) as in Example 1.

EXAMPLE 2   Find all functions g such that

$$g'(x) = 4\sin x - 3x^5 + 6\sqrt[4]{x^3}$$

Solution   We want to find an antiderivative of

$$f(x) = g'(x) = 4\sin x - 3x^5 + 6x^{3/4}$$

Using the formulas in Table 4.34 together with Theorem 4.33, we obtain

$$g(x) = 4(-\cos x) - 3\frac{x^6}{6} + 6\frac{x^{7/4}}{\frac{7}{4}} + C$$

$$= -4\cos x - \frac{x^6}{2} + \frac{24}{7}x^{7/4} + C$$

In applications of calculus it is very common to have a situation as in Example 2, where it is required to find a function, given knowledge about its derivatives. An equation that involves the derivatives of a function is called a **differential equation.** These will be studied in some detail in Section 8.1 and Chapter 15, but for the present we can solve some elementary differential equations. The general solution of a differential equation involves an arbitrary constant (or constants) as in Example 2. However there may be some extra conditions given that will determine the constants and therefore uniquely specify the solution.

EXAMPLE 3   Find f if $f'(x) = x\sqrt{x} + 5(1 + x^2)^{-1}$ and $f(0) = 2$.

Solution   The general antiderivative of

$$f'(x) = x^{3/2} + \frac{5}{1 + x^2}$$

is

$$f(x) = \frac{x^{5/2}}{\frac{5}{2}} + 5\tan^{-1}x + C = \tfrac{2}{5}x^{5/2} + 5\tan^{-1}x + C$$

To determine C we use the fact that $f(0) = 2$:

$$f(0) = 0 + 5\tan^{-1}0 + C = 2$$

Thus we have $C = 2$, so the particular solution is

$$f(x) = \tfrac{2}{5}x^{5/2} + 5\tan^{-1}x + 2$$

EXAMPLE 4   Find f if $f''(x) = 12x^2 + 6x - 4$, $f(0) = 4$, and $f(1) = 1$.

Solution   The general antiderivative of $f''(x) = 12x^2 + 6x - 4$ is

$$f'(x) = 12\frac{x^3}{3} + 6\frac{x^2}{2} - 4x + C = 4x^3 + 3x^2 - 4x + C$$

Using the antidifferentiation rules once more, we find that

$$f(x) = 4\frac{x^4}{4} + 3\frac{x^3}{3} - 4\frac{x^2}{2} + Cx + D = x^4 + x^3 - 2x^2 + Cx + D$$

To determine C and D we use the given conditions that $f(0) = 4$ and $f(1) = 1$. Since $f(0) = 0 + D = 4$, we have $D = 4$. Since

$$f(1) = 1 + 1 - 2 + C + 4 = 1$$

we have $C = -3$. Therefore the required function is

$$f(x) = x^4 + x^3 - 2x^2 - 3x + 4$$

Antidifferentiation is particularly useful in analyzing the motion of an object moving in a straight line. Recall that if the object has position function $s = f(t)$, then the velocity function is $v(t) = s'(t)$. This means that the position function is an antiderivative of the velocity function. Likewise the acceleration function is $a(t) = v'(t)$, so the velocity function is an antiderivative of the acceleration. If the acceleration and the initial values

$s(0)$ and $v(0)$ are known, then the position function can be found by antidifferentiating twice.

EXAMPLE 5   A particle moves in a straight line and has acceleration given by $a(t) = 6t + 4$. Its initial velocity is $v(0) = -6\,\text{cm/s}$ and its initial displacement is $s(0) = 9\,\text{cm}$. Find its position function $s(t)$.

Solution   Since $v'(t) = a(t) = 6t + 4$, antidifferentiation gives

$$v(t) = 6\frac{t^2}{2} + 4t + C = 3t^2 + 4t + C$$

Note that $v(0) = C$. But we are given that $v(0) = -6$, so $C = -6$ and

$$v(t) = 3t^2 + 4t - 6$$

Since $v(t) = s'(t)$, s is the antiderivative of v:

$$s(t) = 3\frac{t^3}{3} + 4\frac{t^2}{2} - 6t + D = t^3 + 2t^2 - 6t + D$$

This gives $s(0) = D$. We are given that $s(0) = 9$, so $D = 9$ and the required position function is

$$s(t) = t^3 + 2t^2 - 6t + 9 \qquad\qquad\bullet$$

An object near the surface of the earth is subject to a gravitational force that produces a downward acceleration denoted by g. For motion close to the earth we may assume that g is constant, its value being about $9.8\,\text{m/s}^2$ (or $32\,\text{ft/s}^2$).

EXAMPLE 6   A ball is thrown upward with a speed of 48 ft/s from the edge of a cliff 432 ft above the ground. Find its height above the ground t seconds later. When does it reach its maximum height? When does it hit the ground?

Solution   The motion is vertical and we choose the positive direction to be upward. At time t the distance above the ground is $s(t)$ and the velocity $v(t)$ is decreasing. Therefore the acceleration must be negative and we have

$$a(t) = \frac{dv}{dt} = -32$$

Taking antiderivatives, we have

$$v(t) = -32t + C$$

To determine C we use the given information that $v(0) = 48$. This gives $48 = 0 + C$, so

$$v(t) = -32t + 48$$

The maximum height is reached when $v(t) = 0$, that is, after 1.5 s. Since $s'(t) = v(t)$, we antidifferentiate again and obtain

$$s(t) = -16t^2 + 48t + D$$

Using the fact that $s(0) = 432$, we have $432 = 0 + D$, and so

$$s(t) = -16t^2 + 48t + 432$$

The expression for $s(t)$ is valid until the ball hits the ground. This happens when $s(t) = 0$, that is, when

$$-16t^2 + 48t + 432 = 0$$

or, equivalently,

$$t^2 - 3t - 27 = 0$$

Using the quadratic formula to solve this equation, we get

$$t = \frac{3 \pm 3\sqrt{13}}{2}$$

We reject the solution with the minus sign since it gives a negative value for t. There-fore the ball hits the ground after $3(1 + \sqrt{13})/2 \approx 6.9$ s.   ●

SECTION 4.8 **Exercises**

In Exercises 1–18 find the most general antiderivative of the given function. (Check your answer by differentiation.)

1. $f(x) = 12x^2 + 6x - 5$ 2. $f(x) = x^3 - 4x^2 + 17$

3. $f(x) = 6x^9 - 4x^7 + 3x^2 + 1$

4. $f(x) = x^{99} - 2x^{49} - 1$

5. $f(x) = \sqrt{x} + \sqrt[3]{x}$ 6. $f(x) = \sqrt[3]{x^2} - \sqrt{x^3}$

7. $f(x) = 6/x^5$ 8. $f(x) = \dfrac{3}{x^2} - \dfrac{5}{x^4}$

9. $f(x) = \sqrt{x} + 1/\sqrt{x}$ 10. $f(x) = x^{2/3} + 2x^{-1/3}$

11. $g(t) = \dfrac{t^3 + 2t^2}{\sqrt{t}}$ 12. $g(t) = \sqrt[5]{t^4} + t^{-6}$

13. $h(x) = \sin x - 2\cos x$

14. $f(t) = \sin t - 2\sqrt{t}$

15. $f(t) = \sec^2 t + t^2$

16. $f(\theta) = \theta + \sec\theta\tan\theta$

17. $f(x) = 2x + 5(1 - x^2)^{-1/2}$

18. $f(x) = \dfrac{x^2 + x + 1}{x}$

In Exercises 19–48 find $f(x)$.

19. $f'(x) = x^4 - 2x^2 + x - 1$

20. $f'(x) = \sin x - \sqrt[5]{x^2}$

21. $f''(x) = x^2 + x^3$ 22. $f''(x) = 60x^4 - 45x^2$

23. $f''(x) = 1$ 24. $f''(x) = \sin x$

25. $f'''(x) = 24x$ 26. $f'''(x) = \sqrt{x}$

27. $f'(x) = 4x + 3, \ f(0) = -9$

28. $f'(x) = 12x^2 - 24x + 1, \ \ f(1) = -2$

29. $f'(x) = 3\sqrt{x} - 1/\sqrt{x}, \ \ f(1) = 2$

30. $f'(x) = 1 + 1/x^2, \ x > 0, \ \ f(1) = 1$

31. $f'(x) = 3\cos x + 5\sin x, \ \ f(0) = 4$

32. $f'(x) = \sin x - 2\sqrt{x}, \ \ f(0) = 0$

33. $f'(x) = 2 + \sqrt[5]{x^3}, \ \ f(1) = 3$

34. $f'(x) = 3x^{-2}, \ \ f(1) = f(-1) = 0$

35. $f'(x) = 2/x, \ x < 0, \ \ f(-1) = 7$

36. $f'(x) = 4 - 3(1 + x^2)^{-1}, \ \ f(\pi/4) = 0$

37. $f''(x) = -8, \ \ f(0) = 6, \ \ f'(0) = 5$

38. $f''(x) = x, \ \ f(0) = -3, \ \ f'(0) = 2$

39. $f''(x) = 20x^3 - 10, \ \ f(1) = 1, \ \ f'(1) = -5$

40. $f''(x) = 12x^2 - 6x + 8, \ \ f(-1) = 5, \ \ f'(-1) = -21$

41. $f''(x) = x^2 + 3\cos x, \ \ f(0) = 2, \ \ f'(0) = 3$

42. $f''(x) = x + \sqrt{x}, \ \ f(1) = 1, \ \ f'(1) = 2$

43. $f''(x) = 6x + 6, \ \ f(0) = 4, \ \ f(1) = 3$

44. $f''(x) = 12x^2 - 6x + 2, \ \ f(0) = 1, \ \ f(2) = 11$

45. $f''(x) = 1/x^3, \ x > 0, \ \ f(1) = 0, \ \ f(2) = 0$

46. $f''(x) = 3e^x + 5\sin x, \ \ f(0) = 1, \ \ f'(0) = 2$

47. $f''(x) = x^{-2}$, $x > 0$, $f(1) = 0$, $f(2) = 0$

48. $f'''(x) = \sin x$, $f(0) = 1$, $f'(0) = 1$, $f''(0) = 1$

49. Given that the graph of f passes through the point $(1, 6)$ and that the slope of its tangent line at $(x, f(x))$ is $2x + 1$, find $f(2)$.

50. Find a function f such that $f'(x) = x^3$ and the line $x + y = 0$ is tangent to the graph of f.

In Exercises 51–56 a particle is moving with the given data. Find the position function of the particle.

51. $v(t) = 3 - 2t$, $s(0) = 4$

52. $v(t) = 3\sqrt{t}$, $s(1) = 5$

53. $a(t) = 3t + 8$, $s(0) = 1$, $v(0) = -2$

54. $a(t) = \cos t + \sin t$, $s(0) = 0$, $v(0) = 5$

55. $a(t) = t^2 - t$, $s(0) = 0$, $s(6) = 12$

56. $a(t) = 10 + 3t - 3t^2$, $s(0) = 0$, $s(2) = 10$

57. A stone is dropped from the upper observation deck (the Space Deck) of the CN Tower, 450 m above the ground.
    (a) Find the distance of the stone above ground level at time t.
    (b) How long does it take to reach the ground?
    (c) With what velocity does it strike the ground?

58. Answer Exercise 57 if the stone is thrown downward with a speed of 5 m/s.

59. Answer Exercise 57 if the stone is thrown upward with a speed of 5 m/s.

60. Show that for motion in a straight line with constant acceleration a, initial velocity v_0, and initial displacement s_0, the displacement after time t is
$$s = \tfrac{1}{2}at^2 + v_0t + s_0$$

61. An object is projected upward with initial velocity v_0 meters per second from a point s_0 meters above the ground. Show that
$$[v(t)]^2 = v_0^2 - 19.6[s(t) - s_0]$$

62. Two balls are thrown upward from the edge of the cliff in Example 6. The first is thrown with a speed of 48 ft/s and the second is thrown 1 s later with a speed of 24 ft/s. Do the balls ever pass each other?

63. A company estimates that the marginal cost (in dollars per item) of producing x items is $1.92 - 0.002x$. If the cost of producing one item is $562, find the cost of producing 100 items.

64. The linear density of a rod of length 1 m is given by $\rho(x) = 1/\sqrt{x}$, in grams per centimeter, where x is measured in centimeters from one end of the rod. Find the mass of the rod.

65. Since raindrops grow as they fall, their surface area increases and therefore the resistance to their falling increases. A raindrop has an initial downward velocity of 10 m/s and its downward acceleration is
$$a = \begin{cases} 9 - 0.9t & \text{if } 0 \le t \le 10 \\ 0 & \text{if } t > 10 \end{cases}$$
If the raindrop is initially 500 m above the ground, how long does it take to fall?

66. A car is traveling at 50 mi/h and the brakes are fully applied, producing a constant deceleration of 40 ft/s^2. What is the distance covered before the car comes to a stop?

67. What constant acceleration is required to increase the speed of a car from 30 mi/h to 50 mi/h in 5 s?

68. A car braked with a constant deceleration of 40 ft/s^2, producing skid marks measuring 160 ft before coming to a stop. How fast was the car traveling when the brakes were first applied?

69. A stone was dropped off a cliff and hit the ground with a speed of 120 ft/s. What is the height of the cliff?

CHAPTER 4

Review

Key Topics

Define, state, or discuss the following.

1. Absolute maximum and minimum values
2. Local maximum and minimum values
3. Extreme Value Theorem
4. Fermat's Theorem
5. Critical number
6. Procedure for finding absolute extreme values of f on $[a, b]$
7. Rolle's Theorem
8. Mean Value Theorem
9. Increasing function; decreasing function; monotonic function
10. Test for Monotonic Functions
11. First Derivative Test
12. Concave upward; concave downward
13. Test for Concavity
14. Point of inflection
15. Second Derivative Test
16. Procedure for curve sketching
17. Slant asymptote
18. First Derivative Test for Absolute Extrema
19. Antiderivative

Exercises

In Exercises 1–14 determine whether the statement is true or false.

1. If $f'(c) = 0$, then f has a local maximum or minimum at c.

2. If f has an absolute minimum value at c, then $f'(c) = 0$.

3. If f is continuous on (a, b), then f attains an absolute maximum value $f(c)$ and an absolute minimum value $f(d)$ at some numbers c and d in (a, b).

4. If f is continuous on $[-1, 1]$ and differentiable on $(-1, 1)$ and $f(-1) = f(1)$, then there is a number c such that $|c| < 1$ and $f'(c) = 0$.

5. If f is continuous on $[1, 6]$ and $f'(x) < 0$ for $x \in (1, 6)$, then f is decreasing on $[1, 6]$.

6. If $f''(c) = 0$, then $(c, f(c))$ is an inflection point of the curve $y = f(x)$.

7. If $f'(x) = g'(x)$ for $0 < x < 1$, then $f(x) = g(x)$ for $0 < x < 1$.

8. If f is monotonic on $[0, 1]$ and monotonic on $[1, 2]$, then f is monotonic on $[0, 2]$.

9. If f and g are increasing on an interval I, then $f + g$ is increasing on I.

10. If f and g are increasing on an interval I, then $f - g$ is increasing on I.

11. If f and g are increasing on an interval I, then fg is increasing on I.

12. If f and g are positive increasing functions on an interval I, then fg is increasing on I.

13. If f is increasing and $f(x) > 0$ on I, then $g(x) = 1/f(x)$ is decreasing on I.

14. The most general antiderivative of $f(x) = x^{-2}$ is $F(x) = (-1/x) + C$.

In Exercises 15–20 find the local and absolute extreme values of the given function on the given interval.

15. $f(x) = x^3 - 12x + 5$, $[-5, 3]$

16. $f(x) = 3x - \ln x$, $[0.1, 1]$

17. $f(x) = \dfrac{x - 2}{x + 2}$, $[0, 4]$

18. $f(x) = \sqrt{x^2 + 4x + 8}$, $[-3, 0]$

19. $f(x) = x - \sqrt{2} \sin x$, $[0, \pi]$

20. $f(x) = 2x + 2\cos x - 4\sin x - \cos 2x$, $[0, \pi]$

In Exercises 21 and 22 sketch the graph of a function that satisfies all of the given conditions.

21. $f(1) = f'(1) = 0$, $\lim_{x \to 2^+} f(x) = \infty$, $\lim_{x \to 2^-} f(x) = -\infty$, $\lim_{x \to 0} f(x) = -\infty$, $\lim_{x \to -\infty} f(x) = \infty$, $\lim_{x \to \infty} f(x) = 0$, $f''(x) > 0$ for $x > 2$, $f''(x) < 0$ for $x < 0$ and for $0 < x < 2$

22. $g(0) = 0$, $g''(x) < 0$ for $x \neq 0$, $\lim_{x \to -\infty} g(x) = \infty$, $\lim_{x \to \infty} g(x) = -\infty$, $\lim_{x \to 0^-} g'(x) = -\infty$, $\lim_{x \to 0^+} g'(x) = \infty$

In Exercises 23–34 discuss the given curve under the headings A–H of Section 4.5.

23. $y = 1 + x + x^3$ **24.** $y = 3x^4 - 4x^3 - 12x^2 + 2$

25. $y = \dfrac{1}{x(x-3)^2}$ **26.** $y = \dfrac{1}{x^2 - x - 6}$

27. $y = x\sqrt{5-x}$ **28.** $y = \dfrac{1}{x} + \dfrac{1}{x+1}$

29. $y = 4x - \tan x,\ -\pi/2 < x < \pi/2$

30. $y = \sin^{-1}(1/x)$ **31.** $y = e^x + e^{-3x}$

32. $y = \ln(x^2 - 1)$ **33.** $y = 2x^2 - \ln x$

34. $y = e^{-1/x^2}$

35. Show that the equation $x^{101} + x^{51} + x - 1 = 0$ has exactly one real root.

36. Suppose that f is continuous on $[0,4]$, $f(0) = 1$, and $2 \le f'(x) \le 5$ for all x in $(0,4)$. Show that $9 \le f(4) \le 21$.

37. By applying the Mean Value Theorem to the function $f(x) = x^{1/5}$ on the interval $[32,33]$, show that
$$2 < \sqrt[5]{33} < 2.0125$$

38. A number x_0 is called a **fixed point** of a function f if $f(x_0) = x_0$. Show that if $f'(x) < 1$ for all x, then f has at most one fixed point.

39. Sketch the graph of a function that satisfies the following conditions: $f(0) = 0$, $f'(-2) = f'(1) = f'(9) = 0$, $\lim_{x \to \infty} f(x) = 0$, $\lim_{x \to 6} f(x) = -\infty$, $f'(x) < 0$ on $(-\infty, -2)$, $(1,6)$, and $(9,\infty)$, $f'(x) > 0$ on $(-2,1)$ and $(6,9)$, $f''(x) > 0$ on $(-\infty,0)$ and $(12,\infty)$, $f''(x) < 0$ on $(0,6)$ and $(6,12)$.

40. Sketch the graph of a function that satisfies the following conditions: $f(0) = 0$, f is continuous and even, $f'(x) = 2x$ if $0 < x < 1$, $f'(x) = -1$ if $1 < x < 3$, and $f'(x) = 1$ if $x > 3$.

41. For what values of the constants a and b is $(1,6)$ a point of inflection of the curve $y = x^3 + ax^2 + bx + 1$?

42. Discuss the curve $y = x^2 + \sin x$ under the headings A–H given in Section 4.5. Use Newton's method when necessary.

43. Show that the shortest distance from the point (x_1, y_1) to the straight line $Ax + By + C = 0$ is
$$\frac{|Ax_1 + By_1 + C|}{\sqrt{A^2 + B^2}}$$

44. Find the point on the hyperbola $xy = 8$ that is closest to the point $(3,0)$.

45. Find the smallest possible area of an isosceles triangle that is circumscribed about a circle of radius r.

46. Find the volume of the largest circular cone that can be inscribed in a sphere of radius r.

47. In $\triangle ABC$, D lies on AB, $CD \perp AB$, $|AD| = |BD| = 4\,\text{cm}$, and $|CD| = 5\,\text{cm}$. Where should a point P be chosen on CD so that the sum $|PA| + |PB| + |PC|$ is a minimum?

48. Do Exercise 47 when $|CD| = 2\,\text{cm}$.

49. A rectangular beam is cut from a cylindrical log of radius R. The strength of the beam is proportional to the width and the square of the depth of a cross-section. Find the dimensions of the strongest such beam.

50. A metal storage tank with volume V is to be constructed in the shape of a right circular cylinder surmounted by a hemisphere. What dimensions will require the least amount of metal?

51. A hockey team plays in an arena with a seating capacity of 15,000 spectators. With ticket prices at \$12, average attendance at a game has been 11,000. A market survey indicates that for each dollar that ticket prices are lowered, the average attendance will increase by 1000. How should the owners of the team set ticket prices to maximize their revenue from ticket sales?

52. A manufacturer determines that the cost of making x units of a commodity is
$$C(x) = 250{,}000 + 0.84x + 0.0002x^2$$
and the demand function is
$$p(x) = 10 - 0.05x$$
(a) What production level will minimize the average cost?
(b) What level of sales will maximize profits?

In Exercises 53–60 find $f(x)$.

53. $f'(x) = x - \sqrt[4]{x}$ **54.** $f'(x) = 2/\sqrt{x^5}$

55. $f'(x) = \dfrac{1+x}{\sqrt{x}},\quad f(1) = 0$

56. $f'(x) = 1 + 2\sin x - \cos x,\quad f(0) = 3$

57. $f'(x) = 2/(1+x^2),\quad f(0) = -1$

58. $f'(x) = e^x - (1/x),\quad f(1) = 0$

59. $f''(x) = x^3 + x,\quad f(0) = -1,\quad f'(0) = 1$

60. $f''(x) = x^4 - 4x^2 + 3x - 2,\quad f(0) = 0,\quad f(1) = 1$

61. A canister is dropped from a helicopter 500 m above the ground. Its parachute does not open, but the canister has been designed to withstand an impact velocity of 100 m/s. Will it burst or not?

62. In an automobile race along a straight road, car A passed car B twice. Prove that at some time during the race their accelerations were equal.

Problems Plus

1. If a rectangle has its base on the x-axis and two vertices on the curve $y = e^{-x^2}$, show that the rectangle has the largest possible area when the two vertices are at the points of inflection of the curve.

2. The figure shows the graph of the *derivative* f' of a function f.
   (a) On what intervals is f increasing or decreasing?
   (b) For what values of x does f have a local maximum or minimum?
   (c) Sketch the graph of f''.
   (d) Sketch a possible graph of f.

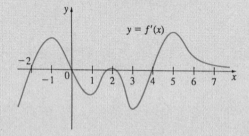

3. If x, y, and z are positive numbers, prove that

$$\frac{(x^2 + 1)(y^2 + 1)(z^2 + 1)}{xyz} \geq 8$$

4. Find the point on the parabola $y = 1 - x^2$ at which the tangent line cuts from the first quadrant the triangle with the smallest area.

5. Find the highest and lowest points on the curve $x^2 + xy + y^2 = 12$.

6. Prove that $\log_2 5$ is an irrational number.

7. Find a function f such that $f(1) = -1$, $f(4) = 7$, and $f'(x) > 3$ for all x, or prove that such a function cannot exist.

8. Find a function f such that $f'(-1) = \frac{1}{2}$, $f'(0) = 0$, and $f''(x) > 0$ for all x, or prove that such a function cannot exist.

9. Find the absolute maximum value of the function $f(x) = \dfrac{1}{1 + |x|} + \dfrac{1}{1 + |x - 2|}$.

10. Find the point where the curves $y = x^3 - 3x + 4$ and $y = 3(x^2 - x)$ are tangent to each other, that is, have a common tangent line. Illustrate by sketching both curves and the common tangent.

11. The figure shows a circle with radius 1 inscribed in the parabola $y = x^2$. Find the center of the circle.

12. Sketch the region in the plane consisting of all points (x, y) such that

$$2xy \le |x - y| \le x^2 + y^2$$

13. Show that

$$\cos\{\arctan[\sin(\operatorname{arccot} x)]\} = \sqrt{\frac{x^2 + 1}{x^2 + 2}}$$

14. Find a condition on the coefficients of a cubic polynomial $P(x) = ax^3 + bx^2 + cx + d$ that will ensure that P is a monotonic function.

15. Suppose that f and g are functions such that $f(0) = 0$, $g(0) = 1$, and $f'(x) = g(x)$, $g'(x) = -f(x)$ for all x.
    (a) Show that $[f(x)]^2 + [g(x)]^2 = 1$ for all x.
    (b) Show that $[f(x) - \sin x]^2 + [g(x) - \cos x]^2 = 0$ for all x.
    (c) Deduce that $f(x) = \sin x$ and $g(x) = \cos x$ for for all x.

16. Sketch the graph of a function f such that $f'(x) < 0$ for all x, $f''(x) > 0$ for $|x| > 1$, $f''(x) < 0$ for $|x| < 1$, and $\lim_{x \to \pm\infty} [f(x) + x] = 0$.

17. Show that $f(x) = [\![x]\!] + \sqrt{x - [\![x]\!]}$ is continuous and increasing on $(-\infty, \infty)$.

18. Draw the graphs of the functions $f(x) = (-1)^{[\![1/x]\!]}$ and $g(x) = x(-1)^{[\![1/x]\!]}$ for $\frac{1}{6} \le |x| \le 2$. Is it possible to define $g(0)$ in such a way that g becomes continuous at 0?

19. For what value of a is it true that

$$\lim_{x \to \infty} \left(\frac{x + a}{x - a}\right)^x = e?$$

20. Sketch the set of all points (x, y) such that $|x + y| \le e^x$.

21. The line $y = mx + b$ intersects the parabola $y = x^2$ in points A and B. Find the point P on the arc AOB of the parabola that maximizes the area of the triangle PAB.

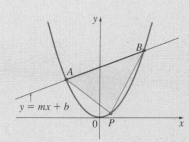

22. Show that $\sin^{-1}(\tanh x) = \tan^{-1}(\sinh x)$.

23. Find $\lim_{x \to \infty} \dfrac{x}{[\![x]\!]}$.

24. Suppose that the roots of a fourth-degree polynomial are consecutive terms of an arithmetic sequence; that is, they are equally spaced. Draw a diagram to make it seem plausible that the roots of its derivative are also consecutive terms of an arithmetic sequence. Then prove it.

5

Integrals

Calculus is the most powerful weapon of thought yet devised by the wit of man.

W.B. Smith

The experimental verification of a theory concerning any natural phenomenon generally rests on the result of an integration.

J.W. Mellor

Chapters 1–4 were concerned with the branch of calculus called differential calculus. We now turn our attention to another branch of calculus, called integral calculus, whose central concept is the definite integral.

In differential calculus, the tangent problem led us to formulate, in terms of limits, the idea of a derivative, which later turned out to be applicable, through velocities and other rates of change, to a variety of applied problems.

In integral calculus, the area problem leads us to formulate, again in terms of limits, the idea of an integral, which will later be used to find volumes, lengths of curves, work, and forces, and to solve problems in chemistry, physics, biology, and economics.

There is a connection between the two branches of calculus. It is called the Fundamental Theorem of Calculus and we see in this chapter that it greatly simplifies the solution of many problems.

Sigma Notation

In finding areas and evaluating integrals we often encounter sums with many terms. A convenient way of writing such sums uses the Greek letter $\sum$ (capital sigma, corresponding to our S) and is called **sigma notation**.

Definition (5.1)

If $a_m, a_{m+1}, \ldots, a_n$ are real numbers and m and n are integers such that $m \leq n$, then

$$\sum_{i=m}^{n} a_i = a_m + a_{m+1} + a_{m+2} + \cdots + a_{n-1} + a_n$$

With function notation, Definition 5.1 can be written as

$$\sum_{i=m}^{n} f(i) = f(m) + f(m+1) + f(m+2) + \cdots + f(n-1) + f(n)$$

Thus the symbol $\sum_{i=m}^{n}$ indicates a summation in which the letter i (called the **index of summation**) takes on the values $m, m+1, \ldots, n$. Other letters can also be used as the index of summation.

315

EXAMPLE 1

(a) $\displaystyle\sum_{i=1}^{4} i^2 = 1^2 + 2^2 + 3^2 + 4^2 = 30$ (b) $\displaystyle\sum_{i=3}^{n} i = 3 + 4 + 5 + \cdots + (n-1) + n$

(c) $\displaystyle\sum_{j=0}^{5} 2^j = 2^0 + 2^1 + 2^2 + 2^3 + 2^4 + 2^5 = 63$

(d) $\displaystyle\sum_{k=1}^{n} \frac{1}{k} = 1 + \frac{1}{2} + \frac{1}{3} + \cdots + \frac{1}{n}$

(e) $\displaystyle\sum_{i=1}^{3} \frac{i-1}{i^2+3} = \frac{1-1}{1^2+3} + \frac{2-1}{2^2+3} + \frac{3-1}{3^2+3} = 0 + \frac{1}{7} + \frac{1}{6}$

(f) $\displaystyle\sum_{i=1}^{4} 2 = 2 + 2 + 2 + 2 = 8$

EXAMPLE 2   Write the sum $2^3 + 3^3 + \cdots + n^3$ in sigma notation.

Solution   There is no unique way of writing a sum in sigma notation. We could write

$$2^3 + 3^3 + \cdots + n^3 = \sum_{i=2}^{n} i^3$$

or $$2^3 + 3^3 + \cdots + n^3 = \sum_{j=1}^{n-1} (j+1)^3$$

or $$2^3 + 3^3 + \cdots + n^3 = \sum_{k=0}^{n-2} (k+2)^3$$

The following theorem gives three simple rules for working with sigma notation.

Theorem (5.2)

If c is any constant (that is, it does not depend on i), then

(a) $\displaystyle\sum_{i=m}^{n} ca_i = c \sum_{i=m}^{n} a_i$

(b) $\displaystyle\sum_{i=m}^{n} (a_i + b_i) = \sum_{i=m}^{n} a_i + \sum_{i=m}^{n} b_i$

(c) $\displaystyle\sum_{i=m}^{n} (a_i - b_i) = \sum_{i=m}^{n} a_i - \sum_{i=m}^{n} b_i$

Proof   To see why these rules are true, all we have to do is write both sides in expanded form. Rule (a) is just the distributive property of real numbers:

$$ca_m + ca_{m+1} + \cdots + ca_n = c(a_m + a_{m+1} + \cdots + a_n)$$

Rule (b) follows from the associative and commutative properties:

$$(a_m + b_m) + (a_{m+1} + b_{m+1}) + \cdots + (a_n + b_n)$$
$$= (a_m + a_{m+1} + \cdots + a_n) + (b_m + b_{m+1} + \cdots + b_n)$$

Rule (c) is proved similarly.

EXAMPLE 3 Find $\displaystyle\sum_{i=1}^{n} 1$.

Solution

$$\sum_{i=1}^{n} 1 = \underbrace{1 + 1 + \cdots + 1}_{n \text{ terms}} = n$$

EXAMPLE 4 Prove the formula for the sum of the first n positive integers:

$$\sum_{i=1}^{n} i = 1 + 2 + 3 + \cdots + n = \frac{n(n+1)}{2}$$

Solution This formula can be proved by mathematical induction (see Appendix A) or by the following method used by the German mathematician Karl Friedrich Gauss (1777–1855) when he was ten years old.

Write the sum S twice, once in the usual order and once in reverse order:

$$
\begin{array}{ccccccccccc}
S = & 1 & + & 2 & + & 3 & + & \cdots & + (n-1) + & n \\
S = & n & + & (n-1) & + & (n-2) & + & \cdots & + \quad 2 \quad + & 1
\end{array}
$$

Adding all columns vertically, we get

$$2S = (n+1) + (n+1) + (n+1) + \cdots + (n+1) + (n+1)$$

On the right-hand side there are n terms, each of which is $n+1$, so

$$2S = n(n+1) \qquad \text{or} \qquad S = \frac{n(n+1)}{2}$$

EXAMPLE 5 Prove the formula for the sum of the squares of the first n positive integers:

$$\sum_{i=1}^{n} i^2 = 1^2 + 2^2 + 3^2 + \cdots + n^2 = \frac{n(n+1)(2n+1)}{6}$$

Solution This formula can be proved by mathematical induction (see Exercise 40) or as follows: Let S be the desired sum. We start with the *telescoping sum* (or collapsing sum)

Most terms cancel in pairs.

$$\sum_{i=1}^{n} [(1+i)^3 - i^3] = (2^3 - 1^3) + (3^3 - 2^3) + (4^3 - 3^3) + \cdots + [(n+1)^3 - n^3]$$
$$= (n+1)^3 - 1^3 = n^3 + 3n^2 + 3n$$

On the other hand, using Theorem 5.2 and Examples 3 and 4, we have

$$\sum_{i=1}^{n} [(1+i)^3 - i^3] = \sum_{i=1}^{n} [3i^2 + 3i + 1]$$

$$= 3 \sum_{i=1}^{n} i^2 + 3 \sum_{i=1}^{n} i + \sum_{i=1}^{n} 1$$

$$= 3S + 3 \frac{n(n+1)}{2} + n$$

$$= 3S + \tfrac{3}{2} n^2 + \tfrac{5}{2} n$$

Thus we have

$$n^3 + 3n^2 + 3n = 3S + \tfrac{3}{2} n^2 + \tfrac{5}{2} n$$

Solving this equation for S, we obtain

$$3S = n^3 + \tfrac{3}{2}n^2 + \tfrac{1}{2}n$$

or $$S = \frac{2n^3 + 3n^2 + n}{6} = \frac{n(n+1)(2n+1)}{6}$$

We list the results of Examples 3, 4, and 5 together with similar results for cubes and fourth powers (see Exercises 41–44) as Theorem 5.3. These formulas are needed for finding areas in the next section.

Theorem (5.3)

Let c be a constant and n a positive integer. Then

(a) $\displaystyle\sum_{i=1}^{n} 1 = n$ (b) $\displaystyle\sum_{i=1}^{n} c = nc$

(c) $\displaystyle\sum_{i=1}^{n} i = \frac{n(n+1)}{2}$ (d) $\displaystyle\sum_{i=1}^{n} i^2 = \frac{n(n+1)(2n+1)}{6}$

(e) $\displaystyle\sum_{i=1}^{n} i^3 = \left[\frac{n(n+1)}{2}\right]^2$

(f) $\displaystyle\sum_{i=1}^{n} i^4 = \frac{n(n+1)(2n+1)(3n^2+3n-1)}{30}$

EXAMPLE 6   Evaluate $\displaystyle\sum_{i=1}^{n} i(4i^2 - 3)$.

Solution   Using Theorem 5.2 and 5.3, we have

$$\sum_{i=1}^{n} i(4i^2 - 3) = \sum_{i=1}^{n}(4i^3 - 3i) = 4\sum_{i=1}^{n} i^3 - 3\sum_{i=1}^{n} i$$

$$= 4\left[\frac{n(n+1)}{2}\right]^2 - 3\,\frac{n(n+1)}{2}$$

$$= \frac{n(n+1)[2n(n+1) - 3]}{2}$$

$$= \frac{n(n+1)(2n^2 + 2n - 3)}{2}$$

EXAMPLE 7   Find $\displaystyle\lim_{n\to\infty} \sum_{i=1}^{n} \frac{3}{n}\left[\left(\frac{i}{n}\right)^2 + 1\right]$.

Solution

$$\lim_{n\to\infty} \sum_{i=1}^{n} \frac{3}{n}\left[\left(\frac{i}{n}\right)^2 + 1\right] = \lim_{n\to\infty} \sum_{i=1}^{n}\left[\frac{3}{n^3}i^2 + \frac{3}{n}\right]$$

$$= \lim_{n\to\infty}\left[\frac{3}{n^3}\sum_{i=1}^{n} i^2 + \frac{3}{n}\sum_{i=1}^{n} 1\right]$$

$$= \lim_{n\to\infty}\left[\frac{3}{n^3}\,\frac{n(n+1)(2n+1)}{6} + \frac{3}{n}\cdot n\right]$$

$$= \lim_{n\to\infty}\left[\frac{1}{2}\cdot\frac{n}{n}\cdot\left(\frac{n+1}{n}\right)\left(\frac{2n+1}{n}\right)+3\right]$$

$$= \lim_{n\to\infty}\left[\frac{1}{2}\cdot1\left(1+\frac{1}{n}\right)\left(2+\frac{1}{n}\right)+3\right]$$

$$= \frac{1}{2}\cdot1\cdot1\cdot2+3 = 4$$

SECTION 5.1 Exercises

In Exercises 1–10 write the sum in expanded form.

1. $\sum_{i=1}^{5} \sqrt{i}$ 2. $\sum_{i=1}^{6} \frac{1}{i+1}$

3. $\sum_{i=4}^{6} 3^i$ 4. $\sum_{i=4}^{6} i^3$

5. $\sum_{k=0}^{4} \frac{2k-1}{2k+1}$ 6. $\sum_{k=5}^{8} x^k$

7. $\sum_{i=1}^{n} i^{10}$ 8. $\sum_{j=n}^{n+3} j^2$

9. $\sum_{j=0}^{n-1} (-1)^j$ 10. $\sum_{i=1}^{n} f(x_i)\,\Delta x_i$

In Exercises 11–20 write the sum in sigma notation.

11. $1+2+3+4+\cdots+10$

12. $\sqrt{3}+\sqrt{4}+\sqrt{5}+\sqrt{6}+\sqrt{7}$

13. $\frac{1}{2}+\frac{2}{3}+\frac{3}{4}+\frac{4}{5}+\cdots+\frac{19}{20}$

14. $\frac{3}{7}+\frac{4}{8}+\frac{5}{9}+\frac{6}{10}+\cdots+\frac{23}{27}$

15. $2+4+6+8+\cdots+2n$

16. $1+3+5+7+\cdots+(2n-1)$

17. $1+2+4+8+16+32$

18. $\frac{1}{1}+\frac{1}{4}+\frac{1}{9}+\frac{1}{16}+\frac{1}{25}+\frac{1}{36}$

19. $x+x^2+x^3+\cdots+x^n$

20. $1-x+x^2-x^3+\cdots+(-1)^n x^n$

In Exercises 21–38 find the value of the sum.

21. $\sum_{i=4}^{8} (3i-2)$ 22. $\sum_{i=3}^{6} i(i+2)$

23. $\sum_{j=1}^{6} 3^{j+1}$ 24. $\sum_{k=0}^{8} \cos k\pi$

25. $\sum_{n=1}^{20} (-1)^n$ 26. $\sum_{i=1}^{100} 4$

27. $\sum_{i=0}^{4} (2^i + i^2)$ 28. $\sum_{i=-2}^{4} 2^{3-i}$

29. $\sum_{i=1}^{n} \left(\frac{1}{i} - \frac{1}{i+1}\right)$ 30. $\sum_{k=1}^{n} (\sqrt{k} - \sqrt{k-1})$

31. $\sum_{i=1}^{n} 2i$ 32. $\sum_{i=1}^{n} (2-5i)$

33. $\sum_{i=1}^{n} (i^2 + 3i + 4)$ 34. $\sum_{i=1}^{n} (3+2i)^2$

35. $\sum_{i=1}^{n} (i+1)(i+2)$ 36. $\sum_{i=1}^{n} i(i+1)(i+2)$

37. $\sum_{i=1}^{n} (i^3 - i - 2)$ 38. $\sum_{k=1}^{n} k^2(k^2 - k + 1)$

39. Prove Formula (b) of Theorem 5.3.

40. Prove Formula (d) of Theorem 5.3 using mathematical induction. (See Appendix A for the Principle of Mathematical Induction.)

41. Prove Formula (e) of Theorem 5.3 using mathematical induction.

42. Prove Formula (e) of Theorem 5.3 using a method similar to that of Example 5 [start with $(1+i)^4 - i^4$].

43. Prove Formula (e) of Theorem 5.3 using the following method published by Abu Bekr Mohammed ibn Alhusain Alkarchi in about A.D. 1010. The figure shows a square $ABCD$ whose sides AB and AD have been divided into segments of length $1, 2, 3, \ldots, n$. Thus the side of the square has length $n(n+1)/2$ so the area is $[n(n+1)/2]^2$. But the area is also the

sum of the areas of the n "gnomons" $G_1, G_2, \ldots, G_n$ shown in the figure. Show that the area of G_i is i^3 and conclude that Formula (e) is true.

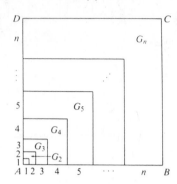

44. Prove Formula (f) of Theorem 5.3 (a) using mathematical induction, (b) using the method of Example 5.

45. Evaluate the following telescoping sums.

(a) $\sum_{i=1}^{n} [i^4 - (i-1)^4]$ (b) $\sum_{i=1}^{100} (5^i - 5^{i-1})$

(c) $\sum_{i=3}^{99} \left(\frac{1}{i} - \frac{1}{i+1}\right)$ (d) $\sum_{i=1}^{n} (a_i - a_{i-1})$

46. Prove the generalized triangle inequality

$$\left| \sum_{i=1}^{n} a_i \right| \le \sum_{i=1}^{n} |a_i|$$

Find the limits in Exercises 47–50.

47. $\lim_{n \to \infty} \sum_{i=1}^{n} \frac{1}{n}\left(\frac{i}{n}\right)^2$ 48. $\lim_{n \to \infty} \sum_{i=1}^{n} \frac{1}{n}\left[\left(\frac{i}{n}\right)^3 + 1\right]$

49. $\lim_{n \to \infty} \sum_{i=1}^{n} \frac{2}{n}\left[\left(\frac{2i}{n}\right)^3 + 5\left(\frac{2i}{n}\right)\right]$

50. $\lim_{n \to \infty} \sum_{i=1}^{n} \frac{3}{n}\left[\left(1 + \frac{3i}{n}\right)^3 - 2\left(1 + \frac{3i}{n}\right)\right]$

51. Prove the formula for the sum of a finite geometric series with first term a and common ratio r:

$$\sum_{i=1}^{n} ar^{i-1} = a + ar + ar^2 + \cdots + ar^{n-1} = \frac{a(r^n - 1)}{r - 1}$$

52. Evaluate $\sum_{i=1}^{n} \frac{3}{2^{i-1}}$.

53. Evaluate $\sum_{i=1}^{n} (2i + 2^i)$.

54. Evaluate $\sum_{i=1}^{m}\left[\sum_{j=1}^{n} (i + j)\right]$.

55. Find the number n such that $\sum_{i=1}^{n} i = 78$.

56. (a) Use the product formula (18a in Appendix B) to show that

$$2 \sin \tfrac{1}{2}x \cos ix = \sin\left(i + \tfrac{1}{2}\right)x - \sin\left(i - \tfrac{1}{2}\right)x$$

(b) Use the identity in part (a) and telescoping sums to prove the formula

$$\sum_{i=1}^{n} \cos ix = \frac{\sin\left(n + \tfrac{1}{2}\right)x - \sin \tfrac{1}{2}x}{2 \sin \tfrac{1}{2}x}$$

where x is not an integer multiple of 2π. Deduce that

$$\sum_{i=1}^{n} \cos ix = \frac{\sin \tfrac{1}{2}nx \cos \tfrac{1}{2}(n+1)x}{\sin \tfrac{1}{2}x}$$

57. Use the method of Exercise 56 to prove the formula

$$\sum_{i=1}^{n} \sin ix = \frac{\sin \tfrac{1}{2}nx \sin \tfrac{1}{2}(n+1)x}{\sin \tfrac{1}{2}x}$$

where x is not an integer multiple of 2π.

SECTION 5.2

Area

We begin by attempting to solve the *area problem*: Find the area of the region S that lies under the curve $y = f(x)$ from a to b. This means that S, illustrated in Figure 5.1, is bounded by the graph of a function f [where $f(x) \ge 0$], the vertical lines $x = a$ and $x = b$, and the x-axis.

In trying to solve the area problem we have to ask ourselves: What is the meaning of the word *area*? This question is easy to answer for regions with straight sides. For a rectangle, the area is defined as the product of the length and the width. The area of a triangle is half the base times the height. The area of a polygon is found by dividing it into triangles (as in Figure 5.2) and adding the areas of the triangles.

Figure 5.1

$S = \{(x, y) \mid a \le x \le b, 0 \le y \le f(x)\}$

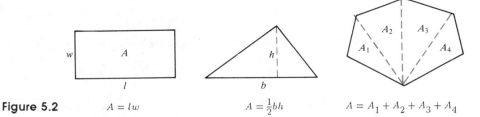

Figure 5.2

$A = lw$ $\qquad A = \frac{1}{2}bh$ $\qquad A = A_1 + A_2 + A_3 + A_4$

However it is not so easy to find the area of a region with curved sides. We all have an intuitive idea of what the area of a region is. But part of the area problem is to make this intuitive idea precise by giving an exact definition of area.

Recall that in defining a tangent we first approximated the slope of the tangent line by slopes of secant lines and then we took the limit of these approximations. We pursue a similar idea for areas. We first approximate the region S by polygons and then we take the limit of the areas of these polygons. The following example will illustrate the procedure.

EXAMPLE 1 Let us try to find the area under the parabola $y = x^2$ from 0 to 1 (the parabolic region illustrated in Figure 5.3). One method of approximating the desired area is to divide the interval $[0, 1]$ into subintervals of equal length and consider the rectangles whose bases are these subintervals and whose heights are the values of the function at the right-hand endpoints of these subintervals. Figure 5.4 shows the approximation of the parabolic region by four, eight, and n rectangles.

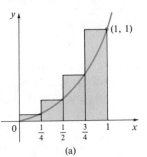

Figure 5.3

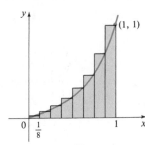
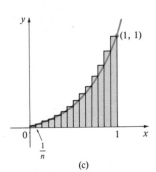
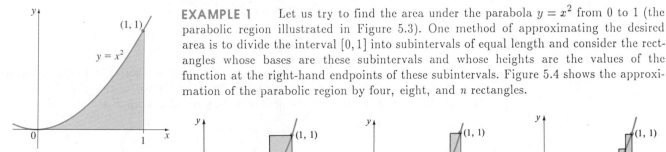

Figure 5.4
(a) $\qquad$ (b) $\qquad$ (c)

Let S_n be the sum of the areas of the n rectangles in Figure 5.4(c). Each rectangle has width $1/n$ and the heights are the values of the function $f(x) = x^2$ at the points $1/n, 2/n, 3/n, \ldots, n/n$; that is, the heights are $(1/n)^2, (2/n)^2, (3/n)^2, \ldots, (n/n)^2$. Thus

$$S_n = \frac{1}{n}\left(\frac{1}{n}\right)^2 + \frac{1}{n}\left(\frac{2}{n}\right)^2 + \frac{1}{n}\left(\frac{3}{n}\right)^2 + \cdots + \frac{1}{n}\left(\frac{n}{n}\right)^2$$

$$= \frac{1}{n}\frac{1}{n^2}(1^2 + 2^2 + 3^2 + \cdots + n^2)$$

$$= \frac{1}{n^3}\sum_{i=1}^{n} i^2$$

Using the formula for the sum of the squares of the first n integers [Equation (d) in Theorem 5.3] we can write

$$S_n = \frac{1}{n^3}\frac{n(n+1)(2n+1)}{6} = \frac{(n+1)(2n+1)}{6n^2}$$

For instance, the sum of the areas of the four shaded rectangles in Figure 5.4(a) is

$$S_4 = \frac{5(9)}{6(16)} = 0.46875$$

and the sum of the areas of the eight rectangles in Figure 5.4(b) is

$$S_8 = \frac{9(17)}{6(64)} = 0.3984375$$

Similarly we find that

$$S_{100} \approx 0.33835 \qquad S_{1000} \approx 0.33383$$

It looks as if S_n is becoming closer to $\frac{1}{3}$ as n increases. In fact,

$$\lim_{n \to \infty} S_n = \lim_{n \to \infty} \frac{(n+1)(2n+1)}{6n^2}$$

$$= \lim_{n \to \infty} \frac{1}{6}\left(\frac{n+1}{n}\right)\left(\frac{2n+1}{n}\right)$$

$$= \lim_{n \to \infty} \frac{1}{6}\left(1 + \frac{1}{n}\right)\left(2 + \frac{1}{n}\right)$$

$$= \frac{1}{6} \cdot 1 \cdot 2 = \frac{1}{3}$$

From Figure 5.4 it appears that, as n increases, S_n becomes a better and better approximation to the area of the parabolic segment. Therefore we *define* the area A to be the limit of the sums of the areas of the approximating rectangles, that is,

$$A = \lim_{n \to \infty} S_n = \frac{1}{3}$$

In applying the idea of Example 1 to the more general region S of Figure 5.1, there is no need to use rectangles of equal width. We start by subdividing the interval $[a, b]$ into n smaller subintervals by choosing partition points $x_0, x_1, x_2, \ldots, x_n$ so that

$$a = x_0 < x_1 < x_2 < \cdots < x_{n-1} < x_n = b$$

Then the n subintervals are

$$[x_0, x_1], [x_1, x_2], [x_2, x_3], \ldots, [x_{n-1}, x_n]$$

This subdivision is called a **partition** of $[a, b]$ and we denote it by P. We use the notation Δx_i for the length of the ith subinterval $[x_{i-1}, x_i]$. Thus

$$\Delta x_i = x_i - x_{i-1}$$

The length of the longest subinterval is denoted by $\|P\|$ and is called the **norm** of P. Thus

$$\|P\| = \max\{\Delta x_1, \Delta x_2, \ldots, \Delta x_n\}$$

Figure 5.5 illustrates one possible partition of $[a, b]$.

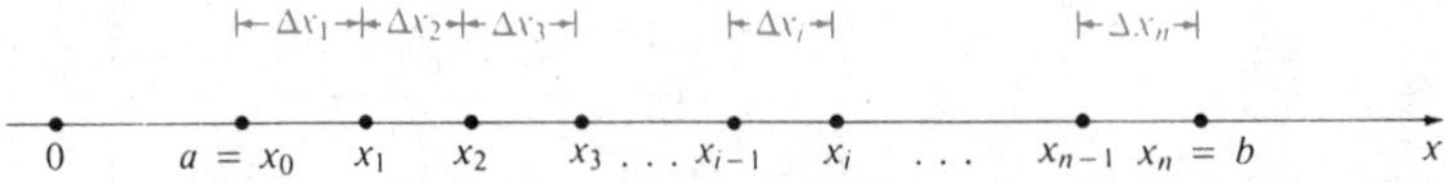

Figure 5.5

By drawing the lines $x = a$, $x = x_1$, $x = x_2, \ldots$, $x = b$, we use the partition P to divide the region S into strips $S_1, S_2, \ldots, S_n$ as in Figure 5.6. Next we approximate these strips S_i by rectangles R_i. To do this we choose a number x_i^* in each subinterval $[x_{i-1}, x_i]$ and construct a rectangle R_i with base Δx_i and height $f(x_i^*)$ as in Figure 5.7.

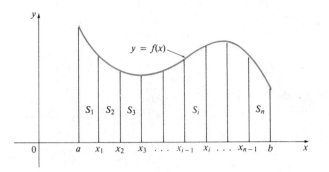

Figure 5.6

Figure 5.7

Each point x_i^* can be anywhere in its subinterval—at the right endpoint (as in Example 1) or at the left endpoint or somewhere in between. The area of the ith rectangle R_i is

$$A_i = f(x_i^*) \, \Delta x_i$$

The n rectangles $R_1, \ldots, R_n$ form a polygonal approximation to the region S. What we think of intuitively as the area of S is approximated by the sum of the areas of these rectangles, which is

(5.4)
$$\sum_{i=1}^{n} A_i = \sum_{i=1}^{n} f(x_i^*) \, \Delta x_i = f(x_1^*) \, \Delta x_1 + \cdots + f(x_n^*) \, \Delta x_n$$

Figure 5.8 shows this approximation for partitions with $n = 2, 4, 8,$ and 12. Notice that this approximation appears to become better and better as the strips become thinner and thinner, that is, as $\|P\| \to 0$. Therefore we define the **area** A of the region S as the limiting value (if it exists) of the areas of the approximating polygons, that is, the limit of the sum (5.4) of the areas of the approximating rectangles. In symbols:

(5.5)
$$A = \lim_{\|P\| \to 0} \sum_{i=1}^{n} f(x_i^*) \, \Delta x_i$$

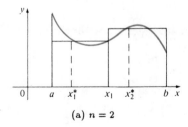

(a) $n = 2$

(b) $n = 4$

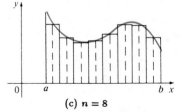

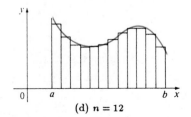

Figure 5.8

(c) $n = 8$

(d) $n = 12$

The preceding discussion and the diagrams in Figure 5.7 and 5.8 show that the definition of area in (5.5) corresponds to our intuitive feeling of what area ought to be.

The limit in (5.5) may or may not exist. It can be shown that if f is continuous, then this limit does exist; that is, the region has an area. [The precise meaning of the limit in Definition 5.5 is that for every $\varepsilon > 0$ there is a corresponding number $\delta > 0$ such that

$$|A - \sum_{i=1}^{n} f(x_i^*)\,\Delta x_i| < \varepsilon \qquad \text{whenever} \qquad \|P\| < \delta$$

In other words, the area can be approximated by a sum of areas of rectangles to within an arbitrary degree of accuracy (ε) by taking the norm of the partition sufficiently small.]

EXAMPLE 2

(a)   If the interval $[0,3]$ is divided into subintervals by the partition P and the set of partition points is $\{0, 0.6, 1.2, 1.6, 2, 2.5, 3\}$, find $\|P\|$.

(b)   If $f(x) = x^2 - 4x + 5$ and x_i^* is chosen to be the left endpoint of the ith subinterval, find the sum of the areas of the approximating rectangles.

(c)   Sketch the approximating rectangles.

Solution

(a)   We are given $x_0 = 0$, $x_1 = 0.6$, $x_2 = 1.2$, $x_3 = 1.6$, $x_4 = 2$, $x_5 = 2.5$, and $x_6 = 3$, so

$$\Delta x_1 = 0.6 - 0 = 0.6$$

$$\Delta x_2 = 1.2 - 0.6 = 0.6$$

$$\Delta x_3 = 1.6 - 1.2 = 0.4$$

$$\Delta x_4 = 2 - 1.6 = 0.4$$

$$\Delta x_5 = 2.5 - 2 = 0.5$$

$$\Delta x_6 = 3 - 2.5 = 0.5$$

[See Figure 5.9(a).] Therefore

$$\|P\| = \max\{0.6, 0.6, 0.4, 0.4, 0.5, 0.5\} = 0.6$$

(b)   Since $x_i^* = x_{i-1}$, the sum of the areas of the approximating rectangles is, by (5.4),

$$\sum_{i=1}^{6} f(x_i^*)\,\Delta x_i = \sum_{i=1}^{6} f(x_{i-1})\,\Delta x_i$$

$$= f(0)\,\Delta x_1 + f(0.6)\,\Delta x_2 + f(1.2)\,\Delta x_3 + f(1.6)\,\Delta x_4 + f(2)\,\Delta x_5 + f(2.5)\,\Delta x_6$$

$$= 5(0.6) + 2.96(0.6) + 1.64(0.4) + 1.16(0.4) + 1(0.5) + 1.25(0.5)$$

$$= 7.021$$

(c)   The graph of f and the approximating rectangles are sketched in Figure 5.9(b).

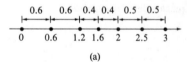
(a)

(b)

Figure 5.9

EXAMPLE 3   Find the area under the parabola $y = x^2 + 1$ from 0 to 2.

Solution   Since $f(x) = x^2 + 1$ is continuous, the limit (5.5) that defines the area must exist for all possible partitions P of the interval $[0,2]$ as long as $\|P\| \to 0$. To simplify things let us take the partition P that divides $[0,2]$ into n subintervals of equal length. (This is called a regular partition.) Then the partition points are

$$x_0 = 0, \quad x_1 = \tfrac{2}{n}, \quad x_2 = \tfrac{4}{n}, \quad \ldots, \quad x_i = \tfrac{2i}{n}, \quad \ldots, \quad x_n = \tfrac{2n}{n} = 2$$

and
$$\Delta x_1 = \Delta x_2 = \cdots = \Delta x_i = \cdots = \Delta x_n = \tfrac{2}{n}$$

so the norm of P is

$$\|P\| = \max\{\Delta x_i\} = \tfrac{2}{n}$$

The point x_i^* can be chosen to be anywhere in the ith subinterval. For the sake of definiteness, let us choose it to be the right-hand endpoint:

$$x_i^* = x_i = \tfrac{2i}{n}$$

Since $\|P\| = 2/n$, the condition $\|P\| \to 0$ is equivalent to $n \to \infty$. So the definition of area (5.5) becomes

$$A = \lim_{\|P\| \to 0} \sum_{i=1}^{n} f(x_i^*) \Delta x_i = \lim_{n \to \infty} \sum_{i=1}^{n} f\left(\tfrac{2i}{n}\right) \tfrac{2}{n}$$

$$= \lim_{n \to \infty} \sum_{i=1}^{n} \left[\left(\tfrac{2i}{n}\right)^2 + 1\right] \tfrac{2}{n} = \lim_{n \to \infty} \sum_{i=1}^{n} \left[\tfrac{8i^2}{n^3} + \tfrac{2}{n}\right]$$

$$= \lim_{n \to \infty} \left[\tfrac{8}{n^3} \sum_{i=1}^{n} i^2 + \tfrac{2}{n} \sum_{i=1}^{n} 1\right] \quad \text{(by Theorem 5.2)}$$

$$= \lim_{n \to \infty} \left[\tfrac{8}{n^3} \cdot \tfrac{n(n+1)(2n+1)}{6} + \tfrac{2}{n} \cdot n\right] \quad \text{(by Theorem 5.3)}$$

$$= \lim_{n \to \infty} \left[\tfrac{4}{3} \cdot 1 \cdot \left(1 + \tfrac{1}{n}\right)\left(2 + \tfrac{1}{n}\right) + 2\right]$$

$$= \tfrac{4}{3} \cdot 1 \cdot 1 \cdot 2 + 2 = \tfrac{14}{3}$$

The sum in this calculation is represented by the areas of the shaded rectangles in Figure 5.10(a). Notice that in this case, with our choice of x_i^* as the right-hand endpoint and since f is increasing, $f(x_i^*)$ is the maximum value of f on $[x_{i-1}, x_i]$ and the required area is approximated at the nth stage by the area of a circumscribed polygon.

We could just as well have chosen x_i^* to be the left-hand endpoint, that is, $x_i^* = x_{i-1} = 2(i-1)/n$. Then $f(x_i^*)$ is the minimum value of f on $[x_{i-1}, x_i]$ and A is approximated by the area of an inscribed polygon [see Figure 5.10(b)]. The calculation with this choice is as follows:

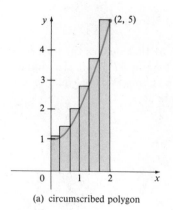

(a) circumscribed polygon

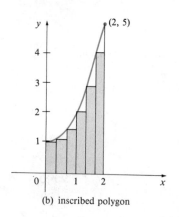

(b) inscribed polygon

Figure 5.10

$$A = \lim_{\|P\| \to 0} \sum_{i=1}^{n} f(x_i^*)\,\Delta x_i$$

$$= \lim_{n \to \infty} \sum_{i=1}^{n} f\!\left(\frac{2(i-1)}{n}\right)\frac{2}{n}$$

$$= \lim_{n \to \infty} \sum_{i=1}^{n} \left\{\left[\frac{2(i-1)}{n}\right]^2 + 1\right\}\frac{2}{n}$$

$$= \lim_{n \to \infty} \sum_{i=1}^{n} \left[\frac{8}{n^3}(i^2 - 2i + 1) + \frac{2}{n}\right]$$

$$= \lim_{n \to \infty} \left[\frac{8}{n^3}\sum_{i=1}^{n} i^2 - \frac{16}{n^3}\sum_{i=1}^{n} i + \frac{8}{n^3}\sum_{i=1}^{n} 1 + \frac{2}{n}\sum_{i=1}^{n} 1\right]$$

$$= \lim_{n \to \infty} \left[\frac{8}{n^3}\frac{n(n+1)(2n+1)}{6} - \frac{16}{n^3}\frac{n(n+1)}{2} + \frac{8}{n^3}n + \frac{2}{n}n\right]$$

$$= \lim_{n \to \infty} \left[\frac{4}{3}\cdot 1\cdot\left(1 + \frac{1}{n}\right)\left(2 + \frac{1}{n}\right) - \frac{8}{n}\left(1 + \frac{1}{n}\right) + \frac{8}{n^2} + 2\right]$$

$$= \frac{4}{3}\cdot 1\cdot 1\cdot 2 - 0\cdot 1 + 0 + 2 = \frac{14}{3}$$

Notice that we have obtained the same answer with the different choice of x_i^*. In fact, we would obtain the same answer if x_i^* was chosen to be the midpoint of $[x_{i-1}, x_i]$ (see Exercise 11) or indeed any other point of this interval. •

EXAMPLE 4   Find the area under the cosine curve from 0 to b, where $0 \le b \le \pi/2$.

Solution   As in the first part of Example 3, we choose a regular partition P so that

$$\|P\| = \Delta x_1 = \Delta x_2 = \cdots = \Delta x_n = \frac{b}{n}$$

and we choose x_i^* to be the right-hand endpoint of the ith subinterval:

$$x_i^* = x_i = \frac{ib}{n}$$

Since $\|P\| = b/n \to 0$ as $n \to \infty$, the area under the cosine curve from 0 to b is

(5.6)
$$A = \lim_{\|P\| \to 0} \sum_{i=1}^{n} f(x_i^*)\,\Delta x_i = \lim_{n \to \infty} \sum_{i=1}^{n} \cos\!\left(i\frac{b}{n}\right)\frac{b}{n}$$

$$= \lim_{n \to \infty} \frac{b}{n} \sum_{i=1}^{n} \cos\!\left(i\frac{b}{n}\right)$$

To evaluate this limit we use the formula of Exercise 56 in Section 5.1:

$$\sum_{i=1}^{n} \cos ix = \frac{\sin\frac{1}{2}nx \cos\frac{1}{2}(n+1)x}{\sin\frac{1}{2}x}$$

with $x = b/n$. Then Equation 5.6 becomes

(5.7)
$$A = \lim_{n \to \infty} \frac{b}{n} \frac{\sin \frac{1}{2}b \cos\left[\frac{(n+1)b}{2n}\right]}{\sin \frac{b}{2n}}$$

Now $\qquad \cos\left[\frac{(n+1)b}{2n}\right] = \cos\left(1 + \frac{1}{n}\right)\frac{b}{2} \to \cos\frac{b}{2} \qquad$ as $n \to \infty$

since cosine is continuous. Letting $t = b/n$ and using Theorem 2.22, we have

$$\lim_{n \to \infty} \frac{b}{n} \cdot \frac{1}{\sin \frac{b}{2n}} = \lim_{t \to 0^+} \frac{t}{\sin \frac{t}{2}} = \lim_{t \to 0^+} 2 \cdot \frac{\frac{t}{2}}{\sin \frac{t}{2}} = 2$$

Putting these limits in Equation 5.7, we obtain

$$A = 2 \sin \frac{b}{2} \cos \frac{b}{2} = \sin b$$

In particular, taking $b = \pi/2$, we have proved that the area under the cosine curve from 0 to $\pi/2$ is $\sin(\pi/2) = 1$ (see Figure 5.11). •

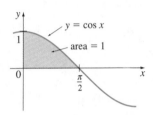

Figure 5.11

Note: The area calculations in Examples 3 and 4 are not easy. We will see in Section 5.5, however, that the Fundamental Theorem of Calculus gives a much easier method for computing these areas.

SECTION 5.2 **Exercises**

In Exercises 1–10 you are given a function f, an interval, partition points, and a description of the point x_i^* within the ith subinterval. (a) Find $\|P\|$. (b) Find the sum of the areas of the approximating rectangles (5.4). (c) Sketch the graph of f and the approximating rectangles.

1. $f(x) = 16 - x^2$, $[0, 4]$,
   $\{0, 1, 2, 3, 4\}$, $x_i^* = $ left endpoint

2. $f(x) = 16 - x^2$, $[0, 4]$,
   $\{0, 1, 2, 3, 4\}$, $x_i^* = $ right endpoint

3. $f(x) = 16 - x^2$, $[0, 4]$,
   $\{0, 1, 2, 3, 4\}$, $x_i^* = $ midpoint

4. $f(x) = 16 - x^2$, $[-4, 4]$,
   $\{-4, -3, -2, -1, 0, 1, 2, 3, 4\}$, $x_i^* = $ midpoint

5. $f(x) = 2x + 1$, $[0, 5]$,
   $\{0, 1, 2, 3, 4, 5\}$, $x_i^* = $ right endpoint

6. $f(x) = 2x + 1$, $[0, 4]$,
   $\{0, 0.5, 1, 2, 4\}$, $x_i^* = $ left endpoint

7. $f(x) = x^3 + 2$, $[-1, 2]$,
   $\{-1, -0.5, 0, 0.5, 1.0, 1.5, 2\}$, $x_i^* = $ right endpoint

8. $f(x) = 1/(x+1)$, $[0, 2]$, $\{0, 0.5, 1.0, 1.5, 2\}$,
   $x_1^* = 0.25$, $x_2^* = 1$, $x_3^* = 1.25$, $x_4^* = 2$

9. $f(x) = 2 \sin x$, $[0, \pi]$, $\{0, \pi/4, \pi/2, 3\pi/4, \pi\}$,
   $x_1^* = \pi/6$, $x_2^* = \pi/3$, $x_3^* = 2\pi/3$, $x_4^* = 5\pi/6$

10. $f(x) = 4 \cos x$, $[0, \pi/2]$,
    $\{0, \pi/6, \pi/4, \pi/3, \pi/2\}$, $x_i^* = $ left endpoint

11. Find the area of Example 3 taking x_i^* to be the midpoint of $[x_{i-1}, x_i]$. Illustrate the approximating rectangles with a sketch like Figure 5.10.

In Exercises 12–14 find the area under the given curve from a to b using subintervals of equal length and taking x_i^* in (5.5) to be the (a) left endpoint, (b) right endpoint, and (c) midpoint of the ith subinterval. In each case sketch the approximating rectangles.

12. $y = 3 - \frac{x}{2}$, $a = -2$, $b = 2$

13. $y = 16 - x^2$, $\quad a = -4$, $\quad b = 4$

14. $y = x^3$, $\quad a = 0$, $\quad b = 1$

In Exercises 15–23 use (5.5) to find the area under the given curve from a to b. Use equal subintervals and take x_i^* to be the right endpoint of the ith subinterval. Sketch the region.

15. $y = 5$, $\quad a = -2$, $\quad b = 2$

16. $y = 2x + 1$, $\quad a = 0$, $\quad b = 5$

17. $y = x^2 + 3x - 2$, $\quad a = 1$, $\quad b = 4$

18. $y = 2x^2 - 4x + 5$, $\quad a = -3$, $\quad b = 2$

19. $y = x^3 + 2x$, $\quad a = 0$, $\quad b = 2$

20. $y = x^3 + 2x^2 + x$, $\quad a = 0$, $\quad b = 1$

21. $y = 1 - 2x^2 + x^4$, $\quad a = -1$, $\quad b = 1$

22. $y = x^4 + 3x + 2$, $\quad a = 0$, $\quad b = 3$

23. Find the area under the curve $y = \sin x$ from 0 to π. [*Hint:* Use equal subintervals and right endpoints, and use Exercise 57 in Section 5.1.]

24. (a) Let A_n be the area of a polygon with n equal sides inscribed in a circle with radius r. By dividing the polygon into n congruent triangles with center angle $2\pi/n$, show that
$$A_n = \tfrac{1}{2} n r^2 \sin(2\pi/n)$$

(b) Show that $\lim_{n \to \infty} A_n = \pi r^2$. [*Hint:* Use Theorem 2.22.]

SECTION 5.3

The Definite Integral

We saw in the preceding section that a limit of the form

(5.8)
$$\lim_{\|P\| \to 0} \sum_{i=1}^{n} f(x_i^*) \, \Delta x_i$$

arises when we compute an area. It turns out that this same type of limit occurs in a wide variety of situations even when f is not necessarily a positive function. In Chapters 6 and 8, we will see that limits of the form (5.8) also arise in finding lengths of curves, volumes of solids, areas of surfaces, centers of mass, fluid pressure, and work, as well as other quantities. We therefore give this type of limit a special name and notation.

Definition of a Definite Integral (5.9)

If f is a function defined on a closed interval $[a, b]$, let P be a partition of $[a, b]$ with partition points $x_0, x_1, \ldots, x_n$, where

$$a = x_0 < x_1 < x_2 < \cdots < x_n = b$$

Choose points x_i^* in $[x_{i-1}, x_i]$ and let $\Delta x_i = x_i - x_{i-1}$ and $\|P\| = \max\{\Delta x_i\}$. Then the **definite integral of f from a to b** is

$$\int_a^b f(x) \, dx = \lim_{\|P\| \to 0} \sum_{i=1}^{n} f(x_i^*) \, \Delta x_i$$

if this limit exists. If the limit does exist, then f is called **integrable** on the interval $[a, b]$.

Note 1: The symbol $\int$ was introduced by Leibniz and is called an **integral sign.** It is an elongated S and was chosen because an integral is a limit of sums. In the notation

$\int_a^b f(x)\,dx$, $f(x)$ is called the **integrand** and a and b are called the **limits of integration;** a is the **lower limit** and b is the **upper limit**. The symbol dx has no meaning by itself; $\int_a^b f(x)\,dx$ is all one symbol. The procedure of calculating an integral is called **integration**.

Note 2: The definite integral $\int_a^b f(x)\,dx$ is a number; it does not depend on x. In fact, we could use any letter in place of x without changing the value of the integral:

$$\int_a^b f(x)\,dx = \int_a^b f(t)\,dt = \int_a^b f(r)\,dr$$

For this reason, x is called a **dummy variable** and the notation $\int_a^b f$ is sometimes used for the definite integral:

$$\int_a^b f = \int_a^b f(x)\,dx$$

Note 3: The sum

(5.10)

$$\sum_{i=1}^{n} f(x_i^*)\,\Delta x_i$$

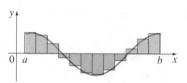

Figure 5.12

that occurs in Definition 5.9 is called a **Riemann sum** after the German mathematician Bernhard Riemann (1826–1866). The definite integral is sometimes called the **Riemann integral**. If f happens to be positive, then the Riemann sum can be interpreted as a sum of areas of approximating rectangles. [Compare (5.10) and (5.4).] If f takes on both positive and negative values, refer to Figure 5.12. The Riemann sum is the sum of the areas of the rectangles that lie above the x-axis and the *negatives* of the areas of the rectangles that lie below the x-axis. (The areas of the red rectangles *minus* the areas of the gray rectangles.)

EXAMPLE 1   Let $f(x) = 1 + 5x$ and consider the partition P of the interval $[-2, 1]$ by means of the set of partition points $\{-2, -1.5, -1, -0.3, 0.2, 1\}$. In this example, $a = -2$, $b = 1$, $n = 5$, and $x_0 = -2$, $x_1 = -1.5$, $x_2 = -1$, $x_3 = -0.3$, $x_4 = 0.2$, and $x_5 = 1$. The lengths of the subintervals are

$$\Delta x_1 = -1.5 - (-2) = 0.5 \qquad\qquad \Delta x_2 = -1 - (-1.5) = 0.5$$

$$\Delta x_3 = -0.3 - (-1) = 0.7 \qquad\qquad \Delta x_4 = 0.2 - (-0.3) = 0.5$$

$$\Delta x_5 = 1 - 0.2 = 0.8$$

Thus the norm of the partition P is

$$\|P\| = \max\{0.5,\,0.5,\,0.7,\,0.5,\,0.8\} = 0.8$$

Suppose we choose $x_1^* = -1.8$, $x_2^* = -1.2$, $x_3^* = -0.3$, $x_4^* = 0$, and $x_5^* = 0.7$. Then the corresponding Riemann sum is

$$\sum_{i=1}^{5} f(x_i^*)\,\Delta x_i = f(-1.8)\,\Delta x_1 + f(-1.2)\,\Delta x_2 + f(-0.3)\,\Delta x_3 + f(0)\,\Delta x_4 + f(0.7)\,\Delta x_5$$

$$= (-8)(0.5) + (-5)(0.5) + (-0.5)(0.7) + 1(0.5) + (4.5)(0.8)$$

$$= -2.75$$

Notice that, in this example, f is not a positive function and so the Riemann sum does not represent a sum of areas of rectangles.

Note 4: An integral need not represent an area. But for *positive* functions, an integral can be interpreted as an area. In fact, comparing Equation 5.5 and Definition 5.9, we see following:

For the special case where $f(x) \geq 0$,

$$\int_a^b f(x)\, dx = \text{the area under the graph of } f \text{ from } a \text{ to } b$$

In general a definite integral can be interpreted as a difference of areas:

$$\int_a^b f(x)\, dx = A_1 - A_2$$

where A_1 is the area of the region above the x-axis and below the graph of f and A_2 is the area of the region below the x-axis and above the graph of f. (This seems reasonable from a comparison of Figures 5.12 and 5.13, but we will be able to see it more clearly in Section 6.1.)

Note 5: The precise meaning of the limit that defines the integral in Definition 5.9 is as follows.

$\int_a^b f(x)\, dx = I$ means that for every $\varepsilon > 0$ there is a corresponding number $\delta > 0$ such that

$$\left| I - \sum_{i=1}^{n} f(x_i^*)\, \Delta x_i \right| < \varepsilon$$

for all partitions P of $[a,b]$ with $\|P\| < \delta$ and for all possible choices of x_i^* in $[x_{i-1}, x_i]$.

Figure 5.13

This means that a definite integral can be approximated to within any desired degree of accuracy by a Riemann sum.

Note 6: In Definition 5.9 we are dealing with a function f defined on an interval $[a,b]$, so we are implicitly assuming that $a < b$. But for some purposes it is useful to extend the definition of $\int_a^b f(x)\, dx$ to the case where $a > b$ or $a = b$ as follows:

If $a > b$, then $\displaystyle\int_a^b f(x)\, dx = -\int_b^a f(x)\, dx.$

If $a = b$, then $\displaystyle\int_a^a f(x)\, dx = 0.$

EXAMPLE 2   Express

$$\lim_{\|P\| \to 0} \sum_{i=1}^{n} \left[(x_i^*)^3 + x_i^* \sin x_i^* \right] \Delta x_i$$

as an integral on the interval $[0, \pi]$.

Solution   Comparing the given limit with the limit in Definition 5.9, we see that they will be identical if we choose

$$f(x) = x^3 + x \sin x$$

We are given that $a = 0$ and $b = \pi$. Therefore, by Definition 5.9, we have

$$\lim_{\|P\| \to 0} \sum_{i=1}^{n} \left[(x_i^*)^3 + x_i^* \sin x_i^* \right] \Delta x_i = \int_0^{\pi} (x^3 + x \sin x) \, dx$$

The question arises: Which functions are integrable? A partial answer is given by the following theorem, which is proved in courses on advanced calculus.

Theorem (5.11)

> If f is either continuous or monotonic on $[a, b]$, then f is integrable on $[a, b]$; that is, the definite integral $\int_a^b f(x) \, dx$ exists.

If f is discontinuous at some points in $[a, b]$, then $\int_a^b f(x) \, dx$ might exist or it might not exist (see Exercises 36 and 37). If f has only a finite number of discontinuities and these are all jump discontinuities, then f is called **piecewise continuous** and it turns out that f is integrable. (See Figure 5.14.)

It can be shown that if f is integrable on $[a, b]$, then f must be a **bounded function** on $[a, b]$; that is, there exists a number M such that $|f(x)| \le M$ for all x in $[a, b]$. Geometrically, this means that the graph of f lies between the horizontal lines $y = M$ and $y = -M$. In particular, if f has an infinite discontinuity at some point in $[a, b]$, then f is not bounded and is therefore not integrable. (See Exercise 36 and Figure 5.15.)

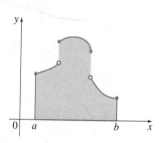

Figure 5.14

Discontinuous integrable function

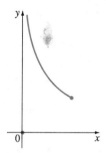

Figure 5.15

Nonintegrable function

If f is integrable on $[a, b]$, then the Riemann sums (5.10) must approach $\int_a^b f$ as $\|P\| \to 0$ no matter how the partitions P are chosen and no matter how the points x_i^* are chosen in $[x_{i-1}, x_i]$. Therefore if it is known beforehand that f is integrable on $[a, b]$ (for instance, if it is known that f is continuous or monotonic), then in calculating the value of an integral we are free to choose partitions P and points x_i^* in any way we like as long as $\|P\| \to 0$. For purposes of calculation, it is often convenient to take P to be a **regular partition**; that is, all the subintervals have the same length Δx. Then

$$\Delta x = \Delta x_1 = \Delta x_2 = \cdots = \Delta x_n = \frac{b - a}{n}$$

and

$$x_0 = a, \; x_1 = a + \Delta x, \; x_2 = a + 2 \Delta x, \; \ldots, \; x_i = a + i \Delta x$$

If we choose x_i^* to be the right endpoint of the ith subinterval, then

$$x_i^* = x_i = a + i \Delta x = a + i \frac{b - a}{n}$$

Since $\|P\| = \Delta x = (b-a)/n$, we have $\|P\| \to 0$ as $n \to \infty$, so Definition 5.9 gives

$$\int_a^b f(x)\,dx = \lim_{\|P\|\to 0} \sum_{i=1}^n f(x_i^*)\,\Delta x$$

$$= \lim_{n\to\infty} \sum_{i=1}^n f\!\left(a + i\frac{b-a}{n}\right)\frac{b-a}{n}$$

Since $(b-a)/n$ does not depend on i, Theorem 5.2 allows us to take it in front of the sigma sign, and we have the following formula for calculating integrals.

Theorem (5.12)

If f is integrable on $[a,b]$, then

$$\int_a^b f(x)\,dx = \lim_{n\to\infty} \frac{b-a}{n} \sum_{i=1}^n f\!\left(a + i\frac{b-a}{n}\right)$$

EXAMPLE 3   Evaluate $\displaystyle\int_1^4 (2x^3 - 5x)\,dx$.

Solution   Here we have $f(x) = 2x^3 - 5x$, $a = 1$, and $b = 4$. Since f is continuous, we know it is integrable and so Theorem 5.12 gives

$$\int_1^4 (2x^3 - 5x)\,dx$$

$$= \lim_{n\to\infty} \frac{3}{n} \sum_{i=1}^n f\!\left(1 + \frac{3i}{n}\right)$$

$$= \lim_{n\to\infty} \frac{3}{n} \sum_{i=1}^n \left[2\left(1 + \frac{3i}{n}\right)^3 - 5\left(1 + \frac{3i}{n}\right)\right]$$

$$= \lim_{n\to\infty} \frac{3}{n} \sum_{i=1}^n \left[-3 + 3\frac{i}{n} + 54\frac{i^2}{n^2} + 54\frac{i^3}{n^3}\right]$$

$$= \lim_{n\to\infty} \left[\frac{3}{n} \sum_{i=1}^n (-3) + \frac{9}{n^2} \sum_{i=1}^n i + \frac{162}{n^3} \sum_{i=1}^n i^2 + \frac{162}{n^4} \sum_{i=1}^n i^3\right]$$

$$= \lim_{n\to\infty} \left\{\frac{3}{n}(-3)n + \frac{9}{n^2}\frac{n(n+1)}{2} + \frac{162}{n^3}\frac{n(n+1)(2n+1)}{6} + \frac{162}{n^4}\left[\frac{n(n+1)}{2}\right]^2\right\}$$

$$= \lim_{n\to\infty} \left\{-9 + \frac{9}{2}\cdot 1 \cdot \left(1 + \frac{1}{n}\right) + 27\cdot 1\cdot\left(1 + \frac{1}{n}\right)\left(2 + \frac{1}{n}\right) + \frac{81}{2}\left[1\left(1 + \frac{1}{n}\right)^2\right]\right\}$$

$$= -9 + \frac{9}{2} + 27\cdot 2 + \frac{81}{2} = 90$$

Note that this integral cannot be interpreted as an area since f takes on both positive and negative values.   •

A much simpler method for evaluating the integral in Example 3 will be given in Section 5.5 after we have proved the Fundamental Theorem of Calculus.

The Midpoint Rule

We often choose x_i^* to be the right endpoint of the ith subinterval because it is convenient for computing the limit. But if the purpose is to find an *approximation* to an integral, it is usually better to choose x_i^* to be the midpoint of the interval, which we denote by $\bar{x}_i$. Any Riemann sum is an approximation to an integral, but if we use midpoints and a regular partition we get the following approximation:

Midpoint Rule

$$\int_a^b f(x)\,dx \approx \sum_{i=1}^n f(\bar{x}_i)\,\Delta x = \Delta x\big[f(\bar{x}_1) + f(\bar{x}_2) + \cdots + f(\bar{x}_n)\big]$$

where

$$\Delta x = \frac{b-a}{n}$$

and

$$\bar{x}_i = \tfrac{1}{2}(x_{i-1} + x_i) = \text{midpoint of } [x_{i-1}, x_i]$$

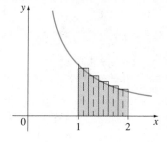

EXAMPLE 4   Use the Midpoint Rule with $n = 5$ to approximate $\displaystyle\int_1^2 \frac{1}{x}\,dx$.

Solution   The partition points are 1, 1.2, 1.4, 1.6, 1.8, and 2.0, so the midpoints of the five intervals are 1.1, 1.3, 1.5, 1.7, and 1.9. The width of the intervals is $\Delta x = (2-1)/5 = \frac{1}{5}$, so the Midpoint Rule gives

$$\int_1^2 \frac{1}{x}\,dx \approx \Delta x[f(1.1) + f(1.3) + f(1.5) + f(1.7) + f(1.9)]$$

$$= \tfrac{1}{5}\left[\frac{1}{1.1} + \frac{1}{1.3} + \frac{1}{1.5} + \frac{1}{1.7} + \frac{1}{1.9}\right]$$

$$\approx 0.691908$$

Since $f(x) = 1/x > 0$ for $1 \le x \le 2$, the integral represents an area and the approximation given by the Midpoint Rule is the sum of the areas of the rectangles shown in Figure 5.16.   ●

At the moment we don't know how accurate the approximation in Example 4 is, but in Section 7.8 we will learn a method for estimating the error involved in using the Midpoint Rule. At that time we will discuss other methods for approximating definite integrals.

SECTION 5.3   **Exercises**

In Exercises 1–6 you are given a function f, an interval, partition points that define a partition P, and points x_i^* in the ith subinterval. (a) Find $\|P\|$. (b) Find the Riemann sum (5.10).

1. $f(x) = 7 - 2x$, $[1,5]$, $\{1, 1.6, 2.2, 3.0, 4.2, 5\}$,
     $x_i^* = \text{midpoint}$

2. $f(x) = 3x - 1$, $[-2, 2]$,
     $\{-2, -1.2, -0.6, 0, 0.8, 1.6, 2\}$, $x_i^* = \text{midpoint}$

3. $f(x) = 2 - x^2$, $[-2, 2]$,
     $\{-2, -1.4, -1, 0, 0.8, 1.4, 2\}$, $x_i^* = \text{right endpoint}$

4. $f(x) = x + x^2$, $[-2, 0]$,
     $\{-2, -1.5, -1, -0.7, -0.4, 0\}$, $x_i^* = \text{left endpoint}$

5. $f(x) = x^3$, $[-1, 1]$, $\{-1, -0.5, 0, 0.5, 1\}$,
     $x_1^* = -1$, $x_2^* = -0.4$, $x_3^* = 0.2$, $x_4^* = 1$

6. $f(x) = \sin x$, $[-\pi/2, \pi]$, $\{-\pi/2, -1, 0, 1, 2, \pi\}$,
     $x_1^* = -1.5$, $x_2^* = -0.5$, $x_3^* = 0.5$, $x_4^* = 1.5$, $x_5^* = 3$

In Exercises 7–10 use the Midpoint Rule with the given value of n to approximate each integral. Round your answers to four decimal places.

7. $\int_0^5 x^3\,dx,\ n=5$ 8. $\int_1^3 \dfrac{1}{2x-7}\,dx,\ n=4$

9. $\int_1^2 \sqrt{1+x^2}\,dx,\ n=10$ 10. $\int_0^{\pi/4} \tan x\,dx,\ n=4$

In Exercises 11–22 use Theorem 5.12 to evaluate the integral.

11. $\int_a^b c\,dx$ *Just set it up*   12. $\int_{-2}^7 (6-2x)\,dx$

13. $\int_1^4 (x^2-2)\,dx$ 14. $\int_0^1 (ax+b)\,dx$

15. $\int_{-3}^0 (2x^2-3x-4)\,dx$ 16. $\int_1^5 (2+3x-x^2)\,dx$

17. $\int_{-1}^1 (t^3-t^2+1)\,dt$ 18. $\int_a^b (Px^2+Qx+R)\,dx$

19. $\int_0^b (x^3+4x)\,dx$ 20. $\int_2^5 (t^3-2t+3)\,dt$

21. $\int_0^2 (x^4-x+1)\,dx$ 22. $\int_0^1 (x^3-5x^4)\,dx$

23. Prove that $\int_a^b x\,dx = \dfrac{b^2-a^2}{2}$.

24. Prove that $\int_a^b x^2\,dx = \dfrac{b^3-a^3}{3}$.

In Exercises 25–28 express the given limit as a definite integral on the given interval.

25. $\lim\limits_{\|P\|\to 0}\sum\limits_{i=1}^n [2(x_i^*)^2-5x_i^*]\,\Delta x_i,\quad [0,1]$

26. $\lim\limits_{\|P\|\to 0}\sum\limits_{i=1}^n \sqrt{x_i^*}\,\Delta x_i,\quad [1,4]$

27. $\lim\limits_{\|P\|\to 0}\sum\limits_{i=1}^n \cos x_i^*\,\Delta x_i,\quad [0,\pi]$

28. $\lim\limits_{\|P\|\to 0}\sum\limits_{i=1}^n \dfrac{\tan x_i}{x_i}\,\Delta x_i,\quad [2,4]$

In Exercises 29–31 express the given limit as a definite integral.

29. $\lim\limits_{n\to\infty}\sum\limits_{i=1}^n \dfrac{i^4}{n^5}$ [Hint: Consider $f(x)=x^4$.]

30. $\lim\limits_{n\to\infty}\dfrac{1}{n}\sum\limits_{i=1}^n \dfrac{1}{1+(i/n)^2}$

31. $\lim\limits_{n\to\infty}\sum\limits_{i=1}^n \left[3\left(1+\dfrac{2i}{n}\right)^5-6\right]\dfrac{2}{n}$

32. Evaluate $\int_1^1 x^2\cos x\,dx$.

33. Given that $\int_4^9 \sqrt{x}\,dx = \frac{38}{3}$, what is $\int_9^4 \sqrt{t}\,dt$?

34. (a) Find an approximation to the integral $\int_0^4 (x^2-3x)\,dx$ using a Riemann sum with right endpoints and $n=8$.

(b) Draw a diagram like Figure 5.12 to illustrate the approximation in part (a).

(c) Evaluate $\int_0^4 (x^2-3x)\,dx$.

(d) Interpret the integral in part (c) as a difference of areas and illustrate with a diagram like Figure 5.13.

35. Which of the following functions are integrable on the interval $[0;2]$?

(a) $f(x)=x^2\sin x$ (b) $f(x)=\sec x$

(c) $f(x)=\begin{cases} x+1 & \text{if } 0\le x<1 \\ 2-x & \text{if } 1\le x\le 2 \end{cases}$

(d) $f(x)=\begin{cases} (x-1)^{-2} & \text{if } x\ne 1 \\ 1 & \text{if } x=1 \end{cases}$

36. Let
$$f(x)=\begin{cases} \dfrac{1}{x} & \text{if } 0<x\le 1 \\ 0 & \text{if } x=0 \end{cases}$$

(a) Show that f is not continuous on $[0,1]$.

(b) Show that f is unbounded on $[0,1]$.

(c) Show that $\int_0^1 f(x)\,dx$ does not exist, that is, f is not integrable on $[0,1]$. [Hint: Show that the first term in the Riemann sum, $f(x_1^*)\Delta x_1$, can be made arbitrarily large.]

37. Let
$$f(x)=\begin{cases} 0 & \text{if } x \text{ is rational} \\ 1 & \text{if } x \text{ is irrational} \end{cases}$$

Show that f is bounded but not integrable on $[a,b]$. (Hint: Show that, no matter how small $\|P\|$ is, some Riemann sums are 0 whereas others are equal to $b-a$.)

38. Evaluate $\int_1^2 x^3\,dx$ using a partition of $[1,2]$ by points of a geometric progression: $x_0=1,\ x_1=2^{1/n}$, $x_2=2^{2/n},\ \dots,\ x_i=2^{i/n},\ \dots,\ x_n=2^{n/n}=2$. Take $x_i^*=x_i$ and use the formula in Exercise 51 in Section 5.1 for the sum of a geometric series.

39. Find $\int_1^2 x^{-2}\,dx$. (Hint: Use a regular partition but choose x_i^* to be the geometric mean of x_{i-1} and x_i ($x_i^*=\sqrt{x_{i-1}x_i}$) and use the identity
$$\frac{1}{m(m+1)}=\frac{1}{m}-\frac{1}{m+1}$$

40. Evaluate $\int_0^1 e^x\,dx$. (Hint: Use Theorem 5.12, sum a geometric series, and then use l'Hospital's Rule.)

Properties of the Definite Integral

In this section we develop some basic properties of integrals that will help us to evaluate integrals in a simple manner.

Properties of the Integral (5.13)

Suppose that all of the following integrals exists. Then

1. $\displaystyle\int_a^b c\,dx = c(b-a)$, where c is any constant

2. $\displaystyle\int_a^b [f(x)+g(x)]\,dx = \int_a^b f(x)\,dx + \int_a^b g(x)\,dx$

3. $\displaystyle\int_a^b c f(x)\,dx = c\int_a^b f(x)\,dx$, where c is any constant

4. $\displaystyle\int_a^b [f(x)-g(x)]\,dx = \int_a^b f(x)\,dx - \int_a^b g(x)\,dx$

5. $\displaystyle\int_a^b f(x)\,dx = \int_a^c f(x)\,dx + \int_c^b f(x)\,dx$

Property 1 was given as Exercise 11 in Section 5.3. It says that the integral of a constant function $f(x) = c$ is the constant times the length of the interval. If $c > 0$ and $a < b$, this is to be expected because $c(b-a)$ is the area of the shaded rectangle in Figure 5.17.

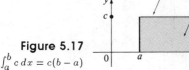

Figure 5.17
$\int_a^b c\,dx = c(b-a)$

Proof of Property 2 Since $\int_a^b (f+g)$ exists, we can compute it using a regular partition and choosing x_i^* to be the right endpoint of the ith subinterval, that is, $x_i^* = x_i$. Using the fact that the limit of a sum is the sum of the limits, we have

$$\int_a^b [f(x)+g(x)]\,dx = \lim_{n\to\infty} \sum_{i=1}^{n} [f(x_i)+g(x_i)]\,\Delta x$$

$$= \lim_{n\to\infty}\left[\sum_{i=1}^{n} f(x_i)\,\Delta x + \sum_{i=1}^{n} g(x_i)\,\Delta x\right] \qquad \text{(by Theorem 5.2)}$$

$$= \lim_{n\to\infty}\sum_{i=1}^{n} f(x_i)\,\Delta x + \lim_{n\to\infty}\sum_{i=1}^{n} g(x_i)\,\Delta x$$

$$= \int_a^b f(x)\,dx + \int_a^b g(x)\,dx$$

Property 2 says that the integral of a sum is the sum of the integrals. Property 3 can be proved in a similar manner (see Exercise 50) and says that the integral of a constant times a function is the constant times the integral of the function. In other words, a

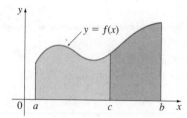

Figure 5.18

constant (but *only* a constant) can be taken in front of an integral sign. Property 4 is proved by writing $f - g = f + (-g)$ and using Properties 2 and 3 with $c = -1$.

Property 5 is somewhat more complicated and is proved in Appendix C, but for the case where $f(x) \geq 0$ and $a < c < b$, it can be seen from the geometric interpretation in Figure 5.18. For positive functions f, $\int_a^b f$ is the total area under $y = f(x)$ from a to b, which is the sum of $\int_a^c f$ (the area from a to c) and $\int_c^b f$ (the area from c to b).

EXAMPLE 1   Use Theorem 5.13 together with the results of Section 5.2 and 5.3 to evaluate the following integrals:

(a) $\displaystyle\int_0^{\pi/2} (x + 3\cos x)\, dx$ (b) $\displaystyle\int_a^b \cos x\, dx$ (c) $\displaystyle\int_{-4}^5 |x|\, dx$

Solution 1

(a)   We know that $\int_0^{\pi/2} x\, dx = \pi^2/8$ (see Exercise 23 in Section 5.3) and also that $\int_0^{\pi/2} \cos x\, dx = 1$ (see Example 4 in Section 5.2). Therefore, using Properties 2 and 3 of integrals, we get

$$\int_0^{\pi/2} (x + 3\cos x)\, dx = \int_0^{\pi/2} x\, dx + 3\int_0^{\pi/2} \cos x\, dx = \frac{\pi^2}{8} + 3$$

(b)   The result of Example 4 in Section 5.2 can be extended to show that $\int_0^b \cos x\, dx = \sin b$. By Property 5 of integrals we have

$$\int_0^b \cos x\, dx = \int_0^a \cos x\, dx + \int_a^b \cos x\, dx$$

and so

$$\int_a^b \cos x\, dx = \int_0^b \cos x\, dx - \int_0^a \cos x\, dx = \sin b - \sin a$$

(c)   Since

$$|x| = \begin{cases} x & \text{if } x \geq 0 \\ -x & \text{if } x < 0 \end{cases}$$

we use Property 5 to split the integral at 0:

$$\int_{-4}^5 |x|\, dx = \int_{-4}^0 |x|\, dx + \int_0^5 |x|\, dx$$

$$= \int_{-4}^0 (-x)\, dx + \int_0^5 x\, dx = -\int_{-4}^0 x\, dx + \int_0^5 x\, dx$$

We know from Exercise 23 in Section 5.3 that

$$\int_a^b x\, dx = \frac{b^2 - a^2}{2}$$

so

$$\int_{-4}^5 |x|\, dx = -\int_{-4}^0 x\, dx + \int_0^5 x\, dx$$

$$= -\tfrac{1}{2}[0^2 - (-4)^2] + \tfrac{1}{2}[5^2 - 0^2]$$

$$= 20.5$$

Notice that Properties 1–5 are true whether $a < b$, $a = b$, or $a > b$. The following properties, however, are true only if $a \leq b$.

Order Properties of the Integral (5.14)

Suppose the following integrals exist and $a \leq b$.

6. If $f(x) \geq 0$ for $a \leq x \leq b$, then $\displaystyle\int_a^b f(x)\,dx \geq 0$.

7. If $f(x) \geq g(x)$ for $a \leq x \leq b$, then $\displaystyle\int_a^b f(x)\,dx \geq \int_a^b g(x)\,dx$.

8. If $m \leq f(x) \leq M$ for $a \leq x \leq b$, then

$$m(b-a) \leq \int_a^b f(x)\,dx \leq M(b-a)$$

9. $\left| \displaystyle\int_a^b f(x)\,dx \right| \leq \displaystyle\int_a^b |f(x)|\,dx$

If $f(x) \geq 0$, then $\int_a^b f$ represents the area under the graph of f, so the geometric interpretation of Property 6 is simply that areas are positive. Property 7 says that a bigger function has a bigger integral.

Proof of Property 6   As in the proof of Property 2, we write

$$\int_a^b f(x)\,dx = \lim_{n\to\infty} \sum_{i=1}^n f(x_i)\,\Delta x$$

Now $f(x_i) \geq 0$ and $\Delta x \geq 0$ so $f(x_i)\,\Delta x \geq 0$ and therefore

$$\sum_{i=1}^n f(x_i)\,\Delta x \geq 0$$

But the limit of nonnegative quantities is nonnegative (from Theorem 1.9) and so $\int_a^b f(x)\,dx \geq 0$.   ●

Proof of Property 7   If $f(x) \geq g(x)$, then $f(x) - g(x) \geq 0$ and so $\int_a^b (f-g) \geq 0$ by Property 6. Then Property 4 gives

$$\int_a^b f - \int_a^b g = \int_a^b (f-g) \geq 0$$

and so $\int_a^b f \geq \int_a^b g$.   ●

Property 8 is illustrated by Figure 5.19 for the case where $f(x) \geq 0$. If f is continuous we could take m and M to be the absolute minimum and maximum values of f on the interval $[a,b]$. In this case Property 8 says that the area under the graph of f is greater than the area of the rectangle with height m and less than the area of the rectangle with height M.

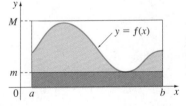

Figure 5.19

Proof of Property 8   Since $m \leq f(x) \leq M$, Property 7 gives

$$\int_a^b m\,dx \leq \int_a^b f(x)\,dx \leq \int_a^b M\,dx$$

Using Property 1 to evaluate the integrals on the left- and right-hand sides, we obtain

$$m(b-a) \le \int_a^b f(x)\,dx \le M(b-a)$$

The proof of Property 9 is left as Exercise 51.

EXAMPLE 2   Show that $\int_1^3 (2x^4 + 1)\,dx \ge \int_1^3 (x^4 + 2)\,dx$.

Solution   In order to use Property 7 we must show that

(5.15)
$$x^4 + 2 \le 2x^4 + 1 \qquad \text{for } 1 \le x \le 3$$

To do this, observe that

$$(2x^4 + 1) - (x^4 + 2) = x^4 - 1 \ge 0 \qquad \text{when } x \ge 1$$

This shows that (5.15) is true, so Property 7 gives

$$\int_1^3 (x^4 + 2)\,dx \le \int_1^3 (2x^4 + 1)\,dx$$

EXAMPLE 3   Use Property 8 to estimate the value of $\int_1^4 \sqrt{x}\,dx$.

Solution   Since $f(x) = \sqrt{x}$ is an increasing function, its absolute minimum on $[1,4]$ is $m = f(1) = 1$ and its absolute maximum on $[1,4]$ is $M = f(4) = \sqrt{4} = 2$. Thus Property 8 gives

$$1(4-1) \le \int_1^4 \sqrt{x}\,dx \le 2(4-1)$$

or

$$3 \le \int_1^4 \sqrt{x}\,dx \le 6$$

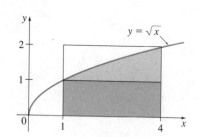

Figure 5.20

The result of Example 3 is illustrated in Figure 5.20. The area under $y = \sqrt{x}$ from 1 to 4 is greater than the area of the lower rectangle and less than the area of the large rectangle.

EXAMPLE 4   Show that $\int_1^4 \sqrt{1 + x^2}\,dx \ge 7.5$.

Solution   The minimum value of $f(x) = \sqrt{1 + x^2}$ on $[0,4]$ is $m = f(0) = 1$ since f is increasing. Thus Property 8 gives

$$\int_1^4 \sqrt{1 + x^2}\,dx \ge 1(4-1) = 3$$

This is not good enough, so instead we use Property 7. Notice that

$$1 + x^2 > x^2 \qquad \Rightarrow \qquad \sqrt{1 + x^2} > \sqrt{x^2} = |x|$$

Since $|x| = x$ for $x > 0$, we have $\sqrt{1 + x^2} > x$ for $1 \le x \le 4$. Thus, by Property 7,

$$\int_1^4 \sqrt{1 + x^2}\,dx \ge \int_1^4 x\,dx = \tfrac{1}{2}(4^2 - 1^2) = 7.5$$

[Here we have used the fact that $\int_a^b x\,dx = (b^2 - a^2)/2$ from Exercise 23 in Section 5.3.]

In Exercises 1–18 use the properties of integrals to evaluate the given integrals. You may assume from Section 5.3 that

$$\int_a^b x \, dx = \tfrac{1}{2}(b^2 - a^2) \qquad \int_a^b x^2 \, dx = \tfrac{1}{3}(b^3 - a^3)$$

$$\int_0^b \cos x \, dx = \sin b$$

1. $\displaystyle\int_2^6 3 \, dx$

2. $\displaystyle\int_{-1}^4 \pi \, dx$

3. $\displaystyle\int_{-4}^{-1} \sqrt{3} \, dx$

4. $\displaystyle\int_{-1}^{-\sqrt{2}} (\sqrt{2} - 1) \, dx$

5. $\displaystyle\int_0^2 (5x + 3) \, dx$

6. $\displaystyle\int_3^6 (4 - 7x) \, dx$

7. $\displaystyle\int_1^4 (2x^2 - 3x + 1) \, dx$

8. $\displaystyle\int_1^3 (x - 2)(x + 3) \, dx$

9. $\displaystyle\int_0^{\pi/3} (1 - 2\cos x) \, dx$

10. $\displaystyle\int_0^1 (5\cos x + 4x) \, dx$

11. $\displaystyle\int_{-2}^2 |x + 1| \, dx$

12. $\displaystyle\int_0^2 |2x - 3| \, dx$

13. $\displaystyle\int_{-2}^5 [\![x]\!] \, dx$

14. $\displaystyle\int_0^2 [\![2x]\!] \, dx$

15. $\displaystyle\int_{-1}^1 f(x) \, dx$, where $f(x) = \begin{cases} -2x & \text{if } -1 \le x < 0 \\ 3x^2 & \text{if } 0 \le x \le 1 \end{cases}$

16. $\displaystyle\int_0^3 g(x) \, dx$, where $g(x) = \begin{cases} 1 - x & \text{if } 0 \le x \le 1 \\ x - 1 & \text{if } 1 < x \le 3 \end{cases}$

17. $\displaystyle\int_0^4 x^2 \, dx + \int_4^{10} x^2 \, dx$

18. $\displaystyle\int_3^4 f(x) \, dx + \int_1^3 f(x) \, dx + \int_4^1 f(x) \, dx$

In Exercises 19–22 write the given sum or difference as a single integral in the form $\int_a^b f(x) \, dx$.

19. $\displaystyle\int_1^3 f(x) \, dx + \int_3^6 f(x) \, dx + \int_6^{12} f(x) \, dx$

20. $\displaystyle\int_5^8 f(x) \, dx + \int_0^5 f(x) \, dx$

21. $\displaystyle\int_2^{10} f(x) \, dx - \int_2^7 f(x) \, dx$

22. $\displaystyle\int_{-3}^5 f(x) \, dx - \int_{-3}^0 f(x) \, dx + \int_5^6 f(x) \, dx$

In Exercises 23–34 use the properties of integrals to verify the given inequality without evaluating the integrals.

23. $\displaystyle\int_0^1 x \, dx \ge \int_0^1 x^2 \, dx$

24. $\displaystyle\int_1^2 x \, dx \le \int_1^2 x^2 \, dx$

25. $\displaystyle\int_2^6 (x^2 - 1) \, dx \ge 0$

26. $\displaystyle\int_{-2}^8 (x^2 - 3x + 4) \, dx \ge 0$

27. $\displaystyle\int_0^{\pi/4} \sin^3 x \, dx \le \int_0^{\pi/4} \sin^2 x \, dx$

28. $\displaystyle\int_0^5 (4x^4 - 3) \, dx \ge \int_0^5 (3x^4 - 4) \, dx$

29. $\displaystyle\int_1^2 \sqrt{5 - x} \, dx \ge \int_1^2 \sqrt{x + 1} \, dx$

30. $\displaystyle\int_4^6 \tfrac{1}{x} \, dx \le \int_4^6 \frac{1}{8 - x} \, dx$

31. $\displaystyle 8 \le \int_2^4 x^2 \, dx \le 32$

32. $\displaystyle \frac{\pi}{6} \le \int_{\pi/6}^{\pi/2} \sin x \, dx \le \frac{\pi}{3}$

33. $\displaystyle 2 \le \int_{-1}^1 \sqrt{1 + x^2} \, dx \le 2\sqrt{2}$

34. $\displaystyle 3 \le \int_1^4 (x^2 - 4x + 5) \, dx \le 15$

In Exercises 35–42 use Property 8 to estimate the value of the integral.

35. $\displaystyle\int_1^3 x^3 \, dx$

36. $\displaystyle\int_0^2 \sqrt{x^3 + 1} \, dx$

37. $\displaystyle\int_1^2 \tfrac{1}{x} \, dx$

38. $\displaystyle\int_{\pi/4}^{\pi/3} \cos x \, dx$

39. $\displaystyle\int_{-3}^0 (x^2 + 2x) \, dx$

40. $\displaystyle\int_{-1}^3 (x^2 - 3x) \, dx$

41. $\displaystyle\int_{-1}^1 \sqrt{1 + x^4} \, dx$

42. $\displaystyle\int_{\pi/4}^{3\pi/4} \sin^2 x \, dx$

In Exercises 43–46 use (5.14), together with the list of integrals in the instructions for Exercises 1–18 to prove the given inequalities.

43. $\displaystyle\int_1^3 \sqrt{x^4 + 1} \, dx \ge \frac{26}{3}$

44. $\displaystyle\int_2^5 \sqrt{x^2 - 1} \, dx \le 10.5$

45. $\displaystyle\int_0^{\pi/2} x \sin x \, dx \le \frac{\pi^2}{8}$

46. $\displaystyle\left| \int_0^\pi x^2 \cos x \, dx \right| \le \frac{\pi^3}{3}$

47. Use Properties 2 and 3 of integrals to show that if c and d are constants, then

$$\int_a^b [cf(x) + dg(x)]\, dx = c \int_a^b f(x)\, dx + d \int_a^b g(x)\, dx$$

48. Extend Property 2 to the sum of any number of functions; that is, given integrable functions $f_1, \ldots, f_n$, prove that

$$\int_a^b \left[\sum_{i=1}^n f_i(x) \right] dx = \sum_{i=1}^n \int_a^b f_i(x)\, dx$$

49. If $f(x) \le 0$ and f is integrable on $[a, b]$, show that $\int_a^b f(x)\, dx \le 0$.

50. Prove Property 3 of Theorem 5.13.

51. Prove Property 9 of Theorem 5.14.
[*Hint:* $-|f(x)| \le f(x) \le |f(x)|$.]

52. Suppose that f is continuous on $[a, b]$ and $f(x) > 0$ for all x in $[a, b]$. Prove that $\int_a^b f(x)\, dx > 0$. (*Hint:* Use the Extreme Value Theorem and Property 8.)

The Fundamental Theorem of Calculus

The Fundamental Theorem of Calculus is appropriately named because it establishes a connection between the two branches of calculus: differential calculus and integral calculus. Differential calculus arose from the tangent problem, whereas integral calculus arose from a seemingly unrelated problem, the area problem. Newton's teacher at Cambridge, Isaac Barrow (1630–1677), discovered that these two problems are actually closely related. In fact, he realized that differentiation and integration are inverse processes. The Fundamental Theorem of Calculus gives the precise inverse relationship between the derivative and the integral. It was Newton and Leibniz who exploited this relationship and used it to develop calculus into a systematic mathematical method. In particular, they saw that the Fundamental Theorem enabled them to compute areas and integrals very easily without having to compute them as limits of sums as we did in Sections 5.2 and 5.3.

In order to motivate the Fundamental Theorem, let f be a continuous function on $[a, b]$ and define a new function g by

$$g(x) = \int_a^x f(t)\, dt \tag{5.16}$$

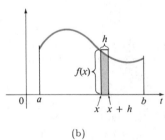

area = $g(x)$

$y = f(t)$

(a)

Observe that g depends only on x, which appears as the variable upper limit in the integral. To make this clear we could use the other notation for an integral and write $g(x) = \int_a^x f$. If x is a fixed number, then the integral $\int_a^x f$ is a definite number. If we then let x vary, the number $\int_a^x f$ also varies and defines a function of x denoted by $g(x)$. For instance, if we take $f(t) = t$ and $a = 0$, then, using Exercise 23 in Section 5.3, we have

$$g(x) = \int_0^x t\, dt = \frac{x^2}{2}$$

Notice that $g'(x) = x$, that is, $g' = f$. In other words, if g is defined as the integral of f by Equation 5.16, then g turns out to be an antiderivative of f, at least in this case.

To see why this might be generally true we consider any continuous function f with $f(x) \ge 0$. Then $g(x) = \int_a^x f$ can be interpreted as the area under the graph of f from a to x, where x can vary from a to b. [Think of g as the "area so far" function; see Figure 5.21(a).]

In order to compute $g'(x)$ from the definition of derivative we first observe that, for $h > 0$, $g(x + h) - g(x)$ is obtained by subtracting areas, so it is the area under the graph

(b)

Figure 5.21

of f from x to $x+h$ [the shaded area in Figure 5.21(b)]. For small h you can see from the figure that this area is approximately equal to the area of the rectangle with height $f(x)$ and width h:

$$g(x+h) - g(x) \approx hf(x) \qquad \frac{g(x+h) - g(x)}{h} \approx f(x)$$

Intuitively, we therefore expect that

$$g'(x) = \lim_{h \to 0} \frac{g(x+h) - g(x)}{h} = f(x)$$

The fact that this is true, even when f is not necessarily positive, is the first part of the Fundamental Theorem of Calculus.

The Fundamental Theorem of Calculus, Part 1 (5.17)

If f is continuous on $[a, b]$, then the function g defined by

$$g(x) = \int_a^x f(t)\,dt \qquad a \le x \le b$$

is continuous on $[a, b]$ and differentiable on (a, b), and $g'(x) = f(x)$.

Proof   If x and $x+h$ are in (a, b), then

$$g(x+h) - g(x) = \int_a^{x+h} f - \int_a^x f$$

$$= \left(\int_a^x f + \int_x^{x+h} f \right) - \int_a^x f \qquad \text{(by Property 5)}$$

$$= \int_x^{x+h} f$$

and so, for $h \ne 0$,

(5.18)
$$\frac{g(x+h) - g(x)}{h} = \frac{1}{h} \int_x^{x+h} f$$

For now let us assume that $h > 0$. Since f is continuous on $[x, x+h]$, the Extreme Value Theorem says that there are numbers u and v in $[x, x+h]$ such that $f(u) = m$ and $f(v) = M$, where m and M are the absolute minimum and maximum values of f on $[x, x+h]$ (see Figure 5.22).
By Property 8 of integrals we have

$$mh \le \int_x^{x+h} f \le Mh$$

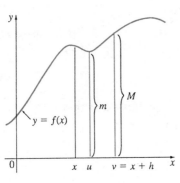

that is,

$$f(u)h \le \int_x^{x+h} f \le f(v)h$$

Figure 5.22

Since $h > 0$, we can divide this inequality by h:

$$f(u) \le \frac{1}{h} \int_x^{x+h} f \le f(v)$$

Now we use Equation 5.18 to replace the middle part of this inequality:

(5.19)
$$f(u) \le \frac{g(x+h) - g(x)}{h} \le f(v)$$

Inequality 5.19 can be proved in a similar manner for the case where $h < 0$ (see Exercise 93).

Now we let $h \to 0$. Then $u \to x$ and $v \to x$ since u and v lie between x and $x + h$. Therefore

$$\lim_{h \to 0} f(u) = \lim_{u \to x} f(u) = f(x) \qquad \lim_{h \to 0} f(v) = \lim_{v \to x} f(v) = f(x)$$

because f is continuous at x. We conclude, from (5.19) and the Squeeze Theorem, that

(5.20)
$$g'(x) = \lim_{h \to 0} \frac{g(x + h) - g(x)}{h} = f(x)$$

If $x = a$ or b, Equation 5.20 can be interpreted as a one-sided limit. Then Theorem 2.8 (modified for one-sided limits) shows that g is continuous on $[a, b]$.   ●

Using the Leibniz notation for derivatives, we can write Theorem 5.17 as

(5.21)
$$\frac{d}{dx} \int_a^x f(t)\, dt = f(x)$$

when f is continuous. Roughly speaking, Equation 5.21 says that if we first integrate f and then differentiate the result, we get back to the original function f.

EXAMPLE 1   Find the derivative of the function $g(x) = \displaystyle\int_0^x \sqrt{1 + t^2}\, dt$.

Solution   Since $f(t) = \sqrt{1 + t^2}$ is continuous, Part 1 of the Fundamental Theorem of Calculus gives

$$g'(x) = \sqrt{1 + x^2}$$

●

EXAMPLE 2   Differentiate $y = \displaystyle\int_1^x \sin \sqrt{u}\, du$.

Solution   Again, Part 1 of the Fundamental Theorem gives

$$\frac{dy}{dx} = \sin \sqrt{x}$$

●

EXAMPLE 3   Find $\dfrac{d}{dx} \displaystyle\int_1^{x^4} \sec t\, dt$.

Solution   Here we have to be careful to use the Chain Rule in conjunction with Part 1 of the Fundamental Theorem. Let $u = x^4$. Then

$$\frac{d}{dx} \int_1^{x^4} \sec t\, dt = \frac{d}{dx} \int_1^{u} \sec t\, dt$$

$$= \frac{d}{du} \left[\int_1^{u} \sec t\, dt \right] \frac{du}{dx} \qquad \text{(by the Chain Rule)}$$

$$= \sec u \, \frac{du}{dx} \qquad \text{(by Theorem 5.17)}$$

$$= \sec(x^4) \cdot 4x^3$$

●

In Section 5.3 we computed integrals from the definition as a limit of Riemann sums and we saw that this procedure is sometimes long and difficult. The second part of the Fundamental Theorem of Calculus, which follows easily from the first part, provides us with a much simpler method for the evaluation of integrals.

The Fundamental Theorem of Calculus, Part 2 (5.22)

If f is continuous on $[a, b]$, then

$$\int_a^b f(x)\,dx = F(b) - F(a)$$

where F is any antiderivative of f, that is, $F' = f$.

Proof   Let $g(x) = \int_a^x f$. We know from Part 1 that $g'(x) = f(x)$; that is, g is an antiderivative of f. If F is any other antiderivative of f on $[a, b]$, then we know from Corollary 4.17 that F and g differ by a constant:

(5.23)
$$F(x) = g(x) + C$$

for $a < x < b$. But both F and g are continuous on $[a, b]$ and so, by taking limits of both sides of Equation 5.23 (as $x \to a^+$ and $x \to b^-$), we see that it also holds when $x = a$ and $x = b$.

If we put $x = a$ in the formula for $g(x)$, we get

$$g(a) = \int_a^a f = 0$$

So, using Equation 5.23 with $x = b$ and $x = a$, we have

$$F(b) - F(a) = [g(b) + C] - [g(a) + C]$$
$$= g(b) - g(a) = g(b) = \int_a^b f$$

Part 2 of the Fundamental Theorem states that if we know an antiderivative F of f, then we can evaluate $\int_a^b f$ simply by subtracting the values of F at the endpoints of the interval $[a, b]$. It is very surprising that $\int_a^b f$, which was defined by a complicated procedure involving all of the values of $f(x)$ for $a \le x \le b$, can be found by knowing the values of $F(x)$ at only two points, a and b.

EXAMPLE 4   Evaluate the integral $\displaystyle\int_{-2}^1 x^3\,dx$.

Solution   The function $f(x) = x^3$ is continuous on $[-2, 1]$ and we know from Section 4.8 that an antiderivative is $F(x) = \frac{1}{4}x^4$, so Part 2 of the Fundamental Theorem gives

$$\int_{-2}^1 x^3\,dx = F(1) - F(-2) = \tfrac{1}{4}(1)^4 - \tfrac{1}{4}(-2)^4 = -\tfrac{15}{4}$$

We use the notation

$$F(x)\big|_a^b = F(b) - F(a)$$

and so the equation of Theorem 5.22 can be written as

$$\int_a^b f(x)\,dx = F(x)\Big|_a^b \qquad \text{where} \qquad F' = f$$

Another common notation is $[F(x)]_a^b$.

EXAMPLE 5   Find the area under the parabola $y = x^2 + 1$ from 0 to 2.

Solution   An antiderivative of $f(x) = x^2 + 1$ is $F(x) = \frac{1}{3}x^3 + x$. The required area A is found using Part 2 of the Fundamental Theorem:

$$A = \int_0^2 (x^2 + 1)\,dx = \frac{x^3}{3} + x\,\Big|_0^2 = \left(\frac{2^3}{3} + 2\right) - \left(\frac{0^3}{3} + 0\right) = \frac{14}{3}$$

Notice that in applying the Fundamental Theorem we used a particular antiderivative F of f. It is not necessary to use the most general antiderivative.   ●

If you compare the calculation in Example 5 with the one in Example 3 in Section 5.2, you will see that the Fundamental Theorem gives us a much shorter method.

EXAMPLE 6   Find the area under the cosine curve from 0 to b, where $0 \le b \le \pi/2$.

Solution   Since an antiderivative of $f(x) = \cos x$ is $F(x) = \sin x$, we have

$$A = \int_0^b \cos x\,dx = \sin x\,\Big|_0^b = \sin b - \sin 0 = \sin b$$ ●

A comparison of Example 6 with Example 4 in Section 5.2 reveals the power of the Fundamental Theorem of Calculus. When the French mathematician Gilles de Roberval first found the area under the sine and cosine curves in 1635, this was a very challenging problem that required a great deal of ingenuity. (Recall that the calculation in Example 4 in Section 5.2 depended on some little-known trigonometric identities.) But in the 1660s and 1670s when the Fundamental Theorem was discovered by Barrow and exploited by Newton and Leibniz, such problems became very easy, as you can see from Example 6.

Notation: Because of the relation given by the Fundamental Theorem between antiderivatives and integrals, the notation $\int f(x)\,dx$ is traditionally used for an antiderivative of f and is called an **indefinite integral**. Thus

(5.24)

$$\int f(x)\,dx = F(x) \qquad \text{means} \qquad F'(x) = f(x)$$

You should distinguish carefully between definite and indefinite integrals. A definite integral $\int_a^b f(x)\,dx$ is a number, whereas an indefinite integral $\int f(x)\,dx$ is a function. The connection between them is given by Part 2 of the Fundamental Theorem. If f is continuous on $[a,b]$, then

(5.25)

$$\int_a^b f(x)\,dx = \int f(x)\,dx\,\Big|_a^b$$

The effectiveness of the Fundamental Theorem depends on having a supply of antiderivatives of functions. We therefore restate the Table of Antidifferentiation Formulas (4.34), together with a few others, in the notation of indefinite integrals. Any formula can be verified by differentiating the function on the right side and obtaining the integrand. For instance,

$$\int \sec^2 x \, dx = \tan x \qquad \text{since} \qquad \frac{d}{dx}(\tan x) = \sec^2 x$$

Table of Indefinite Integrals (5.26)

$$\int c f(x) \, dx = c \int f(x) \, dx$$

$$\int [f(x) + g(x)] \, dx = \int f(x) \, dx + \int g(x) \, dx$$

$$\int x^n \, dx = \frac{x^{n+1}}{n+1} + C \quad (n \neq -1) \qquad \int \frac{1}{x} \, dx = \ln |x| + C$$

$$\int e^x \, dx = e^x + C \qquad \int a^x \, dx = \frac{a^x}{\ln a} + C$$

$$\int \sin x \, dx = -\cos x + C \qquad \int \cos x \, dx = \sin x + C$$

$$\int \sec^2 x \, dx = \tan x + C \qquad \int \csc^2 x \, dx = -\cot x + C$$

$$\int \sec x \tan x \, dx = \sec x + C \qquad \int \csc x \cot x \, dx = -\csc x + C$$

$$\int \frac{1}{x^2 + 1} \, dx = \tan^{-1} x + C \qquad \int \frac{1}{\sqrt{1 - x^2}} \, dx = \sin^{-1} x + C$$

Recall from Theorem 4.33 that the most general antiderivative *on a given interval* is obtained by adding a constant to a particular antiderivative. **We adopt the convention that when a formula for a general indefinite integral is given, it is valid only on an interval.** Thus we write

$$\int \frac{1}{x^2} \, dx = -\frac{1}{x} + C$$

with the understanding that it is valid on the interval $(0, \infty)$ or on the interval $(-\infty, 0)$. This is true despite the fact that the general antiderivative of the function $f(x) = 1/x^2$, $x \neq 0$, is

$$F(x) = \begin{cases} -\frac{1}{x} + C_1 & \text{if } x < 0 \\ -\frac{1}{x} + C_2 & \text{if } x > 0 \end{cases}$$

EXAMPLE 7 Find the general indefinite integral

$$\int (10x^4 + 3 \sec^2 x) \, dx$$

Solution Using our convention and Table 5.26, we have

$$\int (10x^4 + 3 \sec^2 x) \, dx = 10 \frac{x^5}{5} + 3 \tan x + C$$
$$= 2x^5 + 3 \tan x + C$$

Check the answer by differentiating it.

EXAMPLE 8   Evaluate $\int_1^4 (2x^3 - 5x)\,dx$.

Solution   Using Part 2 of the Fundamental Theorem and Table 5.26, we have

$$\int_1^4 (2x^3 - 5x)\,dx = 2\,\frac{x^4}{4} - 5\,\frac{x^2}{2}\Big|_1^4$$

$$= \left(\tfrac{1}{2}\cdot 4^4 - \tfrac{5}{2}\cdot 4^2\right) - \left(\tfrac{1}{2}\cdot 1^4 - \tfrac{5}{2}\cdot 1^2\right)$$

$$= 128 - 40 - \tfrac{1}{2} + \tfrac{5}{2} = 90$$

Compare this calculation with Example 3 in Section 5.3.   ●

EXAMPLE 9 $\int_0^1 e^x\,dx = e^x\big|_0^1 = e^1 - e^0 = e - 1$ ●

EXAMPLE 10

$$\int_1^9 \frac{2t^2 + t^2\sqrt{t} - 1}{t^2}\,dt = \int_1^9 \left(2 + t^{1/2} - t^{-2}\right)dt$$

$$= 2t + \frac{t^{3/2}}{\frac{3}{2}} - \frac{t^{-1}}{-1}\Big|_1^9$$

$$= 2t + \tfrac{2}{3}t^{3/2} + \frac{1}{t}\Big|_1^9$$

$$= \left[2\cdot 9 + \tfrac{2}{3}(9)^{3/2} + \tfrac{1}{9}\right] - \left(2\cdot 1 + \tfrac{2}{3}\cdot 1^{3/2} + \tfrac{1}{1}\right)$$

$$= 18 + 18 + \tfrac{1}{9} - 2 - \tfrac{2}{3} - 1 = 32\tfrac{4}{9}$$ ●

EXAMPLE 11   What is wrong with the following calculation?

$$\int_{-1}^3 \frac{1}{x^2}\,dx = \frac{x^{-1}}{-1}\Big|_{-1}^3 = -\tfrac{1}{3} - 1 = -\tfrac{4}{3}$$

Solution   To start, we notice that this calculation must be wrong because $f(x) = 1/x^2 \geq 0$ and Property 6 of integrals says that $\int_a^b f \geq 0$ when $f \geq 0$. The Fundamental Theorem of Calculus applies to continuous functions. It cannot be applied here because $f(x) = 1/x^2$ is not continuous on $[-1, 3]$. In fact, $f(0)$ is not defined, but even if it were defined [say, by $f(0) = 0$], the integral would not exist because f is unbounded on $[-1, 3]$. In fact, we know that $\lim_{x \to 0}(1/x^2) = \infty$. Thus

$$\int_{-1}^3 \frac{1}{x^2}\,dx \qquad \text{does not exist}$$ ●

Applications of the Fundamental Theorem

Suppose a particle is moving along a straight line with position function $s(t)$, velocity function $v(t)$, and acceleration function $a(t)$. Since $s'(t) = v(t)$, the Fundamental Theorem of Calculus gives

(5.27)

$$\int_{t_1}^{t_2} v(t)\,dt = s(t_2) - s(t_1)$$

The right side of Equation 5.27 is the change of position, or *displacement,* of the particle during the time interval $[t_1, t_2]$. Thus Equation 5.27 enables us to compute the displacement, by integration, if the velocity function is known. Similarly, since $v'(t) = a(t)$, the Fundamental Theorem of Calculus gives

$$\int_{t_1}^{t_2} a(t)\, dt = v(t_2) - v(t_1)$$

If we want to calculate the distance traveled during the time interval, we have to take into account the intervals when $v(t) \geq 0$ (the particle moves to the right) and also the intervals when $v(t) \leq 0$ (the particle moves to the left). In both cases the distance is computed by integrating $|v(t)|$ (the speed). Therefore

(5.28)
$$\text{total distance traveled} = \int_{t_1}^{t_2} |v(t)|\, dt$$

Figure 5.23 shows how both displacement and distance traveled can be interpreted in terms of areas under a velocity curve.   ●

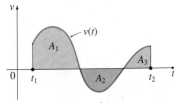

$$\text{displacement} = \int_{t_1}^{t_2} v(t)\, dt$$
$$= A_1 - A_2 + A_3$$

$$\text{distance} = \int_{t_1}^{t_2} |v(t)|\, dt$$
$$= A_1 + A_2 + A_3$$

Figure 5.23

EXAMPLE 12   A particle moves along a line so that its velocity at time t is $v(t) = t^2 - t - 6$ (measured in meters per second).
(a)   Find the displacement of the particle during the time period $1 \leq t \leq 4$.
(b)   Find the distance traveled during this time period.

Solution
(a)   By Equation 5.27, the displacement is

$$s(4) - s(1) = \int_1^4 v(t)\, dt = \int_1^4 (t^2 - t - 6)\, dt$$

$$= \left[\frac{t^3}{3} - \frac{t^2}{2} - 6t \right]_1^4 = -\frac{9}{2}$$

This means that the particle moved 4.5 m to the left.
(b)   Note that $v(t) = t^2 - t - 6 = (t - 3)(t + 2)$ and so $v(t) \leq 0$ on the interval $[1, 3]$ and $v(t) \geq 0$ on $[3, 4]$. Thus, from Equation 5.28, the distance traveled is

$$\int_1^4 |v(t)|\, dt = \int_1^3 (-v(t))\, dt + \int_3^4 v(t)\, dt$$

$$= \int_1^3 (-t^2 + t + 6)\, dt + \int_3^4 (t^2 - t - 6)\, dt$$

$$= \left[-\frac{t^3}{3} + \frac{t^2}{2} + 6t \right]_1^3 + \left[\frac{t^3}{3} - \frac{t^2}{2} - 6t \right]_3^4$$

$$= \frac{61}{6}\, \text{m}$$ ●

The Fundamental Theorem of Calculus can be applied to all of the rates of change in the natural and social sciences that were discussed in Section 2.3. For instance, if the mass of a rod measured from the left end to a point x is $m(x)$ and the linear density is $\rho(x)$, then $m'(x) = \rho(x)$ (see Example 2 in Section 2.3) and so, by the Fundamental Theorem of Calculus, we have

$$m(x_2) - m(x_1) = \int_{x_1}^{x_2} \rho(x)\, dx$$

This enables us to compute the mass of a segment of the rod if the linear density is known.

If the rate of growth dn/dt of a population is known (see Example 6 in Section 2.3), then the Fundamental Theorem of Calculus allows us to compute the population at time t, in terms of the initial population n_0, as

$$n(t) = n_0 + \int_0^t \frac{dn}{dt}\, dt$$

Differentiation and Integration as Inverse Processes

We end this section by bringing the two parts of the Fundamental Theorem together.

The Fundamental Theorem of Calculus (5.29)

Suppose f is continuous on $[a, b]$.

(a)   If $g(x) = \int_a^x f(t)\, dt$, then $g'(x) = f(x)$.

(b) $\int_a^b f(x)\, dx = F(b) - F(a)$, where F is any antiderivative of f, that is, $F' = f$.

We noted that Part 1 can be rewritten as

$$\frac{d}{dx} \int_a^x f(t)\, dt = f(x)$$

which says that if f is integrated and then the result is differentiated, we arrive back at the original function f. Since $F'(x) = f(x)$, Part 2 can be rewritten as

$$\int_a^b F'(x)\, dx = F(b) - F(a)$$

This version says that if we take a function F, first differentiate it, and then integrate the result, we arrive back at the original function F, but in the form $F(b) - F(a)$. Taken together, the two parts of the Fundamental Theorem of Calculus say that differentiation and integration are inverse processes. Each undoes what the other does.

The Fundamental Theorem of Calculus is unquestionably the most important theorem in calculus and, indeed, it ranks as one of the great accomplishments of the human mind. Before it was discovered, from the time of Eudoxus and Archimedes to the time of Galileo and Fermat, problems of finding areas, volumes, and lengths of curves were so difficult that only a genius could meet the challenge. But now, armed with the systematic method that Newton and Leibniz fashioned out of the Fundamental Theorem, we will see in the chapters to come that these challenging problems are accessible to all of us.

SECTION 5.5   **Exercises**

In Exercises 1 and 2 sketch the area represented by $g(x)$. Then find $g'(x)$ in two ways: (a) by using Part 1 of the Fundamental Theorem, and (b) by evaluating the integral using Part 2 and then differentiating.

1. $g(x) = \int_0^x (1 + t^2)\, dt$ **2.** $g(x) = \int_\pi^x (2 + \cos t)\, dt$

In Exercises 3–14 use Part 1 of the Fundamental Theorem of Calculus to find the derivative of the given function.

3. $g(x) = \int_1^x (t^2 - 1)^{20}\, dt$ **4.** $g(x) = \int_{-1}^x \sqrt{t^3 + 1}\, dt$

5. $g(u) = \int_\pi^u \frac{1}{1 + t^4}\, dt$ **6.** $g(t) = \int_0^t \sin(x^2)\, dx$

7. $F(x) = \int_x^2 \cos(t^2)\, dt$

$\left[\text{Hint:} \int_x^2 \cos(t^2)\, dt = -\int_2^x \cos(t^2)\, dt \right]$

8. $F(x) = \int_x^4 (2 + \sqrt{u})^8\, du$

9. $h(x) = \int_2^{1/x} \sin^4 t\, dt$ 10. $h(x) = \int_1^{\sqrt{x}} \frac{s^2}{s^2 + 1}\, ds$

11. $y = \int_{\tan x}^{17} \sin(t^4)\, dt$ 12. $y = \int_{x^2}^{\pi} \frac{\sin t}{t}\, dt$

13. $y = \int_0^{5x+1} \frac{1}{u^2 - 5}\, du$ 14. $y = \int_{-5}^{\sin x} t \cos(t^3)\, dt$

In Exercises 15–60 use Part 2 of the Fundamental Theorem of Calculus to evaluate the integral, or state that it does not exist.

15. $\int_{-1}^8 6\, dx$ 16. $\int_{-2}^4 (3x - 5)\, dx$

17. $\int_0^1 x^{99}\, dx$ 18. $\int_1^2 x^{-2}\, dx$

19. $\int_0^1 (1 - 2x - 3x^2)\, dx$ 20. $\int_1^2 (5x^2 - 4x + 3)\, dx$

21. $\int_{-3}^0 (5y^4 - 6y^2 + 14)\, dy$ 22. $\int_0^1 (y^9 - 2y^5 + 3y)\, dy$

23. $\int_0^4 \sqrt{x}\, dx$ 24. $\int_0^1 x^{3/7}\, dx$

25. $\int_1^3 \left(\frac{1}{t^2} - \frac{1}{t^4} \right) dt$ 26. $\int_1^2 \frac{t^6 - t^2}{t^4}\, dt$

27. $\int_1^2 \frac{x^2 + 1}{\sqrt{x}}\, dx$ 28. $\int_0^2 (x^3 - 1)^2\, dx$

29. $\int_0^1 u(\sqrt{u} + \sqrt[3]{u})\, du$ 30. $\int_{-1}^1 \frac{3}{t^4}\, dt$

31. $\int_{-2}^3 |x^2 - 1|\, dx$ 32. $\int_1^2 \left(x + \frac{1}{x} \right)^2 dx$

33. $\int_3^3 \sqrt{x^5 + 2}\, dx$ 34. $\int_{-1}^2 |x - x^2|\, dx$

35. $\int_{-4}^2 \frac{2}{x^6}\, dx$ 36. $\int_1^{-1} (x - 1)(3x + 2)\, dx$

37. $\int_1^4 \left(\sqrt{t} - \frac{2}{\sqrt{t}} \right) dt$ 38. $\int_1^8 \left(\sqrt[3]{r} + \frac{1}{\sqrt[3]{r}} \right) dr$

39. $\int_{-1}^0 (x + 1)^3\, dx$ 40. $\int_{-5}^{-2} \frac{x^4 - 1}{x^2 + 1}\, dx$

41. $\int_{\pi/4}^{\pi/3} \sin t\, dt$ 42. $\int_0^{\pi/2} (\cos \theta + 2 \sin \theta)\, d\theta$

43. $\int_{\pi/2}^{\pi} \sec x \tan x\, dx$ 44. $\int_{\pi/3}^{\pi/2} \csc x \cot x\, dx$

45. $\int_{\pi/6}^{\pi/3} \csc^2 \theta\, d\theta$ 46. $\int_{\pi/4}^{\pi} \sec^2 \theta\, d\theta$

47. $\int_4^8 \frac{1}{x}\, dx$ 48. $\int_{\ln 3}^{\ln 6} 8e^x\, dx$

49. $\int_8^9 2^t\, dt$ 50. $\int_{-e^2}^{-e} \frac{3}{x}\, dx$

51. $\int_1^{\sqrt{3}} \frac{6}{1 + x^2}\, dx$ 52. $\int_0^{0.5} \frac{dx}{\sqrt{1 - x^2}}$

53. $\int_1^e \frac{x^2 + x + 1}{x}\, dx$ 54. $\int_4^9 \left(\sqrt{x} + \frac{1}{\sqrt{x}} \right)^2 dx$

55. $\int_0^1 \left(\sqrt[4]{x^5} + \sqrt[5]{x^4} \right) dx$ 56. $\int_1^8 \frac{x - 1}{\sqrt[3]{x^2}}\, dx$

57. $\int_{-1}^2 (x - 2|x|)\, dx$ 58. $\int_0^2 (x^2 - |x - 1|)\, dx$

59. $\int_0^2 f(x)\, dx$, where $f(x) = \begin{cases} x^4 & \text{if } 0 \le x < 1 \\ x^5 & \text{if } 1 \le x \le 2 \end{cases}$

60. $\int_{-\pi}^{\pi} f(x)\, dx$, where $f(x) = \begin{cases} x & \text{if } -\pi \le x \le 0 \\ \sin x & \text{if } 0 < x \le \pi \end{cases}$

In Exercises 61–68 find the area of the region that lies beneath the given curve.

61. $y = 4x^2 - 4x + 3,\ 0 \le x \le 2$

62. $y = 1 + x^3,\ 0 \le x \le 4$

63. $y = \sqrt[3]{x},\ 0 \le x \le 27$

64. $y = x^{-4},\ 1 \le x \le 6$

65. $y = \sin x,\ 0 \le x \le \pi$

66. $y = \sec^2 x,\ 0 \le x \le \pi/3$

67. $y = x^{0.8},\ -1 \le x \le 2$

68. $y = 1/x,\ 1 \le x \le 2$

Verify by differentiation that the formulas in Exercises 69–72 are correct.

69. $\int \frac{1}{\sqrt{(a^2 - x^2)^3}}\, dx = \frac{x}{a^2 \sqrt{a^2 - x^2}} + C$

70. $\int \frac{1}{x^2 \sqrt{x^2 + a^2}}\, dx = -\frac{\sqrt{x^2 + a^2}}{a^2 x} + C$

71. $\int \sin^2 x\, dx = \frac{x}{2} - \frac{\sin 2x}{4} + C$

72. $\int x^2 \sin x\, dx = -x^2 \cos x + 2 \int x \cos x\, dx$

In Exercises 73–78 find the general indefinite integral.

73. $\displaystyle\int x\sqrt{x}\,dx$

74. $\displaystyle\int \sqrt{x}\left(x^2 - \frac{1}{x}\right)dx$

75. $\displaystyle\int (2 - \sqrt{x})^2\,dx$

76. $\displaystyle\int (\cos x - 2\sin x)\,dx$

77. $\displaystyle\int (2x + \sec x \tan x)\,dx$

78. $\displaystyle\int \left(x^2 + 1 + \frac{1}{x^2}\right)dx$

In Exercises 79 and 80 the velocity function (in meters per second) is given for a particle moving along a line. Find (a) the displacement and (b) the distance traveled by the particle during the given time interval.

79. $v(t) = 3t - 5,\ 0 \le t \le 3$

80. $v(t) = t^2 - 2t - 8,\ 1 \le t \le 6$

In Exercises 81 and 82 the acceleration function (in m/s^2) and the initial velocity are given for a particle moving along a line. Find (a) the velocity at time t and (b) the distance traveled during the given time interval.

81. $a(t) = t + 4,\quad v(0) = 5,\ 0 \le t \le 10$

82. $a(t) = 2t + 3,\quad v(0) = -4,\ 0 \le t \le 3$

83. The linear density of a rod of length $4\,\text{m}$ is given by $\rho(x) = 9 + 2\sqrt{x}$, measured in kilograms per meter, where x is measured in meters from one end of the rod. Find the total mass of the rod.

84. An animal population is increasing at a rate of $200 + 50t$ per year (where t is measured in years). By how much does the animal population increase between the fourth and tenth years?

In Exercises 85–88 find the derivative of the given function.

85. $\displaystyle g(x) = \int_{2x}^{3x} \frac{u - 1}{u + 1}\,du$

$$\left[\text{Hint: } \int_{2x}^{3x} f(u)\,du = \int_{2x}^{0} f(u)\,du + \int_{0}^{3x} f(u)\,du\right]$$

86. $\displaystyle g(x) = \int_{\tan x}^{x^2} \frac{1}{\sqrt{2 + t^4}}\,dt$

87. $\displaystyle y = \int_{\sqrt{x}}^{x^3} \sqrt{t}\,\sin t\,dt$

88. $\displaystyle y = \int_{\cos x}^{5x} \cos(u^2)\,du$

89. If $F(x) = \displaystyle\int_{1}^{x} f(t)\,dt$, where $f(t) = \displaystyle\int_{1}^{t^2} \frac{\sqrt{1 + u^4}}{u}\,du$, find $F''(2)$.

90. Find the interval on which the curve $y = \displaystyle\int_{0}^{x} \frac{1}{1 + t + t^2}\,dt$ is concave upward.

In Exercises 91 and 92 evaluate the limit by first recognizing the sum as a Riemann sum for a function defined on $[0, 1]$.

91. $\displaystyle\lim_{n\to\infty} \sum_{i=1}^{n} \frac{i^3}{n^4}$

92. $\displaystyle\lim_{n\to\infty} \frac{1}{n}\left(\sqrt{\frac{1}{n}} + \sqrt{\frac{2}{n}} + \sqrt{\frac{3}{n}} + \cdots + \sqrt{\frac{n}{n}}\right)$

93. Justify (5.19) for the case where $h < 0$.

94. If f is continuous and g and h are differentiable functions, find a formula for

$$\frac{d}{dx}\int_{g(x)}^{h(x)} f(t)\,dt$$

95. (a) Show that $1 \le \sqrt{1 + x^3} \le 1 + x^3$ for $x \ge 0$.
    (b) Show that $1 \le \displaystyle\int_{0}^{1} \sqrt{1 + x^3}\,dx \le 1.25$.

96. Let

$$f(x) = \begin{cases} 0 & \text{if } x < 0 \\ x & \text{if } 0 \le x \le 1 \\ 2 - x & \text{if } 1 < x \le 2 \\ 0 & \text{if } x > 2 \end{cases}$$

and

$$g(x) = \int_{0}^{x} f(t)\,dt$$

(a) Find an expression for $g(x)$ similar to the one for $f(x)$.
(b) Sketch the graphs of f and g.
(c) Where is f differentiable? Where is g differentiable?

SECTION 5.6

The Substitution Rule

The Fundamental Theorem of Calculus reduces the problem of integration to the problem of antidifferentiation. But the antidifferentiation formulas in Table 5.26 do not suffice to evaluate integrals such as

$$(5.30) \qquad \int 2x \sqrt{1+x^2}\, dx$$

In such cases the task is simplified by changing from the variable x to a new variable. Suppose that we let u be the quantity under the root sign in (5.30), $u = 1 + x^2$. Then the differential of u is $du = 2x\, dx$. Notice that if the dx in the notation for an integral were to be interpreted as a differential, then the differential $2x\, dx$ would occur in (5.30) so, formally, without justifying our calculation, we could write

Differentials were defined in Section 2.9.

$$(5.31) \qquad \int 2x\sqrt{1+x^2}\, dx = \int \sqrt{1+x^2}\, 2x\, dx$$

$$= \int \sqrt{u}\, du = \tfrac{2}{3} u^{3/2} + C$$

$$= \tfrac{2}{3}(x^2+1)^{3/2} + C$$

But now we could check that we have the correct answer by using the Chain Rule to differentiate the function on the right side of Equation 5.31:

$$\frac{d}{dx}\left[\tfrac{2}{3}(x^2+1)^{3/2} + C \right] = \tfrac{2}{3} \cdot \tfrac{3}{2}(x^2+1)^{1/2} \cdot 2x = 2x\sqrt{x^2+1}$$

In general this method works when we have an integral of the form $\int f(g(x))g'(x)\, dx$. Observe that if $F' = f$, then

$$(5.32) \qquad \int F'(g(x))g'(x)\, dx = F(g(x)) + C$$

because, by the Chain Rule,

$$\frac{d}{dx}[F(g(x))] = F'(g(x))g'(x)$$

If we make the "change of variable" or "substitution" $u = g(x)$, then from Equation 5.32 we have

$$\int F'(g(x))g'(x)\, dx = F(g(x)) + C = F(u) + C = \int F'(u)\, du$$

or, writing $F' = f$, we get

$$\int f(g(x))g'(x)\, dx = \int f(u)\, du$$

Thus we have proved the following rule:

The Substitution Rule (5.33)

If $u = g(x)$ is a differentiable function whose range is an interval I and f is continuous on I, then

$$\int f(g(x))g'(x)\, dx = \int f(u)\, du$$

Notice that the Substitution Rule for integration was proved using the Chain Rule for differentiation. Notice also that if $u = g(x)$, then $du = g'(x)\, dx$, so a way of remembering the Substitution Rule is to think of dx and du in (5.33) as differentials.

Thus the Substitution Rule says: **It is permissible to operate with dx and du after integral signs as if they were differentials.**

EXAMPLE 1   Find $\int x^3 \cos(x^4 + 2)\, dx$.

Solution   We make the substitution $u = x^4 + 2$ because its differential is $du = 4x^3\, dx$, which, apart from the constant factor 4, occurs in the integral. Thus, using $x^3\, dx = du/4$ and the Substitution Rule, we have

$$\int x^3 \cos(x^4 + 2)\, dx = \int \cos u \cdot \tfrac{1}{4}\, du$$

$$= \tfrac{1}{4} \int \cos u\, du$$

$$= \tfrac{1}{4} \sin u + C$$

$$= \tfrac{1}{4} \sin(x^4 + 2) + C$$

Check the answer by differentiating it.

Notice that at the final stage we had to return to the original variable x.

The idea behind the Substitution Rule is to replace a relatively complicated integral by a simpler integral. This is accomplished by changing from the original variable x to a new variable u that is a function of x. Thus in Example 1 we replaced the integral $\int x^3 \cos(x^4 + 2)\, dx$ by the simpler integral $\tfrac{1}{4} \int \cos u\, du$.

The main challenge in using the Substitution Rule is to think of an appropriate substitution. You should try to choose u to be some function in the integrand whose differential also occurs (except for a constant factor). This was the case in Example 1. If that is not possible, try choosing u to be some complicated part of the integrand.

EXAMPLE 2   Evaluate $\int \sqrt{3x + 4}\, dx$.

Solution 1   Let $u = 3x + 4$. Then $du = 3\, dx$, so $dx = du/3$. Thus the Substitution Rule gives

$$\int \sqrt{3x + 4}\, dx = \int \sqrt{u}\, \frac{du}{3} = \tfrac{1}{3} \int u^{1/2}\, du$$

$$= \tfrac{1}{3} \cdot \frac{u^{3/2}}{3/2} + C = \tfrac{2}{9} u^{3/2} + C$$

$$= \tfrac{2}{9}(3x + 4)^{3/2} + C$$

Solution 2   Another possible substitution is $u = \sqrt{3x + 4}$. Then

$$du = \frac{3\, dx}{2\sqrt{3x + 4}} \qquad \text{and} \qquad dx = \tfrac{2}{3}\sqrt{3x + 4}\, du = \tfrac{2}{3} u\, du$$

(Or observe that $u^2 = 3x + 4$, so $2u\, du = 3\, dx$.) Therefore

$$\int \sqrt{3x + 4}\, dx = \int u \cdot \tfrac{2}{3} u\, du = \tfrac{2}{3} \int u^2\, du$$

$$= \tfrac{2}{3} \cdot \frac{u^3}{3} + C = \tfrac{2}{9} u^3 + C$$

$$= \tfrac{2}{9}(3x + 4)^{3/2} + C$$

EXAMPLE 3   Find $\int \dfrac{x}{\sqrt{1 - 4x^2}}\, dx$.

Solution   Let $u = 1 - 4x^2$. Then $du = -8x\,dx$, so $x\,dx = -\frac{1}{8}\,du$ and

$$\int \frac{x}{\sqrt{1-4x^2}}\,dx = -\frac{1}{8}\int \frac{du}{\sqrt{u}} = -\frac{1}{8}\int u^{-1/2}\,du$$

$$= -\frac{1}{8}(2\sqrt{u}) + C$$

$$= -\frac{1}{4}\sqrt{1-4x^2} + C$$

EXAMPLE 4   Calculate $\int \cos 5x\,dx$.

Solution   If we let $u = 5x$, then $du = 5\,dx$, so $dx = \frac{1}{5}\,du$. Therefore

$$\int \cos 5x\,dx = \frac{1}{5}\int \cos u\,du = \frac{1}{5}\sin u + C = \frac{1}{5}\sin 5x + C$$

EXAMPLE 5   Find $\int \sqrt{1+x^2}\,x^5\,dx$.

Solution   An appropriate substitution becomes more obvious if we factor x^5 as $x^4 \cdot x$. Let $u = 1 + x^2$. Then $du = 2x\,dx$, so $x\,dx = du/2$. Also $x^2 = u - 1$, so $x^4 = (u-1)^2$:

$$\int \sqrt{1+x^2}\,x^5\,dx = \int \sqrt{1+x^2}\,x^4 \cdot x\,dx$$

$$= \int \sqrt{u}\,(u-1)^2\frac{du}{2} = \frac{1}{2}\int \sqrt{u}\,(u^2 - 2u + 1)\,du$$

$$= \frac{1}{2}\int \left(u^{5/2} - 2u^{3/2} + u^{1/2}\right)du$$

$$= \frac{1}{2}\left(\frac{2}{7}u^{7/2} - 2\cdot\frac{2}{5}u^{5/2} + \frac{2}{3}u^{3/2}\right) + C$$

$$= \frac{1}{7}(1+x^2)^{7/2} - \frac{2}{5}(1+x^2)^{5/2} + \frac{1}{3}(1+x^2)^{3/2} + C$$

EXAMPLE 6   Calculate $\int \tan x\,dx$.

Solution   First we write tangent in terms of sine and cosine:

$$\int \tan x\,dx = \int \frac{\sin x}{\cos x}\,dx$$

This suggests that we should substitute $u = \cos x$ since then $du = -\sin x\,dx$ and so $\sin x\,dx = -du$:

$$\int \tan x\,dx = \int \frac{\sin x}{\cos x}\,dx = -\int \frac{du}{u}$$

$$= -\ln|u| + C = -\ln|\cos x| + C$$

Since $-\ln|\cos x| = \ln(1/|\cos x|) = \ln|\sec x|$, the result of Example 6 can also be written as

(5.34)

$$\boxed{\int \tan x\,dx = \ln|\sec x| + C}$$

When evaluating a *definite* integral by substitution, there are two possible methods. One method is to evaluate the indefinite integral first and then use the Fundamental

Theorem. For instance, using the result of Example 2, we have

$$\int_0^4 \sqrt{3x+4}\,dx = \int \sqrt{3x+4}\,dx \bigg|_0^4 = \tfrac{2}{9}(3x+4)^{3/2}\bigg|_0^4$$

$$= \tfrac{2}{9}(16)^{3/2} - \tfrac{2}{9}(4)^{3/2} = \tfrac{2}{9}(64-8) = \tfrac{112}{9}$$

Another method, which is usually preferable, is to change the limits of integration when the variable is changed.

The Substitution Rule for Definite Integrals (5.35)

If g' is continuous on $[a,b]$ and f is continuous on the range of g, then

$$\int_a^b f(g(x))g'(x)\,dx = \int_{g(a)}^{g(b)} f(u)\,du$$

Proof

$$\int_a^b f(g(x))g'(x)\,dx = \int f(g(x))g'(x)\,dx \bigg|_{x=a}^{x=b} \qquad \text{(by Equation 5.25)}$$

$$= \int f(u)\,du \bigg|_{x=a}^{x=b} \qquad [\text{(by (5.31)}]$$

$$= \int f(u)\,du \bigg|_{u=g(a)}^{u=g(b)} \qquad [\text{since } u = g(x)]$$

$$= \int_{g(a)}^{g(b)} f(u)\,du \qquad \text{(by Equation 5.25)} \qquad \bullet$$

Rule 5.35 says that when using a substitution in a definite integral, we must put everything in terms of the new variable u—not only x and dx but also the limits of integration. The new limits of integration are the values of u that correspond to $x = a$ and $x = b$.

EXAMPLE 7   Evaluate $\displaystyle\int_0^4 \sqrt{3x+4}\,dx$ using (5.35).

Solution   Using the substitution of Solution 1 of Example 2, we have $u = 3x+4$ and $dx = du/3$. To find the new limits of integration we note that

$$\text{when } x = 0,\ u = 4 \qquad \text{and} \qquad \text{when } x = 4,\ u = 16$$

Therefore

$$\int_0^4 \sqrt{3x+4}\,dx = \tfrac{1}{3}\int_4^{16} \sqrt{u}\,du$$

$$= \tfrac{1}{3} \cdot \tfrac{2}{3}u^{3/2}\bigg|_4^{16}$$

$$= \tfrac{2}{9}(16^{3/2} - 4^{3/2}) = \tfrac{112}{9} \qquad \bullet$$

Observe that when using (5.35) we do not return to the variable x after integrating. We simply evaluate the expression in u between the appropriate values of u.

EXAMPLE 8 Evaluate $\displaystyle\int_1^2 \frac{dx}{(3-5x)^2}$.

Solution Let $u = 3 - 5x$. Then $du = -5\,dx$, so $dx = -du/5$. When $x = 1$, $u = -2$; when $x = 2$, $u = -7$.

$$\int_1^2 \frac{dx}{(3-5x)^2} = -\frac{1}{5}\int_{-2}^{-7} \frac{du}{u^2}$$

$$= -\frac{1}{5}\left[-\frac{1}{u}\right]_{-2}^{-7} = \frac{1}{5u}\Big|_{-2}^{-7}$$

$$= \frac{1}{5}\left(-\frac{1}{7} + \frac{1}{2}\right) = \frac{1}{14}$$

EXAMPLE 9 Calculate $\displaystyle\int_1^e \frac{\ln x}{x}\,dx$.

Solution We let $u = \ln x$ because its differential $du = dx/x$ occurs in the integral. When $x = 1$, $u = \ln 1 = 0$; when $x = e$, $u = \ln e = 1$. Thus

$$\int_1^e \frac{\ln x}{x}\,dx = \int_0^1 u\,du = \frac{u^2}{2}\Big|_0^1 = \frac{1}{2}$$

The next theorem uses the Substitution Rule for Definite Integrals (5.35) to simplify the calculation of integrals of functions that possess symmetry properties.

Integrals of Symmetric Functions (5.36)

Suppose f is continuous on $[-a, a]$.

(a) If f is even $[f(-x) = f(x)]$, then $\int_{-a}^{a} f(x)\,dx = 2\int_0^a f(x)\,dx$.

(b) If f is odd $[f(-x) = -f(x)]$, then $\int_{-a}^{a} f(x)\,dx = 0$.

Proof Using Property 5 of integrals, we have

(5.37)
$$\int_{-a}^{a} f(x)\,dx = \int_{-a}^{0} f(x)\,dx + \int_0^a f(x)\,dx = -\int_0^{-a} f(x)\,dx + \int_0^a f(x)\,dx$$

In the first integral on the right side we make the substitution $u = -x$. Then $du = -dx$ and when $x = -a$, $u = a$. Therefore

(5.38)
$$-\int_0^{-a} f(x)\,dx = -\int_0^{a} f(-u)(-du) = \int_0^a f(-u)\,du$$

(a) If f is even, then $f(-u) = f(u)$ so Equations 5.37 and 5.38 give

$$\int_{-a}^{a} f(x)\,dx = \int_0^a f(-u)\,du + \int_0^a f(x)\,dx$$

$$= \int_0^a f(u)\,du + \int_0^a f(x)\,dx$$

$$= 2\int_0^a f$$

(b) If f is odd, then $f(-u) = -f(u)$ and so

$$\int_{-a}^{a} f(x)\,dx = \int_0^a f(-u)\,du + \int_0^a f(x)\,dx$$

$$= -\int_0^a f(u)\,du + \int_0^a f(x)\,dx = 0$$

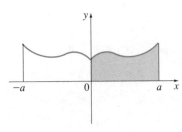

(a) f even, $\int_{-a}^{a} f = 2 \int_{0}^{a} f$

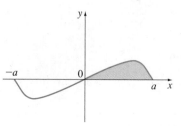

(b) f odd, $\int_{-a}^{a} f = 0$

Figure 5.24

Theorem 5.36 is illustrated by Figure 5.24. For the case where f is positive and even, part (a) says that the area under $y = f(x)$ from $-a$ to a is twice the area from 0 to a because of symmetry. Recall that an integral $\int_{a}^{b} f$ can be expressed as the area above the x-axis and below $y = f(x)$ minus the area below the axis and above the curve. Thus part (b) says the integral is 0 because the areas cancel.

EXAMPLE 10   Since $f(x) = x^6 + 1$ satisfies $f(-x) = f(x)$, it is even and so

$$\int_{-2}^{2} (x^6 + 1)\, dx = 2 \int_{0}^{2} (x^6 + 1)\, dx$$

$$= 2\left[\frac{x^7}{7} + x\right]_{0}^{2} = 2\left(\frac{128}{7} + 2\right) = \frac{284}{7}$$

EXAMPLE 11   Since $f(x) = (\tan x)/(1 + x^2 + x^4)$ satisfies $f(-x) = -f(x)$, it is odd and so

$$\int_{-1}^{1} \frac{\tan x}{1 + x^2 + x^4}\, dx = 0$$

SECTION 5.6   **Exercises**

Evaluate the integrals in Exercises 1–6 by making the given substitution.

1. $\displaystyle\int x(x^2 - 1)^{99}\, dx, \quad u = x^2 - 1$

2. $\displaystyle\int \frac{x^2}{\sqrt{2 + x^3}}\, dx, \quad u = 2 + x^3$

3. $\displaystyle\int \sin 4x\, dx, \quad u = 4x$

4. $\displaystyle\int \frac{dx}{(2x + 1)^2}, \quad u = 2x + 1$

5. $\displaystyle\int \frac{x + 3}{(x^2 + 6x)^2}\, dx, \quad u = x^2 + 6x$

6. $\displaystyle\int \sec a\theta \tan a\theta\, d\theta, \quad u = a\theta$

Evaluate the integrals in Exercises 7–78 if they exist.

7. $\displaystyle\int (2x + 1)(x^2 + x + 1)^3\, dx$

8. $\displaystyle\int x^3(1 - x^4)^5\, dx$

9. $\displaystyle\int \sqrt{x - 1}\, dx$

10. $\displaystyle\int \sqrt[3]{1 - x}\, dx$

11. $\displaystyle\int x^3 \sqrt{2 + x^4}\, dx$

12. $\displaystyle\int x(x^2 + 1)^{3/2}\, dx$

13. $\displaystyle\int \frac{3x - 1}{(3x^2 - 2x + 1)^4}\, dx$

14. $\displaystyle\int \frac{x}{\sqrt{x^2 + 1}}\, dx$

15. $\displaystyle\int \frac{2}{(t + 1)^6}\, dt$

16. $\displaystyle\int \frac{1}{(1 - 3t)^4}\, dt$

17. $\displaystyle\int (1 - 2y)^{1.3}\, dy$

18. $\displaystyle\int \sqrt[5]{3 - 5y}\, dy$

19. $\displaystyle\int \cos 2\theta\, d\theta$

20. $\displaystyle\int \sec^2 3\theta\, d\theta$

21. $\displaystyle\int \frac{x}{\sqrt[4]{x + 2}}\, dx$

22. $\displaystyle\int \frac{x^2}{\sqrt{1 - x}}\, dx$

23. $\displaystyle\int t \sin(t^2)\, dt$

24. $\displaystyle\int \frac{(1 + \sqrt{x})^9}{\sqrt{x}}\, dx$

25. $\displaystyle\int x^3(1 - x^2)^{3/2}\, dx$

26. $\displaystyle\int t^2 \cos(1 - t^3)\, dt$

27. $\displaystyle\int \sin^3 x \cos x\, dx$

28. $\displaystyle\int \frac{\cos \sqrt{x}}{\sqrt{x}}\, dx$

29. $\displaystyle\int \frac{dx}{2x - 1}$

30. $\displaystyle\int \frac{x}{x^2 + 1}\, dx$

31. $\displaystyle\int \frac{(\ln x)^2}{x}\, dx$

32. $\displaystyle\int x e^{x^2}\, dx$

33. $\displaystyle\int e^x(1 + e^x)^{10}\, dx$

34. $\displaystyle\int \frac{\tan^{-1}x}{1+x^2}\,dx$

35. $\displaystyle\int \sec x \tan x \sqrt{1+\sec x}\,dx$

36. $\displaystyle\int \cos^4 x \sin x\,dx$ **37.** $\displaystyle\int \frac{ax+b}{\sqrt{ax^2+2bx+c}}\,dx$

38. $\displaystyle\int \tan^2\theta \sec^2\theta\,d\theta$ **39.** $\displaystyle\int \sin(2x+3)\,dx$

40. $\displaystyle\int \cos(7-3x)\,dx$ **41.** $\displaystyle\int (\sin 3\alpha - \sin 3x)\,dx$

42. $\displaystyle\int \sqrt[3]{x^3+1}\,x^5\,dx$

43. $\displaystyle\int x^a\sqrt{b+cx^{a+1}}\,dx$ $(c \neq 0,\ a \neq -1)$

44. $\displaystyle\int \cos x \cos(\sin x)\,dx$ **45.** $\displaystyle\int \frac{dx}{x \ln x}$

46. $\displaystyle\int e^x \sin(e^x)\,dx$ **47.** $\displaystyle\int \frac{e^x+1}{e^x}\,dx$

48. $\displaystyle\int \frac{e^x}{e^x+1}\,dx$ **49.** $\displaystyle\int \frac{x+1}{x^2+2x}\,dx$

50. $\displaystyle\int \frac{\sin x}{1+\cos^2 x}\,dx$ **51.** $\displaystyle\int \frac{e^x}{e^{2x}+1}\,dx$

52. $\displaystyle\int \frac{dx}{1+9x^2}$ **53.** $\displaystyle\int \frac{1+x}{1+x^2}\,dx$

54. $\displaystyle\int \frac{x^2+1}{x^3+3x+1}\,dx$ **55.** $\displaystyle\int \frac{x}{1+x^4}\,dx$

56. $\displaystyle\int \frac{x}{x+1}\,dx$ **57.** $\displaystyle\int_0^1 (2x-1)^{100}\,dx$

58. $\displaystyle\int_0^{-4} \sqrt{1-2x}\,dx$ **59.** $\displaystyle\int_0^1 (x^4+x)^5(4x^3+1)\,dx$

60. $\displaystyle\int_2^3 \frac{3x^2-1}{(x^3-x)^2}\,dx$ **61.** $\displaystyle\int_1^2 x\sqrt{x-1}\,dx$

62. $\displaystyle\int_0^4 \frac{x}{\sqrt{1+2x}}\,dx$ **63.** $\displaystyle\int_0^1 \cos \pi t\,dt$

64. $\displaystyle\int_0^{\pi/4} \sin 4t\,dt$ **65.** $\displaystyle\int_1^4 \frac{1}{x^2}\sqrt{1+\frac{1}{x}}\,dx$

66. $\displaystyle\int_0^2 \frac{dx}{(2x-3)^2}$ **67.** $\displaystyle\int_0^{\pi/3} \frac{\sin\theta}{\cos^2\theta}\,d\theta$

68. $\displaystyle\int_{-\pi/2}^{\pi/2} \frac{x^2\sin x}{1+x^6}\,dx$ **69.** $\displaystyle\int_0^{13} \frac{dx}{\sqrt[3]{(1+2x)^2}}$

70. $\displaystyle\int_{-\pi/3}^{\pi/3} \sin^5\theta\,d\theta$ **71.** $\displaystyle\int_0^a x\sqrt{a^2-x^2}\,dx$

72. $\displaystyle\int_0^a x\sqrt{x^2+a^2}\,dx$ $(a>0)$

73. $\displaystyle\int_{-a}^a x\sqrt{x^2+a^2}\,dx$ **74.** $\displaystyle\int_0^4 \frac{dx}{(x-2)^3}$

75. $\displaystyle\int_0^3 \frac{dx}{2x+3}$ **76.** $\displaystyle\int_0^1 t^2 2^{-t^3}\,dt$

77. $\displaystyle\int_e^{e^4} \frac{dx}{x\sqrt{\ln x}}$ **78.** $\displaystyle\int_0^{1/2} \frac{\sin^{-1}x}{\sqrt{1-x^2}}\,dx$

In Exercises 79–86 find the area under the given curve.

79. $y = \sqrt{x+1},\ 0 \le x \le 3$

80. $y = \sqrt{2px},\ 0 \le x \le 1$

81. $y = \sin\left(\frac{x}{2}\right),\ 0 \le x \le \frac{\pi}{3}$

82. $y = 3\cos\left(\frac{x}{3}\right),\ -\pi \le x \le \pi$

83. $y = \dfrac{1}{(x+1)^2},\ 0 \le x \le 10$

84. $y = x(x^2+1)^4,\ 1 \le x \le 2$

85. $y = 2e^{-2x},\ 0 \le x \le 1$

86. $y = \dfrac{2}{x-2},\ 3 \le x \le 5$

87. Breathing is cyclic and a full respiratory cycle from the beginning of inhalation to the end of exhalation takes about 5 s. The maximum rate of air flow into the lungs is about 0.5 L/s. This explains, in part, why the function $f(t) = \frac{1}{2}\sin(2\pi t/5)$ has often been used to model the rate of air flow into the lungs. Use this model to find the volume of inhaled air in the lungs at time t.

88. Alabama Instruments Company has set up a production line to manufacture a new calculator. The rate of production of these calculators after t weeks is

$$\frac{dx}{dt} = 5000\left(1 - \frac{100}{(t+10)^2}\right) \quad \text{calculators/week}$$

(Notice that production approaches 5000 per week as time goes on, but the initial production is less because of the workers' unfamiliarity with the new techniques.) Find the number of calculators produced during the third and fourth weeks.

89. If f is continuous on R, prove that

$$\int_a^b f(-x)\,dx = \int_{-b}^{-a} f(x)\,dx$$

90. If f is continuous on R, prove that

$$\int_a^b f(x+c)\,dx = \int_{a+c}^{b+c} f(x)\,dx$$

91. If a and b are positive numbers, show that
$$\int_0^1 x^a(1-x)^b\,dx = \int_0^1 x^b(1-x)^a\,dx$$

92. Use the substitution $u = \pi - x$ to show that
$$\int_0^\pi xf(\sin x)\,dx = \frac{\pi}{2}\int_0^\pi f(\sin x)\,dx$$

93. Show that $\int \cot x\,dx = \ln|\sin x| + C$ (a) by differentiating the right side of the equation and (b) by using the method of Example 6.

94. Use Exercise 92 to evaluate the integral
$$\int_0^\pi \frac{x\sin x}{1+\cos^2 x}\,dx$$

SECTION 5.7

The Logarithm Defined as an Integral

In this section we use the Fundamental Theorem of Calculus to give an alternative treatment of the functions $\ln x, e^x, a^x, \log_a x$, and their derivatives. Instead of starting with a^x and defining $\log_a x$ as its inverse, this time we start by defining $\ln x$ as an integral and then define the exponential function as its inverse. In this section you should bear in mind that we do not use the definitions and results of Sections 3.1, 3.2, 3.4, and 3.5.

The Natural Logarithm

Definition (5.39)

> The **natural logarithmic function** is the function defined by
> $$\ln x = \int_1^x \frac{1}{t}\,dt \qquad x > 0$$

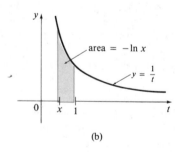

(a)

area $= \ln x$

$y = \frac{1}{t}$

(b)

area $= -\ln x$

$y = \frac{1}{t}$

Figure 5.25

The existence of this function depends on Theorem 5.11, which guarantees that a continuous function is integrable. If $x > 1$, $\ln x$ can be interpreted geometrically as the area under the hyperbola $y = 1/t$ from $t = 1$ to $t = x$ [see Figure 5.25(a)]. For $x = 1$, we have
$$\ln 1 = \int_1^1 \frac{1}{t}\,dt = 0$$

For $0 < x < 1$,
$$\ln x = \int_1^x \frac{1}{t}\,dt = -\int_x^1 \frac{1}{t}\,dt < 0$$

and so $\ln x$ is the negative of the area shown in Figure 5.25(b).

EXAMPLE 1
(a) By comparing areas, show that $\frac{1}{2} < \ln 2 < \frac{3}{4}$.
(b) Use the Midpoint Rule with $n = 10$ to estimate the value of $\ln 2$.

Solution
(a) We can interpret $\ln 2$ as the area under the curve $y = 1/t$ from 1 to 2. From Figure 5.26 we see that this area is larger than the area of rectangle $BCDE$ and smaller than the area of trapezoid $ABCD$. Thus we have
$$\tfrac{1}{2}\cdot 1 < \ln 2 < 1\cdot\tfrac{1}{2}\left(1+\tfrac{1}{2}\right)$$
$$\tfrac{1}{2} < \ln 2 < \tfrac{3}{4}$$

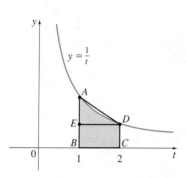

Figure 5.26

(b)   If we use the Midpoint Rule with $f(t) = 1/t$, $n = 10$, and $\Delta t = 0.1$, we get

$$\ln 2 = \int_1^2 \frac{1}{t}\, dt \approx (0.1)[f(1.05) + f(1.15) + \cdots + f(1.95)]$$

$$= (0.1)\left(\frac{1}{1.05} + \frac{1}{1.15} + \cdots + \frac{1}{1.95}\right) \approx 0.693$$ ●

Notice that the integral that defines $\ln x$ is exactly the type of integral discussed in Part 1 of the Fundamental Theorem of Calculus (5.17). In fact, using that theorem, we have

$$\frac{d}{dx}\int_1^x \frac{1}{t}\, dt = \frac{1}{x}$$

and so

(5.40)

$$\boxed{\frac{d}{dx}\ln x = \frac{1}{x}}$$

We now use this differentiation rule to prove the following properties of the logarithm function.

Laws of Logarithms (5.41)

If x and y are positive numbers and r is a rational number, then

(a) $\ln(xy) = \ln x + \ln y$

(b) $\ln\!\left(\frac{x}{y}\right) = \ln x - \ln y$

(c) $\ln(x^r) = r \ln x$

Proof

(a)   Let $f(x) = \ln(ax)$, where a is a positive constant. Then, using Equation 5.40 and the Chain Rule, we have

$$f'(x) = \frac{1}{ax}\frac{d}{dx}(ax) = \frac{1}{ax}\cdot a = \frac{1}{x}$$

Therefore $f(x)$ and $\ln x$ have the same derivative and so they must differ by a constant:

$$\ln(ax) = \ln x + C$$

Putting $x = 1$ in this equation, we get $\ln a = \ln 1 + C = 0 + C = C$. Thus

$$\ln(ax) = \ln x + \ln a$$

If we now replace the constant a by any number y, we have

$$\ln(xy) = \ln x + \ln y$$

(b)   Using law (a) with $x = 1/y$, we have

$$\ln\frac{1}{y} + \ln y = \ln\!\left(\frac{1}{y}\cdot y\right) = \ln 1 = 0$$

and so

$$\ln\frac{1}{y} = -\ln y$$

Using (a) again, we have

$$\ln\left(\frac{x}{y}\right) = \ln\left(x \cdot \frac{1}{y}\right) = \ln x + \ln\frac{1}{y} = \ln x - \ln y$$

The proof of (c) is left to you.   ●

Theorem (5.42)

> (a) $\displaystyle\lim_{x\to\infty} \ln x = \infty$ (b) $\displaystyle\lim_{x\to 0^+} \ln x = -\infty$

Proof

(a)   Using (5.41) law (c) with $x = 2$ and $r = n$ (any positive integer), we have $\ln(2^n) = n\ln 2$. Now $\ln 2 > 0$, so this shows that $\ln(2^n) \to \infty$ as $n \to \infty$. But $\ln x$ is an increasing function since its derivative $1/x > 0$. Therefore $\ln x \to \infty$ as $x \to \infty$.   ●

(b)   If we let $t = 1/x$, then $t \to \infty$ as $x \to 0^+$. Thus, using (a), we have

$$\lim_{x\to 0^+} \ln x = \lim_{t\to\infty} \ln\left(\frac{1}{t}\right) = \lim_{t\to\infty}(-\ln t) = -\infty$$ ●

If $y = \ln x$, $x > 0$, then

$$\frac{dy}{dx} = \frac{1}{x} > 0 \qquad \text{and} \qquad \frac{d^2y}{dx^2} = -\frac{1}{x^2} < 0$$

which shows that $\ln x$ is increasing and concave downward on $(0,\infty)$. Putting this information together with Theorem 5.42, we draw the graph of $y = \ln x$ in Figure 5.27.

Since $\ln 1 = 0$ and $\ln x$ is an increasing continuous function that takes on arbitrarily large values, the Intermediate Value Theorem (1.25) shows that there is a number where $\ln x$ takes on the value 1 (see Figure 5.28). This important number is denoted by e.

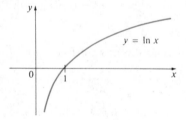

Figure 5.27   Figure 5.28

Definition (5.43)

> e is the number such that $\ln e = 1$

We will show (in Theorem 5.58) that this definition is consistent with our previous definition of e.

The Natural Exponential Function

Since $\ln$ is an increasing function, it is one-to-one and therefore has an inverse function, which we denote by exp. Thus, according to the definition of an inverse function,

(5.44)

and the cancellation equations are

(5.45)

$$\exp(\ln x) = x \quad \text{and} \quad \ln(\exp x) = x$$

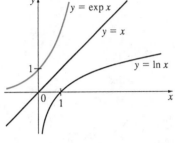

Figure 5.29

In particular, we have

$$\exp(0) = 1 \qquad \text{since } \ln 1 = 0$$

$$\exp(1) = e \qquad \text{since } \ln e = 1$$

We obtain the graph of $y = \exp x$ by reflecting the graph of $y = \ln x$ about the line $y = x$ (see Figure 5.29). The domain of exp is the range of ln, that is, $(-\infty, \infty)$; the range of exp is the domain of ln, that is, $(0, \infty)$.

If r is any rational number, then (5.41) law (c) gives

$$\ln(e^r) = r \ln e = r$$

Therefore, by (5.44), $\qquad \exp(r) = e^r$

Thus $\exp(x) = e^x$ whenever x is a rational number. This leads us to define e^x, even for irrational values of x, by the equation

(5.46)

$$e^x = \exp(x)$$

In other words, for the reasons given, we define e^x to be the inverse of the function $\ln x$. In this notation (5.44) becomes

(5.47)

$$e^x = y \quad \Leftrightarrow \quad \ln y = x$$

and the cancellation equations (5.45) become

(5.48)

$$e^{\ln x} = x \qquad x > 0$$

(5.49)

$$\ln e^x = x \qquad \text{all } x$$

The function $f(x) = e^x$ has domain $(-\infty, \infty)$ and range $(0, \infty)$; that is, $e^x > 0$ for all x. It follows from Theorem 5.42 that

(5.50)

$$\lim_{x \to \infty} e^x = \infty \qquad \text{and} \qquad \lim_{x \to -\infty} e^x = 0$$

We now verify that f has the properties expected of an exponential function.

Laws of Exponents (5.51)

If x and y are real numbers and r is rational, then

(a) $e^{x+y} = e^x e^y$ \qquad (b) $e^{x-y} = \dfrac{e^x}{e^y}$

(c) $(e^x)^r = e^{rx}$

Proof

(a)   From (5.41) and (5.49) we have

$$\ln(e^x e^y) = \ln(e^x) + \ln(e^y) = x + y = \ln(e^{x+y})$$

Since ln is a one-to-one function, it follows that $e^x e^y = e^{x+y}$.

Parts (b) and (c) are proved similarly. As we will see in (5.54), part (c) actually holds when r is any real number.

Theorem (5.52)

$$\frac{d}{dx} e^x = e^x$$

Proof   The function $y = e^x$ is differentiable because it is the inverse function of $y = \ln x$, which we know is differentiable. To find its derivative, we use the inverse function method. Let $y = e^x$. Then $\ln y = x$ and, differentiating this latter equation implicitly with respect to x, we get

$$\frac{1}{y} \frac{dy}{dx} = 1$$

$$\frac{dy}{dx} = y = e^x$$

General Exponential Functions

If $a > 0$ and r is any rational number, then by Equation 5.48 and (5.51) law (c),

$$a^r = (e^{\ln a})^r = e^{r \ln a}$$

Therefore, even for irrational numbers x, we *define*

(5.53)

$$a^x = e^{x \ln a}$$

The function $f(x) = a^x$ is called the **exponential function with base a.** Notice that a^x is positive for all x because e^x is positive for all x. The general laws of exponents follow from Definition 5.53 together with the laws of exponents for e^x.

Laws of Exponents (5.54)

If x and y are real numbers and a, $b > 0$, then

(a) $a^{x+y} = a^x a^y$ (b) $a^{x-y} = \dfrac{a^x}{a^y}$

(c) $(a^x)^y = a^{xy}$ (d) $(ab)^x = a^x b^x$

Proof

(a)   Using Definition 5.53 and (5.51), we have

$$a^{x+y} = e^{(x+y)\ln a} = e^{x \ln a + y \ln a}$$
$$= e^{x \ln a} \, e^{y \ln a} = a^x a^y$$

(c)

$$(a^x)^y = e^{y \ln(a^x)} = e^{y \ln(e^{x \ln a})}$$
$$= e^{yx \ln a} = e^{xy \ln a} = a^{xy}$$ ●

The remaining proofs are left as exercises.

Theorem (5.55)

$$\frac{d}{dx} a^x = a^x \ln a$$

Proof
$$\frac{d}{dx} a^x = \frac{d}{dx}(e^{x \ln a}) = e^{x \ln a}\left(\frac{d}{dx} x \ln a\right)$$

$$= a^x \ln a$$ ●

If $a > 1$, then $\ln a > 0$, so $D_x a^x = a^x \ln a > 0$, which shows that $y = a^x$ is increasing (see Figure 5.30). If $0 < a < 1$, then $\ln a < 0$ and so $y = a^x$ is decreasing (see Figure 5.31).

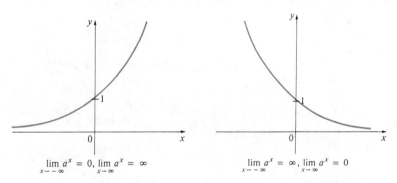

$$\lim_{x \to -\infty} a^x = 0, \ \lim_{x \to \infty} a^x = \infty \qquad\qquad \lim_{x \to -\infty} a^x = \infty, \ \lim_{x \to \infty} a^x = 0$$

Figure 5.30   **Figure 5.31**
$y = a^x, \ a > 1$ $y = a^x, \ 0 < a < 1$

General Logarithmic Functions

If $a > 0$ and $a \neq 1$, then $f(x) = a^x$ is a one-to-one function. Its inverse function is called the **logarithmic function with base a** and is denoted by $\log_a$. Thus

(5.56)

$$\boxed{\log_a x = y \qquad \Leftrightarrow \qquad a^y = x}$$

In particular, we see that

$$\log_e x = \ln x$$

The laws of logarithms are similar to those for the natural logarithm and can be deduced from the laws of exponents (see Exercise 11).

Theorem (5.57)

$$\boxed{\frac{d}{dx} \log_a x = \frac{1}{x \ln a}}$$

Proof   First note that $\log_a x$ is differentiable because it is the inverse function of the differentiable function a^x. Let $y = \log_a x$. Then $a^y = x$, and using Theorem 5.55 to differentiate both sides of the latter equation, we have

$$a^y \ln a \frac{dy}{dx} = 1$$

$$\frac{dy}{dx} = \frac{1}{a^y \ln a} = \frac{1}{x \ln a}$$

An alternative proof is indicated in Exercise 10.

The Number e Expressed as a Limit

In this section we defined e as the number such that $\ln e = 1$. The next theorem shows that this is the same as the number e defined in Section 3.2 (see Equation 3.33).

Theorem (5.58)

$$\boxed{\lim_{x \to 0} (1 + x)^{1/x} = e}$$

Proof   Let $f(x) = \ln x$. Then $f'(x) = 1/x$, so $f'(1) = 1$. But, by the definition of derivative,

$$f'(1) = \lim_{h \to 0} \frac{f(1 + h) - f(1)}{h} = \lim_{x \to 0} \frac{f(1 + x) - f(1)}{x}$$

$$= \lim_{x \to 0} \frac{\ln(1 + x) - \ln 1}{x} = \lim_{x \to 0} \frac{1}{x} \ln(1 + x)$$

$$= \lim_{x \to 0} \ln(1 + x)^{1/x}$$

$$= \ln \left[\lim_{x \to 0} (1 + x)^{1/x} \right] \qquad \text{(since ln is continuous)}$$

Therefore

$$\ln \left[\lim_{x \to 0} (1 + x)^{1/x} \right] = 1$$

It follows that

$$\lim_{x \to 0} (1 + x)^{1/x} = e$$

SECTION 5.7 **Exercises**

In these exercises use only the definitions and results of this section.

1. (a)   By comparing areas, show that
    $$\frac{1}{3} < \ln 1.5 < \frac{5}{12}$$

    (b)   Use the Midpoint Rule with $n = 10$ to estimate $\ln 1.5$.

2. Refer to Example 1. (a) Find the equation of the tangent line to the curve $y = 1/t$ that is parallel to the secant line AD. (b) Use part (a) to show that $\ln 2 > 0.66$.

3. By comparing areas, show that
    $$\frac{1}{2} + \frac{1}{3} + \cdots + \frac{1}{n} < \ln n < 1 + \frac{1}{2} + \frac{1}{3} + \cdots + \frac{1}{n-1}$$

4. (a)   By comparing areas, show that $\ln 2 < 1 < \ln 3$.
    (b)   Deduce that $2 < e < 3$.

5. Prove (5.41) law (c). (*Hint:* Start by showing that both sides of the equation have the same derivative.)

6. Prove (5.51) law (b).

7. Prove (5.51) law (c).

8. Prove (5.54) law (b).

9. Prove (5.54) law (d).

10. (a)   Show that $\log_a x = (\ln x)/\ln a$. [*Hint:* Let $y = \log_a x$ and use (5.56).]
    (b)   Deduce that $D_x \log_a x = 1/(x \ln a)$.

11. Deduce the following laws of logarithms from (5.54):
    (a) $\log_a (xy) = \log_a x + \log_a y$
    (b) $\log_a (x/y) = \log_a x - \log_a y$
    (c) $\log_a (x^y) = y \log_a x$

CHAPTER 5

Review

Key Topics

Define, state, or discuss the following.

1. Sigma notation
2. Partition of $[a, b]$
3. Norm of a partition
4. Area under a curve
5. Riemann sum
6. Definite integral of f from a to b
7. Integrable function
8. Integrand; limits of integration

9. Midpoint Rule
10. Piecewise continuous function
11. Bounded function
12. Properties of the definite integral
13. Fundamental Theorem of Calculus
14. Indefinite integral
15. Substitution Rule
16. Integrals of even or odd functions

Exercises

In Exercises 1–12 determine whether the statement is true or false.

1. $\sum_{i=1}^{n} (a_i + b_i) = \sum_{i=1}^{n} a_i + \sum_{i=1}^{n} b_i$

2. $\sum_{i=1}^{n} a_i b_i = \left(\sum_{i=1}^{n} a_i \right)\left(\sum_{i=1}^{n} b_i \right)$

3. $\frac{d}{dx} \sum_{i=1}^{n} f_i(x) = \sum_{i=1}^{n} \frac{d}{dx} f_i(x)$

4. $\sum_{i=1}^{12} 3 = 36$

5. If f and g are integrable on $[a, b]$, then
    $$\int_a^b [f(x) + g(x)]\, dx = \int_a^b f(x)\, dx + \int_a^b g(x)\, dx$$

6. If f and g are integrable on $[a, b]$, then
    $$\int_a^b [f(x) g(x)]\, dx = \left(\int_a^b f(x)\, dx \right)\left(\int_a^b g(x)\, dx \right)$$

7. If f and g are continuous and $f(x) \geq g(x)$ for $a \leq x \leq b$, then
    $$\int_a^b f(x)\, dx \geq \int_a^b g(x)\, dx$$

8. If f and g are differentiable and $f(x) \geq g(x)$ for $a \leq x \leq b$, then $f'(x) \geq g'(x)$ for $a \leq x \leq b$.

9. $\int_{-1}^{1} \left(x^5 - 6x^9 + \frac{\sin x}{(1 + x^4)^2} \right) dx = 0$

10. If $\int_0^1 f(x)\, dx$ exists, then f is a bounded function on $[0, 1]$.

11. $\int_{-2}^{1} \frac{1}{x^4}\, dx = -\frac{3}{8}$

12. $\int_0^2 (x - x^3)\, dx$ represents the area under the curve $y = x - x^3$ from 0 to 2.

Find the value of the sum in Exercises 13–16.

13. $\sum_{i=2}^{5} \frac{1}{(-10)^i}$

14. $\sum_{j=1}^{5} \sin\left(\frac{j\pi}{2}\right)$

15. $\sum_{k=1}^{n} (2 + k^3)$

16. $\sum_{i=1}^{n} i(i - 2)$

17. Find the Riemann sum for the function $f(x) = 2 + (x - 2)^2$ on the interval $[0, 2]$ using a regular partition with $n = 4$ and choosing x_i^* to be the left endpoint of the ith subinterval. Sketch the graph of f and the approximating rectangles.

18. Do Exercise 17 if x_i^* is the midpoint of the ith subinterval.

Evaluate the integrals in Exercises 19–22 without using the Fundamental Theorem of Calculus.

19. $\int_2^4 (3 - 4x)\, dx$

20. $\int_1^2 (x + 3x^2)\, dx$

21. $\int_0^5 (x^3 - 2x^2)\, dx$

22. $\int_0^b (x^3 + 4x - 1)\, dx$

In Exercises 23–50 evaluate the integral if it exists.

23. $\int_0^5 (x^3 - 2x^2)\, dx$

24. $\int_0^b (x^3 + 4x - 1)\, dx$

25. $\int_0^1 (1 - x^9)\, dx$

26. $\int_0^1 (1 - x)^9\, dx$

27. $\int_1^8 \sqrt[3]{x}\,(x - 1)\, dx$

28. $\int_1^4 \frac{x^2 - x + 1}{\sqrt{x}}\, dx$

29. $\int_0^2 x^2(1 + 2x^3)^3\, dx$

30. $\int_0^4 x\sqrt{16 - 3x}\, dx$

31. $\int_3^{11} \frac{dx}{\sqrt{2x + 3}}$

32. $\int_0^2 \frac{x}{(x^2 - 1)^2}\, dx$

33. $\int_{-2}^{-1} \frac{dx}{(2x + 3)^4}$

34. $\int_{-1}^{1} \frac{x + x^3 + x^5}{1 + x^2 + x^4}\, dx$

35. $\int \frac{x^4}{(2 + x^5)^6}\, dx$

36. $\int (1 - x)\sqrt{2x - x^2}\, dx$

37. $\int \sin \pi x\, dx$

38. $\int \frac{\cos x}{\sqrt{1 + \sin x}}\, dx$

39. $\int \frac{\cos(1/t)}{t^2}\, dt$

40. $\int \csc^2 3t\, dt$

41. $\int \sin x \sec^2(\cos x)\, dx$

42. $\int \frac{x^3}{\sqrt{x^2 + 1}}\, dx$

43. $\int \frac{e^{\sqrt{x}}}{\sqrt{x}}\, dx$

44. $\int \frac{\cos(\ln x)}{x}\, dx$

45. $\int \tan x \ln(\cos x)\, dx$

46. $\int \frac{x}{\sqrt{1 - x^4}}\, dx$

47. $\int \frac{x^3}{1 + x^4}\, dx$

48. $\int \frac{e^x}{(e^x + 1)\ln(e^x + 1)}\, dx$

49. $\int_0^{2\pi} |\sin x|\, dx$

50. $\int_0^8 |x^2 - 6x + 8|\, dx$

Find the derivatives of the functions in Exercises 51–58.

51. $F(x) = \int_1^x \sqrt{1 + t^2 + t^4}\, dt$

52. $F(x) = \int_\pi^x \tan(s^2)\, ds$

53. $g(x) = \int_0^{x^3} \frac{t}{\sqrt{1 + t^3}}\, dt$

54. $g(x) = \int_1^{\cos x} \sqrt[3]{1 - t^2}\, dt$

55. $y = \int_{\sqrt{x}}^{x} \frac{\cos \theta}{\theta}\, d\theta$

56. $y = \int_{2x}^{3x+1} \sin(t^4)\, dt$

57. $f(x) = \int_1^{\sqrt{x}} \frac{e^s}{s}\, ds$

58. $f(x) = \int_{\ln x}^{2x} e^{-t^2}\, dt$

In Exercises 59 and 60 use Property 8 of integrals to estimate the value of the given integral.

59. $\displaystyle\int_1^3 \sqrt{x^2 + 3}\, dx$ **60.** $\displaystyle\int_3^5 \frac{1}{x+1}\, dx$

In Exercises 61–66 use the properties of integrals to verify the given inequalities.

61. $\displaystyle\int_0^\pi \cos^8 x\, dx \le \int_0^\pi \cos^6 x\, dx$

62. $\displaystyle\int_0^{\sqrt{\pi/2}} \sin(x^2)\, dx \le \int_0^{\sqrt{\pi/2}} \cos(x^2)\, dx$

63. $\displaystyle\int_0^1 x^2 \cos x\, dx \le 1/3$

64. $\displaystyle\int_{\pi/4}^{\pi/2} \frac{\sin x}{x}\, dx \le \sqrt{2}/2$

65. $\displaystyle\int_0^1 e^x \cos x\, dx \le e - 1$

66. $\displaystyle\int_0^1 x \sin^{-1} x\, dx \le \pi/4$

67. Show that if n is a positive integer, then
$$\int_0^n [\![x]\!]\, dx = \tfrac{1}{2}n(n-1).$$

68. A particle moves along a line with velocity function $v(t) = t^2 - t$. Find (a) the displacement and (b) the distance traveled by the particle during the time interval $[0, 5]$.

69. Evaluate $\int_{-4}^4 \sqrt{16 - x^2}\, dx$ by interpreting the integral as an area.

70. Let
$$f(x) = \begin{cases} -x - 1 & \text{if } -3 \le x \le 0 \\ -\sqrt{1 - x^2} & \text{if } 0 \le x \le 1 \end{cases}$$

Evaluate $\int_{-3}^1 f(x)\, dx$ by interpreting the integral as a difference of areas.

71. Evaluate the integral $\int_0^1 \sqrt{x}\, dx$ without using the Fundamental Theorem of Calculus. [*Hint:* Use partitions with $x_i^* = x_i = i^2/n^2$.]

72. Find a function f and a value of the constant a such that
$$2 \int_a^x f(t)\, dt = 2 \sin x - 1$$

73. If f' is continuous on $[a, b]$, show that
$$2 \int_a^b f(x) f'(x)\, dx = [f(b)]^2 - [f(a)]^2$$

74. Find $\displaystyle\lim_{h \to 0} \frac{1}{h} \int_2^{2+h} \sqrt{1 + t^3}\, dt$.

75. If f is continuous on $[0, 1]$, prove that
$$\int_0^1 f(x)\, dx = \int_0^1 f(1 - x)\, dx$$

76. Evaluate
$$\lim_{n \to \infty} \frac{1}{n}\left[\left(\frac{1}{n}\right)^9 + \left(\frac{2}{n}\right)^9 + \left(\frac{3}{n}\right)^9 + \cdots + \left(\frac{n}{n}\right)^9 \right]$$

77. Show that the area of the shaded hyperbolic sector in Figure 3.35 is $A(t) = \tfrac{1}{2}t$. [*Hint:* First show that
$$A(t) = \tfrac{1}{2}\sinh t \cosh t - \int_1^{\cosh t} \sqrt{x^2 - 1}\, dx$$
and then verify that $A'(t) = \tfrac{1}{2}$.]

78. Evaluate $\displaystyle\lim_{x \to 0} \frac{1}{x^3} \int_0^x \sin(t^2)\, dt$.

79. If $F(x) = \int_a^b t^x\, dt$, where $a, b > 0$, then, by the Fundamental Theorem,
$$F(x) = \frac{b^{x+1} - a^{x+1}}{x + 1} \qquad x \ne -1$$
$$F(-1) = \ln b - \ln a$$

Use l'Hospital's Rule to show that F is continuous at -1.

Applications Plus

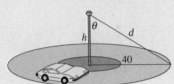

1. A light is to be placed on a pole of height h feet to illuminate a busy traffic circle, which has a radius of 40 ft. The intensity of illumination I at any point P on the circle is directly proportional to the cosine of the angle θ (see figure) and inversely proportional to the square of the distance d from the source.
   (a) How high should the light pole be to maximize I?
   (b) Suppose that the light pole is h feet tall and that a woman is walking away from the base of the pole at the rate of 4 ft/s. At what rate is the intensity of the light at the point on her back 4 ft above the ground decreasing when she reaches the traffic circle?

2. When a foreign object in a person's trachea (windpipe) leads to a cough, the diaphragm thrusts upward causing an increase in pressure in the lungs. This is accompanied by a contraction of the trachea, making a narrower channel for the expelled air to flow through. For a given amount of air to escape in a fixed amount of time, it must move faster through the narrower channel than the wider one. The greater the velocity of the airstream, the greater the force on the foreign object. X rays show that the radius of the circular tracheal tube contracts to two-thirds its normal radius during a cough. It can be shown that the velocity v of the airstream is related to the radius r of the trachea by the equation

$$v(r) = k(r_0 - r)r^2 \text{ cm/s} \qquad \tfrac{1}{2}r_0 \le r \le r_0$$

   where k is a constant and r_0 is the "rest radius" of the trachea. The restriction on r is due to the fact that the tracheal wall stiffens under pressure and a contraction greater than $r_0/2$ is prevented (otherwise we would suffocate).
   (a) Determine the value of r in the interval $[r_0/2, r_0]$ at which v has an absolute maximum. How does this compare with experimental evidence?
   (b) What is the absolute maximum of v on the interval?
   (c) Sketch the graph of v for r in the interval $[0, r_0]$.

3. A high-speed "bullet" train accelerates and decelerates at the rate of 4 ft/s^2. Its maximum cruising speed is 90 mi/h.
   (a) What is the maximum distance the train can travel if it accelerates from rest until it reaches its cruising speed and then runs at that speed for 15 minutes?
   (b) Suppose that the train starts from rest and must come to a complete stop in 15 minutes. What is the maximum distance it can travel under these conditions?

4. A high-tech company purchases a new computing system whose initial value is V. The system will depreciate at the rate $f = f(t)$ and will accumulate maintenance costs at the rate $g = g(t)$, where t is the time measured in months. The company wants to determine the optimum time to replace the system.

   (a) Let
   $$C(t) = \frac{1}{t}\int_0^t [f(s) + g(s)]\,ds$$

   Show that the critical numbers of C occur at the numbers t where $C(t) = f(t) + g(t)$.
   (b) Suppose that
   $$f(t) = \begin{cases} \dfrac{V}{15} - \dfrac{V}{450}t & \text{if } 0 < t \le 30 \\ 0 & \text{if } t > 30 \end{cases} \quad \text{and} \quad g(t) = \frac{Vt^2}{12,900} \text{ if } t > 0$$

Determine the length of time T for the total depreciation $D(t) = \int_0^t f(s)\,ds$ to equal the initial value V.

(c) Determine the absolute minimum of C on $(0, T]$.

(d) Sketch the graphs of C and $f + g$ in the same coordinate system, and verify the result in part (a) in this case.

5. A manufacturing company has a major piece of equipment that depreciates at the (continuous) rate $f = f(t)$, where t is the time measured in months since its last overhaul. Since the company incurs a fixed cost A each time the machine is overhauled, it wants to determine the optimal time T (in months) between overhauls.

(a) Show that $\int_0^t f(s)\,ds$ represents the loss in value of the machine over the period of time t since the last overhaul.

(b) Let $C = C(t)$ be given by

$$C(t) = \frac{1}{t}\left[A + \int_0^t f(s)\,ds \right]$$

What does C represent and why would the company want to minimize C?

(c) Show that C has a minimum value at the numbers $t = T$ where $C(T) = f(T)$.

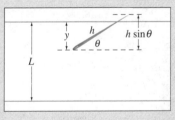

6. Ornithologists have determined that some species of birds tend to avoid flights over large bodies of water during the daylight hours. It is believed that more energy is required to fly over water than land because air generally rises over land and falls over water during the day. A bird with these tendencies is released from an island which is 5 km from the nearest point B on a straight shoreline, flies to a point C on the shoreline, and then flies along the shoreline to its nesting area D. Assume that the bird instinctively chooses a path that will minimize its energy expenditure. Points B and D are 13 km apart.

(a) In general, if it takes 1.4 times as much energy to fly over water as land, to what point C should the bird fly in order to minimize the total energy expended in returning to its nesting area?

(b) Let W denote the energy (in joules) required to fly over water and L the energy required to fly over land. Determine the ratio W/L corresponding to the mimimum expenditure of energy.

(c) What should the value of W/L be in order for the bird to fly directly to its nesting area D? What should the value of W/L be for the bird to fly to B and then along the shoreline to D?

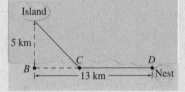

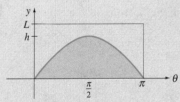

7. In a famous eighteenth-century problem, known as *Buffon's needle problem*, a needle of length h is dropped onto a flat surface (for example, a table or the floor) on which parallel lines L units apart, $L \geq h$, have been drawn. The problem is to determine the probability that the needle will come to rest intersecting one of the lines. Assume that the lines run east-west, parallel to the x-axis in a rectangular coordinate system. Let y be the distance from the "southern" end of the needle to the nearest line to the north. (If the needle's southern end lies on a line, let $y = 0$. If the needle happens to lie east-west, let the "western" end be the "southern" end.) Let θ be the angle that the needle makes with a ray extending eastward from the "southern" end. Then $0 \leq y \leq L$ and $0 \leq \theta < \pi$. Note that the needle intersects one of the lines only when $y < h\sin\theta$. Now, the total set of possibilities for the needle can be identified with the rectangular region $0 \leq y \leq L$, $0 \leq \theta < \pi$, and the proportion of times that the needle intersects a line is the ratio

$$\frac{\text{area under } y = h\sin\theta}{\text{area of rectangle}}$$

This ratio is the probability that the needle intersects a line.
- (a) What is the probability that the needle will intersect a line if $h = L$?
- (b) What is the probability that the needle will intersect a line if $h = L/2$?
- (c) What is the probability that the needle will intersect a line if $h = L/5$?

8. A rectangular beam will be cut from a cylindrical log of radius 10 inches.
   - (a) Show that the beam of maximal cross-sectional area is a square.
   - (b) Four rectangular planks will be cut from the four sections of the log that remain after cutting the square beam. Determine the dimensions of the planks that will have maximal cross-sectional area.
   - (c) Suppose that the strength of a rectangular beam is proportional to the product of its width and the square of its depth. Find the dimensions of the strongest beam that can be cut from the cylindrical log.

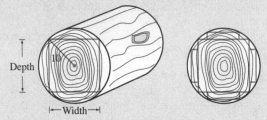

9. Economists use a cumulative distribution called a *Lorenz curve* to describe the distribution of income between households in a given country. Typically, a Lorenz curve is defined on $[0, 1]$, passes through $(0, 0)$ and $(1, 1)$, and is continuous, increasing, and concave upward. The points on this curve are determined by ranking all households by income and then computing the percentage of households whose income is less than or equal to a given percentage of the total income of the country. For example, the point $(a/100, b/100)$ is on the Lorenz curve if the bottom $a\%$ of the households receive less than or equal to $b\%$ of the total income. *Absolute equality* of income distribution would occur if the bottom $a\%$ of the households receive $a\%$ of the income, in which case the Lorenz curve would be the line $y = x$. The area between the Lorenz curve and the line $y = x$ measures how much the income distribution differs from absolute equality. The *coefficient of inequality* is the ratio of the area between the Lorenz curve and the line $y = x$ to the area under $y = x$.

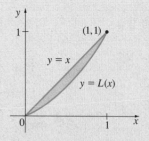

   - (a) Show that the coefficient of inequality is twice the area between the Lorenz curve and the line $y = x$, that is, show that
   $$\text{coefficient of inequality} = 2 \int_0^1 [x - L(x)]\, dx$$
   - (b) The income distribution for a certain country is represented by the Lorenz curve defined by the equation
   $$L(x) = \tfrac{5}{12}x^2 + \tfrac{7}{12}x$$
   What is the percentage of total income received by the bottom 50% of the households? Find the coefficient of inequality.
   - (c) Find the coefficient of inequality for the Lorenz curve defined by the equation
   $$L(x) = \frac{5x^3}{4 + x^2}$$

10. A model rocket is fired vertically upward from rest. Its acceleration for the first three seconds is $a(t) = 60t$ at which time the fuel is exhausted and it becomes a freely "falling" body. After 17 seconds, the rocket's parachute opens, and the (downward) velocity slows linearly to -18 ft/s in 5 s. The rocket then "floats" to the ground at that rate.

(a) Determine the position function s and the velocity function v (for all times t). Sketch the graphs of s and v.

(b) At what time does the rocket reach its maximum height and what is that height?

(c) At what time does the rocket land?

11. If a projectile is fired with an initial velocity v at an angle of inclination θ from the horizontal, then its trajectory, neglecting air resistance, is the parabola

$$y = (\tan \theta)x - \frac{g}{2v^2 \cos^2 \theta} x^2 \qquad 0 \le \theta \le \frac{\pi}{2}$$

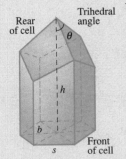

(a) Suppose the projectile is fired from the base of a plane that is inclined at an angle α, $\alpha > 0$, from the horizontal, as shown in the figure. Show that the range of the projectile, measured up the slope, is given by

$$R(\theta) = \frac{2v^2 \cos \theta \sin(\theta - \alpha)}{g \cos^2 \alpha}$$

(b) Determine θ so that R is a maximum.

(c) Suppose the plane is at an angle α *below* the horizontal. Determine the range R in this case and determine the angle at which the projectile should be fired to maximize R.

12. In a beehive, each cell is a regular hexagonal prism, open at one end with a trihedral angle at the other end. It is believed that bees form their cells in such a way as to minimize the surface for a given volume, thus using the least amount of wax in cell construction. Examination of these cells has shown that the measure of the apex angle θ is amazingly consistent. Based on the geometry of the cell, it can be shown that the surface area S is given by

$$S = 6sh - \tfrac{3}{2}s^2 \cot \theta + (3s^2 \sqrt{3}/2) \csc \theta$$

where s, the length of the sides of the hexagon, and h, the height, are constants.

(a) Calculate $dS/d\theta$.

(b) What angle should the bees prefer?

(c) Determine the minimum surface of the area of the cell (in terms of s and h).

Note: Actual measurements of the angle θ in beehives have been made, and the measures of these angles seldom differ from the calculated value by more than $2°$.

Applications of Integration

The calculus is the greatest aid we have to
the appreciation of physical truth in the
broadest sense of the word.

W. F. Osgood

In this chapter we illustrate some of the applications of the definite integral by using it to compute areas between curves, volumes of solids, and the work done by a varying force. The common theme is the following general method that is similar to the one we used to find areas under curves. We break up a quantity Q into a large number of small parts. We then approximate each small part by a quantity of the form $f(x_i^*)\,\Delta x_i$ and thus approximate Q by a Riemann sum. Then we take the limit and express Q as an integral. Finally we evaluate the integral using the Fundamental Theorem of Calculus.

Areas between Curves

So far we have defined and calculated areas of regions that lie under the graphs of functions. In this section we use integrals to find areas of more general regions.

Consider the region S that lies between two curves $y = f(x)$ and $y = g(x)$ and between the vertical lines $x = a$ and $x = b$, where f and g are continuous functions such that $f(x) \geq g(x)$ for all x in $[a, b]$ (see Figure 6.1).

Let P be a partition of $[a, b]$ by points x_i and choose points x_i^* in $[x_{i-1}, x_i]$. Let $\Delta x_i = x_i - x_{i-1}$ and $\|P\| = \max\{\Delta x_i\}$. Figure 6.2 shows the approximating rectangles

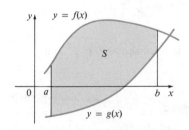

Figure 6.1

$$S = \{(x, y) \mid a \leq x \leq b,$$
$$g(x) \leq y \leq f(x)\}$$

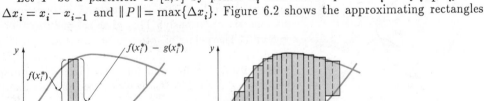

Figure 6.2   (a)   Typical rectangle   (b)   Approximating rectangles

373

with base Δx_i and height $f(x_i^*) - g(x_i^*)$. The Riemann sum

$$\sum_{i=1}^{n} [f(x_i^*) - g(x_i^*)]\,\Delta x_i$$

is therefore an approximation to what we intuitively think of as the area of S.

This approximation appears to become better and better as $\|P\| \to 0$. Therefore we define the **area** A of S as the limiting value of the areas of these approximating rectangles.

(6.1)

$$A = \lim_{\|P\| \to 0} \sum_{i=1}^{n} [f(x_i^*) - g(x_i^*)]\Delta x_i$$

We recognize the limit in (6.1) as a Riemann integral that exists because $f - g$ is continuous. Therefore:

(6.2)

> The area of the region bounded by the curves $y = f(x)$, $y = g(x)$, and the lines $x = a$ and $x = b$, where f and g are continuous and $f(x) \geq g(x)$ for all x in $[a, b]$, is
>
> $$A = \int_a^b [f(x) - g(x)]\,dx$$

Notice that in the special case where $g(x) = 0$, S is the region under the graph of f and our general definition of area (6.1) reduces to our previous definition (5.5).

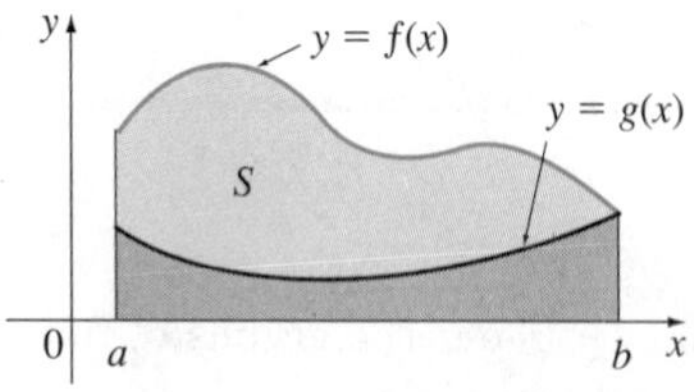

In the case where both f and g are positive, you can see from Figure 6.3 why (6.2) is true:

$$A = [\text{area under } y = f(x)] - [\text{area under } y = g(x)]$$

$$= \int_a^b f(x)\,dx - \int_a^b g(x)\,dx = \int_a^b [f(x) - g(x)]\,dx$$

Figure 6.3

$A = \int_a^b f - \int_a^b g$

EXAMPLE 1   Find the area of the region bounded by the parabolas $y = x^2$ and $y = 2x - x^2$.

Solution   We first find the points of intersection of the parabolas by solving their equations simultaneously. This gives $x^2 = 2x - x^2$ or $2x^2 = 2x$. Thus $x(x-1) = 0$, so $x = 0$ or 1. The points of intersection are $(0, 0)$ and $(1, 1)$ and the region is shown in Figure 6.4.

When using the formula for area (6.2) it is important to ensure that $f(x) \geq g(x)$ when $a \leq x \leq b$. In this case you can see from the diagram that

$$2x - x^2 \geq x^2 \qquad \text{for } 0 \leq x \leq 1$$

and so we choose $f(x) = 2x - x^2$, $g(x) = x^2$, $a = 0$, and $b = 1$. Then (6.2) gives the required area as

$$A = \int_0^1 [(2x - x^2) - x^2]\,dx = \int_0^1 2(x - x^2)\,dx = 2\left[\frac{x^2}{2} - \frac{x^3}{3}\right]_0^1$$

Figure 6.4

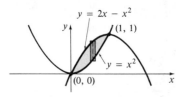

$\int_a^b (\text{upper } y - \text{lower } y)\,dx$

$$= 2\left[\tfrac{1}{2} - \tfrac{1}{3}\right] = \tfrac{1}{3}$$

If we are asked to find the area between curves $y = f(x)$ and $y = g(x)$ where $f(x) \geq g(x)$ for some values of x but $g(x) \geq f(x)$ for other values of x, then we split the given region S into several regions $S_1, S_2, \ldots$ with areas $A_1, A_2, \ldots$ as shown in Figure 6.5.

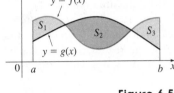

We then define the area of the region S to be the sum of the areas of the smaller regions $S_1, S_2, \ldots$: $A = A_1 + A_2 + \cdots$. Since

$$|f(x) - g(x)| = \begin{cases} f(x) - g(x) & \text{when } f(x) \geq g(x) \\ g(x) - f(x) & \text{when } g(x) \geq f(x) \end{cases}$$

Figure 6.5

we have the following expression for A:

(6.3)

> The area between the curves $y = f(x)$ and $y = g(x)$ and between $x = a$ and $x = b$, is
>
> $$A = \int_a^b |f(x) - g(x)| \, dx$$

When evaluating the integral in (6.3), however, we must still split it into integrals corresponding to $A_1, A_2, \ldots$.

EXAMPLE 2   Find the area of the region bounded by the curves $y = \sin x$, $y = \cos x$, $x = 0$, and $x = \pi/2$.

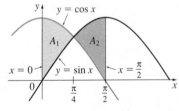

Figure 6.6

Solution   The points of intersection occur when $\sin x = \cos x$, that is, when $x = \pi/4$ (since $0 \leq x \leq \pi/2$). The region is sketched in Figure 6.6. Observe that $\cos x \geq \sin x$ when $0 \leq x \leq \pi/4$ but $\sin x \geq \cos x$ when $\pi/4 \leq x \leq \pi/2$. Therefore the required area is

$$A = \int_0^{\pi/2} |\cos x - \sin x| \, dx$$

$$= A_1 + A_2$$

$$= \int_0^{\pi/4} (\cos x - \sin x) \, dx + \int_{\pi/4}^{\pi/2} (\sin x - \cos x) \, dx$$

$$= \left[\sin x + \cos x\right]_0^{\pi/4} + \left[-\cos x - \sin x\right]_{\pi/4}^{\pi/2}$$

$$= \left(\frac{1}{\sqrt{2}} + \frac{1}{\sqrt{2}} - 0 - 1\right) + \left(-0 - 1 + \frac{1}{\sqrt{2}} + \frac{1}{\sqrt{2}}\right)$$

$$= 2\sqrt{2} - 2$$

In this particular example we could have saved some work by noticing that the region is symmetric about $x = \pi/4$ and so

$$A = 2A_1 = 2\int_0^{\pi/4} (\cos x - \sin x) \, dx$$

EXAMPLE 3   Find the area of the region bounded by the line $y = x - 1$ and the parabola $y^2 = 2x + 6$.

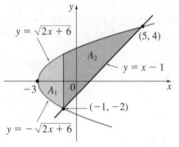

$y = \sqrt{2x + 6}$

A_2

$(5, 4)$

$y = x - 1$

-3 A_1 0 x

$(-1, -2)$

$y = -\sqrt{2x + 6}$

Figure 6.7

Solution   Solving $y^2 = 2x + 6$ and $y = x - 1$, we get $2x + 6 = (x - 1)^2$, which gives $x^2 - 4x - 5 = (x - 5)(x + 1) = 0$. Thus $x = 5$ or -1 and the points of intersection are $(5, 4)$ and $(-1, -2)$. The region is sketched in Figure 6.7. Notice that just because we found the x-coordinates of the points of intersection to be -1 and 5, we cannot simply integrate a difference of functions between -1 and 5 or we miss the area labeled A_1.

In order to use Formula 6.2 we note that the parabola $y^2 = 2x + 6$ defines two functions given by $y = \sqrt{2x + 6}$ (whose graph is the upper half of the parabola) and $y = -\sqrt{2x + 6}$ (the lower half). Since the lower boundary of the region consists of part of a parabola and part of a line, we break up the area as $A = A_1 + A_2$, where the lower and upper boundaries for A_1 are given by

$$y = -\sqrt{2x + 6} \qquad y = \sqrt{2x + 6}$$

and the lower and upper boundaries for A_2 are

$$y = x - 1 \qquad y = \sqrt{2x + 6}$$

Thus the required area is

$$A = A_1 + A_2$$
$$= \int_{-3}^{-1} \left[\sqrt{2x + 6} - (-\sqrt{2x + 6}) \right] dx + \int_{-1}^{5} \left[\sqrt{2x + 6} - (x - 1) \right] dx$$
$$= 2 \int_{-3}^{-1} \sqrt{2x + 6}\, dx + \int_{-1}^{5} \sqrt{2x + 6}\, dx - \int_{-1}^{5} (x - 1)\, dx$$

In the first two integrals we make the substitution $u = 2x + 6$ (so $du = 2\, dx$) and change the limits of integration accordingly:

$$A = 2 \int_{0}^{4} \sqrt{u}\, \frac{du}{2} + \int_{4}^{16} \sqrt{u}\, \frac{du}{2} - \int_{-1}^{5} (x - 1)\, dx$$
$$= \left[\tfrac{2}{3} u^{3/2} \right]_{0}^{4} + \left[\tfrac{1}{3} u^{3/2} \right]_{4}^{16} - \left[\tfrac{x^2}{2} - x \right]_{-1}^{5}$$
$$= \tfrac{2}{3}(8) + \tfrac{1}{3}(64 - 8) - \left(\tfrac{25}{2} - 5 - \tfrac{1}{2} - 1 \right) = 18 \qquad \bullet$$

There is an easier method for solving Example 3. Instead of regarding y as a function of x, let us regard x as a function of y. In general, if a region is bounded by curves with equations $x = f(y)$, $x = g(y)$, $y = c$, and $y = d$, where f and g are continuous and $f(y) \geq g(y)$ for $c \leq y \leq d$ (see Figure 6.8), then its area is

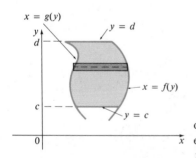

$x = g(y)$

$y = d$

d

$x = f(y)$

c

$y = c$

0 x

Figure 6.8

(6.4)

$$A = \int_{c}^{d} [f(y) - g(y)]\, dy$$

EXAMPLE 4   Use Equation 6.4 to find the area in Example 3.

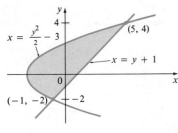

$x = \frac{y^2}{2} - 3$

(5, 4)

$x = y + 1$

(−1, −2)

Figure 6.9

Solution This time we do not have to split the region into two parts. Solving the equation of the parabola for x, we get $x = (y^2/2) - 3$, which represents a single function of y. Also, we write the equation of the line as $x = y + 1$. Notice that

$$y + 1 \geq \frac{y^2}{2} - 3 \qquad \text{for } -2 \leq y \leq 4$$

This means that the left boundary of the region is $x = (y^2/2) - 3$ and the right boundary is $x = y + 1$ (see Figure 6.9). We must integrate between the appropriate y-values, $y = -2$ and $y = 4$. Thus Formula 6.4 gives

$$A = \int_{-2}^{4} \left[(y+1) - \left(\frac{y^2}{2} - 3 \right) \right] dy$$

$$= \int_{-2}^{4} \left[-\frac{y^2}{2} + y + 4 \right] dy$$

$$= -\frac{1}{2} \left(\frac{y^3}{3} \right) + \frac{y^2}{2} + 4y \Big|_{-2}^{4}$$

$$= -\frac{1}{6}(64) + 8 + 16 - \left(\frac{4}{3} + 2 - 8 \right) = 18$$

Recall that the definite integral of a function can be interpreted as an area only when the function is positive. However we can use Formula 6.2 to prove the interpretation of an integral in terms of areas that was given in Section 5.3 as follows: For the case where $f(x) = 0$ and $g(x) < 0$, Formula 6.2 becomes

$$A = \int_a^b [0 - g(x)] \, dx = -\int_a^b g(x) \, dx$$

and so

$$\int_a^b g(x) \, dx = -A$$

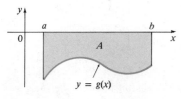

$y = g(x)$

Figure 6.10
$\int_a^b g = -A$

This says that the integral of a negative function can be interpreted as the *negative* of the area of the region *above* its graph between a and b (see Figure 6.10).

In general, if f takes on both positive and negative values, then we can interpret $\int_a^b f$ as a difference of areas:

$$\int_a^b f(x) \, dx = A_1 - A_2$$

where A_1 is the area of the region between a and b that lies above the x-axis and below the graph of f, and A_2 is the area below the x-axis and above the graph of f. For instance, if, as illustrated in Figure 6.11, $f(x) \geq 0$ on $[a,c]$ and $[d,b]$ and $f(x) \leq 0$ on $[c,d]$, then

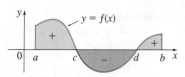

$y = f(x)$

Figure 6.11

$$\int_a^b f = \int_a^c f + \int_c^d f + \int_d^b f$$

$$= \left(\int_a^c f + \int_d^b f \right) - \left(-\int_c^d f \right)$$

$$= A_1 - A_2$$

In Exercises 1–40 sketch the region bounded by the given curves and find the area of the region.

1. $y = x^2 + 3,\quad y = x,\quad x = -1,\quad x = 1$

2. $y = x^4,\quad y = -x - 1,\quad x = -2,\quad x = 0$

3. $y^2 = x,\quad y = x + 5,\quad y = -1,\quad y = 2$

4. $x + y^2 = 0,\quad x = y^2 + 1,\quad y = 0,\quad y = 3$

5. $y = x^2 - 4x,\quad y = 2x$

6. $x^2 + 2x + y = 0,\quad x + y + 2 = 0$

7. $y = x,\quad y = x^2$　　　8. $y = x,\quad y = x^3$

9. $y = x^2,\quad y^2 = x$　　10. $y = x^2,\quad y = x^4$

11. $y = \sqrt{x},\quad y = x/2$

12. $y = \sqrt{x-1},\quad x - 3y + 1 = 0$

13. $y = 4x^2,\quad y = x^2 + 3$

14. $y = x^4 - x^2,\quad y = 1 - x^2$

15. $y = x^2 + 2,\quad y = 2x + 5,\quad x = 0,\quad x = 6$

16. $y = 4 - x^2,\quad y = x + 2,\quad x = -3,\quad x = 0$

17. $y = x^2 + 1,\quad y = 3 - x^2,\quad x = -2,\quad x = 2$

18. $y = x^2 + 2x + 2,\quad y = x + 4,\quad x = -3,\quad x = 2$

19. $y^2 = x,\quad x - 2y = 3$

20. $x + y^2 = 2,\quad x + y = 0$

21. $x = 1 - y^2,\quad x = y^2 - 1$

22. $y = x^3 - 4x^2 + 3x,\quad y = x^2 - x$

23. $y = 2x - x^2,\quad y = x^3$

24. $x = 1 - y^4,\quad x = y^3 - y$

25. $y = x,\quad y = \sin x,\quad x = -\pi/4,\quad x = \pi/2$

26. $y = \cos x,\quad y = \sec^2 x,\quad x = -\pi/4,\quad x = \pi/4$

27. $y = \cos x,\quad y = \sin 2x,\quad x = 0,\quad x = \pi/2$

28. $y = \sin x,\quad y = \sin 2x,\quad x = 0,\quad x = \pi/2$

29. $y = \cos x,\quad y = \sin 2x,\quad x = \pi/2,\quad x = \pi$

30. $y = \sin x,\quad y = \cos 2x,\quad x = 0,\quad x = \pi/4$

31. $y = |x|,\quad y = (x+1)^2 - 7,\quad x = -4$

32. $y = |x - 1|,\quad y = x^2 - 3,\quad x = 0$

33. $x = 3y,\quad x + y = 0,\quad 7x + 3y = 24$

34. $y = x\sqrt{1 - x^2},\quad y = x - x^3$

35. $y = 1/x,\quad y = 1/x^2,\quad x = 1,\quad x = 2$

36. $y = 1/x,\quad x = 0,\quad y = 1,\quad y = 2$

37. $y = x^2,\quad y = 2/(x^2 + 1)$

38. $y = 2^x,\quad y = 5^x,\quad x = -1,\quad x = 1$

39. $y = e^x,\quad y = e^{3x},\quad x = 1$

40. $y = e^x,\quad y = e^{2x}/2,\quad x = 0,\quad x = 1$

In Exercises 41 and 42 find the area of the region bounded by the given curves by two methods:
(a) integrating with respect to x, and
(b) integrating with respect to y.

41. $4x + y^2 = 0,\quad y = 2x + 4$

42. $x + 1 = 2(y - 2)^2,\quad x + 6y = 7$

In Exercises 43 and 44 use calculus to find the area of the triangle with the given vertices.

43. $(0,0),\quad (1,8),\quad (4,3)$

44. $(-2,5),\quad (0,-3),\quad (5,2)$

In Exercises 45 and 46 evaluate the integral and interpret it as the area of a region. Sketch the region.

45. $\displaystyle\int_0^2 |x^2 - x^3|\,dx$　　　46. $\displaystyle\int_0^\pi \left|\sin x - \tfrac{2}{\pi}x\right|\,dx$

In Exercises 47–50 evaluate the integral and interpret it as a difference of areas. Illustrate with a sketch like Figure 6.11.

47. $\displaystyle\int_{-1}^2 x^3\,dx$　　　48. $\displaystyle\int_{-2}^2 (x + x^2)\,dx$

49. $\displaystyle\int_0^\pi \cos x\,dx$　　　50. $\displaystyle\int_{\pi/4}^{5\pi/2} \sin x\,dx$

In Exercises 51 and 52 use the Midpoint Rule with $n = 4$ to approximate the area of the region bounded by the given curves.

51. $y = \sqrt{1 + x^3},\quad y = 1 - x,\quad x = 2$

52. $y = x\tan x,\quad y = x$

53. Find the number b such that the line $y = b$ divides the region bounded by the curves $y = x^2$ and $y = 4$ into two regions with equal area.

54. (a) Find the number a such that the line $x = a$ bisects the area under the curve $y = 1/x^2$, $1 \le x \le 4$.

　　(b) Find the number b such that the line $y = b$ bisects the area in part (a).

SECTION 6.2

Volume

In trying to find the volume of a solid we face the same type of problem as in finding areas. We have an intuitive idea of what volume means, but we must make this idea precise by using calculus to give an exact definition of volume.

We start with a simple type of solid called a **cylinder** (or, more precisely, a *right cylinder*). As illustrated in Figure 6.12(a), a cylinder is bounded by a plane region B_1, called the **base**, and a congruent region B_2 in a parallel plane. The cylinder consists of all points on line segments perpendicular to the base that join B_1 to B_2. If the area of the base is A and the height of the cylinder (the distance from B_1 to B_2) is h, then the volume V of the cylinder is defined as

$$V = Ah$$

In particular, if the base is a circle with radius r, then the cylinder is a circular cylinder with volume $V = \pi r^2 h$ [see Figure 6.12(b)], and if the base is a rectangle with length l and width w, then the cylinder is a rectangular box (also called a rectangular parallelepiped) with volume $V = lwh$ [see Figure 6.12(c)].

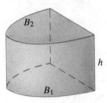 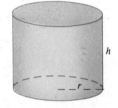 

Figure 6.12   (a) Cylinder $V = Ah$ (b) Circular cylinder $V = \pi r^2 h$ (c) Rectangular box $V = lwh$

Now let S be any solid. The intersection of S with a plane is a plane region that is called a **cross-section** of S. Suppose that the area of the cross-section of S in a plane P_x perpendicular to the x-axis and passing through the point x is $A(x)$, where $a \le x \le b$. (See Figure 6.13. Think of slicing S with a knife through x and computing the area of this slice.) The cross-sectional area $A(x)$ will vary as x increases from a to b.

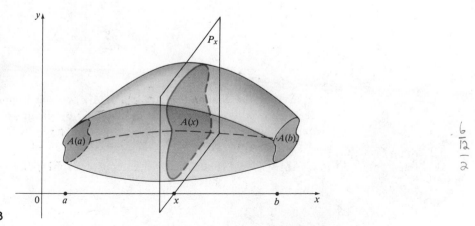

Figure 6.13

Let us take a partition P of the interval $[a,b]$ by points x_i such that $a = x_0 < x_1 < x_2 < \cdots < x_n = b$. The planes P_{x_i} will slice S into smaller "slabs." (Think of slicing a loaf of bread.) If we choose numbers x_i^* in $[x_{i-1}, x_i]$, we can approximate the ith slab S_i (the part of S that lies between the planes $P_{x_{i-1}}$ and P_{x_i}) by a cylinder with base area $A(x_i^*)$ and height $\Delta x_i = x_i - x_{i-1}$ (see Figure 6.14).

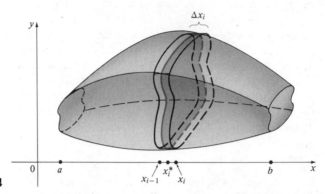

Figure 6.14

The volume of this cylinder is $A(x_i^*)\,\Delta x_i$ so an approximation to our intuitive conception of the volume of the ith slab S_i is

$$V(S_i) \approx A(x_i^*)\,\Delta x_i$$

Adding the volumes of these slabs, we get an approximation to the total volume (that is, what we think of intuitively as the volume):

$$V \approx \sum_{i=1}^{n} A(x_i^*)\,\Delta x_i$$

This approximation appears to become better and better as $\|P\| \to 0$. (Think of the slices as becoming thinner and thinner.) Therefore we *define* the volume as the limit of this sum as $\|P\| \to 0$. But we recognize the limit of a Riemann sum as a definite integral and so we have the following definition:

Definition of Volume (6.5)

Let S be a solid that lies between the planes P_a and P_b. If the cross-sectional area of S in the plane P_x is $A(x)$, where A is an integrable function, then the **volume** of S is

$$V = \lim_{\|P\| \to 0} \sum_{i=1}^{n} A(x_i^*)\Delta x_i = \int_a^b A(x)\,dx$$

In using the volume formula $V = \int_a^b A(x)\,dx$ it is important to remember that $A(x)$ is the area of a moving cross-section obtained by slicing through x perpendicular to the x-axis.

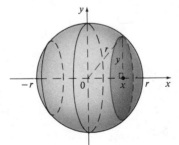

EXAMPLE 1   Show that the volume of a sphere of radius r is

$$V = \tfrac{4}{3}\pi r^3$$

Figure 6.15

Solution   If we place the sphere so that its center is at the origin (see Figure 6.15), then the plane P_x intersects the sphere in a circle whose radius (from the Pythagorean

Theorem) is $y = \sqrt{r^2 - x^2}$. So the cross-sectional area is

$$A(x) = \pi y^2 = \pi(r^2 - x^2)$$

Using the above formula with $a = -r$ and $b = r$, we have

$$V = \int_{-r}^{r} A(x)\,dx = \int_{-r}^{r} \pi(r^2 - x^2)\,dx$$

$$= 2\pi \int_{0}^{r} (r^2 - x^2)\,dx \qquad \text{(The integrand is even.)}$$

$$= 2\pi\left[r^2 x - \frac{x^3}{3}\right]_0^r = 2\pi\left(r^3 - \frac{r^3}{3}\right)$$

$$= \tfrac{4}{3}\pi r^3$$

The sphere in Example 1 is an example of a **solid of revolution** since it can be obtained by revolving a circle about a diameter. In general let S be the solid obtained by revolving the plane region $\mathcal{R}$ bounded by $y = f(x)$, $y = 0$, $x = a$, and $x = b$ about the x-axis (see Figure 6.16).

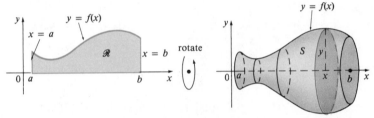

Figure 6.16

Because S is obtained by rotation, a cross-section through x perpendicular to the x-axis is a circular disk with radius $|y| = |f(x)|$ and so the cross-sectional area is

$$A(x) = \pi y^2 = \pi[f(x)]^2$$

Thus, using the basic volume formula $V = \int_a^b A(x)\,dx$, we have the following **formula for a volume of revolution:**

(6.6)
$$\boxed{V = \int_a^b \pi\,[f(x)]^2\,dx}$$

In particular, since the sphere can be obtained by rotating the region under the semicircle

$$y = \sqrt{r^2 - x^2} \qquad -r \le x \le r$$

about the x-axis, Example 1 could be done using Formula 6.6 with $f(x) = \sqrt{r^2 - x^2}$, $a = -r$, and $b = r$.

EXAMPLE 2
(a) Find the volume of the solid obtained by rotating about the x-axis the region under the curve $y = \sqrt{x}$ from 0 to 1.
(b) Illustrate the definition of volume by drawing the approximating cylinders.

Solution

(a)   Taking $a = 0$, $b = 1$, and $f(x) = \sqrt{x}$ in Formula 6.6, we find that the volume of the solid [illustrated in Figure 6.17(b)] is

$$V = \int_0^1 \pi(\sqrt{x})^2\, dx = \pi \int_0^1 x\, dx = \pi \frac{x^2}{2}\bigg|_0^1 = \frac{\pi}{2}$$

(b)   According to the definition of volume in terms of Riemann sums, we have

$$V = \lim_{\|P\| \to 0} \sum_{i=1}^n A(x_i^*)\, \Delta x_i = \lim_{\|P\| \to 0} \sum_{i=1}^n \pi\left(\sqrt{x_i^*}\right)^2 \Delta x_i$$

Taking x_i^* to be the right endpoint of $[x_{i-1}, x_i]$, we can interpret the Riemann sum as the approximating volume of the flat circular cylinders (or disks) shown in Figure 6.17(c). You can see why the use of Formula 6.6 is called the **disk method.** ●

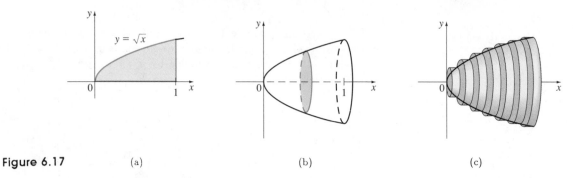

Figure 6.17   (a)   (b)   (c)

Formula 6.6 applies only when the axis of rotation is the x-axis. If, as illustrated in Figure 6.18, the region bounded by the curves $x = g(y)$, $x = 0$, $y = c$, and $y = d$ is rotated about the y-axis, then the corresponding volume of revolution would be

(6.7)

$$V = \int_c^d \pi[g(y)]^2\, dy$$

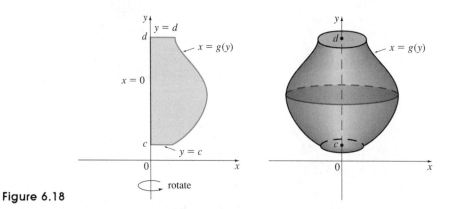

Figure 6.18

EXAMPLE 3   Find the volume of the solid obtained by rotating the region bounded by $y = x^3$, $y = 8$, and $x = 0$ around the y-axis.

Solution   Since the axis of rotation is the y-axis, we write the curve $y = x^3$ in the form $x = \sqrt[3]{y}$. The region and the solid with a typical disk are sketched in Figure 6.19. Using Formula 6.7 with $c = 0$, $d = 8$, and $g(y) = \sqrt[3]{y}$, we have

$$V = \int_0^8 \pi \left(\sqrt[3]{y}\right)^2 dy = \pi \int_0^8 y^{2/3} \, dy$$

$$= \pi \left[\frac{3}{5} y^{5/3}\right]_0^8 = \frac{96\pi}{5}$$

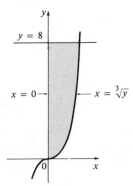

Figure 6.19

EXAMPLE 4   The region $\mathcal{R}$ bounded by the curves $y = x$ and $y = x^2$ is rotated about the x-axis. Find the volume of the resulting solid.

Solution 1   The curves $y = x$ and $y = x^2$ intersect at the points $(0,0)$ and $(1,1)$. The region between them, the solid of rotation, and a cross-section perpendicular to the x-axis are shown in Figure 6.20. A cross-section in the plane P_x has the shape of an annulus (a ring) with inner radius x^2 and outer radius x, so the cross-sectional area is

$$A(x) = \pi x^2 - \pi (x^2)^2 = \pi (x^2 - x^4)$$

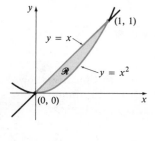
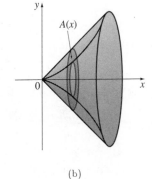
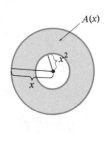

Figure 6.20   (a)   (b)   (c)

Using Definition 6.5, we have

$$V = \int_0^1 A(x) \, dx = \int_0^1 \pi (x^2 - x^4) \, dx$$

$$= \pi \left[\frac{x^3}{3} - \frac{x^5}{5}\right]_0^1 = \frac{2\pi}{15}$$

Solution 2   Another method is to subtract volumes. If V_1 is the volume of the solid obtained by rotating the region under $y = x$ from 0 to 1 around the x-axis and V_2 is the corresponding volume for the curve $y = x^2$, then

$$V = V_1 - V_2$$

Applying Formula 6.6 to calculate each of V_1 and V_2, we have

$$V = \int_0^1 \pi x^2 \, dx - \int_0^1 \pi (x^2)^2 \, dx$$

$$= \pi \int_0^1 (x^2 - x^4) \, dx = \frac{2\pi}{15} \qquad \bullet$$

In general let S be the solid generated when the region bounded by the curves $y = f(x)$, $y = g(x)$, $x = a$, and $x = b$ [where $f(x) \geq g(x)$] is rotated about the x-axis (see Figure 6.21). Then, either by the method of Solution 1 of Example 4 (called the **washer method** because the approximating cylinders have the shape of washers) or by the method of Solution 2 (subtracting volumes), we see that the volume of S is

(6.8)
$$V = \pi \int_a^b \{ [f(x)]^2 - [g(x)]^2 \} \, dx$$

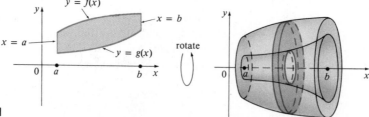

Figure 6.21

EXAMPLE 5   Find the volume of the solid obtained by rotating the region in Example 4 about the line $y = 2$.

Solution   The solid and a cross-section are shown in Figure 6.22. Again a cross-section is an annulus, but this time the inner radius is $2 - x$ and the outer radius is $2 - x^2$. The cross-sectional area is

$$A(x) = \pi (2 - x^2)^2 - \pi (2 - x)^2$$

and so the volume of S is

$$V = \int_0^1 A(x) \, dx = \pi \int_0^1 [(2 - x^2)^2 - (2 - x)^2] \, dx$$

$$= \pi \int_0^1 (x^4 - 5x^2 + 4x) \, dx$$

$$= \pi \left[\frac{x^5}{5} - 5\frac{x^3}{3} + 4\frac{x^2}{2} \right]_0^1 = \frac{8\pi}{15} \qquad \bullet$$

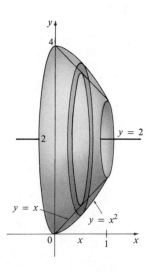

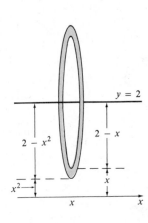

Figure 6.22

EXAMPLE 6   Find the volume of the solid obtained by rotating the region in Example 4 about the y-axis.

Solution   Figure 6.23 shows a cross-section perpendicular to the y-axis. It is an annulus with inner radius y and outer radius $\sqrt{y}$, so the cross-sectional area is

$$A(y) = \pi(\sqrt{y})^2 - \pi y^2$$

and the volume is

$$V = \int_0^1 A(y)\,dy = \pi \int_0^1 [(\sqrt{y})^2 - y^2]\,dy$$

$$= \pi \int_0^1 (y - y^2)\,dy$$

$$= \pi\left[\frac{y^2}{2} - \frac{y^3}{3}\right]_0^1 = \frac{\pi}{6}$$

Figure 6.23

Another method would be to subtract volumes as in Solution 2 of Example 4. Notice that the volume in this example is larger than the volume of revolution about the x-axis in Example 4.

Although we have several volume formulas that are applicable in different situations, it is probably best not to memorize Formulas 6.6, 6.7, or 6.8 but rather to remember the basic volume formula $V = \int_a^b A(x)\,dx$, which applies to any situation. We conclude this section by finding the volumes of three solids that are not solids of revolution.

EXAMPLE 7   A solid has a circular base of radius 1. Parallel cross-sections perpendicular to the base are equilateral triangles. Find the volume of the solid.

Solution   Let us take the circle to be $x^2 + y^2 = 1$. The solid, its base, and a typical cross-section at a distance x from the origin are shown in Figure 6.24. Since B lies on the circle, we have $y = \sqrt{1 - x^2}$ and so the base of the triangle ABC is $|AB| = 2\sqrt{1 - x^2}$. Since the triangle is equilateral, we see from Figure 6.24(c) that its

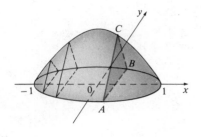

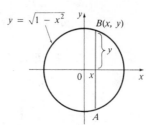

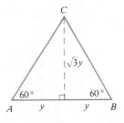

(a) The solid   (b) Its base   (c) A cross-section

Figure 6.24

height is $\sqrt{3}y = \sqrt{3}\sqrt{1 - x^2}$. The cross-sectional area is therefore

$$A(x) = \tfrac{1}{2} \cdot 2\sqrt{1 - x^2} \cdot \sqrt{3}\,\sqrt{1 - x^2} = \sqrt{3}(1 - x^2)$$

and the volume of the solid is

$$V = \int_{-1}^{1} A(x)\,dx = \int_{-1}^{1} \sqrt{3}(1 - x^2)\,dx$$

$$= \sqrt{3}\left[x - \frac{x^3}{3}\right]_{-1}^{1} = \frac{4\sqrt{3}}{3}$$

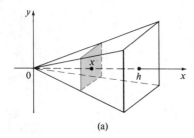

(a)

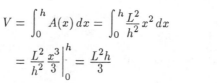

(b)

Figure 6.25

EXAMPLE 8   Find the volume of a pyramid whose base is a square with side L and whose height is h.

Solution   In order to use the formula in Definition 6.5 we place the origin O at the vertex of the pyramid and the x-axis along its central axis as in Figure 6.25(a). Any plane P_x that passes through x and is perpendicular to the x-axis intersects the pyramid in a square with side of length s, say. We can express s in terms of x by observing from the similar triangles in Figure 6.25(b) that

$$\frac{s}{L} = \frac{x}{h}$$

and so $s = Lx/h$. Thus the cross-sectional area is

$$A(x) = s^2 = \frac{L^2}{h^2}x^2$$

Taking $a = 0$ and $b = h$ in Definition 6.5, we see that the volume of the pyramid is

$$V = \int_{0}^{h} A(x)\,dx = \int_{0}^{h} \frac{L^2}{h^2}x^2\,dx$$

$$= \frac{L^2}{h^2}\frac{x^3}{3}\Big|_{0}^{h} = \frac{L^2h}{3}$$

EXAMPLE 9   A wedge is cut out of a circular cylinder of radius 4 by two planes. One plane is perpendicular to the axis of the cylinder. The other intersects the first at an angle of 30° along a diameter of the cylinder. Find the volume of the wedge.

Solution   If we place the x-axis along the diameter where the planes meet, then the base of the solid is a semicircle with equation $y = \sqrt{16 - x^2}$, $-4 \le x \le 4$. A cross-section perpendicular to the x-axis at a distance x from the origin is a triangle ABC as shown

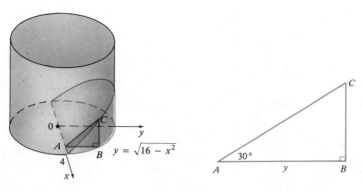

Figure 6.26

in Figure 6.26 with base $y = \sqrt{16 - x^2}$ and height $|BC| = y \tan 30° = \sqrt{16 - x^2}/\sqrt{3}$. Thus the cross-sectional area is

$$A(x) = \tfrac{1}{2}\sqrt{16 - x^2} \cdot \frac{1}{\sqrt{3}}\sqrt{16 - x^2} = \frac{16 - x^2}{2\sqrt{3}}$$

and the volume is

$$V = \int_{-4}^{4} A(x)\,dx = \int_{-4}^{4} \frac{16 - x^2}{2\sqrt{3}}\,dx$$

$$= \frac{1}{\sqrt{3}}\int_{0}^{4} (16 - x^2)\,dx = \frac{1}{\sqrt{3}}\left[16x - \frac{x^3}{3}\right]_{0}^{4}$$

$$= \frac{128}{3\sqrt{3}}$$

For another method see Exercise 63.

SECTION 6.2 **Exercises**

In Exercises 1–12 find the volume of the solid obtained by rotating the region bounded by the given curves about the given axis. Sketch the region, the solid, and a typical disk or "washer."

1. $y = x^2$, $x = 1$, $y = 0$; about the x-axis

2. $y^2 = x^3$, $x = 4$, $y = 0$; about the x-axis

3. $x + y = 1$, $x = 0$, $y = 0$; about the x-axis

4. $y = \sqrt{x - 1}$, $x = 2$, $x = 5$, $y = 0$; about the x-axis

5. $y = x^2$, $y = 4$, $x = 0$, $x = 2$; about the y-axis

6. $x = y - y^2$, $x = 0$; about the y-axis

7. $y = x^2$, $y^2 = x$; about the x-axis

8. $y = x^2 + 1$, $y = 3 - x^2$; about the x-axis

9. $y^2 = x$, $x = 2y$; about the y-axis

10. $y = 2x - x^2$, $y = 0$, $x = 0$, $x = 1$; about the y-axis

11. $y = x^4$, $y = 1$; about $y = 2$

12. $y = x$, $y = 0$, $x = 2$, $x = 4$; about $x = 1$

In Exercises 13–24 refer to the figure and find the volume generated by rotating the given region about the given line.

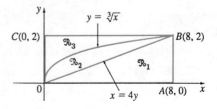

13. $\mathcal{R}_1$ about OA

14. $\mathcal{R}_1$ about OC

15. $\mathcal{R}_1$ about AB

16. $\mathcal{R}_1$ about BC

17. $\mathcal{R}_2$ about OA

18. $\mathcal{R}_2$ about OC

19. $\mathcal{R}_2$ about BC

20. $\mathcal{R}_2$ about AB

21. $\mathcal{R}_3$ about OA

22. $\mathcal{R}_3$ about OC

23. $\mathcal{R}_3$ about BC

24. $\mathcal{R}_3$ about AB

In Exercises 25–32 find the volume of the solid obtained by rotating the region bounded by the given curves about the x-axis.

25. $y = x^2 - 1$, $\quad y = 0$, $\quad x = 0$, $\quad x = 2$

26. $y = -1/x$, $\quad y = 0$, $\quad x = 1$, $\quad x = 3$

27. $y = \sec x$, $\quad y = 1$, $\quad x = -1$, $\quad x = 1$

28. $y = \cos x$, $\quad y = \sin x$, $\quad x = 0$, $\quad x = \pi/4$

29. $y = |x + 2|$, $\quad y = 0$, $\quad x = -3$, $\quad x = 0$

30. $y = [\![x]\!]$, $\quad x = 1$, $\quad x = 6$, $\quad y = 0$

31. $y = e^x$, $\quad y = 0$, $\quad x = 0$, $\quad x = 1$

32. $y = 1/\sqrt{x+1}$, $\quad y = 0$, $\quad x = 0$, $\quad x = 1$

In Exercises 33–38 set up, but do not evaluate, an integral for the volume of the solid obtained by rotating the region bounded by the given curves about the given line.

33. $y = \tan x$, $y = 1$, $x = 0$;   about the x-axis

34. $y = \sqrt{x-1}$, $y = 0$, $x = 5$;   about the y-axis

35. $x - y = 1$, $y = (x - 4)^2 + 1$;   about $y = 7$

36. $y = \cos x$, $y = 0$, $x = 0$, $x = \pi/2$;   about $y = 1$

37. $y = \cos x$, $y = 0$, $x = 0$, $x = \pi/2$;   about $y = -1$

38. $2x + 3y = 6$, $(y - 1)^2 = 4 - x$;   about $x = -5$

In Exercises 39 and 40 sketch and find the volume of the solid obtained by rotating the region under the graph of f about the x-axis.

39. $f(x) = \begin{cases} 3 & \text{if } 0 \le x \le 1 \\ 1 & \text{if } 1 < x < 4 \\ 3 & \text{if } 4 \le x \le 5 \end{cases}$

40. $f(x) = \begin{cases} \frac{1}{2} & \text{if } 0 \le x < 1 \\ x^2 - 2x + 2 & \text{if } 1 \le x \le 2 \end{cases}$

The integrals in Exercises 41–46 represent volumes of solids. Describe the solids.

41. $\pi \displaystyle\int_0^{\pi/4} \tan^2 x \, dx$

42. $\pi \displaystyle\int_1^2 y^6 \, dy$

43. $\pi \displaystyle\int_0^1 (y - y^2) \, dy$

44. $\pi \displaystyle\int_0^4 [16 - (x - 2)^2] \, dx$

45. $\pi \displaystyle\int_0^1 [(5 - 2x^2)^2 - (5 - 2x)^2] \, dx$

46. $\pi \displaystyle\int_{\pi/4}^{\pi/2} [(2 + \sin x)^2 - (2 + \cos x)^2] \, dx$

In Exercises 47–59 find the volume of the described solid S.

47.   A right circular cone with height h and base radius r

48.   A frustum of a right circular cone with height h, lower base radius R, and top radius r

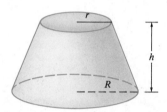

49.   A cap of a sphere with radius r and height h

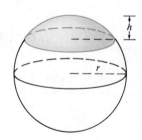

50.   A frustum of a pyramid with square base of side b, square top of side a, and height h

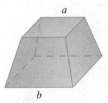

51.   A pyramid with height h and rectangular base with dimensions b and $2b$

52. A pyramid with height h and base an equilateral triangle with side a (a tetrahedron)

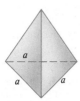

53. A tetrahedron with three mutually perpendicular faces and three mutually perpendicular edges with lengths 3 cm, 4 cm, and 5 cm

54. The base of S is a circular disk with radius r. Parallel cross-sections perpendicular to the base are squares.

55. The base of S is an elliptical region with boundary $9x^2 + 4y^2 = 36$. Cross-sections perpendicular to the x-axis are isosceles right triangles with hypotenuse in the base.

56. The base of S is the parabolic region $\{(x, y) \mid x^2 \leq y \leq 1\}$. Cross-sections perpendicular to the y-axis are equilateral triangles.

57. S has the same base as in Exercise 56 but cross-sections perpendicular to the y-axis are squares.

58. The base of S is the triangular region with vertices $(0,0)$, $(2,0)$, and $(0,1)$. Cross-sections perpendicular to the x-axis are semicircles.

59. S has the same base as in Exercise 58 but cross-sections perpendicular to the x-axis are isosceles triangles with height equal to the base.

60. The base of S is a circular disk with radius r. Parallel cross-sections perpendicular to the base are isosceles triangles with height h and unequal side in the base.
   (a) Set up an integral for the volume of S.
   (b) By interpreting the integral as an area, find the volume of S.

61. (a) Set up an integral for the volume of a solid torus (the donut-shaped solid shown in the figure) with radii r and R.
   (b) By interpreting the integral as an area, find the volume of the torus.

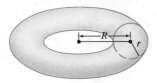

62. Find the volume common to two circular cylinders, each with radius r, if the axes of the cylinders intersect at right angles.

63. Solve Example 9 taking cross-sections to be parallel to the line of intersection of the two planes.

64. Solve Example 9 if the planes intersect at an angle of 45°.

65. A hole of radius r is bored through a cylinder of radius $R > r$ at right angles to the axis of the cylinder. Set up, but do not evaluate, an integral for the volume cut out.

66. A hole of radius r is bored through the center of a sphere of radius $R > r$. Find the volume of the remaining portion of the sphere.

67. (a) Cavalieri's Principle states that if a family of parallel planes gives equal cross-sectional areas for two solids S_1 and S_2, then the volumes of S_1 and S_2 are equal. Prove Cavalieri's Principle.
   (b) Use this principle to find the volume of the oblique cylinder in the figure.

68. A log 10 m long is cut at 1-m intervals and its cross-sectional areas A (at a distance x from the end of the log) are listed in the following table. Use the Midpoint Rule with $n = 5$ to estimate the volume of the log.

x (m)	0	1	2	3	4	5
A (m²)	0.68	0.65	0.64	0.61	0.58	0.59

x (m)	6	7	8	9	10
A (m²)	0.53	0.55	0.52	0.50	0.48

$(3^2)^3$

Volumes by Cylindrical Shells

Some volume problems are very difficult to handle by the methods of the preceding section. For instance, we consider the problem of finding the volume of the solid obtained by rotating about the y-axis the region bounded by $y = x(x-1)^2$ and $y = 0$. If we were to use the "washer method" we would first have to locate the local maximum point (a, b) of $y = x(x-1)^2$ using the methods of Chapter 4. Then we would have to solve the equation $y = x(x-1)^2$ for x in terms of y to obtain the functions $x = g_1(y)$ and $x = g_2(y)$ shown in Figure 6.27. This step would be difficult because it involves the cubic formula. Finally we would find the volume using

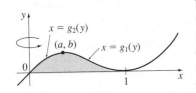

Figure 6.27

$$V = \pi \int_0^b \{[g_1(y)]^2 - [g_2(y)]^2\}\, dx$$

Fortunately there is a method, called the **method of cylindrical shells**, that is easier to use in such a case. Figure 6.28 shows a cylindrical shell with inner radius r_1, outer radius r_2, and height h. Its volume V is calculated by subtracting the volume V_1 of the inner cylinder from the volume V_2 of the outer cylinder:

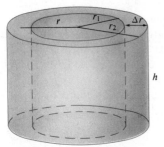

Figure 6.28

$$\begin{aligned}
V &= V_2 - V_1 \\
&= \pi r_2^2 h - \pi r_1^2 h = \pi(r_2^2 - r_1^2)h \\
&= \pi(r_2 + r_1)(r_2 - r_1)h \\
&= 2\pi \frac{r_2 + r_1}{2} h(r_2 - r_1)
\end{aligned}$$

If we let $\Delta r = r_2 - r_1$ (the thickness of the shell) and $r = \frac{1}{2}(r_2 + r_1)$ (the average radius of the shell), then this formula for the volume of a cylindrical shell becomes

(6.9)
$$\boxed{V = 2\pi r h\, \Delta r}$$

and it can be remembered as

$$V = [\text{circumference}][\text{height}][\text{thickness}]$$

Now let S be the solid obtained by rotating about the y-axis the region bounded by $y = f(x)$ [where $f(x) \geq 0$], $y = 0$, $x = a$, and $x = b$, where $b > a \geq 0$ (see Figure 6.29). Let P be a partition of $[a, b]$ by points x_i with $a = x_0 < x_1 < \cdots < x_n = b$ and let x_i^* be the midpoint of $[x_{i-1}, x_i]$, that is, $x_i^* = \frac{1}{2}(x_{i-1} + x_i)$. If the rectangle with base $[x_{i-1}, x_i]$ and height $f(x_i^*)$ is rotated about the y-axis, then the result is a cylindrical shell with average radius x_i^*, height $f(x_i^*)$, and the thickness $\Delta x_i = x_i - x_{i-1}$ (see Figure 6.30), so by

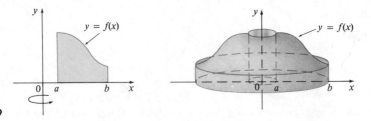

Figure 6.29

Formula 6.9 its volume is

$$V_i = 2\pi x_i^* f(x_i^*)\,\Delta x_i$$

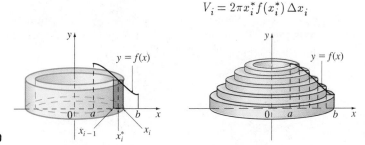

Figure 6.30

Therefore an approximation to the volume V of S is given by the sum of the volumes of these shells:

$$V \approx \sum_{i=1}^{n} V_i = \sum_{i=1}^{n} 2\pi x_i^* f(x_i^*)\,\Delta x_i$$

This approximation appears to become better as $\|P\| \to 0$. But, from the definition of an integral, we know that

$$\lim_{\|P\| \to 0} \sum_{i=1}^{n} 2\pi x_i^* f(x_i^*)\,\Delta x_i = \int_a^b 2\pi x f(x)\,dx$$

Thus the following appears plausible:

(6.10)

> The volume of the solid in Figure 6.29 is
>
> $$V = \int_a^b 2\pi x f(x)\,dx \qquad \text{where } 0 \le a < b$$

The argument using cylindrical shells makes Formula 6.10 seem reasonable but later we will be able to prove it (see Exercise 61 in Section 7.1).

EXAMPLE 1   Find the volume of the solid obtained by rotating about the y-axis the region bounded by $y = x(x-1)^2$ and $y = 0$.

Solution   The region and solid are sketched in Figure 6.31. Formula 6.10 gives the volume of rotation as

$$V = \int_0^1 2\pi x[x(x-1)^2]\,dx$$

$$= 2\pi \int_0^1 (x^4 - 2x^3 + x^2)\,dx$$

$$= 2\pi\left[\frac{x^5}{5} - 2\frac{x^4}{4} + \frac{x^3}{3}\right]_0^1 = \frac{\pi}{15}$$

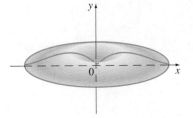

Figure 6.31

Note: Comparing the solution of Example 1 with the remarks at the beginning of this section, we see that the method of cylindrical shells is much easier than the washer method for this problem. We did not have to find the coordinates of the local maximum and we did not have to solve the equation of the curve for x in terms of y. However, in other examples the methods of preceding section may be easier.

EXAMPLE 2   Find the volume of the solid obtained by rotating about the y-axis the region between $y = x$ and $y = x^2$.

Solution   We use Formula 6.10 to find the volume obtained by rotating about the y-axis the region under $y = x$ from 0 to 1 and then we subtract the corresponding volume for the region under $y = x^2$. Thus

$$V = \int_0^1 2\pi x \cdot x \, dx - \int_0^1 2\pi x \cdot x^2 \, dx$$

$$= 2\pi \int_0^1 (x^2 - x^3) \, dx = 2\pi \left[\frac{x^3}{3} - \frac{x^4}{4} \right]_0^1 = \frac{\pi}{6}$$

Comparison with Example 6 in Section 6.2 shows that the washer and shell methods take about the same amount of work in this problem.   ●

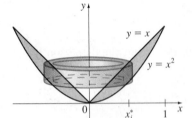

Figure 6.32

Another way of looking at Example 2 is to consider cylindrical shells of height $x_i^* - (x_i^*)^2$ as shown in Figure 6.32. This leads to

$$V = \int_0^1 2\pi x(x - x^2) \, dx$$

In general, by subtracting volumes as in Example 2 or by considering cylindrical shells directly, we see that the volume of the solid generated by rotating about the y-axis the region between the curves $y = f(x)$ and $y = g(x)$ from a to b [where $f(x) \geq g(x)$ and $0 \leq a < b$] is

$$V = \int_a^b 2\pi x[f(x) - g(x)] \, dx$$

The method of cylindrical shells also allows us to compute volumes of revolution about the x-axis. If we interchange the roles of x and y in Formula 6.10, we see that if we rotate the region in Figure 6.33 about the x-axis, then the volume of the resulting solid is

(6.11)
$$V = \int_c^d 2\pi y g(y) \, dy$$

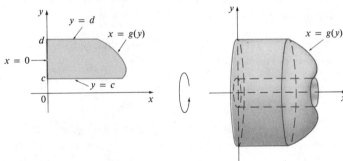

Figure 6.33

EXAMPLE 3   Use cylindrical shells to find the volume of the solid obtained by rotating about the x-axis the region under the curve $y = \sqrt{x}$ from 0 to 1.

Solution   This problem was solved using disks in Example 2 in Section 6.2. To use shells we relabel $y = \sqrt{x}$ in Figure 6.17(a) as $x = y^2$ in Figure 6.34. Then by using For-

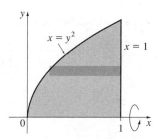

Figure 6.34

mula 6.11 and subtracting volumes, we get

$$V = \int_0^1 2\pi y \cdot 1 \, dy - \int_0^1 2\pi y \cdot y^2 \, dy$$

$$= 2\pi \int_0^1 (y - y^3) \, dy = 2\pi \left[\frac{y^2}{2} - \frac{y^4}{4} \right]_0^1 = \frac{\pi}{2}$$

In this problem the method of Example 2 in Section 6.2 was simpler.

EXAMPLE 4   Find the volume of the solid obtained by rotating the region bounded by $y = x - x^2$ and $y = 0$ about the line $x = 2$.

Solution   Since the axis of rotation is $x = 2$ instead of the y-axis, we cannot use Formula 6.10 so we go back to the basic method of deriving it using cylindrical shells. Figure 6.35 shows a rectangle with base $[x_{i-1}, x_i]$ rotated about $x = 2$ to form a cylindrical shell with average radius $2 - x_i^*$, height $x_i^* - (x_i^*)^2$, and thickness Δx_i, where x_i^* is the midpoint of $[x_{i-1}, x_i]$. By Formula 6.9 the volume of this shell is

$$\underbrace{2\pi(2 - x_i^*)}_{\text{circumference}} \underbrace{[x_i^* - (x_i^*)^2]}_{\text{height}} \underbrace{\Delta x_i}_{\text{thickness}}$$

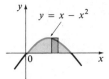
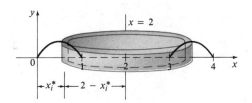

Figure 6.35

and so the volume of the given solid is

$$V = \lim_{\|P\| \to 0} \sum_{i=1}^n 2\pi(2 - x_i^*)[x_i^* - (x_i^*)^2] \Delta x_i$$

$$= \int_0^1 2\pi(2 - x)(x - x^2) \, dx$$

$$= 2\pi \int_0^1 (x^3 - 3x^2 + 2x) \, dx$$

$$= 2\pi \left[\frac{x^4}{4} - x^3 + x^2 \right]_0^1 = \frac{\pi}{2}$$

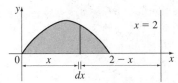

Figure 6.36

The procedure for setting up the integral in Example 4 is often abbreviated heuristically as follows: We think of a very narrow rectangle of width dx as in Figure 6.36. When rotated about the line $x = 2$, it produces a very thin shell of radius $2 - x$, height $x - x^2$, and thickness dx; therefore its volume is $dV = 2\pi(2 - x)(x - x^2) \, dx$. The process of adding a large number of such volumes and taking the limit is abbreviated by writing

$$V = \int dV = \int_0^1 2\pi(2 - x)(x - x^2) \, dx$$

It is customary, particularly among scientists, to use similar reasoning and notation in all applications of integration.

SECTION 6.3   **Exercises**

In Exercises 1–10 use the method of cylindrical shells to find the volume generated by rotating the region bounded by the given curves about the y-axis.

1. $y = x^2$, $y = 0$, $x = 1$, $x = 2$

2. $y = 1/x$, $y = 0$, $x = 1$, $x = 10$

3. $y = \sqrt{4 + x^2}$, $y = 0$, $x = 0$, $x = 4$

4. $y = \sin(x^2)$, $y = 0$, $x = 0$, $x = \sqrt{\pi}$

5. $y = x^2$, $y = 4$, $x = 0$

6. $y^2 = x$, $x = 2y$ 7. $y = x^2 - x^3$, $y = 0$

8. $y = -x^2 + 4x - 3$, $y = 0$

9. $y = x^2 - 6x + 10$, $y = -x^2 + 6x - 6$

10. $y = x - 2$, $y = \sqrt{x - 2}$

In Exercises 11–16 use the method of cylindrical shells to find the volume of the solid obtained by rotating the region bounded by the given curves about the x-axis.

11. $x = \sqrt[4]{y}$, $x = 0$, $y = 16$

12. $x = y^2$, $x = 0$, $y = 2$, $y = 5$

13. $y = x^2$, $y = 9$

14. $y^2 - 6y + x = 0$, $x = 0$

15. $y = \sqrt{x}$, $y = 0$, $x + y = 2$

16. $y = x$, $x = 0$, $x + y = 2$

In Exercises 17–22 use the method of cylindrical shells as in Example 4 to find the volume generated by rotating the region bounded by the given curves about the given axis. Sketch the region and a typical shell.

17. $y = \sqrt{x}$, $y = 0$, $x = 1$, $x = 4$;   about the y-axis

18. $y = x^2$, $y = 0$, $x = -2$, $x = -1$;   about the y-axis

19. $y = x^2$, $y = 0$, $x = 1$, $x = 2$;   about $x = 1$

20. $y = x^2$, $y = 0$, $x = 1$, $x = 2$;   about $x = 4$

21. $y = \sqrt{x - 1}$, $y = 0$, $x = 5$;   about $y = 3$

22. $y = 4x - x^2$, $y = 8x - 2x^2$;   about $x = -2$

In Exercises 23–34 set up, but do not evaluate, an integral for the volume of the solid obtained by rotating the region bounded by the given curves about the given axis.

23. $y = \sin x$, $y = 0$, $x = 2\pi$, $x = 3\pi$;   about the y-axis

24. $y = \dfrac{1}{1 + x^2}$, $y = 0$, $x = 0$, $x = 3$;   about the y-axis

25. $x = \cos y$, $x = 0$, $y = 0$, $y = \pi/4$;   about the x-axis

26. $y = -x^2 + 7x - 10$, $y = x - 2$;   about the x-axis

27. $y = x^4$, $y = \sin(\pi x/2)$;   about $x = -1$

28. $x = 4 - y^2$, $x = 8 - 2y^2$;   about $y = 5$

29. $y = \ln x$, $y = 0$, $x = e$;   about the y-axis

30. $y = e^x$, $y = e^{-x}$, $x = 1$;   about the y-axis

31. $y = e^x$, $x = 0$, $y = \pi$;   about the x-axis

32. $y = e^{-x}$, $y = 0$, $x = -1$, $x = 0$;   about $x = 1$

33. $y = e^x$, $x = 0$, $y = 2$;   about $y = 1$

34. $y = \ln x$, $y = 0$, $x = e$;   about $y = 3$

The integrals in Exercises 35–38 represent volumes of solids. Describe the solids.

35. $\displaystyle\int_0^{\pi/2} 2\pi x \cos x \, dx$ 36. $\displaystyle\int_0^9 2\pi y^{3/2} \, dy$

37. $\displaystyle\int_0^1 2\pi(x^3 - x^7) \, dx$ 38. $\displaystyle\int_0^{\pi} 2\pi(4 - x)\sin^4 x \, dx$

In Exercises 39–44 the region bounded by the given curves is rotated about the given axis. Find the volume of the resulting solid by any method.

39. $y = x^2 + x - 2$, $y = 0$;   about the x-axis

40. $y = x^2 - 3x + 2$, $y = 0$;   about the y-axis

41. $x = 1 - y^2$, $x = 0$;   about the y-axis

42. $y = x\sqrt{1 + x^3}$, $y = 0$, $x = 0$, $x = 2$;   about the y-axis

43. $x^2 + (y - 1)^2 = 1$;   about the y-axis

44. $x^2 + (y - 1)^2 = 1$;   about the x-axis

Use cylindrical shells to find the volumes of the solids in Exercises 45–48.

45. A sphere of radius r

46. The solid torus of Exercise 61 in Section 6.2

47. A right circular cone with height h and base radius r

48. The solid of Exercise 66 in Section 6.2

49. Formula 6.10 is valid only if $a \geq 0$. Show that if $a < b \leq 0$, the volume formula becomes

$$V = -\int_a^b 2\pi x f(x)\, dx$$

50. Find a formula for the volume of the solid obtained when the region under the graph of f from a to b is rotated about the vertical line $x = c$ in the following cases: (a) $c \leq a < b$ and (b) $a < b \leq c$.

51. Use the Midpoint Rule with $n = 4$ to estimate the volume obtained by rotating about the y-axis the region under the curve $y = \tan x$, $0 \leq x \leq \pi/4$.

52. If the region shown in the figure is rotated about the y-axis to form a solid, use the Midpoint Rule with $n = 5$ to estimate the volume of the solid.

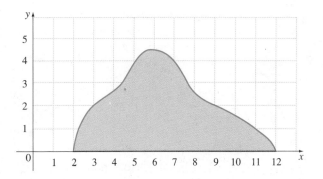

SECTION 6.4

Work

The term **work** is used in everyday language to mean the total amount of effort required to perform a task. In physics it has a technical meaning that depends on the idea of a **force.** Intuitively you can think of a force as describing a push or pull on an object—for example, a horizontal push of a book across a table or the downward pull of the earth's gravity on a ball. In general, if an object moves along a straight line with position function $s(t)$, then the force F on the object (in that same direction) is defined by Newton's Second Law of Motion as the product of its mass m and its acceleration:

(6.12)
$$F = m\frac{d^2 s}{dt^2}$$

In the SI metric system, the mass is measured in kilograms (kg), the displacement in meters (m), the time in seconds (s), and the force in newtons (N = kg-m/s^2). Thus a force of 1 N acting on a mass of 1 kg produces an acceleration of 1 m/s^2. In the British engineering system the fundamental unit is chosen to be the unit of force, which is the pound.

In the case of constant acceleration, the force F is also constant and the work done is defined to be the product of the force F and the distance d that the object moves:

(6.13)
$$W = Fd \qquad \text{work} = \text{force} \times \text{distance}$$

If F is measured in newtons and d in meters, then the unit for W is a newton-meter, which is called a joule (J). If F is measured in pounds and d in feet, then the unit for W is a foot-pound (ft-lb), which is about 1.36 J.

EXAMPLE 1
(a) How much work is done in lifting a 1.2-kg book off the floor to put it on a desk that is 0.7 m high? Use the fact that the acceleration due to gravity is $g = 9.8$ m/s^2.
(b) How much work is done in lifting a 20-lb weight 6 ft off the ground?

Solution

(a)   The force exerted is equal and opposite to that exerted by gravity, so Equation 6.12 gives

$$F = mg = (1.2)(9.8) = 11.76 \, \text{N}$$

and then Equation 6.13 gives the work done as

$$W = Fd = (11.76)(0.7) \approx 8.2 \, \text{J}$$

(b)   Here the force is given as $F = 20 \, \text{lb}$, so the work done is

$$W = Fd = 20 \cdot 6 = 120 \, \text{ft-lb}$$

Notice that in part (b), unlike part (a), we did not have to multiply by g because we were given the *weight* (which is a force) and not the mass of the object.   ●

Equation 6.13 defines work as long as the force is constant, but what happens if the force is variable? Let us suppose that the object moves along the x-axis in the positive direction from $x = a$ to $x = b$ and at each point x between a and b a force $f(x)$ acts on the object, where f is a continuous function. Let P be a partition of $[a, b]$ by points x_i ($i = 1, 2, ..., n$) and let $\Delta x_i = x_i - x_{i-1}$. Choose any x_i^* in the ith subinterval $[x_{i-1}, x_i]$. Then the force at that point is $f(x_i^*)$. If $\|P\|$ is small, then Δx_i is small, and since f is continuous, the values of f do not change very much over the interval $[x_{i-1}, x_i]$. In other words, f is almost constant on the interval and so the work W_i that is done in moving the particle from x_{i-1} to x_i is approximately given by Equation 6.13:

$$W_i \approx f(x_i^*) \, \Delta x_i$$

Thus we can approximate the total work by

(6.14)
$$W \approx \sum_{i=1}^{n} f(x_i^*) \, \Delta x_i$$

It seems that this approximation becomes better as we make $\|P\|$ smaller (and therefore n larger). Therefore we define the **work done in moving the object from a to b** as the limit of this quantity as $\|P\| \to 0$. Since the right side of (6.14) is a Riemann sum, we recognize its limit as being a definite integral, and so

(6.15)
$$W = \lim_{\|P\| \to 0} \sum_{i=1}^{n} f(x_i^*) \, \Delta x_i = \int_a^b f(x) \, dx$$

EXAMPLE 2   When a particle is at a distance x feet from the origin, a force of $x^2 + 2x$ pounds acts on it. How much work is done in moving it from $x = 1$ to $x = 3$?

Solution
$$W = \int_1^3 (x^2 + 2x) \, dx = \frac{x^3}{3} + x^2 \Big|_1^3 = \frac{50}{3}$$

The work done is $16\frac{2}{3}$ ft-lb.   ●

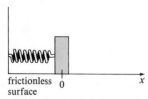

(a) Natural position of spring

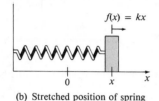

$$f(x) = kx$$

(b) Stretched position of spring

Figure 6.37

Hooke's Law

In the next example we use **Hooke's Law** from physics, which states that the force required to maintain a spring stretched x units beyond its natural length is proportional to x:

$$f(x) = kx$$

where k is a positive constant (called the **spring constant**). Hooke's Law holds provided that x is not too large (see Figure 6.37).

EXAMPLE 3   A force of $40\,\text{N}$ is required to hold a spring that has been stretched from its natural length of $10\,\text{cm}$ to a length of $15\,\text{cm}$. How much work is done in stretching the spring from $15\,\text{cm}$ to $18\,\text{cm}$?

Solution   According to Hooke's Law, the force required to hold the spring stretched x meters beyond its natural length is $f(x) = kx$. In stretching the spring from $10\,\text{cm}$ to $15\,\text{cm}$ the amount stretched is $5\,\text{cm} = 0.05\,\text{m}$. This means that $f(0.05) = 40$, so

$$0.05k = 40 \qquad k = \frac{40}{0.05} = 800$$

Thus $f(x) = 800x$ and the work done in stretching the spring from $15\,\text{cm}$ to $18\,\text{cm}$ is

$$W = \int_{0.05}^{0.08} 800x \, dx = 800 \frac{x^2}{2} \Big|_{0.05}^{0.08}$$
$$= 400[(0.08)^2 - (0.05)^2] = 1.56 \, \text{J}$$

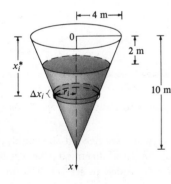

EXAMPLE 4   A tank has the shape of an inverted circular cone with height $10\,\text{m}$ and base radius $4\,\text{m}$. It is filled with water to a height of $8\,\text{m}$. Find the work required to empty the tank by pumping all of the water to the top of the tank. (The density of the water is $1000\,\text{kg/m}^3$.)

Solution   Let us measure depths from the top of the tank by introducing a vertical coordinate line as in Figure 6.38(a). The water extends from a depth of $2\,\text{m}$ to a depth of $10\,\text{m}$ and so we take a partition P of the interval $[2, 10]$ by points x_i with $2 = x_0 < x_1 < \cdots < x_n = 10$ and choose x_i^* in the ith subinterval. This divides the water into n layers. The ith layer is approximated by a circular cylinder with radius r_i and height Δx_i. We can compute r_i from similar triangles using Figure 6.38(b) as follows:

$$\frac{r_i}{10 - x_i^*} = \frac{4}{10} \qquad r_i = \tfrac{2}{5}(10 - x_i^*)$$

Thus an approximation to the volume of the ith layer of water is

$$V_i \approx \pi r_i^2 \Delta x_i = \frac{4\pi}{25}(10 - x_i^*)^2 \Delta x_i$$

and so its mass is

$$m_i = \text{density} \times \text{volume}$$
$$\approx 1000 \cdot \frac{4\pi}{25}(10 - x_i^*)^2 \Delta x_i = 160\pi(10 - x_i^*)^2 \Delta x_i$$

Figure 6.38

The force required to raise this layer must overcome the force of gravity and so

$$F_i = m_i g \approx (9.8)160\pi(10 - x_i^*)^2 \,\Delta x_i$$
$$\approx 1570\pi(10 - x_i^*)^2 \,\Delta x_i$$

Each particle in the layer must travel a distance of approximately x_i^*. The work W_i done to raise this layer to the top is approximately the product of the force F_i and the distance x_i^*:

$$W_i \approx F_i x_i^* \approx 1570\pi x_i^*(10 - x_i^*)^2 \,\Delta x_i$$

To find the total work done in emptying the entire tank, we add the contributions of each of the n layers and then take the limit as $\|P\| \to 0$:

$$W = \lim_{\|P\| \to 0} \sum_{i=1}^{n} 1570\pi x_i^*(10 - x_i^*)^2 \,\Delta x_i$$
$$= \int_{2}^{10} 1570\pi x(10 - x)^2 \, dx$$
$$= 1570\pi \int_{2}^{10} (100x - 20x^2 + x^3) \, dx$$
$$= 1570\pi \left[50x^2 - \frac{20x^3}{3} + \frac{x^4}{4} \right]_{2}^{10}$$
$$= 1570\pi \left(\frac{2048}{3} \right) \approx 3.4 \times 10^6 \text{ J}$$

SECTION 6.4   **Exercises**

1. Find the work done in pushing a car a distance of 8 m while exerting a constant force of 900 N.

2. How much work is done by a weightlifter in raising a 60-kg barbell from the floor to a height of 2 m?

3. A particle is moved along the x-axis by a force that measures $5x^2 + 1$ pounds at a point x feet from the origin. Find the work done in moving the particle from the origin to a distance of 10 ft.

4. When a particle is at a distance x meters from the origin, a force of $\cos(\pi x/3)$ newtons acts on it. How much work is done in moving the particle from $x = 1$ to $x = 2$?

5. A force of 10 lb is required to hold a spring stretched 4 in. beyond its natural length. How much work is done in stretching it from its natural length to 6 in. beyond its natural length?

6. A spring has natural length 20 cm. If a 25-N force is required to keep it stretched it to a length of 30 cm,

how much work is required to stretch it from 20 cm to 25 cm?

7. Suppose that 2 joules of work are needed to stretch a spring from its natural length of 30 cm to a length of 42 cm. How much work is needed to stretch it from 35 cm to 40 cm?

8. If the work required to stretch a spring 1 ft beyond its natural length is 12 ft-lb, how much work is needed to stretch it 9 in. beyond its natural length?

9. How far beyond its natural length will a force of 30 N keep the spring in Exercise 7 stretched?

10. If 6 joules of work are needed to stretch a spring from 10 cm to 12 cm and another 10 joules are needed to stretch it from 12 cm to 14 cm, what is the natural length of the spring?

11. A heavy rope, 50 ft long, weighs 0.5 lb/ft and hangs over the edge of a building 120 ft high. How much work is done in pulling this rope to the top of the building?

12. A uniform cable hanging over the edge of a tall building is 40 ft long and weighs 60 lb. How much work is required to pull 10 ft of the cable to the top?

13. A cable that weighs 2 lb/ft is used to lift 800 lb of coal up a mine shaft 500 ft deep. Find the work done.

14. A bucket that weighs 4 lb and a rope of negligible weight are used to draw water from a well that is 80 ft deep. The bucket starts with 40 lb of water and is pulled up at a rate of 2 ft/s but water leaks out of a hole at a rate of 0.2 lb/s. Find the work done in pulling the bucket to the top of the well.

15. An aquarium 2 m long, 1 m wide, and 1 m deep is full of water. Find the work needed to pump half of the water out of the aquarium.

16. A circular swimming pool has a diameter of 24 ft, the sides are 5 ft high, and the depth of the water is 4 ft. How much work is required to pump all of the water out over the side?

In Exercises 17–20 a tank is full of water. Find the work required to pump the water out of the outlet. In Exercises 19 and 20 use the fact that water weighs 62.5 lb/ft³.

17.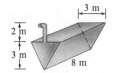

18.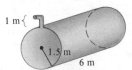

19.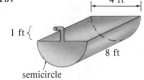
semicircle

20.
hemisphere

21. Solve Exercise 17 if the tank is filled with gasoline that has a density of 680 kg/m³.

22. Solve Exercise 18 if the tank is half full of oil that has a density of 920 kg/m³.

23. When gas expands in a cylinder with radius r, the pressure at any given time is a function of the volume: $P = P(V)$. The force exerted by the gas on the piston (see the figure) is the product of the pressure and the area: $F = \pi r^2 P$. Show that the work done by the gas when the volume expands from volume V_1 to volume V_2 is

$$W = \int_{V_1}^{V_2} P \, dV$$

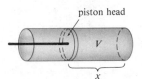
piston head

24. In a steam engine the pressure P and volume V of steam satisfy the equation $PV^{1.4} = k$, where k is a constant. (This is true for adiabatic expansion, that is, expansion in which there is no heat transfer between the cylinder and its surroundings.) Use Exercise 23 to calculate the work done by the engine during a cycle when the steam starts at a pressure of 160 lb/in.² and a volume of 100 in.³ and expands to a volume of 800 in.³.

25. Newton's Law of Gravitation states that two bodies with masses m_1 and m_2 attract each other with a force

$$F = G \frac{m_1 m_2}{r^2}$$

where r is the distance between the bodies and G is the gravitational constant. If one of the bodies is fixed, find the work needed to move the other from $r = a$ to $r = b$.

26. Use Newton's Law of Gravitation to compute the work required to launch a 1000-kg satellite vertically to an orbit 1000 km high. You may assume that the mass of the earth is 5.98×10^{24} kg and is concentrated at the center of the earth. Take the radius of the earth to be 6.37×10^6 m and $G = 6.67 \times 10^{-11}$ N-m²/kg².

SECTION 6.5

Average Value of a Function

It is easy to calculate the average value of finitely many numbers $y_1, y_2, \ldots, y_n$:

$$y_{\text{ave}} = \frac{y_1 + y_2 + \cdots + y_n}{n}$$

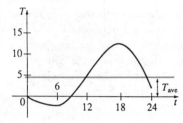

Figure 6.39

But how do we compute the average temperature during a day if there are infinitely many possible temperature readings? Figure 6.39 shows the graph of a temperature function $T(t)$ (where t is measured in hours, T in °C) and a guess at the average temperature, T_{ave}.

In general, let us try to compute the average value of a function $y = f(x)$, $a \le x \le b$. We start by dividing the interval $[a,b]$ into n equal subintervals, each with length $\Delta x = (b-a)/n$. Then we choose points $x_1^*, \dots, x_n^*$ in successive subintervals and calculate the average of the numbers $f(x_1^*), \dots, f(x_n^*)$:

$$\frac{f(x_1^*) + \cdots + f(x_n^*)}{n}$$

(For example, if f represents a temperature function and $n = 24$, this would mean that we take temperature readings every hour and average them.) Since $\Delta x = (b-a)/n$, we can write $n = (b-a)/\Delta x$ and the average value becomes

$$\frac{f(x_1^*) + \cdots + f(x_n^*)}{\dfrac{b-a}{\Delta x}} = \frac{1}{b-a}[f(x_1^*)\,\Delta x + \cdots + f(x_n^*)\,\Delta x]$$

$$= \frac{1}{b-a}\sum_{i=1}^{n} f(x_i^*)\,\Delta x$$

If we let n increase, we would be computing the average value of a large number of closely spaced values. (For example, we would be averaging temperature readings taken every minute or even every second.) The limiting value is

$$\lim_{n \to \infty} \frac{1}{b-a}\sum_{i=1}^{n} f(x_i^*)\,\Delta x = \frac{1}{b-a}\int_a^b f(x)\,dx$$

by the definition of a definite integral.

Therefore we define the **average value of f** on the interval $[a,b]$ as

(6.16)

$$f_{\text{ave}} = \frac{1}{b-a}\int_a^b f(x)\,dx$$

EXAMPLE 1   Find the average value of the function $f(x) = 1 + x^2$ over the interval $[-1,2]$.

Solution   With $a = -1$ and $b = 2$ we have

$$f_{\text{ave}} = \frac{1}{b-a}\int_a^b f(x)\,dx = \frac{1}{2-(-1)}\int_{-1}^2 (1+x^2)\,dx$$

$$= \frac{1}{3}\left[x + \frac{x^3}{3}\right]_{-1}^2 = 2$$

The question arises: Is there a number c at which the value of f is exactly equal to the average value of the function, that is, $f(c) = f_{\text{ave}}$? The following theorem says that this is true for continuous functions.

Mean Value Theorem for Integrals (6.17)

If f is continuous on $[a,b]$, then there exists a number c in $[a,b]$ such that

$$\int_a^b f(x)\,dx = f(c)(b-a)$$

The Mean Value Theorem for Integrals is a consequence of the Mean Value Theorem for derivatives (4.10). The proof is outlined in Exercise 19.

The geometric interpretation of the Mean Value Theorem for Integrals is that, for *positive* functions f, there is a number c such that the rectangle with base $[a, b]$ and height $f(c)$ has the same area as the region under the graph of f from a to b (see Figure 6.40).

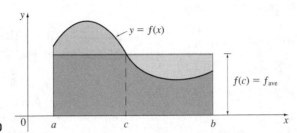

Figure 6.40

EXAMPLE 2　Since $f(x) = 1 + x^2$ is continuous on the interval $[-1, 2]$, the Mean Value Theorem for Integrals says there is a number c in $[-1, 2]$ such that

$$\int_{-1}^{2} (1 + x^2)\,dx = f(c)[2 - (-1)]$$

In this particular case we can find c explicitly. From Example 1 we know that

$$f(c) = f_{\text{ave}} = 2$$

Therefore　　　　　　　　$1 + c^2 = 2 \qquad c^2 = 1$

Thus in this case there happen to be two numbers $c = \pm 1$ in the interval $[-1, 2]$ that work in the Mean Value Theorem for Integrals.　　　　　　　●

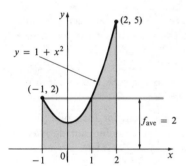

Figure 6.41　Examples 1 and 2 are illustrated by Figure 6.41.

EXAMPLE 3　Show that the average velocity of a car over a time interval $[t_1, t_2]$ is the same as the average of its velocities during the trip.

Solution　If $s(t)$ is the displacement of the car at time t then, by definition, the average velocity of the car over the interval is

$$\frac{\Delta s}{\Delta t} = \frac{s(t_2) - s(t_1)}{t_2 - t_1}$$

On the other hand, the average value of the velocity function on the interval is

$$v_{\text{ave}} = \frac{1}{t_2 - t_1} \int_{t_1}^{t_2} v(t)\,dt = \frac{1}{t_2 - t_1} \int_{t_1}^{t_2} s'(t)\,dt$$

$$= \frac{1}{t_2 - t_1}[s(t_2) - s(t_1)] \qquad \text{(by the Fundamental Theorem)}$$

$$= \frac{s(t_2) - s(t_1)}{t_2 - t_1} = \text{average velocity} \qquad ●$$

SECTION 6.5   **Exercises**

In Exercises 1–8 find the average value of the given function on the given interval.

1. $f(x) = 1 - 2x$, $[0, 3]$ 2. $f(x) = x^2 - 2x$, $[0, 3]$

3. $f(x) = x^2 + 2x - 5$, $[-2, 2]$

4. $f(x) = \sin x$, $[0, \pi]$

5. $f(x) = x^4$, $[-1, 1]$ 6. $f(x) = x^3 - x$, $[1, 3]$

7. $f(x) = \sin^2 x \cos x$, $[-\pi/2, \pi/4]$

8. $f(x) = \sqrt{x}$, $[4, 9]$

In Exercises 9–12, (a) find the average value of f on the given interval, (b) find c such that $f_{\text{ave}} = f(c)$, and (c) sketch the graph of f and a rectangle whose area is the same as the area under the graph of f.

9. $f(x) = 2x$, $[0, 3]$ 10. $f(x) = x^3$, $[-1, 2]$

11. $f(x) = 4 - x^2$, $[0, 2]$ 12. $f(x) = 4x - x^2$, $[0, 3]$

13. The temperature (in °F) in a certain city t hours after 9 A.M. was approximated by the function

$$T(t) = 50 + 14 \sin \frac{\pi t}{12}$$

Find the average temperature during the period from 9 A.M. to 9 P.M.

14. The temperature of a metal rod, 5 m long, is $4x$ (in °C) at a distance x meters from one end of the rod. What is the average temperature of the rod?

15. The linear density in a rod 8 m long is $12/\sqrt{x+1}$ kg/m, where x is measured in meters from one end of the rod. Find the average density of the rod.

16. If a freely falling body starts from rest, then its displacement is given by $s = \frac{1}{2}gt^2$. Let the velocity after a time T be v_T. Show that if we compute the average of the velocities with respect to t we get

$$v_{\text{ave}} = \tfrac{1}{2}v_T$$

but if we compute the average of the velocities with respect to s we get

$$v_{\text{ave}} = \tfrac{2}{3}v_T$$

17. Use the model given in Exercise 87 in Section 5.6 to compute the average volume of inhaled air in the lungs in one respiratory cycle.

18. The velocity v of blood that flows in a blood vessel with radius R and length l at a distance r from the central axis is

$$v(r) = \frac{P}{4\eta l}(R^2 - r^2)$$

where P is the pressure difference between the ends of the vessel and η is the viscosity of the blood (see Example 7 in Section 2.3). Find the average velocity (with respect to r) over the interval $0 \le r \le R$. Compare the average velocity with the maximum velocity.

19. Prove the Mean Value Theorem for Integrals (6.17) by applying the Mean Value Theorem for derivatives (4.10) to the function $F(x) = \int_a^x f(t)\, dt$.

20. If $f_{\text{ave}}[a, b]$ denotes the average value of f on the interval $[a, b]$ and $a < c < b$, show that

$$f_{\text{ave}}[a, b] = \frac{c-a}{b-a} f_{\text{ave}}[a, c] + \frac{b-c}{b-a} f_{\text{ave}}[c, b]$$

CHAPTER 6

════════════════════

Review

Key Topics

Define, state, or discuss the following.

1. Area between curves
2. Volume of a general solid
3. Volume of a solid of revolution: disk (or washer) method

4. Volume of a solid of revolution: shell method
5. Work
6. Average value of a function
7. Mean Value Theorem for Integrals

Exercises

In Exercises 1–8 find the area of the region bounded by the given curves.

1. $y = x^2 - 4x + 3$, $y = 0$

2. $y = 4 + 3x - x^2$, $y = 0$

3. $y = x^2 - 6x$, $y = 12x - 2x^2$

4. $y = x^2 - 6$, $y = 12 - x^2$, $x = -5$, $x = 5$

5. $y = x^3, \quad x = y^3$

6. $x - 2y + 7 = 0, \quad y^2 - 6y - x = 0$

7. $y = \sin x, \quad y = -\cos x, \quad x = 0, \quad x = \pi$

8. $y = x^3, \quad y = x^2 - 4x + 4, \quad x = 0, \quad x = 2$

In Exercises 9–13 find the volume of the solid obtained by rotating the region bounded by the given curves about the given axis.

9. $y = \sqrt{x - 1}, y = 0, x = 3;$ about the x-axis

10. $y = x^3, y = x^2;$ about the x-axis

11. $x + 3 = 4y - y^2, x = 0;$ about the x-axis

12. $y = x^3, y = 8, x = 0;$ about the y-axis

13. $x^2 - y^2 = a^2, x = a + h$ (where $a > 0, h > 0$); about the y-axis

In Exercises 14–16 set up, but do not evaluate, an integral for the volume of the solid obtained by rotating the region bounded by the given curves about the given axis.

14. $y = \cos x, y = 0, x = 3\pi/2, x = 5\pi/2;$ about the y-axis

15. $y = x^3, y = x^2;$ about $y = 1$

16. $y = x^3, y = 8, x = 0;$ about $x = 2$

17. Find the volumes of the solids obtained by rotating the region bounded by curves $y = x$ and $y = x^2$ about (a) the x-axis, (b) the y-axis, and (c) $y = 2$.

18. Let $\mathcal{R}$ be the region in the first quadrant bounded by the curves $y = x^3$ and $y = 2x - x^2$. Calculate the following quantities:
    (a) the area of $\mathcal{R}$
    (b) the volume obtained by rotating $\mathcal{R}$ about the x-axis
    (c) the volume obtained by rotating $\mathcal{R}$ about the y-axis.

The integrals in Exercises 19–22 represent volumes of solids. Describe the solids.

19. $\displaystyle \int_0^{\pi} \pi \sin^2 x \, dx$

20. $\displaystyle \int_0^{\pi} 2\pi x \sin x \, dx$

21. $\displaystyle \int_0^2 2\pi y(4 - y^2) \, dy$

22. $\displaystyle \int_0^1 \pi[(2 - x^2)^2 - (2 - \sqrt{x})^2] \, dx$

23. The base of a solid is a circular disk with radius 3. Find the volume of the solid if parallel cross-sections perpendicular to the base are isosceles right triangles with hypotenuse lying along the base.

24. The base of a solid is the region bounded by the parabolas $y = x^2$ and $y = 2 - x^2$. Find the volume of the solid if the cross-sections perpendicular to the x-axis are squares with one side lying along the base.

25. The height of a monument is 20 m. A horizontal cross-section at a distance x meters from the top is an equilateral triangle with side $x/4$ meters. Find the volume of the monument.

26. (a) The base of solid is a square with vertices at $(1,0)$, $(0,1)$, $(-1,0)$, and $(0,-1)$. Each cross-section perpendicular to the x-axis is a semicircle. Find the volume of the solid.
    (b) Show that by cutting the solid of part (a), it can be rearranged to form a cone. Thus compute its volume more simply.

27. A force of 30 N is required to maintain a spring stretched from its natural length of 12 cm to a length of 15 cm. How much work is done in stretching the spring from 12 cm to 20 cm?

28. A 1600-lb elevator is suspended by a 200-ft cable that weighs 10 lb/ft. How much work is required to raise the elevator from the basement to the third floor, a distance of 30 ft?

29. A tank full of water has the shape of a paraboloid of revolution as in the figure; that is, its shape is obtained by rotating a parabola about a vertical axis. If its height is 4 ft and the radius at the top is 4 ft, find the work required to pump the water out of the tank.

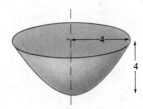

30. Find the average value of the function $f(x) = x^3$ on the interval $[2, 4]$.

31. If f is a continuous function, what is the limit as $h \to 0$ of the average value of f over the interval $[x, x + h]$?

32. (a) Use Newton's method to find the points of intersection of the curves $y = x^2$ and $y = \cos x$ correct to four decimal places.
    (b) Use part (a) to estimate the area of the region bounded by the curves $y = x^2$ and $y = \cos x$.

Problems Plus

1. If $x \sin \pi x = \int_0^{x^2} f(t)\, dt$, where f is a continuous function, find $f(4)$.

2. Suppose the curve $y = f(x)$ passes through the origin and the point $(1,1)$. Find the value of the integral $\int_0^1 f'(x)\, dx$.

3. Show that $\frac{1}{17} \le \int_1^2 \frac{1}{1+x^4}\, dx \le \frac{5}{16}$.

4. An arc PQ of a circle subtends a central angle θ as in the figure. Let $A(\theta)$ be the area between the chord PQ and the arc PQ. Let $B(\theta)$ be the area between the tangent lines PR, QR, and the arc. Find

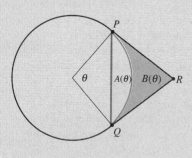

$$\lim_{\theta \to 0^+} \frac{A(\theta)}{B(\theta)}$$

5. Show that

$$\frac{d^n}{dx^n} e^{ax} \sin bx = r^n e^{ax} \sin(bx + n\theta)$$

where $r^2 = a^2 + b^2$ and $\theta = \tan^{-1}(b/a)$.

6. Find a continuous function f such that

$$\int_0^x f(t)\, dt = 3f(x) - 2$$

for all x.

7. A solid is generated by rotating about the x-axis the region bounded by the x-axis, the y-axis, and the curve $y = f(x)$, $x \ge 0$, where f is a positive function. The volume generated by the part of the curve from $x = 0$ to $x = b$ is b^2 for all $b > 0$. Find the function f.

8. Use an integral to estimate the sum $\displaystyle\sum_{i=1}^{10000} \sqrt{i}$.

9. Suppose that a line intersects the parabola $y = x^2$ in two points A and B as shown in the figure. Let C be the point on the parabola where the tangent line is parallel to the line through A and B. Show that the area of the parabolic segment cut off from the parabola by the line is four-thirds the area of triangle ABC.

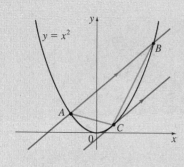

10. Find all functions f such that $f''' = f''$.

11. Evaluate $\displaystyle\lim_{x \to 0} \frac{1}{x} \int_0^x (1 - \tan 2t)^{1/t}\, dt$.

12. A right circular cone with height $1\,\mathrm{m}$ and base radius r is to be separated into three pieces of equal volume by cutting twice parallel to the base. At what heights should the cuts be made?

13. Show that, for $x > 0$,

$$\frac{x}{1+x^2} < \tan^{-1}x < x$$

14. Suppose f is continuous, $f(0) = 0$, $f(1) = 1$, $f'(x) > 0$, and $\int_0^1 f(x)\, dx = \frac{1}{3}$. Find the value of the integral $\int_0^1 f^{-1}(y)\, dy$.

15. If $f(x) = \int_0^{g(x)} \frac{1}{\sqrt{1+t^3}}\, dt$, where $g(x) = \int_0^{\cos x} [1 + \sin(t^2)]\, dt$, find $f'(\pi/2)$.

16. If f is a continuous function such that

$$\int_0^x f(t)\, dt = xe^{2x} + \int_0^x e^{-t} f(t)\, dt$$

for all x, find an explicit formula for $f(x)$.

17. Show that $f(x) = \int_1^x \sqrt{1+t^3}\, dt$ is one-to-one and find $(f^{-1})'(0)$.

18. Find the volume of the solid obtained by rotating the region bounded by the line $y = x$ and the parabola $y = x^2$ about the line $y = x$.

19. Find the interval $[a, b]$ for which the value of the integral $\int_a^b (2 + x - x^2)\, dx$ is a maximum.

20. If $f(x) = \int_0^x x^2 \sin(t^2)\, dt$, find $f'(x)$.

21. (a) Show that $\ln x < x - 1$ for $x > 0$, $x \neq 1$.
    (b) Show that, for $x > 0$, $x \neq 1$,

$$\frac{x-1}{x} < \ln x$$

   (c) Deduce Napier's Inequality:

$$\frac{1}{a} < \frac{\ln a - \ln b}{a - b} < \frac{1}{b}$$

   if $a > b > 0$.

22. Evaluate $\lim\limits_{x \to 3} \left(\frac{x}{x-3} \int_3^x \frac{\sin t}{t}\, dt \right)$.

23. Prove that if f is continuous, then

$$\int_0^x f(u)(x-u)\, du = \int_0^x \left(\int_0^u f(t)\, dt \right) du$$

24. A circular disk of radius r is used in an evaporator and is rotated in a vertical plane. If it is to be partially submerged in the liquid so as to maximize the exposed wetted area of the disk, show that the center of the disk should be positioned at a height $r/\sqrt{1 + \pi^2}$ above the surface of the liquid.

25. Let $f(x) = a_1 \sin x + a_2 \sin 2x + \cdots + a_n \sin nx$, where $a_1, a_2, \ldots, a_n$ are real numbers and n is a positive integer. If it is given that $|f(x)| \leq |\sin x|$ for all x, show that

$$|a_1 + 2a_2 + \cdots + na_n| \leq 1$$

26. Find $\dfrac{d^2}{dx^2} \int_0^x \left(\int_1^{\sin t} \sqrt{1+u^4}\, du \right) dt$.

27. If f is differentiable function such that $\int_0^x f(t)\, dt = [f(x)]^2$ for all x, find f.

7

Techniques of Integration

Common integration is only the memory of
differentiation. The different devices by which
integration is accomplished are changes, not
from the known to the unknown, but from
forms in which memory will not serve us to
those in which it will.

Augustus de Morgan

Because of the Fundamental Theorem of Calculus, we can integrate a function if we know an antiderivative, that is, an indefinite integral. We summarize here the most important integrals that we have learned so far.

$$\int x^n \, dx = \frac{x^{n+1}}{n+1} + C \quad (n \neq -1) \qquad \int \frac{1}{x} \, dx = \ln|x| + C$$

$$\int e^x \, dx = e^x + C \qquad \int a^x \, dx = \frac{a^x}{\ln a} + C$$

$$\int \sin x \, dx = -\cos x + C \qquad \int \cos x \, dx = \sin x + C$$

$$\int \sec^2 x \, dx = \tan x + C \qquad \int \csc^2 x \, dx = -\cot x + C$$

$$\int \sec x \tan x \, dx = \sec x + C \qquad \int \csc x \cot x \, dx = -\csc x + C$$

$$\int \sinh x \, dx = \cosh x + C \qquad \int \cosh x \, dx = \sinh x + C$$

$$\int \tan x \, dx = \ln|\sec x| + C \quad \text{(see Equation 5.34)}$$

$$\int \cot x \, dx = \ln|\sin x| + C \quad \text{(see Exercise 93 in Section 5.6)}$$

$$\int \frac{1}{x^2 + a^2} \, dx = \frac{1}{a} \tan^{-1}\!\left(\frac{x}{a}\right) + C$$

$$\int \frac{1}{\sqrt{a^2 - x^2}} \, dx = \sin^{-1}\!\left(\frac{x}{a}\right) + C$$

In this chapter we develop techniques for using these basic integration formulas to obtain indefinite integrals of more complicated functions. We learned the most important method of integration, the Substitution Rule, in Section 5.6. The other general technique, integration by parts, is presented in Section 7.1. Then we learn methods that are special to particular classes of functions such as trigonometric functions and rational functions.

Integration is not as straightforward as differentiation; there are no rules that absolutely guarantee obtaining an indefinite integral of a function. Therefore in Section 7.6 we discuss a strategy for integration.

Integration by Parts

For every differentiation rule there is a corresponding integration rule. For instance, the Substitution Rule for integration corresponds to the Chain Rule for differentiation. The rule that corresponds to the Product Rule for differentiation is called the rule for integration by parts.

The Product Rule states that if f and g are differentiable functions, then

$$\frac{d}{dx}[f(x)g(x)] = f'(x)g(x) + f(x)g'(x)$$

In the notation for indefinite integrals this equation becomes

$$\int [f'(x)g(x) + f(x)g'(x)]\, dx = f(x)g(x)$$

or

$$\int f'(x)g(x)\, dx + \int f(x)g'(x)\, dx = f(x)g(x)$$

We can rearrange this latter equation as

(7.1)

$$\boxed{\int f(x)g'(x)\, dx = f(x)g(x) - \int f'(x)g(x)\, dx}$$

Formula 7.1 is called **the formula for integration by parts.** It is perhaps easier to remember in the following notation. Let $u = f(x)$ and $v = g(x)$. Then $du = f'(x)\, dx$ and $dv = g'(x)\, dx$, so, by the Substitution Rule, the formula for integration by parts becomes

(7.2)

$$\boxed{\int u\, dv = uv - \int v\, du}$$

EXAMPLE 1   Find $\int x \sin x\, dx$.

Solution using Formula 7.1   Suppose we choose $f(x) = x$ and $g'(x) = \sin x$. Then $f'(x) = 1$ and $g(x) = -\cos x$. (For g we can choose *any* antiderivative of g'.) Thus, using Formula 7.1, we have

$$\int x \sin x\, dx = f(x)g(x) - \int f'(x)g(x)\, dx$$

$$= x(-\cos x) - \int (-\cos x)\, dx$$

$$= -x \cos x + \int \cos x\, dx$$

$$= -x \cos x + \sin x + C$$

It is wise to check the answer by differentiating it. If we do so, we get $x \sin x$, as expected.

Solution using Formula 7.2   Let

It is helpful to use the pattern:
$$u = \square \qquad dv = \square$$

$$u = x \qquad dv = \sin x \, dx$$

$$du = \square \qquad v = \square$$ Then

$$du = dx \qquad v = -\cos x$$

and so

$$\int x \sin x \, dx = \int u \, dv = uv - \int v \, du$$

$$= -x \cos x + \int \cos x \, dx$$

$$= -x \cos x + \sin x + C$$

Note: The object in using integration by parts is to obtain a simpler integral than the one we started with. Thus in Example 1 we started with $\int x \sin x \, dx$ and expressed it in terms of the simpler integral $\int \cos x \, dx$. If we had chosen $u = \sin x$ and $dv = x \, dx$, then $du = \cos x \, dx$ and $v = x^2/2$, so integration by parts gives

$$\int x \sin x \, dx = (\sin x)\frac{x^2}{2} - \frac{1}{2} \int x^2 \cos x \, dx$$

But $\int x^2 \cos x \, dx$ is a more difficult integral than the one we started with. In general, when deciding on a choice for u and dv, we usually try to choose $u = f(x)$ to be a function that becomes simpler when differentiated (or at least not more complicated) as long as $dv = g'(x) \, dx$ can be readily integrated to give v.

EXAMPLE 2   Evaluate $\int \ln x \, dx$.

Solution   Here there is not much choice. Let

$$u = \ln x \qquad dv = dx$$

Then

$$du = \frac{1}{x} \, dx \qquad v = x$$

Integrating by parts, we get

$$\int \ln x \, dx = x \ln x - \int x \, \frac{dx}{x}$$

$$= x \ln x - \int dx = x \ln x - x + C$$

Integration by parts is effective in this example because the derivative of the function $f(x) = \ln x$ is simpler than f.

EXAMPLE 3   Find $\int x^2 e^x \, dx$.

Solution   Let

$$u = x^2 \qquad dv = e^x \, dx$$

Then

$$du = 2x \, dx \qquad v = e^x$$

Integration by parts gives

(7.3)
$$\int x^2 e^x \, dx = x^2 e^x - 2 \int x e^x \, dx$$

The integral that we obtained, $\int x e^x \, dx$, is simpler than the original integral but is still not obvious. Therefore we use integration by parts a second time, this time with $u = x$ and $dv = e^x \, dx$. Then $du = dx$, $v = e^x$, and

$$\int x e^x \, dx = x e^x - \int e^x \, dx$$
$$= x e^x - e^x + C$$

Putting this in Equation 7.3, we get

$$\int x^2 e^x \, dx = x^2 e^x - 2 \int x e^x \, dx$$
$$= x^2 e^x - 2(x e^x - e^x + C)$$
$$= x^2 e^x - 2x e^x + 2e^x + C_1$$

where $C_1 = -2C$.   ●

EXAMPLE 4   Evaluate $\int e^x \sin x \, dx$.

Solution   Let $u = e^x$ and $dv = \sin x \, dx$. Then $du = e^x \, dx$ and $v = -\cos x$, so integration by parts gives

(7.4)
$$\int e^x \sin x \, dx = -e^x \cos x + \int e^x \cos x \, dx$$

The integral that we have obtained, $\int e^x \cos x \, dx$, is no simpler than the original one, but at least it is no more difficult. Having had success in the preceding example integrating by parts twice, we persevere and integrate by parts again. This time we use $u = e^x$ and $dv = \cos x \, dx$. Then $du = e^x$ and $v = \sin x$, and

(7.5)
$$\int e^x \cos x \, dx = e^x \sin x - \int e^x \sin x \, dx$$

At first glance it appears as if we have accomplished nothing because we have arrived at $\int e^x \sin x \, dx$, which is where we started. However if we put Equation 7.5 into Equation 7.4 we get

$$\int e^x \sin x \, dx = -e^x \cos x + e^x \sin x - \int e^x \sin x \, dx$$

This can be regarded as an equation to be solved for the unknown integral. Solving, we obtain

$$2 \int e^x \sin x \, dx = -e^x \cos x + e^x \sin x$$

and, dividing by 2 and adding the constant of integration, we get

Check the answer
by differentiating it.

$$\int e^x \sin x\, dx = \tfrac{1}{2}e^x(\sin x - \cos x) + C$$ •

If we combine the formula for integration by parts with Part 2 of the Fundamental Theorem of Calculus, we can evaluate definite integrals by parts. Evaluating both sides of Formula 7.1 between a and b, assuming f' and g' are continuous, and using the Fundamental Theorem in the form of Equation 5.25, we obtain

(7.6)

$$\int_a^b f(x)g'(x)\, dx = f(x)g(x)\Big|_a^b - \int_a^b f'(x)g(x)\, dx$$

EXAMPLE 5   Calculate $\displaystyle\int_0^1 \tan^{-1}x\, dx$.

Solution   Let

$$u = \tan^{-1}x \qquad\qquad dv = dx$$

Then

$$du = \frac{dx}{1 + x^2} \qquad\qquad v = x$$

So Formula 7.6 gives

$$\int_0^1 \tan^{-1}x\, dx = x\tan^{-1}x\Big|_0^1 - \int_0^1 \frac{x}{1 + x^2}\, dx$$

$$= 1\cdot\tan^{-1}1 - 0\cdot\tan^{-1}0 - \int_0^1 \frac{x}{1 + x^2}\, dx$$

$$= \frac{\pi}{4} - \int_0^1 \frac{x}{1 + x^2}\, dx$$

To evaluate this integral we use the substitution $t = 1 + x^2$ (since u has another meaning in this example). Then $dt = 2x\, dx$ so $x\, dx = dt/2$. When $x = 0$, $t = 1$; when $x = 1$, $t = 2$; so

$$\int_0^1 \frac{x}{1 + x^2}\, dx = \frac{1}{2}\int_1^2 \frac{dt}{t} = \frac{1}{2}\ln|t|\Big|_1^2$$

$$= \tfrac{1}{2}(\ln 2 - \ln 1) = \tfrac{1}{2}\ln 2$$

Therefore

$$\int_0^1 \tan^{-1}x\, dx = \frac{\pi}{4} - \int_0^1 \frac{x}{1 + x^2}\, dx = \frac{\pi}{4} - \frac{\ln 2}{2}$$ •

EXAMPLE 6   Prove the reduction formula

(7.7)

$$\int \sin^n x\, dx = -\frac{1}{n}\cos x\sin^{n-1}x + \frac{n-1}{n}\int \sin^{n-2}x\, dx$$

where $n \geq 2$ is an integer.

Solution   Let

$$u = \sin^{n-1}x \qquad\qquad dv = \sin x\, dx$$

Then $\qquad du = (n-1)\sin^{n-2}x \cos x\, dx \qquad v = -\cos x$

so integration by parts gives

$$\int \sin^n x\, dx = -\cos x \sin^{n-1}x + (n-1)\int \sin^{n-2}x \cos^2 x\, dx$$

Since $\cos^2 x = 1 - \sin^2 x$, we have

$$\int \sin^n x\, dx = -\cos x \sin^{n-1}x + (n-1)\int \sin^{n-2}x\, dx - (n-1)\int \sin^n x\, dx$$

As in Example 4, we solve this equation for the desired integral by taking the last term on the right side to the left side. Thus we have

$$n\int \sin^n x\, dx = -\cos x \sin^{n-1}x + (n-1)\int \sin^{n-2}x\, dx$$

or

$$\int \sin^n x\, dx = -\frac{1}{n} \cos x \sin^{n-1}x + \frac{n-1}{n}\int \sin^{n-2}x\, dx \qquad\qquad \bullet$$

The reduction formula (7.7) is useful because by using it repeatedly we could eventually express $\int \sin^n x\, dx$ in terms of $\int \sin x\, dx$ (if n is odd) or $\int (\sin x)^0\, dx = \int dx$ (if n is even).

SECTION 7.1   **Exercises**

Evaluate the integrals in Exercises 1–36.

1. $\displaystyle\int xe^{2x}\, dx$

2. $\displaystyle\int x \cos x\, dx$

3. $\displaystyle\int x \sin 4x\, dx$

4. $\displaystyle\int x \ln x\, dx$

5. $\displaystyle\int x^2 \cos 3x\, dx$

6. $\displaystyle\int x^2 \sin 2x\, dx$

7. $\displaystyle\int (\ln x)^2\, dx$

8. $\displaystyle\int \sin^{-1}x\, dx$

9. $\displaystyle\int \theta \sin \theta \cos \theta\, d\theta$

10. $\displaystyle\int \theta \sec^2\theta\, d\theta$

11. $\displaystyle\int t^2 \ln t\, dt$

12. $\displaystyle\int t^3 e^t\, dt$

13. $\displaystyle\int e^{2\theta} \sin 3\theta\, d\theta$

14. $\displaystyle\int e^{-\theta} \cos 3\theta\, d\theta$

15. $\displaystyle\int y \sinh y\, dy$

16. $\displaystyle\int y \cosh ay\, dy$

17. $\displaystyle\int_0^1 te^{-t}\, dt$

18. $\displaystyle\int_1^4 \sqrt{t} \ln t\, dt$

19. $\displaystyle\int_0^{\pi/2} x \cos 2x\, dx$

20. $\displaystyle\int_0^1 x^2 e^{-x}\, dx$

21. $\displaystyle\int_0^{1/2} \cos^{-1}x\, dx$

22. $\displaystyle\int_{\pi/4}^{\pi/2} x \csc^2 x\, dx$

23. $\displaystyle\int \sin 3x \cos 5x\, dx$

24. $\displaystyle\int \sin 2x \sin 4x\, dx$

25. $\displaystyle\int \cos x \ln(\sin x)\, dx$

26. $\displaystyle\int x^3 e^{x^2}\, dx$

27. $\displaystyle\int (2x+3)e^x\, dx$

28. $\displaystyle\int x5^x\, dx$

29. $\displaystyle\int \cos(\ln x)\, dx$

30. $\displaystyle\int_1^4 e^{\sqrt{x}}\, dx$

31. $\displaystyle\int_1^4 \ln \sqrt{x}\, dx$

32. $\displaystyle\int \sin(\ln x)\, dx$

33. $\int \sin\sqrt{x}\,dx$ **34.** $\int x^5\cos(x^3)\,dx$

35. $\int x^5 e^{x^2}\,dx$ **36.** $\int x\tan^{-1}x\,dx$

37. (a) Use the reduction formula in Example 6 to show that

$$\int \sin^2 x\,dx = \frac{x}{2} - \frac{\sin 2x}{4} + C$$

(b) Use part (a) and the reduction formula to evaluate $\int \sin^4 x\,dx$.

38. (a) Prove the reduction formula

$$\int \cos^n x\,dx = \frac{1}{n}\cos^{n-1}x\sin x + \frac{n-1}{n}\int \cos^{n-2}x\,dx$$

(b) Use part (a) to evaluate $\int \cos^2 x\,dx$.
(c) Use parts (a) and (b) to evaluate $\int \cos^4 x\,dx$.

39. (a) Use the reduction formula in Example 6 to show that

$$\int_0^{\pi/2} \sin^n x\,dx = \frac{n-1}{n}\int_0^{\pi/2}\sin^{n-2}x\,dx$$

where $n \geq 2$ is an integer.

(b) Use part (a) to evaluate $\int_0^{\pi/2}\sin^3 x\,dx$ and $\int_0^{\pi/2}\sin^5 x\,dx$.

(c) Use part (a) to show that, if n is odd,

$$\int_0^{\pi/2}\sin^n x\,dx = \frac{2\cdot4\cdot6\cdots(n-1)}{3\cdot5\cdot7\cdots n}$$

This formula is called a *Wallis product*.

40. Prove that, if n is even,

$$\int_0^{\pi/2}\sin^n x\,dx = \frac{1\cdot3\cdot5\cdots(n-1)}{2\cdot4\cdot6\cdots n}\frac{\pi}{2}$$

Use integration by parts to prove the reduction formulas in Exercises 41-44.

41. $\int (\ln x)^n\,dx = x(\ln x)^n - n\int (\ln x)^{n-1}\,dx$

42. $\int x^n e^x\,dx = x^n e^x - n\int x^{n-1}e^x\,dx$

43. $\int (x^2+a^2)^n\,dx =$
$$\frac{x(x^2+a^2)^n}{2n+1} + \frac{2na^2}{2n+1}\int (x^2+a^2)^{n-1}\,dx \quad \left(n\neq-\tfrac{1}{2}\right)$$

44. $\int \sec^n x\,dx =$
$$\frac{\tan x\sec^{n-2}x}{n-1} + \frac{n-2}{n-1}\int \sec^{n-2}x\,dx \quad (n\neq1)$$

45. Use Exercise 41 to find $\int (\ln x)^3\,dx$.

46. Use Exercise 42 to find $\int x^4 e^x\,dx$.

In Exercises 47-50 find the area of the region bounded by the given curves.

47. $y=\sin^{-1}x$, $y=0$, $x=0.5$

48. $y=xe^{-x}$, $y=0$, $x=5$

49. $y=\ln x$, $y=0$, $x=e$

50. $y=5\ln x$, $y=x\ln x$

In Exercises 51-54 use the method of cylindrical shells to find the volume generated by rotating the region bounded by the given curves about the given axis.

51. $y=\sin x$, $y=0$, $x=2\pi$, $x=3\pi$; about the y-axis

52. $y=e^x$, $y=e^{-x}$, $x=1$; about the y-axis

53. $y=e^{-x}$, $y=0$, $x=-1$, $x=0$; about $x=1$

54. $y=e^x$, $x=0$, $y=\pi$; about the x-axis

55. A particle that moves along a straight line has velocity $v(t)=t^2 e^{-t}$ meters per second after t seconds. How far will it travel during the first t seconds?

56. If $f(0)=g(0)=0$, show that

$$\int_0^a f(x)g''(x)\,dx =$$
$$f(a)g'(a) - f'(a)g(a) + \int_0^a f''(x)g(x)\,dx$$

57. Use integration by parts to show that

$$\int f(x)\,dx = xf(x) - \int xf'(x)\,dx$$

58. If f and g are inverse functions and f' is continuous, prove that

$$\int_a^b f(x)\,dx = bf(b) - af(a) - \int_{f(a)}^{f(b)} g(y)\,dy$$

[*Hint:* Use Exercise 57 and make the substitution $y=f(x)$.]

59. Use Exercise 58 to evaluate $\int_1^e \ln x\,dx$.

60. In the case where f and g are positive functions and $b>a>0$, draw a diagram to give a geometric interpretation of Exercise 58.

61. We arrived at Formula 6.10, $V=\int_a^b 2\pi xf(x)\,dx$, by using cylindrical shells, but now we can use integration by parts to deduce it from Formula 6.7, at least for the case where f is one-to-one and therefore has an inverse function g. Use the figure to show that

$$V = \pi b^2 d - \pi a^2 c - \int_c^d \pi [g(y)]^2 \, dy$$

Make the substitution $y = f(x)$ and then use integration by parts on the resulting integral to prove Formula 6.10.

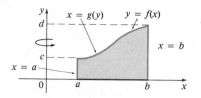

SECTION 7.2

Trigonometric Integrals

In this section we use trigonometric identities to integrate certain combinations of trigonometric functions. We start with powers of sine and cosine.

EXAMPLE 1   Evaluate $\int \cos^3 x \, dx$.

Solution   Here the appropriate identity is $\cos^2 x = 1 - \sin^2 x$. We write

$$\cos^3 x = \cos^2 x \cdot \cos x = (1 - \sin^2 x) \cos x$$

It is useful to have the extra factor of $\cos x$ because if we make the substitution $u = \sin x$, then we have $du = \cos x \, dx$. Thus

$$\int \cos^3 x \, dx = \int \cos^2 x \cdot \cos x \, dx = \int (1 - \sin^2 x) \cos x \, dx$$

$$= \int (1 - u^2) \, du = u - \tfrac{1}{3} u^3 + C$$

$$= \sin x - \tfrac{1}{3} \sin^3 x + C \qquad \bullet$$

The method used in Example 1 suggests the following general strategy to be used in evaluating integrals of the form $\int \sin^m x \cos^n x \, dx$, where $m \geq 0$ and $n \geq 0$ are integers and either m or n is odd.

$\int \sin^m x \cos^n x \, dx$ (7.8)

(a)   If the power of cosine is odd ($n = 2k + 1$), save one cosine factor and use $\cos^2 x = 1 - \sin^2 x$ to express the remaining factors in terms of sine:

$$\int \sin^m x \cos^{2k+1} x \, dx = \int \sin^m x (\cos^2 x)^k \cos x \, dx$$

$$= \int \sin^m x (1 - \sin^2 x)^k \cos x \, dx$$

Then substitute $u = \sin x$.

(b)   If the power of sine is odd ($m = 2k + 1$), save one sine factor and use $\sin^2 x = 1 - \cos^2 x$ to express the remaining factors in terms of cosine:

$$\int \sin^{2k+1} x \cos^n x \, dx = \int (\sin^2 x)^k \cos^n x \sin x \, dx$$

$$= \int (1 - \cos^2 x)^k \cos^n x \sin x \, dx$$

Then substitute $u = \cos x$.

EXAMPLE 2 Find $\int \sin^5 x \cos^2 x \, dx$.

Solution Here the power of sine is odd and so we proceed as in case (b) of (7.8), substituting $u = \cos x$:

$$\int \sin^5 x \cos^2 x \, dx = \int \sin^4 x \cos^2 x \sin x \, dx$$

$$= \int (1 - \cos^2 x)^2 \cos^2 x \sin x \, dx$$

$$= \int (1 - u^2)^2 u^2 \, (-du)$$

$$= -\int (u^2 - 2u^4 + u^6) \, du$$

$$= -\left(\frac{u^3}{3} + 2\frac{u^5}{5} + \frac{u^7}{7}\right) + C$$

$$= -\tfrac{1}{3} \cos^3 x + \tfrac{2}{5} \cos^5 x - \tfrac{1}{7} \cos^7 x + C$$

The advice in (7.8) works if either sine or cosine occurs to an odd power. In the remaining case (both m and n are even) we proceed as follows.

(c) If the powers of both sine and cosine are even, use the half-angle identities

$$\sin^2 x = \tfrac{1}{2}(1 - \cos 2x) \qquad \cos^2 x = \tfrac{1}{2}(1 + \cos 2x)$$

(see Appendix B, Equation 17).

EXAMPLE 3 Evaluate $\int_0^\pi \sin^2 x \, dx$.

Solution Here $m = 2$ and $n = 0$, so we use the half-angle formula for $\sin^2 x$:

$$\int_0^\pi \sin^2 x \, dx = \tfrac{1}{2} \int_0^\pi (1 - \cos 2x) \, dx = \left[\tfrac{1}{2}\left(x - \tfrac{1}{2} \sin 2x\right)\right]_0^\pi$$

$$= \tfrac{1}{2}\left(\pi - \tfrac{1}{2} \sin 2\pi\right) - \tfrac{1}{2}\left(0 - \tfrac{1}{2} \sin 0\right) = \tfrac{1}{2}\pi$$

Notice that we mentally made the substitution $u = 2x$ when integrating $\cos 2x$. Another method for evaluating this integral was given in Exercise 37 in Section 7.1.

EXAMPLE 4 Find $\int \sin^4 x \, dx$.

Solution It would be possible to evaluate this integral using the reduction formula for $\int \sin^n x \, dx$ (Equation 7.7) together with Example 1 (as in Exercise 37 in Section 7.1), but an easier method is to write $\sin^4 x = (\sin^2 x)^2$ and use (c):

$$\int \sin^4 x \, dx = \int (\sin^2 x)^2 \, dx$$

$$= \int \left(\frac{1 - \cos 2x}{2}\right)^2 \, dx$$

$$= \tfrac{1}{4} \int (1 - 2\cos 2x + \cos^2 2x)\, dx$$

Since $\cos^2 2x$ occurs we must use another half-angle formula

$$\cos^2 2x = \tfrac{1}{2}(1 + \cos 4x)$$

This gives

$$\int \sin^4 x\, dx = \tfrac{1}{4} \int [1 - 2\cos 2x + \tfrac{1}{2}(1 + \cos 4x)]\, dx$$

$$= \tfrac{1}{4} \int \left(\tfrac{3}{2} - 2\cos 2x + \tfrac{1}{2}\cos 4x \right) dx$$

$$= \tfrac{1}{4} \left(\tfrac{3}{2}x - \sin 2x + \tfrac{1}{8}\sin 4x \right) + C$$ ●

Integrals of the form $\int \tan^m x \sec^n x\, dx$ can be integrated in the following cases.

$\int \tan^m x \sec^n x\, dx$ (7.9)

(a)   If the power of secant is even ($n = 2k$), save a factor of $\sec^2 x$ and use $\sec^2 x = 1 + \tan^2 x$ to express the remaining factors in terms of $\tan x$:

$$\int \tan^m x \sec^{2k} x\, dx = \int \tan^m x (\sec^2 x)^{k-1} \sec^2 x\, dx$$

$$= \int \tan^m x (1 + \tan^2 x)^{k-1} \sec^2 x\, dx$$

Then substitute $u = \tan x$.

(b)   If the power of tangent is odd ($m = 2k + 1$), save a factor of $\sec x \tan x$ and use $\tan^2 x = \sec^2 x - 1$ to express the remaining factors in terms of $\sec x$:

$$\int \tan^{2k+1} x \sec^n x\, dx = \int (\tan^2 x)^k \sec^{n-1} x \sec x \tan x\, dx$$

$$= \int (\sec^2 x - 1)^k \sec^{n-1} x \sec x \tan x\, dx$$

Then substitute $u = \sec x$.

EXAMPLE 5   Evaluate $\int \tan^6 x \sec^4 x\, dx$.

Solution   Since there is an even power of the secant, we factor off $\sec^2 x$ and substitute $u = \tan x$ so that $du = \sec^2 x\, dx$. Then rest of the integrand is then expressed totally in terms of $\tan x$ by means of the identity $\sec^2 x = 1 + \tan^2 x$:

$$\int \tan^6 x \sec^4 x\, dx = \int \tan^6 x \sec^2 x \sec^2 x\, dx$$

$$= \int \tan^6 x (1 + \tan^2 x) \sec^2 x\, dx$$

$$= \int u^6 (1 + u^2)\, du = \int (u^6 + u^8)\, du$$

$$= \frac{u^7}{7} + \frac{u^9}{9} + C$$

$$= \tfrac{1}{7} \tan^7 x + \tfrac{1}{9} \tan^9 x + C$$ ●

EXAMPLE 6 Find $\int \tan^5 x \sec^7 x \, dx$.

Solution Since there is an odd power of $\tan x$, we factor $\tan x \sec x$ from the integrand and substitute $u = \sec x$ so that $du = \sec x \tan x \, dx$. Since an even power of $\tan x$ remains, we use the identity $\tan^2 x = \sec^2 x - 1$ to express the remainder of the integral totally in terms of $\sec x$:

$$\int \tan^5 x \sec^7 x \, dx = \int \tan^4 x \sec^6 x \sec x \tan x \, dx$$

$$= \int (\sec^2 x - 1)^2 \sec^6 x \sec x \tan x \, dx$$

$$= \int (u^2 - 1)^2 u^6 \, du = \int (u^{10} - 2u^8 + u^6) \, du$$

$$= \frac{u^{11}}{11} - 2\frac{u^9}{9} + \frac{u^7}{7} + C$$

$$= \tfrac{1}{11} \sec^{11} x - \tfrac{2}{9} \sec^9 x + \tfrac{1}{7} \sec^7 x + C$$

If $n = 0$, only $\tan x$ occurs. Here we use $\tan^2 x = \sec^2 x - 1$ and, if necessary, the formula

$$\int \tan x \, dx = \ln |\sec x| + C$$

EXAMPLE 7 Find $\int \tan^3 x \, dx$.

Solution
$$\int \tan^3 x \, dx = \int \tan x \tan^2 x \, dx$$

$$= \int \tan x (\sec^2 x - 1) \, dx$$

$$= \int \tan x \sec^2 x \, dx - \int \tan x \, dx$$

$$= \frac{\tan^2 x}{2} - \ln |\sec x| + C$$

In the first integral we mentally substituted $u = \tan x$ so that $du = \sec^2 x \, dx$.

If n is odd and m is even, we express the integral totally in terms of $\sec x$. Powers of $\sec x$ may require integration by parts.

EXAMPLE 8 Find $\int \sec x \, dx$.

Solution We multiply numerator and denominator by $\sec x + \tan x$:

$$\int \sec x \, dx = \int \sec x \, \frac{\sec x + \tan x}{\sec x + \tan x} \, dx$$

$$= \int \frac{\sec^2 x + \sec x \tan x}{\sec x + \tan x} \, dx$$

If we substitute $u = \sec x + \tan x$, then $du = (\sec x \tan x + \sec^2 x) \, dx$, so the integral becomes $\int (1/u) \, du = \ln |u| + C$. Thus we have

(7.10)

$$\boxed{\int \sec x \, dx = \ln |\sec x + \tan x| + C}$$

The method of Example 8 was admittedly very tricky, but we need Formula 7.10 for our future work.

EXAMPLE 9   Find $\int \sec^3 x \, dx$.

Solution   Here we integrate by parts with

$$u = \sec x \qquad\qquad dv = \sec^2 x \, dx$$

$$du = \sec x \tan x \, dx \qquad\qquad v = \tan x$$

$$\int \sec^3 x \, dx = \sec x \tan x - \int \sec x \tan^2 x \, dx$$

$$= \sec x \tan x - \int \sec x (\sec^2 x - 1) \, dx$$

$$= \sec x \tan x - \int \sec^3 x \, dx + \int \sec x \, dx$$

Using Formula 7.10 and solving for the required integral, we get

$$\int \sec^3 x \, dx = \tfrac{1}{2}\bigl(\sec x \tan x + \ln |\sec x + \tan x|\bigr) + C$$ ●

Integrals such as the one in Example 9 and others of the form (7.9) may seem very special but they occur frequently in applications of integration, as we will see in Chapter 8. Integrals of the form $\int \cot^m x \csc^n x \, dx$ can be found by similar methods because of the identity $1 + \cot^2 x = \csc^2 x$.

(7.11)

> To evaluate the integrals (a) $\int \sin mx \cos nx \, dx$, (b) $\int \sin mx \sin nx \, dx$,
> (c) $\int \cos mx \cos nx \, dx$ use the identities (Appendix B, Equation 18):
>
> (a) $\sin A \cos B = \tfrac{1}{2}[\sin(A - B) + \sin(A + B)]$
> (b) $\sin A \sin B = \tfrac{1}{2}[\cos(A - B) - \cos(A + B)]$
> (c) $\cos A \cos B = \tfrac{1}{2}[\cos(A - B) + \cos(A + B)]$

EXAMPLE 10   Evaluate $\int \sin 4x \cos 5x \, dx$.

Solution   This integral could be evaluated using integration by parts but it is easier to use the identity in (7.11)(a) as follows:

$$\int \sin 4x \cos 5x \, dx = \int \tfrac{1}{2}[\sin(-x) + \sin 9x] \, dx$$

$$= \tfrac{1}{2} \int (-\sin x + \sin 9x) \, dx$$

$$= \tfrac{1}{2}\Bigl(\cos x - \tfrac{1}{9} \cos 9x\Bigr) + C$$ ●

SECTION 7.2 **Exercises**

In Exercises 1–56 evaluate the integral.

1. $\displaystyle\int_0^{\pi/2} \sin^2 3x \, dx$

2. $\displaystyle\int_0^{\pi/2} \cos^2 x \, dx$

3. $\displaystyle\int \cos^4 x \, dx$

4. $\displaystyle\int \sin^3 x \, dx$

5. $\displaystyle\int \sin^3 x \cos^4 x \, dx$

6. $\displaystyle\int \sin^4 x \cos^3 x \, dx$

7. $\displaystyle\int_0^{\pi/4} \sin^4 x \cos^2 x \, dx$

8. $\displaystyle\int_0^{\pi/2} \sin^2 x \cos^2 x \, dx$

9. $\displaystyle\int (1 - \sin 2x)^2 \, dx$

10. $\displaystyle\int \sin\left(x + \frac{\pi}{6}\right) \cos x \, dx$

11. $\displaystyle\int \cos^5 x \sin^5 x \, dx$

12. $\displaystyle\int \sin^6 x \, dx$

13. $\displaystyle\int \cos^6 x \, dx$

14. $\displaystyle\int \sin^5 2x \cos^4 2x \, dx$

15. $\displaystyle\int \sin^5 x \, dx$

16. $\displaystyle\int \sin^4 x \cos^4 x \, dx$

17. $\displaystyle\int \sin^3 x \sqrt{\cos x} \, dx$

18. $\displaystyle\int \frac{\cos^3 x}{\sqrt{\sin x}} \, dx$

19. $\displaystyle\int \frac{\cos^2(\sqrt{x})}{\sqrt{x}} \, dx$

20. $\displaystyle\int x \sin^3(x^2) \, dx$

21. $\displaystyle\int \cos^2 x \tan^3 x \, dx$

22. $\displaystyle\int \cot^5 x \sin^2 x \, dx$

23. $\displaystyle\int \frac{1 - \sin x}{\cos x} \, dx$

24. $\displaystyle\int \frac{dx}{1 - \sin x}$

25. $\displaystyle\int \tan^2 x \, dx$

26. $\displaystyle\int \tan^4 x \, dx$

27. $\displaystyle\int \sec^4 x \, dx$

28. $\displaystyle\int \sec^6 x \, dx$

29. $\displaystyle\int_0^{\pi/4} \tan^4 x \sec^2 x \, dx$

30. $\displaystyle\int_0^{\pi/4} \tan^2 x \sec^4 x \, dx$

31. $\displaystyle\int \tan x \sec^3 x \, dx$

32. $\displaystyle\int \tan^3 x \sec^3 x \, dx$

33. $\displaystyle\int \tan^5 x \, dx$

34. $\displaystyle\int \tan^6 x \, dx$

35. $\displaystyle\int_0^{\pi/3} \tan^5 x \sec x \, dx$

36. $\displaystyle\int_0^{\pi/3} \tan^5 x \sec^3 x \, dx$

37. $\displaystyle\int \tan x \sec^6 x \, dx$

38. $\displaystyle\int \tan^3 x \sec^6 x \, dx$

39. $\displaystyle\int \frac{\sec^2 x}{\cot x} \, dx$

40. $\displaystyle\int \tan^2 x \sec x \, dx$

41. $\displaystyle\int_{\pi/6}^{\pi/2} \cot^2 x \, dx$

42. $\displaystyle\int_{\pi/4}^{\pi/2} \cot^3 x \, dx$

43. $\displaystyle\int \cot^4 x \csc^4 x \, dx$

44. $\displaystyle\int \cot^3 x \csc^4 x \, dx$

45. $\displaystyle\int \csc x \, dx$

46. $\displaystyle\int \csc^3 x \, dx$

47. $\displaystyle\int \frac{\cos^2 x}{\sin x} \, dx$

48. $\displaystyle\int \frac{dx}{\sin^4 x}$

49. $\displaystyle\int \sin 5x \sin 2x \, dx$

50. $\displaystyle\int \sin 3x \cos x \, dx$

51. $\displaystyle\int \cos 3x \cos 4x \, dx$

52. $\displaystyle\int \sin 3x \sin 6x \, dx$

53. $\displaystyle\int \sin x \cos 5x \, dx$

54. $\displaystyle\int \cos x \cos 2x \cos 3x \, dx$

55. $\displaystyle\int \frac{1 - \tan^2 x}{\sec^2 x} \, dx$

56. $\displaystyle\int \frac{\cos x + \sin x}{\sin 2x} \, dx$

57. Find the average value of the function
$f(x) = \sin^2 x \cos^3 x$ on the interval $[-\pi, \pi]$.

58. Evaluate $\int \sin x \cos x \, dx$ by four methods:
(a) the substitution $u = \cos x$, (b) the substitution
$u = \sin x$, (c) the identity $\sin 2x = 2 \sin x \cos x$, and
(d) integration by parts. How do you explain the
different appearances of the answers?

In Exercises 59 and 60 find the area of the region bounded by the given curves.

59. $y = \sin x$, $\quad y = \sin^3 x$, $\quad x = 0$, $\quad x = \pi/2$

60. $y = \sin x$, $\quad y = 2\sin^2 x$, $\quad x = 0$, $\quad x = \pi/2$

In Exercises 61–64 find the volume obtained by rotating the region bounded by the given curves about the given axis.

61. $y = \sin x$, $x = \pi/2$, $x = \pi$, $y = 0$;   about the x-axis

62. $y = \tan^2 x$, $y = 0$, $x = 0$, $x = \pi/4$;   about the x-axis

63. $y = \cos x$, $y = 0$, $x = 0$, $x = \pi/2$;   about $y = -1$

64. $y = \cos x$, $y = 0$, $x = 0$, $x = \pi/2$;   about $y = 1$

65. A particle moves on a straight line with velocity
function $v(t) = \sin \omega t \cos^2 \omega t$. Find its position
function $s = f(t)$ if $f(0) = 0$.

Prove the formulas in Exercises 66–68, where m and n are positive integers.

66. $\displaystyle\int_{-\pi}^{\pi} \sin mx \cos nx \, dx = 0$

67. $\displaystyle\int_{-\pi}^{\pi} \sin mx \sin nx\, dx = \begin{cases} 0 & \text{if } m \neq n \\ \pi & \text{if } m = n \end{cases}$

68. $\displaystyle\int_{-\pi}^{\pi} \cos mx \cos nx\, dx = \begin{cases} 0 & \text{if } m \neq n \\ \pi & \text{if } m = n \end{cases}$

69. If

$$f(x) = \sum_{n=1}^{N} a_n \sin nx$$

$$= a_1 \sin x + a_2 \sin 2x + \cdots + a_N \sin Nx$$

show that the mth coefficient a_m is given by the formula

$$a_m = \frac{1}{\pi} \int_{-\pi}^{\pi} f(x) \sin mx\, dx$$

(The sum is called a *finite Fourier series*.)

SECTION 7.3

═══════════════════

Trigonometric Substitution

In finding the area of a circle or an ellipse, an integral of the form $\int \sqrt{a^2 - x^2}\, dx$ arises, where $a > 0$. If it were $\int x\sqrt{a^2 - x^2}\, dx$, the substitution $u = a^2 - x^2$ would be effective but, as it stands, $\int \sqrt{a^2 - x^2}\, dx$ is more difficult. If we change the variable from x to θ by the substitution $x = a \sin\theta$, then the identity $1 - \sin^2\theta = \cos^2\theta$ allows us to get rid of the root sign because

$$\sqrt{a^2 - x^2} = \sqrt{a^2 - a^2 \sin^2\theta} = \sqrt{a^2(1 - \sin^2\theta)} = \sqrt{a^2 \cos^2\theta} = a\,|\cos\theta|$$

Notice that there is a difference between the substitution $u = a^2 - x^2$ (where the new variable is a function of the old one) and the substitution $x = a \sin\theta$ (where the old variable is a function of the new one).

In general we can make a substitution of the form $x = g(t)$ by using the Substitution Rule in reverse, but only if g has an inverse function, that is, if g is one-to-one. In this case if we replace u by x and x by t in Theorem 5.33, we obtain

$$\int f(x)\, dx = \int f(g(t))g'(t)\, dt$$

This kind of substitution is called *inverse substitution*.

We can make the inverse substitution $x = a \sin\theta$ provided that it defines a one-to-one function. This can be accomplished by restricting θ to lie in the interval $[-\pi/2, \pi/2]$.

In the following table we list trigonometric substitutions that are effective for the given radical expressions because of the given trigonometric identities. In each case the restriction on θ is imposed to ensure that the function that defines the substitution is one-to-one. (These are the same intervals used in Section 3.7 in defining the inverse functions.)

Table of Trigonometric Substitutions (7.12)

Expression	Substitution	Identity
$\sqrt{a^2 - x^2}$	$x = a \sin\theta, \quad -\frac{\pi}{2} \leq \theta \leq \frac{\pi}{2}$	$1 - \sin^2\theta = \cos^2\theta$
$\sqrt{a^2 + x^2}$	$x = a \tan\theta, \quad -\frac{\pi}{2} < \theta < \frac{\pi}{2}$	$1 + \tan^2\theta = \sec^2\theta$
$\sqrt{x^2 - a^2}$	$x = a \sec\theta, \quad 0 \leq \theta < \frac{\pi}{2} \ \text{ or } \ \pi \leq \theta < \frac{3\pi}{2}$	$\sec^2\theta - 1 = \tan^2\theta$

EXAMPLE 1   Evaluate $\displaystyle\int \frac{\sqrt{9-x^2}}{x^2}\,dx$.

Solution   Let $x = 3\sin\theta$, where $-\pi/2 \le \theta \le \pi/2$. Then $dx = 3\cos\theta\,d\theta$ and

$$\sqrt{9-x^2} = \sqrt{9-9\sin^2\theta} = \sqrt{9\cos^2\theta} = 3\,|\cos\theta| = 3\cos\theta$$

(Note that $\cos\theta \ge 0$ because $-\pi/2 \le \theta \le \pi/2$.) Thus the Inverse Substitution Rule gives

$$\int \frac{\sqrt{9-x^2}}{x^2}\,dx = \int \frac{3\cos\theta}{9\sin^2\theta}\,3\cos\theta\,d\theta$$

$$= \int \frac{\cos^2\theta}{\sin^2\theta}\,d\theta = \int \cot^2\theta\,d\theta$$

$$= \int (\csc^2\theta - 1)\,d\theta$$

$$= -\cot\theta - \theta + C$$

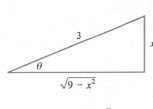

$\sin\theta = \dfrac{x}{3}$

Figure 7.1

Since this is an indefinite integral we must return to the original variable x. This can be done either by using trigonometric identities to express $\cot\theta$ in terms of $\sin\theta = x/3$ or by drawing a diagram, as in Figure 7.1, where θ is interpreted as an angle of a right triangle. Since $\sin\theta = x/3$, we label the opposite side and the hypotenuse as having lengths x and 3. Then the Pythagorean Theorem gives the length of the adjacent side as $\sqrt{9-x^2}$ so we can simply read the value of $\cot\theta$ from the figure:

$$\cot\theta = \frac{\sqrt{9-x^2}}{x}$$

(Although $\theta > 0$ in the diagram, this expression for $\cot\theta$ is valid even when $\theta < 0$.) Since $\sin\theta = x/3$, we have $\theta = \sin^{-1}(x/3)$ and so

$$\int \frac{\sqrt{9-x^2}}{x^2}\,dx = -\frac{\sqrt{9-x^2}}{x} - \sin^{-1}\!\left(\frac{x}{3}\right) + C$$

EXAMPLE 2   Find the area enclosed by the ellipse

$$\frac{x^2}{a^2} + \frac{y^2}{b^2} = 1$$

Solution   Solving the equation of the ellipse for y, we get

$$\frac{y^2}{b^2} = 1 - \frac{x^2}{a^2} = \frac{a^2-x^2}{a^2} \qquad \text{or} \qquad y = \pm\frac{b}{a}\sqrt{a^2-x^2}$$

Because the ellipse is symmetric with respect to both axes, the total area A is four times the area in the first quadrant (see Figure 7.2). The part of the ellipse in the first quadrant is given by the function

$$y = \frac{b}{a}\sqrt{a^2-x^2} \qquad 0 \le x \le a$$

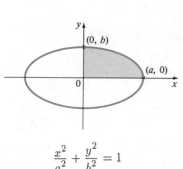

$\dfrac{x^2}{a^2} + \dfrac{y^2}{b^2} = 1$

Figure 7.2   and so

$$\tfrac{1}{4}A = \int_0^a \frac{b}{a}\sqrt{a^2-x^2}\,dx$$

To evaluate this integral we substitute $x = a\sin\theta$. Then $dx = a\cos\theta\,d\theta$. To change the limits of integration we note that when $x = 0$, $\sin\theta = 0$, so $\theta = 0$; when $x = a$, $\sin\theta = 1$, so $\theta = \pi/2$. Also

$$\sqrt{a^2 - x^2} = \sqrt{a^2 - a^2 \sin^2\theta} = \sqrt{a^2 \cos^2\theta} = a\,|\cos\theta| = a\cos\theta$$

since $0 \le \theta \le \pi/2$. Therefore

$$A = 4\frac{b}{a}\int_0^a \sqrt{a^2 - x^2}\,dx = 4\frac{b}{a}\int_0^{\pi/2} a\cos\theta \cdot a\cos\theta\,d\theta$$

$$= 4ab\int_0^{\pi/2} \cos^2\theta\,d\theta = 4ab\int_0^{\pi/2}\tfrac{1}{2}(1 + \cos 2\theta)\,d\theta$$

$$= 2ab\Big[\theta + \tfrac{1}{2}\sin 2\theta\Big]_0^{\pi/2} = 2ab\Big[\frac{\pi}{2} + 0 + 0\Big] = \pi ab$$

We have shown that the area of an ellipse with semiaxes a and b is πab. In particular, taking $a = b = r$, we have proved the famous formula that the area of a circle with radius r is πr^2.   ●

Note: Since the integral in Example 2 was a definite integral, we changed the limits of integration and did not have to convert back to the original variable x.

EXAMPLE 3   Find $\displaystyle\int \frac{1}{x^2\sqrt{x^2 + 4}}\,dx$.

Solution   Let $x = 2\tan\theta$, $-\pi/2 < \theta < \pi/2$. Then $dx = 2\sec^2\theta\,d\theta$ and

$$\sqrt{x^2 + 4} = \sqrt{4(\tan^2\theta + 1)} = \sqrt{4\sec^2\theta} = 2\,|\sec\theta| = 2\sec\theta$$

Thus we have

$$\int \frac{dx}{x^2\sqrt{x^2 + 4}} = \int \frac{2\sec^2\theta\,d\theta}{4\tan^2\theta \cdot 2\sec\theta} = \frac{1}{4}\int \frac{\sec\theta}{\tan^2\theta}\,d\theta$$

To evaluate this trigonometric integral we put everything in terms of $\sin\theta$ and $\cos\theta$:

$$\frac{\sec\theta}{\tan^2\theta} = \frac{1}{\cos\theta} \cdot \frac{\cos^2\theta}{\sin^2\theta} = \frac{\cos\theta}{\sin^2\theta}$$

Therefore, making the substitution $u = \sin\theta$, we have

$$\int \frac{dx}{x^2\sqrt{x^2 + 4}} = \frac{1}{4}\int \frac{\cos\theta}{\sin^2\theta}\,d\theta = \frac{1}{4}\int \frac{du}{u^2}$$

$$= \frac{1}{4}\Big(-\frac{1}{u}\Big) + C = -\frac{1}{4\sin\theta} + C$$

$$= -\frac{\csc\theta}{4} + C$$

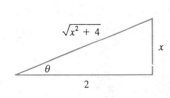

$$\tan\theta = \frac{x}{2}$$

Figure 7.3

We use Figure 7.3 to determine that $\csc\theta = \sqrt{x^2 + 4}/x$ and so

$$\int \frac{dx}{x^2\sqrt{x^2 + 4}} = -\frac{\sqrt{x^2 + 4}}{4x} + C$$ ●

EXAMPLE 4   Find $\displaystyle\int \frac{x}{\sqrt{x^2 + 4}}\,dx$.

Solution   It would be possible to use the trigonometric substitution $x = 2\tan\theta$ here (as in Example 3). But the direct substitution $u = x^2 + 4$ is simpler because then

$$du = 2x\,dx \text{ and}$$

$$\int \frac{x}{\sqrt{x^2+4}}\,dx = \frac{1}{2}\int \frac{du}{\sqrt{u}} = \sqrt{u}+C = \sqrt{x^2+4}+C \qquad \bullet$$

Note: Example 4 illustrates the fact that even when trigonometric substitutions are possible, they may not give the easiest solution. You should look for a simpler method first.

EXAMPLE 5   Evaluate $\displaystyle\int \frac{dx}{\sqrt{x^2-a^2}}$, where $a > 0$.

Solution 1   Let $x = a\sec\theta$ where $0 < \theta < \pi/2$ or $\pi < \theta < 3\pi/2$. Then $dx = a\sec\theta\tan\theta\,d\theta$ and

$$\sqrt{x^2-a^2} = \sqrt{a^2(\sec^2\theta-1)} = \sqrt{a^2\tan^2\theta} = a\,|\tan\theta| = a\tan\theta$$

Therefore

$$\int \frac{dx}{\sqrt{x^2-a^2}} = \int \frac{a\sec\theta\tan\theta}{a\tan\theta}\,d\theta$$

$$= \int \sec\theta\,d\theta = \ln|\sec\theta+\tan\theta|+C$$

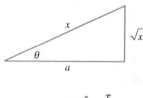 The triangle in Figure 7.4 gives $\tan\theta = \sqrt{x^2-a^2}/a$, so we have

$$\int \frac{dx}{\sqrt{x^2-a^2}} = \ln\left|\frac{x}{a}+\frac{\sqrt{x^2-a^2}}{a}\right|+C$$

$$= \ln\left|x+\sqrt{x^2-a^2}\right|-\ln a+C$$

Figure 7.4   Writing $C_1 = C - \ln a$, we have

(7.13)
$$\int \frac{dx}{\sqrt{x^2-a^2}} = \ln\left|x+\sqrt{x^2-a^2}\right|+C_1$$

Solution 2   For $x > 0$ the hyperbolic substitution $x = a\cosh t$ can also be used. Using the identity $\cosh^2 y - \sinh^2 y = 1$, we have

$$\sqrt{x^2-a^2} = \sqrt{a^2(\cosh^2 t - 1)} = \sqrt{a^2\sinh^2 t} = a\sinh t$$

Since $dx = a\sinh t\,dt$, we obtain

$$\int \frac{dx}{\sqrt{x^2-a^2}} = \int \frac{a\sinh t\,dt}{a\sinh t} = \int dt = t+C$$

Since $\cosh t = x/a$, we have $t = \cosh^{-1}(x/a)$ and

(7.14)
$$\int \frac{dx}{\sqrt{x^2-a^2}} = \cosh^{-1}\left(\frac{x}{a}\right)+C$$

Although Formulas 7.13 and 7.14 look quite different, they are actually equivalent by Formula 3.55. $\bullet$

Note: As Example 5 illustrates, hyperbolic substitutions can be used in place of trigonometric substitutions and sometimes they lead to simpler answers. But we usually use trigonometric substitutions because trigonometric identities are more familiar than hyperbolic identities.

EXAMPLE 6   Find $\displaystyle\int_0^{3\sqrt{3}/2} \frac{x^3}{(4x^2+9)^{3/2}}\,dx$.

Solution   First we note that $(4x^2+9)^{3/2} = \left(\sqrt{4x^2+9}\right)^3$ so trigonometric substitution is appropriate. Although $\sqrt{4x^2+9}$ is not quite one of the expressions in (7.12), it becomes one of them if we make the preliminary substitution $u = 2x$. When we combine this with the tangent substitution we have $x = \frac{3}{2}\tan\theta$, which gives $dx = \frac{3}{2}\sec^2\theta\,d\theta$ and

$$\sqrt{4x^2+9} = \sqrt{9\tan^2\theta + 9} = 3\sec\theta$$

When $x = 0$, $\tan\theta = 0$, so $\theta = 0$; when $x = 3\sqrt{3}/2$, $\tan\theta = \sqrt{3}$, so $\theta = \pi/3$.

$$\int_0^{3\sqrt{3}/2} \frac{x^3}{(4x^2+9)^{3/2}}\,dx = \int_0^{\pi/3} \frac{\frac{27}{8}\tan^3\theta}{27\sec^3\theta}\,\frac{3}{2}\sec^2\theta\,d\theta$$

$$= \frac{3}{16}\int_0^{\pi/3} \frac{\tan^3\theta}{\sec\theta}\,d\theta = \frac{3}{16}\int_0^{\pi/3} \frac{\sin^3\theta}{\cos^2\theta}\,d\theta$$

$$= \frac{3}{16}\int_0^{\pi/3} \frac{1-\cos^2\theta}{\cos^2\theta}\sin\theta\,d\theta$$

Now we substitute $u = \cos\theta$ so that $du = -\sin\theta\,d\theta$. When $\theta = 0$, $u = 1$; when $\theta = \pi/3$, $u = 1/2$. Therefore

$$\int_0^{3\sqrt{3}/2} \frac{x^3}{(4x^2+9)^{3/2}}\,dx = -\frac{3}{16}\int_1^{1/2} \frac{1-u^2}{u^2}\,du = \frac{3}{16}\int_1^{1/2} (1-u^{-2})\,du$$

$$= \frac{3}{16}\left[u + \frac{1}{u}\right]_1^{1/2} = \frac{3}{16}\left[\left(\tfrac{1}{2}+2\right) - (1+1)\right] = \frac{3}{32} \qquad \bullet$$

EXAMPLE 7   Evaluate $\displaystyle\int \frac{x}{\sqrt{3-2x-x^2}}\,dx$.

Solution   We can transform the integrand into a function where trigonometric substitution is appropriate by first completing the square under the root sign:

$$3 - 2x - x^2 = 3 - (x^2 + 2x) = 3 + 1 - (x^2 + 2x + 1)$$
$$= 4 - (x+1)^2$$

This suggests that we make the substitution $u = x + 1$. Then $du = dx$ and $x = u - 1$, so

$$\int \frac{x}{\sqrt{3-2x-x^2}}\,dx = \int \frac{u-1}{\sqrt{4-u^2}}\,du$$

We now substitute $u = 2\sin\theta$, giving $du = 2\cos\theta\,d\theta$ and $\sqrt{4-u^2} = 2\cos\theta$, so

$$\int \frac{x}{\sqrt{3-2x-x^2}}\,dx = \int \frac{2\sin\theta - 1}{2\cos\theta}\,2\cos\theta\,d\theta$$

$$= \int (2\sin\theta - 1)\, d\theta$$

$$= -2\cos\theta - \theta + C$$

$$= -\sqrt{4 - u^2} - \sin^{-1}\left(\frac{u}{2}\right) + C$$

$$= -\sqrt{3 - 2x - x^2} - \sin^{-1}\left(\frac{x+1}{2}\right) + C$$

SECTION 7.3 **Exercises**

In Exercises 1–30 evaluate the integral.

1. $\displaystyle\int_{1/2}^{\sqrt{3}/2} \frac{1}{x^2\sqrt{1-x^2}}\, dx$

2. $\displaystyle\int_0^2 x^3\sqrt{4-x^2}\, dx$

3. $\displaystyle\int \frac{x}{\sqrt{1-x^2}}\, dx$

4. $\displaystyle\int x\sqrt{4-x^2}\, dx$

5. $\displaystyle\int \sqrt{1-4x^2}\, dx$

6. $\displaystyle\int_0^2 \frac{x^3}{\sqrt{x^2+4}}\, dx$

7. $\displaystyle\int_0^3 \frac{dx}{\sqrt{9+x^2}}$

8. $\displaystyle\int \sqrt{x^2+1}\, dx$

9. $\displaystyle\int \frac{dx}{x^3\sqrt{x^2-16}}$

10. $\displaystyle\int \frac{\sqrt{x^2-a^2}\, dx}{x^4}$

11. $\displaystyle\int \frac{\sqrt{9x^2-4}}{x}\, dx$

12. $\displaystyle\int \frac{dx}{x^2\sqrt{16x^2-9}}$

13. $\displaystyle\int \frac{x^2}{(a^2-x^2)^{3/2}}\, dx$

14. $\displaystyle\int \frac{x^2}{\sqrt{5-x^2}}\, dx$

15. $\displaystyle\int \frac{dx}{x\sqrt{x^2+3}}$

16. $\displaystyle\int \frac{x}{(x^2+4)^{5/2}}\, dx$

17. $\displaystyle\int_0^{2/3} x^3\sqrt{4-9x^2}\, dx$

18. $\displaystyle\int_0^3 x^2\sqrt{9-x^2}\, dx$

19. $\displaystyle\int 5x\sqrt{1+x^2}\, dx$

20. $\displaystyle\int \frac{dx}{(4x^2-25)^{3/2}}$

21. $\displaystyle\int \frac{dx}{x^4\sqrt{x^2-2}}$

22. $\displaystyle\int \frac{dx}{(1+x^2)^2}$

23. $\displaystyle\int \sqrt{2x-x^2}\, dx$

24. $\displaystyle\int \frac{dx}{\sqrt{x^2+4x+8}}$

25. $\displaystyle\int \frac{1}{\sqrt{9x^2+6x-8}}\, dx$

26. $\displaystyle\int \frac{x^2}{\sqrt{4x-x^2}}\, dx$

27. $\displaystyle\int \frac{dx}{(x^2+2x+2)^2}$

28. $\displaystyle\int \frac{dx}{(5-4x-x^2)^{5/2}}$

29. $\displaystyle\int e^t\sqrt{9-e^{2t}}\, dt$

30. $\displaystyle\int \sqrt{e^{2t}-9}\, dt$

31. (a) Use trigonometric substitution to show that

$$\int \frac{dx}{\sqrt{x^2+a^2}} = \ln\left(x + \sqrt{x^2+a^2}\right) + C$$

(b) Use the hyperbolic substitution $x = a\sinh t$ to show that

$$\int \frac{dx}{\sqrt{x^2+a^2}} = \sinh^{-1}\left(\frac{x}{a}\right) + C$$

These formulas are connected by Formula 3.54.

32. Evaluate $\displaystyle\int \frac{x^2}{(x^2+a^2)^{3/2}}\, dx$

(a) by the trigonometric substitution and
(b) by the hyperbolic substitution $x = a\sinh t$.

33. Prove the formula $A = \frac{1}{2}r^2\theta$ for the area of a sector of a circle with radius r and central angle θ. (*Hint:* Assume $0 < \theta < \pi/2$ and place the center of the circle at the origin so it has the equation $x^2 + y^2 = r^2$. Then A is the sum of the area of the triangle POQ and the area of the region PQR in the figure.)

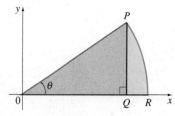

34. Find the area of the region bounded by the hyperbola $9x^2 - 4y^2 = 36$ and the line $x = 3$.

35. A torus is generated by rotating the circle $x^2 + (y-R)^2 = r^2$ about the x-axis. Find the volume enclosed by the torus.

36. Use the result of Example 2 to find the volume enclosed by the ellipsoid

$$\frac{x^2}{a^2} + \frac{y^2}{b^2} + \frac{z^2}{c^2} = 1$$

(See Figure 11.31.)

Integration of Rational Functions by Partial Fractions

In this section we show how to integrate any rational function (a ratio of polynomials) by expressing it as a sum of simpler fractions, called *partial fractions*, that we already know how to integrate.

To illustrate the method, observe that by taking the fractions $2/(x-1)$ and $1/(x+2)$ to a common denominator we obtain

$$\frac{2}{x-1} - \frac{1}{x+2} = \frac{2(x+2)-(x-1)}{(x-1)(x+2)} = \frac{x+5}{x^2+x-2}$$

If we now reverse the procedure we see how to integrate the function on the right side of this equation:

$$\int \frac{x+5}{x^2+x-2}\,dx = \int \left(\frac{2}{x-1} - \frac{1}{x+2}\right)dx$$

$$= 2\ln|x-1| - \ln|x+2| + C$$

To see how the method of partial fractions works in general, let us consider a rational function

$$f(x) = \frac{P(x)}{Q(x)}$$

where P and Q are polynomials. It is possible to express f as a sum of simpler fractions provided that the degree of P is less than the degree of Q. Such a rational function is called *proper*. Recall that if

$$P(x) = a_n x^n + a_{n-1}x^{n-1} + \cdots + a_1 x + a_0$$

where $a_n \neq 0$, then the degree of P is n and we write $\deg(P) = n$.

If f is improper, that is, $\deg(P) \geq \deg(Q)$, then we must take the preliminary step of dividing Q into P (by long division) until a remainder $R(x)$ is obtained with $\deg(R) < \deg(Q)$. The division statement is

(7.15)
$$f(x) = \frac{P(x)}{Q(x)} = S(x) + \frac{R(x)}{Q(x)}$$

where S and R are also polynomials.

As the following example illustrates, sometimes this preliminary step is all that is required.

EXAMPLE 1   Find $\displaystyle\int \frac{x^3+x}{x-1}\,dx$.

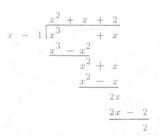

Solution   Since the degree of the numerator is greater than the degree of the denominator, we first perform the long division. This enables us to write

$$\int \frac{x^3+x}{x-1}\,dx = \int \left(x^2+x+2+\frac{2}{x-1}\right)dx$$

$$= \frac{x^3}{3} + \frac{x^2}{2} + 2x + 2\ln|x-1| + C$$

The next step is to factor the denominator $Q(x)$ as far as possible. It can be shown that any polynomial Q can be factored as a product of linear factors (of the form $ax + b$) and irreducible quadratic factors (of the form $ax^2 + bx + c$ where $b^2 - 4ac < 0$). For instance, if $Q(x) = x^4 - 16$, we could factor it as

$$Q(x) = (x^2 - 4)(x^2 + 4) = (x - 2)(x + 2)(x^2 + 4)$$

The third step is to express the proper rational function $R(x)/Q(x)$ (from Equation 7.15) as a sum of **partial fractions** of the form

$$\frac{A}{(ax + b)^i} \quad \text{or} \quad \frac{Ax + B}{(ax^2 + bx + c)^j}$$

A theorem in algebra guarantees that it is always possible to do this. We explain the details for the four cases that occur.

Case 1: **The denominator $Q(x)$ is a product of distinct linear factors.** This means that we can write

$$Q(x) = (a_1x + b_1)(a_2x + b_2)\cdots(a_kx + b_k)$$

where no factor is repeated. In this case the partial fraction theorem states that there exist constants $A_1, A_2, \ldots, A_k$ such that

(7.16)
$$\frac{R(x)}{Q(x)} = \frac{A_1}{a_1x + b_1} + \frac{A_2}{a_2x + b_2} + \cdots + \frac{A_k}{a_kx + b_k}$$

These constants can be determined as in the following example.

EXAMPLE 2   Evaluate $\displaystyle\int \frac{x^2 + 2x - 1}{2x^3 + 3x^2 - 2x}\, dx$.

Solution   Since the degree of the numerator is less than the degree of the denominator, there is no need to divide. We factor the denominator as

$$2x^3 + 3x^2 - 2x = x(2x^2 + 3x - 2) = x(2x - 1)(x + 2)$$

Since there are three distinct linear factors, the partial fraction decomposition of the integrand (7.16) has the form

(7.17)
$$\frac{x^2 + 2x - 1}{x(2x - 1)(x + 2)} = \frac{A}{x} + \frac{B}{2x - 1} + \frac{C}{x + 2}$$

To determine the values of A, B, and C we multiply both sides of this equation by $x(2x - 1)(x + 2)$, obtaining

(7.18)
$$x^2 + 2x - 1 = A(2x - 1)(x + 2) + Bx(x + 2) + Cx(2x - 1)$$

Expanding the right side of Equation 7.18 and writing it in the standard form for polynomials, we get

(7.19)
$$x^2 + 2x - 1 = (2A + B + 2C)x^2 + (3A + 2B - C)x - 2A$$

The polynomials in Equation 7.19 are equal, so their coefficients must be equal. The coefficient of x^2 on the right side, $2A + B + 2C$, must equal the coefficient of x^2 on the left side—namely, 1. Likewise the coefficients of x are equal and the constant terms are

Another method for finding
A, B, and C is given in the
note after this example.

equal. This gives the following system of equations for A, B, and C:

$$2A + B + 2C = 1$$

$$3A + 2B - C = 2$$

$$-2A = -1$$

Solving, we get $A = \frac{1}{2}$, $B = \frac{1}{5}$, and $C = -\frac{1}{10}$, and so

$$\int \frac{x^2 + 2x - 1}{2x^3 + 3x^2 - 2x}\, dx = \int \left[\frac{1}{2}\frac{1}{x} + \frac{1}{5}\frac{1}{2x - 1} - \frac{1}{10}\frac{1}{x + 2}\right] dx$$

$$= \tfrac{1}{2} \ln |x| + \tfrac{1}{10} \ln |2x - 1| - \tfrac{1}{10} \ln |x + 2| + K$$

In integrating the middle term we have made the mental substitution $u = 2x - 1$, which gives $du = 2\, dx$ and $dx = du/2$.   •

Note: There is another way to find the coefficients A, B, and C in Example 2. Equation 7.18 is an identity; it is true for every value of x. Let us choose values of x that simplify the equation. If we put $x = 0$ in Equation 7.18, then the second and third terms on the right side vanish and the equation becomes $-2A = -1$ or $A = \frac{1}{2}$. Likewise $x = \frac{1}{2}$ gives $5B/4 = \frac{1}{4}$ and $x = -2$ gives $10C = -1$, so $B = \frac{1}{5}$ and $C = -\frac{1}{10}$. (You may object that Equation 7.17 is not valid for $x = 0$, $\frac{1}{2}$, or -2, so why should Equation 7.18 be valid for those values? In fact, Equation 7.18 is true for all values of x, even $x = 0$, $\frac{1}{2}$, and -2. See Exercise 73 for the reason.)

EXAMPLE 3   Find $\displaystyle\int \frac{dx}{x^2 - a^2}$, where $a \neq 0$.

Solution   The method of partial fractions gives

$$\frac{1}{x^2 - a^2} = \frac{1}{(x - a)(x + a)} = \frac{A}{x - a} + \frac{B}{x + a}$$

and therefore

$$A(x + a) + B(x - a) = 1$$

Using the method of the note above, we put $x = a$ in this equation and get $A(2a) = 1$, so $A = 1/(2a)$. If we put $x = -a$, we get $B(-2a) = 1$, so $B = -1/(2a)$. Thus

$$\int \frac{dx}{x^2 - a^2} = \frac{1}{2a} \int \left[\frac{1}{x - a} - \frac{1}{x + a}\right] dx$$

$$= \frac{1}{2a}\left[\ln |x - a| - \ln |x + a|\right] + C$$

Since $\ln x - \ln y = \ln(x/y)$, we can write the integral as

(7.20)
$$\boxed{\int \frac{dx}{x^2 - a^2} = \frac{1}{2a} \ln\left|\frac{x - a}{x + a}\right| + C}$$

See Exercises 66–69 for ways of using Formula 7.20.   •

Case 2: $Q(x)$ is a product of linear factors, some of which are repeated. Suppose the first linear factor $(a_1 x + b_1)$ is repeated r times; that is, $(a_1 x + b_1)^r$ occurs in the factorization of $Q(x)$. Then instead of the single term $A_1/(a_1 x + b_1)$ in Equation 7.16, we would use

(7.21)
$$\frac{A_1}{a_1 x + b_1} + \frac{A_2}{(a_1 x + b_1)^2} + \cdots + \frac{A_r}{(a_1 x + b_1)^r}$$

By way of illustration we could write

$$\frac{x^3 - x + 1}{x^2(x-1)^3} = \frac{A}{x} + \frac{B}{x^2} + \frac{C}{x-1} + \frac{D}{(x-1)^2} + \frac{E}{(x-1)^3}$$

but we prefer to work out in detail a simpler example.

EXAMPLE 4   Find $\displaystyle\int \frac{x^4 - 2x^2 + 4x + 1}{x^3 - x^2 - x + 1}\, dx$.

Solution   The first step is to divide. The result of long division is

$$\frac{x^4 - 2x^2 + 4x + 1}{x^3 - x^2 - x + 1} = x + 1 + \frac{4x}{x^3 - x^2 - x + 1}$$

The second step is to factor the denominator $Q(x) = x^3 - x^2 - x + 1$. Since $Q(1) = 0$, we know that $x - 1$ is a factor and we obtain

$$x^3 - x^2 - x + 1 = (x-1)(x^2 - 1) = (x-1)(x-1)(x+1)$$
$$= (x-1)^2(x+1)$$

Since the linear factor $x - 1$ occurs twice, the partial fraction decomposition is

$$\frac{4x}{(x-1)^2(x+1)} = \frac{A}{x-1} + \frac{B}{(x-1)^2} + \frac{C}{x+1}$$

Multiplying by $(x-1)^2(x+1)$, we get

(7.22)
$$4x = A(x-1)(x+1) + B(x+1) + C(x-1)^2$$
$$= (A+C)x^2 + (B - 2C)x + (-A + B + C)$$

Now we equate coefficients:

Another method for finding
the coefficients:
Put $x = 1$ in (7.22): $B = 2$.
Put $x = -1$: $C = -1$.
Put $x = 0$: $A = B + C = 1$.

$$A + C = 0$$
$$B - 2C = 4$$
$$-A + B + C = 0$$

Solving, we obtain $A = 1$, $B = 2$, and $C = -1$, so

$$\int \frac{x^4 - 2x^2 + 4x + 1}{x^3 - x^2 - x + 1}\, dx = \int \left[x + 1 + \frac{1}{x-1} + \frac{2}{(x-1)^2} - \frac{1}{x+1} \right] dx$$

$$= \frac{x^2}{2} + x + \ln|x-1| - \frac{2}{x-1} - \ln|x+1| + K$$

$$= \frac{x^2}{2} + x - \frac{2}{x-1} + \ln\left|\frac{x-1}{x+1}\right| + K$$

Case 3: $Q(x)$ **contains irreducible quadratic factors, none of which is repeated.** If $Q(x)$ has the factor $ax^2 + bx + c$, where $b^2 - 4ac < 0$, then, in addition to the partial fractions in Equation 7.16 and (7.21), the expression for $R(x)/Q(x)$ will have a term of the form

(7.23)
$$\frac{Ax + B}{ax^2 + bx + c}$$

where A and B are constants to be determined. For instance, the function

$$f(x) = \frac{x}{(x - 2)(x^2 + 1)(x^2 + 4)}$$

has a partial fraction decomposition of the form

$$\frac{x}{(x - 2)(x^2 + 1)(x^2 + 4)} = \frac{A}{x - 2} + \frac{Bx + C}{x^2 + 1} + \frac{Dx + E}{x^2 + 4}$$

The term given in (7.23) can be integrated by completing the square and using the formula

(7.24)
$$\boxed{\int \frac{dx}{x^2 + a^2} = \frac{1}{a} \tan^{-1}\left(\frac{x}{a}\right) + C}$$

EXAMPLE 5   Evaluate $\displaystyle\int \frac{2x^2 - x + 4}{x^3 + 4x} \, dx$.

Solution   Since $x^3 + 4x = x(x^2 + 4)$ cannot be factored further, we write

$$\frac{2x^2 - x + 4}{x(x^2 + 4)} = \frac{A}{x} + \frac{Bx + C}{x^2 + 4}$$

Multiplying by $x(x^2 + 4)$, we have

$$2x^2 - x + 4 = A(x^2 + 4) + (Bx + C)x$$
$$= (A + B)x^2 + Cx + 4A$$

Equating coefficients, we obtain

$$A + B = 2 \qquad C = -1 \qquad 4A = 4$$

Thus $A = 1$, $B = 1$, and $C = -1$ and so

$$\int \frac{2x^2 - x + 4}{x^3 + 4x} \, dx = \int \left[\frac{1}{x} + \frac{x - 1}{x^2 + 4} \right] dx$$

In order to integrate the second term we split it into two parts:

$$\int \frac{x - 1}{x^2 + 4} \, dx = \int \frac{x}{x^2 + 4} \, dx - \int \frac{1}{x^2 + 4} \, dx$$

We make the substitution $u = x^2 + 4$ in the first of these integrals so that $du = 2x \, dx$. We evaluate the second integral by means of Formula 7.24 with $a = 2$:

$$\int \frac{2x^2 - x + 4}{x(x^2 + 4)} \, dx = \int \frac{1}{x} \, dx + \int \frac{x}{x^2 + 4} \, dx - \int \frac{1}{x^2 + 4} \, dx$$

$$= \ln|x| + \tfrac{1}{2} \ln(x^2 + 4) - \tfrac{1}{2} \tan^{-1}\left(\frac{x}{2}\right) + K$$

EXAMPLE 6   Evaluate $\displaystyle\int \frac{4x^2 - 3x + 2}{4x^2 - 4x + 3}\, dx$.

Solution   Since the degree of the numerator is not less than the degree of the denominator, we first divide and obtain

$$\frac{4x^2 - 3x + 2}{4x^2 - 4x + 3} = 1 + \frac{x - 1}{4x^2 - 4x + 3}$$

Notice that the quadratic $4x^2 - 4x + 3$ is irreducible because its discriminant is $b^2 - 4ac = -32 < 0$. This means it cannot be factored, so we do not need to use the partial fraction technique.

To integrate the given function we complete the square in the denominator:

$$4x^2 - 4x + 3 = (2x - 1)^2 + 2$$

This suggests that we make the substitution $u = 2x - 1$. Then $du = 2\, dx$ and $x = (u + 1)/2$, so

$$\int \frac{4x^2 - 3x + 2}{4x^2 - 4x + 3}\, dx = \int \left(1 + \frac{x - 1}{4x^2 - 4x + 3} \right) dx$$

$$= x + \frac{1}{2} \int \frac{\frac{1}{2}(u + 1) - 1}{u^2 + 2}\, du$$

$$= x + \frac{1}{4} \int \frac{u - 1}{u^2 + 2}\, du$$

$$= x + \frac{1}{4} \int \frac{u}{u^2 + 2}\, du - \frac{1}{4} \int \frac{1}{u^2 + 2}\, du$$

$$= x + \frac{1}{8}\ln(u^2 + 2) - \frac{1}{4} \cdot \frac{1}{\sqrt{2}} \tan^{-1}\left(\frac{u}{\sqrt{2}}\right) + C$$

$$= x + \frac{1}{8}\ln(4x^2 - 4x + 3) - \frac{1}{4\sqrt{2}} \tan^{-1}\left(\frac{2x - 1}{\sqrt{2}}\right) + C \qquad \bullet$$

Note: Example 6 illustrates the general procedure for integrating a partial fraction of the form

$$\frac{Ax + B}{ax^2 + bx + c} \qquad b^2 - 4ac < 0$$

We complete the square in the denominator and then make a substitution that brings the integral into the form

$$\int \frac{Cu + D}{u^2 + a^2}\, du = C \int \frac{u}{u^2 + a^2}\, du + D \int \frac{1}{u^2 + a^2}\, du$$

Then the first integral is logarithm and the second is expressed in terms of $\tan^{-1}$.

Case 4: $Q(x)$ contains a repeated irreducible quadratic factor. If $Q(x)$ has the factor $(ax^2 + bx + c)^r$, where $b^2 - 4ac < 0$, then instead of the single partial fraction (7.23), the sum

(7.25)
$$\frac{A_1 x + B_1}{ax^2 + bx + c} + \frac{A_2 x + B_2}{(ax^2 + bx + c)^2} + \cdots + \frac{A_r x + B_r}{(ax^2 + bx + c)^r}$$

occurs in the partial fraction decomposition of $R(x)/Q(x)$. Each of terms in (7.25) can be integrated by completing the square and making a tangent substitution.

EXAMPLE 7   Write out the form of the partial fraction decomposition of the function

$$\frac{x^3 + x^2}{x(x-1)(x^2+x+1)(x^2+1)^3}$$

Solution

$$\frac{x^3 + x^2}{x(x-1)(x^2+x+1)(x^2+1)^3} =$$

$$\frac{A}{x} + \frac{B}{x-1} + \frac{Cx+D}{x^2+x+1} + \frac{Ex+F}{x^2+1} + \frac{Gx+H}{(x^2+1)^2} + \frac{Ix+J}{(x^2+1)^3} \qquad \bullet$$

EXAMPLE 8   Evaluate $\displaystyle\int \frac{1 - 3x + 2x^2 - x^3}{x(x^2+1)^2}\, dx.$

Solution   The form of the partial fraction decomposition is

$$\frac{1 - 3x + 2x^2 - x^3}{x(x^2+1)^2} = \frac{A}{x} + \frac{Bx+C}{x^2+1} + \frac{Dx+E}{(x^2+1)^2}$$

Multiplying by $x(x^2+1)^2$, we have

$$
\begin{aligned}
-x^3 + 2x^2 - 3x + 1 &= A(x^2+1)^2 + (Bx+C)x(x^2+1) + (Dx+E)x \\
&= A(x^4 + 2x^2 + 1) + B(x^4 + x^2) + C(x^3 + x) + Dx^2 + Ex \\
&= (A+B)x^4 + Cx^3 + (2A+B+D)x^2 + (C+E)x + A
\end{aligned}
$$

If we equate coefficients we get the system

$$A + B = 0 \qquad C = -1 \qquad 2A + B + D = 2 \qquad C + E = -3 \qquad A = 1$$

which has the solution $A = 1$, $B = -1$, $C = -1$, $D = 1$, and $E = -2$. Thus

$$
\begin{aligned}
\int \frac{1 - 3x + 2x^2 - x^3}{x(x^2+1)^2}\, dx &= \int \left(\frac{1}{x} - \frac{x+1}{x^2+1} + \frac{x-2}{(x^2+1)^2} \right) dx \\
&= \int \frac{dx}{x} - \int \frac{x}{x^2+1}\, dx - \int \frac{dx}{x^2+1} + \int \frac{x\,dx}{(x^2+1)^2} - 2\int \frac{dx}{(x^2+1)^2} \\
&= \ln|x| - \tfrac{1}{2}\ln(x^2+1) - \tan^{-1}x - \frac{1}{2(x^2+1)} - 2\int \frac{dx}{(x^2+1)^2}
\end{aligned}
$$

To evaluate the final integral we substitute $x = \tan\theta$. Then we have $dx = \sec^2\theta\, d\theta$ and $x^2 + 1 = \sec^2\theta$, so

$$\int \frac{dx}{(x^2+1)^2} = \int \frac{\sec^2\theta}{\sec^4\theta}\, d\theta = \int \cos^2\theta\, d\theta$$

$$= \tfrac{1}{2}\int (1 + \cos 2\theta)\, d\theta = \frac{\theta}{2} + \frac{\sin 2\theta}{4} + K$$

$$= \frac{\theta}{2} + \frac{\sin\theta\cos\theta}{2} + K$$

$$= \frac{1}{2}\left(\tan^{-1}x + \frac{x}{\sqrt{x^2+1}} \cdot \frac{1}{\sqrt{x^2+1}} \right) + K \qquad \text{(from Figure 7.5)}$$

$$= \frac{1}{2}\left(\tan^{-1}x + \frac{x}{x^2+1} \right) + K$$

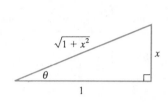

Figure 7.5

Thus

$$\int \frac{1 - 3x + 2x^2 - x^3}{x(x^2+1)^2}\,dx$$

$$= \ln|x| - \frac{1}{2}\ln(x^2+1) - \tan^{-1}x - \frac{1}{2(x^2+1)} - \tan^{-1}x - \frac{x}{x^2+1} + K$$

$$= \ln\frac{|x|}{\sqrt{x^2+1}} - 2\tan^{-1}x - \frac{2x+1}{2(x^2+1)} + K \qquad\bullet$$

Finally we note that sometimes partial fractions can be avoided when integrating a rational function. For instance, although the integral

$$\int \frac{x^2+1}{x(x^2+3)}\,dx$$

could be evaluated by the method of Case 3, it is easier to observe that if

$$u = x(x^2+3) = x^3 + 3x \qquad \text{then} \qquad du = (3x^2+3)\,dx$$

and so

$$\int \frac{x^2+1}{x(x^2+3)}\,dx = \frac{1}{3}\ln|x^3+3x| + C$$

SECTION 7.4 **Exercises**

In Exercises 1–18 write out the form of the partial fraction decomposition of the given function (as in Example 7). Do not determine the numerical values of the coefficients.

1. $\dfrac{1}{(x-1)(x+2)}$

2. $\dfrac{x}{(x+3)(2x-5)}$

3. $\dfrac{x+1}{x^2+2x}$

4. $\dfrac{7}{2x^2+5x-12}$

5. $\dfrac{x^2+3x-4}{(2x-1)^2(2x+3)}$

6. $\dfrac{x^3-x^2}{(x-6)(5x+3)^3}$

7. $\dfrac{1}{x^4-x^3}$

8. $\dfrac{1+x+x^2}{(x+1)(x+2)^2(x+3)^3}$

9. $\dfrac{x^2+1}{x^2-1}$

10. $\dfrac{x^4+x^3-x^2-x+1}{x^3-x}$

11. $\dfrac{x^2-2}{x(x^2+2)}$

12. $\dfrac{x^3-4x^2+2}{(x^2+1)(x^2+2)}$

13. $\dfrac{x^4+x^2+1}{(x^2+1)(x^2+4)^2}$

14. $\dfrac{1+16x}{(2x-3)(x+5)^2(x^2+x+1)}$

15. $\dfrac{x^4}{(x^2+9)^3}$

16. $\dfrac{19x}{(x-1)^3(4x^2+5x+3)^2}$

17. $\dfrac{x^3+x^2+1}{x^4+x^3+2x^2}$

18. $\dfrac{1}{x^6-x^3}$

Evaluate the integrals in Exercises 19–64.

19. $\int \dfrac{x^2}{x+1}\, dx$

20. $\int \dfrac{x}{x-5}\, dx$

21. $\int_2^4 \dfrac{4x-1}{(x-1)(x+2)}\, dx$

22. $\int_3^7 \dfrac{1}{(x+1)(x-2)}\, dx$

23. $\int \dfrac{6x-5}{2x+3}\, dx$

24. $\int \dfrac{1}{(x+a)(x+b)}\, dx$

25. $\int \dfrac{x^2+1}{x^2-x}\, dx$

26. $\int_0^2 \dfrac{x^3+x^2-12x+1}{x^2+x-12}\, dx$

27. $\int_0^1 \dfrac{2x+3}{(x+1)^2}\, dx$

28. $\int \dfrac{1}{(x-1)(x-2)(x-3)}\, dx$

29. $\int \dfrac{1}{x(x+1)(2x+3)}\, dx$

30. $\int \dfrac{3x^2-6x+2}{2x^3-3x^2+x}\, dx$

31. $\int_2^3 \dfrac{6x^2+5x-3}{x^3+2x^2-3x}\, dx$

32. $\int_0^1 \dfrac{x}{x^2+4x+4}\, dx$

33. $\int \dfrac{1}{(x-1)^2(x+4)}\, dx$

34. $\int \dfrac{x^2}{(x-3)(x+2)^2}\, dx$

35. $\int \dfrac{5x^2+3x-2}{x^3+2x^2}\, dx$

36. $\int \dfrac{18-2x-4x^2}{x^3+4x^2+x-6}\, dx$

37. $\int \dfrac{x^2+2x}{x^3+3x^2+4}\, dx$

38. $\int \dfrac{dx}{x^2(x-1)^2}$

39. $\int \dfrac{x^2}{(x+1)^3}\, dx$

40. $\int \dfrac{x^3}{(x+1)^3}\, dx$

41. $\int \dfrac{dx}{x^4-x^2}$

42. $\int \dfrac{2x^3-x}{x^4-x^2+1}\, dx$

43. $\int_0^1 \dfrac{x^3}{x^2+1}\, dx$

44. $\int_0^1 \dfrac{x-1}{x^2+2x+2}\, dx$

45. $\int_0^1 \dfrac{x}{x^2+x+1}\, dx$

46. $\int_{-1/2}^{1/2} \dfrac{4x^2+5x+7}{4x^2+4x+5}\, dx$

47. $\int \dfrac{3x^2-4x+5}{(x-1)(x^2+1)}\, dx$

48. $\int \dfrac{2x+3}{x^3+3x}\, dx$

49. $\int \dfrac{x^2+7x-6}{(x+1)(x^2-4x+7)}\, dx$

50. $\int \dfrac{4x+1}{(x-3)(x^2+6x+12)}\, dx$

51. $\int \dfrac{1}{x^3-1}\, dx$

52. $\int \dfrac{x^3}{x^3+1}\, dx$

53. $\int \dfrac{x^2-2x-1}{(x-1)^2(x^2+1)}\, dx$

54. $\int \dfrac{x^4}{x^4-1}\, dx$

55. $\int \dfrac{3x^3-x^2+6x-4}{(x^2+1)(x^2+2)}\, dx$

56. $\int \dfrac{x^3-2x^2+x+1}{x^4+5x^2+4}\, dx$

57. $\int \dfrac{x-3}{(x^2+2x+4)^2}\, dx$

58. $\int \dfrac{x+1}{(x^2+x+2)^2}\, dx$

59. $\int \dfrac{x^4+1}{x(x^2+1)^2}\, dx$

60. $\int \dfrac{3x^4-2x^3+20x^2-5x+34}{(x-1)(x^2+4)^2}\, dx$

61. $\int \dfrac{8x}{(x^2+4)^3}\, dx$

62. $\int \dfrac{x^2+1}{(x^3+3x)^2}\, dx$

63. $\int \dfrac{(2\sin x-3)\cos x}{\sin^2 x-3\sin x+2}\, dx$

64. $\int \dfrac{\sin x \cos^2 x}{5+\cos^2 x}\, dx$

65. Formula 7.20 can be rewritten as

$$\int \dfrac{dx}{a^2-x^2} = \dfrac{1}{2a}\ln\left|\dfrac{x+a}{x-a}\right| + K$$

Show by a hyperbolic substitution, or by differentiation, that an alternative formula is

$$\int \dfrac{dx}{a^2-x^2} = \begin{cases} \dfrac{1}{a}\tanh^{-1}\left(\dfrac{x}{a}\right) + C & \text{if } |x| < a \\[2mm] \dfrac{1}{a}\coth^{-1}\left(\dfrac{x}{a}\right) + C & \text{if } |x| > a \end{cases}$$

In Exercises 66–69 evaluate the integral by completing the square and using Formula 7.20 or the formula in Exercise 65.

66. $\int \dfrac{dx}{x^2+2x-3}$

67. $\int \dfrac{dx}{x^2-2x}$

68. $\int \dfrac{2x+1}{4x^2+12x-7}\, dx$

69. $\int \dfrac{x}{x^2+x-1}\, dx$

In Exercises 70–72 find the area of the region under the given curve from a to b.

70. $y = \dfrac{1}{x^2-6x+8}$, $a = 5,\ b = 10$

71. $y = \dfrac{x+1}{x-1}$, $a = 2,\ b = 3$

72. $y = \dfrac{x}{x^2-2x+10}$, $a = -1,\ b = 2$

73. Suppose that F, G, and Q are polynomials and

$$\dfrac{F(x)}{Q(x)} = \dfrac{G(x)}{Q(x)}$$

for all x except when $Q(x) = 0$. Prove that $F(x) = G(x)$ for all x. (*Hint:* Use continuity.)

SECTION 7.5

Rationalizing Substitutions

By means of an appropriate substitution, some functions can be changed into rational functions and therefore integrated by the methods of the preceding section. In particular, when an integrand contains an expression of the form $\sqrt[n]{g(x)}$, then the substitution $u = \sqrt[n]{g(x)}$ [or $u = g(x)$] may be effective.

EXAMPLE 1   Evaluate $\displaystyle\int \frac{\sqrt{x+4}}{x}\, dx$.

Solution   Let $u = \sqrt{x+4}$. Then $u^2 = x + 4$, so $x = u^2 - 4$, and $dx = 2u\, du$. Therefore

$$\int \frac{\sqrt{x+4}}{x}\, dx = \int \frac{u}{u^2 - 4}\, 2u\, du$$

$$= 2\int \frac{u^2}{u^2 - 4}\, du \qquad = 2\int \frac{x+4}{u^2-4}\, du = 2\int \frac{u^2-4+4}{u^2-4}$$

$$= 2\int \left(1 + \frac{4}{u^2 - 4}\right) du$$

We can evaluate the integral either by factoring $u^2 - 4$ as $(u-2)(u+2)$ and using partial fractions or by using Formula 7.20 with $a = 2$:

$$\int \frac{\sqrt{x+4}}{x}\, dx = 2\int du + 8\int \frac{du}{u^2 - 4}$$

$$= 2u + 8 \cdot \frac{1}{2 \cdot 2}\ln\left|\frac{u-2}{u+2}\right| + C$$

$$= 2\sqrt{x+4} + 2\ln\left|\frac{\sqrt{x+4}-2}{\sqrt{x+4}+2}\right| + C \qquad \bullet$$

EXAMPLE 2   Find $\displaystyle\int \frac{dx}{\sqrt{x} - \sqrt[3]{x}}$.

Solution   If we were to substitute $u = \sqrt{x}$, then the square root would disappear but a cube root would remain. On the other hand, the substitution $u = \sqrt[3]{x}$ would eliminate the cube root but leave a square root. We can eliminate both roots by means of the substitution $u = \sqrt[6]{x}$. (Note that 6 is the least common multiple of 2 and 3.)

Let $u = \sqrt[6]{x}$. Then $x = u^6$, so $dx = 6u^5\, du$ and $\sqrt{x} = u^3$, $\sqrt[3]{x} = u^2$. Thus

$$\int \frac{dx}{\sqrt{x} - \sqrt[3]{x}} = \int \frac{6u^5\, du}{u^3 - u^2} = 6\int \frac{u^3}{u - 1}\, du$$

$$= 6\int \left(u^2 + u + 1 + \frac{1}{u-1}\right) du \qquad \text{(by long division)}$$

$$= 6\left(\frac{u^3}{3} + \frac{u^2}{2} + u + \ln|u-1|\right) + C$$

$$= 2\sqrt{x} + 3\sqrt[3]{x} + 6\sqrt[6]{x} + 6\ln|\sqrt[6]{x} - 1| + C \qquad \bullet$$

The German mathematician Karl Weierstrass (1815–1897) noticed that the substitution $t = \tan(x/2)$ will convert any rational function of $\sin x$ and $\cos x$ into an ordinary rational function. Let

$$t = \tan\left(\frac{x}{2}\right) \qquad -\pi < x < \pi$$

Then

$$\cos\left(\frac{x}{2}\right) = \frac{1}{\sec\left(\frac{x}{2}\right)} = \frac{1}{\sqrt{1 + \tan^2\left(\frac{x}{2}\right)}} = \frac{1}{\sqrt{1 + t^2}}$$

$$\sin\left(\frac{x}{2}\right) = \cos\left(\frac{x}{2}\right)\tan\left(\frac{x}{2}\right) = \frac{t}{\sqrt{1 + t^2}}$$

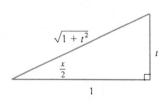

$$\tan\left(\frac{x}{2}\right) = t$$

Figure 7.6

These expressions can also be seen from Figure 7.6. Therefore

$$\sin x = 2\sin\left(\frac{x}{2}\right)\cos\left(\frac{x}{2}\right) = 2\frac{t}{\sqrt{1 + t^2}} \cdot \frac{1}{\sqrt{1 + t^2}} = \frac{2t}{1 + t^2}$$

$$\cos x = \cos^2\left(\frac{x}{2}\right) - \sin^2\left(\frac{x}{2}\right) = \frac{1 - t^2}{1 + t^2}$$

Since $t = \tan(x/2)$ we have $x = 2\tan^{-1}t$, so

$$dx = \frac{2}{1 + t^2}\,dt$$

Thus if we make the Weierstrass substitution $t = \tan(x/2)$, then we have

(7.26)
$$\sin x = \frac{2t}{1 + t^2} \qquad \cos x = \frac{1 - t^2}{1 + t^2} \qquad dx = \frac{2}{1 + t^2}\,dt$$

You can see from (7.26) that the Weierstrass substitution will transform any rational function of $\sin x$ and $\cos x$ into a rational function of t.

EXAMPLE 3   Find $\displaystyle\int \frac{1}{3\sin x - 4\cos x}\,dx$.

Solution   Let $t = \tan(x/2)$. Then, using the expressions in (7.26), we have

$$\int \frac{1}{3\sin x - 4\cos x}\,dx = \int \frac{1}{3\left(\dfrac{2t}{1 + t^2}\right) - 4\left(\dfrac{1 - t^2}{1 + t^2}\right)}\frac{2\,dt}{1 + t^2}$$

$$= 2\int \frac{dt}{3(2t) - 4(1 - t^2)}$$

$$= \int \frac{dt}{2t^2 + 3t - 2}$$

$$= \int \frac{dt}{(2t - 1)(t + 2)}$$

$$= \int \left[\frac{2}{5}\frac{1}{2t - 1} - \frac{1}{5}\frac{1}{t + 2}\right]dt \qquad \text{(using partial fractions)}$$

$$= \tfrac{1}{5}[\ln|2t - 1| - \ln|t + 2|] + C$$

$$= \tfrac{1}{5} \ln \left| \frac{2t-1}{t+2} \right| + C$$

$$= \tfrac{1}{5} \ln \left| \frac{2 \tan\left(\frac{x}{2}\right) - 1}{\tan\left(\frac{x}{2}\right) + 2} \right| + C$$

●

SECTION 7.5 **Exercises**

Evaluate the integrals in Exercises 1–40.

1. $\displaystyle \int_0^1 \frac{1}{1+\sqrt{x}}\, dx$

2. $\displaystyle \int_0^1 \frac{1}{1+\sqrt[3]{x}}\, dx$

3. $\displaystyle \int \frac{\sqrt{x}}{x+1}\, dx$

4. $\displaystyle \int \frac{1}{x\sqrt{x+1}}\, dx$

5. $\displaystyle \int \frac{1}{x-\sqrt[3]{x}}\, dx$

6. $\displaystyle \int \frac{1}{x-\sqrt{x+2}}\, dx$

7. $\displaystyle \int_5^{10} \frac{x^2}{\sqrt{x-1}}\, dx$

8. $\displaystyle \int_1^3 \frac{\sqrt{x-1}}{x+1}\, dx$

9. $\displaystyle \int \frac{1}{\sqrt{1+\sqrt{x}}}\, dx$

10. $\displaystyle \int_{1/3}^3 \frac{\sqrt{x}}{x^2+x}\, dx$

11. $\displaystyle \int \frac{\sqrt{x}+1}{\sqrt{x}-1}\, dx$

12. $\displaystyle \int \frac{\sqrt[3]{x}+1}{\sqrt[3]{x}-1}\, dx$

13. $\displaystyle \int \frac{x^3}{\sqrt[3]{x^2+1}}\, dx$

14. $\displaystyle \int \frac{x}{x^2-\sqrt[3]{x^2}}\, dx$

15. $\displaystyle \int \frac{dx}{\sqrt{x}+\sqrt[3]{x}}$

16. $\displaystyle \int \frac{\sqrt{x}}{\sqrt{x}-\sqrt[3]{x}}\, dx$

17. $\displaystyle \int \frac{1}{\sqrt{x}+\sqrt[4]{x}}\, dx$

18. $\displaystyle \int \frac{1}{\sqrt[3]{x}+\sqrt[4]{x}}\, dx$

19. $\displaystyle \int \frac{1}{(bx+c)\sqrt{x}}\, dx \quad (bc>0)$

20. $\displaystyle \int \frac{1}{(1+\sqrt{x})^3}\, dx$

21. $\displaystyle \int \frac{\sqrt[3]{x-1}}{x}\, dx$

22. $\displaystyle \int \frac{\sqrt[3]{1+\sqrt{x}}}{x}\, dx$

23. $\displaystyle \int \sqrt{\frac{1-x}{x}}\, dx$

24. $\displaystyle \int \sqrt{\frac{x-1}{x}}\, dx$

25. $\displaystyle \int \frac{\cos x}{\sin^2 x + \sin x}\, dx$

26. $\displaystyle \int \frac{\sin x}{\cos^2 x + \cos x - 6}\, dx$

27. $\displaystyle \int \frac{e^{2x}}{e^{2x}+3e^x+2}\, dx$

28. $\displaystyle \int \frac{1}{\sqrt{1+e^x}}\, dx$

29. $\displaystyle \int \sqrt{1-e^x}\, dx$

30. $\displaystyle \int \frac{e^{3x}}{e^{2x}-1}\, dx$

31. $\displaystyle \int \frac{dx}{1-\cos x}$

32. $\displaystyle \int \frac{dx}{3-5\sin x}$

33. $\displaystyle \int_0^{\pi/2} \frac{1}{\sin x + \cos x}\, dx$

34. $\displaystyle \int_{\pi/3}^{\pi/2} \frac{1}{1+\sin x - \cos x}\, dx$

35. $\displaystyle \int \frac{1}{3\sin x + 4\cos x}\, dx$

36. $\displaystyle \int \frac{1}{\sin x + \tan x}\, dx$

37. $\displaystyle \int \frac{1}{2\sin x + \sin 2x}\, dx$

38. $\displaystyle \int \frac{\sec x}{1+\sin x}\, dx$

39. $\displaystyle \int \frac{dx}{a\sin x + b\cos x} \quad (b>0)$

40. $\displaystyle \int \frac{dx}{a^2\sin^2 x + b^2\cos^2 x} \quad (a,b \neq 0)$

41. (a) Use the Weierstrass substitution $t = \tan(x/2)$ to prove the formula
$$\int \sec x\, dx = \ln \frac{|1+\tan(x/2)|}{|1-\tan(x/2)|} + C$$
   (b) Show that
$$\int \sec x\, dx = \ln \left| \tan\left(\frac{\pi}{4} + \frac{x}{2}\right) \right| + C$$
by using the formula for $\tan(x+y)$ and part (a).

42. Find a formula for $\int \csc x\, dx$ similar to the one in Exercise 41(a).

43. Show that the formula in Exercise 41(a) agrees with Formula 7.10.

44. If $a \neq 0$, evaluate the integral
$$\int \frac{dx}{a\sin^2 x + b\sin x \cos x + c\cos^2 x}$$
(*Hint:* Make the substitution $u = \tan x$ and consider separately the cases where $b^2 - 4ac$ is positive, zero, or negative.)

SECTION 7.6

Strategy for Integration

Integration is more challenging than differentiation. In finding the derivative of a function it is obvious which differentiation formula should be applied. But it may not be obvious which technique we should use to integrate a given function.

Until now individual techniques have been applied in each section. For instance, we usually used substitution in Exercises 5.6, integration by parts in Exercises 7.1, and partial fractions in Exercises 7.4. But in this section we present a collection of miscellaneous integrals in random order and the main challenge is to recognize which technique or formula to use. There are no hard and fast rules as to which method applies in a given situation, but we give some advice on strategy that you may find useful.

A prerequisite is a knowledge of the basic integration formulas. In Table 7.27 we collected the integrals from our previous list together with several additional formulas that we have learned in this chapter. Most of them should be memorized. It is useful to know them all, but the ones marked with an asterisk need not be memorized since they are easily derived. Formulas 19 and 20 can be avoided by using fractions, and trigonometric substitutions can be used in place of Formula 21.

Table of Integration Formulas (7.27)

Constants of integration have been omitted.

1. $\int x^n \, dx = \frac{x^{n+1}}{n+1} \quad (n \neq -1)$
2. $\int \frac{1}{x} \, dx = \ln |x|$

3. $\int e^x \, dx = e^x$
4. $\int a^x \, dx = \frac{a^x}{\ln a}$

5. $\int \sin x \, dx = -\cos x$
6. $\int \cos x \, dx = \sin x$

7. $\int \sec^2 x \, dx = \tan x$
8. $\int \csc^2 x \, dx = -\cot x$

9. $\int \sec x \tan x \, dx = \sec x$
10. $\int \csc x \cot x \, dx = -\csc x$

11. $\int \sec x \, dx = \ln |\sec x + \tan x|$
12. $\int \csc x \, dx = \ln |\csc x - \cot x|$

13. $\int \tan x \, dx = \ln |\sec x|$
14. $\int \cot x \, dx = \ln |\sin x|$

15. $\int \sinh x \, dx = \cosh x$
16. $\int \cosh x \, dx = \sinh x$

17. $\int \frac{dx}{x^2 + a^2} = \frac{1}{a} \tan^{-1}\left(\frac{x}{a}\right)$
18. $\int \frac{dx}{\sqrt{a^2 - x^2}} = \sin^{-1}\left(\frac{x}{a}\right)$

*19. $\int \frac{dx}{x^2 - a^2} = \frac{1}{2a} \ln\left|\frac{x-a}{x+a}\right|$

*20. $\int \frac{dx}{a^2 - x^2} = \frac{1}{2a} \ln\left|\frac{x+a}{x-a}\right| = \begin{cases} \frac{1}{a}\tanh^{-1}\left(\frac{x}{a}\right) & \text{if } |x| < a \\ \frac{1}{a}\coth^{-1}\left(\frac{x}{a}\right) & \text{if } |x| > a \end{cases}$

*21. $\int \frac{dx}{\sqrt{x^2 \pm a^2}} = \ln\left|x + \sqrt{x^2 \pm a^2}\right| = \begin{cases} \sinh^{-1}\left(\frac{x}{a}\right) \\ \cosh^{-1}\left(\frac{x}{a}\right) \end{cases}$

Once you are armed with these basic integration formulas, if you do not immediately see how to attack a given integral, you might try the following four-step strategy.

1.   **Simplify the integrand if possible.**   Sometimes the use of algebraic manipulation or trigonometric identities will simplify the integrand and make the method of integration obvious. Here are some examples:

$$\int \sqrt{x}(1+\sqrt{x})\,dx = \int (\sqrt{x}+x)\,dx$$

$$\int \frac{\tan\theta}{\sec^2\theta}\,d\theta = \int \frac{\sin\theta}{\cos\theta}\cos^2\theta\,d\theta$$

$$= \int \sin\theta\cos\theta\,d\theta = \tfrac{1}{2}\int \sin 2\theta\,d\theta$$

$$\int (\sin x + \cos x)^2\,dx = \int (\sin^2 x + 2\sin x\cos x + \cos^2 x)\,dx$$

$$= \int (1 + 2\sin x\cos x)\,dx$$

2.   **Look for an obvious substitution.**   Try to find some function $u = g(x)$ in the integrand whose differential $du = g'(x)\,dx$ also occurs, apart from a constant factor. For instance, in the integral

$$\int \frac{x}{x^2-1}\,dx$$

we notice that if $u = x^2 - 1$, then $du = 2x\,dx$. Therefore we use the substitution $u = x^2 - 1$ instead of the method of partial fractions.

3.   **Classify the integrand according to its form.**   If steps 1 and 2 have not led to the solution, then we take a look at the form of the integrand $f(x)$.
   (a) *Trigonometric functions.* If $f(x)$ is a product of powers of $\sin x$ and $\cos x$, or $\tan x$ and $\sec x$, or $\cot x$ and $\csc x$, then we use the substitutions recommended in Section 7.2. If f is a trigonometric function not of those types but still a rational function of $\sin x$ and $\cos x$, then we use the Weierstrass substitution $t = \tan(x/2)$.
   (b) *Rational functions.* If f is a rational function, we use the procedure of Section 7.4 involving partial fractions.
   (c) *Integration by parts.* If $f(x)$ is a product of power of x (or a polynomial) and a transcendental function (such as a trigonometric, exponential, or logarithmic function), then we try integration by parts, choosing u and dv according to the advice given in Section 7.1. If you look at the functions in Exercises 7.1 you will see that most of them are of the type just described.
   (d) *Radicals.* Particular kinds of substitutions are recommended when certain radicals appear.

   (i) If $\sqrt{\pm x^2 \pm a^2}$ occurs, we use a trigonometric substitution according to Table 7.12.

   (ii) If $\sqrt[n]{ax+b}$ occurs, we use a rationalizing substitution $u = \sqrt[n]{ax+b}$. More generally, this sometimes works for $\sqrt[n]{g(x)}$.

4.   Try again.   If the first three steps have not produced the answer, remember that there are basically only two methods of integration: substitution and parts.

(a)   *Try substitution.*   Even if there is no obvious substitution (step 2), some inspiration or ingenuity (or even desperation) may suggest an appropriate substitution.

(b)   *Try parts.*   Although integration by parts is used most of the time on products of the form described in step 3(c), it is sometimes effective on single functions. Looking at Section 7.1, we see that it worked on $\tan^{-1}x$, $\sin^{-1}x$, and $\ln x$, and these all inverse functions.

(c)   *Manipulate the integrand.*   Algebraic manipulations (perhaps rationalizing the denominator or using trigonometric identities) may be useful in transforming the integral into an easier form. These manipulations may be more substantial than in step 1 and may involve some ingenuity. Here is an example:

$$\int \frac{dx}{1-\cos x} = \int \frac{1}{1-\cos x}\cdot\frac{1+\cos x}{1+\cos x}\,dx = \int \frac{1+\cos x}{1-\cos^2 x}\,dx$$

$$= \int \frac{1+\cos x}{\sin^2 x}\,dx = \int \left(\csc^2 x + \frac{\cos x}{\sin^2 x}\right)dx$$

(d)   *Relate the problem to previous problems.*   When you have built up some experience in integration you may be able to use a method on a given integral that is similar to a method you have already used on a previous integral. Or you may even be able to express the given integral in terms of a previous one. For instance, $\int \tan^2 x \sec x\,dx$ is a challenging integral, but using the identity $\tan^2 x = \sec^2 x - 1$, we can write

$$\int \tan^2 x \sec x\,dx = \int \sec^3 x\,dx - \int \sec x\,dx$$

and if $\int \sec^3 x\,dx$ has previously been evaluated (see Example 9 in Section 7.2), then that calculation can be used in the present problem.

(e)   *Use several methods.*   Sometimes two or three methods are required to evaluate an integral. The evaluation could involve several successive substitutions of different types or it might combine integration by parts with one or more substitutions.

In the following examples we indicate a method of attack but do not fully work out the integral.

EXAMPLE 1 $\displaystyle\int \frac{\tan^3 x}{\cos^3 x}\,dx$

In step 1 we rewrite the integral:

$$\int \frac{\tan^3 x}{\cos^3 x}\,dx = \int \tan^3 x \sec^3 x\,dx$$

The integral is now of the form $\int \tan^m x \sec^n x\,dx$ with m odd, so we can use the advice in (7.9)(b).

Alternatively, if in step 1 we had written

$$\int \frac{\tan^3 x}{\cos^3 x}\, dx = \int \frac{\sin^3 x}{\cos^3 x}\, \frac{1}{\cos^3 x}\, dx = \int \frac{\sin^3 x}{\cos^6 x}\, dx$$

then we could have continued as follows with the substitution $u = \cos x$:

$$\int \frac{\sin^3 x}{\cos^6 x}\, dx = \int \frac{1 - \cos^2 x}{\cos^6 x}\, \sin x\, dx = \int \frac{1 - u^2}{u^6}\, (-du)$$

$$= \int \frac{u^2 - 1}{u^6}\, du = \int (u^{-4} - u^{-6})\, du$$

EXAMPLE 2 $\int e^{\sqrt{x}}\, dx$

According to step 3(d)(ii) we substitute $u = \sqrt{x}$. Then $x = u^2$, so $dx = 2u\, du$ and

$$\int e^{\sqrt{x}}\, dx = 2 \int u e^u\, du$$

The integrand is now a product of u and the transcendental function e^u so it can be integrated by parts.

EXAMPLE 3 $\int \frac{x^5 + 1}{x^3 - 3x^2 - 10x}\, dx$

There is no algebraic simplification or obvious substitution, so steps 1 and 2 do not apply here. The integrand is a rational function so we apply the procedure of Section 7.4, remembering that the first step is to divide.

EXAMPLE 4 $\int \frac{dx}{x\sqrt{\ln x}}$

Here step 2 is all that is needed. We substitute $u = \ln x$ because its differential is $du = dx/x$, which occurs in the integral.

EXAMPLE 5 $\int \sqrt{\frac{1 - x}{1 + x}}\, dx$

Although the rationalizing substitution

$$u = \sqrt{\frac{1 - x}{1 + x}}$$

works here [step 3(d)(ii)], it leads to a very complicated rational function. An easier method is to do some algebraic manipulation [either as step 1 or as step 4(c)]. Multiplying numerator and denominator by $\sqrt{1 - x}$, we have

$$\int \sqrt{\frac{1 - x}{1 + x}}\, dx = \int \frac{1 - x}{\sqrt{1 - x^2}}\, dx$$

$$= \int \frac{1}{\sqrt{1 - x^2}}\, dx - \int \frac{x}{\sqrt{1 - x^2}}\, dx$$

$$= \sin^{-1} x + \sqrt{1 - x^2} + C$$

The question arises: Will our strategy for integration enable us to find the integral of every continuous function? In particular, can we use it to evaluate $\int e^{x^2}\,dx$? The answer is no, at least not in terms of the functions that we are familiar with.

The functions that we have been dealing with in this book are called **elementary functions**. These are the polynomials, rational functions, power functions (x^a), exponential functions (a^x), logarithmic functions, trigonometric and inverse trigonometric functions, hyperbolic and inverse hyperbolic functions, and all functions that can be obtained from these by the five operations of addition, subtraction, multiplication, division, and composition. For instance, the function

$$f(x) = \sqrt{\frac{x^2-1}{x^3+2x-1}} + \ln(\cosh x) - xe^{\sin 2x}$$

is an elementary function.

If f is an elementary function, then f' is an elementary function but $\int f(x)\,dx$ need not be an elementary function. Consider $f(x) = e^{x^2}$. Since f is continuous its integral exists, and if we define the function F by

$$F(x) = \int_0^x e^{t^2}\,dt$$

then we know from Part 1 of the Fundamental Theorem of Calculus that

$$F'(x) = e^{x^2}$$

Thus $f(x) = e^{x^2}$ has an antiderivative F, but it has been proved that F is not an elementary function. This means that no matter how we try, we will never succeed in evaluating $\int e^{x^2}\,dx$ in terms of the functions we know. (In Chapter 10, however, we will see how to express $\int e^{x^2}\,dx$ as an infinite series.) The same can be said of the following integrals:

$$\int \frac{e^x}{x}\,dx \qquad \int \sin(x^2)\,dx \qquad \int \cos(e^x)\,dx$$

$$\int \sqrt{x^3+1}\,dx \qquad \int \frac{1}{\ln x}\,dx \qquad \int \frac{\sin x}{x}\,dx$$

You may rest assured, though, that the integrals in the following exercises are all elementary functions.

SECTION 7.6 **Exercises**

Evaluate the following integrals.

1. $\displaystyle\int \frac{2x+5}{x-3}\,dx$

2. $\displaystyle\int e^{x+e^x}\,dx$

3. $\displaystyle\int \sin^2 x \cos^3 x\,dx$

4. $\displaystyle\int \frac{\sin x - \cos x}{\sin x + \cos x}\,dx$

5. $\displaystyle\int_0^{1/2} \frac{x}{\sqrt{1-x^2}}\,dx$

6. $\displaystyle\int_1^2 x^3 \ln x\,dx$

7. $\displaystyle\int \frac{\sqrt{x-2}}{x+2}\,dx$

8. $\displaystyle\int \frac{x}{(x+2)^2}\,dx$

9. $\displaystyle\int \ln(1+x^2)\,dx$

10. $\displaystyle\int \frac{\sqrt{1+\ln x}}{x \ln x}\,dx$

11. $\displaystyle\int_0^1 \left(1+\sqrt{x}\right)^8\,dx$

12. $\displaystyle\int_0^{\pi/4} \tan^3 x \sec^4 x\,dx$

13. $\displaystyle\int \frac{x}{x^2-2x+2}\,dx$

14. $\displaystyle\int x \sin^{-1} x\,dx$

15. $\displaystyle\int \frac{\sqrt{9-x^2}}{x}\,dx$

16. $\displaystyle\int \frac{x}{x^2+3x+2}\,dx$

17. $\displaystyle\int x^2 \cosh x\,dx$

18. $\displaystyle\int \frac{x^3+x+1}{x^4+2x^2+4x}\,dx$

19. $\displaystyle\int \frac{\cos x}{1+\sin^2 x}\,dx$

20. $\displaystyle\int \cos \sqrt{x}\,dx$

21. $\displaystyle\int_0^1 \cos \pi x \tan \pi x\,dx$

22. $\displaystyle\int \frac{e^{2x}}{1+e^x}\,dx$

23. $\displaystyle\int e^{3x}\cos 5x\,dx$

24. $\displaystyle\int \cos 3x\cos 5x\,dx$

53. $\displaystyle\int (x^2+4x-3)\sin 2x\,dx$

54. $\displaystyle\int \sin x\cos(\cos x)\,dx$

25. $\displaystyle\int \frac{dx}{x^3+x^2+x+1}$

26. $\displaystyle\int x^2\ln(1+x)\,dx$

55. $\displaystyle\int \frac{x}{\sqrt{16-x^4}}\,dx$

56. $\displaystyle\int \frac{x^3}{(x+1)^{10}}\,dx$

27. $\displaystyle\int x^5 e^{-x^3}\,dx$

28. $\displaystyle\int \tan^2 4x\,dx$

57. $\displaystyle\int \cot^3 2x\csc^3 2x\,dx$

58. $\displaystyle\int (x+\sin x)^2\,dx$

29. $\displaystyle\int \frac{1}{\sqrt{9x^2+12x-5}}\,dx$

30. $\displaystyle\int x^2\tan^{-1}x\,dx$

59. $\displaystyle\int \frac{e^{\arctan x}}{1+x^2}\,dx$

60. $\displaystyle\int \frac{dx}{x(x^4+1)}$

31. $\displaystyle\int \sqrt[3]{x}\left(1-\sqrt{x}\right)dx$

32. $\displaystyle\int \frac{dx}{e^x-e^{-x}}$

61. $\displaystyle\int t^3 e^{-2t}\,dt$

62. $\displaystyle\int \frac{\sqrt{t}}{1+\sqrt[3]{t}}\,dt$

33. $\displaystyle\int \frac{x}{x^4+2x^2+10}\,dx$

34. $\displaystyle\int \frac{1}{x+\sqrt[3]{x}}\,dx$

63. $\displaystyle\int \sin x\sin 2x\sin 3x\,dx$

64. $\displaystyle\int_1^3 \left|\ln\!\left(\frac{x}{2}\right)\right|dx$

35. $\displaystyle\int \sin^2 x\cos^4 x\,dx$

36. $\displaystyle\int \frac{1}{\sqrt{5-4x-x^2}}\,dx$

65. $\displaystyle\int \sqrt{\frac{1+x}{1-x}}\,dx$

66. $\displaystyle\int \frac{x\ln x}{\sqrt{x^2-1}}\,dx$

37. $\displaystyle\int \frac{x}{1-x^2+\sqrt{1-x^2}}\,dx$

38. $\displaystyle\int \frac{1+\cos x}{\sin x}\,dx$

67. $\displaystyle\int \frac{x+a}{x^2+a^2}\,dx$

68. $\displaystyle\int \sqrt{1+x-x^2}\,dx$

39. $\displaystyle\int \frac{e^x}{e^{2x}-1}\,dx$

40. $\displaystyle\int \frac{1}{x^3-8}\,dx$

69. $\displaystyle\int \frac{x^4}{x^{10}+16}\,dx$

70. $\displaystyle\int \frac{x+2}{x^2+x+2}\,dx$

41. $\displaystyle\int_{-1}^{1} x^5\cosh x\,dx$

42. $\displaystyle\int_{\pi/4}^{\pi/3} \frac{\ln(\tan x)}{\sin x\cos x}\,dx$

71. $\displaystyle\int x\sec x\tan x\,dx$

72. $\displaystyle\int \frac{x}{x^4-a^4}\,dx$

43. $\displaystyle\int_{-3}^{3} |x^3+x^2-2x|\,dx$

44. $\displaystyle\int_0^{\pi/4} \cos^5\theta\,d\theta$

73. $\displaystyle\int \frac{1}{\sqrt{x+1}+\sqrt{x}}\,dx$

74. $\displaystyle\int \frac{1}{1+2e^x-e^{-x}}\,dx$

45. $\displaystyle\int \cot x\ln(\sin x)\,dx$

46. $\displaystyle\int \frac{1+e^x}{1-e^x}\,dx$

75. $\displaystyle\int \frac{\arctan\sqrt{x}}{\sqrt{x}}\,dx$

76. $\displaystyle\int \frac{\ln(x+1)}{x^2}\,dx$

47. $\displaystyle\int \frac{x}{(x^2+1)(x^2+4)}\,dx$

48. $\displaystyle\int \frac{dx}{4-5\sin x}$

77. $\displaystyle\int \frac{1}{e^{3x}-e^x}\,dx$

78. $\displaystyle\int \frac{1+\cos^2 x}{1-\cos^2 x}\,dx$

49. $\displaystyle\int x^3\sqrt{x+c}\,dx$

50. $\displaystyle\int e^{\sqrt[3]{x}}\,dx$

79. $\displaystyle\int \frac{dx}{x\sqrt{2x-25}}$

80. $\displaystyle\int \frac{\sin 2x}{\sqrt{9-\cos^4 x}}\,dx$

51. $\displaystyle\int \frac{1}{x+4+4\sqrt{x+1}}\,dx$

52. $\displaystyle\int \frac{x^3+1}{x^3-x^2}\,dx$

SECTION 7.7

Using Tables of Integrals

A scientist or engineer often uses tables of integrals when integrating complicated functions. A relatively brief table of 112 integrals is provided in Appendix G. More extensive tables are available in the *CRC Mathematical Tables* (463 entries) or in Gradshteyn and Ryzhik's *Table of Integrals, Series and Products* (New York: Academic Press, 1979), which contains hundreds of pages of integrals. It should be remembered, however, that integrals do not often occur in exactly the form listed in a table. Usually one of the methods of this chapter, such as substitution or integration by parts, is required to transform a given integral into one of the forms in the table.

EXAMPLE 1   The region bounded by the curves $y = \arctan x$, $y = 0$, and $x = 1$ is rotated about the y-axis. Find the volume of the resulting solid.

Solution   Using the method of cylindrical shells, we see that the volume is

$$V = \int_0^1 2\pi x \arctan x \, dx$$

In the section of the Table of Integrals (in Appendix G) entitled *Inverse Trigonometric Forms* we locate Formula 92:

$$\int u \tan^{-1} u \, du = \frac{u^2 + 1}{2} \tan^{-1} u - \frac{u}{2} + C$$

Thus the volume is

$$V = 2\pi \int_0^1 x \tan^{-1} x \, dx = 2\pi \left[\frac{x^2 + 1}{2} \tan^{-1} x - \frac{x}{2} \right]_0^1$$

$$= \pi \left[(x^2 + 1) \tan^{-1} x - x \right]_0^1 = \pi (2 \tan^{-1} 1 - 1)$$

$$= \pi [2(\pi/4) - 1] = \tfrac{1}{2}\pi^2 - \pi$$

EXAMPLE 2   Use the Table of Integrals to find $\displaystyle \int \frac{x^2}{\sqrt{5 - 4x^2}} \, dx$.

Solution   If we look at the section of the table entitled *Forms involving $\sqrt{a^2 - x^2}$*, we see that the closest entry is number 34:

$$\int \frac{x^2}{\sqrt{a^2 - x^2}} \, dx = -\frac{x}{2} \sqrt{a^2 - x^2} + \frac{a^2}{2} \sin^{-1}\left(\frac{x}{a}\right) + C$$

This is not exactly what we have, so we make the substitution $u = 2x$:

$$\int \frac{x^2}{\sqrt{5 - 4x^2}} \, dx = \int \frac{(u/2)^2}{\sqrt{5 - u^2}} \frac{du}{2} = \frac{1}{8} \int \frac{u^2}{\sqrt{5 - u^2}} \, du$$

Then we use Formula 34 with $a^2 = 5$:

$$\int \frac{x^2}{\sqrt{5 - 4x^2}} \, dx = \frac{1}{8} \int \frac{u^2}{\sqrt{5 - u^2}} \, du = \frac{1}{8}\left[-\frac{u}{2} \sqrt{5 - u^2} + \frac{5}{2} \sin^{-1}\left(\frac{u}{\sqrt{5}}\right) \right] + C$$

$$= -\frac{x}{8} \sqrt{5 - 4x^2} + \frac{5}{16} \sin^{-1}\left(\frac{2x}{\sqrt{5}}\right) + C$$

EXAMPLE 3   Use the Table of Integrals to find $\int x^3 \sin x \, dx$.

Solution   We look in the section called *Trigonometric Forms* and use the reduction formula in entry 84 with $n = 3$:

$$\int x^3 \sin x \, dx = -x^3 \cos x + 3 \int x^2 \cos x \, dx$$

Then we use entries 85 and 82:

$$\int x^2 \cos x \, dx = x^2 \sin x - 2 \int x \sin x \, dx$$

$$= x^2 \sin x - 2(\sin x - x \cos x) + C$$

Combining these calculations we get

$$\int x^3 \sin x \, dx = -x^3 \cos x + 3x^2 \sin x + 6x \cos x - 6 \sin x + C$$

●

EXAMPLE 4   Use the Table of Integrals to find $\int x\sqrt{x^2 + 2x + 4} \, dx$.

Solution   Since there are forms involving $\sqrt{a^2 + x^2}$, $\sqrt{a^2 - x^2}$, and $\sqrt{x^2 - a^2}$, but not $\sqrt{ax^2 + bx + c}$, we first complete the square:

$$x^2 + 2x + 4 = (x + 1)^2 + 3$$

Therefore we make the substitution $u = x + 1$:

$$\int x\sqrt{x^2 + 2x + 4} \, dx = \int (u - 1)\sqrt{u^2 + 3} \, du$$

$$= \int u\sqrt{u^2 + 3} \, du - \int \sqrt{u^2 + 3} \, du$$

The first integral is evaluated using the substitution $t = u^2 + 3$:

$$\int u\sqrt{u^2 + 3} \, du = \tfrac{1}{2} \int \sqrt{t} \, dt = \tfrac{1}{2} \cdot \tfrac{2}{3} t^{3/2} = \tfrac{1}{3}(u^2 + 3)^{3/2}$$

For the second integral we use Formula 21 with $a = \sqrt{3}$:

$$\int \sqrt{u^2 + 3} \, du = \tfrac{u}{2}\sqrt{u^2 + 3} + \tfrac{3}{2}\ln\left|u + \sqrt{u^2 + 3}\right|$$

Thus

$$\int x\sqrt{x^2 + 2x + 4} \, dx = \tfrac{1}{3}(x^2 + 2x + 4)^{3/2} - \tfrac{x + 1}{2}\sqrt{x^2 + 2x + 4}$$

$$- \tfrac{3}{2}\ln\left|x + 1 + \sqrt{x^2 + 2x + 4}\right| + C$$

●

SECTION 7.7 **Exercises**

Use the Table of Integrals in Appendix G to evaluate the integrals in Exercises 1–26.

1. $\int e^{-3x} \cos 4x \, dx$

2. $\int \csc^3(x/2) \, dx$

3. $\int \dfrac{\sqrt{9x^2 - 1}}{x^2} \, dx$

4. $\int \dfrac{\sqrt{4 - 3x^2}}{x} \, dx$

5. $\int x^2 e^{3x} \, dx$

6. $\int \dfrac{\sin x \cos x}{\sqrt{1 + \sin x}} \, dx$

7. $\int x \sin^{-1}(x^2) \, dx$

8. $\int x^3 \sin^{-1}(x^2) \, dx$

9. $\int e^x \operatorname{sech}(e^x) \, dx$

10. $\int x^2 \cos 3x \, dx$

11. $\int \sqrt{5 - 4x - x^2} \, dx$

12. $\int \dfrac{x^5}{x^2 + \sqrt{2}} \, dx$

13. $\int \sec^5 x \, dx$

14. $\int \sin^6 2x \, dx$

15. $\int \sin^2 x \cos x \ln(\sin x) \, dx$

16. $\int \dfrac{dx}{e^x(1 + 2e^x)}$

17. $\int \sqrt{2 + 3\cos x} \tan x \, dx$

18. $\int \dfrac{x}{\sqrt{x^2 - 4x}} \, dx$

19. $\int_0^{\pi/2} \cos^5 x \, dx$

20. $\int_0^1 x^4 e^{-x} \, dx$

21. $\displaystyle\int \frac{x^4\,dx}{\sqrt{x^{10}-2}}$ 22. $\displaystyle\int \frac{\sqrt{1-x^2}}{2}\,dx$

23. $\displaystyle\int e^x \ln(1+e^x)\,dx$ 24. $\displaystyle\int x^2 \tan^{-1}x\,dx$

25. $\displaystyle\int \sqrt{e^{2x}-1}\,dx$ 26. $\displaystyle\int e^{\sin x}\sin 2x\,dx$

27. Find the volume of the solid obtained when the region under the curve $y = 1/(1+5x)^2$ from 0 to 1 is rotated about the y-axis.

28. The region under the curve $y = \tan^2 x$ from 0 to $\pi/4$ is rotated about the x-axis. Find the volume of the resulting solid.

29. Verify Formula 53 in the Table of Integrals (a) by differentiation, and (b) by using the substitution $t = a + bu$.

30. Verify Formula 31 (a) by differentiation, and (b) by using the substitution $u = a\sin\theta$.

SECTION 7.8

Approximate Integration

There are two situations in which it is impossible to find the exact value of a definite integral.

The first situation arises from the fact that in order to evaluate $\int_a^b f(x)\,dx$ using the Fundamental Theorem of Calculus we need to know an antiderivative of f. Sometimes, however, it is difficult, or even impossible, to find an antiderivative (see Section 7.6). For example, it is impossible to evaluate the following integrals exactly:

$$\int_0^1 e^{x^2}\,dx \qquad \int_{-1}^1 \sqrt{1+x^3}\,dx$$

The second situation arises when the function is determined from a scientific experiment through instrument readings. There may be no formula for the function (see Example 7).

In both cases there is a need to find approximate values of definite integrals. We already know one such method. Recall that the definite integral is defined as a limit of Riemann sums, so any Riemann sum could be used as an approximation to the integral. In particular, let us take a partition of $[a,b]$ into n subintervals of equal length $\Delta x = (b-a)/n$. Then we have

$$\int_a^b f(x)\,dx \approx \sum_{i=1}^n f(x_i^*)\,\Delta x$$

where x_i^* is any point in the ith subinterval $[x_{i-1}, x_i]$ of the partition. If x_i^* is chosen to be the left endpoint of the interval, then $x_i^* = x_{i-1}$ and we have

(7.28)
$$\int_a^b f(x)\,dx \approx L_n = \sum_{i=1}^n f(x_{i-1})\,\Delta x$$

If $f(x) \ge 0$, then the integral represents an area and (7.28) represents an approximation of this area by the rectangles shown in Figure 7.7(a). If we choose x_i^* to be the right endpoint, then $x_i^* = x_i$ and we have

(7.29)
$$\int_a^b f(x)\,dx \approx R_n = \sum_{i=1}^n f(x_i)\,\Delta x$$

[See Figure 7.7(b).] The approximations L_n and R_n that are defined by Equations 7.28 and 7.29 are called the **left endpoint approximation** and **right endpoint approximation**, respectively.

In Section 5.3 we also considered the case where x_i^* is chosen to be the midpoint $\bar{x}_i$ of the subinterval $[x_{i-1}, x_i]$. Figure 7.7(c) shows the midpoint approximation M_n, which appears to be better than L_n or R_n.

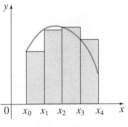

(a) Left endpoint
approximation

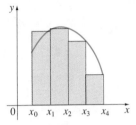

(b) Right endpoint
approximation

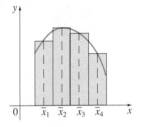

(c) Midpoint
approximation

Figure 7.7

Midpoint Rule (7.30)

$$\int_a^b f(x)\,dx \approx M_n = \Delta x\,[f(\bar{x}_1) + f(\bar{x}_2) + \cdots + f(\bar{x}_n)]$$

where

$$\Delta x = \frac{b-a}{n}$$

and

$$\bar{x}_i = \tfrac{1}{2}(x_{i-1} + x_i) = \text{midpoint of } [x_{i-1}, x_i]$$

Another approximation results from averaging the approximations in Equations 7.28 and 7.29:

$$\int_a^b f(x)\,dx \approx \tfrac{1}{2}\left[\sum_{i=1}^n f(x_{i-1})\,\Delta x + \sum_{i=1}^n f(x_i)\,\Delta x\right]$$

$$= \frac{\Delta x}{2}\left[(f(x_0) + f(x_1)) + (f(x_1) + f(x_2)) + \cdots + (f(x_{n-1}) + f(x_n))\right]$$

$$= \frac{\Delta x}{2}\left[f(x_0) + 2f(x_1) + 2f(x_2) + \cdots + 2f(x_{n-1}) + f(x_n)\right]$$

The Trapezoidal Rule (7.31)

$$\int_a^b f(x)\,dx \approx T_n = \frac{\Delta x}{2}[f(x_0) + 2f(x_1) + 2f(x_2) + \cdots + 2f(x_{n-1}) + f(x_n)]$$

where

$$\Delta x = (b-a)/n \qquad \text{and} \qquad x_i = a + i\,\Delta x$$

The reason for the name Trapezoidal Rule can be seen from Figure 7.8, which illustrates the case where $f(x) \geq 0$. The area of the trapezoid that lies above the ith subinterval is

$$\Delta x\left(\frac{f(x_{i-1}) + f(x_i)}{2}\right) = \frac{\Delta x}{2}[f(x_{i-1}) + f(x_i)]$$

and if we add the areas of all these trapezoids, we get the right side of (7.31).

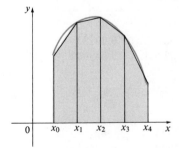

Figure 7.8

Trapezoidal approximation

EXAMPLE 1 Use (a) the Trapezoidal Rule and (b) the Midpoint Rule with $n = 5$ to approximate the integral $\int_1^2 (1/x)\,dx$.

Solution

(a)   With $n = 5$, $a = 1$, and $b = 2$, we have $\Delta x = (2 - 1)/5 = 0.2$, and so the Trapezoidal Rule gives

$$\int_1^2 \frac{1}{x}\,dx \approx T_5 = \frac{0.2}{2}[f(1) + 2f(1.2) + 2f(1.4) + 2f(1.6) + 2f(1.8) + f(2)]$$

$$= 0.1\left[\frac{1}{1} + \frac{2}{1.2} + \frac{2}{1.4} + \frac{2}{1.6} + \frac{2}{1.8} + \frac{1}{2}\right]$$

$$\approx 0.695635$$

(b)   The midpoints of the five intervals are 1.1, 1.3, 1.5, 1.7, and 1.9, so the Midpoint Rule gives

$$\int_1^2 \frac{1}{x}\,dx \approx \Delta x\,[f(1.1) + f(1.3) + f(1.5) + f(1.7) + f(1.9)]$$

$$= \frac{1}{5}\left(\frac{1}{1.1} + \frac{1}{1.3} + \frac{1}{1.5} + \frac{1}{1.7} + \frac{1}{1.9}\right)$$

$$\approx 0.691908$$

In Example 1 we deliberately chose an integral whose value can be computed explicitly so that we can see how accurate the Trapezoidal and Midpoint Rules are. By the Fundamental Theorem of Calculus,

$$\int_1^2 \frac{1}{x}\,dx = \ln x\,\big|_1^2 = \ln 2 = 0.693147\ldots$$

The **error** in using an approximation is defined to be the amount that needs to be added to the approximation to make it exact. From the values in Example 1 we see that the errors in the Trapezoidal and Midpoint approximations for $n = 5$ are

$$E_T \approx -0.002488 \qquad \text{and} \qquad E_M \approx 0.001239$$

In general, we have

$$E_T = \int_a^b f(x)\,dx - T_n \qquad \text{and} \qquad E_M = \int_a^b f(x)\,dx - M_n$$

The following tables show the results of calculations similar to those in Example 1 but for $n = 5$, 10, 20 and for the left and right endpoint approximations as well as the Trapezoidal and Midpoint Rules.

Approximations to $\int_1^2 \frac{1}{x}\,dx$

n	L_n	R_n	T_n	M_n
5	0.745635	0.645635	0.695635	0.691908
10	0.718771	0.668771	0.693771	0.692835
20	0.705803	0.680803	0.693303	0.693069

Corresponding errors

n	E_L	E_R	E_T	E_M
5	−0.052488	0.047512	−0.002488	0.001239
10	−0.025624	0.024376	−0.000624	0.000312
20	−0.012656	0.012344	−0.000156	0.000078

We can make several observations from these tables:

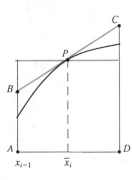

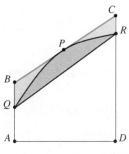

Figure 7.9

1. In all of the methods we get more accurate approximations when we increase the value of n. (But for very large values of n there are so many arithmetic operations that we have to beware of accumulated round-off error.)

2. The errors in the left and right endpoint approximations are opposite in sign and appear to decrease by a factor of about 2 when we double the value of n.

3. The Trapezoidal and Midpoint Rules are much more accurate than the endpoint approximations.

4. The errors in the Trapezoidal and Midpoint Rules are opposite in sign and appear to decrease by a factor of about 4 when we double the value of n.

5. The size of the error in the Midpoint Rule is about half the size of the error in the Trapezoidal Rule.

Figure 7.9 shows why we can expect the Midpoint Rule to be more accurate than the Trapezoidal Rule. The area of a typical rectangle in the Midpoint Rule is the same as the trapezoid $ABCD$ whose upper side is tangent to the graph at P. The area of this trapezoid is closer to the area under the graph than is the area of the trapezoid $AQRD$ used in the Trapezoidal Rule. [The midpoint error (shaded red) is smaller than the trapezoidal error (shaded gray).]

These observations are corroborated in the following error estimates, which are proved in books on numerical analysis. The fact that the estimates depend on the size of the second derivative is not surprising if you look at Figure 7.9, because $f''(x)$ measures how much the graph is curved.

Error Bounds (7.32)

> Suppose $|f''(x)| \leq M$ for $a \leq x \leq b$. If E_T and E_M are the errors in the Trapezoidal and Midpoint Rules, then
>
> $$|E_T| \leq \frac{M(b-a)^3}{12n^2} \quad \text{and} \quad |E_M| \leq \frac{M(b-a)^3}{24n^2}$$

Let us apply this error estimate to the Trapezoidal approximation in Example 1. If $f(x) = 1/x$, then $f'(x) = -1/x^2$ and $f''(x) = 2/x^3$. Since $1 \leq x \leq 2$, we have $1/x \leq 1$, so

$$|f''(x)| = \left|\frac{2}{x^3}\right| \leq \frac{2}{1^3} = 2$$

Therefore, taking $M = 2$, $a = 1$, $b = 2$ and $n = 5$ in the error estimate (7.32), we see that

$$|E_T| \leq \frac{2(2-1)^3}{12(5)^2} \approx 0.006667$$

Comparing this error estimate of 0.006667 with the actual error of about 0.002488, we see that it can happen that the actual error is substantially less than the upper bound for the error given by (7.32).

EXAMPLE 2 How large should we take n in order to guarantee that the Trapezoidal and Midpoint approximations for $\int_1^2 (1/x)\,dx$ are accurate to within 0.0001?

Solution We saw in the preceding calculation that $|f''(x)| \leq 2$ for $1 \leq x \leq 2$, so we can take $M = 2$, $a = 1$, and $b = 2$ in (7.32). For an error less than 0.0001 we should choose n so that

$$\frac{2(1)^3}{12n^2} < 0.0001$$

Solving the inequality for n, we get

$$n^2 > \frac{2}{12(0.0001)}$$

or

$$n > \frac{1}{\sqrt{0.0006}} \approx 40.8$$

Thus, $n = 41$ will ensure the desired accuracy.

For the same accuracy with the Midpoint Rule we require

$$\frac{2(1)^3}{24n} < 0.0001$$

which gives

$$n > \frac{1}{\sqrt{0.0012}} \approx 29$$

EXAMPLE 3

(a)   Use the Trapezoidal Rule with $n = 10$ to approximate the integral $\int_0^1 e^{x^2}\, dx$.

(b)   Give an upper bound for the error involved in this approximation.

Solution

(a)   Since $a = 0$, $b = 1$, and $n = 10$, the Trapezoidal Rule gives

$$\int_0^1 e^{x^2}\, dx \approx \frac{\Delta x}{2}[f(0) + 2f(0.1) + 2f(0.2) + \cdots + 2f(0.9) + f(1)]$$

$$= \frac{0.1}{2}[e^0 + 2e^{0.01} + 2e^{0.04} + 2e^{0.09} + 2e^{0.16} + 2e^{0.25}$$

$$+ 2e^{0.36} + 2e^{0.49} + 2e^{0.64} + 2e^{0.81} + e^1]$$

$$\approx 1.467175$$

(b)   Since $f(x) = e^{x^2}$, we have $f'(x) = 2xe^{x^2}$ and $f''(x) = (2 + 4x^2)e^{x^2}$. Also, since $0 \le x \le 1$, we have $x^2 \le 1$, and so

$$0 \le f''(x) = (2 + 4x^2)e^{x^2} \le 6e$$

Taking $M = 6e$, $a = 0$, $b = 1$ and $n = 10$ in the error estimate (7.32), we see that an upper bound for the error is

$$\frac{6e(1)^3}{12(10)^2} = \frac{e}{200} \approx 0.014$$

Another rule for approximate integration results from using parabolas instead of trapezoids to approximate the area under a curve. As before, we take a partition of $[a, b]$ into n subintervals of equal length $h = \Delta x = (b - a)/n$ but this time we assume that n is an *even* number. Then on each consecutive pair of intervals we approximate the curve $y = f(x) \ge 0$ by a parabola as shown in Figure 7.10. If $y_i = f(x_i)$, then $P_i(x_i, y_i)$ is the

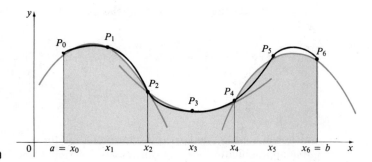

Figure 7.10

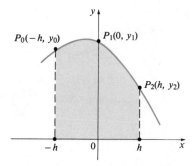

$P_0(-h, y_0)$ $P_1(0, y_1)$

$P_2(h, y_2)$

$-h$ 0 h

Figure 7.11

point on the curve lying above x_i. A typical parabola passes through three consecutive points P_i, P_{i+1}, and P_{i+2}.

In order to simplify our calculations, we first consider the case where $x_0 = -h$, $x_1 = 0$, and $x_2 = h$ (see Figure 7.11). We know that the equation of the parabola through P_0, P_1, and P_2 is of the form $y = Ax^2 + Bx + C$ and so the area under the parabola from $x = -h$ to $x = h$ is

$$\int_{-h}^{h} (Ax^2 + Bx + C)\,dx = A\frac{x^3}{3} + B\frac{x^2}{2} + Cx \Big|_{-h}^{h}$$

$$= A\frac{h^3}{3} + B\frac{h^2}{2} + Ch + A\frac{h^3}{3} - B\frac{h^2}{2} + Ch$$

$$= \frac{h}{3}(2Ah^2 + 6C)$$

But, since the parabola passes through $P_0(-h, y_0)$, $P_1(0, y_1)$, and $P_2(h, y_2)$, we have

$$y_0 = A(-h)^2 + B(-h) + C = Ah^2 - Bh + C$$

$$y_1 = C$$

$$y_2 = Ah^2 + Bh + C$$

and therefore $\qquad\qquad y_0 + 4y_1 + y_2 = 2Ah^2 + 6C$

Thus we can rewrite the area under the parabola as

$$\frac{h}{3}(y_0 + 4y_1 + y_2)$$

Now by shifting this parabola horizontally we do not change the area under it. This means that the area under the parabola through P_0, P_1, and P_2 from $x = x_0$ to $x = x_2$ in Figure 7.10 is still

$$\frac{h}{3}(y_0 + 4y_1 + y_2)$$

Similarly, the area under the parabola through P_2, P_3, and P_4 from $x = x_2$ to $x = x_4$ is

$$\frac{h}{3}(y_2 + 4y_3 + y_4)$$

If we compute the areas under all the parabolas in this manner and add the results, we get

$$\int_a^b f(x)\,dx \approx \frac{h}{3}(y_0 + 4y_1 + y_2) + \frac{h}{3}(y_2 + 4y_3 + y_4) + \cdots + \frac{h}{3}(y_{n-2} + 4y_{n-1} + y_n)$$

$$= \frac{h}{3}(y_0 + 4y_1 + 2y_2 + 4y_3 + 2y_4 + \cdots + 2y_{n-2} + 4y_{n-1} + y_n)$$

Although we have derived this approximation for the case where $f(x) \geq 0$, it is a reasonable approximation for any continuous function f and is called Simpson's Rule after the English mathematician Thomas Simpson (1710–1761). Note the pattern of coefficients:

$$1, 4, 2, 4, 2, 4, 2, \ldots, 4, 2, 4, 1.$$

Simpson's Rule (7.33)

$$\int_a^b f(x)\,dx \approx S_n = \frac{\Delta x}{3}[f(x_0) + 4f(x_1) + 2f(x_2) + 4f(x_3) + \cdots + 2f(x_{n-2})$$
$$+ 4f(x_{n-1}) + f(x_n)]$$

where n is even and $\Delta x = (b - a)/n$.

EXAMPLE 4   Use Simpson's Rule with $n = 10$ to approximate $\int_1^2 (1/x)\,dx$.

Solution   Putting $f(x) = 1/x$, $n = 10$, and $\Delta x = 0.1$ in (7.33), we obtain

$$\int_1^2 \frac{1}{x}\,dx \approx S_{10}$$

$$= \frac{\Delta x}{3}[f(1) + 4f(1.1) + 2f(1.2) + 4f(1.3) + \cdots + 2f(1.8) + 4f(1.9) + f(2)]$$

$$= \frac{0.1}{3}\left[\frac{1}{1} + \frac{4}{1.1} + \frac{2}{1.2} + \frac{4}{1.3} + \frac{2}{1.4} + \frac{4}{1.5} + \frac{2}{1.6} + \frac{4}{1.7} + \frac{2}{1.8} + \frac{4}{1.9} + \frac{1}{2}\right]$$

$$\approx 0.693150$$

Notice that, in Example 4, Simpson's Rule gives us a much better approximation ($S_{10} \approx 0.693150$) to the true value of the integral ($\ln 2 \approx 0.693147\ldots$) than does the Trapezoidal Rule ($T_{10} \approx 0.693771$) or the Midpoint Rule ($M_{10} \approx 0.692835$). It turns out (see Exercise 36) that the approximations in Simpson's Rule are weighted averages of those in the Trapezoidal and Midpoint Rules:

$$S_{2n} = \tfrac{1}{3}T_n + \tfrac{2}{3}M_n$$

(Recall that E_T and E_M have opposite signs and $|E_M|$ is about half the size of $|E_T|$.)

In Exercises 3–6 you are asked to demonstrate, in particular cases, that the error in Simpson's Rule decreases by a factor of about 16 when n is doubled. That is consistent with the appearance of n^4 in the denominator of the following error estimate for Simpson's Rule. It is analogous to the estimate given in (7.32) for the Trapezoidal Rule but it uses the fourth derivative of f.

Error Bound for Simpson's Rule (7.34)

If $|f^{(4)}(x)| \leq M$ for $a \leq x \leq b$, then the error involved in using Simpson's Rule (7.33) is at most

$$\frac{M(b-a)^5}{180n^4}$$

EXAMPLE 5 How large should we take n in order to guarantee that the Simpson's Rule approximation for $\int_1^2 (1/x)\, dx$ is accurate to within 0.0001?

Solution If $f(x) = 1/x$, then $f^{(4)}(x) = 24/x^5$. Since $x \geq 1$, we have $1/x \leq 1$ and so

$$\left| f^{(4)}(x) \right| = \left| \frac{24}{x^5} \right| \leq 24$$

Therefore we can take $M = 24$ in (7.34). Thus for an error less than 0.0001 we should choose n so that

$$\frac{24(1)^5}{180n^4} < 0.0001$$

This gives

$$n^4 > \frac{24}{180(0.0001)}$$

or

$$n > \frac{1}{\sqrt[4]{0.00075}} \approx 6.04$$

Therefore $n = 8$ (n must be even) gives the desired accuracy. (Compare this with Example 2 where we obtained $n = 41$ for the Trapezoidal Rule and $n = 29$ for the Midpoint Rule.) ●

EXAMPLE 6
(a) Use Simpson's Rule with $n = 10$ to approximate the integral $\int_0^1 e^{x^2}\, dx$.

(b) Estimate the error.

Solution
(a) If $n = 10$, then $\Delta x = 0.1$ and Simpson's Rule gives

$$\int_0^1 e^{x^2}\, dx \approx \frac{\Delta x}{3}[f(0) + 4f(0.1) + 2f(0.2) + \cdots + 2f(0.8) + 4f(0.9) + f(1)]$$

$$= \frac{0.1}{3}[e^0 + 4e^{0.01} + 2e^{0.04} + 4e^{0.09} + 2e^{0.16} + 4e^{0.25}$$

$$+ 2e^{0.36} + 4e^{0.49} + 2e^{0.64} + 4e^{0.81} + e^1]$$

$$\approx 1.462681$$

(b) The fourth derivative of $f(x) = e^{x^2}$ is

$$f^{(4)}(x) = (12 + 48x^2 + 16x^4)e^{x^2}$$

and so, since $0 \le x \le 1$, we have

$$0 \le f^{(4)}(x) \le (12 + 48 + 16)e^1 = 76e$$

Therefore, putting $M = 76e$, $a = 0$, $b = 1$, and $n = 10$ in (7.34) we see that the error is at most

$$\frac{76e(1)^5}{180(10)^4} \approx 0.000115$$

(Compare this with Example 3.) Thus, correct to three decimal places, we have

$$\int_0^1 e^{x^2} \, dx \approx 1.463$$

Recall that it is quite possible for y to be a function of x even if no explicit formula is known for y in terms of x. If a scientific experiment establishes values for y corresponding to certain equally spaced values of x, and if there is evidence that the values are not changing rapidly, then the Trapezoidal Rule or Simpson's Rule can still be used to find an approximate value for $\int_a^b y \, dx$, the integral of y with respect to x.

EXAMPLE 7   Suppose the following data were obtained from an experiment:

x	3.0	3.25	3.5	3.75	4.0	4.25	4.5	4.75	5.0
y	6.7	7.4	8.2	9.2	10.4	11.6	12.5	13.3	14.0

Use Simpson's Rule to approximate $\int_3^5 y \, dx$.

Solution   There are $n = 8$ intervals and the interval length is $\Delta x = 0.25$, so Simpson's Rule gives

$$\int_3^5 y \, dx \approx \frac{0.25}{3}[6.7 + 4(7.4) + 2(8.2) + 4(9.2) + 2(10.4)$$
$$+ 4(11.6) + 2(12.5) + 4(13.3) + 14.0]$$
$$\approx 20.7$$

▦ SECTION 7.8 **Exercises**

In Exercises 1 and 2 find the approximations L_n, R_n, T_n, and M_n for $n = 4$, 8, and 16. Then compute the corresponding errors E_L, E_R, E_T, and E_M. (Round your answers to six decimal places.) What observations can you make?

1. $\int_0^1 x^3 \, dx$

2. $\int_0^2 e^x \, dx$

In Exercises 3–6 find the approximations T_n, M_n, and S_n for the given values of n. Then compute the corresponding

errors E_T, E_M, and E_S. (Round your answers to six decimal places.) What observations can you make? In particular, what happens to the errors when n is doubled?*

3. $\int_1^4 \sqrt{x} \, dx$, $n = 6, 12$ 4. $\int_0^1 \frac{1}{1 + x^2} \, dx$, $n = 4, 8$

5. $\int_2^4 \frac{x}{x^2 + 1} \, dx$, $n = 4, 8$ 6. $\int_{-1}^2 x e^x \, dx$, $n = 6, 12$

In Exercises 7–10 use (a) the Trapezoidal Rule and (b) Simpson's Rule to approximate the given integral with the given value of n. (Round your answers to six decimal places.)

7. $\int_{-1}^{1} \sqrt{1+x^3}\,dx, \quad n=8$ 8. $\int_{0}^{1} \cos(x^2)\,dx, \quad n=4$

9. $\int_{\pi/2}^{\pi} \frac{\sin x}{x}\,dx, \quad n=6$ 10. $\int_{0}^{\pi/4} x \tan x\,dx, \quad n=6$

In Exercises 11–20 use (a) the Trapezoidal Rule, (b) the Midpoint Rule, and (c) Simpson's Rule to approximate the given integral with the given value of n. (Round your answers to six decimal places.)

11. $\int_{0}^{1} e^{-x^2}\,dx, \quad n=10$ 12. $\int_{0}^{2} \frac{1}{\sqrt{1+x^3}}\,dx, \quad n=10$

13. $\int_{0}^{1/2} \cos(e^x)\,dx, \quad n=8$ 14. $\int_{2}^{3} \frac{1}{\ln x}\,dx, \quad n=10$

15. $\int_{0}^{1} x^5 e^x\,dx, \quad n=10$ 16. $\int_{0}^{1} \ln(1+e^x)\,dx, \quad n=8$

17. $\int_{1}^{2} e^{1/x}\,dx, \quad n=4$ 18. $\int_{0}^{4} \sqrt{x}\,\sin x\,dx, \quad n=8$

19. $\int_{0}^{3} \frac{1}{1+x^4}\,dx, \quad n=6$ 20. $\int_{2}^{4} \frac{e^x}{x}\,dx, \quad n=10$

21. (a) Find the approximations T_{10} and M_{10} for the integral $\int_{0}^{2} e^{-x^2}\,dx$.
    (b) Estimate the error in the approximations of part (a).

22. (a) Find the approximations T_4, T_8, M_4, and M_8 for $\int_{0}^{1} \cos(x^2)\,dx$.
    (b) Estimate the errors involved in these approximations.

23. (a) Find the approximations T_{10} and S_{10} for $\int_{0}^{1} e^x\,dx$ and the corresponding errors E_T and E_S.
    (b) Compare the actual errors in part (a) with the error estimates given by (7.32) and (7.34).

24. How large do we have to choose n so that the approximations T_n, M_n, and S_n to the integral $\int_{0}^{1} e^x\,dx$ are accurate to within 0.00001?

25. How large do we have to choose n so that the approximations T_n and M_n to the integral $\int_{0}^{1} e^{-x^2}\,dx$ are accurate to within 0.00001?

26. How large should n be to guarantee that the Simpson's Rule approximation to $\int_{0}^{1} e^{x^2}\,dx$ is accurate to within 0.00001?

27. Use the Trapezoidal Rule and the following data to estimate the value of the integral $\int_{1}^{3.2} y\,dx$.

x	1.0	1.2	1.4	1.6	1.8	2.0	2.2	2.4	2.6	2.8	3.0	3.2
y	4.9	5.4	5.8	6.2	6.7	7.0	7.3	7.5	8.0	8.2	8.3	8.3

28. Use Simpson's Rule and the following data to estimate the value of the integral $\int_{2}^{6} y\,dx$.

x	2.0	2.5	3.0	3.5	4.0	4.5	5.0	5.5	6.0
y	9.22	9.01	8.76	8.30	7.52	6.83	7.32	7.69	7.91

29. The speedometer reading (v) on a car was observed at 1-min intervals and recorded in the following chart. Use Simpson's Rule to estimate the distance traveled by the car.

t (min)	0	1	2	3	4	5	6	7	8	9	10
v (mi/h)	40	42	45	49	52	54	56	57	57	55	56

30. The widths (in meters) of a kidney-shaped swimming pool were measured at 2-m intervals as indicated in the figure. Use Simpson's Rule to estimate the area of the pool.

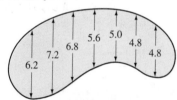

31. Estimate the area under graph in the figure by using (a) the Trapezoidal Rule, (b) the Midpoint Rule, and (c) Simpson's Rule, each with $n=4$.

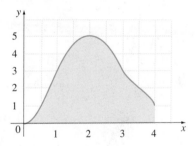

32. A log 10 m long is cut at 1-m intervals and its cross-sectional areas A (at a distance x from the end of the log) are listed in the following table. Use Simpson's Rule to estimate the volume of the log.

x (m)	0	1	2	3	4	5	6	7	8	9	10
A (m^2)	0.68	0.65	0.64	0.61	0.58	0.59	0.53	0.55	0.52	0.50	0.48

33. The region bounded by the curves $y = \sqrt[3]{1 + x^3}$, $y = 0$, $x = 0$, $x = 2$ is rotated about the x-axis. Use Simpson's Rule with $n = 10$ to estimate the volume of the resulting solid.

34. Show that if f is a polynomial of degree 3 or lower, then Simpson's Rule gives the exact value of $\int_a^b f(x)\,dx$.

35. If f is a positive function and $f''(x) < 0$ for $a \le x \le b$, show that
$$T_n < \int_a^b f(x)\,dx < M_n$$

36. Show that $\frac{1}{3}T_n + \frac{2}{3}M_n = S_{2n}$.

Improper Integrals

In defining a definite integral $\int_a^b f(x)\,dx$ we dealt with a function f defined on a finite interval $[a,b]$ and we observed that if the integral exists then f is a bounded function (see Section 5.3). In this section we extend the concept of a definite integral to the case where the interval is infinite and also to the case where f is unbounded (for instance, where f has an infinite discontinuity in $[a,b]$). In either case the integral is called an *improper* integral.

Type 1: Infinite Intervals

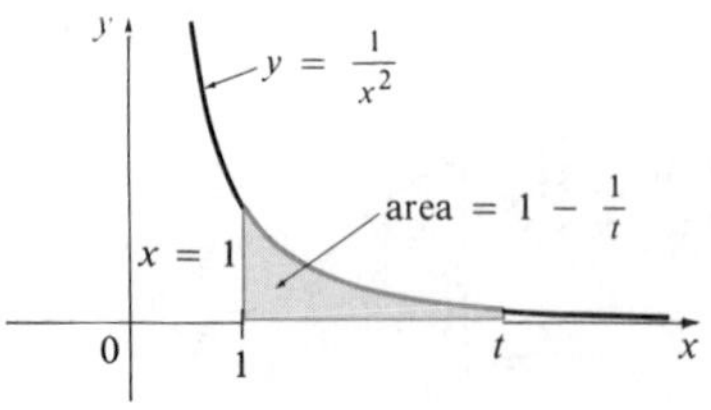

Consider the infinite region S that lies under the curve $y = 1/x^2$, above the x-axis, and to the right of the line $x = 1$. You might think that, since S is infinite in extent, its area must be infinite, but let us take a closer look. The area of the part of S that lies to the left of the line $x = t$ (shaded in Figure 7.12) is

$$A(t) = \int_1^t \frac{1}{x^2}\,dx = -\frac{1}{x}\bigg|_1^t = 1 - \frac{1}{t}$$

Figure 7.12

Notice that $A(t) < 1$ no matter how large t is chosen and

$$\lim_{t\to\infty} A(t) = \lim_{t\to\infty}\left(1 - \frac{1}{t}\right) = 1$$

The area of the shaded region approaches 1 as $t \to \infty$ (see Figure 7.13), so we say that the area of the infinite region S is equal to 1 and we write

$$\int_1^\infty \frac{1}{x^2}\,dx = \lim_{t\to\infty} \int_1^t \frac{1}{x^2}\,dx = 1$$

Using this example as a guide, we define the integral of f (not necessarily a positive function) as the limit of integrals over finite intervals.

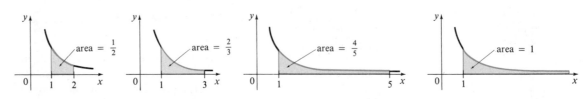

Figure 7.13

Definition of an Improper Integral of Type 1 (7.35)

(a) If $\int_a^t f(x)\,dx$ exists for every number $t \geq a$, then

$$\int_a^\infty f(x)\,dx = \lim_{t\to\infty} \int_a^t f(x)\,dx$$

provided this limit exists (as a finite number).

(b) If $\int_t^b f(x)\,dx$ exists for every number $t \leq b$, then

$$\int_{-\infty}^b f(x)\,dx = \lim_{t\to-\infty} \int_t^b f(x)\,dx$$

provided this limit exists (as a finite number).

The improper integrals in (a) and (b) are called **convergent** if the limit exists and **divergent** if the limit does not exist.

(c) If both $\int_a^\infty f(x)\,dx$ and $\int_{-\infty}^a f(x)\,dx$ are convergent, then we define

$$\int_{-\infty}^\infty f(x)\,dx = \int_{-\infty}^a f(x)\,dx + \int_a^\infty f(x)\,dx$$

In part (c) any real number a can be used (see Exercise 66).

Any of the improper integrals in Definition 7.35 can be interpreted as an area provided that f is a positive function. For instance, in case (a) if $f(x) \geq 0$ and the integral $\int_a^\infty f(x)\,dx$ is convergent, then we define the area of the region $S = \{(x,y) \mid x \geq a,\ 0 \leq y \leq f(x)\}$ in Figure 7.14 to be

$$A(S) = \int_a^\infty f(x)\,dx$$

This is appropriate because $\int_a^\infty f(x)\,dx$ is the limit as $t \to \infty$ of the area under the graph of f from a to t.

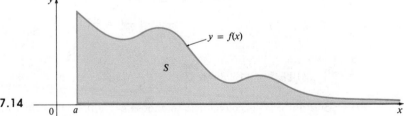

Figure 7.14

EXAMPLE 1 Determine whether the integral $\int_1^\infty (1/x)\,dx$ is convergent or divergent.

Solution According to part (a) of Definition 7.35, we have

$$\int_1^\infty \frac{1}{x}\,dx = \lim_{t\to\infty} \int_1^t \frac{1}{x}\,dx = \lim_{t\to\infty} \ln|x| \Big|_1^t$$

$$= \lim_{t\to\infty} (\ln t - \ln 1) = \lim_{t\to\infty} \ln t = \infty$$

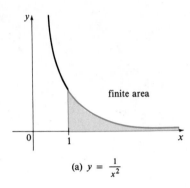

(a) $y = \dfrac{1}{x^2}$

finite area

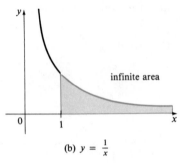

(b) $y = \dfrac{1}{x}$

infinite area

Figure 7.15

The limit does not exist as a finite number and so the improper integral $\int_1^\infty (1/x)\,dx$ is divergent.

Let us compare the result of Example 1 with the example at the beginning of this section:

$$\int_1^\infty \frac{1}{x^2}\,dx \text{ converges} \qquad \int_1^\infty \frac{1}{x}\,dx \text{ diverges}$$

Geometrically, this says that although the curves $y = 1/x^2$ and $y = 1/x$ look very similar for $x > 0$, the region under $y = 1/x^2$ to the right of $x = 1$ [the shaded region in Figure 7.15(a)] has finite area whereas the corresponding region under $y = 1/x$ has infinite area. Note that both $1/x^2$ and $1/x$ approach 0 as $x \to \infty$ but $1/x^2$ approaches 0 faster than $1/x$.

EXAMPLE 2   Evaluate $\displaystyle\int_{-\infty}^0 xe^x\,dx$.

Solution   Using part (b) of Definition 7.35, we have

$$\int_{-\infty}^0 xe^x\,dx = \lim_{t \to -\infty} \int_t^0 xe^x\,dx$$

We integrate by parts with $u = x$, $dv = e^x\,dx$ so that $du = dx$, $v = e^x$:

$$\int_t^0 xe^x\,dx = xe^x \Big|_t^0 - \int_t^0 e^x\,dx$$

$$= -te^t - 1 + e^t$$

We know that $e^t \to 0$ as $t \to -\infty$, and by l'Hospital's Rule we have

$$\lim_{t \to -\infty} te^t = \lim_{t \to -\infty} \frac{t}{e^{-t}} = \lim_{t \to -\infty} \frac{1}{-e^{-t}}$$

$$= \lim_{t \to -\infty} (-e^t) = 0$$

Therefore

$$\int_{-\infty}^0 xe^x\,dx = \lim_{t \to -\infty} (-te^t - 1 + e^t)$$

$$= -0 - 1 + 0 = -1$$

EXAMPLE 3   Evaluate $\displaystyle\int_{-\infty}^\infty \frac{1}{1 + x^2}\,dx$.

Solution   It is convenient to choose $a = 0$ in Definition 7.35(c):

$$\int_{-\infty}^\infty \frac{1}{1 + x^2}\,dx = \int_{-\infty}^0 \frac{1}{1 + x^2}\,dx + \int_0^\infty \frac{1}{1 + x^2}\,dx$$

We must now evaluate the integrals on the right side separately:

$$\int_0^\infty \frac{1}{1 + x^2}\,dx = \lim_{t \to \infty} \int_0^t \frac{dx}{1 + x^2} = \lim_{t \to \infty} \tan^{-1}x \Big|_0^t$$

$$= \lim_{t \to \infty} (\tan^{-1}t - \tan^{-1}0) = \lim_{t \to \infty} \tan^{-1}t$$

$$= \frac{\pi}{2}$$

$$\int_{-\infty}^{0} \frac{1}{1+x^2}\,dx = \lim_{t\to-\infty}\int_{t}^{0}\frac{dx}{1+x^2} = \lim_{t\to-\infty}\tan^{-1}x\Big|_{t}^{0}$$

$$= \lim_{t\to-\infty}\left(\tan^{-1}0 - \tan^{-1}t\right)$$

$$= 0 - \left(-\frac{\pi}{2}\right)$$

$$= \frac{\pi}{2}$$

Since both of these integrals are convergent, the given integral is convergent and

$$\int_{-\infty}^{\infty}\frac{1}{1+x^2}\,dx = \frac{\pi}{2} + \frac{\pi}{2} = \pi$$

Since $1/(1+x^2) > 0$, the given improper integral can be interpreted as the area of the infinite region that lies under the curve $y = 1/(1+x^2)$ and above the x-axis (see Figure 7.16).

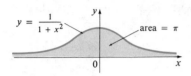

$y = \dfrac{1}{1+x^2}$ area $= \pi$

Figure 7.16

EXAMPLE 4   For what values of p is the integral

$$\int_{1}^{\infty}\frac{1}{x^p}\,dx$$

convergent?

Solution   We know from Example 1 that if $p = 1$ then the integral is divergent, so let us assume that $p \neq 1$. Then

$$\int_{1}^{\infty}\frac{1}{x^p}\,dx = \lim_{t\to\infty}\int_{1}^{t}\frac{1}{x^p}\,dx$$

$$= \lim_{t\to\infty}\frac{x^{-p+1}}{-p+1}\Big|_{x=1}^{x=t}$$

$$= \lim_{t\to\infty}\frac{1}{1-p}\left[\frac{1}{t^{p-1}} - 1\right]$$

If $p > 1$, then $p - 1 > 0$ and so $1/t^{p-1} \to 0$ as $t \to \infty$. Therefore

$$\int_{1}^{\infty}\frac{1}{x^p}\,dx = \frac{1}{p-1} \qquad \text{if } p > 1$$

But if $p < 1$, then $p - 1 < 0$ and so

$$\frac{1}{t^{p-1}} = t^{1-p} \to \infty \qquad \text{as } t \to \infty$$

and the integral diverges.

We summarize the result of Example 4 for future reference:

(7.36)

$$\int_{1}^{\infty}\frac{1}{x^p}\,dx \text{ is convergent if } p > 1 \text{ and divergent if } p \leq 1.$$

Type 2: Discontinuous Integrands

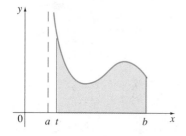

Figure 7.17

Suppose that f is a positive continuous function defined on a finite interval $[a,b)$ but has a vertical asymptote at b. Let S be the unbounded region under the graph of f and above the x-axis between a and b. (For Type 1 integrals, the regions extended indefinitely in a horizontal direction. Here the region is infinite in a vertical direction.) The area of the part of S between a and t (the shaded region in Figure 7.17) is

$$A(t) = \int_a^t f(x)\,dx$$

If it happens that $A(t)$ approaches a definite number A as $t \to b^-$, then we say that the area of the region S is A and we write

$$\int_a^b f(x)\,dx = \lim_{t \to b^-} \int_a^t f(x)\,dx$$

We use this equation to define an improper integral of Type 2 even when f is not a positive function, no matter what type of discontinuity f has at b.

Definition of an Improper Integral of Type 2 (7.37)

(a) If f is continuous on $[a,b)$ and is discontinuous at b, then

$$\int_a^b f(x)\,dx = \lim_{t \to b^-} \int_a^t f(x)\,dx$$

if this limit exists (as a finite number).

(b) If f is continuous on $(a,b]$ and is discontinuous at a, then

$$\int_a^b f(x)\,dx = \lim_{t \to a^+} \int_t^b f(x)\,dx$$

if this limit exists (as a finite number).

The improper integrals in (a) and (b) are called **convergent** if the limit exists and **divergent** if the limit does not exist.

(c) If f has a discontinuity at c, where $a < c < b$, and both $\int_a^c f(x)\,dx$ and $\int_c^b f(x)\,dx$ are convergent, then we define

$$\int_a^b f(x)\,dx = \int_a^c f(x)\,dx + \int_c^b f(x)\,dx$$

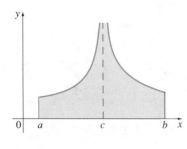

Figure 7.18

Parts (b) and (c) of Definition 7.37 are illustrated in Figures 7.18 and 7.19 for the case where $f(x) \geq 0$ and f has vertical asymptotes at a and c, respectively.

EXAMPLE 5   Find $\displaystyle\int_2^5 \frac{1}{\sqrt{x-2}}\,dx$.

Solution   We note first of all that the given integral is improper because $f(x) = 1/\sqrt{x-2}$ has the vertical asymptote $x = 2$. Since the infinite discontinuity occurs at the left endpoint of $[2,5]$, we use part (b) of Definition 7.37:

$$\int_2^5 \frac{dx}{\sqrt{x-2}} = \lim_{t \to 2^+} \int_t^5 \frac{dx}{\sqrt{x-2}}$$

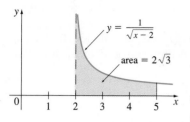

Figure 7.20

$$= \lim_{t \to 2^+} 2\sqrt{x-2} \Big|_t^5$$

$$= \lim_{t \to 2^+} 2(\sqrt{3} - \sqrt{t-2})$$

$$= 2\sqrt{3}$$

Thus the given improper integral is convergent and since the integrand is positive we can interpret the value of the integral as the area of the shaded region in Figure 7.20. ●

EXAMPLE 6   Determine whether $\displaystyle\int_0^{\pi/2} \sec x \, dx$ converges or diverges.

Solution   Note that the given integral is improper because $\displaystyle\lim_{x \to \pi/2^-} \sec x = \infty$. Using part (a) of Definition 7.37 we have

$$\int_0^{\pi/2} \sec x \, dx = \lim_{t \to \pi/2^-} \int_0^t \sec x \, dx$$

$$= \lim_{t \to \pi/2^-} \ln|\sec x + \tan x| \Big|_0^t$$

$$= \lim_{t \to \pi/2^-} [\ln(\sec t + \tan t) - \ln 1]$$

$$= \infty$$

because $\sec t \to \infty$ and $\tan t \to \infty$ as $t \to \pi/2^-$. Thus the given improper integral is divergent. ●

EXAMPLE 7   Evaluate $\displaystyle\int_0^3 \frac{dx}{x-1}$ if possible.

Solution   Observe that the line $x = 1$ is a vertical asymptote of the integrand. Since it occurs in the middle of the interval [0,3] we must use part (c) of Definition 7.37 with $c = 1$:

$$\int_0^3 \frac{dx}{x-1} = \int_0^1 \frac{dx}{x-1} + \int_1^3 \frac{dx}{x-1}$$

where

$$\int_0^1 \frac{dx}{x-1} = \lim_{t \to 1^-} \int_0^t \frac{dx}{x-1} = \lim_{t \to 1^-} \ln|x-1| \Big|_0^t$$

$$= \lim_{t \to 1^-} (\ln|t-1| - \ln|-1|)$$

$$= \lim_{t \to 1^-} \ln(1-t) = -\infty$$

because $1 - t \to 0^+$ as $t \to 1^-$. Thus $\int_0^1 dx/(x-1)$ is divergent. This implies that $\int_0^3 dx/(x-1)$ is divergent. [We do not need to evaluate $\int_1^3 dx/(x-1)$.] ●

Warning: If we had not noticed the asymptote $x = 1$ in Example 7 and had instead confused the integral with an ordinary integral, then we might have made the following erroneous calculation:

$$\int_0^3 \frac{dx}{x-1} = \ln|x-1| \Big|_0^3 = \ln 2 - \ln 1 = \ln 2$$

This is wrong because the integral is improper and must be calculated in terms of limits. (Compare with Example 10 in Section 5.5.)

From now on, whenever you meet the symbol $\int_a^b f(x)\,dx$ you must decide, by looking at the function f on $[a,b]$, whether it is an ordinary definite integral or an improper integral.

EXAMPLE 8   Evaluate $\displaystyle\int_0^1 \ln x\,dx$.

Solution   We know that the function $f(x) = \ln x$ has a vertical asymptote at 0 since $\lim_{x\to 0^+} \ln x = -\infty$. Thus the given integral is improper and we have

$$\int_0^1 \ln x\,dx = \lim_{t\to 0^+} \int_t^1 \ln x\,dx$$

Now we integrate by parts with $u = \ln x$, $dv = dx$, $du = dx/x$, and $v = x$:

$$\int_t^1 \ln x\,dx = x\ln x\,\big|_t^1 - \int_t^1 dx$$

$$= 1\ln 1 - t\ln t - (1 - t)$$

$$= -t\ln t - 1 + t$$

To find the limit of the first term we use l'Hospital's Rule:

$$\lim_{t\to 0^+} t\ln t = \lim_{t\to 0^+} \frac{\ln t}{1/t}$$

$$= \lim_{t\to 0^+} \frac{1/t}{-1/t^2}$$

$$= \lim_{t\to 0^+} (-t) = 0$$

Therefore $$\int_0^1 \ln x\,dx = \lim_{t\to 0^+} (-t\ln t - 1 + t)$$

$$= -0 - 1 + 0 = -1$$

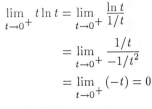

area = 1

$y = \ln x$

Figure 7.21

Figure 7.21 shows the geometric interpretation of this result. The area of the shaded region above $y = \ln x$ and below the x-axis is 1.   ●

A Comparison Test for Improper Integrals

Sometimes it is impossible to find the exact value of an improper integral and yet it is important to know whether it is convergent or divergent. In such cases the following theorem is useful. Although we state it for Type 1 integrals, a similar theorem is true for Type 2 integrals.

Comparison Theorem (7.38)

Suppose that f and g are continuous functions with $f(x) \ge g(x) \ge 0$ for $x \ge a$.

(a)   If $\int_a^\infty f(x)\,dx$ is convergent, then $\int_a^\infty g(x)\,dx$ is convergent.

(b)   If $\int_a^\infty g(x)\,dx$ is divergent, then $\int_a^\infty f(x)\,dx$ is divergent.

We omit the proof of the Comparison Theorem, but Figure 7.22 makes it seem plaus-

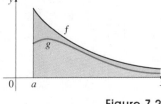

Figure 7.22

ible. If the area under the top curve $y = f(x)$ is finite, then so is the area under the bottom curve $y = g(x)$. And if the area under $y = g(x)$ is infinite, then so is the area under $y = f(x)$.

EXAMPLE 9   Show that $\int_0^\infty e^{-x^2}\, dx$ is convergent.

Solution   We cannot evaluate the integral directly because the antiderivative of e^{-x^2} is not an elementary function (as explained in Section 7.6). We write

$$\int_0^\infty e^{-x^2}\, dx = \int_0^1 e^{-x^2}\, dx + \int_1^\infty e^{-x^2}\, dx$$

and observe that the first integral is just an ordinary definite integral. In the second integral we use the fact that for $x \geq 1$ we have $x^2 \geq x$, so $-x^2 \leq -x$ and therefore $e^{-x^2} \leq e^{-x}$ (see Figure 7.23). The integral of e^{-x} is easy to evaluate:

$$\int_1^\infty e^{-x}\, dx = \lim_{t \to \infty} \int_1^t e^{-x}\, dx = \lim_{t \to \infty} \left(e^{-1} - e^{-t}\right) = e^{-1}$$

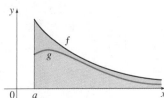

Figure 7.23

Thus, taking $f(x) = e^{-x}$ and $g(x) = e^{-x^2}$ in the Comparison Theorem, we see that $\int_1^\infty e^{-x^2}\, dx$ is convergent. It follows that $\int_0^\infty e^{-x^2}\, dx$ is convergent.   ●

In Example 9 we showed that $\int_0^\infty e^{-x^2}\, dx$ is convergent without computing its value. In Exercise 74 we indicate how to show that its value is approximately 0.89. In probability theory it is important to know the exact value of this improper integral, and in Section 13.4 indirect methods are introduced that give the exact value as $\sqrt{\pi}/2$.

EXAMPLE 10 $\int_1^\infty \dfrac{1 + e^{-x}}{x}\, dx$ is divergent by the Comparison Theorem because

$$\frac{1 + e^{-x}}{x} > \frac{1}{x}$$

and $\int_1^\infty (1/x)\, dx$ is divergent by Example 1 [or (7.36) with $p = 1$].   ●

SECTION 7.9 **Exercises**

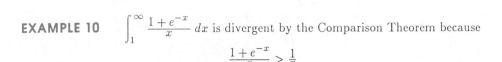

Determine whether the integrals in Exercises 1–46 are convergent or divergent. Evaluate those that are convergent.

1. $\int_2^\infty \dfrac{1}{\sqrt{x+3}}\, dx$

2. $\int_2^\infty \dfrac{1}{(x+3)^{3/2}}\, dx$

3. $\int_{-\infty}^1 \dfrac{1}{(2x-3)^2}\, dx$

4. $\int_{-\infty}^{-1} \dfrac{1}{\sqrt[3]{x-1}}\, dx$

5. $\int_{-\infty}^\infty x\, dx$

6. $\int_{-\infty}^\infty (2x^2 - x + 3)\, dx$

7. $\int_0^\infty e^{-x}\, dx$

8. $\int_{-\infty}^0 e^{3x}\, dx$

9. $\int_{-\infty}^\infty x e^{-x^2}\, dx$

10. $\int_{-\infty}^\infty x^2 e^{-x^3}\, dx$

11. $\int_0^\infty \dfrac{1}{(x+2)(x+3)}\, dx$

12. $\int_0^\infty \dfrac{x}{(x+2)(x+3)}\, dx$

13. $\int_0^\infty \cos x\, dx$

14. $\int_1^\infty \sin \pi x\, dx$

15. $\int_0^\infty \dfrac{5}{2x+3}\, dx$

16. $\int_{-\infty}^3 \dfrac{1}{x^2+9}\, dx$

17. $\int_{-\infty}^1 x e^{2x}\, dx$

18. $\int_0^\infty x e^{-x}\, dx$

19. $\int_1^\infty \dfrac{\ln x}{x}\, dx$

20. $\int_e^\infty \dfrac{1}{x(\ln x)^2}\, dx$

21. $\int_{-\infty}^\infty \dfrac{x}{1+x^2}\, dx$

22. $\int_{-\infty}^\infty e^{-|x|}\, dx$

23. $\displaystyle\int_1^\infty \frac{\ln x}{x^2}\, dx$

24. $\displaystyle\int_{-\infty}^\infty \frac{dx}{x^2 + 4x + 6}$

25. $\displaystyle\int_0^\infty \frac{1}{2^x}\, dx$

26. $\displaystyle\int_1^\infty \frac{\ln x}{x^3}\, dx$

27. $\displaystyle\int_0^3 \frac{1}{\sqrt{x}}\, dx$

28. $\displaystyle\int_0^3 \frac{1}{x\sqrt{x}}\, dx$

29. $\displaystyle\int_{-1}^0 \frac{1}{x^2}\, dx$

30. $\displaystyle\int_1^9 \frac{1}{\sqrt[3]{x-9}}\, dx$

31. $\displaystyle\int_{-2}^3 \frac{1}{x^4}\, dx$

32. $\displaystyle\int_0^2 \frac{1}{4x - 5}\, dx$

33. $\displaystyle\int_4^5 \frac{1}{(5-x)^{2/5}}\, dx$

34. $\displaystyle\int_{\pi/4}^{\pi/2} \sec^2 x\, dx$

35. $\displaystyle\int_{\pi/4}^{\pi/2} \tan^2 x\, dx$

36. $\displaystyle\int_0^{\pi/4} \frac{\cos x}{\sqrt{\sin x}}\, dx$

37. $\displaystyle\int_0^2 \frac{x}{\sqrt{4 - x^2}}\, dx$

38. $\displaystyle\int_0^9 \frac{dx}{(x + 9)\sqrt{x}}$

39. $\displaystyle\int_0^\pi \sec x\, dx$

40. $\displaystyle\int_0^4 \frac{1}{x^2 + x - 6}\, dx$

41. $\displaystyle\int_{-2}^2 \frac{1}{x^2 - 1}\, dx$

42. $\displaystyle\int_{\pi/4}^{3\pi/4} \tan x\, dx$

43. $\displaystyle\int_1^e \frac{1}{x\sqrt[4]{\ln x}}\, dx$

44. $\displaystyle\int_0^2 \frac{x - 3}{2x - 3}\, dx$

45. $\displaystyle\int_0^1 x \ln x\, dx$

46. $\displaystyle\int_0^1 \frac{\ln x}{\sqrt{x}}\, dx$

In Exercises 47–52 sketch the given region and find its area (if the area is finite).

47. $S = \{(x, y) \mid x \le 1, 0 \le y \le e^x\}$

48. $S = \{(x, y) \mid x \ge -2, 0 \le y \le e^{-x/2}\}$

49. $S = \left\{(x, y) \mid 0 \le y \le \dfrac{1}{x^2 - 2x + 5}\right\}$

50. $S = \left\{(x, y) \mid x \ge 0, 0 \le y \le \dfrac{1}{\sqrt{x + 1}}\right\}$

51. $S = \{(x, y) \mid 0 \le x < \pi/2, 0 \le y \le \tan x\}$

52. $S = \left\{(x, y) \mid 3 < x \le 7, 0 \le y \le \dfrac{1}{\sqrt{x - 3}}\right\}$

In Exercises 53–58 use the Comparison Theorem to determine whether the integral is convergent or divergent.

53. $\displaystyle\int_1^\infty \frac{\sin^2 x}{x^2}\, dx$

54. $\displaystyle\int_1^\infty \frac{\sqrt{1 + \sqrt{x}}}{\sqrt{x}}\, dx$

55. $\displaystyle\int_1^\infty \frac{dx}{x + e^{2x}}$

56. $\displaystyle\int_1^\infty \frac{1}{\sqrt{x^3 + 1}}\, dx$

57. $\displaystyle\int_0^{\pi/2} \frac{dx}{x \sin x}$

58. $\displaystyle\int_0^1 \frac{e^{-x}}{\sqrt{x}}\, dx$

59. The integral

$$\int_0^\infty \frac{1}{\sqrt{x}(1 + x)}\, dx$$

is improper for two reasons: the interval $[0, \infty)$ is infinite and the integrand is unbounded near 0. Evaluate it by expressing it as a sum of improper integrals of Types 2 and 1 as follows:

$$\int_0^\infty \frac{1}{\sqrt{x}(1 + x)}\, dx = \int_0^1 \frac{1}{\sqrt{x}(1 + x)}\, dx + \int_1^\infty \frac{1}{\sqrt{x}(1 + x)dx}$$

60. Evaluate

$$\int_2^\infty \frac{1}{x\sqrt{x^2 - 4}}\, dx$$

by the same method as in Exercise 59.

In Exercises 61–63 find the values of p for which the given integral converges and evaluate the integral for those values of p.

61. $\displaystyle\int_0^1 \frac{1}{x^p}\, dx$

62. $\displaystyle\int_e^\infty \frac{1}{x(\ln x)^p}\, dx$

63. $\displaystyle\int_0^1 x^p \ln x\, dx$

64. (a) Evaluate the integral $\int_0^\infty x^n e^{-x}\, dx$ for $n = 0, 1, 2,$ and 3.

(b) Guess the value of $\int_0^\infty x^n e^{-x}\, dx$ when n is an arbitrary positive integer.

(c) Prove your guess using mathematical induction.

65. (a) Show that $\int_{-\infty}^\infty x\, dx$ is divergent.

(b) Show that

$$\lim_{t \to \infty} \int_{-t}^t x\, dx = 0$$

This shows that we cannot define

$$\int_{-\infty}^\infty f(x)\, dx = \lim_{t \to \infty} \int_{-t}^t f(x)\, dx$$

66. If $\int_{-\infty}^\infty f(x)\, dx$ is convergent and a and b are real numbers, show that

$$\int_{-\infty}^a f(x)\, dx + \int_a^\infty f(x)\, dx = \int_{-\infty}^b f(x)\, dx + \int_b^\infty f(x)\, dx$$

67. We know from Example 1 that the region $\mathcal{R} = \{(x, y) \mid x \ge 1, 0 \le y \le 1/x\}$ has infinite area. Show that by rotating $\mathcal{R}$ about the x-axis we obtain

a solid with finite volume.

68. Use the information and data in Exercises 25 and 26 of Section 6.4 to find the work required to propel a 1000-kg satellite out of the earth's gravitational field.

69. Find the *escape velocity* v_0 that is needed to propel a rocket of mass m out of the gravitational field of a planet with mass M and radius R. Use Newton's Law of Gravitation (see Exercise 25 in Section 6.4) and the fact that the initial kinetic energy $\frac{1}{2}mv_0^2$ supplies the needed work.

70. A positive function f is called a *probability density function* if $\int_{-\infty}^{\infty} f(x)\,dx = 1$.
    (a) Show that if $c > 0$, the function f defined by $f(x) = ce^{-cx}$ for $x \geq 0$ and $f(x) = 0$ for $x < 0$ is a probability density function.
    (b) Calculate the *mean*:
    $$\mu = \int_{-\infty}^{\infty} xf(x)\,dx$$
    (c) Calculate the *standard deviation*:
    $$\sigma = \left[\int_{-\infty}^{\infty}(x-\mu)^2 f(x)\,dx\right]^{1/2}$$

71. If $f(t)$ is continuous for $t \geq 0$, the *Laplace transform* of f is the function F defined by
    $$F(s) = \int_0^{\infty} f(t)e^{-st}\,dt$$
    and the domain of F is the set consisting of all numbers s for which the integral converges. Find the Laplace transforms of the following functions:
    (a) $f(t) = 1$ (b) $f(t) = e^t$ (c) $f(t) = t$

72. Show that if $0 \leq f(t) \leq Me^{at}$ for $t \geq 0$, where M and

a are constants, then the Laplace transform $F(s)$ exists for $s > a$.

73. Suppose that $0 \leq f(t) \leq Me^{at}$ and $0 \leq f'(t) \leq Ke^{at}$ for $t \geq 0$, where f' is continuous. If the Laplace transform of $f(t)$ is $F(s)$ and the Laplace transform of $f'(t)$ is $G(s)$, show that
$$G(s) = sF(s) - f(0) \qquad (s > a)$$

74. Estimate the numerical value of $\int_0^{\infty} e^{-x^2}\,dx$ by writing it as the sum of $\int_0^4 e^{-x^2}\,dx$ and $\int_4^{\infty} e^{-x^2}\,dx$. Approximate the first integral by using Simpson's Rule with $n = 8$ and show that the second integral is smaller than $\int_4^{\infty} e^{-4x}\,dx$, which is less than 0.0000001.

75. Show that $\int_0^{\infty} x^2 e^{-x^2}\,dx = \frac{1}{2}\int_0^{\infty} e^{-x^2}\,dx$.

76. Show that $\int_0^{\infty} e^{-x^2}\,dx = \int_0^1 \sqrt{-\ln y}\,dy$ by interpreting the integrals as areas.

77. Find the value of the constant C for which the integral
    $$\int_0^{\infty}\left(\frac{1}{\sqrt{x^2+4}} - \frac{C}{x+2}\right)dx$$
    converges. Evaluate the integral for this value of C.

78. Find the value of the constant C for which the integral
    $$\int_0^{\infty}\left(\frac{x}{x^2+1} - \frac{C}{3x+1}\right)dx$$
    converges. Evaluate the integral for this value of C.

CHAPTER 7

Review

Key Topics

Define, state, or discuss the following.

1. Integration by parts
2. Trigonometric integrals
3. Trigonometric substitution
4. Partial fractions
5. Rationalizing substitutions
6. Strategy for integration
7. The Midpoint Rule
8. The Trapezoidal Rule
9. Simpson's Rule
10. Improper integrals
11. Comparison Theorem

Exercises

(Note: Additional practice in techniques of integration is provided in Exercises 7.6.)

In Exercises 1–8 determine whether the statement is true or false.

1. $\dfrac{x(x^2+4)}{x^2-4}$ can be put in the form $\dfrac{A}{x+2} + \dfrac{B}{x-2}$.

2. $\dfrac{x^2+4}{x(x^2-4)}$ can be put in the form $\dfrac{A}{x} + \dfrac{B}{x+2} + \dfrac{C}{x-2}$.

3. $\dfrac{x^2+4}{x^2(x-4)}$ can be put in the form $\dfrac{A}{x^2} + \dfrac{B}{x-4}$.

4. $\dfrac{x^2-4}{x(x^2+4)}$ can be put in the form $\dfrac{A}{x} + \dfrac{B}{x^2+4}$.

5. $\displaystyle\int_0^4 \dfrac{x}{x^2-1}\,dx = \tfrac{1}{2}\ln 3$

6. $\displaystyle\int_1^\infty \dfrac{1}{x^{\sqrt 2}}\,dx$ is convergent.

7. If f is continuous, then
$$\int_{-\infty}^\infty f(x)\,dx = \lim_{t\to\infty}\int_{-t}^t f(x)\,dx$$

8. The Midpoint Rule is always more accurate than the Trapezoidal Rule.

Evaluate the integrals in Exercises 9–44.

9. $\displaystyle\int \dfrac{x-1}{x+1}\,dx$

10. $\displaystyle\int \dfrac{\sin^3 x}{\cos x}\,dx$

11. $\displaystyle\int \dfrac{(\arctan x)^5}{1+x^2}\,dx$

12. $\displaystyle\int x^2 e^{-3x}\,dx$

13. $\displaystyle\int \dfrac{\cos x}{e^{\sin x}}\,dx$

14. $\displaystyle\int \dfrac{x^2+1}{x-1}\,dx$

15. $\displaystyle\int x^4 \ln x\,dx$

16. $\displaystyle\int \dfrac{\sec^2\theta}{1-\tan\theta}\,d\theta$

17. $\displaystyle\int x\sin(x^2)\,dx$

18. $\displaystyle\int x\sin^2 x\,dx$

19. $\displaystyle\int \dfrac{dx}{2x^2-5x+2}$

20. $\displaystyle\int \dfrac{dt}{\sin^2 t+\cos 2t}$

21. $\displaystyle\int \tan^7 x\sec^3 x\,dx$

22. $\displaystyle\int \dfrac{dx}{\sqrt{8+2x-x^2}}$

23. $\displaystyle\int \dfrac{dx}{\sqrt{1+2x}+3}$

24. $\displaystyle\int x(\tan^{-1}x)^2\,dx$

25. $\displaystyle\int \dfrac{e^{\sqrt x}}{\sqrt x}\,dx$

26. $\displaystyle\int \dfrac{dx}{x^3-2x^2+x}$

27. $\displaystyle\int \ln(x^2+2x+2)\,dx$

28. $\displaystyle\int \dfrac{dx}{x^2\sqrt{1+x^2}}$

29. $\displaystyle\int \dfrac{dx}{x^3+x}$

30. $\displaystyle\int \dfrac{dx}{1+e^x}$

31. $\displaystyle\int \cot^2 x\,dx$

32. $\displaystyle\int \dfrac{dx}{5-3\cos x}$

33. $\displaystyle\int \dfrac{dx}{(x^2-1)^{3/2}}$

34. $\displaystyle\int \sqrt{1+\cos x}\,dx$

35. $\displaystyle\int \dfrac{2x^2+3x+11}{x^3+x^2+3x-5}\,dx$

36. $\displaystyle\int \dfrac{x^3}{(x+1)^{10}}\,dx$

37. $\displaystyle\int \csc^4 4x\,dx$

38. $\displaystyle\int (\arcsin x)^2\,dx$

39. $\displaystyle\int \dfrac{\ln(\ln x)}{x}\,dx$

40. $\displaystyle\int \dfrac{\sin x}{1+\sin x}\,dx$

41. $\displaystyle\int \dfrac{1}{\sqrt{4x^2+4x+5}}\,dx$

42. $\displaystyle\int e^{-x}\sinh x\,dx$

43. $\displaystyle\int (\cos x+\sin x)^2\cos 2x\,dx$

44. $\displaystyle\int \dfrac{e^{2x}}{e^{4x}-7}\,dx$

In Exercises 45–62 evaluate the integral or show that it is divergent.

45. $\displaystyle\int_0^{\pi/2} \cos^3 x\sin 2x\,dx$

46. $\displaystyle\int_{-1}^1 \dfrac{1}{2x+1}\,dx$

47. $\displaystyle\int_0^3 \dfrac{dx}{x^2-x-2}$

48. $\displaystyle\int_0^{\pi/4} \cos^5(2\theta)\,d\theta$

49. $\displaystyle\int_0^1 \dfrac{t^2-1}{t^2+1}\,dt$

50. $\displaystyle\int_2^6 \dfrac{y}{\sqrt{y-2}}\,dy$

51. $\displaystyle\int_0^\infty \dfrac{1}{(x+2)^4}\,dx$

52. $\displaystyle\int_1^4 \dfrac{e^{1/x}}{x^2}\,dx$

53. $\displaystyle\int_1^e \dfrac{dx}{x\sqrt{\ln x}}$

54. $\displaystyle\int_0^\infty \dfrac{dx}{(x+1)^2(x+2)}$

55. $\displaystyle\int_1^4 \dfrac{\sqrt x}{\sqrt x+2}\,dx$

56. $\displaystyle\int_{-3}^3 x\sqrt{1+x^4}\,dx$

57. $\displaystyle\int_{-\infty}^\infty \dfrac{dx}{4x^2+4x+5}$

58. $\displaystyle\int_1^e \dfrac{dx}{x[1+(\ln x)^2]}$

59. $\displaystyle\int_1^2 \dfrac{\sqrt{x^2-1}}{x}\,dx$

60. $\displaystyle\int_{-1}^1 \dfrac{x+1}{\sqrt[3]{x^4}}\,dx$

61. $\displaystyle\int_0^\infty e^{ax}\cos bx\,dx$

62. $\displaystyle\int_1^\infty \dfrac{\tan^{-1}x}{x^2}\,dx$

In Exercises 63–66 use the Table of Integrals in Appendix G to evaluate the integral.

63. $\int e^x \sqrt{1 - e^{2x}}\, dx$ 64. $\int \tan^5 x\, dx$

65. $\int \sqrt{x^2 + x + 1}\, dx$ 66. $\int \dfrac{\cot x}{\sqrt{1 + 2\sin x}}\, dx$

67. Verify Formula 33 in the Table of Integrals (a) by differentiation, and (b) by using a trigonometric substitution.

68. Verify Formula 62 in the Table of Integrals.

In Exercises 69 and 70 use (a) the Trapezoidal Rule (b) the Midpoint Rule, and (c) Simpson's Rule with n = 10 to approximate the given integral. Round your answers to six decimal places.

69. $\int_0^1 \sqrt{1 + x^4}\, dx$ 70. $\int_0^{\pi/2} \sqrt{\sin x}\, dx$

71. Estimate the error involved in Exercise 69(a) and (b).

72. Use Simpson's Rule with $n = 6$ to estimate the area under the curve $y = e^x/x$ from $x = 1$ to $x = 4$.

73. Use the Comparison Theorem to determine whether the integral

$$\int_1^\infty \frac{x^3}{x^5 + 2}\, dx$$

is convergent or divergent.

74. Find the area of the region bounded by the hyperbola $y^2 - x^2 = 1$ and the line $y = 3$.

75. Find the area bounded by the curves $y = \cos x$ and $y = \cos^2 x$ between $x = 0$ and $x = \pi$.

76. Find the area of the region bounded by the curves $y = 1/\!\left(2 + \sqrt{x}\right)$, $y = 1/\!\left(2 - \sqrt{x}\right)$, and $x = 1$.

77. Is it possible to find a number n such that $\int_0^\infty x^n\, dx$ is convergent?

78. If n is a positive integer, prove that

$$\int_0^1 (\ln x)^n\, dx = (-1)^n n!$$

79. If f' is continuous on $[0,\infty)$ and $\lim_{x\to\infty} f(x) = 0$, show that

$$\int_0^\infty f'(x)\, dx = -f(0)$$

80. The magnitude of the repulsive force between two point charges with the same sign, one of size 1 and the other of size q, is

$$F = \frac{q}{4\pi\varepsilon_0 r^2}$$

where r is the distance between the charges and ε_0 is a constant. The *potential* V at a point P due to the charge q is defined to be the work expended in bringing a unit charge to P from infinity along the straight line that joins q and P. Find a formula for V.

81. Use the substitution $u = 1/x$ to show that

$$\int_0^\infty \frac{\ln x}{1 + x^2}\, dx = 0$$

Applications Plus

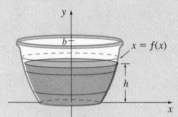

1. A *clepsydra,* or water clock, is a glass container with a small hole in the bottom through which water can flow. The "clock" is calibrated for measuring time by placing markings on the container corresponding to water levels at equally spaced times. Let $x = f(y)$ be continuous on the interval $[0, b]$ and assume that the container is formed by rotating the graph of f about the y-axis. Let V denote the volume of water and h the height of the water level at time t.
   (a) Determine V as a function of h.
   (b) Show that

$$\frac{dV}{dt} = \pi[f(h)]^2 \frac{dh}{dt}$$

   (c) Suppose that A is the area of the hole in the bottom of the container. It follows from Torricelli's Law that the rate of change of the volume of the water is given by

$$\frac{dV}{dt} = kA\sqrt{h}$$

   where k is a negative constant. Determine a formula for the function f such that dh/dt is a constant C. What is the advantage in having $dh/dt = C$?

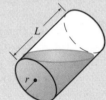

2. A cylindrical glass of radius r and height L is filled with water and then tilted until the water remaining in the glass exactly covers its base.
   (a) Determine a way to "slice" the water into parallel rectangular cross-sections and then *set up* a definite integral for the volume of the water in the glass.
   (b) Determine a way to "slice" the water into parallel cross-sections that are trapezoids and then *set up* a definite integral for the volume of the water.
   (c) Find the volume of water in the glass by evaluating one of the integrals in part (a) or part (b).
   (d) Find the volume of the water in the glass from purely geometric considerations.
   (e) Suppose the glass is tilted until the water exactly covers half the base. In what direction can you "slice" the water into triangular cross-sections? Rectangular cross-sections? Cross-sections that are segments of circles? Find the volume of water in the glass.

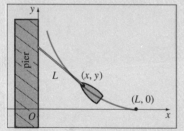

3. A man initially at the point O walks along a pier pulling a row boat by a rope of length L. The man keeps the rope straight and taut. The path followed by the boat is a curve called a *tractrix* and it has the property that the rope is always tangent to the curve (see the figure).
   (a) Show that if the path followed by the boat is the graph of the function $y = f(x)$, then

$$f'(x) = \frac{dy}{dx} = \frac{-\sqrt{L^2 - x^2}}{x}$$

   (b) Determine the function $y = f(x)$.

4. (a) Show that the volume of a segment of height h of a sphere of radius r is

$$V = \tfrac{1}{3}\pi h^2(3r - h)$$

   (b) Show that if a sphere of radius 1 is sliced by a plane at a distance x from the center in such a way that the volume of one segment is twice the volume of the other, then x is a solution of the equation

$$3x^3 - 9x + 2 = 0$$

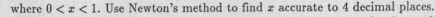

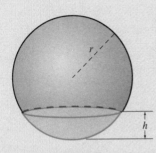

where $0 < x < 1$. Use Newton's method to find x accurate to 4 decimal places.

(c) Using the formula for the volume of a segment of a sphere, it can be shown that the depth x to which a floating sphere of radius r sinks in water is a root of the equation

$$x^3 - 3rx^2 + 4r^3 s = 0$$

where s is the specific gravity of the sphere. Suppose a wooden sphere of radius 0.5 m has specific gravity 0.75. Calculate, to 4 decimal place accuracy, the depth to which the sphere will sink.

(d) A hemispherical bowl has radius 5 in. and water is running into the bowl at the rate of $0.2 \text{ in}^3/\text{s}$.

   (i) How fast is the water level in the bowl rising at the instant the water is 3 in. deep?

   (ii) At a certain instant, the water is 4 in. deep. How long will it take to fill the bowl?

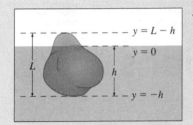

5. Archimedes' Principle states that the buoyant force on an object submerged, or partly submerged, in a fluid is equal to the weight of the fluid the object displaces. Thus, for an object of density ρ_0 floating partly submerged in a fluid of density ρ_f, the buoyant force is given by $F = \rho_f g \int_{-h}^{0} A(y)\,dy$, where g is the acceleration due to gravity and $A(y)$ is the area of a typical cross-section of the object. The weight of the object is given by

$$W = \rho_0 g \int_{-h}^{L-h} A(y)\,dy$$

(a) Show that the percentage of the volume of the object above the liquid is

$$100\,\frac{\rho_f - \rho_0}{\rho_f}$$

(b) The density of ice is 917 kg/m^3 and the density of sea water is 1030 kg/m^3. What percentage of the volume of an iceberg is above water?

(c) A sphere of radius 0.4 m and having negligible weight is floating in a large fresh-water lake. How much work is required to completely submerge the sphere? The density of the water is 1000 kg/m^3.

6. The momentum p of an object is the product of its mass m and its velocity v, that is, $p = mv$. Suppose an object, moving along a straight line, is acted on by a force $F = F(t)$ that is a continuous function of time.

(a) Show that the change in momentum over a time interval $[t_0, t_1]$ is equal to the integral of F from t_0 to t_1, that is, show that

$$p(t_1) - p(t_0) = \int_{t_0}^{t_1} F(t)\,dt$$

This integral is called the *impulse* of the force over time.

(b) A baseball pitcher throws a 90-mi/h fastball to a batter who hits a line drive directly back to the pitcher. Suppose the ball is in contact with the bat for 0.01 s and leaves the bat with velocity 110 mi/h. A baseball weighs 5 oz.

   (i) Determine the change in the ball's momentum.

   (ii) Determine the average force on the bat.

8

Further Applications of Integration

Mathematics takes us still further
from what is human, into the region
of absolute necessity, to which
not only the actual world, but every
possible world, must conform.

Bertrand Russell

We looked at some of the applications of integrals in Chapter 6: areas, volumes, work, and average values. Here we investigate some of many other applications of integration to geometry and physics—the length of a curve, the area of surface, hydrostatic force, centers of mass—as well as quantities of interest in chemistry, biology, and economics.

We start with what is perhaps the most important of all the applications of integration: differential equations. When a scientist uses calculus, more often than not it is to solve a differential equation that has arisen in the description of some physical process.

Differential Equations

A **differential equation** is an equation that contains an unknown function and some of its derivatives. Here are some examples:

(8.1)
$$y' = xy$$

(8.2)
$$y'' + 2y' + y = 0$$

(8.3)
$$\frac{d^3y}{dx^3} + x\frac{d^2y}{dx^2} + \frac{dy}{dx} - 2y = e^{-x}$$

In each of these differential equations y is an unknown function of x. The importance of differential equations lies in the fact that when a scientist or engineer formulates a physical law in mathematical terms, it frequently turns out to be a differential equation.

The **order** of a differential equation is the order of the highest derivative that occurs in the equation. Thus Equations 8.1, 8.2, and 8.3 are of order 1, 2, and 3, respectively.

A function f is called a **solution** of a differential equation if the equation is satisfied when $y = f(x)$ and its derivatives are substituted into the equation. Thus f is a solution of Equation 8.1 if

$$f'(x) = xf(x)$$

for all values of x in some interval.

You can easily verify that both $f(x) = \sin x$ and $g(x) = \cos x$ are solutions of the differential equation

(8.4)
$$y'' + y = 0$$

But when we are asked to *solve* a differential equation we are expected to find all possible solutions of the equation. In Section 15.5 we will show that any solution of Equation 8.4 is of the form

(8.5)
$$y = A \sin x + B \cos x$$

where A and B are constants. So (8.5) is called the **general solution** of the differential equation and particular solutions are obtained by substituting values for the arbitrary constants A and B.

We have already solved some particularly simple differential equations, namely, those of the form

$$y' = f(x)$$

For instance, we know that the general solution of the differential equation

$$y' = x^3$$

is given by
$$y = \frac{x^4}{4} + C$$

where C is an arbitrary constant.

But, in general, solving a differential equation is not an easy matter. There is no systematic technique that will enable us to solve all differential equations. In this section we learn how to solve a certain type of differential equation called a separable equation. Other types of equations will be discussed in Chapter 15.

A **separable equation** is a first-order differential equation that can be written in the form

$$\frac{dy}{dx} = g(x)f(y)$$

The name *separable* comes from the fact that the expression on the right side can be "separated" into a function of x and a function of y. Equivalently, we could write

(8.6)
$$\frac{dy}{dx} = \frac{g(x)}{h(y)}$$

To solve this equation we rewrite it in the differential form

$$h(y)\,dy = g(x)\,dx$$

so that all y's are on one side of the equation and all x's are on the other side. Then we integrate both sides of the equation:

(8.7)
$$\int h(y)\,dy = \int g(x)\,dx$$

Equation 8.7 defines y implicitly as a function of x. In some cases we may be able to solve for y in terms of x.

The justification for the step in Equation 8.7 comes from the Substitution Rule (5.33):

$$\int h(y)\, dy = \int h(y(x)) \frac{dy}{dx}\, dx$$

$$= \int h(y(x)) \frac{g(x)}{h(y(x))}\, dx \qquad \text{(from Equation 8.6)}$$

$$= \int g(x)\, dx$$

EXAMPLE 1   Solve the differential equation $\dfrac{dy}{dx} = \dfrac{6x^2}{2y + \cos y}$.

Solution   Writing the equation in differential form and integrating both sides, we have

$$(2y + \cos y)\, dy = 6x^2\, dx$$

$$\int (2y + \cos y)\, dy = \int 6x^2\, dx$$

(8.8)
$$y^2 + \sin y = 2x^3 + C$$

where C is an arbitrary constant. (We could have used a constant C_1 on the left side and another constant C_2 on the right side. But then we could combine these constants by writing $C = C_2 - C_1$.)

Equation 8.8 gives the general solution implicitly. In this case it is impossible to solve the equation to express y explicitly as a function of x. ●

EXAMPLE 2   Solve the equation $y' = x^2 y$.

Solution   First we rewrite the equation using Leibniz notation:

$$\frac{dy}{dx} = x^2 y$$

If $y \neq 0$, we can rewrite it in differential notation and integrate:

$$\frac{dy}{y} = x^2\, dx \qquad y \neq 0$$

$$\int \frac{dy}{y} = \int x^2\, dx$$

$$\ln |y| = \frac{x^3}{3} + C$$

This defines y implicitly as a function of x. But in this case we can solve explicitly for y as follows:

$$|y| = e^{\ln |y|} = e^{(x^3/3)+C} = e^C e^{x^3/3}$$

$$y = \pm e^C e^{x^3/3}$$

We note that the function $y = 0$ is also a solution of the given differential equation. So we can write the general solution in the form

$$y = Ae^{x^3/3}$$

where A is an arbitrary constant ($A = e^C$ or $A = -e^C$ or $A = 0$).   •

In Examples 1 and 2 we found the general solution of the given differential equation. But in many physical problems it is required to find the particular solution that satisfies a condition of the form $y(x_0) = y_0$. This is called an **initial condition,** and the problem of finding a solution of the differential equation that satisfies the initial condition is called an **initial-value problem.**

EXAMPLE 3   Solve the initial-value problem $xy' = -y$, $x > 0$, $y(4) = 2$.

Solution   We write the differential equation as

$$x\frac{dy}{dx} = -y \qquad \text{or} \qquad \frac{dy}{y} = -\frac{dx}{x}$$

Therefore

$$\int \frac{dy}{y} = -\int \frac{dx}{x}$$

$$\ln|y| = -\ln|x| + C$$

$$|y| = \frac{1}{|x|}e^C$$

$$y = \frac{K}{x}$$

where $K = \pm e^C$ is a constant. To determine K we put $x = 4$ and $y = 2$ in this equation:

$$2 = \frac{K}{4} \qquad K = 8$$

The solution of the initial-value problem is

$$y = \frac{8}{x} \qquad x > 0$$

Figure 8.1 shows the family of solutions $xy = K$ for several values of K (equilateral hyperbolas) and, in particular, the solution that satisfies $y(4) = 2$ [the hyperbola that passes through the point $(4, 2)$].

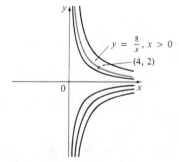

Figure 8.1

EXAMPLE 4   Find the solution of $\dfrac{dy}{dx} = \dfrac{6x^2}{2y + \cos y}$ that satisfies $y(1) = \pi$.

Solution   From Example 1 we know that the general solution is

$$y^2 + \sin y = 2x^3 + C$$

We are given that $y(1) = \pi$, so we substitute $x = 1$ and $y = \pi$ in this equation:

$$\pi^2 + \sin \pi = 2(1)^3 + C$$
$$C = \pi^2 - 2$$

Therefore the solution is given implicitly by

$$y^2 + \sin y = 2x^3 + \pi^2 - 2$$

EXAMPLE 5   Solve $y' = 1 + y^2 - 2x - 2xy^2$, $y(0) = 0$.

Solution   At first glance this does not look like a separable equation, but notice that it is possible to factor the right side as the product of a function of x and a function of y as follows:

$$\frac{dy}{dx} = 1 + y^2 - 2x(1 + y^2) = (1 - 2x)(1 + y^2)$$
$$\int \frac{dy}{1 + y^2} = \int (1 - 2x)\,dx$$
$$\tan^{-1}y = x - x^2 + C$$

Putting $x = 0$ and $y = 0$, we get $C = \tan^{-1} 0 = 0$, so

$$\tan^{-1}y = x - x^2$$

and solving for y we have

$$y = \tan(x - x^2)$$

EXAMPLE 6   A tank contains $20\,\text{kg}$ of salt dissolved in $5000\,\text{L}$ of water. Brine that contains $0.03\,\text{kg}$ of salt per liter of water enters the tank at a rate of $25\,\text{L/min}$. The solution is kept thoroughly mixed and drains from the tank at the same rate. How much salt remains in the tank after half an hour?

Solution   Let $y(t)$ be the amount of salt (in kilograms) after t minutes. We are given that $y(0) = 20$ and we want to find $y(30)$. We do this by finding a differential equation satisfied by $y(t)$. Note that dy/dt is the rate of change of the amount of salt, so

(8.9)
$$\frac{dy}{dt} = (\text{rate in}) - (\text{rate out})$$

where (rate in) is the rate at which salt enters the tank and (rate out) is the rate at which salt leaves the tank. We have

$$\text{rate in} = \left(0.03\frac{\text{kg}}{\text{L}}\right)\left(25\frac{\text{L}}{\text{min}}\right) = 0.75\frac{\text{kg}}{\text{min}}$$

The tank always contains $5000\,\text{L}$ of liquid, so the concentration at time t is $y(t)/5000$ (measured in kilograms per liter). Since the brine flows out at a rate of $25\,\text{L/min}$, we have

$$\text{rate out} = \left(\frac{y(t)\;\text{kg}}{5000\;\text{L}}\right)\left(25\,\frac{\text{L}}{\text{min}}\right) = \frac{y(t)}{200}\frac{\text{kg}}{\text{min}}$$

Thus, from Equation 8.9, we get

$$\frac{dy}{dt} = 0.75 - \frac{y(t)}{200}$$
$$= \frac{150 - y(t)}{200}$$

Solving this separable differential equation, we obtain

$$\int \frac{dy}{150 - y} = \int \frac{dt}{200}$$
$$-\ln|150 - y| = \frac{t}{200} + C$$

Since $y(0) = 20$, we have $-\ln 130 = C$, so

$$-\ln|150 - y| = \frac{t}{200} - \ln 130$$

Therefore $\qquad\qquad |150 - y| = 130e^{-t/200}$

Since $y(t)$ is continuous and $y(0) = 20$ and the right side is never 0, we deduce that $150 - y(t)$ is always positive. Thus $|150 - y| = 150 - y$ and

$$y(t) = 150 - 130e^{-t/200}$$

The amount of salt after 30 min is

$$y(30) = 150 - 130e^{-30/200} \approx 38.1\,\text{kg}$$

EXAMPLE 7   Solve the equation $\dfrac{dy}{dt} = ky$.

Solution   This differential equation was studied in Section 3.6, where it was called the law of natural growth (or decay). Since it is a separable equation we can solve it by the methods of this section as follows:

$$\int \frac{dy}{y} = \int k\,dt \qquad y \neq 0$$
$$\ln|y| = kt + C$$
$$|y| = e^{kt+C} = e^C e^{kt}$$
$$y = Ae^{kt}$$

Where $A\ (= \pm e^C$ or $0)$ is an arbitrary constant.

Logistic Growth

The differential equation of Example 7 is appropriate for modeling population growth ($y' = ky$ says that the rate of growth is proportional to the size of the population) under conditions of unlimited environment and food supply. However, in a restricted environment and with limited food supply, the population cannot exceed a maximal size M at which it consumes its entire food supply. If we make the assumption that the rate of growth of population is jointly proportional to the size of the population (y) and the amount by which y falls short of the maximal size ($M - y$), then we have the equation

(8.10)

$$\frac{dy}{dt} = ky(M - y)$$

where k is a constant. Equation 8.10 is called the **logistic differential equation** and was used by the Dutch mathematical biologist Verhulst in the 1840s to model world population growth.

The logistic equation is separable, so we write it in the form

$$\int \frac{dy}{y(M - y)} = \int k\, dt$$

Using partial fractions, we have

$$\frac{1}{y(M - y)} = \frac{1}{M}\left[\frac{1}{y} + \frac{1}{M - y}\right]$$

and so

$$\frac{1}{M}\left[\int \frac{dy}{y} + \int \frac{dy}{M - y}\right] = \int k\, dt = kt + C$$

$$\frac{1}{M}(\ln|y| - \ln|M - y|) = kt + C$$

But $|y| = y$ and $|M - y| = M - y$ since $0 < y < M$, so we have

$$\ln\frac{y}{M - y} = M(kt + C)$$

$$\frac{y}{M - y} = Ae^{kMt} \qquad (A = e^{MC})$$

If the population at time $t = 0$ is $y(0) = y_0$, then $A = y_0/(M - y_0)$, so

$$\frac{y}{M - y} = \frac{y_0}{M - y_0}e^{kMt}$$

If we solve this equation for y, we get

$$y = \frac{y_0 M e^{kMt}}{M - y_0 + y_0 e^{kMt}} = \frac{y_0 M}{y_0 + (M - y_0)e^{-kMt}}$$

Using the latter expression for y, we see that

$$\lim_{t\to\infty} y(t) = M$$

which is to be expected.

The graph of the logistic growth function is shown in Figure 8.2. At first the graph is concave upward and the growth curve appears to be almost exponential, but then it becomes concave downward and approaches the limiting population M.

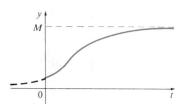

Figure 8.2
Logistic growth function

SECTION 8.1 **Exercises**

Solve the differential equations in Exercises 1–10.

1. $\dfrac{dy}{dx} = y^2$

2. $\dfrac{dy}{dx} = \dfrac{x + \sin x}{3y^2}$

3. $yy' = x$

4. $y' = xy$

5. $x^2 y' + y = 0$

6. $e^{-y} y' + \cos x = 0$

7. $\dfrac{dy}{dx} = \dfrac{x\sqrt{x^2 + 1}}{ye^y}$

8. $y' = \dfrac{\ln x}{xy + xy^3}$

9. $\dfrac{du}{dt} = e^{u + 2t}$

10. $\dfrac{dx}{dt} = 1 + t - x - tx$

In Exercises 11–20 find the solution to the differential equation that satisfies the given initial condition.

11. $\dfrac{dy}{dx} = y^2 + 1, \quad y(1) = 0$

12. $xy' = \sqrt{1 - y^2}, \quad x > 0, \quad y(1) = 0$

13. $e^y y' = \dfrac{3x^2}{1 + y}, \quad y(2) = 0$

14. $\dfrac{dy}{dx} = \dfrac{1 + x}{xy}, \quad x > 0, \quad y(1) = -4$

15. $\dfrac{dy}{dx} = \dfrac{\sin x}{\sin y}, \quad y(0) = \dfrac{\pi}{2}$

16. $\dfrac{dy}{dx} = e^{x - y}, \quad y(0) = 1$

17. $xe^{-t}\dfrac{dx}{dt} = t, \quad x(0) = 1$

18. $x\, dx + 2y\sqrt{x^2 + 1}\, dy = 0, \quad y(0) = 1$

19. $\dfrac{du}{dt} = \dfrac{2t + 1}{2(u - 1)}, \quad u(0) = -1$

20. $\dfrac{dy}{dt} = \dfrac{ty + 3t}{t^2 + 1}, \quad y(2) = 2$

21. Find a function f such that $f'(x) = x^3 f(x)$ and $f(0) = 1$.

22. Find a function g such that $g'(x) = g(x)(1 + g(x))$ and $g(0) = 1$.

23. Find the equation of the curve that satisfies $dy/dx = 4x^3 y$ and whose y-intercept is 7.

24. Find the equation of the curve that passes through the point $(1, 1)$ and whose slope at (x, y) is y^2/x^3.

25. A tank contains 1000 L of brine with 15 kg of dissolved salt. Pure water enters the tank at a rate of 10 L/min. The solution is kept thoroughly mixed and drains from the tank at the same rate. How much salt is in the tank (a) after t minutes and (b) after 20 min?

26. A tank contains 1000 L of pure water. Brine that contains 0.05 kg of salt per liter of water enters the tank at a rate of 5 L/min. Brine that contains 0.04 kg of salt per liter of water enters the tank at a rate of 10 L/min. The solution is kept thoroughly mixed and drains from the tank at a rate of 15 L/min. How much salt is in the tank (a) after t minutes and (b) after an hour?

27. In an elementary chemical reaction, single molecules of two reactants A and B form a molecule of

the product C: $A + B \rightarrow C$. The law of mass action states that the rate of reaction is proportional to the product of the concentrations of A and B:

$$\frac{d[C]}{dt} = k[A][B]$$

(See Example 4 in Section 2.3.) Thus, if the initial concentrations are $[A] = a$ moles/L and $[B] = b$ moles/L and we write $x = [C]$, then we have

$$\frac{dx}{dt} = k(a - x)(b - x)$$

Assuming that $a \neq b$, find x as a function of t.

28. (a) Find $x(t)$ in Exercise 27 assuming that $b = a$.
     (b) How does this expression for $x(t)$ simplify if it is known that $[C] = a/2$ after 20 seconds?

29. The population of the world was about 5 billion in 1986. Using the exponential model for population growth (see Example 7 or Section 3.6) with the recently observed rate of increase of population of 2% per year, find an expression for the population of the world in the year t. Use this model to predict the population of the world in the following years: (a) 2000, (b) 2100, and (c) 2500. The total land surface area of this planet is about 1.8×10^{15} ft^2. How many square feet of land per person will there be in the above years under the exponential model?

30. For a more realistic picture of long-term growth than in Exercise 29, we consider the logistic model for world population growth with $y_0 = 5$ billion in 1986 and an assumed maximum population of $M = 100$ billion. To determine the value of k, use Equation 8.10 and the fact that the population was increasing at a rate of 2% per year when the population was 5 billion. Use this model to predict the population of the world in the following years: (a) 2000, (b) 2100, and (c) 2500. Compare with the results of Exercise 29.

31. One model for the spread of a rumor is that the rate of spread is proportional to the product of the fraction y of the population who have heard the rumor and the fraction who have not heard the rumor.

     (a) Write a differential equation that is satisfied by y.
     (b) Solve the differential equation.
     (c) A small town has 1000 inhabitants. At 8 A.M. 80 people have heard a rumor. By noon half the town has heard it. At what time will 90% of the population have heard the rumor?

32. Biologists stocked a lake with 400 fish and estimated the carrying capacity (the maximum population size

for the fish of that species in that lake) to be 10,000. The number of fish tripled in the first year.
     (a) Assuming that the size of the fish population satisfies the logistic equation, find an expression for the size of the population after t years.
     (b) How long will it take for the population to increase to 5000?

33. Show that the solution to the logistic Equation 8.10 increases most rapidly when $y = M/2$.

34. Another model for a growth function for a limited population is given by the *Gompertz function*, which is a solution of the differential equation

$$\frac{dy}{dt} = c \ln\left(\frac{M}{y}\right) y$$

where c is a constant and M is the maximum size of the population.
     (a) Solve this differential equation.
     (b) Compute $\lim_{t \to \infty} y(t)$.
     (c) Sketch the graph of the Gompertz growth function.

35. In the simple electric circuit shown in the figure, a battery supplies a constant voltage V and produces a current of $I(t)$ amperes at time t. The circuit also contains a resistor with a resistance of R ohms and an inductor with an inductance of L henries. Ohm's Law gives the voltage drop across the resistor as RI; the voltage drop due to the inductor is $L(dI/dt)$. One of Kirchhoff's laws says that the sum of these voltage drops is the applied voltage V, so

$$RI + LI'(t) = V$$

If the switch is closed when $t = 0$, so that $I(0) = 0$, solve this equation to show that

$$I(t) = \frac{V}{R}(1 - e^{-Rt/L})$$

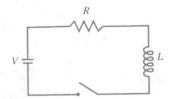

36. When a raindrop falls it increases in size, so its mass at time t is a function of t, $m(t)$. The rate of growth of the mass is $km(t)$ for some positive constant k. When we apply Newton's Law of Motion to the raindrop, we get $(mv)' = gm$, where v is the velocity of the raindrop (directed downward) and g is the acceleration due to gravity. The *terminal velocity* of the raindrop is $\lim_{t \to \infty} v(t)$. Find an expression for the terminal velocity in terms of g and k.

Arc Length

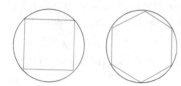

We all have an intuitive idea about what the length of a curve is. But, like the concepts of area and volume, the concept of the length of an arc of a curve requires a careful definition.

For the simple case where the curve is a finite line segment joining the point $P_1(x_1, y_1)$ to the point $P_2(x_2, y_2)$ we know that its length is given by the distance formula

$$|P_1 P_2| = \sqrt{(x_2 - x_1)^2 + (y_2 - y_1)^2}$$

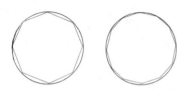

We can also compute the length of a polygon by adding the lengths of the line segments that form the polygon. We define the length of a curve by first approximating it by a polygon and then taking a limit. This process is familiar for the case of a circle where the circumference is the limit of inscribed polygons (see Figure 8.3).

Now suppose that a curve C is defined by the equation $y = f(x)$, where $a \leq x \leq b$. We obtain a polygonal approximation to C by taking a partition P of $[a, b]$ determined by points x_i with $a = x_0 < x_1 < \cdots < x_n = b$. If $y_i = f(x_i)$, then the point $P_i(x_i, y_i)$ lies on C and the polygon with vertices $P_0, P_1, \ldots, P_n$, illustrated in Figure 8.4, is an approximation to C. The length of this polygonal approximation is

Figure 8.3

$$\sum_{i=1}^{n} |P_{i-1} P_i|$$

and this approximation appears to become better as $||P|| \to 0$. (See Figure 8.5 where the arc of the curve between P_{i-1} and P_i has been magnified and approximations with successively smaller $||P||$ are shown.)

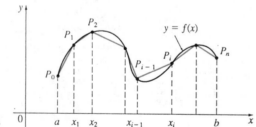

Figure 8.4

Therefore we define the **length** L of the curve C with equation $y = f(x)$, $a \leq x \leq b$, as the limit of the lengths of these inscribed polygons (if the limit exists):

(8.11)

$$L = \lim_{||P|| \to 0} \sum_{i=1}^{n} |P_{i-1} P_i|$$

Notice that the procedure for defining arc length is very similar to the procedure we used for defining area and volume. We divided the curve into a large number of small parts. We then found the approximate lengths of the small parts and added them. Finally we took the limit as $||P|| \to 0$.

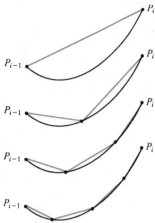

Figure 8.5

The definition of arc length given by Equation 8.11 is not very convenient for computational purposes, but we can derive an integral formula for L in the case where f has a continuous derivative. [Such a function f is called **smooth** because a small change in x produces a small change in $f'(x)$.]

If we let $\Delta y_i = y_i - y_{i-1}$, then

$$|P_{i-1}P_i| = \sqrt{(\Delta x_i)^2 + (\Delta y_i)^2}$$

By applying the Mean Value Theorem (4.10) to f on the interval $[x_{i-1}, x_i]$, we find that there is a number x_i^* between x_{i-1} and x_i such that

$$f(x_i) - f(x_{i-1}) = f'(x_i^*)(x_i - x_{i-1})$$

that is,
$$\Delta y_i = f'(x_i^*)\,\Delta x_i$$

Thus we have

$$|P_{i-1}P_i| = \sqrt{(\Delta x_i)^2 + (\Delta y_i)^2}$$
$$= \sqrt{(\Delta x_i)^2 + [f'(x_i^*)\,\Delta x_i]^2}$$
$$= \sqrt{1 + [f'(x_i^*)]^2}\,\Delta x_i$$

Therefore, by Definition 8.11,

$$L = \lim_{\|P\| \to 0} \sum_{i=1}^{n} |P_{i-1}P_i|$$
$$= \lim_{\|P\| \to 0} \sum_{i=1}^{n} \sqrt{1 + [f'(x_i^*)]^2}\,\Delta x_i$$

We recognize this expression as being equal to

$$\int_a^b \sqrt{1 + [f'(x)]^2}\,dx$$

by the definition of a definite integral. This integral exists because the function $g(x) = \sqrt{1 + [f'(x)]^2}$ is continuous. Thus we have proved the following theorem:

The Arc Length Formula (8.12)

If f' is continuous on $[a, b]$, then the length of the curve $y = f(x)$, $a \le x \le b$, is

$$L = \int_a^b \sqrt{1 + [f'(x)]^2}\,dx$$

If we use the Leibniz notation for derivatives, we can write the arc length formula as follows:

(8.13)

$$L = \int_a^b \sqrt{1 + \left(\frac{dy}{dx}\right)^2}\,dx$$

EXAMPLE 1 Find the length of the arc of the semicubical parabola $y^2 = x^3$ between the points $(1, 1)$ and $(4, 8)$ (see Figure 8.6).

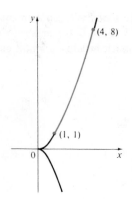

Figure 8.6

Or use Formula 21
in the Table of Integrals

Solution   For the top half of the curve we have

$$y = x^{3/2} \qquad \frac{dy}{dx} = \frac{3}{2}x^{1/2}$$

and so the arc length formula gives

$$L = \int_1^4 \sqrt{1 + \left(\frac{dy}{dx}\right)^2}\, dx = \int_1^4 \sqrt{1 + \frac{9}{4}x}\, dx$$

If we substitute $u = 1 + 9x/4$, then $du = 9\, dx/4$. When $x = 1$, $u = \frac{13}{4}$; when $x = 4$, $u = 10$. Therefore

$$L = \frac{4}{9}\int_{13/4}^{10} \sqrt{u}\, du = \frac{4}{9} \cdot \frac{2}{3}\, u^{3/2}\Big|_{13/4}^{10}$$

$$= \frac{8}{27}\left[10^{3/2} - \left(\frac{13}{4}\right)^{3/2}\right]$$

$$= \frac{80\sqrt{10} - 13\sqrt{13}}{27}$$

If a curve has the equation $x = g(y)$, $c \leq y \leq d$, then by interchanging the roles of x and y in Theorem 8.12 or Equation 8.13, we obtain the following formula for its length:

(8.14)

$$L = \int_c^d \sqrt{1 + [g'(y)]^2}\, dy = \int_c^d \sqrt{1 + \left(\frac{dx}{dy}\right)^2}\, dy$$

EXAMPLE 2   Find the length of the arc of the parabola $y^2 = x$ from $(0,0)$ to $(1,1)$.

Solution   Since $x = y^2$, $dx/dy = 2y$, and Formula 8.14 gives

$$L = \int_0^1 \sqrt{1 + \left(\frac{dx}{dy}\right)^2}\, dy = \int_0^1 \sqrt{1 + 4y^2}\, dy$$

We make the trigonometric substitution $y = \frac{1}{2}\tan\theta$, which gives $dy = \frac{1}{2}\sec^2\theta\, d\theta$ and $\sqrt{1 + 4y^2} = \sqrt{1 + \tan^2\theta} = \sec\theta$. When $y = 0$, $\tan\theta = 0$, so $\theta = 0$; when $y = 1$, $\tan\theta = 2$, so $\theta = \tan^{-1}2 = \alpha$, say. Thus

$$L = \int_0^\alpha \sec\theta \cdot \frac{1}{2}\sec^2\theta\, d\theta = \frac{1}{2}\int_0^\alpha \sec^3\theta\, d\theta$$

$$= \frac{1}{2} \cdot \frac{1}{2}\big[\sec\theta\tan\theta + \ln|\sec\theta + \tan\theta|\big]_0^\alpha \qquad \text{(from Example 9 in Section 7.2)}$$

$$= \frac{1}{4}(\sec\alpha\tan\alpha + \ln|\sec\alpha + \tan\alpha|)$$

Since $\tan\alpha = 2$, we have $\sec^2\alpha = 1 + \tan^2\alpha = 5$, so $\sec\alpha = \sqrt{5}$ and

$$L = \frac{\sqrt{5}}{2} + \frac{\ln(\sqrt{5} + 2)}{4}$$

Because of the presence of the square root sign in Formulas 8.12 and 8.14, the calculation of an arc length often leads to an integral that is very difficult or even impossible to evaluate explicitly. Thus we sometimes have to be content with finding an approximation to the length of a curve as in the following example.

EXAMPLE 3

(a)   Set up an integral for the length of the arc of the hyperbola $xy = 1$ from the point $(1,1)$ to the point $(2,\frac{1}{2})$.

(b)   Use Simpson's Rule with $n = 10$ to estimate the arc length.

Solution

(a)   We have

$$y = \tfrac{1}{x} \qquad \frac{dy}{dx} = -\frac{1}{x^2}$$

and so the arc length is

$$L = \int_1^2 \sqrt{1 + \left(\frac{dy}{dx}\right)^2}\, dx = \int_1^2 \sqrt{1 + \frac{1}{x^4}}\, dx = \int_1^2 \frac{\sqrt{x^4 + 1}}{x^2}\, dx$$

(b)   Using Simpson's Rule (7.33) with $a = 1$, $b = 2$, $n = 10$, $\Delta x = 0.1$, and $f(x) = \sqrt{1 + 1/x^4}$, we have

$$L = \int_1^2 \sqrt{1 + \frac{1}{x^4}}\, dx$$

$$\approx \frac{\Delta x}{3}[f(1) + 4f(1.1) + 2f(1.2) + 4f(1.3) + \cdots + 2f(1.8) + 4f(1.9) + f(2)]$$

$$= \frac{0.1}{3}\left(\sqrt{1 + \frac{1}{1^4}} + 4\sqrt{1 + \frac{1}{(1.1)^4}} + 2\sqrt{1 + \frac{1}{(1.2)^4}} + 4\sqrt{1 + \frac{1}{(1.3)^4}}\right.$$

$$\left. + \cdots + 2\sqrt{1 + \frac{1}{(1.8)^4}} + 4\sqrt{1 + \frac{1}{(1.9)^4}} + \sqrt{1 + \frac{1}{2^4}}\right)$$

$$\approx 1.1321$$

The Arc Length Function

If a smooth curve C has the equation $y = f(x)$, $a \le x \le b$, let $s(x)$ be the distance along C from the initial point $P_0(a, f(a))$ to the point $Q(x, f(x))$. Then s is a function, called the **arc length function,** and, by Formula 8.12,

(8.15)
$$s(x) = \int_a^x \sqrt{1 + [f'(t)]^2}\, dt$$

(We have replaced the dummy variable of integration by t so that x does not have two meanings.) We use Part 1 of the Fundamental Theorem of Calculus to differentiate Equation 8.15 (since the integrand is continuous):

(8.16)
$$\frac{ds}{dx} = \sqrt{1 + [f'(x)]^2} = \sqrt{1 + \left(\frac{dy}{dx}\right)^2}$$

Equation 8.16 shows that the rate of change of s with respect to x is always at least 1 and is equal to 1 when $f'(x)$, the slope of the curve, is 0. The differential of arc length is

(8.17)
$$ds = \sqrt{1 + \left(\frac{dy}{dx}\right)^2}\, dx$$

and this equation is sometimes written in the symmetric form

(8.18)
$$(ds)^2 = (dx)^2 + (dy)^2$$

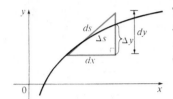

Figure 8.7

The geometric interpretation of Equation 8.18 is shown in Figure 8.7. It can be used as a mnemonic device for remembering both of the formulas 8.13 and 8.14. If we write $L = \int ds$, then from Equation 8.18 either we can solve to get (8.17), which gives (8.13), or we can solve to get

$$ds = \sqrt{1 + \left(\frac{dx}{dy}\right)^2}\, dy$$

which gives (8.14).

EXAMPLE 4   Find the arc length function for the curve $y = x^2 - (\ln x)/8$ taking $P_0(1,1)$ as the starting point.

Solution   If $f(x) = x^2 - (\ln x)/8$, then

$$f'(x) = 2x - \frac{1}{8x}$$

$$1 + [f'(x)]^2 = 1 + \left(2x - \frac{1}{8x}\right)^2 = 1 + 4x^2 - \frac{1}{2} + \frac{1}{64x^2}$$

$$= 4x^2 + \frac{1}{2} + \frac{1}{64x^2} = \left(2x + \frac{1}{8x}\right)^2$$

$$\sqrt{1 + [f'(x)]^2} = 2x + \frac{1}{8x}$$

Thus the arc length function is given by

$$s(x) = \int_1^x \sqrt{1 + [f'(t)]^2}\, dt$$

$$= \int_1^x \left(2t + \frac{1}{8t}\right) dt = t^2 + \frac{1}{8}\ln t \,\Big|_1^x$$

$$= x^2 + \frac{1}{8}\ln x - 1$$

SECTION 8.2   **Exercises**

1. Use the arc length formula (8.13) to find the length of the curve $y = 2x + 1$, $-1 \le x \le 3$. Check your answer by noting that the curve is a line segment and calculating its length by the distance formula.

2. Find the length of the line segment joining the points $(-1, 1)$ and $(2, 2)$ using (a) the formula for the distance between two points, (b) Formula 8.13, and (c) Formula 8.14.

3. Use Formula 8.13 to find the length of the arc of the curve $x^2 = 64y^3$ from $(8,1)$ to $(64,4)$.

4. Use Formula 8.14 to find the arc length in Exercise 3.

In Exercises 5–8 find the length of the arc of the given curve from point A to point B.

5. $y^2 = (x-1)^3$, $A(1,0)$, $B(2,1)$

6. $y = 1 - x^{2/3}$, $A(-8, -3)$, $B(-1,0)$

7. $12xy = 4y^4 + 3$, $A(\frac{7}{12}, 1)$, $B(\frac{67}{24}, 2)$

8. $9y^2 = x(x-3)^2$, $A(0,0)$, $B(4, \frac{2}{3})$

In Exercises 9–20 find the length of the given curve.

9. $y = \frac{1}{3}(x^2 + 2)^{3/2}$, $0 \le x \le 1$

10. $y = \frac{x^3}{6} + \frac{1}{2x}$, $1 \le x \le 2$

11. $y = \frac{x^4}{4} + \frac{1}{8x^2}$, $1 \le x \le 3$

12. $y = \frac{x^2}{2} - \frac{\ln x}{4}$, $2 \le x \le 4$

13. $y = \ln(\cos x)$, $0 \le x \le \pi/4$

14. $y = \ln(\sin x)$, $\pi/6 \le x \le \pi/3$

15. $y = \ln(1 - x^2)$, $0 \le x \le \frac{1}{2}$

16. $y = \ln\left(\dfrac{e^x + 1}{e^x - 1}\right)$, $a \le x \le b$, $a > 0$

17. $y = e^x$, $0 \le x \le 1$ 18. $y = \ln x$, $1 \le x \le \sqrt{3}$

19. $y = \cosh x$, $0 \le x \le 1$ 20. $y^2 = 4x$, $0 \le y \le 2$

In Exercises 21–26 set up, but do not evaluate, an integral for the length of the given curve.

21. $y = x^3$, $0 \le x \le 1$

22. $y = x^4 - x^2$, $-1 \le x \le 2$

23. $y = \sin x$, $0 \le x \le \pi$

24. $y = \tan x$, $0 \le x \le \pi/4$

25. $y = e^x \cos x$, $0 \le x \le \pi/2$

26. $\dfrac{x^2}{a^2} + \dfrac{y^2}{b^2} = 1$

27. Find the arc length function for the curve $y = 2x^{3/2}$ with starting point $P_0(1,2)$.

28. (a) Sketch the curve $y = x^3/3 + 1/(4x)$.
    (b) Find the arc length function for this curve with starting point $P_0(1, \frac{7}{12})$.

29. If a bomb is dropped from an aircraft flying at 200 m/s at an altitude of 4500 m, then the parabolic trajectory of the bomb is described by the equation
$$y = 4500 - \frac{x^2}{8000}$$
until it hits the ground, where y is its height above the ground and x is the horizontal distance traveled in meters. Calculate the distance traveled by the bomb from the time it is dropped to the time it hits the ground. Express your answer correct to the nearest meter.

30. The figure shows a telephone wire hanging between two poles at $x = -b$ and $x = b$. It takes the shape of a catenary with equation $y = a \cosh(x/a)$. Find the length of the wire.

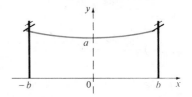

31. Sketch the curve with equation $x^{2/3} + y^{2/3} = 1$ and use symmetry to find its length.

32. (a) Sketch curve $y^3 = x^2$.
    (b) Use Formulas 8.13 and 8.14 to set up two integrals for the arc length from $(0,0)$ to $(1,1)$. Observe that one of these is an improper integral and evaluate both of them.
    (c) Find the length of the arc of this curve from $(-1,1)$ to $(8,4)$.

In Exercises 33–36 use Simpson's Rule with $n = 10$ to estimate the arc length of the given curve.

33. $y = x^3$, $0 \le x \le 1$ 34. $y = x^4$, $0 \le x \le 2$

35. $y = \sin x$, $0 \le x \le \pi$ 36. $y = \tan x$, $0 \le x \le \pi/4$

37. Find the length of the curve $y = \int_1^x \sqrt{t^3 - 1}\, dt$, $1 \le x \le 4$.

38. (a) Evaluate $\int_0^1 \sqrt{1 + 4x^2}\, dx$.
    (b) Identify the integral in part (a) as the length of a curve.
    (c) Without using a calculator, show that
$$\sqrt{2} < \frac{\sqrt{5}}{2} + \frac{\ln(2 + \sqrt{5})}{4}$$

SECTION 8.3

Area of a Surface of Revolution

A surface of revolution is formed when a curve is rotated about a line. Such a surface is the lateral boundary of a solid of a revolution of the type discussed in Sections 6.2 and 6.3.

We want to define the area of a surface of revolution in such a way that it corresponds to our intuition. We can think of peeling away a very thin outer layer of the solid of revolution and laying it out flat so that we can measure its area. Or, if the surface area is A, we can imagine that painting the surface would require the same amount of paint as a flat region with area A.

Let us start with some simple surfaces. The lateral surface area of a circular cylinder with radius r and height h is taken to be $A = 2\pi rh$ because we can imagine cutting the cylinder and unrolling it (as in Figure 8.8) to obtain a rectangle with dimensions $2\pi r$ and h.

Likewise we can take a circular cone with base radius r and slant height l, cut it along the broken line in Figure 8.9, and flatten it to form a sector of a circle with radius l and central angle $\theta = 2\pi r/l$. We know that, in general, the area of a sector of a circle with radius l and angle θ is $\frac{1}{2}l^2\theta$ (see Exercise 33 in Section 7.3) and so in this case it is

$$A = \tfrac{1}{2}l^2\theta = \tfrac{1}{2}l^2\left(\frac{2\pi r}{l}\right) = \pi rl$$

Therefore we define the lateral surface area of a cone to be $A = \pi rl$.

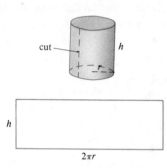

cut — h

h

$2\pi r$

Figure 8.8

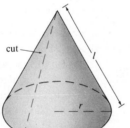

cut —

$2\pi r$

θ l

l

r

Figure 8.9

The area of the band (or frustum of a cone) shown in Figure 8.10 with slant height l and upper and lower radii r_1 and r_2 is found by subtracting the areas of two cones:

(8.19)
$$A = \pi r_2(l_1 + l) - \pi r_1 l_1 = \pi[(r_2 - r_1)l_1 + r_2 l]$$

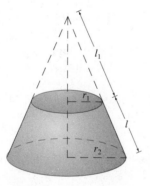

l_1

r_1

l

r_2

Figure 8.10

From similar triangles we have

$$\frac{l_1}{r_1} = \frac{l_1 + l}{r_2}$$

which gives

$$r_2 l_1 = r_1 l_1 + r_1 l \qquad \text{or} \qquad (r_2 - r_1)l_1 = r_1 l$$

Putting this in Equation 8.19, we get

$$A = \pi(r_1 l + r_2 l)$$

(8.20)   or

$$\boxed{A = 2\pi r l}$$

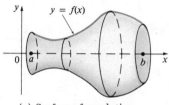

(a) Surface of revolution

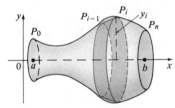

(b) Approximating band

Figure 8.11

where $r = \frac{1}{2}(r_1 + r_2)$ is the average radius of the band.

Now we consider the surface shown in Figure 8.11, which is obtained by rotating the curve $y = f(x)$, $a \le x \le b$, about the x-axis, where f is positive and has a continuous derivative. In order to define its surface area we use a method similar to the one for arc length. We take a partition P of $[a,b]$ by points $a = x_0, x_1, \ldots, x_n = b$, and let $y_i = f(x_i)$ so that the point $P_i(x_i, y_i)$ lies on the curve. The part of the surface between x_{i-1} and x_i is approximated by taking the line segment $P_{i-1}P_i$ and rotating it about the x-axis. The result is a band (a frustum of a cone) with slant height $l = |P_{i-1}P_i|$ and average radius $r = \frac{1}{2}(y_{i-1} + y_i)$ so, by Formula 8.20, its surface area is

$$2\pi \frac{y_{i-1} + y_i}{2} |P_{i-1}P_i|$$

As in the proof of Theorem 8.12, we have

$$|P_{i-1}P_i| = \sqrt{1 + [f'(x_i^*)]^2}\, \Delta x_i$$

where $x_i^* \in [x_{i-1}, x_i]$. When Δx_i is small we have $y_i = f(x_i) \approx f(x_i^*)$ and also $y_{i-1} = f(x_{i-1}) \approx f(x_i^*)$ since f is continuous. Therefore

$$2\pi \frac{y_{i-1} + y_i}{2} |P_{i-1}P_i| \approx 2\pi f(x_i^*) \sqrt{1 + [f'(x_i^*)]^2}\, \Delta x_i$$

and so an approximation to what we think of as the area of the complete surface of revolution is

(8.21)
$$\sum_{i=1}^{n} 2\pi f(x_i^*) \sqrt{1 + [f'(x_i^*)]^2}\, \Delta x_i$$

This approximation appears to become better as $||P|| \to 0$ and, recognizing (8.21) as a Riemann sum for the function $g(x) = 2\pi f(x) \sqrt{1 + [f'(x)]^2}$, we have

$$\lim_{||P|| \to 0} \sum_{i=1}^{n} 2\pi f(x_i^*) \sqrt{1 + [f'(x_i^*)]^2}\, \Delta x_i = \int_a^b 2\pi f(x) \sqrt{1 + [f'(x)]^2}\, dx$$

Therefore, in the case where f is positive and has a continuous derivative, we define the **surface area** of the surface obtained by rotating the curve $y = f(x)$, $a \le x \le b$, about the x-axis as

(8.22)
$$S = \int_a^b 2\pi f(x)\sqrt{1 + [f'(x)]^2}\, dx$$

With the Leibniz notation for derivatives, this formula becomes

(8.23)
$$S = \int_a^b 2\pi y\sqrt{1 + \left(\frac{dy}{dx}\right)^2}\, dx$$

If the curve is described as $x = g(y)$, $c \le y \le d$, then the formula for surface area becomes

(8.24)
$$S = \int_c^d 2\pi y\sqrt{1 + \left(\frac{dx}{dy}\right)^2}\, dy$$

and both Formula 8.23 and Formula 8.24 can be summarized symbolically, using the notation for arc length given in Section 8.2, as

(8.25)
$$S = \int 2\pi y\, ds$$

For rotation about the y-axis, the surface area formula becomes

(8.26)
$$S = \int 2\pi x\, ds$$

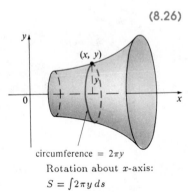

circumference $= 2\pi y$

Rotation about x-axis:

$S = \int 2\pi y\, ds$

circumference

Rotation about y-axis:

$S = \int 2\pi x\, ds$

Figure 8.12

where, as before, we can use either

$$ds = \sqrt{1 + \left(\frac{dy}{dx}\right)^2}\, dx \qquad \text{or} \qquad ds = \sqrt{1 + \left(\frac{dx}{dy}\right)^2}\, dy$$

These formulas can be remembered by thinking of $2\pi y$ or $2\pi x$ as the circumference of a circle traced out by the point (x, y) on the curve as it is rotated about the x-axis or the y-axis, respectively (see Figure 8.12).

EXAMPLE 1   The curve $y = \sqrt{4 - x^2}$, $-1 \le x \le 1$, is an arc of the circle $x^2 + y^2 = 4$. Find the area of the surface obtained by rotating this arc about the x-axis. (The surface is a portion of a sphere of radius 2.)

Solution   We have

$$\frac{dy}{dx} = \tfrac{1}{2}(4 - x^2)^{-1/2}(-2x) = \frac{x}{\sqrt{4 - x^2}}$$

and so, by Formula 8.21, the surface area is

$$S = \int_{-1}^1 2\pi y\sqrt{1 + \left(\frac{dy}{dx}\right)^2}\, dx$$

$$= 2\pi \int_{-1}^1 \sqrt{4 - x^2}\,\sqrt{1 + \frac{x^2}{4 - x^2}}\, dx$$

$$= 2\pi \int_{-1}^{1} \sqrt{4 - x^2} \; \frac{2}{\sqrt{4 - x^2}} \, dx$$

$$= 4\pi \int_{-1}^{1} 1 \, dx = 4\pi(2) = 8\pi \qquad \bullet$$

EXAMPLE 2   The arc of the parabola $y = x^2$ from $(1,1)$ to $(2,4)$ is rotated about the y-axis. Find the area of the resulting surface.

Solution 1   Using

$$y = x^2 \qquad \text{and} \qquad \frac{dy}{dx} = 2x$$

we have, from Formula 8.26,

$$S = \int 2\pi x \, ds = \int_{1}^{2} 2\pi x \sqrt{1 + \left(\frac{dy}{dx}\right)^2} \, dx$$

$$= 2\pi \int_{1}^{2} x\sqrt{1 + 4x^2} \, dx$$

Substituting $u = 1 + 4x^2$, we have $du = 8x \, dx$. Remembering to change the limits of integration, we have

$$S = \frac{\pi}{4} \int_{5}^{17} \sqrt{u} \, du = \frac{\pi}{4}\left[\frac{2}{3}u^{3/2}\right]_{5}^{17}$$

$$= \frac{\pi}{6}\left(17\sqrt{17} - 5\sqrt{5}\right)$$

Solution 2   Using

$$x = \sqrt{y} \qquad \text{and} \qquad \frac{dx}{dy} = \frac{1}{2\sqrt{y}}$$

we have

$$S = \int 2\pi x \, ds = \int_{1}^{4} 2\pi x \sqrt{1 + \left(\frac{dx}{dy}\right)^2} \, dy$$

$$= 2\pi \int_{1}^{4} \sqrt{y}\sqrt{1 + \frac{1}{4y}} \, dy$$

$$= \pi \int_{1}^{4} \sqrt{4y + 1} \, dy$$

$$= \frac{\pi}{4} \int_{5}^{17} \sqrt{u} \, du \qquad \text{(where } u = 1 + 4y\text{)}$$

$$= \frac{\pi}{6}\left(17\sqrt{17} - 5\sqrt{5}\right) \qquad \text{(as in Solution 1)} \qquad \bullet$$

EXAMPLE 3   Find the area of the surface generated by rotating the curve $y = e^x$, $0 \le x \le 1$, about the x-axis.

Solution 1   Using Formula 8.23 with

$$y = e^x \qquad \text{and} \qquad \frac{dy}{dx} = e^x$$

we have

$$S = \int_0^1 2\pi y \sqrt{1 + \left(\frac{dy}{dx}\right)^2}\, dx = 2\pi \int_0^1 e^x \sqrt{1 + e^{2x}}\, dx$$

$$= 2\pi \int_1^e \sqrt{1 + u^2}\, du \qquad \text{(where } u = e^x\text{)}$$

$$= 2\pi \int_{\pi/4}^\alpha \sec^3\theta\, d\theta \qquad \text{(where } u = \tan\theta \text{ and } \alpha = \tan^{-1}e\text{)}$$

Or use Formula 21
in the Table of Integrals

$$= 2\pi \cdot \tfrac{1}{2}\big[\sec\theta\tan\theta + \ln|\sec\theta + \tan\theta|\big]_{\pi/4}^\alpha \qquad \text{(by Example 9 in Section 7.2)}$$

$$= \pi[\sec\alpha\tan\alpha + \ln(\sec\alpha + \tan\alpha) - \sqrt{2} - \ln(\sqrt{2}+1)]$$

Since $\tan\alpha = e$, we have $\sec^2\alpha = 1 + \tan^2\alpha = 1 + e^2$ and

$$S = \pi\Big[e\sqrt{1+e^2} + \ln(e + \sqrt{1+e^2}) - \sqrt{2} - \ln(\sqrt{2}+1)\Big]$$

Solution 2   Using Formula 8.24 with

$$x = \ln y \qquad \text{and} \qquad \frac{dx}{dy} = \frac{1}{y}$$

we have

$$S = \int_1^e 2\pi y \sqrt{1 + \left(\frac{dx}{dy}\right)^2}\, dy$$

$$= 2\pi \int_1^e y\sqrt{1 + \frac{1}{y^2}}\, dy$$

$$= 2\pi \int_1^e \sqrt{y^2 + 1}\, dy$$

which is evaluated as in Solution 1.   ●

In Exercises 1–14 find the area of the surface obtained by rotating the given curve about the x-axis.

1. $y = \sqrt{x}, \quad 4 \le x \le 9$

2. $y^2 = 4x + 4, \quad 0 \le x \le 8$

3. $2y = x + 4, \quad 0 \le x \le 2$

4. $y = x^3, \quad 0 \le x \le 2$

5. $y = x^3 + \frac{1}{12x}, \quad 1 \le x \le 2$

6. $y = \frac{x^2}{4} - \frac{\ln x}{2}, \quad 1 \le x \le 4$

7. $y = \sin x, \quad 0 \le x \le \pi$

8. $y = \cos x, \quad 0 \le x \le \pi/3$

9. $y = \cosh x, \quad 0 \le x \le 1$

10. $2y = 3x^{2/3}, \quad 1 \le x \le 8$

11. $x = \frac{y^4}{2} + \frac{1}{16y^2}, \quad 1 \le y \le 3$

12. $x = \frac{1}{3}(y^2 + 2)^{3/2}, \quad 1 \le y \le 2$

13. $x = 1 + 2y^2, \quad 1 \le y \le 2$

14. $x = 2\ln y, \quad 1 \le y \le \sqrt{3}$

In Exercises 15–22 the given curve is rotated about the y-axis. Find the area of the resulting surface.

15. $y = \sqrt[3]{x}, \quad 1 \le y \le 2$

16. $x = \sqrt{2y - y^2}, \quad 0 \le y \le 1$

17. $y^2 = x^3, \quad 1 \le y \le 8$

18. $4x + 3y = 19, \quad 1 \le x \le 4$

19. $x = e^{2y}, \quad 0 \le y \le \frac{1}{2}$

20. $y = 1 - x^2, \quad 0 \le x \le 1$

21. $x = \dfrac{1}{2\sqrt{2}}(y^2 - \ln y), \quad 1 \le y \le 2$

22. $x = a\cosh(y/a), \quad -a \le y \le a$

In Exercises 23 and 24 use Simpson's Rule with $n = 10$ to find the area of the surface obtained by rotating the given curve about the x-axis.

23. $y = x^4, \quad 0 \le x \le 1$

24. $y = \tan x, \quad 0 \le x \le \pi/4$

25. Find the surface area generated by rotating a loop of the curve $8y^2 = x^2(1 - x^2)$ about the x-axis.

26. If the infinite curve $y = e^{-x}$, $x \ge 0$, is rotated about the x-axis, find the area of the resulting surface.

27. If the region $\mathcal{R} = \{(x, y) \mid x \ge 1, 0 \le y \le 1/x\}$ is rotated about the x-axis, the volume of the resulting solid is finite (see Exercise 67 in Section 7.9). Show that the surface area is infinite.

28. Find the surface area of the torus in Exercise 61 in Section 6.2.

29. The ellipse
$$\frac{x^2}{a^2} + \frac{y^2}{b^2} = 1 \qquad a > b$$

is rotated about the x-axis to form a surface called an ellipsoid. Find the surface area of this ellipsoid.

30. Show that the surface area of a zone of a sphere that lies between two parallel planes is $S = \pi dh$, where d is the diameter of the sphere and h is the distance between the planes. (Notice that S depends only on the distance between the planes and not on their location, provided that both planes intersect the sphere.)

31. Formula 8.22 is valid only when $f(x) \ge 0$. Show that when $f(x)$ is not necessarily positive, the formula for surface area becomes
$$S = \int_a^b 2\pi |f(x)| \sqrt{1 + [f'(x)]^2}\, dx$$

32. If the curve $y = f(x)$, $a \le x \le b$, is rotated about the horizontal line $y = c$, where $f(x) \le c$, find a formula for the area of the resulting surface.

33. Find the area of the surface obtained by rotating the circle $x^2 + y^2 = r^2$ about the line $y = r$.

34. Let L be the length of the curve $y = f(x)$, $a \le x \le b$, where f is positive and has a continuous derivative. Let S_f be the surface area generated by rotating the curve about the x-axis. If c is a positive constant, define $g(x) = f(x) + c$ and let S_g be the corresponding surface area generated by the curve $y = g(x)$, $a \le x \le b$. Express S_g in terms of S_f and L.

SECTION 8.4

Moments and Centers of Mass

The main object of this section is to find the point P on which a thin plate of any given shape balances horizontally as in Figure 8.13. This point is called the **center of mass** (or center of gravity) of the plate.

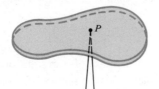

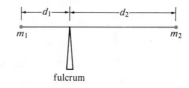

Figure 8.13   **Figure 8.14**

We first consider the simpler situation illustrated in Figure 8.14 where two masses m_1 and m_2 are attached to a rod of negligible mass on opposite sides of a fulcrum and at distances d_1 and d_2 from the fulcrum. The rod will balance if

(8.27)
$$m_1 d_1 = m_2 d_2$$

This is an experimental fact discovered by Archimedes and called the Law of the Lever. (Think of a lighter person balancing a heavier one on a seesaw by sitting farther away from the center.)

Now suppose that the rod lies along the x-axis with m_1 at x_1 and m_2 at x_2 and the center of mass at $\bar{x}$. If we compare Figures 8.14 and 8.15, we see that $d_1 = \bar{x} - x_1$ and $d_2 = x_2 - \bar{x}$ and so Equation 8.27 gives

$$m_1(\bar{x} - x_1) = m_2(x_2 - \bar{x})$$

$$m_1 \bar{x} + m_2 \bar{x} = m_1 x_1 + m_2 x_2$$

(8.28)
$$\bar{x} = \frac{m_1 x_1 + m_2 x_2}{m_1 + m_2}$$

The numbers $m_1 x_1$ and $m_2 x_2$ are called the **moments** of the masses m_1 and m_2 (with respect to the origin), and Equation 8.28 says that the center of mass $\bar{x}$ is obtained by adding the moments of the masses and dividing by the total mass $m = m_1 + m_2$.

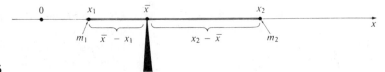

Figure 8.15

In general if we have a system of n particles with masses $m_1, m_2, \ldots, m_n$ located at the points $x_1, x_2, \ldots, x_n$ on the x-axis, it can be shown similarly that the center of mass of the system is located at

(8.29)
$$\bar{x} = \frac{\sum\limits_{i=1}^{n} m_i x_i}{\sum\limits_{i=1}^{n} m_i} = \frac{\sum\limits_{i=1}^{n} m_i x_i}{m}$$

where $m = \sum m_i$ is the total mass of the system, and the sum of the individual moments

(8.30)
$$M = \sum_{i=1}^{n} m_i x_i$$

is called the moment of the system with respect to the origin. Then Equation 8.29 could be rewritten as $m\bar{x} = M$, which says that if the total mass were considered as being concentrated at the center of mass $\bar{x}$, then its moment would be the same as the moment of the system.

EXAMPLE 1   Find the center of mass of a system of four objects with masses 10 g, 45 g, 32 g, and 24 g that are located at the points -4, 1, 3, and 8, respectively, on the x-axis.

Solution   Using Equation 8.29 we have

$$\bar{x} = \frac{10(-4) + 45(1) + 32(3) + 24(8)}{10 + 45 + 32 + 24} = \frac{293}{111}$$

Now we consider a system of n particles with masses $m_1, m_2, \ldots, m_n$ located at the points $(x_1, y_1), (x_2, y_2), \ldots, (x_n, y_n)$ in the xy-plane as shown in Figure 8.16. By analogy with the one-dimensional case we define the **moment of the system about the y-axis** to be

(8.31)
$$M_y = \sum_{i=1}^{n} m_i x_i$$

and the **moment of the system about the x-axis** as

(8.32)
$$M_x = \sum_{i=1}^{n} m_i y_i$$

Then M_y measures the tendency to rotate about the y-axis and M_x measures the tendency to rotate about the x-axis.

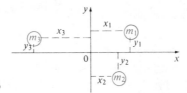

Figure 8.16

As in the one-dimensional case, the coordinates $(\bar{x}, \bar{y})$ of the center of mass are given in terms of the moments by the formulas

(8.33)
$$\bar{x} = \frac{M_y}{m} \qquad \bar{y} = \frac{M_x}{m}$$

where $m = \sum m_i$ is the total mass. Since $m\bar{x} = M_y$ and $m\bar{y} = M_x$ the center of mass $(\bar{x}, \bar{y})$ is the point where a single particle of mass m would have the same moments as the system.

EXAMPLE 2   Find the moments and center of mass of the system of objects that have masses 3, 4, and 8 at the points $(-1, 1)$, $(2, -1)$, and $(3, 2)$.

Solution   We use Equations 8.31 and 8.32 to compute the moments:

$$M_y = 3(-1) + 4(2) + 8(3) = 29$$
$$M_x = 3(1) + 4(-1) + 8(2) = 15$$

Since $m = 3 + 4 + 8 = 15$, we use Equations 8.33 to obtain

$$\bar{x} = \frac{M_y}{m} = \frac{29}{15} \qquad \bar{y} = \frac{M_x}{m} = \frac{15}{15} = 1$$

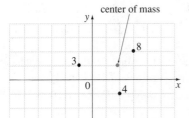

Figure 8.17   Thus the center of mass is $(1\frac{14}{15}, 1)$ (see Figure 8.17). •

Next we consider a flat plate (called a lamina) with uniform density ρ that occupies a region $\mathcal{R}$ of the plane. We wish to locate the center of mass of the plate, which is called the **centroid** of $\mathcal{R}$. In doing so we use the following physical principles: The **symmetry principle** says that if $\mathcal{R}$ is symmetric about a line l, then the centroid of $\mathcal{R}$ lies on l. (If $\mathcal{R}$ is reflected about l, then $\mathcal{R}$ remains the same so its centroid remains fixed. But the only fixed points lie on l.) Thus the centroid of a rectangle is its center. Moments should be defined so that if the entire mass of a region is concentrated at the center of mass,

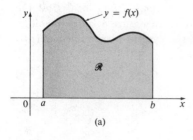

(a)

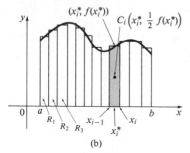

(b)

Figure 8.18

then its moments remain unchanged. Also the moment of the union of two nonoverlapping regions should be the sum of the moments of the individual regions.

First we suppose that the region $\mathcal{R}$ is of the type shown in Figure 8.18(a); that is, $\mathcal{R}$ lies between the lines $x = a$ and $x = b$, above the x-axis, and beneath the graph of f, where f is a continuous function. We take a partition P by points x_i with $a = x_0 < x_1 < \cdots < x_n = b$ and choose x_i^* to be the midpoint of the ith subinterval, that is, $x_i^* = (x_{i-1} + x_i)/2$. This determines the polygonal approximation to $\mathcal{R}$ shown in Figure 8.18(b). The centroid of the ith approximating rectangle is its center $C_i\left(x_i^*, \frac{1}{2}f(x_i^*)\right)$. Its area is $f(x_i^*)\,\Delta x_i$; so its mass is

$$\rho f(x_i^*)\,\Delta x_i$$

The moment of R_i about the y-axis is the product of its mass and the distance from C_i to the y-axis, which is x_i^*. Thus

$$M_y(R_i) = (\rho f(x_i^*)\,\Delta x_i)x_i^* = \rho x_i^* f(x_i^*)\,\Delta x_i$$

Adding these moments, we obtain the moment of the polygonal approximation to $\mathcal{R}$, and then by taking the limit as $\|P\| \to 0$ we obtain the moment of $\mathcal{R}$ itself about the y-axis:

(8.34)
$$M_y = \lim_{\|P\| \to 0} \sum_{i=1}^{n} \rho x_i^* f(x_i^*)\,\Delta x_i = \rho \int_a^b x f(x)\,dx$$

In a similar fashion we compute the moment of R_i about the x-axis as the product of its mass and the distance from C_i to the x-axis:

$$M_x(R_i) = [\rho f(x_i^*)\,\Delta x_i]\tfrac{1}{2}f(x_i^*) = \rho \cdot \tfrac{1}{2}[f(x_i^*)]^2\,\Delta x_i$$

Again we add these moments and take the limit to obtain the moment of $\mathcal{R}$ about the x-axis:

(8.35)
$$M_x = \lim_{\|P\| \to 0} \sum_{i=1}^{n} \rho \cdot \tfrac{1}{2}[f(x_i^*)]^2\,\Delta x_i = \rho \int_a^b \tfrac{1}{2}[f(x)]^2\,dx$$

Just as for systems of particles, the center of mass of the plate is defined so that $m\bar{x} = M_y$ and $m\bar{y} = M_x$. but the mass of the plate is the product of its density and its area:

$$m = \rho A = \rho \int_a^b f(x)\,dx$$

and so
$$\bar{x} = \frac{M_y}{m} = \frac{\rho \displaystyle\int_a^b x f(x)\,dx}{\rho \displaystyle\int_a^b f(x)\,dx} = \frac{\displaystyle\int_a^b x f(x)\,dx}{\displaystyle\int_a^b f(x)\,dx}$$

$$\bar{y} = \frac{M_x}{m} = \frac{\rho \displaystyle\int_a^b \tfrac{1}{2}[f(x)]^2\,dx}{\rho \displaystyle\int_a^b f(x)\,dx} = \frac{\displaystyle\int_a^b \tfrac{1}{2}[f(x)]^2\,dx}{\displaystyle\int_a^b f(x)\,dx}$$

Notice the cancellation of the ρ's. The location of the center of the mass is independent of the density.

In summary, the center of mass of the plate (or the centroid of $\mathcal{R}$) is located at the point $(\bar{x}, \bar{y})$ where

(8.36)

$$\bar{x} = \frac{1}{A} \int_a^b x f(x)\, dx \qquad \bar{y} = \frac{1}{A} \int_a^b \frac{1}{2}[f(x)]^2\, dx$$

EXAMPLE 3   Find the center of mass of a semicircular plate of radius r.

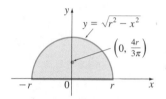

Figure 8.19

Solution   In order to use (8.36) we place the semicircle as in Figure 8.19 so that $f(x) = \sqrt{r^2 - x^2}$ and $a = -r$, $b = r$. Here there is no need to use the formula to calculate $\bar{x}$ because, by the symmetry principle, the center of mass must lie on the y-axis, so $\bar{x} = 0$. The area of the semicircle is $A = \pi r^2/2$, so

$$\bar{y} = \frac{1}{A} \int_{-r}^r \frac{1}{2}[f(x)]^2\, dx$$

$$= \frac{1}{\pi r^2/2} \cdot \frac{1}{2} \int_{-r}^r \left(\sqrt{r^2 - x^2}\right)^2 dx$$

$$= \frac{1}{\pi r^2} \int_{-r}^r (r^2 - x^2)\, dx = \frac{1}{\pi r^2}\left[r^2 x - \frac{x^3}{3}\right]_{-r}^r$$

$$= \frac{1}{\pi r^2} \cdot \frac{4r^3}{3} = \frac{4r}{3\pi}$$

The center of mass is located at the point $(0, 4r/3\pi)$. •

EXAMPLE 4   Find the centroid of the region bounded by the curves $y = \cos x$, $y = 0$, $x = 0$, and $x = \pi/2$.

Solution   The area of the region is

$$A = \int_0^{\pi/2} \cos x\, dx = \sin x \Big|_0^{\pi/2} = 1$$

so Formulas 8.36 give

$$\bar{x} = \frac{1}{A} \int_0^{\pi/2} x f(x)\, dx = \int_0^{\pi/2} x \cos x\ dx$$

$$= x \sin x \Big|_0^{\pi/2} - \int_0^{\pi/2} \sin x\, dx \qquad \text{(by integration by parts)}$$

$$= \frac{\pi}{2} - 1$$

$$\bar{y} = \frac{1}{A} \int_0^{\pi/2} \frac{1}{2}[f(x)]^2\, dx = \frac{1}{2} \int_0^{\pi/2} \cos^2 x\, dx$$

$$= \frac{1}{4} \int_0^{\pi/2} (1 + \cos 2x)\, dx = \frac{1}{4}\left[x + \frac{1}{2}\sin 2x\right]_0^{\pi/2}$$

$$= \frac{\pi}{8}$$

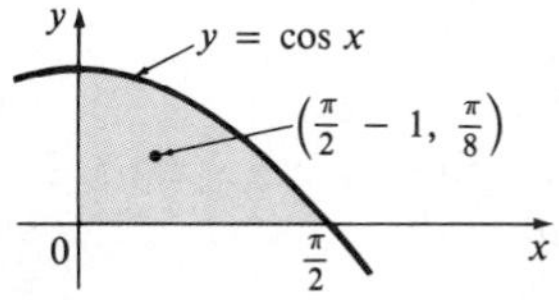

Figure 8.20   The centroid is $((\pi/2) - 1, \pi/8)$ and is shown in Figure 8.20. •

If the region $\mathcal{R}$ lies between two curves $y = f(x)$ and $y = g(x)$ where $f(x) \geq g(x)$, as illustrated in Figure 8.21, then the same sort of argument that led to Formulas 8.36 can be used to show that the centroid of $\mathcal{R}$ is $(\bar{x}, \bar{y})$, where

(8.37)

$$\bar{x} = \frac{1}{A} \int_a^b x[f(x) - g(x)]\, dx$$

$$\bar{y} = \frac{1}{A} \int_a^b \frac{1}{2}\{[f(x)]^2 - [g(x)]^2\}\, dx$$

(See Exercise 31.)

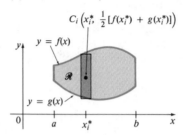

Figure 8.21

EXAMPLE 5   Find the centroid of the region bounded by the line $y = x$ and the parabola $y = x^2$.

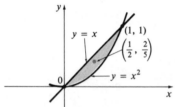

Figure 8.22

Solution   The region is sketched in Figure 8.22. We take $f(x) = x$, $g(x) = x^2$, $a = 0$, and $b = 1$ in Formulas 8.37. First we note that the area of the region is

$$A = \int_0^1 (x - x^2)\, dx = \frac{x^2}{2} - \frac{x^3}{3}\Big|_0^1 = \frac{1}{6}$$

Therefore

$$\bar{x} = \frac{1}{A} \int_0^1 x[f(x) - g(x)]\, dx = \frac{1}{1/6} \int_0^1 x(x - x^2)\, dx$$

$$= 6 \int_0^1 (x^2 - x^3)\, dx = 6\left[\frac{x^3}{3} - \frac{x^4}{4}\right]_0^1 = \frac{1}{2}$$

$$\bar{y} = \frac{1}{A} \int_0^1 \frac{1}{2}\{[f(x)]^2 - [g(x)]^2\}\, dx = \frac{1}{1/6} \int_0^1 \frac{1}{2}(x^2 - x^4)\, dx$$

$$= 3\left[\frac{x^3}{3} - \frac{x^5}{5}\right]_0^1 = \frac{2}{5}$$

The centroid is $\left(\frac{1}{2}, \frac{2}{5}\right)$.   ●

We end this section by showing how centroids can be used in finding volumes of revolution. The following theorem is named after the Greek mathematician Pappus of Alexandria, who lived in the fourth century A.D.

Theorem of Pappus (8.38)

Let $\mathcal{R}$ be a plane region that lies entirely on one side of a line l in the plane. If $\mathcal{R}$ is rotated about l, then the volume of the resulting solid is the product of the area A of $\mathcal{R}$ and the distance d traveled by the centroid of $\mathcal{R}$.

Proof   We give the proof for the special case where the region lies between $y = f(x)$ and $y = g(x)$ as in Figure 8.21 and the line l is the y-axis. Using the method of cylindrical shells (see Section 6.3), we have

$$V = \int_a^b 2\pi x[f(x) - g(x)]\, dx$$

$$= 2\pi \int_a^b x[f(x) - g(x)]\, dx$$

$$= 2\pi(\bar{x}A) \qquad \text{(by Formula 8.37)}$$

$$= (2\pi\bar{x})A = dA$$

where $d = 2\pi\bar{x}$ is the distance traveled by the centroid during one rotation about the y-axis.   ●

EXAMPLE 6   A torus is formed by rotating a circle of radius r about a line in the plane of the circle that is a distance R ($> r$) from the center of the circle. Find the volume of the torus.

Solution   The circle has area $A = \pi r^2$. By the symmetry principle, its centroid is its center and so the distance traveled by the centroid during a rotation is $d = 2\pi R$. Therefore, by the Theorem of Pappus, the volume of the torus is

$$V = dA = (2\pi R)(\pi r^2) = 2\pi^2 r^2 R$$ ●

The method of Example 6 should be compared with the method of Exercise 59 in Section 6.2.

SECTION 8.4 **Exercises**

In Exercises 1–4 the masses m_i are located at the points P_i. Find the moments M_x and M_y and the center of mass of the system.

1. $m_1 = 4,\ m_2 = 8;\quad P_1(-1, 2),\ P_2(2, 4)$

2. $m_1 = 2,\ m_2 = 3,\ m_3 = 1;$
    $P_1(5, 1),\ P_2(3, -2),\ P_3(-2, 4)$

3. $m_1 = 4,\ m_2 = 2,\ m_3 = 5;$
    $P_1(-1, -2),\ P_2(-2, 4),\ P_3(5, -3)$

4. $m_1 = 3,\ m_2 = 3,\ m_3 = 8,\ m_4 = 6;$
    $P_1(0, 0),\ P_2(1, 8),\ P_3(3, -4),\ P_4(-6, -5)$

In Exercises 5–18 find the centroid of the region bounded by the given curves.

5. $y = x^2,\quad y = 0,\quad x = 2$

6. $y = 1 - x^2,\quad y = 0$

7. $y = 3x + 5,\quad y = 0,\quad x = -1,\quad x = 2$

8. $y = \sqrt{x},\quad y = 0,\quad x = 4$

9. $y = \sqrt{x^2 + 1},\quad y = 0,\quad x = 0,\quad x = 1$

10. $y = \dfrac{1}{x - 1},\quad y = 0,\quad x = 2,\quad x = 4$

11. $y = \cos 2x,\quad y = 0,\quad x = -\pi/4,\quad x = \pi/4$

12. $y = \sin x,\quad y = 0,\quad x = 0,\quad x = \pi/2$

13. $y = e^x,\quad y = 0,\quad x = 0,\quad x = 1$

14. $y = \ln x,\quad y = 0,\quad x = e$

15. $y = \sqrt{x},\quad y = x$

16. $y = x^2,\quad y = 8 - x^2$

17. $y = \sin x,\quad y = \cos x,\quad x = 0,\quad x = \pi/4$

18. $y = x,\quad y = 0,\quad y = 1/x,\quad x = 2$

In Exercises 19–22 calculate the moments M_x and M_y and the center of mass of a lamina with the given density and shape.

19. $\rho = 1$

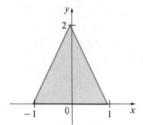

20. $\rho = 2$

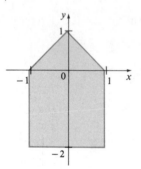

21. $\rho = 4$

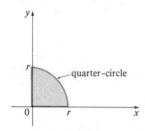

22. $\rho = 5$

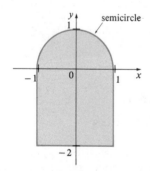

23. Prove that the centroid of any triangle is located at the point of intersection of medians. [*Hint:* Place the axes so that the vertices are $(a, 0)$, $(0, b)$, and $(c, 0)$. Recall that a median is a line segment from a vertex to the midpoint of the opposite side. Recall also that the medians intersect at a point two-thirds of the way from each vertex (along the median) to the opposite side.]

In Exercises 24–27 find the centroid of the region shown, not by integration, but by locating the centroids of the rectangles and triangles (from Exercise 23) and using additivity of moments.

24.

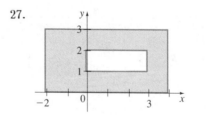

25.

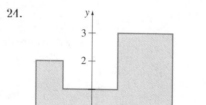

26.

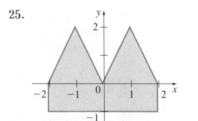

27.

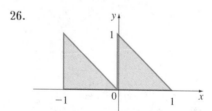

In Exercises 28–30 use the Theorem of Pappus to find the volume of the given solid.

28. A sphere of radius r (Use Example 3.)

29. A cone with height h and base radius r

30. The solid obtained by rotating the quadrilateral with vertices $(0, 0)$, $(1, 4)$, $(7, 4)$, and $(6, 0)$ about the y-axis

31. Prove Formula 8.37.

SECTION 8.5

Hydrostatic Pressure and Force

Divers realize that water pressure increases as they dive deeper. This is because the weight of the water above them increases.

In general, suppose that a thin horizontal plate with area A square meters is submerged in a fluid of density ρ kilograms per cubic meter at a depth d meters below the surface of the fluid as in Figure 8.23. The fluid directly above the plate has volume $V = Ad$ so its mass is $m = \rho V = \rho Ad$. The force exerted by the fluid on the plate is therefore

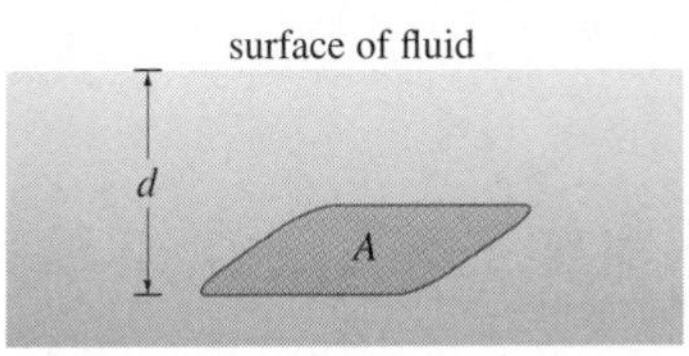

$$F = mg = \rho gdA$$

Figure 8.23

surface of fluid

d

A

where g is the acceleration due to gravity. The pressure P on the plate is defined to be the force per unit area:

$$P = \frac{F}{A} = \rho gd$$

The SI unit for measuring pressure is newton per square meter, which is called a pascal (abbreviation: $1\,\text{N/m}^2 = 1\,\text{Pa}$). Since this is a small unit, the kilopascal (kPa) is often used. For instance, since the density of water is $\rho = 1000\,\text{kg/m}^3$, the pressure at the bottom of a swimming pool 2 m deep is

$$P = \rho gd = 1000\,\text{kg/m}^3 \times 9.8\,\text{m/s}^2 \times 2\,\text{m}$$
$$= 19{,}600\,\text{Pa} = 19.6\,\text{kPa}$$

When using British units, we write $P = \rho gd = \delta d$, where $\delta = \rho g$ is the weight density (as opposed to ρ, which is the mass density). For instance, the weight density of water is $\delta = 62.5\,\text{lb/ft}^3$.

An important principle of fluid pressure is the experimentally verified fact that *at any point in a liquid the pressure is the same in all directions.* (A diver feels the same pressure on both ears and nose.) Thus the pressure in *any* direction at a depth d in a fluid with mass density ρ is given by

(8.39)
$$P = \rho gd = \delta d$$

This helps us determine the hydrostatic force against a *vertical* plate or wall or dam in a fluid. This is not a straightforward problem since the pressure is not constant but increases as the depth increases.

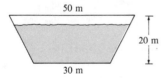

50 m

20 m

30 m

Figure 8.24

EXAMPLE 1   A dam has the shape of the trapezoid shown in Figure 8.24. The height is 20 m and the width is 50 m at the top and 30 m at the bottom. Find the force on the dam due to hydrostatic pressure if the water level is 4 m from the top of the dam.

Solution   We choose a vertical x-axis with origin at the surface of the water as in Figure 8.25(a). The depth of the water is 16 m so we consider a partition P of the interval $[0, 16]$ by points x_i and we choose $x_i^* \in [x_{i-1}, x_i]$. The ith horizontal strip of the dam is approximated by a rectangle with height Δx_i and width w_i where, from similar triangles in Figure 8.25(b),

$$\frac{a}{16 - x_i^*} = \frac{10}{20} \qquad a = \frac{16 - x_i^*}{2} = 8 - \frac{x_i^*}{2}$$

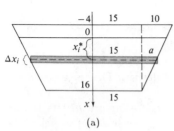

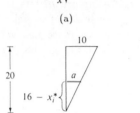

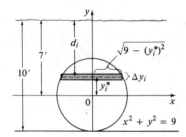

(a)

(b)

Figure 8.25

$$w_i = 2(15 + a) = 2\left(15 + 8 - \frac{x_i^*}{2}\right) = 46 - x_i^*$$

If A_i is the area of the ith strip, then

$$A_i \approx w_i\,\Delta x_i = (46 - x_i^*)\,\Delta x_i$$

If Δx_i is small, then the pressure P_i on the ith strip is almost constant and we can use Equation 8.39 to write

$$P_i \approx 1000g x_i^*$$

The hydrostatic force F_i acting on the ith strip is the product of the pressure and the area:

$$F_i = P_i A_i \approx 1000g x_i^*(46 - x_i^*)\,\Delta x_i$$

Adding these forces and taking the limit as $||P|| \to 0$, we obtain the total hydrostatic force on the dam:

$$F = \lim_{||P|| \to 0} \sum_{i=1}^n 1000g x_i^*(46 - x_i^*)\,\Delta x_i$$

$$= \int_0^{16} 1000g x(46 - x)\,dx$$

$$= 1000(9.8)\int_0^{16} (46x - x^2)\,dx$$

$$= 9800\left[23x^2 - \frac{x^3}{3}\right]_0^{16}$$

$$\approx 4.43 \times 10^7\,\text{N}$$

EXAMPLE 2   Find the hydrostatic force on one end of a cylindrical drum with radius 3 ft if it is submerged in water 10 ft deep.

Solution   In this example it is convenient to choose the axes as in Figure 8.26 so that the origin is placed at the center of the drum. Then the circle has a simple equation, $x^2 + y^2 = 9$. As in Example 1 we divide the circular region into horizontal strips. From the equation of the circle, we see that the length of the ith strip is $2\sqrt{9 - (y_i^*)^2}$ and so its area is

$$A_i = 2\sqrt{9 - (y_i^*)^2}\,\Delta y_i$$

Figure 8.26

The pressure on this strip is approximately

$$\delta d_i = 62.5(7 - y_i^*)$$

and so the force on the strip is approximately

$$\delta d_i A_i = 62.5(7 - y_i^*)2\sqrt{9 - (y_i^*)^2}\,\Delta y_i$$

The total force is obtained by adding the forces on all the strips and taking the limit:

$$F = \lim_{||P|| \to 0} \sum_{i=1}^{n} 62.5(7 - y_i^*)2\sqrt{9 - (y_i^*)^2}\, \Delta y_i$$

$$= 125 \int_{-3}^{3} (7 - y)\sqrt{9 - y^2}\, dy$$

$$= 125 \cdot 7 \int_{-3}^{3} \sqrt{9 - y^2}\, dy - 125 \int_{-3}^{3} y\sqrt{9 - y^2}\, dy$$

The second integral is 0 because the integrand is an odd function (see Theorem 5.36). The first integral can be evaluated using the trigonometric substitution $y = 3\sin\theta$ but it is simpler to observe that it is the area of a semicircular disk with radius 3. Thus

$$F = 875 \int_{-3}^{3} \sqrt{9 - y^2}\, dy = 875 \cdot \tfrac{1}{2}\pi(3)^2$$

$$= \frac{7875\pi}{2} \approx 12,370 \text{ lb}$$

SECTION 8.5 **Exercises**

1. An aquarium 2 m long, 1 m wide, and 1 m deep is full of water. Find (a) the hydrostatic pressure on the bottom of the aquarium, (b) the hydrostatic force on the bottom, and (c) the hydrostatic force on one end of the aquarium.

2. A swimming pool 5 m wide, 10 m long, and 3 m deep is filled with seawater of density 1030 kg/m^3 to a depth of 2.5 m. Find
   (a) the hydrostatic pressure on the bottom of the pool,
   (b) the hydrostatic force on the bottom, and
   (c) the hydrostatic force on one end of the pool.

In Exercises 3–12 a tank contains water. The end of the tank is vertical and has the indicated shape. Find the hydrostatic force against the end of the tank.

3.
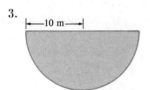

4.
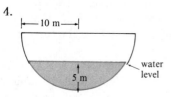

5.

6.
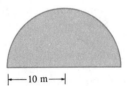

7.

8.
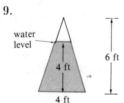

9.
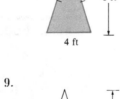

10.

11.
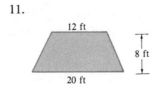

12.
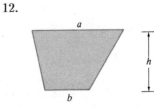

13. A trough is filled with a liquid that has a density of 840 kg/m^3. The ends of the trough are equilateral triangles with sides 8 m and vertex at the bottom. Find the hydrostatic force on one end of the trough.

14. Work Exercise 13 if the height of the liquid is 4 m.

15. A cube with side 20 cm is sitting on the bottom of an aquarium in which the water is 1 m deep. Find the hydrostatic force on (a) the top of the cube and (b) one of the sides of the cube.

16. A vertical dam has a semicircular gate as shown in the figure. Find the hydrostatic force against the gate.

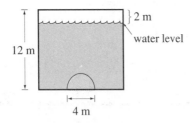

17. A trough 2 m high is filled with water. Its ends are vertical but its sides are 3 m wide and are inclined at an angle of 45° from the vertical. Find the hydrostatic force on one of the sides of the trough.

18. A dam is inclined at an angle of 30° from the vertical and has the shape of an isosceles trapezoid 100 ft

wide at the top and 50 ft wide at the bottom and with a slant height of 70 ft. Find the hydrostatic force on the dam when it is full of water.

19. A swimming pool is 20 ft wide and 40 ft long and its bottom is an inclined plane, the shallow end having a depth of 3 ft and the deep end 9 ft. Find the hydrostatic force on (a) the shallow end, (b) the deep end, (c) one of the sides, and (d) the bottom.

20. Suppose that a plate is immersed vertically in a fluid with density ρ and the width of the plate is $w(x)$ at a depth x meters beneath the surface of the fluid. If the top of the plate is at depth a and the bottom at depth b, show that the hydrostatic force on the plate is

$$F = \int_a^b \rho g x w(x) \, dx$$

21. Use the formula in Exercise 20 to show that

$$F = (\rho g \bar{x}) A$$

where $\bar{x}$ is the x-coordinate of the centroid of the plate and A is its area. This equation shows that the hydrostatic force against a vertical plane region is the same as if the region were horizontal at the depth of the centroid of the region.

22. Use the result of Exercise 21 to give another solution to Exercise 5.

SECTION 8.6

Applications to Economics and Biology

In this section we consider some applications of integration to economics (consumer's surplus, present value of future income) and biology (blood flow, cardiac output). Others are found in the exercises.

Consumer's Surplus

Recall from Section 4.7 that the demand function $p(x)$ is the price that a company has to charge in order to sell x units of a commodity. Usually, selling larger quantities requires lowering prices, so the demand function is a decreasing function. The graph of a typical demand function, called a **demand curve**, is shown in Figure 8.27. If X is the amount of the commodity that is currently available, then $P = p(X)$ is the current selling price.

We partition the interval $[0, X]$ into n subintervals, each of length $\Delta x = X/n$, and let $x_i^* = x_i$ be the right endpoint of the ith subinterval, as in Figure 8.28. If, after the first x_{i-1} units were sold, only a total of x_i units had been available and the price per

Figure 8.27

A demand curve

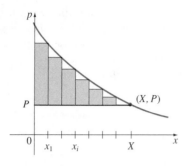

Figure 8.28

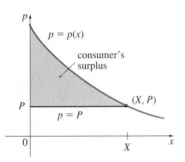

Figure 8.29

unit had been set at $p(x_i)$ dollars, then the additional Δx units could have been sold (but no more). The consumers who would have paid $p(x_i)$ dollars placed a high value on the product; they would have paid what it was worth to them. So in paying only P dollars they have saved an amount of

$$(\text{savings per unit})(\text{number of units}) = [p(x_i) - P]\,\Delta x$$

Considering similar groups of willing consumers for each of the subintervals and adding the savings, we get the total savings:

$$\sum_{i=1}^{n} [p(x_i) - P]\,\Delta x$$

If we let $n \to \infty$, this Riemann sum approaches the integral

(8.40)
$$\int_0^X [p(x) - P]\,dx$$

which economists call the **consumer's surplus** for the commodity.

The consumer's surplus represents the amount of money saved by consumers in purchasing the commodity at price P, corresponding to an amount demanded of X. Figure 8.29 shows the interpretation of the consumer's surplus as the area under the demand curve and above the line $p = P$.

EXAMPLE 1   The demand for a product, in dollars, is $p = 1200 - 0.2x - 0.0001x^2$. Find the consumer's surplus when the sales level is 500.

Solution   Since the number of products sold is $X = 500$, the corresponding price is

$$P = 1200 - (0.2)(500) - (0.0001)(500)^2 = 1075$$

Therefore, from the definition in (8.40), the consumer's surplus is

$$\int_0^{500} [p(x) - P]\,dx = \int_0^{500} (1200 - 0.2x - 0.0001x^2 - 1075)\,dx$$

$$= \int_0^{500} (125 - 0.2x - 0.0001x^2)\,dx$$

$$= 125x - 0.1x^2 - (0.0001)\left(\frac{x^3}{3}\right)\Big|_0^{500}$$

$$= (125)(500) - 0.1(500)^2 - \frac{(0.0001)(500)^3}{3}$$

$$= \$33{,}333.33$$

Present Value of an Income Stream

In Example 4 in Section 3.6 we discussed continuous compounding of interest and found that the value of a dollar after t years, at an annual interest rate r compounded continuously, is e^{rt}. It follows that if A is the amount that will grow to one dollar in t years, then $Ae^{rt} = 1$ and so $A = e^{-rt}$; this is called the *present value* of one dollar. For instance, if the interest rate is 9%, the present value of a dollar to be paid five years from now is $e^{-(0.09)(5)} = e^{-0.45} \approx 64$ cents.

We now suppose that income will be received, not in one lump sum, but over a period of time from $t = a$ to $t = b$ at a rate of $f(t)$ dollars per year at time t. This is referred to as an **income stream**. To find the total present value of this income we partition the interval $[a, b]$ into n subintervals of equal length Δt. Thus, from time $t = t_{i-1}$ to time $t = t_i$ the income received will be about $f(t_i) \Delta t$ dollars, with a present value of

$$e^{-rt_i} f(t_i) \Delta t$$

So an approximation to the present value of the total income is

$$\sum_{i=1}^{n} e^{-rt_i} f(t_i) \Delta t$$

If we let $n \to \infty$, this Riemann sum approaches the integral

(8.41)
$$\int_a^b e^{-rt} f(t) \, dt$$

which is called the **present value of the income stream** $f(t)$ at an interest rate r over the time period from $t = a$ to $t = b$.

EXAMPLE 2   A trust fund pays \$8000 a year for 10 years, starting in 5 years, at a rate of 10% per year compounded continuously.
(a) Find the present value of the trust fund.

(b) Find the value 3 years from now.

Solution
(a) Here the income stream is constant: $f(t) = 8000$. Using (8.41) with $a = 5$, $b = 15$, and $r = 0.1$, we find that the present value of the trust fund is

$$\int_5^{15} e^{-(0.1)t}(8000) \, dt = (8000) \frac{e^{-(0.1)t}}{-0.1} \Big|_5^{15}$$
$$= 80{,}000(e^{-0.5} - e^{-1.5})$$
$$= \$30{,}672.04$$

(b) The value in 3 years from now is

$$(30{,}672.04)e^{(0.1)3} = \$41{,}402.92$$

If we let $b \to \infty$ in Formula 8.41, we get the improper integral

(8.42)
$$\int_a^{\infty} e^{-rt} f(t) \, dt$$

This represents the present value of a **perpetuity**, or perpetual annuity, under which income will be received at a rate of $f(t)$ dollars per year forever. (See Exercises 13 and 14.)

Blood Flow

In Example 7 in Section 2.3 we discussed the law of laminar flow

$$v(r) = \frac{P}{4\eta l}(R^2 - r^2)$$

which gives the velocity v of blood that flows along a blood vessel with radius R and length l at a distance r from the central axis, where P is the pressure difference between the ends of the vessel and η is the viscosity of the blood. Now in order to compute the flux (volume per unit time) we consider radii r_i where

$$0 = r_0 < r_1 < r_2 < \cdots < r_n = R$$

The approximate area of the annulus with inner radius r_{i-1} and outer radius r_i is

$$2\pi r_i \Delta r_i \qquad \text{where } \Delta r_i = r_i - r_{i-1}$$

If Δr_i is small, then the velocity is almost constant throughout this annulus and can be approximated by $v(r_i)$. Thus the volume of blood per unit time that flows across the annulus is approximately

$$(2\pi r_i \Delta r_i)v(r_i) = 2\pi r_i v(r_i)\Delta r_i$$

and the total volume of blood that flows across a cross-section per unit time is approximately

$$\sum_{i=1}^{n} 2\pi r_i v(r_i)\Delta r_i$$

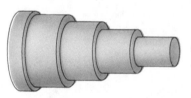

This approximation is illustrated in Figure 8.30. Notice that the velocity (and hence the volume per unit time) increases toward the center of the blood vessel.

The approximation gets better as we take finer subdivisions, that is, as the norm $\|P\| \to 0$. When we take the limit we get the exact value of the *flux* (or *discharge*),

Figure 8.30   which is the volume of blood that passes a cross-section per unit time:

$$\begin{aligned}
F &= \lim_{\|P\| \to 0} \sum_{i=1}^{n} 2\pi r_i v(r_i)\Delta r_i \\
&= \int_0^R 2\pi r v(r)\, dr \\
&= \int_0^R 2\pi r \frac{P}{4\eta l}(R^2 - r^2)\, dr \\
&= \frac{\pi P}{2\eta l}\int_0^R (R^2 r - r^3)\, dr = \frac{\pi P}{2\eta l}\left[R^2 \frac{r^2}{2} - \frac{r^4}{4}\right]_{r=0}^{r=R} \\
&= \frac{\pi P}{2\eta l}\left[\frac{R^4}{2} - \frac{R^4}{4}\right] = \frac{\pi P R^4}{8\eta l}
\end{aligned}$$

The resulting equation

(8.43)
$$F = \frac{\pi P R^4}{4\eta l}$$

is called **Poiseuille's Law** and shows that the flux is proportional to the fourth power of the radius of the blood vessel.

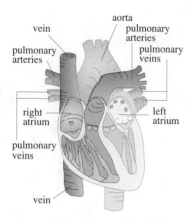

vein

pulmonary arteries

right atrium

pulmonary veins

vein

aorta

pulmonary arteries

pulmonary veins

left atrium

Figure 8.31

Cardiac Output

Figure 8.31 shows the human cardiovascular system. Blood returns from the body through the veins and enters the right atrium of the heart, is pumped into the lungs through the pulmonary arteries for oxygenation, back into the left atrium through the pulmonary veins, and then out to the rest of the body through the aorta. The **cardiac output** of the heart is the volume of blood pumped by the heart per unit time, that is, the rate of flow into the aorta.

The *dye dilution method* is used to measure the cardiac output. Dye is injected into the right atrium and flows through the heart into the aorta. A probe inserted into the aorta measures the concentration of the dye leaving the heart at equally spaced times over a time interval $[0, T]$ until the dye has cleared. Let $c(t)$ be the concentration of the dye at time t. If we partition $[0, T]$ into subintervals of equal length Δt, then the amount of dye that flows past the measuring point during the subinterval from $t = t_{i-1}$ to $t = t_i$ is approximately

$$(\text{concentration})(\text{volume}) = c(t_i)(F \Delta t)$$

where F is the rate of flow that we are trying to determine. Thus the total amount of dye is approximately

$$\sum_{i=1}^{n} c(t_i) F \Delta t = F \sum_{i=1}^{n} c(t_i) \Delta t$$

and, letting $n \to \infty$, we find that the amount of dye is

$$A = F \int_0^T c(t)\, dt$$

Thus the cardiac output is given by

(8.44)
$$F = \frac{A}{\int_0^T c(t)\, dt}$$

where the amount of dye A is known and the integral can be approximated from the concentration readings.

EXAMPLE 3   A 5-mg bolus of dye is injected into a right atrium. The concentration of the dye (in milligrams per liter) is measured in the aorta at one-second intervals as shown in the chart. Estimate the cardiac output.

t	0	1	2	3	4	5	6	7	8	9	10
$c(t)$	0	0.4	2.8	6.5	9.8	8.9	6.1	4.0	2.3	1.1	0

Solution   Here $A = 5$, $\Delta t = 1$, and $T = 10$. We use Simpson's Rule to approximate the integral of the concentration:

$$\int_0^{10} c(t)\, dt \approx \tfrac{1}{3}[0 + 4(0.4) + 2(2.8) + 4(6.5) + 2(9.8) + 4(8.9)$$
$$+ 2(6.1) + 4(4.0) + 2(2.3) + 4(1.1) + 0]$$
$$\approx 41.87$$

Thus Formula 8.44 gives the cardiac output to be

$$F = \frac{A}{\int_0^{10} c(t)\, dt} \approx \frac{5}{41.87}$$
$$\approx 0.12 \text{ L/s} = 7.17 \text{ L/min}$$

SECTION 8.6 **Exercises**

1. The marginal cost function $C'(x)$ was defined to be the derivative of the cost function. (See Sections 2.3 and 4.7.) If the marginal cost of manufacturing x units of a product is $C'(x) = 0.006x^2 - 1.5x + 8$ (measured in dollars per unit) and the fixed start-up cost is $C(0) = \$1,500,000$, use the Fundamental Theorem of Calculus to find the cost of producing the first 2000 units.

2. The marginal revenue from selling x items is $90 - 0.02x$. The revenue from the sale of the first 100 items was \$8800. What was the revenue from the sale of the first 200 items?

3. The marginal cost of producing x units of a certain product is $140 - 0.5x + 0.012x^2$ (in dollars per unit). Find the increase in cost if the production level is raised from 3000 units to 5000 units.

4. The demand function for a certain commodity is $p = 5 - x/10$. Find the consumer's surplus when the sales level is 30. Illustrate by drawing the demand curve and identifying the consumer's surplus as an area.

5. A demand curve is given by $p = 1000/(x + 20)$. Find the consumer's surplus when the selling price is \$20.

6. The **supply function** $p_S(x)$ for a commodity gives the relation between the selling price and the number of units that manufacturers will produce at that price. For a higher price manufacturers will produce more units, so p_S is an increasing function of x. Let X be the amount of the commodity currently produced and $P = p_S(X)$ the current price. Some producers would be willing to make and sell the commodity for a lower selling price and are therefore receiving more than their minimal price. The excess is called the **producer's surplus.** An argument similar to that for consumer's surplus shows that the surplus is given by the integral

$$\int_0^X [P - p_S(x)]\,dx$$

Calculate the producer's surplus for the supply function $p_S(x) = 3 + 0.01x^2$ at the sales level $X = 10$. Illustrate by drawing the supply curve and identifying the producer's surplus as an area.

7. A supply curve is given by $p = 5 + \frac{1}{10}\sqrt{x}$. Find the producer's surplus when the selling price is \$10.

8. For a given commodity and pure competition, the number of units produced and the price per unit are determined as the coordinates of the point of intersection of the supply and demand curves. Given the demand curve $p = 50 - x/20$ and the supply curve $p = 20 + x/10$, find the consumer's surplus and the producer's surplus. Illustrate by sketching the supply and demand curves and identifying the surpluses as areas.

9. A manufacturer has been selling 1000 television sets a week at \$450 each. A market survey indicates that for every \$10 that the price is reduced, the number of sets sold will increase by 100 a week. Find the demand function and calculate the consumer's surplus when the selling price is set at \$400.

10. A trust fund pays \$2000 a year for 5 years, starting immediately. The interest rate is 12% per year compounded continuously. Find the present value of the trust fund.

11. A trust fund starts 10 years from now and pays \$12,000 a year for 15 years. The interest rate is 11% per year compounded continuously. Find the value of the trust fund 5 years from now.

12. A baseball player signs a salary contract whereby he receives a sum that increase continuously and linearly from a starting salary of \$1,000,000 a year and reaches \$3,000,000 a year after 4 years. Thus his salary after t years (in millions of dollars) is $f(t) = 1 + \frac{1}{2}t$. Find the present value of the contract assuming an interest rate of 8% per year compounded continuously.

13. (a) Use (8.42) to show that the present value of a perpetuity under which you and your heirs forever receive an amount A annually is A/r, where r is the annual interest rate, compounded continuously.

    (b) Assuming an interest rate of 10% per year, what is the present value of a perpetuity that pays an annual amount of \$5000?

14. Suppose that a perpetuity pays an amount of $5000 + 1000t$ dollars t years from now. Find the present value.

15. If the amount of capital that a company has at time t is $f(t)$, then the derivative, $f'(t)$, is called the *net investment flow.* Suppose that the net investment flow is $\sqrt{t}$ million dollars per year (where t is measured in years). Find the increase in capital (the capital formation) from the fourth year to the eighth year.

16. An animal population is increasing at a rate of $200 + 50t$ per year (where t is measured in years). By how much does the animal population increase between the fourth and tenth years?

17. Use Poiseuille's Law to calculate the rate of flow in a typical human artery where we can take $\eta = 0.027$, $R = 0.008$ cm, $l = 2$ cm, and $P = 4000$ dynes/cm^2.

18. High blood pressure results from constriction of the arteries. To maintain the same flow rate (flux) the heart has to pump harder, thus increasing the blood pressure. Use Poiseuille's Law to show that if R_0 and P_0 are normal values of the radius and pressure in an artery and the constricted values are R and P, then for the flux to remain constant, P and R are related by the equation

$$\frac{P}{P_0} = \left(\frac{R_0}{R}\right)^4$$

Deduce that if the radius of an artery is reduced to three-fourths of its former value, then the pressure is more than tripled.

19. The dye dilution method is used to measure cardiac output with 8 mg of dye. The dye concentrations are given by $c(t) = \frac{1}{4}t(12 - t)$, $0 \le t \le 12$, where t is measured in seconds. Find the cardiac output.

20. After a 6-mg injection of dye, the readings of dye concentrations at two-second intervals are shown in the table. Use Simpson's Rule to estimate the cardiac output.

t	0	2	4	6	8	10	12	14	16	18	20
$c(t)$	0	2.1	4.5	7.3	5.8	3.6	2.8	1.4	0.6	0.2	0

CHAPTER 8

Review

Key Topics

Define, state, or discuss each of the following:

1. Differential equation
2. Separable equation
3. Logistic growth
4. Arc length
5. Area of a surface of revolution
6. Moments and centroid of a plane region
7. Theorem of Pappus
8. Pressure and force exerted by a fluid
9. Consumer's surplus
10. Present value of an income stream
11. Rate of blood flow
12. Cardiac output

Exercises

Solve the differential equations in Exercises 1–4.

1. $y^2\frac{dy}{dx} = x + \sin x$

2. $\frac{dy}{dx} = \frac{y^2 + 1}{xy}$, $x > 0$

3. $y' = \frac{1}{x^2 y - 2x^2 + y - 2}$

4. $xy' = \sqrt{y}(1 + y)$

In Exercises 5 and 6 find the solution of the differential equation that satisfies the given initial condition.

5. $xyy' = \ln x$, $y(1) = 2$

6. $2u\frac{du}{dt} = te^t$, $u(0) = 1$

In Exercises 7 and 8 find the length of the curve.

7. $3x = 2(y - 1)^{3/2}$, $2 \le y \le 5$

8. $y = \sqrt{x - x^2} + \sin^{-1}\sqrt{x}$

9. (a) Find the length of the curve $y = \frac{x^3}{6} + \frac{1}{2x}$, $1 \le x \le 2$.
    (b) Find the area of the surface obtained by rotating the curve in part (a) about the x-axis.

10. (a) The curve $y = x^2$, $0 \le x \le 1$, is rotated about the y-axis. Find the area of the resulting surface.
    (b) Find the area of the surface obtained by rotating the curve in part (a) about the x-axis.

11. Use Simpson's Rule with $n = 10$ to estimate the length of the arc of the curve $y = 1/x^2$ from $(1, 1)$ to $(2, \frac{1}{4})$.

12. Use Simpson's Rule with $n = 10$ to estimate the area of the surface obtained by rotating the arc of the curve $y = 1/x^2$ from $(1, 1)$ to $(2, \frac{1}{4})$ about the x-axis.

13. Find the surface area generated by rotating the loop of the curve $9ay^2 = x(3a - x)^2$ about the y-axis.

14. If the loop in Exercise 13 is rotated about the x-axis, find the area of the resulting surface.

In Exercises 15 and 16 find the centroid of the region bounded by the given curves.

15. $y = 4 - x^2$, $y = x + 2$

16. $y = 4x - x^2$, $y = 0$

In Exercises 17 and 18 find the centroid of the region shown.

17.

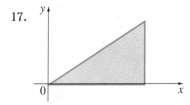

18.

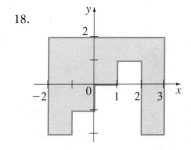

19. Find the volume obtained when the circle of radius 1 with center $(1, 0)$ is rotated about the y-axis.

20. Use the Theorem of Pappus and the fact that the volume of a sphere of radius r is $\frac{4}{3}\pi r^3$ to find the centroid of the semicircular region bounded by the curve $y = \sqrt{r^2 - x^2}$ and the x-axis.

21. A gate in an irrigation canal is in the form of a trapezoid 3 ft wide at the bottom, 5 ft wide at the top, and 2 ft high. It is placed vertically in the canal, with the water extending to its top. Find the hydrostatic force on one side of the gate.

22. A trough is filled with water and its vertical ends have the shape of the parabolic region in the figure. Find the hydrostatic force on one end of the trough.

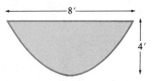

23. The demand function for a commodity is $p = 2000 - 0.1x - 0.01x^2$. Find the consumer's surplus when the sales level is 100.

24. A trust fund starts 8 years from now and pays $10,000 a year for 7 years. The interest rate is 12% per year compounded continuously.
  (a) Find the present value of the trust fund.
  (b) Find the value of the trust fund 5 years from now.
  (c) Find the present value if it lasts not just 7 years but forever.

25. The von Bertalanffy growth model is used to predict the length $L(t)$ of a fish over a period of time. If L_∞ is the largest length for a species, then the hypothesis is that the rate of growth in length is proportional to $L_\infty - L$, the length yet to be achieved.
  (a) Formulate and solve a differential equation to find an expression for $L(t)$.
  (b) For the North Sea haddock it has been determined that $L_\infty = 53$ cm, $L(0) = 10$ cm, and the constant of proportionality is 0.2. What does the expression for $L(t)$ become with these data?

26. A tank contains 100 L of pure water. Brine that contains 0.1 kg of salt per liter enters the tank at a rate of 10 L/min. The solution is kept thoroughly mixed and drains from the tank at the same rate. How much salt is in the tank after 6 minutes?

27. Barbara weighs 60 kg and is on a diet of 1600 calories per day, of which 850 are used up automatically by basal metabolism. She spends about 15 cal/kg/day times her weight doing exercise. If 1 kg of fat contains 10,000 cal and we assume that the storage of calories in the form of fat is 100% efficient, formulate a differential equation and solve it to find her weight as a function of time. Does her weight ultimately approach an equilibrium weight?

Problems Plus

1. (a) Prove that if f is a continuous function, then

$$\int_0^a f(x)\,dx = \int_0^a f(a-x)\,dx$$

   (b) Use part (a) to show that

$$\int_0^{\pi/2} \frac{\sin^n x}{\sin^n x + \cos^n x}\,dx = \frac{\pi}{4}$$

   for all positive numbers n.

2. A student forgot the Product Rule for differentiation and made the mistake of thinking that $(fg)' = f'g'$. However, he was lucky and got the correct answer. The function f that he used was $f(x) = e^{x^2}$ and the domain of his problem was the interval $(\frac{1}{2}, \infty)$. What was the function g?

3. Let f be a function with the property that $f(0) = 1$, $f'(0) = 1$, and $f(a+b) = f(a)f(b)$ for all real numbers a and b. Show that $f'(x) = f(x)$ for all x and deduce that $f(x) = e^x$.

4. Find the centroid of the region enclosed by the loop of the curve $y^2 = x^3 - x^4$.

5. Let a and b be positive numbers. Show that not both of the numbers $a(1-b)$ and $b(1-a)$ can be greater than $\frac{1}{4}$.

6. Evaluate $\displaystyle\int \frac{1}{x^7 - x}\,dx$.

   The straightforward approach would be to start with partial fractions, but that would be unthinkably arduous. Try a substitution.

7. Let f be a continuous function on $[a, b]$. Prove that there exists a number x in $[a, b]$ such that

$$\int_a^x f(t)\,dt = \int_x^b f(t)\,dt$$

8. (a) Show that an observer at height H above the north pole of a sphere of radius r can see a part of the sphere that has area

$$\frac{2\pi r^2 H}{r + H}$$

   (b) Two spheres with radii r and R are placed so that the distance between their centers is d, where $d > r + R$. Where should a light be placed on the line joining the centers of the spheres in order to illuminate the largest total surface?

9. Show that $\displaystyle\int_1^\infty \frac{e^{-x}}{\sqrt{x}}\,dx = 2\int_1^\infty e^{-x^2}\,dx$.

10. Snow began to fall during the morning of February 2 and continued steadily into the afternoon. A snowplow began to clear a street at noon at a constant rate. The plow traveled 6 km from noon to 1 P.M. but only 3 km from 1 P.M. to 2 P.M. When did the snow begin to fall? [*Hints:* To get started, let t be the time measured in hours after noon; let $x(t)$ be the distance traveled by the plow at time t; then the speed of the plow is dx/dt. Let b be the number of hours before noon that it began

to snow. Find an expression for the height of the snow at time t. Then use the given information that the rate of removal R (in m^3/h) is constant.]

11. A function f is defined by $f(x) = 3x - 2x^3$ with the restriction that its domain is $\{x \mid 5x^2 \geq x^4 + 4\}$. Find the maximum value of f.

12. Find all functions f that satisfy the equation $\left(\int f(x) \, dx \right)\left(\int \frac{1}{f(x)} \, dx \right) = -1$.

13. A curve is defined by

$$y = \int_1^x \sqrt{t^2 e^{2t} - 1} \, dt \qquad x \geq 1$$

(a) Where is the curve concave upward?
(b) Find the length of the arc of the curve given by $1 \leq x \leq 2$.

14. A function f is defined by

$$f(x) = \int_0^\pi \cos t \cos(x - t) \, dt \qquad 0 \leq x \leq 2\pi$$

Find the minimum value of f.

15. If $0 < a < b$, find $\displaystyle \lim_{t \to 0} \left\{ \int_0^1 [bx + a(1-x)]^t \, dx \right\}^{1/t}$.

16. Let C be the arc of the curve $y = f(x)$ between the points $P(p, f(p))$ and $Q(q, f(q))$ and let $\mathcal{R}$ be the region bounded by C, by the line $y = mx + b$ (which lies entirely on one side of C), and by the perpendiculars to the line from P and Q.

(a) Show that the area of $\mathcal{R}$ is $\dfrac{1}{1 + m^2} \displaystyle\int_p^q [f(x) - mx - b][1 + mf'(x)] \, dx$.

(b) Find a formula similar to the one in part (a) for the volume of the solid obtained by rotating $\mathcal{R}$ about the line $y = mx + b$.

(c) Find a formula for the area of the surface obtained by rotating C about the line $y = mx + b$.

[*Hint:* The formula in part (a) can be verified by subtracting areas, but it is more instructive to derive it by first approximating the area using rectangles perpendicular to the line, as shown in the figures. This will also help in finding the formulas for parts (b) and (c). Use the figure to help express Δu in terms of Δx.]

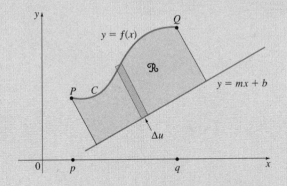
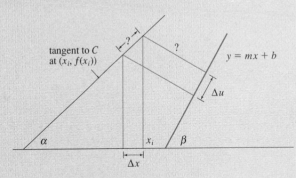

511

9

Parametric Equations and Polar Coordinates

A mathematician, like a painter or poet, is a maker of patterns. If his patterns are more permanent than theirs, it is because they are made with *ideas*.

G.H. Hardy

So far we have described plane curves by giving y as a function of x $[y = f(x)]$ or x as a function of y $[x = g(y)]$ or by giving a relation between x and y that defines y implicitly as a function of x $[f(x,y) = 0]$. In this chapter we discuss two new methods for describing curves.

Some curves, such as the cycloid, are best handled when both x and y are given in terms of a third variable t called a parameter $[x = f(t),\ y = g(t)]$. Other curves, such as the cardioid, have their most convenient description when we use a new coordinate system, called the polar coordinate system.

Curves Defined by Parametric Equations

Suppose that x and y are both given as continuous functions of a third variable t (called a **parameter**) by the equations

(9.1)
$$x = f(t) \qquad y = g(t)$$

(called **parametric equations**). Each value of t determines a point (x, y), which we can plot in a coordinate plane. As t varies, the point $(x, y) = (f(t), g(t))$ varies and traces out a curve C. If we interpret t as time and $(x, y) = (f(t), g(t))$ as the position of a particle at time t, then we can imagine the particle moving along the curve C.

t	x	y
-2	8	-1
-1	3	0
0	0	1
1	-1	2
2	0	3
3	3	4
4	8	5

EXAMPLE 1 Sketch and identify the curve defined by the parametric equations $x = t^2 - 2t$ and $y = t + 1$.

Solution Each value of t gives a point on the curve, as shown in the table. For instance, if $t = 0$, then $x = 0$, $y = 1$ and so the corresponding point is $(0, 1)$. In Figure 9.1 we plot the points (x, y) determined by several values of the parameter.

A particle whose position is given by the parametric equations moves along the curve in the direction of the arrows as t increases. It appears from Figure 9.1 that the curve traced out by the particle may be a parabola. This can be confirmed by eliminating the parameter t as follows. We obtain $t = y - 1$ from the second equation and substitute

513

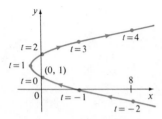

Figure 9.1

into the first equation. This gives

$$x = (y-1)^2 - 2(y-1) = y^2 - 4y + 3$$

and so the curve represented by the given parametric equations is the parabola $x = y^2 - 4y + 3$. ●

EXAMPLE 2   What curve is represented by the parametric equations $x = \cos t$ and $y = \sin t$, $0 \le t \le 2\pi$?

Solution   We can eliminate t by noting that

$$x^2 + y^2 = \cos^2 t + \sin^2 t = 1$$

Thus the point (x, y) moves on the unit circle $x^2 + y^2 = 1$. Notice that in this example the parameter t can be interpreted as the angle shown in Figure 9.2. As t increases from 0 to 2π, the point $(x, y) = (\cos t, \sin t)$ moves once around the circle in the counterclockwise direction starting from the point $(1, 0)$. ●

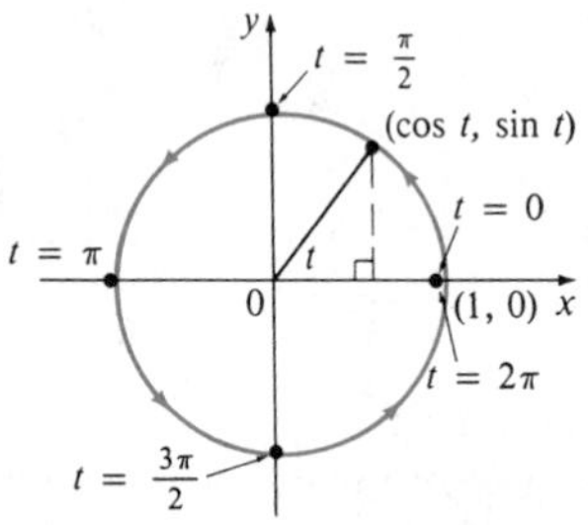

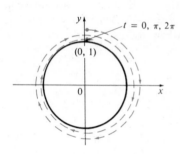

Figure 9.2

EXAMPLE 3   What curve is represented by the parametric equations $x = \sin 2t$ and $y = \cos 2t$, $0 \le t \le 2\pi$?

Solution   Again we have

$$x^2 + y^2 = \sin^2 2t + \cos^2 2t = 1$$

so the parametric equations again represent the unit circle $x^2 + y^2 = 1$. But as t increases from 0 to 2π, the point $(x, y) = (\sin 2t, \cos 2t)$ starts at $(0, 1)$ and moves *twice* around the circle in the clockwise direction as indicated in Figure 9.3. ●

Figure 9.3

EXAMPLE 4   Sketch the curve with parametric equations $x = \sin t$ and $y = \sin^2 t$.

Solution   Observe that $y = x^2$ and so the point (x, y) moves on the parabola $y = x^2$. But note also that, since $-1 \le \sin t \le 1$, we have $-1 \le x \le 1$, so the parametric equations represent only the part of the parabola for which $-1 \le x \le 1$. Since $\sin t$ is periodic, the point $(x, y) = (\sin t, \sin^2 t)$ moves back and forth infinitely often along the parabola from $(-1, 1)$ to $(1, 1)$ (see Figure 9.4). ●

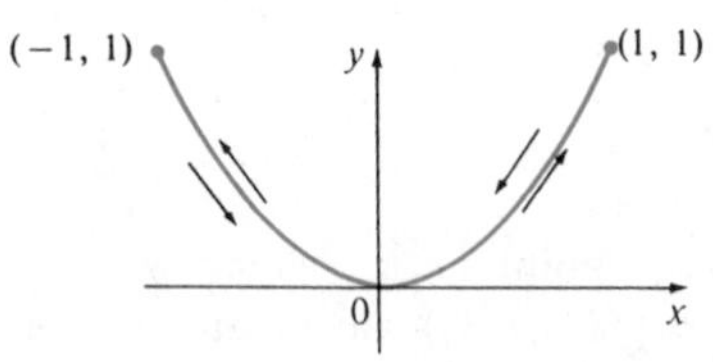

Figure 9.4

EXAMPLE 5   The curve traced out by a point P on the circumference of a circle as the circle rolls along a straight line is called a **cycloid** (see Figure 9.5). If the circle has radius r and rolls along the x-axis and if one position of P is the origin, find parametric equations for the cycloid.

Figure 9.5

Solution   We choose as parameter the angle of rotation θ of the circle ($\theta = 0$ when P is at the origin). When the circle has rotated through θ radians, the distance it has rolled from the origin is

$$|OT| = \text{arc } PT = r\theta$$

and so the center of the circle is $C(r\theta, r)$. Let the coordinates of P be (x, y). Then from

Figure 9.6 we see that

$$x = |OT| - |PQ| = r\theta - r\sin\theta = r(\theta - \sin\theta)$$

$$y = |TC| - |QC| = r - r\cos\theta = r(1 - \cos\theta)$$

Therefore the parametric equations of the cycloid are

(9.2) $$x = r(\theta - \sin\theta) \qquad y = r(1 - \cos\theta) \qquad \theta \in R$$

One arch of the cycloid comes from one rotation of the circle and so is described by $0 \le \theta \le 2\pi$. Although Equations 9.2 were derived from Figure 9.6, which illustrates the case where $0 < \theta < \pi/2$, it can be seen that these equations are still valid for other values of θ (see Exercise 33).

Although it is possible to eliminate the parameter θ from Equations 9.2, the resulting Cartesian equation in x and y is very complicated and not as convenient to work with as the parametric equations.   ●

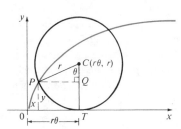

Figure 9.6

One of the first people to study the cycloid was Galileo, who proposed that bridges be built in the shape of cycloids and who tried to find the area under one arch of a cycloid. Later this curve arose in connection with the *Brachistochrone problem:* Find the curve along which a particle will slide in the shortest time (under the influence of gravity) from a point A to a lower point B not directly beneath A. The Swiss mathematician John Bernoulli, who posed this problem in 1696, showed that among all possible curves that join A to B as in Figure 9.7, the particle will take the least time sliding from A to B if the curve is an inverted arch of a cycloid.

The Dutch physicist Huygens had already shown that the cycloid is also the solution to the *Tautochrone problem;* that is, no matter where a particle P is placed on an inverted cycloid, it takes the same time to slide to the bottom (see Figure 9.8). Huygens proposed that pendulum clocks (which he invented) should swing in cycloidal arcs because then the pendulum would take the same time to make a complete oscillation whether it swings through a wide or a small arc.

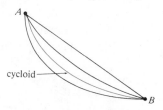

Figure 9.7

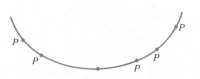

Figure 9.8

SECTION 9.1   **Exercises**

In Exercises 1–20, (a) sketch the curve represented by the given parametric equations, and (b) eliminate the parameter to find the Cartesian equation of the curve.

1. $x = 1 - t, y = 2 + 3t$

2. $x = 2t - 1, y = 2 - t, -3 \le t \le 3$

3. $x = 3t^2, y = 2 + 5t, 0 \le t \le 2$

4. $x = 2t - 1, y = t^2 - 1$

5. $x = \sqrt{t}, y = 1 - t$ 6. $x = t^2, y = t^3$

7. $x = \sin\theta, y = \cos\theta, 0 \le \theta \le \pi$

8. $x = 3\cos\theta, y = 2\sin\theta, 0 \le \theta \le 2\pi$

9. $x = \sin^2\theta, y = \cos^2\theta$

10. $x = \sec\theta, y = \tan\theta, -\pi/2 < \theta < \pi/2$

11. $x = e^t, y = e^t$

12. $x = e^t, y = e^{-t}$

13. $x = \cos^2 t, y = \cos^4 t$

14. $x = \cos t, y = \cos 2t$

15. $x = \cos^2\theta, y = \sin\theta$

16. $x = \dfrac{1 - t^2}{1 + t^2}, y = \dfrac{2t}{1 + t^2}$

17. $x = e^t, y = \sqrt{t}, 0 \le t \le 1$

18. $x = \dfrac{1 - t}{1 + t}, y = t^2, 0 \le t \le 1$

19. $x = \cosh t, y = \sinh t$

20. $x = 4\sinh t, y = 3\cosh t$

In Exercises 21–26 describe the motion of a particle with position (x, y) as t varies in the given time interval.

21. $x = \cos\pi t, y = \sin\pi t, 1 \le t \le 2$

22. $x = 2 + \cos t, y = 3 + \sin t, 0 \le t \le 2\pi$

23. $x = 8t - 3, y = 2 - t, 0 \le t \le 1$

24. $x = \cos^2 t, y = \cos t, 0 \le t \le 4\pi$

25. $x = 2\sin t, y = 3\cos t, 0 \le t \le 2\pi$

26. $x = \sin t, y = \csc t, \pi/6 \le t \le 1$

In Exercises 27–30 sketch the curve represented by the given parametric equations.

27. $x = 3(t^2 - 3), y = t^3 - 3t$

28. $x = t\cos t, y = t\sin t$

29. $x = \tan\theta + \sin\theta, y = \cos\theta$

30. $x = \dfrac{3t}{1 + t^3}, y = \dfrac{3t^2}{1 + t^3}$

31. Show that the parametric equations
$x = x_1 + (x_2 - x_1)t$ and $y = y_1 + (y_2 - y_1)t$,
$0 \le t \le 1$, describe the line segment that joins the
points $P_1(x_1, y_1)$ and $P_2(x_2, y_2)$.

32. If a projectile is fired with an initial velocity of v_0
meters per second at an angle α above the horizontal, then its position after t seconds is given by the
parametric equations

$$x = (v_0\cos\alpha)t \qquad y = (v_0\sin\alpha)t - \tfrac{1}{2}gt^2$$

where g is the acceleration due to gravity ($9.8\,\text{m/s}^2$).
(a) If a gun is fired with $\alpha = 30°$ and $v_0 = 500\,\text{m/s}$,
     when will the bullet hit the ground? How far
     from the gun will it hit the ground?

(b) What is the maximum height reached by the
     bullet ?
(c) Show that the path is parabolic by eliminating
     the parameter.

33. Derive Equations 9.2 for the case where $\pi/2 < \theta < \pi$.

34. Let P be a point at a distance d from the center of a
circle of radius r. The curve traced out by P as the
circle rolls along a straight line is called a **trochoid**.
(Think of the motion of a point on a spoke of a bi-
cycle wheel.) The cycloid is the special case of a
trochoid with $d = r$. Using the same parameter θ as
for the cycloid and assuming the line is the x-axis
and $\theta = 0$ when P is at one of its lowest points, show
that the parametric equations of the trochoid are

$$x = r\theta - d\sin\theta \qquad y = r - d\cos\theta$$

Sketch the trochoid for the cases $d < r$ and $d > r$.

35. Find parametric equations for the set of all points P
determined as shown in the figure, using the angle θ
as the parameter. Then eliminate the parameter and
identify the curve.

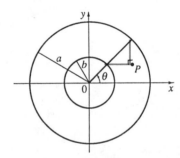

36. Find parametric equations for the set of all points P
determined as shown in the figure, using the angle θ
as a parameter. The line segment AB is tangent to
the larger circle.

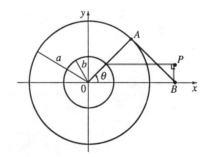

37. (a) A circle C of radius b rolls on the inside of a
         larger circle with center O and radius a. The
         curve traced out by a fixed point P on C is
         called a **hypocycloid**. If the initial position of P
         is $(a, 0)$ and the parameter θ is chosen as in the

figure, show that the parametric equations of the hypocycloid are

$$x = (a - b)\cos\theta + b\cos\left(\frac{a-b}{b}\theta\right)$$

$$y = (a - b)\sin\theta - b\sin\left(\frac{a-b}{b}\theta\right)$$

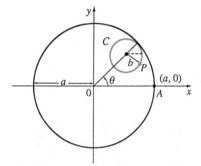

(b) If $b = a/4$, the hypocycloid is called a **hypocycloid of four cusps** or an **astroid**. Show that in this case the parametric equations reduce to

$$x = a\cos^3\theta \qquad y = a\sin^3\theta$$

and sketch the curve.

38. If the circle C of Exercise 37 rolls on the outside of larger circle, the curve traced out by P is called an **epicycloid**. (The special case where $a = b$ is called a **cardioid**.) Find the parametric equations for the epicycloid. Sketch the epicycloid for the case $b = a/4$.

39. A curve, called a **witch of Maria Agnesi**, consists of all points P determined as shown in the figure. Show that parametric equations for this curve can be written as

$$x = 2a\cot\theta \qquad y = 2a\sin^2\theta$$

Sketch the curve.

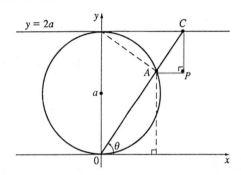

40. Find parametric equations for the set of all points P determined as shown in the figure so that $|OP| = |AB|$. Sketch the curve. (This curve is called the **cissoid of Diocles** after the Greek scholar Diocles, who introduced the cissoid as a graphical method for constructing the edge of a cube whose volume is twice that of a given cube.)

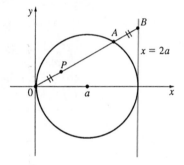

SECTION 9.2

Tangents and Areas

In this section we adapt our previous methods for finding tangents and areas for curves of the form $y = F(x)$ and apply them to curves given by parametric equations.

In the preceding section we saw that some curves defined by parametric equations $x = f(t)$, $y = g(t)$ can also be expressed, by eliminating the parameter, in the form $y = F(x)$. (See Exercise 38 for general conditions under which this is possible.) If we substitute $x = f(t)$ and $y = g(t)$ in the equation $y = F(x)$, we get

$$g(t) = F(f(t))$$

and so, if g, F, and f are differentiable, the Chain Rule gives

$$g'(t) = F'(f(t))f'(t) = F'(x)f'(t)$$

If $f'(t) \neq 0$, we can solve for $F'(x)$:

(9.3)
$$F'(x) = \frac{g'(t)}{f'(t)}$$

Since the slope of the tangent to the curve $y = F(x)$ at $(x, F(x))$ is $F'(x)$, Equation 9.3 enables us to find tangents to parametric curves without having to eliminate the parameter. Using Leibniz notation we can rewrite Equation 9.3 in the easily remembered form

(9.4)
$$\frac{dy}{dx} = \frac{\dfrac{dy}{dt}}{\dfrac{dx}{dt}} \qquad \text{if} \quad \frac{dx}{dt} \neq 0$$

It can be seen from Equation 9.4 that the curve has a horizontal tangent when $dy/dt = 0$ (provided that $dx/dt \neq 0$) and it has a vertical tangent when $dx/dt = 0$ (provided that $dy/dt \neq 0$). This information is useful when sketching parametric curves.

As we know from Chapter 4, it is also useful to consider d^2y/dx^2. This can again be found from Equation 9.4 as follows:

$$\frac{d^2y}{dx^2} = \frac{d}{dx}\left(\frac{dy}{dx}\right) = \frac{\dfrac{d}{dt}\left(\dfrac{dy}{dx}\right)}{\dfrac{dx}{dt}}$$

EXAMPLE 1

(a)   Find dy/dx and d^2y/dx^2 for the cycloid $x = r(\theta - \sin\theta)$, $y = r(1 - \cos\theta)$. (See Example 5 in Section 9.1.)
(b)   Find the tangent to the cycloid at the point where $\theta = \pi/3$.
(c)   At what points is the tangent horizontal?
(d)   Discuss the concavity.

Solution

(a)
$$\frac{dy}{dx} = \frac{\dfrac{dy}{d\theta}}{\dfrac{dx}{d\theta}} = \frac{r\sin\theta}{r(1-\cos\theta)} = \frac{\sin\theta}{1-\cos\theta}$$

$$\frac{d}{d\theta}\left(\frac{dy}{dx}\right) = \frac{d}{d\theta}\left(\frac{\sin\theta}{1-\cos\theta}\right) = \frac{\cos\theta(1-\cos\theta) - \sin\theta\sin\theta}{(1-\cos\theta)^2}$$

$$= \frac{\cos\theta - 1}{(1-\cos\theta)^2} = -\frac{1}{1-\cos\theta}$$

$$\frac{d^2y}{dx^2} = \frac{\dfrac{d}{d\theta}\left(\dfrac{dy}{dx}\right)}{\dfrac{dx}{d\theta}} = \frac{-\dfrac{1}{1-\cos\theta}}{r(1-\cos\theta)} = -\frac{1}{r(1-\cos\theta)^2}$$

(b)   When $\theta = \pi/3$, we have

$$x = r\left(\frac{\pi}{3} - \sin\frac{\pi}{3}\right) = r\left(\frac{\pi}{3} - \frac{\sqrt{3}}{2}\right) \qquad y = r\left(1 - \cos\frac{\pi}{3}\right) = \frac{r}{2}$$

and
$$\frac{dy}{dx} = \frac{\sin(\pi/3)}{1-\cos(\pi/3)} = \frac{\sqrt{3}/2}{1-\frac{1}{2}} = \sqrt{3}$$

Therefore the slope of the tangent is $\sqrt{3}$ and its equation is

$$y - \frac{r}{2} = \sqrt{3}\left(x - \frac{r\pi}{3} + \frac{r\sqrt{3}}{2}\right) \quad \text{or} \quad \sqrt{3}x - y = r\left(\frac{\pi}{\sqrt{3}} - 2\right)$$

This tangent is sketched in Figure 9.9.

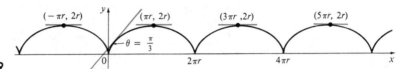

Figure 9.9

(c)   The tangent is horizontal when $dy/dx = 0$, which occurs when $\sin\theta = 0$ and $1 - \cos\theta \neq 0$, that is, $\theta = (2n-1)\pi$, n an integer. The corresponding point on the cycloid is $((2n-1)\pi r, 2r)$. (It can be shown that the cycloid has vertical tangents when $\theta = 2n\pi$. See Exercise 28.)

(d)   From part (a) we have $d^2y/dx^2 = -1/[r(1-\cos\theta)^2]$, which shows that $d^2y/dx^2 < 0$ except when $\cos\theta = 1$. Thus the cycloid is concave downward on the intervals $(2n\pi, 2(n+1)\pi)$.   ●

EXAMPLE 2   A curve C is defined by the parametric equations $x = t^2$ and $y = t^3 - 3t$.
(a)   Show that C has two tangents at the point $(3, 0)$ and find their equations.
(b)   Find the points on C where the tangent is horizontal or vertical.
(c)   Determine where the curve rises and falls and where it is concave upward or downward.
(d)   Sketch the curve.

Solution
(a)   Notice that $y = t^3 - 3t = t(t^2 - 3) = 0$ when $t = 0$ or $t = \pm\sqrt{3}$. Therefore the point $(3, 0)$ on C arises from two values of the parameter, $t = \sqrt{3}$ and $t = -\sqrt{3}$. This indicates that C crosses itself at $(3, 0)$. Since

$$\frac{dy}{dx} = \frac{\dfrac{dy}{dt}}{\dfrac{dx}{dt}} = \frac{3t^2 - 3}{2t} = \frac{3}{2}\left(t - \frac{1}{t}\right)$$

the slope of the tangent when $t = \pm\sqrt{3}$ is $dy/dx = \pm 6/(2\sqrt{3}) = \pm\sqrt{3}$ so the equations of the tangents at $(3, 0)$ are

$$y = \sqrt{3}(x - 3) \quad \text{and} \quad y = -\sqrt{3}(x - 3)$$

(b) C has a vertical tangent when $dx/dt = 2t = 0$, that is, $t = 0$. The corresponding point on C is $(0, 0)$. C has a horizontal tangent when $dy/dt = 3t^2 - 3 = 0$, that is, $t = \pm 1$. The corresponding points on C are $(1, -2)$ and $(1, 2)$.

(c)   Since

$$\frac{dx}{dt} = 2t \quad \text{and} \quad \frac{dy}{dt} = 3(t-1)(t+1)$$

we can summarize the parameter intervals in which the curve rises and falls as in the following table.

	$t < -1$	$-1 < t < 0$	$0 < t < 1$	$t > 1$
dx/dt	$-$	$-$	$+$	$+$
dy/dt	$+$	$-$	$-$	$+$
x	$\leftarrow$	$\leftarrow$	$\rightarrow$	$\rightarrow$
y	$\uparrow$	$\downarrow$	$\downarrow$	$\uparrow$
curve	$\nwarrow$	$\swarrow$	$\searrow$	$\nearrow$

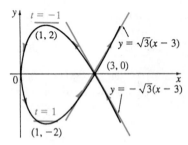

Figure 9.10

To determine concavity we calculate the second derivative:

$$\frac{d^2 y}{dx^2} = \frac{\dfrac{d}{dt}\left(\dfrac{dy}{dx}\right)}{\dfrac{dx}{dt}} = \frac{\frac{3}{2}\left(1 + \dfrac{1}{t^2}\right)}{2t} = \frac{3(t^2 + 1)}{4t^3}$$

Thus the curve is concave upward when $t > 0$ and concave downward when $t < 0$.

(d)   Using the information from parts (b) and (c), we sketch C in Figure 9.10.

We know that the area under a curve $y = F(x)$ from a to b is given by $A = \int_a^b F(x)\, dx$, where $F(x) \geq 0$. If the curve is given by parametric equations $x = f(t)$ and $y = g(t)$, $\alpha \leq t \leq \beta$, then we can adapt the earlier formula by using the Substitution Rule for Definite Integrals (5.35) as follows:

$$A = \int_a^b y\, dx = \int_\alpha^\beta g(t) f'(t)\, dt \qquad \left[\text{or} \int_\beta^\alpha g(t) f'(t)\, dt \right]$$

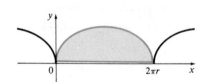

Figure 9.11

EXAMPLE 3   Find the area under one arch of the cycloid $x = r(\theta - \sin\theta)$, $y = r(1 - \cos\theta)$ (see Figure 9.11).

Solution   One arch of the cycloid is given by $0 \leq \theta \leq 2\pi$. Using the Substitution Rule with $y = r(1 - \cos\theta)$ and $dx = r(1 - \cos\theta)\, d\theta$, we have

$$A = \int_0^{2\pi r} y\, dx = \int_0^{2\pi} r(1 - \cos\theta) r(1 - \cos\theta)\, d\theta$$

$$= r^2 \int_0^{2\pi} (1 - \cos\theta)^2\, d\theta = r^2 \int_0^{2\pi} (1 - 2\cos\theta + \cos^2\theta)\, d\theta$$

$$= r^2 \int_0^{2\pi} \left[1 - 2\cos\theta + \tfrac{1}{2}(1 + \cos 2\theta) \right] d\theta$$

$$= r^2 \left[\tfrac{3}{2}\theta - 2\sin\theta + \tfrac{1}{4}\sin 2\theta \right]_0^{2\pi}$$

$$= r^2 \left(\tfrac{3}{2} \cdot 2\pi \right) = 3\pi r^2$$

The result of Example 3 says that the area under one arch of the cycloid is three times the area of the rolling circle that generates the cycloid (see Example 5 in Section 9.1). Galileo guessed this result but it was first proved by the French mathematician Roberval and the Italian mathematician Torricelli.

SECTION 9.2 Exercises

In Exercises 1–6 find the equation of the tangent to the given curve at the point corresponding to the given value of the parameter.

1. $x = t^2 + t,\ y = t^2 - t;\quad t = 0$

2. $x = 1 - t^3,\ y = t^2 - 3t + 1;\quad t = 1$

3. $x = t^2 + t,\ y = \sqrt{t};\quad t = 4$

4. $x = \ln t,\ y = te^t;\quad t = 1$

5. $x = 2\sin\theta,\ y = 3\cos\theta;\quad \theta = \pi/4$

6. $x = t\sin t,\ y = t\cos t;\quad t = \pi$

In Exercises 7–10 find the equation of the tangent to the given curve at the given point by two methods: (a) without eliminating the parameter and (b) by first eliminating the parameter.

7. $x = 1 - t,\ y = 1 - t^2;\quad (1, 1)$

8. $x = 2t + 3,\ y = t^2 + 2t;\quad (5, 3)$

9. $x = 5\cos t,\ y = 5\sin t;\quad (3, 4)$

10. $x = t^3,\ y = t^2;\quad (1, 1)$

In Exercises 11–18 find dy/dx and d^2y/dx^2.

11. $x = t^2 + t,\ y = t^2 + 1$

12. $x = t^3 + t^2 + 1,\ y = 1 - t^2$

13. $x = \sqrt{t + 1},\ y = t^2 - 3t$

14. $x = t^4 - t^2 + t,\ y = \sqrt[3]{t}$

15. $x = \sin \pi t,\ y = \cos \pi t$

16. $x = t + 2\cos t,\ y = \sin 2t$

17. $x = e^{-t},\ y = te^{2t}$

18. $x = 1 + t^2,\ y = t\ln t$

In Exercises 19–24 find the points on the given curve where the tangent is horizontal or vertical. Then use an analysis of the intervals in which the curve rises and falls, as in Example 2, to sketch the curve.

19. $x = 1 - 2\cos t,\ y = 2 + 3\sin t$

20. $x = t^3 - 3t^2,\ y = t^3 - 3t$

21. $x = t(t^2 - 3),\ y = 3(t^2 - 3)$

22. $x = \sin 2t,\ y = \sin t$

23. $x = \dfrac{3t}{1 + t^3},\ y = \dfrac{3t^2}{1 + t^3}$

24. $x = a(\cos\theta - \cos^2\theta),\ y = a(\sin\theta - \sin\theta\cos\theta)$

25. Show that the curve $x = \cos t,\ y = \sin t \cos t$ has two tangents at $(0, 0)$ and find their equations. Sketch the curve.

26. At what point does the curve $x = 1 - 2\cos^2 t$, $y = (\tan t)(1 - 2\cos^2 t)$ cross itself? Find the equations of both tangents at that point.

27. At what point does the curve of Exercise 21 cross itself? Find the equations of both tangents at that point.

28. ✓ Show that the cycloid $x = r(\theta - \sin\theta)$, $y = r(1 - \cos\theta)$ has a vertical tangent at the origin by showing that $dy/dx \to \infty$ as $\theta \to 0^+$.

29. (a) Find the slope of the tangent line to the trochoid $x = r\theta - d\sin\theta,\ y = r - d\cos\theta$ in terms of θ. (See Exercise 34 in Section 9.1.)

   (b) Show that if $d < r$, then the trochoid does not have a vertical tangent.

30. (a) Find the slope of the tangent to the astroid $x = a\cos^3\theta,\ y = a\sin^3\theta$ in terms of θ. (See Exercise 37 in Section 9.1.)

   (b) At what points is the tangent horizontal or vertical?

   (c) At what points does the tangent have slope 1 or -1?

31. At what points on the curve $x = t^3 + 4t,\ y = 6t^2$ is the tangent parallel to the line with equations $x = -7t,\ y = 12t - 5$?

32. Find the equations of the tangents to the curve $x = 3t^2 + 1,\ y = 2t^3 + 1$ that pass through the point $(4, 3)$.

33. Use the parametric equations of an ellipse, $x = a\cos\theta,\ y = b\sin\theta,\ 0 \le \theta \le 2\pi$, to find the area that it encloses.

34. Find the area bounded by the curve $x = t - 1/t$, $y = t + 1/t$ and the line $y = 2.5$.

35. Find the area bounded by the curve $x = \cos t$, $y = e^t, 0 \le t \le \pi/2$, and the lines $y = 1$ and $x = 0$.

36. Find the area of the region enclosed by the astroid $x = a\cos^3\theta,\ y = a\sin^3\theta$. (See Exercise 37 in Section 9.1.)

37. Find the area under one arch of the trochoid of Exercise 34 in Section 9.1 for the case where $d < r$.

38. If f' is continuous and $f'(t) \ne 0$ for $a \le t \le b$, show that the parametric curve $x = f(t),\ y = g(t)$, $a \le t \le b$, can be put in the form $y = F(x)$.

39. A string is wound around a circle and then unwound while being held taut. The curve traced by the point P at the end of the string is called the **involute** of the circle. If the circle has radius r and center O and the initial position of P is $(r,0)$, and if the parameter θ is chosen as in the figure, show that the parametric equations of the involute are

$$x = r(\cos\theta + \theta\sin\theta) \qquad y = r(\sin\theta - \theta\cos\theta)$$

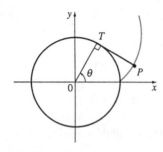

40. A cow is tied to a silo with radius r by a rope just long enough to reach the opposite side of the silo. Find the area available for grazing by the cow.

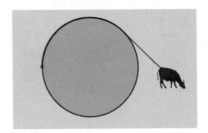

SECTION 9.3

Arc Length and Surface Area

We already know how to find the length L of a curve C given in the form $y = F(x)$, $a \le x \le b$. Formula 8.13 says that if F' is continuous, then

(9.5)
$$L = \int_a^b \sqrt{1 + \left(\frac{dy}{dx}\right)^2}\, dx$$

Suppose that C can also be described by the parametric equations $x = f(t)$ and $y = g(t)$, $\alpha \le t \le \beta$, where $dx/dt = f'(t) > 0$. This means that C is traversed once, from left to right, as t increases from α to β and $f(\alpha) = a$, $f(\beta) = b$. Putting Formula 9.4 into Formula 9.5 and using the Substitution Rule, we obtain

$$L = \int_a^b \sqrt{1 + \left(\frac{dy}{dx}\right)^2}\, dx = \int_\alpha^\beta \sqrt{1 + \left(\frac{dy/dt}{dx/dt}\right)^2}\, \frac{dx}{dt}\, dt$$

Since $dx/dt > 0$, we have

(9.6)
$$L = \int_\alpha^\beta \sqrt{\left(\frac{dx}{dt}\right)^2 + \left(\frac{dy}{dt}\right)^2}\, dt$$

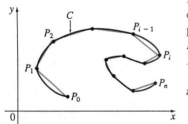

Figure 9.12

Even if C cannot be expressed in the form $y = F(x)$, Formula 9.6 is still valid but we obtain it by polygonal approximations as follows. Let P be a partition of $[\alpha, \beta]$ by points t_i with $\alpha = t_0 < t_1 < \cdots < t_n = \beta$. Let $x_i = f(t_i)$, $y_i = g(t_i)$, $\Delta x_i = x_i - x_{i-1}$, and $\Delta y_i = y_i - y_{i-1}$. Then the point $P_i(x_i, y_i)$ lies on C and the polygon with vertices P_0, $P_1, \ldots, P_n$ approximates C (see Figure 9.12).

As in Section 8.2, we define the length L of C to be the limit of the lengths of these approximating polygons as $\|P\| \to 0$:

$$L = \lim_{\|P\| \to 0} \sum_{i=1}^n |P_{i-1}P_i|$$

The Mean Value Theorem, when applied to f on the interval $[t_{i-1}, t_i]$, gives a number t_i^* in (t_{i-1}, t_i) such that

$$f(t_i) - f(t_{i-1}) = f'(t_i^*)(t_i - t_{i-1})$$

that is,
$$\Delta x_i = f'(t_i^*)\,\Delta t_i$$

Similarly, when applied to g, the Mean Value Theorem gives a number t_i^{**} in (t_{i-1}, t_i) such that

$$\Delta y_i = g'(t_i^{**})\,\Delta t_i$$

Therefore
$$
\begin{aligned}
|P_{i-1}P_i| &= \sqrt{(\Delta x_i)^2 + (\Delta y_i)^2} \\
&= \sqrt{[f'(t_i^*)\,\Delta t_i]^2 + [g'(t_i^{**})\,\Delta t_i]^2} \\
&= \sqrt{[f'(t_i^*)]^2 + [g'(t_i^{**})]^2}\,\Delta t_i
\end{aligned}
$$

and so

(9.7)
$$L = \lim_{\|P\| \to 0} \sum_{i=1}^{n} \sqrt{[f'(t_i^*)]^2 + [g'(t_i^{**})]^2}\,\Delta t_i$$

The sum in (9.7) resembles a Riemann sum for the function $\sqrt{[f'(t)]^2 + [g'(t)]^2}$ but it is not exactly a Riemann sum because $t_i^* \neq t_i^{**}$ in general. Nevertheless if f' and g' are continuous, it can be shown that the limit in (9.7) is the same as if t_i^* and t_i^{**} were equal, namely,

$$L = \int_\alpha^\beta \sqrt{[f'(t)]^2 + [g'(t)]^2}\,dt$$

Thus, using Leibniz notation, we have the following result, which has the same form as (9.6).

Theorem (9.8)

If a curve C is described by the parametric equations $x = f(t)$, $y = g(t)$, $\alpha \leq t \leq \beta$, where f' and g' are continuous on $[\alpha, \beta]$ and C is traversed exactly once as t increases from α to β, then the length of C is

$$L = \int_\alpha^\beta \sqrt{\left(\frac{dx}{dt}\right)^2 + \left(\frac{dy}{dt}\right)^2}\,dt$$

Notice that the formula in Theorem 9.8 is consistent with the general formulas $L = \int ds$ and $(ds)^2 = (dx)^2 + (dy)^2$ of Section 8.2.

EXAMPLE 1 If we use the representation of the unit circle given in Example 2 in Section 9.1,

$$x = \cos t \qquad y = \sin t \qquad 0 \leq t \leq 2\pi$$

then $dx/dt = -\sin t$ and $dy/dt = \cos t$, so Theorem 9.8 gives

$$
\begin{aligned}
L &= \int_0^{2\pi} \sqrt{\left(\frac{dx}{dt}\right)^2 + \left(\frac{dy}{dt}\right)^2}\,dt = \int_0^{2\pi} \sqrt{\sin^2 t + \cos^2 t}\,dt \\
&= \int_0^{2\pi} dt = 2\pi
\end{aligned}
$$

as expected. If, on the other hand, we use the representation given in Example 3 in Section 9.1,

$$x = \sin 2t \qquad y = \cos 2t \qquad 0 \le t \le 2\pi$$

then $dx/dt = 2\cos 2t$ and $dy/dt = -2\sin 2t$, and the integral in Theorem 9.8 gives

$$\int_0^{2\pi} \sqrt{\left(\frac{dx}{dt}\right)^2 + \left(\frac{dy}{dt}\right)^2}\, dt = \int_0^{2\pi} \sqrt{4\cos^2 2t + 4\sin^2 2t}\, dt = \int_0^{2\pi} 2\, dt = 4\pi$$

Notice that the integral gives twice the arc length of the circle because as t increases from 0 to 2π, the point $(\sin 2t, \cos 2t)$ traverses the circle twice. In general, when finding the length of a curve C from a parametric representation, we have to be careful to ensure that C is traversed only once as t increases from α to β.   •

EXAMPLE 2   Find the length of one arch of the cycloid $x = r(\theta - \sin\theta)$, $y = r(1 - \cos\theta)$.

Solution   From Example 5 in Section 9.1 we see that one arch is described by the parameter interval $0 \le \theta \le 2\pi$. Since

$$\frac{dx}{d\theta} = r(1 - \cos\theta) \qquad \text{and} \qquad \frac{dy}{d\theta} = r\sin\theta$$

we have

$$L = \int_0^{2\pi} \sqrt{\left(\frac{dx}{d\theta}\right)^2 + \left(\frac{dy}{d\theta}\right)^2}\, d\theta = \int_0^{2\pi} \sqrt{r^2(1 - \cos\theta)^2 + r^2\sin^2\theta}\, d\theta$$

$$= \int_0^{2\pi} \sqrt{r^2(1 - 2\cos\theta + \cos^2\theta + \sin^2\theta)}\, d\theta = r\int_0^{2\pi} \sqrt{2(1 - \cos\theta)}\, d\theta$$

In order to evaluate this integral we use the identity $\sin^2 x = \frac{1}{2}(1 - \cos 2x)$ with $\theta = 2x$, which gives $1 - \cos\theta = 2\sin^2(\theta/2)$. Since $0 \le \theta \le 2\pi$, we have $0 \le \theta/2 \le \pi$ and so $\sin(\theta/2) \ge 0$. Therefore

$$\sqrt{2(1 - \cos\theta)} = \sqrt{4\sin^2\!\left(\frac{\theta}{2}\right)} = 2\left|\sin\!\left(\frac{\theta}{2}\right)\right| = 2\sin\!\left(\frac{\theta}{2}\right)$$

and so

$$L = 2r\int_0^{2\pi} \sin\!\left(\frac{\theta}{2}\right) d\theta = 2r\left[-2\cos\!\left(\frac{\theta}{2}\right)\right]_0^{2\pi}$$

$$= 2r[2 + 2] = 8r$$

          •

The result of Example 2 says that the length of one arch of a cycloid is eight times the radius of the generating circle. This was first proved in 1658 by Sir Christopher Wren, who later became the architect of St. Paul's Cathedral in London.

Surface Area

In the same way as for arc length, we can adapt Formula 8.23 to obtain a formula for surface area. If the curve given by the parametric equations $x = f(t)$, $y = g(t)$, $\alpha \le t \le \beta$, is rotated about the x-axis, where f', g' are continuous and $g(t) \ge 0$, then the area of the resulting surface is given by

(9.9)
$$S = \int_\alpha^\beta 2\pi y \sqrt{\left(\frac{dx}{dt}\right)^2 + \left(\frac{dy}{dt}\right)^2}\, dt$$

The general symbolic formulas $S = \int 2\pi y\, ds$ and $S = \int 2\pi x\, ds$ (Formulas 8.25 and

8.26) are still valid, but for parametric curves we use

$$ds = \sqrt{\left(\frac{dx}{dt}\right)^2 + \left(\frac{dy}{dt}\right)^2} \, dt$$

EXAMPLE 3   Show that the surface area of a sphere of radius r is $4\pi r^2$.

Solution   The sphere is obtained by rotating the semicircle

$$x = r\cos t \qquad y = r\sin t, \qquad 0 \le t \le \pi$$

about the x-axis. Therefore, from Formula 9.9, we get

$$S = \int_0^\pi 2\pi r \sin t \sqrt{(-r\sin t)^2 + (r\cos t)^2} \, dt$$

$$= 2\pi \int_0^\pi r \sin t \sqrt{r^2(\sin^2 t + \cos^2 t)} \, dt$$

$$= 2\pi r^2 \int_0^\pi \sin t \, dt = 2\pi r^2 (-\cos t)\Big|_0^\pi = 4\pi r^2$$ ●

EXAMPLE 4   Find the area of the surface generated by rotating one arch of the cycloid $x = r(\theta - \sin\theta)$, $y = r(1 - \cos\theta)$ about the x-axis.

Solution   Using Formula 9.9 and the same identity as in Example 2, we have

$$S = \int_0^{2\pi} 2\pi y \sqrt{\left(\frac{dx}{d\theta}\right)^2 + \left(\frac{dy}{d\theta}\right)^2} \, d\theta$$

$$= \int_0^{2\pi} 2\pi r(1 - \cos\theta)\sqrt{r^2(1 - \cos^2\theta) + r^2 \sin^2\theta} \, d\theta$$

$$= 2\pi r^2 \int_0^{2\pi} (1 - \cos\theta)\sqrt{2(1 - \cos\theta)} \, d\theta = 2\pi r^2 \int_0^{2\pi} 2\sin^2\left(\frac{\theta}{2}\right) 2\sin\left(\frac{\theta}{2}\right) d\theta$$

$$= 8\pi r^2 \int_0^{2\pi} \left[1 - \cos^2\left(\frac{\theta}{2}\right)\right] \sin\left(\frac{\theta}{2}\right) d\theta = 16\pi r^2 \int_0^\pi (\sin t - \cos^2 t \sin t) \, dt$$

$$= 16\pi r^2 \left[-\cos t + \tfrac{1}{3}\cos^3 t\right]_0^\pi = \frac{64\pi r^2}{3}$$ ●

The problem in Example 4 was solved by Blaise Pascal (1623–1662) without using the Fundamental Theorem of Calculus.

SECTION 9.3   **Exercises**

In Exercises 1–4 set up, but do not evaluate, an integral that represents the length of the curve.

1. $x = t^3$, $y = t^4$, $0 \le t \le 1$

2. $x = t^2$, $y = 1 + 4t$, $0 \le t \le 2$

3. $x = t\sin t$, $y = t\cos t$, $0 \le t \le \pi/2$

4. $x = e^{-t}$, $y = te^{2t}$, $-1 \le t \le 1$

Find the lengths of the curves in Exercises 5–12.

5. $x = 1 + 2\sin \pi t$, $y = 3 - 2\cos \pi t$, $0 \le t \le 1$

6. $x = t^3$, $y = t^2$, $0 \le t \le 4$

7. $x = 5t^2 + 1$, $y = 4 - 3t^2$, $0 \le t \le 2$

8. $x = 3t - t^3$, $y = 3t^2$, $0 \le t \le 2$

9. $x = e^t \cos t$, $y = e^t \sin t$, $0 \le t \le \pi$

10. $x = a(\cos\theta + \theta\sin\theta)$, $y = a(\sin\theta - \theta\cos\theta)$, $0 \le \theta \le \pi$

11. $x = 2 - 3\sin^2\theta$, $y = \cos 2\theta$, $0 \le \theta \le \pi/2$

12. $x = e^t - t$, $y = 4e^{t/2}$, $0 \le t \le 1$

13. Use Simpson's Rule with $n = 10$ to estimate the length of the curve $x = \ln t$, $y = e^{-t}$, $1 \le t \le 2$.

14. In Exercise 39 in Section 9.1 you were asked to derive the parametric equations $x = 2a\cot\theta$, $y = 2a\sin^2\theta$ for the curve called the witch of Maria Agnesi. Use Simpson's Rule with $n = 4$ to estimate the length of the arc of this curve given by $\pi/4 \le \theta \le \pi/2$.

In Exercises 15 and 16 find the distance traveled by a particle with position (x, y) as t varies in the given time interval. Compare with the length of the curve.

15. $x = \sin^2\theta$, $y = \cos^2\theta$, $0 \le \theta \le 3\pi$

16. $x = \cos^2 t$, $y = \cos t$, $0 \le t \le 4\pi$

17. Show that the total length of the ellipse $x = a\sin\theta$, $y = b\cos\theta$, $a > b > 0$, is

$$L = 4a \int_0^{\pi/2} \sqrt{1 - e^2\sin^2\theta}\, d\theta$$

where e is the eccentricity of the ellipse ($e = c/a$ where $c = \sqrt{a^2 - b^2}$).

18. Find the total length of the astroid $x = a\cos^3\theta$, $y = a\sin^3\theta$.

19. Set up, but do not evaluate, an integral that represents the area of the surface obtained by rotating the curve $x = t^3$, $y = t^4$, $0 \le t \le 1$, about the x-axis.

20. If the arc of the curve in Exercise 14 is rotated about x-axis, estimate the area of the resulting surface using Simpson's Rule with $n = 4$.

In Exercises 21–26 find the area of the surface obtained by rotating the given curve about the x-axis.

21. $x = t^2 + 2/t$, $y = 8\sqrt{t}$, $1 \le t \le 9$

22. $x = t^3$, $y = t^2$, $0 \le t \le 1$

23. $x = e^t \cos t$, $y = e^t \sin t$, $0 \le t \le \pi/2$

24. $x = 3t - t^3$, $y = 3t^2$, $0 \le t \le 1$

25. $x = 2\cos\theta - \cos 2\theta$, $y = 2\sin\theta - \sin 2\theta$

26. $x = a\cos^3\theta$, $y = a\sin^3\theta$, $0 \le \theta \le \pi/2$

In Exercises 27 and 28 find the surface area generated by rotating the given curve about y-axis.

27. $x = 3t^2$, $y = 2t^3$, $0 \le t \le 5$

28. $x = e^t - t$, $y = 4e^{t/2}$, $0 \le t \le 1$

29. Find the surface area of the ellipsoid obtained by rotating the ellipse $x = a\cos\theta$, $y = b\sin\theta$ $(a > b)$ about (a) the x-axis and (b) the y-axis.

30. Use Formula 9.4 to derive Formula 9.9 from Formula 8.23 for the case where the curve can be represented in the form $y = F(x)$, $a \le x \le b$.

31. The **curvature** at a point P of a curve is defined as

$$\kappa = \left|\frac{d\phi}{ds}\right|$$

where ϕ is the angle of inclination of the tangent line at P, as shown in the figure. Thus the curvature is the absolute value of the rate of change of ϕ with respect to arc length. It can be regarded as a measure of the rate of change of direction of the curve at P and will be studied in greater detail in Chapter 11.

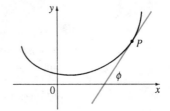

(a) For a parametric curve $x = x(t)$, $y = y(t)$, derive the formula

$$\kappa = \frac{|\dot{x}\ddot{y} - \ddot{x}\dot{y}|}{(\dot{x}^2 + \dot{y}^2)^{3/2}}$$

where the dots indicate derivatives with respect to t, so $\dot{x} = dx/dt$. [*Hint:* Use $\phi = \tan^{-1}(dy/dx)$ and Equation 9.4 to find $d\phi/dt$. Then use the Chain Rule to find $d\phi/ds$.]

(b) By regarding a curve $y = f(x)$ as the parametric curve $x = x$, $y = f(x)$, with parameter x, show that the formula in part (a) becomes

$$\kappa = \frac{|d^2y/dx^2|}{[1 + (dy/dx)^2]^{3/2}}$$

32. (a) Use the formula in Exercise 31(b) to find the curvature of the parabola $y = x^2$ at the point $(1, 1)$.

    (b) At what point does this parabola have maximum curvature?

33. Use the formula in Exercise 31(a) to find the curvature of the cycloid $x = \theta - \sin\theta$, $y = 1 - \cos\theta$ at the top of one of its arches.

34. (a) Show that the curvature at each point of a straight line is $\kappa = 0$.

    (b) Show that the curvature at each point of a circle of radius r is $\kappa = 1/r$.

SECTION 9.4

Polar Coordinates

A coordinate system represents a point in the plane by an ordered pair of numbers called coordinates. So far we have been using Cartesian coordinates, which are directed distances from two perpendicular axes. In this section we describe a coordinate system introduced by Newton, called the **polar coordinate system**, which is more convenient for many purposes.

We choose a point in the plane that is called the **pole** (or origin) and labeled O. Then we draw a ray (half-line) starting at O called the **polar axis**. This axis is usually drawn horizontally to the right and corresponds to the positive x-axis in Cartesian coordinates.

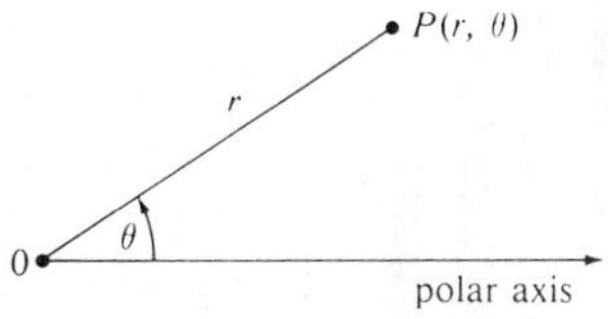

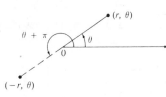

Figure 9.13

If P is any other point in the plane, let r be the distance from O to P and let θ be the angle (usually measured in radians) between the polar axis and the line OP as in Figure 9.13. Then the point P is represented by the ordered pair (r,θ) and r,θ are called **polar coordinates** of P. We use the convention that an angle is positive if measured in the counterclockwise direction from the polar axis and negative in the clockwise direction. If $P = O$, then $r = 0$ and we agree that $(0,\theta)$ represents the pole for any value of θ.

We extend the meaning of polar coordinates (r,θ) to the case where r is negative by agreeing that, as in Figure 9.14, the points $(-r,\theta)$ and (r,θ) lie on the same line through O and at the same distance $|r|$ from O, but on opposite sides of O. If $r > 0$, the point (r,θ) lies in the same quadrant as θ; if $r < 0$, it lies in the quadrant on the opposite side of the pole. Notice that $(-r,\theta)$ represents the same point as $(r,\theta + \pi)$.

Figure 9.14

EXAMPLE 1   Plot the points whose polar coordinates are (a) $(1,5\pi/4)$, (b) $(2,3\pi)$, (c) $(2,-2\pi/3)$, and (d) $(-3,3\pi/4)$.

Solution   The points are plotted in Figure 9.15. In part (d) the point $(-3,3\pi/4)$ is located three units from the pole in the fourth quadrant because the angle $3\pi/4$ is in the second quadrant and $r = -3$ is negative.

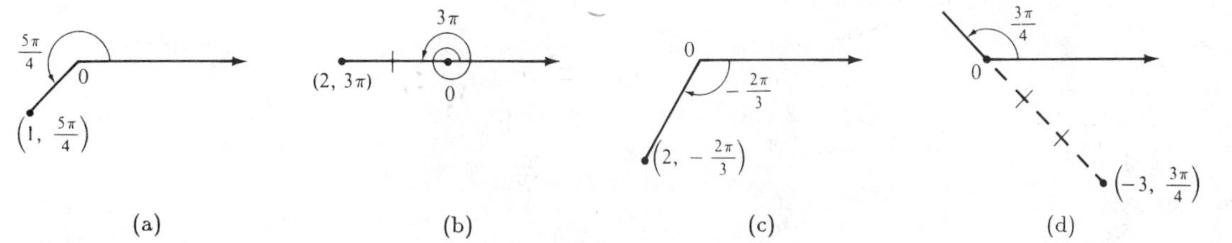

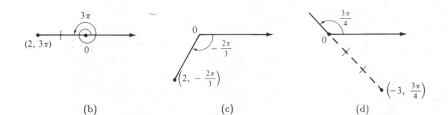

(a)   (b)   (c)   (d)

Figure 9.15

In the Cartesian coordinate system every point has only one representation, but in the polar coordinate system each point has many representations. For instance, the point $(1,5\pi/4)$ in Example 1(a) could be written as $(1,-3\pi/4)$ or $(1,13\pi/4)$ or $(-1,\pi/4)$ (see Figure 9.16).

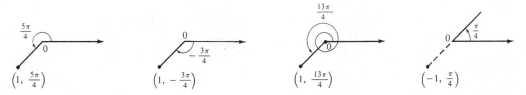

$\left(1,\dfrac{5\pi}{4}\right)$ $\left(1,-\dfrac{3\pi}{4}\right)$ $\left(1,\dfrac{13\pi}{4}\right)$ $\left(-1,\dfrac{\pi}{4}\right)$

Figure 9.16

In fact, since a complete counterclockwise rotation is given by an angle 2π, the point represented by polar coordinates (r, θ) is also represented by

$$(r, \theta + 2n\pi) \qquad \text{and} \qquad (-r, \theta + (2n+1)\pi)$$

where n is any integer.

Figure 9.17

The connection between polar and Cartesian coordinates can be seen from Figure 9.17 where the pole corresponds to the origin and the polar axis coincides with the positive x-axis.

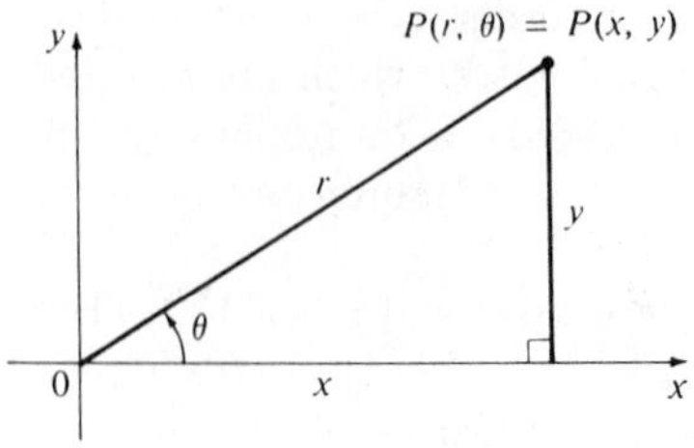

If the point P has Cartesian coordinates (x, y) and polar coordinates (r, θ), then, from the figure, we have

$$\cos\theta = \frac{x}{r} \qquad \sin\theta = \frac{y}{r}$$

(9.10)   and so

$$x = r\cos\theta \qquad y = r\sin\theta$$

Although Equations 9.10 were deduced from Figure 9.17, which illustrates the case where $r > 0$ and $0 < \theta < \pi/2$, these equations are valid for all values of r and θ. (See the general definition of $\sin\theta$ and $\cos\theta$ in Appendix B.)

Equations 9.10 allow us to find the Cartesian coordinates of a point when the polar coordinates are known. To find r and θ when x and y are known we use the equations

(9.11)

$$r^2 = x^2 + y^2 \qquad \tan\theta = \frac{y}{x}$$

which can be deduced from Equations 9.10 or simply read from Figure 9.17.

EXAMPLE 2   Convert the point $(2, \pi/3)$ from polar to Cartesian coordinates.

Solution   Since $r = 2$ and $\theta = \pi/3$, Equations 9.10 give

$$x = r\cos\theta = 2\cos\frac{\pi}{3} = 2 \cdot \frac{1}{2} = 1$$

$$y = r\sin\theta = 2\sin\frac{\pi}{3} = 2 \cdot \frac{\sqrt{3}}{2} = \sqrt{3}$$

Therefore the point is $(1, \sqrt{3})$ in Cartesian coordinates. $\qquad \bullet$

EXAMPLE 3   Represent the point with Cartesian coordinates $(1, -1)$ in terms of polar coordinates.

Solution   If we choose r to be positive, then Equations 9.11 give

$$r = \sqrt{x^2 + y^2} = \sqrt{1^2 + (-1)^2} = \sqrt{2}$$

$$\tan\theta = \frac{y}{x} = -1$$

Since the point $(1, -1)$ lies in the fourth quadrant, we can choose $\theta = -\pi/4$ or $\theta = 7\pi/4$. Thus one possible answer is $(\sqrt{2}, -\pi/4)$. Another is $(\sqrt{2}, 7\pi/4)$.

Note: Equations 9.11 do not uniquely determine θ when x and y are given because, as θ increases through the interval $0 \leq \theta \leq 2\pi$, each value of $\tan\theta$ occurs twice. Therefore, in converting from Cartesian to polar coordinates, it is not good enough just to find r and θ that satisfy Equations 9.11. As in Example 3, we must choose θ so that the point (r, θ) lies in the correct quadrant.

The **graph of a polar equation** $r = f(\theta)$, or more generally $F(r, \theta) = 0$, consists of all points P that have at least one polar representation (r, θ) whose coordinates satisfy the equation.

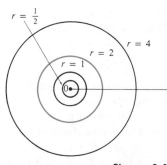

Figure 9.18

EXAMPLE 4   What curve is represented by the polar equation $r = 2$?

Solution   The curve consists of all points (r, θ) with $r = 2$. Since r represents the distance from the point to the pole, the curve $r = 2$ represents the circle with center O and radius 2. In general the equation $r = a$ represents a circle with center O and radius $|a|$ (see Figure 9.18).

EXAMPLE 5   Sketch the polar curve $\theta = 1$.

Solution   This curve consists of all points (r, θ) such that the polar angle θ is 1 radian. It is the straight line that passes through O and makes an angle of 1 radian with the polar axis (see Figure 9.19). Notice that the points $(r, 1)$ on the line with $r > 0$ are in the first quadrant, whereas those with $r < 0$ are in the third quadrant.

EXAMPLE 6
(a)   Sketch the curve with polar equation $r = 2\cos\theta$.
(b)   Find a Cartesian equation for this curve.

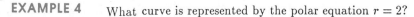

Figure 9.19

Solution

(a)   In Figure 9.20 we find the values of r for some convenient values of θ and plot the corresponding points (r, θ). Then we join these points to sketch the curve, which appears to be a circle. We have used only values of θ between 0 and π because if we let θ increase beyond π we obtain the same points again.

θ	$r = 2\cos\theta$
0	2
$\pi/6$	$\sqrt{3}$
$\pi/4$	$\sqrt{2}$
$\pi/3$	1
$\pi/2$	0
$2\pi/3$	-1
$3\pi/4$	$-\sqrt{2}$
$5\pi/6$	$-\sqrt{3}$
π	-2

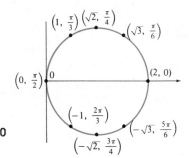

Figure 9.20
$r = 2\cos\theta$

(b)   To convert the given equation into a Cartesian equation we use Equations 9.10 and 9.11. From $x = r\cos\theta$ we have $\cos\theta = x/r$, so the equation $r = 2\cos\theta$ becomes $r = 2x/r$, which gives

$$2x = r^2 = x^2 + y^2 \qquad \text{or} \qquad x^2 + y^2 - 2x = 0$$

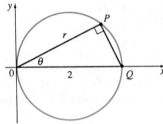

Completing the square, we obtain

$$(x-1)^2 + y^2 = 1$$

which is the equation of a circle with center $(1,0)$ and radius 1.   ●

Figure 9.21 shows a geometrical illustration that the circle in Example 6 has the equation $r = 2\cos\theta$. The angle OPQ is a right angle (Why?) and so $r/2 = \cos\theta$.

Figure 9.21

EXAMPLE 7   Sketch the curve $r = 1 + \sin\theta$.

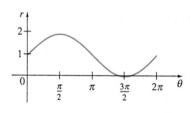

Figure 9.22

$r = 1 + \sin\theta$ in Cartesian coordinates, $0 < \theta \le 2\pi$

Solution   Instead of plotting points as in Example 6, we first sketch the graph of $r = 1 + \sin\theta$ in *Cartesian* coordinates in Figure 9.22 by shifting the sine curve up one unit. This will enable us to read at a glance the values of r that correspond to increasing values of θ. For instance, we see that as θ increases from 0 to $\pi/2$, r (the distance from O) increases from 1 to 2, so we sketch the corresponding part of the polar curve in Figure 9.23(a). As θ increases from $\pi/2$ to π, Figure 9.22 shows that r decreases from 2 to 1, so we sketch the next part of the curve in Figure 9.23(b). As θ increases from π to $3\pi/2$, r decreases from 1 to 0 as shown in part (c). Finally, as θ increases from $3\pi/2$ to 2π, r increases from 0 to 1 as shown in part (d). If we let θ increase beyond 2π or decrease beyond 0, we would simply retrace our path. Putting together the parts of the curve from Figure 9.23(a)–(d), we sketch the complete curve in Figure 9.23(e). It is called a **cardioid** because it is shaped like a heart.

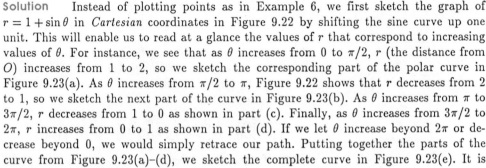

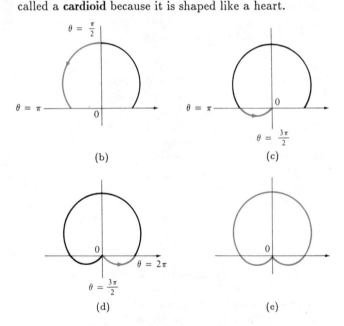

Figure 9.23

Stages in sketching the cardioid
$r = 1 + \sin\theta$

EXAMPLE 8   Sketch the curve $r = \cos 2\theta$.

Solution   As in Example 7, we first sketch $r = \cos 2\theta$, $0 \le \theta \le 2\pi$, in Cartesian coordinates in Figure 9.24. As θ increases from 0 to $\pi/4$, Figure 9.24 shows that r decreases from 1 to 0 and so we draw the corresponding portion of the polar curve in Figure 9.25 (indicated by a single arrow). As θ increases from $\pi/4$ to $\pi/2$, r goes from 0 to -1. This means that the distance from O increases from 0 to 1, but instead of being in the second quadrant this portion of the polar curve (indicated by a double arrow) lies on the opposite side of the pole in the fourth quadrant. The remainder of the curve is drawn in a similar fashion, with the arrows and numbers indicating the order in which the portions are traced out. The resulting curve has four loops and is called a **four-leaved rose.**

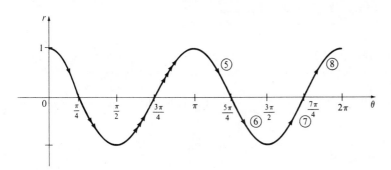

Figure 9.24

$r = \cos 2\theta$ in Cartesian coordinates

Figure 9.25

Four-leaved rose $r = \cos 2\theta$

In sketching polar curves it is sometimes helpful to take advantage of symmetry. The following three rules are explained by Figure 9.26.

(a) If a polar equation is unchanged when θ is replaced by $-\theta$, the curve is symmetric about the polar axis.

(b) If the equation is unchanged when r is replaced by $-r$, the curve is symmetric about the pole.

(c) If the equation is unchanged when θ is replaced by $\pi - \theta$, the curve is symmetric about the vertical line $\theta = \pi/2$.

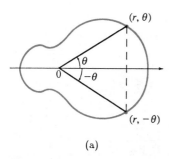

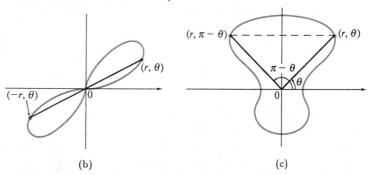

(a)   (b)   (c)

Figure 9.26

The curves in Examples 6 and 8 are symmetric about the polar axis since $\cos(-\theta) = \cos\theta$. The curves in Examples 7 and 8 are symmetric about $\theta = \pi/2$ because $\sin(\pi - \theta) = \sin\theta$ and $\cos 2(\pi - \theta) = \cos 2\theta$. The four-leaved rose is also symmetric about the pole. These symmetry properties could have been used in sketching the curves. For instance, in Example 6 we need only have plotted points for $0 \le \theta \le \pi/2$ and then reflected in the polar axis to obtain the complete circle.

Tangents to Polar Curves

To find a tangent line to a polar curve $r = f(\theta)$ we regard θ as a parameter and write its parametric equations as

$$x = r\cos\theta = f(\theta)\cos\theta \qquad y = r\sin\theta = f(\theta)\sin\theta$$

Then, using the method for finding slopes of parametric curves (Equation 9.4), we have

(9.12)
$$\frac{dy}{dx} = \frac{\dfrac{dy}{d\theta}}{\dfrac{dx}{d\theta}} = \frac{\dfrac{dr}{d\theta}\sin\theta + r\cos\theta}{\dfrac{dr}{d\theta}\cos\theta - r\sin\theta}$$

We locate horizontal tangents by finding the points where $dy/d\theta = 0$ (provided that $dx/d\theta \neq 0$). Likewise we locate vertical tangents at the points where $dx/d\theta = 0$ (provided that $dy/d\theta \neq 0$).

Notice that if we are looking for tangent lines at the pole, then $r = 0$ and Equation 9.12 simplifies to

$$\frac{dy}{dx} = \tan\theta \qquad \text{if} \quad \frac{dr}{d\theta} \neq 0$$

For instance, in Example 8 we found that $r = \cos 2\theta = 0$ when $\theta = \pi/4$ or $3\pi/4$. This means that the lines $\theta = \pi/4$ and $\theta = 3\pi/4$ (or $y = x$ and $y = -x$) are tangent lines to $r = \cos 2\theta$ at the origin.

EXAMPLE 9
(a) For the cardioid $r = 1 + \sin\theta$ of Example 7, find the slope of the tangent line when $\theta = \pi/3$.
(b) Find the points on the cardioid where the tangent line is horizontal or vertical.

Solution   Using Equation 9.12 with $r = 1 + \sin\theta$, we have

$$\frac{dy}{dx} = \frac{\dfrac{dr}{d\theta}\sin\theta + r\cos\theta}{\dfrac{dr}{d\theta}\cos\theta - r\sin\theta} = \frac{\cos\theta\sin\theta + (1 + \sin\theta)\cos\theta}{\cos\theta\cos\theta - (1 + \sin\theta)\sin\theta}$$

$$= \frac{\cos\theta(1 + 2\sin\theta)}{1 - 2\sin^2\theta - \sin\theta} = \frac{\cos\theta(1 + 2\sin\theta)}{(1 + \sin\theta)(1 - 2\sin\theta)}$$

(a) The slope of the tangent at the point where $\theta = \pi/3$ is

$$\frac{dy}{dx}\bigg|_{\theta=\pi/3} = \frac{\cos(\pi/3)(1 + 2\sin(\pi/3))}{(1 + \sin(\pi/3))(1 - 2\sin(\pi/3))}$$

$$= \frac{\frac{1}{2}(1 + \sqrt{3})}{\left(1 + \dfrac{\sqrt{3}}{2}\right)(1 - \sqrt{3})} = \frac{1 + \sqrt{3}}{(2 + \sqrt{3})(1 - \sqrt{3})}$$

$$= \frac{1 + \sqrt{3}}{-1 - \sqrt{3}} = -1$$

(b) Observe that

$$\frac{dy}{d\theta} = \cos\theta(1 + 2\sin\theta) = 0 \qquad \text{when } \theta = \frac{\pi}{2}, \frac{3\pi}{2}, \frac{7\pi}{6}, \frac{11\pi}{6}$$

$$\frac{dx}{d\theta} = (1 + \sin\theta)(1 - 2\sin\theta) = 0 \qquad \text{when } \theta = \frac{3\pi}{2}, \frac{\pi}{6}, \frac{5\pi}{6}$$

Therefore there are horizontal tangents at the points $(2, \pi/2)$, $(\frac{1}{2}, 7\pi/6)$, $(\frac{1}{2}, 11\pi/6)$ and vertical tangents at $(\frac{3}{2}, \pi/6)$ and $(\frac{3}{2}, 5\pi/6)$. When $\theta = 3\pi/2$, both $dy/d\theta$ and $dx/d\theta$ are 0, so we must be careful. Using L'Hospital's Rule, we have

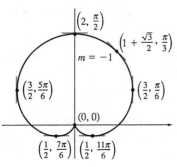

Figure 9.27

Tangent lines for $r = 1 + \sin\theta$

$$\lim_{\theta \to 3\pi/2^-} \frac{dy}{dx} = -\frac{1}{3} \lim_{\theta \to 3\pi/2^-} \frac{\cos\theta}{1 + \sin\theta}$$

$$= -\frac{1}{3} \lim_{\theta \to 3\pi/2^-} \frac{-\sin\theta}{\cos\theta} = \infty$$

By symmetry,

$$\lim_{\theta \to 3\pi/2^+} \frac{dy}{dx} = -\infty$$

Thus there is a vertical tangent line at the pole (see Figure 9.27).

Note: Instead of having to remember Equation 9.12, we could employ the method used to derive it. For instance, in Example 9 we could have written

$$x = r\cos\theta = (1 + \sin\theta)\cos\theta = \cos\theta + \tfrac{1}{2}\sin 2\theta$$

$$y = r\sin\theta = (1 + \sin\theta)\sin\theta = \sin\theta + \sin^2\theta$$

$$\frac{dy}{dx} = \frac{dy/d\theta}{dx/d\theta} = \frac{\cos\theta + 2\sin\theta\cos\theta}{-\sin\theta + \cos 2\theta} = \frac{\cos\theta + \sin 2\theta}{-\sin\theta + \cos 2\theta}$$

SECTION 9.4 **Exercises**

In Exercises 1–8 plot the points whose polar coordinates are given. Then find two other pairs of polar coordinates of these points, one with $r > 0$ and one with $r < 0$.

1. $(1, \pi/2)$
2. $(3, 0)$
3. $(-1, \pi/5)$
4. $(2, -\pi/7)$
5. $(4, -2\pi/3)$
6. $(3, 2)$
7. $(-1, \pi)$
8. $(-2, 3\pi/2)$

In Exercises 9–16 plot the points whose polar coordinates are given. Then find the Cartesian coordinates of these points.

9. $(\sqrt{2}, \pi/4)$
10. $(2, 2\pi/3)$
11. $(1.5, 3\pi/2)$
12. $(4, 3\pi)$
13. $(4, -7\pi/6)$
14. $(-4, 5\pi/4)$
15. $(-1, \pi/3)$
16. $(-2, -5\pi/6)$

In Exercises 17–20 the Cartesian coordinates of a point are given. Find the polar coordinates (r, θ) of the given point, where $r > 0$ and $0 \le \theta < 2\pi$.

17. $(-1, 1)$
18. $(-1, -\sqrt{3})$
19. $(2\sqrt{3}, -2)$
20. $(3, 4)$

In Exercises 21–26 sketch the region in the plane consisting of points whose polar coordinates satisfy the given conditions.

21. $r > 1$
22. $0 \le \theta \le \pi/3$
23. $0 \le r \le 2, \ \pi/2 \le \theta \le \pi$

24. $1 \le r < 3, \ -\pi/4 \le \theta \le \pi/4$
25. $3 < r < 4, \ -\pi/2 \le \theta \le \pi$
26. $-1 \le r \le 1, \ \pi/4 \le \theta \le 3\pi/4$
27. Find the distance between the points with polar coordinates $(1, \pi/6)$ and $(3, 3\pi/4)$.
28. Find a formula for the distance between the points with polar coordinates (r_1, θ_1) and (r_2, θ_2).

In Exercises 29–34 find a Cartesian equation for the curve described by the given polar equation.

29. $r\sin\theta = 2$
30. $r = 2\sin\theta$
31. $r = \dfrac{1}{1 - \cos\theta}$
32. $r = \dfrac{5}{3 - 4\sin\theta}$
33. $r^2 = \sin 2\theta$
34. $r^2 = \theta$

In Exercises 35–40 find a polar equation for the curve represented by the given Cartesian equation.

35. $y = 5$
36. $y = x + 1$
37. $x^2 + y^2 = 25$
38. $x^2 = 4y$
39. $2xy = 1$
40. $x^2 - y^2 = 1$

In Exercises 41–72 sketch the curve whose polar equation is given.

41. $r = 5$
42. $r = -1$
43. $\theta = 3\pi/4$
44. $\theta = -\pi/4$

45. $r = 2\sin\theta$ 46. $r = -4\sin\theta$

47. $r = -\cos\theta$

48. $r = 2\sin\theta + 2\cos\theta$

49. $r = \cos\theta - \sin\theta$

50. $r = 2(1 - \sin\theta)$

51. $r = 3(1 - \cos\theta)$

52. $r = 1 + \cos\theta$

53. $r = \theta,\ \theta \geq 0$ (spiral)

54. $r = \theta/2,\ -4\pi \leq \theta \leq 4\pi$

55. $r = 1/\theta$ (reciprocal spiral)

56. $r = e^{\theta}$ (logarithmic spiral)

57. $r = 1 - 2\cos\theta$ (limaçon)

58. $r = 2 + \cos\theta$ (limaçon)

59. $r = 3 + 2\sin\theta$ (limaçon)

60. $r = 3 - 4\sin\theta$ (limaçon)

61. $r = -3\cos 2\theta$

62. $r = \sin 2\theta$

63. $r = \sin 3\theta$ (three-leaved rose)

64. $r = 2\cos 3\theta$ (three-leaved rose)

65. $r = 2\cos 4\theta$ (eight-leaved rose)

66. $r = \sin 4\theta$ (eight-leaved rose)

67. $r = \sin 5\theta$ (five-leaved rose)

68. $r = -\cos 5\theta$ (five-leaved rose)

69. $r^2 = 4\cos 2\theta$ (lemniscate)

70. $r^2 = \sin 2\theta$ (lemniscate)

71. $r = 2\cos(3\theta/2)$

72. $r^2\theta = 1$ (lituus)

73. Show that the polar curve $r = 4 + 2\sec\theta$ (called a **conchoid**) has the line $x = 2$ as a vertical asymptote by showing that $\lim_{r\to\pm\infty} x = 2$. Use this fact to help sketch the conchoid.

74. Show that the curve $r = 2 - \csc\theta$ (also a conchoid) has the line $y = -1$ as a horizontal asymptote by showing that $\lim_{r\to\pm\infty} y = -1$. Use this fact to help sketch the conchoid.

75. Show that the curve $r = \sin\theta\tan\theta$ (called a **cissoid of Diocles**) has the line $x = 1$ as a vertical asymptote. Show also that the curve lies entirely within the vertical strip $0 \leq x < 1$. Use these facts to help sketch the cissoid.

76. Sketch the curve $(x^2 + y^2)^3 = 4x^2y^2$.

In Exercises 77–84 find the slope of the tangent line to the given polar curve at the point given by the value of θ.

77. $r = 3\cos\theta,\ \theta = \pi/3$

78. $r = \cos\theta + \sin\theta,\ \theta = \pi/4$

79. $r = \theta,\ \theta = \pi/2$

80. $r = \ln\theta,\ \theta = e$

81. $r = 1 + \cos\theta,\ \theta = \pi/6$

82. $r = 2 + 4\cos\theta,\ \theta = \pi/6$

83. $r = \sin 3\theta,\ \theta = \pi/3$

84. $r = \sin 3\theta,\ \theta = \pi/6$

In Exercises 85–90 find the points on the given curve where the tangent line is horizontal or vertical.

85. $r = 3\cos\theta$ 86. $r = \cos\theta + \sin\theta$

87. $r = \cos 2\theta$ 88. $r^2 = \sin 2\theta$

89. $r = 1 + \cos\theta$ 90. $r = e^{\theta}$

91. Show that the polar equation $r = a\sin\theta + b\cos\theta$, where $ab \neq 0$, represents a circle and find its center and radius.

92. Show that the curves $r = a\sin\theta$ and $r = a\cos\theta$ intersect at right angles.

93. Let P be any point (except the origin) on the curve $r = f(\theta)$. If ψ is the angle between the tangent line at P and the radial line OP, show that

$$\tan\psi = \frac{r}{dr/d\theta}$$

(*Hint:* Observe that $\psi = \phi - \theta$ in the figure.)

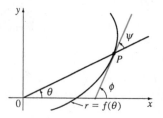

SECTION 9.5

Areas and Lengths in Polar Coordinates

In this section we develop the formula for the area of a region whose boundary is given by a polar equation. We need to use the formula for the area of a sector of a circle

(9.13)
$$A = \tfrac{1}{2}r^2\theta$$

where, as in Figure 9.28, r is the radius and θ is the radian measure of the central angle. Formula 9.13 can be proved as in Exercise 33 in Section 7.3 or by using the fact that the area of a sector is proportional to its central angle and so $A = (\theta/2\pi)\pi r^2 = \tfrac{1}{2}r^2\theta$.

Let $\mathcal{R}$ be the region, illustrated in Figure 9.29, bounded by the polar curve $r = f(\theta)$ and the rays $\theta = a$ and $\theta = b$, where f is a positive continuous function and $0 < b - a \leq 2\pi$. Now let P be a partition of $[a,b]$ by numbers θ_i such that $a = \theta_0 < \theta_1 < \cdots < \theta_n = b$. The rays $\theta = \theta_i$ then divide $\mathcal{R}$ into n smaller regions with central angles $\Delta\theta_i = \theta_i - \theta_{i-1}$. If we choose θ_i^* in the ith subinterval $[\theta_{i-1}, \theta_i]$, then the area ΔA_i of the ith region is approximated by the area of the sector of a circle with central angle $\Delta\theta_i$ and radius $f(\theta_i^*)$ (see Figure 9.30).

Figure 9.28

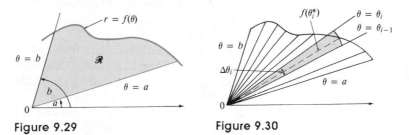

Figure 9.29   **Figure 9.30**

Thus from Formula 9.13 we have

$$\Delta A_i \approx \tfrac{1}{2}[f(\theta_i^*)]^2 \Delta\theta_i$$

and so an approximation to the total area A of $\mathcal{R}$ is

(9.14)
$$A \approx \sum_{i=1}^{n} \tfrac{1}{2}[f(\theta_i^*)]^2 \Delta\theta_i$$

It appears from Figure 9.30 that the approximation in (9.14) will improve as $\|P\| \to 0$. But the sums in (9.14) are Riemann sums for the function $g(\theta) = \tfrac{1}{2}[f(\theta)]^2$, so

$$\lim_{\|P\| \to 0} \sum_{i=1}^{n} \tfrac{1}{2}[f(\theta_i^*)]^2 \Delta\theta_i = \int_a^b \tfrac{1}{2}[f(\theta)]^2 \, d\theta$$

It therefore appears plausible (and can in fact be proved) that the formula for the area A of the polar region $\mathcal{R}$ is

(9.15)
$$A = \int_a^b \tfrac{1}{2}[f(\theta)]^2 \, d\theta$$

Formula 9.15 is often written as

(9.16)
$$A = \int_a^b \tfrac{1}{2}r^2 \, d\theta$$

with the understanding that $r = f(\theta)$. Note the similarity between Formulas 9.13 and 9.16.

In applying Formula 9.15 or 9.16 it is helpful to think of the area as being swept out by a rotating ray through O that starts with angle a and ends with angle b.

EXAMPLE 1　　Find the area enclosed by one loop of the four-leaved rose $r = \cos 2\theta$.

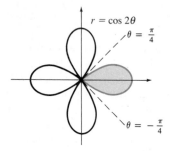

$r = \cos 2\theta$

$\theta = \dfrac{\pi}{4}$

$\theta = -\dfrac{\pi}{4}$

Figure 9.31

Solution　　The curve $r = \cos 2\theta$ was sketched in Example 8 in Section 9.4. Notice from Figure 9.31 that the region enclosed by the right loop is swept out by a ray that rotates from $\theta = -\pi/4$ to $\theta = \pi/4$. Therefore Formula 9.16 gives

$$A = \int_{-\pi/4}^{\pi/4} \tfrac{1}{2}r^2 \, d\theta = \tfrac{1}{2}\int_{-\pi/4}^{\pi/4} \cos^2 2\theta \, d\theta$$

$$= \tfrac{1}{2}\int_{-\pi/4}^{\pi/4} \tfrac{1}{2}(1 + \cos 4\theta) \, d\theta$$

$$= \tfrac{1}{4}\Big[\theta + \tfrac{1}{4}\sin 4\theta\Big]_{-\pi/4}^{\pi/4} = \frac{\pi}{8}$$

EXAMPLE 2　　Find the area of the region that lies inside the circle $r = 3\sin\theta$ and outside the cardioid $r = 1 + \sin\theta$.

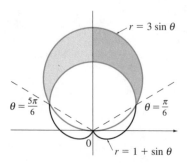

$r = 3\sin\theta$

$\theta = \dfrac{5\pi}{6}$

$\theta = \dfrac{\pi}{6}$

0

$r = 1 + \sin\theta$

Figure 9.32

Solution　　The cardioid (see Example 7 in Section 9.4) and the circle are sketched in Figure 9.32 and the desired region is shaded. The values of a and b in Formula 9.16 are determined by finding the points of intersection of the two curves. They intersect when $3\sin\theta = 1 + \sin\theta$, which gives $\sin\theta = \tfrac{1}{2}$, so $\theta = \pi/6$, $5\pi/6$. The desired area can be found by subtracting the area inside the cardioid between $\theta = \pi/6$ and $\theta = 5\pi/6$ from the area inside the circle from $\pi/6$ to $5\pi/6$. Thus

$$A = \tfrac{1}{2}\int_{\pi/6}^{5\pi/6} (3\sin\theta)^2 \, d\theta - \tfrac{1}{2}\int_{\pi/6}^{5\pi/6} (1 + \sin\theta)^2 \, d\theta$$

Since the region is symmetric about the vertical axis $\theta = \pi/2$, we can write

$$A = 2\left[\tfrac{1}{2}\int_{\pi/6}^{\pi/2} 9\sin^2\theta \, d\theta - \tfrac{1}{2}\int_{\pi/6}^{\pi/2} (1 + 2\sin\theta + \sin^2\theta) \, d\theta\right]$$

$$= \int_{\pi/6}^{\pi/2} (8\sin^2\theta - 1 - 2\sin\theta) \, d\theta$$

$$= \int_{\pi/6}^{\pi/2} (3 - 4\cos 2\theta - 2\sin\theta) \, d\theta$$

$$= 3\theta - 2\sin 2\theta + 2\cos\theta \,\big|_{\pi/6}^{\pi/2} = \pi$$

Example 2 illustrates the procedure for finding the area of the region bounded by two polar curves. In general let $\mathcal{R}$ be a region, as illustrated in Figure 9.33, that is bounded by curves with polar equations $r = f(\theta)$, $r = g(\theta)$, $\theta = a$, and $\theta = b$, where $f(\theta) \ge g(\theta) \ge 0$ and $0 < b - a \le 2\pi$. The area A of $\mathcal{R}$ is found by subtracting the area inside $r = g(\theta)$ from the area inside $r = f(\theta)$, so using Formula 9.15, we have

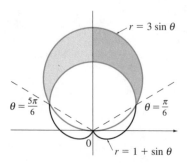

$r = f(\theta)$

$\mathcal{R}$

$\theta = b$

$r = g(\theta)$

$\theta = a$

0

Figure 9.33

$$A = \int_a^b \tfrac{1}{2}[f(\theta)]^2 \, d\theta - \int_a^b \tfrac{1}{2}[g(\theta)]^2 \, d\theta$$

$$= \tfrac{1}{2}\int_a^b \big([f(\theta)]^2 - [g(\theta)]^2\big) \, d\theta$$

Caution: The fact that a single point has many representations in polar coordinates sometimes makes it difficult to find all of the points of intersection of two polar curves. For instance, it is obvious from Figure 9.32 that the circle and the cardioid have three points of intersection; however, in Example 2 we solved the equations $r = 3\sin\theta$, $r = 1 + \sin\theta$ and found only two such points, $(3/2, \pi/6)$ and $(3/2, 5\pi/6)$. The origin is also a point of intersection but we could not find it by solving the equations of the curves because the origin has no single representation in polar coordinates that satisfies both equations. Notice that, when represented as $(0,0)$ or $(0,\pi)$, the origin satisfies $r = 3\sin\theta$ and so it lies on the circle; when represented as $(0, 3\pi/2)$, it satisfies $r = 1 + \sin\theta$ and so it lies on the cardioid. Think of two points moving along the curves as the parameter value θ increases from 0 to 2π. On one curve the origin is reached at $\theta = 0$ and $\theta = \pi$; on the other curve it is reached at $\theta = 3\pi/2$. The points do not collide at the origin because they reach the origin at different times, but the curves intersect there nonetheless.

Thus, to find *all* points of intersection of two curves, it is essential to draw the graphs of both curves.

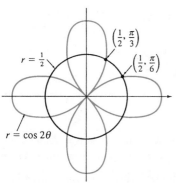

EXAMPLE 3   Find all points of intersection of the curves $r = \cos 2\theta$ and $r = 1/2$.

Solution   If we solve the equations $r = \cos 2\theta$ and $r = 1/2$, we get $\cos 2\theta = 1/2$ and therefore $2\theta = \pi/3, 5\pi/3, 7\pi/3, 11\pi/3$. Thus the values of θ between 0 and 2π that satisfy both equations are $\theta = \pi/6, 5\pi/6, 7\pi/6, 11\pi/6$. We have found four points of intersection: $(1/2, \pi/6)$, $(1/2, 5\pi/6)$, $(1/2, 7\pi/6)$, $(1/2, 11\pi/6)$.

However, you can see from Figure 9.34 that there are four other points of intersection—namely, $(1/2, \pi/3)$, $(1/2, 2\pi/3)$, $(1/2, 4\pi/3)$, and $(1/2, 5\pi/3)$. These can be found using symmetry or by noticing that another equation of the circle is $r = -1/2$ and solving the equations $r = \cos 2\theta$ and $r = -1/2$.   •

Figure 9.34

Arc Length

To find the length of a polar curve $r = f(\theta)$, $a \le \theta \le b$, we regard θ as a parameter and use Equations 9.10 to write the parametric equations of the curve as

$$x = r\cos\theta = f(\theta)\cos\theta \qquad y = r\sin\theta = f(\theta)\sin\theta$$

Using the Product Rule and differentiating with respect to θ, we obtain

$$\frac{dx}{d\theta} = \frac{dr}{d\theta}\cos\theta - r\sin\theta \qquad \frac{dy}{d\theta} = \frac{dr}{d\theta}\sin\theta + r\cos\theta$$

so, using $\cos^2\theta + \sin^2\theta = 1$, we have

$$\left(\frac{dx}{d\theta}\right)^2 + \left(\frac{dy}{d\theta}\right)^2 = \left(\frac{dr}{d\theta}\right)^2\cos^2\theta - 2r\frac{dr}{d\theta}\cos\theta\sin\theta + r^2\sin^2\theta$$
$$+ \left(\frac{dr}{d\theta}\right)^2\sin^2\theta + 2r\frac{dr}{d\theta}\sin\theta\cos\theta + r^2\cos^2\theta$$
$$= \left(\frac{dr}{d\theta}\right)^2 + r^2$$

Assuming that f' is continuous, we can use Theorem 9.8 to write the arc length as

$$L = \int_a^b \sqrt{\left(\frac{dx}{d\theta}\right)^2 + \left(\frac{dy}{d\theta}\right)^2}\, d\theta$$

Therefore the length of a curve with polar equation $r = f(\theta)$, $a \le \theta \le b$, is

(9.17)

$$L = \int_a^b \sqrt{r^2 + \left(\frac{dr}{d\theta}\right)^2}\, d\theta$$

EXAMPLE 4   Find the length of the cardioid $r = 1 + \sin\theta$.

Solution   This cardioid was sketched in Figure 9.23. Notice that it is symmetric about the vertical line $\theta = \pi/2$ so we can compute its total length as being twice the length of its right half:

$$L = 2\int_{-\pi/2}^{\pi/2} \sqrt{r^2 + \left(\frac{dr}{d\theta}\right)^2}\, d\theta = 2\int_{-\pi/2}^{\pi/2}\sqrt{(1+\sin\theta)^2 + \cos^2\theta}\, d\theta$$

$$= 2\int_{-\pi/2}^{\pi/2}\sqrt{2 + 2\sin\theta}\, d\theta = 2\sqrt{2}\int_{-\pi/2}^{\pi/2}\sqrt{1+\sin\theta}\,\frac{\sqrt{1-\sin\theta}}{\sqrt{1-\sin\theta}}\, d\theta$$

$$= 2\sqrt{2}\int_{-\pi/2}^{\pi/2}\frac{\sqrt{1-\sin^2\theta}}{\sqrt{1-\sin\theta}}\, d\theta = 2\sqrt{2}\int_{-\pi/2}^{\pi/2}\frac{\cos\theta}{\sqrt{1-\sin\theta}}\, d\theta$$

$$= 2\sqrt{2}\left[-2\sqrt{1-\sin\theta}\right]_{-\pi/2}^{\pi/2} = 8$$

Note: In the solution of Example 4 we used the fact that $\sqrt{\cos^2\theta} = \cos\theta$ when $-\pi/2 \le \theta \le \pi/2$ because $\cos\theta \ge 0$ for these values of θ. In general we must write $\sqrt{\cos^2\theta} = |\cos\theta|$. If we had not taken advantage of symmetry, then we would have had to write

$$L = \int_0^{2\pi}\sqrt{r^2 + \left(\frac{dr}{d\theta}\right)^2}\, d\theta$$

or

$$L = \int_{-\pi}^{\pi}\sqrt{r^2 + \left(\frac{dr}{d\theta}\right)^2}\, d\theta$$

which leads to

(9.18)

$$L = \sqrt{2}\int_0^{2\pi}\frac{\sqrt{\cos^2\theta}}{\sqrt{1-\sin\theta}}\, d\theta = \sqrt{2}\int_0^{2\pi}\frac{|\cos\theta|}{\sqrt{1-\sin\theta}}\, d\theta$$

If the absolute value symbol had been forgotten we would have obtained the false equation

$$L = \sqrt{2}\int_0^{2\pi}\frac{\cos\theta}{\sqrt{1-\sin\theta}}\, d\theta = -2\sqrt{2}\left[\sqrt{1-\sin\theta}\right]_0^{2\pi} = 0$$

The integral in Equation 9.18 can be evaluated by observing that $\cos\theta \ge 0$ on the intervals $[0, \pi/2]$ and $[3\pi/2, 2\pi]$ but $\cos\theta \le 0$ on the interval $[\pi/2, 3\pi/2]$. Thus

$$L = \sqrt{2}\int_0^{\pi/2}\frac{\cos\theta}{\sqrt{1-\sin\theta}}\, d\theta - \sqrt{2}\int_{\pi/2}^{3\pi/2}\frac{\cos\theta}{\sqrt{1-\sin\theta}}\, d\theta$$

$$+ \sqrt{2}\int_{3\pi/2}^{2\pi}\frac{\cos\theta}{\sqrt{1-\sin\theta}}\, d\theta$$

When these integrals are worked out, it is found that $L = 8$, but the work involved is much greater than in the solution using symmetry.

In Exercises 1–8 find the area of the region that is bounded by the given curve and lies in the given sector.

1. $r = \theta, \quad 0 \le \theta \le \pi$

2. $r = e^{\theta}, \quad -\pi/2 \le \theta \le \pi/2$

3. $r = 2\cos\theta, \quad 0 \le \theta \le \pi/6$

4. $r = 3\sin\theta, \quad \pi/4 \le \theta \le 3\pi/4$

5. $r = \theta^2, \quad \pi/2 \le \theta \le 3\pi/2$

6. $r = 1/\theta, \quad \pi/6 \le \theta \le 5\pi/6$

7. $r = \sin 2\theta, \quad 0 \le \theta \le \pi/6$

8. $r = \cos 3\theta, \quad -\pi/12 \le \theta \le \pi/12$

In Exercises 9–18 sketch the given curve and find the area that it encloses.

9. $r = 5\sin\theta$

10. $r = 2\cos\theta$

11. $r = 1 + \sin\theta$

12. $r = 4(1 - \cos\theta)$

13. $r^2 = 4\cos 2\theta$

14. $r^2 = \sin 2\theta$

15. $r = 4 - \sin\theta$

16. $r = 3 - \cos\theta$

17. $r = \sin 4\theta$

18. $r = \sin 3\theta$

In Exercises 19–24 find the area of the region enclosed by one loop of the given curve.

19. $r = \cos 3\theta$

20. $r = 3\sin 2\theta$

21. $r = \sin 5\theta$

22. $r = 2\cos 4\theta$

23. $r = 1 + 2\sin\theta$
    (inner loop)

24. $r = 2 + 3\cos\theta$
    (inner loop)

In Exercises 25–30 find the area of the region that lies inside the first curve and outside the second curve.

25. $r = 1 - \cos\theta, \quad r = \frac{3}{2}$

26. $r = 1 - \sin\theta, \quad r = 1$

27. $r = 4\sin\theta, \quad r = 2$

28. $r = 3\cos\theta, \quad r = 2 - \cos\theta$

29. $r = 3\cos\theta, \quad r = 1 + \cos\theta$

30. $r = 1 + \cos\theta, \quad r = 3\cos\theta$

In Exercises 31–36 find the area of the region that lies inside both curves.

31. $r = \sin\theta, \quad r = \cos\theta$

32. $r = \sin 2\theta, \quad r = \sin\theta$

33. $r = \sin 2\theta, \quad r = \cos 2\theta$

34. $r^2 = 2\sin 2\theta, \quad r = 1$

35. $r = 3 + 2\sin\theta, \quad r = 2$

36. $r = a\sin\theta, \quad r = b\cos\theta, \quad a > 0, b > 0$

37. Find the area inside the larger loop and outside the smaller loop of the limaçon $r = \frac{1}{2} + \cos\theta$.

38. Find the area inside the larger loop and outside the smaller loop of the limaçon $r = 3 + 4\sin\theta$.

In Exercises 39–44 find all points of intersection of the given curves.

39. $r = \sin\theta, \quad r = \cos\theta$

40. $r = 2, \quad r = 2\cos 2\theta$

41. $r = \cos\theta, \quad r = 1 - \cos\theta$

42. $r = \cos 3\theta, \quad r = \sin 3\theta$

43. $r = \sin\theta, \quad r = \sin 2\theta$

44. $r^2 = \sin 2\theta, \quad r^2 = \cos 2\theta$

In Exercises 45–52 find the length of the given polar curve.

45. $r = 5\cos\theta, \quad 0 \le \theta \le 3\pi/4$

46. $r = e^{-\theta}, \quad 0 \le \theta \le 3\pi$

47. $r = 2^{\theta}, \quad 0 \le \theta \le 2\pi$ 48. $r = \theta, \quad 0 \le \theta \le 2\pi$

49. $r = \theta^2, \quad 0 \le \theta \le 2\pi$ 50. $r = 1 + \cos\theta$

51. $r = \cos^2(\theta/4)$ 52. $r = \cos^2(\theta/2)$

53. Use Simpson's Rule with $n = 4$ to estimate the length of one loop of the four-leaved rose $r = \cos 2\theta$.

54. Use Simpson's Rule with $n = 4$ to estimate the length of the loop of the conchoid $r = 4 + 2\sec\theta$.

55. (a) Use Formula 9.9 to show that the area of the surface generated by rotating the polar curve $r = f(\theta), a \le \theta \le b$ (where f' is continuous and $0 \le a < b \le \pi$), about the polar axis is

$$S = \int_a^b 2\pi r\sin\theta\sqrt{r^2 + \left(\frac{dr}{d\theta}\right)^2}\,d\theta$$

(b) Use the formula in part (a) to find the surface area generated by rotating the lemniscate $r^2 = \cos 2\theta$ about the polar axis.

56. (a) Find a formula for the area of the surface generated by rotating the polar curve $r = f(\theta)$, $a \le \theta \le b$ (where f' is continuous and $0 \le a < b \le \pi$), about the line $\theta = \pi/2$.

(b) Find the surface area generated by rotating the lemniscate $r^2 = \cos 2\theta$ about the line $\theta = \pi/2$.

Conic Sections

In Section 3 of Review and Preview we sketched parabolas, ellipses, and hyperbolas starting with their equations. In this section we give geometric definitions of these curves and derive their standard equations. They are called **conic sections**, or **conics**, because they result from intersecting a cone with a plane as in Figure 9.35.

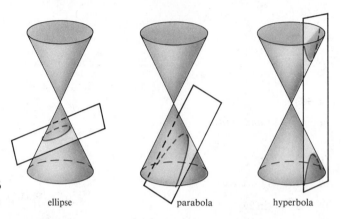

Figure 9.35

Conics   ellipse   parabola   hyperbola

Parabolas

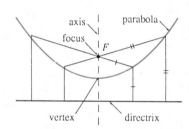

Figure 9.36

A **parabola** is the set of points in a plane that are equidistant from a fixed point F (called the **focus**) and a fixed line (called the **directrix**). This definition is illustrated by Figure 9.36. Notice that the point halfway between the focus and the directrix lies on the parabola; it is called the **vertex**. The line through the focus perpendicular to the directrix is called the **axis** of the parabola.

In the 16th century, Galileo showed that the path of a projectile that is shot into the air at an angle to the ground is a parabola. Since then, parabolic shapes have been used in designing automobile headlights, reflecting telescopes, and suspension bridges. (See problem 9 on page 254 for the reflection property of parabolas that makes them so useful.)

We obtain a particularly simple equation for a parabola if we place its vertex at the origin O and its directrix parallel to the x-axis as in Figure 9.37. If the focus is the point $(0, p)$, then the directrix has the equation $y = -p$. If $P(x, y)$ is any point on the parabola, then the distance from P to the focus is

$$|PF| = \sqrt{x^2 + (y - p)^2}$$

and the distance from P to the directrix is $|y + p|$. (Figure 9.37 illustrates the case where $p > 0$.) The defining property of a parabola is that these distances are equal:

$$\sqrt{x^2 + (y - p)^2} = |y + p|$$

We get an equivalent equation by squaring and simplifying:

$$x^2 + (y - p)^2 = |y + p|^2 = (y + p)^2$$

$$x^2 + y^2 - 2py + p^2 = y^2 + 2py + p^2$$

$$x^2 = 4py$$

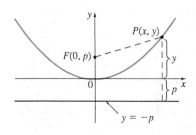

Figure 9.37

(9.19)

> The equation of a parabola with focus $(0, p)$ and directrix $y = -p$ is
>
> $$x^2 = 4py$$

If we write $a = 1/(4p)$, then the standard equation of a parabola (9.19) becomes $y = ax^2$ as in Section 3 of Review and Preview. It opens upward if $p > 0$ and downward if $p < 0$ (see Figure 9.38). The graph is symmetric with respect to the y-axis since (9.19) is unchanged when x is replaced by $-x$.

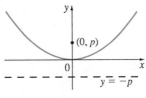 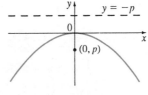

Figure 9.38   (a) $x^2 = 4py$, $p > 0$ (b) $x^2 = 4py$, $p < 0$

If we interchange x and y in (9.19), we obtain

(9.20)

$$y^2 = 4px$$

which is the equation of a parabola with focus $(p, 0)$ and directrix $x = -p$. (Interchanging x and y amounts to reflecting about the diagonal line $y = x$.) The parabola opens to the right if $p > 0$ and to the left if $p < 0$ (see Figure 9.39). In both cases the graph is symmetric with respect to the x-axis, which is the axis of the parabola.

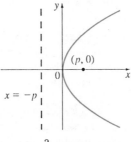 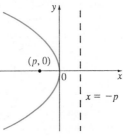

Figure 9.39   (a) $y^2 = 4px$, $p > 0$ (b) $y^2 = 4px$, $p < 0$

EXAMPLE 1   Find the focus and directrix of the parabola $y^2 + 10x = 0$ and sketch the graph.

Solution   If we write the equation as $y^2 = -10x$ and compare with Equation 9.20, we see that $4p = -10$, so $p = -\frac{5}{2}$. Thus the focus is $(p, 0) = (-\frac{5}{2}, 0)$ and the directrix is $x = \frac{5}{2}$. The sketch is shown in Figure 9.40.   ●

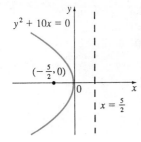

Ellipses

Figure 9.40   An **ellipse** is the set of points in a plane the sum of whose distances from two fixed points F_1 and F_2 is a constant (see Figure 9.41). These two fixed points are called the **foci** (plural of **focus**). One of Kepler's laws is that the orbits of the planets in the solar system are ellipses with the sun at one focus.

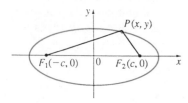

Figure 9.41

In order to obtain the simplest equation for an ellipse, we place the foci on the x-axis at the points $(-c, 0)$ and $(c, 0)$ as in Figure 9.42 so that the origin is halfway between the foci. Let the sum of the distances from a point on the ellipse to the foci be $2a > 0$. Then $P(x, y)$ will be a point on the ellipse when

$$|PF_1| + |PF_2| = 2a$$

that is,

$$\sqrt{(x+c)^2 + y^2} + \sqrt{(x-c)^2 + y^2} = 2a$$

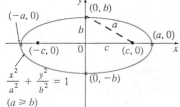

or

$$\sqrt{(x-c)^2 + y^2} = 2a - \sqrt{(x+c)^2 + y^2}$$

Squaring both sides, we have

$$x^2 - 2cx + c^2 + y^2 = 4a^2 - 4a\sqrt{(x+c)^2 + y^2} + x^2 + 2cx + c^2 + y^2$$

Figure 9.42

which simplifies to

$$a\sqrt{(x+c)^2 + y^2} = a^2 + cx$$

We square again:

$$a^2(x^2 + 2cx + c^2 + y^2) = a^4 + 2a^2cx + c^2x^2$$

which becomes

$$(a^2 - c^2)x^2 + a^2y^2 = a^2(a^2 - c^2)$$

From triangle F_1F_2P in Figure 9.42 we see that $2c < 2a$, so $c < a$ and therefore $a^2 - c^2 > 0$. For convenience let $b^2 = a^2 - c^2$. Then the equation of the ellipse becomes $b^2x^2 + a^2y^2 = a^2b^2$ or, if both sides are divided by a^2b^2,

(9.21)

$$\frac{x^2}{a^2} + \frac{y^2}{b^2} = 1$$

Since $b^2 = a^2 - c^2 < a^2$, it follows that $b < a$. The x-intercepts are found by setting $y = 0$. Then $x^2/a^2 = 1$, or $x^2 = a^2$, so $x = \pm a$. The corresponding points $(a, 0)$ and $(-a, 0)$ are called the **vertices** of the ellipse and the line segment joining the vertices is called the **major axis**. To find the y-intercepts we set $x = 0$ and obtain $y^2 = b^2$, so $y = \pm b$. Equation 9.21 is unchanged if x is replaced by $-x$ or y is replaced by $-y$, so the ellipse is symmetric about both axes. Notice that if the foci coincide, then $c = 0$, so $a = b$ and the ellipse becomes a circle with radius $r = a = b$.

We summarize this discussion as follows (see also Figure 9.43):

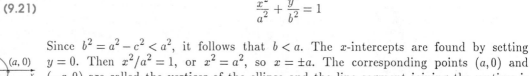

Figure 9.43

(9.22)

The ellipse

$$\frac{x^2}{a^2} + \frac{y^2}{b^2} = 1 \qquad a \geq b > 0$$

has foci $(\pm c, 0)$, where $c^2 = a^2 - b^2$, and vertices $(\pm a, 0)$.

If the foci of an ellipse are located on the y-axis at $(0, \pm c)$, then we can find its equation by interchanging x and y in (9.22) (see Figure 9.44).

(9.23)

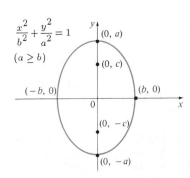

$$\frac{x^2}{b^2} + \frac{y^2}{a^2} = 1$$

$(a \geq b)$

Figure 9.44

> The ellipse
>
> $$\frac{x^2}{b^2} + \frac{y^2}{a^2} = 1 \qquad a \geq b > 0$$
>
> has foci $(0, \pm c)$, where $c^2 = a^2 - b^2$, and vertices $(0, \pm a)$.

EXAMPLE 2 Sketch the graph of $9x^2 + 16y^2 = 144$ and locate the foci.

Solution Divided both sides of the equation by 144:

$$\frac{x^2}{16} + \frac{y^2}{9} = 1$$

The equation is now in the standard form for an ellipse, so we have $a^2 = 16$, $b^2 = 9$, $a = 4$, and $b = 3$. The x-intercepts are ± 4; the y-intercepts are ± 3. Also $c^2 = a^2 - b^2 = 7$, so $c = \sqrt{7}$ and the foci are $(\pm\sqrt{7}, 0)$. The graph is sketched in Figure 9.45.

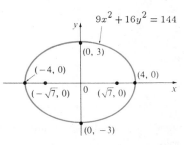

Figure 9.45

EXAMPLE 3 Find an equation of the ellipse with foci $(0, \pm 2)$ and vertices $(0, \pm 3)$.

Solution Using the notation of (9.23) we have $c = 2$ and $a = 3$. Then we obtain $b^2 = a^2 - c^2 = 9 - 4 = 5$, so the equation of the ellipse is

$$\frac{x^2}{5} + \frac{y^2}{9} = 1$$

Another way of writing the equation is $9x^2 + 5y^2 = 45$.

Like parabolas, ellipses have an interesting reflection property that has practical consequences. If a source of light or sound is placed at one focus of a surface with elliptical cross-sections, then all the light or sound is reflected off the surface to the other focus (see Exercise 49). This principle is used in *lithotripsy*, the new treatment for kidney stones. A reflector with elliptical cross-section is placed in such a way that the kidney stone is at one focus. High-intensity sound waves generated at the other focus are reflected to the stone and destroy it without damaging surrounding tissue. The patient is spared the trauma of surgery and recovers within a few days.

Hyperbolas

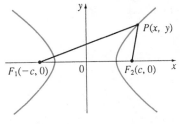

Figure 9.46

P is on the hyperbola when
$$|PF_1| - |PF_2| = \pm 2a$$

A **hyperbola** is the set of all points in a plane the difference of whose distances from two fixed points F_1 and F_2 (the foci) is a constant. This definition is illustrated in Figure 9.46.

Notice that the definition of a hyperbola is similar to that of an ellipse; the only change is that the sum of distances has become a difference of distances. In fact, the derivation of the equation of a hyperbola is also similar to the one given earlier for an ellipse. It is left as Exercise 42 to show that when the foci are on the x-axis at $(\pm c, 0)$ and the difference of distances is $|PF_1| - |PF_2| = \pm 2a$, then the equation of the hyperbola is

(9.24)

$$\frac{x^2}{a^2} - \frac{y^2}{b^2} = 1$$

where $c^2 = a^2 + b^2$. Notice that the x-intercepts are again $\pm a$ and the points $(a, 0)$ and $(-a, 0)$ are the **vertices** of the hyperbola. But if we put $x = 0$ in Equation 9.24 we get $y^2 = -b^2$, which is impossible, so there is no y-intercept. The hyperbola is symmetric with respect to both axes.

To analyze the hyperbola further, we look at equation 9.24 and obtain

$$\frac{x^2}{a^2} = 1 + \frac{y^2}{b^2} \geq 1$$

This shows that $x^2 \geq a^2$, so $|x| = \sqrt{x^2} \geq a$. Therefore we have $x \geq a$ or $x \leq -a$. This means that the hyperbola consists of two parts, called its branches.

In drawing a hyperbola it is useful to draw first its **asymptotes**, which are the lines $y = (b/a)x$ and $y = -(b/a)x$ shown in Figure 9.47. Both branches of the hyperbola approach the asymptotes; that is, they come arbitrarily close to the asymptotes. [See Exercise 81 in Section 4.5 where $y = (b/a)x$ is shown to be a slant asymptote.]

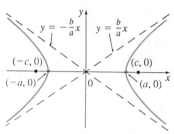

Figure 9.47
$$\frac{x^2}{a^2} - \frac{y^2}{b^2} = 1$$

(9.25)

The hyperbola

$$\frac{x^2}{a^2} - \frac{y^2}{b^2} = 1$$

has foci $(\pm c, 0)$, where $c^2 = a^2 + b^2$, vertices $(\pm a, 0)$, and asymptotes $y = \pm(b/a)x$.

If the foci of a hyperbola are on the y-axis, then by reversing the roles of x and y we obtain the following information, which is illustrated in Figure 9.48.

(9.26)

The hyperbola

$$\frac{y^2}{a^2} - \frac{x^2}{b^2} = 1$$

has foci $(0, \pm c)$, where $c^2 = a^2 + b^2$, vertices $(0, \pm a)$, and asymptotes $y = \pm(a/b)x$.

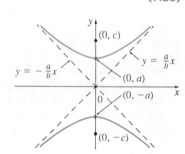

Figure 9.48
$$\frac{y^2}{a^2} - \frac{x^2}{b^2} = 1$$

EXAMPLE 4   Find the foci and asymptotes of the hyperbola $9x^2 - 16y^2 = 144$ and sketch its graph.

Solution   If we divide both sides of the equation by 144, it becomes

$$\frac{x^2}{16} - \frac{y^2}{9} = 1$$

which is of the form given in (9.25) with $a = 4$ and $b = 3$. Since $c^2 = 16 + 9 = 25$, the foci are $(\pm 5, 0)$. The asymptotes are the lines $y = \frac{3}{4}x$ and $y = -\frac{3}{4}x$. The graph is shown in Figure 9.49. ●

EXAMPLE 5   Find the foci and equation of the hyperbola with vertices $(0, \pm 1)$ and asymptote $y = 2x$.

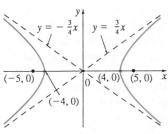

Figure 9.49
$9x^2 - 16y^2 = 144$

Solution   From (9.26) and the given information we see that $a = 1$ and $a/b = 2$. Thus $b = a/2 = \frac{1}{2}$ and $c^2 = a^2 + b^2 = \frac{5}{4}$. The foci are $(0, \pm\sqrt{5}/2)$ and the equation of the hyperbola is

$$y^2 - 4x^2 = 1$$

Shifted Conics

As in Section 3 in Review and Preview we shift conics by taking the standard equations 9.19, 9.20 9.22, 9.23, 9.25, and 9.26 and replacing x and y by $x - h$ and $y - k$.

EXAMPLE 6   Find an equation of the ellipse with foci $(2, -2)$, $(4, -2)$ and vertices $(1, -2)$, $(5, -2)$.

Solution   The major axis is the line segment that joins the vertices $(1, -2)$ and $(5, -2)$, which has length 4, so $a = 2$. The distance between the foci is 2, so $c = 1$. Thus $b^2 = a^2 - c^2 = 3$. Since the center of the ellipse is $(3, -2)$ we replace x and y in (9.22) by $x - 3$ and $y + 2$ to obtain

$$\frac{(x-3)^2}{4} + \frac{(y+2)^2}{3} = 1$$

as the equation of the given ellipse.

EXAMPLE 7   Sketch the conic

$$9x^2 - 4y^2 - 72x + 8y + 176 = 0$$

and find its foci.

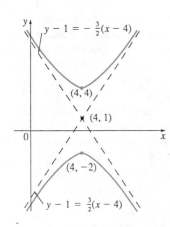

Figure 9.50
$9x^2 - 4y^2 - 72x + 8y + 176 = 0$

Solution   We complete the squares as follows:

$$4(y^2 - 2y) - 9(x^2 - 8x) = 176$$
$$4(y^2 - 2y + 1) - 9(x^2 - 8x + 16) = 176 + 4 - 144$$
$$4(y-1)^2 - 9(x-4)^2 = 36$$
$$\frac{(y-1)^2}{9} - \frac{(x-4)^2}{4} = 1$$

This is in the form (9.26) except that x and y are replaced by $x - 4$ and $y - 1$. Thus $a^2 = 9$, $b^2 = 4$, and $c^2 = 13$. The hyperbola is shifted four units to the right and one unit upward. The foci are $(4, 1 + \sqrt{13})$ and $(4, 1 - \sqrt{13})$ and the vertices are $(4, 4)$ and $(4, -2)$. The asymptotes are $y - 1 = \pm\frac{3}{2}(x - 4)$. The hyperbola is sketched in Figure 9.50.

SECTION 9.6 **Exercises**

In Exercises 1–8 find the vertex, focus, and directrix of the parabola and sketch the graph.

1. $x^2 = -8y$

2. $x = -5y^2$

3. $y^2 = x$

4. $2x^2 = y$

5. $x + 1 = 2(y - 3)^2$

6. $x^2 - 6x + 8y = 7$

7. $2x + y^2 - 8y + 12 = 0$

8. $x^2 + 12x - y + 39 = 0$

In Exercises 9–20 find the vertices and foci of the conic and sketch the graph. In the case of a hyperbola, find the asymptotes.

9. $\dfrac{x^2}{16} + \dfrac{y^2}{4} = 1$ 10. $\dfrac{x^2}{4} + \dfrac{y^2}{25} = 1$

11. $25x^2 + 9y^2 = 225$ 12. $x^2 + 4y^2 = 4$

13. $\dfrac{x^2}{144} - \dfrac{y^2}{25} = 1$ 14. $\dfrac{y^2}{25} - \dfrac{x^2}{144} = 1$

15. $9y^2 - x^2 = 9$ 16. $x^2 - y^2 = 1$

17. $9x^2 - 18x + 4y^2 = 27$

18. $16x^2 - 9y^2 + 64x - 90y = 305$

19. $2y^2 - 3x^2 - 4y + 12x + 8 = 0$

20. $x^2 + 2y^2 - 6x + 4y + 7 = 0$

In Exercises 21–38 find an equation for the conic that satisfies the given conditions.

21.   Parabola, focus $(0,3)$, directrix $y = -3$

22.   Parabola, focus $(-2,0)$, directrix $x = 2$

23.   Parabola, focus $(3,0)$, directrix $x = 1$

24.   Parabola, focus $(1,-1)$, directrix $y = 5$

25.   Parabola, vertex $(0,0)$, axis the x-axis, passing through $(1,-4)$

26.   Parabola, vertical axis, passing through $(-2,3)$, $(0,3)$, and $(1,9)$

27.   Ellipse, foci $(\pm 1, 0)$, vertices $(\pm 2, 0)$

28.   Ellipse, foci $(0, \pm 4)$, vertices $(0, \pm 5)$

29.   Ellipse, foci $(3, \pm 1)$, vertices $(3, \pm 3)$

30.   Ellipse, foci $(\pm 1, 2)$, length of major axis 6

31.   Ellipse, center $(2,2)$, focus $(0,2)$, vertex $(5,2)$

32.   Ellipse, foci $(\pm 2, 0)$, passing through $(2,1)$

33.   Hyperbola, foci $(0, \pm 3)$, vertices $(0, \pm 1)$

34.   Hyperbola, foci $(\pm 6, 0)$, vertices $(\pm 4, 0)$

35.   Hyperbola, foci $(1,3)$ and $(7,3)$, vertices $(2,3)$ and $(6,3)$

36.   Hyperbola, foci $(2,-2)$ and $(2,8)$, vertices $(2,0)$ and $(2,6)$

37.   Hyperbola, vertices $(\pm 3, 0)$, asymptotes $y = \pm 2x$

38.   Hyperbola, foci $(2,2)$ and $(6,2)$, asymptotes $y = x - 2$ and $y = 6 - x$

39.   The point in a lunar orbit nearest the surface of the moon is called *perilune* and the point farthest from the surface is called *apolune*. The Apollo 11 spacecraft was placed in an elliptical lunar orbit with perilune altitude 110 km and apolune altitude 314 km (above the moon). Find an equation of this ellipse if the radius of the moon is 1728 km and the center of the moon is at one focus.

40.   A cross-section of a parabolic reflector is shown in the figure. The bulb is located at the focus and the opening at the focus is 10 cm.
     (a)   Find an equation of the parabola.
     (b)   Find the diameter of the opening $|CD|$, 11 cm from the vertex.

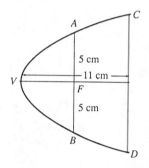

41.   In the LORAN (LOng RAnge Navigation) radio navigation system two radio stations located at A and B transmit simultaneous signals to a ship or an aircraft located at P. The onboard computer converts the time difference in receiving these signals into a distance difference $|PA| - |PB|$, and this, according to the definition of a hyperbola, locates the ship or aircraft on one branch of a hyperbola (see the figure). Suppose that station B is located 400 mi due east of station A on a coastline. A ship received the signal from B 1200 microseconds before it received the signal from A.

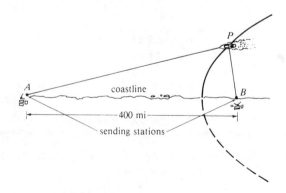

(a) Assuming that radio signals travel at a speed of 980 ft/microsecond, find the equation of the hyperbola on which the ship lies.

(b) If the ship is due north of B, how far off the coastline is the ship?

42. Use the definition of a hyperbola to derive Equation 9.24 for a hyperbola with foci $(\pm c, 0)$ and vertices $(\pm a, 0)$.

43. Show that the function defined by the upper branch of the hyperbola $y^2/a^2 - x^2/b^2 = 1$ is concave upward.

44. Find an equation for the ellipse with foci $(1, 1)$, $(-1, -1)$ and major axis of length 4.

45. What type of curve is represented by the equation

$$\frac{x^2}{k} + \frac{y^2}{k - 16} = 1$$

in the following cases: (a) $k > 16$, (b) $0 < k < 16$, and (c) $k < 0$? (d) Show that all of the curves in parts (a) and (b) have the same foci, no matter what the value of k is.

46. (a) Show that the equation of the tangent line to the parabola $y^2 = 4px$ at the point (x_0, y_0) can be written as

$$y_0 y = 2p(x + x_0)$$

(b) What is the x-intercept of this tangent line? Use this fact to draw the tangent line.

47. Use Simpson's Rule with $n = 10$ to estimate the length of the ellipse $x^2 + 4y^2 = 4$.

48. The planet Pluto travels in an elliptical orbit around the sun (at one focus). The length of the major axis is 1.18×10^{10} km and the length of the minor axis is 1.14×10^{10} km. Use Simpson's Rule with $n = 10$ to estimate the distance traveled by the planet during one complete orbit around the sun.

49. Let $P(x_1, y_1)$ be a point on the ellipse $x^2/a^2 + y^2/b^2 = 1$ with foci F_1 and F_2 and let α and β be the angles between the lines PF_1, PF_2, and the ellipse as in the figure. Prove that $\alpha = \beta$. This ex-

plains how whispering galleries and lithotripsy work. Sound coming from one focus is reflected and passes through the other focus. [*Hint:* Use the formula in Exercise 64(b) in Section 2 of Review and Preview to show that $\tan \alpha = \tan \beta$.]

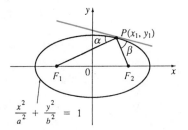

50. Let $P(x_1, y_1)$ be a point on the hyperbola $x^2/a^2 - y^2/b^2 = 1$ with foci F_1 and F_2 and let α, β be the angles between the lines PF_1 and PF_2 and the hyperbola as in the figure. Prove that $\alpha = \beta$. (This is the "reflection property" of the hyperbola. It shows that light aimed at a focus F_2 of a hyperbolic mirror is reflected toward the other focus F_1.)

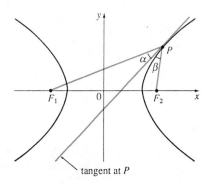

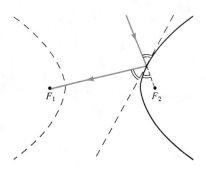

SECTION 9.7

Conic Sections in Polar Coordinates

In the preceding section we defined the parabola in terms of a focus and directrix but we defined the ellipse and hyperbola in terms of two foci. In this section we give a more unified treatment of all three types of conic sections in terms of a focus and directrix. Furthermore, if we place the focus at the origin, then a conic section has a simple polar equation. In Chapter 11 we will use the polar equation of an ellipse to derive Kepler's laws of planetary motion.

Theorem (9.27)

> Let F be a fixed point (called the **focus**) and l be a fixed line (called the **directrix**) in a plane. Let e be a fixed positive number (called the **eccentricity**). The set of all points P in the plane such that
>
> $$\frac{|PF|}{|Pl|} = e$$
>
> (that is, the ratio of **the distance from** F to the distance from l is the constant e) is a conic section. The conic is
>
> (a)   an ellipse if $e < 1$
>
> (b)   a parabola if $e = 1$
>
> (c)   a hyperbola if $e > 1$

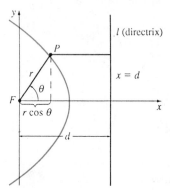

Figure 9.51

Proof   Notice that if the eccentricity is $e = 1$, then $|PF| = |Pl|$ and so the given condition simply becomes the definition of a parabola as in Section 9.6.

Let us place the focus F at the origin and the directrix parallel to the y-axis and d units to the right. Thus the directrix has equation $x = d$ and is perpendicular to the polar axis. If the point P has polar coordinates (r, θ), we see from Figure 9.51 that

$$|PF| = r \qquad |Pl| = d - r\cos\theta$$

Thus the condition $|PF|/|Pl| = e$, or $|PF| = e|Pl|$, becomes

(9.28)
$$r = e(d - r\cos\theta)$$

If we square both sides of this polar equation and convert to rectangular coordinates, we get

$$x^2 + y^2 = e^2(d - x)^2 = e^2(d^2 - 2\,dx + x^2)$$

or
$$(1 - e^2)x^2 + 2de^2x + y^2 = e^2d^2$$

After completing the square, we have

(9.29)
$$\left(x + \frac{e^2d}{1-e^2}\right)^2 + \frac{y^2}{1-e^2} = \frac{e^2d^2}{(1-e^2)^2}$$

If $e < 1$, we recognize Equation 9.29 as being the equation of an ellipse. In fact, it is of the form

$$\frac{(x-h)^2}{a^2} + \frac{y^2}{b^2} = 1$$

(9.30) where
$$h = -\frac{e^2 d}{1-e^2} \qquad a^2 = \frac{e^2 d^2}{(1-e^2)^2} \qquad b^2 = \frac{e^2 d^2}{1-e^2}$$

In Section 9.6 we found that the foci of an ellipse are at a distance c from the center, where

(9.31)
$$c^2 = a^2 - b^2 = \frac{e^4 d^2}{(1-e^2)^2}$$

This shows that
$$c = \frac{e^2 d}{1-e^2} = -h$$

and confirms that the focus as defined in Theorem 9.27 means the same as the focus defined in Section 9.6. It also follows from the Equations 9.30 and 9.31 that the eccentricity is given by

$$e = \frac{c}{a}$$

If $e > 1$, then $1 - e^2 < 0$ and we see that Equation 9.29 represents a hyperbola. Just as we did before, we could rewrite Equation 9.29 in the form

$$\frac{(x-h)^2}{a^2} - \frac{y^2}{b^2} = 1$$

and see that
$$e = \frac{c}{a} \qquad \text{where } c^2 = a^2 + b^2 \qquad \bullet$$

By solving Equation 9.28 for r, we see that the polar equation of the conic shown in Figure 9.51 can be written as

$$r = \frac{ed}{1 + e\cos\theta}$$

If the directrix is chosen to be to the left of the focus as $x = -d$, or if the directrix is chosen to be parallel to the polar axis as $y = \pm d$, then the polar equation of the conic is given by the following theorem, which is illustrated by Figure 9.52. (See Exercises 21–23.)

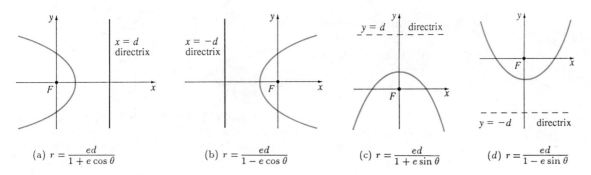

(a) $r = \dfrac{ed}{1 + e\cos\theta}$ (b) $r = \dfrac{ed}{1 - e\cos\theta}$ (c) $r = \dfrac{ed}{1 + e\sin\theta}$ (d) $r = \dfrac{ed}{1 - e\sin\theta}$

Figure 9.52

Polar equations of conics

Theorem (9.32)

A polar equation of the form

$$r = \frac{ed}{1 \pm e\cos\theta} \qquad \text{or} \qquad r = \frac{ed}{1 \pm e\sin\theta}$$

represents a conic section with eccentricity e. The conic is an ellipse if $e < 1$, a parabola if $e = 1$, or a hyperbola if $e > 1$.

EXAMPLE 1   Find a polar equation for a parabola that has its focus at the origin and whose directrix is the line $y = -6$.

Solution   Using Theorem 9.32 with $e = 1$ and $d = 6$, and using part (d) of Figure 9.52, we see that the equation of the parabola is

$$r = \frac{6}{1 - \sin\theta}$$

EXAMPLE 2   A conic is given by the polar equation

$$r = \frac{10}{3 - 2\cos\theta}$$

Find the eccentricity, identify the conic, locate the directrix, and sketch the conic.

Solution   Dividing numerator and denominator by 3, we write the equation as

$$r = \frac{\frac{10}{3}}{1 - \frac{2}{3}\cos\theta}$$

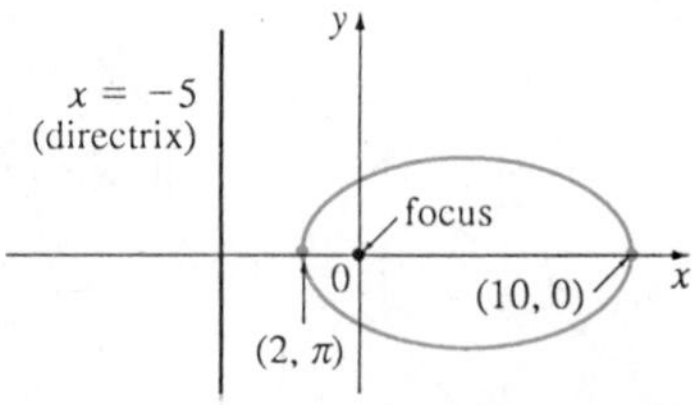

From Theorem 9.32 we see that this represents an ellipse with $e = \frac{2}{3}$. Since $ed = \frac{10}{3}$, we have

$$d = \frac{\frac{10}{3}}{e} = \frac{\frac{10}{3}}{\frac{2}{3}} = 5$$

so the directrix has Cartesian equation $x = -5$. When $\theta = 0$, $r = 10$; when $\theta = \pi$, $r = 2$. So the vertices have polar coordinates $(10, 0)$, $(2, \pi)$. The ellipse is sketched in Figure 9.53.

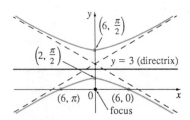

Figure 9.53

$$r = \frac{10}{3 - 2\cos\theta}$$

EXAMPLE 3   Sketch the conic $r = \dfrac{12}{2 + 4\sin\theta}$.

Solution   Writing the equation in the form

$$r = \frac{6}{1 + 2\sin\theta}$$

we see that the eccentricity is $e = 2$ and the equation therefore represents a hyperbola. Since $ed = 6$, $d = 3$ and the directrix has equation $y = 3$. The vertices occur when $\theta = \pi/2$ and $3\pi/2$, so they are $(2, \pi/2)$ and $(-6, 3\pi/2) = (6, \pi/2)$. It is also useful to plot the x-intercepts. These occur when $\theta = 0$, π and in both cases $r = 6$. For additional accuracy we could draw the asymptotes. Note that $r \to \pm\infty$ when $2 + 4\sin\theta \to 0^+$ or 0^- and $2 + 4\sin\theta = 0$ when $\sin\theta = -\frac{1}{2}$. Thus the asymptotes are parallel to the rays $\theta = 7\pi/6$ and $\theta = 11\pi/6$. The hyperbola is sketched in Figure 9.54.

Figure 9.54

$$r = \frac{12}{2 + 4\sin\theta}$$

In Figure 9.55 we sketch a number of conics to demonstrate the effect of varying the eccentricity e. Notice that when e is close to 0 the ellipse is nearly circular, whereas it becomes more elongated as $e \to 1^-$. When $e = 1$, of course, the conic is a parabola.

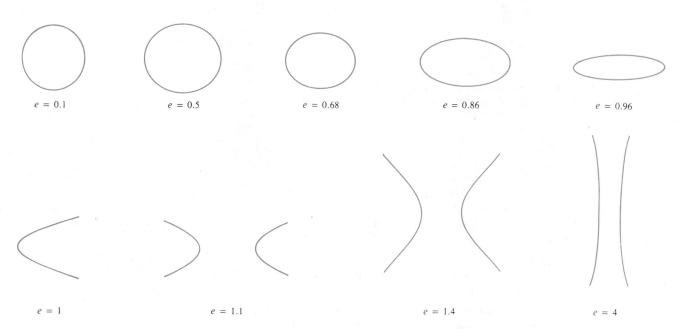

$e = 0.1$ $e = 0.5$ $e = 0.68$ $e = 0.86$ $e = 0.96$

$e = 1$ $e = 1.1$ $e = 1.4$ $e = 4$

Figure 9.55

SECTION 9.7 **Exercises**

In Exercises 1–8 write a polar equation of a conic with the focus at the origin and the given data.

1. Ellipse, eccentricity $\frac{2}{3}$, directrix $x = 3$

2. Hyperbola, eccentricity $\frac{4}{3}$, directrix $x = -3$

3. Parabola, directrix $y = 2$

4. Ellipse, eccentricity $\frac{1}{2}$, directrix $y = -4$

5. Hyperbola, eccentricity 4, directrix $r = 5 \sec \theta$

6. Ellipse, eccentricity 0.6, directrix $r = 2 \csc \theta$

7. Parabola, vertex at $(5, \pi/2)$

8. Ellipse, eccentricity 0.4, vertex at $(2, 0)$

In Exercises 9–20, (a) find the eccentricity, (b) identify the conic, (c) give an equation of the directrix, and (d) sketch the conic.

9. $r = \dfrac{4}{1 + 3 \cos \theta}$

10. $r = \dfrac{8}{3 + 3 \cos \theta}$

11. $r = \dfrac{2}{1 - \cos \theta}$

12. $r = \dfrac{10}{3 - 2 \sin \theta}$

13. $r = \dfrac{6}{2 + \sin \theta}$

14. $r = \dfrac{5}{2 - 3 \sin \theta}$

15. $r = \dfrac{1}{4 - 3 \cos \theta}$

16. $r = \dfrac{6}{1 + 5 \cos \theta}$

17. $r = \dfrac{7}{2 - 5 \sin \theta}$

18. $r = \dfrac{8}{3 + \cos \theta}$

19. $r = \dfrac{5}{2 + 2 \sin \theta}$

20. $r = \dfrac{1}{1 - \sin \theta}$

21. Show that a conic with focus at the origin, eccentricity e, and directrix $x = -d$ has polar equation $r = ed/(1 - e \cos \theta)$.

22. Show that a conic with focus at the origin, eccentricity e, and directrix $y = d$ has polar equation $r = ed/(1 + e \sin \theta)$.

23. Show that a conic with focus at the origin, eccentricity e, and directrix $y = -d$ has polar equation $r = ed/(1 - e \sin \theta)$.

24. Show that the parabolas $r = c/(1 + \cos \theta)$ and $r = d/(1 - \cos \theta)$ intersect at right angles.

25. (a) Show that the polar equation of an ellipse with directrix $x = -d$ can be written in the form

$$r = \frac{a(1 - e^2)}{1 - e \cos \theta}$$

(b) Find an approximate polar equation for the elliptical orbit of the earth around the sun (at one focus) given that the eccentricity is about 0.017 and the length of the major axis is about 2.99×10^8 km.

26. (a) The planets move around the sun in elliptical orbits with the sun at one focus. The positions of a planet that are closest to and farthest from the sun are called *perihelion* and *aphelion*, respectively. Use Exercise 25(a) to show that the perihelion distance from a planet to the sun is $a(1 - e)$ and the aphelion distance is $a(1 + e)$.

(b) Use the data of Exercise 25(b) to find the distances from the earth to the sun at perihelion and at aphelion.

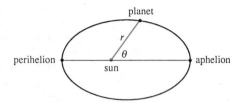

27. The planet Mercury travels in an elliptical orbit with eccentricity 0.206. Its minimum distance from the sun is 4.6×10^7 km. Use the results of Exercise 26(a) to find its maximum distance from the sun.

28. The distance from the planet Pluto to the sun is 4.43×10^9 km at perihelion and 7.37×10^9 km at aphelion. Use Exercise 26 to find the eccentricity of Pluto's orbit.

CHAPTER 9

Review

Key Topics

Define, state, or discuss the following:

1. Parametric equations of a curve
2. Slope of a tangent line to a parametric curve
3. Area under a parametric curve
4. Arc length of a parametric curve
5. Surface area for a rotated parametric curve
6. Polar coordinate system
7. Relation between rectangular and polar coordinates
8. Graph of a polar equation
9. Slope of a tangent line to a polar curve

10. Area formula in polar coordinates
11. Length of a polar curve
12. Definition, focus, directrix, vertex, and axis of a parabola
13. Definition, foci, and vertices of an ellipse
14. Definition, foci, and vertices of a hyperbola
15. Equations of conics in rectangular coordinates
16. Eccentricity
17. Equations of conics in polar coordinates

Exercises

In Exercises 1–4 sketch the given parametric curve and eliminate the parameter to find the Cartesian equation of the curve.

1. $x = 1 - t^2$, $y = 1 - t$, $-1 \le t \le 1$
2. $x = t^2 + 1$, $y = t^2 - 1$
3. $x = 1 + \sin t$, $y = 2 + \cos t$
4. $x = 1 + \cos t$, $y = 1 + \sin^2 t$

In Exercises 5–12 sketch the curve whose polar equation is given.

5. $r = 1 + 3 \cos \theta$
6. $r = 3 - \sin \theta$
7. $r^2 = \sec 2\theta$
8. $r = \tan \theta$
9. $r = 2 \cos^2(\theta/2)$
10. $r = 2 \cos(\theta/2)$
11. $r = \dfrac{1}{1 + \cos \theta}$
12. $r = \dfrac{5}{1 - 3 \sin \theta}$

In Exercises 13 and 14 find a polar equation for the curve represented by the given Cartesian equation.

13. $x^2 + y^2 = 4x$ 14. $x + y^2 = 0$

In Exercises 15–18 find the slope of the tangent line to the given curve at the point corresponding to the given value of the parameter.

15. $x = t^2 + 2t$, $y = t^3 - t$; $t = 1$

16. $x = te^t$, $y = 1 + \sqrt{1+t}$; $t = 0$

17. $r = 0$; $\theta = \pi/4$

18. $r = 3 - 2\sin\theta$; $\theta = \pi/2$

In Exercises 19 and 20 find dy/dx and d^2y/dx^2.

19. $x = t\cos t$, $y = t\sin t$

20. $x = t^6 + t^3$, $y = t^4 + t^2$

21. At what points does the curve $x = 2a\cos t - a\cos 2t$, $y = 2a\sin t - a\sin 2t$ have vertical or horizontal tangents? Use this information to help sketch the curve.

22. Find the area enclosed by the curve in Exercise 21.

23. Find the area enclosed by the curve $r^2 = 9\cos 5\theta$.

24. Find the area enclosed by the inner loop of the curve $r = 1 - 3\sin\theta$.

25. Find the points of intersection of the curves $r = 2$ and $r = 4\cos\theta$.

26. Find the points of intersection of the curves $r = \cot\theta$ and $r = 2\cos\theta$.

27. Find the area of the region that lies inside both of the circles $r = 2\sin\theta$ and $r = \sin\theta + \cos\theta$.

28. Find the area of the region that lies inside the curve $r = 2 + \cos 2\theta$ but outside the curve $r = 2 + \sin\theta$.

Find the lengths of the curves in Exercises 29–32.

29. $x = 3t^2$, $y = 2t^3$, $0 \le t \le 2$

30. $x = \cos t + \ln\tan(t/2)$, $y = \sin t$, $\pi/2 \le t \le 3\pi/4$

31. $r = 1/\theta$, $\pi \le \theta \le 2\pi$

32. $r = \sin^3(\theta/3)$, $0 \le \theta \le \pi$

In Exercises 33 and 34 find the area of the surface obtained by rotating the given curve about the x-axis.

33. $x = 4\sqrt{t}$, $y = \dfrac{t^3}{3} + \dfrac{1}{2t^2}$, $1 \le t \le 4$

34. $x = \cos t + \ln\tan(t/2)$, $y = \sin t$, $\pi/2 \le t \le 3\pi/4$

In Exercises 35–38 find the foci and vertices and sketch the graph.

35. $\dfrac{x^2}{9} + \dfrac{y^2}{8} = 1$ 36. $4x^2 - y^2 = 16$

37. $6y^2 + x - 36y + 55 = 0$

38. $25x^2 + 4y^2 + 50x - 16y = 59$

39. Find an equation of the parabola with focus $(0,6)$ and directrix $y = 2$.

40. Find an equation of the hyperbola with foci $(0,\pm 5)$ and vertices $(0,\pm 2)$.

41. Find an equation of the hyperbola with foci $(\pm 3, 0)$ and asymptotes $2y = \pm x$.

42. Find an equation of the ellipse with foci $(3,\pm 2)$ and major axis with length 8.

43. Find an equation for the ellipse that shares a vertex and a focus with the parabola $x^2 + y = 100$ and that has its other focus at the origin.

44. Show that if m is any real number, then there are exactly two lines of slope m that are tangent to the ellipse $x^2/a^2 + y^2/b^2 = 1$ and their equations are

$$y = mx \pm \sqrt{a^2m^2 + b^2}.$$

45. Find a polar equation for the ellipse with focus at the origin, eccentricity $\frac{1}{3}$, and directrix with equation $r = 4\sec\theta$.

46. Show that the angles between the polar axis and the asymptotes of the hyperbola $r = ed/(1 - e\cos\theta)$, $e > 1$, are given by $\cos^{-1}(\pm 1/e)$.

47. In the figure the circle of radius a is stationary, and for every θ, the point P is the midpoint of the segment QR. The curve traced out by P for $0 < \theta < \pi$ is called the **longbow curve**. Find parametric equations for this curve.

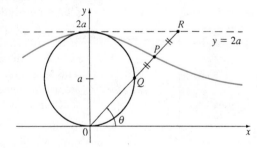

Applications Plus

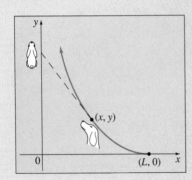

1. A dog sees a rabbit running in a straight line across an open field and gives chase. In a rectangular coordinate system, assume:

   (i) The rabbit is at the origin and the dog is at the point $(L, 0)$ at the instant the dog first sees the rabbit.

   (ii) The rabbit runs up the y-axis and the dog always runs straight for the rabbit.

   (iii) The dog runs at the same speed as the rabbit.

   (a) Show that the dog's path is the graph of the function $y = f(x)$, where y satisfies the differential equation

   $$x \frac{d^2y}{dx^2} = \sqrt{1 + \left(\frac{dy}{dx}\right)^2}$$

   (b) Determine the solution of the equation in part (a) that satisfies the initial conditions $y = y' = 0$ when $x = L$. (*Hint:* Let $z = dy/dx$ in the differential equation and solve the resulting first order equation to find z; then integrate z to find y.)

   (c) Does the dog ever catch the rabbit?

2. Replace hypothesis (iii) in Problem 1 by

   (iii)′ The dog runs twice as fast as the rabbit.

   (a) Show that the dog's path is the graph of the function $y = f(x)$, where y satisfies the differential equation

   $$2x \frac{d^2y}{dx^2} = \sqrt{1 + \left(\frac{dy}{dx}\right)^2}$$

   (b) Determine the solution of the differential equation in part (a) that satisfies the initial conditions $y = y' = 0$ when $x = L$.

   (c) Show that the dog has run twice as far as the rabbit at the instant the dog catches him.

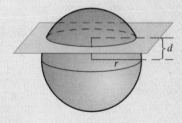

3. If a sphere of radius r is sliced by a plane whose distance from the center of the sphere is d, then the sphere is divided into two pieces called *segments of one base*. The corresponding surfaces are called *spherical zones of one base*.

   (a) Determine the surface areas of the two spherical zones indicated in the figure.

   (b) Determine the approximate area of the Arctic Ocean by assuming that it is approximately circular in shape, with center at the North Pole and "circumference" at 75° north latitude. Use $r = 3960$ mi for the radius of the earth.

   (c) A sphere of radius r is inscribed in a right circular cylinder of radius r. Two planes perpendicular to the central axis of the cylinder and a distance h apart cut off a *spherical zone of two bases* on the sphere. Show that the surface area of the spherical zone equals the surface area of the region that the two planes cut off on the cylinder.

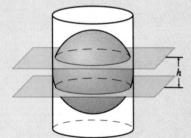

   (d) The *Torrid Zone* is the region on the surface of the earth that is between the tropic of Cancer (23.45° north latitude) and the Tropic of Capricorn (23.45° south latitude). What is the area of the Torrid Zone?

4. In a second order chemical reaction, one molecule of a substance A interacts with one molecule of a substance B to form one molecule of a third substance C. The reaction takes place only if the two molecules collide and are "energetic" enough to supply the necessary activation energy. It is assumed that the frequency with which this happens is proportional to the product of the concentrations of the substances A and B. If $x = x(t)$ denotes the concentration of the substance C at time t, then the

mathematical model for the reaction is

$$\frac{dx}{dt} = k(a - x)(b - x)$$

where a and b are the initial concentrations of A and B, respectively, and k is a positive constant. (See Exercise 27 in Section 8.1.) This model takes into account only the "forward reaction" of A and B combining to form C. There is also a "reverse reaction" in which C dissociates back into A and B. It is assumed that the mechanism for this dissociation is the collision of sufficiently energetic C molecules and so the rate of dissociation is proportional to x^2. Thus, the mathematical model that includes the reverse reaction is

$$\frac{dx}{dt} = k(a - x)(b - x) - rx^2$$

where a and b are the initial concentrations of A and B, and k and r are the reaction rates for the forward and reverse reactions. Assume that $k > r$.
(a) Suppose that $x(0) = 0$. Determine the concentration $x = x(t)$ of the substance C at any time t.
(b) Determine $\lim_{t \to \infty} x(t)$.

5. Let ε be a positive number. A differential equation of the form

$$\frac{dy}{dt} = ky^{1+\varepsilon}$$

where k is a positive constant, is called a *doomsday equation* because the "growth" term $ky^{1+\varepsilon}$ is larger than that for natural growth (that is, ky).
(a) Determine that solution that satisfies the initial condition $y(0) = y_0$.
(b) Show that there is a finite time $t = T$ such that $\lim_{t \to T} y(t) = \infty$.
(c) An especially prolific breed of rabbits has the growth term $ky^{1.01}$. There are 2 such rabbits initially and 16 rabbits after three months. When is doomsday?

6. In a *seasonal-growth model*, a periodic function of time is introduced to account for seasonal variations in the rate of growth. Such variations could, for example, be caused by seasonal changes in the availability of food.
(a) Determine the solution of the seasonal-growth model

$$\frac{dy}{dt} = k\cos(rt - \phi)y \qquad y(0) = y_0$$

where k, r, and ϕ are positive constants. Sketch the graph of the solution. What can you say about $\lim_{t \to \infty} y(t)$?
(b) Determine the solution of the seasonal-growth model

$$\frac{dy}{dt} = k\cos^2(rt - \phi)y \qquad y(0) = y_0$$

where k, r, and ϕ are positive constants. Sketch the graph of the solution. What can you say about $\lim_{t \to \infty} y(t)$ in this case?

7. Two points in the plane A and B, with B "below" A, are joined by a thin wire. A bead is allowed to slide without friction down the wire from A to B. For convenience, assume that the point A is at the origin in a rectangular coordinate system with the positive y-axis in the downward direction. A classical problem,

posed by Johann Bernoulli in 1696, and solved by Newton, Leibniz, and Johann and Jakob Bernoulli, is to determine the path from A to B along which the bead will fall in the shortest time. This problem is called the *brachistrochrone problem*. Using principles from physics and optics, it can be shown that if $y = f(x)$ is the path giving the shortest time of descent, then y must satisfy the differential equation

$$y\left[1 + \left(\frac{dy}{dx}\right)^2\right] = C$$

where C is a constant. Solving this equation for dy/dx and separating the variables leads to the equation

$$\sqrt{\frac{y}{C-y}}\, dy = dx$$

(a) Introduce an auxiliary variable θ by means of the equation

$$\tan\theta = \sqrt{\frac{y}{C-y}}$$

Show that $y = C\sin^2\theta = \frac{1}{2}C(1 - \cos 2\theta)$ and $dx = C(1 - \cos 2\theta)\, d\theta$. Determine x as a function of θ, $x = g(\theta)$. Use the fact that the curve passes through the origin to evaluate the constant of integration.

(b) Identify the curve given parametrically by $x = g(\theta)$, $y = \frac{1}{2}C(1 - \cos 2\theta)$.

Note: As discussed in Section 9.1, it can also be shown that this curve has the property that, no matter where the bead starts its fall between A and B, it will take the same length of time for it to reach B. This is known as the *tautochrone property* of the curve.

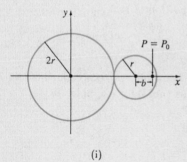

(i)

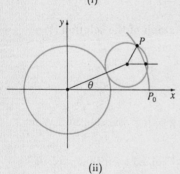

(ii)

8. A circle C of radius $2r$ has its center at the origin. A circle of radius r rolls without slipping in the counterclockwise direction around C. A point P is located on a fixed radius of the rolling circle at a distance b from its center, $0 < b < r$. See figures (i) and (ii). Let L be the line from the center of C to the center of the rolling circle and let θ be the angle that L makes with the positive x-axis.

(a) Using θ as a parameter, show that the parametric equations of the path traced out by P are:

$$x = b\cos 3\theta + 3r\cos\theta \qquad y = b\sin 3\theta + 3r\sin\theta$$

Note: If $b = 0$, the path is a circle of radius $3r$; if $b = r$, the path is an *epicycloid*. The path traced out by P for $0 < b < r$ is called an *epitrochoid*.

(b) Suppose $0 < b < r$. Sketch the graph of the curve.

(c) Show that an equilateral triangle can be inscribed in the epitrochoid and that its centroid is on the circle of radius b centered at the origin.

Note: This is the principle of the Wankel rotary engine. When the equilateral triangle rotates with its vertices on the epitrochoid, its centroid sweeps out a circle whose center is at the center of the curve.

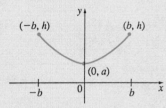

9. When a flexible cable of uniform density is suspended between two fixed points and hangs of its own weight, the shape $y = f(x)$ of the cable must satisfy a differential equation of the form

$$\frac{d^2y}{dx^2} = k\sqrt{1 + \left(\frac{dy}{dx}\right)^2}$$

where k is a positive constant. Consider the cable indicated in the figure.
(a) Let $z = dy/dx$ in the differential equation. Solve the resulting first order differential equation (in z), and then integrate to find y.
(b) Determine the length of the cable.

10. Consider a model for population growth involving two species in which one species has an ample food supply and the other species feeds on the first species. In this situation, the first species is called the *prey* and the second species is called the *predators*. Examples might be rabbits and wolves in an isolated forest, food fish and sharks, aphids and ladybugs, and so forth. Let $x = x(t)$ denote the number of prey and $y = y(t)$ the number of predators at time t. In the absence of predators, the ample food supply will support exponential growth of the prey, that is, $dx/dt = kx$, $k > 0$. Similarly, in the absence of prey, the predator population will decline exponentially: $dy/dt = -ry$, $r > 0$. However, with both species present, assume that the principal cause of death among the prey is being eaten by a predator and that the birth and survival rate for the predators depends on their available food supply, namely, the prey. Assume, further, that the two species encounter each other by chance at a rate that is proportional to the product of the sizes of the two populations. These assumptions lead to the system of differential equations

$$\frac{dx}{dt} = kx - axy \qquad \frac{dy}{dt} = -ry + bxy$$

where a, b, k, and r are positive constants. These equations are known as *predator-prey equations*, or *Lotka-Volterra equations*. Although it is not possible to find elementary general solutions for this system of equations, one can find relationships between x and y. Since the number of predators depends on the number of prey available for food, assume that y is a function of x. Then

$$\frac{dy}{dx} = \frac{dy/dt}{dx/dt} = \frac{-ry + bxy}{kx - axy} \qquad (1)$$

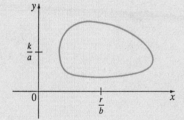

(a) Show that the constant functions $x(t) = r/b$, $y(t) = k/a$ are solutions of the predator-prey equations.
(b) Determine the general solution of the differential equation (1) and express your answer in terms of exponentials.
(c) Show that the populations x and y are bounded by analyzing the general solution found in part (b).

A typical solution curve for (1) is shown in the figure.

10

Infinite Sequences and Series

Our minds are finite, and yet even in those circumstances of finitude, we are surrounded by possibilities that are infinite, and the purpose of human life is to grasp as much as we can out of that infinitude.

Alfred North Whitehead

Infinite sequences and series were briefly introduced in the Preview of Calculus (Section 7 in Review and Preview) in connection with Zeno's paradoxes and the decimal representation of numbers. Their importance in calculus stems from Newton's idea of representing functions as sums of infinite series. For instance, in finding areas he often integrated a function by first expressing it as a series and then integrating each term of the series. We pursue this idea in Section 10.9 in order to integrate such functions as e^{-x^2}. (Recall that we have previously been unable to do this.) Many of the functions that arise in mathematical physics and chemistry, such as Bessel functions, are defined as sums of series, so it is important to be familiar with the basic concepts of convergence of infinite sequences and series.

Sequences

A **sequence** is a set of numbers written in a definite order:

$$a_1, a_2, a_3, a_4, \ldots, a_n, \ldots$$

The number a_1 is called the *first term*, a_2 is called the *second term*, and in general a_n is the *nth term*. We will deal exclusively with infinite sequences and so each term a_n will have a successor a_{n+1}.

Notice that for every positive integer n there is a corresponding number a_n and so a sequence can be regarded as a function whose domain is the set of positive integers. But we usually write a_n instead of the function notation $f(n)$ for the value of the function at the number n.

Notation: The sequence $\{a_1, a_2, a_3, \ldots\}$ is denoted by

$$\{a_n\} \qquad \text{or} \qquad \{a_n\}_{n=1}^{\infty}$$

EXAMPLE 1   Some sequences can be defined by giving a formula for the nth term. In the following examples we give three descriptions of the sequence: one by using the above notation, another by using the defining formula, and a third by writing out the terms of the sequence.

(a) $\left\{\dfrac{n}{n+1}\right\}_{n=1}^{\infty}$ $a_n = \dfrac{n}{n+1}$ $\left\{\dfrac{1}{2}, \dfrac{2}{3}, \dfrac{3}{4}, \dfrac{4}{5}, \ldots, \dfrac{n}{n+1}, \ldots\right\}$

(b) $\left\{\dfrac{(-1)^n(n+1)}{3^n}\right\}$ $a_n = \dfrac{(-1)^n(n+1)}{3^n}$ $\left\{-\dfrac{2}{3}, \dfrac{3}{9}, -\dfrac{4}{27}, \dfrac{5}{81}, \ldots, \dfrac{(-1)^n(n+1)}{3^n}, \ldots\right\}$

(c) $\left\{\sqrt{n-3}\right\}_{n=3}^{\infty}$ $a_n = \sqrt{n-3},\ n \geq 3$ $\{0, 1, \sqrt{2}, \sqrt{3}, \ldots, \sqrt{n-3}, \ldots\}$

(d) $\left\{\cos\dfrac{n\pi}{6}\right\}_{n=0}^{\infty}$ $a_n = \cos\dfrac{n\pi}{6},\ n \geq 0$ $\left\{1, \dfrac{\sqrt{3}}{2}, \dfrac{1}{2}, 0, \ldots, \cos\dfrac{n\pi}{6}, \ldots\right\}$ ●

EXAMPLE 2   Here are some sequences that do not have a simple defining equation.

(a) The sequence $\{p_n\}$, where p_n is the population of the world as of January 1 in the year n.

(b) If we let a_n be the digit in the nth decimal place of the number e, then $\{a_n\}$ is a well-defined sequence whose first few terms are

$$\{7, 1, 8, 2, 8, 1, 8, 2, 8, 4, 5, \ldots\}$$

(c) The **Fibonacci sequence** $\{f_n\}$ is defined recursively by the conditions

$$f_1 = 1 \qquad f_2 = 1 \qquad f_n = f_{n-1} + f_{n-2} \qquad (n \geq 3)$$

Each term is the sum of the two preceding terms. The first few terms are

$$\{1, 1, 2, 3, 5, 8, 13, 21, \ldots\}$$

This sequence arose when the 13th-century Italian mathematician known as Fibonacci solved a problem concerning breeding of rabbits (see Exercise 71).   ●

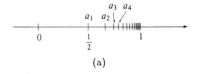

(a)

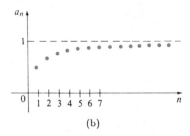

(b)

Figure 10.1

A sequence such as the one in Example 1(a), $a_n = n/(n+1)$, can be pictured either by plotting its terms on a number line as in Figure 10.1(a) or by plotting its graph as in Figure 10.1(b). From Figure 10.1(a) or (b) it appears that the terms of the sequence $a_n = n/(n+1)$ are approaching 1 as n becomes large. In fact, the difference

$$1 - \frac{n}{n+1} = \frac{1}{n+1}$$

can be made as small as we like by taking n sufficiently large. We indicate this by writing

$$\lim_{n\to\infty} \frac{n}{n+1} = 1$$

In general, the notation

$$\lim_{n\to\infty} a_n = L$$

means that the terms of the sequence $\{a_n\}$ can be made arbitrarily close to L by taking

n sufficiently large. Notice that the following precise definition of the limit of a sequence is very similar to the definition of a limit of a function at infinity given in Section 1.6.

Definition (10.1)

A sequence $\{a_n\}$ has the **limit** L and we write

$$\lim_{n \to \infty} a_n = L \qquad \text{or} \qquad a_n \to L \text{ as } n \to \infty$$

if for every $\varepsilon > 0$ there is a corresponding integer N such that

$$|a_n - L| < \varepsilon \qquad \text{whenever} \qquad n > N$$

If $\lim_{n \to \infty} a_n$ exists, we say the sequence **converges** (or is **convergent**). Otherwise we say the sequence **diverges** (or is **divergent**).

Definition 10.1 is illustrated by Figure 10.2 in which the terms a_1, a_2, a_3, ... are plotted on a number line. No matter how small an interval $(L - \varepsilon, L + \varepsilon)$ is chosen, there exists an N such that all terms of the sequence from a_{N+1} onward must lie in that interval.

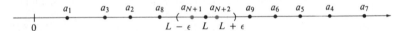

Figure 10.2

Another illustration of Definition 10.1 is given in Figure 10.3. The points on the graph of $\{a_n\}$ must lie between the horizontal lines $y = L + \varepsilon$ and $y = L - \varepsilon$ if $n > N$. This picture must be valid no matter how small ε is chosen, but usually a smaller ε requires a larger N.

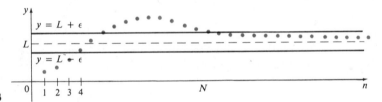

Figure 10.3

Comparison of Definitions 10.1 and 1.30 shows that the only difference between $\lim_{n \to \infty} a_n = L$ and $\lim_{x \to \infty} f(x) = L$ is that n is required to be an integer. Thus we have the following theorem, which is illustrated by Figure 10.4.

Theorem (10.2)

If $\lim_{x \to \infty} f(x) = L$ and $f(n) = a_n$ when n is an integer, then $\lim_{n \to \infty} a_n = L$.

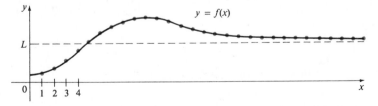

Figure 10.4

In particular, since we know that $\lim_{x \to \infty} (1/x^r) = 0$ where $r > 0$ (Theorem 1.29), we have

(10.3)
$$\lim_{n\to\infty} \frac{1}{n^r} = 0 \qquad \text{if } r > 0$$

The analogue of Definition 1.38 is the following:

Definition (10.4)

$\lim_{n\to\infty} a_n = \infty$ means that for every positive number M there is an integer N such that

$$a_n > M \qquad \text{whenever} \qquad n > N$$

If $\lim_{n\to\infty} a_n = \infty$, then the sequence $\{a_n\}$ is divergent but in a special way. We say that $\{a_n\}$ diverges to ∞.

The Limit Laws given in Section 1.3 also hold for limits of sequences. For instance, if $\{a_n\}$ and $\{b_n\}$ are convergent sequences, then

$$\lim_{n\to\infty} (a_n + b_n) = \lim_{n\to\infty} a_n + \lim_{n\to\infty} b_n$$

and
$$\lim_{n\to\infty} (a_n b_n) = \left(\lim_{n\to\infty} a_n\right)\left(\lim_{n\to\infty} b_n\right)$$

The Squeeze Theorem can also be adapted for sequences as follows. (The proof is similar to the proof of Theorem 1.10 in Appendix C.)

The Squeeze Theorem for Sequences (10.5)

If $a_n \le b_n \le c_n$ for $n \ge n_0$ and

$$\lim_{n\to\infty} a_n = \lim_{n\to\infty} c_n = L$$

then $\lim_{n\to\infty} b_n = L$

Another useful fact about limits of sequences is given by the following theorem, whose proof is requested in Exercise 73.

Theorem (10.6)

If $\lim_{n\to\infty} |a_n| = 0$, then $\lim_{n\to\infty} a_n = 0$.

EXAMPLE 3   Find $\lim_{n\to\infty} \frac{n}{n+1}$.

Solution   The method is similar to the one we used in Section 1.6: divide numerator and denominator by the highest power of n:

$$\lim_{n\to\infty} \frac{n}{n+1} = \lim_{n\to\infty} \frac{1}{1 + \frac{1}{n}}$$
$$= \frac{1}{1+0} = 1$$

Here we have used Equation 10.3 with $r = 1$.   ●

EXAMPLE 4   Calculate $\lim_{n\to\infty} \frac{\ln n}{n}$.

Solution   Notice that both numerator and denominator approach infinity as $n \to \infty$. We cannot apply l'Hospital's Rule directly since it applies not to sequences but to functions of a real variable. However we can apply l'Hospital's Rule to the related function $f(x) = (\ln x)/x$ and obtain

$$\lim_{x \to \infty} \frac{\ln x}{x} = \lim_{x \to \infty} \frac{1/x}{1} = 0$$

Therefore by Theorem 10.2 we have

$$\lim_{n \to \infty} \frac{\ln n}{n} = 0$$ ●

EXAMPLE 5   Determine whether the sequence $a_n = (-1)^n$ is convergent or divergent.

Solution   If we write out the terms of the sequence, we obtain

$$\{-1, 1, -1, 1, -1, 1, -1, \ldots\}$$

The graph of this sequence is shown in Figure 10.5. Since the terms oscillate between 1 and -1 infinitely often, a_n does not approach any number. Thus $\lim_{n \to \infty} (-1)^n$ does not exist; that is, the sequence $\{(-1)^n\}$ is divergent.

Figure 10.5

●

EXAMPLE 6   Evaluate $\lim_{n \to \infty} \dfrac{(-1)^n}{n}$ if it exists.

Solution

$$\lim_{n \to \infty} \left| \frac{(-1)^n}{n} \right| = \lim_{n \to \infty} \frac{1}{n} = 0$$

Therefore, by Theorem 10.6,

$$\lim_{n \to \infty} \frac{(-1)^n}{n} = 0$$ ●

EXAMPLE 7   Discuss the convergence of the sequence $a_n = n!/n^n$, where $n! = 1 \cdot 2 \cdot 3 \cdots \cdot n$.

Solution   Both numerator and denominator approach infinity as $n \to \infty$ but here there is no corresponding function for use with l'Hospital's Rule ($x!$ is not defined when x is not an integer). Let us write out a few terms to get a feeling for what happens to a_n as n gets large:

$$a_1 = 1 \qquad a_2 = \frac{1 \cdot 2}{2 \cdot 2} \qquad a_3 = \frac{1 \cdot 2 \cdot 3}{3 \cdot 3 \cdot 3}$$

(10.7)
$$a_n = \frac{1 \cdot 2 \cdot 3 \cdots \cdot n}{n \cdot n \cdot n \cdots \cdot n}$$

It appears that the terms are decreasing and perhaps approach 0. To confirm this, observe from Equation 10.7 that

$$a_n = \frac{1}{n}\left(\frac{2 \cdot 3 \cdot \cdots \cdot n}{n \cdot n \cdot \cdots \cdot n}\right)$$

so

$$0 < a_n \leq \frac{1}{n}$$

We know that $1/n \to 0$ as $n \to \infty$. Therefore, $a_n \to 0$ as $n \to \infty$ by the Squeeze Theorem.   ●

EXAMPLE 8   For what values of r is the sequence $\{r^n\}$ convergent?

Solution   We know from (3.3) that $\lim_{x \to \infty} a^x = \infty$ for $a > 1$ and $\lim_{x \to \infty} a^x = 0$ for $0 < a < 1$. Therefore, putting $a = r$ and using Theorem 10.2, we have

$$\lim_{n \to \infty} r^n = \begin{cases} \infty & \text{if } r > 1 \\ 0 & \text{if } 0 < r < 1 \end{cases}$$

It is obvious that

$$\lim_{n \to \infty} 1^n = 1 \quad \text{and} \quad \lim_{n \to \infty} 0^n = 0$$

If $-1 < r < 0$, then $0 < |r| < 1$, so

$$\lim_{n \to \infty} |r^n| = \lim_{n \to \infty} |r|^n = 0$$

and therefore $\lim_{n \to \infty} r^n = 0$ by Theorem 10.6. If $r \leq -1$, then $\{r^n\}$ diverges as in Example 5.   ●

The results of Example 8 are summarized for future use as follows:

(10.8)

> The sequence $\{r^n\}$ is convergent if $-1 < r \leq 1$ and divergent for all other values of r.
>
> $$\lim_{n \to \infty} r^n = \begin{cases} 0 & \text{if } -1 < r < 1 \\ 1 & \text{if } r = 1 \end{cases}$$

Definition (10.9)

> A sequence $\{a_n\}$ is called **increasing** if $a_n \leq a_{n+1}$ for all $n \geq 1$, that is, $a_1 \leq a_2 \leq a_3 \leq \cdots$. It is called **decreasing** if $a_n \geq a_{n+1}$ for all $n \geq 1$. It is called **monotonic** if it is either increasing or decreasing.

EXAMPLE 9   The sequence $\left\{\frac{3}{n+5}\right\}$ is decreasing because

$$\frac{3}{n+5} > \frac{3}{n+6}$$

for all $n \geq 1$. (The right side is smaller because it has a larger denominator.)   ●

EXAMPLE 10   Show that the sequence $a_n = \dfrac{n}{n^2 + 1}$ is decreasing.

Solution 1   We must show that $a_{n+1} \le a_n$, that is,

$$\frac{n+1}{(n+1)^2 + 1} \le \frac{n}{n^2 + 1}$$

This inequality is equivalent to the one we get by cross-multiplication:

$$\frac{n+1}{(n+1)^2 + 1} \le \frac{n}{n^2 + 1} \quad \Leftrightarrow \quad (n+1)(n^2 + 1) \le n[(n+1)^2 + 1]$$

$$\Leftrightarrow \quad n^3 + n^2 + n + 1 \le n^3 + 2n^2 + 2n$$

$$\Leftrightarrow \quad 1 \le n^2 + n$$

It is obvious that $n^2 + n \ge 1$ is true for $n \ge 1$. Therefore $a_{n+1} \le a_n$ and so $\{a_n\}$ is decreasing.

Solution 2   Consider the function $f(x) = \dfrac{x}{x^2 + 1}$:

$$f'(x) = \frac{x^2 + 1 - 2x^2}{(x^2 + 1)^2} = \frac{1 - x^2}{(x^2 + 1)^2} < 0 \qquad \text{when } x^2 > 1$$

Thus f is decreasing on $[1, \infty)$ and so $f(n) > f(n+1)$. Therefore $\{a_n\}$ is decreasing.   ●

Definition (10.10)

A sequence $\{a_n\}$ is **bounded above** if there is a number M such that

$$a_n \le M \qquad \text{for all } n \ge 1$$

It is **bounded below** if there is a number m such that

$$m \le a_n \qquad \text{for all } n \ge 1$$

If it is bounded above and below, $\{a_n\}$ is called **bounded**.

For instance, the sequence $a_n = n$ is bounded below ($a_n > 0$) but not above. The sequence $a_n = n/(n+1)$ is bounded because $0 < a_n < 1$ for all n.

We know that not all bounded sequences are convergent [$a_n = (-1)^n$ satisfies $-1 \le a_n \le 1$ but is divergent from Example 5] and not all monotonic sequences are convergent ($a_n = n \to \infty$). But if a sequence is both bounded *and* monotonic, then it must be convergent. This fact is proved as Theorem 10.11 but intuitively you can understand why it is true by looking at Figure 10.6. If $\{a_n\}$ is increasing and $a_n \le M$ for all n, then the terms are forced to crowd together and approach some number L.

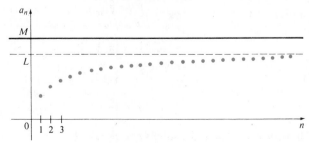

Figure 10.6

The proof of Theorem 10.11 is based on the **Completeness Axiom** for the set R of real numbers, which says that if S is a nonempty set of real numbers that has an upper bound M ($x \leq M$ for all x in S), then S has a **least upper bound** b. (This means that b is an upper bound for S, but if M is any other upper bound, then $b \leq M$.) The Completeness Axiom is an expression of the fact that there are no gaps or holes in the real number line.

Theorem (10.11)

> Every bounded, monotonic sequence is convergent.

Proof   Suppose $\{a_n\}$ is an increasing sequence. Since $\{a_n\}$ is bounded, the set $S = \{a_n \mid n \geq 1\}$ has an upper bound. By the Completeness Axiom it has a least upper bound L. Given $\varepsilon > 0$, $L - \varepsilon$ is *not* an upper bound for S (since L is the *least* upper bound). Therefore

$$a_N > L - \varepsilon \qquad \text{for some integer } N$$

But the sequence is increasing so $a_n \geq a_N$ for every $n > N$. Thus if $n > N$ we have

$$a_n > L - \varepsilon$$

so
$$0 \leq L - a_n < \varepsilon$$

Since $a_n \leq L$. Thus

$$\left| L - a_n \right| < \varepsilon \qquad \text{whenever} \qquad n > N$$

so $\lim_{n \to \infty} a_n = L$.

A similar proof (using the greatest lower bound) works if $\{a_n\}$ is decreasing.   ●

The proof of Theorem 10.11 shows that a sequence that is increasing and bounded above is convergent. (Likewise a decreasing sequence that is bounded below is convergent.) This fact is used many times in dealing with infinite series.

SECTION 10.1 **Exercises**

In Exercises 1–10 write the first five terms of the sequence.

1. $a_n = \dfrac{n}{2n+1}$

2. $a_n = \dfrac{4n-3}{3n+4}$

3. $a_n = \dfrac{(-1)^{n-1}n}{2^n}$

4. $a_n = \left(-\dfrac{2}{3}\right)^n$

5. $a_n = \dfrac{1 \cdot 3 \cdot 5 \cdot \, \cdots \, \cdot (2n-1)}{n!}$

6. $\left\{ \dfrac{(-7)^{n+1}}{n!} \right\}$

7. $\left\{ \sin \dfrac{n\pi}{2} \right\}$

8. $\{\cos n\pi\}$

9. $a_1 = 1, \ a_{n+1} = \dfrac{1}{1+a_n}$

10. $a_1 = 0, \ a_2 = 1, \ a_n = a_{n-1} - a_{n-2}$

In Exercises 11–18 find a formula for the general term a_n of the given sequence assuming that the pattern of the first few terms continues.

11. $\left\{ \dfrac{1}{2}, \dfrac{1}{4}, \dfrac{1}{8}, \dfrac{1}{16}, \ldots \right\}$

12. $\left\{ \dfrac{1}{2}, \dfrac{1}{4}, \dfrac{1}{6}, \dfrac{1}{8}, \ldots \right\}$

13. $\{1, 4, 7, 10, \ldots\}$

14. $\left\{ \dfrac{3}{16}, \dfrac{4}{25}, \dfrac{5}{36}, \dfrac{6}{49}, \ldots \right\}$

15. $\{-1, 2, -6, 24, \ldots\}$

16. $\left\{ \dfrac{3}{2}, -\dfrac{9}{4}, \dfrac{27}{8}, -\dfrac{81}{16}, \ldots \right\}$

17. $\left\{ \dfrac{2}{3}, -\dfrac{3}{5}, \dfrac{4}{7}, -\dfrac{5}{9}, \ldots \right\}$

18. $\{0, 2, 0, 2, 0, 2, \ldots\}$

$$\frac{1}{0} = \infty \quad , \quad \frac{1}{\infty} = 0$$

19. Illustrate Example 6 by drawing the graph of the sequence.

20. Draw the graph of the sequence

$$\left\{(-1)^n \frac{n+1}{n}\right\}$$

Is the sequence convergent or divergent?

In Exercises 21–56 determine whether the given sequence converges or diverges. If it converges, find the limit.

21. $a_n = \dfrac{1}{4n^2}$

22. $a_n = 4\sqrt{n}$

23. $a_n = \dfrac{n^2 - 1}{n^2 + 1}$

24. $a_n = \dfrac{4n - 3}{3n + 4}$

25. $a_n = \dfrac{n^2}{n + 1}$

26. $a_n = \dfrac{\sqrt[3]{n} + \sqrt[4]{n}}{\sqrt{n} + \sqrt[5]{n}}$

27. $a_n = (-1)^n \dfrac{n^2}{1 + n^3}$

28. $a_n = (-1)^{n-1} \dfrac{n^4}{1 + n^2 + n^3}$

29. $a_n = 1/5^n$

30. $a_n = 2 + \left(-\dfrac{2}{\pi}\right)^n$

31. $a_n = \cos\left(\dfrac{n\pi}{2}\right)$

32. $a_n = \sin\left(\dfrac{n\pi}{2}\right)$

33. $\left\{\dfrac{\pi^n}{3^n}\right\}$

34. $\left\{\arctan\left(\dfrac{2n}{2n+1}\right)\right\}$

35. $\{\arctan 2n\}$

36. $\left\{\dfrac{\sin n}{\sqrt{n}}\right\}$

37. $\left\{\dfrac{3 + (-1)^n}{n^2}\right\}$

38. $\left\{\dfrac{n!}{(n+2)!}\right\}$

39. $\left\{\dfrac{\ln(n^2)}{n}\right\}$

40. $\left\{(-1)^n \sin\left(\dfrac{1}{n}\right)\right\}$

41. $\{\sqrt{n+2} - \sqrt{n}\}$

42. $\left\{\dfrac{\ln(2 + e^n)}{3n}\right\}$

43. $a_n = n2^{-n}$

44. $a_n = \ln(n+1) - \ln n$

45. $a_n = n^{-1/n}$

46. $a_n = (1 + 3n)^{1/n}$

47. $a_n = \dfrac{\cos^2 n}{2^n}$

48. $a_n = \dfrac{n \cos n}{n^2 + 1}$

49. $a_n = \dfrac{1}{n^2} + \dfrac{2}{n^2} + \cdots + \dfrac{n}{n^2}$

50. $a_n = \left(\sqrt{n+1} - \sqrt{n}\right)\sqrt{n + \frac{1}{2}}$

51. $a_n = \dfrac{n!}{2^n}$

52. $a_n = \dfrac{(-3)^n}{n!}$

53. $a_n = \dfrac{n^3}{n!}$

54. $a_n = \sqrt[n]{3^n + 5^n}$

55. $a_n = \dfrac{1 \cdot 3 \cdot 5 \cdot \cdots \cdot (2n-1)}{(2n)^n}$

56. $a_n = \dfrac{1 \cdot 3 \cdot 5 \cdot \cdots \cdot (2n-1)}{n!}$

57. For what values of r is the sequence $\{nr^n\}$ convergent?

58. (a) If $\{a_n\}$ is convergent, show that $\lim_{n\to\infty} a_{n+1} = \lim_{n\to\infty} a_n$.

(b) A sequence $\{a_n\}$ is defined by $a_1 = 1$ and $a_{n+1} = 1/(1 + a_n)$ for $n \geq 1$. Assuming that $\{a_n\}$ is convergent, find its limit.

In Exercises 59–66 determine whether the given sequence is increasing, decreasing, or not monotonic.

59. $a_n = \dfrac{1}{3n + 5}$

60. $a_n = \dfrac{1}{5^n}$

61. $a_n = \dfrac{n - 2}{n + 2}$

62. $a_n = \dfrac{3n + 4}{2n + 5}$

63. $a_n = \cos\left(\dfrac{n\pi}{2}\right)$

64. $a_n = 3 + \dfrac{(-1)^n}{n}$

65. $a_n = \dfrac{n}{n^2 + n - 1}$

66. $a_n = \dfrac{\sqrt{n+1}}{5n + 3}$

67. Find the limit of the sequence

$$\left\{\sqrt{2},\ \sqrt{2\sqrt{2}},\ \sqrt{2\sqrt{2\sqrt{2}}},\ \ldots\right\}$$

68. A sequence $\{a_n\}$ is given by

$$a_1 = \sqrt{2},\ a_{n+1} = \sqrt{2 + a_n}$$

(a) By induction, or otherwise, show that $\{a_n\}$ is increasing and bounded above by 3. Apply Theorem 10.11 to show that $\lim_{n\to\infty} a_n$ exists.

(b) Find $\lim_{n\to\infty} a_n$.

69. Repeat Exercise 68 for the sequence defined by $a_1 = 1$, $a_{n+1} = 3 - 1/a_n$.

70. (a) Let $a_1 = a$, $a_2 = f(a)$, $a_3 = f(a_2) = f(f(a))$, $\ldots$, $a_{n+1} = f(a_n)$, where f is a continuous function. If $\lim_{n\to\infty} a_n = L$, show that $f(L) = L$.

(b) Illustrate part (a) by taking $f(x) = \cos x$, $a = 1$, and estimating the value of L to five decimal places.

71. (a) Fibonacci posed the following problem: Suppose that rabbits live forever and that every month each pair produces a new pair that becomes productive at age 2 months. If we start with one newborn pair, how many pairs of rabbits will there be in the nth month? Show that the answer is f_n, where $\{f_n\}$ is the Fibonacci sequence defined in Example 2(c).

(b)   Let $a_n = f_{n+1}/f_n$ and show that $a_{n-1} = 1 + 1/a_{n-2}$. Assuming that $\{a_n\}$ is convergent, find its limit.

72.   Use Definition 10.1 directly to prove that $\lim_{n\to\infty} r^n = 0$ when $|r| < 1$.

73.   Prove Theorem 10.6. (*Hint:* Use either Definition 10.1 or the Squeeze Theorem.)

74.   Let $a_n = \left(1 + \frac{1}{n}\right)^n$.

    (a)   Show that if $0 \le a < b$, then
$$\frac{b^{n+1} - a^{n+1}}{b - a} < (n+1)b^n$$

    (b)   Deduce that $b^n[(n+1)a - nb] < a^{n+1}$.

    (c)   Put $a = 1 + 1/(n+1)$ and $b = 1 + 1/n$ in part (b) to show that $\{a_n\}$ is increasing.

    (d)   Put $a = 1$ and $b = 1 + 1/(2n)$ in part (b) to show that $a_{2n} < 4$.

    (e)   Use parts (c) and (d) to show that $a_n < 4$ for all n.

    (f)   Use Theorem 10.11 to show that $\lim_{n\to\infty}(1 + 1/n)^n$ exists. (The limit is e. See Equation 3.34.)

75.   Let a and b be positive numbers with $a > b$. Let a_1 be their arithmetic mean and b_1 their geometric mean:

$$a_1 = \frac{a+b}{2} \qquad b_1 = \sqrt{ab}$$

Repeat this process so that, in general,

$$a_{n+1} = \frac{a_n + b_n}{2} \qquad b_{n+1} = \sqrt{a_n b_n}$$

    (a)   Use mathematical induction to show that

$$a_n > a_{n+1} > b_{n+1} > b_n$$

    (b)   Deduce that both $\{a_n\}$ and $\{b_n\}$ are convergent.

    (c)   Show that $\lim_{n\to\infty} a_n = \lim_{n\to\infty} b_n$

Gauss called the common value of these limits the *arithmetic-geometric mean* of the numbers a and b.

76.   (a)   Show that if $\lim_{n\to\infty} a_{2n} = L$ and $\lim_{n\to\infty} a_{2n+1} = L$, then $\{a_n\}$ is convergent and $\lim_{n\to\infty} a_n = L$.

    (b)   If $a_1 = 1$ and $a_{n+1} = 1 + \dfrac{1}{1 + a_n}$ find the first eight terms of the sequence $\{a_n\}$. Then use part (a) to show that $\lim_{n\to\infty} a_n = \sqrt{2}$. This gives the *continued fraction expansion*

$$\sqrt{2} = 1 + \cfrac{1}{2 + \cfrac{1}{2 + \cdots}}$$

SECTION 10.2

Series

If we try to add the terms of an infinite sequence $\{a_n\}_{n=1}^{\infty}$ we get an expression of the form

$$(10.12) \qquad a_1 + a_2 + a_3 + \cdots + a_n + \cdots$$

which is called an **infinite series** (or just a **series**) and is denoted, for short, by the symbol

$$\sum_{n=1}^{\infty} a_n \qquad \text{or} \qquad \sum a_n$$

But does it make sense to talk about the sum of infinitely many terms?

It would be impossible to find a finite sum for the series

$$1 + 2 + 3 + 4 + 5 + \cdots + n + \cdots$$

because if we start adding the terms we get the cumulative sums 1, 3, 6, 10, 15, 21, ... and, after the nth term, $n(n+1)/2$, which becomes very large as n increases.

However, if we start to add the terms of the series

$$\frac{1}{2} + \frac{1}{4} + \frac{1}{8} + \frac{1}{16} + \frac{1}{32} + \frac{1}{64} + \cdots + \frac{1}{2^n} + \cdots$$

we get $\frac{1}{2}, \frac{3}{4}, \frac{7}{8}, \frac{15}{16}, \frac{31}{32}, \frac{63}{64}, \ldots, 1 - 1/2^n, \ldots$. As we add more and more terms, these partial sums become closer and closer to 1. (See also Figure 82 in Section 7 of Review and Preview.) In fact, by adding sufficiently many terms of the series we can make the partial sums as close as we like to 1. So it seems reasonable to say that the sum of this infinite series is 1 and to write

$$\sum_{i=1}^{\infty} \frac{1}{2^n} = \frac{1}{2} + \frac{1}{4} + \frac{1}{8} + \frac{1}{16} + \cdots + \frac{1}{2^n} + \cdots = 1$$

We use a similar idea to determine whether or not a general series (10.12) has a sum. We consider the **partial sums**

$$s_1 = a_1$$
$$s_2 = a_1 + a_2$$
$$s_3 = a_1 + a_2 + a_3$$
$$s_4 = a_1 + a_2 + a_3 + a_4$$

and, in general,

$$s_n = a_1 + a_2 + a_3 + \cdots + a_n = \sum_{i=1}^{n} a_i$$

These partial sums form a new sequence $\{s_n\}$, which may or may not have a limit. If $\lim_{n \to \infty} s_n = s$ exists (as a finite number) then, as in the preceding example, we call it the sum of the infinite series $\sum a_n$.

Definition (10.13)

> Given a series $\sum_{n=1}^{\infty} a_n = a_1 + a_2 + a_3 + \cdots$, let s_n denote its nth partial sum:
>
> $$s_n = \sum_{i=1}^{n} a_i = a_1 + a_2 + \cdots + a_n$$
>
> If the sequence $\{s_n\}$ is convergent and $\lim_{n \to \infty} s_n = s$ exists as a real number, then the series $\sum a_n$ is called **convergent** and we write
>
> $$a_1 + a_2 + \cdots + a_n + \cdots = s \qquad \text{or} \qquad \sum_{n=1}^{\infty} a_n = s$$
>
> The number s is called the **sum** of the series. Otherwise the series is called **divergent**.

Thus when we write $\sum_{n=1}^{\infty} a_n = s$ we mean that by adding sufficiently many terms of the series we can get as close as we like to the number s.

Notice that

$$\sum_{n=1}^{\infty} a_n = \lim_{n \to \infty} \sum_{i=1}^{n} a_i$$

EXAMPLE 1   An important example of an infinite series is the **geometric series**

$$a + ar + ar^2 + ar^3 + \cdots + ar^{n-1} + \cdots = \sum_{n=1}^{\infty} ar^{n-1} \qquad a \neq 0$$

Each term is obtained from the preceding one by multiplying by the common ratio r. (We have already considered the special case where $a = \frac{1}{2}$ and $r = \frac{1}{2}$.)

If $r = 1$, then $s_n = a + a + \cdots + a = na \to \pm\infty$. Since $\lim_{n \to \infty} s_n$ does not exist, the geometric series diverges in this case.

If $r \neq 1$, we have

$$s_n = a + ar + ar^2 + \cdots + ar^{n-1}$$

and

$$rs_n = ar + ar^2 + \cdots + ar^{n-1} + ar^n$$

Subtracting these equations, we get

$$s_n - rs_n = a - ar^n$$

(10.14)
$$s_n = \frac{a(1 - r^n)}{1 - r}$$

If $-1 < r < 1$, we know from (10.8) that $r^n \to 0$ as $n \to \infty$, so

$$\lim_{n \to \infty} s_n = \lim_{n \to \infty} \frac{a(1 - r^n)}{1 - r} = \frac{a}{1 - r} - \frac{a}{1 - r} \lim_{n \to \infty} r^n = \frac{a}{1 - r}$$

Thus when $|r| < 1$ the geometric series is convergent and its sum is $a/(1 - r)$.

If $r \leq -1$ or $r > 1$, the sequence $\{r^n\}$ is divergent by (10.8) and so, by Equation 10.14, $\lim_{n \to \infty} s_n$ does not exist. Therefore the geometric series diverges in those cases. ●

We summarize the results of Example 1 as follows:

(10.15)

> The geometric series
>
> $$\sum_{n=1}^{\infty} ar^{n-1} = a + ar + ar^2 + \cdots$$
>
> is convergent if $|r| < 1$ and its sum is
>
> $$\sum_{n=1}^{\infty} ar^{n-1} = \frac{a}{1 - r} \qquad |r| < 1$$
>
> If $|r| \geq 1$, the geometric series is divergent.

EXAMPLE 2   Find the sum of the geometric series

$$5 - \tfrac{10}{3} + \tfrac{20}{9} - \tfrac{40}{27} + \cdots$$

Solution   The first term is $a = 5$ and the common ratio is $r = -\tfrac{2}{3}$. Since $|r| = \tfrac{2}{3} < 1$, the series is convergent by (10.15) and its sum is

$$5 - \tfrac{10}{3} + \tfrac{20}{9} - \tfrac{40}{27} + \cdots = \frac{5}{1 - \left(-\tfrac{2}{3}\right)} = \frac{5}{\tfrac{5}{3}} = 3$$ ●

EXAMPLE 3   Is the series $\sum_{n=1}^{\infty} 2^{2n}3^{1-n}$ convergent or divergent?

Solution

$$\sum_{n=1}^{\infty} 2^{2n}3^{1-n} = \sum_{n=1}^{\infty} \frac{4^n}{3^{n-1}} = \sum_{n=1}^{\infty} 4\left(\tfrac{4}{3}\right)^{n-1}$$

We recognize this series as a geometric series with $a = 4$ and $r = \tfrac{4}{3}$. Since $r > 1$, the series diverges by (10.15). ●

EXAMPLE 4   Write the number $2.3\overline{17} = 2.3171717\ldots$ as a ratio of integers.

Solution

$$2.3171717\ldots = 2.3 + \frac{17}{10^3} + \frac{17}{10^5} + \frac{17}{10^7} + \cdots$$

After the first term we have a geometric series with $a = 17/10^3$ and $r = 1/10^2$. Therefore

$$2.3\overline{17} = 2.3 + \frac{\dfrac{17}{10^3}}{1 - \dfrac{1}{10^2}}$$

$$= 2.3 + \frac{\dfrac{17}{1000}}{\dfrac{99}{100}}$$

$$= \frac{23}{10} + \frac{17}{990} = \frac{1147}{495}$$

EXAMPLE 5   Find the sum of the series $\sum\limits_{n=0}^{\infty} x^n$, where $|x| < 1$.

Solution   Notice that this series starts with $n = 0$ and so the first term is $x^0 = 1$. Thus

$$\sum_{n=0}^{\infty} x^n = 1 + x + x^2 + x^3 + x^4 + \cdots$$

This is a geometric series with $a = 1$ and $r = x$. Since $|r| = |x| < 1$, it converges and (10.15) gives

(10.16)
$$\sum_{n=0}^{\infty} x^n = \frac{1}{1-x}$$

EXAMPLE 6   Show that the series $\sum\limits_{n=1}^{\infty} \dfrac{1}{n(n+1)}$ is convergent and find its sum.

Solution   This is not a geometric series so we go back to the definition of a convergent series and compute the partial sums

$$s_n = \sum_{i=1}^{n} \frac{1}{i(i+1)} = \frac{1}{1\cdot 2} + \frac{1}{2\cdot 3} + \frac{1}{3\cdot 4} + \cdots + \frac{1}{n(n+1)}$$

We can simplify this expression if we use the partial fraction decomposition

$$\frac{1}{i(i+1)} = \frac{1}{i} - \frac{1}{i+1}$$

(see Section 7.4). Thus we have

$$s_n = \sum_{i=1}^{n} \frac{1}{i(i+1)} = \sum_{i=1}^{n} \left(\frac{1}{i} - \frac{1}{i+1}\right)$$

Terms cancel in pairs (telescoping sum)

$$= \left(1 - \frac{1}{2}\right) + \left(\frac{1}{2} - \frac{1}{3}\right) + \left(\frac{1}{3} - \frac{1}{4}\right) + \cdots + \left(\frac{1}{n} - \frac{1}{n+1}\right)$$

$$= 1 - \frac{1}{n+1}$$

and so

$$\lim_{n\to\infty} s_n = \lim_{n\to\infty} \left(1 - \frac{1}{n+1}\right) = 1 - 0 = 1$$

Therefore the given series is convergent and

$$\sum_{n=1}^{\infty} \frac{1}{n(n+1)} = 1$$

EXAMPLE 7   Show that the **harmonic series**

$$\sum_{n=1}^{n} \frac{1}{n} = 1 + \frac{1}{2} + \frac{1}{3} + \frac{1}{4} + \cdots$$

is divergent.

Solution

$$s_1 = 1$$

$$s_2 = 1 + \frac{1}{2}$$

$$s_4 = 1 + \frac{1}{2} + \left(\frac{1}{3} + \frac{1}{4}\right) > 1 + \frac{1}{2} + \left(\frac{1}{4} + \frac{1}{4}\right) = 1 + \frac{2}{2}$$

$$s_8 = 1 + \frac{1}{2} + \left(\frac{1}{3} + \frac{1}{4}\right) + \left(\frac{1}{5} + \frac{1}{6} + \frac{1}{7} + \frac{1}{8}\right)$$

$$> 1 + \frac{1}{2} + \left(\frac{1}{4} + \frac{1}{4}\right) + \left(\frac{1}{8} + \frac{1}{8} + \frac{1}{8} + \frac{1}{8}\right)$$

$$= 1 + \frac{1}{2} + \frac{1}{2} + \frac{1}{2} = 1 + \frac{3}{2}$$

$$s_{16} = 1 + \frac{1}{2} + \left(\frac{1}{3} + \frac{1}{4}\right) + \left(\frac{1}{5} + \cdots + \frac{1}{8}\right) + \left(\frac{1}{9} + \cdots + \frac{1}{16}\right)$$

$$> 1 + \frac{1}{2} + \left(\frac{1}{4} + \frac{1}{4}\right) + \left(\frac{1}{8} + \cdots + \frac{1}{8}\right) + \left(\frac{1}{16} + \cdots + \frac{1}{16}\right)$$

$$= 1 + \frac{1}{2} + \frac{1}{2} + \frac{1}{2} + \frac{1}{2} = 1 + \frac{4}{2}$$

Similarly $s_{32} > 1 + \frac{5}{2}$, $s_{64} > 1 + \frac{6}{2}$, and in general

$$s_{2^n} > 1 + \frac{n}{2}$$

This shows that $s_{2^n} \to \infty$ as $n \to \infty$ and so $\{s_n\}$ is divergent. Therefore the harmonic series diverges.

Theorem (10.17)

> If the series $\sum_{n=1}^{\infty} a_n$ is convergent, then $\lim_{n \to \infty} a_n = 0$.

Proof   Let $s_n = a_1 + a_2 + \cdots + a_n$. Then $a_n = s_n - s_{n-1}$. Since $\sum a_n$ is convergent, the sequence $\{s_n\}$ is convergent. Let $\lim_{n \to \infty} s_n = s$. Since $n - 1 \to \infty$ as $n \to \infty$, we also have $\lim_{n \to \infty} s_{n-1} = s$. Therefore

$$\lim_{n \to \infty} a_n = \lim_{n \to \infty} (s_n - s_{n-1}) = \lim_{n \to \infty} s_n - \lim_{n \to \infty} s_{n-1}$$

$$= s - s = 0$$

Note 1:   With any *series* $\sum a_n$ we associate two *sequences:* the sequence $\{s_n\}$ of its partial sums and the sequence $\{a_n\}$ of its terms. If $\sum a_n$ is convergent, then the limit of the sequence $\{s_n\}$ is s and, as Theorem 10.17 asserts, the limit of the sequence $\{a_n\}$ is 0.

Note 2:   The converse of Theorem 10.17 is not true in general. If $\lim_{n\to\infty} a_n = 0$, we cannot conclude that $\sum a_n$ is convergent. Observe that for the harmonic series $\sum 1/n$ we have $a_n = 1/n \to 0$ as $n \to \infty$, but we showed in Example 7 that $\sum 1/n$ is divergent.

The Test for Divergence (10.18)

> If $\lim\limits_{n\to\infty} a_n$ does not exist or if $\lim\limits_{n\to\infty} a_n \neq 0$, then the series $\sum\limits_{n=1}^{\infty} a_n$ is divergent.

Proof   This follows immediately from Theorem 10.17.

EXAMPLE 8   Show that the series $\sum\limits_{n=1}^{\infty} \dfrac{n^2}{5n^2 + 4}$ diverges.

Solution

$$\lim_{n\to\infty} a_n = \lim_{n\to\infty} \frac{n^2}{5n^2 + 4}$$

$$= \lim_{n\to\infty} \frac{1}{5 + 4/n^2} = \frac{1}{5} \neq 0$$

So the series diverges by the Test for Divergence.

Note 3:   If we find that $\lim_{n\to\infty} a_n \neq 0$, we know that $\sum a_n$ is divergent. If we find that $\lim_{n\to\infty} a_n = 0$, we know *nothing* about the convergence or divergence of $\sum a_n$. Remember the warning in Note 2: If $\lim_{n\to\infty} a_n = 0$, the series $\sum a_n$ might converge or it might diverge.

Theorem (10.19)

> If $\sum a_n$ and $\sum b_n$ are convergent series, then so are the series $\sum c a_n$ (where c is a constant), $\sum (a_n + b_n)$, and $\sum (a_n - b_n)$, and
>
> (a) $\sum\limits_{n=1}^{\infty} c a_n = c \sum\limits_{n=1}^{\infty} a_n$
>
> (b) $\sum\limits_{n=1}^{\infty} (a_n + b_n) = \sum\limits_{n=1}^{\infty} a_n + \sum\limits_{n=1}^{\infty} b_n$
>
> (c) $\sum\limits_{n=1}^{\infty} (a_n - b_n) = \sum\limits_{n=1}^{\infty} a_n - \sum\limits_{n=1}^{\infty} b_n$

Proof
(a)   This proof is left as Exercise 53.
(b)   Let

$$s_n = \sum_{i=1}^{n} a_i \qquad s = \sum_{n=1}^{\infty} a_n \qquad t_n = \sum_{i=1}^{n} b_i \qquad t = \sum_{n=1}^{\infty} b_n$$

The nth partial sum for the series $\sum (a_n + b_n)$ is

$$u_n = \sum_{i=1}^{n} (a_i + b_i)$$

and, using Theorem 5.2, we have

$$\lim_{n \to \infty} u_n = \lim_{n \to \infty} \sum_{i=1}^{n} (a_i + b_i) = \lim_{n \to \infty} \left(\sum_{i=1}^{n} a_i + \sum_{i=1}^{n} b_i \right)$$

$$= \lim_{n \to \infty} \sum_{i=1}^{n} a_i + \lim_{n \to \infty} \sum_{i=1}^{n} b_i$$

$$= \lim_{n \to \infty} s_n + \lim_{n \to \infty} t_n = s + t$$

Therefore $\sum (a_n + b_n)$ is convergent and its sum is

$$\sum_{n=1}^{\infty} (a_n + b_n) = s + t = \sum_{n=1}^{\infty} a_n + \sum_{n=1}^{\infty} b_n$$

(c) This equation is proved like part (b) or it can be deduced from parts (a) and (b).

●

EXAMPLE 9   Find the sum of the series $\displaystyle\sum_{n=1}^{\infty} \left(\frac{3}{n(n+1)} + \frac{1}{2^n} \right)$.

Solution   The series $\sum 1/2^n$ is a geometric series with $a = \frac{1}{2}$ and $r = \frac{1}{2}$, so

$$\sum_{n=1}^{\infty} \frac{1}{2^n} = \frac{\frac{1}{2}}{1 - \frac{1}{2}} = 1$$

In Example 6 we found that

$$\sum_{n=1}^{\infty} \frac{1}{n(n+1)} = 1$$

So, by Theorem 10.19, the given series is convergent and

$$\sum_{n=1}^{\infty} \left(\frac{3}{n(n+1)} + \frac{1}{2^n} \right) = 3 \sum_{n=1}^{\infty} \frac{1}{n(n+1)} + \sum_{n=1}^{\infty} \frac{1}{2^n}$$

$$= 3 \cdot 1 + 1 = 4$$

●

Note 4:   A finite number of terms cannot affect the convergence of a series. For instance, suppose that we were able to show that the series

$$\sum_{n=4}^{\infty} \frac{n}{n^3 + 1}$$

is convergent. Since

$$\sum_{n=1}^{\infty} \frac{n}{n^3 + 1} = \frac{1}{2} + \frac{2}{9} + \frac{3}{28} + \sum_{n=4}^{\infty} \frac{n}{n^3 + 1}$$

it follows that the entire series $\sum_{n=1}^{\infty} n/(n^3 + 1)$ is convergent. Similarly if it is known that the series $\sum_{n=N+1}^{\infty} a_n$ converges, then the full series

$$\sum_{n=1}^{\infty} a_n = \sum_{n=1}^{N} a_n + \sum_{n=N+1}^{\infty} a_n$$

is also convergent.

SECTION 10.2 **Exercises**

In Exercises 1–36 determine whether the given series is convergent or divergent. If it is convergent, find its sum.

1. $4 + \frac{8}{5} + \frac{16}{25} + \frac{32}{125} + \cdots$ 2. $1 - \frac{1}{2} + \frac{1}{4} - \frac{1}{8} + \cdots$

3. $\frac{2}{3} - \frac{2}{9} + \frac{2}{27} - \frac{2}{81} + \cdots$

4. $\frac{1}{2^6} + \frac{1}{2^8} + \frac{1}{2^{10}} + \frac{1}{2^{12}} + \cdots$

5. $\frac{1}{36} + \frac{1}{30} + \frac{1}{25} + \frac{6}{125} + \cdots$

6. $-\frac{81}{100} + \frac{9}{10} - 1 + \frac{10}{9} - \cdots$

7. $\sum_{n=1}^{\infty} 2\left(\frac{3}{4}\right)^{n-1}$ 8. $\sum_{n=1}^{\infty} \left(-\frac{3}{\pi}\right)^{n-1}$

9. $\sum_{n=1}^{\infty} 5\left(\frac{e}{3}\right)^{n}$ 10. $\sum_{n=1}^{\infty} \frac{1}{e^{2n}}$

11. $\sum_{n=0}^{\infty} \frac{5^n}{8^n}$ 12. $\sum_{n=0}^{\infty} \frac{4^{n+1}}{5^n}$

13. $\sum_{n=1}^{\infty} 3^{-n} 8^{n+1}$ 14. $\sum_{n=1}^{\infty} 3^{1-n} 5^{n/2}$

15. $\sum_{n=1}^{\infty} \frac{(-2)^{n+3}}{5^{n-2}}$ 16. $\sum_{n=1}^{\infty} (-1)^{n-1} \frac{3^{2n}}{2^{3n+1}}$

17. $\sum_{n=1}^{\infty} \frac{n}{n+1}$ 18. $\sum_{n=4}^{\infty} \frac{3}{n(n-1)}$

19. $\sum_{n=1}^{\infty} \frac{1}{2n}$ 20. $\sum_{n=1}^{\infty} \frac{n^2}{3(n+1)(n+2)}$

21. $\sum_{n=1}^{\infty} \frac{1}{(3n-2)(3n+1)}$ 22. $\sum_{n=1}^{\infty} \left(\frac{1}{2^{n-1}} + \frac{2}{3^{n-1}}\right)$

23. $\sum_{n=1}^{\infty} [2(0.1)^n + (0.2)^n]$ 24. $\sum_{n=1}^{\infty} \left(\frac{1}{n} + 2^n\right)$

25. $\sum_{n=1}^{\infty} \frac{n}{\sqrt{1+n^2}}$ 26. $\sum_{n=1}^{\infty} \frac{1}{4n^2 - 1}$

27. $\sum_{n=1}^{\infty} \frac{1}{n(n+2)}$ 28. $\sum_{n=1}^{\infty} \ln\left(\frac{n}{2n+5}\right)$

29. $\sum_{n=1}^{\infty} \frac{3^n + 2^n}{6^n}$ 30. $\sum_{n=1}^{\infty} \frac{2n+1}{n^2(n+1)^2}$

31. $\sum_{n=1}^{\infty} \left[\sin\left(\frac{1}{n}\right) - \sin\left(\frac{1}{n+1}\right)\right]$

32. $\sum_{n=1}^{\infty} \frac{1}{5 + 2^{-n}}$

33. $\sum_{n=1}^{\infty} \arctan n$ 34. $\sum_{n=1}^{\infty} \frac{1}{n(n+1)(n+2)}$

35. $\sum_{n=1}^{\infty} \ln \frac{n}{n+1}$ 36. $\sum_{n=1}^{\infty} \ln \frac{n^2-1}{n^2}$

Express the numbers in Exercises 37–42 as ratio of integers.

37. $0.\overline{5} = 0.5555\ldots$ 38. $0.\overline{15} = 0.15151515\ldots$

39. $0.\overline{307} = 0.307307307307\ldots$

40. $1.1\overline{23}$

41. $0.123\overline{456}$ 42. $4.\overline{1570}$

In Exercises 43–48 find the values of x for which the given series converges. Find the sum of the series for those values of x.

43. $\sum_{n=0}^{\infty} (x-3)^n$ 44. $\sum_{n=0}^{\infty} 3^n x^n$

45. $\sum_{n=2}^{\infty} \frac{x^n}{5^n}$ 46. $\sum_{n=0}^{\infty} \frac{1}{x^n}$

47. $\sum_{n=0}^{\infty} 2^n \sin^n x$ 48. $\sum_{n=0}^{\infty} \tan^n x$

49. A rubber ball is dropped from a height of 4 m and bounces back to half its height after each fall. If it continues to bounce indefinitely, find the total distance it travels.

50. A right triangle ABC is given with $\angle A = \theta$ and $|AC| = b$. CD is drawn perpendicular to AB, DE is drawn perpendicular to BC, $EF \perp AB$, and this process is continued indefinitely as in the figure. Find the total length of all the perpendiculars

$$|CD| + |DE| + |EF| + |FG| + \cdots$$

in terms of b and θ.

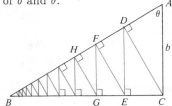

51. What is wrong with the following calculation?

$$0 = 0 + 0 + 0 + \cdots$$
$$= (1-1) + (1-1) + (1-1) + \cdots$$
$$= 1 - 1 + 1 - 1 + 1 - 1 + \cdots$$
$$= 1 + (-1+1) + (-1+1) + (-1+1) + \cdots$$
$$= 1 + 0 + 0 + 0 + \cdots = 1$$

(Guido Ubaldus thought that this proved the existence of God because "something has been created out of nothing.")

52. Suppose that $\sum_{n=1}^{\infty} a_n$ $(a_n \neq 0)$ is known to be a convergent series. Prove that $\sum_{n=1}^{\infty} 1/a_n$ is a divergent series.

53. Prove Theorem 10.19 part (a).

54. If $\sum a_n$ is divergent and $c \neq 0$, show that $\sum ca_n$ is divergent.

55. If $\sum a_n$ is convergent and $\sum b_n$ is divergent, show that the series $\sum (a_n + b_n)$ is divergent. (*Hint:* Argue by contradiction.)

56. If $\sum a_n$ and $\sum b_n$ are both divergent, is $\sum (a_n + b_n)$ necessarily divergent?

57. Consider the series $\sum_{n=1}^{\infty} \dfrac{n}{(n+1)!}$.
   (a) Find the partial sums s_1, s_2, s_3, and s_4. Do you recognize the denominators? Use the pattern to guess a formula for s_n.
   (b) Use mathematical induction to prove your guess.

   (c) Show that the given infinite series is convergent and find its sum.

58. In the figure there are infinitely many circles approaching the vertices of an equilateral triangle, each circle touching other circles and sides of the triangles. If the triangle has sides of length 1, find the total area occupied by the circles.

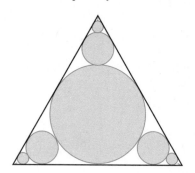

The Integral Test

In general it is difficult to find the exact sum of a series. We were able to accomplish this for geometric series and the series $\sum 1/[n\,(n+1)]$ because in those cases there was a simple formula for the nth partial sum s_n. But usually it is not easy to compute $\lim_{n\to\infty} s_n$. Therefore in the next few sections we develop several tests that enable us to say whether a series is convergent or divergent without explicitly finding its sum. Our first test involves improper integrals.

The Integral Test (10.20)

Suppose f is a continuous, positive, decreasing function on $[1,\infty)$ and let $a_n = f(n)$. Then the series $\sum_{n=1}^{\infty} a_n$ is convergent if and only if the improper integral $\int_1^{\infty} f(x)\,dx$ is convergent. In other words:

(a) If $\displaystyle\int_1^{\infty} f(x)\,dx$ is convergent, then $\displaystyle\sum_{n=1}^{\infty} a_n$ is convergent.

(b) If $\displaystyle\int_1^{\infty} f(x)\,dx$ is divergent, then $\displaystyle\sum_{n=1}^{\infty} a_n$ is divergent.

Proof   The basic idea behind the Integral Test can be seen by looking at Figures 10.7 and 10.8. The area of the first shaded rectangle in Figure 10.7 is the value of f at the right endpoint of $[1,2]$, that is, $f(2) = a_2$. So comparing the areas of the shaded rectangles with the area under $y = f(x)$ from 1 to n, we see that

(10.21)
$$a_2 + a_3 + \cdots + a_n \leq \int_1^n f(x)\,dx$$

(Notice that this inequality depends on the fact that f is decreasing.) Likewise Figure 10.8 shows that

(10.22)

$$\int_1^n f(x)\,dx \le a_1 + a_2 + \cdots + a_{n-1}$$

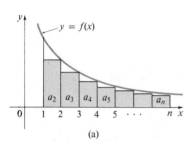

(a)   If $\int_1^\infty f(x)\,dx$ is convergent, then (10.21) gives

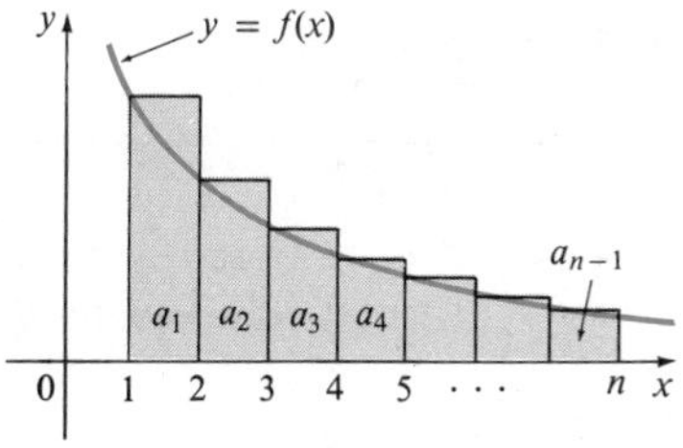

$$\sum_{i=2}^n a_i \le \int_1^n f(x)\,dx \le \int_1^\infty f(x)\,dx$$

since $f(x) \ge 0$. Therefore

$$s_n = a_1 + \sum_{i=2}^n a_i \le a_1 + \int_1^\infty f(x)\,dx = M, \text{ say}$$

Figure 10.7   Since $s_n \le M$ for all n, the sequence $\{s_n\}$ is bounded above. Also

$$s_{n+1} = s_n + a_{n+1} \ge s_n$$

since $a_{n+1} = f(n+1) \ge 0$. Thus $\{s_n\}$ is an increasing bounded sequence and so it is convergent by Theorem 10.11. This means that $\sum a_n$ is convergent.

(b)   If $\int_1^\infty f(x)\,dx$ is divergent, then $\int_1^n f(x)\,dx \to \infty$ as $n \to \infty$ because $f(x) \ge 0$. But (10.22) gives

$$\int_1^n f(x)\,dx \le \sum_{i=1}^{n-1} a_i = s_{n-1}$$

Figure 10.8

and so $s_{n-1} \to \infty$. This implies that $s_n \to \infty$ and so $\sum a_n$ diverges.   ●

Note: When using the Integral Test it is not necessary to start the series or the integral at $n = 1$. For instance, in testing the series

$$\sum_{n=4}^\infty \frac{1}{(n-3)^2} \qquad \text{we use} \qquad \int_4^\infty \frac{1}{(x-3)^2}\,dx$$

Also, it is not necessary that f be always decreasing. What is important is that f be *ultimately* decreasing, that is, decreasing for x larger than some number N. Then $\sum_{n=N}^\infty a_n$ is convergent, so $\sum_{n=1}^\infty a_n$ is convergent by Note 4 of Section 10.2.

EXAMPLE 1   Test the series $\displaystyle\sum_{n=1}^\infty \frac{1}{n^2+1}$ for convergence or divergence.

Solution   The function $f(x) = 1/(x^2+1)$ is continuous, positive, and decreasing on $[1,\infty)$ so we use the Integral Test:

$$\int_1^\infty \frac{1}{x^2+1}\,dx = \lim_{t\to\infty} \int_1^t \frac{1}{x^2+1}\,dx = \lim_{t\to\infty} \tan^{-1}x\,\Big|_1^t$$

$$= \lim_{t\to\infty}\left(\tan^{-1}t - \frac{\pi}{4}\right) = \frac{\pi}{2} - \frac{\pi}{4} = \frac{\pi}{4}$$

Thus $\int_1^\infty 1/(x^2+1)\,dx$ is a convergent integral and so, by the Integral Test, the series $\sum 1/(n^2+1)$ is convergent.   ●

EXAMPLE 2   For what values of p is the series $\displaystyle\sum_{n=1}^\infty \frac{1}{n^p}$ convergent?

Solution   If $p < 0$, then $\lim_{n\to\infty}(1/n^p) = \infty$. If $p = 0$, then $\lim_{n\to\infty}(1/n^p) = 1$. In either case $\lim_{n\to\infty}(1/n^p) \ne 0$, so the given series diverges by the Test for Divergence (10.18).

If $p > 0$, then the function $f(x) = 1/x^p$ is clearly continuous, positive, and decreasing on $[1, \infty)$. We found in Chapter 7 [see (7.36)] that

$$\int_1^\infty \frac{1}{x^p}\, dx \text{ converges if } p > 1 \text{ and diverges if } p \le 1$$

It follows from the Integral Test that the series $\sum 1/n^p$ converges if $p > 1$ and diverges if $0 < p \le 1$. (For $p = 1$, this series is the harmonic series discussed in Example 7 in Section 10.2.)   ●

The series in Example 2 is called the **p-series.** It is important in the rest of this chapter so we summarize the results of Example 2 for future reference as follows:

(10.23) | The p-series $\sum_{n=1}^\infty \dfrac{1}{n^p}$ is convergent if $p > 1$ and divergent if $p \le 1$.

EXAMPLE 3   The series $\sum 1/n^2$ is convergent because it is a p-series with $p = 2 > 1$. The exact sum of this series was found by the Swiss mathematician Leonhard Euler (1707–1783) to be

$$\sum_{n=1}^\infty \frac{1}{n^2} = \frac{\pi^2}{6}$$

but the proof of this fact is beyond the scope of this book.   ●

Note: We should *not* infer from the Integral Test that the sum of the series is equal to the value of the integral. In fact, we know from Example 3 and from Section 7.9 that

$$\sum_{n=1}^\infty \frac{1}{n^2} = \frac{\pi^2}{6}$$

while

$$\int_1^\infty \frac{1}{x^2}\, dx = 1$$

Therefore, in general,

$$\sum_{n=1}^\infty a_n \neq \int_1^\infty f(x)\, dx$$

EXAMPLE 4   The series $\sum 1/\sqrt{n}$ is divergent because it can be rewritten as $\sum 1/n^{1/2}$, which is a p-series with $p = \frac{1}{2} < 1$.   ●

EXAMPLE 5   Determine whether the series $\sum_{n=1}^\infty \dfrac{\ln n}{n}$ converges or diverges.

Solution   The function $f(x) = (\ln x)/x$ is positive and continuous for $x > 1$ since the logarithm function is continuous. But it is not obvious whether or not f is decreasing so we compute its derivative:

$$f'(x) = \frac{(1/x)x - \ln x}{x^2} = \frac{1 - \ln x}{x^2}$$

Thus $f'(x) < 0$ when $\ln x > 1$, that is, $x > e$. It follows that f is decreasing when $x > e$ and so we can apply the Integral Test:

$$\int_1^\infty \frac{\ln x}{x}\, dx = \lim_{t\to\infty} \int_1^t \frac{\ln x}{x}\, dx = \lim_{t\to\infty} \frac{(\ln x)^2}{2}\Big|_1^t$$

$$= \lim_{t\to\infty} \frac{(\ln t)^2}{2} = \infty$$

Since this improper integral is divergent, the series $\sum (\ln n)/n$ is also divergent by the Integral Test.

SECTION 10.3 Exercises

Test the series in Exercises 1–22 for convergence or divergence.

1. $\displaystyle\sum_{n=1}^\infty \frac{2}{\sqrt[3]{n}}$

2. $\displaystyle\sum_{n=1}^\infty \left(\frac{2}{n\sqrt{n}} + \frac{3}{n^3}\right)$

3. $\displaystyle\sum_{n=5}^\infty \frac{1}{n^{1.0001}}$

4. $\displaystyle\sum_{n=1}^\infty n^{-0.99}$

5. $\displaystyle\sum_{n=5}^\infty \frac{1}{(n-4)^2}$

6. $\displaystyle\sum_{n=1}^\infty \frac{1}{2n+3}$

7. $\displaystyle\sum_{n=1}^\infty \frac{1}{\sqrt{n+1}}$

8. $\displaystyle\sum_{n=2}^\infty \frac{1}{n^2-1}$

9. $\displaystyle\sum_{n=1}^\infty n e^{-n^2}$

10. $\displaystyle\sum_{n=1}^\infty \frac{n}{2^n}$

11. $\displaystyle\sum_{n=1}^\infty \frac{n}{n^2+1}$

12. $\displaystyle\sum_{n=2}^\infty \frac{1}{2n^2-n-1}$

13. $\displaystyle\sum_{n=2}^\infty \frac{1}{n\ln n}$

14. $\displaystyle\sum_{n=1}^\infty \frac{1}{4n^2+1}$

15. $\displaystyle\sum_{n=1}^\infty \frac{\arctan n}{1+n^2}$

16. $\displaystyle\sum_{n=2}^\infty \frac{1}{n(\ln n)^2}$

17. $\displaystyle\sum_{n=1}^\infty \frac{\ln n}{n^2}$

18. $\displaystyle\sum_{n=1}^\infty \left(\frac{\ln n}{n}\right)^2$

19. $\displaystyle\sum_{n=1}^\infty \frac{\sin\left(\frac{1}{n}\right)}{n^2}$

20. $\displaystyle\sum_{n=3}^\infty \frac{1}{n\ln n \ln(\ln n)}$

21. $\displaystyle\sum_{n=1}^\infty \frac{1}{n^2+2n+2}$

22. $\displaystyle\sum_{n=1}^\infty \operatorname{sech}^2 n$

In Exercises 23–26 find the values of p for which the given series is convergent.

23. $\displaystyle\sum_{n=2}^\infty \frac{1}{n(\ln n)^p}$

24. $\displaystyle\sum_{n=3}^\infty \frac{1}{n\ln n[\ln(\ln n)]^p}$

25. $\displaystyle\sum_{n=1}^\infty n(1+n^2)^p$

26. $\displaystyle\sum_{n=1}^\infty \frac{\ln n}{n^p}$

27. The Riemann zeta-function ζ is defined by

$$\zeta(x) = \sum_{n=1}^\infty \frac{1}{n^x}$$

and is used in number theory to study the distribution of prime numbers. What is the domain of ζ?

28. (a) Use (10.21) to show that if s_n is the nth partial sum of the harmonic series, then

$$s_n \le 1 + \ln n$$

  (b) The harmonic series diverges, but very slowly. Use part (a) to show that the sum of the first million terms is less than 15 and the sum of the first billion terms is less than 22.

29. According to the proof of the Integral Test, the sum of the series $\sum_{n=1}^\infty n^{-1.001}$ must lie between what two values?

30. Use the following steps to show that the sequence

$$t_n = 1 + \tfrac{1}{2} + \tfrac{1}{3} + \cdots + \tfrac{1}{n} - \ln n$$

has a limit. (The value of the limit is denoted by γ and is called Euler's constant.)

  (a) Draw a picture like Figure 10.8 with $f(x) = 1/x$ and interpret t_n as an area [or use (10.22)] to show that $t_n > 0$ for all n.

  (b) Interpret

$$t_n - t_{n+1} = [\ln(n+1) - \ln n] - \frac{1}{n+1}$$

  as a difference of areas to show that $t_n - t_{n+1} > 0$. Therefore $\{t_n\}$ is a decreasing sequence.

  (c) Use Theorem 10.11 to show that $\{t_n\}$ is convergent.

SECTION 10.4

The Comparison Tests

In the comparison tests the idea is to compare a given series with a series that is known to be convergent or divergent.

The Comparison Test (10.24)

Suppose that $\sum a_n$ and $\sum b_n$ are series with positive terms.

(a) If $\sum b_n$ is convergent and $a_n \le b_n$ for all n, then $\sum a_n$ is also convergent.

(b) If $\sum b_n$ is divergent and $a_n \ge b_n$ for all n, then $\sum a_n$ is also divergent.

Proof

(a) Let

$$s_n = \sum_{i=1}^{n} a_i \qquad t_n = \sum_{i=1}^{n} b_i \qquad t = \sum_{n=1}^{\infty} b_n$$

Since both series have positive terms, the sequences $\{s_n\}$ and $\{t_n\}$ are increasing $(s_{n+1} = s_n + a_{n+1} \ge s_n)$. Also $t_n \to t$, so $t_n \le t$ for all n. Since $a_i \le b_i$, we have $s_n \le t_n$. Thus $s_n \le t$ for all n. This means that $\{s_n\}$ is increasing and bounded above and therefore converges by Theorem 10.11. Thus $\sum a_n$ converges.

(b) If $\sum b_n$ is divergent, then $t_n \to \infty$ (since $\{t_n\}$ is increasing). But $a_i \ge b_i$ so $s_n \ge t_n$. Thus $s_n \to \infty$. Therefore $\sum a_n$ diverges.   ●

In using the Comparison Test we must, of course, have some known series $\sum b_n$ for purposes of comparison. Most of the time we use either a p-series [$\sum 1/n^p$ converges if $p > 1$ and diverges if $p \le 1$; see (10.23)] or a geometric series [$\sum ar^{n-1}$ converges if $|r| < 1$ and diverges if $|r| \ge 1$; see (10.15)].

EXAMPLE 1   Determine whether the series $\sum_{n=1}^{\infty} \dfrac{5}{2n^2 + 4n + 3}$ converges or diverges.

Solution   For large n the dominant term in the denominator is $2n^2$ so we compare the given series with the series $\sum 5/(2n^2)$. Observe that

$$\frac{5}{2n^2 + 4n + 3} < \frac{5}{2n^2}$$

since the left side has a bigger denominator. (In the notation of Theorem 10.24, a_n is the left side and b_n is the right side.) We know that

$$\sum_{n=1}^{\infty} \frac{5}{2n^2} = \frac{5}{2} \sum_{n=1}^{\infty} \frac{1}{n^2}$$

is convergent (p-series with $p = 2 > 1$). Therefore

$$\sum_{n=1}^{\infty} \frac{5}{2n^2 + 4n + 3}$$

is convergent by part (a) of the Comparison Test.   ●

Although the condition $a_n \le b_n$ or $a_n \ge b_n$ in the Comparison Test was given for all n, we need only verify that it holds for $n \ge N$, where N is some fixed integer, because the convergence of a series is not affected by a finite number of terms. This is illustrated in the next example.

EXAMPLE 2   Test the series $\sum_{n=1}^{\infty} \frac{\ln n}{n}$ for convergence or divergence.

Solution   This series was tested (using the Integral Test) in Example 5 in Section 10.3, but it is also possible to test it by comparing it with the harmonic series. Observe that $\ln n > 1$ for all $n \geq 3$ and so

$$\frac{\ln n}{n} > \frac{1}{n} \qquad n \geq 3$$

We know that $\sum 1/n$ is divergent (p-series with $p = 1$). Thus the given series is divergent by the Comparison Test.   ●

EXAMPLE 3   Test the series $\sum_{n=1}^{\infty} \frac{1}{2^n + 1}$ for convergence or divergence.

Solution   Notice that

$$\frac{1}{2^n + 1} < \frac{1}{2^n} = \left(\frac{1}{2}\right)^n \qquad n \geq 1$$

The series $\sum (1/2)^n$ is convergent (geometric series with $r = 1/2$) and so the given series converges by the Comparison Test.   ●

Note: The terms of the series being tested must be smaller than those of a convergent series or larger than those of a divergent series. If the terms are larger than the terms of a convergent series or smaller than those of a divergent series, then the Comparison Test does not apply. For instance, suppose that in Example 3 we had been given the similar series

$$\sum_{n=1}^{\infty} \frac{1}{2^n - 1}$$

The inequality

$$\frac{1}{2^n - 1} > \frac{1}{2^n}$$

is useless as far as the Comparison Test is concerned because $\sum b_n = \sum (1/2)^n$ is convergent and $a_n > b_n$. Nonetheless we have the feeling that $\sum 1/(2^n - 1)$ ought to be convergent since it is very similar to the convergent geometric series $\sum (1/2)^n$. In such cases the following test can be used.

The Limit Comparison Test
(10.25)

Suppose that $\sum a_n$ and $\sum b_n$ are series with positive terms.

(a)   If $\lim_{n \to \infty} \frac{a_n}{b_n} = c > 0$, then either both series converge or both diverge.

(b)   If $\lim_{n \to \infty} \frac{a_n}{b_n} = 0$ and $\sum b_n$ converges, then $\sum a_n$ also converges.

(c)   If $\lim_{n \to \infty} \frac{a_n}{b_n} = \infty$ and $\sum b_n$ diverges, then $\sum a_n$ also diverges.

Proof   To prove part (a) we take $\varepsilon = c/2$ in Definition 10.1 and see that, since $\lim_{n \to \infty}(a_n/b_n) = c$, there is an integer N such that

$$\left| \frac{a_n}{b_n} - c \right| < \frac{c}{2} \qquad \text{when} \qquad n > N$$

Thus

$$\frac{c}{2} < \frac{a_n}{b_n} < \frac{3c}{2} \qquad \text{when} \qquad n > N$$

(10.26)   and so

$$\left(\frac{c}{2} \right) b_n < a_n < \left(\frac{3c}{2} \right) b_n \qquad \text{when} \qquad n > N$$

If $\sum b_n$ converges, so does $\sum (3c/2)b_n$. The right half of (10.26) then shows that $\sum_N^\infty a_n$ converges by the Comparison Test. It follows that $\sum_1^\infty a_n$ converges. If $\sum b_n$ diverges, so does $\sum (c/2)b_n$ and the left half of (10.26) together with part (b) of the Comparison Test shows that $\sum a_n$ diverges. The proofs of parts (b) and (c) are similar to that of part (a) and are left as Exercises 42 and 43.   ●

EXAMPLE 4   Test the series $\sum_{n=1}^\infty \dfrac{1}{2^n - 1}$ for convergence or divergence.

Solution   We use the Limit Comparison Test with

$$a_n = \frac{1}{2^n - 1} \qquad b_n = \frac{1}{2^n}$$

$$\lim_{n \to \infty} \frac{a_n}{b_n} = \lim_{n \to \infty} \frac{2^n}{2^n - 1} = \lim_{n \to \infty} \frac{1}{1 - 1/2^n} = 1$$

Since this limit exists and $\sum 1/2^n$ is a convergent geometric series, the given series converges by the Limit Comparison Test.   ●

EXAMPLE 5   Solve Example 2 using the Limit Comparison Test.

Solution   Taking $a_n = (\ln n)/n$ and $b_n = 1/n$, we have

$$\lim_{n \to \infty} \frac{a_n}{b_n} = \lim_{n \to \infty} \frac{\frac{\ln n}{n}}{\frac{1}{n}} = \lim_{n \to \infty} \ln n = \infty$$

We know that the harmonic series $\sum 1/n$ is divergent so by part (c) of the Limit Comparison Test, $\sum (\ln n)/n$ diverges.   ●

EXAMPLE 6   Determine whether the series $\sum_{n=1}^\infty \dfrac{2n^2 + 3n}{\sqrt{5 + n^7}}$ converges or diverges.

Solution   The dominant part of the numerator is $2n^2$ and the dominant part of the denominator is $\sqrt{n^7} = n^{7/2}$. This suggests taking

$$a_n = \frac{2n^2 + 3n}{\sqrt{5 + n^7}} \qquad b_n = \frac{2n^2}{n^{7/2}} = \frac{2}{n^{3/2}}$$

$$\lim_{n \to \infty} \frac{a_n}{b_n} = \lim_{n \to \infty} \frac{2n^2 + 3n}{\sqrt{5 + n^7}} \cdot \frac{n^{3/2}}{2} = \lim_{n \to \infty} \frac{2n^{7/2} + 3n^{5/2}}{2\sqrt{5 + n^7}}$$

$$= \lim_{n \to \infty} \frac{2 + \frac{3}{n}}{2\sqrt{\frac{5}{n^7} + 1}} = \frac{2 + 0}{2\sqrt{0 + 1}} = 1$$

Since $\sum b_n = 2 \sum 1/n^{3/2}$ is convergent (p-series, $p = \frac{3}{2} > 1$), the given series converges by the Limit Comparison Test.

Notice that in testing many series we find a suitable comparison series $\sum b_n$ by keeping only the highest powers in the numerator and denominator.

SECTION 10.4 **Exercises**

In Exercises 1–38 determine whether the given series converges or diverges.

1. $\displaystyle\sum_{n=1}^{\infty} \frac{1}{n^3 + n^2}$

2. $\displaystyle\sum_{n=1}^{\infty} \frac{3}{4^n + 5}$

3. $\displaystyle\sum_{n=1}^{\infty} \frac{3}{n2^n}$

4. $\displaystyle\sum_{n=2}^{\infty} \frac{1}{\sqrt{n} - 1}$

5. $\displaystyle\sum_{n=0}^{\infty} \frac{1 + 5^n}{4^n}$

6. $\displaystyle\sum_{n=1}^{\infty} \frac{\sin^2 n}{n\sqrt{n}}$

7. $\displaystyle\sum_{n=1}^{\infty} \frac{3}{n(n+3)}$

8. $\displaystyle\sum_{n=1}^{\infty} \frac{1}{n^n}$

9. $\displaystyle\sum_{n=1}^{\infty} \frac{1}{\sqrt{n^3 + 1}}$

10. $\displaystyle\sum_{n=1}^{\infty} \frac{1}{\sqrt{n(n+1)(n+2)}}$

11. $\displaystyle\sum_{n=2}^{\infty} \frac{\sqrt{n}}{n - 1}$

12. $\displaystyle\sum_{n=1}^{\infty} \frac{1}{\sqrt[3]{n(n+1)(n+2)}}$

13. $\displaystyle\sum_{n=1}^{\infty} \frac{n-1}{n^3 + 1}$

14. $\displaystyle\sum_{n=1}^{\infty} \frac{n}{(n+1)2^n}$

15. $\displaystyle\sum_{n=1}^{\infty} \frac{3 + \cos n}{3^n}$

16. $\displaystyle\sum_{n=1}^{\infty} \frac{5n}{2n^2 - 5}$

17. $\displaystyle\sum_{n=1}^{\infty} \frac{4}{(n+1)(2n+1)}$

18. $\displaystyle\sum_{n=1}^{\infty} \frac{n}{(n+1)(n+2)(n+3)}$

19. $\displaystyle\sum_{n=1}^{\infty} \frac{n}{\sqrt{n^5 + 4}}$

20. $\displaystyle\sum_{n=1}^{\infty} \frac{\arctan n}{n^4}$

21. $\displaystyle\sum_{n=1}^{\infty} \frac{2^n}{1 + 3^n}$

22. $\displaystyle\sum_{n=1}^{\infty} \frac{1 + 2^n}{1 + 3^n}$

23. $\displaystyle\sum_{n=1}^{\infty} \frac{1}{1 + \sqrt{n}}$

24. $\displaystyle\sum_{n=3}^{\infty} \frac{1}{n^2 - 4}$

25. $\displaystyle\sum_{n=1}^{\infty} \frac{n^2 + 1}{n^4 + 1}$

26. $\displaystyle\sum_{n=1}^{\infty} \frac{3n^3 - 2n^2}{n^4 + n^2 + 1}$

27. $\displaystyle\sum_{n=1}^{\infty} \frac{n^2 - n + 2}{\sqrt[4]{n^{10} + n^5 + 3}}$

28. $\displaystyle\sum_{n=1}^{\infty} \frac{n^2 - 3n}{\sqrt[3]{n^{10} - 4n^2}}$

29. $\displaystyle\sum_{n=1}^{\infty} \frac{n+1}{n2^n}$

30. $\displaystyle\sum_{n=1}^{\infty} \frac{2n^2 + 7n}{3^n(n^2 + 5n - 1)}$

31. $\displaystyle\sum_{n=1}^{\infty} \frac{\ln n}{n^3}$

32. $\displaystyle\sum_{n=2}^{\infty} \frac{1}{\ln n}$

33. $\displaystyle\sum_{n=1}^{\infty} \frac{1}{n!}$

34. $\displaystyle\sum_{n=1}^{\infty} \frac{n!}{(2n)!}$

35. $\displaystyle\sum_{n=1}^{\infty} \frac{n!}{n^2}$

36. $\displaystyle\sum_{n=1}^{\infty} \frac{n!}{n^n}$

37. $\displaystyle\sum_{n=1}^{\infty} \sin\!\left(\frac{1}{n}\right)$

38. $\displaystyle\sum_{n=1}^{\infty} \frac{1}{n^{1 + 1/n}}$

39. The meaning of the decimal representation of a number $0.d_1 d_2 d_3 \ldots$ (where the digit d_i is one of the numbers 0, 1, 2, ..., 9) is that

$$0.d_1 d_2 d_3 d_4 \ldots = \frac{d_1}{10} + \frac{d_2}{10^2} + \frac{d_3}{10^3} + \frac{d_4}{10^4} + \cdots$$

Show that this series always converges.

40. For what values of p does the series $\sum_{n=2}^{\infty} 1/(n^p \ln n)$ converge?

41. Prove that if $a_n \geq 0$ and $\sum a_n$ converges, then $\sum a_n^2$ also converges.

42. Prove part (b) of the Limit Comparison Test.

43. Prove part (c) of the Limit Comparison Test.

44. Given an example of a pair of series $\sum a_n$ and $\sum b_n$ with positive terms where $\lim_{n\to\infty} (a_n/b_n) = 0$, $\sum b_n$ diverges, but $\sum a_n$ converges. [Compare with part (b) of Theorem 10.25.]

45. Show that if $a_n > 0$ and $\lim_{n\to\infty} n a_n \neq 0$, then $\sum a_n$ is divergent.

46. Show that if $a_n > 0$ and $\sum a_n$ is convergent, then $\sum \ln(1 + a_n)$ is convergent.

Alternating Series

An **alternating series** is a series whose terms are alternately positive and negative. Here are two examples:

$$1 - \frac{1}{2} + \frac{1}{3} - \frac{1}{4} + \frac{1}{5} - \frac{1}{6} + \cdots = \sum_{n=1}^{\infty} \frac{(-1)^{n-1}}{n}$$

$$-\frac{1}{2} + \frac{2}{3} - \frac{3}{4} + \frac{4}{5} - \frac{5}{6} + \frac{6}{7} - \cdots = \sum_{n=1}^{\infty} (-1)^n \frac{n}{n+1}$$

The following test says that if the terms of an alternating series decrease to 0 in absolute value, then the series converges.

The Alternating Series Test (10.27)

If the alternating series

$$\sum_{n=1}^{\infty} (-1)^{n-1} a_n = a_1 - a_2 + a_3 - a_4 + a_5 - a_6 + \cdots \qquad (a_n > 0)$$

satisfies

(a) $a_{n+1} \le a_n$ for all n

(b) $\lim_{n \to \infty} a_n = 0$

then the series is convergent.

Before giving the proof of Theorem 10.27 let us take a look at Figure 10.9, which gives a picture of the idea behind the proof. We first plot $s_1 = a_1$ on a number line. To find s_2 we subtract a_2, so s_2 is to the left of s_1. Then to find s_3 we add a_3, so s_3 is to the right of s_2. But, since $a_3 < a_2$, s_3 is to the left of s_1. Continuing in this manner, we see that the partial sums oscillate back and forth. Since $a_n \to 0$, the successive steps are becoming smaller and smaller. The even partial sums $s_2, s_4, s_6, \ldots$ are increasing and the odd partial sums $s_1, s_3, s_5, \ldots$ are decreasing and it seems plausible that both are converging to some number s. Therefore in the following proof we consider the even and odd partial sums separately.

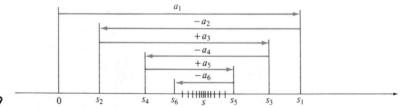

Figure 10.9

Proof of the Alternating Series Test   We first consider the even partial sums:

$$s_2 = a_1 - a_2 \ge 0 \qquad\qquad \text{since } a_2 \le a_1$$

$$s_4 = s_2 + (a_3 - a_4) \ge s_2 \qquad\qquad \text{since } a_4 \le a_3$$

In general

$$s_{2n} = s_{2n-2} + (a_{2n-1} - a_{2n}) \geq s_{2n-2} \qquad \text{since } a_{2n} \leq a_{2n-1}$$

Thus

$$0 \leq s_2 \leq s_4 \leq s_6 \leq \cdots \leq s_{2n} \leq \cdots$$

But we can also write

$$s_{2n} = a_1 - (a_2 - a_3) - (a_4 - a_5) - \cdots - (a_{2n-2} - a_{2n-1}) - a_{2n}$$

Every term in brackets is positive, so $s_{2n} \leq a_1$ for all n. Therefore the sequence $\{s_{2n}\}$ of even partial sums is increasing and bounded above. It is therefore convergent by Theorem 10.11. Let us call its limit s, that is,

$$\lim_{n \to \infty} s_{2n} = s$$

Now we compute the limit of the odd partial sums:

$$\begin{aligned}
\lim_{n \to \infty} s_{2n+1} &= \lim_{n \to \infty} (s_{2n} + a_{2n+1}) \\
&= \lim_{n \to \infty} s_{2n} + \lim_{n \to \infty} a_{2n+1} \\
&= s + 0 \qquad \text{[by condition (b) of (10.27)]} \\
&= s
\end{aligned}$$

Since both the even and odd partial sums converge to s, we have $\lim_{n \to \infty} s_n = s$ and so the series is convergent. ●

EXAMPLE 1   The alternating harmonic series

$$1 - \frac{1}{2} + \frac{1}{3} - \frac{1}{4} + \cdots = \sum_{n=1}^{\infty} \frac{(-1)^{n-1}}{n}$$

satisfies

(a) $a_{n+1} < a_n$ because $\frac{1}{n+1} < \frac{1}{n}$

(b) $\lim_{n \to \infty} a_n = \lim_{n \to \infty} \frac{1}{n} = 0$

so the series is convergent by the Alternating Series Test. It can be shown that its sum is $\ln 2$ (see Exercise 39). ●

EXAMPLE 2   The series $\sum_{n=1}^{\infty} \frac{(-1)^n 3n}{4n - 1}$ is alternating but

$$\lim_{n \to \infty} a_n = \lim_{n \to \infty} \frac{3n}{4n - 1} = \lim_{n \to \infty} \frac{3}{4 - \frac{1}{n}} = \frac{3}{4}$$

so condition (b) is not satisfied. In fact, the limit of the nth term of the series does not exist, so the series diverges by the Test for Divergence. ●

EXAMPLE 3   Test the series $\sum_{n=1}^{\infty} (-1)^{n+1} \frac{n^2}{n^3 + 1}$ for convergence or divergence.

Solution   The given series is alternating so we try to verify conditions (a) and (b) of Theorem 10.27.

Unlike the situation in Example 1, it is not obvious that the sequence given by $a_n = n^2/(n^3 + 1)$ is decreasing. However, if we consider the related function $f(x) = x^2/(x^3 + 1)$ we find that

$$f'(x) = \frac{x(2 - x^3)}{(x^3 + 1)^2}$$

Since we are considering only positive x, we see that $f'(x) < 0$ if $2 - x^3 < 0$, that is, $x > \sqrt[3]{2}$. Thus f is decreasing on the interval $[\sqrt[3]{2}, \infty)$. This means that $f(n + 1) < f(n)$ and therefore $a_{n+1} < a_n$ when $n \geq 2$. (The inequality $a_2 < a_1$ can be verified directly but all that really matters is that the sequence $\{a_n\}$ is eventually decreasing.)

Condition (b) is readily verified:

$$\lim_{n \to \infty} a_n = \lim_{n \to \infty} \frac{n^2}{n^3 + 1} = \lim_{n \to \infty} \frac{\frac{1}{n}}{1 + \frac{1}{n^3}} = 0$$

Thus the given series is convergent by the Alternating Series Test.   ●

Note: Instead of verifying condition (a) of the Alternating Series Test by computing a derivative as in Example 3, it is possible to verify that $a_{n+1} < a_n$ directly by using the technique of Solution 1 of Example 10 in Section 10.1.

A partial sum s_n of any convergent series can be used as an approximation to the total sum s but this is not of much use unless we can estimate the accuracy of the approximation. The error involved in using $s \approx s_n$ is $|s - s_n|$. The next theorem says that for series that satisfy the conditions of the Alternating Series Test, the error is smaller than a_{n+1}, which is the magnitude of the first neglected term.

Theorem (10.28)

If $s = \sum (-1)^{n-1} a_n$ is the sum of an alternating series that satisfies

(a) $0 \leq a_{n+1} \leq a_n$ and (b) $\lim_{n \to \infty} a_n = 0$, then

$$|s - s_n| \leq a_{n+1}$$

Proof   The idea is similar to the one for the proof of the Alternating Series Test. (Indeed the result of Theorem 10.28 can be seen geometrically by looking at Figure 10.9.) We have

$$s - s_n = (-1)^n a_{n+1} + (-1)^{n+1} a_{n+2} + (-1)^{n+2} a_{n+3} + \cdots$$
$$= (-1)^n [a_{n+1} - a_{n+2} + a_{n+3} - \cdots]$$

and so

$$|s - s_n| = (a_{n+1} - a_{n+2}) + (a_{n+3} - a_{n+4}) + \cdots$$
$$= a_{n+1} - (a_{n+2} - a_{n+3}) - (a_{n+4} - a_{n+5}) - \cdots$$

Every term in brackets is positive, so $|s - s_n| \leq a_{n+1}$.   ●

EXAMPLE 4   Find the sum of the series $\sum_{n=0}^{\infty} \frac{(-1)^n}{n!}$ correct to three decimal places. (By definition, $0! = 1$.)

Solution   We first observe that the series is convergent by the Alternating Series Test because

(a) $\quad \dfrac{1}{(n+1)!} = \dfrac{1}{n!(n+1)} < \dfrac{1}{n!}$

(b) $\quad 0 < \dfrac{1}{n!} < \dfrac{1}{n} \to 0 \quad$ so $\dfrac{1}{n!} \to 0$ as $n \to \infty$

To get a feel for how many terms we need to use in our approximation, let us write out the first few terms of the series:

$$s = \frac{1}{0!} - \frac{1}{1!} + \frac{1}{2!} - \frac{1}{3!} + \frac{1}{4!} - \frac{1}{5!} + \frac{1}{6!} - \frac{1}{7!} + \cdots$$

$$= 1 - 1 + \frac{1}{2} - \frac{1}{6} + \frac{1}{24} - \frac{1}{120} + \frac{1}{720} - \frac{1}{5040} + \cdots$$

Notice that

$$a_7 = \frac{1}{5040} < \frac{1}{5000} = 0.0002$$

 and

$$s_6 = 1 - 1 + \frac{1}{2} - \frac{1}{6} + \frac{1}{24} - \frac{1}{120} + \frac{1}{720} \approx 0.368056$$

By Theorem 10.28 we know that

$$|s - s_6| \le a_7 < 0.0002$$

This error of less than 0.0002 does not affect the third decimal place, so we have

$$s \approx 0.368$$

correct to three decimal places.

In Section 10.11 we will prove that $e^x = \sum_{n=0}^{\infty} x^n/n!$ for all x, so what we have obtained in this example is actually an approximation to the number e^{-1}. ●

Note: The rule that the error (in using s_n to approximate s) is smaller than the first neglected term is, in general, valid only for alternating series that satisfy the conditions of Theorem 10.28. The rule does not apply to other types of series. (See the example in Appendix D.)

SECTION 10.5 **Exercises**

In Exercises 1–26 test the given series for convergence or divergence.

1. $\dfrac{3}{5} - \dfrac{3}{6} + \dfrac{3}{7} - \dfrac{3}{8} + \dfrac{3}{9} - \cdots$

2. $-5 - \dfrac{5}{2} + \dfrac{5}{5} - \dfrac{5}{8} + \dfrac{5}{11} - \dfrac{5}{14} + \cdots$

3. $-\dfrac{1}{2} + \dfrac{2}{3} - \dfrac{3}{4} + \dfrac{4}{5} - \dfrac{5}{6} + \dfrac{6}{7} - \cdots$

4. $\dfrac{1}{\ln 2} - \dfrac{1}{\ln 3} + \dfrac{1}{\ln 4} - \dfrac{1}{\ln 5} + \dfrac{1}{\ln 6} - \cdots$

5. $\displaystyle\sum_{n=1}^{\infty} \dfrac{(-1)^{n-1}}{n^2}$

6. $\displaystyle\sum_{n=1}^{\infty} \dfrac{(-1)^n}{\sqrt{n+3}}$

7. $\displaystyle\sum_{n=1}^{\infty} (-1)^{n+1} \dfrac{n}{5n+1}$

8. $\displaystyle\sum_{n=2}^{\infty} \dfrac{(-1)^{n-1}}{n \ln n}$

9. $\displaystyle\sum_{n=1}^{\infty} (-1)^n \dfrac{n}{n^2+1}$

10. $\displaystyle\sum_{n=1}^{\infty} (-1)^n \dfrac{n^2}{n^2+1}$

11. $\displaystyle\sum_{n=1}^{\infty} (-1)^{n-1} \dfrac{\sqrt{n}}{n+4}$

12. $\displaystyle\sum_{n=1}^{\infty} (-1)^{n+1} \dfrac{n}{2^n}$

13. $\displaystyle\sum_{n=2}^{\infty} (-1)^n \dfrac{n}{\ln n}$

14. $\displaystyle\sum_{n=1}^{\infty} (-1)^{n-1} \dfrac{\ln n}{n}$

15. $\displaystyle\sum_{n=1}^{\infty} (-1)^{n+1} \dfrac{n+10}{n(n+1)}$

16. $\displaystyle\sum_{n=1}^{\infty} (-1)^{n-1} \dfrac{(n+9)(n+10)}{n(n+1)}$

17. $\displaystyle\sum_{n=1}^{\infty} \dfrac{\cos n\pi}{n^{3/4}}$

18. $\displaystyle\sum_{n=1}^{\infty} \dfrac{\sin\left(\frac{n\pi}{2}\right)}{n!}$

19. $\displaystyle\sum_{n=1}^{\infty} (-1)^n \sin\!\left(\frac{\pi}{n}\right)$ 20. $\displaystyle\sum_{n=1}^{\infty} (-1)^n \cos\!\left(\frac{\pi}{n}\right)$

21. $\displaystyle\sum_{n=1}^{\infty} \frac{(-1)^n}{\sqrt[n]{n}}$ 22. $\displaystyle\sum_{n=1}^{\infty} \frac{(-1)^n}{|n - 10\pi|}$

23. $\displaystyle\sum_{n=1}^{\infty} (-1)^n \frac{n(n+1)}{(n+2)^3}$ 24. $\displaystyle\sum_{n=1}^{\infty} (-1)^n \frac{n^n}{n!}$

25. $\displaystyle\sum_{n=1}^{\infty} \frac{(-3)^n}{n^2}$ 26. $\displaystyle\sum_{n=2}^{\infty} \frac{(-1)^{n-1}}{\sqrt[3]{\ln n}}$

27. Show that the series $\sum (-1)^{n-1} a_n$, where $a_n = 1/n$ if n is odd and $a_n = 1/n^2$ if n is even, is divergent. Why does the Alternating Series Test not apply?

For what values of p are the series in Exercises 28–30 convergent?

28. $\displaystyle\sum_{n=1}^{\infty} \frac{(-1)^{n-1}}{n^p}$ 29. $\displaystyle\sum_{n=1}^{\infty} \frac{(-1)^n}{n + p}$

30. $\displaystyle\sum_{n=1}^{\infty} (-1)^{n-1} \frac{(\ln n)^p}{n}$

In Exercises 31–38 approximate the sum of the given series to the indicated accuracy.

31. $\displaystyle\sum_{n=1}^{\infty} \frac{(-1)^{n-1}}{n^2}$ (error < 0.01)

32. $\displaystyle\sum_{n=1}^{\infty} \frac{(-1)^{n+1}}{n^4}$ (error < 0.001)

33. $\displaystyle\sum_{n=0}^{\infty} \frac{(-2)^n}{n!}$ (error < 0.01)

34. $\displaystyle\sum_{n=0}^{\infty} \frac{(-1)^n n}{4^n}$ (error < 0.002)

35. $\displaystyle\sum_{n=1}^{\infty} \frac{(-1)^{n-1}}{(2n-1)!}$ (four decimal places)

36. $\displaystyle\sum_{n=0}^{\infty} \frac{(-1)^n}{(2n)!}$ (four decimal places)

37. $\displaystyle\sum_{n=0}^{\infty} \frac{(-1)^n}{2^n n!}$ (four decimal places)

38. $\displaystyle\sum_{n=1}^{\infty} \frac{(-1)^{n-1}}{n^6}$ (five decimal places)

39. Use the following steps to show that
$$\sum_{n=1}^{\infty} \frac{(-1)^{n-1}}{n} = \ln 2$$

Let h_n and s_n be the partial sums of the harmonic and alternating harmonic series.
(a) Show that $s_{2n} = h_{2n} - h_n$.
(b) From Exercise 30 in Section 10.3 we have $h_n - \ln n \to \gamma$ as $n \to \infty$ and therefore $h_{2n} - \ln(2n) \to \gamma$ as $n \to \infty$. Use these facts together with part (a) to show that $s_{2n} \to \ln 2$ as $n \to \infty$.

SECTION 10.6

Absolute Convergence and the Ratio and Root Tests

Given any series $\sum a_n$, we can consider the corresponding series

$$\sum_{n=1}^{\infty} |a_n| = |a_1| + |a_2| + |a_3| + \cdots$$

whose terms are the absolute values of the terms of the original series.

Definition (10.29)

A series $\sum a_n$ is called **absolutely convergent** if the series of absolute values $\sum |a_n|$ is convergent.

Notice that if $\sum a_n$ is a series with positive terms, then $|a_n| = a_n$ and so absolute convergence is the same as convergence.

EXAMPLE 1   The series

$$\sum_{n=1}^{\infty} \frac{(-1)^{n-1}}{n^2} = 1 - \frac{1}{2^2} + \frac{1}{3^2} - \frac{1}{4^2} + \cdots$$

is absolutely convergent because

$$\sum_{n=1}^{\infty} \left| \frac{(-1)^{n-1}}{n^2} \right| = \sum_{n=1}^{\infty} \frac{1}{n^2} = 1 + \frac{1}{2^2} + \frac{1}{3^2} + \frac{1}{4^2} + \cdots$$

is a convergent p-series $(p = 2)$.

EXAMPLE 2   We know that the alternating harmonic series

$$\sum_{n=1}^{\infty} \frac{(-1)^{n-1}}{n} = 1 - \tfrac{1}{2} + \tfrac{1}{3} - \tfrac{1}{4} + \cdots$$

is convergent (see Example 1 in Section 10.5) but it is not absolutely convergent because the corresponding series of absolute values is

$$\sum_{n=1}^{\infty} \left| \frac{(-1)^{n-1}}{n} \right| = \sum_{n=1}^{\infty} \frac{1}{n} = 1 + \tfrac{1}{2} + \tfrac{1}{3} + \tfrac{1}{4} + \cdots$$

which is the harmonic series (p–series, $p = 1$) and is therefore divergent.

Definition (10.30)
> A series $\sum a_n$ is called **conditionally convergent** if it is convergent but not absolutely convergent.

Example 2 shows that the alternating harmonic series is conditionally convergent. Thus it is possible for a series to be convergent but not absolutely convergent. However the next theorem shows that absolute convergence implies convergence.

Theorem (10.31)
> If a series $\sum a_n$ is absolutely convergent, then it is convergent.

Proof   Observe that the inequality

$$-|a_n| \le a_n \le |a_n|$$

is true because a_n is either $-|a_n|$ or $|a_n|$. If we now add $|a_n|$ to each side of this inequality, we get

$$0 \le a_n + |a_n| \le 2|a_n|$$

Let $b_n = a_n + |a_n|$. Then $0 \le b_n \le 2|a_n|$. If $\sum a_n$ is absolutely convergent, then $\sum |a_n|$ is convergent, so $\sum 2|a_n|$ is convergent by Theorem 10.19 part (a). Therefore $\sum b_n$ is convergent by the Comparison Test. Since $a_n = b_n - |a_n|$,

$$\sum a_n = \sum b_n - \sum |a_n|$$

is convergent by Theorem 10.19 part (c).

EXAMPLE 3   Determine whether the series

$$\sum_{n=1}^{\infty} \frac{\cos n}{n^2} = \frac{\cos 1}{1^2} + \frac{\cos 2}{2^2} + \frac{\cos 3}{3^2} + \cdots$$

is convergent or divergent.

Solution   This series has both positive and negative terms but it is not alternating. (The first term is positive, the next three are negative, and the following three are positive. The signs change irregularly.) We can apply the Comparison Test to the series of absolute values

$$\sum_{n=1}^{\infty} \left| \frac{\cos n}{n^2} \right| = \sum_{n=1}^{\infty} \frac{|\cos n|}{n^2}$$

Since $|\cos n| \leq 1$ for all n we have

$$\frac{|\cos n|}{n^2} \leq \frac{1}{n^2}$$

We know that $\sum 1/n^2$ is convergent (p-series, $p = 2$) and therefore $\sum |\cos n|/n^2$ is convergent by the Comparison Test. Thus the given series $\sum (\cos n)/n^2$ is absolutely convergent and therefore convergent by Theorem 10.31.   ●

The following test is very useful in determining whether a given series is absolutely convergent.

The Ratio Test (10.32)

(a)   If $\lim_{n \to \infty} \left| \frac{a_{n+1}}{a_n} \right| = L < 1$, then the series $\sum_{n=1}^{\infty} a_n$ is absolutely convergent (and therefore convergent).

(b)   If $\lim_{n \to \infty} \left| \frac{a_{n+1}}{a_n} \right| = L > 1$ or $\lim_{n \to \infty} \left| \frac{a_{n+1}}{a_n} \right| = \infty$, then the series $\sum_{n=1}^{\infty} a_n$ is divergent.

Proof
(a)   The idea is to compare the given series with a convergent geometric series. Since $L < 1$ we can choose a number r such that $L < r < 1$. Since

$$\lim_{n \to \infty} \left| \frac{a_{n+1}}{a_n} \right| = L \qquad \text{and} \qquad L < r$$

the ratio $|a_{n+1}/a_n|$ will eventually be less than r, that is, there exists an integer N such that

$$\left| \frac{a_{n+1}}{a_n} \right| < r \qquad \text{whenever} \qquad n \geq N$$

or equivalently

(10.33) $$|a_{n+1}| < |a_n| r \qquad \text{whenever} \qquad n \geq N$$

Putting n successively equal to N, $N+1$, $N+2$, ... in (10.33), we obtain

$$|a_{N+1}| < |a_N|r$$

$$|a_{N+2}| < |a_{N+1}|r < |a_N|r^2$$

$$|a_{N+3}| < |a_{N+2}|r < |a_N|r^3$$

and, in general,

(10.34) $$|a_{N+k}| < |a_N|r^k \qquad \text{for all } k \geq 1$$

Now the series

$$\sum_{k=1}^{\infty} |a_N|r^k = |a_N|r + |a_N|r^2 + |a_N|r^3 + \cdots$$

is convergent since it is a geometric series with $0 < r < 1$. So the inequality (10.34) together with the Comparison Test shows that the series

$$\sum_{n=N+1}^{\infty} |a_n| = \sum_{k=1}^{\infty} |a_{N+k}| = |a_{N+1}| + |a_{N+2}| + |a_{N+3}| + \cdots$$

is also convergent. It follows that the series $\sum_{n=1}^{\infty} |a_n|$ is convergent. (Recall that a finite number of terms cannot affect convergence.) Therefore $\sum a_n$ is absolutely convergent.

(b)   If $|a_{n+1}/a_n| \to L > 1$ or $|a_{n+1}/a_n| \to \infty$, then the ratio $|a_{n+1}/a_n|$ will eventually be greater then 1; that is, there exists an integer N such that

$$\left| \frac{a_{n+1}}{a_n} \right| > 1 \qquad \text{whenever } n \geq N$$

This means that $|a_{n+1}| > |a_n|$ when $n \geq N$ and so

$$\lim_{n \to \infty} a_n \neq 0$$

Therefore $\sum a_n$ diverges by the Test for Divergence.   ●

Note: If $\lim_{n \to \infty} |a_{n+1}/a_n| = 1$, the Ratio Test gives no information. For instance, for the convergent series $\sum 1/n^2$ we have

$$\left| \frac{a_{n+1}}{a_n} \right| = \frac{\dfrac{1}{(n+1)^2}}{\dfrac{1}{n^2}} = \frac{n^2}{(n+1)^2} = \frac{1}{\left(1 + \dfrac{1}{n}\right)^2} \to 1 \qquad \text{as } n \to \infty$$

whereas for the divergent series $\sum 1/n$ we have

$$\left| \frac{a_{n+1}}{a_n} \right| = \frac{\dfrac{1}{n+1}}{\dfrac{1}{n}} = \frac{n}{n+1} = \frac{1}{1 + \dfrac{1}{n}} \to 1 \qquad \text{as } n \to \infty$$

Therefore if $\lim_{n \to \infty} |a_{n+1}/a_n| = 1$, the series $\sum a_n$ might converge or it might diverge. In this case the Ratio Test fails and we must use some other test.

EXAMPLE 4   Test the series $\sum_{n=1}^{\infty} (-1)^n \frac{n^3}{3^n}$ for absolute convergence.

Solution   We use the Ratio Test with $a_n = (-1)^n n^3 / 3^n$:

$$\left| \frac{a_{n+1}}{a_n} \right| = \left| \frac{\frac{(-1)^{n+1}(n+1)^3}{3^{n+1}}}{\frac{(-1)^n n^3}{3^n}} \right| = \frac{(n+1)^3}{3^{n+1}} \cdot \frac{3^n}{n^3}$$

$$= \frac{1}{3}\left(\frac{n+1}{n}\right)^3 = \frac{1}{3}\left(1 + \frac{1}{n}\right)^3 \to \frac{1}{3} < 1$$

Thus, by the Ratio Test, the given series is absolutely convergent and therefore convergent.   ●

EXAMPLE 5   Test the convergence of the series $\sum_{n=1}^{\infty} \frac{n^n}{n!}$.

Solution   Since the terms $a_n = n^n / n!$ are positive we do not need the absolute value signs.

$$\frac{a_{n+1}}{a_n} = \frac{(n+1)^{n+1}}{(n+1)!} \cdot \frac{n!}{n^n} = \frac{(n+1)(n+1)^n}{(n+1)n!} \cdot \frac{n!}{n^n}$$

$$= \left(\frac{n+1}{n}\right)^n = \left(1 + \frac{1}{n}\right)^n \to e \qquad \text{as } n \to \infty$$

(see Equation 3.34). Since $e > 1$, the given series is divergent by the Ratio Test.   ●

Note: Although the Ratio Test works in Example 5, an easier method is to use the Test for Divergence. Since

$$a_n = \frac{n^n}{n!} = \frac{n \cdot n \cdot n \cdots n}{1 \cdot 2 \cdot 3 \cdots n} \geq n$$

it follows that a_n does not approach 0 as $n \to \infty$. Therefore the given series is divergent by the Test for Divergence.

The following test is convenient to apply when nth powers occur. Its proof is similar to the proof of the Ratio Test and is left as Exercise 40.

The Root Test (10.35)

(a) If $\lim_{n\to\infty} \sqrt[n]{|a_n|} = L < 1$, then the series $\sum_{n=1}^{\infty} a_n$ is absolutely convergent (and therefore convergent).

(b) If $\lim_{n\to\infty} \sqrt[n]{|a_n|} = L > 1$ or $\lim_{n\to\infty} \sqrt[n]{|a_n|} = \infty$, then the series $\sum_{n=1}^{\infty} a_n$ is divergent.

If $\lim_{n \to \infty} \sqrt[n]{|a_n|} = 1$, then the Root Test gives no information. The series $\sum a_n$ could converge or diverge. (If $L = 1$ in the Ratio Test, do not try the Root Test because L will again be 1.)

EXAMPLE 6   Test the convergence of the series $\sum\limits_{n=1}^{\infty} \left(\dfrac{2n+3}{3n+2}\right)^n$.

Solution
$$a_n = \left(\frac{2n+3}{3n+2}\right)^n$$

$$\sqrt[n]{|a_n|} = \frac{2n+3}{3n+2} = \frac{2 + \frac{3}{n}}{3 + \frac{2}{n}} \to \frac{2}{3} < 1$$

Thus the given series converges by the Root Test.   •

Rearrangements

The question of whether a given convergent series is absolutely convergent or conditionally convergent has a bearing on the question of whether infinite sums behave like finite sums.

If we rearrange the order of the terms in a finite sum, then of course the value of the sum remains unchanged. But this is not always the case for an infinite series. By a **rearrangement** of an infinite series $\sum a_n$ we mean a series obtained by simply changing the order of the terms. For instance, a rearrangement of $\sum a_n$ could start as follows:

$$a_1 + a_2 + a_5 + a_3 + a_4 + a_{15} + a_6 + a_7 + a_{20} + \cdots$$

It turns out that if $\sum a_n$ is an absolutely convergent series with sum s, then any rearrangement of $\sum a_n$ has the same sum s. However any conditionally convergent series can be rearranged to give a different sum. To illustrate this fact let us consider the alternating harmonic series

$$1 - \tfrac{1}{2} + \tfrac{1}{3} - \tfrac{1}{4} + \tfrac{1}{5} - \tfrac{1}{6} + \tfrac{1}{7} - \tfrac{1}{8} + \cdots = \ln 2 \tag{1}$$

(See Exercise 39 in Section 10.5.) If we multiply this series by $\tfrac{1}{2}$, we get

$$\tfrac{1}{2} - \tfrac{1}{4} + \tfrac{1}{6} - \tfrac{1}{8} + \cdots = \tfrac{1}{2}\ln 2$$

Inserting zeros between the terms of this series, we have

Adding these zeros does not affect the sum of the series; each term in the sequence of partial sums is repeated, but the limit is the same.

$$0 + \tfrac{1}{2} + 0 - \tfrac{1}{4} + 0 + \tfrac{1}{6} + 0 - \tfrac{1}{8} + \cdots = \tfrac{1}{2}\ln 2 \tag{2}$$

Now we add the series in Equations 1 and 2 using Theorem 10.19:

$$1 + \tfrac{1}{3} - \tfrac{1}{2} + \tfrac{1}{5} + \tfrac{1}{7} - \tfrac{1}{4} + \cdots = \tfrac{3}{2}\ln 2 \tag{3}$$

Notice that the series in (3) contains the same terms as in (1), but rearranged so that one negative term occurs after each pair of positive terms. The sums of these series, however, are different. In fact, Riemann proved that if $\sum a_n$ is a **conditionally convergent series and r is any real number whatsoever, then there is a rearrangement of $\sum a_n$ that has a sum equal to r.** A proof of this fact is outlined in Exercise 42.

SECTION 10.6 **Exercises**

In Exercises 1–36 determine whether the given series is absolutely convergent, conditionally convergent, or divergent.

1. $\displaystyle\sum_{n=1}^{\infty} \frac{(-1)^{n-1}}{n\sqrt{n}}$

2. $\displaystyle\sum_{n=1}^{\infty} \frac{(-1)^n}{\sqrt{n}}$

3. $\displaystyle\sum_{n=1}^{\infty} \frac{(-3)^n}{n^3}$

4. $\displaystyle\sum_{n=0}^{\infty} \frac{(-3)^n}{n!}$

5. $\displaystyle\sum_{n=1}^{\infty} \frac{(-1)^{n+1}}{2n+1}$

6. $\displaystyle\sum_{n=1}^{\infty} \frac{(-1)^{n-1}}{n^2+1}$

7. $\displaystyle\sum_{n=1}^{\infty} \frac{(-1)^{n-1}}{(2n-1)!}$

8. $\displaystyle\sum_{n=1}^{\infty} e^{-n} n!$

9. $\displaystyle\sum_{n=1}^{\infty} (-1)^n \frac{n}{n^2+4}$

10. $\displaystyle\sum_{n=1}^{\infty} (-1)^{n-1} \frac{\sqrt{n}}{n+1}$

11. $\displaystyle\sum_{n=1}^{\infty} (-1)^n \frac{2n}{3n-4}$

12. $\displaystyle\sum_{n=1}^{\infty} (-1)^n \frac{2^n}{n^2+1}$

13. $\displaystyle\sum_{n=1}^{\infty} \frac{\sin 2n}{n^2}$

14. $\displaystyle\sum_{n=1}^{\infty} \frac{(-1)^n \arctan n}{n^3}$

15. $\displaystyle\sum_{n=1}^{\infty} \frac{(-2)^n}{n 3^{n+1}}$

16. $\displaystyle\sum_{n=1}^{\infty} \frac{(-1)^{n+1} 5^{n-1}}{(n+1)^2 4^{n+2}}$

17. $\displaystyle\sum_{n=1}^{\infty} \frac{(n+1) 5^n}{n 3^{2n}}$

18. $\displaystyle\sum_{n=1}^{\infty} \frac{8-n^3}{n!}$

19. $\displaystyle\sum_{n=2}^{\infty} \frac{(-1)^n}{\ln n}$

20. $\displaystyle\sum_{n=1}^{\infty} \frac{\cos\left(\frac{n\pi}{6}\right)}{n\sqrt{n}}$

21. $\displaystyle\sum_{n=1}^{\infty} \frac{n!}{(-10)^n}$

22. $\displaystyle\sum_{n=1}^{\infty} \frac{n!}{n^n}$

23. $\displaystyle\sum_{n=1}^{\infty} \frac{\cos\left(\frac{n\pi}{3}\right)}{n!}$

24. $\displaystyle\sum_{n=2}^{\infty} \frac{(-1)^n}{(\ln n)^n}$

25. $\displaystyle\sum_{n=1}^{\infty} \frac{(-n)^n}{5^{2n+3}}$

26. $\displaystyle\sum_{n=2}^{\infty} \frac{(-1)^n}{n \ln n}$

27. $\displaystyle\sum_{n=1}^{\infty} \left(\frac{1-3n}{3+4n}\right)^n$

28. $\displaystyle\sum_{n=1}^{\infty} \frac{(-2)^n n^2}{(n+2)!}$

29. $1 - \dfrac{2!}{1\cdot 3} + \dfrac{3!}{1\cdot 3\cdot 5} - \dfrac{4!}{1\cdot 3\cdot 5\cdot 7} + \cdots$

$\quad + \dfrac{(-1)^{n-1} n!}{1\cdot 3\cdot 5\cdot \cdots \cdot (2n-1)} + \cdots$

30. $\dfrac{1}{3} + \dfrac{1\cdot 4}{3\cdot 5} + \dfrac{1\cdot 4\cdot 7}{3\cdot 5\cdot 7} + \dfrac{1\cdot 4\cdot 7\cdot 10}{3\cdot 5\cdot 7\cdot 9} + \cdots$

$\quad + \dfrac{1\cdot 4\cdot 7\cdot \cdots \cdot (3n-2)}{3\cdot 5\cdot 7\cdot \cdots \cdot (2n+1)} + \cdots$

31. $\displaystyle\sum_{n=1}^{\infty} \frac{2\cdot 4\cdot 6\cdot \cdots \cdot (2n)}{n!}$

32. $\displaystyle\sum_{n=1}^{\infty} (-1)^n \frac{2^n n!}{5\cdot 8\cdot 11\cdot \cdots \cdot (3n+2)}$

33. $\displaystyle\sum_{n=1}^{\infty} \frac{(n+2)!}{n! 10^n}$

34. $\displaystyle\sum_{n=1}^{\infty} \frac{(n!)^2}{(2n)!}$

35. $\displaystyle\sum_{n=1}^{\infty} \frac{\sin(3n)\cdot n^2}{(1.1)^n}$

36. $\displaystyle\sum_{n=1}^{\infty} \frac{(-1)^n}{(\arctan n)^n}$

37. For which of the following series is the Ratio Test inconclusive (that is, fails to give a definite answer)?

(a) $\displaystyle\sum_{n=1}^{\infty} \frac{1}{n^3}$ 　　 (b) $\displaystyle\sum_{n=1}^{\infty} \frac{n}{2^n}$

(c) $\displaystyle\sum_{n=1}^{\infty} \frac{(-3)^{n-1}}{\sqrt{n}}$ 　　 (d) $\displaystyle\sum_{n=1}^{\infty} \frac{\sqrt{n}}{1+n^2}$

38. (a) Show that $\sum_{n=0}^{\infty} x^n/n!$ converges for all x.
　　(b) Deduce that $\lim_{n\to\infty} x^n/n! = 0$ for all x.

39. Prove that if $\sum a_n$ is absolutely convergent, then

$$\left| \sum_{n=1}^{\infty} a_n \right| \le \sum_{n=1}^{\infty} |a_n|$$

40. Prove the Root Test (10.35). [*Hint for part (a):* Take any number r such that $L < r < 1$ and use the fact that there is an integer N such that $\sqrt[n]{|a_n|} < r$ when $n \ge N$.]

41. Given any series $\sum a_n$, we define a series $\sum a_n^+$ whose terms are all the positive terms of $\sum a_n$ and a series $\sum a_n^-$ whose terms are all the negative terms of $\sum a_n$. To be specific, we let

$$a_n^+ = \frac{a_n + |a_n|}{2} \qquad a_n^- = \frac{a_n - |a_n|}{2}$$

Notice that if $a_n > 0$, then $a_n^+ = a_n$ and $a_n^- = 0$, whereas if $a_n < 0$, then $a_n^- = a_n$ and $a_n^+ = 0$.

(a) If $\sum a_n$ is absolutely convergent, show that both of the series $\sum a_n^+$ and $\sum a_n^-$ are convergent.

(b) If $\sum a_n$ is conditionally convergent, show that both of the series $\sum a_n^+$ and $\sum a_n^-$ are divergent.

42. Prove that if $\sum a_n$ is a conditionally convergent series and r is any real number, then there is a rearrangement of $\sum a_n$ whose sum is r. (*Hints:* Use the notation of Exercise 41. Take just enough positive terms a_n^+ so that their sum is greater than r. Then add just enough negative terms a_n^- so that the cumulative sum is less than r. Continue in this manner and use Theorem 10.17.)

SECTION 10.7

Strategy for Testing Series

We now have several ways of testing a series for convergence or divergence; the problem is to decide which test to use on which series. In this respect testing series is similar to integrating functions. Again there are no hard and fast rules about which test to apply to a given series, but you may find the following advice of some use.

It is not wise to have a list of the tests to be applied in a specific order until one finally works. That would be a waste of time and effort. Instead, as with integration, the main strategy is to classify the series according to its *form*.

1. If the series is of the form $\sum 1/n^p$, it is a *p*-series, which we know to be convergent if $p > 1$ and divergent if $p \leq 1$.

2. If the series has the form $\sum ar^{n-1}$ or $\sum ar^n$, it is a geometric series, which converges if $|r| < 1$ and diverges if $|r| \geq 1$. Some preliminary algebraic manipulation may be required to bring the series into this form.

3. If the series has a form that is similar to a *p*-series or a geometric series, then one of the comparison tests should be considered. In particular, if a_n is a rational function or algebraic function of n (involving roots of polynomials), then the series should be compared to a *p*-series. Notice that most of the series in Exercises 10.4 have this form. (The value of p should be chosen as in Section 10.4 by keeping only the highest powers of n in the numerator and denominator.) The comparison tests apply only to series with positive terms, but if $\sum a_n$ has some negative terms then we can apply the Comparison Test to $\sum |a_n|$ and test for absolute convergence.

4. If you can see at a glance that $\lim_{n \to \infty} a_n \neq 0$, then the Test for Divergence should be used.

5. If the series is of the form $\sum (-1)^{n-1} a_n$ or $\sum (-1)^n a_n$, then the Alternating Series Test is an obvious possibility.

6. Series that involve factorials or other products (including a constant raised to the *n*th power) are often conveniently tested using the Ratio Test. Bear in mind that $|a_{n+1}/a_n| \to 1$ as $n \to \infty$ for all *p*-series and therefore all rational or algebraic functions of n. Thus the Ratio Test should not be used for such series.

7. If a_n is of the form $(b_n)^n$, then the Root Test may be useful.

8. If $a_n = f(n)$, where $\int_1^\infty f(x)\,dx$ is easily evaluated, then the Integral Test is effective (assuming the hypotheses of this test are satisfied).

In the following examples we do not work out all the details but simply indicate which tests should be used.

EXAMPLE 1 $\displaystyle\sum_{n=1}^{\infty} \frac{n-1}{2n+1}$

Since $a_n \to \frac{1}{2} \neq 0$ as $n \to \infty$, we should use the Test for Divergence.   ●

EXAMPLE 2 $\displaystyle\sum_{n=1}^{\infty} \frac{\sqrt{n^3+1}}{3n^3+4n^2+2}$

Since a_n is an algebraic function of n, we compare the given series with a p-series. The comparison series is $\sum b_n$, where

$$b_n = \frac{\sqrt{n^3}}{3n^3} = \frac{n^{3/2}}{3n^3} = \frac{1}{3n^{3/2}}$$

EXAMPLE 3 $\displaystyle\sum_{n=1}^{\infty} ne^{-n^2}$

Since the integral $\displaystyle\int_{1}^{\infty} xe^{-x^2}\, dx$ is easily evaluated, we use the Integral Test. The Ratio Test also works.

EXAMPLE 4 $\displaystyle\sum_{n=1}^{\infty} (-1)^n \frac{n^3}{n^4 + 1}$

Since the series is alternating, we use the Alternating Series Test.

EXAMPLE 5 $\displaystyle\sum_{n=1}^{\infty} \frac{2^n}{n!}$

Since the series involves $n!$, we use the Ratio Test.

EXAMPLE 6 $\displaystyle\sum_{n=1}^{\infty} \frac{1}{2 + 3^n}$

Since the series is closely related to the geometric series $\sum 1/3^n$, we use the Comparison Test.

SECTION 10.7 **Exercises**

Test the following series for convergence or divergence.

1. $\displaystyle\sum_{n=1}^{\infty} \frac{\sqrt{n}}{n^2 + 1}$

2. $\displaystyle\sum_{n=1}^{\infty} \cos n$

3. $\displaystyle\sum_{n=1}^{\infty} \frac{4^n}{3^{2n-1}}$

4. $\displaystyle\sum_{i=1}^{\infty} \frac{i^4}{4^i}$

5. $\displaystyle\sum_{n=2}^{\infty} \frac{(-1)^n}{(\ln n)^2}$

6. $\displaystyle\sum_{n=1}^{\infty} n^2 e^{-n^3}$

7. $\displaystyle\sum_{k=1}^{\infty} k^{-1.7}$

8. $\displaystyle\sum_{n=0}^{\infty} \frac{10^n}{n!}$

9. $\displaystyle\sum_{n=1}^{\infty} \frac{n}{e^n}$

10. $\displaystyle\sum_{m=1}^{\infty} \frac{2m}{8m - 5}$

11. $\displaystyle\sum_{n=2}^{\infty} \frac{n^3 + 1}{n^4 - 1}$

12. $\displaystyle\sum_{n=1}^{\infty} \left(\frac{n^2 + 1}{2n^2 + 1}\right)^n$

13. $\displaystyle\sum_{n=2}^{\infty} \frac{2}{n(\ln n)^3}$

14. $\displaystyle\sum_{n=1}^{\infty} \frac{\sqrt{n}}{e^{\sqrt{n}}}$

15. $\displaystyle\sum_{n=1}^{\infty} \frac{3^n n^2}{n!}$

16. $\displaystyle\sum_{n=1}^{\infty} \frac{3}{4n - 5}$

17. $\displaystyle\sum_{n=1}^{\infty} \frac{3^n}{5^n + n}$

18. $\displaystyle\sum_{k=1}^{\infty} \frac{k + 5}{5^k}$

19. $\displaystyle\sum_{n=0}^{\infty} \frac{n!}{2 \cdot 5 \cdot 8 \cdots (3n + 2)}$

20. $\displaystyle\sum_{n=1}^{\infty} \frac{(-1)^n n}{(n + 1)(n + 2)}$

21. $\displaystyle\sum_{i=1}^{\infty} \frac{1}{\sqrt{i(i + 1)}}$

22. $\displaystyle\sum_{n=1}^{\infty} \frac{n^2}{\sqrt{n^5 + n^2 + 2}}$

23. $\displaystyle\sum_{n=1}^{\infty} (-1)^n 2^{1/n}$

24. $\displaystyle\sum_{n=1}^{\infty} \frac{\cos\left(\frac{n}{2}\right)}{n^2 + 4n}$

25. $\displaystyle\sum_{n=1}^{\infty} (-1)^n \frac{\ln n}{\sqrt{n}}$

26. $\displaystyle\sum_{n=1}^{\infty} \frac{\tan\left(\frac{1}{n}\right)}{n}$

27. $\displaystyle\sum_{n=0}^{\infty} (-\pi)^n$

28. $\displaystyle\sum_{n=1}^{\infty} \frac{\sqrt[3]{n} + 1}{n(\sqrt{n} + 1)}$

29. $\displaystyle\sum_{n=1}^{\infty} \frac{(-2)^{2n}}{n^n}$

30. $\displaystyle\sum_{n=1}^{\infty} \frac{2^{3n-1}}{n^2 + 1}$

31. $\displaystyle\sum_{k=1}^{\infty} \frac{k \ln k}{(k + 1)^3}$

32. $\displaystyle\sum_{n=1}^{\infty} \frac{e^{1/n}}{n^2}$

33. $\displaystyle\sum_{n=1}^{\infty} \frac{2^n}{(2n + 1)!}$

34. $\displaystyle\sum_{j=1}^{\infty} (-1)^j \frac{\sqrt{j}}{j + 5}$

35. $\displaystyle\sum_{n=1}^{\infty} \frac{\tan^{-1} n}{n\sqrt{n}}$

36. $\displaystyle\sum_{n=1}^{\infty} \frac{(2n)^n}{n^{2n}}$ **37.** $\displaystyle\sum_{n=1}^{\infty} \frac{1}{1+\left(\frac{3}{\pi}\right)^n}$ **39.** $\displaystyle\sum_{n=1}^{\infty} \left(\sqrt[n]{2}-1\right)^n$ **40.** $\displaystyle\sum_{n=1}^{\infty} \left(\sqrt[n]{2}-1\right)$

38. $\displaystyle\sum_{n=2}^{\infty} \frac{1}{(\ln n)^{\ln n}}$

SECTION 10.8

Power Series

A **power series** is a series of the form

(10.36)
$$\sum_{n=0}^{\infty} a_n x^n = a_0 + a_1 x + a_2 x^2 + a_3 x^3 + \cdots$$

where x is a variable and the a_n's are constants called the **coefficients** of the series. For each fixed x, the series (10.36) is a series of constants that we can test for convergence or divergence. A power series may converge for some values of x and diverge for other values of x. The sum of the series is a function

$$f(x) = a_0 + a_1 x + a_2 x^2 + \cdots + a_n x^n + \cdots$$

whose domain is the set of all x for which the series converges. Notice that f resembles a polynomial. The only difference is that f has infinitely many terms.

For instance, if we take $a_n = 1$ for all n, the power series becomes the geometric series

$$\sum_{n=0}^{\infty} x^n = 1 + x + x^2 + x^3 + \cdots + x^n + \cdots = \frac{1}{1-x}$$

which converges when $-1 < x < 1$ and diverges when $|x| \geq 1$ (see Equation 10.16).

More generally, a series of the form

(10.37)
$$\sum_{n=0}^{\infty} a_n (x-c)^n = a_0 + a_1(x-c) + a_2(x-c)^2 + \cdots$$

is called a **power series in $(x-c)$** or a **power series centered at c** or a **power series about c**. Notice that in writing out the term corresponding to $n = 0$ in Equations 10.36 and 10.37 we have adopted the convention that $(x-c)^0 = 1$ even when $x = c$. Notice also that when $x = c$ all of the terms are 0 for $n \geq 1$ and so the power series (10.37) always converges when $x = c$.

EXAMPLE 1   For what values of x is the series $\displaystyle\sum_{n=0}^{\infty} n! x^n$ convergent?

Solution   We use the Ratio Test. Since we have been using a_n to mean the coefficient of x^n in this section, we use u_n to denote the nth term of the series. Thus $u_n = n! x^n$ and, if $x \neq 0$, we have

$$\lim_{n\to\infty} \left| \frac{u_{n+1}}{u_n} \right| = \lim_{n\to\infty} \left| \frac{(n+1)! x^{n+1}}{n! x^n} \right|$$

$$= \lim_{n\to\infty} (n+1)|x| = \infty$$

By the Ratio Test, the series diverges when $x \neq 0$. Thus the given series converges only when $x = 0$.   ●

EXAMPLE 2   For what values of x does the series $\sum_{n=1}^{\infty} \dfrac{(x-3)^n}{n}$ converge?

Solution   Let $u_n = (x-3)^n/n$. Then

$$\left|\frac{u_{n+1}}{u_n}\right| = \left|\frac{(x-3)^{n+1}}{n+1} \cdot \frac{n}{(x-3)^n}\right|$$

$$= \frac{1}{1+\frac{1}{n}}|x-3| \to |x-3| \qquad \text{as } n \to \infty$$

By the Ratio Test, the given series is absolutely convergent, and therefore convergent, when $|x-3| < 1$ and divergent when $|x-3| > 1$. Now

$$|x-3| < 1 \qquad \Leftrightarrow \qquad -1 < x-3 < 1 \qquad \Leftrightarrow \qquad 2 < x < 4$$

so the series converges when $2 < x < 4$ and diverges when $x < 2$ or $x > 4$.

The Ratio Test gives no information when $|x-3| = 1$ so we must consider $x = 2$ and $x = 4$ separately. If we put $x = 4$ in the series, it becomes $\sum 1/n$, the harmonic series, which is divergent. If $x = 2$, the series is $\sum (-1)^n/n$, which converges by the Alternating Series Test. Thus the given power series converges for $2 \le x < 4$.   •

We will see that the main use of a power series is that it provides a way to represent some of the most important functions that arise in mathematics, physics, and chemistry. In particular, the sum of the power series in the next example is called a **Bessel function**, after the German astronomer Friedrich Bessel (1784–1846), and the function given in Exercise 33 is another example of a Bessel function. In fact, Bessel functions first arose in solving Kepler's equation, which describes planetary motion. Since then they have been applied in many different physical situations, such as the temperature distribution in a circular plate.

EXAMPLE 3   What is the domain of the Bessel function of order 0 defined by

$$J_0(x) = \sum_{n=0}^{\infty} \frac{(-1)^n x^{2n}}{2^{2n}(n!)^2}$$

Solution   Let $u_n = (-1)^n x^{2n}/[2^{2n}(n!)^2]$. Then

$$\left|\frac{u_{n+1}}{u_n}\right| = \left|\frac{(-1)^{n+1} x^{2(n+1)}}{2^{2(n+1)}[(n+1)!]^2} \cdot \frac{2^{2n}(n!)^2}{(-1)^n x^{2n}}\right|$$

$$= \frac{x^2}{4(n+1)^2} \to 0 < 1 \qquad \text{for all } x$$

Thus, by the Ratio Test, the given series converges for all values of x. In other words, the domain of the Bessel function J_0 is $(-\infty, \infty) = R$.   •

For the power series that we have looked at so far, the set of values of x for which the series is convergent has always turned out to be an interval [a finite interval for the geometric series and in Example 2, the infinite interval $(-\infty, \infty)$ in Example 3, and a collapsed interval $[0,0] = \{0\}$ in Example 1]. The following theorem, proved in Appendix C, says that this is true in general.

Theorem (10.38)

For a given power series $\sum_{n=0}^{\infty} a_n(x-c)^n$ there are only three possibilities:

(a) The series converges only when $x = c$.

(b) The series converges for all x.

(c) There is a positive number R such that the series converges if $|x - c| < R$ and diverges if $|x - c| > R$.

The number R in case (c) is called the **radius of convergence** of the power series. By convention, the radius of convergence is $R = 0$ in case (a) and $R = \infty$ in case (b). The **interval of convergence** of a power series is the interval that consists of all values of x for which the series converges. In case (a) the interval consists of just a single point c. In case (b) the interval is $(-\infty, \infty)$. In case (c) note that the inequality $|x - c| < R$ can be rewritten as $c - R < x < c + R$. When x is an *endpoint* of the interval, that is, $x = c \pm R$, anything can happen—the series might converge at one or both endpoints or it might diverge at both endpoints. Thus in case (c) there are four possibilities for the interval of convergence:

$$(c - R, c + R) \quad (c - R, c + R] \quad [c - R, c + R) \quad [c - R, c + R]$$

The situation is illustrated in Figure 10.10.

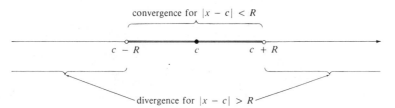

Figure 10.10

We summarize here the radius and interval of convergence for the examples already considered in this section.

	Series	Radius of convergence	Interval of convergence
Geometric series	$\sum_{n=0}^{\infty} x^n$	$R = 1$	$(-1, 1)$
Example 1	$\sum_{n=0}^{\infty} n! x^n$	$R = 0$	$\{0\}$
Example 2	$\sum_{n=1}^{\infty} \dfrac{(x-3)^n}{n}$	$R = 1$	$[2, 4)$
Example 3	$\sum_{n=0}^{\infty} \dfrac{(-1)^n x^{2n}}{2^{2n}(n!)^2}$	$R = \infty$	$(-\infty, \infty)$

In general the Ratio Test (or sometimes the Root Test) should be used to determine the radius of convergence R. The Ratio and Root Tests always fail when x is an endpoint of the interval of convergence, so the endpoints should be checked using some other test.

EXAMPLE 4   Find the radius of convergence and interval of convergence of the series

$$\sum_{n=0}^{\infty} \frac{(-3)^n x^n}{\sqrt{n+1}}$$

Solution   Let $u_n = (-3)^n x^n / \sqrt{n+1}$. Then

$$\left| \frac{u_{n+1}}{u_n} \right| = \left| \frac{(-3)^{n+1} x^{n+1}}{\sqrt{n+2}} \cdot \frac{\sqrt{n+1}}{(-3)^n x^n} \right|$$

$$= 3 \sqrt{\frac{1 + (1/n)}{1 + (2/n)}} |x| \to 3|x| \qquad \text{as } n \to \infty$$

By the Ratio Test, the given series converges if $3|x| < 1$ and diverges if $3|x| > 1$. Thus it converges if $|x| < \frac{1}{3}$ and diverges if $|x| > \frac{1}{3}$. This means that the radius of convergence is $R = \frac{1}{3}$.

We know the series converges in the interval $(-\frac{1}{3}, \frac{1}{3})$ but we must now test for convergence at the endpoints of this interval. If $x = -\frac{1}{3}$, the series becomes

$$\sum_{n=0}^{\infty} \frac{(-3)^n \left(-\frac{1}{3}\right)^n}{\sqrt{n+1}} = \sum_{n=0}^{\infty} \frac{1}{\sqrt{n+1}} = \frac{1}{\sqrt{1}} + \frac{1}{\sqrt{2}} + \frac{1}{\sqrt{3}} + \frac{1}{\sqrt{4}} + \cdots$$

which diverges. (Use the Integral Test or observe that it is a p-series with $p = \frac{1}{2} < 1$.) If $x = \frac{1}{3}$, the series is

$$\sum_{n=0}^{\infty} \frac{(-3)^n \left(\frac{1}{3}\right)^n}{\sqrt{n+1}} = \sum_{n=0}^{\infty} \frac{(-1)^n}{\sqrt{n+1}}$$

which converges by the Alternating Series Test. Therefore the given power series converges when $-\frac{1}{3} < x \le \frac{1}{3}$, so the interval of convergence is $(-\frac{1}{3}, \frac{1}{3}]$.   ●

EXAMPLE 5   Find the radius of convergence and interval of convergence of the series

$$\sum_{n=0}^{\infty} \frac{n(x+2)^n}{3^{n+1}}$$

Solution   If $u_n = n(x+2)^n / 3^{n+1}$, then

$$\left| \frac{u_{n+1}}{u_n} \right| = \left| \frac{(n+1)(x+2)^{n+1}}{3^{n+2}} \cdot \frac{3^{n+1}}{n(x+2)^n} \right|$$

$$= \left(1 + \frac{1}{n}\right) \frac{|x+2|}{3} \to \frac{|x+2|}{3} \qquad \text{as } n \to \infty$$

Using the Ratio Test, we see that the series converges if $|x+2|/3 < 1$ and it diverges if $|x+2|/3 > 1$. So it converges if $|x+2| < 3$ and diverges if $|x+2| > 3$. Thus the radius of convergence is $R = 3$.

The inequality $|x+2| < 3$ can be written as $-5 < x < 1$, so we test the series at the endpoints -5 and 1. When $x = -5$, the series is

$$\sum_{n=0}^{\infty} \frac{n(-3)^n}{3^{n+1}} = \frac{1}{3} \sum_{n=0}^{\infty} (-1)^n n$$

which diverges by the Test for Divergence [$(-1)^n n$ does not converge to 0]. When $x = 1$, the series is

$$\sum_{n=0}^{\infty} \frac{n(3)^n}{3^{n+1}} = \frac{1}{3} \sum_{n=0}^{\infty} n$$

which also diverges by the Test for Divergence. Thus the series converges only when $-5 < x < 1$ so the interval of convergence is $(-5, 1)$. •

SECTION 10.8 Exercises

1. If $\sum_{n=0}^{\infty} a_n 4^n$ is convergent, does it follow that the following series are convergent?

   (a) $\displaystyle\sum_{n=0}^{\infty} a_n(-2)^n$ (b) $\displaystyle\sum_{n=0}^{\infty} a_n(-4)^n$

2. Suppose that $\sum_{n=0}^{\infty} a_n x^n$ converges when $x = -4$ and diverges when $x = 6$. What can be said about the convergence or divergence of the following series?

   (a) $\displaystyle\sum_{n=0}^{\infty} a_n$ (b) $\displaystyle\sum_{n=0}^{\infty} a_n 8^n$

   (c) $\displaystyle\sum_{n=0}^{\infty} a_n(-3)^n$ (d) $\displaystyle\sum_{n=0}^{\infty} (-1)^n a_n 9^n$

Find the radius of convergence and interval of convergence of the series in Exercises 3–32.

3. $\displaystyle\sum_{n=0}^{\infty} \frac{x^n}{n+2}$ 4. $\displaystyle\sum_{n=1}^{\infty} \frac{(-1)^n x^n}{\sqrt[3]{n}}$ 5. $\displaystyle\sum_{n=0}^{\infty} n x^n$

6. $\displaystyle\sum_{n=1}^{\infty} \frac{x^n}{n^2}$ 7. $\displaystyle\sum_{n=0}^{\infty} \frac{x^n}{n!}$ 8. $\displaystyle\sum_{n=1}^{\infty} n^n x^n$

9. $\displaystyle\sum_{n=1}^{\infty} \frac{(-1)^n x^n}{n 2^n}$ 10. $\displaystyle\sum_{n=1}^{\infty} n 5^n x^n$ 11. $\displaystyle\sum_{n=0}^{\infty} \frac{3^n x^n}{(n+1)^2}$

12. $\displaystyle\sum_{n=0}^{\infty} \frac{n^2 x^n}{10^n}$ 13. $\displaystyle\sum_{n=2}^{\infty} \frac{x^n}{\ln n}$

14. $\displaystyle\sum_{n=0}^{\infty} \sqrt{n}(3x+2)^n$ 15. $\displaystyle\sum_{n=0}^{\infty} \frac{n}{4^n}(2x-1)^n$

16. $\displaystyle\sum_{n=1}^{\infty} \frac{(-1)^n x^{2n-1}}{(2n-1)!}$ 17. $\displaystyle\sum_{n=1}^{\infty} (-1)^n \frac{(x-1)^n}{\sqrt{n}}$

18. $\displaystyle\sum_{n=1}^{\infty} \frac{(x-4)^n}{n 5^n}$ 19. $\displaystyle\sum_{n=1}^{\infty} \frac{(x-2)^n}{n^n}$

20. $\displaystyle\sum_{n=0}^{\infty} \frac{(-3)^n (x-1)^n}{\sqrt{n+1}}$ 21. $\displaystyle\sum_{n=0}^{\infty} \frac{2^n(x-3)^n}{n+3}$

22. $\displaystyle\sum_{n=1}^{\infty} \frac{(x+1)^n}{n(n+1)}$

23. $\displaystyle\sum_{n=0}^{\infty} \frac{n}{(n^2+1)4^n}(x+10)^n$

24. $\displaystyle\sum_{n=0}^{\infty} \frac{n!}{10^n}(x-\pi)^n$ 25. $\displaystyle\sum_{n=1}^{\infty} \left(\frac{n}{2}\right)^n (x+6)^n$

26. $\displaystyle\sum_{n=1}^{\infty} \frac{n x^n}{1 \cdot 3 \cdot 5 \cdot \cdots \cdot (2n-1)}$

27. $\displaystyle\sum_{n=1}^{\infty} \frac{(2x-1)^n}{n^3}$ 28. $\displaystyle\sum_{n=2}^{\infty} (-1)^n \frac{(2x+3)^n}{n \ln n}$

29. $\displaystyle\sum_{n=0}^{\infty} \frac{n}{\sqrt{n+1}}(x-e)^n$ 30. $\displaystyle\sum_{n=2}^{\infty} \frac{x^n}{(\ln n)^n}$

31. $\displaystyle\sum_{n=1}^{\infty} \frac{n! x^n}{(2n)!}$

32. $\displaystyle\sum_{n=1}^{\infty} \frac{2 \cdot 4 \cdot 6 \cdot \cdots \cdot (2n)}{1 \cdot 3 \cdot 5 \cdot \cdots \cdot (2n-1)} x^n$

33. The function J_1 defined by

$$J_1(x) = \sum_{n=0}^{\infty} \frac{(-1)^n x^{2n+1}}{n!(n+1)! 2^{2n+1}}$$

is called the *Bessel function of order* 1. Find its domain.

34. The function A is defined by

$$A(x) = 1 + \frac{x^3}{2 \cdot 3} + \frac{x^6}{2 \cdot 3 \cdot 5 \cdot 6} + \frac{x^9}{2 \cdot 3 \cdot 5 \cdot 6 \cdot 8 \cdot 9} + \cdots$$

is called the *Airy function* after the English mathematician and philosopher Sir George Airy (1801–1892). Find the domain of the Airy function.

35. A function f is defined by

$$f(x) = 1 + 2x + x^2 + 2x^3 + x^4 + \cdots$$

that is, its coefficients are $a_{2n} = 1$ and $a_{2n+1} = 2$ for all $n \ge 0$. Find the interval of convergence of the series and find an explicit formula for $f(x)$.

36. If $f(x) = \sum_{n=0}^{\infty} a_n x^n$, where $a_{n+4} = a_n$ for all $n \ge 0$, find the interval of convergence of the series and a formula for $f(x)$.

37. Show that if $\lim_{n \to \infty} \sqrt[n]{|a_n|} = a$, then the radius of convergence of the power series $\sum a_n x^n$ is $R = 1/a$.

38. Suppose that the radius of convergence of the power series $\sum a_n x^n$ is R. What is the radius of convergence of the power series $\sum a_n x^{2n}$?

Taylor and Maclaurin Series

The sum of a power series is a function $f(x) = \sum_{n=0}^{\infty} a_n(x - c)^n$ whose domain is the interval of convergence of the series. We would like to be able to differentiate and integrate such functions, and the following theorem says that we can do so by differentiating or integrating each individual term in the series, just as we would for a polynomial. This is called **term-by-term differentiation and integration.** The proof is lengthy and is therefore omitted.

Theorem (10.39)

If the power series $\sum a_n(x - c)^n$ has radius of convergence $R > 0$, then the function f defined by

$$f(x) = a_0 + a_1(x - c) + a_2(x - c)^2 + \cdots = \sum_{n=0}^{\infty} a_n(x - c)^n$$

is differentiable (and therefore continuous) on the interval $(c - R, c + R)$ and

(a) $\quad f'(x) = a_1 + 2a_2(x - c) + 3a_3(x - c)^2 + \cdots = \sum_{n=1}^{\infty} na_n(x - c)^{n-1}$

(b) $\quad \displaystyle\int f(x)\, dx = C + a_0(x - c) + a_1\frac{(x - c)^2}{2} + a_2\frac{(x - c)^3}{3} + \cdots$

$$= C + \sum_{n=0}^{\infty} a_n \frac{(x - c)^{n+1}}{n + 1}$$

The radii of convergence of the power series in Equations (a) and (b) are both R.

Note 1: Equations (a) and (b) can be rewritten in the form

(c) $\quad \displaystyle\frac{d}{dx}\left[\sum_{n=0}^{\infty} a_n(x - c)^n\right] = \sum_{n=0}^{\infty} \frac{d}{dx}[a_n(x - c)^n]$

(d) $\quad \displaystyle\int\left[\sum_{n=0}^{\infty} a_n(x - c)^n\right] dx = \sum_{n=0}^{\infty} \int a_n(x - c)^n\, dx$

We know that, for finite sums, the derivative of a sum is the sum of the derivatives and the integral of a sum is the sum of the integrals. Equations (c) and (d) assert that the same is true for infinite sums provided that we are dealing with *power series*. (For other types of series of functions the situation is not as simple; see Exercise 70.)

Note 2: Although Theorem 10.39 says that the radius of convergence remains the same when a power series is differentiated or integrated, that does not mean that the *interval* of convergence remains the same. It may happen that the original series converges at an endpoint, whereas the differentiated series diverges there (see Exercise 69).

Note 3: The idea of differentiating a power series term by term is the basis for a powerful method for solving differential equations. We will discuss this method in Chapter 15.

EXAMPLE 1   In Example 3 in Section 10.8 we saw that the Bessel function

$$J_0(x) = \sum_{n=0}^{\infty} \frac{(-1)^n x^{2n}}{2^{2n}(n!)^2}$$

is defined for all x. Thus, by Theorem 10.39, J_0 is differentiable for all x and its derivative is found by term-by-term differentiation as follows:

$$J_0'(x) = \sum_{n=1}^{\infty} \frac{(-1)^n 2nx^{2n-1}}{2^{2n}(n!)^2}$$

EXAMPLE 2   Let us apply Theorem 10.39 to the geometric series

$$\frac{1}{1-x} = 1 + x + x^2 + \cdots = \sum_{n=0}^{\infty} x^n \qquad |x| < 1$$

(See Equation 10.16.) Differentiating each side of this equation, we have

$$\frac{1}{(1-x)^2} = 1 + 2x + 3x^2 + \cdots = \sum_{n=1}^{\infty} nx^{n-1} \qquad |x| < 1$$

Integrating each side gives

$$-\ln(1-x) = \int \frac{1}{1-x}\, dx = C + x + \frac{x^2}{2} + \frac{x^3}{3} + \cdots$$
$$= C + \sum_{n=0}^{\infty} \frac{x^{n+1}}{n+1} = C + \sum_{n=1}^{\infty} \frac{x^n}{n} \qquad |x| < 1$$

To determine the value of C we put $x = 0$ in this equation and obtain $-\ln(1-0) = C$. Thus $C = 0$ and

$$\ln(1-x) = -x - \frac{x^2}{2} - \frac{x^3}{3} - \cdots = -\sum_{n=1}^{\infty} \frac{x^n}{n} \qquad |x| < 1$$

In particular, putting $x = \frac{1}{2}$ and using the fact that $\ln(\frac{1}{2}) = -\ln 2$, we see that

$$\ln 2 = \frac{1}{2} + \frac{1}{8} + \frac{1}{24} + \frac{1}{64} + \cdots = \sum_{n=1}^{\infty} \frac{1}{n2^n}$$

Now let us suppose that f is any function that can be represented by a power series

(10.40) $$f(x) = a_0 + a_1(x-c) + a_2(x-c)^2 + a_3(x-c)^3 + a_4(x-c)^4 + \cdots \qquad |x-c| < R$$

and let us try to determine what the coefficients a_n must be in terms of f. To begin with, notice that if we put $x = c$ in Equation 10.40, then all terms after the first one are 0 and we get

$$f(c) = a_0$$

If we apply Theorem 10.39 to Equation 10.40, we obtain

(10.41) $$f'(x) = a_1 + 2a_2(x-c) + 3a_3(x-c)^2 + 4a_4(x-c)^3 + \cdots \qquad |x-c| < R$$

and substitution of $x = c$ in Equation 10.41 gives

$$f'(c) = a_1$$

Now we apply Theorem 10.39 a second time, this time to Equation 10.41, and obtain

(10.42) $$f''(x) = 2a_2 + 2\cdot 3a_3(x-c) + 3\cdot 4a_4(x-c)^2 + \cdots \qquad |x-c| < R$$

Again we put $x = c$ in Equation 10.42. The result is

$$f''(c) = 2a_2$$

Let us apply the procedure one more time. Differentiation of the series in Equation 10.42 yields

(10.43) $\qquad f'''(x) = 2 \cdot 3a_3 + 2 \cdot 3 \cdot 4a_4(x-c) + 3 \cdot 4 \cdot 5a_5(x-c)^2 + \cdots \qquad |x-c| < R$

and putting $x = c$ in Equation 10.43 gives

$$f'''(c) = 2 \cdot 3a_3 = 3!a_3$$

By now you can see the pattern. If we continue to differentiate and put $x = c$, we obtain

$$f^{(n)}(c) = 2 \cdot 3 \cdot 4 \cdots \cdot na_n = n!a_n$$

Solving this equation for the nth coefficient a_n, we get

$$a_n = \frac{f^{(n)}(c)}{n!}$$

This formula remains valid even for $n = 0$ if we adopt the conventions that $0! = 1$ and $f^{(0)} = f$. Thus we have proved the following theorem:

Theorem (10.44)

If f has a power series representation (expansion) at c, that is, if

$$f(x) = \sum_{n=0}^{\infty} a_n(x-c)^n \qquad |x-c| < R$$

then its coefficients are given by the formula

$$a_n = \frac{f^{(n)}(c)}{n!}$$

Putting this formula for a_n back into the series, we see that *if f has a power series expansion at c, then it must be of the form*

(10.45)
$$f(x) = \sum_{n=0}^{\infty} \frac{f^{(n)}(c)}{n!}(x-c)^n$$
$$= f(c) + \frac{f'(c)}{1!}(x-c) + \frac{f''(c)}{2!}(x-c)^2 + \frac{f'''(c)}{3!}(x-c)^3 + \cdots$$

The series in Equation 10.45 is called the **Taylor series of the function f at c** after the English mathematician Brook Taylor (1685–1731). For the special case where $c = 0$ the Taylor series becomes

(10.46)
$$f(x) = \sum_{n=0}^{\infty} \frac{f^{(n)}(0)}{n!}x^n = f(0) + \frac{f'(0)}{1!}x + \frac{f''(0)}{2!}x^2 + \cdots$$

This case arises frequently enough that it is given the special name **Maclaurin series** in honor of the Scottish mathematician Colin Maclaurin (1698–1746).

Note: We have shown that *if* f can be represented as a power series about c (such functions are called **analytic at** c), then f is equal to its Taylor series. Theorem 10.39 shows that analytic functions are infinitely differentiable at c; that is, they have derivatives of all orders at c. However not all infinitely differentiable functions are analytic. Exercise 71 gives an example of an infinitely differentiable function that is not analytic at 0. This function is therefore not equal to its Taylor series.

EXAMPLE 3   Find the Taylor series of the function $f(x) = e^x$ at 0 and at 1 and the associated radii of convergence.

Solution   If $f(x) = e^x$, then $f^{(n)}(x) = e^x$, so $f^{(n)}(0) = e^0 = 1$ for all n. Therefore the Taylor series for f at 0 (that is, the Maclaurin series) is

$$e^x = \sum_{n=0}^{\infty} \frac{f^{(n)}(0)}{n!} x^n = \sum_{n=0}^{\infty} \frac{x^n}{n!} = 1 + \frac{x}{1!} + \frac{x^2}{2!} + \frac{x^3}{3!} + \cdots$$

To find the radius of convergence we let $u_n = x^n/n!$. Then

$$\left| \frac{u_{n+1}}{u_n} \right| = \left| \frac{x^{n+1}}{(n+1)!} \cdot \frac{n!}{x^n} \right| = \frac{|x|}{n+1} \to 0 < 1$$

so, by the Ratio Test, the series converges for all x and the radius of convergence is $R = \infty$.

We have proved that if e^x has a power series expansion at 0, then

(10.47)
$$\boxed{\; e^x = \sum_{n=0}^{\infty} \frac{x^n}{n!} \quad \text{for all } x \;}$$

(In Section 10.11 we prove that Equation 10.47 is true without the prior assumption that e^x can be expanded as a power series.)

In particular if we put $x = 1$ in Equation 10.47, we obtain the following expression for the number e as a sum of an infinite series:

(10.48)
$$\boxed{\; e = \sum_{n=0}^{\infty} \frac{1}{n!} = 1 + \frac{1}{1!} + \frac{1}{2!} + \frac{1}{3!} + \cdots \;}$$

To find the Taylor series of $f(x) = e^x$ at 1 we observe that $f^{(n)}(1) = e^1 = e$ and so, putting $c = 1$ in Equation 10.45, we have

$$e^x = \sum_{n=0}^{\infty} \frac{f^{(n)}(1)}{n!} (x-1)^n = \sum_{n=0}^{\infty} \frac{e}{n!} (x-1)^n$$

Again it can be verified that the radius of convergence is $R = \infty$.   ●

EXAMPLE 4   Find the Maclaurin series for $\sin x$. For what values of x does it converge?

Solution　　We arrange our computation in the following manner:

$$f(x) = \sin x \qquad\qquad f(0) = 0$$
$$f'(x) = \cos x \qquad\qquad f'(0) = 1$$
$$f''(x) = -\sin x \qquad\qquad f''(0) = 0$$
$$f'''(x) = -\cos x \qquad\qquad f'''(0) = -1$$
$$f^{(4)}(x) = \sin x \qquad\qquad f^{(4)}(0) = 0$$

Since the derivatives repeat in a cycle of four, we can write the Maclaurin series as follows:

$$\sin x = f(0) + \frac{f'(0)}{1!}x + \frac{f''(0)}{2!}x^2 + \frac{f'''(0)}{3!}x^3 + \cdots$$

$$= x - \frac{x^3}{3!} + \frac{x^5}{5!} - \frac{x^7}{7!} + \cdots = \sum_{n=0}^{\infty} (-1)^n \frac{x^{2n+1}}{(2n+1)!}$$

If $u_n = (-1)^n x^{2n+1}/(2n+1)!$, then

$$\left| \frac{u_{n+1}}{u_n} \right| = \left| \frac{(-1)^{n+1}x^{2n+3}}{(2n+3)!} \cdot \frac{(2n+1)!}{(-1)^n x^{2n+1}} \right| = \frac{x^2}{(2n+3)(2n+2)} \to 0 < 1$$

so the series converges for all x. Thus, under the assumption that the sine function has a power series expansion, we have shown that

(10.49)

$$\boxed{\begin{array}{c} \sin x = x - \dfrac{x^3}{3!} + \dfrac{x^5}{5!} - \dfrac{x^7}{7!} + \cdots \\[2mm] = \displaystyle\sum_{n=0}^{\infty} (-1)^n \dfrac{x^{2n+1}}{(2n+1)!} \qquad \text{for all } x \end{array}}$$

EXAMPLE 5　　Find the Maclaurin series for $\cos x$.

Solution　　We could proceed directly as in Example 4 but it is easier to use Theorem 10.39 to differentiate the Maclaurin series for $\sin x$ given by Equation 10.49:

$$\cos x = \frac{d}{dx}(\sin x) = \frac{d}{dx}\left(x - \frac{x^3}{3!} + \frac{x^5}{5!} - \frac{x^7}{7!} + \cdots\right)$$

$$= 1 - \frac{3x^2}{3!} + \frac{5x^4}{5!} - \frac{7x^6}{7!} + \cdots = 1 - \frac{x^2}{2!} + \frac{x^4}{4!} - \frac{x^6}{6!} + \cdots$$

Since the Maclaurin series for $\sin x$ converges for all x, Theorem 10.39 tells us that the differentiated series for $\cos x$ also converges for all x. Thus

(10.50)

$$\boxed{\begin{array}{c} \cos x = 1 - \dfrac{x^2}{2!} + \dfrac{x^4}{4!} - \dfrac{x^6}{6!} + \cdots \\[2mm] = \displaystyle\sum_{n=0}^{\infty} (-1)^n \dfrac{x^{2n}}{(2n)!} \qquad \text{for all } x \end{array}}$$

EXAMPLE 6   Find the Maclaurin series for the function $f(x) = x \cos x$.

Solution   Instead of computing derivatives and substituting in Equation 10.46, it is easier to multiply the series for $\cos x$ (Equation 10.50) by x:

$$x \cos x = x \sum_{n=0}^{\infty} (-1)^n \frac{x^{2n}}{(2n)!} = \sum_{n=0}^{\infty} (-1)^n \frac{x^{2n+1}}{(2n)!}$$

Notice that we have used Theorem 10.19 part (a).

EXAMPLE 7   Find the Taylor series for $\ln x$ at 1.

Solution   Arranging our work in columns as in Example 4, we have

$$f(x) = \ln x \qquad\qquad f(1) = \ln 1 = 0$$
$$f'(x) = x^{-1} \qquad\qquad f'(1) = 1$$
$$f''(x) = -x^{-2} \qquad\qquad f''(1) = -1$$
$$f'''(x) = 2x^{-3} \qquad\qquad f'''(1) = 2$$
$$f^{(4)}(x) = -2 \cdot 3 x^{-4} \qquad\qquad f^{(4)}(1) = -2 \cdot 3$$
$$\vdots \qquad\qquad\qquad \vdots$$
$$f^{(n)}(x) = (-1)^{n-1}(n-1)! x^{-n} \qquad\qquad f^{(n)}(1) = (-1)^{n-1}(n-1)!$$

So, using Equation 10.45 with $c = 1$, we see that the Taylor expansion of $\ln x$ is

$$\ln x = f(1) + \frac{f'(1)}{1!}(x-1) + \frac{f''(1)}{2!}(x-1)^2 + \frac{f'''(1)}{3!}(x-1)^3 + \cdots$$

$$= (x-1) - \frac{1}{2!}(x-1)^2 + \frac{2!}{3!}(x-1)^3 - \frac{3!}{4!}(x-1)^4 + \cdots$$

$$= (x-1) - \frac{(x-1)^2}{2} + \frac{(x-1)^3}{3} - \frac{(x-1)^4}{4} + \cdots$$

$$= \sum_{n=1}^{\infty} (-1)^{n-1} \frac{(x-1)^n}{n}$$

If $u_n = (-1)^{n-1}(x-1)^n/n$, then

$$\left| \frac{u_{n+1}}{u_n} \right| = \left| \frac{(-1)^n (x-1)^{n+1}}{n+1} \cdot \frac{n}{(-1)^{n-1}(x-1)^n} \right|$$

$$= \frac{|x-1|}{1+\frac{1}{n}} \to |x-1| \qquad \text{as } n \to \infty$$

By the Ratio Test, the series converges if $|x-1| < 1$ and diverges if $|x-1| > 1$. So we have convergence if $0 < x < 2$. When $x = 0$ the series is the negative of the harmonic series, which diverges. When $x = 2$ the series is the alternating harmonic series, which converges (and represents $\ln 2$ by Exercise 39 in Section 10.5). Thus, if $\ln x$ has a power series expansion, then

$$\ln x = \sum_{n=1}^{\infty} (-1)^{n-1} \frac{(x-1)^n}{n} \qquad 0 < x \leq 2$$

EXAMPLE 8   Find the Maclaurin series for $f(x) = \dfrac{1}{1+x^2}$.

Solution   We could proceed as in Examples 4 and 7 but another method is to use a geometric series. Replacing x by $-x^2$ in Equation 10.16 or Example 2, we have

$$\frac{1}{1+x^2} = \frac{1}{1-(-x^2)} = \sum_{n=0}^{\infty} (-x^2)^n$$

$$= \sum_{n=0}^{\infty} (-1)^n x^{2n} = 1 - x^2 + x^4 - x^6 + x^8 - \cdots$$

Because this is a geometric series, it converges when $|-x^2| < 1$, this is, $x^2 < 1$, or $|x| < 1$.

         ●

EXAMPLE 9   Find the Maclaurin series for $f(x) = \tan^{-1}x$.

Solution   For this function it is extremely arduous to compute $f^{(n)}(0)$ directly, so we observe that $f'(x) = 1/(1+x^2)$ and find the required series by integrating the Maclaurin series for $1/(1+x^2)$ found in Example 8. Theorem 10.39 part (b) gives

$$\tan^{-1}x = \int \frac{1}{1+x^2}\,dx = \int (1 - x^2 + x^4 - x^6 + \cdots)\,dx$$

$$= C + x - \frac{x^3}{3} + \frac{x^5}{5} - \frac{x^7}{7} + \cdots$$

To find C we put $x = 0$ and obtain $C = \tan^{-1}(0) = 0$. Therefore

$$\tan^{-1}x = x - \frac{x^3}{3} + \frac{x^5}{5} - \frac{x^7}{7} + \cdots = \sum_{n=0}^{\infty} (-1)^n \frac{x^{2n+1}}{2n+1}$$

Since the radius of convergence of the series for $1/(1+x^2)$ is 1, the radius of convergence of this Maclaurin series for $\tan^{-1}x$ is also 1.

         ●

Note: In Examples 5, 6, 8, and 9 we can be sure that the series we obtained by indirect methods are indeed the Taylor or Maclaurin series of the given functions because Theorem 10.44 asserts that, no matter how a power series representation $f(x) = \sum a_n(x-c)^n$ is obtained, it is always true that $a_n = f^{(n)}(c)/n!$. In other words, the coefficients are uniquely determined.

    One reason that Taylor series are important is that they enable us to integrate functions that we could not previously handle. In fact, in the introduction to this chapter we mentioned that Newton often integrated functions by first expressing them as power series and then integrating them term by term. The function $f(x) = e^{-x^2}$ cannot be integrated in the usual way because its antiderivative is not an elementary function (see Section 7.6). In the following example we use Newton's idea to integrate this function.

EXAMPLE 10

(a)   Evaluate $\int e^{-x^2}\,dx$ as an infinite series.

(b)   Evaluate $\int_0^1 e^{-x^2}\,dx$ correct to within an error of 0.001.

Solution

    (a)   First we find the Maclaurin series for $f(x) = e^{-x^2}$. Although it is possible to use the direct method, let us find it simply by replacing x by $-x^2$ in Equation 10.47. Thus

$$e^{-x^2} = \sum_{n=0}^{\infty} \frac{(-x^2)^n}{n!} = \sum_{n=0}^{\infty} (-1)^n \frac{x^{2n}}{n!} = 1 - \frac{x^2}{1!} + \frac{x^4}{2!} - \frac{x^6}{3!} + \cdots$$

Now we integrate term by term using Theorem 10.39:

$$\int e^{-x^2}\,dx = \int \left(1 - \frac{x^2}{1!} + \frac{x^4}{2!} - \frac{x^6}{3!} + \cdots + (-1)^n \frac{x^{2n}}{n!} + \cdots\right)dx$$

$$= C + x - \frac{x^3}{3 \cdot 1!} + \frac{x^5}{5 \cdot 2!} - \frac{x^7}{7 \cdot 3!} + \cdots + (-1)^n \frac{x^{2n+1}}{(2n+1)n!} + \cdots$$

This series converges for all x because the original series for e^{-x^2} converges for all x.

(b)   The Fundamental Theorem of Calculus gives

$$\int_0^1 e^{-x^2}\,dx = \left[x - \frac{x^3}{3 \cdot 1!} + \frac{x^5}{5 \cdot 2!} - \frac{x^7}{7 \cdot 3!} + \frac{x^9}{9 \cdot 4!} - \cdots\right]_0^1$$

$$= 1 - \frac{1}{3} + \frac{1}{10} - \frac{1}{42} + \frac{1}{216} - \cdots$$

$$\approx 0.7475$$

Since this series is alternating, Theorem 10.28 shows that the error involved in this approximation is less than

$$\frac{1}{11 \cdot 5!} = \frac{1}{1320} < 0.001$$ •

Multiplication and Division of Power Series

If power series are added or subtracted, they behave like polynomials (Theorem 10.19 shows this). In fact, as the following example illustrates, they can also be multiplied and divided like polynomials. We find only the first few terms because the calculations for the later terms become tedious and the initial terms are the most important ones.

EXAMPLE 11   Find the first three nonzero terms in the Maclaurin series for (a) $e^x \sin x$ and (b) $\tan x$.

Solution
(a) From Equations 10.47 and 10.49 we have

$$e^x \sin x = \left(1 + \frac{x}{1!} + \frac{x^2}{2!} + \frac{x^3}{3!} + \cdots\right)\left(x - \frac{x^3}{3!} + \cdots\right)$$

We multiply these expressions, collecting like terms just as for polynomials:

$$
\begin{array}{ccccccccc}
1 & + & x & + & \frac{1}{2}x^2 & + & \frac{1}{6}x^3 & + & \cdots \\
 & & x & & & - & \frac{1}{6}x^3 & + & \cdots \\
\hline
 & & x & + & x^2 & + & \frac{1}{2}x^3 & + & \frac{1}{6}x^4 & + & \cdots \\
 & & & & & - & \frac{1}{6}x^3 & - & \frac{1}{6}x^4 & - & \cdots \\
\hline
 & & x & + & x^2 & + & \frac{1}{3}x^3 & + & \cdots
\end{array}
$$

Thus

$$e^x \sin x = x + x^2 + \frac{1}{3}x^3 + \cdots$$

(b) From Equations 10.49 and 10.50 we have

$$\tan x = \frac{\sin x}{\cos x} = \frac{x - \frac{x^3}{3!} + \frac{x^5}{5!} - \cdots}{1 - \frac{x^2}{2!} + \frac{x^4}{4!} - \cdots}$$

We use a procedure like long division:

$$
\begin{array}{r}
x + \frac{1}{3}x^3 + \frac{2}{15}x^5 + \cdots \\
1 - \frac{1}{2}x^2 + \frac{1}{24}x^4 - \cdots \overline{\smash{\big)}\; x - \frac{1}{6}x^3 + \frac{1}{120}x^5 - \cdots} \\
x - \frac{1}{2}x^3 + \frac{1}{24}x^5 - \cdots \\
\hline
\frac{1}{3}x^3 - \frac{1}{30}x^5 + \cdots \\
\frac{1}{3}x^3 - \frac{1}{6}x^5 + \cdots \\
\hline
\frac{2}{15}x^5 + \cdots
\end{array}
$$

Thus

$$\tan x = x + \tfrac{1}{3}x^3 + \tfrac{2}{15}x^5 + \cdots$$

Although we have not attempted to justify the formal manipulations used in Example 11, they are legitimate. There is a theorem that states that if $f(x) = \sum a_n x^n$ and $g(x) = \sum b_n x^n$ both converge for $|x| < R$ and the series are multiplied as if they were polynomials, then the resulting series also converges for $|x| < R$ and represents $f(x)g(x)$. For division we require $b_0 \neq 0$ and the resulting series converges for sufficiently small $|x|$.

SECTION 10.9 **Exercises**

In the following exercises, assume that all of the functions possess power series expansions.

In Exercises 1–12 use the direct method of Examples 3, 4, and 7 to find Taylor series of f at the given value of c. Also find the associated radius of convergence.

1. $f(x) = \cos x$, $c = 0$
2. $f(x) = \sin 2x$, $c = 0$
3. $f(x) = \sin x$, $c = \pi/4$
4. $f(x) = \cos x$, $c = -\pi/4$
5. $f(x) = \dfrac{1}{(1+x)^2}$, $c = 0$
6. $f(x) = \dfrac{x}{1-x}$, $c = 0$
7. $f(x) = 1/x$, $c = 1$
8. $f(x) = \sqrt{x}$, $c = 4$
9. $f(x) = e^x$, $c = 3$
10. $f(x) = \ln x$, $c = 2$
11. $f(x) = \sinh x$, $c = 0$
12. $f(x) = \cosh x$, $c = 0$

In Exercises 13–24 find the Maclaurin series of f and its radius of convergence by using a geometric series or by differentiating or integrating a geometric series.

13. $f(x) = \dfrac{1}{1+x}$
14. $f(x) = \ln(1+x)$

15. $f(x) = \dfrac{1}{(1+x)^2}$
16. $f(x) = \dfrac{x}{1-x}$
17. $f(x) = \dfrac{1}{1+4x^2}$
18. $f(x) = \tan^{-1}(2x)$
19. $f(x) = \dfrac{1}{4+x^2}$
20. $f(x) = \dfrac{1}{x^4+16}$
21. $f(x) = \dfrac{1}{1-x^2}$
22. $f(x) = \dfrac{1+x^2}{1-x^2}$
23. $f(x) = \ln\left(\dfrac{1+x}{1-x}\right)$
24. $f(x) = \dfrac{2}{3x+4}$

In Exercises 25–42 use any method to find the Maclaurin series of f and its radius of convergence. In particular, you may use the known series 10.47, 10.49, and 10.50.

25. $f(x) = e^{3x}$
26. $f(x) = \sin 2x$
27. $f(x) = x^2 \cos x$
28. $f(x) = \cos(x^3)$
29. $f(x) = x\sin(x/2)$
30. $f(x) = xe^{-x}$
31. $f(x) = \sin^2 x$ [Hint: Use $\sin^2 x = \tfrac{1}{2}(1 - \cos 2x)$.]

32. $f(x) = \cos^2 x$

33. $f(x) = \begin{cases} \frac{\sin x}{x} & \text{if } x \neq 0 \\ 1 & \text{if } x = 0 \end{cases}$

34. $f(x) = \begin{cases} \frac{1 - \cos x}{x^2} & \text{if } x \neq 0 \\ \frac{1}{2} & \text{if } x = 0 \end{cases}$

35. $f(x) = \sqrt{1 + x}$ **36.** $f(x) = \dfrac{1}{\sqrt{1 + 2x}}$

37. $f(x) = \dfrac{1}{\sqrt[3]{1 - x}}$ **38.** $f(x) = (1 + x)^{2/3}$

39. $f(x) = (1 + x)^{-3}$ **40.** $f(x) = 2^x$

41. $f(x) = \ln(5 + x)$ **42.** $f(x) = \log_{10}(1 + x)$

43. Find the Maclaurin series for $\ln(1 + x)$ and use it to calculate $\ln 1.1$ correct to five decimal places.

44. Use the Maclaurin series for $\sin x$ to compute $\sin 3°$ correct to five decimal places.

In Exercises 45–50 evaluate the given indefinite integral as an infinite series.

45. $\displaystyle\int \sin(x^2)\,dx$ **46.** $\displaystyle\int \frac{\sin x}{x}\,dx$

47. $\displaystyle\int \frac{1}{1 + x^4}\,dx$ **48.** $\displaystyle\int e^{x^3}\,dx$

49. $\displaystyle\int \sqrt{x^3 + 1}\,dx$ **50.** $\displaystyle\int \frac{x}{1 + x^5}\,dx$

In Exercises 51–56 use series to approximate the given definite integral to within the indicated accuracy.

51. $\displaystyle\int_0^1 \sin(x^2)\,dx$ (three decimal places)

52. $\displaystyle\int_0^{0.5} \cos(x^2)\,dx$ (three decimal places)

53. $\displaystyle\int_0^{0.5} \frac{dx}{1 + x^6}$ (four decimal places)

54. $\displaystyle\int_0^{0.1} \frac{dx}{\sqrt{1 + x^3}}$ (error $< 10^{-8}$)

55. $\displaystyle\int_0^{0.5} x^2 e^{-x^2}\,dx$ (error < 0.001)

56. $\displaystyle\int_0^1 \cos(x^3)\,dx$ (four decimal places)

57. (a) Show that J_0 (the Bessel function of order 0 given in Example 1) satisfies the differential equation
$$x^2 J_0''(x) + x J_0'(x) + x^2 J_0(x) = 0$$

(b) Evaluate $\int_0^1 J_0(x)\,dx$ correct to three decimal places.

58. The Bessel function of order 1 is defined by
$$J_1(x) = \sum_{n=0}^{\infty} \frac{(-1)^n x^{2n+1}}{n!(n + 1)!\,2^{2n+1}}$$

(a) Show that J_1 satisfies the differential equation
$$x^2 J_1''(x) + x J_1'(x) + (x^2 - 1)J_1(x) = 0$$

(b) Show that $J_0'(x) = -J_1(x)$.

In Exercises 59–64 find the sum of the series.

59. $\displaystyle\sum_{n=0}^{\infty} (-1)^n \frac{x^{4n}}{n!}$ **60.** $\displaystyle\sum_{n=0}^{\infty} \frac{(-1)^n \pi^{2n}}{6^{2n}(2n)!}$

61. $\displaystyle\sum_{n=0}^{\infty} \frac{(-1)^n \pi^{2n+1}}{4^{2n+1}(2n + 1)!}$ **62.** $\displaystyle\sum_{n=2}^{\infty} \frac{x^{3n+1}}{n!}$

63. $\displaystyle\sum_{n=0}^{\infty} \frac{x^{n+1}}{(n + 1)!}$ **64.** $\displaystyle\sum_{n=0}^{\infty} \frac{x^n}{2^n(n + 1)!}$

In Exercises 65–68 use multiplication or division of power series to find the first three nonzero terms in the Maclaurin series for the given function.

65. $y = e^{-x^2} \cos x$ **66.** $y = \sec x$

67. $y = \dfrac{\ln(1 - x)}{e^x}$ **68.** $y = e^x \ln(1 - x)$

69. Let
$$f(x) = \sum_{n=1}^{\infty} \frac{x^n}{n^2}$$
Find the intervals of convergence for f, f', and f''.

70. Let $f_n(x) = (\sin nx)/n^2$. Show that the series $\sum f_n(x)$ converges for all values of x but the series of derivatives $\sum f_n'(x)$ diverges when $x = 2n\pi$, n an integer. For what values of x does the series $\sum f_n''(x)$ converge?

71. Show that the function defined in Exercise 84 in Section 3.9 is not equal to its Maclaurin series.

72. (a) Starting with the geometric series $\displaystyle\sum_{n=0}^{\infty} x^n$, find the sum of the series $\displaystyle\sum_{n=1}^{\infty} n x^{n-1}$ for $|x| < 1$.

(b) Find the sums of the following series.

(i) $\displaystyle\sum_{n=1}^{\infty} n x^n$, $|x| < 1$ (ii) $\displaystyle\sum_{n=1}^{\infty} \frac{n}{2^n}$

(c) Find the sums of the following series.

(i) $\displaystyle\sum_{n=2}^{\infty} n(n - 1)x^n$, $|x| < 1$ (ii) $\displaystyle\sum_{n=2}^{\infty} \frac{n^2 - n}{2^n}$

(iii) $\displaystyle\sum_{n=1}^{\infty} \frac{n^2}{2^n}$

The Binomial Series

You may be acquainted with the Binomial Theorem, which states that if a and b are any real numbers and k is a positive integer, then

$$(a+b)^k = a^k + ka^{k-1}b + \frac{k(k-1)}{2!}a^{k-2}b^2 + \frac{k(k-1)(k-2)}{3!}a^{k-3}b^3$$
$$+ \cdots + \frac{k(k-1)(k-2)\cdots(k-n+1)}{n!}a^{k-n}b^n$$
$$+ \cdots + kab^{k-1} + b^k$$

The traditional notation for the binomial coefficients is

$$\binom{k}{0} = 1 \qquad \binom{k}{n} = \frac{k(k-1)(k-2)\cdots(k-n+1)}{n!} \qquad n = 1, 2, \ldots, k$$

which enables us to write the Binomial Theorem in the abbreviated form

$$(a+b)^k = \sum_{n=0}^{k} \binom{k}{n}a^{k-n}b^n$$

In particular, if we put $a = 1$ and $b = x$, we get

(10.51)
$$(1+x)^k = \sum_{n=0}^{k} \binom{k}{n}x^n$$

One of Newton's accomplishments was to extend the Binomial Theorem (Equation 10.51) to the case where k is no longer a positive integer. In this case the expression for $(1+x)^k$ is no longer a finite sum; it becomes an infinite series.

Assuming that $(1+x)^k$ can be expanded as a power series, we compute its Maclaurin series in the usual way:

$$f(x) = (1+x)^k \qquad\qquad f(0) = 1$$
$$f'(x) = k(1+x)^{k-1} \qquad\qquad f'(0) = k$$
$$f''(x) = k(k-1)(1+x)^{k-2} \qquad\qquad f''(0) = k(k-1)$$
$$f'''(x) = k(k-1)(k-2)(1+x)^{k-3} \qquad\qquad f'''(0) = k(k-1)(k-2)$$
$$\vdots \qquad\qquad\qquad\qquad \vdots$$
$$f^{(n)}(x) = k(k-1)\cdots(k-n+1)(1+x)^{k-n} \qquad f^{(n)}(0) = k(k-1)\cdots(k-n+1)$$

$$(1+x)^k = \sum_{n=0}^{\infty} \frac{f^{(n)}(0)}{n!}x^n = \sum_{n=0}^{\infty} \frac{k(k-1)\cdots(k-n+1)}{n!}x^n$$

If u_n is the nth term of this series, then

$$\left|\frac{u_{n+1}}{u_n}\right| = \left|\frac{k(k-1)\cdots(k-n+1)(k-n)x^{n+1}}{(n+1)!} \cdot \frac{n!}{k(k-1)\cdots(k-n+1)x^n}\right|$$

$$= \frac{|k-n|}{n+1}|x| = \frac{\left|1-\dfrac{k}{n}\right|}{1+\dfrac{1}{n}}|x| \to |x| \qquad \text{as } n \to \infty$$

Thus, by the Ratio Test, the binomial series converges if $|x| < 1$ and diverges if $|x| > 1$.

The Binomial Series (10.52)

If k is any real number and $|x| < 1$, then

$$(1+x)^k = 1 + kx + \frac{k(k-1)}{2!}x^2 + \frac{k(k-1)(k-2)}{3!}x^3 + \cdots$$

$$= \sum_{n=0}^{\infty} \binom{k}{n}x^n$$

where

$$\binom{k}{n} = \frac{k(k-1)\cdots(k-n+1)}{n!} \quad (n \geq 1) \qquad \binom{k}{0} = 1$$

We have proved (10.52) under the assumption that $(1+x)^k$ has a power series expansion. For a proof without that assumption see Exercise 21.

Although the binomial series always converges when $|x| < 1$, the question of whether or not it converges at the endpoints, ± 1, depends on the value of k. It turns out that the series converges at 1 if $-1 < k \leq 0$ and at both endpoints if $k \geq 0$. Notice that if k is a positive integer and $n > k$, then the expression for $\binom{k}{n}$ contains a factor $(k-k)$, so $\binom{k}{n} = 0$ for $n > k$. This means that the series terminates and reduces to the ordinary Binomial Theorem (Equation 10.51) when k is a positive integer.

Although, as we have seen, the binomial series is just a special case of the Maclaurin series, it occurs frequently and so it is worth remembering.

EXAMPLE 1   Expand $\dfrac{1}{(1+x)^2}$ as a power series.

Solution   We use the binomial series with $k = -2$. The binomial coefficient is

$$\binom{-2}{n} = \frac{(-2)(-3)(-4)\cdots(-2-n+1)}{n!}$$

$$= \frac{(-1)^n 2 \cdot 3 \cdot 4 \cdot \cdots \cdot n(n+1)}{n!} = (-1)^n(n+1)$$

and so, when $|x| < 1$,

$$\frac{1}{(1+x)^2} = (1+x)^{-2} = \sum_{n=0}^{\infty} \binom{-2}{n}x^n$$

$$= \sum_{n=0}^{\infty} (-1)^n(n+1)x^n$$

EXAMPLE 2   Find the Maclaurin series for the function $f(x) = 1/\sqrt{4-x}$ and its radius of convergence.

Solution   As given, $f(x)$ is not quite of the form $(1+x)^k$ so we rewrite it as follows:

$$\frac{1}{\sqrt{4-x}} = \frac{1}{\sqrt{4\left(1-\frac{x}{4}\right)}} = \frac{1}{2\sqrt{1-\frac{x}{4}}} = \frac{1}{2}\left(1-\frac{x}{4}\right)^{-1/2}$$

Using the binomial series with $k = -\frac{1}{2}$ and with x replaced by $-x/4$, we have

$$\frac{1}{\sqrt{4-x}} = \frac{1}{2}\left(1 - \frac{x}{4}\right)^{-1/2} = \frac{1}{2}\sum_{n=0}^{\infty}\binom{-\frac{1}{2}}{n}\left(-\frac{x}{4}\right)^n$$

$$= \frac{1}{2}\left[1 + \left(-\frac{1}{2}\right)\left(-\frac{x}{4}\right) + \frac{\left(-\frac{1}{2}\right)\left(-\frac{3}{2}\right)}{2!}\left(-\frac{x}{4}\right)^2 + \frac{\left(-\frac{1}{2}\right)\left(-\frac{3}{2}\right)\left(-\frac{5}{2}\right)}{3!}\left(-\frac{x}{4}\right)^3 \right.$$

$$\left. + \cdots + \frac{\left(-\frac{1}{2}\right)\left(-\frac{3}{2}\right)\left(-\frac{5}{2}\right)\cdots\left(-\frac{1}{2} - n + 1\right)}{n!}\left(-\frac{x}{4}\right)^n + \cdots \right]$$

$$= \frac{1}{2}\left(1 + \frac{1}{8}x + \frac{1\cdot3}{2!\,8^2}x^2 + \frac{1\cdot3\cdot5}{3!\,8^3}x^3 + \cdots + \frac{1\cdot3\cdot5\cdot\cdots\cdot(2n-1)}{n!\,8^n}x^n + \cdots\right)$$

We know from (10.52) that this series converges when $|-x/4| < 1$, that is, $|x| < 4$, so the radius of convergence is $R = 4$.

SECTION 10.10　Exercises

In Exercises 1–12 use the binomial series to expand the given function as a power series. State the radius of convergence.

1. $\sqrt{1+x}$

2. $\dfrac{1}{(1+x)^3}$

3. $\dfrac{1}{(1+2x)^4}$

4. $\sqrt[3]{1+x^2}$

5. $\dfrac{x}{\sqrt{1-x}}$

6. $\dfrac{1}{\sqrt{2+x}}$

7. $\dfrac{1}{\sqrt[3]{8+x}}$

8. $(4+x)^{3/2}$

9. $\sqrt[4]{1-x^4}$

10. $\dfrac{x^2}{\sqrt{1-x^3}}$

11. $\left(\dfrac{x}{1-x}\right)^5$

12. $\sqrt[5]{x-1}$

13. (a) Use the binomial series to expand $1/\sqrt{1-x^2}$.
    (b) Use part (a) to find the Maclaurin series for $\sin^{-1}x$.

14. (a) Use the binomial series to expand $1/\sqrt{1+x^2}$.
    (b) Use part (a) to find the Maclaurin series for $\sinh^{-1}x$.

15. (a) Expand $1/\sqrt{1+x}$ as a power series.
    (b) Use part (a) to estimate $1/\sqrt{1.1}$ correct to three decimal places.

16. (a) Expand $\sqrt[3]{8+x}$ as a power series.
    (b) Use part (a) to estimate $\sqrt[3]{8.2}$ correct to four decimal places.

17. (a) Expand $f(x) = x/(1-x)^2$ as a power series.
    (b) Use part (a) to find the sum of the series

    $$\sum_{n=1}^{\infty}\frac{n}{2^n}$$

18. (a) Expand $f(x) = (x+x^2)/(1-x)^3$ as a power series.
    (b) Use part (a) to find the sum of the series

    $$\sum_{n=1}^{\infty}\frac{n^2}{2^n}$$

19. (a) Use the binomial series to find the Maclaurin series of $f(x) = \sqrt{1+x^2}$.
    (b) Use part (a) to evaluate $f^{(10)}(0)$.

20. (a) Use the binomial series to find the Maclaurin series of $f(x) = 1/\sqrt{1+x^3}$.
    (b) Use part (a) to evaluate $f^{(9)}(0)$.

21. Use the following steps to prove (10.52).
    (a) Let $g(x) = \sum_{n=0}^{\infty}\binom{k}{n}x^n$. Differentiate this series to show that

    $$g'(x) = \frac{kg(x)}{1+x} \qquad -1 < x < 1$$

    (b) Let $h(x) = (1+x)^{-k}g(x)$ and show that $h'(x) = 0$.
    (c) Deduce that $g(x) = (1+x)^k$.

SECTION 10.11

Approximation by Taylor Polynomials

Recall that in Section 10.9 we were able to express some functions f as the sum of their Taylor series:

$$f(x) = \sum_{n=0}^{\infty} \frac{f^{(n)}(c)}{n!}(x-c)^n$$

As with any series, the partial sums are approximations to the total sum of the series. In the case of the Taylor series the partial sums are

(10.53)
$$T_n(x) = \sum_{i=0}^{n} \frac{f^{(i)}(c)}{i!}(x-c)^i$$

$$= f(c) + \frac{f'(c)}{1!}(x-c) + \frac{f''(c)}{2!}(x-c)^2 + \cdots + \frac{f^{(n)}(c)}{n!}(x-c)^n$$

Notice that T_n is a polynomial of degree n called the **nth-degree Taylor polynomial of f at c**. If f is the sum of its Taylor series, then $T_n(x) \to f(x)$ as $n \to \infty$ and so T_n can be used as an approximation to f: $f(x) \approx T_n(x)$. It is useful to be able to approximate a function by a polynomial because polynomials are the simplest of functions. We can often gain information about a function by looking at its Taylor polynomials T_n.

We know from Example 3 in Section 10.9 that the Taylor series for $f(x) = e^x$ at $c = 0$ is

$$1 + \frac{x}{1!} + \frac{x^2}{2!} + \frac{x^3}{3!} + \cdots + \frac{x^n}{n!} + \cdots$$

so its first three Taylor polynomials at 0 (Maclaurin polynomials) are

$$T_1(x) = 1 + x \qquad T_2(x) = 1 + x + \frac{x^2}{2!} \qquad T_3(x) = 1 + x + \frac{x^2}{2!} + \frac{x^3}{3!}$$

The graphs of the exponential function and these three Taylor polynomials are drawn in Figure 10.11.

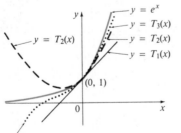

Figure 10.11 $y = T_3(x)$

The graph of T_1 is the tangent line to $y = e^x$ at $(0, 1)$, which is the best linear approximation to e^x near $(0, 1)$. The graph of T_2 is the parabola $y = 1 + x + x^2/2$ and the graph of T_3 is the cubic curve $y = 1 + x + x^2/2 + x^3/6$, which is a closer fit to the exponential curve $y = e^x$ than T_2. The next Taylor polynomial T_4 would be an even better approximation, and so on. In general, the derivatives of T_n at c agree with those of f up to and including derivatives of order n (see Exercise 50).

When using a Taylor polynomial T_n to approximate a function f, we have to ask the question: How good an approximation is it? or How large should we take n to be in order to achieve a desired accuracy? The following theorem can be used to answer these questions.

Taylor's Formula (10.54)

If f has $n+1$ derivatives in an interval I that contains the number c, then for x in I there is a number z strictly between x and c such that

$$f(x) = f(c) + \frac{f'(c)}{1!}(x-c) + \frac{f''(c)}{2!}(x-c)^2 + \cdots + \frac{f^{(n)}(c)}{n!}(x-c)^n + R_n(x)$$

where
$$R_n(x) = \frac{f^{(n+1)}(z)}{(n+1)!}(x-c)^{n+1}$$

Note 1: For the special case $n = 0$, if we put $x = b$, $c = a$, and $z = c$, we get $f(b) = f(a) + f'(c)(b-a)$, which is the Mean Value Theorem. In fact, Theorem 10.54 can be proved by a method similar to the proof of the Mean Value Theorem. The proof is given at the end of this section.

Note 2: Using the notation (Equation 10.53) for Taylor polynomials, we can write Taylor's Formula as

$$f(x) = T_n(x) + R_n(x)$$

Notice that the remainder term

(10.55)
$$R_n(x) = \frac{f^{(n+1)}(z)}{(n+1)!}(x-c)^{n+1}$$

is very similar to the terms in the Taylor series except that $f^{(n+1)}$ is evaluated at z instead of at c. All we can say about the number z is that it lies somewhere between x and c. The expression for $R_n(x)$ in Equation 10.55 is known as **Lagrange's form of the remainder term.**

The **error** in the approximation $f(x) \approx T_n(x)$ is

$$|f(x) - T_n(x)| = |R_n(x)|$$

and this can often be estimated by using Equation 10.55.

Note 3: For the special case $n = 1$, this approximation is the same as the approximation by differentials (see Section 2.9) since

$$f(x) \approx T_1(x) = f(c) + f'(c)(x-c) = f(c) + f'(c)\,dx = f(c) + dy$$

but now we have an expression for the error $|R_1(x)|$.

EXAMPLE 1

(a)   Write Taylor's Formula for the case where $f(x) = \ln(1+x)$, $c = 0$, $n = 5$.
(b)   Use part (a) to find an approximate value for $\ln(1.2)$ and to give an estimate for the error involved in this approximation.

Solution

(a)   We arrange our work in columns just as with Taylor series:

$$f(x) = \ln(1 + x) \qquad\qquad f(0) = \ln 1 = 0$$
$$f'(x) = (1 + x)^{-1} \qquad\qquad f'(0) = 1$$
$$f''(x) = -(1 + x)^{-2} \qquad\qquad f''(0) = -1$$
$$f'''(x) = 2(1 + x)^{-3} \qquad\qquad f'''(0) = 2$$
$$f^{(4)}(x) = -6(1 + x)^{-4} \qquad\qquad f^{(4)}(0) = -6$$
$$f^{(5)}(x) = 24(1 + x)^{-5} \qquad\qquad f^{(5)}(0) = 24$$
$$f^{(6)}(x) = -120(1 + x)^{-6}$$

Therefore Taylor's Formula gives

$$\ln(1 + x) = f(0) + \frac{f'(0)}{1!}x + \frac{f''(0)}{2!}x^2 + \frac{f'''(0)}{3!}x^3 + \frac{f^{(4)}(0)}{4!}x^4 + \frac{f^{(5)}(0)}{5!}x^5 + R_5(x)$$

$$= x - \frac{1}{2!}x^2 + \frac{2}{3!}x^3 - \frac{6}{4!}x^4 + \frac{24}{5!}x^5 + R_5(x)$$

$$= x - \frac{x^2}{2} + \frac{x^3}{3} - \frac{x^4}{4} + \frac{x^5}{5} + R_5(x)$$

where
$$R_5(x) = \frac{f^{(6)}(z)}{6!}x^6 = -\frac{120}{(1+z)^6}\frac{x^6}{6!} = -\frac{x^6}{6(1+z)^6}$$

and z is a number between 0 and x.

(b)   From part (a) we have the approximation

$$\ln(1 + x) \approx T_5(x) = x - \frac{x^2}{2} + \frac{x^3}{3} - \frac{x^4}{4} + \frac{x^5}{5}$$

In particular, with $x = 0.2$, this approximation becomes

$$\ln(1.2) \approx 0.2 - \frac{(0.2)^2}{2} + \frac{(0.2)^3}{3} - \frac{(0.2)^4}{4} + \frac{(0.2)^5}{5}$$

$$\approx 0.18233067$$

The error in this approximation is

$$|R_5(0.2)| = \frac{(0.2)^6}{6(1+z)^6} \qquad \text{where } 0 < z < 0.2$$

Since $z > 0$ we have

$$\frac{1}{1+z} < 1 \qquad \text{so} \qquad \frac{1}{(1+z)^6} < 1$$

and
$$|R_5(0.2)| = \frac{(0.2)^6}{6(1+z)^6} < \frac{(0.2)^6}{6} = \frac{0.000064}{6} < 0.000011$$

so the error in this approximation is less than 0.000011.

EXAMPLE 2

(a)   Approximate the function $f(x) = \sqrt[3]{x}$ by a Taylor polynomial of degree 2 at $c = 8$.

(b)   How accurate is this approximation when $7 \le x \le 9$?

12th

Solution

(a)
$$f(x) = \sqrt[3]{x} = x^{1/3} \qquad\qquad f(8) = 2$$
$$f'(x) = \tfrac{1}{3}x^{-2/3} \qquad\qquad f'(8) = \tfrac{1}{12}$$
$$f''(x) = -\tfrac{2}{9}x^{-5/3} \qquad\qquad f''(8) = -\tfrac{1}{144}$$
$$f'''(x) = \tfrac{10}{27}x^{-8/3}$$

Thus Taylor's Formula becomes

$$\sqrt[3]{x} = f(8) + \frac{f'(8)}{1!}(x-8) + \frac{f''(8)}{2!}(x-8)^2 + R_2(x)$$

$$= 2 + \tfrac{1}{12}(x-8) - \tfrac{1}{288}(x-8)^2 + R_2(x)$$

The desired approximation is

$$\sqrt[3]{x} \approx T_2(x) = 2 + \tfrac{1}{12}(x-8) - \tfrac{1}{288}(x-8)^2$$

(b)   The remainder term is

$$R_2(x) = \frac{f'''(z)}{3!}(x-8)^3 = \frac{10}{27}z^{-8/3}\frac{(x-8)^3}{3!} = \frac{5(x-8)^3}{81z^{8/3}}$$

where z lies between 8 and x. In order to estimate the error $|R_2(x)|$ we note that if $7 \le x \le 9$, then $-1 \le x - 8 \le 1$, so $|x-8| \le 1$ and therefore $|x-8|^3 \le 1$. Also, since $z > 7$, we have

$$z^{8/3} > 7^{8/3} > 179$$

and so
$$|R_2(x)| = \frac{5|x-8|^3}{81z^{8/3}} < \frac{5 \cdot 1}{81 \cdot 179} < 0.0004$$

Thus if $7 \le x \le 9$, the approximation in part (a) is accurate to within 0.0004.
Even without a calculator we could have made the cruder estimate

$$z^{8/3} > 7^{8/3} > 7^2 = 49$$

which gives the error estimate

$$|R_2(x)| < \frac{5 \cdot 1}{81 \cdot 49} < 0.002$$

EXAMPLE 3   Estimate the value of $\sqrt[4]{e}$ to within an accuracy of 0.0001.

Solution   Since $\sqrt[4]{e} = e^{1/4}$, we use Taylor's Formula with $f(x) = e^x$ and $c = 0$. Notice that this example is different from Examples 1 and 2 in that we are given the prescribed accuracy but not the value of n. So we start by writing Taylor's Formula for general n. Since $f^{(n)}(x) = e^x$ for all n, we obtain

$$e^x = 1 + \frac{x}{1!} + \frac{x^2}{2!} + \frac{x^3}{3!} + \cdots + \frac{x^n}{n!} + R_n(x)$$

where
$$R_n(x) = \frac{f^{(n+1)}(z)}{(n+1)!}x^{n+1} = \frac{e^z}{(n+1)!}x^{n+1}$$

and z lies between 0 and x. Since we are approximating $e^{1/4}$, we take $x = \tfrac{1}{4}$. Then $0 < z < \tfrac{1}{4}$ so, using the fact that $e < 3$, we have

$$e^z < e^{1/4} < 3^{1/4} = \sqrt[4]{3} < 2$$

and

$$\left| R_n\!\left(\tfrac{1}{4}\right) \right| = \frac{e^z}{(n+1)!}\left(\tfrac{1}{4}\right)^{n+1} < \frac{2}{(n+1)!4^{n+1}} = \frac{1}{2\cdot 4^n(n+1)!}$$

For $n = 3$ we have

$$\left| R_3\!\left(\tfrac{1}{4}\right) \right| < \frac{1}{2\cdot 4^3\cdot 4!} = \frac{1}{3072} < 0.0004$$

This is not good enough, so we try $n = 4$ instead:

$$\left| R_4\!\left(\tfrac{1}{4}\right) \right| < \frac{1}{2\cdot 4^4\cdot 5!} = \frac{1}{61,440} < 0.00002$$

Therefore we choose $n = 4$ and the approximation becomes

$$\sqrt[4]{e} \approx T_4\!\left(\tfrac{1}{4}\right) = 1 + \frac{1}{4} + \frac{1}{4^2\cdot 2!} + \frac{1}{4^3\cdot 3!} + \frac{1}{4^4\cdot 4!}$$

$$\approx 1.28402$$

EXAMPLE 4   What is the maximum possible error in using the approximation

$$\sin x \approx x - \frac{x^3}{3!} + \frac{x^5}{5!}$$

when $-0.3 \le x \le 0.3$? Use this approximation to find $\sin 12°$ correct to six decimal places.

Solution   If $f(x) = \sin x$, then $f^{(6)}(0) = -\sin 0 = 0$, so we can use Taylor's Formula with $n = 6$ to write

$$\sin x = x - \frac{x^3}{3!} + \frac{x^5}{5!} + R_6(x)$$

where
$$R_6(x) = \frac{f^{(7)}(z)}{7!}x^7 = -\cos z\frac{x^7}{7!}$$

If $-0.3 \le x \le 0.3$, then $|x| \le 0.3$ so

$$\left| R_6(x) \right| = |\cos z|\frac{|x|^7}{7!} \le \frac{|x|^7}{7!} \le \frac{(0.3)^7}{7!} < 0.00000005$$

This means that the maximum possible error is less than 0.00000005.
   To find $\sin 12°$ we first convert to radian measure.

$$\sin 12° = \sin\!\left(\frac{12\pi}{180}\right) = \sin\!\left(\frac{\pi}{15}\right)$$

$$\approx \frac{\pi}{15} - \left(\frac{\pi}{15}\right)^3\frac{1}{3!} + \left(\frac{\pi}{15}\right)^5\frac{1}{5!}$$

$$\approx 0.20791169$$

Thus, correct to six decimal places, $\sin 12° \approx 0.207912$.

If we had been asked to approximate $\sin 72°$ instead of $\sin 12°$ in Example 4, it would have been wise to use the Taylor polynomials at $c = \pi/3$ (instead of $c = 0$) because they are better approximations to $\sin x$ for values of x close to $\pi/3$. Notice that $72°$ is close to $60°$ (or $\pi/3$ radians) and the derivatives of $\sin x$ are easy to compute at $\pi/3$.

Figure 10.12 shows the graphs of the Taylor polynomial approximations

$$T_1(x) = x \qquad\qquad\qquad T_3(x) = x - \frac{x^3}{3!}$$

$$T_5(x) = x - \frac{x^3}{3!} + \frac{x^5}{5!} \qquad\qquad T_7(x) = x - \frac{x^3}{3!} + \frac{x^5}{5!} - \frac{x^7}{7!}$$

to the sine curve. You can see that as n increases, $T_n(x)$ is a good approximation to $\sin x$ on a larger and larger interval.

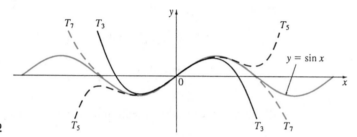

Figure 10.12

Figures 10.13 and 10.14 illustrate some of the Taylor approximations to the functions $y = \ln x$ and $y = \cos x$, $x \ge 0$.

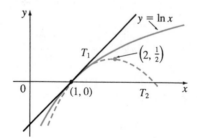

Figure 10.13

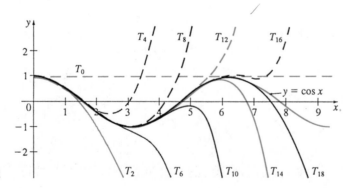

Figure 10.14

One use of the type of calculation done in Examples 1–4 occurs in calculators and computers. For instance, when you press the sin or e^x key on your calculator, or when a computer programmer uses a subroutine for a trigonometric or exponential or Bessel function, in many machines a polynomial approximation is calculated. The polynomial is often a Taylor polynomial that has been modified so that the error is spread more evenly throughout an interval.

Another use of Taylor's Formula occurs when we try to show that a function is equal to its Taylor series according to the following principle.

Theorem (10.56)

If f has derivatives of all orders and, in the notation of Taylor's Formula,

$$\lim_{n \to \infty} R_n(x) = \lim_{n \to \infty} \frac{f^{(n+1)}(z)}{(n+1)!}(x-c)^{n+1} = 0$$

for $|x - c| < R$, then f is equal to its Taylor series on the interval $|x - c| < R$; that is, f is analytic at c.

Proof If T_n is the nth Taylor polynomial of f at c, then $f(x) = T_n(x) + R_n(x)$, so

$$\lim_{n \to \infty} T_n(x) = \lim_{n \to \infty} [f(x) - R_n(x)] = f(x) - \lim_{n \to \infty} R_n(x) = f(x)$$

But $T_n(x)$ is just the nth partial sum of the Taylor series. Therefore f is the sum of its Taylor series. ●

In applying Theorem 10.56 it is often helpful to make use of the following fact:

(10.57)

$$\lim_{n \to \infty} \frac{x^n}{n!} = 0 \qquad \text{for every real number } x$$

This is true because we know from Example 3 in Section 10.9 that the series $\sum x^n/n!$ converges for all x and so its nth term approaches 0.

EXAMPLE 5 Prove that $\cos x = \sum_{n=0}^{\infty} (-1)^n \frac{x^{2n}}{(2n)!}$ without assuming that $\cos x$ has a power series expansion.

Solution Using the remainder term with $c = 0$, we have

$$R_n(x) = \frac{f^{(n+1)}(z)}{(n+1)!} x^{n+1}$$

where $f(x) = \cos x$ and z lies between 0 and x. But $f^{(n+1)}(z)$ is $\pm\sin z$ or $\pm\cos z$. In any case $|f^{(n+1)}(z)| \le 1$ and so

$$|R_n(x)| = \frac{|f^{(n+1)}(z)|}{(n+1)!} |x^{n+1}| \le \frac{|x|^{n+1}}{(n+1)!}$$

By Equation 10.57 the right side of this inequality approaches 0 as $n \to \infty$, so $|R_n(x)| \to 0$ by the Squeeze Theorem. It follows that $R_n(x) \to 0$ as $n \to \infty$, so $\cos x$ is equal to its Taylor series (which we found in Example 5 in Section 10.9) by Theorem 10.56.   ●

EXAMPLE 6   Prove that e^x is equal to its Taylor series.

Solution   If $f(x) = e^x$, then $f^{(n+1)}(x) = e^x$, so the remainder term in Taylor's Formula is

$$R_n(x) = \frac{e^z}{(n+1)!} x^{n+1}$$

where z lies between 0 and x. (Note, however, that z depends on n.) If $x > 0$, then $0 < z < x$, so $e^z < e^x$. Therefore

$$0 < R_n(x) = \frac{e^z}{(n+1)!} x^{n+1} < e^x \frac{x^{n+1}}{(n+1)!} \to 0$$

by Equation 10.57, so $R_n(x) \to 0$ as $n \to \infty$ by the Squeeze Theorem. If $x < 0$, then $x < z < 0$, so $e^z < e^0 = 1$ and

$$|R_n(x)| < \frac{|x|^{n+1}}{(n+1)!} \to 0$$

Again $R_n(x) \to 0$. Thus, by Theorem 10.56, e^x is equal to its Taylor series, that is,

$$e^x = \sum_{n=0}^{\infty} \frac{x^n}{n!} \qquad \text{for all } x$$

●

EXAMPLE 7   Show that $e^x > 1 + x$ for $x > 0$.

Solution   Although we were able to prove this inequality earlier (see Exercise 53(a) in Section 4.3), we now have a much simpler method. In Example 6 we proved that

$$e^x = 1 + x + \frac{x^2}{2!} + \frac{x^3}{3!} + \cdots$$

for all x. For $x > 0$, all terms on the right side are positive. It follows that for $x > 0$,

$$e^x = 1 + x + \frac{x^2}{2!} + \frac{x^3}{3!} + \cdots > 1 + x$$

In fact, it follows just as easily, that

$$e^x > 1 + x + \frac{x^2}{2!} + \cdots + \frac{x^n}{n!}$$

for $x > 0$ and $n > 0$. (Compare this with Exercise 53(c) in Section 4.3).   ●

Proof of Taylor's Formula   Let $R_n(x) = f(x) - T_n(x)$, where T_n is the nth-degree Taylor polynomial of f at c. The idea for the proof is the same as that for the Mean Value Theorem: We apply Rolle's Theorem to a specially constructed function. We think of x as a constant, $x \neq c$, and we define a function g on I by

$$g(t) = f(x) - f(t) - f'(t)(x - t) - \frac{f''(t)}{2!}(x - t)^2 - \cdots$$

$$- \frac{f^{(n)}(t)}{n!}(x - t)^n - R_n(x)\frac{(x - t)^{n+1}}{(x - c)^{n+1}}$$

Then
$$g(x) = f(x) - f(x) - 0 - \cdots - 0 = 0$$

$$g(c) = f(x) - [T_n(x) + R_n(x)] = f(x) - f(x) = 0$$

Thus by Rolle's Theorem (applied to g on the interval from c to x), there is a number z between x and c such that $g'(z) = 0$. If we differentiate the expression for g, then most terms will cancel. We leave it to you to verify that the expression for $g'(t)$ simplifies to

$$g'(t) = -\frac{f^{(n+1)}(t)}{n!}(x-t)^n + (n+1)R_n(x)\frac{(x-t)^n}{(x-c)^{n+1}}$$

Thus we have

$$g'(z) = -\frac{f^{(n+1)}(z)}{n!}(x-z)^n + (n+1)R_n(x)\frac{(x-z)^n}{(x-c)^{n+1}} = 0$$

and so
$$R_n(x) = \frac{f^{(n+1)}(z)}{(n+1)!}(x-c)^{n+1}$$

Therefore $f(x) = T_n(x) + R_n(x)$, where $R_n(x)$ has the form stated in Taylor's Formula.

There are many applications of Taylor's Formula. In this section we have seen that it is useful in approximating functions and in showing that a function is equal to its Taylor series. Another use is explored in Exercise 51, which shows how quickly the approximations in Newton's method converge to a root of an equation.

SECTION 10.11 **Exercises**

In Exercises 1–12 find the Taylor polynomial $T_n(x)$ for the given function f at the number c.

1. $f(x) = 1 + 2x + 3x^2 + 4x^3$, $c = -1$, $n = 4$

2. $f(x) = x^3 - 1$, $c = 1$, $n = 3$

3. $f(x) = \sin x$, $c = \pi/6$, $n = 3$

4. $f(x) = \cos x$, $c = 2\pi/3$, $n = 4$

5. $f(x) = \tan x$, $c = 0$, $n = 4$

6. $f(x) = \tan x$, $c = \pi/4$, $n = 4$

7. $f(x) = e^x \sin x$, $c = 0$, $n = 3$

8. $f(x) = e^x \cos 2x$, $c = 0$, $n = 3$

9. $f(x) = \sqrt{x}$, $c = 9$, $n = 3$

10. $f(x) = 1/\sqrt[3]{x}$, $c = 8$, $n = 3$

11. $f(x) = \ln \sin x$, $c = \pi/2$, $n = 3$

12. $f(x) = \sec x$, $c = \pi/3$, $n = 3$

In Exercises 13 and 14 find the Taylor polynomials $T_n(x)$ at c for the given function. Then sketch the graphs of f and these approximating polynomials.

13. $f(x) = \cos x$, $c = 0$, $n = 1, 2, 3, 4$

14. $f(x) = 1/x$, $c = 1$, $n = 1, 2, 3$

In Exercises 15–26, (a) write Taylor's Formula (with remainder term) for the function f at c with the given value of n, and (b) use the remainder term to estimate the accuracy of the approximation $f(x) \approx T_n(x)$ when x lies in the given interval.

15. $f(x) = \sqrt{1+x}$, $c = 0$, $n = 1$, $0 \le x \le 0.1$

16. $f(x) = 1/x$, $c = 1$, $n = 3$, $0.8 \le x \le 1.2$

17. $f(x) = \sin x$, $c = \pi/4$, $n = 5$, $0 \le x \le \pi/2$

18. $f(x) = \cos x$, $c = \pi/3$, $n = 4$, $0 \le x \le 2\pi/3$

19. $f(x) = 1/(1+2x)^4$, $c = 0$, $n = 3$, $|x| \le 0.1$

20. $f(x) = \sqrt[3]{1+x^2}$, $c = 0$, $n = 2$, $|x| \le 0.5$

21. $f(x) = \tan x$, $c = 0$, $n = 3$, $0 \le x \le \pi/6$

22. $f(x) = \ln \cos x$, $c = 0$, $n = 3$, $0 \le x \le \pi/4$

23. $f(x) = e^{x^2}$, $c = 0$, $n = 3$, $0 \le x \le 0.1$

24. $f(x) = \cosh x$, $c = 0$, $n = 5$, $|x| \le 1$

25. $f(x) = x^{3/4}$, $c = 16$, $n = 3$, $15 \le x \le 17$

26. $f(x) = \ln x$, $c = 4$, $n = 3$, $3 \le x \le 5$

In Exercises 27–38 use Taylor's Formula to estimate the given number with the indicated accuracy.

27. $e^{0.1}$,   error < 0.00001

28. $\sqrt[3]{e}$,   to four decimal places

29. $\sqrt[5]{1.1}$,   to four decimal places

30. $\sqrt{4.08}$,   to four decimal places

31. $\ln 1.4$,   error < 0.001

32. $1/1.09$,   error < 0.0001

33. $\sin(0.5)$,   error < 0.0001

34. $\cos(0.4)$,   error < 0.0001

35. $\cos 10°$,   to five decimal places

36. $\sin 15°$,   to five decimal places

37. $\sin 35°$,   to five decimal places

38. $\cos 69°$,   to five decimal places

In Exercise 39 and 40 use the remainder term in Taylor's Formula to estimate the range of values of x for which the given approximation is accurate to within the stated error.

39. $\sin x \approx x - \dfrac{x^3}{6}$,   error < 0.01

40. $\cos x \approx 1 - \dfrac{x^2}{2} + \dfrac{x^4}{24}$,   error < 0.005

In Exercises 41–44 prove that the given function f has the stated Taylor series representation without assuming that f possesses a power series expansion.

41. $\sin x = \displaystyle\sum_{n=0}^{\infty} \dfrac{(-1)^n x^{2n+1}}{(2n+1)!}$

42. $\sin x = \displaystyle\sum_{n=0}^{\infty} \dfrac{(-1)^{n(n-1)/2}}{\sqrt{2}} \dfrac{\left(x - \frac{\pi}{4}\right)^n}{n!}$

43. $\sinh x = \displaystyle\sum_{n=0}^{\infty} \dfrac{x^{2n+1}}{(2n+1)!}$ 44. $\cosh x = \displaystyle\sum_{n=0}^{\infty} \dfrac{x^{2n}}{(2n)!}$

45.   Show that $(1+x)^n > 1 + nx$ for all $x > 0$ and $n > 1$.

46.   Show that $\cosh x \geq 1 + \frac{1}{2}x^2$ for all x.

47.   The limit

$$\lim_{x \to 0} \frac{\sin x - x + \frac{1}{6}x^3}{x^5}$$

could be evaluated using l'Hospital's Rule. Instead, use a series to evaluate it.

48.   Use the result of Example 7 in Section 10.9 to evaluate the limit

$$\lim_{x \to 1} \frac{\ln x}{x - 1}$$

49.   Show that $y = T_1(x)$ is the equation of the tangent to $y = f(x)$ at $(c, f(c))$.

50.   Show that T_n and f have the same derivatives at c up to order n.

51.   In Section 2.10 we considered Newton's method for approximating a root r of the equation $f(x) = 0$, and from an initial approximation x_0 we obtained successive approximations x_1, x_2, ..., where

$$x_{n+1} = x_n - \frac{f(x_n)}{f'(x_n)}$$

Use Taylor's Formula with $n = 1$, $c = x_n$, and $x = r$ to show that if $f''(x)$ exists on an interval I containing r, x_n, and x_{n+1}, and $|f''(x)| \leq M$, $|f'(x)| \geq K$ for all $x \in I$, then

$$|x_{n+1} - r| \leq \frac{M}{2K}|x_n - r|^2$$

[This means that if x_n is accurate to d decimal places, then x_{n+1} is accurate to about $2d$ decimal places. More precisely, if the error at stage n is at most 10^{-m}, then the error at stage $n + 1$ is at most $(M/2K)10^{-2m}$.]

52.   If f has derivatives of all orders on an interval $I = (c - R, c + R)$ and these derivatives have a common bound M [$|f^{(n)}(x)| \leq M$ for all x in I and all $n = 1, 2, 3, ...$], prove that f is analytic at c.

53.   Use the following outline to prove that e is an irrational number.
(a)   If e were rational, it would be of the form $e = p/q$, where p and q are positive integers and $q > 2$. Use Taylor's Formula to write

$$\frac{p}{q} = e = 1 + \frac{1}{1!} + \frac{1}{2!} + \cdots + \frac{1}{q!} + \frac{e^z}{(q+1)!}$$

$$= s_q + \frac{e^z}{(q+1)!}$$

where $0 < z < 1$.
(b)   Show that $q!(e - s_q)$ is an integer.
(c)   Show that $0 < q!(e - s_q) < 1$.
(d)   Use part (b) and (c) to deduce that e is irrational.

CHAPTER 10

Review

Key Topics

Define, state, or discuss the following:

1. Sequence
2. Limit of a sequence
3. Convergent sequence; divergent sequence
4. Increasing, decreasing, and monotonic sequences
5. Bounded sequence
6. Completeness Axiom
7. Convergence of bounded, monotonic sequences
8. Series
9. Partial sums
10. Convergent series; divergent series
11. Sum of a series
12. Geometric series
13. Harmonic series
14. Test for Divergence
15. Integral Test
16. Convergence of a p-series
17. Comparison Test
18. Limit Comparison Test
19. Alternating Series Test
20. Absolute convergence
21. Conditional convergence
22. Relation between convergence and absolute convergence
23. Ratio Test
24. Root Test
25. Power series
26. Radius of convergence
27. Interval of convergence
28. Differentiation and integration of power series
29. Taylor series
30. Maclaurin series
31. Maclaurin series for e^x, $\sin x$, $\cos x$
32. Binomial series
33. Taylor polynomial
34. Taylor's Formula
35. Convergence of Taylor series

Exercises

In Exercises 1–14 determine whether the statement is true or false.

1. If $\lim_{n\to\infty} a_n = 0$, then $\sum a_n$ is convergent.

2. If $\sum a_n 6^n$ is convergent, then $\sum a_n(-2)^n$ is convergent.

3. If $\sum a_n 6^n$ is convergent, then $\sum a_n(-6)^n$ is convergent.

4. If $\sum a_n x^n$ diverges when $x = 6$, then it diverges when $x = 10$.

5. The Ratio Test can be used to determine whether $\sum 1/n^3$ converges.

6. The Ratio Test can be used to determine whether $\sum 1/n!$ converges.

7. If $0 \le a_n \le b_n$ and $\sum b_n$ diverges, then $\sum a_n$ diverges.

8. $\sum_{n=0}^{\infty} \dfrac{(-1)^n}{n!} = \dfrac{1}{e}$

9. $1^x + 2^x + 3^x + \cdots$ is a power series.

10. If f has infinitely many derivatives on $(-\infty, \infty)$, then $f(x) = \sum_{n=0}^{\infty} \dfrac{f^{(n)}(0)}{n!} x^n$ for all x.

11. If $-1 < \alpha < 1$, then $\lim_{n\to\infty} \alpha^n = 0$.

12. If $\sum a_n$ is divergent, then $\sum |a_n|$ is divergent.

13. If $f(x) = 2x - x^2 + \frac{1}{3}x^3 - \cdots$ converges for all x, then $f'''(0) = 2$.

14. If $\{a_n\}$ and $\{b_n\}$ are divergent, then $\{a_n + b_n\}$ is divergent.

In Exercises 15–22 determine whether the sequence is convergent or divergent. If it is convergent, find the limit.

15. $a_n = \dfrac{n}{2n+5}$

16. $a_n = 5 - (0.9)^n$

17. $a_n = 2n + 5$

18. $a_n = n/\ln n$

19. $a_n = \sin n$

20. $a_n = (\sin n)/n$

21. $\{(1 + 3/n)^{4n}\}$

22. $\{(-10)^n/n!\}$

In Exercises 23–34 determine whether the series is convergent or divergent.

23. $\sum_{n=1}^{\infty} \dfrac{n^2}{n^3+1}$

24. $\sum_{n=1}^{\infty} \dfrac{n+n^2}{n+n^4}$

25. $\sum_{n=1}^{\infty} \dfrac{(-1)^n}{\sqrt[4]{n}}$

26. $\sum_{n=1}^{\infty} \dfrac{n^2}{3^n}$

27. $\displaystyle\sum_{n=1}^{\infty} \left(\frac{n}{3n+1}\right)^n$

28. $\displaystyle\sum_{n=1}^{\infty} \sqrt{\frac{n-1}{n}}$

29. $\displaystyle\sum_{n=1}^{\infty} \frac{\sin n}{1+n^2}$

30. $\displaystyle\sum_{n=2}^{\infty} \frac{1}{n(\ln n)^2}$

31. $\displaystyle\sum_{n=1}^{\infty} \frac{1\cdot 3\cdot 5\cdots(2n-1)}{5^n n!}$

32. $\displaystyle\sum_{n=1}^{\infty} (-1)^{n+1}\frac{\ln n}{\sqrt{n}}$

33. $\displaystyle\sum_{n=1}^{\infty} \frac{4^n}{n3^n}$

34. $\displaystyle\sum_{n=1}^{\infty} \frac{\sqrt{n+1}-\sqrt{n-1}}{n}$

In Exercises 35–38 determine whether the series is conditionally convergent, absolutely convergent, or divergent.

35. $\displaystyle\sum_{n=1}^{\infty} (-1)^{n-1}n^{-1/3}$

36. $\displaystyle\sum_{n=1}^{\infty} (-1)^{n-1}n^{-3}$

37. $\displaystyle\sum_{n=1}^{\infty} \frac{(-1)^n(n+1)3^n}{2^{2n+1}}$

38. $\displaystyle\sum_{n=1}^{\infty} \frac{(-1)^n\sqrt{n}}{\ln n}$

In Exercises 39–42 find the sum of the series.

39. $\displaystyle\sum_{n=1}^{\infty} \frac{2^{2n+1}}{5^n}$

40. $\displaystyle\sum_{n=1}^{\infty} \frac{1}{n(n+3)}$

41. $\displaystyle\sum_{n=1}^{\infty} [\tan^{-1}(n+1)-\tan^{-1}n]$

42. $\displaystyle\sum_{n=0}^{\infty} \frac{(-1)^n x^n}{2^{2n}n!}$

43. Express the repeating decimal $1.2345345345\ldots$ as a fraction.

44. For what values of x does the series $\displaystyle\sum_{n=1}^{\infty} (\ln x)^n$ converge?

45. Find the sum of the series $\displaystyle\sum_{n=1}^{\infty} \frac{(-1)^{n+1}}{n^5}$ correct to four decimal places.

46. (a) Show that the series $\displaystyle\sum_{n=1}^{\infty} \frac{n^n}{(2n)!}$ is convergent.

   (b) Deduce that $\displaystyle\lim_{n\to\infty} \frac{n^n}{(2n)!} = 0$.

47. Prove that if the series $\displaystyle\sum_{n=1}^{\infty} a_n$ is absolutely convergent, then the series

$$\sum_{n=1}^{\infty} \left(\frac{n+1}{n}\right) a_n$$

   is also absolutely convergent.

In Exercises 48–51 find the radius of convergence and interval of convergence of the given series.

48. $\displaystyle\sum_{n=0}^{\infty} \frac{(-3)^n x^{2n}}{n+1}$

49. $\displaystyle\sum_{n=1}^{\infty} \frac{x^n}{3^n n^3}$

50. $\displaystyle\sum_{n=1}^{\infty} \frac{(x+1)^n}{n^n}$

51. $\displaystyle\sum_{n=0}^{\infty} \frac{2^n(x-3)^n}{\sqrt{n+3}}$

52. Find the radius of convergence of the series

$$\sum_{n=1}^{\infty} \frac{(2n)!}{(n!)^2} x^n$$

53. Find the Taylor series of $f(x)=\sin x$ at $c=\pi/6$.

54. Find the Taylor series of $f(x)=\cos x$ at $c=\pi/3$.

In Exercises 55–62 find the Maclaurin series for f and its radius of convergence. You may use either the direct method (definition of a Maclaurin series) or known series such as geometric series, binomial series, or the Maclaurin series for e^x and $\sin x$.

55. $f(x)=\dfrac{x^2}{1+x}$

56. $f(x)=\sqrt{1-x^2}$

57. $f(x)=\ln(1-x)$

58. $f(x)=xe^{2x}$

59. $f(x)=\sin(x^4)$

60. $f(x)=10^x$

61. $f(x)=1/\sqrt[4]{16-x}$

62. $f(x)=(1-3x)^{-5}$

63. Evaluate $\displaystyle\int \frac{e^x}{x}\,dx$ as an infinite series.

64. Use series to approximate $\displaystyle\int_0^1 \sqrt{1-x^4}\,dx$ correct to three decimal places.

In Exercises 65 and 66 use Taylor's Formula to estimate the given number with an error less than 0.0001.

65. $1/\sqrt[4]{e}$

66. $\ln(0.95)$

In Exercises 67 and 68 write out Taylor's Formula (with remainder term) for f at c with the given value of n. Then use the remainder term to estimate the accuracy of the approximation $f(x)\approx T_n(x)$ when x lies in the given interval.

67. $f(x)=\sqrt{x}$, $\quad c=1$, $\quad n=3$, $\quad 0.9\le x\le 1.1$

68. $f(x)=\sec x$, $\quad c=0$, $\quad n=2$, $\quad 0\le x\le\dfrac{\pi}{6}$

69. Use series to evaluate the limit $\displaystyle\lim_{x\to\infty} x^2\left(1-e^{-1/x^2}\right)$.

70. Suppose that $f(x)=\sum_{n=0}^{\infty} a_n x^n$ for all x.
   (a) If f is an odd function, show that
$$a_0=a_2=a_4=\cdots=0$$
   (b) If f is an even function, show that
$$a_1=a_3=a_5=\cdots=0$$

71. If $f(x)=e^{x^2}$, show that $f^{(2n)}(0)=\dfrac{(2n)!}{n!}$.

72. If $f(x)=\sum_{m=0}^{\infty} a_m x^m$ has positive radius of convergence and $e^{f(x)}=\sum_{n=0}^{\infty} b_n x^n$, show that
$$nb_n=\sum_{i=1}^{n} i a_i b_{n-i} \qquad n\ge 1$$

Problems Plus

1. If $f(x) = \sin(x^3)$, find $f^{(15)}(0)$.

2. A function f is defined by

$$f(x) = \lim_{n \to \infty} \frac{x^{2n} - 1}{x^{2n} + 1}$$

Where is f continuous?

3. To construct the **snowflake curve** start with an equilateral triangle with side of length 1. Step 1 in the construction is to divide each side into three equal parts, construct an equilateral triangle on the middle part, and then delete the middle part (see the figure). Step 2 is to repeat step 1 for each side of the resulting polygon. This process is repeated at each succeeding step. The snowflake curve is the curve that results from repeating this process indefinitely.

   (a) Let s_n, l_n, and p_n represent the number of sides, the length of a side, and the total length of the nth approximating curve (the curve obtained after step n of the construction), respectively. Find formulas for s_n, l_n, and p_n.

   (b) Show that $p_n \to \infty$ as $n \to \infty$.

   (c) Sum an infinite series to find the area enclosed by the snowflake curve. Parts (b) and (c) show that the snowflake curve is infinitely long but encloses only a finite area.

4. (a) Find the highest and lowest points on the curve $x^4 + y^4 = x^2 + y^2$.

   (b) Sketch the curve. (Notice that it is symmetric with respect to both axes and both of the lines $y = \pm x$, so it suffices to consider $y \geq x \geq 0$ initially.)

   (c) Find the area enclosed by the curve. (You may want to use polar coordinates for this part.)

5. (a) Show that $\tan \frac{1}{2}x = \cot \frac{1}{2}x - 2\cot x$.

   (b) Find the sum of the series

$$\sum_{n=1}^{\infty} \frac{1}{2^n} \tan \frac{x}{2^n}$$

6. A curve is defined by the parametric equations

$$x = \int_1^t \frac{\cos u}{u}\, du \qquad y = \int_1^t \frac{\sin u}{u}\, du$$

Find the length of the arc of the curve from the origin to the nearest point where there is a vertical tangent line.

7. If $a_1 = \cos\theta$, $-\pi/2 \leq \theta \leq \pi/2$, $b_1 = 1$, and

$$a_{n+1} = \tfrac{1}{2}(a_n + b_n) \qquad b_{n+1} = \sqrt{b_n a_{n+1}}$$

show that

$$\lim_{n \to \infty} a_n = \lim_{n \to \infty} b_n = \frac{\sin\theta}{\theta}$$

8. Find the sum of the series $\displaystyle\sum_{n=0}^{\infty} \frac{(x+2)^n}{(n+3)!}$.

9. Let

$$u = 1 + \frac{x^3}{3!} + \frac{x^6}{6!} + \frac{x^9}{9!} + \cdots$$

$$v = x + \frac{x^4}{4!} + \frac{x^7}{7!} + \frac{x^{10}}{10!} + \cdots$$

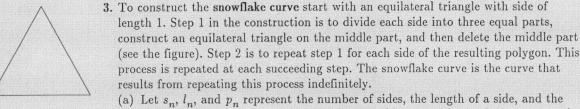

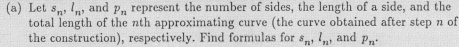

627

$$w = \frac{x^2}{2!} + \frac{x^5}{5!} + \frac{x^8}{8!} + \cdots$$

Show that $u^3 + v^3 + w^3 - 3uvw = 1$.

10. A curve called the **folium of Descartes** is defined by the parametric equations

$$x = \frac{3t}{1+t^3} \qquad y = \frac{3t^2}{1+t^3}$$

(a) Show that if (a, b) lies on the curve, then so does (b, a); that is, the curve is symmetric with respect to the line $y = x$. Where does the curve intersect this line?
(b) Find the points on the curve where the tangent lines are horizontal or vertical.
(c) Show that the Cartesian equation of this curve is $x^3 + y^3 = 3xy$.
(d) Show that the line $y = -x - 1$ is a slant asymptote.
(e) Sketch the curve.
(f) Find the area enclosed by the loop of this curve.

11. Find the interval of convergence of $\sum_{n=1}^{\infty} n^3 x^n$ and find its sum.

12. Let $\{P_n\}$ be a sequence of points determined as in the figure. Thus $|AP_1| = 1$, $|P_n P_{n+1}| = 2^{n-1}$, and angle $AP_n P_{n+1}$ is a right angle. Find $\lim_{n \to \infty} \angle P_n A P_{n+1}$.

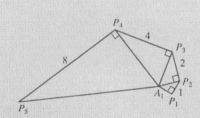

13. (a) Show that, for $xy \neq 1$,

$$\arctan x - \arctan y = \arctan \frac{x-y}{1+xy}$$

if the left-hand side lies between $-\pi/2$ and $\pi/2$.
(b) Show that

$$\arctan \frac{120}{119} - \arctan \frac{1}{239} = \frac{\pi}{4}$$

(c) Deduce the following formula of John Machin (1680–1751):

$$4 \arctan \frac{1}{5} - \arctan \frac{1}{239} = \frac{\pi}{4}$$

(d) Use the Maclaurin series for arctan to show that

$$0.197395560 < \arctan \frac{1}{5} < 0.197395562$$

(e) Show that

$$0.004184075 < \arctan \frac{1}{239} < 0.004184077$$

(f) Deduce that, correct to seven decimal places,

$$\pi \approx 3.1415927$$

Machin used this method in 1706 to find π correct to 100 decimal places. In this century, with the aid of computers, the value of π has been computed to increasingly greater accuracy. In 1989 Gregory and David Chudnovsky of Columbia University used supercomputers to find the value of π correct to more than a billion decimal places!

14. (a) Prove a formula similar to the one in Problem 13(a) but involving arccot instead of arctan.
   (b) Find the sum of the series

$$\sum_{n=0}^{\infty} \operatorname{arccot}(n^2 + n + 1)$$

15. If $0 < a \le b \le c$, show that

$$\lim_{n \to \infty} (a^n + b^n + c^n)^{1/n} = c$$

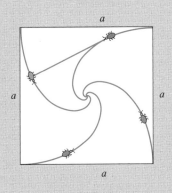

16. Four bugs are placed at the four corners of a square with side length a. The bugs crawl counterclockwise at the same speed and each bug crawls directly toward the next bug at all times. They approach the center of the square along spiral paths.
   (a) Find the polar equation of a bug's path assuming the pole is at the center of the square. (Use the fact that the line joining one bug to the next will be tangent to the bug's path.)
   (b) Find the distance traveled by a bug by the time it meets the other bugs at the center.

17. If the value of x^x at $x = 0$ is taken to be 1, show that

$$\int_0^1 x^x \, dx = \sum_{n=1}^{\infty} \frac{(-1)^n}{n^n}$$

18. For which numbers c is it true that $\cosh x \le e^{cx^2}$ for all x?

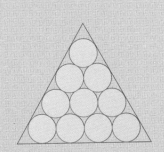

19. Suppose that circles of equal diameter are tightly packed in n rows inside an equilateral triangle. (The figure illustrates the case $n = 4$.) If A is the area of the triangle and A_n is the total area occupied by the n rows of circles, show that

$$\lim_{n \to \infty} \frac{A_n}{A} = \frac{\pi}{2\sqrt{3}}$$

20. If $a_0 + a_1 + a_2 + \cdots + a_k = 0$, show that

$$\lim_{n \to \infty} \left(a_0 \sqrt{n} + a_1 \sqrt{n+1} + a_2 \sqrt{n+2} + \cdots + a_k \sqrt{n+k} \right) = 0$$

If you don't see how to prove this, try the problem-solving strategy of analogy (see the front endpapers of this book). Try the special cases $k = 1$ and $k = 2$ first. If you can see how to prove the assertion for these cases, then you will probably see how to prove it in general.

21. Let $f(x) = x \sin(\pi/x)$, $-1 \le x \le 1$, $x \ne 0$, and $f(0) = 0$.
   (a) Show that f is continuous on $(-1, 1)$.
   (b) Sketch the graph of f, showing the local maxima and minima.
   (c) Use your graph from part (b) to show that the length of the graph from $x = 1/n$ to $x = 1/(n-1)$ is greater than $2/(2n-1)$. Deduce that the graph of f has infinite length. (This shows that it is possible for a continuous function on a finite interval to have a graph that is infinitely long.)

Appendixes

A

Review of Algebra

Arithmetic Operations

The real numbers have the following properties:

$$a + b = b + a \qquad ab = ba \qquad \text{(Commutative Law)}$$

$$(a + b) + c = a + (b + c) \qquad (ab)c = a(bc) \qquad \text{(Associative Law)}$$

$$a(b + c) = ab + ac \qquad \text{(Distributive Law)}$$

In particular, putting $a = -1$ in the Distributive Law, we get

$$-(b + c) = (-1)(b + c) = (-1)b + (-1)c$$

and so

$$-(b + c) = -b - c$$

EXAMPLE 1

(a) $(3xy)(-4x) = 3(-4)x^2y = -12x^2y$

(b) $2t(7x + 2tx - 11) = 14tx + 4t^2x - 22t$

(c) $4 - 3(x - 2) = 4 - 3x + 6 = 10 - 3x$

If we use the Distributive Law three times, we get

$$(a+b)(c+d) = (a+b)c + (a+b)d = ac + bc + ad + bd$$

This says that we multiply two factors by multiplying each term in one factor by each term in the other factor and adding the products. Schematically, we have

$$(a + b)(c + d)$$

In the case where $c = a$ and $d = b$, we have

$$(a+b)^2 = a^2 + ba + ab + b^2$$

(1) or

$$(a+b)^2 = a^2 + 2ab + b^2$$

Similarly, we obtain

(2)

$$(a-b)^2 = a^2 - 2ab + b^2$$

EXAMPLE 2

(a) $(2x+1)(3x-5) = 6x^2 + 3x - 10x - 5 = 6x^2 - 7x - 5$

(b) $(x+6)^2 = x^2 + 12x + 36$

(c) $3(x-1)(4x+3) - 2(x+6) = 3(4x^2 - x - 3) - 2x - 12$
$$= 12x^2 - 3x - 9 - 2x - 12$$
$$= 12x^2 - 5x - 21$$

Fractions

To add two fractions with the same denominator, we use the Distributive Law:

$$\frac{a}{b} + \frac{c}{b} = \frac{1}{b} \times a + \frac{1}{b} \times c = \frac{1}{b}(a+c) = \frac{a+c}{b}$$

Thus it is true that

$$\frac{a+c}{b} = \frac{a}{b} + \frac{c}{b}$$

But remember to avoid the following common error:

$$\frac{a}{b+c} \ne \frac{a}{b} + \frac{a}{c}$$

(For instance, take $a = b = c = 1$ to see the error.)

To add two fractions with different denominators, we use a common denominator:

$$\frac{a}{b} + \frac{c}{d} = \frac{ad + bc}{bd}$$

We multiply such fraction as follows:

$$\frac{a}{b} \cdot \frac{c}{d} = \frac{ac}{bd}$$

In particular, it is true that

$$\frac{-a}{b} = -\frac{a}{b} = \frac{a}{-b}$$

To divide two fractions, we invert and multiply:

$$\frac{\frac{a}{b}}{\frac{c}{d}} = \frac{a}{b} \times \frac{d}{c} = \frac{ad}{bc}$$

EXAMPLE 3

(a) $\quad \dfrac{x+3}{x} = \dfrac{x}{x} + \dfrac{3}{x} = 1 + \dfrac{3}{x}$

(b) $\quad \dfrac{3}{x-1} + \dfrac{x}{x+2} = \dfrac{3(x+2) + x(x-1)}{(x-1)(x+2)} = \dfrac{3x+6+x^2-x}{x^2+x-2}$

$$= \dfrac{x^2+2x+6}{x^2+x-2}$$

(c) $\quad \dfrac{s^2t}{u} \cdot \dfrac{ut}{-2} = \dfrac{s^2t^2u}{-2u} = -\dfrac{s^2t^2}{2}$

(d) $\quad \dfrac{\frac{x}{y}+1}{1-\frac{y}{x}} = \dfrac{\frac{x+y}{y}}{\frac{x-y}{x}} = \dfrac{x+y}{y} \times \dfrac{x}{x-y} = \dfrac{x(x+y)}{y(x-y)} = \dfrac{x^2+xy}{xy-y^2}$

Factoring

We have used the Distributive Law to expand certain algebraic expressions. We sometimes need to reverse this process (again using the Distributive Law) by factoring an expression as a product of simpler ones. The easiest situation occurs when the expression has a common factor as follows:

$$\longrightarrow \text{Expanding} \longrightarrow$$
$$3x(x-2) = 3x^2 - 6x$$
$$\longleftarrow \text{Factoring} \longleftarrow$$

To factor a quadratic of the form $x^2 + bx + c$ we note that

$$(x + r)(x + s) = x^2 + (r + s)x + rs$$

so we need to choose numbers r and s so that $r + s = b$ and $rs = c$.

EXAMPLE 4   Factor $x^2 + 5x - 24$.

Solution   The two integers that add to give 5 and multiply to give -24 are -3 and 8. Therefore

$$x^2 + 5x - 24 = (x - 3)(x + 8)$$ ●

EXAMPLE 5   Factor $2x^2 - 7x - 4$.

Solution   Even though the coefficient of x^2 is not 1, we can still look for factors of the form $2x + r$ and $x + s$, where $rs = -4$. Experimentation reveals that

$$2x^2 - 7x - 4 = (2x + 1)(x - 4)$$ ●

Some special quadratics can be factored by using Equations 1 or 2 (from right to left) or by using the formula for a difference of squares.

(3)
$$a^2 - b^2 = (a - b)(a + b)$$

The analogous formula for a difference of cubes is

(4)
$$a^3 - b^3 = (a - b)(a^2 + ab + b^2)$$

which you can verify by expanding the right side. For a sum of cubes we have

(5)
$$a^3 + b^3 = (a + b)(a^2 - ab + b^2)$$

EXAMPLE 6

(a) $x^2 - 6x + 9 = (x - 3)^2$ (Equation 2; $a = x$, $b = 3$)

(b) $4x^2 - 25 = (2x - 5)(2x + 5)$ (Equation 3; $a = 2x$, $b = 5$)

(c) $x^3 + 8 = (x + 2)(x^2 - 2x + 4)$ (Equation 5; $a = x$, $b = 2$)   ●

EXAMPLE 7   Simplify $\dfrac{x^2 - 16}{x^2 - 2x - 8}$.

Solution   Factoring numerator and denominator, we have

$$\frac{x^2 - 16}{x^2 - 2x - 8} = \frac{(x - 4)(x + 4)}{(x - 4)(x + 2)} = \frac{x + 4}{x + 2}$$ ●

To factor polynomials of degree 3 or more, we sometimes use the following fact.

The Factor Theorem (6)

If P is a polynomial and $P(b) = 0$, then $x - b$ is a factor of $P(x)$.

EXAMPLE 8   Factor $x^3 - 3x^2 - 10x + 24$.

Solution   Let $P(x) = x^3 - 3x^2 - 10x + 24$. If $P(b) = 0$, where b is an integer, then b is a factor of 24. Thus the possibilities for b are ± 1, ± 2, ± 3, ± 4, ± 6, ± 8, ± 12, ± 24. We find that $P(1) = 12$, $P(-1) = 30$, $P(2) = 0$. By the Factor Theorem, $x - 2$ is a factor. Instead of substituting further, we use long division as follows:

$$
\begin{array}{r}
x^2 \quad - \quad x \quad - \quad 12 \\
x - 2 \overline{\smash{\big)}\, x^3 \quad - \quad 3x^2 \quad - \quad 10x \quad + \quad 24} \\
\underline{x^3 \quad - \quad 2x^2} \\
- x^2 \quad - \quad 10x \\
\underline{- x^2 \quad + \quad 2x} \\
- 12x \quad + \quad 24 \\
\underline{- 12x \quad + \quad 24}
\end{array}
$$

Therefore

$$
\begin{aligned}
x^3 - 3x^2 - 10x + 24 &= (x - 2)(x^2 - x - 12) \\
&= (x - 2)(x + 3)(x - 4)
\end{aligned}
$$

Completing the Square

Completing the square is a useful techique for graphing parabolas (as in Example 2 in Section 6 in Review and Preview) or integrating rational functions (as in Example 6 in Section 7.4). Completing the square means rewriting a quadratic $ax^2 + bx + c$ in the form $a(x + p)^2 + q$ and can be accomplished by:

1.   Factoring the number a from the terms involving x.

2.   Adding and subtracting the square of half the coefficient of x.

In general, we have

$$
ax^2 + bx + c = a\left[x^2 + \frac{b}{a}x\right] + c
$$

$$
= a\left[x^2 + \frac{b}{a}x + \left(\frac{b}{2a}\right)^2 - \left(\frac{b}{2a}\right)^2\right] + c
$$

$$
= a\left(x + \frac{b}{2a}\right)^2 + \left(c - \frac{b^2}{4a}\right)
$$

EXAMPLE 9   Rewrite $x^2 + x + 1$ by completing the square.

Solution   The square of half the coefficient of x is $\frac{1}{4}$. Thus

$$
x^2 + x + 1 = x^2 + x + \tfrac{1}{4} - \tfrac{1}{4} + 1 = \left(x + \tfrac{1}{2}\right)^2 + \tfrac{3}{4}
$$

EXAMPLE 10

$$2x^2 - 12x + 11 = 2[x^2 - 6x] + 11 = 2[x^2 - 6x + 9 - 9] + 11$$
$$= 2[(x-3)^2 - 9] + 11 = 2(x-3)^2 - 7$$

Quadratic Formula

By completing the square as above we can obtain the following formula for the roots of a quadratic equation.

The Quadratic Formula (7)

The roots of the quadratic equation $ax^2 + bx + c = 0$ are

$$x = \frac{-b \pm \sqrt{b^2 - 4ac}}{2a}$$

EXAMPLE 11   Solve the equation $5x^2 + 3x - 3 = 0$.

Solution   With $a = 5$, $b = 3$, $c = -3$, the quadratic formula gives the solutions

$$x = \frac{-3 \pm \sqrt{3^2 - 4(5)(-3)}}{2(5)} = \frac{-3 \pm \sqrt{69}}{10}$$

The quantity $b^2 - 4ac$ that appears in the quadratic formula is called the **discriminant**. There are three possibilities:

1.   If $b^2 - 4ac > 0$, the equation has two real roots.

2.   If $b^2 - 4ac = 0$, the roots are equal.

3.   If $b^2 - 4ac < 0$, the equation has no real root. (The roots are complex.)

These three cases correspond to the fact that the number of times the parabola $y = ax^2 + bx + c$ crosses the x-axis is 2, 1, or 0 (see Figure 1). In case (3) the quadratic $ax^2 + bx + c$ cannot be factored and is called **irreducible**.

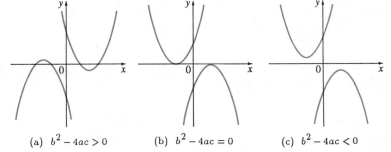

Figure 1

Possible graphs of
$y = ax^2 + bx + c$

(a) $b^2 - 4ac > 0$ (b) $b^2 - 4ac = 0$ (c) $b^2 - 4ac < 0$

EXAMPLE 12   The quadratic $x^2 + x + 2$ is irreducible because its discriminant is negative:

$$b^2 - 4ac = 1^2 - 4(1)(2) = -7 < 0$$

Therefore it is impossible to factor $x^2 + x + 2$.

The Binomial Theorem

Recall the binomial expansion from Equation 1:

$$(a+b)^2 = a^2 + 2ab + b^2$$

If we multiply both sides by $(a+b)$ and simplify, we get the binomial expansion

(8)

$$(a+b)^3 = a^3 + 3a^2b + 3ab^2 + b^3$$

Repeating this procedure, we get

$$(a+b)^4 = a^4 + 4a^3b + 6a^2b^2 + 4ab^3 + b^4$$

In general, we have the following formula (which is a special case of the Binomial Theorem of Section 10.10):

The Binomial Theorem (9)

If k is a positive integer, then

$$(a+b)^k = a^k + ka^{k-1}b + \frac{k(k-1)}{1\cdot 2}a^{k-2}b^2 + \frac{k(k-1)(k-2)}{1\cdot 2\cdot 3}a^{k-3}b^3$$
$$+ \cdots + \frac{k(k-1)\cdots(k-n+1)}{1\cdot 2\cdot 3\cdot\,\cdots\,\cdot n}a^{k-n}b^n$$
$$+ \cdots + kab^{k-1} + b^k$$

EXAMPLE 13   Expand $(x-2)^5$.

Solution   Using the Binomial Theorem with $a = x$, $b = -2$, $k = 5$, we have

$$(x-2)^5 = x^5 + 5x^4(-2) + \frac{5\cdot 4}{1\cdot 2}x^3(-2)^2 + \frac{5\cdot 4\cdot 3}{1\cdot 2\cdot 3}x^2(-2)^3 + 5x(-2)^4 + (-2)^5$$
$$= x^5 - 10x^4 + 40x^3 - 80x^2 + 80x - 32$$

Radicals

The most commonly occurring radicals are square roots. The symbol $\sqrt{}$ means "the positive square root of." Thus

$$x = \sqrt{a} \quad \text{means} \quad x^2 = a \quad \text{and} \quad x \geq 0$$

Since $a = x^2 \geq 0$, the symbol $\sqrt{a}$ makes sense only when $a \geq 0$. Here are two rules for working with square roots:

(10)

$$\sqrt{ab} = \sqrt{a}\sqrt{b} \qquad \sqrt{\frac{a}{b}} = \frac{\sqrt{a}}{\sqrt{b}}$$

However there is no similar rule for the square root of a sum. In fact, you should remember to avoid the following common error:

$$\sqrt{a + b} \neq \sqrt{a} + \sqrt{b}$$

(For instance, take $a = 9$ and $b = 16$ to see the error.)

EXAMPLE 14

(a) $\dfrac{\sqrt{18}}{\sqrt{2}} = \sqrt{\dfrac{18}{2}} = \sqrt{9} = 3$

(b) $\sqrt{x^2 y} = \sqrt{x^2}\sqrt{y} = |x|\sqrt{y}$

Notice that $\sqrt{x^2} = |x|$ because $\sqrt{}$ indicates the positive square root. (see Section 1 of Review and Preview.)

In general, if n is a positive integer,

$$x = \sqrt[n]{a} \quad \text{means} \quad x^n = a$$

If n is even, then $a \geq 0$ and $x \geq 0$.

Thus $\sqrt[3]{-8} = -2$ because $(-2)^3 = -8$, but $\sqrt[4]{-8}$ and $\sqrt[6]{-8}$ are not defined. The following rules are valid:

$$\sqrt[n]{ab} = \sqrt[n]{a}\sqrt[n]{b} \qquad \sqrt[n]{\dfrac{a}{b}} = \dfrac{\sqrt[n]{a}}{\sqrt[n]{b}}$$

EXAMPLE 15 $\quad \sqrt[3]{x^4} = \sqrt[3]{x^3 x} = \sqrt[3]{x^3}\,\sqrt[3]{x} = x\sqrt[3]{x}$

To **rationalize** a numerator or denominator that contains an expression such as $\sqrt{a} - \sqrt{b}$, we multiply both the numerator and the denominator by the conjugate radical $\sqrt{a} + \sqrt{b}$. Then we can take advantage of the formula for a difference of squares:

$$(\sqrt{a} - \sqrt{b})(\sqrt{a} + \sqrt{b}) = (\sqrt{a})^2 - (\sqrt{b})^2 = a - b$$

EXAMPLE 16 Rationalize the numerator in the expression $\dfrac{\sqrt{x+4} - 2}{x}$.

Solution We multiply the numerator and the denominator by the conjugate radical $\sqrt{x+4} + 2$:

$$\dfrac{\sqrt{x+4} - 2}{x} = \left(\dfrac{\sqrt{x+4} - 2}{x}\right)\left(\dfrac{\sqrt{x+4} + 2}{\sqrt{x+4} + 2}\right) = \dfrac{(x+4) - 4}{x(\sqrt{x+4} + 2)}$$

$$= \dfrac{x}{x(\sqrt{x+4} + 2)} = \dfrac{1}{\sqrt{x+4} + 2}$$

Exponents

Let a be any positive number and let n be a positive integer. Then, by definition,

1. $a^n = \underbrace{a \cdot a \cdot \cdots \cdot a}_{n \text{ factors}}$

2. $a^0 = 1$

3. $a^{-n} = \dfrac{1}{a^n}$

4. $a^{1/n} = \sqrt[n]{a}$

    $a^{m/n} = \sqrt[n]{a^m} = (\sqrt[n]{a})^m \qquad m \text{ is any integer}$

Laws of Exponents (11)

Let a and b be positive numbers and let r and s be any rational numbers (that is, ratios of integers). Then

1. $a^r \times a^s = a^{r+s}$ 2. $\dfrac{a^r}{a^s} = a^{r-s}$

3. $(a^r)^s = a^{rs}$ 4. $(ab)^r = a^r b^r$

5. $\left(\dfrac{a}{b}\right)^r = \dfrac{a^r}{b^r} \qquad b \neq 0$

In words, these five laws can be stated as follows:

1. To multiply two powers of the same number, we add the exponents.

2. To divide two powers of the same number, we subtract the exponents.

3. To raise a power to a new power, we multiply the exponents.

4. To raise a product to a power, we raise each factor to the power.

5. To raise a quotient to a power, we raise both numerator and denominator to the power.

EXAMPLE 17

(a) $\quad 2^8 \times 8^2 = 2^8 \times (2^3)^2 = 2^8 \times 2^6 = 2^{14}$

(b) $\quad \dfrac{x^{-2} - y^{-2}}{x^{-1} + y^{-1}} = \dfrac{\dfrac{1}{x^2} - \dfrac{1}{y^2}}{\dfrac{1}{x} + \dfrac{1}{y}} = \dfrac{\dfrac{y^2 - x^2}{x^2 y^2}}{\dfrac{y + x}{xy}} = \dfrac{y^2 - x^2}{x^2 y^2} \cdot \dfrac{xy}{y + x}$

$\qquad\qquad = \dfrac{(y - x)(y + x)}{xy(y + x)} = \dfrac{y - x}{xy}$

(c) $\quad 4^{3/2} = \sqrt{4^3} = \sqrt{64} = 8 \qquad$ Alternative solution: $4^{3/2} = (\sqrt{4})^3 = 2^3 = 8$

(d) $\quad \dfrac{1}{\sqrt[3]{x^4}} = \dfrac{1}{x^{4/3}} = x^{-4/3}$

(e) $\quad \left(\dfrac{x}{y}\right)^3 \left(\dfrac{y^2 x}{z}\right)^4 = \dfrac{x^3}{y^3} \cdot \dfrac{y^8 x^4}{z^4} = x^7 y^5 z^{-4}$

Mathematical Induction

The principle of mathematical induction is useful when proving a statement S_n about the positive integer n. For instance, if S_n is the statement

$$(ab)^n = a^n b^n$$

then S_1 says that $ab = ab$

S_2 says that $(ab)^2 = a^2 b^2$

and so on.

**Principle of
Mathematical Induction**

Let S_n be a statement about the positive integer n. Suppose that

1. S_1 is true.

2. S_{k+1} is true whenever S_k is true.

Then S_n is true for all positive integers n.

This is reasonable because, since S_1 is true, it follows from condition 2 (with $k = 1$) that S_2 is true. Then, using condition 2 with $k = 2$, we see that S_3 is true. Again using condition 2, this time with $k = 3$, we have that S_4 is true. This procedure can be followed indefinitely.

In using the principle of mathematical induction, there are three steps.

Step 1:   Prove that S_n is true when $n = 1$.

Step 2:   Assume that S_n is true when $n = k$ and deduce that S_n is true when $n = k + 1$.

Step 3:   Conclude that S_n is true for all n by the principle of mathematical induction.

EXAMPLE 18   If a and b are real numbers, prove that $(ab)^n = a^n b^n$ for every positive integer n.

Solution   Let S_n be the given statement.

1. S_1 is true because $(ab)^1 = ab = a^1 b^1$.

2.   Assume that S_k is true, that is, $(ab)^k = a^k b^k$. Then

$$(ab)^{k+1} = (ab)^k (ab) = a^k b^k ab$$
$$= (a^k a)(b^k b) = a^{k+1} b^{k+1}$$

This says that S_{k+1} is true.

3.   Therefore, by the principle of mathematical induction, S_n is true for all n; that is, $(ab)^n = a^n b^n$ for every positive integer n.   ●

EXAMPLE 19   Prove that, for every positive integer n,

$$1 + 2 + 3 + \cdots + n = \frac{n(n+1)}{2}$$

Let S_n be the given statement.

1. S_1 is true because

$$1 = \frac{1(1+1)}{2}$$

2. Assume that S_k is true, that is,

$$1 + 2 + \cdots + k = \frac{k(k+1)}{2}$$

Then

$$1 + 2 + \cdots + (k+1) = (1 + 2 + \cdots + k) + (k+1)$$

$$= \frac{k(k+1)}{2} + k + 1$$

$$= \frac{k(k+1) + 2(k+1)}{2}$$

$$= \frac{(k+1)(k+2)}{2}$$

Thus $\qquad\qquad 1 + 2 + \cdots + (k+1) = \frac{(k+1)[(k+1)+1]}{2}$

which shows that S_{k+1} is true.

3. Therefore, S_n is true for all n by mathematical induction, that is,

$$1 + 2 + \cdots + (n+1) = \frac{n(n+1)}{2}$$

for every positive integer n.

APPENDIX A Exercises

In Exercises 1–16 expand and simplify.

1. $(-6ab)(0.5ac)$

2. $-(2x^2 y)(-xy^4)$

3. $2x(x-5)$

4. $(4-3x)x$

5. $-2(4-3a)$

6. $8 - (4+x)$

7. $4(x^2 - x + 2) - 5(x^2 - 2x + 1)$

8. $5(3t-4) - (t^2 + 2) - 2t(t-3)$

9. $(4x-1)(3x+7)$

10. $x(x-1)(x+2)$

11. $(2x-1)^2$

12. $(2+3x)^2$

13. $y^4(6-y)(5+y)$

14. $(t-5)^2 - 2(t+3)(8t-1)$

15. $(1+2x)(x^2 - 3x + 1)$

16. $(1+x-x^2)^2$

In Exercises 17–28 perform the indicated operations and simplify.

17. $\dfrac{2+8x}{2}$

18. $\dfrac{9b-6}{3b}$

19. $\dfrac{1}{x+5} + \dfrac{2}{x-3}$

20. $\dfrac{1}{x+1} + \dfrac{1}{x-1}$

21. $u + 1 + \dfrac{u}{u+1}$

22. $\dfrac{2}{a^2} - \dfrac{3}{ab} + \dfrac{4}{b^2}$

23. $\dfrac{x/y}{z}$

24. $\dfrac{x}{y/z}$

25. $\left(\dfrac{-2r}{s}\right)\left(\dfrac{s^2}{-6t}\right)$

26. $\dfrac{a}{bc} \div \dfrac{b}{ac}$

27. $\dfrac{1 + \dfrac{1}{c-1}}{1 - \dfrac{1}{c-1}}$

28. $1 + \dfrac{1}{1 + \dfrac{1}{1+x}}$

In Exercises 29–48 factor the given expression.

29. $2x + 12x^3$

30. $5ab - 8abc$

31. $x^2 + 7x + 6$

32. $x^2 - x - 6$

33. $x^2 - 2x - 8$

34. $2x^2 + 7x - 4$

35. $9x^2 - 36$

36. $8x^2 + 10x + 3$

37. $6x^2 - 5x - 6$

38. $x^2 + 10x + 25$

39. $t^3 + 1$

40. $4t^2 - 9s^2$

41. $4t^2 - 12t + 9$

42. $x^3 - 27$

43. $x^3 + 2x^2 + x$ 44. $x^3 - 4x^2 + 5x - 2$

45. $x^3 + 3x^2 - x - 3$ 46. $x^3 - 2x^2 - 23x + 60$

47. $x^3 + 5x^2 - 2x - 24$ 48. $x^3 - 3x^2 - 4x + 12$

In Exercises 49–54 simplify the given expression.

49. $\dfrac{x^2 + x - 2}{x^2 - 3x + 2}$ 50. $\dfrac{2x^2 - 3x - 2}{x^2 - 4}$

51. $\dfrac{x^2 - 1}{x^2 - 9x + 8}$ 52. $\dfrac{x^3 + 5x^2 + 6x}{x^2 - x - 12}$

53. $\dfrac{1}{x + 3} + \dfrac{1}{x^2 - 9}$ 54. $\dfrac{x}{x^2 + x - 2} - \dfrac{2}{x^2 - 5x + 4}$

In Exercises 55–60 complete the square.

55. $x^2 + 2x + 5$ 56. $x^2 - 16x + 80$

57. $x^2 - 5x + 10$ 58. $x^2 + 3x + 1$

59. $4x^2 + 4x - 2$ 60. $3x^2 - 24x + 50$

In Exercises 61–68 solve the given equation.

61. $x^2 + 9x - 10 = 0$ 62. $x^2 - 2x - 8 = 0$

63. $x^2 + 9x - 1 = 0$ 64. $x^2 - 2x - 7 = 0$

65. $3x^2 + 5x + 1 = 0$ 66. $2x^2 + 7x + 2 = 0$

67. $x^3 - 2x + 1 = 0$ 68. $x^3 + 3x^2 + x - 1 = 0$

Which of the quadratics in Exercises 69–72 are irreducible?

69. $2x^2 + 3x + 4$ 70. $2x^2 + 9x + 4$

71. $3x^2 + x - 6$ 72. $x^2 + 3x + 6$

In Exercises 73–76 use the Binomial Theorem to expand the given expressions.

73. $(a + b)^6$ 74. $(a + b)^7$

75. $(x^2 - 1)^4$ 76. $(3 + x^2)^5$

In Exercises 77–82 simplify the given radicals.

77. $\sqrt{32}\sqrt{2}$ 78. $\dfrac{\sqrt[3]{-2}}{\sqrt[3]{54}}$

79. $\dfrac{\sqrt[4]{32x^4}}{\sqrt[4]{2}}$ 80. $\sqrt{xy}\sqrt{x^3y}$

81. $\sqrt{16a^4b^3}$ 82. $\dfrac{\sqrt[5]{96a^6}}{\sqrt[5]{3a}}$

In Exercises 83–100 use the Laws of Exponents to rewrite and simplify the given expression.

83. $3^{10} \times 9^8$ 84. $2^{16} \times 4^{10} \times 16^6$

85. $\dfrac{x^9(2x)^4}{x^3}$ 86. $\dfrac{a^n \times a^{2n+1}}{a^{n-2}}$

87. $\dfrac{a^{-3}b^4}{a^{-5}b^5}$ 88. $\dfrac{x^{-1} + y^{-1}}{(x+y)^{-1}}$

89. $3^{-1/2}$ 90. $96^{1/5}$

91. $125^{2/3}$ 92. $64^{-4/3}$

93. $(2x^2y^4)^{3/2}$ 94. $(x^{-5}y^3z^{10})^{-3/5}$

95. $\sqrt[5]{y^6}$ 96. $(\sqrt[4]{a})^3$

97. $\dfrac{1}{(\sqrt{t})^5}$ 98. $\dfrac{\sqrt[8]{x^5}}{\sqrt[4]{x^3}}$

99. $\sqrt[4]{\dfrac{t^{1/2}\sqrt{st}}{s^{2/3}}}$ 100. $\sqrt[4]{r^{2n+1}} \times \sqrt[4]{r^{-1}}$

In Exercises 101–108 rationalize the given expression.

101. $\dfrac{\sqrt{x} - 3}{x - 9}$ 102. $\dfrac{(1/\sqrt{x}) - 1}{x - 1}$

103. $\dfrac{x\sqrt{x} - 8}{x - 4}$ 104. $\dfrac{\sqrt{2+h} + \sqrt{2-h}}{h}$

105. $\dfrac{2}{3 - \sqrt{5}}$ 106. $\dfrac{1}{\sqrt{x} - \sqrt{y}}$

107. $\sqrt{x^2 + 3x + 4} - x$ 108. $\sqrt{x^2 + x} - \sqrt{x^2 - x}$

In Exercises 109–116 state whether or not the given equation is true for all values of the variable.

109. $\sqrt{x^2} = x$ 110. $\sqrt{x^2 + 4} = |x| + 2$

111. $\dfrac{16 + a}{16} = 1 + \dfrac{a}{16}$ 112. $\dfrac{1}{x^{-1} + y^{-1}} = x + y$

113. $\dfrac{x}{x + y} = \dfrac{1}{1 + y}$ 114. $\dfrac{2}{4 + x} = \dfrac{1}{2} + \dfrac{2}{x}$

115. $(x^3)^4 = x^7$

116. $6 - 4(x + a) = 6 - 4x - 4a$

In Exercises 117–126 n represents a positive integer. Use mathematical induction to prove the given statement.

117. $2^n > n$ 118. $3^n > 2n$

119. $(1 + x)^n \geq 1 + nx$ (where $x \geq -1$)

120. If $0 \leq a < b$, then $a^n < b^n$.

121. $7^n - 1$ is divisible by 6.

122. $\left(\dfrac{a}{b}\right)^n = \dfrac{a^n}{b^n}$

123. $1 + 3 + 5 + \cdots + (2n - 1) = n^2$

124. $2 + 6 + 12 + \cdots + n(n+1) = \dfrac{n(n+1)(n+2)}{3}$

125. $\dfrac{1}{2} + \dfrac{1}{6} + \dfrac{1}{12} + \cdots + \dfrac{1}{n(n+1)} = \dfrac{n}{n+1}$

126. $a + ar + ar^2 + \cdots + ar^{n-1} = \dfrac{a(1 - r^n)}{1 - r}$ $(r \neq 1)$

B

Review of Trigonometry

Angles

Angles can be measured in degrees or radians (abbreviated as rad). The angle given by a complete revolution contains 360°, which is the same as 2π rad. Therefore

(1)

$$\boxed{\pi \text{ rad} = 180°}$$

and

(2)

$$1 \text{ rad} = \left(\frac{180}{\pi}\right)° \approx 57.3° \qquad 1° = \frac{\pi}{180} \text{ rad} \approx 0.017 \text{ rad}$$

EXAMPLE 1
(a) Find the radian measure of 60°.   (b) Express $5\pi/4$ rad in degrees.

Solution
(a) From Equation 1 or 2 we see that to convert from degrees to radians we multiply by $\pi/180$. Therefore

$$60° = 60\left(\frac{\pi}{180}\right) = \frac{\pi}{3} \text{ rad}$$

(b) To convert from radians to degrees we multiply by $180/\pi$. Thus

$$\frac{5\pi}{4} \text{ rad} = \frac{5\pi}{4}\left(\frac{180}{\pi}\right) = 225°$$

A14

In calculus we use radians to measure angles except when otherwise indicated. The following table gives the correspondence between degree and radian measures of some common angles.

Degrees	0°	30°	45°	60°	90°	120°	135°	150°	180°	270°	360°
Radians	0	$\dfrac{\pi}{6}$	$\dfrac{\pi}{4}$	$\dfrac{\pi}{3}$	$\dfrac{\pi}{2}$	$\dfrac{2\pi}{3}$	$\dfrac{3\pi}{4}$	$\dfrac{5\pi}{6}$	π	$\dfrac{3\pi}{2}$	2π

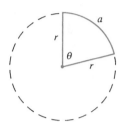

Figure 1

Figure 1 shows a sector of a circle with central angle θ and radius r subtending an arc with length a. Since the length of the arc is proportional to the size of the angle, and since the entire circle has circumference $2\pi r$ and central angle 2π, we have

$$\frac{\theta}{2\pi} = \frac{a}{2\pi r}$$

Solving this equation for θ and for a, we obtain

(3)
$$\boxed{\theta = \frac{a}{r}} \qquad \boxed{a = r\theta}$$

Remember that Equations 3 are valid only when θ is measured in radians.

In particular, putting $a = r$ in Equation 3, we see that an angle of 1 rad is the angle subtended at the center of a circle by an arc equal in length to the radius of the circle (see Figure 2).

Figure 2

EXAMPLE 2
(a) If the radius of a circle is 5 cm, what angle is subtended by an arc of 6 cm?
(b) If a circle has radius 3 cm, what is the length of an arc subtended by a central angle of $3\pi/8$ rad?

Solution
(a) Using Equation 3 with $a = 6$ and $r = 5$, we see that the angle is

$$\theta = \frac{6}{5} = 1.2 \, \text{rad}$$

(b) With $r = 3$ cm and $\theta = 3\pi/8$ rad, the arc length is

$$a = r\theta = 3\left(\frac{3\pi}{8}\right) = \frac{9\pi}{8} \, \text{cm}$$

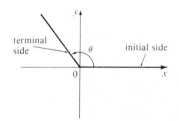

Figure 3
$\theta \geq 0$

The **standard position** of an angle occurs when we place its vertex at the origin of a coordinate system and its initial side on the positive x-axis as in Figure 3. A **positive** angle is obtained by rotating the initial side counterclockwise until it coincides with the terminal side. Likewise, **negative** angles are obtained by clockwise rotation as in Figure 4. Figure 5 shows several examples of angles in standard position. Notice that different

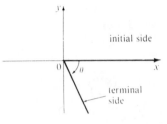

Figure 4

$\theta < 0$

Figure 5

Angles in standard position

angles can have the same terminal side. For instance, the angles $3\pi/4, -5\pi/4$, and $11\pi/4$ have the same initial and terminal sides because

$$\frac{3\pi}{4} - 2\pi = -\frac{5\pi}{4} \qquad \frac{3\pi}{4} + 2\pi = \frac{11\pi}{4}$$

and 2π rad represents a complete revolution.

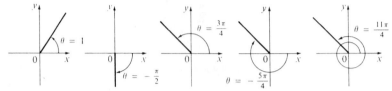

The Trigonometric Functions

For an acute angle θ the six trigonometric functions are defined as ratios of lengths of sides of a right triangle as follows (see Figure 6):

(4)

$$\sin\theta = \frac{\text{opp}}{\text{hyp}} \qquad \csc\theta = \frac{\text{hyp}}{\text{opp}}$$

$$\cos\theta = \frac{\text{adj}}{\text{hyp}} \qquad \sec\theta = \frac{\text{hyp}}{\text{adj}}$$

$$\tan\theta = \frac{\text{opp}}{\text{adj}} \qquad \cot\theta = \frac{\text{adj}}{\text{opp}}$$

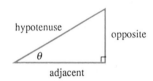

Figure 6

This definition does not apply to obtuse or negative angles, so for a general angle θ in standard position we let $P(x, y)$ be any point on the terminal side of θ and we let r be the distance $|OP|$ as in Figure 7. Then we define

(5)

$$\sin\theta = \frac{y}{r} \qquad \csc\theta = \frac{r}{y}$$

$$\cos\theta = \frac{x}{r} \qquad \sec\theta = \frac{r}{x}$$

$$\tan\theta = \frac{y}{x} \qquad \cot\theta = \frac{x}{y}$$

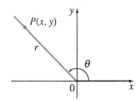

Figure 7

Since division by 0 is not defined, $\tan\theta$ and $\sec\theta$ are undefined when $x = 0$ and $\csc\theta$ and $\cot\theta$ are undefined when $y = 0$. Notice that the definitions in (4) and (5) are consistent when θ is an acute angle.

If θ is a number, the convention is that $\sin\theta$ means the sine of the angle whose *radian* measure is θ. For example, the expression $\sin 3$ implies that we are dealing with an angle of 3 rad. When finding a calculator approximation to this number we must remember to set our calculator in radian mode and then we obtain

$$\sin 3 \approx 0.14112$$

If we want to know the sine of the angle 3° we would write $\sin 3°$ and, with our calculator in degree mode, we find that

$$\sin 3° \approx 0.05234$$

The exact trigonometric ratios for certain angles can be read from the triangles in Figure 8. For instance,

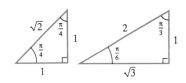

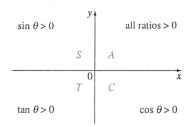

$$\sin \frac{\pi}{4} = \frac{1}{\sqrt{2}} \qquad \sin \frac{\pi}{6} = \frac{1}{2} \qquad \sin \frac{\pi}{3} = \frac{\sqrt{3}}{2}$$

$$\cos \frac{\pi}{4} = \frac{1}{\sqrt{2}} \qquad \cos \frac{\pi}{6} = \frac{\sqrt{3}}{2} \qquad \cos \frac{\pi}{3} = \frac{1}{2}$$

Figure 8

$$\tan \frac{\pi}{4} = 1 \qquad \tan \frac{\pi}{6} = \frac{1}{\sqrt{3}} \qquad \tan \frac{\pi}{3} = \sqrt{3}$$

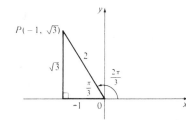

Figure 9

The signs of the trigonometric functions for angles in each of the four quadrants can be remembered by means of the rule "All Students Take Calculus" in Figure 9.

EXAMPLE 3   Find the exact trigonometric ratios for $\theta = 2\pi/3$.

Solution   From Figure 10 we see that a point on the terminal line for $\theta = 2\pi/3$ is $P(-1, \sqrt{3})$. Therefore, taking

$$x = -1 \qquad y = \sqrt{3} \qquad r = 2$$

in the definitions of the trigonometric ratios, we have

$$\sin \frac{2\pi}{3} = \frac{\sqrt{3}}{2} \qquad \cos \frac{2\pi}{3} = -\frac{1}{2} \qquad \tan \frac{2\pi}{3} = -\sqrt{3}$$

$$\csc \frac{2\pi}{3} = \frac{2}{\sqrt{3}} \qquad \sec \frac{2\pi}{3} = -2 \qquad \cot \frac{2\pi}{3} = -\frac{1}{\sqrt{3}}$$

●

Figure 10

The following table gives some values of $\sin \theta$ and $\cos \theta$ found by the method of Example 3.

θ	0	$\frac{\pi}{6}$	$\frac{\pi}{4}$	$\frac{\pi}{3}$	$\frac{\pi}{2}$	$\frac{2\pi}{3}$	$\frac{3\pi}{4}$	$\frac{5\pi}{6}$	π	$\frac{3\pi}{2}$	2π
$\sin \theta$	0	$\frac{1}{2}$	$\frac{1}{\sqrt{2}}$	$\frac{\sqrt{3}}{2}$	1	$\frac{\sqrt{3}}{2}$	$\frac{1}{\sqrt{2}}$	$\frac{1}{2}$	0	-1	0
$\cos \theta$	1	$\frac{\sqrt{3}}{2}$	$\frac{1}{\sqrt{2}}$	$\frac{1}{2}$	0	$-\frac{1}{2}$	$-\frac{1}{\sqrt{2}}$	$-\frac{\sqrt{3}}{2}$	-1	0	1

EXAMPLE 4   If $\cos \theta = \frac{2}{5}$ and $0 < \theta < \pi/2$, find the other five trigonometric functions of θ.

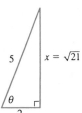

Solution   Since $\cos \theta = \frac{2}{5}$, we can label the hypotenuse as having length 5 and the adjacent side as having length 2 in Figure 11. If the opposite side has length x, then the Pythagorean Theorem gives $x^2 + 4 = 25$ and so $x^2 = 21$, $x = \sqrt{21}$. We can now use the **Figure 11**   diagram to write the other five trigonometric functions:

$$\sin \theta = \frac{\sqrt{21}}{5} \qquad \tan \theta = \frac{\sqrt{21}}{2}$$

$$\csc \theta = \frac{5}{\sqrt{21}} \qquad \sec \theta = \frac{5}{2} \qquad \cot \theta = \frac{2}{\sqrt{21}}$$

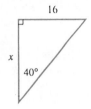

Figure 12

EXAMPLE 5   Use a calculator to approximate the value of x in Figure 12.

Solution   From the diagram we see that

$$\tan 40° = \frac{16}{x}$$

Therefore

$$x = \frac{16}{\tan 40°} \approx 19.07$$

Trigonometric Identities

A trigonometric identity is a relationship among the trigonometric functions. The most elementary are the following, which are immediate consequences of the definitions of the trigonometric functions.

(6)

$$\csc \theta = \frac{1}{\sin \theta} \qquad \sec \theta = \frac{1}{\cos \theta} \qquad \cot \theta = \frac{1}{\tan \theta}$$

$$\tan \theta = \frac{\sin \theta}{\cos \theta} \qquad \cot \theta = \frac{\cos \theta}{\sin \theta}$$

For the next identity we refer to Figure 7. The distance formula (or, equivalently, the Pythagorean Theorem) tells us that $x^2 + y^2 = r^2$. Therefore

$$\sin^2\theta + \cos^2\theta = \frac{y^2}{r^2} + \frac{x^2}{r^2} = \frac{x^2 + y^2}{r^2} = \frac{r^2}{r^2} = 1$$

We have therefore proved one of the most useful of all trigonometric identities:

(7)

$$\sin^2\theta + \cos^2\theta = 1$$

If we now divide both sides of Equation 7 by $\cos^2\theta$ and use Equations 6, we get

(8)

$$\tan^2\theta + 1 = \sec^2\theta$$

Similarly, if we divide both sides of Equation 7 by $\sin^2\theta$, we get

(9)

$$1 + \cot^2\theta = \csc^2\theta$$

The identities

(10a)

(10b)

$$\sin(-\theta) = -\sin\theta$$

$$\cos(-\theta) = \cos\theta$$

show that sin is an odd function and cos is an even function. (See Section 4 of Review and Preview.) They are easily proved by drawing a diagram showing θ and $-\theta$ in standard position (see Exercise 39).

Since the angles θ and $\theta + 2\pi$ have the same terminal side, we have

(11)

$$\sin(\theta + 2\pi) = \sin\theta \qquad \cos(\theta + 2\pi) = \cos\theta$$

These identities show that the sine and cosine functions are periodic with period 2π.

The remaining trigonometric identities are all consequences of two basic identities called the **addition formulas:**

(12a)

(12b)

$$\sin(x + y) = \sin x \cos y + \cos x \sin y$$

$$\cos(x + y) = \cos x \cos y - \sin x \sin y$$

The proofs of these addition formulas are indicated in Exercises 85, 86, and 87.

By substituting $-y$ for y in Equations 12a and 12b and using Equations 10a and 10b we obtain the following **subtraction formulas:**

(13a)

(13b)

$$\sin(x - y) = \sin x \cos y - \cos x \sin y$$

$$\cos(x - y) = \cos x \cos y + \sin x \sin y$$

Then, by dividing the formulas in Equations 12 or Equations 13, we obtain the corresponding formulas for $\tan(x \pm y)$:

(14a)

$$\tan(x + y) = \frac{\tan x + \tan y}{1 - \tan x \tan y}$$

(14b)

$$\tan(x - y) = \frac{\tan x - \tan y}{1 + \tan x \tan y}$$

If we put $y = x$ in the addition formulas (12) we get the **double angle formulas:**

(15a)

(15b)

$$\sin 2x = 2\sin x \cos x$$

$$\cos 2x = \cos^2 x - \sin^2 x$$

Then, by using the identity $\sin^2 x + \cos^2 x = 1$, we obtain the following alternate forms of the double angle formulas for $\cos 2x$:

(16a)

(16b)

$$\cos 2x = 2\cos^2 x - 1$$

$$\cos 2x = 1 - 2\sin^2 x$$

If we now solve these equations for $\cos^2 x$ and $\sin^2 x$ we get the following **half-angle formulas**, which are useful in integral calculus:

(17a)

(17b)

$$\cos^2 x = \frac{1 + \cos 2x}{2}$$

$$\sin^2 x = \frac{1 - \cos 2x}{2}$$

Finally we state the **product formulas**, which can be deduced from Equations 12 and 13.

(18a)

(18b)

(18c)

$$\sin x \cos y = \tfrac{1}{2}[\sin(x+y) + \sin(x-y)]$$

$$\cos x \cos y = \tfrac{1}{2}[\cos(x+y) + \cos(x-y)]$$

$$\sin x \sin y = \tfrac{1}{2}[\cos(x-y) - \cos(x+y)]$$

There are many other trigonometric identities but those we have stated are the ones used most often in calculus. If you forget any of them, remember that they can all be deduced from Equations 12a and 12b.

EXAMPLE 6　　Find all values of x in the interval $[0, 2\pi]$ such that $\sin x = \sin 2x$.

Solution　　Using the double angle formula (15a), we rewrite the given equation as

$$\sin x = 2\sin x \cos x \quad \text{or} \quad \sin x(1 - 2\cos x) = 0$$

Therefore there are two possibilities:

$$\sin x = 0 \qquad \text{or} \qquad 1 - 2\cos x = 0$$

$$x = 0, \ \pi, \ 2\pi \qquad\qquad\qquad \cos x = \tfrac{1}{2}$$

$$x = \frac{\pi}{3}, \frac{5\pi}{3}$$

There are five solutions to the given equation: 0, $\pi/3$, π, $5\pi/3$, and 2π.　　　　●

Graphs of the Trigonometric Functions

The graph of the function $f(x) = \sin x$, shown in Figure 13(a), is obtained by plotting points for $0 \le x \le 2\pi$ and then using the periodic nature of the function (from Equation 11) to complete the graph. Notice that the zeros of the sine function occur at the integer multiples of π, that is,

$$\sin x = 0 \qquad \text{when } x = n\pi \qquad n \text{ an integer}$$

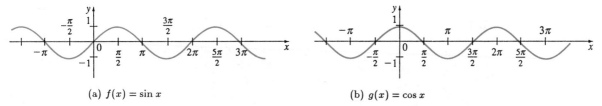

(a) $f(x) = \sin x$ (b) $g(x) = \cos x$

Figure 13

Because of the identity

$$\cos x = \sin\left(x + \frac{\pi}{2}\right)$$

(which can be verified using Equation 12a) the graph of cosine is obtained by shifting the graph of sine by an amount $\pi/2$ to the left [see Figure 13(b)]. Note that for both the sine and cosine functions the domain is $(-\infty, \infty)$ and the range is the closed interval $[-1, 1]$. Thus, for all values of x, we have

$$-1 \leq \sin x \leq 1 \qquad -1 \leq \cos x \leq 1$$

The graphs of the remaining four trigonometric functions are shown in Figure 14 and their domains are indicated there. Notice that tangent and cotangent have range $(-\infty, \infty)$, whereas cosecant and secant have range $(-\infty, -1] \cup [1, \infty)$. All four functions are periodic: tangent and cotangent have period π, whereas cosecant and secant have period 2π.

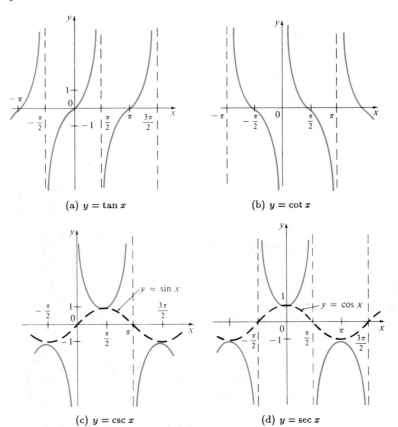

(a) $y = \tan x$ (b) $y = \cot x$

Figure 14   (c) $y = \csc x$ (d) $y = \sec x$

APPENDIX B **Exercises**

In Exercises 1–6 convert from degrees to radians.

1. 210°   2. 300°   3. 9°
4. −315°   5. 900°   6. 36°

In Exercises 7–12 convert from radians to degrees.

7. 4π 8. $-\dfrac{7\pi}{2}$ 9. $\dfrac{5\pi}{12}$

10. $\dfrac{8\pi}{3}$ 11. $-\dfrac{3\pi}{8}$ 12. 5

13. Find the length of a circular arc subtended by an angle of $\pi/12$ rad if the radius of the circle is 36 cm.

14. If a circle has radius 10 cm, what is the length of the arc subtended by a central angle of 72°?

15. A circle has radius 1.5 m. What angle is subtended at the center of the circle by an arc 1 m long?

16. Find the radius of a circular sector with angle $3\pi/4$ and arc length 6 cm.

In Exercises 17–22 draw, in standard position, the angle whose measure is given.

17. 315°   18. −150°

19. $-\dfrac{3\pi}{4}$ rad   20. $\dfrac{7\pi}{3}$ rad

21. 2 rad   22. −3 rad

In Exercises 23–28 find the exact trigonometric ratios for the angles whose radian measures are given.

23. $\dfrac{3\pi}{4}$ 24. $\dfrac{4\pi}{3}$

25. $\dfrac{9\pi}{2}$ 26. -5π

27. $\dfrac{5\pi}{6}$ 28. $\dfrac{11\pi}{4}$

In Exercises 29–34 find the remaining trigonometric ratios.

29. $\sin\theta = \dfrac{3}{5}, \quad 0 < \theta < \dfrac{\pi}{2}$

30. $\tan\alpha = 2, \quad 0 < \alpha < \dfrac{\pi}{2}$

31. $\sec\phi = -1.5, \quad \dfrac{\pi}{2} < \phi < \pi$

32. $\cos x = -\dfrac{1}{3}, \quad \pi < x < \dfrac{3\pi}{2}$

33. $\cot\beta = 3, \quad \pi < \beta < 2\pi$

34. $\csc\theta = -\dfrac{4}{3}, \quad \dfrac{3\pi}{2} < \theta < 2\pi$

In Exercises 35–38 find, correct to five decimal places, the length of the side labeled x.

35.

36.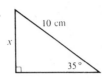

37.

38.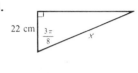

39. (a) Prove Equation 10a.
     (b) Prove Equation 10b.

40. (a) Prove Equation 14a.
     (b) Prove Equation 14b.

41. (a) Prove Equation 18a.
     (b) Prove Equation 18b.
     (c) Prove Equation 18c.

Prove the identities in Exercises 42–58.

42. $\cos\left(\dfrac{\pi}{2} - x\right) = \sin x$

43. $\sin\left(\dfrac{\pi}{2} + x\right) = \cos x$

44. $\sin(\pi - x) = \sin x$

45. $\sin\theta \cot\theta = \cos\theta$

46. $(\sin x + \cos x)^2 = 1 + \sin 2x$

47. $\sec y - \cos y = \tan y \sin y$

48. $\tan^2\alpha - \sin^2\alpha = \tan^2\alpha \sin^2\alpha$

49. $\cot^2\theta + \sec^2\theta = \tan^2\theta + \csc^2\theta$

50. $2 \csc 2t = \sec t \csc t$

51. $\tan 2\theta = \dfrac{2\tan\theta}{1 - \tan^2\theta}$

52. $\dfrac{1}{1 - \sin\theta} + \dfrac{1}{1 + \sin\theta} = 2\sec^2\theta$

53. $\sin x \sin 2x + \cos x \cos 2x = \cos x$

54. $\sin^2 x - \sin^2 y = \sin(x + y)\sin(x - y)$

55. $\dfrac{\sin\phi}{1 - \cos\phi} = \csc\phi + \cot\phi$

56. $\tan x + \tan y = \dfrac{\sin(x + y)}{\cos x \cos y}$

57. $\sin 3\theta + \sin\theta = 2\sin 2\theta \cos\theta$

58. $\cos 3\theta = 4\cos^3\theta - 3\cos\theta$

If $\sin x = \frac{1}{3}$ and $\sec y = \frac{5}{4}$, where x and y lie between 0 and $\pi/2$, evaluate the expressions in Exercises 59–64.

59. $\sin(x + y)$ **60.** $\cos(x + y)$

61. $\cos(x - y)$ **62.** $\sin(x - y)$

63. $\sin 2y$ **64.** $\cos 2y$

In Exercises 65–72 find all values of x in the interval $[0, 2\pi]$ that satisfy the given equation.

65. $2 \cos x - 1 = 0$ **66.** $3 \cot^2 x = 1$

67. $2 \sin^2 x = 1$ **68.** $|\tan x| = 1$

69. $\sin 2x = \cos x$ **70.** $2 \cos x + \sin 2x = 0$

71. $\sin x = \tan x$ **72.** $2 + \cos 2x = 3 \cos x$

In Exercises 73–76 find all values of x in the interval $[0, 2\pi]$ that satisfy the given inequality.

73. $\sin x \le \frac{1}{2}$ **74.** $2 \cos x + 1 > 0$

75. $-1 < \tan x < 1$ **76.** $\sin x > \cos x$

In Exercises 77–82 graph the given functions by starting with the graphs in Figures 12 and 13 and applying the transformations of Section 6 of Review and Preview where appropriate.

77. $y = \cos\left(x - \frac{\pi}{3}\right)$ **78.** $y = \tan 2x$

79. $y = \frac{1}{3}\tan\left(x - \frac{\pi}{2}\right)$ **80.** $y = 1 + \sec x$

81. $y = |\sin x|$ **82.** $y = 2 + \sin\left(x + \frac{\pi}{4}\right)$

83. Prove the **Law of Cosines:** If a triangle has sides with lengths a, b, and c, and θ is the angle between the sides with lengths a and b, then

$$c^2 = a^2 + b^2 - 2ab \cos \theta$$

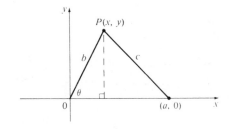

(*Hint:* Introduce a coordinate system so that θ is in standard position as in the figure. Express x and y in terms of θ and then use the distance formula to compute c.)

84. In order to find the distance $|AB|$ across a small inlet, a point C is located as in the figure and the following measurements were recorded: $\angle C = 103°$, $|AC| = 820$ m, $|BC| = 910$ m. Use the Law of Cosines to find the required distance.

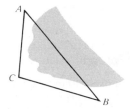

85. Use the figure to prove the subtraction formula

$$\cos(\alpha - \beta) = \cos \alpha \cos \beta + \sin \alpha \sin \beta$$

[*Hint:* Compute c^2 in two ways (using the Law of Cosines from Exercise 83 and also using the distance formula) and compare the two expressions.]

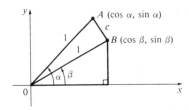

86. Use the formula in Exercise 85 to prove the addition formula for cosine (12b).

87. Use the subtraction formula for cosine and the identities

$$\cos\left(\frac{\pi}{2} - \theta\right) = \sin \theta \qquad \sin\left(\frac{\pi}{2} - \theta\right) = \cos \theta$$

to prove the subtraction formula for the sine function.

Proofs of Theorems

In this appendix we present proofs of several theorems that were stated in the main body of the text. These theorems are numbered according to their original statement.

Limit Laws (1.6)

Suppose that c is a constant and the limits

$$\lim_{x \to a} f(x) = L \quad \text{and} \quad \lim_{x \to a} g(x) = M$$

exist. Then

1. $\lim_{x \to a} [f(x) + g(x)] = L + M$

2. $\lim_{x \to a} [f(x) - g(x)] = L - M$

3. $\lim_{x \to a} [cf(x)] = cL$

4. $\lim_{x \to a} [f(x)g(x)] = LM$

5. $\lim_{x \to a} \dfrac{f(x)}{g(x)} = \dfrac{L}{M} \quad \text{if } M \neq 0$

Proof of Law 4   Let $\varepsilon > 0$ be given. We want to find $\delta > 0$ such that

$$\left| f(x)g(x) - LM \right| < \varepsilon \quad \text{whenever } 0 < |x - a| < \delta$$

In order to get terms that contain $|f(x) - L|$ and $|g(x) - M|$ we add and subtract $Lg(x)$ as follows:

$$|f(x)g(x) - LM| = |f(x)g(x) - Lg(x) + Lg(x) - LM|$$

$$= |(f(x) - L)g(x) + L(g(x) - M)|$$

$$\leq |(f(x) - L)g(x)| + |L(g(x) - M)| \qquad \text{(Triangle Inequality)}$$

$$= |f(x) - L|\,|g(x)| + |L|\,|g(x) - M|$$

We want to make each of these terms less than $\varepsilon/2$.

Since $\lim_{x \to a} g(x) = M$, there is a number $\delta_1 > 0$ such that

$$|g(x) - M| < \frac{\varepsilon}{2(1 + |L|)} \qquad \text{whenever } 0 < |x - a| < \delta_1$$

Also, there is a number $\delta_2 > 0$ such that if $0 < |x - a| < \delta_2$, then

$$|g(x) - M| < 1$$

and therefore

$$|g(x)| = |g(x) - M + M| \leq |g(x) - M| + |M| < 1 + |M|$$

Since $\lim_{x \to a} f(x) = L$, there is a number $\delta_3 > 0$ such that

$$|f(x) - L| < \frac{\varepsilon}{2(1 + |M|)} \qquad \text{whenever } 0 < |\,x - a| \leq \delta_3$$

Let $\delta = \min\{\delta_1, \delta_2, \delta_3\}$. If $0 < |x - a| < \delta$, then we have $0 < |x - a| < \delta_1$, $0 < |x - a| < \delta_2$, and $0 < |x - a| < \delta_3$, so we can combine the inequalities to obtain

$$|f(x)g(x) - LM| \;\leq\; |f(x) - L|\,|g(x)| + |L|\,|g(x) - M|$$

$$< \frac{\varepsilon}{2(1 + |M|)}\,(1 + |M|) + |L|\,\frac{\varepsilon}{2(1 + |L|)}$$

$$< \frac{\varepsilon}{2} + \frac{\varepsilon}{2} = \varepsilon$$

This shows that $\lim_{x \to a} f(x)g(x) = LM$.

Proof of Law 3   If we take $g(x) = c$ in Law 4, we get

$$\lim_{x \to a} [cf(x)] = \lim_{x \to a} [g(x)f(x)]$$

$$= \lim_{x \to a} g(x) \cdot \lim_{x \to a} f(x)$$

$$= \lim_{x \to a} c \cdot \lim_{x \to a} f(x)$$

$$= c \lim_{x \to a} f(x) \qquad \text{(by Law 7)}$$

Proof of Law 2   Using Law 1 and Law 3 with $c = -1$, we have

$$\lim_{x \to a} [f(x) - g(x)] = \lim_{x \to a} [f(x) + (-1)g(x)]$$

$$= \lim_{x \to a} f(x) + \lim_{x \to a} (-1)g(x)$$

$$= \lim_{x \to a} f(x) + (-1) \lim_{x \to a} g(x)$$

$$= \lim_{x \to a} f(x) - \lim_{x \to a} g(x)$$

Proof of Law 5   First let us show that

$$\lim_{x \to a} \frac{1}{g(x)} = \frac{1}{M}$$

To do this we must show that, given $\varepsilon > 0$, there exists $\delta > 0$ such that

$$\left| \frac{1}{g(x)} - \frac{1}{M} \right| < \varepsilon \qquad \text{whenever } 0 < |x - a| < \delta$$

Observe that

$$\left| \frac{1}{g(x)} - \frac{1}{M} \right| = \frac{|M - g(x)|}{|Mg(x)|}$$

We know that we can make the numerator small. But we also need to know that the denominator is not small when x is near a. Since $\lim_{x \to a} g(x) = M$, there is a number $\delta_1 > 0$ such that, whenever $0 < |x - a| < \delta_1$, we have

$$|g(x) - M| < \frac{|M|}{2}$$

and therefore

$$|M| = |M - g(x) + g(x)| \le |M - g(x)| + |g(x)|$$
$$< \frac{|M|}{2} + |g(x)|$$

This shows that

$$|g(x)| > \frac{M}{2} \qquad \text{whenever } 0 < |x - a| < \delta_1$$

and so, for these values of x,

$$\frac{1}{|Mg(x)|} = \frac{1}{|M||g(x)|} < \frac{1}{|M|} \cdot \frac{2}{|M|} = \frac{2}{M^2}$$

Also, there exists $\delta_2 > 0$ such that

$$|g(x) - M| < \frac{M^2}{2} \varepsilon \qquad \text{whenever } 0 < |x - a| < \delta_2$$

Let $\delta = \min\{\delta_1, \delta_2\}$. Then, for $0 < |x - a| < \delta$, we have

$$\left| \frac{1}{g(x)} - \frac{1}{M} \right| = \frac{|M - g(x)|}{|Mg(x)|} < \frac{2}{M^2} \frac{M^2}{2} \varepsilon = \varepsilon$$

It follows that $\lim_{x \to a} 1/g(x) = 1/M$. Finally, using Law 4, we obtain

$$\lim_{x \to a} \frac{f(x)}{g(x)} = \lim_{x \to a} f(x) \left(\frac{1}{g(x)} \right)$$
$$= \lim_{x \to a} f(x) \cdot \lim_{x \to a} \frac{1}{g(x)} = L \cdot \frac{1}{M} = \frac{L}{M}$$

Theorem (1.9)

> If $f(x) \leq g(x)$ for all x in an open interval that contains a (except possibly at a) and
>
> $$\lim_{x \to a} f(x) = L \quad \text{and} \quad \lim_{x \to a} g(x) = M$$
>
> then $L \leq M$.

Proof   We use the method of proof by contradiction. Suppose, if possible, that $L > M$. Limit Law 2 says that

$$\lim_{x \to a} [g(x) - f(x)] = M - L$$

Therefore, for any $\varepsilon > 0$, there exists $\delta > 0$ such that

$$\big|[g(x) - f(x)] - (M - L)\big| < \varepsilon \qquad \text{whenever} \qquad 0 < |x - a| < \delta$$

In particular, taking $\varepsilon = L - M$ (note that $L - M > 0$ by hypothesis), we have a number $\delta > 0$ such that

$$\big|[g(x) - f(x)] - (M - L)\big| < L - M \qquad \text{whenever } 0 < |x - a| < \delta$$

Since $a \leq |a|$ for any number a, we have

$$[g(x) - f(x)] - (M - L) < L - M \qquad \text{whenever } 0 < |x - a| < \delta$$

which simplifies to

$$g(x) < f(x) \qquad \text{whenever } 0 < |x - a| < \delta$$

But this contradicts $f(x) \leq g(x)$. Thus the inequality $L > M$ must be false. Therefore $L \leq M$.   ●

The Squeeze Theorem (1.10)

> If $f(x) \leq g(x) \leq h(x)$ for all x in an open interval that contains a (except possibly at a) and
>
> $$\lim_{x \to a} f(x) = \lim_{x \to a} h(x) = L$$
>
> then
> $$\lim_{x \to a} g(x) = L$$

Proof   Let $\varepsilon > 0$ be given. Since $\lim_{x \to a} f(x) = L$, there is a number $\delta_1 > 0$ such that

$$|f(x) - L| < \varepsilon \qquad \text{whenever } 0 < |x - a| < \delta_1$$

that is, $$L - \varepsilon < f(x) < L + \varepsilon \qquad \text{whenever } 0 < |x - a| < \delta_1$$

Since $\lim_{x \to a} h(x) = L$, there is a number $\delta_2 > 0$ such that

$$|h(x) - L| < \varepsilon \qquad \text{whenever } 0 < |x - a| < \delta_2$$

that is, $\qquad L - \varepsilon < h(x) < L + \varepsilon \qquad \text{whenever } 0 < |x - a| < \delta_2$

Let $\delta = \min\{\delta_1, \delta_2\}$. If $0 < |x - a| < \delta$, then $0 < |x - a| < \delta_1$ and $0 < |x - a| < \delta_2$, so

$$L - \varepsilon \; < f(x) \le g(x) \le h(x) < L + \varepsilon$$

In particular,

$$L - \varepsilon < g(x) < L + \varepsilon$$

and so $|g(x) - L| < \varepsilon$. Therefore $\lim_{x \to a} g(x) = L$.   ●

Theorem (1.23)

> If f is continuous at b and $\lim_{x \to a} g(x) = b$, then
>
> $$\lim_{x \to a} f(g(x)) = f(b)$$

Proof   Let $\varepsilon > 0$ be given. We want to find a number $\delta > 0$ such that

$$|f(g(x)) - f(b)| < \varepsilon \qquad \text{whenever } 0 < |x - a| < \delta$$

Since f is continuous at b, we have

$$\lim_{y \to a} f(y) = f(b)$$

and so there exists $\delta_1 > 0$ such that

$$|f(y) - f(b)| < \varepsilon \qquad \text{whenever } |y - b| < \delta_1$$

Since $\lim_{x \to a} g(x) = b$, there is a number $\delta > 0$ such that

$$|g(x) - b| < \delta_1 \qquad \text{whenever } 0 < |x - a| < \delta$$

Combining these two statements, we see that whenever $0 < |x - a| < \delta$ we have $|g(x) - b| < \delta_1$, which implies that $|f(g(x)) - f(b)| < \varepsilon$. Therefore we have proved that $\lim_{x \to a} f(g(x)) = f(b)$.   ●

The proof of the following result was promised in the proof of Theorem 2.22.

Theorem

> If $0 < \theta < \frac{\pi}{2}$, then $\theta \le \tan \theta$.

Proof   Figure 1 shows a sector of a circle with center O, central angle θ, and radius 1. Then

$$|AD| = |OA| \tan \theta = \tan \theta$$

We approximate the arc AB by an inscribed polygon consisting of n equal line segments and we look at a typical segment PQ. We extend the lines OP and OQ to meet AD in the points R and S. Then we draw $RT \parallel PQ$ as in Figure 1. Observe that

$$\angle RTO = \angle PQO < 90°$$

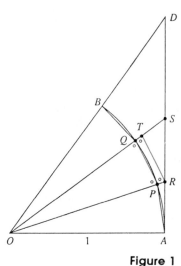

Figure 1

and so $\angle RTS > 90°$. Therefore we have

$$|PQ| < |RT| < |RS|$$

If we add n such inequalities, we get

$$L_n < |AD| = \tan\theta$$

where L_n is the length of the inscribed polygon. Thus, by Theorem 1.9, we have

$$\lim_{n \to \infty} L_n \le \tan\theta$$

But the arc length was defined in Equation 8.11 as the limit of the lengths of inscribed polygons, so

$$\theta = \lim_{n \to \infty} L_n \le \tan\theta \qquad \bullet$$

Theorem (3.17)

> If f is a one-to-one continuous function defined on an interval (a, b), then its inverse function f^{-1} is also continuous.

Proof   First we show that if f is both one-to-one and continuous on (a, b), then it must be either increasing or decreasing on (a, b). If it were neither increasing nor decreasing, then there would exist numbers x_1, x_2, and x_3 in (a, b) with $x_1 < x_2 < x_3$ such that $f(x_2)$ does not lie between $f(x_1)$ and $f(x_3)$. There are two possibilities: either (1) $f(x_3)$ lies between $f(x_1)$ and $f(x_2)$ or (2) $f(x_1)$ lies between $f(x_2)$ and $f(x_3)$. (Draw a picture.) In case (1) we apply the Intermediate Value Theorem to the continuous function f to get a number c between x_1 and x_2 such that $f(c) = f(x_3)$. In case (2) the Intermediate Value Theorem gives a number c between x_2 and x_3 such that $f(c) = f(x_1)$. In either case we have contradicted the fact that f is one-to-one.

Let us assume, for the sake of definiteness, that f is increasing on (a, b). We take any number y_0 in the domain of f^{-1} and we let $f^{-1}(y_0) = x_0$; that is, x_0 is the number in (a, b) such that $f(x_0) = y_0$. To show that f^{-1} is continuous at y_0 we take any $\varepsilon > 0$ such that the interval $(x_0 - \varepsilon, x_0 + \varepsilon)$ is contained in the interval (a, b). Since f is increasing, it maps the numbers in the interval $(x_0 - \varepsilon, x_0 + \varepsilon)$ onto the numbers in the interval $(f(x_0 - \varepsilon), f(x_0 + \varepsilon))$ and f^{-1} reverses the correspondence. If we let δ denote the smaller of the numbers $\delta_1 = y_0 - f(x_0 - \varepsilon)$ and $\delta_2 = f(x_0 + \varepsilon) - y_0$, then the interval $(y_0 - \delta, y_0 + \delta)$ is contained in the interval $(f(x_0 - \varepsilon), f(x_0 + \varepsilon))$ and so is mapped into the interval $(x_0 - \varepsilon, x_0 + \varepsilon)$ by f^{-1}. (See the arrow diagram in Figure 2.) We have therefore found a number $\delta > 0$ such that

$$|f^{-1}(y) - f^{-1}(y_0)| < \varepsilon \qquad \text{whenever} \quad |y - y_0| < \delta$$

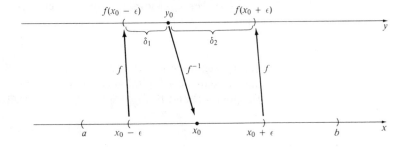

Figure 2

This shows that $\lim_{y \to y_0} f^{-1}(y) = f^{-1}(y_0)$ and so f^{-1} is continuous at any number y_0 in its domain.

In order to give the promised proof of l'Hospital's Rule we first need a generalization of the Mean Value Theorem, which is due to the French mathematician, Augustin Cauchy (1789–1867).

Cauchy's Mean Value Theorem

Suppose that the functions f and g are continuous on $[a,b]$ and differentiable on (a,b) and $g'(x) \neq 0$ for all x in (a,b). Then there is a number c in (a,b) such that

$$\frac{f'(c)}{g'(c)} = \frac{f(b) - f(a)}{g(b) - g(a)}$$

Notice that if we take the special case where $g(x) = x$, then $g'(c) = 1$ and Cauchy's Mean Value Theorem is just the ordinary Mean Value Theorem (4.10). Furthermore, Cauchy's Theorem can be proved in a manner very similar to that of Theorem 4.10. You can verify that all we have to do is change the function h given by (4.14) to the function

$$h(x) = f(x) - f(a) - \frac{f(b) - f(a)}{g(b) - g(a)} [g(x) - g(a)]$$

and apply Rolle's Theorem as before.

L'Hospital's Rule (3.60)

Suppose f and g are differentiable and $g'(x) \neq 0$ on an open interval I that contains a (except possibly at a). Suppose that

$$\lim_{x \to a} f(x) = 0 \qquad \text{and} \qquad \lim_{x \to a} g(x) = 0$$

or that

$$\lim_{x \to a} f(x) = \pm\infty \qquad \text{and} \qquad \lim_{x \to a} g(x) = \pm\infty$$

(In other words, we have an indeterminate from of type $0/0$ or ∞/∞.) Then

$$\lim_{x \to a} \frac{f(x)}{g(x)} = \lim_{x \to a} \frac{f'(x)}{g'(x)}$$

if the limit on the right side exists (or is ∞ or $-\infty$).

Proof of L'Hospital's Rule   We are assuming that $\lim_{x \to a} f(x) = 0$ and $\lim_{x \to a} g(x) = 0$. Let

$$L = \lim_{x \to a} \frac{f'(x)}{g'(x)}$$

We must show that $\lim_{x \to a} f(x)/g(x) = L$. Define

$$F(x) = \begin{cases} f(x) & \text{if } x \neq a \\ 0 & \text{if } x = a \end{cases} \qquad G(x) = \begin{cases} g(x) & \text{if } x \neq a \\ 0 & \text{if } x = a \end{cases}$$

Then F is continuous on I since f is continuous on $\{x \in I \mid x \neq a\}$ and

$$\lim_{x \to a} F(x) = \lim_{x \to a} f(x) = 0 = F(a)$$

Likewise G is continuous on I. Let $x \in I$ and $x > a$. Then F and G are continuous on $[a, x]$ and differentiable on (a, x) and $G' \neq 0$ there (since $F' = f'$ and $G' = g'$). Therefore by Cauchy's Mean Value Theorem there is a number y such that $a < y < x$ and

$$\frac{F'(y)}{G'(y)} = \frac{F(x) - F(a)}{G(x) - G(a)} = \frac{F(x)}{G(x)}$$

Here we have used the fact that, by definition, $F(a) = 0$ and $G(a) = 0$. Now let $x \to a^+$. Then $y \to a^+$ (since $a < y < x$) so

$$\lim_{x \to a^+} \frac{f(x)}{g(x)} = \lim_{x \to a^+} \frac{F(x)}{G(x)} = \lim_{y \to a^+} \frac{F'(y)}{G'(y)} = \lim_{y \to a^+} \frac{f'(y)}{g'(y)} = L$$

A similar argument shows that the left-hand limit is also L. Therefore

$$\lim_{x \to a} \frac{f(x)}{g(x)} = L$$

This proves l'Hospital's Rule for the case where a is finite.

If a is infinite, we let $t = 1/x$. Then $t \to 0^+$ as $x \to \infty$, so we have

$$\lim_{x \to \infty} \frac{f(x)}{g(x)} = \lim_{t \to 0^+} \frac{f\left(\frac{1}{t}\right)}{g\left(\frac{1}{t}\right)}$$

$$= \lim_{t \to 0^+} \frac{f'\left(\frac{1}{t}\right)\frac{-1}{t^2}}{g'\left(\frac{1}{t}\right)\frac{-1}{t^2}} \qquad \text{(by l'Hospital's Rule for finite } a\text{)}$$

$$= \lim_{t \to 0^+} \frac{f'\left(\frac{1}{t}\right)}{g'\left(\frac{1}{t}\right)} = \lim_{x \to \infty} \frac{f'(x)}{g'(x)}$$

Property 5 of Integrals (5.13)

$$\int_a^b f(x)\,dx = \int_a^c f(x)\,dx + \int_c^b f(x)\,dx$$

if all of these integrals exist.

Proof We first assume that $a < c < b$. Since we are assuming that $\int_a^b f(x)\,dx$ exists, we can compute it as a limit of Riemann sums using only partitions P that include c as one of the partition points. If P is such a partition, let P_1 be the corresponding partition of $[a, c]$ determined by those partition points of P that lie in $[a, c]$. Similarly, P_2 will denote the corresponding partition of $[c, b]$. Note that $\|P_1\| \leq \|P\|$ and $\|P_2\| \leq \|P\|$. Thus, if $\|P\| \to 0$, it follows that $\|P_1\| \to 0$ and $\|P_2\| \to 0$. If $\{x_i \mid 1 \leq i \leq n\}$ is the set of partition points for P and $n = k + m$, where k is the number of subintervals in $[a, c]$ and m is the number of subintervals in $[c, b]$, then $\{x_i \mid 1 \leq i \leq k\}$ is the set of partition points for P_1. If we write $t_j = x_{k+j}$ for the partition points to the right of c, then $\{t_j \mid 1 \leq j \leq m\}$ is the set of partition points for P_2. Thus we have

$$a = x_0 < x_1 < \cdots < x_k < x_{k+1} < \cdots < x_n = b$$

$$c < t_1 < \cdots < t_m = b$$

Choosing $x_i^* = x_i$ and letting $\Delta t_j = t_j - t_{j-1}$, we compute $\int_a^b f(x)\, dx$ as follows:

$$\int_a^b f(x)\, dx = \lim_{\|P\| \to 0} \sum_{i=1}^n f(x_i)\, \Delta x_i$$

$$= \lim_{\|P\| \to 0} \left[\sum_{i=1}^k f(x_i)\, \Delta x_i + \sum_{i=k+1}^n f(x_i)\, \Delta x_i \right]$$

$$= \lim_{\|P\| \to 0} \left[\sum_{i=1}^k f(x_i)\, \Delta x_i + \sum_{j=1}^m f(t_j)\, \Delta t_j \right]$$

$$= \lim_{\|P_1\| \to 0} \sum_{i=1}^k f(x_i)\, \Delta x_i + \lim_{\|P_2\| \to 0} \sum_{j=1}^m f(t_j)\, \Delta t_j$$

$$= \int_a^c f(x)\, dx + \int_c^b f(t)\, dt$$

Now suppose that $c < a < b$. By what we have already proved, we have

$$\int_c^b f(x)\, dx = \int_c^a f(x)\, dx + \int_a^b f(x)\, dx$$

Therefore
$$\int_a^b f(x)\, dx = -\int_c^a f(x)\, dx + \int_c^b f(x)\, dx$$

$$= \int_a^c f(x)\, dx + \int_c^b f(x)\, dx$$

(See Note 6 in Section 5.3.) The proofs are similar for the remaining four orderings of a, b, and c.   ●

In order to prove Theorem 10.38 we first need the following results:

Theorem

1. If a power series $\sum a_n x^n$ converges when $x = b$ (where $b \neq 0$), then it converges whenever $|x| < |b|$.

2. If a power series $\sum a_n x^n$ diverges when $x = d$ (where $d \neq 0$), then it diverges whenever $|x| > |d|$.

Proof of 1   Suppose that $\sum a_n b^n$ converges. Then, by Theorem 10.17, we have $\lim_{n \to \infty} a_n b^n = 0$. According to Definition 10.1 with $\varepsilon = 1$, there is a positive integer N such that $|a_n b^n| < 1$ whenever $n \geq N$. Thus, for $n \geq N$, we have

$$|a_n x^n| = \left| \frac{a_n b^n x^n}{b^n} \right| = |a_n b^n| \left| \frac{x}{b} \right|^n < \left| \frac{x}{b} \right|^n$$

If $|x| < |b|$, then $|x/b| < 1$, so $\sum |x/b|^n$ is a convergent geometric series. Therefore, by the Comparison Test, the series $\sum_{n=N}^{\infty} |a_n x^n|$ is convergent. Thus the series $\sum a_n x^n$ is absolutely convergent and therefore convergent. •

Proof of 2   Suppose that $\sum a_n d^n$ diverges. If x is any number such that $|x| > |d|$, then $\sum a_n x^n$ cannot converge because, by part 1, the convergence of $\sum a_n x^n$ would imply the convergence of $\sum a_n d^n$. Therefore $\sum a_n x^n$ diverges whenever $|x| > |d|$. •

Theorem | For a power series $\sum a_n x^n$ there are only three possibilities:

1. The series converges only when $x = 0$.

2. The series converges for all x.

3. There is a positive number R such that the series converges if $|x| < R$ and diverges if $|x| > R$.

Proof   Suppose that neither case (1) nor case (3) is true. Then there are nonzero numbers b and d such that $\sum a_n x^n$ converges for $x = b$ and diverges for $x = d$. Thus the set $S = \{x \mid \sum a_n x^n \text{ converges}\}$ is not empty. By the preceding theorem, the series diverges if $|x| > |d|$, so $|x| \leq |d|$ for all $x \in S$. This says that $|d|$ is an upper bound for the set S. Thus, by the Completeness Axiom (see Section 10.1), S has a least upper bound R. If $|x| > R$, then $x \notin S$, so $\sum a_n x^n$ diverges. If $|x| < R$, then $|x|$ is not an upper bound for S and so there exists $b \in S$ such that $b > |x|$. Since $b \in S$, $\sum a_n b^n$ converges, so by the preceding theorem $\sum a_n x^n$ converges. •

Theorem (10.38) | For a power series $\sum a_n (x - c)^n$ there are only three possibilities:

1. The series converges only when $x = c$.

2. The series converges for all x.

3. There is a positive number R such that the series converges if $|x - c| < R$ and diverges if $|x - c| > R$.

Proof   If we make the change of variable $u = x - c$, then the power series becomes $\sum a_n u^n$ and we can apply the preceding theorem to this series. In case (3) we have convergence for $|u| < R$ and divergence for $|u| > R$. Thus we have convergence for $|x - c| < R$ and divergence for $|x - c| > R$. •

D

Lies My Calculator or Computer Told Me

A wide variety of pocket-size calculating devices are currently marketed. Some can run programs prepared by the user; some have pre-programmed packages for frequently used calculus procedures, including the display of curve sketches. All have certain limitations in common: a limited range of magnitude (usually less than 10^{100} for calculators) and a bound on accuracy (typically eight to thirteen digits).

A calculator usually comes with an owner's manual. Read it! The manual will tell you about further limitations (for example, for angles when entering trigonometric functions) and perhaps how to overcome them.

Program packages for microcomputers (even the most fundamental ones, which realize arithmetical operations and elementary functions) often suffer from hidden flaws. You will be made aware of some of them in the examples below, and you are encouraged to experiment using the ideas we present.

Preliminary Experiments with Your Calculator or Computer

To have a first look at the limitations and quality of your calculator, make it compute $2 \div 3$. Of course, the answer is not a terminating decimal so it cannot be represented exactly on your calculator. If the last displayed digit is 6 rather than 7, then your calculator approximates $\frac{2}{3}$ by truncating instead of rounding, so be prepared for slightly greater loss of accuracy in longer calculations.

Now multiply the result by 3; that is, calculate $(2 \div 3) \times 3$. If the answer is 2, then subtract 2 from the result, thereby calculating $(2 \div 3) \times 3 - 2$. Instead of obtaining 0 as the answer, you might obtain a small negative number, which depends on the construction of the circuits. (The calculator keeps, in this case, a few "spare" digits that are remembered but not shown.) This is all right because, as previously mentioned, the finite number of digits makes it impossible to represent $2 \div 3$ exactly.

A similar situation occurs when you calculate $(\sqrt{6})^2 - 6$. If you do not obtain 0, the order of magnitude of the result will tell you how many digits the calculator uses internally.

Next, try to compute $(-1)^5$ using the y^x key. Many calculators will indicate an error because they are built to attempt $e^{5\ln(-1)}$. One way to overcome this is to use the fact that $(-1)^k = \cos k\pi$ when k is an integer.

Calculators are usually constructed to operate in the decimal number system. In contrast, some microcomputer packages of arithmetical programs operate in a number system with base other than 10 (typically 2 or 16). Here the list of unwelcome tricks your device can play on you is even longer, since not all terminating decimal numbers are represented exactly. A recent implementation of the BASIC language shows (in double precision) examples of incorrect conversion from one number system into another, for example,

$$8 \times 0.1 \overset{?}{=} 0.79999\ 99999\ 99999\ 9$$

while
$$19 \times 0.1 \overset{?}{=} 1.90000\ 00000\ 00001$$

Yet another implementation, apparently free of the above anomalies, will not calculate standard functions in double precision. For example, the number $\pi = 4 \times \tan^{-1} 1$, whose representation with sixteen decimal digits should be 3.14159 26535 89793, appears as 3.14159 29794 31152; this is off by more than 3×10^{-7}. What is worse, the cosine function is programmed so badly that its "cos" $0 = 1 + 2^{-23}$. (Can you invent a situation when this could ruin your calculations?) These or similar defects exist in other programming languages, too.

A word of caution is also due about graphing calculators. If you have one, graph a function with a discontinuity at a point, such as $y = \coth(x - 1)$ near $x = 1$. Experimenting with different scales may produce a steep line segment from the bottom to the top of the screen, and this near-vertical line may disappear with a change of scale. You will be the judge of whether to accept this line or not.

The Perils of Subtraction

You might have observed that subtraction of two numbers that are close to each other is a tricky operation. The difficulty is similar to this thought exercise: Imagine that you walk blindfolded 100 steps forward and then turn around and walk 99 steps. Are you sure that you end up exactly one step from where you started?

The name of this phenomenon is "loss of significant digits." To illustrate, let us calculate

$$8721\sqrt{3} - 10,681\sqrt{2}$$

The approximations from my calculator are

$$8721\sqrt{3} \approx 15105.21509 \qquad \text{and} \qquad 10,681\sqrt{2} \approx 15105.21506$$

and so we get $8721\sqrt{3} - 10,681\sqrt{2} \approx 0.00003$. Even with three spare digits exposed, the difference comes out as 0.00003306. As you can see, the two ten-digit numbers agree in nine digits that, after subtraction, become zeros before the first nonzero digit. To make things worse, the formerly small errors in the square roots become more visible. In this particular example we can use rationalization to write

$$8721\sqrt{3} - 10,681\sqrt{2} = \frac{1}{8721\sqrt{3} + 10,681\sqrt{2}}$$

(work out the details!) and now the loss of significant digits does not occur:

$$\frac{1}{8721\sqrt{3} + 10,681\sqrt{2}} \approx 0.00003310115 \quad \text{to seven digits}$$

(It would take too much space to explain why all seven digits are reliable; the subject called *numerical analysis* deals with these and similar situations.) See Exercise 7 for another instance of restoring lost digits.

Now you can see why in Exercise 18 in Section 1.2 the guess at the limit was bound to go wrong: $\tan x$ becomes so close to x that the values will eventually agree in all digits that the calculator is capable of carrying. Similarly, if you start with just about any continuous function f and try to guess the value of

$$f'(x) = \lim_{h \to 0} \frac{f(x+h) - f(x)}{h}$$

long enough using a calculator, you will end up with a zero, despite all the rules in Section 2.2!

Where Calculus Is More Powerful than Calculators and Computers

One of the secrets of success of calculus in overcoming the difficulties connected with subtraction is symbolic manipulation. For instance, $(a+b) - a$ is always b, although the calculated value may be different. Try it with $a = 10^7$ and $b = \sqrt{2} \times 10^{-5}$. Another powerful tool is the use of inequalities; a good example is the Squeeze Theorem (1.10) as demonstrated in Example 12 in Section 1.3. Yet another method for avoiding computational difficulties is provided by the Mean Value Theorem (4.10) (see Exercise 5) and its consequences, such as l'Hospital's Rule (3.60) (which helps solve the aforementioned Exercise 18 in Section 1.2 and others) and Taylor's Formula (10.54).

The limitations of calculators and computers are further illustrated by infinite series. It is a common misconception that a series can be summed by adding terms until there is "practically nothing to add" and "the error is less than the first neglected term." The latter statement is true for certain alternating series (Theorem 10.28) but not in general; a modified version is true for another class of series (Exercise 10). As an example to refute these misconceptions, let us consider the series

$$\sum_{n=1}^{\infty} \frac{1}{n^{1.001}}$$

which is a convergent p-series ($p = 1.001 > 1$). Suppose we were to try to sum this series correct to eight decimal places, by adding terms until they are less than 5 in the ninth decimal place. In other words, we stop when

$$\frac{1}{n^{1.001}} < 0.00000\,0005$$

that is, when $n = N = 196,217,284$. (This would require a high-speed computer and increased precision.) After going to all this trouble, we would end up with the approximating partial sum

$$S_N = \sum_{n=1}^{N} \frac{1}{n^{1.001}} < 19.5$$

But from the proof of the Integral Test (see Figure 10.8 in Section 10.3) we have

$$\sum_{n=1}^{\infty} \frac{1}{n^{1.001}} > \int_{1}^{\infty} \frac{dx}{x^{1.001}} = 1000$$

Thus the machine result represents less than 2% of the correct answer!

Suppose that we then wanted to add a huge number of terms of this series, say 10^{100} terms, in order to approximate the infinite sum more closely. (This number 10^{100}, called a *googol*, is outside the range of pocket calculators and is much larger than the number of elementary particles in our solar system.) If we were to add 10^{100} terms of the above series (only in theory; a million years is less than 10^{26} microseconds) we would still obtain a sum of less than 207 compared with the true sum of more than 1000. (This estimate of 207 is obtained by using a more precise form of the Integral Test, known as the Euler-Maclaurin Formula, and only then using a calculator. The formula provides a way to accelerate the convergence of this and other series.)

If the two preceding approaches did not give the right information about the accuracy of the partial sums, what does? A suitable inequality satisfied by the remainder of the series, as you can see from Exercise 6.

Symbolic Manipulation

It would be erroneous to leave the impression that most of the symbolic manipulation done in a calculus course *cannot* be done correctly on a computer or calculator. New software is capable of executing many of the techniques detailed in this course and less capable, but still powerful, pocket calculators are able to do many symbolic operations as well as plot and graphically display data. Calculators like the HP series and software for microcomputers, such as Maple®, Calculus T/L®, Derive®, and Mathematica®, are all useful and powerful calculus tools. Just as the best chess playing programs are capable of beating all but the experts, the best mathematical software is much more capable and efficient than most students at its specific programmed activity—such as computing limits, taking derivatives, finding coefficients, or computing indefinite integrals.

Despite considerable progress in recent years, this field is still in its infancy. Computers and software are conspicuously worse at doing mathematics than most mathematicians. There is much mathematical reasoning and computation that people can do, sometimes easily, that computers can't yet do, and probably a great deal that computing devices will never be able to do. Software is *not* very useful for converting real problems into executable sequences of calculations, and computers are very weak at symbolic reasoning (doing proofs) or providing explanations of what the calculations have accomplished. These creative aspects of mathematics are still well beyond what we usually think of as computable operations. On the other hand, students usually have difficulty with these skills as well.

Maple® *and Calculus* by J.S. Devitt (Brooks/Cole, 1991) describes some of the considerable capabilities of the Maple computer algebra system and its usefulness to calculus students. For example, Maple would be very effective at solving almost all the problems at the end of Appendix E, on rotations of axes; however, it would be considerably less useful in solving Problems Plus 3, 7, and 12 at the end of Chapter 8, or Exercise 74 in Section 10.1. Even if computer algebra systems could handle these types of problems, there would still be great pedagogical value in practicing symbolic manipulation by hand, for the same reason that it will always be important to know how to do arithmetic without a calculator.

In short, computers and calculators are not replacements for mathematical thought. They are just replacements for some kinds of mathematical labor, either numerical or symbolic. There are, and always will be, mathematical problems untreatable by a calculator or computer, regardless of its size and speed. A calculator or computer does stretch the human capacity for handling numbers and symbols, but there is still considerable scope and necessity for "thinking before doing."

APPENDIX D Exercises

1. Guess the value of

    $$\lim_{x \to 0}\left(\frac{1}{\sin^2 x} - \frac{1}{x^2}\right)$$

    and determine when to stop guessing before the loss of significant digits destroys your results. (The answer will depend on your calculator.) Then find the precise answer using an appropriate calculus method.

2. Guess the value of

    $$\lim_{x \to 0}\frac{\ln(1+h)}{h}$$

    and determine when to stop guessing before the loss of significant digits destroys your results. This time the detrimental subtraction takes place inside the machine. Explain how (assuming that the Taylor series with center $c = 1$ is used to approximate $\ln x$). Then find the precise answer using an appropriate calculus method.

3. Even innocent-looking calculus problems can lead to numbers beyond the calculator range. Show that the maximum value of the function

    $$f(x) = \frac{x^{25}}{(1.0001)^x}$$

    is greater than 10^{124}. (*Hint:* Use logarithms.) What is the limit of $f(x)$ as $x \to \infty$?

4. What is a numerically reliable expression to replace $\sqrt{1 - \cos x}$, especially when x is a small number? Refer to the identities in Appendix B for help. (Recall that some computer packages would signal an unnecessary error condition, or even switch to complex arithmetic, when $x = 0$.)

5. Try to evaluate

    $$D = \ln \ln(10^9 + 1) - \ln \ln(10^9)$$

    on your calculator. These numbers are so close together that you will likely obtain 0 or just a few digits of accuracy. However we can use the Mean

    Value Theorem to achieve much greater accuracy.
    (a) Let $f(x) = \ln \ln x$, $a = 10^9$, and $b = 10^9 + 1$. Then the Mean Value Theorem gives

    $$f(b) - f(a) = f'(c)(b - a) = f'(c)$$

    where $a < c < b$. Since f' is decreasing, we have $f'(a) > f'(c) > f'(b)$. Use this to estimate the value of D.
    (b) Use the Mean Value Theorem a second time to discover why quantities $f'(a)$ and $f'(b)$ in part (a) are so close to each other.

6. For the series $\sum_{n=1}^{\infty} n^{-1.001}$, studied in the text, exactly how many terms do we need (in theory) to make the error less than 5 in the ninth decimal place? You can use the inequalities from the proof of the Integral Test:

    $$\int_{N+1}^{\infty} f(x)\,dx < \sum_{n=N+1}^{\infty} f(n) < \int_{N}^{\infty} f(x)\,dx$$

7. Archimedes found an approximation to 2π by considering the perimeter p of a regular 96-gon inscribed in a circle of radius 1. His formula, in modern notation, is

    $$p = 96\sqrt{2 - \sqrt{2 + \sqrt{2 + \sqrt{2 + \sqrt{3}}}}}$$

    (a) Carry out the calculations and compare with the value of p from more accurate sources, say $p = 192 \sin(\pi/96)$. How many digits did you lose?
    (b) Perform rationalization to avoid subtraction of approximate numbers, and count the exact digits again.

8. This exercise is related to Exercise 2. Suppose that your computing device has an excellent program for the exponential function $\exp(x) = e^x$ but a poor program for $\ln x$. Use the identity

    $$\ln a = b + \ln\left(1 + \frac{a - e^b}{e^b}\right)$$

and Taylor's Formula (10.54) to improve the accuracy of $\ln x$.

9. The cubic equation

$$x^3 + px + q = 0$$

where we assume for simplicity that $p > 0$, has a classical solution formula for the real root, called Cardano's formula:

$$x = \frac{1}{3}\left[\left(\frac{27q + \sqrt{729q^2 + 108p^3}}{2}\right)^{\frac{1}{3}} + \left(\frac{27q - \sqrt{729q^2 + 108p^3}}{2}\right)^{\frac{1}{3}}\right]$$

For a user of a pocket-size calculator, as well as for an inexperienced programmer, the solution presents several stumbling blocks. First, the second radicand is negative and the fractional power key or routine may not handle it. Next, even if we fix the negative radical problem, when q is small in magnitude and p is of moderate size, the small number x is the difference of two numbers close to $\sqrt{p/3}$.

(a) Show that all these troubles are avoided by the formula

$$x = \frac{-9q}{a^{2/3} + 3p + 9p^2 a^{-2/3}}$$

where

$$a = \frac{27|q| + \sqrt{729q^2 + 108p^3}}{2}$$

Hint: Use the factorization formula

$$A + B = \frac{A^3 + B^3}{A^2 - AB + B^2}.$$

(b) Evaluate

$$u = \frac{4}{(2 + \sqrt{5})^{2/3} + 1 + (2 + \sqrt{5})^{-2/3}}$$

and if the result is simple, relate it to part (a), that is, restore the cubic equation whose root is u written in this form.

10. (a) A series convergent by the Ratio Test (10.32) offers an error estimate similar to the "first neglected term" rule (10.28). When the index N is large enough to make

$$\left|\frac{a_{n+1}}{a_n}\right| \leq r < 1 \quad \text{for every } n \geq N$$

(and we know that it will happen if the limit of the quotient is < 1), we can use this constant r to write

$$|s_n - s| = |a_{n+1} + a_{n+2} + \cdots|$$
$$\leq |a_{n+1}| + |a_{n+2}| + \cdots$$
$$\leq |a_n| \cdot \frac{r}{1 - r}$$

where s_n is the partial sum and s is the sum. (This estimate becomes especially simple when $r \leq \frac{1}{2}$, making $r/(1-r) \leq 1$: "the error is not more than the last used term.") Find the place in the proof of the Ratio Test where you can draw this conclusion.

(b) Consider the power series

$$f(x) = \sum_{n=1}^{\infty} \frac{x^n}{100^n + 1}$$

It is easy to show that its radius of convergence is $r = 100$. The series will converge rather slowly at $x = 99$: find out how many terms will make the error less than 5×10^{-7}.

(c) There is a way to speed up the convergence of the series in part (b). Show that

$$f(x) = \frac{x}{100 - x} - f\left(\frac{x}{100}\right)$$

and find the number of the terms of this transformed series that leads to an error less than 5×10^{-7}. [*Hint:* Compare with the series $\sum_{n=1}^{\infty}(x^n/100^n)$ whose sum you know.]

11. The positive numbers

$$a_n = \int_0^1 e^{1-x} x^n \, dx$$

can, in theory, be calculated from a reduction formula obtained by integration by parts: $a_0 = e - 1$, $a_n = na_{n-1} - 1$. Prove, using $1 \leq e^{1-x} \leq e$ and the Squeeze Theorem, that $\lim_{n\to\infty} a_n = 0$. Then try to calculate a_{20} from the reduction formula using your calculator. What went wrong?

The initial term $a_0 = e - 1$ cannot be represented exactly in a calculator. Let us call c the approximation of $e - 1$ that we can enter. Verify from the reduction formula (by observing the pattern after a few steps) that

$$a_n = \left[c - \left(\frac{1}{1!} + \frac{1}{2!} + \cdots + \frac{1}{n!}\right)\right]n!$$

and recall from Equation 10.48 that

$$\frac{1}{1!} + \frac{1}{2!} + \cdots + \frac{1}{n!}$$

converges to $e - 1$ as $n \to \infty$. The expression in square brackets converges to $c - (e - 1)$, a nonzero number, which gets multiplied by a fast-growing factor $n!$. We conclude that even if all further calculations (after entering a_0) were performed without errors, the initial inaccuracy would cause the computed sequence $\{a_n\}$ to diverge.

12. (a) A consolation after the catastrophic outcome of Exercise 11: If we rewrite the reduction formula to read

$$a_{n-1} = \frac{1 + a_n}{n}$$

we can use the inequality used in the squeeze argument to obtain improvements of the approximations of a_n. Try a_{20} again using this reverse approach.

(b) We used the reversed reduction formula to calculate quantities for which we have elementary formulas. To see that the idea is even more powerful, develop it for the integrals

$$\int_0^1 x^{n-\theta} e^{1-x} \, dx$$

where θ is a constant, $0 < \theta < 1$, and $n = 0, 1, \ldots$. For such θ, the integrals are no longer elementary (not solvable in "finite terms"), but the numbers can be calculated quickly. Find the integrals for the particular choice $\theta = \frac{1}{3}$ and $n = 0, 1, \ldots, 5$ to five digits of accuracy.

13. An advanced calculator has a key for a peculiar function:

$$E(x) = \begin{cases} 1 & \text{if } x = 0 \\ \dfrac{e^x - 1}{x} & \text{if } x \neq 0 \end{cases}$$

After so many warnings about the subtraction of close numbers, you may appreciate that the definition $\sinh x = \frac{1}{2}(e^x - e^{-x})$ gives inaccurate results for small x, where $\sinh x$ is close to x. Show that the use of the accurately evaluated function $E(x)$ helps restore the accuracy of $\sinh x$ for small x.

Rotation of Axes

In Section 9.6 we studied conic sections with equations of the form

$$Ax^2 + Cy^2 + Dx + Ey + F = 0$$

In this appendix we show that the general second-degree equation

(1)
$$Ax^2 + Bxy + Cy^2 + Dx + Ey + F = 0$$

can be analyzed by rotating the axes so as to eliminate the term Bxy.

In Figure 1 the x and y axes have been rotated about the origin through an acute angle θ to produce the X and Y axes. Thus a given point P has coordinates (x, y) in the first coordinate system (X, Y) in the new coordinate system. To see how X and Y are related to x and y we observe from Figure 2 that

$$X = r \cos \phi \qquad\qquad Y = r \sin \phi$$
$$x = r \cos (\theta + \phi) \qquad\qquad y = r \sin (\theta + \phi)$$

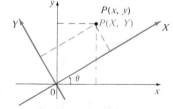

Figure 1

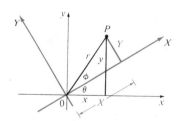

Figure 2

The addition formula for the cosine function then gives

$$x = r\cos(\theta + \phi) = r(\cos\theta\cos\phi - \sin\theta\sin\phi)$$
$$= (r\cos\phi)\cos\theta - (r\sin\phi)\sin\theta = X\cos\theta - Y\sin\theta$$

A similar computation gives y in terms of X and Y and so we have the following formulas:

(2)
$$x = X\cos\theta - Y\sin\theta \qquad y = X\sin\theta + Y\cos\theta$$

By solving Equations 2 for X and Y we obtain

(3)
$$X = x\cos\theta + y\sin\theta \qquad Y = -x\sin\theta + y\cos\theta$$

EXAMPLE 1   If the axes are rotated through $60°$, find the XY-coordinates of the point whose xy-coordinates are $(2, 6)$.

Solution   Using Equations 3 with $x = 2$, $y = 6$, and $\theta = 60°$, we have

$$X = 2\cos 60° + 6\sin 60° = 1 + 3\sqrt{3}$$
$$Y = -2\sin 60° + 6\cos 60° = -\sqrt{3} + 3$$

The XY-coordinates are $(1 + 3\sqrt{3}, 3 - \sqrt{3})$.

Now let us try to determine an angle θ such that the term Bxy in Equation 1 disappears when the axes are rotated through the angle θ. If we substitute from Equations 2 in Equation 1, we get

$$A(X\cos\theta - Y\sin\theta)^2 + B(X\cos\theta - Y\sin\theta)(X\sin\theta + Y\cos\theta)$$
$$+ C(X\sin\theta + Y\cos\theta)^2 + D(X\cos\theta - Y\sin\theta)$$
$$+ E(X\sin\theta + Y\cos\theta) + F = 0$$

Expanding and collecting terms, we obtain an equation of the form

(4)
$$A'X^2 + B'XY + C'Y^2 + D'X + E'Y + F = 0$$

where the coefficient B' of XY is

$$B' = 2(C - A)\sin\theta\cos\theta + B(\cos^2\theta - \sin^2\theta)$$
$$= (C - A)\sin 2\theta + B\cos 2\theta$$

To eliminate the XY term we choose θ so that $B' = 0$, that is,

$$(A - C)\sin 2\theta = B\cos 2\theta$$

(5) or
$$\cot 2\theta = \frac{A - C}{B}$$

EXAMPLE 2   Show that the graph of the equation $xy = 1$ is a hyperbola.

Solution   Notice that the equation $xy = 1$ is in the form of Equation 1 where $A = 0$, $B = 1$, and $C = 0$. According to Equation 5, the xy term will be eliminated if we choose θ so that

$$\cot 2\theta = \frac{A - C}{B} = 0$$

This will be true if $2\theta = \pi/2$, that is, $\theta = \pi/4$. Then $\cos\theta = \sin\theta = 1/\sqrt{2}$ and Equations 2 become

$$x = \frac{X}{\sqrt{2}} - \frac{Y}{\sqrt{2}} \qquad y = \frac{X}{\sqrt{2}} + \frac{Y}{\sqrt{2}}$$

Substituting these expressions into the original equation gives

$$\left(\frac{X}{\sqrt{2}} - \frac{Y}{\sqrt{2}}\right)\left(\frac{X}{\sqrt{2}} + \frac{Y}{\sqrt{2}}\right) = 1 \qquad \text{or} \qquad \frac{X^2}{2} - \frac{Y^2}{2} = 1$$

We recognize this as a hyperbola with vertices $(\pm\sqrt{2}, 0)$ in the XY-coordinate system. The asymptotes are $Y = \pm X$ in the XY-system, which correspond to the coordinate axes in the xy-system (see Figure 3).   ●

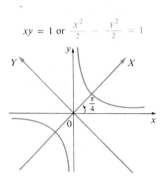

$xy = 1$ or $\dfrac{X^2}{2} - \dfrac{Y^2}{2} = 1$

Figure 3

EXAMPLE 3   Identify and sketch the curve

$$73x^2 + 72xy + 52y^2 + 30x - 40y - 75 = 0$$

Solution   This equation is in the form of Equation 1 with $A = 73$, $B = 72$, and $C = 52$. Thus

$$\cot 2\theta = \frac{A - C}{B} = \frac{73 - 52}{72} = \frac{7}{24}$$

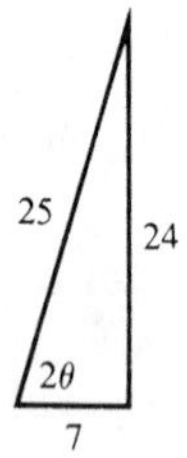

From the triangle in Figure 4 we see that

$$\cos 2\theta = \frac{7}{25}$$

Figure 4

The values of $\cos\theta$ and $\sin\theta$ can then be computed from the half-angle formulas:

$$\cos\theta = \sqrt{\frac{1 + \cos 2\theta}{2}} = \sqrt{\frac{1 + \frac{7}{25}}{2}} = \frac{4}{5}$$

$$\sin\theta = \sqrt{\frac{1 - \cos 2\theta}{2}} = \sqrt{\frac{1 - \frac{7}{25}}{2}} = \frac{3}{5}$$

The rotation Equations 2 become

$$x = \tfrac{4}{5}X - \tfrac{3}{5}Y \qquad y = \tfrac{3}{5}X + \tfrac{4}{5}Y$$

Substituting into the given equation, we have

$$73\left(\tfrac{4}{5}X - \tfrac{3}{5}Y\right)^2 + 72\left(\tfrac{4}{5}X - \tfrac{3}{5}Y\right)\left(\tfrac{3}{5}X + \tfrac{4}{5}Y\right) + 52\left(\tfrac{3}{5}X + \tfrac{4}{5}Y\right)^2$$

$$+ 30\left(\tfrac{4}{5}X - \tfrac{3}{5}Y\right) - 40\left(\tfrac{3}{5}X + \tfrac{4}{5}Y\right) - 75 = 0$$

which simplifies to $\qquad$ $4X^2 + Y^2 - 2Y = 3$

Completing the square gives

$$4X^2 + (Y-1)^2 = 4 \qquad \text{or} \qquad X^2 + \frac{(Y-1)^2}{4} = 1$$

and we recognize this as being an ellipse whose center is $(0,1)$ in XY-coordinates. Since $\theta = \cos^{-1}\left(\frac{4}{5}\right) \approx 37°$, we can sketch the graph in Figure 5.

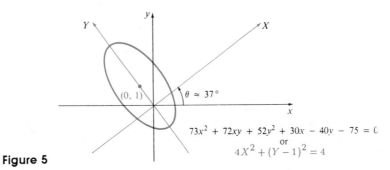

Figure 5

In Exercises 1–4 find the XY-coordinates of the given point if the axes are rotated through the given angle.

1. $(1, 4), 30°$
2. $(4, 3), 45°$
3. $(-2, 4), 60°$
4. $(1, 1), 15°$

In Exercises 5–12 use rotation of axes to identify and sketch the given curve.

5. $x^2 - 2xy + y^2 - x - y = 0$
6. $x^2 - xy + y^2 = 1$
7. $x^2 + xy + y^2 = 1$
8. $\sqrt{3}\,xy + y^2 = 1$
9. $97x^2 + 192xy + 153y^2 = 225$
10. $3x^2 - 12\sqrt{5}\,xy + 6y^2 + 9 = 0$
11. $2\sqrt{3}\,xy - 2y^2 - \sqrt{3}x - y = 0$
12. $16x^2 - 8\sqrt{2}\,xy + 2y^2 + (8\sqrt{2} - 3)x - (6\sqrt{2} + 4)y = 7$

13. (a) Use rotation of axes to show that the equation
$$36x^2 + 96xy + 64y^2 + 20x - 15y + 25 = 0$$
represents a parabola.
   (b) Find the XY-coordinates of the focus. Then find the xy-coordinates of the focus.
   (c) Find an equation of the directrix in the xy-coordinate system.

14. (a) Use rotation of axes to show that the equation
$$2x^2 - 72xy + 23y^2 - 80x - 60y = 125$$
represents a hyperbola.
   (b) Find the XY-coordinates of the foci. Then find the xy-coordinates of the foci.
   (c) Find the xy-coordinates of the vertices.
   (d) Find the equations of the asymptotes in the xy-coordinate system.
   (e) Find the eccentricity of the hyperbola.

15. Suppose that a rotation changes Equation 1 into Equation 4. Show that
$$A' + C' = A + C$$

16. Suppose that a rotation changes Equation 1 into Equation 4. Show that
$$(B')^2 - 4A'C' = B^2 - 4AC$$

17. Use Exercise 16 to show that Equation 1 represents (a) a parabola if $B^2 - 4AC = 0$, (b) an ellipse if $B^2 - 4AC < 0$, and (c) a hyperbola if $B^2 - 4AC > 0$ except in degenerate cases when it reduces to a point, a line, a pair of lines, or no graph at all.

18. Use Exercise 17 to determine the type of curve in Exercises 9–12.

Complex Numbers

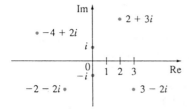

Figure 1

Complex numbers as points in the Argand plane

A **complex number** can be represented by an expression of the form $a + bi$, where a and b are real numbers and i is a symbol with the property that $i^2 = -1$. The complex number $a + bi$ can also be represented by the ordered pair (a, b) and plotted as a point in a plane (called the Argand plane) as in Figure 1. Thus the complex number $i = 0 + 1 \cdot i$ is identified with the point $(0, 1)$.

The **real part** of the complex number $a + bi$ is the real number a and the **imaginary part** is the real number b. Thus the real part of $4 - 3i$ is 4 and the imaginary part is -3. Two complex numbers $a + bi$ and $c + di$ are **equal** if $a = c$ and $b = d$, that is, their real parts are equal and their imaginary parts are equal. In the Argand plane the x-axis is called the real axis and the y-axis is called the imaginary axis.

The sum and difference of two complex numbers are defined by adding or subtracting their real parts and their imaginary parts, respectively:

$$(a + bi) + (c + di) = (a + c) + (b + d)i$$
$$(a + bi) - (c + di) = (a - c) + (b - d)i$$

For instance,

$$(1 - i) + (4 + 7i) = (1 + 4) + (-1 + 7)i = 5 + 6i$$

The product of complex numbers is defined so that the usual commutative and distributive laws hold:

$$(a + bi)(c + di) = a(c + di) + (bi)(c + di)$$
$$= ac + adi + bci + bdi^2$$

A46

Since $i^2 = -1$, this becomes

$$(a + bi)(c + di) = (ac - bd) + (ad + bc)i$$

EXAMPLE 1

$$(-1 + 3i)(2 - 5i) = -(2 - 5i) + 3i(2 - 5i)$$
$$= -2 + 5i + 6i - 15(-1) = 13 + 11i$$

Division of complex numbers is much like rationalizing the denominator of a rational expression. For the complex numbers $z = a + bi$, we define its **complex conjugate** to be $\bar{z} = a - bi$. To find the quotient of two complex numbers we multiply numerator and denominator by the complex conjugate of the denominator.

EXAMPLE 2   Express the number $\dfrac{-1 + 3i}{2 + 5i}$ in the form $a + bi$.

Solution   We multiply numerator and denominator by the complex conjugate of $2 + 5i$, namely $2 - 5i$, and we take advantage of the result of Example 1:

$$\frac{-1 + 3i}{2 + 5i} = \frac{-1 + 3i}{2 + 5i} \cdot \frac{2 - 5i}{2 - 5i} = \frac{13 + 11i}{2^2 + 5^2} = \frac{13}{29} + \frac{11}{29}i$$

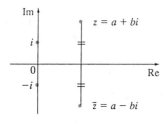

Figure 2

The geometric interpretation of the complex conjugate is shown in Figure 2: $\bar{z}$ is the reflection of z in the real axis. We list some of the properties of the complex conjugate in the following table. The proofs follow from the definition and are requested in Exercise 18.

Properties of Conjugates

$$\overline{z + w} = \bar{z} + \bar{w} \qquad \overline{zw} = \bar{z}\bar{w} \qquad \overline{z^n} = \bar{z}^n$$

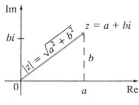

Figure 3

The **modulus**, or **absolute value**, $|z|$ of a complex number $z = a + bi$ is its distance from the origin. From Figure 3 we see that if $z = a + bi$, then

$$|z| = \sqrt{a^2 + b^2}$$

Notice that

$$z\bar{z} = (a + bi)(a - bi) = a^2 + abi - abi - b^2i^2 = a^2 + b^2$$

and so

$$z\bar{z} = |z|^2$$

This explains why the division procedure in Example 2 works in general:

$$\frac{z}{w} = \frac{z\bar{w}}{w\bar{w}} = \frac{z\bar{w}}{|w|^2}$$

Since $i^2 = -1$, we can think of i as a square root of -1. But we also have $(-i)^2 = i^2 = -1$ and so $-i$ is also a square root of -1. We say that i is the **principal square root** of -1 and write $\sqrt{-1} = i$. In general, if c is any positive number we write

$$\sqrt{-c} = \sqrt{c}\, i$$

With this convention the usual derivation and formula for the roots of the quadratic equation $ax^2 + bx + c = 0$ are valid even when $b^2 - 4ac < 0$:

$$x = \frac{-b \pm \sqrt{b^2 - 4ac}}{2a}$$

EXAMPLE 3   Find the roots of the equation $x^2 + x + 1 = 0$.

Solution   Using the quadratic formula, we have

$$x = \frac{-1 \pm \sqrt{1^2 - 4 \cdot 1}}{2} = \frac{-1 \pm \sqrt{-3}}{2} = \frac{-1 \pm \sqrt{3}i}{2}$$ ●

We observe that the solutions of the equation in Example 3 are complex conjugates of each other. In general, the solutions of any quadratic equation $ax^2 + bx + c = 0$ with real coefficients a, b, and c are always complex conjugates. (If z is real, $\overline{z} = z$, so z is its own conjugate.)

We have seen that if we allow complex numbers as solutions, then every quadratic equation has a solution. More generally, it is true that every polynomial equation

$$a_n x^n + a_{n-1} x^{n-1} + \cdots + a_1 x + a_0 = 0$$

of degree at least one has a solution among the complex numbers. This fact is known as the Fundamental Theorem of Algebra and was proved by Gauss.

Polar Form

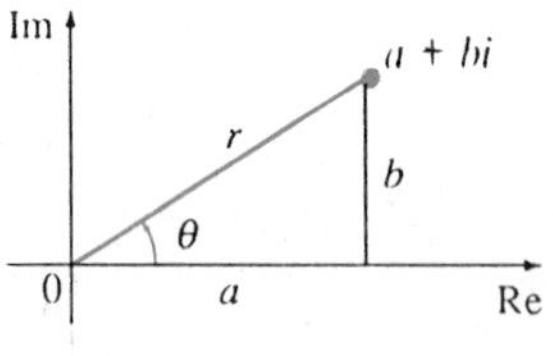

Figure 4

We know that any complex number $z = a + bi$ can be considered as a point (a, b) and that any such point can be represented by polar coordinates (r, θ) with $r \geq 0$. In fact,

$$a = r \cos \theta \qquad b = r \sin \theta$$

as in Figure 4. Therefore we have

$$z = a + bi = (r \cos \theta) + (r \sin \theta)i$$

Thus we can write any complex number z in the form

$$\boxed{z = r (\cos \theta + i \sin \theta)}$$

where $r = |z| = \sqrt{a^2 + b^2}$ and $\tan \theta = \frac{b}{a}$

The angle θ is called the **argument** of z and we write $\theta = \arg(z)$. Note that $\arg(z)$ is not unique; any two arguments of z differ by an integer multiple of 2π.

EXAMPLE 4   Write the following numbers in polar form:

(a) $z = 1 + i$ (b) $w = \sqrt{3} - i$

Solution

(a)   We have $r = |z| = \sqrt{1+1} = \sqrt{2}$ and $\tan\theta = 1$, so we can take $\theta = \pi/4$. Therefore the polar form is

$$z = \sqrt{2}\left(\cos\frac{\pi}{4} + i\sin\frac{\pi}{4}\right)$$

(b)   Here we have $r = |w| = \sqrt{3+1} = 2$ and $\tan\theta = -1/\sqrt{3}$. Since w is lies in the fourth quadrant, we take $\theta = -\pi/6$ and

$$w = 2\left[\cos\left(-\frac{\pi}{6}\right) + i\sin\left(-\frac{\pi}{6}\right)\right]$$

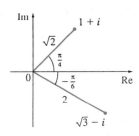

Figure 5   The numbers z and w are shown in Figure 5.

The polar form of complex numbers gives insight into multiplication and division. Let

$$z_1 = r_1(\cos\theta_1 + i\sin\theta_1) \qquad z_2 = r_2(\cos\theta_2 + i\sin\theta_2)$$

be two complex numbers written in polar form. Then

$$z_1 z_2 = r_1 r_2 (\cos\theta_1 + i\sin\theta_1)(\cos\theta_2 + i\sin\theta_2)$$

$$= r_1 r_2[(\cos\theta_1\cos\theta_2 - \sin\theta_1\sin\theta_2) + i(\sin\theta_1\cos\theta_2 + \cos\theta_1\sin\theta_2)]$$

Therefore, using the addition formulas for cosine and sine, we have

(1)
$$\boxed{z_1 z_2 = r_1 r_2[\cos(\theta_1 + \theta_2) + i\sin(\theta_1 + \theta_2)]}$$

This formula says that *to multiply two complex numbers we multiply the moduli and add the arguments.* (See Figure 6.)

A similar argument using the subtraction formulas for sine and cosine shows that *to divide two complex numbers we divide the moduli and subtract the arguments.*

$$\boxed{\frac{z_1}{z_2} = \frac{r_1}{r_2}[\cos(\theta_1 - \theta_2) + i\sin(\theta_1 - \theta_2)], \qquad z_2 \neq 0}$$

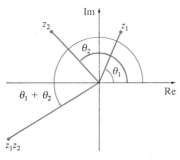

Figure 6

In particular, taking $z_1 = 1$ and $z_2 = z$, we have the following, which is illustrated in Figure 7.

$$\boxed{\text{If } z = r(\cos\theta + i\sin\theta), \text{ then } \frac{1}{z} = \frac{1}{r}(\cos\theta - i\sin\theta)}$$

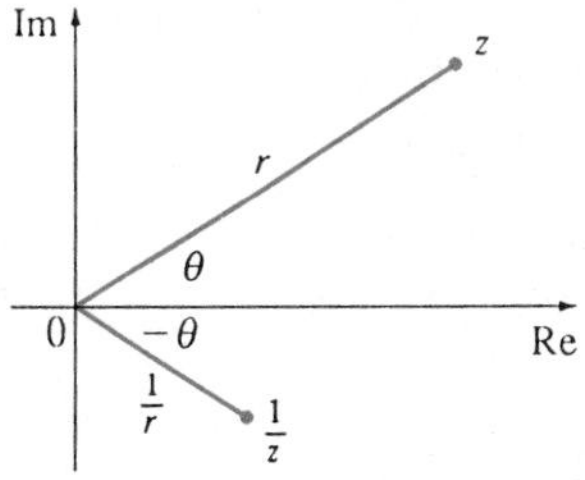

EXAMPLE 5   Find the product of the complex numbers $1 + i$ and $\sqrt{3} - i$ in polar form.

Solution   From Example 4 we have

Figure 7
$$1 + i = \sqrt{2}\left(\cos\frac{\pi}{4} + i\sin\frac{\pi}{4}\right) \quad \text{and} \quad \sqrt{3} - i = 2\left[\cos\left(-\frac{\pi}{6}\right) + i\sin\left(-\frac{\pi}{6}\right)\right]$$

So by Equation 1,

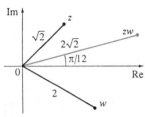

Figure 8

$$(1+i)(\sqrt{3}-i) = 2\sqrt{2}\left[\cos\left(\frac{\pi}{4}-\frac{\pi}{6}\right)+i\sin\left(\frac{\pi}{4}-\frac{\pi}{6}\right)\right]$$

$$= 2\sqrt{2}\left(\cos\frac{\pi}{12}+i\sin\frac{\pi}{12}\right)$$

This is illustrated in Figure 8.

Repeated use of Formula 1 shows how to compute powers of a complex number. If

$$z = r(\cos\theta + i\sin\theta)$$

then
$$z^2 = r^2(\cos 2\theta + i\sin 2\theta)$$

and
$$z^3 = zz^2 = r^3(\cos 3\theta + i\sin 3\theta)$$

In general, we obtain the following result, named after the French mathematician Abraham De Moivre (1667–1754).

De Moivre's Theorem (2)

> If $z = r(\cos\theta + i\sin\theta)$ and n is a positive integer, then
>
> $$z^n = [r(\cos\theta + i\sin\theta)]^n = r^n(\cos n\theta + i\sin n\theta)$$

This says that *to take the nth power of a complex number we take the nth power of the modulus and multiply the argument by n.*

EXAMPLE 6   Find $(\frac{1}{2}+\frac{1}{2}i)^{10}$.

Solution   Since $\frac{1}{2}+\frac{1}{2}i = \frac{1}{2}(1+i)$, it follows from Example 4(a) that $\frac{1}{2}+\frac{1}{2}i$ has the polar form

$$\frac{1}{2}+\frac{1}{2}i = \frac{\sqrt{2}}{2}\left(\cos\frac{\pi}{4}+i\sin\frac{\pi}{4}\right)$$

So by De Moivre's Theorem,

$$\left(\frac{1}{2}+\frac{1}{2}i\right)^{10} = \left(\frac{\sqrt{2}}{2}\right)^{10}\left(\cos\frac{10\pi}{4}+i\sin\frac{10\pi}{4}\right)$$

$$= \frac{2^5}{2^{10}}\left(\cos\frac{5\pi}{2}+i\sin\frac{5\pi}{2}\right) = \frac{1}{32}i$$

De Moivre's Theorem can also be used to find the nth roots of complex numbers. An nth root of the complex number z is a complex number w such that

$$w^n = z$$

Writing these two numbers in trigonometric form as

$$w = s(\cos\phi + i\sin\phi) \qquad \text{and} \qquad z = r(\cos\theta + i\sin\theta)$$

and using De Moivre's Theorem, we get

$$s^n(\cos n\phi + i\sin n\phi) = r(\cos\theta + i\sin\theta)$$

The equality of these two complex numbers shows that

$$s^n = r \qquad \text{or} \qquad s = r^{1/n}$$

and $\qquad\qquad\qquad \cos n\phi = \cos\theta \qquad \text{and} \qquad \sin n\phi = \sin\theta$

From the fact that sine and cosine have period 2π it follows that

$$n\phi = \theta + 2k\pi \qquad \text{or} \qquad \phi = \frac{\theta + 2k\pi}{n}$$

Thus

$$w = r^{1/n}\left[\cos\left(\frac{\theta + 2k\pi}{n}\right) + i\sin\left(\frac{\theta + 2k\pi}{n}\right)\right]$$

Since this expression gives a different value of w for $k = 0, 1, 2, \ldots, n-1$, we have the following:

Roots of a Complex Number (3)

Let $z = r(\cos\theta + i\sin\theta)$ and let n be a positive integer. Then z has the n distinct nth roots

$$w_k = r^{1/n}\left[\cos\left(\frac{\theta + 2k\pi}{n}\right) + i\sin\left(\frac{\theta + 2k\pi}{n}\right)\right]$$

where $k = 0, 1, 2, \ldots, n-1$.

Notice that each of the nth roots of z has modulus $|w_k| = r^{1/n}$. Thus all the nth roots of z lie on the circle of radius $r^{1/n}$ in the complex plane. Also, since the argument of each successive nth root exceeds the argument of the previous root by $2\pi/n$, we see that the nth roots of z are equally spaced on this circle.

EXAMPLE 7   Find the six sixth roots of $z = -8$ and graph these roots in the complex plane.

Solution   In trigonometric form $z = 8(\cos\pi + i\sin\pi)$. Applying Equation 3 with $n = 6$, we get

$$w_k = 8^{1/6}\left(\cos\frac{\pi + 2k\pi}{6} + i\sin\frac{\pi + 2k\pi}{6}\right)$$

We get six sixth roots of -8 by taking $k = 0, 1, 2, 3, 4, 5$ in this formula:

$$w_0 = 8^{1/6}\left(\cos\frac{\pi}{6} + i\sin\frac{\pi}{6}\right) = \sqrt{2}\left(\frac{\sqrt{3}}{2} + \frac{1}{2}i\right)$$

$$w_1 = 8^{1/6}\left(\cos\frac{\pi}{2} + i\sin\frac{\pi}{2}\right) = \sqrt{2}i$$

$$w_2 = 8^{1/6}\left(\cos\frac{5\pi}{6} + i\sin\frac{5\pi}{6}\right) = \sqrt{2}\left(-\frac{\sqrt{3}}{2} + \frac{1}{2}i\right)$$

$$w_3 = 8^{1/6}\left(\cos\frac{7\pi}{6} + i\sin\frac{7\pi}{6}\right) = \sqrt{2}\left(-\frac{\sqrt{3}}{2} - \frac{1}{2}i\right)$$

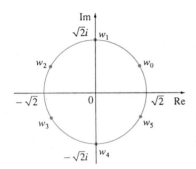

Figure 9

The six sixth roots of $z = -8$

$$w_4 = 8^{1/6}\left(\cos\frac{3\pi}{2} + i\sin\frac{3\pi}{2}\right) = -\sqrt{2}i$$

$$w_5 = 8^{1/6}\left(\cos\frac{11\pi}{6} + i\sin\frac{11\pi}{6}\right) = \sqrt{2}\left(\frac{\sqrt{3}}{2} - \frac{1}{2}i\right)$$

All these points lie on the circle of radius $\sqrt{2}$ as shown in Figure 9.

Complex Exponentials

We also need to give a meaning to the expression e^z when $z = x + iy$ is a complex number. The theory of infinite series as developed in Chapter 10 can be extended to the case where the terms are complex numbers. Using the Taylor series for e^x (10.47) as our guide, we define

(4)
$$e^z = \sum_{n=0}^{\infty} \frac{z^n}{n!} = 1 + z + \frac{z^2}{2!} + \frac{z^3}{3!} + \cdots$$

and it turns out that this complex exponential function has the same properties as the real exponential function. In particular, it is true that

(5)
$$e^{z_1 + z_2} = e^{z_1} e^{z_2}$$

If we put $z = iy$, where y is a real number, in Equation 4, and use the facts that

$$i^2 = -1, \quad i^3 = i^2 i = -i, \quad i^4 = 1, \quad i^5 = i, \cdots$$

we get
$$e^{iy} = 1 + iy + \frac{(iy)^2}{2!} + \frac{(iy)^3}{3!} + \frac{(iy)^4}{4!} + \frac{(iy)^5}{5!} + \cdots$$

$$= 1 + iy - \frac{y^2}{2!} - i\frac{y^3}{3!} + \frac{y^4}{4!} + i\frac{y^5}{5!} + \cdots$$

$$= \left(1 - \frac{y^2}{2!} + \frac{y^4}{4!} - \frac{y^6}{6!} + \cdots\right) + i\left(y - \frac{y^3}{3!} + \frac{y^5}{5!} - \cdots\right)$$

$$= \cos y + i\sin y$$

Here we have used the Taylor series for $\cos y$ and $\sin y$ (Equations 10.50 and 10.49). The result is a famous formula called **Euler's formula:**

(6)
$$\boxed{e^{iy} = \cos y + i\sin y}$$

Comparing Euler's formula with Equation 5, we get

(7)
$$e^{x+iy} = e^x e^{iy} = e^x(\cos y + i\sin y)$$

EXAMPLE 8   Evaluate: (a) $e^{i\pi}$ (b) $e^{-1+i\pi/2}$

Solution

(a)   From Euler's equation (6) we have

$$e^{i\pi} = \cos\pi + i\sin\pi = -1 + i(0) = -1$$

(b)   Using Equation 7 we get

$$e^{-1+i\pi/2} = e^{-1}\left(\cos\frac{\pi}{2} + i\sin\frac{\pi}{2}\right) = \frac{1}{e}[0 + i(1)] = \frac{i}{e}$$

•

Finally we note that Euler's equation provides us with an easier method of proving De Moivre's Theorem:

$$[r(\cos\theta + i\sin\theta)]^n = (re^{i\theta})^n = r^n e^{in\theta} = r^n(\cos n\theta + i\sin n\theta)$$

APPENDIX F   **Exercises**

In Exercises 1–14 evaluate the expression and write your answer in the form $a + bi$.

1. $(3 + 2i) + (7 - 3i)$ 2. $(1 + i) - (2 - 3i)$

3. $(3 - i)(4 + i)$ 4. $(4 - 7i)(1 + 3i)$

5. $\overline{12 + 7i}$ 6. $\overline{2i(\frac{1}{2} - i)}$

7. $\dfrac{2 + 3i}{1 - 5i}$ 8. $\dfrac{5 - i}{3 + 4i}$

9. $\dfrac{1}{1 + i}$ 10. $\dfrac{3}{4 - 3i}$

11. i^3 12. i^{100}

13. $\sqrt{-25}$ 14. $\sqrt{-3}\sqrt{-12}$

In Exercises 15–17 find the complete conjugate and the modulus of the number.

15. $3 + 4i$ 16. $\sqrt{3} - i$ 17. $-4i$

18. Prove the following properties of complex numbers.
   (a) $\overline{z + w} = \overline{z} + \overline{w}$ (b) $\overline{zw} = \overline{z}\,\overline{w}$
   (c) $\overline{z^n} = \overline{z}^n$, where n is a positive integer
       (*Hint:* Write $z = a + bi$, $w = c + di$.)

In Exercises 19–24 find all solutions of the given equation.

19. $4x^2 + 9 = 0$ 20. $x^4 = 1$

21. $x^2 - 8x + 17 = 0$ 22. $x^2 - 4x + 5 = 0$

23. $z^2 + z + 2 = 0$ 24. $z^2 + \frac{1}{2}z + \frac{1}{4} = 0$

In Exercises 25–28 write the number in polar form with argument between 0 and 2π.

25. $-3 + 3i$ 26. $1 - \sqrt{3}i$

27. $3 + 4i$ 28. $8i$

In Exercises 29–32 find polar forms for zw, z/w, and $1/z$ by first putting z and w into polar form.

29. $z = \sqrt{3} + i$, $w = 1 + \sqrt{3}i$

30. $z = 4\sqrt{3} - 4i$, $w = 8i$

31. $z = 2\sqrt{3} - 2i$, $w = -1 + i$

32. $z = 4(\sqrt{3} + i)$, $w = -3 - 3i$

In Exercises 33–36 find the indicated power using De Moivre's Theorem.

33. $(1 + i)^{20}$ 34. $(1 - \sqrt{3}i)^5$

35. $(2\sqrt{3} + 2i)^5$ 36. $(1 - i)^8$

In Exercises 37–40 find the indicated roots. Sketch the roots in the complex plane.

37. The eighth roots of 1

38. The fifth roots of 32

39. The cube roots of i

40. The cube roots of $1 + i$

In Exercises 41–46 write the number in the form $a + bi$.

41. $e^{i\pi/2}$ 42. $e^{2\pi i}$ 43. $e^{i3\pi/4}$

44. $e^{-i\pi}$ 45. $e^{2 + i\pi}$ 46. $e^{1 + 2i}$

47. Use De Moivre's Theorem with $n = 3$ to express $\cos 3\theta$ and $\sin 3\theta$ in terms of $\cos\theta$ and $\sin\theta$.

48. Use Euler's formula to prove the following formulas for $\cos x$ and $\sin x$

$$\cos x = \frac{e^{ix} + e^{-ix}}{2}$$

$$\sin x = \frac{e^{ix} - e^{-ix}}{2i}$$

49. If $u(x) = f(x) + ig(x)$ is a complex-valued function of a real variable x and the real and imaginary parts $f(x)$ and $g(x)$ are differentiable functions of x, then the derivative of u is defined to be $u'(x) = f'(x) + ig'(x)$. Use this together with Equation 7 to prove that if $F(x) = e^{rx}$ then $F'(x) = re^{rx}$ when $r = a + bi$ is a complex number.

Table of Integrals

Basic Forms

1. $\displaystyle\int u\,dv = uv - \int v\,du$

2. $\displaystyle\int u^n\,du = \frac{1}{n+1}u^{n+1} + C, \quad n \neq -1$

3. $\displaystyle\int \frac{du}{u} = \ln|u| + C$

4. $\displaystyle\int e^u\,du = e^u + C$

5. $\displaystyle\int a^u\,du = \frac{1}{\ln a}a^u + C$

6. $\displaystyle\int \sin u\,du = -\cos u + C$

7. $\displaystyle\int \cos u\,du = \sin u + C$

8. $\displaystyle\int \sec^2 u\,du = \tan u + C$

9. $\displaystyle\int \csc^2 u\,du = -\cot u + C$

10. $\displaystyle\int \sec u \tan u\,du = \sec u + C$

11. $\displaystyle\int \csc u \cot u\,du = -\csc u + C$

12. $\displaystyle\int \tan u\,du = \ln|\sec u| + C$

13. $\displaystyle\int \cot u\,du = \ln|\sin u| + C$

14. $\displaystyle\int \sec u\,du = \ln|\sec u + \tan u| + C$

15. $\displaystyle\int \csc u\,du = \ln|\csc u - \cot u| + C$

16. $\displaystyle\int \frac{du}{\sqrt{a^2 - u^2}} = \sin^{-1}\frac{u}{a} + C$

17. $\displaystyle\int \frac{du}{a^2 + u^2} = \frac{1}{a}\tan^{-1}\frac{u}{a} + C$

18. $\displaystyle\int \frac{du}{u\sqrt{u^2 - a^2}} = \frac{1}{a}\sec^{-1}\frac{u}{a} + C$

19. $\displaystyle\int \frac{du}{a^2 - u^2} = \frac{1}{2a}\ln\left|\frac{u+a}{u-a}\right| + C$

20. $\displaystyle\int \frac{du}{u^2 - a^2} = \frac{1}{2a}\ln\left|\frac{u-a}{u+a}\right| + C$

Forms involving $\sqrt{a^2+u^2}$

21. $\displaystyle \int \sqrt{a^2+u^2}\,du = \frac{u}{2}\sqrt{a^2+u^2} + \frac{a^2}{2}\ln\left|u+\sqrt{a^2+u^2}\right| + C$

22. $\displaystyle \int u^2\sqrt{a^2+u^2}\,du = \frac{u}{8}(a^2+2u^2)\sqrt{a^2+u^2} - \frac{a^4}{8}\ln\left|u+\sqrt{a^2+u^2}\right| + C$

23. $\displaystyle \int \frac{\sqrt{a^2+u^2}}{u}\,du = \sqrt{a^2+u^2} - a\ln\left|\frac{a+\sqrt{a^2+u^2}}{u}\right| + C$

24. $\displaystyle \int \frac{\sqrt{a^2+u^2}}{u^2}\,du = -\frac{\sqrt{a^2+u^2}}{u} + \ln\left|u+\sqrt{a^2+u^2}\right| + C$

27. $\displaystyle \int \frac{du}{u\sqrt{a^2+u^2}} = -\frac{1}{a}\ln\left|\frac{\sqrt{a^2+u^2}+a}{u}\right| + C$

25. $\displaystyle \int \frac{du}{\sqrt{a^2+u^2}} = \ln\left|u+\sqrt{a^2+u^2}\right| + C$

28. $\displaystyle \int \frac{du}{u^2\sqrt{a^2+u^2}} = -\frac{\sqrt{a^2+u^2}}{a^2 u} + C$

26. $\displaystyle \int \frac{u^2\,du}{\sqrt{a^2+u^2}} = \frac{u}{2}\sqrt{a^2+u^2} - \frac{a^2}{2}\ln\left|u+\sqrt{a^2+u^2}\right| + C$

29. $\displaystyle \int \frac{du}{(a^2+u^2)^{3/2}} = \frac{u}{a^2\sqrt{a^2+u^2}} + C$

Forms involving $\sqrt{a^2-u^2}$

30. $\displaystyle \int \sqrt{a^2-u^2}\,du = \frac{u}{2}\sqrt{a^2-u^2} + \frac{a^2}{2}\sin^{-1}\frac{u}{a} + C$

31. $\displaystyle \int u^2\sqrt{a^2-u^2}\,du = \frac{u}{8}(2u^2-a^2)\sqrt{a^2-u^2} + \frac{a^4}{8}\sin^{-1}\frac{u}{a} + C$

32. $\displaystyle \int \frac{\sqrt{a^2-u^2}}{u}\,du = \sqrt{a^2-u^2} - a\ln\left|\frac{a+\sqrt{a^2-u^2}}{u}\right| + C$

33. $\displaystyle \int \frac{\sqrt{a^2-u^2}}{u^2}\,du = -\frac{1}{u}\sqrt{a^2-u^2} - \sin^{-1}\frac{u}{a} + C$

34. $\displaystyle \int \frac{u^2\,du}{\sqrt{a^2-u^2}} = -\frac{u}{2}\sqrt{a^2-u^2} + \frac{a^2}{2}\sin^{-1}\frac{u}{a} + C$

35. $\displaystyle \int \frac{du}{u\sqrt{a^2-u^2}} = -\frac{1}{a}\ln\left|\frac{a+\sqrt{a^2-u^2}}{u}\right| + C$

36. $\displaystyle \int \frac{du}{u^2\sqrt{a^2-u^2}} = -\frac{1}{a^2 u}\sqrt{a^2-u^2} + C$

37. $\displaystyle \int (a^2-u^2)^{3/2}\,du = -\frac{u}{8}(2u^2-5a^2)\sqrt{a^2-u^2} + \frac{3a^4}{8}\sin^{-1}\frac{u}{a} + C$

38. $\displaystyle \int \frac{du}{(a^2-u^2)^{3/2}} = \frac{u}{a^2\sqrt{a^2-u^2}} + C$

Forms involving $\sqrt{u^2-a^2}$

39. $\displaystyle \int \sqrt{u^2-a^2}\,du = \frac{u}{2}\sqrt{u^2-a^2} - \frac{a^2}{2}\ln\left|u+\sqrt{u^2-a^2}\right| + C$

40. $\displaystyle \int u^2\sqrt{u^2-a^2}\,du = \frac{u}{8}(2u^2-a^2)\sqrt{u^2-a^2} - \frac{a^4}{8}\ln\left|u+\sqrt{u^2-a^2}\right| + C$

41. $\displaystyle\int \frac{\sqrt{u^2-a^2}}{u}\,du = \sqrt{u^2-a^2} - a\cos^{-1}\frac{a}{u} + C$

42. $\displaystyle\int \frac{\sqrt{u^2-a^2}}{u^2}\,du = -\frac{\sqrt{u^2-a^2}}{u} + \ln|u+\sqrt{u^2-a^2}| + C$

43. $\displaystyle\int \frac{du}{\sqrt{u^2-a^2}} = \ln|u+\sqrt{u^2-a^2}| + C$

44. $\displaystyle\int \frac{u^2\,du}{\sqrt{u^2-a^2}} = \frac{u}{2}\sqrt{u^2-a^2} + \frac{a^2}{2}\ln|u+\sqrt{u^2-a^2}| + C$

45. $\displaystyle\int \frac{du}{u^2\sqrt{u^2-a^2}} = \frac{\sqrt{u^2-a^2}}{a^2 u} + C$

46. $\displaystyle\int \frac{du}{(u^2-a^2)^{3/2}} = -\frac{u}{a^2\sqrt{u^2-a^2}} + C$

Forms involving $a+bu$

47. $\displaystyle\int \frac{u\,du}{a+bu} = \frac{1}{b^2}\left(a+bu - a\ln|a+bu|\right) + C$

48. $\displaystyle\int \frac{u^2\,du}{a+bu}$

$\qquad = \dfrac{1}{2b^3}\left[(a+bu)^2 - 4a(a+bu) + 2a^2\ln|a+bu|\right] + C$

49. $\displaystyle\int \frac{du}{u(a+bu)} = \frac{1}{a}\ln\left|\frac{u}{a+bu}\right| + C$

50. $\displaystyle\int \frac{du}{u^2(a+bu)} = -\frac{1}{au} + \frac{b}{a^2}\ln\left|\frac{a+bu}{u}\right| + C$

51. $\displaystyle\int \frac{u\,du}{(a+bu)^2} = \frac{a}{b^2(a+bu)} + \frac{1}{b^2}\ln|a+bu| + C$

52. $\displaystyle\int \frac{du}{u(a+bu)^2} = \frac{1}{a(a+bu)} - \frac{1}{a^2}\ln\left|\frac{a+bu}{u}\right| + C$

53. $\displaystyle\int \frac{u^2\,du}{(a+bu)^2} = \frac{1}{b^3}\left(a+bu - \frac{a^2}{a+bu} - 2a\ln|a+bu|\right) + C$

54. $\displaystyle\int u\sqrt{a+bu}\,du = \frac{2}{15b^2}(3bu-2a)(a+bu)^{3/2} + C$

55. $\displaystyle\int \frac{u\,du}{\sqrt{a+bu}} = \frac{2}{3b^2}(bu-2a)\sqrt{a+bu}$

56. $\displaystyle\int \frac{u^2\,du}{\sqrt{a+bu}} = \frac{2}{15b^3}(8a^2+3b^2u^2-4abu)\sqrt{a+bu}$

57. $\displaystyle\int \frac{du}{u\sqrt{a+bu}} = \frac{1}{\sqrt{a}}\ln\left|\frac{\sqrt{a+bu}-\sqrt{a}}{\sqrt{a+bu}+\sqrt{a}}\right| + C, \text{ if } a>0$

$\qquad = \dfrac{2}{\sqrt{-a}}\tan^{-1}\sqrt{\dfrac{a+bu}{-a}} + C, \quad \text{ if } a<0$

58. $\displaystyle\int \frac{\sqrt{a+bu}}{u}\,du = 2\sqrt{a+bu} + a\int \frac{du}{u\sqrt{a+bu}}$

59. $\displaystyle\int \frac{\sqrt{a+bu}}{u^2}\,du = -\frac{\sqrt{a+bu}}{u} + \frac{b}{2}\int \frac{du}{u\sqrt{a+bu}}$

60. $\displaystyle\int u^n\sqrt{a+bu}\,du$

$\qquad = \dfrac{2}{b(2n+3)}\left[u^n(a+bu)^{3/2} - na\int u^{n-1}\sqrt{a+bu}\,du\right]$

61. $\displaystyle\int \frac{u^n\,du}{\sqrt{a+bu}} = \frac{2u^n\sqrt{a+bu}}{b(2n+1)} - \frac{2na}{b(2n+1)}\int \frac{u^{n-1}\,du}{\sqrt{a+bu}}$

62. $\displaystyle\int \frac{du}{u^n\sqrt{a+bu}}$

$\qquad = -\dfrac{\sqrt{a+bu}}{a(n-1)u^{n-1}} - \dfrac{b(2n-3)}{2a(n-1)}\int \dfrac{du}{u^{n-1}\sqrt{a+bu}}$

Trigonometric Forms

63. $\displaystyle\int \sin^2 u\,du = \frac{1}{2}u - \frac{1}{4}\sin 2u + C$

64. $\displaystyle\int \cos^2 u\,du = \frac{1}{2}u + \frac{1}{4}\sin 2u + C$

65. $\displaystyle\int \tan^2 u\,du = \tan u - u + C$

66. $\displaystyle\int \cot^2 u\,du = -\cot u - u + C$

67. $\displaystyle\int \sin^3 u\,du = -\frac{1}{3}(2+\sin^2 u)\cos u + C$

68. $\displaystyle\int \cos^3 u\,du = \frac{1}{3}(2+\cos^2 u)\sin u + C$

69. $\displaystyle\int \tan^3 u\,du = \frac{1}{2}\tan^2 u + \ln|\cos u| + C$

70. $\displaystyle\int \cot^3 u\,du = -\frac{1}{2}\cot^2 u - \ln|\sin u| + C$

71. $\int \sec^3 u \, du = \frac{1}{2} \sec u \tan u + \frac{1}{2} \ln |\sec u + \tan u| + C$

80. $\int \cos au \cos bu \, du = \frac{\sin(a-b)u}{2(a-b)} + \frac{\sin(a+b)u}{2(a+b)} + C$

72. $\int \csc^3 u \, du = -\frac{1}{2} \csc u \cot u + \frac{1}{2} \ln |\csc u - \cot u| + C$

81. $\int \sin au \cos bu \, du = -\frac{\cos(a-b)u}{2(a-b)} - \frac{\cos(a+b)u}{2(a+b)} + C$

73. $\int \sin^n u \, du = -\frac{1}{n} \sin^{n-1} u \cos u + \frac{n-1}{n} \int \sin^{n-2} u \, du$

82. $\int u \sin u \, du = \sin u - u \cos u + C$

74. $\int \cos^n u \, du = \frac{1}{n} \cos^{n-1} u \sin u + \frac{n-1}{n} \int \cos^{n-2} u \, du$

83. $\int u \cos u \, du = \cos u + u \sin u + C$

75. $\int \tan^n u \, du = \frac{1}{n-1} \tan^{n-1} u - \int \tan^{n-2} u \, du$

84. $\int u^n \sin u \, du = -u^n \cos u + n \int u^{n-1} \cos u \, du$

76. $\int \cot^n u \, du = \frac{-1}{n-1} \cot^{n-1} u - \int \cot^{n-2} u \, du$

85. $\int u^n \cos u \, du = u^n \sin u - n \int u^{n-1} \sin u \, du$

77. $\int \sec^n u \, du = \frac{1}{n-1} \tan u \sec^{n-2} u + \frac{n-2}{n-1} \int \sec^{n-2} u \, du$

86. $\int \sin^n u \cos^m u \, du$

78. $\int \csc^n u \, du = \frac{-1}{n-1} \cot u \csc^{n-2} u + \frac{n-2}{n-1} \int \csc^{n-2} u \, du$

$$= -\frac{\sin^{n-1} u \cos^{m+1} u}{n+m} + \frac{n-1}{n+m} \int \sin^{n-2} u \cos^m u \, du$$

79. $\int \sin au \sin bu \, du = \frac{\sin(a-b)u}{2(a-b)} - \frac{\sin(a+b)u}{2(a+b)} + C$

$$= \frac{\sin^{n+1} u \cos^{m-1} u}{n+m} + \frac{m-1}{n+m} \int \sin^n u \cos^{m-2} u \, du$$

Inverse Trigonometric Forms

87. $\int \sin^{-1} u \, du = u \sin^{-1} u + \sqrt{1-u^2} + C$

93. $\int u^n \sin^{-1} u \, du = \frac{1}{n+1} \left[u^{n+1} \sin^{-1} u - \int \frac{u^{n+1} \, du}{\sqrt{1-u^2}} \right],$
$$n \neq -1$$

88. $\int \cos^{-1} u \, du = u \cos^{-1} u - \sqrt{1-u^2} + C$

89. $\int \tan^{-1} u \, du = u \tan^{-1} u - \frac{1}{2} \ln(1+u^2) + C$

94. $\int u^n \cos^{-1} u \, du = \frac{1}{n+1} \left[u^{n+1} \cos^{-1} u + \int \frac{u^{n+1} \, du}{\sqrt{1-u^2}} \right],$
$$n \neq -1$$

90. $\int u \sin^{-1} u \, du = \frac{2u^2-1}{4} \sin^{-1} u + \frac{u\sqrt{1-u^2}}{4} + C$

91. $\int u \cos^{-1} u \, du = \frac{2u^2-1}{4} \cos^{-1} u - \frac{u\sqrt{1-u^2}}{4} + C$

95. $\int u^n \tan^{-1} u \, du = \frac{1}{n+1} \left[u^{n+1} \tan^{-1} u - \int \frac{u^{n+1} \, du}{1+u^2} \right],$
$$n \neq -1$$

92. $\int u \tan^{-1} u \, du = \frac{u^2+1}{2} \tan^{-1} u - \frac{u}{2} + C$

Exponential and Logarithmic Forms

96. $\int u e^{au} \, du = \frac{1}{a^2} (au - 1) e^{au} + C$

98. $\int e^{au} \sin bu \, du = \frac{e^{au}}{a^2 + b^2} (a \sin bu - b \cos bu) + C$

97. $\int u^n e^{au} \, du = \frac{1}{a} u^n e^{au} - \frac{n}{a} \int u^{n-1} e^{au} \, du$

99. $\int e^{au} \cos bu \, du = \frac{e^{au}}{a^2 + b^2} (a \cos bu + b \sin bu) + C$

100. $\int \ln u \, du = u \ln u - u + C$

101. $\int u^n \ln u \, du = \frac{u^{n+1}}{(n+1)^2} [(n+1) \ln u - 1] + C$

102. $\int \frac{1}{u \ln u} \, du = \ln |\ln u| + C$

Hyperbolic Forms

103. $\int \sinh u \, du = \cosh u + C$

104. $\int \cosh u \, du = \sinh u + C$

105. $\int \tanh u \, du = \ln \cosh u + C$

106. $\int \coth u \, du = \ln |\sinh u| + C$

107. $\int \operatorname{sech} u \, du = \tan^{-1} |\sinh u| + C$

108. $\int \operatorname{csch} u \, du = \ln |\tanh \tfrac{1}{2}u| + C$

109. $\int \operatorname{sech}^2 u \, du = \tanh u + C$

110. $\int \operatorname{csch}^2 u \, du = -\coth u + C$

111. $\int \operatorname{sech} u \tanh u \, du = -\operatorname{sech} u + C$

112. $\int \operatorname{csch} u \coth u \, du = -\operatorname{csch} u + C$

Forms involving $\sqrt{2au - u^2}$

113. $\int \sqrt{2au - u^2} \, du = \frac{u-a}{2} \sqrt{2au - u^2} + \frac{a^2}{2} \cos^{-1}\left(\frac{a-u}{a}\right) + C$

114. $\int u\sqrt{2au - u^2} \, du = \frac{2u^2 - au - 3a^2}{6} \sqrt{2au - u^2} + \frac{a^3}{2} \cos^{-1}\left(\frac{a-u}{a}\right) + C$

115. $\int \frac{\sqrt{2au - u^2}}{u} \, du = \sqrt{2au - u^2} + a \cos^{-1}\left(\frac{a-u}{a}\right) + C$

116. $\int \frac{\sqrt{2au - u^2}}{u^2} \, du = -\frac{2x\sqrt{2au - u^2}}{u} - \cos^{-1}\left(\frac{a-u}{a}\right) + C$

117. $\int \frac{du}{\sqrt{2au - u^2}} = \cos^{-1}\left(\frac{a-u}{a}\right) + C$

118. $\int \frac{u \, du}{\sqrt{2au - u^2}} = -\sqrt{2au - u^2} + a \cos^{-1}\left(\frac{a-u}{a}\right) + C$

119. $\int \frac{u^2 \, du}{\sqrt{2au - u^2}} = -\frac{u + 3a}{2} \sqrt{2au - u^2} + \frac{3a^2}{2} \cos^{-1}\left(\frac{a-u}{a}\right) + C$

120. $\int \frac{du}{u\sqrt{2au - u^2}} = -\frac{\sqrt{2au - u^2}}{au} + C$

H

Answers to Odd-Numbered Exercises

Review and Preview

Exercises 1 (page 12)

1. 18 **3.** π **5.** $5 - \sqrt{5}$ **7.** $2 - x$

9. $|x+1| = \begin{cases} x+1 & \text{for } x \geq -1 \\ -x-1 & \text{for } x < -1 \end{cases}$ **11.** $x^2 + 1$

13. $(-2, \infty)$

15. $[-1, \infty)$

17. $(3, \infty)$

19. $(2, 6)$

21. $(0, 1]$

23. $[-1, \frac{1}{2})$

25. $[2, 3]$

27. $(-\infty, 1) \cup (2, \infty)$

29. $[-1, \frac{1}{2}]$

31. $(-\infty, \infty)$

33. $(-\sqrt{3}, \sqrt{3})$

35. $(-\infty, 1]$

37. $(-1, 0) \cup (1, \infty)$

39. $(-\infty, 0) \cup (\frac{1}{4}, \infty)$

41. $(-2, 0) \cup (2, \infty)$

43. $(-\infty, 5) \cup (16, \infty)$

45. $(-\infty, -1] \cup [1, \infty)$

47. $10 \leq C \leq 35$

49. (a) $T = 20 - 10h$, $0 \leq h \leq 12$ (b) $-30°C \leq T \leq 20°C$
51. $\pm\frac{3}{2}$ **53.** $2, -\frac{4}{3}$ **55.** $(-3, 3)$ **57.** $(3, 5)$
59. $(-\infty, -7] \cup [-3, \infty)$ **61.** $[1.3, 1.7]$
63. $[-4, -1] \cup [1, 4]$ **65.** $(\frac{1}{2}, \infty)$ **67.** $(-1, \infty)$
69. $x \geq (a+b)c/(ab)$ **71.** $x > (c-b)/a$

Exercises 2 (page 19)

1. 5 **3.** $\sqrt{74}$ **5.** $2\sqrt{37}$ **7.** 2 **9.** $-\frac{9}{2}$

17.

19.

11. Parabola　　**13.** Ellipse　　**15.** Hyperbola

  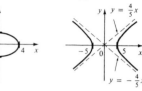

21. $y = 6x - 15$　　**23.** $2x - 3y + 19 = 0$
25. $5x + y = 11$　　**27.** $y = 3x - 2$　　**29.** $y = 3x - 3$
31. $y = 5$　　**33.** $x + 2y + 11 = 0$　　**35.** $5x - 2y + 1 = 0$
37. $m = -\frac{1}{3},\ b = 0$　　　　**39.** $m = 0,\ b = -2$

 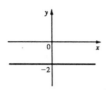

17. Ellipse　　**19.** Parabola　　**21.** Hyperbola

41. $m = \frac{3}{4},\ b = -3$　　**43.**

 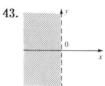

23. Hyperbola　　**25.** Ellipse　　**27.** Parabola

45.　　　　　**47.**

29. Parabola　　**31.** Ellipse　　**33.**

49.　　　　　**51.**

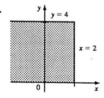 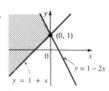

35. $y = x^2 - 2x$　　**37.**　　**39.**

 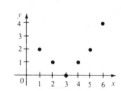

53. $(0, -4)$　　**55.** (a) $(4, 9)$　(b) $(3.5, -3)$　　**57.** $(1, -2)$
59. $y = x - 3$　　**61.** (b) $4x - 3y - 24 = 0$

63. (a) $d = 48t$　　　　　(b)

(c) $m = 48$, speed in mi/h

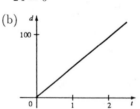

Exercises 3 (page 27)

1. $(x - 3)^2 + (y + 1)^2 = 25$　　**3.** $x^2 + y^2 = 65$
5. $(2, -5),\ 4$　　**7.** $(-\frac{1}{2}, 0),\ \frac{1}{2}$　　**9.** $(\frac{1}{4}, -\frac{1}{4}),\ \sqrt{10}/4$

Exercises 4 (page 34)

1. $0,\ 12,\ \frac{3}{4},\ 7 - 3\sqrt{5},\ a^2 - 3a + 2,\ a^2 + 3a + 2$
3. $-\frac{1}{3},\ -3,\ \frac{1 - \pi}{1 + \pi},\ \frac{1 - a}{1 + a},\ \frac{2 - a}{a},\ \frac{1 + a}{1 - a}$
5. $-(h^2 + 3h + 2),\ x + h - x^2 - 2xh - h^2,\ 1 - 2x - h$
7.

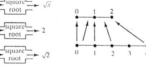

9. $\{0, 1, 2, 4\}$

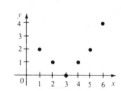

11. $[-2, 3],\ [-6, 14]$
13. $\{x \mid x \neq \frac{5}{3}\} = (-\infty, \frac{5}{3}) \cup (\frac{5}{3}, \infty),$
　　$\{y \mid y \neq 0\} = (-\infty, 0) \cup (0, \infty)$

15. $[\frac{5}{2}, \infty)$, $[0, \infty)$ **17.** $\{x \mid |x| \le 1\} = [-1, 1]$, $[0, 1]$

19. $\{x \mid x \ne \pm 1\} = (-\infty, -1) \cup (-1, 1) \cup (1, \infty)$

21. $\{x \mid x \le 0 \text{ or } x \ge 6\} = (-\infty, 0] \cup [6, \infty)$

23. $[0, \pi)$ **25.** $(-\infty, \infty)$

27. $(-\infty, \infty)$ **29.** $(-\infty, \infty)$ **31.** $(-\infty, \infty)$

33. $(-\infty, \infty)$ **35.** $(-\infty, \infty)$ **37.** $(-\infty, 0]$

39. $[-2, 2]$ **41.** $\{x \mid x \ne 0\}$ **43.** $(-\infty, \infty)$

45. $(-\infty, \infty)$ **47.** $(-\infty, 0) \cup (0, \infty)$ **49.** $(-\infty, 1) \cup (1, \infty)$

51. $(-\infty, \infty)$ **53.** $(-\infty, \infty)$ **55.** $(-\infty, \infty)$

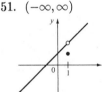 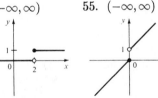

57. $(-\infty, \infty)$ **59.** $(-\infty, \infty)$ **61.** $(-\infty, \infty)$

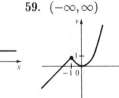

63. Yes, $[-3, 2]$, $[-2, 2]$ **65.** No

67. $f(x) = -\frac{7}{6}x - \frac{4}{3}$, $-2 \le x \le 4$

69. $f(x) = 1 - \sqrt{-x}$

71. $A(L) = 10L - L^2$, $0 < L < 10$

73. $A(x) = \sqrt{3}x^2/4$, $x > 0$

75. $S(x) = x^2 + (8/x)$, $x > 0$

77. $V(x) = 4x^3 - 64x^2 + 240x$, $0 < x < 6$

79. Even   **81.** Neither   **83.** Odd

 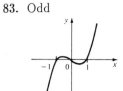

Exercises 5 (page 40)

1. $(f + g)(x) = x^2 + 5$, $(-\infty, \infty)$; $(f - g)(x) = x^2 - 2x - 5$, $(-\infty, \infty)$; $(fg)(x) = x^3 + 4x^2 - 5x$, $(-\infty, \infty)$; $(f/g)(x) = (x^2 - x)/(x + 5)$, $(-\infty, -5) \cup (-5, \infty)$

3. $(f + g)(x) = \sqrt{1 + x} + \sqrt{1 - x}$, $[-1, 1]$; $(f - g)(x) = \sqrt{1 + x} - \sqrt{1 - x}$, $[-1, 1]$; $(fg)(x) = \sqrt{1 - x^2}$, $[-1, 1]$; $(f/g)(x) = \sqrt{(1 + x)/(1 - x)}$, $[-1, 1)$

5. $(f + g)(x) = \sqrt{x} + \sqrt[3]{x}$, $[0, \infty)$; $(f - g)(x) = \sqrt{x} - \sqrt[3]{x}$, $[0, \infty)$; $(fg)(x) = x^{5/6}$, $[0, \infty)$; $(f/g)(x) = \sqrt[6]{x}$, $(0, \infty)$

7. $\{x \mid -3 \le x \le 4, x \ne \pm \sqrt{2}\}$ $= [-3, -\sqrt{2}) \cup (-\sqrt{2}, \sqrt{2}) \cup (\sqrt{2}, 4]$

9. 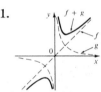   **11.**

13. $(f \circ g)(x) = 8x + 1$, $(-\infty, \infty)$; $(g \circ f)(x) = 8x + 11$, $(-\infty, \infty)$; $(f \circ f)(x) = 4x + 9$, $(-\infty, \infty)$; $(g \circ g)(x) = 16x - 5$, $(-\infty, \infty)$

15. $(f \circ g)(x) = 3(6x^2 + 7x + 2)$, $(-\infty, \infty)$; $(g \circ f)(x) = 6x^2 - 3x + 2$, $(-\infty, \infty)$; $(f \circ f)(x) = 8x^4 - 8x^3 + x$, $(-\infty, \infty)$; $(g \circ g)(x) = 9x + 8$, $(-\infty, \infty)$

17. $(f \circ g)(x) = 1/(x^3 + 2x)$, $\{x \mid x \ne 0\}$; $(g \circ f)(x) = (1/x^3) + (2/x)$, $\{x \mid x \ne 0\}$; $(f \circ f)(x) = x$, $\{x \mid x \ne 0\}$; $(g \circ g)(x) = x^9 + 6x^7 + 12x^5 + 10x^3 + 4x$, $(-\infty, \infty)$

19. $(f \circ g)(x) = \sqrt[3]{1 - \sqrt{x}}$, $[0, \infty)$; $(g \circ f)(x) = 1 - \sqrt[6]{x}$, $[0, \infty)$; $(f \circ f)(x) = \sqrt[9]{x}$, $(-\infty, \infty)$; $(g \circ g)(x) = 1 - \sqrt{1 - \sqrt{x}}$, $[0, 1]$

21. $(f \circ g)(x) = (3x - 4)/(3x - 2)$, $\{x \mid x \ne 2, \frac{2}{3}\}$; $(g \circ f)(x) = -(x + 2)/(3x)$, $\{x \mid x \ne 0, -\frac{1}{2}\}$; $(f \circ f)(x) = (5x + 4)/(4x + 5)$, $\{x \mid x \ne -\frac{1}{2}, -\frac{5}{4}\}$; $(g \circ g)(x) = x/(4 - x)$, $\{x \mid x \ne 2, 4\}$

23. $(f \circ g \circ h)(x) = \sqrt{x - 1} - 1$

25. $(f \circ g \circ h)(x) = (\sqrt{x} - 5)^4 + 1$

27. $g(x) = x - 9$, $f(x) = x^5$

29. $g(x) = x^2$, $f(x) = x/(x + 4)$

31. $h(x) = x^2$, $g(x) = x + 1$, $f(x) = 1/x$

33. $A = 3600\pi t^2$ **35.** $g(x) = x^2 + x - 1$
37. Domain of $f \circ f$ is $\{x \mid x \neq 0\}$

Exercises 6 (page 47)

1.

3.

5.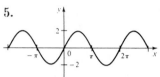

7.

9.

11.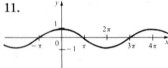

13.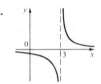

15.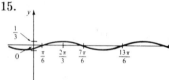

17.

19.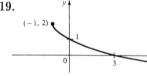

21.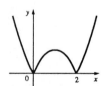

23.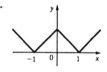

25. (a) The portion of the graph of $y = f(x)$ to the right of the y-axis is reflected in the y-axis.
  (b)

Review Exercises for Review and Preview (page 54)

1. $(-\infty, 1]$ **3.** $(-10, 4)$ **5.** $(-\infty, 2) \cup (3, \infty)$
7. 10   **9.** $(x - 2)^2 + (y - 1)^2 = 9$
11. Center $(-1, 4)$, radius 3
13. $5x - 3y = 13$ **15.** $x + 2y = 8$
17. $10x - 6y = 15$

19. Parabola   **21.** Hyperbola   **23.** Ellipse

25. $3, 1 + 2\sqrt{2}, 1 + \sqrt{-x - 1}, 1 + \sqrt{x^2 - 1}, x + 2\sqrt{x - 1}$
27. $\{x \mid x \neq (-1 \pm \sqrt{5})/2\}$
29.   **31.**   **33.**

35.    **37.**

39. (a) $(f + g)(x) = x^2 + x + 2, (-\infty, \infty)$
  (b) $(f/g)(x) = x^2/(x + 2), (-\infty, -2) \cup (-2, \infty)$
  (c) $(f \circ g)(x) = (x + 2)^2, (-\infty, \infty)$
  (d) $(g \circ f)(x) = x^2 + 2, (-\infty, \infty)$
41. $f(x) = \sqrt{x}, g(x) = x^2 + x + 9$
43. $A = C^2/(4\pi), C > 0$

Chapter 1

Exercises 1.1 (page 60)

1. (a) (i) 4 (ii) 3.5 (iii) 3.1 (iv) 3.01 (v) 3.001 (vi) 2
  (vii) 2.5 (viii) 2.9 (ix) 2.99 (x) 2.999   (b) 3
  (c) $y = 3x$
3. (a) (i) 0.236068 (ii) 0.242641 (iii) 0.248457
  (iv) 0.249844 (v) 0.249984 (vi) 0.267949
  (vii) 0.258343 (viii) 0.251582 (ix) 0.250156
  (x) 0.250016   (b) $\frac{1}{4}$ (c) $x - 4y + 4 = 0$
5. (a) (i) $-32\,\text{ft/s}$ (ii) $-25.6\,\text{ft/s}$ (iii) $-24.8\,\text{ft/s}$
  (iv) $-24.16\,\text{ft/s}$ (b) $-24\,\text{ft/s}$
7. (a) (i) $3\,\text{m/s}$ (ii) $2\,\text{m/s}$ (iii) $1.5\,\text{m/s}$ (iv) $1.1\,\text{m/s}$
  (b) $1\,\text{m/s}$ (c), (d)

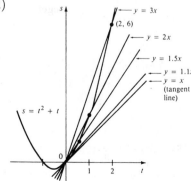

Exercises 1.2 (page 67)

1. (a) 3   (b) 2   (c) -2 (d) Does not exist
   (e) 1   (f) -1 (g) -1 (h) -1 (i) -3
3. (a) 2   (b) -1 (c) 1   (d) 1   (e) 2
   (f) Does not exist
5. (a)   (b) (i) 1 (ii) 1 (iii) 1

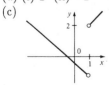

7. $-1, -5.75, -12.718, -14.761198, -14.976012, -53,$
   $-30.25, -17.522, -15.241202, -15.024012; -15$
9. $0.806452, 0.641026, 0.510204, 0.409836, 0.369004,$
   $0.336689, 0.165563, 0.193798, 0.229358, 0.274725,$
   $0.302115, 0.330022; \frac{1}{3}$
11. $-0.003884, -0.003941, -0.003988, -0.003994,$
    $-0.003999, -0.004124, -0.004061, -0.004012,$
    $-0.004006, -0.004001; -0.004$
13. $0.459698, 0.489670, 0.493369, 0.496261, 0.498336,$
    $0.499583, 0.499896, 0.499996; \frac{1}{2}$
15. $2.0, 2.593742, 2.704814, 2.716924, 2.718146, 2.718268,$
    $2.718280, 2.718282, 2.718282, 2.718282; 2.71828$
17. (a) $0.998000, 0.638259, 0.358484, 0.158680, 0.038851,$
    $0.008928, 0.001465; 0$ (b) $0.000572, -0.000614,$
    $-0.000907, -0.000978, -0.000993, -0.001000;$
    $-0.001 = -\frac{1}{1000}$

Exercises 1.3 (page 77)

1. 75   3. 60   5. $\frac{1}{2}$ 7. 2   9. -3 11. -2
13. $\frac{2}{5}$
15. (a) 5   (b) 9   (c) 2   (d) $-\frac{1}{3}$ (e) $-\frac{3}{8}$
    (f) 0   (g) Does not exist   (h) $-\frac{6}{11}$
17. Does not exist   19. 1   21. -3 23. 1
25. -10 27. 4   29. $-\frac{1}{5}$ 31. -2 33. 6
35. $-\sqrt{2}/4$ 37. 108   39. $-\frac{1}{2}$ 41. $\frac{2}{3}$ 43. 3
47. 5   49. -1.5 51. -1 53. Does not exist
55. 9   57. Does not exist   59. 9   61. 0   63. 0
65. 0   69. (a) 1, 1   (b) Yes, 1   (c)
71. (a) (i) $n - 1$ (ii) n
    (b) a is not an integer
73. (a) (i) 2 (ii) -2 (b) No
    (c)

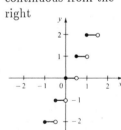

81. $c/3$

Exercises 1.4 (page 85)

1. (a) $|x - 3| < \frac{1}{60}$ (b) $|x - 3| < \frac{1}{600}$

Exercises 1.5 (page 94)

1. (a) $-5, -3, -1, 3, 5, 8, 10$ (b) Left, left, neither,
   neither, neither, right, neither
13. $f(2)$ undefined   15. $f(1)$ undefined

17. $\lim\limits_{x \to 1} f(x)$ does not exist   19. $\lim\limits_{x \to -3} f(x) \neq f(-3)$

    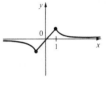

21. R 23. $\{x \mid x \neq -1\}$ 25. $(-1, \infty)$ 27. R
29. R 31. $(-\infty, -5) \cup [2, \infty)$ 33. R
37. 0, continuous from   39. Continuous at all points
    the right

41. 0, 1, continuous from   43. $\{n/2 \mid n$ an integer$\}$,
    right at both   continuous from the
                                  right

45. $\frac{1}{3}$ 57. (b) $(-1.33, -1.32)$ 59. Never continuous

Exercises 1.6 (page 103)

1. 0   3. -2 5. $\frac{1}{2}$ 7. $\frac{1}{6}$ 9. 0   11. 0
13. 2   15. 0   17. -1 19. 0   21. 0   23. $-\frac{1}{2}$
25. $y = 2$ 27. $y = 1, y = -1$ 29. 4
31. $0, 0.5, 1.0, 1.125, 1.0, 0.78125, 0.5625, 0.3828125,$
    $0.25, 0.158203125, 0.09765625, 0.00038147,$
    $2.2204 \times 10^{-12}, 7.8886 \times 10^{-27}; 0$

Exercises 1.7 (page 111)

1. (a) ∞ (b) $-\infty$ (c) $-\infty$ (d) ∞ (e) $-\infty$ (f) 2 (g) -2
(h) $x = -9$, $x = -4$, $x = 3$, $x = 7$ (i) $y = -2$, $y = 2$
3. $-\infty$ 5. ∞ 7. $-\infty$ 9. $-\infty$ 11. ∞
13. $-\infty$ 15. ∞ 17. $-\infty$ 19. $-\infty$ 21. ∞
23. $-\infty$ 25. 0   27. ∞ 29. $-\infty$
31. $x = 1$, $x = -1$; $y = 1$ 33. $x = 0$, $y = 1$
35. $-\infty, -\infty$ 37. $\infty, -\infty$

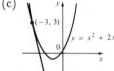

39. $\left| x - (-3) \right| < 0.1$

Exercises 1.8 (page 118)

1. (a) (i) -4 (ii) -4 (b) $4x + y + 9 = 0$
(c)

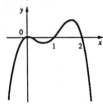

3. $10x - y + 13 = 0$ 5. $x - 4y + 3 = 0$
7. (a) $-2/(a+3)^2$ (b) (i) $-\frac{1}{2}$ (ii) $-\frac{2}{9}$ (iii) $-\frac{1}{8}$
9. -24 ft/s   11. $12a^2 + 6$, 18 m/s, 54 m/s, 114 m/s
13. (a) (i) $-1.2°$/h (ii) $-1.25°$/h (iii) $-1.3°$/h (b) $-1.6°$/h
15. (a) (i) \$20.25/unit (ii) \$20.05/unit   (b) \$20/unit

Review Exercises for Chapter 1 (page 120)

1. False   3. True   5. False   7. True   9. True
11. $\sqrt{6}$ 13. $\frac{1}{2}$ 15. 2   17. 0   19. $-\frac{1}{8}$
21. -1 23. 0   25. ∞ 27. ∞ 29. 0
31. $\frac{1}{2}$ 33. 2   35. $-\infty$ 37. ∞
43. (a) (i) 3 (ii) 0 (iii) Does not exist (iv) 0 (v) 0
(vi) 0   (b) At 0 and 3   (c)
45. R
49. (a) -8
(b) $8x + y = 17$
51. (a) (i) 3 m/s (ii) 2.75 m/s (iii) 2.625 m/s
(iv) 2.525 m/s (b) 2.5 m/s
53. 0

Chapter 2

Exercises 2.1 (page 130)

1. 7, $7x - y - 12 = 0$ 3. $1/(2\sqrt{a+7})$, $x - 6y + 16 = 0$
5. -2 m/s   7. $1 - 4a$ 9. $-1/(2a-1)^2$

11. $1/(3-a)^{3/2}$ 13. $1/(3a^{2/3})$
15. $f(x) = \sqrt{x}$, $a = 1$ 17. $f(x) = x^9$, $a = 1$
19. $f(x) = \sin x$, $a = \pi/2$ 21. $f'(x) = 5$, R, R
23. $f'(x) = 3x^2 - 2x + 2$, R, R
25. $f'(x) = 1 + (2/x^2)$, $\{x \mid x \neq 0\}$, $\{x \mid x \neq 0\}$
27. $g'(x) = 1/\sqrt{1+2x}$, $[-\frac{1}{2}, \infty)$, $(-\frac{1}{2}, \infty)$
29. $G'(x) = -10/(2+x)^2$, $\{x \mid x \neq -2\}$, $\{x \mid x \neq -2\}$
31. $f'(x) = 4x^3$, R, R 33. 1, $2x$, $3x^2$, nx^{n-1}, $5x^4$

35.   37.   39.

41.   43.

45. $f'(x) = \begin{cases} 1 & \text{if } x > 6 \\ -1 & \text{if } x < 6 \end{cases}$ or $f'(x) = \dfrac{x-6}{|x-6|}$

47. (a)   (b) All x (c) $f'(x) = 2|x|$

49. (a) $-5, 5$ 51. Does not exist

Exercises 2.2 (page 140)

1. $f'(x) = 2x - 10$ 3. $V'(r) = 4\pi r^2$
5. $F'(x) = 12288x^2$ 7. $Y'(t) = -54t^{-10}$
9. $g'(x) = 2x - (2/x^3)$ 11. $h'(x) = -3/(x-1)^2$
13. $G'(s) = (2s+1)(s^2+2) + (s^2+s+1)(2s)$
$[\,= 4s^3 + 3s^2 + 6s + 2\,]$
15. $y' = \frac{3}{2}\sqrt{x} + (2/\sqrt{x}) - 3/(2x\sqrt{x})$ 17. $y' = \sqrt{5}/(2\sqrt{x})$
19. $y' = -(4x^3 + 2x)/(x^4 + x^2 + 1)^2$
21. $y' = 2ax + b$
23. $y' = (-3t^2 + 14t + 23)/(t^2 + 5t - 4)^2$
25. $y' = 1 + 2/(5\sqrt[5]{x^3})$ 27. $u' = \sqrt{2}\,x^{\sqrt{2}-1}$
29. $v' = \frac{3}{2}\sqrt{x} - 5/(2x^3\sqrt{x})$ 31. $f'(x) = 2cx/(x^2 + c)^2$
33. $f'(x) = 2x^4(x^3 - 5)/(x^3 - 2)^2$
35. $s' = (1 + 2t - 2t^2)/[t^2(t+1)^2]$
37. $P'(x) = na_n x^{n-1} + (n-1)a_{n-1}x^{n-2}$
$+ \cdots + 2a_2 x + a_1$
39. $y = 4$ 41. $x - 2y + 2 = 0$ 43. (4, 8)
45. (1, 0), $(-\frac{1}{3}, \frac{32}{27})$ 47. 2, $(-2 \pm \sqrt{3}, (1 \mp \sqrt{3})/2)$

51. $x - 4y - 14 = 0$ **53.** $12x + y + 98 = 0$

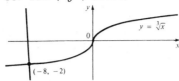

55. $(-\frac{1}{4}, \frac{1}{256})$ **57.** (a) -16 (b) $-\frac{20}{9}$ (c) 20

61. $y' = (x^4 + x + 1)(2x - 3)/(2\sqrt{x})$
$\qquad + \sqrt{x}\,[(4x^3 + 1)(2x - 3) + 2(x^4 + x + 1)]$

63. No

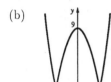 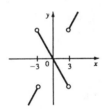

65. (a) Not differentiable at 3 or -3

$$f'(x) = \begin{cases} 2x & \text{if } |x| > 3 \\ -2x & \text{if } |x| < 3 \end{cases}$$

(b)

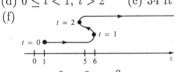

Exercises 2.3 (page 150)

1. (a) $2t - 6$ (b) -2 ft/s   (c) $t = 3$ (d) $t > 3$
(e) 10 ft   (f)

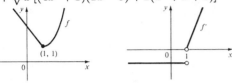

3. (a) $6t^2 - 18t + 12$ (b) 0 ft/s   (c) $t = 1, 2$
(d) $0 \le t < 1,\ t > 2$ (e) 34 ft
(f)

5. (a) $(1 - t^2)/(t^2 + 1)^2$ (b) $-\frac{3}{25}$ ft/s   (c) $t = 1$
(d) $0 \le t < 1$ (e) $\frac{13}{17}$ ft   (f)

7. $t = 4$ s   **9.** (a) (i) 91   (ii) 76.51   (iii) 75.1501   (b) 75
11. (a) 7200π cm^2/s (b) 21600π cm^2/s (c) 36000π cm^2/s
13. (a) 8π ft^2/ft   (b) 16π ft^2/ft   (c) 24π ft^2/ft
15. (a) 6 kg/m   (b) 12 kg/m   (c) 18 kg/m
17. (a) 4.75 A   (b) 5 A   **19.** (a) $dV/dP = -C/P^2$
21. (a) $a^2 k/(akt + 1)^2$ **23.** -0.04
25. ≈ -92.6 (cm/s)/cm
27. $C'(x) = 1.5 + 0.004x$; $1.90, 1.902$
29. $C'(x) = 3 + 0.02x + 0.0006x^2$; $11, 11.07$

Exercises 2.4 (page 157)

1. 1   **3.** $(\sqrt{3} - 1)/2$ **5.** $2\sqrt{2}/(3\pi)$ **7.** 0   **9.** 5
11. $\sin 1$ **13.** $1/\pi$ **15.** 0   **17.** $\frac{1}{2}$ **19.** 9
25. $\cos x - \sin x$ **27.** $-\csc x \cot^2 x - \csc^3 x$
29. $(x\sec^2 x - \tan x)/x^2$
31. $(\sin x + \cos x + x \sin x - x \cos x)/(1 + \sin 2x)$
33. $x^{-4}\sin x(-3\tan x + x + x\sec^2 x)$
35. $(2x \tan x + x^2)/\sec x$ **37.** $4x - 2y = \pi - 2$
39. $(2n + 1)\pi \pm \pi/3$, n an integer   **43.** $\frac{1}{2}$ **45.** $\frac{1}{2}$
47. $\frac{1}{2}$ **49.** 1
51. (a) $\sec^2 x = 1/\cos^2 x$ (b) $\sec x \tan x = (\sin x)/\cos^2 x$
(c) $\cos x - \sin x = (\cot x - 1)/\csc x$

Exercises 2.5 (page 164)

1. $4u(x + 1)$, 48   **3.** $3u^2(1 - 1/x^2)$, 0
5. $F'(x) = 10(x^2 + 4x + 6)^4(x + 2)$
7. $G'(x) = 6(3x - 2)^9(5x^2 - x + 1)^{11}(85x^2 - 51x + 9)$
9. $f'(t) = -16(2t^2 - 6t + 1)^{-9}(2t - 3)$
11. $g'(x) = (2x - 7)/(2\sqrt{x^2 - 7x})$
13. $h'(t) = \frac{3}{2}(t - 1/t)^{1/2}(1 + 1/t^2)$
15. $F'(y) = 39(y - 6)^2/(y + 7)^4$
17. $f'(z) = -\frac{2}{5}(2z - 1)^{-6/5}$
19. $y' = 8(2x - 5)^3(8x^2 - 5)^{-4}(-4x^2 + 30x - 5)$
21. $y' = 3\sec^2 3x$ **23.** $y' = -3x^2 \sin(x^3)$
25. $y' = -12\cos x \sin x(1 + \cos^2 x)^5$
27. $y' = -\sin(\tan x)\sec^2 x$ **29.** $y' = 0$
31. $y' = -(1/3)\csc(x/3)\cot(x/3)$
33. $y' = 3\sin x \cos x(\sin x - \cos x)$
35. $y' = -\cos(1/x)/x^2$ **37.** $y' = 4(\cos 2x)/(1 - \sin 2x)^2$
39. $y' = 6x^2 \tan(x^3)\sec^2(x^3)$
41. $y' = \sin[2(1 - \sqrt{x})/(1 + \sqrt{x})]/[\sqrt{x}(1 + \sqrt{x})^2]$
43. $y' = 0$ **45.** $y' = [1 + 1/(2\sqrt{x})]/(2\sqrt{x + \sqrt{x}})$
47. $f'(x) = 9[x^3 + (2x - 1)^3]^2(9x^2 - 8x + 2)$
49. $p'(t) = -2[(1 + 2/t)^{-1} + 3t]^{-3}[2(t + 2)^{-2} + 3]$
51. $y' = \cos(\tan \sqrt{\sin x})(\sec^2 \sqrt{\sin x})[1/(2\sqrt{\sin x})](\cos x)$
53. $y = 0$ **55.** $3x + 16y = 44$ **57.** $4x + y = \pi + 1$
59. $((\pi/2) + 2n\pi, 3), ((3\pi/2) + 2n\pi, -1)$, n any integer
61. 28
63. $f'(x) = 2x(\sec^2 3x)(1 + 3x \tan 3x)$,
$\quad \mathrm{dom}(f) = \mathrm{dom}(f') = \left\{x \mid x \ne \dfrac{(2n - 1)\pi}{6},\ n \text{ any integer}\right\}$
65. $f'(x) = -(\sin \sqrt{x})/(4\sqrt{x}\sqrt{\cos \sqrt{x}})$,
$\quad \mathrm{dom}(f) = \{x \mid 0 \le x \le \pi^2/4$ or $[(4n - 1)\pi/2]^2 \le x$
$\qquad \le [(4n + 1)\pi/2]^2$ for some $n = 1, 2, \ldots\}$,
$\quad \mathrm{dom}(f') = \{x \mid 0 < x < \pi^2/4$ or $[(4n - 1)\pi/2]^2 < x$
$\qquad < [(4n + 1)\pi/2]^2$ for some $n = 1, 2, \ldots\}$
67. $(5\pi/2)\cos(10\pi t)$ cm/s
69. (a) $(7\pi/54)\cos(2\pi t/5.4)$ (b) 0.16
71. (a) On $(0, \infty)$ (b) $G'(x) = h'(\sqrt{x})/(2\sqrt{x})$

73. (a) $F'(x) = -\sin x\, f'(\cos x)$
　　(b) $G'(x) = -\sin(f(x))f'(x)$
79. $f'(x) = x/|x|$
81. $h'(x) = |2x - 1| + 2x(2x - 1)/|2x - 1|$

Exercises 2.6 (page 170)

1. (a) $y' = -(2x + y + 3)/x$
　　(b) $y = (5/x) - x - 3$, $y' = -(5/x^2) - 1$
3. (a) $y' = -y^2/x^2$
　　(b) $y = x/(3x - 1)$, $y' = -1/(3x - 1)^2$
5. (a) $y' = (2x - y)/(x + 4y)$
　　(b) $y = \left(-x \pm \sqrt{9x^2 + 24}\right)/4$,
　　　　$y' = \left(-1 \pm 9x/\sqrt{9x^2 + 24}\right)/4$
7. $y' = (y - 2x)/(3y^2 - x)$
9. $y' = (18x - x^{-2/3}y^{1/3})/(12y + x^{1/3}y^{-2/3})$
11. $y' = -x^3/y^3$
13. $y' = (3x\sqrt{x^2 + y^2} - 2y)/(2x - 3y\sqrt{x^2 + y^2})$
15. $y' = (y/x) + 2(x - y)^2$ [or $(3x^2 + 1 - 2xy)/(x^2 + 2)$]
17. $y' = [\sin(x - y) + y \cos x]/[\sin(x - y) - \sin x]$
19. $y' = -y/x$
21. $dx/dy = (1 - 4y^3 - 2x^2y - x^4)/(2xy^2 + 4yx^3)$
23. $-\frac{1}{6}$　　25. $5x + 4y + 16 = 0$　　27. $y = x$
29. $9x + 13y - 40 = 0$　　31. $(\pm 5\sqrt{3}/4, \pm 5/4)$
33. $(x_0x/a^2) - (y_0y/b^2) = 1$

39. 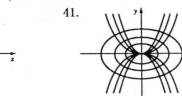　　41.

43. $(-1, -1)$, $(1, 1)$

Exercises 2.7 (page 175)

1. $f'(x) = 4x^3 - 9x^2 + 16$, $f''(x) = 12x^2 - 18x$
3. $h'(x) = x/\sqrt{x^2 + 1}$, $h''(x) = 1/(x^2 + 1)^{3/2}$
5. $F'(s) = 24(3s + 5)^7$, $F''(s) = 504(3s + 5)^6$
7. $y' = 1/(1 - x)^2$, $y'' = 2/(1 - x)^3$
9. $y' = -\frac{3}{2}x(1 - x^2)^{-1/4}$, $y'' = \frac{3}{4}(1 - x^2)^{-5/4}(x^2 - 2)$
11. $H'(t) = 6\tan^2(2t - 1)\sec^2(2t - 1)$,
　　$H''(t) = 24\tan(2t - 1)\sec^4(2t - 1)$
　　　　　　　　$+ 24\tan^3(2t - 1)\sec^2(2t - 1)$
13. $F'(r) = (\sec\sqrt{r}\,\tan\sqrt{r})/(2\sqrt{r})$,
　　$F''(r) = \dfrac{\sec\sqrt{r}\tan^2\sqrt{r} + \sec^3\sqrt{r} - r^{-1/2}\sec\sqrt{r}\tan\sqrt{r}}{4r}$
15. 0　　17. $\frac{375}{8}(5t - 1)^{-5/2}$
19. $1/\sqrt{2}$, $3/(4\sqrt{2})$, $27/(16\sqrt{2})$, $405/(64\sqrt{2})$　　21. -80
23. $-2x/y^5$　　25. $64/(3x + y)^3$

27. $f'(x) = 1 - 2x + 3x^2 - 4x^3 + 5x^4 - 6x^5$,
　　$f''(x) = -2 + 6x - 12x^2 + 20x^3 - 30x^4$,
　　$f'''(x) = 6 - 24x + 60x^2 - 120x^3$,
　　$f^{(4)}(x) = -24 + 120x - 360x^2$, $f^{(5)}(x) = 120 - 720x$,
　　$f^{(6)}(x) = -720$, $f^{(n)}(x) = 0$ for $7 \le n \le 73$
29. $n!$　　31. $(-1)^n(n + 2)!/(6x^{n+3})$　　33. $-2^{50}\cos 2x$
35. (a) $v(t) = 3t^2 - 3$, $a(t) = 6t$　(b) $6\,\text{m/s}^2$
　　(c) $a(1) = 6\,\text{m/s}^2$
37. (a) $v(t) = 2At + B$, $a(t) = 2A$　　(b) $2A\,\text{m/s}^2$
　　(c) $2A\,\text{m/s}^2$
39. (a) $t = 0, 2$
　　(b) $s(0) = 2\,\text{m}$, $v(0) = 0\,\text{m/s}$, $s(2) = -14\,\text{m}$,
　　　　$v(2) = -16\,\text{m/s}$
41. (a) $v(t) = A\omega\cos\omega t$, $a(t) = -A\omega^2\sin\omega t$
43. $P(x) = x^2 - x + 3$
47. $f''(x) = 6xg'(x^2) + 4x^3g''(x^2)$
49. $f''(x) = [\sqrt{x}g''(\sqrt{x}) - g'(\sqrt{x})]/(4x\sqrt{x})$

Exercises 2.8 (page 179)

1. $dV/dt = 3x^2\,dx/dt$　　3. -1　　5. $1/(50\pi)\,\text{cm/min}$
7. (a) $\frac{25}{3}\,\text{ft/s}$　(b) $\frac{10}{3}\,\text{ft/s}$　　9. $250\sqrt{3}\,\text{mi/h}$
11. $65\,\text{mi/h}$　　13. $\frac{720}{13} \approx 55.4\,\text{km/h}$　　15. $-1.6\,\text{cm/min}$
17. $(10000 + 800000\pi/9) \approx 2.89 \times 10^5\,\text{cm}^3/\text{min}$
19. $\frac{10}{3}\,\text{cm/min}$　　21. $6/(5\pi)\,\text{ft/min}$　　23. $0.3\,\text{m}^2/\text{s}$
25. $80\,\text{cm}^3/\text{min}$　　27. $1650/\sqrt{31} \approx 296\,\text{km/h}$
29. $\sqrt{2}/5\,\text{rad/s}$

Exercises 2.9 (page 185)

1. $dy = 5x^4\,dx$　　3. $dy = [(2x^3 + x)/\sqrt{x^4 + x^2 + 1}]\,dx$
5. $dy = 7\,dx/(2x + 3)^2$　　7. $dy = 2\cos 2x\,dx$
9. (a) $dy = -2x\,dx$　　(b) -5
11. (a) $dy = 6x(x^2 + 5)^2\,dx$　　(b) 10.8
13. (a) $dy = -(3x + 2)^{-4/3}\,dx$　　(b) 0.0025
15. (a) $dy = -\sin x\,dx$　　(b) -0.025
17. $\Delta y = 1.25$, $dy = 1$　　　19. $\Delta y = 1.44$, $dy = 1.6$

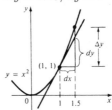　　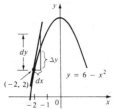

21. $\Delta y = 77, 33.25, 5.882, 0.571802$; $dy = 57, 28.5, 5.7$,
　　0.57; $\Delta y - dy = 20, 4.75, 0.182, 0.001802$
23. $6 + \frac{1}{120} \approx 6.0083$　　25. $6 + \frac{1}{54} \approx 6.0185$　　27. 0.099
29. 0.857　　31. $1 + (\pi/45) \approx 1.07$　　33. $L(x) = 3x - 2$
35. $L(x) = \frac{1}{2} - \frac{1}{16}x$
41. $\sqrt{1 - x} \approx 1 - \frac{1}{2}x$,
　　$\sqrt{0.9} \approx 0.95$,
　　$\sqrt{0.99} \approx 0.995$

43. (a) $270 \, \text{cm}^3$ (b) $36 \, \text{cm}^2$
45. (a) $84/\pi \approx 27 \, \text{cm}^2$ (b) $\frac{1}{84} \approx 0.012$
47. (a) $V \approx 2\pi r h \, \Delta r$ (b) $\pi (\Delta r)^2 h$

Exercises 2.10 (page 189)

1. -0.6860 **3.** 1.5850 **5.** 2.165737 **7.** 1.618034
9. 3.992020 **11.** 1.895494
13. $-2.114908, 0.254102, 1.860806$ **15.** $1, -0.569840$
17. $0, 1.109144, 3.698154$ **19.** (b) 31.622777

Review Exercises for Chapter 2 (page 190)

1. False   **3.** False   **5.** True   **7.** False   **9.** True
11. $f'(x) = 3x^2 + 5$ **13.** $f'(x) = -5/(2\sqrt{3-5x})$
15. $y' = 2(7x+18)(x+2)^7(x+3)^5$
17. $y' = (9-2x)/(9-4x)^{3/2}$
19. $y' = (1-2xy^3)/(3x^2y^2 + 6y + 4)$
21. $y' = 7/(8\sqrt[8]{x})$ **23.** $y' = 8/(8-3x)^2$
25. $y' = \frac{1}{5}(x \tan x)^{-4/5}(\tan x + x \sec^2 x)$
27. $y' = 2x/(2y+1)$
29. $y' = 2(2x-5)/[(x-2)^2(x-3)^2]$
31. $y' = -(\sec^2 \sqrt{1-x})/(2\sqrt{1-x})$
33. $y' = \cos\left(\tan\sqrt{1+x^3}\right)\left(\sec^2\sqrt{1+x^3}\right)\left(3x^2/\left(2\sqrt{1+x^3}\right)\right)$
35. $y' = -6x \csc^2(3x^2+5)$
37. $y' = (mx \cos mx - \sin mx)/x^2$
39. $y' = -\sin(2\tan x)\sec^2 x$
41. $y' = -[x(x^3+7)^5/\sqrt{7-x^2}] + 15x^2(x^3+7)^4\sqrt{7-x^2}$
43. -120 **45.** $-5x^4/y^{11}$ **47.** $3x + 2y - 8 = 0$
49. $4x - y + \sqrt{3} - (4\pi/3) = 0$
51. $(\pi/4, \sqrt{2}), (5\pi/4, -\sqrt{2})$
53. (a) $v(t) = 3t^2 - 12, \, a(t) = 6t$
   (b) Upward when $t > 2$, downward when $0 \le t < 2$
   (c) 23
57. (a) 2   (b) 44   **59.** $f'(x) = 2xg(x) + x^2 g'(x)$
61. $f'(x) = 2g(x)g'(x)$ **63.** $f'(x) = g'(g(x))g'(x)$
65. $h'(x) = (f'(x)[g(x)]^2 + g'(x)[f(x)]^2)/[f(x) + g(x)]^2$
67. $h'(x) = f'(g(\sin 4x))g'(\sin 4x)(\cos 4x)(4)$
69. $4 \, \text{kg/m}$ **71.** $\frac{4}{3} \, \text{cm}^2/\text{min}$ **73.** $13 \, \text{ft/s}$
75. $400 \, \text{ft/h}$ **77.** 0.8
79. $L(x) = 1 + x, \sqrt[3]{1+3x} \approx 1 + x, \sqrt[3]{1.03} \approx 1.01$
81. 0.724492 **83.** $(0.9340, -2.0634)$ **85.** 192

Problems Plus (page 192)

1. $(\pm\sqrt{3}/2, \frac{1}{4})$ **3.** True for all x **5.**

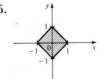

7. -4
11. (a) $[-1, 2]$
   (b) $-1/\left(8\sqrt{3-x}\sqrt{2-\sqrt{3-x}}\sqrt{1-\sqrt{2-\sqrt{3-x}}}\right)$

13. (a)

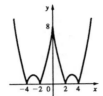

       (b) $0, \pm 2, \pm 4$

17. (a) 0   (b) 4   (c) $\frac{1}{2}$
21. (b) (i) $53°$ (or $127°$)   (ii) $63°$ (or $117°$)

Chapter 3

Exercises 3.1 (page 198)

1.

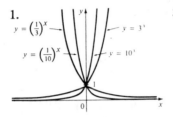

3.

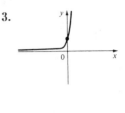

5.

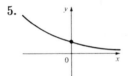

7.

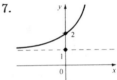

9.

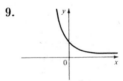

11.

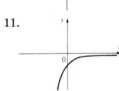

13.

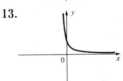

15.

17. ∞ **19.** 0   **21.** ∞ **23.** 0   **25.** ∞
27. ∞ **29.** 1   **31.** ∞ **33.** 1   **35.** ∞
37. (a) 0   (b) ∞ (c) $\frac{1}{4}$

Exercises 3.2 (page 202)

1. (a)

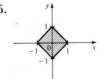

(b) $1.4870, 1.3959, 1.3873, 1.3864$;   slope of secant line
(c) 1.39 (d) Slope of tangent line to $y = 4^x$ at $(0, 1)$

3. $f'(x) = e^{\sqrt{x}}/(2\sqrt{x})$ **5.** $y' = e^{2x}(2x+1)$

7. $h'(t) = -e^t/(2\sqrt{1-e^t})$ **9.** $y' = e^{x\cos x}(\cos x - x\sin x)$

11. $y' = e^{-1/x}/x^2$ **13.** $y' = 3e^{3x-2}\sec^2(e^{3x-2})$

15. $y' = (3e^{3x} + 2e^{4x})/(1+e^x)^2$ **17.** $y' = exe^{-1}$

19. $x + e^\pi y = \pi$

21. $y' = 1 + (x+1)e^x/\sin(x-y)$ **25.** $r = 1, -6$

27. $256e^{-2x}$

29. (b) c = initial velocity   (c) $t = (\ln 2)/k$

31.

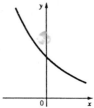

33. 1   **35.** 0   **37.** 0

Exercises 3.3 (page 209)

1. No   **3.** Yes   **5.** No   **7.** Yes   **9.** Yes   **11.** No

13. $f^{-1}(x) = (x-7)/4$ **15.** $f^{-1}(x) = (5x-1)/(2x+3)$

17. $f^{-1}(x) = (x^2-2)/5, x \geq 0$

19. (b) $\frac{1}{2}$ (c) $g(x) = (x-1)/2$, domain $= R =$ range

(e)

21. (b) $\frac{1}{12}$ (c) $g(x) = \sqrt[3]{x}$, domain $= R =$ range

(e)

23. (b) $-\frac{1}{2}$

(c) $g(x) = \sqrt{9-x}$, domain $= [0,9]$, range $= [0,3]$

(e)

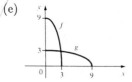

25. 1   **27.** $1/\pi$ **29.** 1   **31.** $\frac{3}{2}$

Exercises 3.4 (page 214)

1. 6   **3.** $\frac{1}{3}$ **5.** -3 **7.** $\sqrt{2}$ **9.** 2   **11.** 1

13. 15   **15.** $\log_5(ab/c)$ **17.** $3\ln 2$

19. $\ln[\sqrt[3]{x}/(2x+3)^4]$

21.

23.

25.

27.

29.

31.

33.

35. 8   **37.** $4\ln 2$ **39.** $(e^3+1)/2$ **41.** $\log_3 m - 2$

43. 40   **45.** 3   **47.** e^e **49.** $\log_3(\log_2 5)$ **51.** 4

53. 25.0855   **55.** -0.3319 **61.** 8.3   **63.** $-\infty$

65. ∞ **67.** $-\infty$ **69.** 0   **71.** $(-\infty, 1), (-\infty, \infty)$

73. $(1, \infty), (-\infty, \infty)$ **75.** $y = e^x - 3$ **77.** $y = (\ln x)^2$

79. $y = \log_{10}[x/(1-x)]$ **81.** (b) $y = \frac{1}{2}(e^x - e^{-x})$

83. $\log_9 82$ **87.** 1, $\log_2 6$

Exercises 3.5 (page 222)

1. $f'(x) = 1/(x+1), (-1, \infty), (-1, \infty)$

3. $f'(x) = -\tan x$,

$\{x \mid (4n-1)\pi/2 < x < (4n+1)\pi/2, n = 0, \pm1, \pm2, ...\}$,

same

5. $f'(x) = (-1-2x)/(2-x-x^2), (-2,1), (-2,1)$

7. $f'(x) = 2x\ln(1-x^2) - 2x^3/(1-x^2), (-1,1), (-1,1)$

9. $f'(x) = (\log_3 e)2x/(x^2-4), |x| > 2, |x| > 2$

11. $y' = 1 + \ln x, y'' = 1/x$

13. $y' = 1/(x\ln 10), y'' = -1/(x^2\ln 10)$

15. $f'(x) = (2 + \ln x)/(2\sqrt{x})$

17. $g'(x) = -2a/(a^2-x^2)$ **19.** $F'(x) = 1/(2x)$

21. $f'(t) = (\log_2 e)(4t^3 - 2t)/(t^4 - t^2 + 1)$

23. $h'(y) = (3/y) + \cot y$

25. $g'(u) = -2/[u(1+\ln u)^2]$ **27.** $y' = 3\cot x(\ln\sin x)^2$

29. $y' = [1 + x^2(1-2\ln x)]/[x(1+x^2)^2]$

31. $y' = -6/[5(x^2-1)]$ **33.** $y' = (3x-2)/[x(x-1)]$

35. $F'(x) = e^x(\ln x + 1/x)$ **37.** $f'(t) = -\pi^{-t}\ln\pi$

39. $h'(t) = 3t^2 - 3^t\ln 3$ **41.** $y' = (\ln 2)(\ln 3)2^{3^x}3^x$

43. $y' = -x/(1+x)$

45. $y' = x^{\sin x}[\cos x\ln x + (\sin x)/x]$

47. $y' = e^x x^{e^x}(\ln x + 1/x)$

49. $y' = (\ln x)^x(\ln \ln x + 1/\ln x)$

51. $y' = 0$ 53. $y' = x^{1/x}(1 - \ln x)/x^2$

55. $y' = x^{x^2+1}(1 + 2\ln x)$ 57. 0   59. $x - ey = e$

61. $x - 2y + \ln 4 = 0$ 63. $y' = 2x/(x^2 + y^2 - 2y)$

65. $f^{(n)}(x) = (-1)^{n-1}(n-1)!/(x-1)^n$

67. $y' = (3x - 7)^4(8x^2 - 1)^3[12/(3x - 7) + 48x/(8x^2 - 1)]$

69. $y' = \dfrac{(x+1)^4(x-5)^3}{(x-3)^8}\left[\dfrac{4}{x+1} + \dfrac{3}{(x-5)} - \dfrac{8}{x-3}\right]$

71. $y' = \dfrac{e^x\sqrt{x^5+2}}{(x+1)^4(x^2+3)^2}\left[1 + \dfrac{5x^4}{2(x^5+2)} - \dfrac{4}{x+1} - \dfrac{4x}{x^2+3}\right]$

73. $\frac{1}{3}$

Exercises 3.6 (page 227)

1. (a) 100×2^{3t} (b) $\approx 1.07 \times 10^{11}$
   (c) $(\ln 100)/(3\ln 2) \approx 2.2$ h

3. (a) $500e^{(\ln 16)t/3}$ (b) $\approx 20{,}159$
   (c) $(3\ln 60)/\ln 16 \approx 4.4$ h

5. (a) 506,157   (b) 683,241

7. (a) $Ce^{-0.0005t}$ (b) $-2000\ln 0.9 \approx 211$ s

9. (a) $50 \times 2^{-t/0.00014}$ (b) $\approx 1.57 \times 10^{-20}$ mg
   (c) $\approx 4.5 \times 10^{-5}$ s   11. ≈ 2500 years

13. $21 + 12e^{-t/10}$ 15. (a) $\approx 137°F$ (b) ≈ 116 min

17. (a) ≈ 64.5 kPa   (b) ≈ 39.9 kPa

19. (a) \$4615.87   (b) \$4658.91   (c) \$4697.04
    (d) \$4703.11   (e) \$4704.67   (f) \$4704.94

21. (a) $450e^{-0.4} \approx 302$ kg   (b) ≈ 30.4 min

Exercises 3.7 (page 234)

1. π 3. $\pi/3$ 5. $\pi/4$ 7. $5\pi/6$ 9. $-\pi/4$

11. $-\pi/6$ 13. 0.7   15. 10   17. $\frac{1}{2}$ 19. $\frac{3}{5}$

21. $-\pi/4$ 23. $\frac{119}{169}$ 25. $(\sqrt{5} + 4\sqrt{2})/9$

29. $x/\sqrt{x^2+1}$ 37. $g'(x) = 3x^2/(1 + x^6)$

39. $y' = 2x/\sqrt{1 - x^4}$ 41. $F'(x) = a/(a^2 + x^2)$

43. $H'(x) = 1 + 2x\arctan x$

45. $g'(t) = -4/\sqrt{t^4 - 16t^2}$

47. $G'(t) = -1/\sqrt{2(-2t^2 + 3t - 1)}$

49. $y' = x/[|x|(1 + x^2)]$ 51. $y' = \cos x/(1 + \sin^2 x)$

53. $y' = \pi/[2\sqrt{1 - x^2}(\cos^{-1}x)^2]$

55. $y' = -1/[(1 + x^2)(\tan^{-1}x)^2]$

57. $y' = 2x\cot^{-1}3x - 3x^2/(1 + 9x^2)$

59. $y' = \sqrt{a^2 - b^2}/(a + b\cos x)$

61. $g'(x) = 3/\sqrt{-9x^2 - 6x}$, $[-\frac{2}{3}, 0]$, $(-\frac{2}{3}, 0)$

63. $S'(x) = 1/[(1 + x^2)\sqrt{1 - (\tan^{-1}x)^2}]$, $[-\tan 1, \tan 1]$, $(-\tan 1, \tan 1)$

65. $G'(x) = -1/[2x\sqrt{(x^2 - 1)\csc^{-1}x}]$, $|x| \geq 1$, $|x| > 1$

67. $U'(t) = 2^{\arctan t}(\ln 2)/(1 + t^2)$, $(-\infty, \infty)$, $(-\infty, \infty)$

69. $\pi/6$ 71. $-\pi/2$ 73. π 75. π 77. $\pi^2/4$

79. 0   81. $\frac{1}{4}$ rad/s

83.   87. (a) Yes   (b) No

Exercises 3.8 (page 240)

17. $\coth x = \frac{5}{4}$, $\operatorname{sech} x = \frac{3}{5}$, $\cosh x = \frac{5}{3}$, $\sinh x = \frac{4}{3}$,
    $\operatorname{csch} x = \frac{3}{4}$

19. (a) 1   (b) -1 (c) ∞ (d) $-\infty$ (e) 0
    (f) 1   (g) ∞ (h) $-\infty$ (i) 0

27. $f'(x) = 3\operatorname{sech}^2 3x$ 29. $h'(x) = 4x^3\sinh(x^4)$

31. $G'(x) = 2x\operatorname{sech} x - x^2\operatorname{sech} x\tanh x$

33. $H'(t) = e^t\operatorname{sech}^2(e^t)$

35. $y' = x^{\cosh x}[\sinh x\ln x + (\cosh x)/x]$

37. $y' = 2x/\sqrt{x^4 - 1}$ 39. $y' = \ln(\operatorname{sech} 4x) - 4x\tanh 4x$

41. $y' = a/(a^2 - x^2)$ 43. $y' = -4/(x\sqrt{x^8 + 1})$

45. $y' = -1/(x\sqrt{x^2 + 1})$ 47. $(\ln(1 + \sqrt{2}), \sqrt{2})$

Exercises 3.9 (page 247)

1. $\frac{1}{4}$ 3. $\frac{3}{2}$ 5. 1   7. ∞ 9. $\frac{1}{2}$ 11. 0   13. ∞

15. $1/(3a^{2/3})$ 17. $\frac{1}{2}$ 19. 0   21. $\frac{1}{2}$ 23. ∞

25. 0   27. $\frac{2}{3}$ 29. α 31. $\frac{2}{3}$ 33. -2 35. 0

37. $\frac{1}{2}$ 39. 0   41. 0   43. 0   45. 1   47. ∞

49. 0   51. 0   53. 0   55. 1   57. e 59. e^{-2}

61. e^3 63. 1   65. 1   67. $1/e$ 69. 0   71. $\frac{1}{2}$

73. $-\infty$ 75. 4   77. $\frac{1}{24}$

Review Exercises for Chapter 3 (page 249)

1.   3.

5. 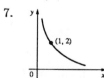   7.

9. $\ln 5$ 11. $\ln 10$ 13. $e^{2/\pi}$

15. $\tan^{-1}4 + n\pi$, n an integer

17. $(\log_{10}e)(2x-1)/(x^2-x)$

19. $\dfrac{\sqrt{x+1}(2-x)^5}{(x+3)^7}\left[\dfrac{1}{2(x+1)} - \dfrac{5}{(2-x)} - \dfrac{7}{x+3}\right]$

21. $(1+c^2)e^{cx}\sin x$ 23. $2\tan x$ 25. $e^{-1/x}(1+1/x)$

27. $\dfrac{(\cos^{-1}x)^{\sin^{-1}x}[\cos^{-1}x\ln(\cos^{-1}x) - \sin^{-1}x]}{(\cos^{-1}x)\sqrt{1-x^2}}$

29. e^{x+e^x} 31. $-1/x - 1/[x(\ln x)^2]$

33. $7^{\sqrt{2x}}(\ln 7)/\sqrt{2x}$ 35. $e^{\sec^{-1}x}(1+1/\sqrt{x^2-1})$

37. $3\tanh 3x$ 39. $\cosh x/\sqrt{\sinh^2 x - 1}$
41. $\cot x - \sin x\cos x$ 43. $1/[\sqrt{x}(x+1)]$
45. $1/(x^4+x^2+1)$ 47. $2^x(\ln 2)^n$
51. $3x - 2y + \ln 4 = 0$ 53. $(-3, 0)$ 55. ∞ 57. $-\infty$
59. ∞ 61. 0 63. $\pi/3$ 65. $-1/(2\pi)$ 67. 0
69. $-\frac{1}{3}$ 71. 0 73. 0 75. 1

77. (a) $1000e^{(\ln 9)t/2} = 1000 \times 3^t$ (b) $27,000$
    (c) $(\ln 2)/\ln 3$ h
79. (a) $\$14,641.00$ (b) $\$14,774.55$ (c) $\$14,845.06$
    (d) $\$14,893.54$ (e) $\$14,917.44$ (f) $\$14,918.25$
81. (a) $C_0 e^{-kt}$ (b) ≈ 100 h
83. $v(t) = -Ae^{-ct}[c\cos(\omega t + \delta) + \omega\sin(\omega t + \delta)]$,
    $a(t) = Ae^{-ct}[(c^2 - \omega^2)\cos(\omega t + \delta) + 2c\omega\sin(\omega t + \delta)]$
85. $f'(x) = e^{g(x)}g'(x)$ 87. $f'(x) = g'(x)/g(x)$
89. $f'(x) = [g'(e^x)e^x]/g(e^x)$ 91. $\frac{2}{3}$

Applications Plus (page 252)

1. $(100 - 50\sqrt{2}, 150 - 100\sqrt{2})$, about 29 m east and
   9 m north of the origin

3. (a) $(r/k) - [(r/k) - C_0]e^{-kt}$
   (b) r/k; the concentration increases to this limiting
       concentration

5. $3/(\sqrt[3]{2} - 1) \approx 11\frac{1}{2}$ h

7. (a) 9.8 h   (b) $31,900\pi \approx 100,000$ ft^2; ≈ 6283 ft^2/h
   (c) ≈ 5.1 h

Chapter 4

Exercises 4.1 (page 262)

1. Absolute maximum at e, local maximum at b and e,
   absolute minimum at d, local minimum at d and s.

3.
No maximum, absolute
minimum
$f(-1) = -1$

5.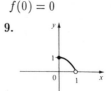
Absolute maximum
$f(-2) = 2$,
local and absolute minimum
$f(0) = 0$

7.
No maximum or minimum

9.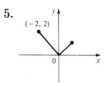
Absolute maximum $f(0) = 1$,
no minimum

11.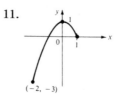
Local and absolute maximum
$f(0) = 1$, absolute minimum
$f(-2) = -3$

13.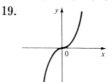
No maximum or minimum

15.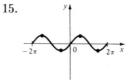
Local and absolute maximum
$f(\pi/2) = f(-3\pi/2) = 1$,
local and absolute minimum
$f(3\pi/2) = f(-\pi/2) = -1$

17.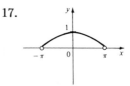
Local and absolute
maximum
$f(0) = 1$, no minimum

19.
No maximum or minimum

21.
No maximum, absolute
minimum
$f(0) = f(2) = 0$

23. $\frac{1}{3}$ 25. ± 1 27. None   29. $(-1 \pm \sqrt{5})/2$
31. 0 33. $-2, 0$ 35. ± 1 37. $0, 4, \frac{8}{7}$
39. 2 41. $n\pi/4$ (n any integer)   43. $1/e$
45. $f(3) = 5, f(1) = 1$ 47. $f(5) = 66, f(2) = -15$
49. $f(1) = 9, f(-2) = 0$
51. $f(-3) = 47, f(\pm\sqrt{2}) = -2$ 53. $f(2) = 5, f(1) = 3$
55. $f(-32) = 16, f(0) = 0$ 57. $f(-1) = 1, f(1) = -1$

59. $f(\pi/4) = \sqrt{2}$, $f(0) = 1$

61. $f(1) = 1/e$, $f(0) = 0$

69.

Exercises 4.2 (page 268)

1. $\pm 1/\sqrt{3}$ **3.** $\pi/2$ **7.** $\frac{3}{2}$ **9.** $\pm\sqrt{7/3}$ **11.** $\sqrt{2}$

13. $1 + (7/3)^{3/2}$ **15.** f is not differentiable at 1

23. No   **27.** No

Exercises 4.3 (page 274)

1. (a) Increasing on $(-\infty, -\frac{1}{2}]$, decreasing on $[-\frac{1}{2}, \infty)$

(b) Local maximum $f(-\frac{1}{2}) = 20.25$

(c)

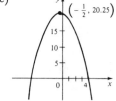

3. (a) Increasing on $(-\infty, \infty)$

(b) No maximum or minimum

(c)

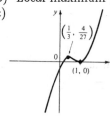

5. (a) Increasing on $(-\infty, \frac{1}{3}]$, decreasing on $[\frac{1}{3}, 1]$, increasing on $[1, \infty)$

(b) Local maximum $f(\frac{1}{3}) = \frac{4}{27}$, minimum $f(1) = 0$

(c)

7. (a) Increasing on $(-\infty, -1]$ and $[0, 1]$, decreasing on $[-1, 0]$ and $[1, \infty)$

(b) Local maximum $f(\pm 1) = 1$, local minimum $f(0) = 0$

(c)

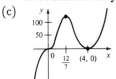

9. (a) Decreasing on $(-\infty, -1]$, increasing on $[-1, \infty)$

(b) Local minimum $f(-1) = -2$

(c)

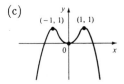

11. (a) Increasing on $(-\infty, \frac{12}{7}]$ and $[4, \infty)$, decreasing on $[\frac{12}{7}, 4]$

(b) Local maximum $f(\frac{12}{7}) = 12^3 \cdot 16^4/7^7$, local minimum $f(4) = 0$

(c)

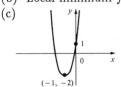

13. (a) Increasing on $(-\infty, 4]$, decreasing on $[4, 6]$

(b) Local maximum $f(4) = 4\sqrt{2}$

(c)

15. (a) Decreasing on $(-\infty, -\frac{1}{6}]$, increasing on $[-\frac{1}{6}, \infty)$

(b) Local minimum $f(-\frac{1}{6}) = -5/6^{1.2}$

(c)

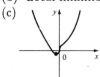

17. (a) Increasing on $[0, \frac{3}{4}]$, decreasing on $[\frac{3}{4}, 1]$

(b) Local maximum $f(\frac{3}{4}) = 3\sqrt{3}/16$

(c)

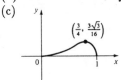

19. (a) Decreasing on $[0, \pi/3]$ and $[5\pi/3, 2\pi]$, increasing on $[\pi/3, 5\pi/3]$

(b) Local maximum $f(5\pi/3) = (5\pi/3) + \sqrt{3}$, minimum $f(\pi/3) = (\pi/3) - \sqrt{3}$

(c)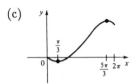

21. (a) Decreasing on $[0, \pi/4]$, $[\pi/2, 3\pi/4]$, $[\pi, 5\pi/4]$, $[3\pi/2, 7\pi/4]$, increasing on $[\pi/4, \pi/2]$, $[3\pi/4, \pi]$, $[5\pi/4, 3\pi/2]$, $[7\pi/4, 2\pi]$
(b) Local maximum $f(\pi/2) = f(\pi) = f(3\pi/2) = 1$, minimum $f(\pi/4) = f(3\pi/4) = f(5\pi/4) = f(7\pi/4) = \frac{1}{2}$
(c)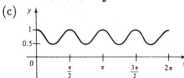

23. Increasing on $(-\infty, (-2 - \sqrt{7})/3]$ and $[(-2 + \sqrt{7})/3, \infty)$, decreasing on $[(-2 - \sqrt{7})/3, (-2 + \sqrt{7})/3]$
25. Decreasing on $(-\infty, -2]$, increasing on $[-2, \infty)$
27. Decreasing on $(-\infty, -1]$, increasing on $[-1, \infty)$
29. Increasing on $(0, e^2]$, decreasing on $[e^2, \infty)$
31. Absolute maximum $f(1) = 2$, absolute minimum $f(-1) = -12$, no local extremum
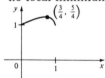

33. Local and absolute maximum $f(\frac{3}{4}) = \frac{5}{4}$, absolute minimum $f(0) = f(1) = 1$, no local minimum
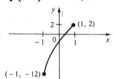

35. Local and absolute maximum $g(1) = \frac{1}{2}$, local and absolute minimum $g(-1) = -\frac{1}{2}$
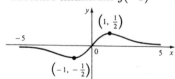

45. $f(x) = \frac{1}{9}(2x^3 + 3x^2 - 12x + 7)$

47.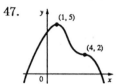

Exercises 4.4 (page 280)

1. (a) Increasing on $(-\infty, -1/\sqrt{3}]$ and $[1/\sqrt{3}, \infty)$, decreasing on $[-1/\sqrt{3}, 1/\sqrt{3}]$
(b) Local maximum $f(-1/\sqrt{3}) = 2/(3\sqrt{3})$, local minimum $f(1/\sqrt{3}) = -2/(3\sqrt{3})$
(c) CD on $(-\infty, 0)$, CU on $(0, \infty)$
(d) 0
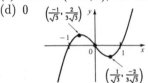

3. (a) Increasing on $(-\infty, -\frac{1}{3}]$ and $[1, \infty)$, decreasing on $[-\frac{1}{3}, 1]$
(b) Local maximum $f(-\frac{1}{3}) = \frac{32}{27}$, local minimum $f(1) = 0$
(c) CD on $(-\infty, \frac{1}{3})$, CU on $(\frac{1}{3}, \infty)$
(d) $\frac{1}{3}$
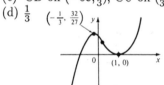

5. (a) Decreasing on $(-\infty, \frac{1}{4}]$, increasing on $[\frac{1}{4}, \infty)$
(b) Local minimum $g(\frac{1}{4}) = -\frac{27}{256}$
(c) CU on $(-\infty, \frac{1}{2})$ and $(1, \infty)$, CD on $(\frac{1}{2}, 1)$
(d) $\frac{1}{2}$, 1
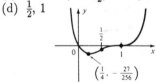

7. (a) Increasing on $(-\infty, -1]$ and $[1, \infty)$, decreasing on $[-1, 1]$
(b) Local maximum $h(-1) = 5$, minimum $h(1) = 1$
(c) CD on $(-\infty, -1/\sqrt{2})$ and $(0, 1/\sqrt{2})$, CU on $(-1/\sqrt{2}, 0)$ and $(1/\sqrt{2}, \infty)$
(d) 0, $\pm 1/\sqrt{2}$
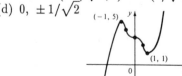

9. (a) Decreasing on $(-\infty, \infty)$
   (b) No maximum or minimum
   (c) CD on $(-\infty, 0)$, CU on $(0, \infty)$
   (d) 0

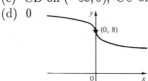

11. (a) Increasing on $(-\infty, \infty)$
    (b) No maximum or minimum
    (c) CD on $(-\infty, 0)$, CU on $(0, \infty)$
    (d) 0

13. (a) Increasing on $(-\infty, -3]$ and $[-1, \infty)$, decreasing on $[-3, -1]$
    (b) Local maximum $Q(-3) = 0$, minimum $Q(-1) = -\sqrt[3]{4}$
    (c) CU on $(-\infty, -3)$ and $(-3, 0)$, CD on $(0, \infty)$
    (d) 0

15. (a) Increasing on $[n\pi, (2n+1)\pi/2]$, decreasing on remaining intervals
    (b) Maximum $f((2n+1)\pi/2) = 1$, minimum $f(n\pi) = 0$
    (c) CD on $(n\pi + (\pi/4), n\pi + (3\pi/4))$, CU on $(n\pi - (\pi/4), n\pi + (\pi/4))$
    (d) $n\pi \pm \pi/4$

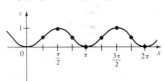

17. (a) Increasing on $(-\infty, \infty)$
    (b) No maximum or minimum
    (c) CD on $(2n\pi, (2n+1)\pi)$, CU on $((2n+1)\pi, (2n+2)\pi)$
    (d) $n\pi$

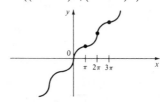

19. $((-1-\sqrt{5})/2, (-1+\sqrt{5})/2)$ 21. $(2, \infty)$

23. $(-2, \infty)$ 25. $(e^{8/3}, \infty)$

27. 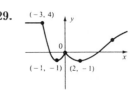   29.

Exercises 4.5 (page 289)

(*abbreviations:* VA, *vertical asymptote;* HA, *horizontal asymptote;* IP, *inflection point*)

1. (a) R (b) y-intercept 1   (c) None   (d) None
   (e) Increasing on $[\frac{1}{3}, 3]$, decreasing on $(-\infty, \frac{1}{3}]$ and $[3, \infty)$
   (f) Local minimum $f(\frac{1}{3}) = \frac{14}{27}$; local maximum $f(3) = 10$
   (g) CD on $(\frac{5}{3}, \infty)$, CU on $(-\infty, \frac{5}{3})$, IP $(\frac{5}{3}, \frac{142}{27})$
   (h)

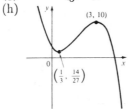

3. (a) R (b) y-intercept 0; x-intercepts 0, $\pm\sqrt{6}$
   (c) About y-axis   (d) None
   (e) Increasing on $[-\sqrt{3}, 0]$ and $[\sqrt{3}, \infty)$, decreasing on $(-\infty, -\sqrt{3}]$ and $[0, \sqrt{3}]$
   (f) Local minima $f(\pm\sqrt{3}) = -9$, maximum $f(0) = 0$
   (g) CU on $(-\infty, -1)$ and $(1, \infty)$, CD on $(-1, 1)$, IP $(1, -5)$ and $(-1, -5)$
   (h)

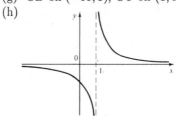

5. (a) $\{x \mid x \neq 1\}$ (b) y-intercept -1
   (c) None   (d) VA $x = 1$, HA $y = 0$
   (e) Decreasing on $(-\infty, 1)$ and $(1, \infty)$
   (f) No maximum or minimum
   (g) CD on $(-\infty, 1)$, CU on $(1, \infty)$, no IP
   (h)

7. (a) $\{x \mid x \neq \pm 3\}$ (b) y-intercept $-\frac{1}{9}$
   (c) About y-axis   (c) VA $x = \pm 3$, HA $y = 0$
   (e) Increasing on $(-\infty, -3)$ and $(-3, 0]$,
       decreasing on $[0, 3)$ and $(3, \infty)$
   (f) Local maximum $f(0) = -\frac{1}{9}$
   (g) CU on $(-\infty, -3)$ and $(3, \infty)$, CD on $(-3, 3)$, no IP
   (h)

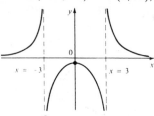

9. (a) $\{x \mid x \neq \frac{3}{2}\}$ (b) x-intercept 0, y-intercept 0
   (c) None   (d) HA $y = 0$, VA $x = \frac{3}{2}$
   (e) Decreasing on $(-\infty, -\frac{3}{2}]$ and $(\frac{3}{2}, \infty)$
       increasing on $[-\frac{3}{2}, \frac{3}{2})$
   (f) Local minimum $f(-\frac{3}{2}) = -\frac{1}{24}$
   (g) CD on $(-\infty, -3)$, CU on $(-3, \frac{3}{2})$ and $(\frac{3}{2}, \infty)$,
       IP$(-3, -\frac{1}{27})$
   (h)

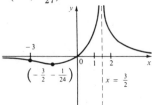

11. (a) $\{x \mid x \neq -3\}$
    (b) x-intercept 3, y-intercept -1
    (c) None (d) VA $x = -3$, HA $y = 1$
    (e) Increasing on $(-\infty, -3)$ and $(-3, \infty)$
    (f) No maximum or minimum
    (g) CU on $(-\infty, -3)$, CD on $(-3, \infty)$, no IP
    (h)

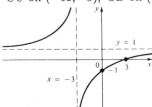

13. (a) $\{x \mid x \neq 1, -2\}$ (b) y-intercept $-\frac{1}{2}$
    (c) None   (d) VA $x = 1$, $x = -2$, HA $y = 0$
    (e) Increasing on $(-\infty, -2)$ and $(-2, -\frac{1}{2}]$,
        decreasing on $[-\frac{1}{2}, 1)$ and $(1, \infty)$
    (f) Local maximum $f(-\frac{1}{2}) = -\frac{4}{9}$
    (g) CU on $(-\infty, -2)$ and $(1, \infty)$, CD on $(-2, 1)$, no IP
    (h)

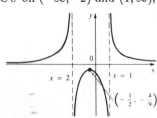

15. (a) $\{x \mid x \neq \pm 1\}$ (b) y-intercept 1
    (c) About y-axis   (d) VA $x = \pm 1$, HA $y = -1$
    (e) Decreasing on $(-\infty, -1)$ and $(-1, 0]$, increasing on
        $[0, 1)$ and $(1, \infty)$
    (f) Local minimum $f(0) = 1$
    (g) CD on $(-\infty, -1)$ and $(1, \infty)$, CU on $(-1, 1)$, no IP
    (h)

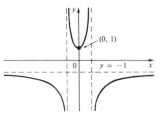

17. (a) $\{x \mid x \neq 0, \pm 1\}$ (b) None   (c) About $(0, 0)$
    (d) VA $x = -1$, $x = 0$, $x = 1$; HA $y = 0$
    (e) Decreasing on $(-\infty, -1)$, $(-1, -1/\sqrt{3}]$, $[1/\sqrt{3}, 1)$,
        and $(1, \infty)$, increasing on $[-1/\sqrt{3}, 0)$ and
        $(0, 1/\sqrt{3}]$
    (f) Local minimum $f(-1/\sqrt{3}) = 3\sqrt{3}/2$, maximum
        $f(1/\sqrt{3}) = -3\sqrt{3}/2$
    (g) CD on $(-\infty, -1)$, $(0, 1)$; CU on $(-1, 0)$, $(1, \infty)$
    (h)

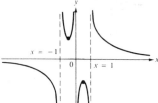

19. (a) $(-\infty, 2]$ (b) x-intercept 2, y-intercept $\sqrt{2}$
    (c) None   (d) None   (e) Decreasing on $(-\infty, 2]$
    (f) No local maximum or minimum
    (g) CD on $(-\infty, 2)$, no IP
    (h)

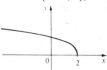

21. (a) $[0, \infty)$ (b) x-intercept 0, y-intercept 0
    (c) None   (d) None   (e) Increasing on $[0, \infty)$
    (f) No local maximum or minimum
    (g) CD on $(0, \infty)$, no IP
    (h)

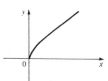

23. (a) R (b) y-intercept 1   (c) None
    (d) HA $y = 0$ (e) Decreasing on $(-\infty, \infty)$
    (f) No local maximum or minimum

(g) CU on $(-\infty, \infty)$, no IP

(h)

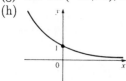

25. (a) $\{x \mid |x| \geq 5\} = (-\infty, -5] \cup [5, \infty)$
    (b) x-intercepts ± 5 (c) About y-axis
    (d) None
    (e) Decreasing on $(-\infty, -5]$, increasing on $[5, \infty)$
    (f) No local maximum or minimum
    (g) CD on $(-\infty, -5)$ and $(5, \infty)$, no IP
    (h)

27. (a) $\{x \mid |x| \geq 3\} = (-\infty, -3] \cup [3, \infty)$
    (b) x-intercepts ± 3 (c) About $(0,0)$ (d) None
    (e) Increasing on $(-\infty, -3]$ and $[3, \infty)$
    (f) No local maximum or minimum
    (g) CU on $(-3\sqrt{3/2}, -3)$ and $(3\sqrt{3/2}, \infty)$,
        CD on $(-\infty, -3\sqrt{3/2})$ and $(3, 3\sqrt{3/2})$,
        IP when $x = \pm 3\sqrt{3/2}$
    (h)

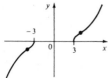

29. (a) $\{x \mid |x| \leq 1, \ x \neq 0\} = [-1, 0) \cup (0, 1]$
    (b) x-intercepts ± 1 (c) About $(0,0)$
    (d) VA $x = 0$ (e) Decreasing on $[-1, 0)$ and $(0, 1]$
    (f) No local maximum or minimum
    (g) CU on $(-1, -\sqrt{2/3})$ and $(0, \sqrt{2/3})$,
        CD on $(-\sqrt{2/3}, 0)$ and $(\sqrt{2/3}, 1)$,
        IP $(\pm \sqrt{2/3}, \pm 1/\sqrt{2})$
    (h)

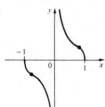

31. (a) $(-\infty, \infty)$ (b) x-intercepts $0, -27$, y-intercept 0
    (c) None   (d) None
    (e) Increasing on $(-\infty, -8]$ and $[0, \infty)$,
        decreasing on $[-8, 0]$
    (f) Local minimum $f(0) = 0$, local maximum
        $f(-8) = 4$

(g) CD on $(-\infty, 0)$ and $(0, \infty)$

(h)

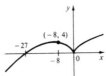

33. (a) R
    (b) y-intercept 1,
        x-intercepts $n\pi + (\pi/4)$ (n an integer)
    (c) Period 2π (d) None
    (e) Decreasing on $[2n\pi - \frac{\pi}{4}, 2n\pi + \frac{3\pi}{4}]$,
        increasing on $[2n\pi + \frac{3\pi}{4}, 2n\pi + \frac{7\pi}{4}]$
    (f) Local minimum $f(2n\pi + \frac{3\pi}{4}) = -\sqrt{2}$,
        local maximum $f(2n\pi - \frac{\pi}{4}) = \sqrt{2}$
    (g) CU on $(2n\pi + (\pi/4), 2n\pi + (5\pi/4))$,
        CD on $(2n\pi - (3\pi/4), 2n\pi + (\pi/4))$,
        IP $(n\pi + (\pi/4), 0)$
    (h)

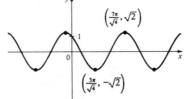

35. (a) $\{x \mid x \neq (2n+1)\pi/2\}$
    (b) y-intercept 0, x-intercept $n\pi$
    (c) About $(0,0)$ and period 2π
    (d) VA $x = (2n+1)\pi/2$
    (e) Decreasing on $((2n-1)\pi/2, (2n+1)\pi/2)$
    (f) No maximum or minimum
    (g) CU on $((2n-1)\pi/2, n\pi)$, CD on
        $(n\pi, (2n+1)\pi/2)$, IP $(n\pi, 0)$
    (h)

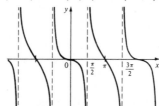

37. (a) $(-\pi/2, \pi/2)$ (b) x-intercept 0, y-intercept 0
    (c) About y-axis   (d) VA $x = \pm\frac{\pi}{2}$
    (e) Decreasing on $(-\pi/2, 0]$, increasing on $[0, \pi/2)$
    (f) Local minimum $f(0) = 0$
    (g) CU on $(-\pi/2, \pi/2)$, no IP

(h)

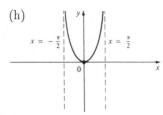

39. (a) $(0, 3\pi)$ (c) None   (d) None
     (e) Decreasing on $(0, \pi/3]$ and $[5\pi/3, 7\pi/3]$,
        increasing on $[\pi/3, 5\pi/3]$ and $[7\pi/3, 3\pi)$
     (f) Local minimum $f(\pi/3) = (\pi/6) - (\sqrt{3}/2)$,
        $f(7\pi/3) = (7\pi/6) - (\sqrt{3}/2)$,
        maximum $f(5\pi/3) = (5\pi/6) + (\sqrt{3}/2)$
     (g) CU on $(0, \pi)$ and $(2\pi, 3\pi)$, CD on $(\pi, 2\pi)$,
        IP $(\pi, \pi/2)$, $(2\pi, \pi)$
     (h)

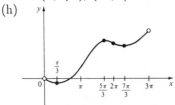

41. (a) $(-\infty, \infty)$ (b) y-intercept 2
     (c) About y-axis, period 2π (d) None
     (e) Increasing on $[(2n-1)\pi, 2n\pi]$,
        decreasing on $[2n\pi, (2n+1)\pi]$
     (f) Maximum $f(2n\pi) = 2$,
        minimum $f((2n+1)\pi) = -2$
     (g) CD on $(2n\pi - (2\pi/3), 2n\pi + (2\pi/3))$,
        CU on remaining intervals,
        IP when $x = 2n\pi \pm (2\pi/3)$
     (h)

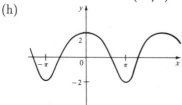

43. (a) R (b) y-intercept 0, x-intercept $n\pi$
     (c) About $(0, 0)$, period 2π (d) None
     (e) Increasing on $[-\pi, -2\pi/3]$ and $[2\pi/3, \pi]$,
        decreasing on $[-2\pi/3, 2\pi/3]$
     (f) Local maximum $f(-2\pi/3) = 3\sqrt{3}/2$,
        local minimum $f(2\pi/3) = -3\sqrt{3}/2$
     (g) CD on $(-\pi, -\alpha)$ and $(0, \alpha)$, where $\cos \alpha = \frac{1}{4}$,
        CU on $(-\alpha, 0)$ and (α, π), IP when $x = 0, \pm \alpha, \pi$
     (h)

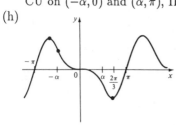

45. (a) R (b) y-intercept 1   (c) About y-axis
     (d) Horizontal asymptote $y = 0$
     (e) Increasing on $(-\infty, 0]$, decreasing on $[0, \infty)$
     (f) Local maximum $f(0) = 1$
     (g) CU on $(-\infty, -1\sqrt{2})$ and $(1/\sqrt{2}, \infty)$,
        CD on $(-1/\sqrt{2}, 1/\sqrt{2})$, IP $(\pm 1/\sqrt{2}, 1/\sqrt{e})$
     (h)

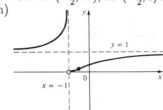

47. (a) $\{x \,|\, x \neq -1\}$ (b) y-intercept $f(0) = 1/e$
     (c) None
     (d) Horizontal asymptote $y = 1$,
        vertical asymptote $x = -1$
     (e) Increasing on $(-\infty, -1)$ and $(-1, \infty)$
     (f) No maximum or minimum
     (g) CU on $(-\infty, -1)$ and $(-1, -\frac{1}{2})$,
        CD on $(-\frac{1}{2}, \infty)$, IP $(-\frac{1}{2}, 1/e^2)$
     (h)

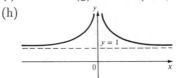

49. (a) $\{x \,|\, x \neq 0\}$ (b) None   (c) About y-axis
     (d) Horizontal asymptote $y = 1$,
        vertical asymptote $x = 0$
     (e) Increasing on $(-\infty, 0)$, decreasing on $(0, \infty)$
     (f) None   (g) CU on $(-\infty, 0)$ and $(0, \infty)$, no IP
     (h)

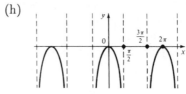

51. (a) $\sum_{n=-\infty}^{\infty} (2n\pi - \pi/2, 2n\pi + \pi/2)$
     (b) x-intercepts $2n\pi$, y-intercept 0
     (c) About y-axis and period 2π
     (d) Vertical asymptotes $x = (2n+1)\pi/2$
     (e) Increasing on $(2n\pi - \pi/2, 2n\pi]$, decreasing on
        $[2n\pi, 2n\pi + \pi/2)$
     (f) Local maximum $f(2n\pi) = 0$
     (g) CD on $(2n\pi - \pi/2, 2n\pi + \pi/2)$, no IP
     (h)

53. (a) R (b) x-intercept 0, y-intercept 0
(c) About $(0,0)$ (d) None
(e) Increasing on $(-\infty, \infty)$ (f) None
(g) CU on $(-\infty, 0)$, CD on $(0, \infty)$, IP $(0,0)$
(h)

55. (a) R (b) x-intercept 0, y-intercept 0
(c) About y-axis   (d) None
(e) Decreasing on $(-\infty, 0]$, increasing on $[0, \infty)$
(f) Local minimum $f(0) = 0$
(g) CD on $(-\infty, -1)$ and $(1, \infty)$, CU on $(-1, 1)$,
IP $(\pm 1, \ln 2)$
(h)

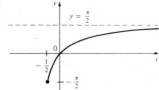

57. (a) $[-\frac{1}{2}, \infty)$ (b) x-intercept 0, y-intercept 0
(c) None   (d) Horizontal asymptote $y = \pi/2$
(e) Increasing on $[-\frac{1}{2}, \infty)$
(f) No local maximum or minimum
(g) CD on $(-\frac{1}{2}, \infty)$, no IP
(h)

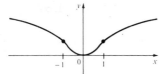

59. (a) R (b) x-intercept 0, y-intercept 0
(c) None   (d) Horizontal asymptote $y = 0$
(e) Increasing on $(-\infty, 1]$, decreasing on $[1, \infty)$
(f) Local maximum $f(1) = 1/e$
(g) CD on $(-\infty, 2)$, CU on $(2, \infty)$, IP $(2, 2/e^2)$
(h)

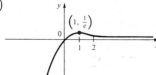

61. (a) $(0, \infty)$ (b) x-intercept 1
(c) None   (d) None
(e) Decreasing on $(0, 1/e]$, increasing on $[1/e, \infty)$
(f) Local minimum $f(1/e) = -1/e$
(g) CU on $(0, \infty)$, no IP

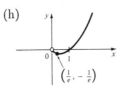

$\left(\frac{1}{e}, -\frac{1}{e}\right)$

63. (a) $(0, \infty)$ (b) x-intercept 1
(c) None   (d) None
(e) Decreasing on $(0, 1/\sqrt{e}]$, increasing on $[1/\sqrt{e}, \infty)$
(f) Local minimum $f(1/\sqrt{e}) = -1/(2e)$
(g) CD on $(0, e^{-3/2})$, CU on $(e^{-3/2}, \infty)$,
IP $(e^{-3/2}, -3/(2e^3))$
(h)

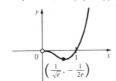

$\left(\frac{1}{\sqrt{e}}, -\frac{1}{2e}\right)$

65. (a) R (b) x-intercept 0, y-intercept 0
(c) About $(0,0)$
(d) Horizontal asymptote $y = 0$
(e) Decreasing on $(-\infty, -1/\sqrt{2}]$ and $[1/\sqrt{2}, \infty)$,
increasing on $[-1/\sqrt{2}, 1/\sqrt{2}]$
(f) Local minimum $f(-1/\sqrt{2}) = -1/\sqrt{2e}$,
local maximum $f(1/\sqrt{2}) = 1/\sqrt{2e}$
(g) CD on $(-\infty, -\sqrt{3/2})$ and $(0, \sqrt{3/2})$,
CU on $(-\sqrt{3/2}, 0)$ and $(\sqrt{3/2}, \infty)$,
IP $(\pm\sqrt{3/2}, \pm\sqrt{3/2}e^{-3/2})$ and $(0,0)$
(h)

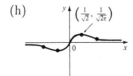

67. (a) $\{x \mid x \neq 0\}$ (b) None   (c) None
(d) Horizontal asymptote $y = 0$,
vertical asymptote $x = 0$
(e) Decreasing on $(-\infty, 0)$ and $(0, 1]$, increasing
on $[1, \infty)$ (f) Local minimum $f(1) = e$
(g) CD on $(-\infty, 0)$, CU on $(0, \infty)$, no IP
(h)

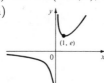

69. (a) $\{x \mid x \neq 0\}$ (b) None
(c) None   (d) Vertical asymptote $x = 0$
(e) Increasing on $(-\infty, 0)$ and $[1, \infty)$,
decreasing on $(0, 1]$

(f) Local minimum $f(1) = e$

(g) CD on $(-\infty, 0)$, CU on $(0, \infty)$, no IP

(h)

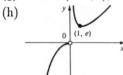

71. (a) $(-1, \infty)$ (b) x-intercept 0, y-intercept 0
    (c) None   (d) Vertical asymptote $x = -1$
    (e) Decreasing on $(-1, 0]$, increasing on $[0, \infty)$
    (f) Local minimum $f(0) = 0$
    (g) CU on $(-1, \infty)$, no IP
    (h)

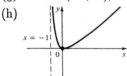

73. (a) $\{x \mid x \neq \pm 1\}$ (b) x-intercept 0, y-intercept 0
    (c) About the origin
    (d) VA $x = \pm 1$, slant asymptote $y = x$
    (e) Increasing on $(-\infty, -\sqrt{3}]$ and $[\sqrt{3}, \infty)$,
        decreasing on $[-\sqrt{3}, -1)$, $(-1, 1)$, and $(1, \sqrt{3}]$
    (f) Local maximum $f(-\sqrt{3}) = -3\sqrt{3}/2$,
        minimum $f(\sqrt{3}) = 3\sqrt{3}/2$
    (g) CD on $(-\infty, -1)$ and $(0, 1)$,
        CU on $(-1, 0)$ and $(1, \infty)$, IP $(0, 0)$
    (h)

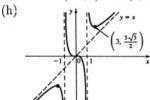

75. (a) $\{x \mid x \neq 0\}$ (b) No intercepts
    (c) About $(0, 0)$
    (d) VA $x = 0$, slant asymptote $y = x$
    (e) Increasing on $(-\infty, -2]$ and $[2, \infty)$,
        decreasing on $[-2, 0)$ and $(0, 2]$
    (f) Local maximum $f(-2) = -4$,
        minimum $f(2) = 4$
    (g) CD on $(-\infty, 0)$, CU on $(0, \infty)$, no IP
    (h)

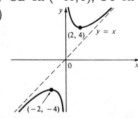

77. (a) $\{x \mid x \neq 1\}$
    (b) y-intercept -1, x-intercepts $(1 \pm \sqrt{5})/2$
    (c) None   (d) VA $x = 1$, slant asymptote $y = -x$
    (e) Decreasing on $(-\infty, 1)$ and $(1, \infty)$

(f) No maximum or minimum

(g) CD on $(-\infty, 1)$, CU on $(1, \infty)$, no IP

(h)

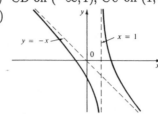

79. (a) R (b) y-intercept 1
    (c) None   (d) Slant asymptote $y = x$
    (e) Increasing on $(-\infty, \infty)$ (f) None
    (g) CU on $(-\infty, \infty)$
    (h)

Exercises 4.6 (page 295)

1. 50, 50   3. 10, 10   7. 14,062.5 ft^2 9. 4000 cm^3
11. \$191.28   13. $(1.2, -0.6)$ 15. $(1, \pm\sqrt{5})$
17. Square, side $\sqrt{2}r$ 19. $L/2$, $\sqrt{3}L/4$
21. Base $\sqrt{3}r$, height $3r/2$ 23. $4\pi r^3/(3\sqrt{3})$
25. $\pi r^2(1 + \sqrt{5})$ 27. 24 cm, 36 cm
29. (a) All of the wire for the square
    (b) $40\sqrt{3}/(9 + 4\sqrt{3})$ m for the square
31. Height = radius = $\sqrt[3]{V/\pi}$ cm
33. $V = 2\pi R^3/(9\sqrt{3})$ 37. Row directly to B
39. $10\sqrt[3]{3}/(1 + \sqrt[3]{3})$ ft from the stronger source
45. At a distance $5 - 2\sqrt{5}$ from A 47. $x = 6$ in
49. $\pi/6$ 51. $(L + W)^2/2$

Exercises 4.7 (page 302)

1. (a) \$1,035,000, \$1035, \$2025/unit   (b) 100   (c) \$225
3. (a) \$2330.71, \$2.33, \$4.07/unit   (b) 159   (c) \$1.07
5. (a) \$188.25, \$0.19, \$0.28/unit   (b) 400   (c) \$0.15
7. 400   9. 16,667   11. 672   13. 100
15. (a) $p(x) = 19 - (x/3000)$ (b) \$9.50
17. (a) $p(x) = 550 - (x/10)$ (b) \$175   (c) \$100

Exercises 4.8 (page 307)

1. $4x^3 + 3x^2 - 5x + C$
3. $(3x^{10}/5) - (x^8/2) + x^3 + x + C$
5. $(2x^{3/2}/3) + (3x^{4/3}/4) + C$
7. $-3/(2x^4) + C_1$ if $x > 0$, $-3/(2x^4) + C_2$ if $x < 0$
9. $(2x\sqrt{x}/3) + 2\sqrt{x} + C$ 11. $(2t^{7/2}/7) + (4t^{5/2}/5) + C$
13. $-\cos x - 2\sin x + C$
15. $\tan t + (t^3/3) + C_n$, $(2n - 1)\pi/2 < t < (2n + 1)\pi/2$
17. $x^2 + 5\sin^{-1}x + C$
19. $(x^5/5) - (2x^3/3) + (x^2/2) - x + C$

21. $(x^4/12) + (x^5/20) + Cx + D$ 23. $(x^2/2) + Cx + D$
25. $x^4 + (Cx^2/2) + Dx + E$ 27. $2x^2 + 3x - 9$
29. $2x^{3/2} - 2\sqrt{x} + 2$ 31. $3 \sin x - 5 \cos x + 9$
33. $2x + \frac{5}{8}x^{8/5} + \frac{3}{8}$ 35. $2 \ln|x| + 7$
37. $-4x^2 + 5x + 6$ 39. $x^5 - 5x^2 + 5$
41. $(x^4/12) - 3 \cos x + 3x + 5$ 43. $x^3 + 3x^2 - 5x + 4$
45. $f(x) = 1/(2x) + (x/4) - (3/4)$
47. $-\ln x + (\ln 2)x - \ln 2$
49. 10   51. $s(t) = 3t - t^2 + 4$
53. $s(t) = (t^3/2) + 4t^2 - 2t + 1$
55. $s(t) = (t^4/12) - (t^3/6) - 10t$
57. (a) $s(t) = 450 - 4.9t^2$ (b) $\sqrt{450/4.9} \approx 9.58$ s
    (c) $-9.8\sqrt{450/4.9} \approx -93.9$ m/s
59. (a) $s(t) = 450 + 5t - 4.9t^2$
    (b) $(5 + \sqrt{8845})/9.8 \approx 10.1$ s   (c) ≈ -94.0 m/s
63. \$742.08   65. $\frac{130}{11} \approx 11.8$ s   67. $\frac{88}{15}$ ft/s^2
69. 225 ft

Review Exercises for Chapter 4 (page 309)

1. False   3. False   5. True   7. False   9. True
11. False   13. True
15. Absolute and local maximum $f(-2) = 21$, local
    minimum $f(2) = -11$, absolute minimum
    $f(-5) = -60$
17. Absolute maximum $f(4) = \frac{1}{3}$, absolute minimum
    $f(0) = -1$
19. Local and absolute minimum $f(\frac{\pi}{4}) = \frac{\pi}{4} - 1$, absolute
    maximum $f(\pi) = \pi$

21.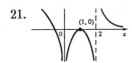

23. (a) R (b) y-intercept 1   (c) None   (d) None
    (e) Increasing on $(-\infty, \infty)$ (f) None
    (g) CD on $(-\infty, 0)$, CU on $(0, \infty)$, IP$(0, 1)$
    (h)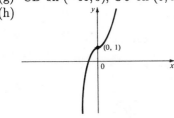

25. (a) $\{x \mid x \neq 0, 3\} = (-\infty, 0) \cup (0, 3) \cup (3, \infty)$
    (b) None   (c) None
    (d) HA $y = 0$, VA $x = 0$, $x = 3$
    (e) Decreasing on $(-\infty, 0), (0, 1], (3, \infty)$,
        increasing on $[1, 3)$
    (f) Local minimum $f(1) = \frac{1}{4}$
    (g) CU on $(0, 3), (3, \infty)$, CD on $(-\infty, 0)$

(h)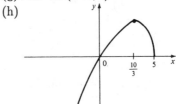

27. (a) $(-\infty, 5]$
    (b) x-intercepts 0, 5, y-intercept 0   (c) None
    (d) None
    (e) Increasing on $(-\infty, \frac{10}{3}]$, decreasing on $[\frac{10}{3}, 5]$
    (f) Local maximum $f(\frac{10}{3}) = 10\sqrt{5}/(3\sqrt{3})$
    (g) CD on $(-\infty, 5)$
    (h)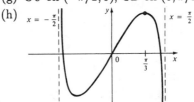

29. (a) $(-\pi/2, \pi/2)$ (b) y-intercept 0
    (c) About $(0, 0)$ (d) VA $x = \pm\pi/2$
    (e) Increasing on $[-\pi/3, \pi/3]$,
        decreasing on $(-\pi/2, -\pi/3]$ and $[\pi/3, \pi/2)$
    (f) Local maximum $f(\frac{\pi}{3}) = (4\pi/3) - \sqrt{3}$,
        minimum $f(-\frac{\pi}{3}) = -(4\pi/3) + \sqrt{3}$
    (g) CU on $(-\pi/2, 0)$, CD on $(0, \pi/2)$, IP$(0, 0)$
    (h)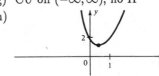

31. (a) R (b) y-intercept 2   (c) None   (d) None
    (e) Decreasing on $(-\infty, \frac{1}{4}\ln 3]$, increasing on
        $[\frac{1}{4}\ln 3, \infty)$
    (f) Local minimum $f(\frac{1}{4}\ln 3) = 3^{1/4} + 3^{-3/4}$
    (g) CU on $(-\infty, \infty)$, no IP
    (h)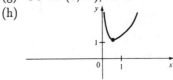

33. (a) $(0, \infty)$ (b) None   (c) None
    (d) Vertical asymptote $x = 0$
    (e) Decreasing on $(0, \frac{1}{2}]$, increasing on $[\frac{1}{2}, \infty)$
    (f) Local minimum $f(\frac{1}{2}) = \frac{1}{2} + \ln 2$
    (g) CU on $(0, \infty)$, no IP
    (h)

39.

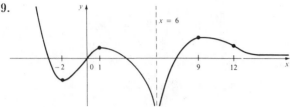

41. $a = -3$, $b = 7$ **45.** $3\sqrt{3}r^2$
47. $4/\sqrt{3}$ cm from D
49. Width $2R/\sqrt{3}$, depth $2\sqrt{2}R/\sqrt{3}$
51. $11.50   **53.** $(x^2/2) - (4x^{5/4}/5) + C$
55. $2\sqrt{x} + (2x^{3/2}/3) - (8/3)$ **57.** $2\tan^{-1}x - 1$
59. $(x^5/20) + (x^3/6) + x - 1$ **61.** No

Problems Plus (page 312)

5. $(-2, 4)$ $(2, -4)$ **7.** Does not exist   **9.** $\frac{4}{3}$
11. $(0, \frac{5}{4})$ **19.** $\frac{1}{2}$ **21.** $(m/2, m^2/4)$ **23.** 1

Chapter 5

Exercises 5.1 (page 319)

1. $\sqrt{1} + \sqrt{2} + \sqrt{3} + \sqrt{4} + \sqrt{5}$ **3.** $3^4 + 3^5 + 3^6$
5. $-1 + \frac{1}{3} + \frac{3}{5} + \frac{5}{7} + \frac{7}{9}$ **7.** $1^{10} + 2^{10} + 3^{10} + \cdots + n^{10}$
9. $1 - 1 + 1 - 1 + \cdots + (-1)^{n-1}$ **11.** $\sum_{i=1}^{10} i$
13. $\sum_{i=1}^{19} \frac{i}{i+1}$ **15.** $\sum_{i=1}^{n} 2i$ **17.** $\sum_{i=0}^{5} 2^i$ **19.** $\sum_{i=1}^{n} x^i$
21. 80   **23.** 3276   **25.** 0   **27.** 61   **29.** $n/(n+1)$
31. $n(n+1)$ **33.** $n(n^2 + 6n + 17)/3$
35. $n(n^2 + 6n + 11)/3$ **37.** $n(n^3 + 2n^2 - n - 10)/4$
45. (a) n^4 (b) $5^{100} - 1$ (c) $\frac{97}{300}$ (d) $a_n - a_0$
47. $\frac{1}{3}$ **49.** 14   **53.** $2^{n+1} + n^2 + n - 2$ **55.** 12

Exercises 5.2 (page 327)

1. (a) 1   (b) 50   (c)

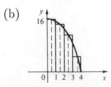

3. (a) 1   (b) 43   (b)

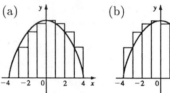

5. (a) 1   (b) 35   (c)

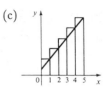

7. (a) 0.5   (b) 12.1875   (c)

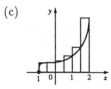

9. (a) $\pi/4$ (b) $\pi(1 + \sqrt{3})/2$ (c)

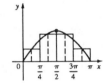

11.

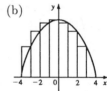

13. $\frac{256}{3}$

(a)   (b)   (c)

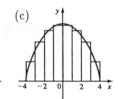

15. 20

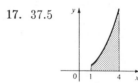 **17.** 37.5

19. 8

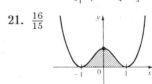 **21.** $\frac{16}{15}$

23. 2

Exercises 5.3 (page 333)

1. (a) 1.2   (b) 4   **3.** (a) 1   (b) 2.336
5. (a) 0.5   (b) -0.028 **7.** 153.125   **9.** 1.8100
11. $(b - a)c$ **13.** 15   **15.** 19.5   **17.** $\frac{4}{3}$
19. $\frac{1}{4}b^4 + 2b^2$ **21.** $\frac{32}{5}$ **25.** $\int_0^1 (2x^2 - 5x)\,dx$
27. $\int_0^\pi \cos x\,dx$ **29.** $\int_0^1 x^4\,dx$ **31.** $\int_1^3 (3x^5 - 6)\,dx$
33. $-\frac{38}{3}$ **35.** (a) and (c) are integrable   **39.** $\frac{1}{2}$

Exercises 5.4 (page 339)

1. 12 **3.** $3\sqrt{3}$ **5.** 16 **7.** 22.5 **9.** $(\pi/3) - \sqrt{3}$
11. 5 **13.** 7 **15.** 2 **17.** $\frac{1000}{3}$ **19.** $\int_{1}^{12} f(x)\, dx$
21. $\int_{7}^{10} f(x)\, dx$ **35.** $2 \le \int_{1}^{3} x^3\, dx \le 54$
37. $\frac{1}{2} \le \int_{1}^{2} dx/x \le 1$ **39.** $-3 \le \int_{-3}^{0}(x^2 + 2x)\, dx \le 9$
41. $2 \le \int_{-1}^{1}\sqrt{1 + x^4}\, dx \le 2\sqrt{2}$

Exercises 5.5 (page 348)

1. $g'(x) = 1 + x^2$

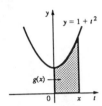

3. $g'(x) = (x^2 - 1)^{20}$ **5.** $g'(u) = 1/(1 + u^4)$
7. $F'(x) = -\cos(x^2)$ **9.** $h'(x) = -\sin^4(1/x)/x^2$
11. $y' = -\sin(\tan^4 x)\sec^2 x$
13. $y' = 5/(25x^2 + 10x - 4)$ **15.** 54 **17.** $\frac{1}{100}$
19. -1 **21.** 231 **23.** $\frac{16}{3}$ **25.** $\frac{28}{81}$
27. $6(3\sqrt{2} - 2)/5$ **29.** $\frac{29}{35}$ **31.** $\frac{28}{3}$ **33.** 0
35. Does not exist **37.** $\frac{2}{3}$ **39.** $\frac{1}{4}$ **41.** $(\sqrt{2} - 1)/2$
43. Does not exist **45.** $2\sqrt{3}/3$ **47.** $\ln 2$
49. $2^8/\ln 2$ **51.** $\pi/2$ **53.** $\frac{1}{2}e^2 + e - \frac{1}{2}$ **55.** 1
57. -3.5 **59.** 10.7 **61.** $\frac{26}{3}$ **63.** $\frac{243}{4}$ **65.** 2
67. $(1 + 2^{1.8})/1.8$ **73.** $\frac{2}{5}x^{5/2} + C$
75. $4x - \frac{8}{3}x^{3/2} + \frac{1}{2}x^2 + C$ **77.** $x^2 + \sec x + C$
79. (a) $-\frac{3}{2}$ m (b) $\frac{41}{6}$ m
81. (a) $v(t) = \frac{1}{2}t^2 + 4t + 5$ (b) $\frac{1250}{3}$ m **83.** $46\frac{2}{3}$ kg
85. $g'(x) = -2(2x - 1)/(2x + 1) + 3(3x - 1)/(3x + 1)$
87. $y' = 3x^{7/2}\sin(x^3) - (\sin\sqrt{x})/(2\sqrt[4]{x})$
89. $\sqrt{257}$ **91.** $\frac{1}{4}$

Exercises 5.6 (page 356)

1. $(x^2 - 1)^{100}/200 + C$ **3.** $-\frac{1}{4}\cos 4x + C$
5. $-1/[2(x^2 + 6x)] + C$ **7.** $(x^2 + x + 1)^4/4 + C$
9. $\frac{2}{3}(x - 1)^{3/2} + C$ **11.** $(2 + x^4)^{3/2}/6 + C$
13. $-1/[6(3x^2 - 2x + 1)^3] + C$ **15.** $-2/[5(t + 1)^5] + C$
17. $-(1 - 2y)^{2.3}/4.6 + C$ **19.** $\frac{1}{2}\sin 2\theta + C$
21. $\frac{4}{7}(x + 2)^{7/4} - \frac{8}{3}(x + 2)^{3/4} + C$ **23.** $-\frac{1}{2}\cos(t^2) + C$
25. $\frac{1}{7}(1 - x^2)^{7/2} - \frac{1}{5}(1 - x^2)^{5/2} + C$ **27.** $\frac{1}{4}\sin^4 x + C$
29. $\frac{1}{2}\ln|2x - 1| + C$ **31.** $(\ln x)^3/3 + C$
33. $(1 + e^x)^{11}/11 + C$ **35.** $\frac{2}{3}(1 + \sec x)^{3/2} + C$
37. $\sqrt{ax^2 + 2bx + c} + C$ **39.** $-\frac{1}{2}\cos(2x + 3) + C$
41. $(\sin 3\alpha)x + \frac{1}{3}\cos 3x + C$

43. $2(b + cx^{a+1})^{3/2}/[3c(a + 1)] + C$ **45.** $\ln|\ln x| + C$
47. $x - e^{-x} + C$ **49.** $\frac{1}{2}\ln|x^2 + 2x| + C$
51. $\tan^{-1}(e^x) + C$ **53.** $\tan^{-1}x + \frac{1}{2}\ln(1 + x^2) + C$
55. $\frac{1}{2}\tan^{-1}(x^2) + C$ **57.** $\frac{1}{101}$ **59.** $\frac{32}{3}$ **61.** $\frac{16}{15}$
63. 0 **65.** $(4\sqrt{2}/3) - (5\sqrt{5}/12)$ **67.** 1 **69.** 3
71. $a^3/3$ **73.** 0 **75.** $\frac{1}{2}\ln 3$ **77.** 2 **79.** $\frac{14}{3}$
81. $2 - \sqrt{3}$ **83.** $\frac{10}{11}$ **85.** $1 - e^{-2}$
87. $[5/(4\pi)][1 - \cos(2\pi t/5)]$ L

Exercises 5.7 (page 365)

1. (b) 0.405

Review Exercises for Chapter 5 (page 365)

1. True **3.** True **5.** True **7.** True **9.** True
11. False **13.** 0.00909 **15.** $2n + n^2(n + 1)^2/4$
17. 7.75

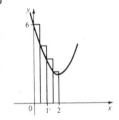

19. -18 **21.** $\frac{875}{12}$ **23.** $\frac{875}{12}$ **25.** $\frac{9}{10}$ **27.** $\frac{1209}{28}$
29. 3480 **31.** 2 **33.** Does not exist
35. $-1/[25(2 + x^5)^5] + C$ **37.** $-(\cos \pi x)/\pi + C$
39. $-\sin(1/t) + C$ **41.** $-\tan(\cos x) + C$
43. $2e^{\sqrt{x}} + C$ **45.** $-\frac{1}{2}[\ln(\cos x)]^2 + C$
47. $\frac{1}{4}\ln(1 + x^4) + C$ **49.** 4
51. $F'(x) = \sqrt{1 + x^2 + x^4}$ **53.** $g'(x) = 3x^5/\sqrt{1 + x^9}$
55. $y' = (2\cos x - \cos\sqrt{x})/(2x)$ **57.** $e^{\sqrt{x}}/(2x)$
59. $4 \le \int_{1}^{3}\sqrt{x^2 + 3}\, dx \le 4\sqrt{3}$ **69.** 8π **71.** $\frac{2}{3}$

Applications Plus (page 368)

1. (a) $\sqrt{800} \approx 28$ ft
(b) $dI/dt = -480k(h - 4)/[(h - 4)^2 + 1600]^{5/2}$,
where k is the constant of proportionality
3. (a) 22.9125 mi (b) 21.675 mi
5. (b) Average expenditure over $[0, t]$.
Minimize average expenditure.
7. (a) $2/\pi$ (b) $1/\pi$ (c) $2/(5\pi)$
9. (b) At most 40%, $\frac{5}{36}$ (c) $-4 + 20\ln 1.25 \approx 0.46$
11. (b) $\frac{1}{4}\pi + \frac{1}{2}\alpha$
(c) $2v^2\cos\theta\sin(\theta + \alpha)/[g\cos^2\alpha]$; $\frac{1}{4}\pi - \frac{1}{2}\alpha$

Chapter 6

Exercises 6.1 (page 378)

1. $\frac{20}{3}$
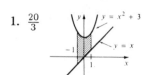

3. 16.5
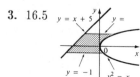

5. 36
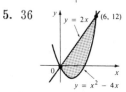

7. $\frac{1}{6}$
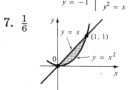

9. $\frac{1}{3}$
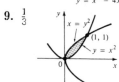

11. $\frac{4}{3}$
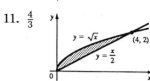

13. 4
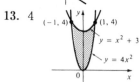

15. 36
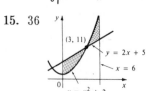

17. 8
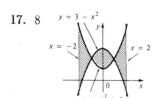

19. $\frac{32}{3}$
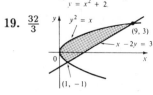

21. $\frac{8}{3}$
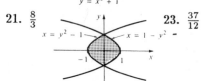

23. $\frac{37}{12}$
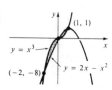

25. $\frac{5}{32}\pi^2 + (1/\sqrt{2}) - 2$
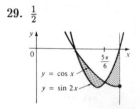

27. $\frac{1}{2}$
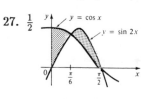

29. $\frac{1}{2}$

31. 34
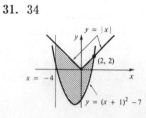

33. 12
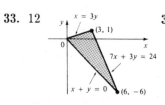

35. $\ln 2 - \frac{1}{2}$
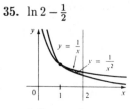

37. $\pi - \frac{2}{3}$
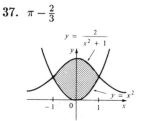

39. $\frac{1}{3}e^3 - e + \frac{2}{3}$
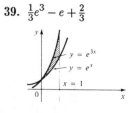

41. 9   **43.** 14.5   **45.** 1.5
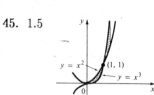

47. 3.75
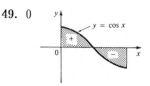

49. 0
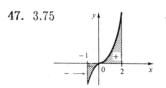

51. 3.220   **53.** $4^{2/3}$

Exercises 6.2 (page 387)

1.
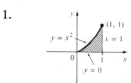

3. $\pi/3$
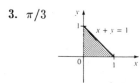

5. 8π
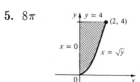

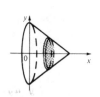

7. $3\pi/10$

9. $64\pi/15$

11. $208\pi/45$

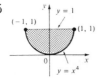

13. $32\pi/3$ **15.** $128\pi/3$ **17.** $128\pi/15$
19. $112\pi/15$ **21.** $64\pi/5$ **23.** $16\pi/5$ **25.** $46\pi/15$
27. $2\pi(\tan 1 - 1)$ **29.** 3π **31.** $\pi(e^2 - 1)/2$
33. $\pi \int_0^{\pi/4} (1 - \tan^2 x)\, dx$
35. $\pi \int_3^6 \{[6 - (x-4)^2]^2 - (8-x)^2\}\, dx$
37. $\pi \int_0^{\pi/2} [(1 + \cos x)^2 - 1]\, dx$
39. 21π

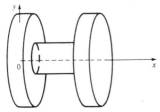

41. Solid obtained by rotating the region under
  $y = \tan x$ from 0 to $\pi/4$ about the x-axis
43. Solid obtained by rotating the region between $x = y$
  and $x = \sqrt{y}$ about the y-axis
45. Solid obtained by rotating the region bounded by
  $y = 2x$ and $y = 2x^2$ about the line $y = 5$
  [or the region bounded by $y = 5 - 2x$ and $y = 5 - 2x^2$
  about the x-axis]
47. $\pi r^2 h/3$ **49.** $\pi h^2[r - (h/3)]$ **51.** $2b^2 h/3$
53. 10 cm^3 **55.** 24 **57.** 2 **59.** $\frac{1}{3}$
61. (a) $8\pi R \int_0^r \sqrt{r^2 - y^2}\, dy$ (b) $2\pi^2 r^2 R$
65. $8 \int_0^r \sqrt{R^2 - y^2} \sqrt{r^2 - y^2}\, dy$ **67.** (b) $\pi r^2 h$

Exercises 6.3 (page 394)

1. $15\pi/2$ **3.** $16\pi(5\sqrt{5} - 1)/3$ **5.** 8π **7.** $\pi/10$
9. 16π **11.** $4096\pi/9$ **13.** $1944\pi/5$ **15.** $5\pi/6$
17. $124\pi/5$

19. $17\pi/6$

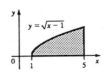 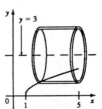

21. 24π

 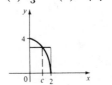

23. $\int_{2\pi}^{3\pi} 2\pi x \sin x\, dx$ **25.** $\int_0^{\pi/4} 2\pi y \cos y\, dy$
27. $\int_0^1 2\pi(x+1)[\sin(\pi x/2) - x^4]\, dx$ **29.** $\int_1^e 2\pi x \ln x\, dx$
31. $\int_1^{\pi} 2\pi y \ln y\, dy$ **33.** $\int_1^2 2\pi(y-1) \ln y\, dy$
35. Solid obtained by rotating the region under $y = \cos x$
  from 0 to $\pi/2$ about the y-axis
37. Solid obtained by rotating the region in the first
  quadrant bounded by $y = x^2$ and $y = x^6$ about the
  y-axis
39. $81\pi/10$ **41.** $16\pi/15$ **43.** $4\pi/3$ **45.** $\frac{4}{3}\pi r^3$
47. $\frac{1}{3}\pi r^2 h$ **49.** 1.142

Exercises 6.4 (page 398)

1. 7200 J   **3.** $\frac{5030}{3}$ ft-lb   **5.** $\frac{15}{4}$ ft-lb   **7.** $\frac{25}{24} \approx 1.04$ J
9. 10.8 cm   **11.** 625 ft-lb   **13.** $650{,}000$ ft-lb
15. $\approx 2.45 \times 10^3$ J   **17.** $\approx 1.06 \times 10^6$ J
19. $\approx 5.8 \times 10^3$ ft-lb   **21.** $\approx 7.20 \times 10^5$ J
25. $Gm_1 m_2[(1/a) - (1/b)]$

Exercises 6.5 (page 402)

1. -2 **3.** $-\frac{11}{3}$ **5.** $\frac{1}{5}$ **7.** $(\sqrt{2} + 4)/(9\pi)$
9. (a) 3 (b) $\frac{3}{2}$ (c)   **11.** (a) $\frac{8}{3}$ (b) $2/\sqrt{3}$ (c)

13. $(50 + 28/\pi)°$F $\approx 59°$ F   **15.** 6 kg/m
17. $5/(4\pi) \approx 0.4$ L

Review Exercises for Chapter 6 (page 402)

1. $\frac{4}{3}$ **3.** 108 **5.** 1 **7.** $2\sqrt{2}$ **9.** 2π
11. $16\pi/3$ **13.** $4\pi(2ah + h^2)^{3/2}/3$
15. $\int_0^1 \pi[(1 - x^3)^2 - (1 - x^2)^2]\, dx$
17. (a) $2\pi/15$ (b) $\pi/6$ (c) $8\pi/15$
19. Solid obtained by rotating the region under
  $y = \sin x$ from 0 to π about the x-axis

21. Solid obtained by rotating the region in the first quadrant bounded by $x = 4 - y^2$ and the axes about the x-axis
23. 36   **25.** $125\sqrt{3}/3$ m^3 **27.** 3.2 J
29. $8000\pi/3$ ft-lb   **31.** $f(x)$

Problems Plus (page 404)

1. $\pi/2$ **7.** $f(x) = \sqrt{2x/\pi}$ **11.** $1/e^2$ **15.** -1
17. $1/\sqrt{2}$ **19.** $[-1, 2]$ **27.** $f(x) = \frac{1}{2}x$ or $f(x) = 0$

Chapter 7

Exercises 7.1 (page 412)

1. $(xe^{2x}/2) - (e^{2x}/4) + C$
3. $-\frac{1}{4}x \cos 4x + \frac{1}{16} \sin 4x + C$
5. $\frac{1}{3}x^2 \sin 3x + \frac{2}{9}x \cos 3x - \frac{2}{27} \sin 3x + C$
7. $x(\ln x)^2 - 2x \ln x + 2x + C$
9. $\frac{1}{8}(\sin 2\theta - 2\theta \cos 2\theta) + C$ **11.** $t^3(3 \ln t - 1)/9 + C$
13. $e^{2\theta}(2 \sin 3\theta - 3 \cos 3\theta)/13 + C$
15. $y \cosh y - \sinh y + C$ **17.** $1 - 2/e$ **19.** $-\frac{1}{2}$
21. $(\pi + 6 - 3\sqrt{3})/6$
23. $(5 \sin 3x \sin 5x + 3 \cos 3x \cos 5x)/16 + C$
25. $\sin x(\ln \sin x - 1) + C$ **27.** $(2x + 1)e^x + C$
29. $x(\cos \ln x + \sin \ln x)/2 + C$ **31.** $4 \ln 2 - \frac{3}{2}$
33. $2(\sin \sqrt{x} - \sqrt{x} \cos \sqrt{x}) + C$
35. $e^{x^2}[(x^4/2) - x^2 + 1] + C$
37. $-\frac{1}{4} \cos x \sin^3 x + (3x/8) - \frac{3}{16} \sin 2x + C$ **39.** (b) $\frac{2}{3}$, $\frac{8}{15}$
45. $x[(\ln x)^3 - 3(\ln x)^2 + 6 \ln x - 6] + C$
47. $(\pi + 6\sqrt{3} - 12)/12$ **49.** 1   **51.** $10\pi^2$ **53.** $2\pi e$
55. $2 - e^{-t}(t^2 + 2t + 2)$ m   **59.** 1

Exercises 7.2 (page 419)

1. $\pi/4$ **3.** $\frac{3}{8}x + \frac{1}{4} \sin 2x + \frac{1}{32} \sin 4x + C$
5. $\frac{1}{7} \cos^7 x - \frac{1}{5} \cos^5 x + C$ **7.** $(3\pi - 4)/192$
9. $(3x/2) + \cos 2x - \frac{1}{8} \sin 4x + C$
11. $\frac{1}{6} \sin^6 x - \frac{1}{4} \sin^8 x + \frac{1}{10} \sin^{10} x + C$
    $[\text{or} -\frac{1}{6} \cos^6 x + \frac{1}{4} \cos^8 x - \frac{1}{10} \cos^{10} x + C_1]$
13. $\frac{1}{8}[(5x/2) + 2 \sin 2x + \frac{3}{8} \sin 4x - \frac{1}{6} \sin^3 2x] + C$
15. $-\frac{1}{5} \cos^5 x + \frac{2}{3} \cos^3 x - \cos x + C$
17. $[\frac{2}{7} \cos^3 x - \frac{2}{3} \cos x] \sqrt{\cos x} + C$
19. $\sqrt{x} + \frac{1}{2} \sin 2\sqrt{x} + C$ **21.** $\frac{1}{2} \cos^2 x - \ln |\cos x| + C$
23. $\ln(1 + \sin x) + C$ **25.** $\tan x - x + C$
27. $\tan x + \frac{1}{3} \tan^3 x + C$ **29.** $\frac{1}{5}$ **31.** $\frac{1}{3} \sec^3 x + C$
33. $\frac{1}{4} \sec^4 x - \tan^2 x + \ln |\sec x| + C$ **35.** $\frac{38}{15}$
37. $\frac{1}{6} \sec^6 x + C$ [or $\frac{1}{2} \tan^2 x + \frac{1}{2} \tan^4 x + \frac{1}{6} \tan^6 x + C_1$]
39. $\frac{1}{2} \tan^2 x + C$ **41.** $\sqrt{3} - (\pi/3)$
43. $-\frac{1}{5} \cot^5 x - \frac{1}{7} \cot^7 x + C$ **45.** $\ln |\csc x - \cot x| + C$
47. $\ln |\csc x - \cot x| + \cos x + C$
49. $\frac{1}{2}[\frac{1}{3} \sin 3x - \frac{1}{7} \sin 7x] + C$ **51.** $\frac{1}{2}[\sin x + \frac{1}{7} \sin 7x] + C$

53. $\frac{1}{2}[\frac{1}{4} \cos 4x - \frac{1}{6} \cos 6x] + C$ **55.** $\frac{1}{2} \sin 2x + C$ **57.** 0
59. $\frac{1}{3}$ **61.** $\pi^2/4$ **63.** $2\pi + (\pi^2/4)$
63. $s = (1 - \cos^3 \omega t)/(3\omega)$

Exercises 7.3 (page 425)

1. $2/\sqrt{3}$ **3.** $-\sqrt{1 - x^2} + C$
5. $\frac{1}{4} \sin^{-1}(2x) + \frac{1}{2}x \sqrt{1 - 4x^2} + C$ **7.** $\ln(1 + \sqrt{2})$
9. $\left[\sqrt{x^2 - 16}/(32 x^2)\right] + \frac{1}{128} \sec^{-1}(x/4) + C$
11. $\sqrt{9x^2 - 4} - 2 \sec^{-1}(3x/2) + C$
13. $\left(x/\sqrt{a^2 - x^2}\right) - \sin^{-1}(x/a) + C$
15. $(1/\sqrt{3}) \ln \left|(\sqrt{x^2 + 3} - \sqrt{3})/x\right| + C$ **17.** $\frac{64}{1215}$
19. $\frac{5}{3}(1 + x^2)^{3/2} + C$
21. $\frac{1}{4}\left[\left(\sqrt{x^2 - 2}/x\right) - (x^2 - 2)^{3/2}/(3x^3)\right] + C$
23. $\frac{1}{2}\left[\sin^{-1}(x - 1) + (x - 1)\sqrt{2x - x^2}\right] + C$
25. $\frac{1}{3} \ln \left|3x + 1 + \sqrt{9x^2 + 6x - 8}\right| + C$
27. $\frac{1}{2}[\tan^{-1}(x + 1) + (x + 1)/(x^2 + 2x + 2)] + C$
29. $\frac{1}{2}\left[e^t \sqrt{9 - e^{2t}} + 9 \sin^{-1}(e^t/3)\right] + C$ **35.** $2\pi^2 Rr^2$

Exercises 7.4 (page 433)

1. $\dfrac{A}{x - 1} + \dfrac{B}{x + 2}$ **3.** $\dfrac{A}{x} + \dfrac{B}{x + 2}$
5. $\dfrac{A}{2x - 1} + \dfrac{B}{(2x - 1)^2} + \dfrac{C}{2x + 3}$ **7.** $\dfrac{A}{x} + \dfrac{B}{x^2} + \dfrac{C}{x^3} + \dfrac{D}{x - 1}$
9. $1 + \dfrac{A}{x - 1} + \dfrac{B}{x + 1}$ **11.** $\dfrac{A}{x} + \dfrac{Bx + C}{x^2 + 2}$
13. $\dfrac{Ax + B}{x^2 + 1} + \dfrac{Cx + D}{x^2 + 4} + \dfrac{Ex + F}{(x^2 + 4)^2}$
15. $\dfrac{Ax + B}{x^2 + 9} + \dfrac{Cx + D}{(x^2 + 9)^2} + \dfrac{Ex + F}{(x^2 + 9)^3}$
17. $\dfrac{A}{x} + \dfrac{B}{x^2} + \dfrac{Cx + D}{x^2 + x + 2}$ **19.** $\dfrac{x^2}{2} - x + \ln |x + 1| + C$
21. $\ln 3 + 3 \ln 6 - 3 \ln 4 = \ln \frac{81}{8}$
23. $3x - 7 \ln |2x + 3| + C$
25. $x - \ln |x| + 2 \ln |x - 1| + C = x + \ln((x - 1)^2/|x|) + C$
27. $2 \ln 2 + \frac{1}{2}$ **29.** $\frac{1}{3} \ln |x| - \ln |x + 1| + \frac{2}{3} \ln |2x + 3| + C$
31. $4 \ln 6 - 3 \ln 5$
33. $-1/[5(x - 1)] + \frac{1}{25} \ln |(x + 4)/(x - 1)| + C$
35. $2 \ln |x| + 3 \ln |x + 2| + (1/x) + C$
37. $\frac{1}{3} \ln |x^3 + 3x^2 + 4| + C$
39. $\ln |x + 1| + 2/(x + 1) - 1/[2(x + 1)^2] + C$
41. $(1/x) + \frac{1}{2} \ln |(x - 1)/(x + 1)| + C$ **43.** $(1 - \ln 2)/2$
45. $\ln \sqrt{3} - [\pi/(6\sqrt{3})]$
47. $\ln(x - 1)^2 + \ln \sqrt{x^2 + 1} - 3 \tan^{-1} x + C$

49. $-\ln|x+1| + \ln(x^2 - 4x + 7) + \dfrac{5}{\sqrt{3}} \tan^{-1}\!\left(\dfrac{x-2}{\sqrt{3}}\right) + C$

51. $\frac{1}{3}\ln|x-1| - \frac{1}{6}\ln(x^2+x+1) - \dfrac{1}{\sqrt{3}}\tan^{-1}\!\left(\dfrac{2x+1}{\sqrt{3}}\right) + C$

53. $\ln|x-1| - \ln\sqrt{x^2+1} + [1/(x-1)] + \tan^{-1}x + C$

55. $\frac{3}{2}\ln(x^2+1) - 3\tan^{-1}x + \sqrt{2}\,\tan^{-1}(x/\sqrt{2}) + C$

57. $\dfrac{-1}{2(x^2+2x+4)} - \dfrac{2}{3\sqrt{3}}\tan^{-1}\!\left(\dfrac{x+1}{\sqrt{3}}\right) - \dfrac{2(x+1)}{3(x^2+2x+4)} + C$

59. $\ln|x| + [1/(x^2+1)] + C$ **61.** $-2/(x^2+4)^2 + C$

63. $\ln|\sin^2 x - 3\sin x + 2| + C$ **67.** $\frac{1}{2}\ln|(x-2)/x| + C$

69. $\frac{1}{2}\ln|x^2+x-1| - \dfrac{1}{2\sqrt{5}}\ln\left|\dfrac{2x+1-\sqrt{5}}{2x+1+\sqrt{5}}\right| + C$

71. $1 + 2\ln 2$

Exercises 7.5 (page 437)

1. $2(1 - \ln 2)$ **3.** $2(\sqrt{x} - \tan^{-1}\sqrt{x}) + C$

5. $\frac{3}{2}\ln|x^{2/3} - 1| + C$ **7.** $\frac{1676}{15}$

9. $\frac{4}{3}\left(\sqrt{1+\sqrt{x}}\right)^3 - 4\sqrt{1+\sqrt{x}} + C$

11. $x + 4\sqrt{x} + 4\ln|\sqrt{x} - 1| + C$

13. $\frac{3}{10}(x^2+1)^{5/3} - \frac{3}{4}(x^2+1)^{2/3} + C$

15. $2\sqrt{x} - 3\sqrt[3]{x} + 6\sqrt[6]{x} - 6\ln(\sqrt[6]{x} + 1) + C$

17. $2\sqrt{x} - 4\sqrt[4]{x} + 4\ln(1 + \sqrt[4]{x}) + C$

19. $(2/\sqrt{bc})\tan^{-1}\sqrt{bx/c} + C$

21. $3\sqrt[3]{x-1} - \ln|1 + \sqrt[3]{x-1}|$
$+ \frac{1}{2}\ln[(x-1)^{2/3} - (x-1)^{1/3} + 1] - \sqrt{3}\tan^{-1}\!\left(\dfrac{2\sqrt[3]{x-1} - 1}{\sqrt{3}}\right) + C$

23. $\sqrt{x - x^2} - \tan^{-1}\sqrt{(1-x)/x} + C$

25. $\ln|(\sin x)/(1 + \sin x)| + C$

27. $\ln[(e^x + 2)^2/(e^x + 1)] + C$

29. $2\sqrt{1 - e^x} + \ln\left[\left(1 - \sqrt{1-e^x}\right)/\left(1 + \sqrt{1-e^x}\right)\right] + C$

31. $C - \cot(x/2)$ $[\text{or} -\cot x - \csc x + C]$

33. $-\sqrt{2}\ln(\sqrt{2} - 1) = (1/\sqrt{2})\ln(3 + 2\sqrt{2})$

35. $\frac{1}{5}\ln|(2\tan(x/2) + 1)/(\tan(x/2) - 2)| + C$

37. $\frac{1}{4}\ln|\tan(x/2)| + \frac{1}{8}\tan^2(x/2) + C$

39. $\dfrac{1}{\sqrt{a^2+b^2}}\ln\left|\dfrac{b\tan(x/2) - a + \sqrt{a^2+b^2}}{b\tan(x/2) - a - \sqrt{a^2+b^2}}\right| + C$

Exercises 7.6 (page 442)

1. $2x + 11\ln|x-3| + C$ **3.** $\frac{1}{3}\sin^3 x - \frac{1}{5}\sin^5 x + C$

5. $1 - (\sqrt{3}/2)$ **7.** $2\sqrt{x-2} - 4\tan^{-1}(\sqrt{x-2}/2) + C$

9. $x\ln(1 + x^2) - 2x + 2\tan^{-1}x + C$ **11.** $\frac{4097}{45}$

13. $\frac{1}{2}\ln(x^2 - 2x + 2) + \tan^{-1}(x - 1) + C$

15. $3\ln\left|\left(3 - \sqrt{9 - x^2}\right)/x\right| + \sqrt{9 - x^2} + C$

17. $(x^2 + 2)\sinh x - 2x\cosh x + C$ **19.** $\tan^{-1}(\sin x) + C$

21. $2/\pi$ **23.** $e^{3x}(5\sin 5x + 3\cos 5x)/34 + C$

25. $\frac{1}{2}\left[\ln|x+1| - \frac{1}{2}\ln(x^2+1) + \tan^{-1}x\right] + C$

27. $-\frac{1}{3}(x^3 + 1)e^{-x^3} + C$

29. $\frac{1}{3}\ln\left|3x + 2 + \sqrt{9x^2 + 12x - 5}\right| + C$

31. $\frac{3}{4}x^{4/3} - \frac{6}{11}x^{11/6} + C$ **33.** $\frac{1}{6}\tan^{-1}[(x^2+1)/3] + C$

35. $\frac{1}{16}\left(x - \frac{1}{4}\sin 4x + \frac{1}{3}\sin^3 2x\right) + C$

37. $-\ln\left(1 + \sqrt{1 - x^2}\right) + C$

39. $\frac{1}{2}\ln|(e^x - 1)/(e^x + 1)| + C$ **41.** 0 **43.** $\frac{86}{3}$

45. $\frac{1}{2}(\ln\sin x)^2 + C$ **47.** $\frac{1}{6}\ln[(x^2+1)/(x^2+4)] + C$

49. $\frac{3}{7}(x+c)^{7/3} - \frac{3}{4}c(x+c)^{4/3} + C$

51. $3\ln(\sqrt{x+1} + 3) - \ln(\sqrt{x+1} + 1) + C$

53. $\frac{1}{2}(x+2)\sin 2x - \frac{1}{4}(2x^2 + 8x - 7)\cos 2x + C$

55. $\frac{1}{2}\sin^{-1}(x^2/4) + C$ **57.** $\frac{1}{6}\csc^3 2x - \frac{1}{10}\csc^5 2x + C$

59. $e^{\arctan x} + C$ **61.** $-\frac{1}{8}e^{-2t}(4t^3 + 6t^2 + 6t + 3) + C$

63. $\frac{1}{24}\cos 6x - \frac{1}{16}\cos 4x - \frac{1}{8}\cos 2x + C$

65. $\sin^{-1}x - \sqrt{1 - x^2} + C$

67. $\ln\sqrt{x^2 + a^2} + \tan^{-1}(x/a) + C$

69. $\frac{1}{20}\tan^{-1}(x^5/4) + C$

71. $x\sec x - \ln|\sec x + \tan x| + C$

73. $\frac{2}{3}[(x+1)^{3/2} - x^{3/2}] + C$

75. $2\sqrt{x}\tan^{-1}\sqrt{x} - \ln(1 + x) + C$

77. $e^{-x} + \frac{1}{2}\ln|(e^x - 1)/(e^x + 1)| + C$

79. $\frac{2}{5}\tan^{-1}(\sqrt{2x - 25}/5) + C$

Exercises 7.7 (page 445)

1. $\frac{1}{25}e^{-3x}(-3\cos 4x + 4\sin 4x) + C$

3. $\left(-\sqrt{9x^2 - 1}/x\right) + 3\ln\left|3x + \sqrt{9x^2 - 1}\right| + C$

5. $e^{3x}(9x^2 - 6x + 2)/27 + C$

7. $\frac{1}{2}\left[x^2\sin^{-1}(x^2) + \sqrt{1 - x^4}\right] + C$

9. $\tan^{-1}(\sinh e^x) + C$

11. $\frac{1}{2}(x+2)\sqrt{5 - 4x - x^2} + \frac{9}{2}\sin^{-1}[(x+2)/3] + C$

13. $\frac{1}{4}\tan x\sec^3 x + \frac{3}{8}\tan x\sec x + \frac{3}{8}\ln|\sec x + \tan x| + C$

15. $\frac{1}{9}\sin^3 x[3\ln(\sin x) - 1] + C$

17. $-2\sqrt{2 + 3\cos x} - \sqrt{2}\ln\left|\dfrac{\sqrt{2 + 3\cos x} - \sqrt{2}}{\sqrt{2 + 3\cos x} + \sqrt{2}}\right| + C$

19. $\frac{8}{15}$ **21.** $\frac{1}{5}\ln\left|x^5 + \sqrt{x^{10} - 2}\right| + C$

23. $(1 + e^x)\ln(1 + e^x) - e^x + C$

25. $\sqrt{e^{2x}-1}-\cos^{-1}(e^{-x})+C$ **27.** $(2\pi/25)\left(\ln 6 - \frac{5}{6}\right)$

Exercises 7.8 (page 454)

1.

n	L_n	R_n	T_n	M_n
4	0.140625	0.390625	0.265625	0.242188
8	0.191406	0.316406	0.253906	0.248047
16	0.219727	0.282227	0.250977	0.249512

n	E_L	E_R	E_T	E_M
4	0.109375	-0.140625	-0.015625	0.007812
8	0.058594	-0.066406	-0.003906	0.001953
16	0.030273	-0.032227	-0.000977	0.000488

Observations same as after Example 1.

3.

n	T_n	M_n	S_n
6	4.661488	4.669245	4.666563
12	4.665367	4.667316	4.666659

n	E_T	E_M	E_S
6	0.005178	-0.002578	0.000104
12	0.001300	-0.000649	0.000007

Observations same as after Example 1.

5.

n	T_n	M_n	S_n
4	0.613313	0.611173	0.611868
8	0.612243	0.611710	0.611886

n	E_T	E_M	E_S
4	-0.001425	0.000715	0.000020
8	-0.000355	0.000178	0.000002

7. (a) 1.913972   (b) 1.934766
9. (a) 0.481672   (b) 0.481172
11. (a) 0.746211   (b) 0.747131   (c) 0.746825
13. (a) 0.132465   (b) 0.132857   (c) 0.132727
15. (a) 0.409140   (b) 0.388849   (c) 0.395802
17. (a) 2.031893   (b) 2.014207   (c) 2.020651
19. (a) 1.098004   (b) 1.098709   (c) 1.109031
21. (a) $T_{10} \approx 0.881839$, $M_{10} \approx 0.882202$
    (b) $|E_T| < 0.01\overline{3}$, $|E_M| < 0.00\overline{6}$
23. (a) $T_{10} \approx 1.719713$, $E_T \approx -0.001432$,
     $S_{10} \approx 1.718283$, $E_S \approx -0.000001$
    (b) $|E_T| < 0.002266$, $|E_S| < 0.0000016$

25. $n = 130$ for T_n, $n = 92$ for M_n **27.** 15.4
29. 8.6 mi   **31.** (a) 11.5   (b) 12   (c) $11.\overline{6}$
33. 12.3251

Exercises 7.9 (page 463)

1. Divergent   **3.** $\frac{1}{2}$ **5.** Divergent   **7.** 1   **9.** 0
11. $-\ln\frac{2}{3}$ **13.** Divergent   **15.** Divergent
17. $e^2/4$ **19.** Divergent   **21.** Divergent   **23.** 1
25. $1/\ln 2$ **27.** $2\sqrt{3}$ **29.** Divergent
31. Divergent   **33.** $\frac{5}{3}$ **35.** Divergent   **37.** 2
39. Divergent   **41.** Divergent   **43.** $\frac{4}{3}$ **45.** $-\frac{1}{4}$
47. e **49.** $\pi/2$

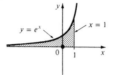   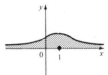

51. Infinite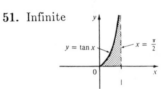

53. Convergent   **55.** Convergent   **57.** Divergent
59. π **61.** $1/(1-p)$, $p < 1$
63. $-1/(p+1)^2$, $p > -1$ **69.** $\sqrt{2GM/R}$
71. (a) $F(s) = 1/s$, $s > 0$ (b) $F(s) = 1/(s-1)$, $s > 1$
    (c) $F(s) = 1/s^2$, $s > 0$ **77.** $C = 1$, $\ln 2$

Review Exercises for Chapter 7 (page 466)

1. False   **3.** False   **5.** False   **7.** False
9. $x - 2\ln|x+1| + C$ **11.** $\frac{1}{6}(\arctan x)^6 + C$
13. $-1/e^{\sin x} + C$ **15.** $\frac{1}{25}x^5(5\ln x - 1) + C$
17. $-\frac{1}{2}\cos(x^2) + C$ **19.** $\frac{1}{3}\ln|(x-2)/(2x-1)| + C$
21. $\frac{1}{9}\sec^9 x - \frac{3}{7}\sec^7 x + \frac{3}{5}\sec^5 x - \frac{1}{3}\sec^3 x + C$
23. $\sqrt{1+2x} - 3\ln(\sqrt{1+2x}+3) + C$ **25.** $2e^{\sqrt{x}} + C$
27. $(x+1)\ln(x^2+2x+2) - 2x + 2\tan^{-1}(x+1) + C$
29. $\ln|x| - \frac{1}{2}\ln(x^2+1) + C$ **31.** $-\cot x - x + C$
33. $-x/\sqrt{x^2-1} + C$
35. $2\ln|x-1| - \frac{1}{2}\tan^{-1}[(x+1)/2] + C$
37. $-\frac{1}{12}(\cot^3 4x + 3\cot 4x) + C$
39. $\ln x[\ln(\ln x) - 1] + C$
41. $\frac{1}{2}\ln\left(2x+1+\sqrt{4x^2+4x+5}\right) + C$
43. $\frac{1}{2}\sin 2x - \frac{1}{8}\cos 4x + C$ **45.** $\frac{2}{5}$ **47.** Divergent
49. $1 - (\pi/2)$ **51.** $\frac{1}{24}$ **53.** 2   **55.** $8\ln\frac{4}{3} - 1$
57. $\pi/4$ **59.** $\sqrt{3} - (\pi/3)$
61. $-a/(a^2+b^2)$ if $a < 0$, divergent if $a \geq 0$

63. $\frac{1}{2}\left[e^x\sqrt{1-e^{2x}}+\sin^{-1}(e^x)\right]+C$

65. $\frac{1}{4}(2x+1)\sqrt{x^2+x+1}+\frac{3}{8}\ln\left|x+\frac{1}{2}+\sqrt{x^2+x+1}\right|+C$

69. (a) 1.090608   (b) 1.088840   (c) 1.089429

71. 0.0067   **73.** Convergent   **75.** 2   **77.** No

Applications Plus (page 468)

1. (a) $V=\int_0^h \pi[f(y)]^2\,dy$
(c) $f(y)=\sqrt{kA/(\pi C)}\,y^{1/4}$, Advantage: the markings on the container are equally spaced

3. (b) $f(x)=L\ln\left[\left(L+\sqrt{L^2-x^2}\right)\big/x\right]-\sqrt{L^2-x^2}$

5. (b) $\approx 11\%$ (c) $\approx 1.05\times 10^3$ J

Chapter 8

Exercises 8.1 (page 478)

1. $y=-1/(x+C)$ or $y=0$ **3.** $x^2-y^2=C$

5. $y=Ce^{1/x}$ **7.** $(y-1)e^y=\frac{1}{3}(x^2+1)^{3/2}+C$

9. $u=-\ln\left(C-\frac{1}{2}e^{2t}\right)$ **11.** $y=\tan(x-1)$

13. $ye^y=x^3-8$ **15.** $\cos y=\cos x-1$

17. $x=\sqrt{2(t-1)e^t+3}$ **19.** $u=1-\sqrt{t^2+t+4}$

21. $f(x)=e^{x^4/4}$ **23.** $y=7e^{x^4}$

25. (a) $15e^{-t/100}$ kg   (b) $15e^{-0.2}\approx 12.3$ kg

27. $x=ab(e^{(b-a)kt}-1)/(be^{(b-a)kt}-a)$

29. $5\times 10^9 e^{0.02(t-1986)}$

(a) 6.6 billion   (b) 49 billion   (c) 146 trillion
(a) $\approx 270{,}000$ (b) $\approx 37{,}000$ (c) ≈ 12

31. (a) $dy/dt=ky(1-y)$
(b) $y=y_0/[y_0+(1-y_0)e^{-kt}]$ (c) 3:36 P.M.

Exercises 8.2 (page 484)

1. $4\sqrt{5}$ **3.** $(577^{3/2}-145^{3/2})/216$

5. $(13\sqrt{13}-8)/27$ **7.** $\frac{59}{24}$ **9.** $\frac{4}{3}$ **11.** $\frac{181}{9}$

13. $\ln(\sqrt{2}+1)$ **15.** $\ln 3-\frac{1}{2}$

17. $\sqrt{1+e^2}-\sqrt{2}+\ln(\sqrt{1+e^2}-1)-1-\ln(\sqrt{2}-1)$

19. $\sinh 1$ **21.** $\int_0^1\sqrt{1+9x^4}\,dx$

23. $\int_0^\pi\sqrt{1+\cos^2 x}\,dx$ **25.** $\int_0^{\pi/2}\sqrt{1+e^{2x}(1-\sin 2x)}\,dx$

27. $s(x)=\frac{2}{27}\left[(1+9x)^{3/2}-10\sqrt{10}\right]$ **29.** 7798 m

31. 6   **33.** 1.548   **35.** 3.820   **37.** 12.4

Exercises 8.3 (page 490)

1. $\pi(37\sqrt{37}-17\sqrt{17})/6$ **3.** $5\sqrt{5}\,\pi$ **5.** $12289\pi/192$

7. $2\pi\left[\sqrt{2}+\ln\left(\sqrt{2}+1\right)\right]$ **9.** $\pi\left[1+\frac{1}{4}(e^2-e^{-2})\right]$

11. $5813\pi/30$ **13.** $\pi(65\sqrt{65}-17\sqrt{17})/24$

15. $\pi(145\sqrt{145}-10\sqrt{10})/27$

17. $\pi(3200\sqrt{10}-\frac{494}{5}\sqrt{13})/243$

19. $\frac{1}{4}\pi\left[2e\sqrt{1+4e^2}-2\sqrt{5}+\ln\left(\dfrac{2e+\sqrt{1+4e^2}}{2+\sqrt{5}}\right)\right]$

21. $\pi[21-8\ln 2-(\ln 2)^2]/8$ **23.** ≈ 3.44 **25.** $\pi/4$

29. $2\pi\left[b^2+a^2 b\sin^{-1}\left(\sqrt{a^2-b^2}\big/a\right)\big/\sqrt{a^2-b^2}\right]$

33. $4\pi^2 r^2$

Exercises 8.4 (page 497)

1. $40, 12, \left(1,\frac{10}{3}\right)$ **3.** $-15, 17\left(\frac{17}{11},-\frac{15}{11}\right)$

5. $(1.5,1.2)$ **7.** $\left(\frac{11}{13},\frac{49}{13}\right)$

9. $\left(\dfrac{\frac{2}{3}(2\sqrt{2}-1)}{\sqrt{2}+\ln(1+\sqrt{2})},\dfrac{4}{3[\sqrt{2}+\ln(1+\sqrt{2})]}\right)$ **11.** $\left(0,\frac{1}{8}\pi\right)$

13. $(1/(e-1),(e+1)/4)$ **15.** $(0.4,0.5)$

17. $\left((\pi\sqrt{2}-4)/[4(\sqrt{2}-1)],1/[4(\sqrt{2}-1)]\right)$

19. $\frac{4}{3}, 0, \left(0,\frac{2}{3}\right)$ **21.** $-\frac{44}{3}, 0, \left(0,-\frac{11}{15}\right)$ **25.** $\left(0,\frac{1}{12}\right)$

27. $\left(\frac{9}{10},\frac{3}{2}\right)$ **29.** $\frac{1}{3}\pi r^2 h$

Exercises 8.5 (page 501)

1. (a) 9.8 kPa   (b) 1.96×10^4 N   (c) 4.90×10^3 N

3. 6.5×10^6 N   **5.** $1000g\pi r^3$ N   **7.** 3.00×10^3 lb

9. 1.56×10^3 lb   **11.** 3.47×10^4 lb   **13.** 5.27×10^5 N

15. (a) 314 N   (b) 353 N   **17.** 8.32×10^4 N

19. (a) 5.63×10^3 lb   (b) 5.06×10^4 lb
(c) 4.88×10^4 lb   (d) 3.03×10^5 lb

Exercises 8.6 (page 507)

1. \$14,516,000   **3.** \$388,280,000   **5.** \$316.29

7. \$4166.67   **9.** \$112,500   **11.** \$50,852.36

13. (b) \$50,000   **15.** $16(2\sqrt{2}-1)/3\approx$ \$9.75 million

17. 1.19×10^{-4} cm^3/s   **19.** $\frac{1}{9}$ L/s

Review Exercises for Chapter 8 (page 508)

1. $y=\sqrt[3]{(3x^2/2)-3\cos x+K}$

3. $y=2\pm\sqrt{2\tan^{-1}x+C}$ **5.** $y=\sqrt{(\ln x)^2+4}$

7. $\frac{2}{3}(5\sqrt{5}-2\sqrt{2})$ **9.** (a) $\frac{17}{12}$ (b) $\frac{47}{16}\pi$ **11.** 1.297

13. $56\sqrt{3}\pi a^2/5$ **15.** $(-\frac{1}{2},\frac{12}{5})$ **17.** $(2,\frac{2}{3})$ **19.** $2\pi^2$

21. 458 lb   **23.** \$7166.67

25. (a) $L(t)=L_\infty-(L_\infty-L(0))e^{-kt}$
(b) $L(t)=53-43e^{-0.2t}$

27. $50+10e^{-3t/2000}$, 50 kg

Problems Plus (page 510)

11. 10   **13.** (a) On $(-1, \infty)$ (b) e^2
15. $(b^b a^{-a})^{1/(b-a)} e^{-1}$

Chapter 9

Exercises 9.1 (page 515)

1. (a)
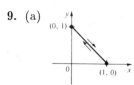

(b) $3x + y = 5$

3. (a)
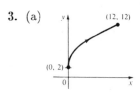

(b) $x = \frac{3}{25}(y - 2)^2$, $2 \le y \le 12$

5. (a)
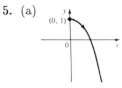

(b) $y = 1 - x^2$, $x \ge 0$

7. (a)
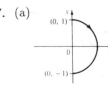

(b) $x^2 + y^2 = 1$, $x \ge 0$

9. (a)
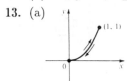

(b) $x + y = 1$, $0 \le x \le 1$

11. (a)
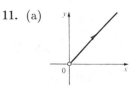

(b) $y = x$, $x > 0$

13. (a)

(b) $y = x^2$, $0 \le x \le 1$

15. (a)
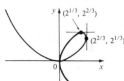

(b) $x = 1 - y^2$, $-1 \le y \le 1$

17. (a)
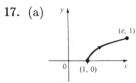

(b) $y = \sqrt{\ln x}$, $1 \le x \le e$,
or $x = e^{y^2}$, $0 \le y \le 1$

19. (a)
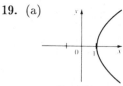

(b) $x^2 - y^2 = 1$, $x \ge 1$

21. Moves counterclockwise along the circle $x^2 + y^2 = 1$ from $(-1, 0)$ to $(1, 0)$
23. Moves along the line $x + 8y = 13$ from $(-3, 2)$ to $(5, 1)$
25. Moves once clockwise around the ellipse $(x^2/4) + (y^2/9) = 1$, starting and ending at $(0, 3)$

27.
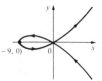

29.
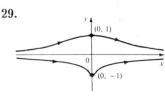

35. $x = a \cos\theta$, $y = b \sin\theta$; $(x^2/a^2) + (y^2/b^2) = 1$, ellipse

39.
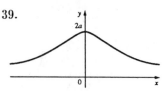

Exercises 9.2 (page 521)

1. $x + y = 0$ **3.** $x - 36y + 52 = 0$
5. $3x + 2y = 6\sqrt{2}$ **7.** $y = 1$ **9.** $3x + 4y = 25$
11. $2t/(2t + 1)$, $2/(2t + 1)^3$
13. $2(2t - 3)\sqrt{t + 1}$, $2(6t + 1)$ **15.** $-\tan\pi t$, $-\sec^3\pi t$
17. $-e^{3t}(2t + 1)$, $e^{4t}(6t + 5)$
19. Horizontal at $(1, 5)$, $(1, -1)$;   **21.** Horizontal at $(0, -9)$; vertical at $(-1, 2)$, $(3, 2)$ vertical at $(\pm 2, -6)$

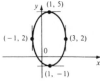

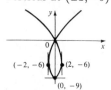

23. Horizontal at $(0, 0)$, $(2^{1/3}, 2^{2/3})$;   **25.** $y = x$, $y = -x$ vertical at $(0, 0)$, $(2^{2/3}, 2^{1/3})$

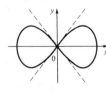

27. $(0, 0)$, $y = \pm\sqrt{3}x$ **29.** (a) $d \sin\theta/(r - d\cos\theta)$
31. $(-5, 6)$, $(-\frac{208}{27}, \frac{32}{3})$ **33.** πab **35.** $(e^{\pi/2} - 1)/2$
37. $2\pi r^2 + \pi d^2$

Exercises 9.3 (page 525)

1. $\int_0^1 \sqrt{9t^4 + 16t^6}\, dt$ **3.** $\int_0^{\pi/2} \sqrt{1 + t^2}\, dt$ **5.** 2π
7. $4\sqrt{34}$ **9.** $\sqrt{2}(e^\pi - 1)$ **11.** $\sqrt{13}$ **13.** 0.7314
15. $6\sqrt{2}$, $\sqrt{2}$ **19.** $2\pi\int_0^1 t^4\sqrt{9t^4 + 16t^6}\, dt$
21. $47{,}104\pi/15$ **23.** $2\sqrt{2}\pi(2e^\pi + 1)/5$ **25.** $128\pi/5$
27. $24\pi(949\sqrt{26} + 1)/5$
29. (a) $2\pi[b^2 + (ab/e)\sin^{-1}e]$, $e = \dfrac{\sqrt{a^2 - b^2}}{a}$ = eccentricity
   (b) $2\pi\{a^2 + (b^2/2e)\ln[(1 + e)/(1 - e)]\}$
33. $\frac{1}{4}$

Exercises 9.4 (page 533)

1.

$(1, 5\pi/2), (-1, 3\pi/2)$

3.

$(1, 6\pi/5), (-1, 11\pi/5)$

5.
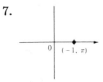
$(4, 4\pi/3), (-4, \pi/3)$

7.

$(1, 0)(-1, 3\pi)$

9.

$(1, 1)$

11.

$(0, -1.5)$

13.
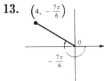
$(-2\sqrt{3}, 2)$

15.

$\left(-\frac{1}{2}, -\sqrt{3}/2\right)$

17. $(\sqrt{2}, 3\pi/4)$ **19.** $(4, 11\pi/6)$

21.

23.

25.

27. $\frac{1}{2}\sqrt{40 + 6\sqrt{6} - 6\sqrt{2}}$ **29.** $y = 2$

31. $y^2 = 2x + 1$ **33.** $(x^2 + y^2)^2 = 2xy$

35. $r\sin\theta = 5$ **37.** $r = 5$ **39.** $r^2 = \csc 2\theta$

41.

43.

45.

47.

49.

51.
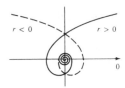

53.

55.

57.

59.
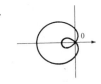

61.

63.

65.
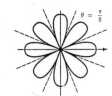

67.

69.

71.

73.

75.

77. $1/\sqrt{3}$ **79.** $-2/\pi$ **81.** -1 **83.** $\sqrt{3}$

85. Horizontal at $(3/\sqrt{2}, \pi/4), (-3/\sqrt{2}, 3\pi/4)$; vertical at $(3, 0), (0, \pi/2)$

87. Horizontal at $(1, 3\pi/2)$, $(1, \pi/2)$, $(\frac{2}{3}, \alpha)$, $(\frac{2}{3}, \pi - \alpha)$, $(\frac{2}{3}, \pi + \alpha)$, $(\frac{2}{3}, 2\pi - \alpha)$, where $\alpha = \sin^{-1}(1/\sqrt{6})$; vertical at $(1, 0)$, $(1, \pi)$, $(\frac{2}{3}, 3\pi/2 - \alpha)$, $(\frac{2}{3}, 3\pi/2 + \alpha)$, $(\frac{2}{3}, \pi/2 - \alpha)$, $(\frac{2}{3}, \pi/2 + \alpha)$

89. Horizontal at $(\frac{3}{2}, \pi/3)$, $(\frac{3}{2}, 5\pi/3)$ and the pole; vertical at $(2, 0)$, $(\frac{1}{2}, 2\pi/3)$, $(\frac{1}{2}, 4\pi/3)$

91. Center $(b/2, a/2)$, radius $\sqrt{a^2 + b^2}/2$

Exercises 9.5 (page 539)

1. $\pi^3/6$ **3.** $(\pi/6) + (\sqrt{3}/4)$ **5.** $121\pi^5/160$
7. $(4\pi - 3\sqrt{3})/96$ **9.** $25\pi/4$

11. $3\pi/2$ **13.** 4

15. $33\pi/2$ **17.** $\pi/2$

19. $\pi/12$ **21.** $\pi/20$ **23.** $\pi - (3\sqrt{3}/2)$
25. $(9\sqrt{3}/8) - (\pi/4)$ **27.** $(4\pi/3) + 2\sqrt{3}$ **29.** π
31. $(\pi - 2)/8$ **33.** $(\pi/2) - 1$
35. $(19\pi/3) - (11\sqrt{3}/2)$ **37.** $(\pi + 3\sqrt{3})/4$
39. $(1/\sqrt{2}, \pi/4)$ and the pole
41. $(\frac{1}{2}, \pi/3)$, $(\frac{1}{2}, 5\pi/3)$ and the pole
43. $(\sqrt{3}/2, \pi/3)$, $(\sqrt{3}/2, 2\pi/3)$, and the pole
45. $15\pi/4$ **47.** $\sqrt{1 + (\ln 2)^2}(4^\pi - 1)/\ln 2$
49. $\frac{8}{3}[(\pi^2 + 1)^{3/2} - 1]$ **51.** $\frac{16}{3}$ **53.** 2.4
55. (b) $2\pi(2 - \sqrt{2})$

Exercises 9.6 (page 545)

1. $(0, 0)$, $(0, -2)$, $y = 2$ **3.** $(0, 0)$, $(\frac{1}{4}, 0)$, $x = -\frac{1}{4}$

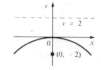

5. $(-1, 3)$, $(-\frac{7}{8}, 3)$, $x = -\frac{9}{8}$ **7.** $(2, 4)$, $(\frac{3}{2}, 4)$, $x = \frac{5}{2}$

9. $(\pm 4, 0)$, $(\pm 2\sqrt{3}, 0)$ **11.** $(0, \pm 5)$, $(0, \pm 4)$

13. $(\pm 12, 0)$, $(\pm 13, 0)$,   **15.** $(0, \pm 1)$, $(0, \pm \sqrt{10})$, $y = \pm \frac{5}{12}x$ $y = \pm x/3$

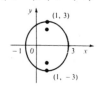

17. $(1, \pm 3)$, $(1, \pm\sqrt{5})$

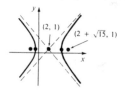

19. $(2 \pm \sqrt{6}, 1)$, $(2 \pm \sqrt{15}, 1)$, $y - 1 = \pm(\sqrt{6}/2)(x - 2)$

21. $x^2 = 12y$ **23.** $y^2 = 4(x - 2)$ **25.** $y^2 = 16x$
27. $(x^2/4) + (y^2/3) = 1$ **29.** $[(x - 3)^2/8] + (y^2/9) = 1$
31. $[(x - 2)^2/9] + [(y - 2)^2/5] = 1$ **33.** $y^2 - (x^2/8) = 1$
35. $[(x - 4)^2/4] - [(y - 3)^2/5] = 1$
37. $(x^2/9) - (y^2/36) = 1$
39. $(x^2/3,763,600) + (y^2/3,753,196) = 1$
41. (a) $(121x^2/1,500,625) - (121y^2/3,339,375) = 1$
    (b) ≈ 248 mi
45. (a) Ellipse   (b) Hyperbola   (c) No curve
47. 9.69

Exercises 9.7 (page 551)

1. $r = 6/(3 + 2\cos\theta)$ **3.** $r = 2/(1 + \sin\theta)$
5. $r = 20/(1 + 4\cos\theta)$ **7.** $r = 10/(1 + \sin\theta)$

9. (a) 3   (b) Hyperbola
   (c) $x = \frac{4}{3}$
   (d)
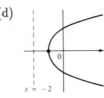

11. (a) 1   (b) Parabola
    (c) $x = -2$
    (d)
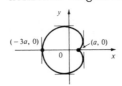

21. Vertical tangent at $(3a/2, \pm\sqrt{3}a/2)$, $(-3a, 0)$;
    horizontal tangent at $(a, 0)$, $(-a/2, \pm 3\sqrt{3}a/2)$
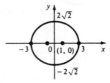

13. (a) $\frac{1}{2}$ (b) Ellipse
    (c) $y = 6$
    (d)

15. (a) $\frac{3}{4}$ (b) Ellipse
    (c) $x = -\frac{1}{3}$
    (d)

23. 18   25. $(2, \pm\pi/3)$ 27. $(\pi - 1)/2$
29. $2(5\sqrt{5} - 1)$

31. $\dfrac{2\sqrt{\pi^2 + 1} - \sqrt{4\pi^2 + 1}}{2\pi} + \ln\left(\dfrac{2\pi + \sqrt{4\pi^2 + 1}}{\pi + \sqrt{\pi^2 + 1}}\right)$

33. $471{,}295\pi/1024$
35. $(\pm 1, 0)$, $(\pm 3, 0)$ 37. $(-\frac{25}{24}, 3)$, $(-1, 3)$
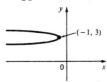

17. (a) $\frac{5}{2}$ (b) Hyperbola
    (c) $y = -\frac{7}{5}$
    (d)

19. (a) 1   (b) Parabola
    (c) $y = \frac{5}{2}$
    (d)

39. $x^2 = 8(y - 4)$ 41. $5x^2 - 20y^2 = 36$
43. $(x^2/25) + (8y - 399)^2/(160{,}801) = 1$
45. $r = 4/(3 + \cos\theta)$
47. $x = a(\cot\theta + \sin\theta\cos\theta)$, $y = a(1 + \sin^2\theta)$

25. (b) $r = (1.49 \times 10^8)/(1 - 0.017\cos\theta)$
27. 7.0×10^7 km

Review Exercises for Chapter 9 (page 552)

1.

$x = 2y - y^2$, $0 \le y \le 2$

3.
$(x - 1)^2 + (y - 2)^2 = 1$

5.

7.

9.

11.

13. $r = 4\cos\theta$ 15. $\frac{1}{2}$ 17. $(4 + \pi)/(4 - \pi)$
19. $(\sin t + t\cos t)/(\cos t - t\sin t)$, $(t^2 + 2)/(\cos t - t\sin t)^3$

Applications Plus (page 554)

1. (b) $f(x) = (x^2 - L^2)/(4L) - (L/2)\ln(x/L)$ (c) No
3. (a) $2\pi r(r \pm d)$ (b) $\approx 3{,}360{,}000$ mi^2
   (d) $\approx 78{,}400{,}000$ mi^2
5. (a) $y(t) = y_0/(1 - \varepsilon y_0^\varepsilon kt)^{1/\varepsilon}$ (b) $T = 1/(\varepsilon y_0^\varepsilon k)$
   (c) ≈ 12.15 years
7. (a) $x = C(\theta - \frac{1}{2}\sin 2\theta)$ (b) Cycloid
9. (a) $y = (1/k)\cosh kx + a - 1/k$ or
      $y = (1/k)\cosh kx - (1/k)\cosh kb + h$ (b) $(2/k)\sinh kb$

Chapter 10

Exercises 10.1 (page 566)

1. $\left\{\frac{1}{3}, \frac{2}{5}, \frac{3}{7}, \frac{4}{9}, \frac{5}{11}, \cdots\right\}$ 3. $\left\{\frac{1}{2}, -\frac{2}{4}, \frac{3}{8}, -\frac{4}{16}, \frac{5}{32}, \cdots\right\}$

5. $\left\{1, \frac{3}{2}, \frac{5}{2}, \frac{35}{8}, \frac{63}{8}, \cdots\right\}$ 7. $\{1, 0, -1, 0, 1, \cdots\}$

9. $\left\{1, \frac{1}{2}, \frac{2}{3}, \frac{3}{5}, \frac{5}{8}, \cdots\right\}$ 11. $a_n = 1/2^n$ 13. $a_n = 3n - 2$

15. $a_n = (-1)^n n!$ 17. $a_n = (-1)^{n+1}(n+1)/(2n+1)$

19.
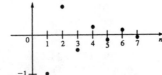

21. 0   **23.** 1   **25.** Diverges (to ∞)   **27.** 0   **29.** 0
31. Diverges   **33.** Diverges (to ∞)   **35.** $\pi/2$
37. 0   **39.** 0   **41.** 0   **43.** 0   **45.** 1   **47.** 0
49. $\frac{1}{2}$ **51.** Diverges (to ∞)   **53.** 0   **55.** 0
57. $-1 < r < 1$ **59.** Decreasing   **61.** Increasing
63. Not monotonic   **65.** Decreasing   **67.** 2
69. (b) $(3 + \sqrt{5})/2$ **71.** (b) $(1 + \sqrt{5})/2$

Exercises 10.2 (page 575)

1. $\frac{20}{3}$ **3.** $\frac{1}{2}$ **5.** Divergent   **7.** 8   **9.** $5e/(3 - e)$
11. $\frac{8}{3}$ **13.** Divergent   **15.** $\frac{400}{7}$ **17.** Divergent
19. Divergent   **21.** $\frac{1}{3}$ **23.** $\frac{17}{36}$ **25.** Divergent
27. $\frac{3}{4}$ **29.** $\frac{3}{2}$ **31.** $\sin 1$ **33.** Divergent
35. Divergent   **37.** $\frac{5}{9}$ **39.** $\frac{307}{999}$ **41.** $41{,}111/333{,}000$
43. $2 < x < 4$, $1/(4 - x)$ **45.** $-5 < x < 5$, $x^2/[5(5 - x)]$
47. $|x - n\pi| < \pi/6$, n any integer, $1/(1 - 2\sin x)$
49. $12\,\mathrm{m}$ **51.** The series is divergent
57. (a) $\frac{1}{2}, \frac{5}{6}, \frac{23}{24}, \frac{119}{120}$; $[(n+1)! - 1]/(n+1)!$ (c) 1

Exercises 10.3 (page 579)

(*abbreviations:* C, *convergent;* D, *divergent*)

1. D   **3.** C   **5.** C   **7.** D   **9.** C   **11.** D
13. D   **15.** C   **17.** C   **19.** C   **21.** C
23. $p > 1$ **25.** $p < -1$ **27.** $(1, \infty)$
29. Between 1000 and 1001

Exercises 10.4 (page 583)

1. C   **3.** C   **5.** D   **7.** C   **9.** C   **11.** D
13. C   **15.** C   **17.** C   **19.** C   **21.** C   **23.** D
25. C   **27.** D   **29.** C   **31.** C   **33.** C   **35.** D
37. D

Exercises 10.5 (page 587)

1. C   **3.** D   **5.** C   **7.** D   **9.** C   **11.** C
13. D   **15.** C   **17.** C   **19.** C   **21.** D   **23.** C
25. D   **27.** $\{a_n\}$ is not decreasing
29. p is not a negative integer   **31.** 0.82
33. 0.13 (or 0.137)   **35.** 0.8415   **37.** 0.6065

Exercises 10.6 (page 594)

(*abbreviations:* AC, *absolutely convergent;* CC, *conditionally convergent*)

1. AC   **3.** D   **5.** CC   **7.** AC   **9.** CC   **11.** D
13. AC   **15.** AC   **17.** AC   **19.** CC   **21.** D
23. AC   **25.** D   **27.** AC   **29.** AC   **31.** D
33. AC   **35.** AC   **37.** (a) and (d)

Exercises 10.7 (page 596)

1. C   **3.** C   **5.** C   **7.** C   **9.** C   **11.** D
13. C   **15.** C   **17.** C   **19.** C   **21.** D   **23.** D

25. C   **27.** D   **29.** C   **31.** C   **33.** C   **35.** C
37. D   **39.** C

Exercises 10.8 (page 601)

1. (a) Yes   (b) No   **3.** $1, [-1, 1)$ **5.** $1, (-1, 1)$
7. $\infty, (-\infty, \infty)$ **9.** $2, (-2, 2]$ **11.** $\frac{1}{3}, [-\frac{1}{3}, \frac{1}{3}]$
13. $1, [-1, 1)$ **15.** $2, (-\frac{3}{2}, \frac{5}{2})$ **17.** $1, (0, 2]$
19. $\infty, (-\infty, \infty)$ **21.** $0.5, [2.5, 3.5)$ **23.** $4, [-14, -6)$
25. $0, \{-6\}$ **27.** $\frac{1}{2}, [0, 1]$ **29.** $1, (e-1, e+1)$
31. $\infty, (-\infty, \infty)$ **33.** $(-\infty, \infty)$
35. $(-1, 1)$, $f(x) = (1 + 2x)/(1 - x^2)$

Exercises 10.9 (page 610)

1. $\displaystyle\sum_{n=0}^{\infty} (-1)^n \frac{x^{2n}}{(2n)!}$, $R = \infty$

3. $\displaystyle\sum_{n=0}^{\infty} \frac{(-1)^{n(n-1)/2}}{\sqrt{2}} \frac{(x - (\pi/4))^n}{n!}$, $R = \infty$

5. $\displaystyle\sum_{n=0}^{\infty} (-1)^n (n+1)x^n$, $R = 1$

7. $\displaystyle\sum_{n=0}^{\infty} (-1)^n (x-1)^n$, $R = 1$ **9.** $\displaystyle\sum_{n=0}^{\infty} \frac{e^3}{n!}(x-3)^n$, $R = \infty$

11. $\displaystyle\sum_{n=0}^{\infty} \frac{x^{2n+1}}{(2n+1)!}$, $R = \infty$ **13.** $\displaystyle\sum_{n=0}^{\infty} (-1)^n x^n$, $R = 1$

15. $\displaystyle\sum_{n=0}^{\infty} (-1)^n (n+1)x^n$, $R = 1$

17. $\displaystyle\sum_{n=0}^{\infty} (-1)^n 4^n x^{2n}$, $R = \frac{1}{2}$ **19.** $\displaystyle\sum_{n=0}^{\infty} \frac{(-1)^n}{4^{n+1}} x^{2n}$, $R = 2$

21. $\displaystyle\sum_{n=0}^{\infty} x^{2n}$, $R = 1$ **23.** $\displaystyle\sum_{n=0}^{\infty} \frac{2x^{2n+1}}{2n+1}$, $R = 1$

25. $\displaystyle\sum_{n=0}^{\infty} \frac{3^n x^n}{n!}$, $R = \infty$ **27.** $\displaystyle\sum_{n=0}^{\infty} \frac{(-1)^n x^{2n+2}}{(2n)!}$, $R = \infty$

29. $\displaystyle\sum_{n=0}^{\infty} \frac{(-1)^n x^{2n+2}}{2^{2n+1}(2n+1)!}$, $R = \infty$

31. $\displaystyle\sum_{n=1}^{\infty} \frac{(-1)^{n+1} 2^{2n-1} x^{2n}}{(2n)!}$, $R = \infty$

33. $\displaystyle\sum_{n=0}^{\infty} \frac{(-1)^n x^{2n}}{(2n+1)!}$, $R = \infty$

35. $1 + \dfrac{x}{2} + \displaystyle\sum_{n=2}^{\infty} (-1)^{n-1} \frac{1 \cdot 3 \cdot 5 \cdot \cdots \cdot (2n-3)}{2^n n!} x^n$, $R = 1$

37. $1 + \displaystyle\sum_{n=1}^{\infty} \frac{1 \cdot 4 \cdot 7 \cdot \cdots \cdot (3n-2)}{3^n n!} x^n$, $R = 1$

39. $\displaystyle\sum_{n=0}^{\infty} \frac{(-1)^n}{2}(n+1)(n+2)x^n$, $R = 1$

41. $\ln 5 + \displaystyle\sum_{n=1}^{\infty} \frac{(-1)^{n-1} x^n}{5^n n}$, $R = 5$

43. $\sum\limits_{n=1}^{\infty} (-1)^{n-1}\dfrac{x^n}{n}$, 0.09531

45. $C + \sum\limits_{n=0}^{\infty} \dfrac{(-1)^n x^{4n+3}}{(4n+3)(2n+1)!}$ 47. $C + \sum\limits_{n=0}^{\infty} \dfrac{(-1)^n x^{4n+1}}{4n+1}$

49. $C + x + \dfrac{x^4}{8} + \sum\limits_{n=2}^{\infty} (-1)^{n-1}\dfrac{1\cdot 3\cdot 5\cdot\,\cdots\,\cdot(2n-3)}{2^n n!(3n+1)}x^{3n+1}$

51. 0.310   53. 0.4989   55. 0.0354   57. (b) 0.920

59. e^{-x^4} 61. $1/\sqrt{2}$ 63. $e^x - 1$

65. $1 - \dfrac{3}{2}x^2 + \dfrac{25}{24}x^4$ 67. $-x + \dfrac{1}{2}x^2 - \dfrac{1}{3}x^3$

69. $[-1,1]$, $[-1,1)$, $(-1,1)$

Exercises 10.10 (page 614)

1. $1 + \dfrac{x}{2} + \sum\limits_{n=2}^{\infty} (-1)^{n-1}\dfrac{1\cdot 3\cdot 5\cdot\,\cdots\,\cdot(2n-3)}{2^n n!}x^n$, $R = 1$

3. $\sum\limits_{n=0}^{\infty} (-1)^n \dfrac{(n+1)(n+2)(n+3)2^n}{6}x^n$, $R = \dfrac{1}{2}$

5. $x + \sum\limits_{n=1}^{\infty} \dfrac{1\cdot 3\cdot 5\cdot\,\cdots\,\cdot(2n-1)}{2^n n!}x^{n+1}$, $R = 1$

7. $\dfrac{1}{2} + \dfrac{1}{2}\sum\limits_{n=1}^{\infty} \dfrac{(-1)^n 1\cdot 4\cdot 7\cdot\,\cdots\,\cdot(3n-2)}{24^n n!}x^n$, $R = 8$

9. $1 - \dfrac{x^4}{4} - \sum\limits_{n=2}^{\infty} \dfrac{3\cdot 7\cdot 11\cdot\,\cdots\,\cdot(4n-5)}{4^n n!}x^{4n}$, $R = 1$

11. $\sum\limits_{n=0}^{\infty} \dfrac{(n+1)(n+2)(n+3)(n+4)}{24}x^{n+5}$, $R = 1$

13. (a) $1 + \sum\limits_{n=1}^{\infty} \dfrac{1\cdot 3\cdot 5\cdot\,\cdots\,\cdot(2n-1)}{2^n n!}x^{2n}$

   (b) $x + \sum\limits_{n=1}^{\infty} \dfrac{1\cdot 3\cdot 5\cdot\,\cdots\,\cdot(2n-1)}{2^n n!}\dfrac{x^{2n+1}}{2n+1}$

15. (a) $1 + \sum\limits_{n=1}^{\infty} (-1)^n\dfrac{1\cdot 3\cdot 5\cdot\,\cdots\,\cdot(2n-1)}{2^n n!}x^n$ (b) 0.953

17. (a) $\sum\limits_{n=1}^{\infty} nx^n$ (b) 2

19. (a) $1 + \dfrac{x^2}{2} + \sum\limits_{n=2}^{\infty} (-1)^{n-1}\dfrac{1\cdot 3\cdot 5\cdot\,\cdots\,\cdot(2n-3)}{2^n n!}x^{2n}$

   (b) 99,225

Exercises 10.11 (page 623)

1. $-2 + 8(x+1) - 9(x+1)^2 + 4(x+1)^3$

3. $\dfrac{1}{2} + \dfrac{\sqrt{3}}{2}\left(x - \dfrac{\pi}{6}\right) - \dfrac{1}{4}\left(x - \dfrac{\pi}{6}\right)^2 - \dfrac{\sqrt{3}}{12}\left(x - \dfrac{\pi}{6}\right)^3$

5. $x + \dfrac{1}{3}x^3$ 7. $x + x^2 + \dfrac{1}{3}x^3$

9. $3 + \dfrac{1}{6}(x-9) - \dfrac{1}{216}(x-9)^2 + \dfrac{1}{3888}(x-9)^3$

11. $-\dfrac{1}{2}\left(x - \dfrac{\pi}{2}\right)^2$

13. $T_1(x) = 1$, $T_2(x) = 1 - \dfrac{x^2}{2}$, $T_3(x) = 1 - \dfrac{x^2}{2}$,

   $T_4(x) = 1 - \dfrac{x^2}{2} + \dfrac{x^4}{24}$

15. (a) $\sqrt{1+x} = 1 + \dfrac{x}{2} - \dfrac{1}{8(1+z)^{3/2}}x^2$ (b) 0.00125

17. (a) $\dfrac{1}{\sqrt{2}} + \dfrac{1}{\sqrt{2}}\left(x - \dfrac{\pi}{4}\right) - \dfrac{1}{2\sqrt{2}}\left(x - \dfrac{\pi}{4}\right)^2 - \dfrac{1}{6\sqrt{2}}\left(x - \dfrac{\pi}{4}\right)^3$

   $+ \dfrac{1}{24\sqrt{2}}\left(x - \dfrac{\pi}{4}\right)^4 + \dfrac{1}{120\sqrt{2}}\left(x - \dfrac{\pi}{4}\right)^5 - \dfrac{\sin z}{720}\left(x - \dfrac{\pi}{4}\right)^6$

   (b) 0.00033

19. (a) $\dfrac{1}{(1+2x)^4} = 1 - 8x + 40x^2 - 160x^3 + \dfrac{560}{(1+2z)^8}x^4$

   (b) 0.34

21. (a) $\tan x = x + \dfrac{x^3}{3} + \dfrac{\sec^2 z\,\tan^3 z + 2\sec^4 z\,\tan z}{3}x^4$

   (b) 0.06

23. (a) $e^{x^2} = 1 + x^2 + \dfrac{e^{z^2}(3 + 12z^2 + 4z^4)}{6}x^4$

   (b) 0.00006

25. (a) $x^{3/4} = 8 + \dfrac{3}{8}(x-16) - \dfrac{3}{1024}(x-16)^2$

   $+ \dfrac{15}{196,608}(x-16)^3 - \dfrac{135(x-16)^4}{6144z^{13/4}}$

   (b) 0.0000034

27. 1.10517   29. 1.0192   31. 0.336   33. 0.4794

35. 0.98481   37. 0.57358   39. $|x| \leq 1$ 47. $\dfrac{1}{120}$

Review Exercises for Chapter 10 (page 625)

1. False   3. False   5. False   7. False   9. False

11. True   13. True   15. Convergent, $\dfrac{1}{2}$

17. Divergent   19. Divergent   21. Convergent, e^{12}

23. D   25. C   27. C   29. C   31. C   33. D

35. Conditionally convergent

37. Absolutely convergent   39. 8   41. $\pi/4$

43. $\dfrac{4111}{3330}$ 45. 0.9721   49. 3, $[-3,3]$ 51. $[2.5, 3.5]$

53. $\dfrac{1}{2} + \dfrac{\sqrt{3}}{2}\left(x - \dfrac{\pi}{6}\right) - \dfrac{1}{2}\dfrac{1}{2!}\left(x - \dfrac{\pi}{6}\right)^2 - \dfrac{\sqrt{3}}{2}\dfrac{1}{3!}\left(x - \dfrac{\pi}{6}\right)^3 + \cdots$

   $= \dfrac{1}{2}\sum\limits_{n=0}^{\infty} (-1)^n\left[\dfrac{1}{(2n)!}\left(x - \dfrac{\pi}{6}\right)^{2n} + \dfrac{\sqrt{3}}{(2n+1)!}\left(x - \dfrac{\pi}{6}\right)^{2n+1}\right]$

55. $\sum\limits_{n=0}^{\infty} (-1)^n x^{n+2}$, 1   57. $-\sum\limits_{n=1}^{\infty} \dfrac{x^n}{n}$, 1

59. $\sum\limits_{n=0}^{\infty} (-1)^n \dfrac{x^{8n+4}}{(2n+1)!}$, ∞

61. $\dfrac{1}{2}\left(1 + \dfrac{x}{2^6} + \dfrac{1\cdot 5}{2!}\dfrac{x^2}{2^{12}} + \dfrac{1\cdot 5\cdot 9}{3!}\dfrac{x^3}{2^{18}} + \cdots\right.$

   $\left. + \dfrac{1\cdot 5\cdot 9\cdot\,\cdots\,\cdot(4n-3)}{n!2^{6n}}x^n + \cdots\right)$, 16

63. $\ln|x| + \sum\limits_{n=1}^{\infty} \dfrac{x^n}{n\cdot n!} + C$ 65. 0.7788

67. $\sqrt{x} = 1 + \frac{1}{2}(x-1) - \frac{1}{8}(x-1)^2 + \frac{1}{16}(x-1)^3$
$$- \frac{5}{128z^{7/2}}(x-1)^4, \ 0.000006$$

69. 1

Problems Plus (page 628)

1. $15!/5! = 10,897,286,400$
3. (a) $s_n = 3 \cdot 4^n$, $l_n = 1/3^n$, $p_n = 4^n/3^{n-1}$ (c) $2\sqrt{3}/5$
5. (b) 0 if $x = 0$, $(1/x) - \cot x$ if $x \neq n\pi$
11. $(-1, 1)$, $(x^3 + 4x^2 + x)/(1-x)^4$
21. (b)

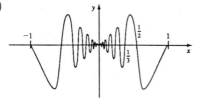

Appendices

Appendix A (page A12)

1. $-3a^2bc$ **3.** $2x^2 - 10x$ **5.** $-8 + 6a$
7. $-x^2 + 6x + 3$ **9.** $12x^2 + 25x - 7$ **11.** $4x^2 - 4x + 1$
13. $30y^4 + y^5 - y^6$ **15.** $2x^3 - 5x^2 - x + 1$ **17.** $1 + 4x$
19. $(3x+7)/(x^2+2x-15)$ **21.** $(u^2+3u+1)/(u+1)$
23. $x/(yz)$ **25.** $rs/(3t)$ **27.** $c/(c-2)$
29. $2x(1+6x^2)$ **31.** $(x+1)(x+6)$
33. $(x+2)(x-4)$ **35.** $9(x-2)(x+2)$
37. $(3x+2)(2x-3)$ **39.** $(t+1)(t^2-t+1)$
41. $(2t-3)^2$ **43.** $x(x+1)^2$
45. $(x-1)(x+1)(x+3)$ **47.** $(x-2)(x+3)(x+4)$
49. $(x+2)/(x-2)$ **51.** $(x+1)/(x-8)$
53. $(x-2)/(x^2-9)$ **55.** $(x+1)^2 + 4$
57. $(x-\frac{5}{2})^2 + \frac{15}{4}$ **59.** $(2x+1)^2 - 3$ **61.** $1, -10$

63. $(-9 \pm \sqrt{85})/2$ **65.** $(-5 \pm \sqrt{13})/6$

67. $1, (-1 \pm \sqrt{5})/2$ **69.** Irreducible
71. Not irreducible (two real roots)
73. $a^6 + 6a^5b + 15a^4b^2 + 20a^3b^3 + 15a^2b^4 + 6ab^5 + b^6$
75. $x^8 - 4x^6 + 6x^4 - 4x^2 + 1$ **77.** 8   **79.** $2|x|$
81. $4a^2b\sqrt{b}$ **83.** 3^{26} **85.** $16x^{10}$ **87.** a^2b^{-1}
89. $1/\sqrt{3}$ **91.** 25   **93.** $2\sqrt{2}|x|^3y^6$ **95.** $y^{6/5}$
97. $t^{-5/2}$ **99.** $t^{1/4}/s^{1/24}$ **101.** $1/(\sqrt{x}+3)$

103. $(x^2+4x+16)/(x\sqrt{x}+8)$ **105.** $(3+\sqrt{5})/2$

107. $(3x+4)/(\sqrt{x^2+3x+4}+x)$ **109.** False
111. True   **113.** False   **115.** False

Appendix B (page A22)

1. $7\pi/6$ **3.** $\pi/20$ **5.** 5π **7.** $720°$ **9.** $75°$
11. $-67.5°$ **13.** 3π cm   **15.** $\frac{2}{3}$ rad $= \left(\frac{120}{\pi}\right)°$
17.   **19.**

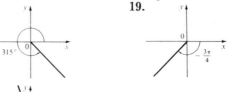

21.

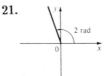

23. $\sin(3\pi/4) = 1/\sqrt{2}$, $\cos(3\pi/4) = -1/\sqrt{2}$,
$\tan(3\pi/4) = -1$, $\csc(3\pi/4) = \sqrt{2}$, $\sec(3\pi/4) = -\sqrt{2}$,
$\cot(3\pi/4) = -1$
25. $\sin(9\pi/2) = 1$, $\cos(9\pi/2) = 0$, $\csc(9\pi/2) = 1$,
$\cot(9\pi/2) = 0$, $\tan(9\pi/2)$ and $\sec(9\pi/2)$ undefined
27. $\frac{1}{2}, -\sqrt{3}/2, -1/\sqrt{3}, 2, -2/\sqrt{3}, -\sqrt{3}$
29. $\cos\theta = \frac{4}{5}$, $\tan\theta = \frac{3}{4}$, $\csc\theta = \frac{5}{3}$, $\sec\theta = \frac{5}{4}$, $\cot\theta = \frac{4}{3}$
31. $\sin\phi = \sqrt{5}/3$, $\cos\phi = -\frac{2}{3}$, $\tan\phi = -\sqrt{5}/2$,
$\csc\phi = 3/\sqrt{5}$, $\cot\phi = -2/\sqrt{5}$
33. $\sin\beta = -1/\sqrt{10}$, $\cos\beta = -3/\sqrt{10}$, $\tan\beta = \frac{1}{3}$,
$\csc\beta = -\sqrt{10}$, $\sec\beta = -\sqrt{10}/3$ **35.** 5.73576 cm
37. 24.62147 cm   **59.** $(4+6\sqrt{2})/15$
61. $(3+8\sqrt{2})/15$ **63.** $\frac{24}{25}$ **65.** $\pi/3, 5\pi/3$
67. $\pi/4, 3\pi/4, 5\pi/4, 7\pi/4$ **69.** $\pi/6, \pi/2, 5\pi/6, 3\pi/2$
71. $0, \pi, 2\pi$ **73.** $0 \leq x \leq \pi/6$ and $5\pi/6 \leq x \leq 2\pi$
75. $0 \leq x < \pi/4, 3\pi/4 < x < 5\pi/4, 7\pi/4 < x \leq 2\pi$

77.

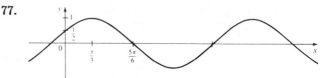

79.

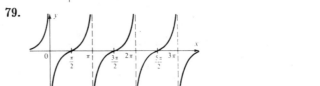

81.

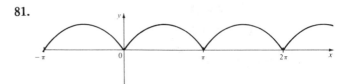

Appendix D (page A39)

1. $\frac{1}{3}$ 3. 0   5. (a) $4.82549424 \times 10^{-11}$
9. (b) 1, $x^3 + 3x - 4 = 0$

Appendix E (page A45)

1. $((\sqrt{3}+4)/2, (4\sqrt{3}-1)/2)$ 3. $(2\sqrt{3}-1, \sqrt{3}+2)$
5. $X = \sqrt{2}Y^2$, parabola   7. $3X^2 + Y^2 = 2$, ellipse

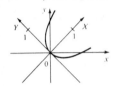
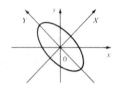

9. $X^2 + (Y^2/9) = 1$,
   ellipse

11. $(X-1)^2 - 3Y^2 = 1$,
    hyperbola

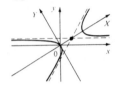
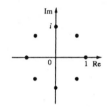

13. (a) $Y - 1 = 4X^2$ (b) $\left(0, \frac{17}{16}\right), \left(-\frac{17}{20}, \frac{51}{80}\right)$
    (c) $64x - 48y + 75 = 0$

Appendix F (page A53)

1. $10 - i$ 3. $13 - i$ 5. $12 - 7i$ 7. $-\frac{1}{2} + \frac{1}{2}i$
9. $\frac{1}{2} - \frac{1}{2}i$ 11. $-i$ 13. $5i$ 15. $3 - 4i, 5$ 17. $4i, 4$
19. $\pm\frac{3}{2}i$ 21. $4 \pm i$ 23. $-\frac{1}{2} \pm (\sqrt{7}/2)i$
25. $3\sqrt{2}[\cos(3\pi/4) + i\sin(3\pi/4)]$
27. $5\{\cos[\tan^{-1}(\frac{4}{3})] + i\sin[\tan^{-1}(\frac{4}{3})]\}$
29. $4[\cos(\pi/2) + i\sin(\pi/2)], \cos(-\pi/6) + i\sin(-\pi/6),$
    $\frac{1}{2}[\cos(-\pi/6) + i\sin(-\pi/6)]$
31. $4\sqrt{2}[\cos(7\pi/12) + i\sin(7\pi/12)],$
    $(2\sqrt{2})[\cos(13\pi/12) + i\sin(13\pi/12)],$
    $\frac{1}{4}[\cos(\pi/6) + i\sin(\pi/6)]$
33. -1024 35. $-512\sqrt{3} + 512i$
37. $\pm 1, \pm i, (1/\sqrt{2})(\pm 1 \pm i)$ 39. $\pm(\sqrt{3}/2) + \frac{1}{2}i, -i$

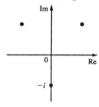

41. i 43. $(-1/\sqrt{2}) + (1/\sqrt{2})i$ 45. $-e^2$
47. $\cos 3\theta = \cos^3\theta - 3\cos\theta\sin^2\theta,$
    $\sin 3\theta = 3\cos^2\theta\sin\theta - \sin^3\theta$

Index

Algebra

Exponents and Radicals

$x^m x^n = x^{m+n}$

$\dfrac{x^m}{x^n} = x^{m-n}$

$(x^m)^n = x^{mn}$

$x^{-n} = \dfrac{1}{x^n}$

$(xy)^n = x^n y^n$

$\left(\dfrac{x}{y}\right)^n = \dfrac{x^n}{y^n}$

$x^{1/n} = \sqrt[n]{x}$

$\sqrt[n]{xy} = \sqrt[n]{x}\,\sqrt[n]{y}$

$x^{m/n} = \sqrt[n]{x^m} = (\sqrt[n]{x})^m$

$\sqrt[m]{\sqrt[n]{x}} = \sqrt[nm]{x} = \sqrt[mn]{x}$

$\sqrt[n]{\dfrac{x}{y}} = \dfrac{\sqrt[n]{x}}{\sqrt[n]{y}}$

Special Products

$(x + y)^2 = x^2 + 2xy + y^2$

$(x - y)^2 = x^2 - 2xy + y^2$

$(x + y)^3 = x^3 + 3x^2 y + 3xy^2 + y^3$

$(x - y)^3 = x^3 - 3x^2 y + 3xy^2 - y^3$

Factoring Polynomials

$x^2 - y^2 = (x + y)(x - y)$

$x^2 + 2xy + y^2 = (x + y)^2$

$x^2 - 2xy + y^2 = (x - y)^2$

$x^3 + y^3 = (x + y)(x^2 - xy + y^2)$

$x^3 - y^3 = (x - y)(x^2 + xy + y^2)$

Quadratic Formula

If $ax^2 + bx + c = 0$, then

$$x = \frac{-b \pm \sqrt{b^2 - 4ac}}{2a}$$

Inequalities and Absolute Value

If $a < b$ and $b < c$, then $a < c$.

If $a < b$, then $a + c < b + c$.

If $a < b$ and $c > 0$, then $ca < cb$.

If $a < b$ and $c < 0$, then $ca > cb$.

If $a > 0$, then

$|x| = a$ means $x = a$ or $x = -a$

$|x| < a$ means $-a < x < a$

$|x| > a$ means $x > a$ or $x < -a$

Geometry

Geometric Formulas

Formulas for area A, circumference C, and volume V:

Triangle

$A = \frac{1}{2}bh$

Circle

$A = \pi r^2$

$C = 2\pi r$

Sphere

$V = \frac{4}{3}\pi r^3$

$A = 4\pi r^2$

Cylinder

$V = \pi r^2 h$

Cone

$V = \frac{1}{3}\pi r^2 h$

Distance and Midpoint Formulas

Distance between $P_1(x_1, y_1)$ and $P_2(x_2, y_2)$:

$$d = \sqrt{(x_2 - x_1)^2 + (y_2 - y_1)^2}$$

Midpoint of $\overline{P_1 P_2}$: $\left(\dfrac{x_1 + x_2}{2}, \dfrac{y_1 + y_2}{2}\right)$

Lines

Slope of line through $P_1(x_1, y_1)$ and $P_2(x_2, y_2)$:

$$m = \frac{y_2 - y_1}{x_2 - x_1}$$

Point-slope equation of line through $P_1(x_1, y_1)$ with slope m:

$$y - y_1 = m(x - x_1)$$

Slope-intercept equation of line with slope m and y-intercept b:

$$y = mx + b$$

Circles

Equation of the circle with center (h, k) and radius r:

$$(x - h)^2 + (y - k)^2 = r^2$$